187.⁰⁰

A.M. Joussen · T.W. Gardner · B. Kirchhof · S.J. Ryan (Eds.)
Retinal Vascular Disease

A.M. Joussen · T.W. Gardner · B. Kirchhof · S.J. Ryan (Eds.)

Retinal Vascular Disease

With 565 Figures in 1043 Parts and 330 Tables

 Springer

ANTONIA M. JOUSSEN, MD, PhD
Professor of Ophthalmology
Department of Ophthalmology, University of Düsseldorf
Moorenstraße 5, 40225 Düsseldorf, Germany

THOMAS W. GARDNER, MD, MS
Professor of Ophthalmology
Department of Ophthalmology
Penn State College of Medicine
500 University
Drive Box 850
Hershey, PA 17033, USA

BERND KIRCHHOF, MD
Professor of Ophthalmology
Department of Vitreoretinal Surgery
Center for Ophthalmology
University of Cologne
Joseph-Stelzmann-Straße 9, 50931 Cologne, Germany

STEPHEN J. RYAN, MD
Professor of Ophthalmology
Doheny Eye Institute, University of Southern California
1450 San Pablo Street, Los Angeles, CA 90033, USA

ISBN 978-3-540-29541-9   Springer-Verlag Berlin Heidelberg New York

Library of Congress Control Number: 2007931904

Springer is a part of Springer Science+Business Media
http://www.springer.com

© Springer-Verlag Berlin Heidelberg 2007

Printed in Germany

Editor: Marion Philipp, Heidelberg, Germany
Desk Editor: Martina Himberger, Heidelberg, Germany
Production Editor: Joachim W. Schmidt, München, Germany

Cover design: eStudio Calamar, Spain

Typesetting: FotoSatz Pfeifer GmbH, D-82166 Gräfelfing
Printed on acid-free paper – 24/3150 – 5 4 3 2 1 0

Foreword

Angiogenesis inhibitors comprise a new class of drugs that have recently received FDA approval for use in age-related macular degeneration. They are currently in clinical trials for the treatment of diabetic retinopathy. These new drugs emerged after more than three decades of cancer research. This journey was driven by the hypothesis that tumor growth is angiogenesis dependent, and sustained by experimental demonstrations that antiangiogenic therapy could become a fourth modality to help control neoplastic disease.

At the dawning of this new field of angiogenesis research, animal models of corneal neovascularization became the test tube for discovery of angiogenesis regulatory molecules. It seems more than a coincidence that the two implantable polymers still employed by scientists today for slow release of angiogenesis regulatory molecules into the avascular cornea as a bioassay, originated from soft contact lenses or from a wearable device to treat glaucoma. In this sense, two distinct specialties of medicine, oncology and ophthalmology, are now linked. Angiogenesis has become the organizing principle.

Over the years, numerous ophthalmologists have studied in a cancer biology lab and have gone on to distinguished careers in ophthalmology. Without their contributions it is unlikely that an oncologist's armamentarium would today contain approved anti-cancer drugs that inhibit angiogenesis directly or indirectly.

It is as exciting that these new drugs have begun to increase survival for the three common cancers, colon, breast and lung, as it is that eyesight has been improved in patients with age-related macular degeneration ("A very effective treatment for neovascular macular degeneration," E.M. Stone, *N Engl J Med* 2006 355:1493).

The editors and authors of this book have brought the field of retinal vascular disease up to date, as the principles of antiangiogenic therapy are rapidly being translated to clinical practice. This book is also very valuable because it inspires the reader to think about possible future improvements in the management of retinal vascular diseases. Can biomarkers in the blood or urine be developed to detect recurrence of retinal neovascularization before symptoms or before detection by ophthalmoscopy? Can oral angiogenesis inhibitors maintain suppression of neovascular macular degeneration after sight has been improved by repeated intravitreal injections of antiangiogenic drugs? Because platelets are now known to carry high concentrations of angiogenesis regulatory molecules stored and segregated in alpha granules, will it become possible to therapeutically instruct platelets to release antiangiogenic proteins? Beyond these questions it is possible to anticipate that angiogenesis research will continue to bring new insights into the biology and molecular mechanisms of retinal vascular disease.

Judah Folkman, MD

Preface

Great progress has been made in the treatment of vascular disease in recent years, and we hope and expect this is a prelude to even greater progress in the near future. The advent of clinically applicable anti-VEGF therapies is only the "tip of the iceberg" of future additions to the clinical armamentarium. However, there is no unique formula for angiogenesis. Our understanding of complex disease specific interactions is dependent on a knowledge of the basic mechanisms involved in the vascular reactions specific to different disease entities. Despite these new treatments and the explosion of knowledge on the basic mechanisms of vascular biology, vascular disease of the retina remains to date a major cause of blindness in all age groups.

The topics covered by this book range from fundamental concepts of molecular biology to basic clinical appearance, specific pathology and treatment of retinal vascular disease.

In the first part, the current thinking in vascular biology is discussed in relation to retinal vascular disease. The emphasis is on general pathogenic concepts including ischemia, inflammation and their associated pathology. Experimental approaches are reviewed as well as animal models which might help in the investigation of the diseases.

The second part includes modern diagnostic features and general treatment strategies. Diagnostic procedures are discussed with respect to their relevance for clinical decision making. Similarly, treatment procedures are described stepwise by schematic graphics.

The third part describes each disease in a comprehensive manner including demographics, clinical course and treatment. Clinical image series including illustrated single case follow-ups are given major emphasis and provide an atlas like presentation. The book includes topics which are not currently found in other textbooks of retinal disease including case reports and clinical follow-ups. Furthermore, all the treatment procedures are explained in detail to facilitate their use by ophthalmologists in training.

More than 50 experts in the field contributed to this state of the art review of basic and clinical science, which aims to enhance our understanding of retinal vascular disease and to help the clinician in the evaluation of current and future treatment approaches. The authors are internationally recognized leaders in clinical ophthalmology, including the areas of medical retina, vitreoretinal surgery, and uveitis. In a unique way leaders and experts in their fields of molecular mechanisms and general concepts of vascular surgery have contributed to this book even though their previous major focus has not necessarily been on ophthalmic disease.

Nevertheless, the field of vascular biology and retinal vascular disease is so broad and the evolution of knowledge so rapid that the work cannot be comprehensive and the information given only resembles the current knowledge at the time of printing.

As a multiauthored text, there are many literary styles, and the editors have not sacrificed the originality and the style of the individual authors. Although there will be some aspects that are discussed in more than one chapter, the unique interpretation by each author justifies some overlap and is an attractive feature.

The editors gratefully acknowledge the support of the contributing authors who, in addition to their large clinical load and scientific research efforts, found the time to make such a large contribution to the completion of this project. In particular, Andrew P. Schachat, MD, and his team at The Wilmer Eye Institute were of invaluable help.

At Springer, Marion Philipp and Martina Himberger helped to create this book and its unique combination of basic science and clinical application.

We hope that *Retinal Vascular Disease* will help to inspire clinicians and scientists in the future to making further efforts and advances in this field.

Düsseldorf, Hershey, Cologne, Los Angeles

August 2007

Antonia M. Joussen
Thomas W. Gardner
Bernd Kirchhof
Stephen J. Ryan

Contents

Section I: Pathogenesis of Retinal Vascular Disease

**1   Functional Anatomy, Fine Structure and Basic Pathology
of the Retinal Vasculature**
D.B. Archer, T.A. Gardiner, A.W. Stitt   3
1.1   Anatomical Organization of the Retinal Vasculature   3
1.1.1   Microvascular Arrangement   3
1.1.2   Nature of the Retinal Vasculature   6
1.2   Responses of the Retina and Its Vasculature to Stress and Disease:
Histological and Pathological Consequences   7
1.2.1   Hemodynamic Changes   7
1.2.2   Oxygen Saturation Changes   11
1.2.3   Occlusion – Ischemia   12
1.2.4   Repair and Remodeling   13
1.2.5   Metabolic Stresses   14
1.2.6   Trauma   17
1.2.7   Drug Toxicity   18
1.2.8   Inflammation   18
1.2.9   Retinal-Choroidal Tumors   20
1.2.10 Primary Neuropile Atrophy and Degeneration   22
1.2.11 Remote Effects of Retinal Vascular Pathology   22
References   22

2   Retinal Vascular Development
M.I. Dorrell, M. Friedlander, L.E.H. Smith   24
2.1   Introduction   24
2.1.1   General Vascular Development   24
2.1.2   Basis of Clinical Identification of Blood Vessels   24
2.1.3   Major Cellular Components of Vessel Formation   25
2.2   Endothelial Cells   25
2.2.1   Vascular Heterogeneity (Morphological Classification of Vessels)   25
2.3   Mural Cells   25
2.4   Vascular Patterning   26
2.5   Retinal Vascular Development   27
2.5.1   The Role of Astrocytes   29
2.5.2   The Role of Subcellular Endothelial Processes   31
2.6   Development of the Deep Retinal Vascular Plexuses   32
2.7   Vascular Maturation   32
2.8   Vascular Pruning Mechanisms   34
2.9   Mouse Retinal Vascular Development as a Model for General
Vascular Development   34
2.10   Use of Retinal Vascular Development as Models for Clinical Ocular
Neovascularization   35

2.10.1   Mouse Retinal Angiogenesis Model 35
2.10.2   Oxygen-Induced Retinopathy 35
References 35

3   Retinal Angiogenesis and Growth Factors

3.1   General Concepts of Angiogenesis and Vasculogenesis
         C. RUIZ DE ALMODOVAR, A. NY, P. CARMELIET 38
3.1.1   General Introduction 38
3.1.2   Angiogenic Disorders...................................... 38
3.1.3   Modes of Vessel Growth 40
3.1.4   Vasculogenesis 40
3.1.4.1 Role of Endothelial Progenitors in the Embryo 41
3.1.4.2 Role of Endothelial Progenitors in the Adult 42
3.1.4.3 The Endothelial/Hematopoietic Connection – An Emerging
         Theme 44
3.1.4.4 Arterial, Venous and Lymphatic Cell Fate Specification ... 44
3.1.4.5 Tissue-Specific EC Differentiation 45
3.1.5   Angiogenesis 46
3.1.5.1 Vascular Permeability and Extracellular Matrix Degradation ... 46
3.1.5.2 Endothelial Budding and Sprouting 48
3.1.5.3 Vascular Lumen Formation 49
3.1.5.4 Guided Navigation of Vessels 50
3.1.5.5 Vessel Maintenance 51
3.1.6   Arteriogenesis 52
3.1.6.1 Smooth Muscle Progenitor Cells 52
3.1.6.2 Smooth Muscle Cell Recruitment, Growth and Differentiation ... 53
3.1.7   Therapeutic Implications 54
References 56

3.2   Vascular Endothelial Growth Factor in Retinal Vascular Disease
         G.L. KING, K. SUZUMA, J.K. SUN 66
3.2.1   VEGF Regulation and Receptors 66
3.2.1.1 VEGFR2, PKC, PI3-kinase 66
3.2.2   Vascular Endothelial Growth Factor 67
3.2.3   VEGF and Systemic Diseases 68
3.2.4   VEGF and Retinal Vascular Disease 68
3.2.4.1 VEGF and Other Growth Factors 68
3.2.4.2 VEGF and Diabetic Retinopathy 69
3.2.4.3 Hypertension As an Aggravating Factor in Diabetes-Induced
         Activation of VEGF 70
3.2.4.4 VEGF in Neovascularization Secondary to Retinal Vascular
         Occlusions 70
References 70

**3.3   Involvement of the Ephrin/Eph System in Angioproliferative
          Ocular Diseases**
         H. AGOSTINI, G. MARTIN 73
3.3.1   First Studies: Ephrins in Retinotectal Projection 73
3.3.2   Ephrins in Vascular Development 74
3.3.3   Ephrins and Ocular Angiogenesis 75
3.3.4   Ephrins in Retinal and Subretinal Animal Models 75
3.3.5   Therapeutic Potential 76
References 76

**4   Hematopoietic Stem Cells in Vascular Development
and Ocular Neovascularization**
N. SENGUPTA, M.B. GRANT, S. CABALLERO, M.E. BOULTON 78
4.1   Background 78
4.2   Developmental Origins of HSCs 78
4.2.1 HSCs Lack Regional Patterning 79
4.3   Defining the Adult HSC 79
4.3.1 HSC Self-Renewal 79
4.3.2 HSC Pluripotency/Plasticity 81
4.4   The HSC Niche 82
4.4.1 Molecular Mechanisms for Maintenance in the Niche 82
4.5   HSC Mobilization 82
4.5.1 The SDF-1/CXCR4 Axis 84
4.6   Surface Marker Expression – HSC Identification 84
4.7   Surface Marker Expression – HSC Isolation 84
4.8   Methodologies 85
4.8.1 Extraction of HSC from Donor Mice 85
4.8.2 Reconstitution of Bone Marrow-Ablated Recipient Mice 85
4.8.3 EPC Culture 86
4.9   Mouse Models of HSC Involvement in Ocular Neovascularization 88
4.9.1 Preretinal Neovascularization 88
4.9.2 Iris Neovascularization 89
4.9.3 Choroidal Neovascularization 90
4.10   Conclusion 90
References 91

**5   Inflammation as a Stimulus for Vascular Leakage
and Proliferation**
A.M. JOUSSEN, A.P. ADAMIS 97
5.1   Evidence for Inflammation in the Pathogenesis of Diabetic Retinopathy   97
5.1.1 Upregulation of Inflammatory Mediators in Diabetic Retinopathy 97
5.1.2 Diabetic Retinal Pathology Can Be Inhibited by Anti-inflammatory
       Agents 98
5.1.3 VEGF Is a Key Mediator of Inflammatory Changes in the Diabetic
       Retina: General Considerations 99
5.1.4 Angiopoietin-1 Regulates Vascular Permeability and Expression of
       Inflammatory Mediators in Diabetic Retinopathy 99
5.2   Cellular and Molecular Mechanisms Mediating Diabetes-Associated
       Vascular Damage and the Neovascularizing Response to Ischemia 101
5.2.1 Leukocytes Mediate Retinal Vascular Remodeling During Develop-
       ment and Vaso-obliteration in Disease 101
5.2.2 VEGF and Leukocyte Invasion Are Important Factors in Regulating
       Both Ischemia-Mediated Ocular Neovascularization and Vascular
       Damage in Diabetic Retinopathy 102
5.2.3 Regulation of Ischemia-Mediated Retinal Vascularization 102
5.2.4 Diabetes Associated Vascular Damage Is Accelerated by Leukostasis
       and Fas-FasL-Mediated Apoptosis 103
5.3   Conclusions 103
References 104

6   The Neuronal Influence on Retinal Vascular Pathology
A.J. BARBER, H.D. VAN GUILDER, M.J. GASTINGER 108
6.1   Introduction 108
6.2   The Phenotype of the Retinal Vasculature Is Determined by the
       Tissue It Serves 109

6.3   Diabetes Causes a Loss of the Blood-Retinal Barrier Phenotype ... 109
6.4   The Increase in Vascular Permeability Is Likely Due to an Increase
        in the Expression of VEGF 109
6.5   VEGF Originates from the Neural Tissue of the Retina
        in Diabetes 110
6.6   Injury and Neurodegeneration in the Brain Are Accompanied by
        Increased VEGF Expression and Vascular Abnormalities 110
6.7   Diabetes Increases Neural Apoptosis, Leading to a Cumulative
        Reduction in the Thickness of the Inner Layers of the Retina 111
6.8   Specific Neurons Are Lost by Apoptosis in Diabetes 113
6.9   Neurons in the Retina Have Morphological Characteristics of
        Neurodegeneration 113
6.10   The Glial Cells of the Retina React to Diabetes As if an Injury
        Has Occurred 114
6.11   There is a Loss of Function in Diabetic Retinopathy that Begins
        soon After the Onset of Diabetes 115
6.12   Conclusions 115
6.13   Methodology 115
References 116

7   Hypoxia in the Pathogenesis of Retinal Disease
        V. POULAKI 121
7.1   Introduction 121
7.2   The HIF Pathway and Its Role in Hypoxia Signaling.............. 121
7.2.1   Structure of the HIF Transcription Factor Complex and Regulation
        of Its Activity 122
7.2.2   Downstream Targets of HIF 123
7.2.3   Activation of HIF by Non-hypoxic Stimuli 124
7.2.4   VHL and Its Role in Retinal Angiogenesis 124
7.3   VEGF 125
7.4   Retinal Hypoxia and Retinopathy of Prematurity 126
7.4.1   Introduction 126
7.4.2   Pathophysiology of Retinopathy of Prematurity 126
7.5   Retinal Hypoxia and Diabetic Retinopathy 129
7.5.1   Introduction 129
7.5.2   Pathophysiology of Diabetic Retinopathy 129
7.6   Retinal Hypoxia and Vascular Occlusive Disease 131
7.7   Other Diseases 132
7.7.1   Cystoid Macular Edema 132
7.7.2   Retinal Degeneration 132
7.7.3   Glaucoma 133
7.7.4   Retinal Detachment 134
7.8   Conclusions 134
References 134

8   Blood Retinal Barrier

8.1   Blood-Retinal Barrier, Retinal Vascular Leakage
        and Macular Edema
        B.E. PHILLIPS, D.A. ANTONETTI 139
8.1.1   Introduction 139
8.1.2   Characteristics of the Blood-Retinal Barrier 139
8.1.2.1 Blood-Retinal Barrier Physiology 139
8.1.2.2 Fenestrations 140
8.1.2.3 Endocytosis and Facilitated Diffusion 140
8.1.2.4 Water Transport 140

8.1.2.5   Tight Junctions 141
8.1.2.6   P-Glycoprotein 141
8.1.3   Diabetic Retinopathy 142
8.1.3.1   Permeability and Macular Edema 142
8.1.3.2   Vascular Endothelial Growth Factor 142
8.1.3.3   Apoptosis and Cell Loss 143
8.1.3.4   Routes of Paracellular Flux 143
8.1.4   The Structure and Role of the Tight Junctions 144
8.1.4.1   Tight Junction 144
8.1.4.2   Polarity and Tight Junction Assembly 144
8.1.4.3   Tight Junction Structure 144
8.1.4.4   Occludin 145
8.1.4.5   Claudins 146
8.1.4.6   Zonula Occludens 147
8.1.5   Summary 147
References 147

8.2   MRI Studies of Blood-Retinal Barrier: New Potential for Translation of Animal Results to Human Application

          B.A. BERKOWITZ 154
8.2.1   Introduction 154
8.2.2   Magnetic Resonance Imaging 155
8.2.3   Conclusions 164
References 164

9   Retinal Blood Flow

          L. SCHMETTERER, G. GARHÖFER 167
9.1   Anatomy 167
9.2   Quantification of Retinal Blood Flow 168
9.3   Blood Flow Regulation 170
9.4   Blood Flow in Retinal Vascular Disease 173
References 174

10   Genetic Approach to Retinal Vascular Disease

10.1   Gene Therapy for Proliferative Ocular Disease

          T.J. MCFARLAND, J.T. STOUT 175
10.1.1   Introduction 175
10.1.2   Delivery Systems 175
10.1.2.1 Retroviral 177
10.1.2.2 Adenovirus 177
10.1.2.3 AAV 178
10.1.2.4 Non-viral 179
10.1.2.5 Physical 179
10.1.3   Target Diseases 180
10.1.3.1 Retinopathy of Prematurity 180
10.1.3.2 Age Related Macular Degeneration 181
10.1.3.3 Proliferative Diabetic Retinopathy 181
10.1.3.4 Retinal Vaso-occlusive Disorders 181
10.1.4   Fundamentals of Gene Therapy 182
10.1.4.1 In Vitro 182
10.1.4.2 Ex Vivo Versus In Vivo 182
10.1.4.3 Somatic Cells Versus Stem Cells 183
10.1.5   Genes as Drugs/Clinical Trials 183
10.1.6   Ethical/Safety Concerns 184
References 184

10.2   Norrin and Its Role During Angiogenesis of the Retina
M. Scholz, E.R. Tamm 186
10.2.1   Introduction 186
10.2.2   Norrie Disease 186
10.2.3   Norrie Disease Mutant Mice 186
10.2.4   In Vivo Overexpression of Norrin 187
10.2.5   Conclusion 188
References 188

Section II: General Concepts in the Diagnosis and Treatment of Retinal Vascular Disease

11   A Practical Guide to Fluorescein Angiography

H. Heimann, S. Wolf 193
11.1   History 193
11.2   Concept 194
11.3   Performing Fluorescein Angiography.................... 194
11.4   Interpretation of Fluorescein Angiography 195
11.5   Quantitative Evaluation of Fluorescein Angiography 203
11.6   Avoiding Unnecessary Angiographies 203
References 204

12   Optical Coherence Tomography in the Diagnosis of Retinal Vascular Disease

A. Walsh, S. Sadda 205
12.1   Overview 205
12.2   Evaluation 205
12.2.1   Optical Coherence Tomography 206
12.2.2   Biomicroscopic Examination 207
12.2.3   Stereoscopic Fundus Photography 208
12.2.4   Fluorescein Angiography 208
12.2.5   Ultrasound 209
12.3   Diagnosis 209
12.3.1   Intraretinal Edema 209
12.3.2   Cystoid Macular Edema 210
12.3.3   Serous Retinal Detachment 212
12.3.4   Vitreomacular Traction Syndrome 213
12.3.5   Miscellaneous Findings 214
12.4   Management 216
12.4.1   Focal and Panretinal Photocoagulation 217
12.4.2   Pharmacologic Therapy 217
12.4.3   Surgery 218
12.5   Future Directions 219
12.5.1   Current OCT Limitations 219
12.5.2   Clinical Applications 221
12.5.3   The Future of OCT Hardware 222
12.5.4   The Future of OCT Software 223
References 224

13   General Concepts in Laser Treatment for Retinal Vascular Disease

F. Rüfer, J. Roider 228
13.1   History of Laser Treatment 228
13.2   Laser Sources 228
13.2.1   Histological Findings After Light and Laser Coagulation 229

13.2.2 Mechanisms of Treatment 231
13.3   Standards and Indications for Panretinal Laser Coagulation 232
13.3.1 Full Scatter Panretinal Laser Coagulation 232
13.3.2 Mild Scatter Panretinal Laser Coagulation 233
13.4   Focal Laser Application 234
13.5   Subthreshold Laser Coagulation for Retinal Disease 234
References 236

14 The Role of Photodynamic Therapy in Retinal Vascular Disease

B. Jurklies, N. Bornfeld 239

14.1   Introduction 240
14.2   Photodynamic Therapy 240
14.2.1 Effects of Light on Biological Tissue 240
14.2.2 Differences Between PDT and Laser Coagulation 242
14.2.3 Mechanisms of Photodynamic Therapy 242
14.3   Verteporfin in PDT 244
14.3.1 Characteristics 244
14.3.2 Effects of PDT in Animal Experiments 244
14.3.3 Effects of PDT in (Normal) Human Tissue 245
14.3.4 Toxic Effects and Adverse Effects of PDT with Verteporfin 247
14.4   Current Treatment Recommendations 248
14.5   Verteporfin in Retinal Vascular Disease 248
14.5.1 Retinal Capillary Hemangioma 248
14.5.2 Vasoproliferative Tumor 251
14.5.3 Parafoveal Telangiectasis 251
14.6   Conclusions 252
References 252

15 Cryosurgery in Retinal Vascular Disease

B. Kirchhof, A.M. Joussen 256

15.1   Technique of Cryotherapy and Equipment 257
15.2   Indications of Cryosurgery in Retinal Vascular Disease 258
15.2.1 Cryotherapy for Retinopathy of Prematurity 258
15.2.2 Cryotherapy for Diabetic Retinopathy and Retinal Ischemic
        Disease 258
15.2.3 Cryotherapy for Vascular Abnormalities and Exudative
        Retinopathies: Coats' Disease, FEVR and Small Peripheral
        Hemangiomas 258
References 259

16 Vitrectomy in Retinal Vascular Disease: Surgical Principles

A.M. Joussen, B. Kirchhof 260

16.1   Introduction 260
16.2   Prerequisites for Vitreoretinal Surgery in Retinal Vascular Disease:
        Microscope Requirements and Wide-Angle Viewing Systems 260
16.3   Surgical Techniques 261
16.3.1 Three-Port Vitrectomy 261
16.3.2 "Chromovitrectomy" 262
16.3.3 Heavy Liquids 262
16.3.4 Vitreous Tamponades 263
16.3.5 Endolaser Coagulation 266
16.3.6 Lens Management and Compartmentalization 266
16.4   Indications for Vitreoretinal Surgery in Retinal Vascular Disease .. 267
16.4.1 Destruction of Ischemic Retina and Pathological Vessels 267

16.4.2   Removal of Dense Vitreous Hemorrhages 267
16.4.3   Release of Traction 267
References 270

17   Treatment of Rubeotic Secondary Glaucoma
         T. SCHLOTE, K.U. BARTZ-SCHMIDT 274
17.1   Definition and Classification 274
17.2   Prognosis 275
17.3   Treatment 275
17.3.1   Early Rubeotic Secondary Glaucoma (Open Angle Type) 275
17.3.2   Advanced Rubeotic Secondary Glaucoma (Angle Closure Type) 278
17.3.3   Blind Painful Eye with Rubeotic Secondary Glaucoma 279
17.3.4   Treatment Options Under Investigation 280
References 281

18   Intravitreal Injections: Guidelines to Minimize the Risk
         of Endophthalmitis
         I.U. SCOTT, H.W. FLYNN, JR 283
18.1   Introduction 283
18.2   Risk of Endophthalmitis Associated with the Intravitreal
         Injection Procedure 283
18.3   Injection Procedure Guidelines 284
18.4   Guidelines for Follow-up 286
18.5   Non-infectious Endophthalmitis 287
18.6   Summary 287
References 287

Section III: Pathology, Clinical Course and Treatment
              of Retinal Vascular Diseases

19   Grading of Diabetic Retinopathy
         M. LARSEN, W. SOLIMAN 291
19.01   Early Treatment Diabetic Retinopathy Study Grading –
         An Extension of the Modified Airlie House Classification 292
19.02   EURODIAB Grading 293
19.03   International Clinical Diabetic Retinopathy Severity Scale 301
References 301

19.1   Nonproliferative Diabetic Retinopathy

19.1.1 Nonproliferative Stages of Diabetic Retinopathy: Animal Models
          and Pathogenesis
          T.S. KERN, S. MOHR 303
19.1.1.1 Early Stages of Diabetic Retinopathy: Histopathology 303
19.1.1.2 Animal Models of the Early Stages of Diabetic Retinopathy 305
19.1.1.3 Insulin Therapy to Inhibit Development of Retinopathy 306
19.1.1.4 Is Diabetic Retinopathy a Chronic Inflammatory Disease? 307
19.1.1.5 Therapies 307
19.1.1.6 Summary 311
References 311

19.1.2 Pharmacological Approach and Current Clinical Studies

19.1.2.1 Protein Kinase C Inhibitors
           A. GIRACH, D.S. FONG 317
19.1.2.1.1 Introduction 317

19.1.2.1.2 Protein Kinase C Family of Isoenzymes 317
19.1.2.1.3 PKC Activation and Diabetic Retinopathy 318
19.1.2.1.4 PKC-β Inhibitor – Ruboxistaurin 318
19.1.2.1.5 PKC-DRS Trial 320
19.1.2.1.6 PKC-DMES (MBBK) Trial 320
19.1.2.1.7 PKC-DRS2 (MBCM) Trial 321
19.1.2.1.8 Safety of Ruboxistaurin 322
19.1.2.1.9 Conclusion 323
References 323

19.1.2.2   Somatostatin Analogues
          G.E. Lang 324
19.1.2.2.1 Pathogenesis of Diabetic Retinopathy 324
19.1.2.2.2 Somatostatin and Somatostatin Analogues 324
19.1.2.2.3 Current Clinical Use of Octreotide 327
References 329

19.2   Proliferative Diabetic Retinopathy

19.2.1   A Surgical Approach to Proliferative Diabetic Retinopathy
          H. Helbig 330
19.2.1.1 Rationale for Surgery in Diabetic Retinopathy 330
19.2.1.2 Indications for Surgery in Diabetic Retinopathy: A Practical
          Approach 330
19.2.1.3 Surgical Principles 338
References 340

19.2.2   Laser Coagulation of Proliferative Diabetic Retinopathy
          W. Soliman, M. Larsen 342
19.2.2.1 History of Photocoagulation 342
19.2.2.2 Mechanisms of Action 343
19.2.2.3 Clinical Trials and Indications for Retinal Photocoagulation
          Treatment 343
19.2.2.4 Clinical Practice 344
19.2.2.5 Preparations for Photocoagulation: Information and Consent .. 345
19.2.2.6 Consent to Treatment 348
19.2.2.7 Photocoagulation Protocol 348
19.2.2.8 Complications of Photocoagulation for PDR 351
References 351

19.3   Diabetic Macular Edema

19.3.1   Therapeutic Approaches to (Diabetic) Macular Edema
          A.M. Joussen 353
19.3.1.1 Macular Edema as a Result of Various Disease Mechanisms 353
19.3.1.2 Diagnosis and Current Imaging Modalities 359
19.3.1.3 Treatment of Macular Edema 360
19.3.1.4 Discussion: Open Questions and Technical Aspects 370
19.3.1.5 Summary 371
References 371

19.3.2   Pegaptanib for Diabetic Macular Edema
          A.P. Adamis, B. Katz 377
19.3.2.1 Background 377
19.3.2.2 Phase 2 Trial – Intravitreous Pegaptanib as a Treatment for
          DME 380
19.3.2.3 Conclusions 383
References 383

19.3.3   **Ranibizumab for the Treatment of Diabetic Macular Edema**
         D.V. Do, Quan Dong Nguyen, S.M. Shah, J.A. Haller 386
19.3.3.1 Introduction 386
19.3.3.2 Pathogenesis 386
19.3.3.3 Vascular Endothelial Growth Factor 386
19.3.3.4 Ranibizumab 387
19.3.3.5 Ranibizumab in Diabetic Macular Edema 387
19.3.3.6 Summary 390
References 390

20   Retinopathy of Prematurity

20.1   Retinopathy of Prematurity: Pathophysiology of Disease

         L.E.H. Smith 392
20.1.1   History of Retinopathy of Prematurity 392
20.1.2   ROP: Disruption of Normal Vascular Development 392
20.1.3   Pathogenesis: Two Phases of ROP 393
20.1.4   ROP: Phase I 393
20.1.5   ROP: Phase II 393
20.1.6   Mouse Model of ROP 393
20.1.7   Oxygen Regulated Factors: Vascular Endothelial Growth Factor
         in ROP 393
20.1.8   VEGF is Critical to Phase II of ROP 395
20.1.9   VEGF in Phase I of ROP 395
20.1.10 VEGF Role in Retinal Vessel Loss 395
20.1.11 VEGF Role in Cessation of Normal Vascular Development 395
20.1.12 Other Growth Factors in ROP 396
20.1.13 IGF-1 Deficiency in the Preterm Infant 396
20.1.14 Growth Hormone and IGF-1 in Phase II of ROP 396
20.1.15 IGF-1 and VEGF Interaction 396
20.1.16 Low Levels of IGF-I and Phase I of ROP 397
20.1.17 Clinical Studies: Low IGF-1 is Associated with the Degree of
         ROP 397
20.1.18 Low IGF-1 is Associated with Decreased Vascular Density 398
20.1.19 IGF-1 and ROP 398
20.1.20 IGF-1 and Brain Development 399
20.1.21 Conclusion: A Rationale for the Evolution of ROP 399
20.1.22 Possible Medical Intervention to Prevent ROP 399
References 399

20.2   Clinical Course and Treatment

         C. Jandeck, M.H. Foerster 403
20.2.1   Historical Developments and Epidemiology 403
20.2.2   Explanation of Important Terms, The International Classification
         and the Cryo-ROP Study 404
20.2.2.1 Terms Used to Describe Age 404
20.2.2.2 International Classification 404
20.2.2.3 Cryo-ROP Study 406
20.2.3   Guidelines for Screening 406
20.2.3.1 Examination Technique 408
20.2.4   Indications for Coagulation Treatment 408
20.2.5   Incidence of ROP and Incidence of Treatment 409
20.2.6   Zone I Disease 409
20.2.7   Treatment Modalities 409
20.2.7.1 ROP Stage 3+ 409
20.2.7.2 Treatment Principles 410

20.2.7.3 Treatment in Stages 4 and 5 411
20.2.8   Conservative Therapy 411
20.2.8.1 Vitamin E 411
20.2.8.2 STOP-ROP Study 411
20.2.8.3 Light-ROP Study 411
20.2.8.4 Surfactant 411
20.2.9   Late Changes 412
20.2.9.1 Risk of Myopia 412
20.2.9.2 Visual Acuity 412
20.2.9.3 Strabismus 412
20.2.9.4 Glaucoma 413
20.2.9.5 Regressive Late Changes in the Retina 413
20.2.9.6 Differential Diagnosis 413
20.2.9.7 Outlook 413
20.2.9.8 Future Treatment Possibilities 414
          References 414

20.3   **Surgical Management of Retinopathy of Prematurity**
          P. Quiram, M. Lai, M. Trese 418
20.3.1   Introduction: Preoperative Evaluation and Timing of Surgical
          Intervention 418
20.3.2   Eyes with Peripheral Ablation 419
20.3.3   Eyes Without Peripheral Ablation 419
20.3.4   Treatment of Retinal Detachment 419
20.3.4.1 Scleral Buckling 419
20.3.4.2 Lens-Sparing Vitrectomy 419
20.3.4.3 Management of Stage 4B Retinopathy of Prematurity 421
20.3.4.4 Management of Stage 5 Retinopathy of Prematurity 421
20.3.4.5 Follow-up Care 422
          References 422

21   **Vascular Occlusive Disease**

21.1   **Plasma Proteins – Possible Risk Factors for Retinal Vascular
          Occlusive Disease**
          E. Tourville, A.P. Schachat 424
21.1.1   Introduction 424
21.1.2   Homocysteine Metabolism and the Role of MTHFR 426
21.1.2.1 Homocysteine 427
21.1.2.2 Methylenetetrahydrofolate Reductase Gene 677 CT Polymorphism:
          The Thermolabile Form (TT MTHFR) 427
21.1.2.3 Literature Evidence on Homocysteine 427
21.1.2.4 Literature Evidence Concerning MTHFR 429
21.1.3   Antiphospholipid Antibodies (APA): Lupus Anticoagulant and
          Anticardiolipin Antibodies 430
21.1.4   Disorders of Coagulation and Anti-coagulation 432
21.1.4.1 Factor V Leiden (FVL) and Activated Protein C (aPC)
          Resistance 432
21.1.4.2 Natural Anticoagulants Deficiency: Protein C, S and
          Antithrombin III 436
21.1.4.3 Prothrombin G20210A Gene Mutation 437
21.1.4.4 Lipoprotein A 438
21.1.4.5 Other Factors 438
21.1.5   Conclusions 439
          References 440

21.2   Central Retinal Vein Occlusion
         L.L. HANSEN 443
21.2.1   Background 443
21.2.2   Epidemiology 443
21.2.3   Etiology and Pathogenesis 444
21.2.3.1 Systemic Risk Factors 444
21.2.3.2 Local Risk Factors 445
21.2.3.3 Pathogenesis of CRVO 446
21.2.4   Clinical Features 447
21.2.4.1 Symptoms and Funduscopic Features 448
21.2.4.2 Classification of CRVO 450
21.2.4.3 Diagnosis 452
21.2.4.4 Differential Diagnosis 453
21.2.4.5 Natural Course 454
21.2.5   Treatment 456
21.2.5.1 Early Treatment 457
21.2.5.2 Late Treatment 461
References 461

21.3   Branch Retinal Vein Occlusion
         H. HOERAUF 467
21.3.1   History, Epidemiology and Classification 467
21.3.2   Anatomy and Histopathology 469
21.3.3   Pathogenesis 471
21.3.4   Clinical Appearance and Symptoms 472
21.3.4.1 Early Findings 472
21.3.4.2 Late Findings and Complications 476
21.3.5   Clinical Evaluation and Diagnostic Methods 478
21.3.5.1 Visual Function and Perimetry 478
21.3.5.2 Angiographic Features 479
21.3.5.3 Optical Coherence Tomography 480
21.3.5.4 General Medical Examination and Laboratory Parameters 481
21.3.6   Natural Course 481
21.3.7   Associated Systemic Disorders and Risk Factors 483
21.3.7.1 Systemic Risk Factors 483
21.3.7.2 Local Risk Factors 484
21.3.7.3 Hematologic Risk Factors 484
21.3.8   Treatment 485
21.3.8.1 Improvement of Hemodynamic Properties and Secondary
         Prevention 485
21.3.8.2 Reduction of Macular Edema 486
21.3.8.3 Treatment of Neovascular Related Complications 495
21.3.9   Conclusions 498
21.3.9.1 Recommendations for Therapy 498
References 499

21.4   Retinal Arterial Occlusion
         M. BURTON, Z. GREGOR 507
21.4.1   Epidemiology 507
21.4.2   Mechanisms of Retinal Arterial Obstruction 507
21.4.2.1 Emboli 507
21.4.2.2 Thrombosis 508
21.4.2.3 Thrombophilia 508
21.4.2.4 Vasculitis 509
21.4.2.5 Infectious 509
21.4.2.6 Other Causes 509

21.4.2.7 Retinal Arterial Occlusion in the Young 510
21.4.3   Clinical Features 510
21.4.3.1 Anatomical Classification 510
21.4.3.2 History 510
21.4.3.3 Examination 510
21.4.4   Differential Diagnosis 512
21.4.5   Systemic Clinical Assessment 513
21.4.6   Investigations 513
21.4.6.1 Investigations to Confirm the Diagnosis of Retinal Arterial
          Occlusion 513
21.4.6.2 Investigations of the Cause of Retinal Arterial Occlusion 513
21.4.7   Pathology 514
21.4.8   Management 514
21.4.8.1 Lie Patient Flat 514
21.4.8.2 Ocular Massage 514
21.4.8.3 Anterior Chamber Paracentesis 514
21.4.8.4 Pharmacological Reduction of Intraocular Pressure 515
21.4.8.5 Pharmacological Vasodilatation 515
21.4.8.6 Carbogen 515
21.4.8.7 Hyperbaric Oxygen 515
21.4.8.8 Steroids for Temporal Arteritis 515
21.4.8.9 Thrombolysis 515
21.4.8.10 Secondary Prevention 516
21.4.9   Outcome and Follow-up 516
References 516

21.5   **The Ocular Ischemic Syndrome**
          G.C. BROWN, M.M. BROWN 519
21.5.1   Pathophysiology 519
21.5.2   Demography 520
21.5.3   Clinical Features 520
21.5.4   Ancillary Diagnostic Studies 524
21.5.5   Systemic Associations 524
21.5.6   Therapeutic Modalities 525
21.5.7   Differential Diagnosis 526
References 526

22   **Vascular Abnormalities**

22.1   **Idiopathic Juxtafoveolar Retinal Telangiectasis**
          D. PAULEIKHOFF, B. PADGE 528
22.1.1   History 528
22.1.2   Clinical Course of the Disease 528
22.1.3   Electron Microscopic and Light Microscopic Changes 532
22.1.4   Natural Course of IJRT 532
22.1.5   Association with Systemic Diseases and Differential Diagnosis ... 532
22.1.6   Therapy 533
References 533

22.2   **Congenital Arteriovenous Communications and Wyburn-Mason
          Syndrome**
          A. WESSING 535
22.2.1   History 535
22.2.2   Classification 536
22.2.3   Clinical Features 536
22.2.4   Fluorescein Angiography 538

22.2.5   Differential Diagnosis 538
22.2.6   Natural Course 539
22.2.7   Histopathology 539
22.2.8   Systemic Involvement 539
22.2.9   Genetics 540
22.2.10   Therapy 540
References 540

22.3   **Retinal Arterial Macroaneurysms**
          S. Bopp 543
22.3.1   Clinical Presentation 543
22.3.1.1 Typical Clinical Findings 543
22.3.1.2 Special Clinical Findings 544
22.3.1.3 Clinical Symptoms 544
22.3.2   Epidemiologic Data and Risk Factors 545
22.3.3   Pathogenesis and Pathomorphology 545
22.3.4   Natural History 545
22.3.5   Terminology and Classification 547
22.3.6   Diagnosis and Imaging 547
22.3.7   Differential Diagnosis 548
22.3.8   Therapy 549
22.3.8.1 Exudative Complications 549
22.3.8.2 Hemorrhagic Complications 550
22.3.8.3 Pearls and Pitfalls of Surgery for Hemorrhagic RAMs 552
22.3.9   Conclusions and Practical Recommendations 557
22.3.10   Personal Approach to Symptomatic RAM 558
References 558

22.4   **Coats' Disease**
          A. Schueler, N. Bornfeld 561
22.4.1   Introduction 561
22.4.2   Classification 562
22.4.3   Differential Diagnosis 563
22.4.4   Treatment 564
22.4.5   Future Prospects 565
References 565

22.5   **Familial Exudative Vitreoretinopathy**
          A.M. Joussen, B. Kirchhof 567
22.5.1   History 567
22.5.2   Special Pathological Features 567
22.5.3   Genetics and Molecular Mechanisms 570
22.5.4   Clinical Course of the Disease 571
22.5.5   Differential Diagnosis 574
22.5.6   Treatment Recommendations Including Follow-up 576
22.5.6.1 General Recommendations for Laser Photocoagulation in FEVR 576
22.5.6.2 Indications for Vitrectomy in FEVR 577
22.5.6.3 General Recommendations for Vitrectomy in FEVR 577
22.5.6.4 Surgical Approach to Falciform Detachment in FEVR 579
22.5.7   Treatment Options Under Investigation 580
References 580

23   **Vasculopathy After Treatment of Choroidal Melanoma**
          B. Damato 582
23.1   Uveal Melanoma 582
23.2   Radiotherapy 583

23.2.1 Radiotherapy Techniques 583
23.2.2 Radiation Vasculopathy 583
23.2.3 Treatment of Radiation Vasculopathy 584
23.3 Phototherapy 588
23.3.1 Phototherapy Techniques 588
23.3.2 Vasculopathy After Phototherapy 588
23.3.3 Treatment of Vasculopathy After Phototherapy 589
23.4 Local Resection 589
23.4.1 Local Resection Techniques 589
23.4.2 Vasculopathy After Local Resection 589
23.4.3 Treatment of Vasculopathy After Local Resection 590
23.5 Conclusion 590
References 591

24 Vasculopathies with Acute Systemic Diseases

24.1 Purtscher's Retinopathy

D.V. Do, A.P. Schachat 592
24.1.1 History 592
24.1.2 Special Pathologic Features 592
24.1.3 Clinical Course of the Disease 593
24.1.4 Differential Diagnosis 593
24.1.5 Treatment Recommendations Including Follow-up 594
References 594

24.2 Terson Syndrome

F. Kuhn, R. Morris, V. Mester 595
24.2.1 History 595
24.2.2 Pathophysiology and Pathoanatomy 595
24.2.2.1 The Origin of the Intraocular Blood 595
24.2.2.2 The Type and Location of the Intraocular Blood 596
24.2.3 Epidemiology 597
24.2.4 Significance 598
24.2.4.1 Natural History of Vitreous Hemorrhage in Eyes with Terson
         Syndrome 598
24.2.4.2 Systemic Significance 598
24.2.4.3 Human Significance 598
24.2.5 Diagnosis 598
24.2.5.1 Differential Diagnosis 599
24.2.6 Treatment 599
24.2.6.1 Indication and Counseling 599
24.2.6.2 Surgery 599
24.2.6.3 Complications 599
24.2.6.4 Follow-up 600
24.2.7 Summary 600
References 600

24.3 Disseminated Intravascular Coagulopathy

M.M. Lai, A. Schachat 602
24.3.1 Introduction 602
24.3.2 History 602
24.3.3 Pathologic Features 603
24.3.4 Clinical Course 603
24.3.5 Differential Diagnosis 604
24.3.6 Treatment and Follow-up 605
References 605

24.4 Bone Marrow Transplant Associated Retinopathy
H. TABANDEH, N. RAFIEI, A.P. SCHACHAT 606
24.4.1 Introduction 606
24.4.2 Pathophysiology 607
24.4.3 Bone Marrow Transplant Associated Retinopathy 608
24.4.3.1 Microvascular Retinopathy 608
24.4.3.2 Retinal Pigment Epitheliopathy and Choroidopathy 609
24.4.3.3 Hematologic Complications 609
24.4.3.4 Optic Neuropathy 610
24.4.3.5 Infectious Complications 610
24.4.3.6 Other Complications 610
24.4.4 Pathologic Features 611
24.4.5 Management 611
References 611

25 Inflammatory Vascular Disease

25.1 Eales' Disease
S. GADKARI 613
25.1.1 History 613
25.1.2 Epidemiology 613
25.1.3 Clinical Features 613
25.1.3.1 Signs of Inflammation 614
25.1.3.2 Signs of Ischemia 614
25.1.3.3 Signs of Neovascularization and Its Sequelae 616
25.1.3.4 Central Eales' 619
25.1.4 Natural Course 620
25.1.5 Differential Diagnosis 620
25.1.6 Systemic Associations Described in Eales' Disease 620
25.1.7 Attempts at Classification 621
25.1.8 Pathology 621
25.1.9 Etiopathology 621
25.1.10 Management 622
25.1.10.1 Fundus Fluorescein Angiography 622
25.1.10.2 Medical Treatment 622
25.1.10.3 Role of Laser Treatment 622
25.1.10.4 Vitreoretinal Surgery 623
References 626

25.2 Ocular Manifestations of Systemic Lupus Erythematosus
J.T. ROSENBAUM, F. MACKENSEN 628
25.2.1 Epidemiology and Disease Criteria for SLE 628
25.2.2 Frequency and Prognostic Value of Ocular Findings in SLE 629
25.2.3 Special Pathological Features and Molecular Mechanisms 630
25.2.4 Retinopathy in SLE patients: Clinical Picture and Course
of Disease 630
25.2.5 Differential Diagnosis 630
25.2.6 Treatment Recommendations, Follow-up and Recommendations
for Ophthalmologic Screening of SLE Patients 632
References 633

25.3 Behçet's Disease
M. ZIERHUT, N. STÜBIGER, I. KÖTTER, C. DEUTER 635
25.3.1 Epidemiology and Definition of Behçet's Disease 635
25.3.1.1 Definition 635

25.3.1.2 Epidemiology 635
25.3.2 Special Pathological Features 636
25.3.3 Genetics and Molecular Mechanisms 638
25.3.4 Course of the Disease 639
25.3.4.1 Extraocular Manifestations 639
25.3.4.2 Ocular Manifestations 640
25.3.5 Differential Diagnosis 642
25.3.6 Treatment 643
25.3.6.1 Corticosteroids 643
25.3.6.2 Immunosuppressive and Cytotoxic Agents 643
25.3.6.3 Novel Medical Treatment Approaches (Biologicals) 644
25.3.6.4 Surgical Treatment 644
25.3.6.5 Recommendations for the Clinician 645
References 645

25.4   **Vasculitis in Multiple Sclerosis**
          M.D. BECKER, U. WIEHLER, D.W. MILLER 650
25.4.1 History 650
25.4.2 Epidemiology 650
25.4.3 Pathogenesis 650
25.4.4 Intermediate Uveitis and Retinal Vasculitis as the Clinical
        Manifestations in MS-Associated Uveitis 651
25.4.5 Differential Diagnosis 653
25.4.6 Therapeutic Options 653
25.4.6.1 Corticosteroids and Immunosuppression 653
25.4.6.2 Immunomodulatory Therapy with Interferon for Cystoid
          Macular Edema 653
25.4.6.3 Laser Photocoagulation 654
25.4.6.4 Treatment Strategy 654
References 655

25.5   **Sarcoidosis**
          S. SIVAPRASAD, N. OKHRAVI, S. LIGHTMAN 657
25.5.1 Epidemiology 657
25.5.2 Etiology 657
25.5.3 Molecular Mechanisms 657
25.5.4 Pathology 657
25.5.5 Special Pathological Findings 658
25.5.5.1 Systemic Features 658
25.5.5.2 Ocular Sarcoidosis 658
25.5.5.3 Ocular Sarcoidosis in Children 662
25.5.6 Diagnosis of Sarcoidosis 663
25.5.6.1 Serum Angiotensin Converting Enzyme 663
25.5.6.2 Chest Radiographs 663
25.5.7 Clinical Course and Prognosis 663
25.5.7.1 Ocular Disease 663
25.5.7.2 Systemic Disease 664
25.5.8 Treatment of Ocular Sarcoidosis 664
25.5.9 Treatment of Systemic Sarcoidosis 664
25.5.10 Conclusion 665
References 665

25.6   **Necrotizing Vasculitis**
          J.L. DAVIS 668
25.6.1 Polyarteritis Nodosa and Microscopic Polyangiitis 668
25.6.1.1 Synonyms and Related Conditions 668

25.6.1.2 Histopathology 668
25.6.1.3 Systemic Course of PAN 668
25.6.1.4 Ocular Manifestations of PAN 669
25.6.2 Churg-Strauss Syndrome 670
25.6.2.1 Synonyms and Related Conditions 670
25.6.2.2 Histopathology 670
25.6.2.3 Systemic Course of Disease 670
25.6.2.4 Ocular Manifestations 670
25.6.3 Wegener Granulomatosis 670
25.6.3.1 Synonyms and Related Conditions 670
25.6.3.2 Histopathology 672
25.6.3.3 Systemic Course of Disease 672
25.6.3.4 Ocular Manifestations 672
25.6.4 Differential Diagnosis 673
25.6.5 Treatment 673
25.6.6 Prognosis 673
References 673

25.7 Systemic Immunosuppression in Retinal Vasculitis and Rheumatic Diseases

J.J. HUANG, C.S. FOSTER 675
25.7.1 Introduction 675
25.7.2 Diagnosis, Imaging and Electrophysiology 676
25.7.2.1 Laboratory Tests 677
25.7.3 Treatment 678
25.7.3.1 Corticosteroids 678
25.7.3.2 Nonsteroidal Anti-inflammatory Drugs 678
25.7.3.3 Immunosuppressive Agents 679
25.7.4 Immunosuppressive Therapy in Children 685
25.7.5 Combination Therapy 686
25.7.6 Conclusion 686
References 686

26 Hypertensive Retinopathies

26.1 General Basics of Hypertensive Retinopathy

S. WOLF 688
26.1.1 Pathophysiology of the Retinal Vessels in Arterial Hypertension .. 688
26.1.2 Fundus Changes in Arterial Hypertension 689
26.1.3 Fundus Changes in Hypertensive Retinopathy 689
26.1.4 Clinical Diagnoses in Hypertensive Retinopathy 689
26.1.5 Classification of Fundus Changes in Arterial Hypertension 689

26.2 Pregnancy-Induced Hypertension (Preeclampsia/Eclampsia)

T.R. KLESERT, A.P. SCHACHAT 691
26.2.1 History 691
26.2.2 Hypertension in Pregnancy 691
26.2.3 Systemic Complications of PIH 692
26.2.4 Ocular and Neurologic Manifestations 693
26.2.5 Pathophysiology and Epidemiology 694
26.2.6 Diagnostic Evaluation 695
26.2.7 Treatment 696
26.2.8 Clinical Course and Outcomes 697
References 698

27 Sickle Cell Retinopathy and Hemoglobinopathies

27.1 Histopathology of Sickle Cell Retinopathy
G.A. LUTTY 700
27.1.1 Introduction 700
27.1.2 Nonproliferative Changes 700
27.1.2.1 Retinal Vessel Occlusions and Remodeling 700
27.1.2.2 Causes of Vaso-occlusions 705
27.1.2.3 Hemorrhage, Schisis Cavities, Iridescent Spots, and Black Sunbursts 706
27.1.3 Proliferative Retinopathy 707
27.1.4 Choroidopathy 709
27.1.5 Conclusions 710
References 710

27.2 Retinal Vascular Disease in Sickle Cell Patients
J.C. VAN MEURS, A.C. BIRD, S.M. DOWNES 712
27.2.1 Introduction to Sickle Cell Disease 713
27.2.1.1 Normal Hemoglobin and Sickle Hemoglobin 713
27.2.1.2 Determinants of HbS Polymerization and Patient Categories 713
27.2.1.3 Combinations with Thalassemia 714
27.2.1.4 Current Concepts in the Pathogenesis of Vaso-occlusion 714
27.2.1.5 Pathophysiology of Ocular Vaso-occlusion 714
27.2.2 Retinal Vascular Disease in Sickle Patients 714
27.2.2.1 Introduction 714
27.2.2.2 Database 714
27.2.3 Choroid 715
27.2.3.1 Choriocapillary 715
27.2.4 Retinal Vessel Occlusions 715
27.2.4.1 Major Retinal Vessel Occlusions 715
27.2.4.2 Retinal Vein Occlusions 716
27.2.5 Retinal Hemorrhages, Salmon Patches, Iridescent Spots and Black Sunbursts 716
27.2.5.1 Reports 716
27.2.5.2 Course 717
27.2.5.3 Relevance to Function 718
27.2.6 Capillary Occlusions of the Posterior Pole 718
27.2.6.1 Natural History 718
27.2.6.2 Relevance of These Findings to Function 718
27.2.7 Peripheral Vessel Occlusions 718
27.2.7.1 Reports 718
27.2.7.2 Natural History 719
27.2.7.3 Relevance of These Findings to Function 719
27.2.8 Proliferative Sickle Retinopathy 719
27.2.8.1 Classification of Sickle Cell Retinopathy 719
27.2.8.2 Reports of Proliferative Sickle Retinopathy 719
27.2.8.3 Non-peripheral Proliferative Lesions 721
27.2.8.4 Natural History of Proliferative Sickle Retinopathy 721
27.2.8.5 Clinical Correlations 723
27.2.8.6 Differential Diagnosis of Proliferative Sickle Retinopathy 723
27.2.9 Prophylactic Treatment of PSR.......................... 725
27.2.9.1 Reports 725
27.2.10 Epiretinal Membranes 726
27.2.10.1 Etiology.. 726
27.2.10.2 Relevance to Function 726
27.2.11 Macular Holes 726

27.2.12 Treatment of Vitreous Hemorrhages and Retinal Detachments .. 727
27.2.12.1 Anterior Segment Ischemia 727
27.2.12.2 Surgical Reports 727
27.2.12.3 Comment 728
27.2.12.4 Measures to Consider in Vitreoretinal Surgery 728
27.2.13 Blindness Caused by Sickle Cell Disease 728
27.2.13.1 Reports 728
References 732

28 Vascular Tumors of the Retina

28.1 Histopathology of Retinal Vascular Tumors and Selected
Vascular Lesions
M.A. CHANG, W.R. GREEN 735
28.1.1 Cavernous Hemangioma 735
28.1.2 Capillary Hemangioma 736
28.1.3 Retinal Vasoproliferative Tumors 739
28.1.4 Combined Hamartoma of the Retinal Pigment Epithelium
and Retina 741
28.1.5 Racemose Hemangioma 743
28.1.6 Retinal Angiomatous Proliferation 744
References 746

28.2 Retinal Capillary Hemangioma
C.L. SHIELDS, J.A. SHIELDS 749
28.2.1 General Considerations 749
28.2.2 Definition and Incidence 749
28.2.3 Clinical Features 750
28.2.4 Differential Diagnosis 751
28.2.5 Pathology and Pathogenesis 751
28.2.6 Diagnostic Approaches 752
28.2.6.1 Fluorescein Angiography 753
28.2.6.2 Indocyanine Green Angiography 753
28.2.6.3 Ultrasonography 753
28.2.6.4 Optical Coherence Tomography 753
28.2.6.5 Color Doppler Imaging 754
28.2.6.6 Computed Tomography 754
28.2.6.7 Magnetic Resonance Imaging 754
28.2.7 Systemic Evaluation 754
28.2.8 Management 754
28.2.8.1 Ocular 754
28.2.8.2 Systemic 758
28.2.9 Prognosis 758
28.2.10 Summary 758
References 758

28.3 Cavernous Hemangioma
B. JURKLIES, N. BORNFELD 760
28.3.1 Introduction 760
28.3.2 History 760
28.3.3 Pathological Features 761
28.3.4 Clinical Findings and Characteristics of Cavernous Hemangioma 761
28.3.5 Differential Diagnosis 763
28.3.6 Treatment 763
28.3.7 Conclusions 763
References 764

28.4 **Vasoproliferative Retinal Tumor**

B. DAMATO, J. ELIZALDE, H. HEIMANN 766

28.4.1 History 766

28.4.2 Histological Features 766

28.4.3 Pathogenesis 766

28.4.4 Clinical Features 767

28.4.5 Differential Diagnosis 768

28.4.6 Treatment 768

28.4.7 Prognosis 770

References 770

Subject Index 771

List of Contributors

Anthony P. Adamis, MD
Jerini Ophthalmic, 111 West 50th Street, 7th floor
New York NY 10020, USA

Hansjürgen Agostini, MD
Department of Ophthalmology, University of
Freiburg, Killianstrasse 5, 79106 Freiburg, Germany

Lloyd P. Aiello, MD, PhD
Joslin Diabetes Center, One Joslin Place, Boston
MA 02115, USA

Carmen Ruiz de Almodovar, PhD
Center for Transgene Technology and Gene
Therapy, Flanders Interuniversity Institute for
Biotechnology, KULeuven, Campus Gasthuisberg
Herestraat 49, 3000, Leuven, Belgium

David A. Antonetti, PhD
Cellular & Molecular Physiology, Penn State College
of Medicine, 500 University Drive
Hershey, PA 17033, USA

Desmond B. Archer, FRCS, FRCSOph
Department of Ophthalmology, Queen's University
of Belfast, Institute of Clinical Science, Royal
Victoria Hospital, Belfast, BT12 6BA, Northern
Ireland, UK

Alistair J. Barber, PhD
Ophthalmology C4800 H166, Penn State College of
Medicine, 500 University Drive, Hershey, PA 17033
USA

Karl-Ulrich Bartz-Schmidt, MD
Department of Ophthalmology, University of
Tübingen, Schleichstr. 12 – 16, 72076 Tübingen
Germany

Matthias D. Becker, MD, PhD, FEBO
Department of Ophthalmology, University of
Heidelberg, Im Neuenheimer Feld 400
69120 Heidelberg, Germany

Bruce Berkowitz, PhD
Department of Anatomy and Cell Biology/
Department of Ophthalmology, Wayne State
University School of Medicine
540 E. Canfield, Detroit, MI 48201, USA

Alan C. Bird, FRCS, FRCOphth
Moorfields Eye Hospital London, City Road
London, EC1V 2PD, UK

Silvia Bopp, MD
Augenklinik Universitätsallee, Parkallee 301/
Universitätsallee, 28213 Bremen, Germany

Norbert Bornfeld, MD
Department of Vitreoretinal Surgery, University of
Essen, Hufelandstrasse 55, 45122 Essen, Germany

Mike Boulton, PhD
Department of Ophthalmology & Visual Sciences
The University of Texas Medical Branch, 301
University Blvd., Galveston, TX 77555-1106, USA

Gary C. Brown, MD, MBA
Retina Service, Wills Eye Hospital, Jefferson
Medical College, 840 Walnut Street, Philadelphia
PA 19107, USA; and the Center for Value-Based
Medicine, Box 355, Flourtown, PA 19031, USA

Melissa M. Brown, MD, MN, MBA
Department of Ophthalmology, University of
Pennsylvania, Philadelphia, PA; and the Center for
Value-Based Medicine, Box 355, Flourtown
PA 19031, USA

Matthew Burton, MA, PhD, MRCP, MRCOphth
International Centre for Eye Health, Department of
Infectious & Tropical Diseases, London School of
Hygiene and Tropical Medicine, Keppel Street
London, WC1E 7HT, UK

Sergio Caballero, PhD
University of Florida, Academic Research Bldg.
ARB5-250, 1600 SW Archer Rd.
Gainesville FL 32618, USA

Peter Carmeliet, MD, PhD
Center for Transgene Technology and Gene
Therapy, Flanders Interuniversity Institute for
Biotechnology, KULeuven, Campus Gasthuisberg
Herestraat 49, 3000, Leuven, Belgium

Margaret A. Chang, MD, MS
Eye Pathology Laboratory, The Wilmer Eye Institute
and Department of Pathology, Johns Hopkins
Medical Institution, 600 N. Wolfe Street, Baltimore
MD 21287-9248, USA

Bertil Damato, FRCS, FRCOphth
Ophthalmology Department, Link 8Z, Royal
Liverpool University Hospital, Prescot Street
Liverpool, L7 8XP, UK

Janet L. Davis, MD
Bascom Palmer Eye Institute, 900 N.W. 17th Street
Miami, FL 33136, USA

Christoph Deuter, MD
Department of Ophthalmology, University of
Tübingen, Schleichstr. 12–16, 72076 Tübingen
Germany

Diana V. Do, MD
The Wilmer Eye Institute, The Johns Hopkins
University School of Medicine, 600 N. Wolfe Street
Baltimore, MD 21287, USA

Michael I. Dorrell, PhD
Department of Cell Biology, The Scripps Research
Institute, 10550 N. Torrey Pines Rd., La Jolla
CA 92124, USA

Susan M. Downes, FRCOphth
Oxford Eye Hospital, Radcliffe Infirmary
Oxford, OX2 6HE, UK

Javier Elizalde, MD
Ocular Oncology Service, Vitreoretinal Surgery
Centro de Oftalmología Barraquer, Muntaner 314
08021 Barcelona, Spain

Harry W. Flynn, Jr, MD
Bascom Palmer Eye Institute, 900 N.W. 17th Street
Miami, FL 33136, USA

Michael H. Foerster, MD
Department of Ophthalmology, Campus Benjamin
Franklin, Humboldt University Berlin
Hindenburgdamm 30, 12200 Berlin, Germany

Donald Fong, MD, MPH
Ophthalmology, Kaiser Permanente Southern
California, 100 S. Los Robles, Pasadena, CA 91101
USA

C. Stephen Foster, MD, PhD, FACS, FACR
Massachusetts Eye Research and Surgery Institute
5 Cambridge Center, 8th Floor, Cambridge
MA 02142, USA

Martin Friedlander, MD, PhD
Department of Cell Biology, The Scripps Research
Institute, 10550 N. Torrey Pines Rd., La Jolla
CA 92037, USA

Salil Gadkari, MS, FRCS, MRCOpth
Gadkari Eye Center, Opp Karve Road, Telephone
Exchange, Nal Stop, Pune 411004 Pune
Maharashtra, India

Tom Gardiner, PhD
Department of Ophthalmology, Queen's University
of Belfast, Institute of Clinical Science, Royal
Victoria Hospital, Belfast, BT12 6BA, Northern
Ireland, UK

Thomas W. Gardner, MD, MS
Department of Ophthalmology, Penn State College
of Medicine, 500 University Drive Box 850, Hershey
PA 17033, USA

Gerhard Garhöfer, MD
Department of Clinical Pharmacology, Währinger
Gürtel 18–20, 1090 Vienna, Austria

Matthew J. Gastinger, PhD
Ophthalmology C4800 H166, Penn State College of
Medicine, 500 University Drive, Hershey, PA 17033
USA

Aniz Girach, MD
PKC Product Team, Eli Lilly & Co., Erl Wood
Manor, Sunninghill Road, Windlesham, Surrey
GU20 6PH, UK

Maria B. Grant, MD
Department of Pharmacology and Therapeutics
University of Florida, Academic Research Bldg.
ARB5-250, 1600 SW Archer Rd., Gainesville
FL 32618, USA

W. Richard Green, MD, PhD
Eye Pathology Laboratory, The Wilmer Eye Institute
and Department of Pathology, Johns Hopkins
Medical Institution, 600 N. Wolfe Street, Baltimore
MD 21287-9248, USA

Zdenek Gregor, FRCS, FRCOphth
Moorfields Eye Hospital, City Road
London, EC1V 2PD, UK

Julia A. Haller, MD
The Wilmer Eye Institute, The John Hopkins
University School of Medicine, Maumenee 709
600 N. Wolfe Street, Baltimore, MD 21287–9277
USA

Lutz L. Hansen, MD
Department of Ophthalmology, University of
Freiburg, Killianstrasse 5, 79106 Freiburg, Germany

Heinrich Heimann, MD
St. Paul's Eye Unit, Royal Liverpool University
Hospital, Prescot Street, Liverpool, L7 8XP, UK

Horst Helbig, MD
Department of Ophthalmology, University of
Regensburg, Franz-Josef-Strauss-Allee 11
93053 Regensburg, Germany

Hans Hoerauf, MD
Department of Ophthalmology, University of
Göttingen, Robert-Koch-Str. 40, 37075 Göttingen
Germany

John J. Huang, MD
Yale Eye Center, 330 Cedar Street, P.O. Box 208061
New Haven, CT 06520-8061, USA

Claudia Jandeck, MD
Department of Ophthalmology
Campus Benjamin Franklin
Charité Universitätsmedizin Berlin, Germany

Antonia M. Joussen, MD, PhD
Department of Ophthalmology, University of
Düsseldorf, Moorenstr. 5, 40225 Düsseldorf
Germany

Bernhard Jurklies, MD
Department of Vitreoretinal Surgery, University of
Essen, Hufelandstrasse 55, 45122 Essen, Germany

William G. Kaelin, Jr, MD
Dana-Farber Cancer Institute, 44 Binney Street
Mayer 457, Boston, MA 02115, USA

C. Ronald Kahn, MD, PhD
Joslin Diabetes Center, One Joslin Place, Boston
MA 02215, USA

Barrett Katz, MD, MBA
Weill Medical College, Cornell University, New
York-Presbyterian Hospital, 525 East 68th St., New
York, NY 10021, USA; and Fovea Pharmaceuticals
SA, 12 rue Jean-Antoine de Baif, 75013 Paris, France

Timothy S. Kern, PhD
Department of Medicine, Division of Clinical and
Molecular Endocrinology, Center for Diabetes
Research, Case Western Reserve University, 10900
Euclid Ave., Cleveland, OH 44106-4951, USA

George L. King, MD
Joslin Diabetes Center, Harvard Medical School
One Joslin Place, Boston, MA 02215, USA

Bernd Kirchhof, MD
Department of Vitreoretinal Surgery, Center for
Ophthalmology, University of Cologne, Joseph-
Stelzmann-Strasse 9, 50931 Cologne, Germany

Todd Klesert, MD, PhD
Wilmer Eye Institute, Johns Hopkins Hospital
600 N. Wolfe Street, Baltimore, MD 21287, USA

Ina Kötter, MD
Department of Internal Medicine, University of
Tübingen, Ottfried-Müller-Str. 10, 72076 Tübingen
Germany

Ferenc Kuhn, MD, PhD
Helen Keller Foundation for Research and
Education, University of Alabama at Birmingham
1201 11th Avenue South, Birmingham, AL 35205
USA

Michael Lai, MD, PhD
Associated Retinal Consultants, William Beaumont
Hospital Medical Building, Suite 632, 3535 W.
Thirteen Mile Road, Royal Oak, MI 48073, USA

Gabriele E. Lang, MD
Department of Ophthalmology, University of Ulm
Prittwitzstr. 43, 89075 Ulm, Germany

Michael Larsen, MD, DMSc
Department of Ophthalmology, Herlev Hospital
University of Copenhagen, Copenhagen, Denmark

Susan Lightman, FRCP, FRCOphth, PhD
Department of Clinical Ophthalmology, Institute of
Ophthalmology, Moorfields Eye Hospital London
City Road, London, EC1V 2PD, UK

Gerard A. Lutty, PhD
The Wilmer Eye Institute, Johns Hopkins Hospital
600 N. Wolfe Street, Baltimore, MD 21287-9115
USA

Friederike Mackensen, MD
Department of Ophthalmology, University of
Heidelberg, Im Neuenheimer Feld 400
69120 Heidelberg, Germany

Gottfried Martin, PhD
Department of Ophthalmology, University of
Freiburg, Killianstrasse 5, 79106 Freiburg, Germany

Trevor J. McFarland, BS
Casey Eye Institute, Oregon Health and Science
University, 3375 SW Terwilliger Blvd., Portland
OR 97239, USA

Viktoria Mester, MD
General Authority of Health Services, for the
Emirates of Abu Dhabi Middle Region, Mafraq
Hospital, PO Box 2951, Abu Dhabi, United Arab
Emirates

Daniel W. Miller, MD
Department of Ophthalmology, University of
Heidelberg, Im Neuenheimer Feld 400
69120 Heidelberg, Germany

Susanne Mohr, PhD
Department of Ophthalmology, Biomedical
Research Building, 434, School of Medicine, Case
Western Reserve University, 10900 Euclid Ave.
Cleveland, OH 44106–4951, USA

Robert Morris, MD
Helen Keller Foundation for Research and
Education, University of Alabama at Birmingham
1201 11th Avenue South, Birmingham, AL 35205, USA

Quan Don Nguyen, MD, MSc
The Wilmer Eye Institute, The Johns Hopkins
Hospital, 600 N. Wolfe Street, Baltimore, MD 21287
USA

Annelii Ny, PhD
Center for Transgene Technology and Gene
Therapy, Flanders Interuniversity Institute for
Biotechnology, KULeuven, Campus Gasthuisberg
Herestraat 49, 3000, Leuven, Belgium

Narciss Okhravi, BSc, FRCOphth, PhD
Department of Clinical Ophthalmology, Institute of
Ophthalmology, Moorfields Eye Hospital London
City Road, London, EC1V 2PD, UK

Björn Padge, MD
Department of Ophthalmology, St. Franziskus-
Hospital, Hohenzollernring 74, 48415 Münster
Germany

Daniel Pauleikhoff, MD
Department of Ophthalmology, St. Franziskus-
Hospital, Hohenzollernring 74, 48415 Münster
Germany

Brett E. Phillips, PhD
Cellular & Molecular Physiology, Penn State College
of Medicine, 500 University Drive, Hershey
PA 17033, USA

Vasiliki Poulaki, MD, PhD
Massachusetts Eye and Ear Infirmary, Harvard
Medical School, 243 Charles Street, Boston
MA 02114, USA

Polly Quiram, MD, PhD
Associated Retinal Consultants, William Beaumont
Hospital Medical Building, Suite 632, 3535 W.
Thirteen Mile Road, Royal Oak, MI 48073, USA

Nastaran Rafiei, MD
Repartment of Neurology, University of California –
Irvine California, USA

Narsing A. Rao, MD
Department of Ophthalmology, Doheny Eye Institute
University of Southern California, Los Angeles
California, USA

Johann Roider, MD
Department of Ophthalmology, University of
Schleswig-Holstein, Kiel Campus, Hegewischstr. 2
24105 Kiel, Germany

James T. Rosenbaum, MD
Casey Eye Institute, Oregon Health & Science
University, 3375 SW Terwilliger Blvd., Portland
OR 97239-4146, USA

Florian Rüfer, MD
Department of Ophthalmology, University of
Schleswig-Holstein, Kiel Campus, Hegewischstr. 2
24105 Kiel, Germany

Stephen J. Ryan, MD
Doheny Eye Institute, University of Southern
California, 1450 San Pablo Street, Los Angeles
CA 90033, USA

Srinivas R. Sadda, MD
Department of Ophthalmology, Doheny Eye
Institute, University of Southern California
1450 San Pablo Street, Los Angeles, CA 90033
USA

Andrew P. Schachat, MD
The Cole Eye Institute, The Cleveland Clinic
Foundation, 9500 Euclid Avenue, Cleveland
Ohio 44195 USA

Torsten Schlote, MD, PhD
Augenzentrum der Klinik Pallas, Louis-Giroud-
Strasse 20, 4600 Olten, Switzerland

Leopold Schmetterer, PhD
Department of Clinical Pharmacology, Währinger
Gürtel 18–20, 1090 Vienna, Austria

Michael Scholz, PhD
Institute of Anatomy, University of Erlangen
Universitätsstrasse 19, 91054 Erlangen, Germany

Andreas Schüler, MD
Augenklinik Universitätsallee, Parkallee 301/
Universitätsallee, 28213 Bremen, Germany

Ingrid U. Scott, MD, MPH
Department of Ophthalmology, Penn State College
of Medicine, 500 University Drive, HU19, Hershey
PA 17033-0850, USA

Nilanjana Sengupta, PhD
University of Florida, Academic Research Bldg.
ARB5-250, 1600 SW Archer Rd, Gainesville
FL 32618, USA

Syed Mahmood Shah, MD
The Wilmer Eye Institute, The Johns Hopkins
Hospital, 600 N. Wolfe Street, Baltimore, MD 21287
USA

Carol L. Shields, MD
Ocular Oncology Service, Department of Ocular
Oncology, Wills Eye Hospital, 840 Walnut Street
Philadelphia, PA 19107; Thomas Jefferson
University, Philadelphia, Pennsylvania, USA

Jerry A. Shields, MD, PhD
Ocular Oncology Service, Department of Ocular
Oncology, Wills Eye Hospital, 840 Walnut Street
Philadelphia, PA 19107; Thomas Jefferson
University, Philadelphia, Pennsylvania, USA

Sobha Sivaprasad, MS, DNB, FRCS
Department of Clinical Ophthalmology, Institute of
Ophthalmology, Moorfields Eye Hospital London
City Road, London, EC1V 2PD, UK

Lois E.H. Smith, MD, PhD
Department of Ophthalmology, Children's Hospital
Harvard Medical School, 300 Longwood Avenue
Boston, MA 02115, USA

Wael Soliman, MD
Department of Ophthalmology, Herlev Hospital
University of Copenhagen, Herlev Ringvej 75
2730 Herlev, Denmark

Alan W. Stitt, PhD
Department of Ophthalmology, Queen's University
of Belfast, Institute of Clinical Science, Royal
Victoria Hospital, Belfast, BT12 6BA, Northern
Ireland, UK

J. Timothy Stout, MD, PhD
Casey Eye Institute, 3375 SW Terwilliger Blvd.
Portland, OR 97237-4197, USA

Nicole Stübiger, MD
Department of Ophthalmology, University of
Tübingen, Schleichstr. 12 – 16, 72076 Tübingen
Germany

Jennifer K. Sun, MD
Section of Eye Research, Joslin Diabetes Center
Harvard Medical School, One Joslin Place, Boston
MA 02215, USA

Kiyoshi Suzuma, MD
Department of Ophthalmology and Visual Sciences
Graduate School of Medicine, Kyoto University
54 Kawaracho Shogoin, Sakyo-ku Kyoto 606-8507
Japan

Homayoun Tabandeh, MD, MS, FRCP, FRCOphth
Retina-Vitreous Associates Medical Group
8641 Wilshire Blvd., #210 Beverly Hills, CA 90211
USA

Ernst R. Tamm, MD
Institute of Human Anatomy and Embryology
University of Regensburg, Universitätsstr. 31,
93053 Regensburg, Germany

Eric Tourville, MD
Department of Ophthalmology, The Wilmer Eye
Institute, Johns Hopkins University, 600 N. Wolfe
Street, Baltimore, MD 21287, USA

Michael T. Trese, MD
Associated Retinal Consultants, William Beaumont
Hospital Medical Building, Suite 344, 3535 W.
Thirteen Mile Road, Royal Oak, MI 48073, USA

Heather D. VanGuilder, BSc
Ophthalmology C4800 H166, Penn State College of
Medicine, 500 University Drive, Hershey, PA 17033
USA

Jan van Meurs, MD, PhD
The Rotterdam Eye Hospital, Schiedamsevest 180
3011, BH Rotterdam, The Netherlands

Alexander Walsh, MD
Department of Ophthalmology, Doheny Eye
Institute, University of Southern California, 1450
San Pablo Street, Los Angeles, CA 90033, USA

Achim Wessing, MD
Horster Strasse 115, 45968 Gladbeck, Germany

Ute Wiehler, MD
Department of Ophthalmology, University of
Heidelberg, Im Neuenheimer Feld 400,
69120 Heidelberg, Germany

Sebastian Wolf, MD, PhD
Department of Ophthalmology, Inselspital
University of Bern, 3010 Bern, Switzerland

Manfred Zierhut, MD
Department of Ophthalmology, University of
Tübingen, Schleichstr. 12 – 16, 72076 Tübingen
Germany

Section I
Pathogenesis of Retinal Vascular Disease

1 Functional Anatomy, Fine Structure and Basic Pathology of the Retinal Vasculature

D.B. ARCHER, T.A. GARDINER, A.W. STITT

Core Messages
- The retinal vasculature is arranged in a three-dimensional network. Abnormalities of retinal circulation are key pointers to retinal dysfunction and disease and may highlight perturbations of systemic circulation such as diabetes, hypertension and sickle cell disease
- There are specific responses of the retina and its vasculature to stress and disease, combining hemodynamic changes, oxygen saturation changes, occlusion and ischemia, and repair and remodeling. Furthermore, there are specific reactions to trauma, radiational damage, drug toxicity, and inflammation

1.1 Anatomical Organization of the Retinal Vasculature

1.1.1 Microvascular Arrangement

The densely cellular retina with its intricate arrangement of neurons requires highly specialized circulations to meet its demanding metabolic requirements without compromising its extracellular space, which is a highly defined microenvironment conducive to neurotransmission, phototransduction and the complex interaction of metabolites, growth factors and vasoactive agents. The retinal circulation, which supplies the inner retina, is observed ophthalmoscopically as a regular geometrically arranged network of vessels and their three-dimensional complexity reflects the cellular density of the retinal neuropile. The caliber of directly viewed vessels is determined by the size of the red cell column, as the vessel walls and peripheral plasma layer are virtually transparent. The vessels accordingly appear wider during fluorescein angiography as dye mixes with the luminal plasma. As the vessel walls sclerose with age, stress or disease processes they become visible, due to reflected light on ophthalmoscopy, and obscure the red cell column to varying degrees. Abnormalities of the retinal circulation are key pointers to retinal dysfunction and disease and frequently highlight perturbations of the systemic circulation, e.g., diabetes, hypertension, and sickle cell disease.

The central retinal artery is a direct branch of the ophthalmic artery arising from bifurcations adjacent

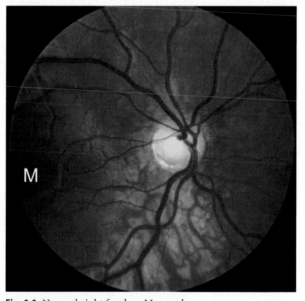

Fig. 1.1. Normal right fundus. *M* macula

to the optic disk to form a unique, intraretinal, end-artery microvascular system (Fig. 1.1). The development of the blood supply is intimately linked with embryological progress of the neural retina and other intraocular structures, such as the lens [29], developing a complex network in unison with oxygen demands and local tissue gradients of vasoactive growth factors such as vascular endothelial growth factor (VEGF) [7, 27, 36]. Oxygen tension in the mammalian inner retina has been identified as a key regulator of retinal cell differentiation and microvascular permeability, growth and survival by altering expression of VEGF [2, 23, 25], controlled, in part, by cellular oxygen and activation of the transcription factor hypoxia inducible factor-1α (HIF-1α) [9]. In the mature retinal microvasculature, this oxygen-regulated control of capillary density is well illustrated by the appearance of a so-called "capillary-free zone" adjacent to arterial walls where oxygen tension is high and local expression of VEGF is low [8] (Figs. 1.2, 1.3).

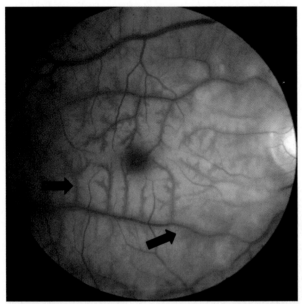

Fig. 1.2. Right fundus – central retinal artery obstruction: immediate periarteriolar retina survives (*arrows*) due to arteriolar oxygen "leak" and is transparent; ischemic retina is edematous and opaque

Cilioretinal arteries may also contribute to the retinal circulation and these originate from posterior ciliary vessels. The cilioretinal system typically supplies the macular region or part of the macular region of the neuropile and it is notable that these vessels adopt a "retinal structure" typified by tight junctions and barrier properties (Fig. 1.4).

Branch retinal "arteries" lack an internal elastic lamina and, anatomically speaking, are arterioles. These arterioles are ~200 μm in diameter and in the peripheral retina bifurcate to third and fourth orders and finally to pre-capillary arterioles (Fig. 1.5) that lack a pre-capillary sphincter that is a feature of the rodent retinal vasculature. However, this arrangement is altered in the central retina where small pre-capillary arterioles emerge abruptly from the larger radial arteries. These vessels merge seamlessly into highly complex capillary networks consisting of two

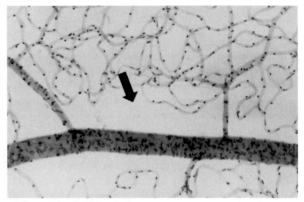

Fig. 1.3. Trypsin digest preparation shows the capillary-free zone adjacent to a retinal artery (*arrow*)

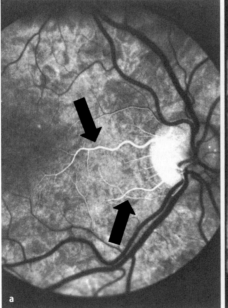
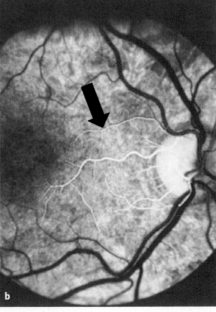

Fig. 1.4. a Cilioretinal artery system (*arrows*) supplies right nasal macula and fills coincident with choroidal circulation and before retinal arteries on fluorescein angiography. **b** Cilioretinal vain (*arrow*)

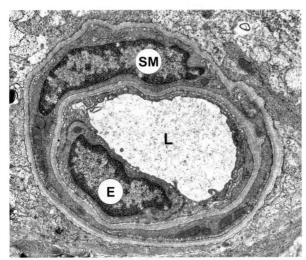

Fig. 1.5. A retina pre-capillary arteriole has capillary dimensions but is enveloped by a single continuous layer of smooth muscle (*SM*). *E* endothelial cell, *L* lumen

to four plexi, reflecting the thickness of inner retinal neuropile and metabolic demands of the neurons. The peripapillary retina has four layers of capillaries while the macular and peripheral retinas have three and one to two respectively. There is a capillary-free zone at the fovea, where the inner retinal neurons and their processes show lateral displacement to allow unobstructed passage of light to the midget central cones for accurate resolution of visual images (Figs. 1.6, 1.7).

Each capillary unit is ~10–15 μm in diameter and consists of a continuous endothelium surrounded by pericytes, and both cell types are in direct communication via gap junctions and exchange paracrine signals through their shared basement membrane (BM) (Fig. 1.8). This BM acts as an important regulatory matrix for passage and sequestration of vasoactive agents and pro-survival growth factors in humans. Retinal pericytes occur in a 1:1 ratio with endothelial cells, which is a unique feature of this microvasculature and translates to an even greater pericyte coverage than in brain capillaries [10]. These cells have been shown to possess contractile properties [6] and, although of less importance than the smooth muscle cells of the upstream pre-capillary arterioles, may exert a significant influence on retinal hemodynamics and fine control of blood flow through the capillaries.

The capillary flow drains into the venular system, which is localized in the deeper retina (inner plexiform layer), and eventually to the retinal veins which, characteristically, lack a well-developed smooth muscle covering and have a larger diameter

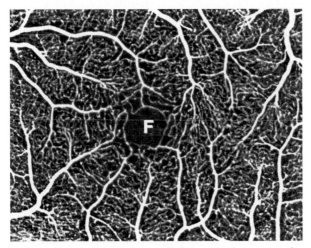

Fig. 1.6. Fluorescein angiogram of left macula shows avascular foveola (*F*)

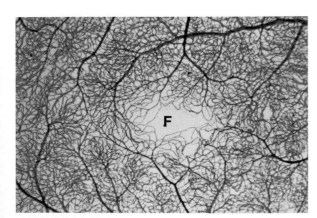

Fig. 1.7. Trypsin digest of right macula and avascular foveola (*F*)

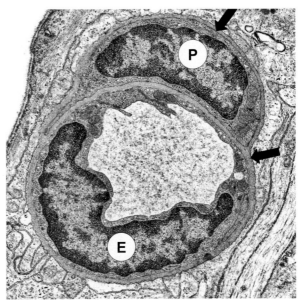

Fig. 1.8. Electron micrograph of typical retinal capillary showing endothelial cell (*E*) and pericyte (*P*) within the cohesive basement membrane (*arrows*)

1

than arteries (30–300 µm). The central retinal vein lies within the optic nerve head and is drained by the ophthalmic vein and cavernous sinus. It is notable that in a number of species the retinal vasculature shows arteriovenous (A-V) crossings where the arterial and venous partners share a common basement membrane (Figs. 1.9, 1.10). At such crossings there is intimate juxtaposition of arterial and venous vascular cells (smooth muscle and endothelial cells) and opportunity for cellular "cross talk"; however, the significance of this is not known. Although the possible physiological regulatory advantages of A-V crossings have not been identified, there is no doubt concerning their role in retinal vascular pathology:

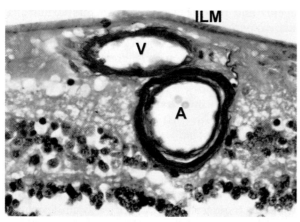

Fig. 1.9. A hematoxylin and eosin stained histological section of an arteriovenous crossing from human retina shows that the vein (*V*) lies between its arterial partner (*A*) and the internal limiting membrane (*ILM*)

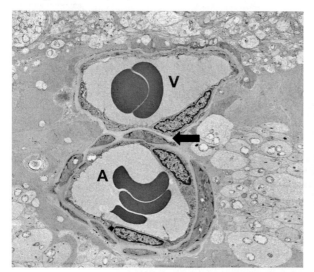

Fig. 1.10. Electron micrograph of an arteriovenous crossing from the retinal vasculature of a dog reveals the shared basement membrane (*arrow*) and juxtaposition of the vascular smooth muscle of the artery (*A*) and the endothelium of the vein (*V*)

Excessive extracellular matrix accumulation in the arterial wall, with age or in response to hemodynamic stress, such as in hypertension, narrows and distorts the venous lumen with aberrations of flow, e.g., eddying, turbulence, and stasis, and increases the possibility of thrombosis and occlusion of the affected vein.

1.1.2 Nature of the Retinal Vasculature

While the retinal vasculature is a classic end-artery system it lacks any obvious autonomic nerve supply and blood flow into the capillary beds is autoregulated in response to the local metabolic needs of the retinal parenchyma. This is achieved, in large part, by the sensitivity of the component smooth muscle cells in the retinal arteries and arterioles to endothelial-generated vasodilators and constrictors, such as nitric oxide (NO), endothelins and bradykinin [30]. The structure of the pre-capillary arterioles and arteries with a highly organized smooth muscle covering, in which component cells are coordinated and directly linked, facilitates sensitive control of luminal diameter and effective autoregulation of retinal blood flow against the high tissue pressure, i.e., intraocular pressure of up to 40–50 mm Hg.

An important normal physiological function of the retinal vasculature is maintenance of the inner blood-retinal barrier (iBRB), which prevents nonspecific permeation of the retinal neuropile by macromolecules yet facilitates exchange of respiratory gases, amino acids, salts, sugars and some peptides. Furthermore, retinal capillaries possess an array of abluminal anionic pumps contained within plasmolemmal caveolae [20] (Fig. 1.11) that assist the removal of excess fluid and waste products from the

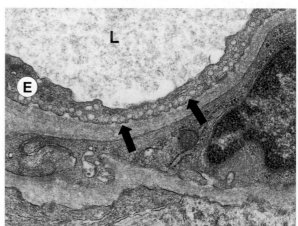

Fig. 1.11. The abluminal plasma membranes of the retinal vascular endothelium are covered by elaborate arrays of caveolae (*arrows*), in contrast to the luminal surface, where these structures are relatively sparse and occur either singly or in small groups. *E* endothelium, *L* lumen

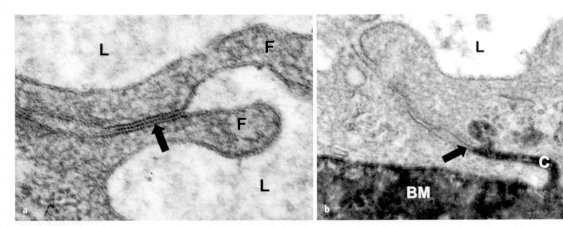

Fig. 1.12. a Electron micrograph of a tight junction between retinal vascular endothelial cells shows fusion of the outer membrane leaflets (*arrow*) to produce the typical pentalaminar arrangement of three electron-dense and two electron-lucent layers. Protruding cytoplasmic folds (*F*) are a regular feature at the luminal aspect of tight junctions in the retinal vascular endothelium. *L* lumen. **b** The position of the tight junction (*arrow*) between contiguous retinal vascular endothelial cells is illustrated in this electron micrograph of a retinal vessel from a rat eye in which an intravitreal injection of horseradish peroxidase serves to stain the extracellular space of the retina, including the vascular basement membranes (*BM*) and intercellular cleft (*C*). An adjacent caveolus on the lateral plasma membrane of one cell is also filled by the tracer. *L* lumen

extracellular space into the retinal circulation. The endothelial cells of the retinal vessels form a continuous, non-fenestrated monolayer, with each cell being fused to juxtaposed neighbors by *zonulae occludens* tight junctions that maintain barrier function (Fig. 1.12a, b). This property of the intraretinal circulation is structurally and functionally analogous to the blood brain barrier and is thought to be maintained in part by the surrounding neuroglia [15] and pericytes that promote tight junction integrity and the non-fenestrated phenotype of the retinal vascular endothelium [3]. When these cell relationships are disrupted as in some pathological situations (diabetes, radiation-retinopathy), the barrier properties of the retinal vessels may be compromised or lost. In such situations the endothelial cells can lose tight-junction integrity and may even assume a fenestrated phenotype (see later in chapter).

In regard to cell turnover and the cell cycle, the endothelial cells and pericytes of the retinal circulation represent highly stable populations with the vast majority of the cells in G0 phase of the cell cycle; however, they are able to reenter the cycle in response to vascular injury, cell loss or angiogenic stimuli. Under physiological conditions there is a low but measurable labeling index with DNA precursors such as tritiated thymidine (^{3}H-T), reflecting an ongoing need for cell replacement. In rats this index is 0.33% for retinal endothelial cells and 0.16% for pericytes [31] (Fig. 1.13). Interestingly in diabetes, where there is accelerated vascular cell death [21], the endothelial cells respond by an increased turnover (^{3}H-T labeling index 0.91%), although the pericytes remained quiescent.

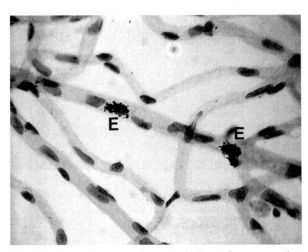

Fig. 1.13. Autoradiogram of a retinal digest from a rat that received a continuous infusion of ^{3}H-thymidine for 8 days. The nuclei of two neighboring endothelial cells (*E*) within the same vessel are heavily decorated with silver grains

1.2 Responses of the Retina and Its Vasculature to Stress and Disease: Histological and Pathological Consequences

1.2.1 Hemodynamic Changes

Increased arteriolar intraluminal pressure induces reactive vessel narrowing probably by stretch-activated calcium channels[22]. This may be focal in nature and have a downstream influence on arterioles and capillary beds. Severe and sustained hypertension and associative arteriolar narrowing (Fig. 1.14) may lead to occlusion of pre-arteriolar branches, subsequent to endothelial damage and insudation of plasma into the vessel wall. Occlusion of downstream retinal vessels may cause impaired axoplasmic transport due to loss of efficient oxidative phosphorylation and ATP depletion in such regions. Clinically such dysfunction is manifested as cotton-wool spots in the inner retina, and electron microscopy reveals swollen axons containing cytoid bodies in the nerve fiber layer [13]. Cytoid bodies are heterogeneous aggregates of vesicular and lysosomal debris, disordered bundles of microfilaments and dense accumulations of mitochondria in a "log-jam" like event resulting from the failure of multiple microtubule-linked molecular motors where their course traversed the area of ischemia (Fig. 1.15a, b).

Normalization of intravascular pressure results in recovery of competence and sometimes local reorganization of the affected microvasculature; but the legacy of focal capillary fallout often persists in the form of microaneurysms, persistent inner retina-exudates and focal reactive microgliosis. Fibrosis of the vascular wall and varying degrees of obscuration of the columns of red blood cells ("copper/silver wiring" effects) can occur and this is often linked to occlusion of downstream capillaries and non-perfusion (Fig. 1.16).

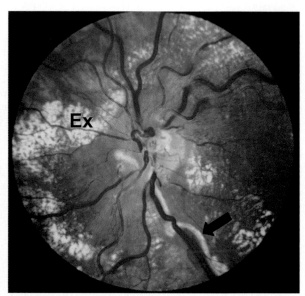

Fig. 1.16. Left fundus – hypertension – retinal vein obstruction – exudation (*Ex*) and "silver wiring" (*arrow*) of inferotemporal artery

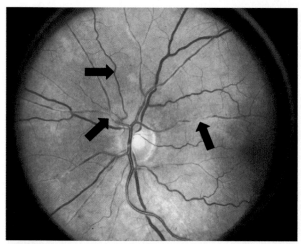

Fig. 1.14. Left fundus patient with accelerated hypertension shows widespread focal narrowing of peripapillary arterioles (*arrows*)

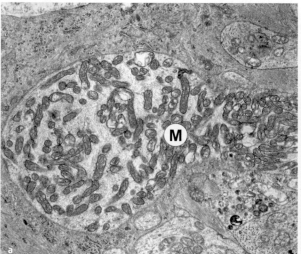

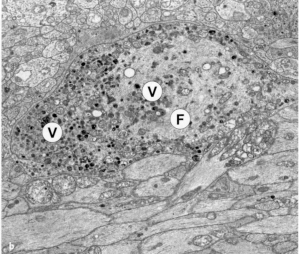

Fig. 1.15. a The nerve fiber layer of a rat with experimental radiation retinopathy shows a distended axon filled with mitochondria (*M*) indicating impaired axoplasmic transport in an area of retinal ischemia. **b** A cytoid body in a swollen axon adjacent to that depicted in **a** contains disorganized aggregates of filamentous material (*F*) and heterogeneous vesicular bodies (*V*)

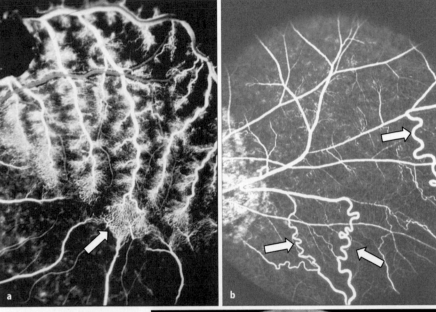

Fig. 1.17. a Experimental branch vein occlusion: Venovenous collaterals (*arrow*) across horizontal raphe at macula. **b** The tortuous courses of mature venous collaterals (*arrows*) follow low resistance pathways in remodeled capillary beds; peripheral retina of fundus shown in **a. c** Fundus image from patient with hemispheric vein obstruction with papillary venous collaterals linking superior and inferior circulations (*arrow*)

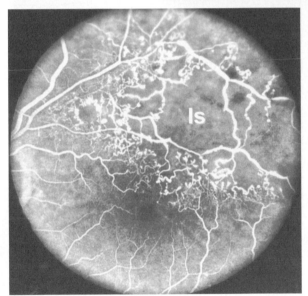

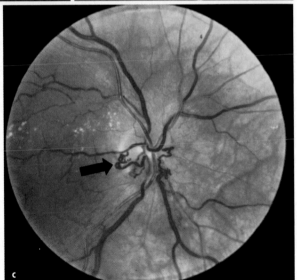

Fig. 1.18. Left 3rd order superior branch vein occlusion. Areas of inner retinal ischemia (*Is*) are bordered by dilated, incompetent, capillaries

Sudden increases in venous intraluminal pressure during central/branch vein occlusion/venular obstruction cause an acute rise in venous pressure and induce dilatation, flow stasis and transudation of fluid/blood into the perivascular retina (Fig. 1.16). In such disorders, blood is shunted into the non-obstructed circulation (central vein–ciliary circulation) via competent capillary/venular collaterals which border the regions of the occluded circulation (Fig. 1.17a–c). Venous macroaneurysms may develop and varying degrees of capillary collapse persist according to the

order of vein occlusion, residual perfusion pressure (arterial sufficiency) and efficiency of formed collaterals (Fig. 1.18). Upon achieving a steady state, retinal veins become sclerosed to varying degrees (refer to Fig. 1.16), reflecting persistent stresses. Retinal arterial macroaneurysms may reflect longstanding arteriolar stress in patients with hypertension (Fig. 1.19).

A reduction in intraocular pressure may cause macular edema with cystoid degenerative changes and secondary atrophic alterations at the outer retina. Similarly, a reduction in retinal perfusion pressure, often linked to carotid/ophthalmic artery insufficiency, can have similar retinal manifestations and in extreme circumstances there may be retrograde filling of arteries from fellow veins (Figs. 1.20, 1.21). Where intraluminal pressure falls below critical closing pressure the tone of the arteriolar wall cannot be resisted and the downstream capillary bed collapses (Fig. 1.22). If this situation persists the endothelial cells can become "fibrin locked." Endothelial cells deprived of their circulation and nutrition die and only acellular BMs persist.

In the face of venous obstruction and compromised arteriolar flow some residual perfusion of capillary beds can be maintained by collateral formation, reopened/reendothelized vessels and occasionally reversal of blood flow from veins to grossly compromised arterial circulations (Figs. 1.20, 1.21). Where vascular occlusions are significant, areas of retinal non-perfusion persist, usually with bordering areas of imperfectly perfused (hypoxic) retina. If the area of abnormally perfused retina is sufficiently large, the ischemic/hypoxic/metabolically compromised retina can produce a range of angiogenic growth factors and leads to pre-retinal/optic disk or even iris and anterior chamber angle neovasculari-

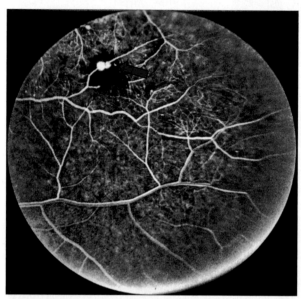

Fig. 1.19. Arterial macroaneurysm (*arrow*) at site of focal ischemia in left fundus of a patient with hypertension

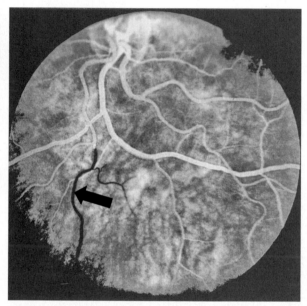

Fig. 1.21. Angiogram shows gross stasis in affected arteriole (*arrow*)

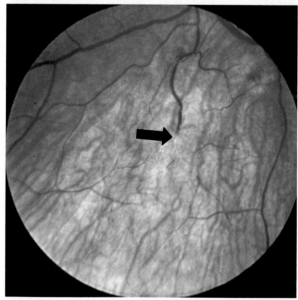

Fig. 1.20. Occluded inferior artery (*arrow*) fills in retrograde fashion from venous circulation and contains desaturated blood

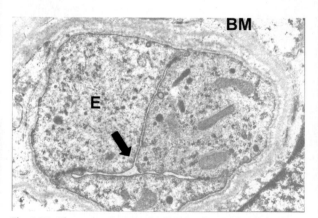

Fig. 1.22. Electron micrograph showing luminal collapse in a non-perfused retinal vessel. The processes of three apparently normal endothelial cells (*E*) are represented and fill the luminal space, which is reduced to a T-shaped slit (*arrow*). *BM* basement membrane

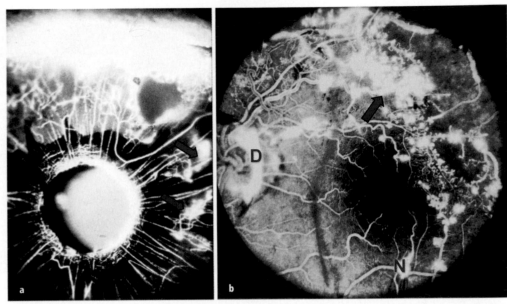

Fig. 1.23. a Multiple foci of iris neovascularization (*arrows*) in a patient with ischemic central retinal vein obstruction. **b** Left superotemporal branch vein occlusion. Extensive areas of inner retinal ischemia induce extensive new vessel formation at the optic disk (*D*), regions bordering non-perfused retina (*arrow*) and from "normal" retinal venules in the inferior macula (*N*)

zation (Fig. 1.23a). When the angiogenic stimulus is sufficiently strong pre-retinal new vessels can arise from unaffected, normal venules (Fig. 1.23b). Systemic changes in blood rheology such as alterations in plasma viscosity, an enhanced pro-coagulative state or physical changes in shape, rigidity and oxygen carrying capacity of red cells can greatly alter laminar flow, increase shear forces within the retinal circulation and cause vascular endothelial damage and microvascular occlusion, e.g., sickle cell disease.

The unique branching pattern of the retinal circulation is engineered to maintain smooth laminar flow of a very dynamic circulation in relatively small vessels. Alterations in branching angles following vascular obstruction and formation of collaterals can induce eddying and turbulence of flow and compromise the vascular endothelium. Key pathological changes to the intraretinal microvasculature following blood flow changes include vascular dilatation, tortuosity, shunt formation and varying degrees of closure of capillary bed. Lack of capillary perfusion or impaired oxygen carrying capacity has a profound influence on the inner retina similar to that outlined above.

1.2.2 Oxygen Saturation Changes

The retina can readily adapt to physiological variations in oxygen saturation or other blood gases to maintain adequate oxygenation of the neuropile. This efficient autoregulation of retinal blood flow

maintains retinal homeostasis over a considerable range of blood oxygen levels and evidence suggests that this is normally altered diurnally in response to oxygen usage by the retinal photoreceptors; not only in the intraretinal microvasculature, but also in the choroidal circulation [26]. Reduced oxygen saturation rapidly affects the metabolically demanding retinal neuropile, which can release metabolites such as adenosine and lactate, and this triggers local vasodilatation and increased blood flow as a direct consequence of vasogenic agents (e.g., NO). This is counterbalanced by integrated release of vasoconstrictors (e.g., endothelin-1).

When levels of oxygenation reach extreme levels this can have a profound influence on the retina and its microcirculation. Sustained hyperoxia has an exaggerated effect on immature retinal vessels with vascular closure and death of growth factor sensitive retinal vascular cells (covered elsewhere). By the same token, hypoxia due to circulatory failure/obstruction/stasis results in persistent vasodilation, vascular tortuosity and microvascular incompetence due to adverse hemodynamic events and hypoxia induced release of growth factors (e.g., VEGF) (Fig. 1.24a, b). Chronic hypoxia induces vascular endothelial cell proliferation to revascularize metabolically deprived retina and a perturbation of this response causes pre-retinal neovascularization with aphysiological, incompetent, fragile and unsupported vessels ramifying in the pre-retinal and vitreous spaces.

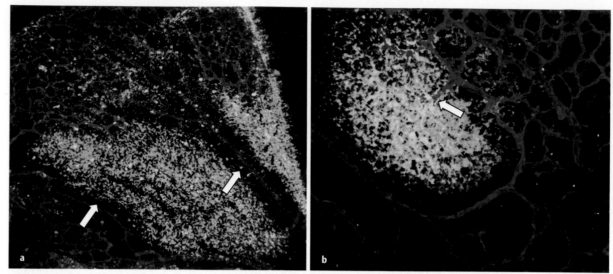

Fig. 1.24. a Ischemic/hypoxic retina in a 12-day-old mouse with oxygen-induced retinopathy shows deposition of the hypoxia-sensitive drug pimonidazole in hypoxic ischemic retina (*green fluorescence*). Note exclusion of the drug adducts in the oxygenated zone around the retinal arterioles as they traverse the ischemic central retina (*arrows*). The drug-protein adducts were detected by immunofluorescent staining and confocal microscopy and the vessels are stained orange. **b** Intraretinal angiogenic sprouts (*arrow*) invade a hypoxic area of the retina shown in **a**

1.2.3 Occlusion – Ischemia

Collapse of the downstream circulation in occlusive disorders is characterized by dilated capillaries, venovenous shunts, microaneurysms, adventitial sclerosis and areas of capillary closure where the intraluminal pressure falls below the critical closing pressure of the affected vessel. In the case of central/branch retinal artery occlusion, involutionary sclerosis of the compromised artery is common. Whatever the cause of the occlusion and the nature of the vessel, this hypoxic insult has a profound influence on the wellbeing of the affected retinal neuropile (apoptosis, necrosis, glial scar). If severe or uncompensated, these disorders can lead to focal retina ischemia (microinfarct, cotton-wool spot and disordered axoplasmic transport; Fig. 1.25) and significant damage to the neuropile in the form of macular edema, cystoid degeneration, focal atrophy of macular photoreceptors, glial cell abnormalities and pathophysiological changes in the retinal pigment epithelium (RPE). The disordered metabolism of the hypoxic retina, particularly the accumulation of growth factors (e.g., VEGF), has a profound effect on the residual vasculature, e.g., retinal arterioles traversing ischemic retina dilate, stain with dye on angiography and show varying degrees of incompetence (Fig. 1.26). Pre-retinal neovascularization reflects a "wounding" response to non-perfusion and the persistence of viable but metabolically disadvantaged retinal cells. Where there is acute infarction followed

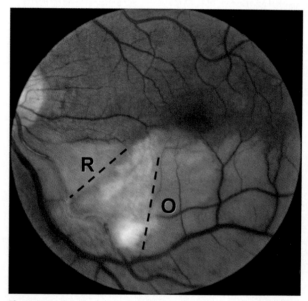

Fig. 1.25. Left inferotemporal artery occlusion producing retinal edema within the ischemic retina. *Demarcation lines* denote orthograde (*O*) and retrograde (*R*) obstruction of axoplasmic flow

recovery of sufficient circulation to service the residual atrophic gliotic retina (central retinal/branch artery occlusion) retinal neovascularization is not a feature.

1.2.4 Repair and Remodeling

Microvascular repair and remodeling are a feature of acute and chronic vaso-occlusion where there is continuing stasis, hypoxia and variations in tissue perfusion pressure. This is particularly notable in vein occlusions, radiation retinopathy and in the presence of cotton-wool spots as retinal hemodynamics equilibrate and compromised cells undergo apoptosis or necrosis. Capillaries dilate or attenuate and microaneurysms form and subse-quently show a pattern of sclerosis or recanalization.

A limited degree of intraretinal neovascularization occurs where redundant and acellular basement membrane tubes are recanalized and connect with residual radicals of the existing circulation (Fig. 1.27). Some new intraretinal channels form independently of degenerate or defunct vessels but are exceptionally slow growing and probably do not contribute significantly to the revitalization of the defunct or ailing retinal parenchyma (Fig. 1.28a, b).

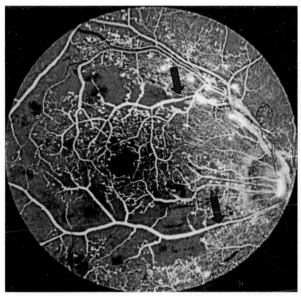

Fig. 1.26. Angiogram proliferative/ischemic diabetic retinopathy. Arterioles traversing areas of ischemic retina dilate, show. endothelial staining with dye and incompetence (*arrows*). On entering perfused retina, arterioles show more normal features

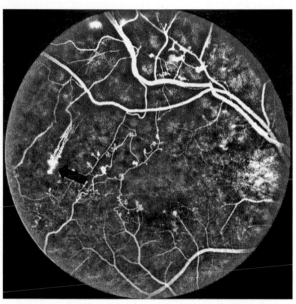

Fig. 1.27. Right superotemporal branch vein occlusion. A collection of new, competent, intraretinal vessels invade non-perfused retina in linear fashion. The growing front leaks dye (*arrow*), possibly as the vessels gain the pre-retinal space

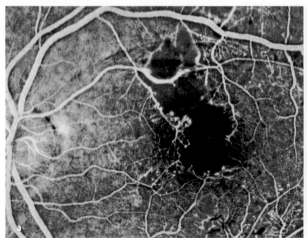

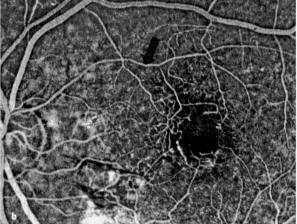

Fig. 1.28. a Radiation retinopathy of the left eye. Acute phase shows focal ischemia and microvascular disorganization in the posterior fundus. Note dilated incompetent arteriole crossing an ischemic area (*arrow*). **b** Radiation retinopathy of region depicted in **a**; 1 year later showing limited revascularization of ischemic superior macula. The superior macular arteriole has now normal caliber and competence (*arrow*), reflecting reduction in hypoxic retina

1

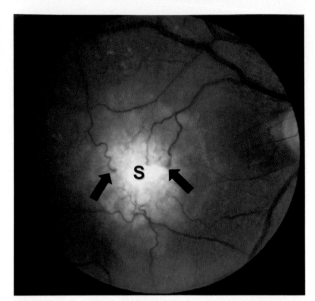

Fig. 1.29. Advanced age-related macular degeneration with disciform scar (*S*). Large macular venules (*arrows*) penetrate the scar to communicate with the choroidal circulation

When a critical threshold of hypoxia (plus associated metabolites and growth factors) is reached, intraretinal new vessels, usually from the vicinity of vein or major venules, breach the internal limiting membrane and proliferate unimpeded by the physical constraints of the densely cellular inner retina, at the vitreoretinal interface (Fig. 1.23b). The form and orientation of these new vessels is determined by concentration gradients of growth factors, e.g., VEGF, TGFβ, and availability of angiogenic stem cells. At sites of abundant scar tissue, e.g., disciform degeneration of the macula (age-related macular degeneration), the retinal circulation may establish connections (anastomoses) with the choroidal circulation (Fig. 1.29).

1.2.5 Metabolic Stresses

The retinal microvasculature can be influenced by a range of systemic disorders such as galactosemia, Fabry's disease, hyperlipidemia and homocystinuria. Many of these are associated with venous stasis and a variety of occlusive events. Perhaps the most common and well-studied systemic disorder that has a profound influence on the retinal vessel condition is diabetes. Type 1 and Type 2 diabetes is characterized by failure to regulate blood glucose, but these patients also experience hypertension and dyslipidemia that is linked to their disease. Together these metabolic insults result in considerable histopathological alterations to the retina and its microvasculature.

Diabetic retinopathy is often regarded as a quintessential disease of the intraretinal vasculature, but it should be appreciated that there is also a subtle concurrent neuropathy associated with this disorder. Neural defects, which may be linked to oxygenation and vascular function, include early alterations in the electroretinograph (ERG) [37], decreased color and contrast sensitivity, neuronal/glial abnormalities and eventual depletion of ganglion cells [12]. Microscopical choroidal vascular changes may also occur; however, it remains uncertain whether these influence vision.

Retinal vascular dysfunction commences soon after the onset of diabetes and is characterized by impaired autoregulation in the microvasculature which may be an important factor in the initiation and progression of the vascular lesions in diabetic retinopathy [1, 17]. Most clinical hemodynamic studies in diabetes conclude that increased blood flow and impaired autoregulation are features of diabetic retinopathy [30]. Although autoregulatory mechanisms are blunted in these vessels during diabetic retinopathy, they do still show a significant regulatory response to oxygen [14]. Hemodynamic abnormalities associated with diabetes are almost certainly accentuated by the increases in the retinal capillary and venular pressures and diameters that are observed in diabetic patients [30] and which possibly represent a direct consequence of arteriolar dysfunction.

The lesions that are manifest in the retinal vasculature in postmortem specimens from diabetic patients and in appropriate long-term diabetic animal models include retinal capillary BM thickening, which is thought to be a direct consequence of increased synthesis of BM components (e.g., collagen IV, fibronectin and laminin) and/or reduced degradation by catabolic enzymes [18, 28, 32, 34] (Fig. 1.30a, b). It remains uncertain if BM thickening is of primary or secondary importance in the development of diabetic retinopathy but it has been speculated that such matrix modifications may contribute to impaired endothelial–pericyte communication, defects in capillary autoregulation or inappropriate cell interaction with constituent BM proteins [4, 24, 35]. Pericyte loss is a hallmark of diabetic retinopathy and is manifest in trypsin digest preparations by the appearance of pericyte ghosts (Fig. 1.31a, b). Also in the retinal capillary unit, the demise of the endothelial cells follows soon after pericyte loss with formation of acellular capillaries (Fig. 1.31c) and this is more evident in the vessels adjacent to the arterial side of the circulation, often in close association with microaneurysms (Fig. 1.32).

Death of vascular smooth muscle cells (VSMC) also occurs in diabetic retinal arteries and arterioles

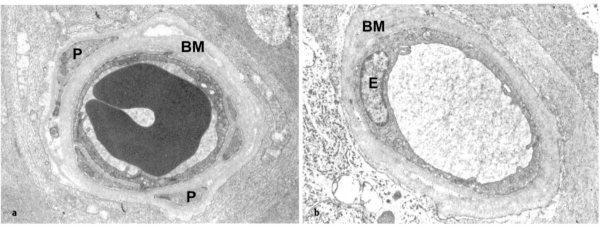

Fig. 1.30. a Retinal capillary from a 4-year diabetic dog shows an abnormally thickened basement membrane (*BM*) but with viable pericyte processes (*P*). **b** Capillary from same diabetic retina as that depicted in **a** shows a grossly thickened BM but no viable pericyte processes. In absence of pericytes the unsupported endothelium (*E*) assumes a smooth round or oval profile

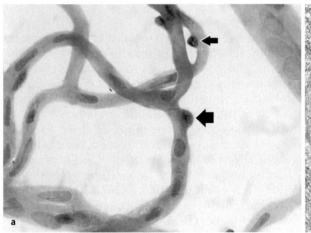

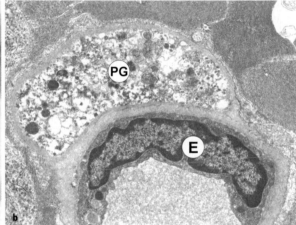

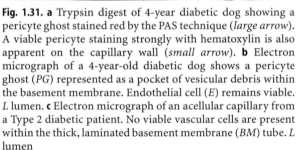

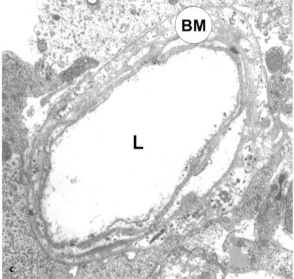

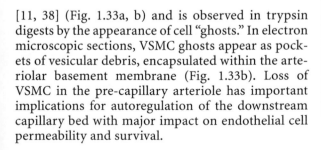

Fig. 1.31. a Trypsin digest of 4-year diabetic dog showing a pericyte ghost stained red by the PAS technique (*large arrow*). A viable pericyte staining strongly with hematoxylin is also apparent on the capillary wall (*small arrow*). **b** Electron micrograph of a 4-year-old diabetic dog shows a pericyte ghost (*PG*) represented as a pocket of vesicular debris within the basement membrane. Endothelial cell (*E*) remains viable. *L* lumen. **c** Electron micrograph of an acellular capillary from a Type 2 diabetic patient. No viable vascular cells are present within the thick, laminated basement membrane (*BM*) tube. *L* lumen

[11, 38] (Fig. 1.33a, b) and is observed in trypsin digests by the appearance of cell "ghosts." In electron microscopic sections, VSMC ghosts appear as pockets of vesicular debris, encapsulated within the arteriolar basement membrane (Fig. 1.33b). Loss of VSMC in the pre-capillary arteriole has important implications for autoregulation of the downstream capillary bed with major impact on endothelial cell permeability and survival.

1

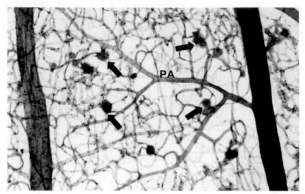

Fig. 1.32. Trypsin digest specimen from elderly Type 2 diabetic patient shows massive loss of smooth muscle from a pre-capillary arteriole (*PA*), numerous microaneurysms and acellular vessels in the capillary bed immediately downstream of this vessel (*arrows*)

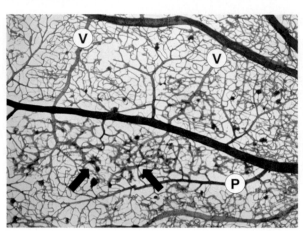

Fig. 1.34. Trypsin digest preparation from Type 2 diabetic showing vascular remodeling with formation of IRMA (*arrows*) between pre-capillary arterioles (*P*) and adjacent postcapillary venules (*V*) that bridge regions of confluent capillary loss in the periarterial zone

The microaneurysm is a hallmark of retinal microvascular disease in diabetic patients. Using the ophthalmoscope, microaneurysms may appear as dark red or white spots (occluded) in the fundus while fluorescein angiography typically outlines perfused microaneurysms as discrete hyperfluorescent spots. Clinicopathological studies of diabetic retinopathy in which exact correlations were made between fluorescein angiograms and trypsin digest preparations have confirmed that regions of capillary acellularity corresponded to non-perfused microvasculature angiographically [5, 16, 19] often downstream from areas where microaneurysms were abundant (Fig. 1.31). One of the earliest abnormalities that predisposes to formation of microaneurysms during diabetic retinopathy is the loss of pericytes [33, 40, 41], and it is likely that localized increases in hydrostatic pressure could account for capillary wall stretching at weak points and subsequent microaneurysm formation [33]. Uncontrolled hydrostatic pressure in capillary

beds could result from, or at least be greatly exacerbated by, selective loss of smooth muscle cells in the arteries and pre-capillary arterioles. The effects of intraluminal pressure in microaneurysm formation are further underlined by the presence of similar lesions in other, non-diabetic, conditions such as hypertension [39] where high capillary pressure results from failure of autoregulation, obstruction of venous return or development of high-flow shunts.

Increasing closure of capillaries may be linked with cotton-wool spots in the neural retina and also the occurrence of so-called intraretinal microvascular abnormalities (IRMA). These striking lesions are represented in trypsin digests by wide caliber multicellular channels within the capillary bed [5, 16]. IRMA contain large numbers of endothelial-like cells and occur in association with acellular capillaries close to the arterial side of the circulation (Fig. 1.34). They could reflect increasing retinal ischemia and an attempt to revascularize hypoxic neuropile, possibly to form shunt-like channels.

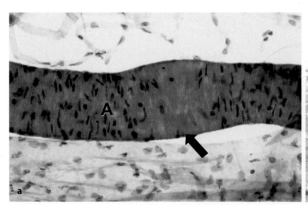

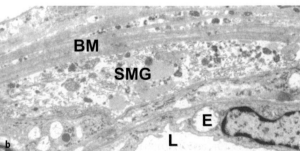

Fig. 1.33. a Trypsin digest specimen of a 7-year-old diabetic dog shows focal loss of smooth muscle (*arrow*) from a radial arteriole (*A*) in the central retina. **b** Electron micrograph showing a retinal arteriole from a 68-year-old diabetic male. A smooth muscle cell "ghost" (*SMG*) is represented as a pocket of cell debris within the basement membrane (*BM*). *E* endothelium, *L* lumen

As a direct consequence of progressive retinal microvascular degeneration during diabetes, the inner retina would be expected to experience widespread hypoxia-related insults, largely experienced by the metabolically demanding retinal neurons in this area of the retina. Hypoxia increases expression of VEGF and other peptide growth factors that have an important modulatory role in development of macular edema and pre-retinal neovascularization (refer to Fig. 1.26).

1.2.6 Trauma

Traumatic injury to the eye, whatever form it may take, can lead to many pathological manifestations in the retinal microvasculature (Table 1.1).

1.2.6.1 Radiational Damage

Damage to the eye can occur as a result of exposure to radiation of various types. Laser induced injury can cause acute coagulation of retinal microvascular networks and incite a local and/or a remote attraction of inflammatory cells (Fig. 1.39).

Light damage to the photoreceptors has been shown to induce invasion of the outer retina by retinal capillaries. Likewise exposure to ionizing radiation (>20 Gy) causes damage to the photoreceptors but also the retinal microvascular component cells – presumably as a result of damage to the nuclear DNA. It is curious that endothelial cells show more drop-out than pericytes (Fig. 1.40) and this is probably related to the replicative turnover of these cells. The latent period typical of radiation retinopathy, with respect

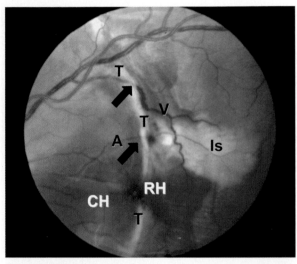

Fig. 1.35. Elliptical chorioretinal tear (*T*) with choroidal (*CH*) and retinal hemorrhage (*RH*). A superior macular arteriole (*A*) is damaged (*lower arrow*) with opaque ischemic retina (*Is*) in its distribution. A macular venule (*V*) has also been obstructed (*upper arrow*)

to the retinal microvasculature, is linked to the initial survival period of damaged endothelial cells. Typical lesions are capillary fallout, pre-capillary arteriolar occlusion, microvascular incompetence, exudation and neovascularization where there is extensive capillary damage (Fig. 1.41). Sometimes in patients with radiation retinopathy, capillary collaterals and shunts develop, but where outer retina is lost. The BM of the retinal vessels becomes greatly expanded with separation of the glial interface. In such vessels the component endothelial cells may become fenestrated.

Table 1.1. Mechanical trauma to the retina

Mechanical trauma type	Retinal vascular pathology	Neural retina pathology
Direct	Retinal rupture complicated by arteriolar occlusion and infarction (Fig. 1.35)	Avulsion optic nerve characterized by central retina artery obstruction and formation of cilioretinal communications
	Formation of retinal choroidal shunts at sites of chorioretinal tears	
Indirect	Acute rise in intravascular pressure, with secondary closure of pre-capillary arterioles, microinfarction, retinal hemorrhage, exudates and/or edema	Legacy of intraretinal ischemia, residual capillary abnormalities, shunts, microaneurysms and secondary damage to outer retina, especially macula
Head/chest injuries (Purtscher retinopathy)	Residual capillary abnormalities, shunts, microaneurysms (Fig. 1.36a, b)	
Traction	Forces exerted cause distortion of anatomy and alterations in microvascular competence, e.g., fibroglial scars (Fig. 1.37)	Disturbance of normal retinal neuroglial vascular cell relationships
	Contusional and traction forces causing occlusion of retinal vessels at sites of tears and retinal detachment (Fig. 1.38a, b)	Impaired perfusion of retina surrounding tear (Fig. 1.36a, b)

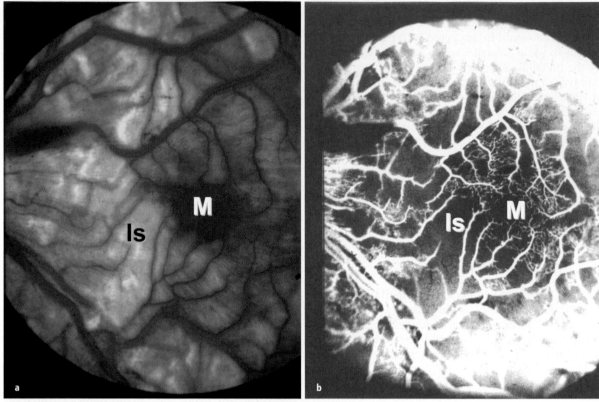

Fig. 1.36. a Purtscher retinopathy left fundus following closed head injury characterized by extensive venous engorgement and inner retinal ischemia (*Is*). **b** Angiography confirms extensive non-perfusion of inner retina (*Is*). *M* macula

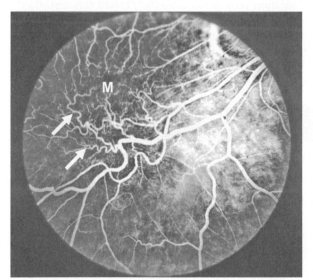

Fig. 1.37. Fibroglial scar right inferotemporal macula. Disorganization macular circulation, microaneurysms, distorted vessels (*arrows*). *M* macula

1.2.7 Drug Toxicity

Systemic delivery of therapeutic agents can often influence the retinal blood vessels. For example, oua-bain can produce severe and extensive atrophy of the outer retina and, as in radiation retinopathy, this may induce movement of retinal capillaries into the subretinal space. Another example is quinine or chloroquine, traditionally used for protection against malaria. If used in high doses for prolonged periods, these drugs lead to subtle closure of the capillary beds in the retina, secondary to atrophy of the neural retina and the retinal pigment epithelium.

1.2.8 Inflammation

The retinal microvasculature is central to retinal inflammatory processes by transporting inflammatory cells to sites of disease and removing unwanted products of inflammation and cell breakdown from the extravascular space by anionic pump mechanisms. There can be leakage by the retinal vessels which is often observable on fluorescein angiography but without overt retinal thickening. Subclinical inflammatory processes such as postcataract maculopathy (Fig. 1.42) and early diabetic retinopathy may result in intraretinal fluid accumulation and retinal thickening.

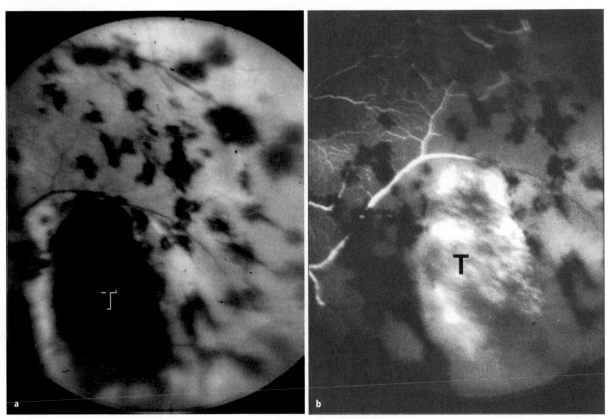

Fig. 1.38. a A contusional retinal tear (*T*) retina has produced edema, hemorrhage and pigment epithelial defects. **b** Angiography defines the extent of retinal capillary non-perfusion in vicinity of tear

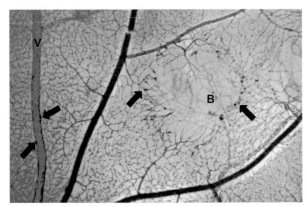

Fig. 1.39. Trypsin digest autoradiogram of experimental photocoagulation burn (*B*) following intravitreal injection of tritiated thymidine to mark dividing cells. The involved microvasculature is acellular and represented by pale tubes of residual basement membrane. Darkly stained mononuclear and endothelial cells surrounding the burn are decorated with silver grains (*arrows*). Labeled monocytes (*arrows*) are also present in the wall of a remote venule (*V*)

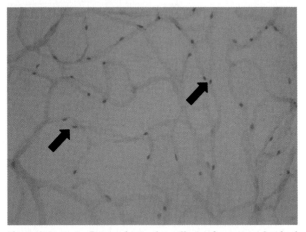

Fig. 1.40. Trypsin digest of retinal capillaries from a rat that had received 20 Gy of X-radiation to the eye. A confluent group of capillaries show numerous darkly stained pericyte nuclei (*arrows*) but no surviving endothelial cells. Electron microscopy confirmed absence of endothelium in such vessels

1

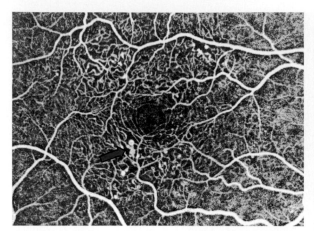

Fig. 1.41. Radiation retinopathy: Right macula shows dilated, incompetent microvasculature, focal capillary fallout and microaneurysms (*arrow*)

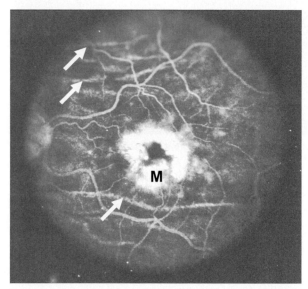

Fig. 1.42. Postsurgical cystoid macular edema. Incompetent capillaries pool dye, in the inner plexiform layer at the macula (*M*): choroidal folds (*arrows*) reflect hypotony

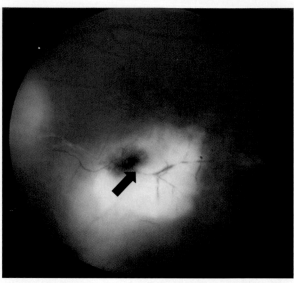

Fig. 1.43. Toxoplasmosis retinitis in left inferior macula. Retinal vessels traversing the inflammatory focus are thrombosed (*arrow*)

Established retinal perivasculitis is associated with focal damage to vessel walls (usually venous) and leakage of serum proteins and lipids into the extravascular space, e.g., sarcoidosis and multiple sclerosis. Following resolution of the initiating inflammatory process, veins may show residual alterations in caliber and collaterals can remain, perhaps in association with areas of capillary closure, especially if the obstruction during perivasculitis has been severe. Such retinal microvascular pathology is also common in conditions such as severe retinitis toxoplasmosis (Fig. 1.43) or herpesvirus infections, which again can lead to capillary occlusion at the site of acute inflammation. Extensive retinal phlebitis and occlusion associated with Eales' disease and sarcoidosis may precipitate pre-retinal neovascularization/vitreous hemorrhage.

1.2.9 Retinal-Choroidal Tumors

Pathological evaluation shows that the retinal microvasculature may play an important role in growth and metastasis of retinal and optic disk tumors. This microcirculation also reacts to localized release of vasoactive agents released from large, proximal choroidal tumors.

Optic disk tumors: these structures attract retinal vessels from superficial capillaries of the optic nerve head (Fig. 1.44). The associated retinal vessels enlarge as the metabolic requirements of the expanding tumor increase. In some circumstances, the normally "tight" vessels can become fenestrated and lose tight junction integrity, probably in response to high concentrations of VEGF in the microenvironment. The accompanying breakdown of iBRB accounts for exudates into the neural retina and vitreous.

Highly vascularized retinal tumors such as angiomata recruit their blood supply from the adjacent retinal microvasculature. As tumor oxygen requirements increase, retinal feeder vessels develop to provide the necessary nutritional support. The tumor vessels are fenestrated, irregular and incompetent, while larger hyperdynamic feeder vessels (within the neural retina) are typically competent. Some vascular tumors with "in-built shunting systems" have "arterialized" venous systems. Large vascular tumors are grossly incompetent and lipid/protein rich

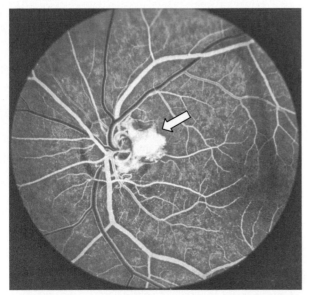

Fig. 1.44. Von-Hippel Lindau disease. A papillary angioma (*arrow*) communicates with optic disk circulation and is incompetent to dye

exudates infiltrate the neuropile and may accumulate in the subretinal space and detach the retina. Vitreous hemorrhage may occur as a late phenomenon.

Neuronal/glial tumors also recruit retinal vessels which develop complex tumor circulations as the lesion slowly enlarges. This intra-tumor vasculature is typically incompetent and the resulting exudates can infiltrate adjacent retina. Occasionally, intravitreal hemorrhage can occur. Retinoblastomas are very fast growing tumors and can often outstrip their vascular supply. They become grossly hypoxic and subsequently produce large intravitreal concentrations of angiogenic growth factors that can induce pre-retinal neovascularization in neighboring retina (Fig. 1.45a–d).

A similar phenomenon can occur in choroid tumors where breakdown of the iBRB and neovascularization can occur. Sometimes large choroidal tumors can cause compressional effects on adjacent retina and these manifest as retinal hemorrhage and capillary perfusion deficits.

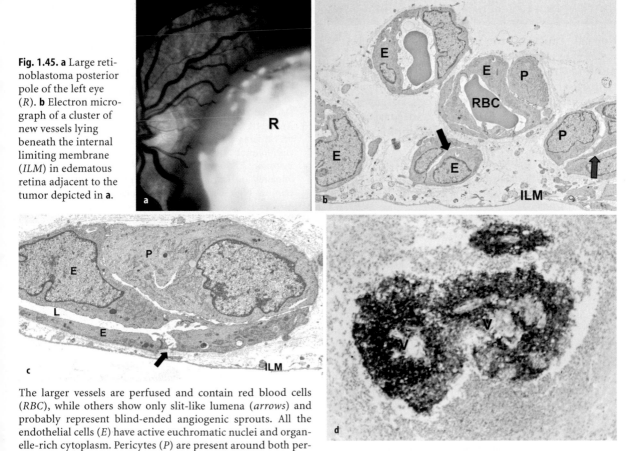

Fig. 1.45. a Large retinoblastoma posterior pole of the left eye (*R*). **b** Electron micrograph of a cluster of new vessels lying beneath the internal limiting membrane (*ILM*) in edematous retina adjacent to the tumor depicted in **a**.

The larger vessels are perfused and contain red blood cells (*RBC*), while others show only slit-like lumena (*arrows*) and probably represent blind-ended angiogenic sprouts. All the endothelial cells (*E*) have active euchromatic nuclei and organelle-rich cytoplasm. Pericytes (*P*) are present around both perfused vessels and sprouts. **c** Sprout-like retinal capillary from region depicted in **b** shows the presence of fenestration in the endothelium (*arrow*). *E* endothelial cells, *P* pericyte, *L* lumen, *ILM* internal limiting membrane. **d** Retinoblastoma in a perivascular location within a largely necrotic tumor shows expression of mRNA for VEGF (*blue-purple*) in a tissue section stained by in situ hybridization. *V* vessels

1

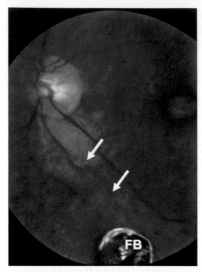

Fig. 1.46. Retained intraocular metallic foreign body left inferior fundus. Retinal vessels in the vicinity of the foreign body (*FB*) are grossly attenuated or absent (*arrows*)

1.2.10 Primary Neuropile Atrophy and Degeneration

Normal function of the retinal circulation is completely dependent on intimate cell-cell communication with neural and glial elements of the retina. Neuroglial degenerative disorders can deprive the retinal microvascular component cells of trophic factors necessary for functionality and survival, e.g., NFGR, PDGF, VEGF, and BDGF. In retinitis pigmentosa, trauma, toxic retinopathy, and loss of retinal parenchyma is associated with retinal capillary cell attrition, closure of capillary beds, narrowing of supply vessels and involutional sclerosis of larger radicals (Fig. 1.46).

1.2.11 Remote Effects of Retinal Vascular Pathology

Retinal hypoxia may lead to elevated vitreous levels of angiogenic factors such as VEGF and these can have a pathogenic influence on the other ocular structures. Iris neovascularization is a clear example of this phenomenon. Vitrectomy can also cause localized vasoproliferation, especially in diabetic retinopathy where the compromised retinal circulation may develop fibrovascular responses at the site of the vitrectomy port. Complications include traction, retinal detachment, proliferative vitreoretinopathy and iris neovascularization.

Peripheral retinal vasoproliferation, as in the case of retinopathy of prematurity, can also impact on the iris, ciliary body, lens and angle anterior chamber with inflammatory and traction effects.

In summary the retinal vasculature shows a common range of responses to diverse pathological stimuli. This normally quiescent system demonstrates a remarkable ability to recover from obstructive events induced by factors ranging from radiation injury and metabolic imbalance to immune cell inundation and gross hemodynamic insult. In such situations it is physiological recanalisation and other forms of vascular remodelling that predominante. However, when this regenerative pattern is altered by chronic hypoxia/ischemia more threatening aberrations are manifest: change of endothelial phenotype, chronic neuroretinal oedema and vitreo-retinal neovascularization. In recent years advances made possible through molecular cell biology have provided us with new perspectives on the morphological changes induced by disease in the retinal vasculature. We have also gained an increased awareness of the unity of the neuro-vascular complex of the retina: a uniquue branch of the central nervous system imposing an equally specialised phenotype on the vasculature that serves it.

References

1. Alder VA, Su EN, et al. (1998) Overview of studies on metabolic and vascular regulatory changes in early diabetic retinopathy. Aust N Z J Ophthalmol 26(2):141–148
2. Alon T, Hemo I, et al. (1995) Vascular endothelial growth factor acts as a survival factor for newly formed retinal vessels and has implications for retinopathy of prematurity. Nat Med 1(10):1024–1028
3. Antonetti DA, Lieth E, et al. (1999) Molecular mechanisms of vascular permeability in diabetic retinopathy. Semin Ophthalmol 14(4):240–248
4. Beltramo E, Pomero F, et al. (2002) Pericyte adhesion is impaired on extracellular matrix produced by endothelial cells in high hexose concentrations. Diabetologia 45(3):416–419
5. Bresnick GH, Davis MD, et al. (1977) Clinicopathologic correlations in diabetic retinopathy. II. Clinical and histologic appearances of retinal capillary microaneurysms. Arch Ophthalmol 95(7):1215–1220
6. Chakravarthy U, Gardiner TA (1999) Endothelium-derived agents in pericyte function/dysfunction. Prog Retin Eye Res 18(4):511–527
7. Chan-Ling T, Gock B, et al. (1995) The effect of oxygen on vasoformative cell division. Evidence that 'physiological hypoxia' is the stimulus for normal retinal vasculogenesis. Invest Ophthalmol Vis Sci 36(7):1201–1214
8. Claxton S, Fruttiger M (2003) Role of arteries in oxygen induced vaso-obliteration. Exp Eye Res 77(3):305–311
9. Dery MA, Michaud MD, et al. (2005) Hypoxia-inducible factor 1: regulation by hypoxic and non-hypoxic activators. Int J Biochem Cell Biol 37(3):535–540
10. Frank RN, Dutta S, et al. (1987) Pericyte coverage is greater in the retinal than in the cerebral capillaries of the rat. Invest Ophthalmol Vis Sci 28(7):1086–1091
11. Gardiner TA, Stitt AW, et al. (1994) Selective loss of vascular smooth muscle cells in the retinal microcirculation of diabetic dogs. Br J Ophthalmol 78(1):54–60

12. Gardner TW, Antonetti DA, et al. (2000) New insights into the pathophysiology of diabetic retinopathy: potential cell-specific therapeutic targets. Diabetes Technol Ther 2(4):601–608

13. Garner A, Ashton N (1979) Pathogenesis of hypertensive retinopathy: a review. J R Soc Med 72(5):362–365

14. Grunwald JE, Riva CE, et al. (1984) Altered retinal vascular response to 100% oxygen breathing in diabetes mellitus. Ophthalmology 91(12):1447–1452

15. Janzer RC, Raff MC (1987) Astrocytes induce blood-brain barrier properties in endothelial cells. Nature 325(6101):253–257

16. Kohner EM, Henkind P (1970) Correlation of fluorescein angiogram and retinal digest in diabetic retinopathy. Am J Ophthalmol 69(3):403–414

17. Kohner EM, Patel V, et al. (1995) Role of blood flow and impaired autoregulation in the pathogenesis of diabetic retinopathy. Diabetes 44(6):603–607

18. Ljubimov AV, Burgeson RE, et al. (1996) Basement membrane abnormalities in human eyes with diabetic retinopathy. J Histochem Cytochem 44(12):1469–1479

19. Gardiner TA, Archer DB, Curtis TM, Stitt AW (2007) Arteriolar involvement in the microvascular lesions of diabetic retinopathy: implications for pathogenesis. Micorcirculation 14(1):25–38

20. Minshall RD, Sessa WC, et al. (2003) Caveolin regulation of endothelial function. Am J Physiol Lung Cell Mol Physiol 285(6):L1179–1183

21. Mizutani M, Kern TS, et al. (1996) Accelerated death of retinal microvascular cells in human and experimental diabetic retinopathy. J Clin Invest 97(12):2883–2890

22. Muraki K, Iwata Y, Katanosaka Y, Ito T, Ohya S, Shigekawa M, Imaizumi Y (2003) TRPV2 is a component of osmotically sensitive cation channels in murine aortic myocytes. Circ Res 93(9):829–38

23. Nomura M, Yamagishi S, et al. (1995) Possible participation of autocrine and paracrine vascular endothelial growth factors in hypoxia-induced proliferation of endothelial cells and pericytes. J Biol Chem 270(47):28316–28324

24. Padayatti PS, Jiang C, et al. (2001) High concentrations of glucose induce synthesis of argpyrimidine in retinal endothelial cells. Curr Eye Res 23(2):106–115

25. Pierce EA, Avery RL, et al. (1995) Vascular endothelial growth factor/vascular permeability factor expression in a mouse model of retinal neovascularization. Proc Natl Acad Sci U S A 92(3):905–909

26. Polska E, Polak K, et al. (2004) Twelve hour reproducibility of choroidal blood flow parameters in healthy subjects. Br J Ophthalmol 88(4):533–537

27. Provis JM (2001) Development of the primate retinal vasculature. Prog Retin Eye Res 20(6):799–821

28. Roy S, Maiello M, et al. (1994) Increased expression of basement membrane collagen in human diabetic retinopathy. J Clin Invest 93(1):438–442

29. Saint-Geniez M, D'Amore PA (2004) Development and pathology of the hyaloid, choroidal and retinal vasculature. Int J Dev Biol 48(8–9):1045–1058

30. Schmetterer L, Wolzt M (1999) Ocular blood flow and associated functional deviations in diabetic retinopathy. Diabetologia 42(4):387–405

31. Sharma NK, Gardiner TA, et al. (1985) A morphologic and autoradiographic study of cell death and regeneration in the retinal microvasculature of normal and diabetic rats. Am J Ophthalmol 100(1):51–60

32. Stitt AW, Anderson HR, et al. (1994) Diabetic retinopathy: quantitative variation in capillary basement membrane thickening in arterial or venous environments. Br J Ophthalmol 78(2):133–137

33. Stitt AW, Gardiner TA, et al. (1995) Histological and ultrastructural investigation of retinal microaneurysm development in diabetic patients. Br J Ophthalmol 79(4):362–367

34. Stitt AW, Gardiner TA, et al. (2002) The AGE inhibitor pyridoxamine inhibits development of retinopathy in experimental diabetes. Diabetes 51(9):2826–2832

35. Stitt AW, Hughes SJ, et al. (2004) Substrates modified by advanced glycation end-products cause dysfunction and death in retinal pericytes by reducing survival signals mediated by platelet-derived growth factor. Diabetologia 47(10):1735–1746

36. Stone J, Itin A, et al. (1995) Development of retinal vasculature is mediated by hypoxia-induced vascular endothelial growth factor (VEGF) expression by neuroglia. J Neurosci 15(7):4738–4747

37. Tzekov R, Arden GB (1999) The electroretinogram in diabetic retinopathy. Surv Ophthalmol 44(1):53–60

38. vom Hagen F, Feng Y, et al. (2005) Early loss of arteriolar smooth muscle cells: more than just a pericyte loss in diabetic retinopathy. Exp Clin Endocrinol Diabetes 113(10):573–576

39. Wong TY, McIntosh R (2005) Hypertensive retinopathy signs as risk indicators of cardiovascular morbidity and mortality. Br Med Bull 73–74:57–70

40. Yanoff M (1966) Diabetic retinopathy. N Engl J Med 274(24):1344–1349

41. Yanoff M (1969) Ocular pathology of diabetes mellitus. Am J Ophthalmol 67(1):21–38

I 1

2 Retinal Vascular Development

M.I. DORRELL, M. FRIEDLANDER, L.E.H. SMITH

Core Messages

- The retinal vasculature is highly ordered and the formation of the normal vascular plexuses involves the complex interactions of numerous cell types, including endothelial cells, glial cells, pericytes, and myeloid cells
- Understanding the process of normal vascular development relates to understanding pathological neovascularization
- Endothelial cell guidance during retinal vascular development is mediated by an astrocytic template, filopodial extensions from the developing endothelial cells, and cell-cell adhesion
- Retinal vascular development models can be used effectively to test the activity of clinical and pre-clinical compounds on neovascularization

2.1 Introduction

2.1.1 General Vascular Development

The circulatory system evolved so that nutrients and chemicals required for cellular function can be efficiently transferred from central organs to the extremities. Because of its importance to the growth and survival of other tissues, the circulatory system forms during early stages of development, and its correct development and early function is absolutely critical for survival of the embryo. During development, blood vessel formation occurs by three processes, the initial formation of vessels from yolk sacs during early embryogenesis, and by the distinct processes of vasculogenesis and angiogenesis during subsequent development [33, 35].

Yolk Sacs. Angiogenic clusters containing hematopoietic cells (blood precursors) at the center and angioblasts (endothelial cell precursors) lining the periphery. These cells, originally differentiating from mesoderm cells, form the earliest identifiable blood vessels.

Vasculogenesis. The assembly of vessels from separate endothelial precursor cells (angioblasts) as they differentiate into mature endothelial cells.

Angiogenesis. The formation of new blood vessels from preexisting capillaries. Differentiated endothelial cells are induced to proliferate, thus facilitating the sprouting of new vessels from existing vessels.

Clinically, blood vessels of the cardiovascular system are identified mainly on the basis of the tissue in which they are found, their position relative to other vessels, the lumen size, and the pattern of branching. This system of identification is possible because of the consistency with which vascular patterns are formed from individual to individual, and even across various mammalian species.

2.1.2 Basis of Clinical Identification of Blood Vessels

Essentials

- The tissue that the vessels supply (i.e., retinal vasculature, brain vasculature, etc.)
- Position within the vascular tree (i.e., arterial, venous, or capillary)
- Lumen size
- Branching patterns

Historically, research investigating the vascular development has primarily focused on the creation of the basic structural unit, the endothelial lined tube. However, as numerous molecules are continuously added to the growing list of neovascular factors [6, 15, 35], the complexity of this process is becoming evident. This complexity highlights the limitations of a conceptual framework that defines vascular development solely in terms of creating the blood vessel unit. Instead, a complete analysis of the morphogenetic events that occur beyond endothelial cell development is required to fully understand neovascular processes. These events include vessel branching, vessel guidance, recruitment of vascular-associated non-endothelial cells (mural cells), and specification of vessel identity. As a whole, these processes are critical to neovascularization of the eye, both during development and neovascular disease progression.

2.1.3 Major Cellular Components of Vessel Formation

Essentials
- Endothelial cells: form the vessel wall creating the lumen through which blood flows
- Mural cells: perivascular cells that associate with the vessels. Critical for maturation and maintenance of the functioning vasculature (e.g., pericytes, perivascular macrophages, astroglial cells)
- Hematopoietic cells: blood lineage cells that flow through the vessel lumen. In addition to their normal roles, these cells can play important roles during initiation, maturation, and regression of blood vessels.

2.2 Endothelial Cells

Endothelial cells, the cells that form the actual blood vessel wall, are generally the major cell type referred to during discussions of neovascularization. During neovascular development, these cells must undergo proliferation, migration, and final maturation to form a new mature vessel. As embryonic blood vessels become established and various tissues differentiate, endothelial cells continue to undergo tissue-specific changes, generating functionally distinct vascular beds. These differences are primarily defined by the types of junctions that form between adjacent endothelial cells in the vessel wall. For example, lymphatic vessels have a discontinuous or even partially absent basement membrane [22] and endothelial cells in the kidney are fenestrated so that waste can be cleared from the blood stream and efficiently removed from the body. The choroidal vasculature of the eye is also fenestrated. In the brain and to a lesser extent in the retina, tight junctions form connections between the endothelial cells resulting in the formation of the blood-brain and blood-retinal barriers. This barrier helps regulate the neural microenvironment by protecting the brain or retina from fluctuations in plasma composition, and by preventing circulating agents and small molecules

from entering the tissue and disrupting neural function [1]. Many other differences between the vasculature from different tissues also become apparent during development, with each specific morphology evolutionarily tailored to the different vascular requirements for each tissue.

2.2.1 Vascular Heterogeneity (Morphological Classification of Vessels)

Essentials
- Type and complexity of interendothelial adhesions:
  - *Continuous.* Tight junctions between adjacent endothelial cells in a vessel. Limited materials can pass through the vessel wall.
    Examples: retina, heart, lung, brain, skeletal muscle
  - *Fenestrated.* Fewer tight junctions between adjacent endothelial cells. Vessels are more permeable to cells and larger molecules. Examples: choroidal plexus, intestinal villi, endocrine and exocrine glands
  - *Discontinuous.* Incomplete membrane wall. These types of vessels are often found in "filtration tissues." Examples: hepatic sinusoids in the liver
- Size of vessels (arteries or veins with large lumens vs. smaller capillaries)
- Shape of individual vessels (straight vs. tortuous)
- Patterning of the vascular plexus (hexagonal patterning, planar vs. three-dimensional, etc.)
- Presence or absence of diaphragms or valves to prevent misdirection of blood flow [8]

2.3 Mural Cells

Once a new vessel has formed, whether by vasculogenesis or angiogenesis, it must undergo maturation to become a stable, functional vessel. This involves mural cell recruitment and, ultimately, remodeling of the vascular bed. Perivascular mural cells are closely associated with vessels and lie just external to the endothelial cells. Just as vessel morphology is tissue specific, the type of associated mural cells, their morphology, and their function, also vary depending on the location of the associated vessel. For example, in arteries and larger vessels, smooth muscle cells form a multilayered sheath around the vessel wall. Smaller arterioles and venules also recruit smooth muscle cells, but mainly as a continuous

2

sheet around the vessels. Unlike larger arteries, these smaller vessels contain a basement membrane layer between the endothelial cells and the associated smooth muscle cells. In general, pericytes mainly associate with, and extend long processes along, capillaries and postcapillary venules [9, 20]. The type of associated mural cells also differs by tissue. For example, astrocytes remain closely associated with the brain and retinal vasculature while Schwann glial cells are associated with the peripheral vasculature during development [27]. Appropriate recruitment of the various mural cells is important for stabilization and maturation of new vessels during developmental neovascularization. Vessels lacking mural cells have been found to be more susceptible to apoptosis and degeneration during vascular remodeling processes than vessels that have appropriately associated mural cells [3]. Mural cells are also important for maintaining vascular quiescence after vascular development is complete [28]. In fact, the loss of mural cell association has been correlated with a number of neovascular ocular diseases including diabetic retinopathy.

2.4 Vascular Patterning

The appropriate organization and development of vascular plexuses is critical for proper tissue function. The loss of these guidance mechanisms can lead to abnormal vascularization and a variety of pathological conditions. This is particularly true for the development of the retina, a highly organized neuronal tissue with perfectly layered neuronal cells and synaptic plexuses. The retina consists of many parallel, anatomically equipotent microcircuits derived from millions of neuronal cells. Each of the approximately 55 distinct cell types (for mammalian species) have different, important functions [26] and must be organized in such a way that visual information can be passed from the photoreceptors in the back of the retina, through various neurons in the inner nuclear layer, to the ganglion cells in the front of the retina, and eventually on to the brain. Thus, the neurons and their glial helper cells must be organized perfectly during development so that functional synapses can form between appropriate neurons within the correct retinal regions (Fig. 2.1).

Similarly, the retinal vasculature must also develop appropriately and must localize to specific retinal layers. Incorrect vascular organization can lead to the disruption of retinal function. In most higher order mammals, three planar vascular plexuses are formed during development; the superficial plexus forms within the ganglion cell layer at the front of the retina, and the deep and intermediate vascular plex-

uses form at the outer and inner edges of the inner nuclear layer respectively (Fig. 2.1). If any of the three vascular layers fail to form, neuronal degeneration can occur due to hypoxic stress. If the vasculature forms but the layers are disorganized, or if vessels become located within regions of the retina that normally remain avascular, such as the photoreceptor layer or the macula, these vessels can disrupt the neuronal synapses, again leading to the disruption of visual function [41]. For proof of the importance of proper vessel formation in the retina, one needs only to look at the devastating diseases that lead to a loss of vision because of abnormal neovascularization [5, 23]. Many of the ocular diseases that can cause a catastrophic loss of vision due to complications from abnormal neovascularization are reviewed extensively in this book.

As our knowledge of the processes that mediate neovascularization progresses, it is becoming increasingly clear that there are common mechanisms observed in developmental and pathological angiogenesis. In either circumstance, proliferation of endothelial cells must be initiated. The endothelial cells must then degrade the extracellular matrix, undergo guided migration, and ultimately mature into functional vessels. With such complex mechanisms to ensure proper regulation of neovascularization, it is not surprising that many of the factors involved in promoting vascular development are also implicated in pathological vascularization. Thus, by investigating the mechanisms that mediate vascular development, novel insights are also gained regarding potential treatments of vascular diseases. Differences between developmental and pathological angiogenesis are also important. In fact, these differences may prove to be even more useful for the development of novel anti-angiogenic treatments since factors that are solely implicated in pathological angiogenesis would be logical therapeutic targets. Disruption of their function would not affect vascular development or other important angiogenic processes such as wound healing, and thus they are likely to have few negative affects on the established vasculature.

It is reasonable to think that many vascular diseases arise from a loss of complex regulation mechanisms associated with normal developmental neovascularization. In fact, many of the problems associated with abnormal angiogenesis do not arise from the growth of new blood vessels per se, but from abnormalities in the neovessels such as leakiness and hemorrhaging. This is particularly true for the pathology of ocular diseases with associated neovascularization. For example, many of the characteristics of neovascularization that occur during retina development are mirrored in diabetic retinopathy. However, key problems begin to arise as neovessels

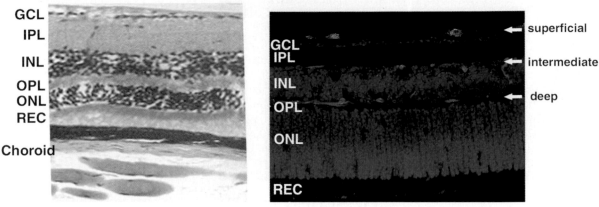

Fig. 2.1. The retina is a highly ordered, striated tissue with distinct neuronal and vascular layers. The many different neuronal subtypes are seen in fluorescently stained mouse retinal cross-sections demonstrating the normal development of three planar vascular plexuses (*red* blood vessels, *blue* DAPI staining of cell nuclei, *GCL* ganglion cell layer, *IPL* inner plexiform layer, *INL* inner nuclear layer, *OPL* outer plexiform layer, *ONL* outer nuclear layer, *REC* photoreceptor segments)

associated with diabetic retinopathy fail to recruit and maintain appropriate mural cell association. The apoptotic loss of pericytes surrounding the retinal vasculature occurs during early stages of diabetic retinopathy. This is likely to lead to reduced vessel stability, the onset of endothelial cell apoptosis, and subsequent retinal ischemia. New vessels that form as a result of this ischemia fail to properly recruit mural cells. Because these neovessels lack pericyte and normal basement membrane coverage, they do not mature or develop normally, leading to the observed vascular leakiness associated with retinal edema and vision loss [14]. Thus, understanding the developmental mechanisms that facilitate proper formation of mature vascular systems is important not only because of the parallels between developmental and pathological neovascularization, but also so similar mechanisms may be employed clinically to stabilize pathological vessels and treat diseases associated with abnormal, leaky vessels. To this extent, in various mouse models of retinal degeneration, bone marrow-derived cells have been used to stabilize degenerating vasculature [32]. Interestingly, stabilization of the degenerating vasculature also appears to facilitate rescue of degenerating cones (Fig. 2.2) [32]. In addition to their potential utility for the treatment of degenerative diseases, such therapeutic methods may also be useful for the treatment of neovascular eye diseases such as age related macular degeneration (ARMD) and diabetic retinopathy. In contrast to simply blocking the growth of new vessels using angiostatics, similar bone marrow-derived cells may be able to stabilize neovessels, allowing the underlying hypoxic drive to be relieved while eliminating the major clinical problems associated with neovessels [29].

2.5 Retinal Vascular Development

Along with the development of any tissue, the development of the retina requires the concomitant formation of a complex vascular system. In the human fetus, formation of the ocular vasculature occurs mainly during the second and third trimesters in utero and the retinal vasculature is one of the last vascular systems to form. During early development of the eye, the retina is initially supplied by the choroid and the hyaloidal vessels. However, due to the natural regression of the hyaloidal vasculature (Fig. 2.3), as well as the thickening of the retina due to the final development and differentiation of the neuronal layers, the retina eventually finds itself in a hypoxic environment and must begin to establish its own vascular system [36].

During development, retinal blood vessels form three vascular plexuses in a highly reproducible manner, thereby resulting in the formation of distinct vascular and avascular zones (Fig. 2.4). Mammalian retinas generally vascularize in similar fashions during development, albeit on different time scales; human retinal developmental vascularization begins during the second trimester in utero and continues, becoming complete only after birth. In most non-primate species, the retina is avascular at birth and developmental neovascularization occurs during the first few weeks postnatally. Another important distinction between primates and non-primates is the development of the macula, the highly dense region of cone photoreceptors where central vision occurs. This region, which forms in primates, remains avascular throughout development and into adulthood. The abnormal vascularization or leakage of fluid into the macular region is a major cause of devastating loss of vision.

2

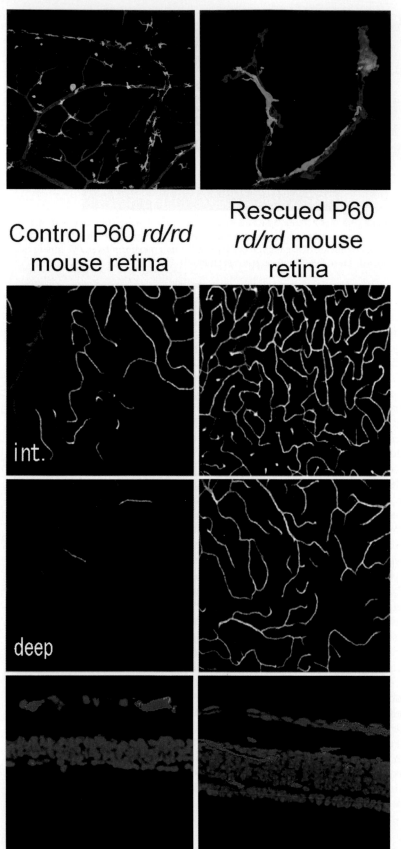

Control P60 *rd/rd* mouse retina

Rescued P60 *rd/rd* mouse retina

int.

deep

Fig. 2.2. Bone marrow-derived progenitor cells associate with the retinal vasculature and have vasculo- and neurotrophic properties. **a** Lin-bone marrow-derived progenitor cells can target to, and associate with, the developing retinal vasculature in mice. **b** These cells can rescue both vascular (*top and middle panels*) and neuronal cells (*bottom panel;* notice the rescue of DAPI stained cells in the INL and ONL) normally observed in models of retinal degeneration. (Adapted from [32])

Fig. 2.3. Visualization of the natural regression of the hyaloid vasculature using live imaging in mice. **a** Live, in vivo imaging can be used to visualize the various ocular vascular plexuses including the iris vasculature at the front of the eye (*red*), the hyaloidal vessels in the vitreal cavity (*blue*), and the retinal vessels at the back of the eye (*green*) (adapted from [36]). **b** Using in vivo imaging, natural regression of the hyaloidal vasculature can be followed demonstrating the existence of a full hyaloidal network in 1-week-old mice (*left*), a regressing hyaloidal vasculature in 16-day-old mice (*middle*), and a nearly fully regressed hyaloidal vasculature in 22-day-old mice (*right*)

The superficial vascular plexus within the ganglion cell layer is the first to form, with the early stages of development beginning at gestational age 14 to 16 weeks in humans [17, 34]. This plexus grows radially from the optic nerve head toward the retinal periphery (Fig. 2.4A). Currently there is conflicting evidence as to whether the superficial plexus forms by the process of vasculogenesis or angiogenesis. While there is no evidence of angioblasts ahead of the migrating vascular front in rodent species [16, 17], results from the few studies investigating human retinal development suggest that vasculogenesis may be involved [7].

2.5.1 The Role of Astrocytes

Astrocytic neuroglia play a critical role during retinal vascular development. Only species with vascularized retinas are known to express retinal astrocytes, and within the retinas of those species, astrocytes are only observed in regions where vascularization occurs. For example, astrocytes do not develop in the macula, which also remains avascular. Astrocytes emerge from the optic nerve head prior to neovascularization. As the retina becomes hypoxic, the retinal glial cells, consisting of astrocytes within the ganglion cell layer and Müller cells whose processes extend throughout the neural retina, respond to hypoxia by

2

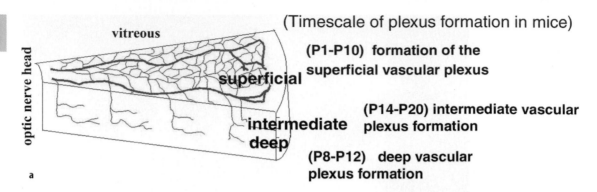

Fig. 2.4. Normal development of the mouse retinal vasculature. **a** The superficial plexus forms after birth in mice as vessels emerge from the optic nerve head and migrate out toward the retinal periphery during the first 10 days. By 3 weeks after birth, a normal adult-like vasculature has formed. **b** During development of the mouse retinal vasculature, three vascular plexuses are formed. During the 1st week after birth the superficial plexus forms within the ganglion cell layer (*GCL*), the intermediate plexus forms at the inner edge of the inner nuclear layer (*INL*) by the 2nd week after birth, and the deep plexus forms at the outer edge of the INL by the 3rd week. This vascular development is similar to the formation of the human retinal vasculature during the 3rd trimester in utero. (Images adapted from [11])

secreting growth factors, thus initiating proliferation of endothelial cells and vascularization of the retina [40]. Vascular endothelial growth factor (VEGF) plays an important role during retinal vascular development. Astrocytes secrete VEGF, which the endothelial cells respond to through VEGF receptors on the endothelial cell surface. Other growth factors such as insulin-like growth factor-1 (IGF-1) and fibroblast growth factors (FGFs) are also likely to participate in the initiation of retinal vascular development.

In addition to their important roles as initiators of retinal vascular development, astrocytes also play a critical role during guidance and maintenance of the neovascular plexus. During formation of the superficial plexus, retinal vessels are intimately associated with astrocytes [37, 40]. These neovessels develop in

direct contact with the astrocytes, migrating along the preexisting astrocytic template (Fig. 2.5). Recent evidence has demonstrated that $VEGF_{165}$ [39] and R-cadherin mediated cell-cell adhesion [11] are critical for normal astrocytic guidance of the superficial vessels. There are three main VEGF-A isoforms, $VEGF_{122}$, $VEGF_{165}$, and $VEGF_{188}$, each named for protein size. $VEGF_{165}$ and $VEGF_{188}$ both have heparin-sulfate binding motifs that are eliminated in the $VEGF_{122}$ splice variant. This limits the permeability of $VEGF_{165}$ and $VEGF_{188}$, causing these molecules to remain in the extracellular milieu proximate to the cells from which they were expressed. Thus, as $VEGF_{165}$ is secreted by the astrocytes in response to retinal hypoxia, the limited diffusion properties result in a VEGF pattern similar to the underlying

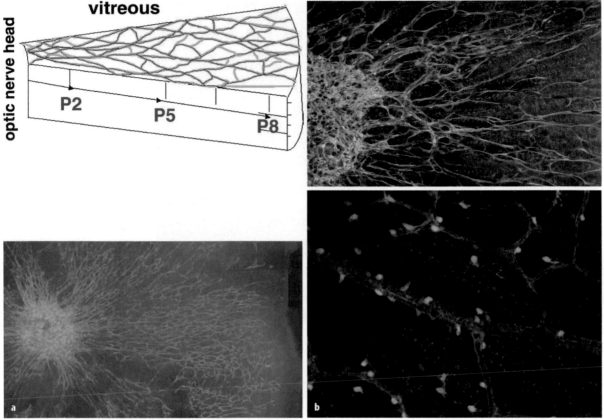

Fig. 2.5. Retinal vessels develop in close association with retinal astrocytes. **a** Retinal astrocytes (GFAP, *green*) emerge from the optic nerve head ahead of the retinal vessels (*red*), which subsequently grow along the preexisting astrocytic template (**b**). Mature retinal vessels (*red*) are wrapped by astrocytes (*green*) helping to form the blood-retinal barrier (*blue* stains the underlying nerve fiber layer). (Images adapted from [11])

astrocytes. Subsequent interactions between VEGF and the endothelial-VEGF receptors lead to the initial endothelial cell-astrocyte association.

Continuous interactions between endothelial cells and the underlying astrocytic template are mediated by adherens junctions. R-cadherin molecules on the astrocyte cell surface directly interact with adherence molecules on the endothelial cells (presumably other cadherin molecules), mediating the lasting endothelial cell-astrocyte association. When R-cadherin function is blocked, endothelial cells are no longer able to migrate along the astrocytic template and the superficial vasculature fails to develop normally [11]. Thus, through specific expression of growth factors and adhesion molecules, the astrocytes form a template that guides vascular development and patterning in the retina. As the vasculature matures, astrocytes begin to wrap around the newly formed vessels and this vessel-astrocyte association remains as an important aspect of the blood-retinal barrier throughout the adult life (Fig. 2.5).

2.5.2 The Role of Subcellular Endothelial Processes

Endothelial cells must be able to respond specifically to guidance cues expressed by the astrocytic template. This is mediated by growth factor receptors and specific adhesion molecules expressed at the endothelial cell surface. However, since $VEGF_{165}$ has limited diffusion capabilities and subsequent adherence junctions are between membrane-bound molecules, the endothelial cells must have a method of accessing these molecular guidance cues. To this extent, specialized endothelial cells at the tips of the migrating vascular front extend long filopodial-like processes that can access and initially respond to environmental cues (Fig. 2.6) [11, 18]. These processes mediate response to conditions within the regions ahead of the migrating vascular front, and thus mediate response to the astrocytic-guidance cues. Although not directly proven, evidence strongly suggests that the filopodia that fall along underlying astrocytes are stabilized by interactions between VEGF and its receptor, as well as the formation of R-

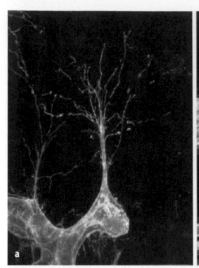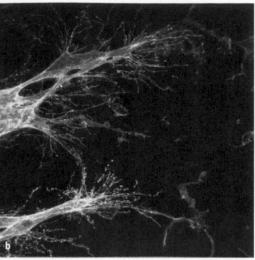

Fig. 2.6. Filopodial processes are extended from the tips of retinal endothelial cells at the developing vascular front. Isolectin *Griffonia simplicifolia* staining of the developing retinal vasculature allows visualization of the filopodial processes. (Images adapted from [11])

cadherin mediated adherence junctions. Longer, more established filopodial extensions tend to colocalize with underlying astrocytes [11, 18]. These stabilized filopodia then remain, initiating vascular growth and migration in that direction, while other non-stabilized filopodia are retracted. In this manner, the endothelial cells are guided along the astrocytic template during retinal development.

2.6 Development of the Deep Retinal Vascular Plexuses

As the retina continues to expand due to the final development and differentiation of neurons within the inner nuclear layer, the superficial vessels will eventually branch and begin formation of the deep and intermediate vascular plexuses. These vascular branches sprout perpendicular to the superficial plexus and dive toward the outer edge of the inner nuclear layer where they anastomose laterally and form a planar microvascular plexus. Eventually a third, intermediate vascular plexus will form at the inner edge of the inner nuclear layer from sprouts off the branches between the superficial and deep plexus (Fig. 2.4). While debate continues as to whether the superficial plexus forms by vasculogenesis or angiogenesis, general consensus is that the deep and intermediate plexuses fom solely by the process of angiogenesis.

The general principles that mediate formation of the deep vascular plexuses are similar to those that initiate and guide formation of the superficial plexus. A cytokine gradient is created toward the outer retina, mainly through the expression of growth facrs by retinal pigment epithelium (RPE), photoreceptors and Müller cells in response to the growing hypoxia created by neuronal growth, differentiation, and activity. This gradient is sensed by endothelial cells

in the superficial plexus initiating new growth toward the deep plexus [10]. Filopodia-like sprouts are again observed extended ahead of vessels migrating toward the deep plexus, suggesting that these endothelial cells can also respond to specific environmental guidance cues (Fig. 2.7). Indeed, R-cadherin expression is also present in the zones where the deep and intermediate plexus forms, just prior to, and throughout neovascularization of these regions. In mouse models, if R-cadherin mediated adhesion is blocked, the vessels are no longer guided to the appropriate layers and instead migrate directly past the inner nuclear layer into the normally avascular photoreceptor layer [11]. Thus, similar to the superficial plexus, mechanisms involving specific expression of growth factors and cell-cell adhesion molecules are also important for initiation and guidance of the deep retinal vascular plexuses. Bone marrow derived cells also rely on similar cues to target to sites of retinal neovascularization, even ahead of the developing endogenous vessels, further supporting the existence of preexisting guidance cues [12]. Normally, these Lin⁻ bone marrow-derived cells will target to astrocytes and the developing vasculature (Fig. 2.2). However, when R-cadherin mediated adhesion is blocked, many of these cells lose targeting abilities and instead migrate through the retina and mistarget to the subretinal space [12] similar to the mistargeting of the endogenous vasculature (Fig. 2.7).

2.7 Vascular Maturation

As the characteristic retinal plexuses are formed, they must undergo final maturation before vascular development is complete. This involves appropriate mural cell recruitment, and remodeling of the vascular plexus. Mural cell recruitment occurs almost con-

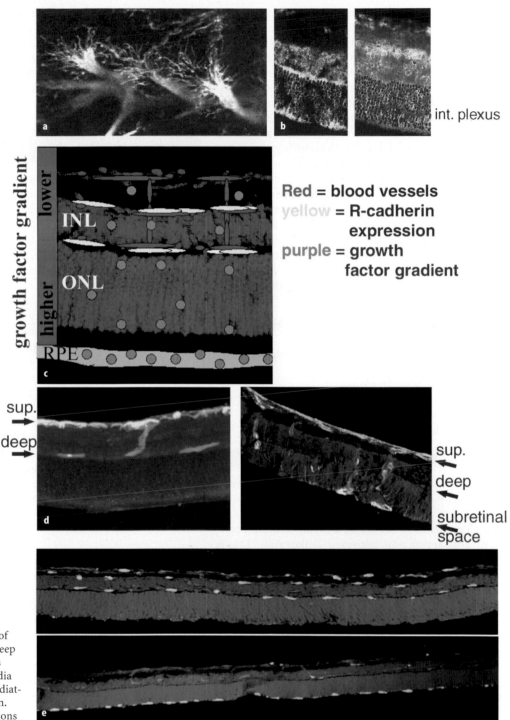

Fig. 2.7. Formation of the characteristic deep vascular plexuses is mediated by filopodia and R-cadherin mediated cell-cell adhesion.
a Filopodial extensions are observed at the endothelial tips as vessels branch and migrate from the superficial plexus to the deep vascular plexuses. **b** R-cadherin expression correlates spatially and temporally with deep vascular plexus formation. **c** As vessels branch from the superficial plexus and migrate toward the deep retina, normal R-cadherin guidance cues guide the vessels to the normal deep vascular plexus. **d** The normal deep retinal plexus forms at the outer edge of the inner nuclear layer (*left*), but when R-cadherin adhesion is blocked (*right panel*), these guidance cues are lost and the deep vascular plexus fails to form normally resulting in the abnormal formation of subretinal vessels (*green* astrocytes, *red* blood vessels, *blue* DAPI nuclei). **e** Lin-bone marrow derived progenitor cells (*green*) also use R-cadherin mediated adhesion to target to the three vascular plexuses (*top*). When R-cadherin adhesion is blocked, targeting of the bone marrow-derived progenitors is also lost and these cells become abnormally localized within the subretinal space (bottom). (**a, b** images adapted from [11]; **c** image adapted from [10]; **d, e** image adapted from [12])

comitantly with neovascular formation [9]. As the new vessels grow, endothelial cells secrete platelet derived growth factor (PDGF). Mural cells respond to this signal through receptors on their cell surface and are thereby recruited to the neovessel surface. The lack of functional PDGF is embryonic lethal due to numerous microvascular aneurysms thought to be caused by a lack of pericyte association with the developing vasculature [24]. Similar to the importance of VEGF localization during the endothelial cell guidance, PDGF localization is also important for appropriate mural cell recruitment. PDGF also has a heparin binding motif that prevents its diffusion away from the neovessels. This helps localize the recruited mural cells to the neovessel surface. Removal of the heparin-binding domain from PDGF results in a lack of pericyte recruitment in the developing retinal vasculature and these vessels eventually regress leading to several vascular abnormalities.

Deletion of the PDGF Heparin Binding Motif Results In:
- Loss of mural cell recruitment in the developing retina (fewer pericytes)
- 1elayed vascular development
- Abnormal vessels with irregular branching
- Excessive remodeling
- Vascular leakiness

2.8 Vascular Pruning Mechanisms

The newly formed retinal vascular plexuses are initially highly dense networks of vessels that must be pruned during the later stages of vascular maturation. This pruning occurs as a result of the recruitment and activity of activated leukocytes [21]. A brief period of hyperoxia results from the initial overly dense vascular networks. This causes upregulation of intercellular cell adhesion molecule 1 (ICAM-1) on the lumen surface of the endothelial cells. CD18 molecules on circulating leukocytes adhere to ICAM-1 on the sides of the vessels leading to extravasation and activation of these cells. Through Fas-L mediated apoptosis, these cells then cause the regulated cell death of certain endothelial cells resulting in the mature, pruned retinal vasculature observed in a normal, healthy adult.

Vascular Pruning Activities:
- Mural cell recruitment during vascular development
- Hyperoxia caused by the dense neovascular plexuses formed during early retinal vascular development
- Upregulation of ICAM-1 on the endothelial cells' lumen surface

- Recruitment and activation of circulating leukocytes through CD18/ICAM-1 interactions
- Regulated endothelial cell death leading to a mature, pruned vascular network. Selective mural cell association may help stabilize certain vessels and regulate the extent of vascular regression.

2.9 Mouse Retinal Vascular Development as a Model for General Vascular Development

Investigating the mechanisms of vascular development can be quite difficult due to the fact that most vascular systems develop during embryogenesis. In vitro assays, while important, often fail to replicate the intricate nature of neovascularization. For in vivo studies, the importance of specific factors has historically been determined by analyzing the effect of genetic knock-downs. However, this is complicated by natural compensatory mechanisms that often result in little or no phenotypic differences, even when factors known to be important for vascular development are deleted. Also, it is very difficult to administer exogenous compounds to developing fetuses in a controlled manner, making it difficult to assess the roles of particular factors using molecular agonists or antagonists.

The retina, particularly in non-primate species, has several advantages that overcome many of the experimental complications normally associated with in vivo studies of developmental vascularization, allowing researchers to study the molecular events of developmental neovascularization in its natural context.

Essentials
- In non-primates, retinal vascular development occurs postnatally
- Direct intravitreal injections of exogenous molecules are possible
- Retinal vascular development is highly organized
  - Allows even subtle alterations in vascular patterning to be observed more easily
  - Mediates specific studies of vascular guidance mechanisms
- Retinal vascular development is temporally consistent
  - Facilitates specific studies regarding different phases of retinal vascular development (superficial vs. deep plexus, etc.)
  - Facilitates analysis of the effects of exogenous pro- or anti-angiogenic molecules

- Utilization of vascular targeting bone marrow-derived progenitor cells facilitates analysis of the effects of specific knock-ins or knock-downs
  - Faster and more simple than engineering and producing various transgenic mouse colonies
  - Use of siRNA or other non-genetic methods of altering protein expression is applicable to multiple species

2.10 Use of Retinal Vascular Development as Models for Clinical Ocular Neovascularization

Because of the inherent advantages, and the similarities between developmental and pathological angiogenesis, many aspects of retinal vascular development have been adapted as models for clinical ocular neovascularization. These models are used for various preclinical studies ranging from identifying mechanisms associated with vascular retinopathies, to testing potential angiostatics for clinical use.

2.10.1 Mouse Retinal Angiogenesis Model

In the neonatal mouse, the retinal vasculature develops during the first 3 weeks after birth, with the superficial plexus forming during the first 10 days. The deep vascular plexus begins formation during the 2nd week with sprouts budding from the superficial plexus around postnatal day 8 (P8) and a fully dense, albeit not yet fully remodeled, plexus in place by P12. By injecting various angiostatic compounds at P7–P8, either systemically by intravenous injection, or locally by intravitreal or subretinal injection, and assessing the effects on the formation of the deep vascular plexus, the activity of these compounds can be analyzed relatively easily in vivo (Fig. 2.8). Importantly, by analyzing the effects on the already established superficial plexus, one can also assess whether these compounds may adversely affect pre-established normal retinal vessels. This model has been used extensively to test various angiostatic and combinations of angiostatic molecules [13, 30].

2.10.2 Oxygen-Induced Retinopathy

By altering the oxygen levels to which the neonatal mouse (or rat) pups are exposed, one can also manipulate the retinal vascular development in a way that mimics many aspects of clinical disease. In several animal species, including the kitten, the beagle puppy, the rat, and the mouse [25], exposing newborn animals to hyperoxia (or to alternating hyperoxia and hypoxia) prompts regression or delay of retinal vascular development, followed by abnormal neovascularization upon return to normal oxygen levels (Fig. 2.8). These models mirror the events that occur during retinopathy of prematurity (ROP), a condition involving pathological neovascularization that can affect premature infants [2, 19, 38]. Retinopathy of prematurity will be discussed extensively in later chapters. In recent years, the use of this model has been extended to the general study of ischemic vasculopathies and related anti-angiogenic interventions, and it is now used extensively in both basic and applied research environments.

References

1. Abbott NJ (2002) Astrocyte-endothelial interactions and blood-brain barrier permeability. J Anat 200:629–38
2. Ashton N (1966) Oxygen and the growth and development of retinal vessels. In vivo and in vitro studies. The XX Francis I. Proctor Lecture. Am J Ophthalmol 62:412–35
3. Augustin HG, Braun K, Telemenakis I, Modlich U, Kuhn W (1995) Ovarian angiogenesis. Phenotypic characterization of endothelial cells in a physiological model of blood vessel growth and regression. Am J Pathol 147:339–51
4. Banin E, et al. (2006) T2-TrpRS inhibits preretinal neovascularization and enhances physiological vascular regrowth in OIR as assessed by a new method of quantification. Invest Ophthalmol Vis Sci 47(5):2125–34
5. Campochiaro PA (2000) Retinal and choroidal neovascularization. J Cell Physiol 184:301–10
6. Carmeliet P (2000) Mechanisms of angiogenesis and arteriogenesis. Nat Med 6:389–95
7. Chan-Ling T, McLeod DS, Hughes S, Baxter L, Chu Y, Hasegawa T, Lutty GA (2004) Astrocyte-endothelial cell relationships during human retinal vascular development. Invest Ophthalmol Vis Sci 45:2020–32
8. Cleaver O, Melton DA (2003) Endothelial signaling during development. Nat Med 9:661–8
9. Darland DC, D'Amore PA (2001) Cell-cell interactions in vascular development. Curr Top Dev Biol 52:107–49
10. Dorrell MI, Friedlander M (2006) Mechanisms of endothelial cell guidance and vascular patterning in the developing mouse retina. Prog Retin Eye Res 25:277–95
11. Dorrell MI, Aguilar E, Friedlander M (2002) Retinal vascular development is mediated by endothelial filopodia, a preexisting astrocytic template and specific R-cadherin adhesion. Invest Ophthalmol Vis Sci 43:3500–10
12. Dorrell MI, Otani A, Aguilar E, Moreno SK, Friedlander M (2004) Adult bone marrow-derived stem cells utilize R-cadherin to target sites of neovascularization in the developing retina. Blood 103:3420
13. Dorrell MI, et al. (2007) Combination angiostatic therapy completely inhibits ocular and tumor angiogenesis. Proc Natl Acad Sci U S A 104(3):967–72
14. Dosso AA, Leuenberger PM, Rungger-Brandle E (1999) Remodeling of retinal capillaries in the diabetic hypertensive rat. Invest Ophthalmol Vis Sci 40(10):2405–10
15. Folkman J, D'Amore PA (1996) Blood vessel formation: what is its molecular basis? Cell 87:1153–5

2

Inhibition levels

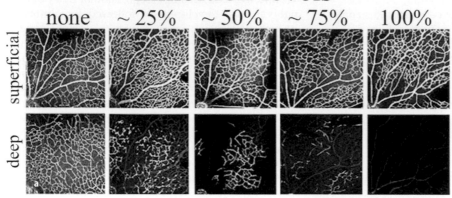

none   ~ 25%   ~ 50%   ~ 75%   100%

superficial

deep

a

Mouse OIR Model

P10
(in hyperoxia)

P12 (return
to normoxia)

P15 (initial
revascularization)

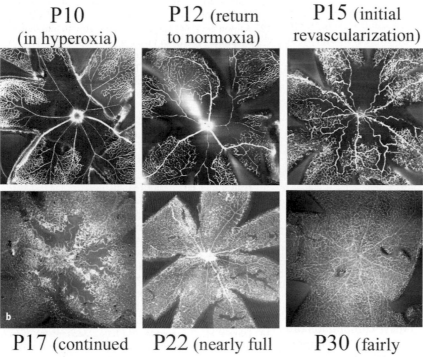

b

P17 (continued
revascularization,
many pathological
tuft formations)

P22 (nearly full
revascularization,
few tufts remain)

P30 (fairly
normal retinal
vasculature

Fig. 2.8. Clinical utility of mouse models of developmental retinal vascularization. **a** Anti-angiogenic factors can be injected at postnatal day 7, just before formation of the deep vascular plexus, and the effect on the formation of the deep vascular plexus can be assessed 5 days later. This allows analysis of angiostatic efficacy by visualizing the resulting deep vascular plexus as well as analysis of toxicity by visualizing the previously formed superficial vascular plexus. **b** By incubating animals in hyperoxia during vascular development, the normal vascular development is disrupted. As these animals with an underdeveloped retinal vasculature are returned to normoxia, a hypoxic situation ensues. This leads to abnormal pathological revascularization in a process that mirrors the human condition retinopathy of prematurity and models ischemic retinopathies

16. Fruttiger M (2002) Development of the mouse retinal vasculature: angiogenesis versus vasculogenesis. Invest Ophthalmol Vis Sci 43:522–7
17. Gariano RF (2003) Cellular mechanisms in retinal vascular development. Prog Retin Eye Res 22:295–306
18. Gerhardt H, Golding M, Fruttiger M, Ruhrberg C, Lundkvist A, Abramsson A, Jeltsch M, Mitchell C, Alitalo K, Shima D, Betsholtz C (2003) VEGF guides angiogenic sprouting utilizing endothelial tip cell filopodia. J Cell Biol 161:1163–77

19. Hellstrom A, Perruzzi C, Ju M, Engstrom E, Hard AL, Liu JL, Albertsson-Wikland K, Carlsson B, Niklasson A, Sjodell L, LeRoith D, Senger DR, Smith LE (2001) Low IGF-I suppresses VEGF-survival signaling in retinal endothelial cells: direct correlation with clinical retinopathy of prematurity. Proc Natl Acad Sci U S A 98:5804–8
20. Hirschi KK, D'Amore PA (1996) Pericytes in the microvasculature. Cardiovasc Res 32:687–98
21. Ishida S, Yamashiro K, Usui T, Kaji Y, Ogura Y, Hida T, Honda Y, Oguchi Y, Adamis AP (2003) Leukocytes mediate reti-

nal vascular remodeling during development and vaso-obliteration in disease. Nat Med 9:781–8

22. Leak LV (1970) Electron microscopic observations on lymphatic capillaries and the structural components of the connective tissue-lymph interface. Microvasc Res 2:361–91

23. Lee P, Wang CC, Adamis AP (1998) Ocular neovascularization: an epidemiologic review. Surv Ophthalmol 43:245–69

24. Lindahl P, Johansson BR, Leveen P, Betsholtz C (1997) Pericyte loss and microaneurysm formation in PDGF-B-deficient mice. Science 277:242–5

25. Madan A, Penn JS (2003) Animal models of oxygen-induced retinopathy. Front Biosci 8:d1030–43

26. Masland RH (2001) The fundamental plan of the retina. Nat Neurosci 4:877–86

27. Mukouyama YS, Shin D, Britsch S, Taniguchi M, Anderson DJ (2002) Sensory nerves determine the pattern of arterial differentiation and blood vessel branching in the skin. Cell 109:693–705

28. Orlidge A, D'Amore PA (1987) Inhibition of capillary endothelial cell growth by pericytes and smooth muscle cells. J Cell Biol 105:1455–62

29. Otani A, Friedlander M (2005) Retinal vascular regeneration. Semin Ophthalmol 20:43–50

30. Otani A, Slike BM, Dorrell MI, Hood J, Kinder K, Ewalt KL, Cheresh D, Schimmel P, Friedlander M (2002) A fragment of human TrpRS as a potent antagonist of ocular angiogenesis. Proc Natl Acad Sci U S A 99:178–83

31. Otani A, et al. (2002) Bone marrow-derived stem cells target retinal astrocytes and can promote or inhibit retinal angiogenesis. Nat Med 8(9):1004–10

32. Otani A, et al. (2004) Rescue of retinal degeneration by intra-vitreally injected adult bone marrow-derived lineage-negative hematopoietic stem cells. J Clin Invest 114(6):765–74

33. Patan S (2000) Vasculogenesis and angiogenesis as mechanisms of vascular network formation, growth and remodeling. J Neurooncol 50:1–15

34. Provis JM (2001) Development of the primate retinal vasculature. Prog Retin Eye Res 20:799–821

35. Risau W (1997) Mechanisms of angiogenesis. Nature 386:671–4

36. Ritter MR, et al. (2005) Three-dimensional in vivo imaging of the mouse intraocular vasculature during development and disease. Invest Ophthalmol Vis Sci 46(9):3021–6

37. Sandercoe TM, Madigan MC, Billson FA, Penfold PL, Provis JM (1999) Astrocyte proliferation during development of the human retinal vasculature. Exp Eye Res 69:511–23

38. Smith LE (2002) Pathogenesis of retinopathy of prematurity. Acta Paediatr Suppl 91:26–8

39. Stalmans I, Ng YS, Rohan R, Fruttiger M, Bouche A, Yuce A, Fujisawa H, Hermans B, Shani M, Jansen S, Hicklin D, Anderson DJ, Gardiner T, Hammes HP, Moons L, Dewerchin M, Collen D, Carmeliet P, D'Amore PA (2002) Arteriolar and venular patterning in retinas of mice selectively expressing VEGF isoforms. J Clin Invest 109:327–36

40. Stone J, Itin A, Alon T, Pe'er J, Gnessin H, Chan-Ling T, Keshet E (1995) Development of retinal vasculature is mediated by hypoxia-induced vascular endothelial growth factor (VEGF) expression by neuroglia. J Neurosci 15:4738–47

41. Wechsler-Reya RJ, Barres BA (1997) Retinal development: communication helps you see the light. Curr Biol 7:R433–6

3   Retinal Angiogenesis and Growth Factors

3.1 General Concepts of Angiogenesis and Vasculogenesis

C. Ruiz de Almodovar, A. Ny, P. Carmeliet

Core Messages

- Vasculogenesis in the embryo is a result of mesoderm-derived endothelial precursor cells
- There is a tissue specific endothelial cell differentiation resulting in heterogeneous tight-junction formation and fenestration properties. The tissue specific differentiation of endothelial cells is regulated by a variety of growth factors
- Angiogenesis is the formation of blood vessels from existing ones and includes spouting, bridging, and intussusceptive growth from preexisting vessels, remodeling and pruning

3.1.1 General Introduction

The vasculature is the first organ to arise during development. Blood vessels run through virtually every organ in the body, ensuring metabolic homeostasis by supplying oxygen and nutrients and removing waste products. Consequently, a dysfunction of blood vessels compromises normal organ performance. This in turn may lead to congenital or acquired diseases, disability or even death. The lymphatic system develops in parallel but secondary to the blood vascular system. It serves an essential function in absorbing and transporting tissue fluid and extravasated proteins and cells back to the venous circulation. Understanding the principles of how blood and lymph vessels form and which angiogenic factors are involved might provide novel attractive opportunities for treatment of angiogenic disorders.

3.1.2 Angiogenic Disorders

Dysregulation of vessel growth, either because of an excess or an insufficient number of vessels, has a major impact on our health and contributes to the pathogenesis of many disorders. The first identified and best known angiogenic disorders are cancer, arthritis, psoriasis and blinding retinopathy [90, 162]. However, there are numerous other inflammatory, allergic, infectious, traumatic, metabolic or hormonal disorders, which are characterized by excessive vessel growth including atherosclerosis, restenosis, transplant arteriopathy, warts, allergic dermatitis, scar keloids, peritoneal adhesions, synovitis, osteomyelitis, asthma, nasal polyps, choroideal and intraocular disorders, retinopathy of prematurity, diabetic retinopathy, leukomalacia, AIDS, endometriosis, uterine bleeding, ovarian cysts, ovarian hyperstimulation, liver cirrhosis, nasal polyps and the list is still growing (Table 3.1.1) [41, 43, 45]. In obesity, adipose tissue may also show excessive vessel growth. A high fat diet induces an angiogenic gene program in fat [191] and angiogenic factors stimulate adipogenesis, while treatment of obese mice with anti-angiogenic agents results in weight reduction and adipose tissue loss [276]. Viral and bacterial pathogens carry angiogenic genes of their own [218], or induce the expression of angiogenic genes in the host [122]. The human herpesvirus 8 transforms endothelial cells (ECs) and causes Kaposi's sarcoma in HIV-1 infected AIDS patients. Infectious diseases, such as chronic airway inflammation, are also angiogenic [21].

Vessel pruning is a physiological mechanism to match perfusion with metabolic demand when the nascent vasculature consists of too many vessels. However, vessel regression also contributes to the pathogenesis of numerous disorders (Table 3.1.2). For instance, lower levels of the primary angiogenic factor, vascular endothelial growth factor (VEGF), cause organ dysfunction in pregnant women with preeclampsia [213]. A progressive loss of the microvasculature underlies many age-related diseases. In the skin, age-dependent reductions in vessel density and maturation cause vessel fragility leading to the development of purpura, telangiectasia, pallor, angioma and venous lake formation [54]. In old age, a pro-

Table 3.1.1. Diseases characterized or caused by abnormal or excessive angiogenesis

Organ	Disease in mice or humans
Numerous organs	Cancer (activation of oncogenes; loss of tumor suppressors) and metastasis; infectious diseases (pathogens express angiogenic genes [218], induce angiogenic programs [122] or transform ECs [22, 345]); vasculitis and angiogenesis in autoimmune diseases such as systemic sclerosis, multiple sclerosis, Sjögren's syndrome [166, 235, 309]
Vasculature	Vascular malformations (Tie-2 mutation [334]); DiGeorge syndrome (low VEGF/Nrp-1 expression [305]); hereditary hemorrhagic telangiectasia (mutation of endoglin or ALK [183, 331]); cavernous hemangioma (loss of Cx37/40 [298]); cutaneous hemangioma (VG5Q mutation [178, 321]); transplant arteriopathy and atherosclerosis [154, 163, 228]
Skin	Psoriasis (high VEGF and Tie2 [188, 340, 351]), warts [122], allergic dermatitis (high VEGF and PlGF [3, 242]), scar keloids [107, 355], pyogenic granulomas, blistering disease [32], Kaposi's sarcoma in AIDS patients [22], systemic sclerosis [75].
Adipose tissue	Obesity (angiogenesis induced by fat diet); weight loss by angiogenesis inhibitors [276]; anti-VEGFR2 inhibits preadipocyte differentiation via effects on ECs [93]; adipocytokines stimulate angiogenesis [292]
Eye	Persistent hyperplastic vitreous syndrome (loss of Ang-2 [94, 119] or VEGF164 [304]); diabetic retinopathy; retinopathy of prematurity [36]; choroidal neovascularization [36] (TIMP-3 mutation [256])
Bone, joints	Arthritis and synovitis [10, 176, 313, 318], osteomyelitis [125], osteophyte formation [202]; HIV-induced bone marrow angiogenesis [249]
Lung	Primary pulmonary hypertension (BMPR2 mutation; somatic EC mutations [140, 337, 359]); asthma [17], nasal polyps [111]; rhinitis [167]; chronic airway inflammation [21], cystic fibrosis [297]
Gastrointestinal tract	Inflammatory bowel disease (ulcerative colitis [169]), liver cirrhosis [83, 216, 346]
Reproductive system	Endometriosis [114, 139], uterine bleeding, ovarian cysts [1], ovarian hyperstimulation [184]
Kidney	Diabetic nephropathy [283, 354]

gressive loss of the microvasculature has been implicated in nephropathy [157], bone loss [210], and impaired reendothelialization after arterial injury [101]. Diabetes, atherosclerosis and hyperlipidemia also impair vessel growth [319, 328, 343], whereas hypertension causes microvascular rarefaction [29]. Reduced angiogenic signaling also causes pulmonary fibrosis [171] and emphysema [159]. Insufficient vascular growth or function may also significantly impact on the nervous system, as the extent of angiogenesis correlated with survival in stroke patients [172] and insufficient VEGF levels causes motor neuron degeneration, reminiscent of amyotrophic lateral sclerosis [239]. Amyloid-β, a molecule believed to play a central role in the pathogenesis of Alzheimer's disease and plaque formation, has negative effects on the cerebral vasculature [67].

Dysregulation of lymphatic vessels is also known to give rise to severe diseases, of which lymphedema and lymphatic metastasis are the most common. In lymphedema, the transport ability of lymphatic vessels is impaired and this leads to fluid accumulation in the tissue. The effects are manifested as chronic and disabling swelling, tissue fibrosis, adipose degeneration, poor immune function, susceptibility to infections and impaired wound healing [5, 238, 270].

Primary lymphedemas are rare genetic disorders while secondary lymphedema results from various types of damage to the lymphatic vessels caused by for example radiation therapy, surgery or parasite infections. At present, there is no cure to this disease and only symptomatic use of supportive stockings to reduce the swellings can be offered. Cancer metastasis today is one of the leading causes of mortality in our society. Excessive proliferation of lymphatic vessels is mainly linked to cancer progression and tumor cell metastasis to lymph nodes and represents a major criterion for evaluating the prognosis of patients.

Thus, as an excess or impairment of blood and lymphatic vessel growth contributes to the pathogenesis of numerous disorders, it is important to define the molecular basis of these defects and how vessels grow, fail to grow or grow excessively. The development of compounds or strategies, stimulating or inhibiting the growth of blood vessels (a process called "angiogenesis") and lymphatic vessels (a process termed "lymphangiogenesis") offers unprecedented opportunities for treating lymphedema and cancer among many other diseases. We will therefore discuss some key principles of the various steps of vessel growth in the embryo and adult.

Table 3.1.2. Diseases characterized or caused by insufficient angiogenesis or vessel regression

Organ	Disease in mice or humans	Angiogenic mechanism
Nervous system	Alzheimer's disease	Vasoconstriction, microvascular degeneration and cerebral angiopathy due to EC toxicity by amyloid-β [68, 366]
	Amyotrophic lateral sclerosis; diabetic neuropathy	Impaired perfusion and neuroprotection, causing motoneuron or axon degeneration due to insufficient VEGF production [15, 179, 239, 307, 308]
	Stroke	Correlation of survival with angiogenesis in brain [172]; stroke due to arteriopathy (Notch-3 mutations [155])
Vasculature	Diabetes	Characterized by impaired collateral growth [343], and angiogenesis in ischemic limbs [269], but enhanced retinal neovascularization secondary to pericyte dropout [35]
	Hypertension	Microvessel rarefaction due to impaired vasodilation or angiogenesis [29, 173, 279]
	Atherosclerosis	Characterized by impaired collateral vessel development [328]
	Restenosis	Impaired reendothelialization after arterial injury [101]
Heart	Ischemic heart disease, cardiac failure	Imbalance in capillary-to-cardiomyocyte fiber ratio due to reduced VEGF levels [173, 295]
Gastrointestinal tract	Gastric or oral ulcerations	Delayed healing due to production of angiogenesis inhibitors by pathogens (*Helicobacter pylori*) [150, 164]
	Crohn's disease	Characterized by mucosal ischemia [123, 169]
Bone	Osteoporosis, impaired bone fracture healing	Impaired bone formation due to age-dependent decline of VEGF-driven angiogenesis [210]; angiogenesis inhibitors prevent fracture healing [360]; osteoporosis due to low VEGF [255]; healing of fracture non-union is impaired by insufficient angiogenesis [125]
Skin	Hair loss	Retarded hair growth by angiogenesis inhibitors [358]
	Skin purpura, telangiectasia, and venous lake formation	Age-dependent reduction of vessel number and maturation (SMC dropout) due to EC telomere shortening [357]
	Systemic sclerosis, lupus	Insufficient compensatory angiogenic response [204]
Reproductive system	Preeclampsia	EC dysfunction, resulting in organ failure, thrombosis and hypertension due to deprivation of VEGF by soluble Flt1 [189, 213]
	Menorrhagia (uterine bleeding)	Fragility of SMC-poor vessels due to low Ang-1 production [129]
Lung	Neonatal respiratory distress syndrome (RDS)	Insufficient lung maturation and surfactant production in premature mice with low HIF-2/VEGF [61]; low VEGF levels in human neonates also correlate with RDS [326]
	Pulmonary fibrosis, emphysema	Alveolar EC apoptosis upon VEGF inhibition [159, 214, 317]
Kidney	Nephropathy (ageing; metabolic syndrome); glomerulosclerosis; tubulointerstitial fibrosis	Characterized by vessel dropout, microvasculopathy and EC dysfunction (low VEGF; high TSP1) [99, 157, 198]; recovery of glomerular/peritubular ECs in glomerulonephritis, thrombotic microangiopathy and nephrotoxicity is VEGF-dependent [282]

3.1.3 Modes of Vessel Growth

In the developing embryo, as well as in adult tissues, there are key events and distinct mechanisms that exist to establish and maintain a functional vascular network. Endothelial progenitor cells (EPCs) arising from various embryonic regions or from adult bone marrow (BM) can form vessels in a process referred to as vasculogenesis (Fig. 3.1.1). *Angiogenesis* denotes the process in which budding from preexisting vessels gives rise to sprouts of new blood vessels, while *arteriogenesis* refers to the stabilization of these new sprouts by mural cells such as pericytes and smooth muscle cells – arteriogenesis is critical for the new vasculature to become stable, mature and functional (Fig. 3.1.1). Collateral vessel growth represents the formation of collateral bridges between arterial networks and remodeling of preexisting vessels after occlusion of a main artery – this type of vessel growth is of major therapeutic importance. A fine-tuned interplay between molecular markers in a spatial and temporal manner is necessary for these essential events to occur. We will now discuss these individual steps in more detail.

3.1.4 Vasculogenesis

Vasculogenesis in the embryo is different from that after birth. In the embryo, mesoderm-derived endothelial precursor cells give rise to the first embryonic

Embryonal Vasculogenesis   **Angiogenesis**
                                                Arteriogenesis

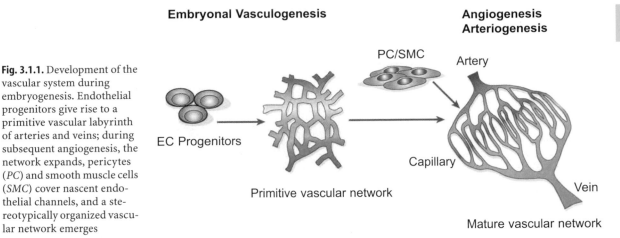

Fig. 3.1.1. Development of the vascular system during embryogenesis. Endothelial progenitors give rise to a primitive vascular labyrinth of arteries and veins; during subsequent angiogenesis, the network expands, pericytes (*PC*) and smooth muscle cells (*SMC*) cover nascent endothelial channels, and a stereotypically organized vascular network emerges

blood vessels (vasculogenesis). In the adult, EPCs originating from the bone marrow can enter the blood circulation and are recruited to sites of neovascularization (adult vasculogenesis). These two different events will be discussed separately in this chapter. In addition, we will discuss the exciting recent insights that hematopoietic progenitors also contribute to the formation of new vessels in the embryo and adult.

3.1.4.1 Role of Endothelial Progenitors in the Embryo

In the embryo, blood vessels emerge through recruitment of separate mesodermal precursors at distinct locations in the mesoderm. In amniotes in particular, the first blood vessels arise in the extraembryonic mesoderm of the yolk sac, when mesenchymal cells aggregate to form blood islands. The cells on the outer boundaries give rise to precursors of ECs, while the inner cell aggregates form primitive hematopoietic cells. Vascular progenitor cells that contribute to the formation of the major intraembryonic vascular system also derive from the intraembryonic mesoderm, and differentiate to form the dorsal aorta, the cardinal veins and the vitelline plexus. Within the embryo proper, the different mesodermal compartments vary in their vasculogenic capacity, the splanchnopleural and the paraxial mesoderm being the richest in endothelial precursor cells. Grafting experiments in quail and chick embryos suggest the existence of two distinct lineages of endothelial progenitor cells [244]. A first lineage, derived from the paraxial mesoderm, is known to have solely angioblastic capacity. A second bipotential hemangioblastic precursor cell line is derived from the splanchnopleural mesoderm, and differentiates to both endothelial and hematopoietic cells. Although no ultimate proof has been provided, the close prox-

imity of differentiating hematopoietic and ECs at sites of both extra- and intraembryonic vasculogenesis [66] suggests the existence of a bipotential mesodermal precursor cell for both systems, the *hemangioblast* [59, 264]. Lymphatic vessels arise via transdifferentiation of venous ECs, but may also develop from lymphangioblast. The latest has been demonstrated in birds such as chick and quail [281, 350] and recently also in the amphibian *Xenopus laevis* [231] (Fig. 3.1.2).

A common origin of blood cells and ECs is further suggested by molecular links between the embryonic precursors of endothelial and hematopoietic lineages, which share expression of various markers, including VEGF receptor-2 (VEGFR-2; also known as fetal liver kinase 1, Flk-1), CD34, AC133, PECAM- 1, c-Kit and Sca-1. In addition, in vitro experiments have shown that a transient population of so-called blast colony forming cells (BL-CFC) can be derived from embryonic stem cell cultures [161, 230]. BL-CFCs are responsive to VEGF and can produce both endothelial and hematopoietic cells. Strikingly, expression of mesodermal genes precedes the expression of genes marking early stage endothelial and hematopoietic development in these embryonic stem cell lines [82, 161], thus recapitulating the gene expression sequence observed in the yolk sac in vivo. Moreover, isolation of single cells expressing VEGFR-2 that can give rise to both endothelial and hematopoietic cells in vitro strongly suggests the existence of a common progenitor for the two lineages [161], even though it does not rule out the possibility that this progenitor is actually a more primitive multipotent precursor.

The molecular players that determine the early steps of hemangioblast differentiation are not completely elucidated yet. However, several genes have so far been implicated in this process, including

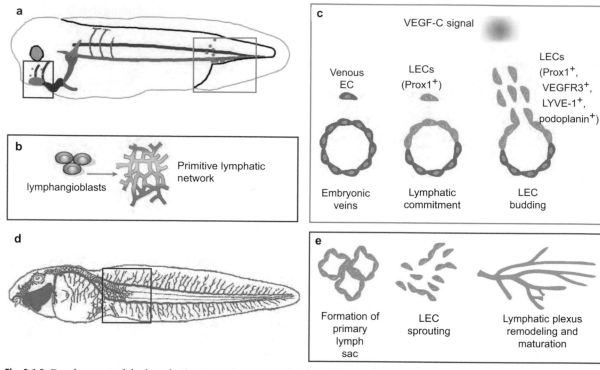

Fig. 3.1.2. Development of the lymphatic system in *Xenopus laevis*. **a** Scheme of vascular development in tadpoles, veins in *blue*, arteries in *red* and lymphatic vessels in *green*. At stage 33/34, lymphangioblasts give rise to the lymphatic endothelial cells (*LECs*) in the rostral lymph sac (*inset*, **b**), LECs in the trunk region detach from the posterior cardinal vein, form the ventral caudal lymphatic vessel and migrate dorsally to form the dorsal lymphatic vessel (*inset*, **c**). **b** Lymphangioblasts form a primitive lymphatic network in the rostral lymph sac. **c** Transdifferentiation of veins giving rise to lymphatic vessels in the trunk area. **d** Modification of drawing by M. Hoyer [137] displaying the maturing lymphatic network in the tadpole (*inset*, **e**). **e** The formation of the primary lymph sacs is followed by the LEC sprouting and finally by the remodeling and maturation of the lymphatic plexus

Ets-1, Fli-1, Vezf-1, VEGFR-2 and members of the GATA-, Hox-, the transcription factor Tal-1 (T-cell acute leukemia, also known as Scl) and inhibitor of differentiation (Id) protein families. VEGFR-2 is expressed on endothelial precursors, and development of any blood vessels or hematopoietic cells is defective in VEGFR-2 deficient embryos [288, 289]. Loss of VEGF, by contrast, induces severe vascular defects, but EC differentiation still occurs to a certain extent [46]. Whether this implies that VEGF-C, which also binds VEGFR-2, additionally controls hemangioblast differentiation remains to be further determined. VEGF also regulates arterial versus venous specification (see below). Tal-1 is involved in early cell fate determination of the hemangioblast, most likely by synergizing with VEGFR-2 [79, 336]. In contrast, targeted inactivation of VEGF receptor-1, VEGFR-1 (also fms-like tyrosine kinase 1, Flt-1), does not prevent hemangioblast differentiation, but leads to vascular disorganization, most likely due to an uncontrolled excess in endothelial progenitor cells [91, 160].

Endodermal signals may also regulate vasculogenesis in the adjacent mesoderm. Recent studies in amphibian and avian embryos suggest that the endoderm regulates the assembly of angioblasts to vascular tubes, rather than the specification of the hemangioblast lineage, and that Indian Hedgehog signaling is the key mediator involved in this interaction [338, 339]. Genetic studies in zebrafish show, however, that the endoderm provides a substratum for EC migration and that it is involved in regulating the timing of this process, but that it is not essential for the direction of migration neither for the formation of the vascular cord and lumen of vessels [153].

3.1.4.2 Role of Endothelial Progenitors in the Adult

Neovascularization in the adult has long been solely attributed to the process of sprouting angiogenesis. The isolation of putative endothelial progenitor cells from circulating mononuclear cells in the peripheral blood of adult humans has revealed that EPCs may home in from the bone marrow to sites of ongoing physiological or pathological neovascularization (Fig. 3.1.3). Apart from the bone marrow, there

Adult Vasculogenesis   Incorporation of EPCs

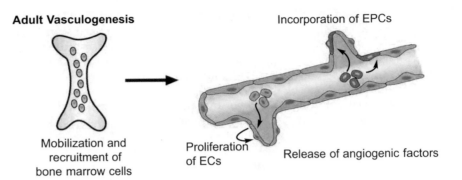

Mobilization and   Proliferation
recruitment of   of ECs   Release of angiogenic factors
bone marrow cells

Fig. 3.1.3. Vasculogenesis in the adult. Bone marrow progenitor cells (EPCs and myeloid cells) are recruited from the bone marrow and incorporated into nascent vessels or stimulate new vessel growth by releasing proangiogenic factors and inducing the proliferation of resident endothelial cells

might be alternative sources of endothelial progenitor cells. Indeed, multipotent adult progenitor cells (MAPCs) with angioblastic potency have been identified in the bone marrow [261], while tissue specific stem cells might also exist. In skeletal muscle, myoendothelial progenitors might differentiate locally into muscle or ECs [315]. Unlike mature ECs, EPCs are characterized by a greater capacity to proliferate [195] and by the fact that they have not yet acquired mature endothelial markers. EPCs should be distinguished from circulating ECs (CECs), which are sloughed off due to shedding from the existing vasculature and enter the circulation as a result of traumatic and infectious vascular injury or tumor growth. EPCs express EC surface antigens such as VEGFR-2, CD34, vascular endothelial cadherin (VE-cadherin) and AC133 among other markers [257]. AC133 is lost upon differentiation of the EPCs into more mature ECs [146]. Since mature myelomonocytic cells have lost surface expression of AC133, this marker also provides an effective means to distinguish EPCs from mature cells of myelomonocytic origin [259].

EPCs in the bone marrow likely reside in close association with hematopoietic stem cells and stromal cells in bone marrow niches. Though still largely undefined, these cells are possibly involved in promoting local EPC proliferation and transmigration across the bone marrow/blood barrier via secretion of VEGF, PlGF (placental growth factor, a homologue of VEGF) and other angiogenic factors [324]. Mobilization of EPCs from the bone marrow, as well as their recruitment to sites of adult vasculogenesis, involves a number of similar cues that regulate EC sprouting (angiogenesis), such as VEGF [11], fibroblast growth factor 2 (FGF-2), PlGF [49, 257], the recently discovered platelet-derived growth factor family member PDGF-CC [192], and angiopoietin-1 (Ang-1) [124]. In addition to stimulating proliferation of ECs, the chemokine stromal cell-derived factor (SDF-1), a chemoattractant for hematopoietic progenitor cells (HPCs), also induces migration of EPCs, which

express CXCR4, a receptor for SDF-1 [229, 250, 353], while inhibition of SDF-1 blocks EPC recruitment to sites of neovascularization [117].

It remains an outstanding question to what extent and how precisely EPCs contribute to vascular growth. Apart from differentiating to mature ECs that become incorporated as building blocks in the endothelial layer of nascent vessels [250], mononuclear cells might, together with accessory cells derived from the bone marrow, create a proangiogenic microenvironment to facilitate neovascularization. CD34-expressing cells mobilized from the bone marrow stimulate vascularization in myocardial infarcts both via vasculogenic in situ vessel formation and via stimulation of angiogenic sprouting from resident endothelium by secretion of angiogenic growth factors [168]. Recently, a new role has been described for SDF-1 as a retention factor for angiocompetent cells recruited from the bone marrow to the site of neovascularization [115].

The relative numeric contribution of bone marrow-derived EPCs to adult organ and tumor neovascularization is highly variable. In different experimental settings of pathological angiogenesis, incorporation of EPCs into the growing vasculature has been reported to be remarkably high [98] or negligibly low [258, 365]. Apart from differences in the genetic background of mouse strains used for these studies, the variability might also reflect remarkable differences of spontaneous and xenografted tumors in their dependence on bone marrow derived endothelial precursors [277]. Mathematical models have been suggested to calculate – and possibly predict – the contribution of EPCs to tumor neovascularization [306].

Despite these unresolved questions, the concept of postnatal vasculogenesis offers challenging clinical perspectives for the treatment of cardiovascular disorders and cancer. Mobilization of endothelial progenitor cells from the bone marrow is enhanced in patients with ischemic conditions [294], and levels of circulating endothelial progenitor cells have been

3 |

introduced as a valuable clinical parameter for cardiovascular risk assessment [131]. In tumor bearing mice, EPC levels in peripheral blood correlate with the anti-angiogenic effect of angiogenesis inhibitors on tumor angiogenesis and growth [131], suggesting that EPCs (and CECs) may be useful biomarkers for dose finding and monitoring the effect of anti-angiogenic treatment in cancer. Intriguingly, a recent study has shown that in multiple myeloma (MM) patients with 13q14 deletion, CECs carried the same chromosome aberration as the neoplastic plasma cells, suggesting that not only EPCs or host-derived ECs, but also MM-derived CECs, may contribute to tumor vasculogenesis [265].

3.1.4.3 The Endothelial/Hematopoietic Connection – An Emerging Theme

In the embryo, hematopoietic stem cells (HSCs) migrate into avascular areas and attract sprouting ECs by releasing angiogenic factors, such as angiopoietin-1 [314]. In the adult, bone marrow-derived hematopoietic cells, expressing markers such as Sca-1, c-Kit, CXCR4 and/or VEGFR1, become recruited, often together with EPCs, to tumors or ischemic tissues in response to VEGF and PlGF [115, 240, 257]. These angiocompetent cells extravasate around nascent vessels, where they are retained by SDF-1, and stimulate growth of resident vessels by releasing angiogenic factors such as VEGF, PlGF and angiopoietin-2 (Ang-2) [15, 34, 237]. In other cases, these cells function as hemangioblasts, producing both hematopoietic and endothelial progenitors which give rise to new blood vessels [257]. Furthermore, in response to PlGF released by tumor cells, VEGFR-1 expressing hematopoietic bone marrow progenitors home in to tumor-specific premetastatic sites, where they recruit tumor cells and EPCs; anti-VEGFR1 antibodies prevent the formation of such premetastatic niches [263].

3.1.4.4 Arterial, Venous and Lymphatic Cell Fate Specification

Arteries and veins have evolved as anatomically distinct but closely interconnected blood vessel types. The structural differences between arteries and veins were attributed to different flow dynamics and distinct physiological requirements. But, evidence has recently emerged that molecular differences between arterial and venous ECs already exist, even before blood vessels are formed, and that complex genetic pathways are responsible for this arterial versus venous specification. The expression of the ligand ephrinB2 in arteries and of the Eph receptor tyrosine kinase EphB4 in veins occurs before the onset of circulation [103, 104]. This indicates that while ephrins are essential for proper distinction between arterial and venous cells, they are not required for the initial fate decision that distinguishes arterial and venous endothelial progenitors.

Lineage tracking in zebrafish embryos indicates that angioblasts in the lateral posterior mesoderm receive signals from the notochord and the ventral endoderm, and become restricted to the aorta or trunk vein [363]. Studies in zebrafish and *Xenopus* indicate that sonic hedgehog (Shh), produced by the notochord, specifies arterial EC fate [181]. Indeed, formation of the aorta is impaired in zebrafish embryos mutant for *sonic you* (*syu*), the zebrafish homolog of Shh [31, 56] or after morpholino knockdown of Shh [181]. Shh induces the expression in the adjacent somites of VEGF, which, in turn, drives arterial differentiation of angioblasts. In the chick, the early extraembryonic blood islands contain a mixture of subpopulations of cells expressing neuropilins 1 and 2 (Nrp-1 and Nrp-2), which subsequently become lineage markers of arteries and veins respectively [128]. This suggests that even early angioblasts may already be committed to either the arterial or venous lineage. Further evidence that VEGF has a role stems from findings that, when released from Schwann cells, it induces arterial specification of vessels, tracking alongside these nerves [224] and that Nrp-1, a receptor selective for the VEGF165 isoform, is expressed in arterial beds [222, 223, 304]. VEGF also determines arterial EC specification after birth in the heart and retina, where the matrix-binding VEGF188 isoform is critical for arterial development [304, 335].

The Notch pathway acts downstream of VEGF in arterial EC specification [181]. Notch signaling is initiated when the Notch receptors (Notch-1 – 4) are activated by their ligands Jagged-1, Jagged-2 and Delta-like-1, -3 and -4 [7]. During vascular development, defects in Notch signaling disrupt normal arterial-venous differentiation, resulting in loss of artery-specific markers (e.g., ephrin-B2) and ectopic expression of venous markers (e.g., VEGFR-3/Flt-4) in the aorta [180]. Conversely, overactivation of Notch induces ectopic arterial markers in veins, thereby suppressing vein differentiation. Mutation in Delta-4 (dll4), which is specifically expressed in developing arterial endothelium, leads to defective development of the dorsal aorta, with development of arteriovenous shunts [77, 95]. This is associated with downregulation of arterial markers and upregulation of venous markers in the dorsal aorta. Furthermore, Hey2, a transcription factor induced by Notch signaling, confers features of arterial EC gene expression to venous ECs, upregulating arterial-specific genes (ADHA1, EVA1, and keratin 7), while supn pressing vein-specific genes (GDF, lefty-1 and lefty-2) [57]. The hairy-related transcription factor *gridlock,* which is a downstream target of Notch signaling, is required for the early assignment of arterial endothelial identity [364]. Zebrafish lacking this protein show a disrupted assembly of the aorta in the posterior part of the body [363, 364], while overexpression of gridlock causes a similar disruption of the vein without affecting the artery [363]. Thus, activation of Notch signaling and its downstream response genes seem to be a requirement for specification of arterial cell fate with venous fate being the "default" state. Although arterial differentiation has been studied in more detail, little is known about venous cell fate specification. Recent insights have shown that the orphan nuclear receptor, COUP-TFII, is expressed specifically in venous endothelium and that mutation leads to activation of arterial markers in veins [361]. Together, these results suggest that an active pathway promoting venous fate exists and that COUP-TFII may participate suppressing Notch signaling.

All these genetic findings appear to refute the hypothesis that hemodynamic forces alone are responsible for arteriovenous differentiation. However, even after ECs attain a specific arterial or venous phenotype late in embryonic development, this genetic program still remains remarkably plastic [222]. Indeed, manipulation of circulatory flow has shown that flow could change gene expression and cell fate [145, 182]. Thus, vascular cell identity is refined by an interplay of hemodynamic forces and circulatory flow patterns in combination with an underlying genetic programming of arterial versus venous fate.

The lymphatic vessels do not start developing until the blood vasculature has been established. The events involved in the formation of the lymphatic vessels resemble the ones occurring in the development of blood vessels, such as the initiation of a primitive capillary plexus and remodeling to generate a mature vascular network. The earliest known event in the lymphatic vascular development occurs in the mouse embryo at stage E10.5 and is initiated by polarized expression of the homeobox transcription factor Prox-1 in a subset of venous ECs, a process known as lymphatic commitment [348] (Fig. 3.1.2). It is still not clear what is initiating Prox-1 expression, but a recent study has reported that interleukin-3 regulates Prox-1 expression in blood endothelial cells (BECs) and lymphatic endothelial cells (LECs) [113]. Next, the Prox-1 positive ECs bud off from the vein while changing their fate to become committed LECs overexpressing lymphatic endothelial specific genes (such as VEGFR3, lymphatic vessel endothelial hyaluronan receptor-1, LYVE-1, and

podoplanin) and downregulating blood vascular endothelial specific genes. The committed ECs migrate along a VEGF-C gradient emanating from nearby mesenchymal cells [158] and form primitive lymph sacs from which new lymphatic capillaries will sprout further to form primary lymphatic plexus. Following the formation of lymph sacs, the blood and the lymphatic structures develop separately and remain connected only at specific sites to allow lymph to return to the blood circulation (Fig. 3.1.2). The cornea of animals such as mouse or rabbit is being used to study lymphangiogenesis. In normal conditions the cornea is reported to lack both blood and lymphatic vessels, but under certain pathological conditions (see Box 1 below) the vessels are described to invade the cornea.

Much less is known about the subsequent remodeling and maturation of lymphatic vessels when they form a superficial lymphatic capillary network and deeper collecting lymphatic trunks. Recently, however, studies have revealed that the forkhead transcription factor, FOXC2, controls the late stages of lymphatic vascular development [158]. FOXC2 is highly expressed in developing lymphatic vessels and lymphatic valves; this gene is also mutated in the human hereditary disease lymphedema distichiasis [65, 158]. Mice deficient for *Foxc2* have lymphatic vessels that exhibit disorganized patterning, lack valves and acquire abnormal covering of smooth muscle cells resulting in backflow. A similar effect has been observed in patients diagnosed with lymphedema distichiasis [30, 158], which suggests that FOXC2 might control the maturation of the lymphatic vascular development and the formation of the lymphatic valves.

3.1.4.5 Tissue-Specific EC Differentiation

Endothelial cells in different organs acquire highly specialized properties, which permit these cells to optimally perform specific functions within each tissue and organ [275]. For instance, ECs in the brain are tightly linked to each other and are surrounded by numerous periendothelial cells, which constitute a barrier that protects brain cells from potentially toxic blood-derived molecules. The development of the blood-brain barrier requires interactions between astroglial cells that express glial fibrillary acidic protein, pericytes and adequate angiotensinogen levels [271]. The tight junctional complex between ECs consists of numerous integral membrane and cytosolic proteins from the families of cadherins, occludins, claudins and membrane-associated guanylate kinase homologous proteins [327]. In contrast, vessels in endocrine glands lack these tight junctions. Their endothelium is rather discontinuous and fenestrated, allowing high volume molecular and ion transport as well as hormone trafficking. Overall, the factors that regulate acquisition of specific endothelial properties are largely unknown. However, it appears that the interaction with the host environmental extracellular matrix, in concert with VEGF, plays a major role [267, 268]. Besides vascular cell heterogeneity in distinct organs, ECs within the same organ can even be heterogeneous. In the heart, for instance, ECs in distinct locations of the coronary vascular tree differ in their expression of the endothelial constitutive nitric oxide synthase isoform [9], brain-derived neurotrophic factor [76] or adhesion molecules [69, 70]. Even within a single vessel, ECs may have distinct cell fates. For example, three types of ECs, each with a distinct cell fate, build the intersegmental vessels in the zebrafish embryo [58].

Recently, genetic studies in mice, zebrafish and *Xenopus* are starting to unravel the transcriptional code that determines EC fate [31, 193]. This code involves bHLH transcriptional activators (hypoxia-inducible factor-2α, stem cell leukemia factor, Tfeb) [39] as well as Id repressors as demonstrated by the perturbation of developmental and tumor associated angiogenesis in mice lacking Id-1/3. Several members of the homeobox transcription factors (HOX) family are expressed in the vascular system during development and have also been implicated in EC fate determination [110]. Within the lymphatic system, LYVE-1 is predominantly expressed in the lymphatic capillary endothelium, while podoplanin is expressed both in the capillaries and valved collecting lymphatic vessels [209]. However, there are other mechanisms determining endothelial heterogeneity and organ-specific angiogenesis. For instance, the activity of VEGF and Ang-1 varies in different tissues. Low permeability tumors overexpress Ang-1 and/or underexpress VEGF or PlGF, whereas tumors with high permeability lack Ang-1, but overexpress Ang-2 [149]. Another example is the effect of Ang-1, which stimulates angiogenesis in the skin but suppresses vascular growth in the heart [312, 335].

An exciting new development is the discovery of tissue-specific vascular endothelial growth factors. A striking example of this novel class of cues is the endocrine gland-derived VEGF (EG-VEGF), which selectively affects EC growth and differentiation (fenestration) in endocrine glands [185, 186]. Other organ-specific angiogenic factors include prokineticin-2 and Bv8 in endocrine glands [184], Bves and fibulin-2 in the heart [342], and glial derived neurotrophic factor in the brain. That ECs in different tissues are distinct is further suggested by their considerably different response to anti-angiogenic factors. Indeed, ECs in endocrine glands rapidly lose their fenestrations and even become apoptotic in response to VEGF inhibitors, resulting in a 70% loss of the microvasculature in these organs [156]. By contrast, the microvasculature in other organs is much more resistant to such pruning in response to anti-VEGF therapy [156, 316]. Malignant cells also induce ECs in tumor vessels to acquire a distinct fate and express unique markers ("vascular zip codes") that are absent or barely detectable in quiescent blood vessels of normal tissue [273, 274]. Tumor cells also change the responsiveness of ECs to cues – for instance, epidermal growth factor (EGF) upregulates its receptors in tumor-associated vessels and thereby makes these ECs responsive to the mitogenic activity of EGF [20] – a finding of significant therapeutic relevance.

3.1.5 Angiogenesis

After vasculogenesis, the nascent primitive vascular labyrinth expands and becomes remodeled into a more complex, hierarchically and stereotypically organized network of larger vessels, ramifying into smaller vessels. This process involves sprouting, bridging and intussusceptive growth from preexisting vessels, navigation and guidance, remodeling and pruning. Many experimental models have been developed to study angiogenesis. Among them we have highlighted the most commonly used to study angiogenesis in the eye in Box 2. We will now discuss this process as the following orderly series of events (Figs. 3.1.1, 3.1.4). The avascular areas in the embryo release angiogenic cues that diffuse into the nearby tissues and activate ECs to induce extracellular matrix (ECM) degradation. ECs then proliferate and navigate toward these cues forming a sprouting. Finally, arteriogenesis takes place, whereby the newly formed blood vessel are stabilized by smooth muscle cells and pericytes.

3.1.5.1 Vascular Permeability and Extracellular Matrix Degradation

Water and nutrients move from blood to tissues across the walls of capillaries and venules. The wall of blood vessels is composed of ECs and mural cells, namely pericytes and smooth muscle cells, which are embedded in an ECM. The expression of cell adhesion molecules such as VE-cadherin in adherent junctions and claudins, as well as occludins and JAM-1 in tight junctions of quiescent vessels, provides mechanical strength and tightness to the vessel wall and establishes a permeability barrier. Between vascular cells, an interstitial matrix of collagen-I and elastin provides both viscoelasticity and strength to the vessel wall. The ECM is responsible for the contacts between ECs and the surrounding tissue, and

while anti-IL-8 antibodies block tumor growth [221]. Further, emerging evidence indicates that SDF-1 stimulates angiogenesis via direct effects on ECs, as well as via recruitment of bone marrow-derived EPCs and HPCs both in ischemic and malignant tissues [34, 51]; antagonists of SDF-1α block tumor growth [117].

Integrins also regulate proliferation of ECs. They are heterodimeric cell surface receptors of specific ECM molecules which, by bi-directionally transmitting signals between the outside and inside of vascular cells, assist vascular cells to build new vessels in coordination with their surroundings [136, 142]. The αvβ3 and αvβ5 integrins have long been considered to positively regulate the angiogenic switch [285], because their pharmacological antagonists which are currently being evaluated in clinical trials suppress pathological (i.e., tumor) angiogenesis [215]. Furthermore, a combination of antibodies against α1β1 and α2β1 integrins reduces tumor vascularization [285]. However, genetic deletion studies suggest that vascular integrins inhibit, not stimulate, tumor angiogenesis [262]. This inhibitory activity may be attributable to the ability of these integrins to suppress VEGFR-2 mediated EC survival, trans-dominantly block other integrins, or mediate the anti-angiogenic activity of angiogenesis inhibitors such as tumstatin, endostatin, and canstatin [42, 121, 187, 207, 262, 310]. Thus, while these genetic insights do not invalidate the promising (pre)clinical results obtained with integrin-antagonists for cancer treatment, a better understanding of whether and in which conditions integrins play positive or negative roles in tumor angiogenesis is desirable.

Epidermal growth factor (EGF) is a mitogen for epithelial cells and overexpressed in various tumors. While it does not regulate vascular development, it has been implicated in tumor angiogenesis. Indeed, EGF induces the expression of its own receptors in ECs and is mitogenic for EGFR-positive ECs [20]. In addition, EGF indirectly stimulates tumor angiogenesis by inducing the release of VEGF and the expression of VEGF receptors in tumor vessels [330]. Another growth factor, hepatocyte growth factor (HGF), stimulates angiogenesis when exogenously administered [152]. Other molecules are capable of stimulating EC growth in vitro or angiogenesis in experimental models, but their endogenous role in angiogenesis during development or disease often remains incompletely determined. Some examples include erythropoietin, IGF-1, neutropeptide-Y, leptin, tissue factor, interleukins and others [43].

Angiogenic sprouting is a complex process, requiring a finely tuned balance between activators and inhibitors. Some of the endogenous angiogenesis inhibitors that are currently being evaluated for clinical use include angiostatin [233], endostatin [234], antithrombin III, interferon-β, leukemia inhibitory factor and platelet factor 4 [41], tumstatin, PEX and vasostatin [41, 43, 232].

3.1.5.3 Vascular Lumen Formation

Sprouting ECs assemble into solid cords, which then undergo tubulogenesis to form vessels with a central lumen (Fig. 3.1.4). Little is known about how lumen formation is regulated in vivo. Gene targeting studies revealed that VEGF coordinatedly regulates vessel size and guidance. VEGF exists in different isoforms, with distinct affinities for the extracellular matrix. Thus, VEGF121 is diffusible, VEGF189 binds to the matrix, whereas VEGF165 has an intermediate profile (in mice, all VEGF isoforms are shorter by one residue). By virtue of their distinct affinities, the isoforms produce a gradient, with VEGF120 acting over a long range and VEGF188 over a short range [105, 272]. In the mouse retina, a gradient of matrix-bound VEGF produced by astrocytes guides endothelial tip cells, as alteration of the gradient by either loss- or gain-of-function manipulations led to loss or increased branching of vessels, respectively [105]. Further evidence for a role of VEGF gradients in tip cell guidance was deduced from the analysis of three mouse lines (the VEGF[120], VEGF[164], and VEGF[188] lines), each engineered to express a single VEGF isoform. VEGF[164] mice are normal, but VEGF[120] or VEGF[188] mice exhibit serious vascular remodeling defects [47, 304]. Vessels in VEGF[120] mutants are enlarged, stunted and exhibit reduced branching. Their tip cell filopodia extend chaotically in all directions, which is thought to cause lumen enlargement at the expense of directed branch formation and elongation. These defects presumably result from replacement of the normal VEGF gradient by a non-directional deposition of the highly diffusible VEGF120. In VEGF[188] mice, a shortage of diffusible VEGF causes the opposite phenotype, i.e., supernumerary branches at the expense of luminal enlargement.

In combination with VEGF, Ang-1 also augments lumen diameter [312]. Egfl7, a recently identified endothelial-derived secreted factor, is expressed at high levels in the developing vasculature. Knockdown of Egfl7 in zebrafish embryos revealed that Eglf7 specifically blocks vascular tubulogenesis. In Eglf7 knockdown embryos, angioblasts fail to separate and retain extensive tight junctions, which suggests that Eglf7 provides a permissive substrate that favors motility over stable attachment, thereby allowing the local movement of angioblasts that is required for tube formation [248]. Other molecules involved in the control of lumen formation are the

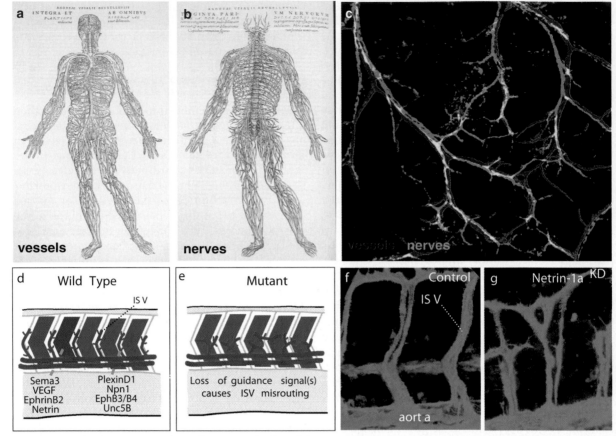

Fig. 3.1.5. Vessel and nerve patterning. **a, b** Drawings by the Belgian anatomist Andreas Vesalius [333], highlighting the similarities in the arborization of the vascular and nervous systems. **c** Vessel (*red*) and nerves (*green*) track together toward their targets (reproduced with permission from ref. [128]). **d** Schematic representation of the zebrafish embryo trunk, showing the somites (*red*) producing the indicated guidance cues and ISVs (*blue*) producing the indicated receptors. **e** In the absence of these guidance cues, the ISVs are misrouted. **f** Stereotyped pattern of ISVs in zebrafish embryos (the endothelial cells express an eGFP transgene). **g** After knockdown (KD) of *netrin1a*, the ISVs are misguided

integrins α5,β1 and α$_v$β$_3$ [23], likely because of their interaction with the surrounding ECM. Finally, thrombosondin-1 (TSP-1) and tubedown (tbdn)-1 suppress vascular lumen formation [100]. Several studies have indicated that the endoderm is required for proper vascular tubulogenesis [243, 339]; however, it has been shown recently that the endoderm is dispensable for vascular tube formation in zebrafish [153], raising the question of whether temporal and spatial expression patterns of molecular cues regulating angioblast migration and lumen formation have evolved differently.

3.1.5.4 Guided Navigation of Vessels

During evolution, organisms have come to perform more specialized tasks, requiring an increased supply of nutrients by blood vessels. Wiring of blood vessels into functional circuits is therefore of utmost importance. The complexity of this task is under-scored by the high degree of orderly patterning of the vascular networks. Five centuries ago, Andreas Vesalius illustrated the parallels in the stereotyped branching patterns of vessels and nerves (Fig. 3.1.5). Today, evidence is emerging that vessels, which arose later in evolution than nerves, coopted several of the organizational principles and molecular mechanisms that evolved to wire up the nervous system. The choreographed morphogenesis of both networks suggests that they are directed by genetically programmed mechanisms. Specialized endothelial "tip" cells are present at the forefront of navigating vessels and share many similarities with axonal growth cones [105]. They extend and retract numerous filopodia in saltatory fashion to explore their environment, suggesting that they direct the extension of vessel sprouts. The key function of the tip cells appears to be to "pave the path" for the subjacent "stalk" ECs. Tip cells proliferate minimally, whereas stalk cells proliferate extensively while migrating in the wake of the tip cell, thus permitting extension of the nascent vessel.

Guidance of embryonic vessels requires local guidance cues that instruct them to navigate along specific paths to reach their correct targets. Nerves and vessels face similar challenges in finding their trajectories, which are staked out with multiple checkpoints that divide navigation over a long trajectory into a series of shorter decision-making events [13]. Axons and vessels often take advantage of one another to follow the same path. In some cases, vessels produce signals (such as artemin and neurotrophin 3) that attract axons to track alongside the pioneer vessel [135, 175]. Conversely, nerves may also produce signals such as VEGF to guide blood vessels [224]. Very recent evidence reveals that the same cues that control axon guidance also function to pattern blood vessels. Four families of axon guidance cues, acting over a short (cell- or matrix-associated signals) or long range (secreted diffusible signals), have been identified: netrins and their DCC and Unc5 receptors, semaphorins and their neuropilin and plexin receptors, Slits and their Robos receptors and ephrins and their Eph receptors (reviewed in [44, 73, 138]). A role for Unc5b and Netrin1a in vessel guidance was identified by analysis of the developing intersegmental vessels (ISVs) in zebrafish embryos. Pathfinding of these vessels is stereotyped and believed to be genetically programmed by an interaction of attractive and repulsive cues. In control embryos, ISVs sprout from the dorsal aorta to the dorsolateral roof of the neural tube. After knockdown of Unc5b or Netrin1a, ISVs exhibit supernumerary, often ectopically located, filopodial extensions [199], and the dorsal trajectory of most ISVs is irregular, with numerous extra branches, deviating from the normal stereotyped path (Fig. 3.1.5). These findings suggest a tight control of ISV navigation by netrin family members.

Crosstalk between semaphorins and their receptors (e.g., neuropilins and plexins) is also necessary for ISVs to select the appropriate branching site along the dorsal aorta and to follow the pathway along the somites boundaries [116, 219, 286, 296, 325]. A vascular specific Robo homologue, Robo4, has been identified [247]. In vitro, Slit2 is able to repel ECs and Robo4 may mediate this effect [247]. A Robo4 knockdown study in zebrafish showed that some Robo4-expressing ISVs failed to sprout from the dorsal aorta or arrested midway through their dorsal migration path, whereas others deviated from their normal dorsal trajectory [24]. Repulsive ephrin-Eph signals provide short-range guidance cues for vessels to navigate through tissue boundaries. For instance, ephrinB2 repels EphB3/Eph-B4-expressing ISVs from entering somites [2, 236, 344].

Understanding this process in the primitive zebrafish embryo may have therapeutic implications beyond those for understanding better vascular biology, as many of these guidance signals are excessively expressed in tumors, which characteristically develop a chaotic, misguided vasculature [13].

3.1.5.5 Vessel Maintenance

When vessels sprout, they initially consist of naked EC channels. Once assembled in new vessels, these ECs become quiescent and survive for years. The importance of endothelial survival is demonstrated by the finding that diminished endothelial survival causes vascular dysfunction and regression in the embryo as well as in the adult [16, 48, 102]. The molecular mechanisms enabling a confluent endothelium to maintain its physiological function in various vascular beds for long periods of time are still unclear. However, some insights have been obtained from in vivo and in vitro studies. For instance, Ang-1 promotes, while Ang-2 suppresses, endothelial survival in the absence of angiogenic stimuli, and contributes to the regression of tumor vessels [96, 134, 312]. Hemodynamic forces are also essential for the maintenance of the vascular integrity in different vascular beds, as physiological shear stress reduces endothelial turnover and abrogates TNF-α mediated endothelial apoptosis [74]. Flow is critical too for maintaining vessel branches, as hypoperfused sprouts often regress.

But perhaps the most critical survival factor for quiescent ECs in the adult is VEGF. Thus, when VEGF levels are reduced, for instance by exposure of premature babies to hyperoxia, retinal vessels regress [6, 217]. The recent clinical experience with VEGF inhibitors has revealed that these anti-angiogenic agents may cause rare but important adverse effects, such as thrombosis, hypertension, bleeding and renal dysfunction [141]. Some of these adverse effects of anti-VEGF therapy can be explained by the requirement of threshold levels of VEGF for the survival and maintenance of quiescent vessels in healthy organs. For instance, the thrombotic risk may be related to the reduced release of fibrinolytic components [252], the increased release of fibrinolytic inhibitors and pro-coagulants [203], the reduced release of nitric oxide (inhibitor of platelet aggregation and vasospasms) [356], and to EC dysfunction resulting from deprivation of VEGF vessel maintenance signals [78, 102]. The hypertension is likely attributable to reduced vasodilation by NO, and possibly to pruning of normal vessels [279], while the proteinuria and glomerulonephritis may be related to the maintenance role of VEGF in podocyte functioning [80]. Bleeding in centrally located cavitary

3

necrotic lung tumors is likely due to vessel disintegration. As mentioned above, VEGF inhibitors cause the microvasculature to regress by 70% in endocrine organs, further highlighting the importance of VEGF as a maintenance cue for quiescent vessels in healthy organs [156, 316].

3.1.6 Arteriogenesis

Establishment of a functional vascular network requires that nascent vessels – formed by vasculogenesis and angiogenesis – mature into durable, stable, non-leaky and functional vessels. This stabilization requires recruitment of mural periendothelial and smooth muscle cells (SMCs), generation of an extracellular matrix and specialization of the vessel wall for structural support and regulation of vessel function – a process termed arteriogenesis (Fig. 3.1.6). Endothelial channels are covered by multiple layers of SMCs in large vessels in proximal parts of the vasculature and by single pericytes around smaller distal vessels. There are several sources of mural cells.

They can differentiate from neural crest cells, endothelial and mesenchymal cells and also originate from bone marrow derived progenitors. The coverage of vessels by mural cells (pericytes and SMCs) not only regulates EC proliferation, survival, differentiation and hemostatic control, but also assists ECs in acquiring specialized functions to accommodate various needs in different vascular beds (see above). Moreover, interstitial matrix components, generated by mural cells, interconnect ECs and provide blood vessels with viscoelastic properties.

3.1.6.1 Smooth Muscle Progenitor Cells

Smooth muscle cells (SMCs) provide structural integrity of the vessel wall. These cells contract spontaneously or in response to agonists, and regulate tissue perfusion. A striking feature is the considerable heterogeneous origin of SMCs, both during development and in the adult. As mentioned above, smooth muscle cells may differentiate from ECs, from mesenchymal cells, or from bone marrow precursors or

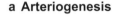

a Arteriogenesis

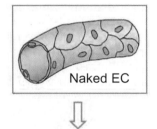

Naked EC

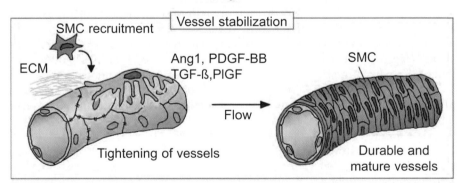

b Collateral vessel growth

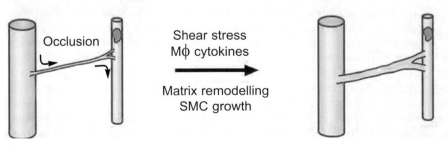

Fig. 3.1.6. Vessel maturation and stabilization. **a** When vessels sprout, they initially consist of naked endothelial cell channels. These nascent vessels become stabilized by recruitment of SMC and PC (a process called arteriogenesis), deposition of extracellular matrix and tightening of cell junctions. Blood flow plays a critical role in making these vessels durable. **b** Upon vascular occlusion, for instance by a thrombus, preexisting collateral vessels expand to ensure blood circulation. Recruitment of macrophages and monocytes to the shear stress-activated endothelium plays a critical role in this process

macrophages. The first SMCs in the dorsal aorta and the SMC-like myofibroblasts in the prospective cardiac valves transdifferentiate from the endothelium [108, 227]. Cardiac neural crest cells are the source of smooth muscle cells of the large thoracic blood vessels and the proximal coronary arteries [63, 108]. Mural cells from the distal coronary arteries are recruited from the epicardial layer [108], whereas coronary vein SMCs are derived from the atrial myocardium [71].

In the adult, SMCs arise from dividing preexisting SMCs or from BM-derived SMC progenitors, as exemplified in heterotypic cardiac or aortic transplantation in mice [132, 278]. In addition, fibroblasts can differentiate into myofibroblasts, which in turn differentiate into vascular SMCs in response to biochemical or mechanical cues. In humans, SMC progenitors have been identified in the mononuclear fraction of the peripheral blood [300]. However, the numeric contribution of BM-derived SMCs to vessel growth or maturation still remains debatable. Whereas, in some animal studies, BM-derived SMCs made a major contribution (10–50%) to neointima formation after vessel transplantation, balloon injury and primary atherosclerosis [37, 260, 280], other reports demonstrated a more modest (1–10%) contribution of BM-derived SMCs during these events [133, 190]. Moreover, SMCs were shown to transdifferentiate from circulating EPCs and mature ECs [92, 300].

3.1.6.2 Smooth Muscle Cell Recruitment, Growth and Differentiation

Recruitment of mural cells to nascent vessels is achieved by the involvement of several regulatory pathways. The PDGF family comprises four family members (e.g., PDGF-A to D) which bind, with distinct selectivity, the receptor tyrosine kinases PDGFR-α and -β, expressed on ECs and SMCs. PDGF-BB and its receptor PDGFR-β play essential roles in the stabilization of nascent blood vessels. By releasing PDGF-BB, ECs stimulate growth and differentiation of PDGFR-β-positive mesenchymal progenitors and recruit them around nascent vessels. Absence or insufficient recruitment of periendothelial cells in mice embryos lacking PDGF-B or PDGFR-β increases endothelial growth and permeability, enlarges vessel size and enhances fragility, which results in bleeding, impaired perfusion and hypoxia [127, 197]. The subsequent increase in VEGF levels further aggravates vascular permeability, edema and promotes hemangioma formation. Similar neovascularization occurs in the retina of diabetic subjects, when their pericytes are killed by toxic metabolites.

Unlike in normal tissues, vessels in tumors are covered by fewer pericytes [241]. These mural cells differentiate from pools of c-Kit$^+$Sca-1$^+$VEGFR1$^+$ perivascular progenitor cells, which are mobilized from the bone marrow in response to PDGF-BB [302]. When PDGFs are overexpressed, tumor vessels are covered by more mural cells and tumor growth is accelerated [241]. Conversely, when PDGFRβ signaling is inhibited, fewer pericytes are recruited, tumor vessels are dilated and EC apoptosis is increased. Recently, Song et al. described that a subset of progenitor perivascular cells expressing PDGFR-β is recruited from the bone marrow to perivascular sites in tumors contributing in this way to regulation of vessel stability and vascular survival in tumors [302]. Combined administration of receptor tyrosine kinase inhibitors (RTKIs), targeting VEGFRs and PDGFRβ, increases the anti-angiogenic effect, even in the often-intractable late-stage of solid tumors [27]. PDGRβ inhibitors also destabilize the larger SMC-covered vessels, which supply bulk flow to tumors, and thereby render them more susceptible to EC-specific inhibitors.

Also critical for vessel maintenance and stabilization are the Ang/Tie-2 and PlGF/VEGFR-1 signaling systems. Ang-1 tightens vessels by affecting junctional molecules, including PECAM, VE-cadherin and occludins [97, 320]. Furthermore Ang-1, by counteracting Ang-2 activity, recruits pericytes and promotes the interaction between nascent endothelial channels and mural cells by serving as an adhesive ECM-associated and α5-binding protein [38, 352]. A precise balance of Tie-2 signals thus seems critical, as a hereditary dysfunction of Tie-2 in humans induces vascular malformations, characterized by enlarged vessels with reduced smooth muscle cell coverage [334]. PlGF, via binding to VEGFR-1, directly affects SMCs and fibroblasts which express VEGFR-1, but may also indirectly influence SMC proliferation and migration through cytokine release from activated ECs [202]. Through these effects, PlGF recruits SMCs around nascent vessels, thereby stabilizing them into mature, durable and non-leaky vessels [143, 202, 334].

TGF-β1, a multifunctional cytokine, promotes vessel maturation by stimulation of ECM production and by inducing differentiation of mesenchymal cells to mural cells [53, 251]. It is expressed in a number of different cell types, including endothelial and periendothelial cells and, depending on the context and concentration, both pro- and anti-angiogenic properties have been ascribed to TGF-β1 [109]. Inactivation of the gene encoding the transcription factor MEF2C leads to embryonic lethality due to a severe vascular disorder characterized by impaired smooth muscle cell differentiation [5].

3

Collateral vessel growth is the process of enlargement of preexisting collateral arterioles upon occlusion of a supply artery in the myocardium and peripheral limbs (Fig. 3.1.6). The mechanisms of collateral growth differ from the ones of angiogenesis [126]. Because of an increase in collateral blood flow and shear stress, ECs are activated and secrete cytokines (i.e., MCP-1, GM-CSF, ICAM-1, TGF-β1, TNF-α) to recruit monocytes [33, 332, 341]. Monocytes then infiltrate and proteolytically remodel the vessel wall enabling SMCs to migrate and divide. While VEGF alone appears to affect capillary angiogenesis more efficiently than collateral vessel growth [126, 144], PlGF is able to enhance collateral growth, by recruiting monocytes and by stimulating EC and SMC growth [202, 253]. Coadministration of VEGF with additional molecules such as PDGF, PlGF or Ang-1 may therefore enhance its therapeutic potential for the treatment of ischemic heart and limb disease [52].

3.1.7 Therapeutic Implications

Over the last decade, intensive efforts have been undertaken to develop therapeutic strategies to promote revascularization of ischemic tissues or to inhibit angiogenesis in cancer, ocular, joint or skin disorders. Over 500 million people worldwide have been estimated to benefit from either pro- or anti-angiogenic therapy. Unfortunately, clinical trials testing the pro-angiogenic potential of VEGF or FGF have not met with the expected results [299]. While some of this failure is attributable to suboptimal delivery strategies, stimulating the growth of durable and functional vessels is a more formidable challenge than previously anticipated. Novel strategies, involving transplantation of bone marrow-derived cells or the delivery of molecules capable of stimulating the growth of not only distal capillaries but also of proximal collateral conduit vessels, will be required in the future [89, 299].

Most previous efforts have thus been focused on developing anti-angiogenic agents, blocking the activity of VEGF. The first two VEGF antagonists, an anti-VEGF antibody (Avastin) [88] and a VEGF165 aptamer (Macugen, Eyetech), have been recently approved by the FDA for treatment of ocular and malignant disease. Avastin provides an overall survival benefit in colorectal, breast and lung cancer patients when combined with conventional chemotherapy [141], while monotherapy with the multi-targeted receptor tyrosine kinase inhibitors (RTKIs) Sorafenib (Bayer and Onyx) or Sutent (Pfizer), which target ECs as well as cancer, mural, stromal and hematopoietic cells, demonstrates clinical benefit in certain cancers [266, 349].

However, a number of outstanding questions still remain to be addressed in the future. For instance, despite promising success, cancer patients receiving a single class of angiogenesis inhibitors still die – does this suggest that this strategy is insufficient or does it evoke resistance and, if so, how can we avoid resistance? Can we develop more reliable biomarkers to monitor the efficacy of anti-angiogenic therapy? In addition, adverse effects have been reported – what are the molecular mechanisms? Combined treatment of anti-angiogenic agents with distinct complementary mechanisms of action, targeting other angiogenic molecules and/or targeting not only ECs but also other pro-angiogenic cell types, may thus offer advantages of increased efficacy – at least if toxicity is not a concern (see below) [45]. Another advantage is that such combinations may give the tumor less chance to escape from anti-angiogenic treatment. Exploring strategies to delay, minimize or even avoid resistance to anti-angiogenic agents might further increase the benefit of anti-angiogenic treatments [45].

As anti-angiogenic agents will hopefully be delivered earlier and earlier to more and more patients for less advanced and life-threatening disease, likely in combination with additional medications, safety of anti-angiogenic treatment is a topic of emerging importance. Given that VEGF/VEGFR-2 targeting agents may provoke unwanted toxicity [211], our improved understanding of why the PlGF/VEGR-1 pathway is effective only in pathological but not in physiological angiogenesis [14, 49, 202] should therefore be relevant for the design of safe therapeutic strategies. Thus, development of effective, safe angiogenesis inhibitors will become an increasingly important issue in the future.

Methodology A: Ocular lymphangiogenesis
Within the eye structures, blood vessels are known to be present in the retina but so far the presence of lymphatic vessels has not been described. The cornea, on the other hand, has been reported to lack both blood and lymphatic vessels. However, under certain pathological conditions, such as inflammatory disease, herpetic keratitis, and corneal melanoma, blood vessels have been described to invade the cornea. Experimental corneal neovascularization assays in rabbits have also demonstrated the presence of lymphatic vessels in the cornea, ultrastructurally and by drainage of ink [60]. Moreover, by using antibodies against lymphatic markers (LYVE-1 and podoplanin) lymphatic vessels were reported to be present in human corneas exhibiting neovascularization secondary to keratitis, transplant rejection, trauma and limbal insufficiency [64]. The cornea hence seems to be an excellent model organ with which to study angiogenesis and lymphangiogenesis processes as well as the possible interaction between them (see Box 2 and Fig. 3.1.7). By using this

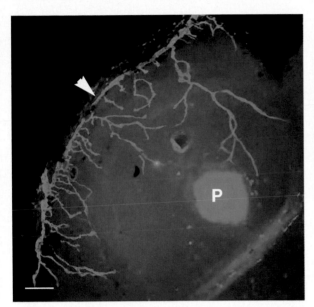

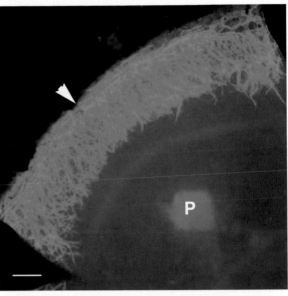

Fig. 3.1.7. Corneal lymphangiogenesis. Mouse FGF 12.5 ng (*P*) stimulates corneal lymphangiogenesis from the peripheral limbal vasculature (*arrowhead*). The vessels are viewed under fluorescent microscopy after labeling lymphatic vessels (*green*) with antibodies against lymphatic marker LYVE-1. *Scale bar* 100 μm

Fig. 3.1.8. Corneal angiogenesis. Mouse VEGF 300 ng (*P*) stimulates corneal angiogenesis from the peripheral limbal vasculature (*arrowhead*). The vessels are viewed under fluorescent microscopy (*green*) after labeling blood vessels with antibodies against CD31. *Scale bar* 100 μm

model in mice, studies have shown that the implantation of fibroblast growth factor 2 (FGF2) in the cornea specifically induces a lymphangiogenic response [55, 174]. These results bring more understanding to the processes of what signals an angiogenic versus a lymphangiogenic response during pathological conditions in the cornea and might eventually lead to better therapeutic treatments. Tumor metastasis has been described in some eye tumor types such as corneal melanoma. Since tumor cells are able to metastasize to other organs not only by entering into the blood circulation but also by invading the lymph circulation, to prevent metastasis, apart from inhibiting angiogenesis, it is of great importance to inhibit lymphatic vessel growth within the cornea.

Methodoology B: In vivo models for the study of ocular angiogenesis

In vivo angiogenesis assays are used to test the efficacy of pro- and anti-angiogenic agents [12]. Some of the most commonly used models for the study of angiogenesis and vasculogenesis in the eye are the retinopathy of prematurity (ROP) model, the cornea pocket assay and the choroidal neovascularization (CNV) model. ROP is a condition that occurs in preterm infants who are born at a time when lung development and retinal vascular development are not complete [301]. It is characterized by an unregulated growth of retinal blood vessels caused by abnormal sprouting of existing vessels. A ROP rat model is well established to study retinal neovascularization and angiogenesis in vivo [205]. The cornea pocket assay

was first developed in rabbit eyes [106] but has been adapted to mice [225, 226]. Because the cornea itself is avascular, this assay is an easy method with which to study angiogenesis as well as lymphangiogenesis (see Box 1) in vivo. Briefly, a pocket is made in the mouse cornea, and test tissues, tumors or potential angiogenic factors, when introduced into this pocket, elicit the ingrowth of new vessels from the peripheral limbal vasculature (Fig. 3.1.8). In CNV the majority of the damage to the retina occurs at Bruch's membrane when new blood vessels begin to grow from the choroid up to the retinal pigment epithelial (RPE) cells and beneath the retina. The newly formed blood vessels are permeable, which allows blood and fluid to leak into the retina. The leakage causes the retina to swell, which impairs the function and leads to poor central vision and distortion. The CNV may evolve into a scar, which can create a permanent blind spot in the central vision. Among the animal models for the study of CNV, the most commonly used is the laser-induced CNV model in mouse or rat [322]. Since CNV represents new blood vessel growth, this experimental model is also commonly used to study angiogenesis and vasculogenesis processes [290]. The role of different molecules in neovascularization can for instance be elegantly studied by overexpressing or blocking experiments after inducing CNV in this animal model [50, 291].

Mice lacking the macrophage chemoattractant Ccl-2 or its receptor Ccr-2 could be used as a model for the study of CNV since it has recently been shown that they spontaneously develop cardinal features of age-related macular degeneration (AMD), including CNV, RPE and photoreceptor atrophy [8].

3

References

1. Abd El Aal DE, Mohamed SA, Amine AF, Meki AR (2005) Vascular endothelial growth factor and insulin-like growth factor-1 in polycystic ovary syndrome and their relation to ovarian blood flow. Eur J Obstet Gynecol Reprod Biol 118:219–24

2. Adams RH, et al. (1999) Roles of ephrinB ligands and EphB receptors in cardiovascular development: demarcation of arterial/venous domains, vascular morphogenesis, and sprouting angiogenesis. Genes Dev 13:295–306

3. Agha-Majzoub R, Becker RP, Schraufnagel DE, Chan LS (2005) Angiogenesis: the major abnormality of the keratin-14 IL-4 transgenic mouse model of atopic dermatitis. Microcirculation 12:455–76

4. Ahmad SA, et al. (2001) The effects of angiopoietin-1 and -2 on tumor growth and angiogenesis in human colon cancer. Cancer Res 61:1255–9

5. Alitalo K, Carmeliet P (2002) Molecular mechanisms of lymphangiogenesis in health and disease. Cancer Cell 1:219–27

6. Alon T, et al. (1995) Vascular endothelial growth factor acts as a survival factor for newly formed retinal vessels and has implications for retinopathy of prematurity. Nat Med 1:1024–8

7. Alva JA, Iruela-Arispe ML (2004) Notch signaling in vascular morphogenesis. Curr Opin Hematol 11:278–83

8. Ambati J, et al. (2003) An animal model of age-related macular degeneration in senescent Ccl-2- or Ccr-2-deficient mice. Nat Med 9:1390–7

9. Andries LJ, Brutsaert DL, Sys SU (1998) Nonuniformity of endothelial constitutive nitric oxide synthase distribution in cardiac endothelium. Circ Res 82:195–203

10. Arima K, et al. (2005) RS3PE syndrome presenting as vascular endothelial growth factor associated disorder. Ann Rheum Dis 64:1653–5

11. Asahara T, et al. (1999) VEGF contributes to postnatal neovascularization by mobilizing bone marrow-derived endothelial progenitor cells. EMBO J 18:3964–72

12. Auerbach R, Lewis R, Shinners B, Kubai L, Akhtar N (2003) Angiogenesis assays: a critical overview. Clin Chem 49:32–40

13. Autiero M, De Smet F, Claes F, Carmeliet P (2005) Role of neural guidance signals in blood vessel navigation. Cardiovasc Res 65:629–38

14. Autiero M, et al. (2003) Role of PlGF in the intra- and intermolecular cross talk between the VEGF receptors Flt1 and Flk1. Nat Med 9:936–43

15. Azzouz M, et al. (2004) VEGF delivery with retrogradely transported lentivector prolongs survival in a mouse ALS model. Nature 429:413–7

16. Baffert F, et al. (2005) Cellular changes in normal blood capillaries undergoing regression after inhibition of VEGF signaling. Am J Physiol Heart Circ Physiol 290:H547–H559

17. Bai TR, Knight DA (2005) Structural changes in the airways in asthma: observations and consequences. Clin Sci (Lond) 108:463–77

18. Bajou K, et al. (1998) Absence of host plasminogen activator inhibitor 1 prevents cancer invasion and vascularization. Nat Med 4:923–8

19. Bajou K, et al. (2001) The plasminogen activator inhibitor PAI-1 controls in vivo tumor vascularization by interaction with proteases, not vitronectin. Implications for antiangiogenic strategies. J Cell Biol 152:777–84

20. Baker CH, et al. (2002) Blockade of epidermal growth factor receptor signaling on tumor cells and tumor-associated endothelial cells for therapy of human carcinomas. Am J Pathol 161:929–38

21. Baluk P, et al. (2005) Pathogenesis of persistent lymphatic vessel hyperplasia in chronic airway inflammation. J Clin Invest 115:247–57

22. Barillari G, Ensoli B (2002) Angiogenic effects of extracellular human immunodeficiency virus type 1 Tat protein and its role in the pathogenesis of AIDS-associated Kaposi's sarcoma. Clin Microbiol Rev 15:310–26

23. Bayless KJ, Salazar R, Davis GE (2000) RGD-dependent vacuolation and lumen formation observed during endothelial cell morphogenesis in three-dimensional fibrin matrices involves the alpha(v)beta(3) and alpha(5)beta(1) integrins. Am J Pathol 156:1673–83

24. Bedell VM, et al. (2005) roundabout4 is essential for angiogenesis in vivo. Proc Natl Acad Sci U S A 102:6373–8

25. Berchem G, et al. (2002) Cathepsin-D affects multiple tumor progression steps in vivo: proliferation, angiogenesis and apoptosis. Oncogene 21:5951–5

26. Bergers G, et al. (2000) Matrix metalloproteinase-9 triggers the angiogenic switch during carcinogenesis. Nat Cell Biol 2:737–44

27. Bergers G, Song S, Meyer-Morse N, Bergsland E, Hanahan D (2003) Benefits of targeting both pericytes and endothelial cells in the tumor vasculature with kinase inhibitors. J Clin Invest 111:1287–95

28. Bernardini G, et al. (2003) Analysis of the role of chemokines in angiogenesis. J Immunol Methods 273:83–101

29. Boudier HA (1999) Arteriolar and capillary remodelling in hypertension. Drugs 58: Spec No 1:37–40

30. Brice G, et al. (2002) Analysis of the phenotypic abnormalities in lymphoedema-distichiasis syndrome in 74 patients with FOXC2 mutations or linkage to 16q24. J Med Genet 39:478–83

31. Brown LA, et al. (2000) Insights into early vasculogenesis revealed by expression of the ETS-domain transcription factor Fli-1 in wild-type and mutant zebrafish embryos. Mech Dev 90:237–52

32. Brown LF, et al. (1995) Increased expression of vascular permeability factor (vascular endothelial growth factor) in bullous pemphigoid, dermatitis herpetiformis, and erythema multiforme. J Invest Dermatol 104:744–9

33. Buschmann IR, et al. (2001) GM-CSF: a strong arteriogenic factor acting by amplification of monocyte function. Atherosclerosis 159:343–56

34. Butler JM, et al. (2005) SDF-1 is both necessary and sufficient to promote proliferative retinopathy. J Clin Invest 115:86–93

35. Caldwell RB, et al. (2005) Vascular endothelial growth factor and diabetic retinopathy: role of oxidative stress. Curr Drug Targets 6:511–24

36. Campochiaro PA (2004) Ocular neovascularisation and excessive vascular permeability. Expert Opin Biol Ther 4:1395–402

37. Caplice NM, et al. (2003) Smooth muscle cells in human coronary atherosclerosis can originate from cells administered at marrow transplantation. Proc Natl Acad Sci U S A 100:4754–9

38. Carlson TR, Feng Y, Maisonpierre PC, Mrksich M, Morla AO (2001) Direct cell adhesion to the angiopoietins mediated by integrins. J Biol Chem 276:26516–25

39. Carmeliet P (1999) Developmental biology. Controlling the cellular brakes. Nature 401:657–8

40. Carmeliet P (2000) Fibroblast growth factor-1 stimulates

branching and survival of myocardial arteries: a goal for therapeutic angiogenesis? Circ Res 87:176–8

41. Carmeliet P (2000) Mechanisms of angiogenesis and arteriogenesis. Nat Med 6:389–95
42. Carmeliet P (2002) Integrin indecision. Nat Med 8:14–6
43. Carmeliet P (2003) Angiogenesis in health and disease. Nat Med 9:653–60
44. Carmeliet P (2003) Blood vessels and nerves: common signals, pathways and diseases. Nat Rev Genet 4:710–20
45. Carmeliet P (2005) Angiogenesis in life, disease and medicine. Nature 438:932–6
46. Carmeliet P, et al. (1996) Abnormal blood vessel development and lethality in embryos lacking a single VEGF allele. Nature 380:435–9
47. Carmeliet P, et al. (1999) Impaired myocardial angiogenesis and ischemic cardiomyopathy in mice lacking the vascular endothelial growth factor isoforms VEGF164 and VEGF188 [see comments]. Nat Med 5:495–502
48. Carmeliet P, et al. (1999) Targeted deficiency or cytosolic truncation of the VE-cadherin gene in mice impairs VEGF-mediated endothelial survival and angiogenesis. Cell 98:147–57
49. Carmeliet P, et al. (2001) Synergism between vascular endothelial growth factor and placental growth factor contributes to angiogenesis and plasma extravasation in pathological conditions. Nat Med 7:575–83
50. Castro MR, Lutz D, Edelman JL (2004) Effect of COX inhibitors on VEGF-induced retinal vascular leakage and experimental corneal and choroidal neovascularization. Exp Eye Res 79:275–85
51. Ceradini DJ, et al. (2004) Progenitor cell trafficking is regulated by hypoxic gradients through HIF-1 induction of SDF-1. Nat Med 10:858–64
52. Chae JK, et al. (2000) Coadminstration of angiopoietin-1 and vascular endothelial growth factor enhances collateral vascularization. Arterioscler Thromb Vasc Biol 20: 2573–8
53. Chambers RC, Leoni P, Kaminski N, Laurent GJ, Heller RA (2003) Global expression profiling of fibroblast responses to transforming growth factor-beta1 reveals the induction of inhibitor of differentiation-1 and provides evidence of smooth muscle cell phenotypic switching. Am J Pathol 162:533–46
54. Chang E, Yang J, Nagavarapu U, Herron GS (2002) Aging and survival of cutaneous microvasculature. J Invest Dermatol 118:752–8
55. Chang LK, et al. (2004) Dose-dependent response of FGF-2 for lymphangiogenesis. Proc Natl Acad Sci U S A 101: 11658–63
56. Chen JN, et al. (1996) Mutations affecting the cardiovascular system and other internal organs in zebrafish. Development 123:293–302
57. Chi JT, et al. (2003) Endothelial cell diversity revealed by global expression profiling. Proc Natl Acad Sci U S A 100:10623–8
58. Childs S, Chen JN, Garrity DM, Fishman MC (2002) Patterning of angiogenesis in the zebrafish embryo. Development 129:973–82
59. Choi K (2002) The hemangioblast: a common progenitor of hematopoietic and endothelial cells. J Hematother Stem Cell Res 11:91–101
60. Collin HB (1970) Lymphatic drainage of 131-I-albumin from the vascularized cornea. Invest Ophthalmol 9:146–55
61. Compernolle V, et al. (2002) Loss of HIF-2alpha and inhibition of VEGF impair fetal lung maturation, whereas treatment with VEGF prevents fatal respiratory distress in premature mice. Nat Med 8:702–10

62. Coussens LM, et al. (1999) Inflammatory mast cells up-regulate angiogenesis during squamous epithelial carcinogenesis. Genes Dev 13:1382–97
63. Creazzo TL, Godt RE, Leatherbury L, Conway SJ, Kirby ML (1998) Role of cardiac neural crest cells in cardiovascular development. Annu Rev Physiol 60:267–86
64. Cursiefen C, et al. (2002) Lymphatic vessels in vascularized human corneas: immunohistochemical investigation using LYVE-1 and podoplanin. Invest Ophthalmol Vis Sci 43:2127–35
65. Dagenais SL, et al. (2004) Foxc2 is expressed in developing lymphatic vessels and other tissues associated with lymphedema-distichiasis syndrome. Gene Expr Patterns 4:611–9
66. de Bruijn MF, Speck NA, Peeters MC, Dzierzak E (2000) Definitive hematopoietic stem cells first develop within the major arterial regions of the mouse embryo. EMBO J 19:2465–74
67. de la Torre JC (2002) Alzheimer's disease: how does it start? J Alzheimers Dis 4:497–512
68. de la Torre JC (2004) Alzheimer's disease is a vasocognopathy: a new term to describe its nature. Neurol Res 26:517–24
69. Derhaag JG, Duijvestijn AM, Emeis JJ, Engels W, van Breda Vriesman PJ (1996) Production and characterization of spontaneous rat heart endothelial cell lines. Lab Invest 74:437–51
70. Derhaag JG, Duijvestijn AM, Van Breda Vriesman PJ (1997) Heart EC respond heterogeneous on cytokine stimulation in ICAM-1 and VCAM-1, but not in MHC expression. A study with 3 rat heart endothelial cell (RHEC) lines. Endothelium 5:307–19
71. Dettman RW, Denetclaw W, Jr, Ordahl CP, Bristow J (1998) Common epicardial origin of coronary vascular smooth muscle, perivascular fibroblasts, and intermyocardial fibroblasts in the avian heart. Dev Biol 193:169–81
72. Devy L, et al. (2002) The pro- or antiangiogenic effect of plasminogen activator inhibitor 1 is dose dependent. FASEB J 16:147–54
73. Dickson BJ (2002) Molecular mechanisms of axon guidance. Science 298:1959–64
74. Dimmeler S, Haendeler J, Rippmann V, Nehls M, Zeiher AM (1996) Shear stress inhibits apoptosis of human endothelial cells. FEBS Lett 399:71–4
75. Distler O, et al. (2004) Uncontrolled expression of vascular endothelial growth factor and its receptors leads to insufficient skin angiogenesis in patients with systemic sclerosis. Circ Res 95:109–16
76. Donovan MJ, et al. (2000) Brain derived neurotrophic factor is an endothelial cell survival factor required for intramyocardial vessel stabilization. Development 127:4531–40
77. Duarte A, et al. (2004) Dosage-sensitive requirement for mouse Dll4 in artery development. Genes Dev 18:2474–8
78. Eliceiri BP, et al. (1999) Selective requirement for Src kinases during VEGF-induced angiogenesis and vascular permeability. Mol Cell 4:915–24
79. Ema M, et al. (2003) Combinatorial effects of Flk1 and Tal1 on vascular and hematopoietic development in the mouse. Genes Dev 17:380–93
80. Eremina V, et al. (2003) Glomerular-specific alterations of VEGF-A expression lead to distinct congenital and acquired renal diseases. J Clin Invest 111:707–16
81. Etoh T, et al. (2001) Angiopoietin-2 is related to tumor angiogenesis in gastric carcinoma: possible in vivo regulation via induction of proteases. Cancer Res 61:2145–53
82. Fehling HJ, et al. (2003) Tracking mesoderm induction and

its specification to the hemangioblast during embryonic stem cell differentiation. Development 130:4217–27

83. Fernandez M, et al. (2005) Inhibition of VEGF receptor-2 decreases the development of hyperdynamic splanchnic circulation and portal-systemic collateral vessels in portal hypertensive rats. J Hepatol 43:98–103

84. Ferrara N (2000) Vascular endothelial growth factor and the regulation of angiogenesis. Recent Prog Horm Res 55:15–35; discussion: 35–6

85. Ferrara N (2002) VEGF and the quest for tumour angiogenesis factors. Nat Rev Cancer 2:795–803

86. Ferrara N, et al. (1996) Heterozygous embryonic lethality induced by targeted inactivation of the VEGF gene. Nature 380:439–42

87. Ferrara N, Gerber HP, LeCouter J (2003) The biology of VEGF and its receptors. Nat Med 9:669–76

88. Ferrara N, Hillan KJ, Gerber HP, Novotny W (2004) Discovery and development of bevacizumab, an anti-VEGF antibody for treating cancer. Nat Rev Drug Discovery 3:391–400

89. Ferrara N, Kerbel R (2005) Angiogenesis as a therapeutic target. Nature 438:967–974

90. Folkman J (2001) Angiogenesis-dependent diseases. Semin Oncol 28:536–42

91. Fong GH, Rossant J, Gertsenstein M, Breitman ML (1995) Role of the Flt-1 receptor tyrosine kinase in regulating the assembly of vascular endothelium. Nature 376:66–70

92. Frid MG, Kale VA, Stenmark KR (2002) Mature vascular endothelium can give rise to smooth muscle cells via endothelial-mesenchymal transdifferentiation: in vitro analysis. Circ Res 90:1189–96

93. Fukumura D, et al. (2003) Paracrine regulation of angiogenesis and adipocyte differentiation during in vivo adipogenesis. Circ Res 93:e88–97

94. Gale NW, et al. (2002) Angiopoietin-2 is required for postnatal angiogenesis and lymphatic patterning, and only the latter role is rescued by Angiopoietin-1. Dev Cell 3:411–23

95. Gale NW, et al. (2004) Haploinsufficiency of delta-like 4 ligand results in embryonic lethality due to major defects in arterial and vascular development. Proc Natl Acad Sci U S A 101:15949–54

96. Gale NW, Yancopoulos GD (1999) Growth factors acting via endothelial cell-specific receptor tyrosine kinases: VEGFs, angiopoietins, and ephrins in vascular development. Genes Dev 13:1055–66

97. Gamble JR, et al. (2000) Angiopoietin-1 is an antipermeability and anti-inflammatory agent in vitro and targets cell junctions. Circ Res 87:603–7

98. Garcia-Barros M, et al. (2003) Tumor response to radiotherapy regulated by endothelial cell apoptosis. Science 300: 1155–9

99. Gealekman O, et al. (2004) Endothelial dysfunction as a modifier of angiogenic response in Zucker diabetic fat rat: amelioration with Ebselen. Kidney Int 66:2337–47

100. Gendron RL, Adams LC, Paradis H (2000) Tubedown-1, a novel acetyltransferase associated with blood vessel development. Dev Dyn 218:300–15

101. Gennaro G, Menard C, Michaud SE, Rivard A (2003) Age-dependent impairment of reendothelialization after arterial injury: role of vascular endothelial growth factor. Circulation 107:230–3

102. Gerber HP, et al. (1999) VEGF is required for growth and survival in neonatal mice. Development 126:1149–59

103. Gerety SS, Anderson DJ (2002) Cardiovascular ephrinB2 function is essential for embryonic angiogenesis. Development 129:1397–410

104. Gerety SS, Wang HU, Chen ZF, Anderson DJ (1999) Symmetrical mutant phenotypes of the receptor EphB4 and its specific transmembrane ligand ephrin-B2 in cardiovascular development. Mol Cell 4:403–14

105. Gerhardt H, et al. (2003) VEGF guides angiogenic sprouting utilizing endothelial tip cell filopodia. J Cell Biol 161:1163–77

106. Gimbrone MA, Jr, Cotran RS, Leapman SB, Folkman J (1974) Tumor growth and neovascularization: an experimental model using the rabbit cornea. J Natl Cancer Inst 52:413–27

107. Gira AK, Brown LF, Washington CV, Cohen C, Arbiser JL (2004) Keloids demonstrate high-level epidermal expression of vascular endothelial growth factor. J Am Acad Dermatol 50:850–3

108. Gittenberger-de Groot AC, DeRuiter MC, Bergwerff M, Poelmann RE (1999) Smooth muscle cell origin and its relation to heterogeneity in development and disease. Arterioscler Thromb Vasc Biol 19:1589–94

109. Gohongi T, et al. (1999) Tumor-host interactions in the gallbladder suppress distal angiogenesis and tumor growth: involvement of transforming growth factor beta1. Nat Med 5:1203–8

110. Gorski DH, Walsh K (2003) Control of vascular cell differentiation by homeobox transcription factors. Trends Cardiovasc Med 13:213–20

111. Gosepath J, Brieger J, Lehr HA, Mann WJ (2005) Expression, localization, and significance of vascular permeability/vascular endothelial growth factor in nasal polyps. Am J Rhinol 19:7–13

112. Green CJ, et al. (2001) Placenta growth factor gene expression is induced by hypoxia in fibroblasts: a central role for metal transcription factor-1. Cancer Res 61:2696–703

113. Groger M, et al. (2004) IL-3 induces expression of lymphatic markers Prox-1 and podoplanin in human endothelial cells. J Immunol 173:7161–9

114. Groothuis PG, Nap AW, Winterhager E, Grummer R (2005) Vascular development in endometriosis. Angiogenesis 8:147–156

115. Grunewald M, et al. (2003) VEGF-induced adult neovascularization: recruitment, retention, and role of accessory cells. Cell 124:175–89

116. Gu C, et al. (2005) Semaphorin 3E and plexin-D1 control vascular pattern independently of neuropilins. Science 307:265–8

117. Guleng B, et al. (2005) Blockade of the stromal cell-derived factor-1/CXCR4 axis attenuates in vivo tumor growth by inhibiting angiogenesis in a vascular endothelial growth factor-independent manner. Cancer Res 65:5864–71

118. Guo DQ, et al. (2000) Tumor necrosis factor employs a protein-tyrosine phosphatase to inhibit activation of KDR and vascular endothelial cell growth factor-induced endothelial cell proliferation. J Biol Chem 275:11216–21

119. Hackett SF, et al. (2000) Angiopoietin 2 expression in the retina: upregulation during physiologic and pathologic neovascularization. J Cell Physiol 184:275–84

120. Haigh JJ, Gerber HP, Ferrara N, Wagner EF (2000) Conditional inactivation of VEGF-A in areas of collagen2a1 expression results in embryonic lethality in the heterozygous state. Development 127:1445–53

121. Hamano Y, et al. (2003) Physiological levels of tumstatin, a

fragment of collagen IV alpha3 chain, are generated by MMP-9 proteolysis and suppress angiogenesis via alphaV beta3 integrin. Cancer Cell 3:589–601

122. Harada K, Lu S, Chisholm DM, Syrjanen S, Schor AM (2000) Angiogenesis and vasodilation in skin warts. Association with HPV infection. Anticancer Res 20:4519–23

123. Hatoum OA, Binion DG, Gutterman DD (2005) Paradox of simultaneous intestinal ischaemia and hyperaemia in inflammatory bowel disease. Eur J Clin Invest 35:599–609

124. Hattori K, et al. (2001) Vascular endothelial growth factor and angiopoietin-1 stimulate postnatal hematopoiesis by recruitment of vasculogenic and hematopoietic stem cells. J Exp Med 193:1005–14

125. Hausman MR, Rinker BD (2004) Intractable wounds and infections: the role of impaired vascularity and advanced surgical methods for treatment. Am J Surg 187:44S–55S

126. Helisch A, Schaper W (2003) Arteriogenesis: the development and growth of collateral arteries. Microcirculation 10:83–97

127. Hellstrom M, Kalen M, Lindahl P, Abramsson A, Betsholtz C (1999) Role of PDGF-B and PDGFR-beta in recruitment of vascular smooth muscle cells and pericytes during embryonic blood vessel formation in the mouse. Development 126:3047–55

128. Herzog Y, Guttmann-Raviv N, Neufeld G (2005) Segregation of arterial and venous markers in subpopulations of blood islands before vessel formation. Dev Dyn 232:1047–55

129. Hewett P, et al. (2002) Down-regulation of angiopoietin-1 expression in menorrhagia. Am J Pathol 160:773–80

130. Heymans S, et al. (1999) Inhibition of plasminogen activators or matrix metalloproteinases prevents cardiac rupture but impairs therapeutic angiogenesis and causes cardiac failure. Nat Med 5:1135–42

131. Hill JM, et al. (2003) Circulating endothelial progenitor cells, vascular function, and cardiovascular risk. N Engl J Med 348:593–600

132. Hillebrands JL, et al. (2001) Origin of neointimal endothelium and alpha-actin-positive smooth muscle cells in transplant arteriosclerosis. J Clin Invest 107:1411–22

133. Hillebrands JL, Klatter FA, van Dijk WD, Rozing J (2002) Bone marrow does not contribute substantially to endothelial-cell replacement in transplant arteriosclerosis. Nat Med 8:194–5

134. Holash J, et al. (1999) Vessel cooption, regression, and growth in tumors mediated by angiopoietins and VEGF. Science 284:1994–8

135. Honma Y, et al. (2002) Artemin is a vascular-derived neurotropic factor for developing sympathetic neurons. Neuron 35:267–82

136. Hood JD, Cheresh DA (2002) Role of integrins in cell invasion and migration. Nat Rev Cancer 2:91–100

137. Hoyer M (1908) Untersuchungen über das Lymphgefässsystem der Froschlarven. Teil II. Bull Acad Cracov 451–464

138. Huber AB, Kolodkin AL, Ginty DD, Cloutier JF (2003) Signaling at the growth cone: ligand-receptor complexes and the control of axon growth and guidance. Annu Rev Neurosci 26:509–63

139. Hull ML, et al. (2003) Antiangiogenic agents are effective inhibitors of endometriosis. J Clin Endocrinol Metab 88:2889–99

140. Humbert M, Trembath RC (2002) Genetics of pulmonary hypertension: from bench to bedside. Eur Respir J 20:741–9

141. Hurwitz H, et al. (2004) Bevacizumab plus irinotecan, fluorouracil, and leucovorin for metastatic colorectal cancer. N Engl J Med 350:2335–42

142. Hynes RO (2002) A reevaluation of integrins as regulators of angiogenesis. Nat Med 8:918–21

143. Ishida A, et al. (2001) Expression of vascular endothelial growth factor receptors in smooth muscle cells. J Cell Physiol 188:359–68

144. Isner JM (2002) Myocardial gene therapy. Nature 415:234–9

145. Isogai S, Lawson ND, Torrealday S, Horiguchi M, Weinstein BM (2003) Angiogenic network formation in the developing vertebrate trunk. Development 130:5281–90

146. Iwami Y, Masuda H, Asahara T (2004) Endothelial progenitor cells: past, state of the art, and future. J Cell Mol Med 8:488–97

147. Jackson C (2002) Matrix metalloproteinases and angiogenesis. Curr Opin Nephrol Hypertens 11:295–9

148. Jain RJ, Duda DG, Clark JW, Loeffler JS (2005) General review: Lessons from phase III clinical trials on anti-VEGF therapy for cancer. Nature Clin Pract Oncol 3:24–40

149. Jain RK, Munn LL (2000) Leaky vessels? Call Ang1! Nat Med 6:131–2

150. Jenkinson L, Bardhan KD, Atherton J, Kalia N (2002) Helicobacter pylori prevents proliferative stage of angiogenesis in vitro: role of cytokines. Dig Dis Sci 47:1857–62

151. Jesmin S, et al. (2005) Age-related changes in cardiac expression of VEGF and its angiogenic receptor KDR in stroke-prone spontaneously hypertensive rats. Mol Cell Biochem 272:63–73

152. Jiang WG, et al. (2005) Hepatocyte growth factor, its receptor, and their potential value in cancer therapies. Crit Rev Oncol Hematol 53:35–69

153. Jin SW, Beis D, Mitchell T, Chen JN, Stainier DY (2005) Cellular and molecular analyses of vascular tube and lumen formation in zebrafish. Development 132:5199–209

154. Kahlon R, Shapero J, Gotlieb AI (1992) Angiogenesis in atherosclerosis. Can J Cardiol 8:60–4

155. Kalimo H, Ruchoux MM, Viitanen M, Kalaria RN (2002) CADASIL: a common form of hereditary arteriopathy causing brain infarcts and dementia. Brain Pathol 12:371–84

156. Kamba T, et al. (2005) VEGF-dependent plasticity of fenestrated capillaries in the normal adult microvasculature. Am J Physiol Heart Circ Physiol 290:H560–H576

157. Kang DH, et al. (2001) Impaired angiogenesis in the aging kidney: vascular endothelial growth factor and thrombospondin-1 in renal disease. Am J Kidney Dis 37:601–11

158. Karkkainen MJ, et al. (2004) Vascular endothelial growth factor C is required for sprouting of the first lymphatic vessels from embryonic veins. Nat Immunol 5:74–80

159. Kasahara Y, et al. (2000) Inhibition of VEGF receptors causes lung cell apoptosis and emphysema. J Clin Invest 106:1311–9

160. Kearney JB, et al. (2002) Vascular endothelial growth factor receptor Flt-1 negatively regulates developmental blood vessel formation by modulating endothelial cell division. Blood 99:2397–407

161. Kennedy M, et al. (1997) A common precursor for primitive erythropoiesis and definitive haematopoiesis. Nature 386:488–93

162. Kerbel R, Folkman J (2002) Clinical translation of angiogenesis inhibitors. Nat Rev Cancer 2:727–39

163. Khurana R, et al. (2005) Placental growth factor promotes atherosclerotic intimal thickening and macrophage accumulation. Circulation 111:2828–2836

164. Kim JS, Kim JM, Jung HC, Song IS (2004) *Helicobacter pylori* down-regulates the receptors of vascular endothelial growth factor and angiopoietin in vascular endothelial cells: implications in the impairment of gastric ulcer healing. Dig Dis Sci 49:778–86

165. Kimura H, et al. (2000) Hypoxia response element of the human vascular endothelial growth factor gene mediates transcriptional regulation by nitric oxide: control of hypoxia-inducible factor-1 activity by nitric oxide. Blood 95:189–97

166. Kirk SL, Karlik SJ (2003) VEGF and vascular changes in chronic neuroinflammation. J Autoimmun 21:353–63

167. Kirmaz C, et al. (2004) Increased expression of angiogenic markers in patients with seasonal allergic rhinitis. Eur Cytokine Netw 15:317–22

168. Kocher AA, et al. (2001) Neovascularization of ischemic myocardium by human bone-marrow-derived angioblasts prevents cardiomyocyte apoptosis, reduces remodeling and improves cardiac function. Nat Med 7:430–6

169. Konno S, et al. (2004) Altered expression of angiogenic factors in the VEGF-Ets-1 cascades in inflammatory bowel disease. J Gastroenterol 39:931–9

170. Kostoulas G, Lang A, Nagase H, Baici A (1999) Stimulation of angiogenesis through cathepsin B inactivation of the tissue inhibitors of matrix metalloproteinases. FEBS Lett 455:286–90

171. Koyama S, et al. (2002) Decreased level of vascular endothelial growth factor in bronchoalveolar lavage fluid of normal smokers and patients with pulmonary fibrosis. Am J Respir Crit Care Med 166:382–5

172. Krupinski J, Kaluza J, Kumar P, Kumar S, Wang JM (1994) Role of angiogenesis in patients with cerebral ischemic stroke. Stroke 25:1794–8

173. Kubis N, et al. (2002) Decreased arteriolar density in endothelial nitric oxide synthase knockout mice is due to hypertension, not to the constitutive defect in endothelial nitric oxide synthase enzyme. J Hypertens 20:273–80

174. Kubo H, et al. (2002) Blockade of vascular endothelial growth factor receptor-3 signaling inhibits fibroblast growth factor-2-induced lymphangiogenesis in mouse cornea. Proc Natl Acad Sci U S A 99:8868–73

175. Kuruvilla R, et al. (2004) A neurotrophin signaling cascade coordinates sympathetic neuron development through differential control of TrkA trafficking and retrograde signaling. Cell 118:243–55

176. Lainer DT, Brahn E (2005) New antiangiogenic strategies for the treatment of proliferative synovitis. Expert Opin Investig Drugs 14:1–17

177. Lambert V, et al. (2001) Influence of plasminogen activator inhibitor type 1 on choroidal neovascularization. FASEB J 15:1021–7

178. Lambrechts D, Carmeliet P (2004) Medicine: genetic spotlight on a blood defect. Nature 427:592–4

179. Lambrechts D, et al. (2003) VEGF is a modifier of amyotrophic lateral sclerosis in mice and humans and protects motoneurons against ischemic death. Nat Genet 34:383–94

180. Lawson ND, et al. (2001) Notch signaling is required for arterial-venous differentiation during embryonic vascular development. Development 128:3675–83

181. Lawson ND, Vogel AM, Weinstein BM (2002) Sonic hedgehog and vascular endothelial growth factor act upstream of the Notch pathway during arterial endothelial differentiation. Dev Cell 3:127–36

182. le Noble F, et al. (2004) Flow regulates arterial-venous differentiation in the chick embryo yolk sac. Development 131:361–75

183. Lebrin F, Deckers M, Bertolino P, Ten Dijke P (2005) TGF-beta receptor function in the endothelium. Cardiovasc Res 65:599–608

184. LeCouter J, et al. (2001) Identification of an angiogenic mitogen selective for endocrine gland endothelium. Nature 412:877–84

185. LeCouter J, Lin R, Ferrara N (2002) Endocrine gland-derived VEGF and the emerging hypothesis of organ-specific regulation of angiogenesis. Nat Med 8:913–7

186. LeCouter J, Lin R, Ferrara N (2004) EG-VEGF: a novel mediator of endocrine-specific angiogenesis, endothelial phenotype, and function. Ann N Y Acad Sci 1014:50–7

187. Lee JW, Juliano R (2004) Mitogenic signal transduction by integrin- and growth factor receptor-mediated pathways. Mol Cells 17:188–202

188. Leong TT, Fearon U, Veale DJ (2005) Angiogenesis in psoriasis and psoriatic arthritis: clues to disease pathogenesis. Curr Rheumatol Rep 7:325–9

189. Levine RJ, et al. (2004) Circulating angiogenic factors and the risk of preeclampsia. N Engl J Med 350:672–83

190. Li J, et al. (2001) Vascular smooth muscle cells of recipient origin mediate intimal expansion after aortic allotransplantation in mice. Am J Pathol 158:1943–7

191. Li J, Yu X, Pan W, Unger RH (2002) Gene expression profile of rat adipose tissue at the onset of high-fat-diet obesity. Am J Physiol Endocrinol Metab 282:E1334–41

192. Li X, et al. (2005) Revascularization of ischemic tissues by PDGF-CC via effects on endothelial cells and their progenitors. J Clin Invest 115:118–27

193. Liao W, Ho CY, Yan YL, Postlethwait J, Stainier DY (2000) Hhex and scl function in parallel to regulate early endothelial and blood differentiation in zebrafish. Development 127:4303–13

194. Lin Q, et al. (1998) Requirement of the MADS-box transcription factor MEF2C for vascular development. Development 125:4565–74

195. Lin Y, Weisdorf DJ, Solovey A, Hebbel RP (2000) Origins of circulating endothelial cells and endothelial outgrowth from blood. J Clin Invest 105:71–7

196. Lindahl P, et al. (1999) Role of platelet-derived growth factors in angiogenesis and alveogenesis. Curr Top Pathol 93:27–33

197. Lindahl P, Hellstrom M, Kalen M, Betsholtz C (1998) Endothelial-perivascular cell signaling in vascular development: lessons from knockout mice. Curr Opin Lipidol 9:407–11

198. Long DA, Mu W, Price KL, Johnson RJ (2005) Blood vessels and the aging kidney. Nephron Exp Nephrol 101:e95–9

199. Lu X, et al. (2004) The netrin receptor UNC5B mediates guidance events controlling morphogenesis of the vascular system. Nature 432:179–86

200. Luttun A, Autiero M, Tjwa M, Carmeliet P (2004) Genetic dissection of tumor angiogenesis: are PlGF and VEGFR-1 novel anti-cancer targets? Biochim Biophys Acta 1654:79–94

201. Luttun A, Dewerchin M, Collen D, Carmeliet P (2000) The role of proteinases in angiogenesis, heart development, restenosis, atherosclerosis, myocardial ischemia, and stroke: insights from genetic studies. Curr Atheroscler Rep 2:407–16

202. Luttun A, et al. (2002) Revascularization of ischemic tis-

sues by PlGF treatment, and inhibition of tumor angiogenesis, arthritis and atherosclerosis by anti-Flt1. Nat Med 8:831–40

203. Ma L, et al. (2005) In vitro procoagulant activity induced in endothelial cells by chemotherapy and antiangiogenic drug combinations: modulation by lower-dose chemotherapy. Cancer Res 65:5365–73

204. Mackiewicz Z, et al. (2002) Increased but imbalanced expression of VEGF and its receptors has no positive effect on angiogenesis in systemic sclerosis skin. Clin Exp Rheumatol 20:641–6

205. Madan A, Penn JS (2003) Animal models of oxygen-induced retinopathy. Front Biosci 8:d1030–43

206. Maes C, et al. (2002) Impaired angiogenesis and endochondral bone formation in mice lacking the vascular endothelial growth factor isoforms VEGF164 and VEGF188. Mech Dev 111:61–73

207. Magnon C, et al. (2005) Canstatin acts on endothelial and tumor cells via mitochondrial damage initiated through interaction with alphavbeta3 and alphavbeta5 integrins. Cancer Res 65:4353–61

208. Maisonpierre PC, et al. (1997) Angiopoietin-2, a natural antagonist for Tie2 that disrupts in vivo angiogenesis. Science 277:55–60

209. Makinen T, et al. (2005) PDZ interaction site in ephrinB2 is required for the remodeling of lymphatic vasculature. Genes Dev 19:397–410

210. Martinez P, Esbrit P, Rodrigo A, Alvarez-Arroyo MV, Martinez ME (2002) Age-related changes in parathyroid hormone-related protein and vascular endothelial growth factor in human osteoblastic cells. Osteoporos Int 13: 874–81

211. Marx GM, et al. (2002) Unexpected serious toxicity with chemotherapy and antiangiogenic combinations: time to take stock! J Clin Oncol 20:1446–8

212. Mattot V, et al. (2002) Loss of the VEGF(164) and VEGF(188) isoforms impairs postnatal glomerular angiogenesis and renal arteriogenesis in mice. J Am Soc Nephrol 13:1548–60

213. Maynard SE, et al. (2003) Excess placental soluble fms-like tyrosine kinase 1 (sFlt1) may contribute to endothelial dysfunction, hypertension, and proteinuria in preeclampsia. J Clin Invest 111:649–58

214. McGrath-Morrow SA, Cho C, Zhen L, Hicklin DJ, Tuder RM (2005) Vascular endothelial growth factor receptor 2 blockade disrupts postnatal lung development. Am J Respir Cell Mol Biol 32:420–7

215. McNeel DG, et al. (2005) Phase I trial of a monoclonal antibody specific for alphavbeta3 integrin (MEDI-522) in patients with advanced malignancies, including an assessment of effect on tumor perfusion. Clin Cancer Res 11:7851–60

216. Medina J, et al. (2005) Evidence of angiogenesis in primary biliary cirrhosis: an immunohistochemical descriptive study. J Hepatol 42:124–31

217. Meeson AP, Argilla M, Ko K, Witte L, Lang RA (1999) VEGF deprivation-induced apoptosis is a component of programmed capillary regression. Development 126:1407–15

218. Meyer M, et al. (1999) A novel vascular endothelial growth factor encoded by Orf virus, VEGF-E, mediates angiogenesis via signalling through VEGFR-2 (KDR) but not VEGFR-1 (Flt-1) receptor tyrosine kinases. EMBO J 18:363–74

219. Miao HQ, et al. (1999) Neuropilin-1 mediates collapsin-1/semaphorin III inhibition of endothelial cell motility: functional competition of collapsin-1 and vascular endothelial growth factor-165. J Cell Biol 146:233–42

220. Miao RQ, Agata J, Chao L, Chao J (2002) Kallistatin is a new inhibitor of angiogenesis and tumor growth. Blood 100:3245–52

221. Mizukami Y, et al. (2005) Induction of interleukin-8 preserves the angiogenic response in HIF-1alpha-deficient colon cancer cells. Nat Med 11:992–7

222. Moyon D, Pardanaud L, Yuan L, Breant C, Eichmann A (2001) Plasticity of endothelial cells during arterial-venous differentiation in the avian embryo. Development 128: 3359–70

223. Mukouyama YS, Gerber HP, Ferrara N, Gu C, Anderson DJ (2005) Peripheral nerve-derived VEGF promotes arterial differentiation via neuropilin 1-mediated positive feedback. Development 132:941–52

224. Mukouyama YS, Shin D, Britsch S, Taniguchi M, Anderson DJ (2002) Sensory nerves determine the pattern of arterial differentiation and blood vessel branching in the skin. Cell 109:693–705

225. Muthukkaruppan V, Auerbach R (1979) Angiogenesis in the mouse cornea. Science 205:1416–8

226. Muthukkaruppan VR, Kubai L, Auerbach R (1982) Tumor-induced neovascularization in the mouse eye. J Natl Cancer Inst 69:699–708

227. Nakajima Y, Mironov V, Yamagishi T, Nakamura H, Markwald RR (1997) Expression of smooth muscle alpha-actin in mesenchymal cells during formation of avian endocardial cushion tissue: a role for transforming growth factor beta3. Dev Dyn 209:296–309

228. Nakano T, et al. (2005) Angiogenesis and lymphangiogenesis and expression of lymphangiogenic factors in the atherosclerotic intima of human coronary arteries. Hum Pathol 36:330–40

229. Neuhaus T, et al. (2003) Stromal cell-derived factor 1alpha (SDF-1alpha) induces gene-expression of early growth response-1 (Egr-1) and VEGF in human arterial endothelial cells and enhances VEGF induced cell proliferation. Cell Prolif 36:75–86

230. Nishikawa SI, Nishikawa S, Hirashima M, Matsuyoshi N, Kodama H (1998) Progressive lineage analysis by cell sorting and culture identifies FLK1+VE-cadherin+ cells at a diverging point of endothelial and hemopoietic lineages. Development 125:1747–57

231. Ny A, et al. (2005) A genetic *Xenopus laevis* tadpole model to study lymphangiogenesis. Nat Med 11:998–1004

232. Nyberg P, Xie L, Kalluri R (2005) Endogenous inhibitors of angiogenesis. Cancer Res 65:3967–79

233. O'Reilly MS, et al. (1994) Angiostatin: a novel angiogenesis inhibitor that mediates the suppression of metastases by a Lewis lung carcinoma. Cell 79:315–28

234. O'Reilly MS, et al. (1997) Endostatin: an endogenous inhibitor of angiogenesis and tumor growth. Cell 88:277–85

235. Ohno A, et al. (2004) Dermatomyositis associated with Sjogren's syndrome: VEGF involvement in vasculitis. Clin Neuropathol 23:178–82

236. Oike Y, et al. (2002) Regulation of vasculogenesis and angiogenesis by EphB/ephrin-B2 signaling between endothelial cells and surrounding mesenchymal cells. Blood 100:1326–33

237. Okamoto R, et al. (2005) Hematopoietic cells regulate the angiogenic switch during tumorigenesis. Blood 105:2757–63

238. Oliver G, Detmar M (2002) The rediscovery of the lym-

3

phatic system: old and new insights into the development and biological function of the lymphatic vasculature. Genes Dev 16:773–83

239. Oosthuyse B, et al. (2001) Deletion of the hypoxia-response element in the vascular endothelial growth factor promoter causes motor neuron degeneration. Nat Genet 28:131–8

240. Orimo A, et al. (2005) Stromal fibroblasts present in invasive human breast carcinomas promote tumor growth and angiogenesis through elevated SDF-1/CXCL12 secretion. Cell 121:335–48

241. Ostman A (2004) PDGF receptors-mediators of autocrine tumor growth and regulators of tumor vasculature and stroma. Cytokine Growth Factor Rev 15:275–86

242. Oura H, et al. (2003) A critical role of placental growth factor in the induction of inflammation and edema formation. Blood 101:560–7

243. Palis J, McGrath KE, Kingsley PD (1995) Initiation of hematopoiesis and vasculogenesis in murine yolk sac explants. Blood 86:156–63

244. Pardanaud L, et al. (1996) Two distinct endothelial lineages in ontogeny, one of them related to hemopoiesis. Development 122:1363–71

245. Park JE, Chen HH, Winer J, Houck KA, Ferrara N (1994) Placenta growth factor. Potentiation of vascular endothelial growth factor bioactivity, in vitro and in vivo, and high affinity binding to Flt-1 but not to Flk-1/KDR. J Biol Chem 269:25646–54

246. Park JE, Keller GA, Ferrara N (1993) The vascular endothelial growth factor (VEGF) isoforms: differential deposition into the subepithelial extracellular matrix and bioactivity of extracellular matrix-bound VEGF. Mol Biol Cell 4:1317–26

247. Park KW, et al. (2003) Robo4 is a vascular-specific receptor that inhibits endothelial migration. Dev Biol 261:251–67

248. Parker LH, et al. (2004) The endothelial-cell-derived secreted factor Egfl7 regulates vascular tube formation. Nature 428:754–8

249. Patsouris E, et al. (2004) Increased microvascular network in bone marrow of HIV-positive haemophilic patients. HIV Med 5:18–25

250. Peichev M, et al. (2000) Expression of VEGFR-2 and AC133 by circulating human CD34(+) cells identifies a population of functional endothelial precursors. Blood 95:952–8

251. Pepper MS (1997) Transforming growth factor-beta: vasculogenesis, angiogenesis, and vessel wall integrity. Cytokine Growth Factor Rev 8:21–43

252. Pepper MS, Rosnoblet C, Di Sanza C, Kruithof EK (2001) Synergistic induction of t-PA by vascular endothelial growth factor and basic fibroblast growth factor and localization of t-PA to Weibel-Palade bodies in bovine microvascular endothelial cells. Thromb Haemost 86:702–9

253. Pipp F, et al. (2003) VEGFR-1-selective VEGF homologue PlGF is arteriogenic: evidence for a monocyte-mediated mechanism. Circ Res 92:378–85

254. Plank MJ, Sleeman BD, Jones PF (2004) The role of the angiopoietins in tumour angiogenesis. Growth Factors 22:1–11

255. Pufe T, et al. (2003) The role of vascular endothelial growth factor in glucocorticoid-induced bone loss: evaluation in a minipig model. Bone 33:869–76

256. Qi JH, et al. (2003) A novel function for tissue inhibitor of metalloproteinases-3 (TIMP3): inhibition of angiogenesis by blockage of VEGF binding to VEGF receptor-2. Nat Med 9:407–15

257. Rafii S, Lyden D (2003) Therapeutic stem and progenitor cell transplantation for organ vascularization and regeneration. Nat Med 9:702–12

258. Rajantie I, et al. (2004) Adult bone marrow-derived cells recruited during angiogenesis comprise precursors for periendothelial vascular mural cells. Blood 104:2084–6

259. Rehman J, Li J, Orschell CM, March KL (2003) Peripheral blood "endothelial progenitor cells" are derived from monocyte/macrophages and secrete angiogenic growth factors. Circulation 107:1164–9

260. Religa P, et al. (2002) Smooth-muscle progenitor cells of bone marrow origin contribute to the development of neointimal thickenings in rat aortic allografts and injured rat carotid arteries. Transplantation 74:1310–5

261. Reyes M, et al. (2002) Origin of endothelial progenitors in human postnatal bone marrow. J Clin Invest 109:337–46

262. Reynolds LE, et al. (2002) Enhanced pathological angiogenesis in mice lacking beta3 integrin or beta3 and beta5 integrins. Nat Med 8:27–34

263. Riba RD, et al. (2005) VEGFR1+ haematopoietic bone marrow progenitors initiate the premestatic niche. Nature 438:820–827

264. Ribatti D, Vacca A, Nico B, Ria R, Dammacco F (2002) Cross-talk between hematopoiesis and angiogenesis signaling pathways. Curr Mol Med 2:537–43

265. Rigolin GM, et al. (2005) Neoplastic circulating endothelial cells in multiple myeloma with 13q14 deletion. Blood 107:2531–2535

266. Rini BI, Sosman JA, Motzer RJ (2005) Therapy targeted at vascular endothelial growth factor in metastatic renal cell carcinoma: biology, clinical results and future development. BJU Int 96:286–90

267. Risau W (1995) Differentiation of endothelium. FASEB J 9:926–33

268. Risau W (1998) Development and differentiation of endothelium. Kidney Int Suppl 67:S3–6

269. Rivard A, et al. (1999) Rescue of diabetes-related impairment of angiogenesis by intramuscular gene therapy with adeno-VEGF. Am J Pathol 154:355–63

270. Rockson SG (2002) Preclinical models of lymphatic disease: the potential for growth factor and gene therapy. Ann N Y Acad Sci 979:64–75; discussion 76–9

271. Rubin LL, Staddon JM (1999) The cell biology of the blood-brain barrier. Annu Rev Neurosci 22:11–28

272. Ruhrberg C, et al. (2002) Spatially restricted patterning cues provided by heparin-binding VEGF-A control blood vessel branching morphogenesis. Genes Dev 16:2684–98

273. Ruoslahti E (2002) Specialization of tumour vasculature. Nat Rev Cancer 2:83–90

274. Ruoslahti E (2004) Vascular zip codes in angiogenesis and metastasis. Biochem Soc Trans 32:397–402

275. Ruoslahti E, Rajotte D (2000) An address system in the vasculature of normal tissues and tumors. Annu Rev Immunol 18:813–27

276. Rupnick MA, et al. (2002) Adipose tissue mass can be regulated through the vasculature. Proc Natl Acad Sci U S A 99:10730–5

277. Ruzinova MB, et al. (2003) Effect of angiogenesis inhibition by Id loss and the contribution of bone-marrow-derived endothelial cells in spontaneous murine tumors. Cancer Cell 4:277–89

278. Saiura A, Sata M, Hirata Y, Nagai R, Makuuchi M (2001) Circulating smooth muscle progenitor cells contribute to atherosclerosis. Nat Med 7:382–3

279. Sane DC, Anton L, Brosnihan KB (2004) Angiogenic growth factors and hypertension. Angiogenesis 7:193–201

280. Sata M, et al. (2002) Hematopoietic stem cells differentiate into vascular cells that participate in the pathogenesis of atherosclerosis. Nat Med 8:403–9

281. Schneider M, Othman-Hassan K, Christ B, Wilting J (1999) Lymphangioblasts in the avian wing bud. Dev Dyn 216: 311–9

282. Schrijvers BF, Flyvbjerg A, De Vriese AS (2004) The role of vascular endothelial growth factor (VEGF) in renal pathophysiology. Kidney Int 65:2003–17

283. Schrijvers BF, Flyvbjerg A, Tilton RG, Lameire NH, Vriese AS (2005) A neutralizing VEGF antibody prevents glomerular hypertrophy in a model of obese type 2 diabetes, the Zucker diabetic fatty rat. Nephrol Dial Transplant 21(2):324–329

284. Senger DR, et al. (1983) Tumor cells secrete a vascular permeability factor that promotes accumulation of ascites fluid. Science 219:983–5

285. Senger DR, et al. (2002) The alpha(1)beta(1) and alpha(2)-beta(1) integrins provide critical support for vascular endothelial growth factor signaling, endothelial cell migration, and tumor angiogenesis. Am J Pathol 160: 195–204

286. Serini G, et al. (2003) Class 3 semaphorins control vascular morphogenesis by inhibiting integrin function. Nature 424:391–7

287. Shaked Y, et al. (2005) Genetic heterogeneity of the vasculogenic phenotype parallels angiogenesis: Implications for cellular surrogate marker analysis of antiangiogenesis. Cancer Cell 7:101–11

288. Shalaby F, et al. (1995) Failure of blood-island formation and vasculogenesis in Flk-1-deficient mice. Nature 376: 62–6

289. Shalaby F, et al. (1997) A requirement for Flk1 in primitive and definitive hematopoiesis and vasculogenesis. Cell 89:981–90

290. Shen J, et al. (2005) Suppression of ocular neovascularization with siRNA targeting VEGF receptor 1. Gene Ther 29:1–10

291. Sheu SJ, et al. (2005) Suppression of choroidal neovascularization by intramuscular polymer-based gene delivery of vasostatin. Exp Eye Res 81(6):673–9

292. Shibata R, et al. (2004) Adiponectin stimulates angiogenesis in response to tissue ischemia through stimulation of amp-activated protein kinase signaling. J Biol Chem 279:28670–4

293. Shim WS, et al. (2002) Angiopoietin 1 promotes tumor angiogenesis and tumor vessel plasticity of human cervical cancer in mice. Exp Cell Res 279:299–309

294. Shintani S, et al. (2001) Mobilization of endothelial progenitor cells in patients with acute myocardial infarction. Circulation 103:2776–9

295. Shiojima I, et al. (2005) Disruption of coordinated cardiac hypertrophy and angiogenesis contributes to the transition to heart failure. J Clin Invest 115:2108–18

296. Shoji W, Isogai S, Sato-Maeda M, Obinata M, Kuwada JY (2003) Semaphorin3a1 regulates angioblast migration and vascular development in zebrafish embryos. Development 130:3227–36

297. Shute J, Marshall L, Bodey K, Bush A (2003) Growth factors in cystic fibrosis – when more is not enough. Paediatr Respir Rev 4:120–7

298. Simon AM, McWhorter AR (2002) Vascular abnormalities in mice lacking the endothelial gap junction proteins connexin37 and connexin40. Dev Biol 251:206–20

299. Simons M (2005) Angiogenesis: where do we stand now? Circulation 111:1556–66

300. Simper D, Stalboerger PG, Panetta CJ, Wang S, Caplice NM (2002) Smooth muscle progenitor cells in human blood. Circulation 106:1199–204

301. Smith LE (2003) Pathogenesis of retinopathy of prematurity. Semin Neonatol 8:469–73

302. Song S, Ewald AJ, Stallcup W, Werb Z, Bergers G (2005) PDGFRbeta+ perivascular progenitor cells in tumours regulate pericyte differentiation and vascular survival. Nat Cell Biol 7:870–9

303. Sounni NE, et al. (2002) MT1-MMP expression promotes tumor growth and angiogenesis through an up-regulation of vascular endothelial growth factor expression. FASEB J 16:555–64

304. Stalmans I, et al. (2002) Arteriolar and venular patterning in retinas of mice selectively expressing VEGF isoforms. J Clin Invest 109:327–36

305. Stalmans I, et al. (2003) VEGF: a modifier of the del22q11 (DiGeorge) syndrome? Nat Med 9:173–82

306. Stoll BR, Migliorini C, Kadambi A, Munn LL, Jain RKA (2003) Mathematical model of the contribution of endothelial progenitor cells to angiogenesis in tumors: implications for antiangiogenic therapy. Blood 102:2555–61

307. Storkebaum E, Carmeliet P (2004) VEGF: a critical player in neurodegeneration. J Clin Invest 113:14–8

308. Storkebaum E, et al. (2005) Treatment of motoneuron degeneration by intracerebroventricular delivery of VEGF in a rat model of ALS. Nat Neurosci 8:85–92

309. Storkebaum E, Lambrechts D, Carmeliet P (2004) VEGF: once regarded as a specific angiogenic factor, now implicated in neuroprotection. Bioessays 26:943–54

310. Sudhakar A, et al. (2005) Human alpha1 type IV collagen NC1 domain exhibits distinct antiangiogenic activity mediated by alpha1beta1 integrin. J Clin Invest 115: 2801–10

311. Suri C, et al. (1996) Requisite role of angiopoietin-1, a ligand for the TIE2 receptor, during embryonic angiogenesis. Cell 87:1171–80

312. Suri C, et al. (1998) Increased vascularization in mice overexpressing angiopoietin-1. Science 282:468–71

313. Szekanecz Z, Gaspar L, Koch AE (2005) Angiogenesis in rheumatoid arthritis. Front Biosci 10:1739–53

314. Takakura N, et al. (2000) A role for hematopoietic stem cells in promoting angiogenesis. Cell 102:199–209

315. Tamaki T, et al. (2002) Identification of myogenic-endothelial progenitor cells in the interstitial spaces of skeletal muscle. J Cell Biol 157:571–7

316. Tang K, Breen EC, Gerber HP, Ferrara NM, Wagner PD (2004) Capillary regression in vascular endothelial growth factor-deficient skeletal muscle. Physiol Genomics 18:63–9

317. Tang K, Rossiter HB, Wagner PD, Breen EC (2004) Lung-targeted VEGF inactivation leads to an emphysema phenotype in mice. J Appl Physiol 97:1559–66; discussion: 1549

318. Taylor PC, Sivakumar B (2005) Hypoxia and angiogenesis in rheumatoid arthritis. Curr Opin Rheumatol 17:293–8

319. Tepper OM, et al. (2002) Human endothelial progenitor cells from type II diabetics exhibit impaired proliferation, adhesion, and incorporation into vascular structures. Circulation 106:2781–6

320. Thurston G, et al. (2000) Angiopoietin-1 protects the adult vasculature against plasma leakage. Nat Med 6:460–3

321. Tian XL, et al. (2004) Identification of an angiogenic factor that when mutated causes susceptibility to Klippel-Trenaunay syndrome. Nature 427:640–5

322. Tobe T, et al. (1998) Targeted disruption of the FGF2 gene does not prevent choroidal neovascularization in a murine model. Am J Pathol 153:1641–6

323. Tong RT, et al. (2004) Vascular normalization by vascular endothelial growth factor receptor 2 blockade induces a pressure gradient across the vasculature and improves drug penetration in tumors. Cancer Res 64:3731–6

324. Tordjman R, et al. (2001) Erythroblasts are a source of angiogenic factors. Blood 97:1968–74

325. Torres-Vazquez J, et al. (2004) Semaphorin-plexin signaling guides patterning of the developing vasculature. Dev Cell 7:117–23

326. Tsao PN, et al. (2005) Vascular endothelial growth factor in preterm infants with respiratory distress syndrome. Pediatr Pulmonol 39:461–5

327. Tsukita S, Furuse M (1999) Occludin and claudins in tight-junction strands: leading or supporting players? Trends Cell Biol 9:268–73

328. Van Belle E, et al. (1997) Hypercholesterolemia attenuates angiogenesis but does not preclude augmentation by angiogenic cytokines. Circulation 96:2667–74

329. van Bruggen N, et al. (1999) VEGF antagonism reduces edema formation and tissue damage after ischemia/reperfusion injury in the mouse brain. J Clin Invest 104:1613–20

330. van Cruijsen H, Giaccone G, Hoekman K (2005) Epidermal growth factor receptor and angiogenesis: Opportunities for combined anticancer strategies. Int J Cancer 117:883–8

331. van den Driesche S, Mummery CL, Westermann CJ (2003) Hereditary hemorrhagic telangiectasia: an update on transforming growth factor beta signaling in vasculogenesis and angiogenesis. Cardiovasc Res 58:20–31

332. van Royen N, et al. (2002) Exogenous application of transforming growth factor beta 1 stimulates arteriogenesis in the peripheral circulation. FASEB J 16:432–4

333. Vesalius A (1543) De Humani Corporis Fabrica (The fabric of the human body). Oporinus, Basel

334. Vikkula M, et al. (1996) Vascular dysmorphogenesis caused by an activating mutation in the receptor tyrosine kinase TIE2. Cell 87:1181–90

335. Visconti RP, Richardson CD, Sato TN (2002) Orchestration of angiogenesis and arteriovenous contribution by angiopoietins and vascular endothelial growth factor (VEGF). Proc Natl Acad Sci U S A 99:8219–24

336. Visvader JE, Fujiwara Y, Orkin SH (1998) Unsuspected role for the T-cell leukemia protein SCL/tal-1 in vascular development. Genes Dev 12:473–9

337. Voelkel NF, et al. (2002) Janus face of vascular endothelial growth factor: the obligatory survival factor for lung vascular endothelium controls precapillary artery remodeling in severe pulmonary hypertension. Crit Care Med 30:S251–6

338. Vokes SA, et al. (2004) Hedgehog signaling is essential for endothelial tube formation during vasculogenesis. Development 131:4371–80

339. Vokes SA, Krieg PA (2002) Endoderm is required for vascular endothelial tube formation, but not for angioblast specification. Development 129:775–85

340. Voskas D, et al. (2005) A cyclosporine-sensitive psoriasis-like disease produced in Tie2 transgenic mice. Am J Pathol 166:843–55

341. Voskuil M, et al. (2003) Modulation of collateral artery growth in a porcine hindlimb ligation model using MCP-1. Am J Physiol Heart Circ Physiol 284:H1422–8

342. Wada AM, Reese DE, Bader DM (2001) Bves: prototype of a new class of cell adhesion molecules expressed during coronary artery development. Development 128:2085–93

343. Waltenberger J (2001) Impaired collateral vessel development in diabetes: potential cellular mechanisms and therapeutic implications. Cardiovasc Res 49:554–60

344. Wang HU, Chen ZF, Anderson DJ (1998) Molecular distinction and angiogenic interaction between embryonic arteries and veins revealed by ephrin-B2 and its receptor Eph-B4. Cell 93:741–53

345. Wang HW, et al. (2004) Kaposi sarcoma herpesvirus-induced cellular reprogramming contributes to the lymphatic endothelial gene expression in Kaposi sarcoma. Nat Genet 36:687–93

346. Ward NL, Van Slyke P, Sturk C, Cruz M, Dumont DJ (2004) Angiopoietin 1 expression levels in the myocardium direct coronary vessel development. Dev Dyn 229:500–9

347. Weis SM, Cheresh DA (2005) Pathophysiological consequences of VEGF-induced vascular permeability. Nature 437:497–504

348. Wigle JT, Oliver G (1999) Prox1 function is required for the development of the murine lymphatic system. Cell 98:769–78

349. Wilhelm SM, et al. (2004) BAY 43–9006 exhibits broad spectrum oral antitumor activity and targets the RAF/MEK/ERK pathway and receptor tyrosine kinases involved in tumor progression and angiogenesis. Cancer Res 64:7099–109

350. Wilting J, et al. (2001) Development of the avian lymphatic system. Microsc Res Tech 55:81–91

351. Xia YP, et al. (2003) Transgenic delivery of VEGF to mouse skin leads to an inflammatory condition resembling human psoriasis. Blood 102:161–8

352. Xu Y, Yu Q (2001) Angiopoietin-1, unlike angiopoietin-2, is incorporated into the extracellular matrix via its linker peptide region. J Biol Chem 276:34990–8

353. Yamaguchi J, et al. (2003) Stromal cell-derived factor-1 effects on ex vivo expanded endothelial progenitor cell recruitment for ischemic neovascularization. Circulation 107:1322–8

354. Yamamoto Y, et al. (2004) Tumstatin peptide, an inhibitor of angiogenesis, prevents glomerular hypertrophy in the early stage of diabetic nephropathy. Diabetes 53:1831–40

355. Yang GP, Lim IJ, Phan TT, Lorenz HP, Longaker MT (2003) From scarless fetal wounds to keloids: molecular studies in wound healing. Wound Repair Regen 11:411–8

356. Yang R, et al. (1996) Effects of vascular endothelial growth factor on hemodynamics and cardiac performance. J Cardiovasc Pharmacol 27:838–44

357. Yang ZJ, et al. (2002) Role of vascular endothelial growth factor in neuronal DNA damage and repair in rat brain following a transient cerebral ischemia. J Neurosci Res 70:140–9

358. Yano K, Brown LF, Detmar M (2001) Control of hair growth and follicle size by VEGF-mediated angiogenesis. J Clin Invest 107:409–17

359. Yeager ME, Halley GR, Golpon HA, Voelkel NF, Tuder RM (2001) Microsatellite instability of endothelial cell growth and apoptosis genes within plexiform lesions in primary pulmonary hypertension. Circ Res 88:E2–E11

360. Yin G, et al. (2002) Endostatin gene transfer inhibits joint angiogenesis and pannus formation in inflammatory arthritis. Mol Ther 5:547–54

361. You LR, et al. (2005) Suppression of Notch signalling by the COUP-TFII transcription factor regulates vein identity. Nature 435:98–104

362. Yousef GM, Diamandis EP (2002) Expanded human tissue kallikrein family – a novel panel of cancer biomarkers. Tumour Biol 23:185–92

363. Zhong TP, Childs S, Leu JP, Fishman MC (2001) Gridlock signalling pathway fashions the first embryonic artery. Nature 414:216–20

364. Zhong TP, Rosenberg M, Mohideen MA, Weinstein B, Fishman MC (2000) gridlock, an HLH gene required for assembly of the aorta in zebrafish. Science 287:1820–4

365. Ziegelhoeffer T, et al. (2004) Bone marrow-derived cells do not incorporate into the adult growing vasculature. Circ Res 94:230–8

366. Zlokovic BV (2005) Neurovascular mechanisms of Alzheimer's neurodegeneration. Trends Neurosci 28:202–8

Carmen Ruiz de Almodovar and Annelii Ny contributed equally to this work.

3 |

3.2 Vascular Endothelial Growth Factor in Retinal Vascular Disease

G.L. KING, K. SUZUMA, J.K. SUN

Core Messages

- Vascular endothelial growth factor (VEGF) is an angiogenic and vasopermeability factor induced under hypoxic conditions
- Although VEGF-induced neovascularization plays a beneficial role in resolving tissue ischemia in some forms of non-ocular pathology, such as myocardial infarction, VEGF-related

neovascularization in the eye can lead to severe visual loss
- In ocular tissues, VEGF, in conjunction with other cytokines and hormones, is critically involved in the pathogenesis of vascular disease, including neovascularization secondary to diabetic retinopathy and vascular occlusions

3.2.1 VEGF Regulation and Receptors

Essentials

- 48 kDa homodimeric glycoprotein
- Activated by hypoxia, glucose, ROS, PKC, AGE
- Five member molecular family
- Three VEGF receptors

3.2.1.1 VEGFR2, PKC, PI3-kinase

Vascular endothelial growth factor (VEGF), or vascular permeability factor (VPF), is a 48 kDa homodimeric glycoprotein that functions as an endothelial cell-specific mitogen and vasopermeability factor [18, 27]. The expression of VEGF is potentiated in

response to hypoxia [3], high glucose and protein kinase C (PKC) activation [40], advanced glycation end-products (AGE) [20], reactive oxygen species (ROS), activated oncogenes, and a variety of cytokines [21]. Activation of VEGF induces endothelial cell proliferation, increases vascular permeability, promotes cell migration, and inhibits apoptosis [21].

The VEGF molecular family consists of five members: placenta growth factor (PlGF), VEGF-A, VEGF-B, VEGF-C, and VEGF-D. Each of the VEGF factors may interact with one or more of three VEGF receptors (Fig. 3.2.1).

The tyrosine-kinase receptors for VEGF have been identified as Flt (VEGFR1), KDR/Flk (VEGFR2), and Flt4 (VEGFR3). KDR/Flk (VEGFR2) is reported to play a dominant role in VEGF signaling in endothelial cells. Activation of this signaling pathway leads to the tyrosine phosphorylation of phospholipase C γ (PLCγ), elevation of diacylglycerol (DAG) levels, activation of several PKC isoforms and mitogen-activated protein kinase (MAP-kinase), and

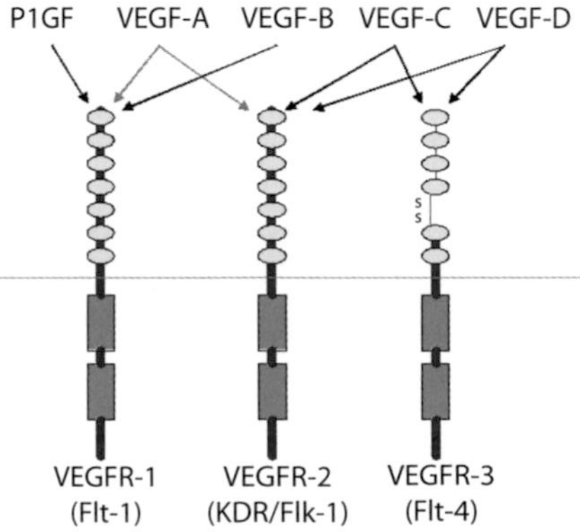

Fig. 3.2.1. VEGF family and receptors

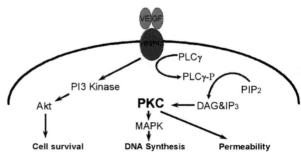

Fig. 3.2.2. Mechanism of VEGF action

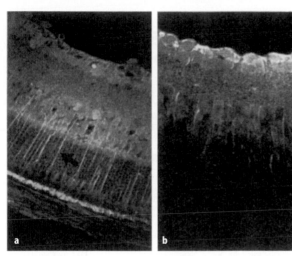

Fig. 3.2.3. Confocal immunohistochemistry of VEGF/VPF. Sections from the eyes of mice at the ages indicated were analyzed by immunohistochemistry with anti-VEGF/VPF antibodies and viewed by confocal microscopy. **a** P17, after 120 h of hypoxia, stained with anti-VEGF/VPF antibodies. Note Müller cell processes spanning retina (*arrow*). **b** Higher power view of **a** showing Müller cell foot plates in inner retina. **a** ×21, **b** ×42. (Fig. 5 from [25])

the activation of the PI3-kinase–Akt pathway (Fig. 3.2.2) [34, 41]. These actions subsequently lead to increased vasopermeability and endothelial cell proliferation as well as cell migration and inhibition of apoptosis [21].

3.2.2 Vascular Endothelial Growth Factor

Essentials
- Involved in formation of retinal neovascularization
- Animal models of retinal ischemia
- Human ischemic retinal disease

Hypoxia stimulates VEGF production in almost all tissue types, including ocular tissues [3]. Data from both animal models and human patients have provided evidence that VEGF plays an active role in the development of retinal neovascularization.

In an animal (mouse) model of ischemic retinopathy, VEGF expression is increased, especially in cells of the inner nuclear layer of the retina that have been identified morphologically as Müller cells (Fig. 3.2.3). Inhibition of VEGF by soluble VEGF receptor-IgG chimeric proteins in the same model markedly reduces retinal neovascularization without significant retinal toxicity (Fig. 3.2.4).

The fact that VEGF is increased in ocular fluids from human patients with active neovascularization secondary to ischemic diseases such as diabetic retinopathy and retinal-vein occlusion provides further supportive evidence that VEGF plays a role in mediating active intraocular neovascularization (Fig. 3.2.5). Increased levels of VEGF are found in ocular fluids from patients with active proliferative diabetic retinopathy, but not in patients with quiescent or non-proliferative disease.

The following model has been suggested for the initiation and control of ischemic ocular neovascularization. The expression of VEGF is increased in ischemic tissue, either bound to local cell surface or basement membrane (VEGF isoforms 189 or 286) or freely diffusible within the vitreous and aqueous cavity (VEGF isoforms 121 or 165) [3]. Neovascularization induced by VEGF's direct action on endothelial cells can arise at areas of increased exposure to VEGF, such as at the optic nerve head, along the vascular arcades, or at the pupillary border. Because the degree of retinal ischemia is directly proportional to VEGF production, a relative reduction in retinal ischemia as a result of reperfusion or from retinal tissue death (e.g., from laser photocoagulation) may reduce VEGF production and thereby result in neovascular regression.

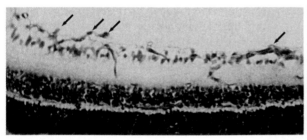

Fig. 3.2.4. Soluble VEGF receptor-IgG chimeric proteins reduce histologically evident ischemia-induced retinal neovascularization. Retinal ischemia was induced in C57BL/6J mice as described in Fig. 2. The right eye of each mouse was injected with 250 ng of human CD4-IgG control chimeric protein on P12 and P14 (*left*). The left eye received intravitreal injections of 250 ng of human Flt-IgG chimera at the same times (*right*). Paraffin-embedded, periodic acid/Schiff reagent, and hematoxylin-stained 6-μm serial sections were obtained. Typical findings from corresponding retinal locations from both eyes of the same mouse are shown and are representative of all animals studied. Vascular cell nuclei internal to the inner limiting membrane represent areas of retinal neovascularization and are indicated with *arrows*. No vascular cell nuclei anterior to the internal limiting membrane are observed in normal, unmanipulated animals. ×50. (Fig. 4 from [4])

3

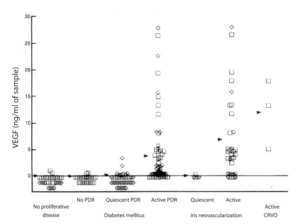

Fig. 3.2.5. Concentrations of immunoreactive VEGF in ocular fluids from patients undergoing intraocular surgery. Aqueous (*squares*), vitreous (*diamonds*), and mean (*arrowheads*). VEGF concentrations are shown. Values of zero or below on the *y*-axis denote concentrations below 0.05 ng/ml. *PDR* proliferative diabetic retinopathy, *CRVO* central-retinal-vein occlusion. (Fig. 2 from [2])

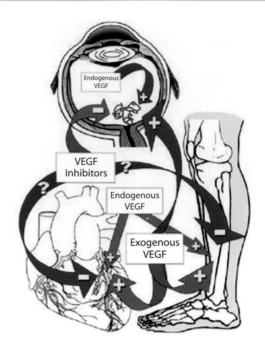

Fig. 3.2.6. Schematic representation of the VEGF agonist/antagonist paradox. Principal organs affected by microvascular (ocular) and macrovascular (cardiac and peripheral vascular) complications in diabetes are shown in relation to their predicted response to VEGF inhibitors and VEGF agonists from either endogenous or exogenous sources. *Blue arrows* represent potential beneficial effects. *Red arrows* reflect possible adverse actions. + stimulation of angiogenesis at the site, – inhibition of angiogenesis at the site. (Fig. 1 from [10])

3.2.3 VEGF and Systemic Diseases

Essentials
- Divergent roles throughout the body
- Role in ischemic heart disease
- Importance of therapeutic selectivity for target organ

Although this chapter focuses primarily on the ocular effects of VEGF activation, it is important to acknowledge the widely divergent roles that VEGF plays throughout the body. In ischemic ocular diseases, neovascularization leads to severe visual loss such as neovascular glaucoma and vitreous hemorrhage. In contrast, VEGF-stimulated collateral blood vessel formation helps to preserve myocardial function during coronary arterial obstruction. Several observations, including the fact that direct VEGF gene transfer therapies have proved to be effective in coronary heart disease as well as peripheral vascular disease, suggest that VEGF plays a significant role in this adaptive process [10]. Thus, it is possible that systemic treatment aimed at one disease may theoretically exacerbate another disease (Fig. 3.2.6). Although anti-VEGF therapies appear promising as a means of reducing neovascular complications from ischemic ocular diseases, they also have the potential to decrease collateral vascular formation and thereby increase macrovascular complications associated with myocardial infarction and peripheral limb ischemia. Because of this consideration, anti-VEGF

therapies for ocular disease must be meticulously designed to have limited activity in other organs.

3.2.4 VEGF and Retinal Vascular Disease

Essentials
- Interaction with other growth factors
- Role in proliferative diabetic retinopathy and macular edema
- Hypertension independently increases VEGF
- Role in central and branch retinal vein occlusions

3.2.4.1 VEGF and Other Growth Factors

In the past, multiple growth factors in the eye, including insulin-like growth factors (IGF), transforming growth factor (TGF), and fibroblast growth factors (FGF), were postulated to play a primary role in the development of ocular complications of diabetes. It is certain that many of these factors have actions that affect both the retinal and choroidal vas-

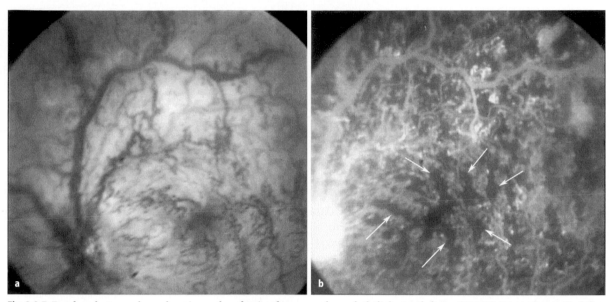

Fig. 3.2.7. Fundus photography and angiography of retina from vascular endothelial growth factor (VEGF)-injected eye. **a** Red-free fundus photograph of a VEGF-injected eye shows dilation and tortuosity of retinal vessels in the posterior pole. 35-degree field photograph. **b** Fluorescein angiogram of the same eye shown in **a** 54 s after dye injection shows non-perfusion of regions throughout the posterior pole including macula (*arrows*). There is leakage from the optic nerve and the retinal vessels end in bulbous microaneurysm-like structures. (Fig. 1 from [35])

culature. However, the fact that many of these growth factors are not regulated by tissue oxygenation suggests that they may play a subordinate role in coordinating vascularization in the retina. Recent studies suggest that many of their effects may be mediated through interactions with VEGF, a molecule whose production is regulated by tissue oxygenation, although other factors may also increase the expression of VEGF (Table 3.2.1).

Table 3.2.1. Interaction of VEGF and other growth factors

	Angio-genesis	VEGF expression	References
IGF	+	+	[28, 17]
Insulin	+	+	[14]
FGF	++	+	[30, 5]
TGF	++	+	[24]
CTGF	+	?	[34, 6]
Estrogen	+	+	[33]
Leptin	+	+	[31]
Angiopoietin	+	+	[32]
Erythropoietin	+	+	[39, 15]
PEDF	–	?	[9]

CTGF connective tissue growth factor, *PEDF* pigment epithelial-derived factor

3.2.4.2 VEGF and Diabetic Retinopathy

Patients with diabetic retinopathy develop proliferative complications (secondary to neovascularization) and/or macular edema (secondary to increased vascular permeability). Recent work has implicated VEGF in human eye diseases characterized by both these conditions [1, 2, 12, 36]. Levels of ocular VEGF are tightly correlated with both growth and permeability of new vessels. Furthermore, introduction of VEGF into normal primate eyes induces similar pathological processes to those seen in diabetic retinopathy (Fig. 3.2.7). In patients with proliferative diabetic retinopathy in which tissue hypoxia promotes neovascularization, levels of VEGF are elevated in ocular tissues [2]. These elevated levels of VEGF decline when treatment with panretinal laser photocoagulation induces regression of neovascularization.

These observations suggest that VEGF may be an important therapeutic target in ocular neovascularization and macular edema secondary to diabetes. Results from a recent phase 2 trial demonstrate that patients with diabetic macular edema (DME) who are treated with an intravitreal anti-VEGF agent, pegaptanib sodium (Macugen), had better visual acuity outcomes, were more likely to show a reduction in macular edema, and were less likely to need focal laser photocoagulation for their edema than patients treated with a sham injection [8]. Addi-

3

tional case series data suggest that other anti-VEGF agents, including bevacizumab, may have efficacy against diabetic neovascularization [29]. Currently, phase 2 and phase 3 clinical trials are ongoing using anti-VEGF agents to treat both diabetic neovascularization and DME.

VEGF expression may also be increased in earlier stages of diabetic retinopathy when clear documentation of intraretinal hypoxia has not yet been clearly documented. Thus, other metabolites of glucose induced toxins such as oxidants, glycated proteins or activation of signaling pathways such as protein kinase C (PKC) have also been suggested to increase capillary permeability and to contribute to macular edema. Recently, the results of clinical trials using a selective inhibitors of the PKC β isoforms (ruboxistaurin) have shown that ruboxistaurin to preserves vision in diabetic patients with vision threatening macular edema. The therapeutic effects of this PKC β isoform inhibitor may be mediated by its effects by on both VEGF dependent and independent pathways.

3.2.4.3 Hypertension As an Aggravating Factor in Diabetes-Induced Activation of VEGF

Epidemiological studies clearly identify hypertension as an independent risk factor for diabetic retinopathy. Hypertension increases the risk of retinopathy progression and the development of proliferative diabetic retinopathy (PDR) [16]. Patients with higher ranges of blood pressure are threefold more likely to develop PDR [26]. The incidence of macular edema is similarly increased with the concomitant presence of hypertension [11]. Patients with hypertension are 3.2-fold more likely to develop diffuse macular edema [19].

Severe hypertension itself can induce a retinopathy characterized by increased retinal vascular leakage [37, 38]. It is also known to independently increase levels of VEGF in the eye. The mechanism by which hypertension induces VEGF expression is by increasing mechanical cyclic stretch in vascular walls, an action that may also involve the enhancement of angiotensin actions. It is possible that the additional induction of VEGF is the reason that systemic hypertension exacerbates coexistent diabetic retinopathy and other ocular diseases that lead to neovascularization and increased vascular permeability.

3.2.4.4 VEGF in Neovascularization Secondary to Retinal Vascular Occlusions

Patients with both branch and central retinal vascular occlusions are at risk for development of retinal neovascularization that is similar to that observed in diabetic retinopathy. As in diabetes, hypoxic induction of VEGF is likely the link between retinal ischemia in these clinical conditions and the development of neovascularization. Clinical studies analyzing VEGF concentrations in aqueous humor have established that there is a correlation between aqueous VEGF concentrations and the extent of non-perfusion in patients with central and branch retinal vein occlusions (CRVO and BRVO) [7, 22]. Increased VEGF levels are also correlated with the onset and persistence of neovascularization of the iris in cases of ischemic CRVO [7], and with increasing vascular permeability and severity of macular edema in cases of BRVO [22, 23]. Initial clinical studies appear to suggest that intravitreal anti-VEGF agents such as bevacizumab may be effective in improving clinical outcomes in CRVO patients, specifically in regard to decreasing macular edema and improving visual acuity [13].

References

1. Adamis AP, Miller JW, Bernal MT, D'Amico DJ, Folkman J, Yeo TK, Yeo KT (1994) Increased vascular endothelial growth factor levels in the vitreous of eyes with proliferative diabetic retinopathy. Am J Ophthalmol 118:445–450
2. Aiello LP, Avery RL, Arrigg PG, Keyt BA, Jampel HD, Shah ST, Pasquale LR, Thieme H, Iwamoto MA, Park JE, Nguyen HV, Aiello LM, Ferrara N, King GL (1994) Vascular endothelial growth factor in ocular fluid of patients with diabetic retinopathy and other retinal disorders. N Engl J Med 331:1480–1487
3. Aiello LP, Northrup JM, Keyt BA, Takagi H, Iwamoto MA (1995) Hypoxic regulation of vascular endothelial growth factor in retinal cells. Arch Ophthalmol 113:1538–1544
4. Aiello LP, Pierce EA, Foley ED, Takagi H, Chen H, Riddle L, Ferrara N, King GL, Smith LE (1995) Suppression of retinal neovascularization in vivo by inhibition of vascular endothelial growth factor (VEGF) using soluble VEGF-receptor chimeric proteins. Proc Natl Acad Sci U S A 92:10457–10461
5. Asahara T, Bauters C, Zheng LP, Takeshita S, Bunting S, Ferrara N, Symes JF, Isner JM (1995) Synergistic effect of vascular endothelial growth factor and basic fibroblast growth factor on angiogenesis in vivo. Circulation 92:II365–371
6. Babic AM, Chen CC, Lau LF (1999) Fisp12/mouse connective tissue growth factor mediates endothelial cell adhesion and migration through integrin alphavbeta3, promotes endothelial cell survival, and induces angiogenesis in vivo. Mol Cell Biol 19:2958–2966
7. Boyd SR, Zachary I, Chakravarthy U, Allen GJ, Wisdom GB, Cree IA, Martin JF, Hykin PG (2002) Correlation of increased vascular endothelian growth factor with neovascularization and permeability in ischaemic central vein occlusion. Arch Ophthalmol 120:1644–1650
8. Cunningham ET, Jr, Adamis AP, Altaweel M, Aiello LP, Bressler NM, D'Amico DJ, Goldbaum M, Guyer DR, Katz B, Patel M, Schwartz SD (2005) A phase II randomized double-masked trial of pegaptanib, an anti-vascular endothelial

growth factor aptamer, for diabetic macular edema. Ophthalmology 112:1747–1757

9. Dawson DW, Volpert OV, Gillis P, Crawford SE, Xu H, Benedict W, Bouck NP (1999) Pigment epithelium-derived factor: a potent inhibitor of angiogenesis. Science 285:245–248

10. Duh E, Aiello LP (1999) Vascular endothelial growth factor and diabetes: the agonist versus antagonist paradox. Diabetes 48:1899–1906

11. El-Asrar AM, Al-Rubeaan KA, Al-Amro SA, Kangave D, Moharram OA (1998) Risk factors for diabetic retinopathy among Saudi diabetics. Int Ophthalmol 22:155–161

12. Funatsu H, Yamashita H, Noma H, Mimura T, Yamashita T, Hori S (2002) Increased levels of vascular endothelial growth factor and interleukin-6 in the aqueous humor of diabetics with macular edema. Am J Ophthalmol 133:70–77

13. Iturralde D, Spaide RF, Meyerle CB, Klancnik JM, Yannuzzi LA, Fisher YL, Sorenson J, Slakter JS, Freund KB, Cooney M, Fine HF (2006) Intravitreal bevacizumab (Avastin) treatment of macular edema in central retinal vein occlusion: a short-term study. Retina 26:279–284

14. Jiang ZY, He Z, King BL, Kuroki T, Opland DM, Suzuma K, Suzuma I, Ueki K, Kulkarni RN, Kahn CR, King GL (2003) Characterization of multiple signaling pathways of insulin in the regulation of vascular endothelial growth factor expression in vascular cells and angiogenesis. J Biol Chem 278:31964–31971

15. Kertesz N, Wu J, Chen TH, Sucov HM, Wu H (2004) The role of erythropoietin in regulating angiogenesis. Dev Biol 276:101–110

16. Klein R, Klein BE, Moss SE, Cruickshanks KJ (1998) The Wisconsin Epidemiologic Study of diabetic Retinopathy: XVII. The 14-year incidence and progression of diabetic retinopathy and associated risk factors in type 1 diabetes. Ophthalmology 105:1801–1815

17. Kondo T, Vicent D, Suzuma K, Yanagisawa M, King GL, Holzenberger M, Kahn CR (2003) Knockout of insulin and IGF-1 receptors on vascular endothelial cells protects against retinal neovascularization. J Clin Invest 111:1835–1842

18. Leung DW, Cachianes G, Kuang WJ, Goeddel DV, Ferrara N (1989) Vascular endothelial growth factor is a secreted angiogenic mitogen. Science 246:1306–1309

19. Lopes de Faria JM, Jalkh AE, Trempe CL, McMeel JW (1999) Diabetic macular edema: risk factors and concomitants. Acta Ophthalmol Scand 77:170–175

20. Lu M, Kuroki M, Amano S, Tolentino M, Keough K, Kim I, Bucala R, Adamis AP (1998) Advanced glycation end products increase retinal vascular endothelial growth factor expression. J Clin Invest 101:1219–1224

21. Neufeld G, Cohen T, Gengrinovitch S, Poltorak Z (1999) Vascular endothelial growth factor (VEGF) and its receptors. FASEB J 13:9–22

22. Noma H, Funatsu H, Yamasaki M, Tsukamoto H, Mimura T, Sone T, Jian K, Sakamoto I, Nakano K, Yamashita H, Minamoto A, Mishima HK (2005) Pathogenesis of macular edema with branch retinal vein occlusion and intraocular levels of vascular endothelial growth factor and interleukin-6. Am J Ophthalmol 140:256–261

23. Noma H, Minamoto A, Funatsu H, Tsukamoto H, Nakano K, Yamashita H, Mishima HK (2006) Intravitreal levels of vascular endothelial growth factor and interleukin-6 are correlated with macular edema in branch retinal vein occlusion. Graefes Arch Clin Exp Ophthalmol 244:309–315

24. Pertovaara L, Kaipainen A, Mustonen T, Orpana A, Ferrara N, Saksela O, Alitalo K (1994) Vascular endothelial growth factor is induced in response to transforming growth factor-beta in fibroblastic and epithelial cells. J Biol Chem 269:6271–6274

25. Pierce EA, Avery RL, Foley ED, Aiello LP, Smith LE (1995) Vascular endothelial growth factor/vascular permeability factor expression in a mouse model of retinal neovascularization. Proc Natl Acad Sci U S A 92:905–909

26. Roy MS (2000) Diabetic retinopathy in African Americans with type 1 diabetes: The New Jersey 725: II. Risk factors. Arch Ophthalmol 118:105–115

27. Senger DR, Perruzzi CA, Feder J, Dvorak HF (1986) A highly conserved vascular permeability factor secreted by a variety of human and rodent tumor cell lines. Cancer Res 46:5629–5632

28. Smith LE, Shen W, Perruzzi C, Soker S, Kinose F, Xu X, Robinson G, Driver S, Bischoff J, Zhang B, Schaeffer JM, Senger DR (1999) Regulation of vascular endothelial growth factor-dependent retinal neovascularization by insulin-like growth factor-1 receptor. Nat Med 5:1390–1395

29. Spaide RF, Fisher YL (2006) Intravitreal bevacizumab (Avastin) treatment of proliferative diabetic retinopathy complicated by vitreous hemorrhage. Retina 26:275–278

30. Stavri GT, Zachary IC, Baskerville PA, Martin JF, Erusalimsky JD (1995) Basic fibroblast growth factor upregulates the expression of vascular endothelial growth factor in vascular smooth muscle cells. Synergistic interaction with hypoxia. Circulation 92:11–14

31. Suganami E, Takagi H, Ohashi H, Suzuma K, Suzuma I, Oh H, Watanabe D, Ojima T, Suganami T, Fujio Y, Nakao K, Ogawa Y, Yoshimura N (2004) Leptin stimulates ischemia-induced retinal neovascularization: possible role of vascular endothelial growth factor expressed in retinal endothelial cells. Diabetes 53:2443–2448

32. Suri C, Jones PF, Patan S, Bartunkova S, Maisonpierre PC, Davis S, Sato TN, Yancopoulos GD (1996) Requisite role of angiopoietin-1, a ligand for the TIE2 receptor, during embryonic angiogenesis. Cell 87:1171–1180

33. Suzuma I, Mandai M, Takagi H, Suzuma K, Otani A, Oh H, Kobayashi K, Honda Y (1999) 17{beta}-estradiol increases VEGF receptor-2 and promotes DNA synthesis in retinal microvascular endothelial cells. Invest Ophthalmol Vis Sci 40:2122–2129

34. Suzuma K, Naruse K, Suzuma I, Takahara N, Ueki K, Aiello LP, King GL (2000) Vascular endothelial growth factor induces expression of connective tissue growth factor via KDR, Flt1, and phosphatidylinositol 3-kinase-Akt-dependent pathways in retinal vascular cells. J Biol Chem 275:40725–40731

35. Tolentino MJ, McLeod DS, Taomoto M, Otsuji T, Adamis AP, Lutty GA (2002) Pathologic features of vascular endothelial growth factor-induced retinopathy in the nonhuman primate. Am J Ophthalmol 133:373–385

36. Tolentino MJ, Miller JW, Gragoudas ES, Jakobiec FA, Flynn E, Chatzistefanou K, Ferrara N, Adamis AP (1996) Intravitreous injections of vascular endothelial growth factor produce retinal ischemia and microangiopathy in an adult primate. Ophthalmology 103:1820–1828

37. Tso MO, Jampol LM (1982) Pathophysiology of hypertensive retinopathy. Ophthalmology 89:1132–1145

38. Walsh JB (1982) Hypertensive retinopathy: description, classification, and prognosis. Ophthalmology 89:1127–1131

39. Watanabe D, Suzuma K, Matsui S, Kurimoto M, Kiryu J, Kita M, Suzuma I, Ohashi H, Ojima T, Murakami T, Kobayashi T, Masuda S, Nagao M, Yoshimura N, Takagi H (2005) Erythropoietin as a retinal angiogenic factor in proliferative diabetic retinopathy. N Engl J Med 353:782–792

40. Williams B, Gallacher B, Patel H, Orme C (1997) Glucose-induced protein kinase C activation regulates vascular permeability factor mRNA expression and peptide production by human vascular smooth muscle cells in vitro. Diabetes 46:1497–1503

41. Xia P, Aiello LP, Ishii H, Jiang ZY, Park DJ, Robinson GS, Takagi H, Newsome WP, Jirousek MR, King GL (1996) Characterization of vascular endothelial growth factor's effect on the activation of protein kinase C, its isoforms, and endothelial cell growth. J Clin Invest 98:2018–2026

3.3 Involvement of the Ephrin/Eph System in Angioproliferative Ocular Diseases

H. Agostini, G. Martin

Core Messages

- Ephrin and their Eph receptors both are membrane bound and signal upon cell-cell contact
- Ephrin/Ephs are essential for axonal guidance during neuronal development and are involved in angiogenic cell recruitment
- EphrinB2 is expressed in arteries, EphB4 in veins
- Expression of ephrinB2, EphB2, and EphB3 is increased in diabetic retinopathy, and expression of EphA7 in RPE is increased in age-related macular degeneration (AMD)
- The potential for a therapeutic use of recombinant ephrins or Ephs remains to be determined

3.3.1 First Studies: Ephrins in Retinotectal Projection

The ephrins and Eph receptors are a large subgroup of the receptor tyrosine kinases (RTK, Fig. 3.3.1). They are bound to the cell membrane and mediate signals of cell contact and interaction between cells. This way they provide positional information during cell growth and cell migration, especially if they are expressed in a gradient. Their role in a developing neuronal network was intensively studied in axon guidance (reviewed in [15, 17, 19, 20, 24, 27, 28]). The differentiating murine retina and its projection in the brain is a well studied example of the biological function of the ephrin/Eph system during axonal growth. The essentials are summarized in Fig. 3.3.2.

Topographic mapping of axons of retinal ganglion cells (RGCs) at the neuronal target site occurs along

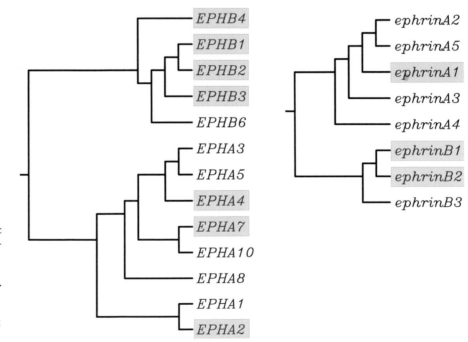

Fig. 3.3.1. Human Eph receptors and ephrins. Angiogenic ephrins are highlighted. EphrinAs interact with EphA receptors, and ephrinBs interact with EphB receptors. Only EphA4 interacts with EphB receptors, too. The sequence of EphA6 is not yet fully determined

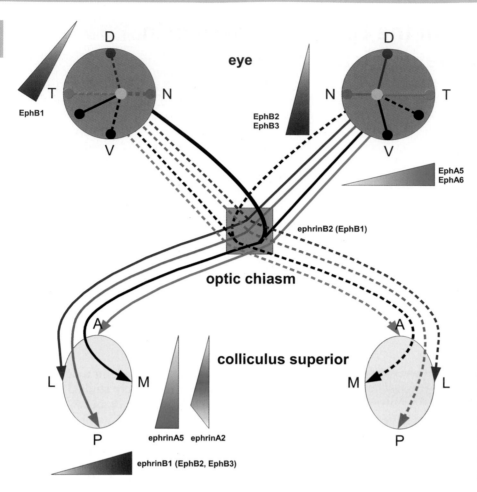

Fig. 3.3.2. Ephrins in mouse retinotectal projection. Retinal ganglion cells form axons which grow through the optic nerve head to the optic chiasm where they are projected contralaterally or ipsilaterally to the colliculus superior. Ephrin gradients (indicated by *graded triangles*) in the optic chiasm or the colliculus superior guide the growing axons depending on their load of Eph receptors. Note that the portion of the non-crossing temporal axons is much smaller in mice than in humans. The known interactors of the ephrins are indicated in *parentheses*

two orthogonal axes: the nasal-temporal axis of the retina maps along the anterior-posterior axis (A-P) of the corpus geniculatum laterale (man) or colliculus superior (mouse) and the dorsal-ventral retinal axis along the lateral-medial axis (L-M). During eye development in the embryo, RGCs express EphA5 and EphA6 in a high temporal to low nasal gradient [11]. At the superior colliculus (SC) in the midbrain, they reach an area with a low anterior to high posterior gradient expression of ephrinA5 and ephrinA2 [12]. The more Eph an RGC expresses on the axon tip the earlier it generates a signal upon interaction with the ephrin, resulting in a stop of growth.

Similarly, EphB2 and EphB3 are expressed in a high ventral to low dorsal gradient in the mouse retina. This corresponds to a low lateral to high medial gradient of ephrinB1 in the SC resulting in lateral innervation of dorsal RGCs in the SC and in medial innervation of ventral RGCs. EfnB1 seems to be attractive rather than repulsive for EphB2 or EphB3 expressing RGCs. Mutants without EphB2 or EphB3 show additional innervation of ventral RGCs in the lateral SC [14].

The EphB1 receptor is expressed in a high ventral to low dorsal gradient on RGCs during axon pathfinding in the mouse. At the optic chiasm, axons of the ventral RGCs are repelled by ephrinB2 to their own side while axons from dorsal RGCs cross to the other side. Loss of EphB1, but not EphB2 or EphB3, in mutant mice reduced ipsilateral projection as did injection into the chiasm of soluble EphB4-Fc, which blocks ephrinB2 [35].

In humans, the retinotectal development seems to follow comparable molecular mechanisms [18]. However, species specific differences exist. As binocular vision is much more prominent in humans than in mice, EphB1 is expressed in the whole temporal retina compared to the small border in mice, resulting in a higher part of ipsilateral projection. The center of the EphA5 and EphA6 expression is shifted to the foveal area, and this shift is related to the lateral orientation of the mouse eyes.

3.3.2 Ephrins in Vascular Development

The pattern of the peripheral nervous system is similar to the vascular system. Therefore, it was asked if the factors involved in axon guidance are also

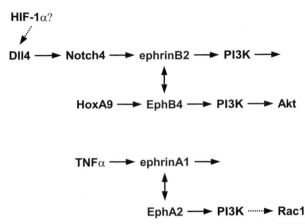

Fig. 3.3.3. Signal transduction of angiogenic ephrins. The *EphB4* gene is activated by *HoxA9* while the *ephrinB2* gene is activated by *Notch4* and the *ephrinA1* gene by *TNFα*. One way of *EphB4* signaling is through *PI3K* and *Akt* while signal transduction of *ephrinB2* also uses *PI3K* in a different way. The same is true for *EphA2* signaling. Data are not yet complete

involved in angiogenesis. Indeed, all major groups of axon guidance factors like ephrin/Eph, netrin/DCC/Unc5, semaphorin/neuropilin/plexin, and slit/robo were also found to be involved in angiogenesis [6, 10, 34]. VEGF, which is produced by nerve cells, guides arterial formation through VEGF receptors expressed on endothelial cells [25].

The role of ephrins in angiogenesis has been reviewed [2, 3, 8]. EphrinB2 is expressed in arterial angioblast and endothelial and perivascular mesenchymal cells, while EphB4 is expressed in endothelial cells of veins. The knockout mutants have severe defects in vascular development and die at embryonic stages [33]. Similar defects are observed in EphB2/EphB3 double knockout mutants, while the single EphB2 or EphB3 mutants have no vascular phenotype [1]. EphrinB1, EphB1, EphB2, and EphB3 are expressed both in arteries and veins. EphB4 signals through PI3K and Akt [30] and is activated by *HoxA9* [5]. EphrinB2 is activated by *Notch4* and *Dll4* in cell culture experiments with human endothelial cells [16, 29] (Fig. 3.3.3).

EphrinA1 is expressed in vascular tissues during embryonic development [23]. Its gene is activated by TNFα [7]. The ephrinA1 receptor, EphA2, has no expression in embryonic vasculature. Mutants are viable and fertile but show reduced angiogenesis in adult animals, for example in wound healing or tumor growth [4]. EphA2 signals through PI3K and Rac1 GTPase, which is a rather common intracellular signaling pathway [3] (Fig. 3.3.3). Inhibition of both results in reduced tumor growth, and both are involved in tumor vascularization [26].

3.3.3 Ephrins and Ocular Angiogenesis

Immunohistochemical staining for ephrinB2, EphB2, and EphB3 expression was found in endothelial cells of fibroproliferative membranes from human eyes [32]. Membranes were derived from patients with proliferative diabetic retinopathy and from retinopathy of prematurity. The staining pattern was equal for all factors and correlated with that of factor VIII, a marker for endothelial cells. Although detectable in pathological specimens, the biological functions of these factors are not yet clear in this context.

EphA7 was found in cells cultured from choroidal neovascularization membranes (CNV-RPE) as well as in paraffin sections of CNV membranes by immunohistochemistry [21]. It is thought that EphA7 is involved in the transdifferentiation process of RPE cells into CNV matrix cells, especially during late stages of age-related macular degeneration (AMD).

Human retinal endothelial cells (HRECs) express ephrinB2 but only low levels of EphB4 [31]. Stimulation of ephrinB2 by a soluble EphB4/Fc fusion protein resulted in phosphorylation of PI3K, Src, and ERK1/2, indicating that MAPK signaling may be involved.

These data show that ephrins are present in normal and pathologic tissues of the eye. But at the moment, their role in pathogenesis needs further clarification.

3.3.4 Ephrins in Retinal and Subretinal Animal Models

EfnB2 as well as its inducer Dll4 is not expressed in retinal arteries of young mice if they are treated with mild hypoxia (10% oxygen) directly after birth for up to 6 days [9]. This corresponds with a 2.5-fold increase of ephrinB2 expression during treatment of 7-day-old mice with 75% oxygen [22]. EphrinB2 may be induced by oxygen via HIF1.

EphrinB2 and EphB4 were found to be expressed in choroidal endothelial cells and in laser-induced choroidal neovascularizations (CNV) in rat [13]. CNV formation was inhibited by intravitreal injection of monomeric sEphB4 consisting of the extracellular part of EphB4 only. A system for testing retinal neovascularization is the mouse model of oxygen-induced retinopathy. Seven-day-old mice are placed in 75% oxygen and returned to normal air after 5 days. During the next 5 days, they develop retinal neovascularization prompted by relative hypoxia. This retinal neovascularization is enhanced after intravitreal injection of dimeric EphB4-Fc or ephrinB2-Fc and reduced by injection of monomeric sEphB4 or s-ephrinB2 [22]. Phosphorylated forms of Eph receptors are detected in preretinal blood vessel tufts in histological eye sections stained with a pan-anti-phospho-Eph antibody [36]. Expression of EphB4 as well as of VEGF, VEGF-R1, and VEGF-R2, but not of ephrinB2, is enhanced 2 days after return of the mice to normal oxygen. This indicates that EphB4 and ephrinB2 influence angiogenesis by independent signaling pathways.

3.3.5 Therapeutic Potential

Ephrins are involved in late stages of capillary formation. Since ephrins themselves have no known direct proliferative effect on endothelial cells, intervention at the level of ephrin action is expected to have an indirect effect on angiogenesis by changing cell recruitment.

In order to manipulate the Ephrin/Eph system it is helpful to visualize the mode of how cell-bound ephrins or Ephs are activated. Upon binding of a corresponding ligand, clustering is thought to be essential before the signaling cascade is activated. So far, two different experimental approaches with therapeutic potential were pursued by different groups by modulating angiogenesis with monomeric or dimeric ephrinB2 and EphB4, respectively. Monomeric ligands or receptors are thought to inhibit clustering whereas dimeric recombinant proteins could be potential activators of cluster formation. The results are still conflicting. Monomeric EphB4 seems to be the strongest inhibitor of vessel formation followed by monomeric ephrinB2. With regard to the dimeric forms both inhibition and enhancement of angiogenesis are described. Further studies will be necessary before a clear recommendation for clinical use of ephrin/Eph inhibitors can be given.

References

1. Adams RH, Wilkinson GA, Weiss C, Diella F, Gale NW, Deutsch U, Risau W, Klein R (1999) Roles of ephrinB ligands and EphB receptors in cardiovascular development: demarcation of arterial/venous domains, vascular morphogenesis, and sprouting angiogenesis. Genes Dev 13:295–306
2. Augustin HG, Reiss Y (2003) EphB receptors and ephrinB ligands: regulators of vascular assembly and homeostasis. Cell Tissue Res 314:25–31
3. Brantley-Sieders DM, Chen J (2004) Eph receptor tyrosine kinases in angiogenesis: from development to disease. Angiogenesis 7:17–28
4. Brantley-Sieders DM, Caughron J, Hicks D, Pozzi A, Ruiz JC, Chen J (2004) EphA2 receptor tyrosine kinase regulates endothelial cell migration and vascular assembly through phosphoinositide 3-kinase-mediated Rac1 GTPase activation. J Cell Sci 117:2037–2049
5. Bruhl T, Urbich C, Aicher D, Acker-Palmer A, Zeiher AM, Dimmeler S (2004) Homeobox A9 transcriptionally regulates the EphB4 receptor to modulate endothelial cell migration and tube formation. Circ Res 94:743–751
6. Carmeliet P, Tessier-Lavigne M (2005) Common mechanisms of nerve and blood vessel wiring. Nature 436:193–200
7. Cheng N, Chen J (2001) Tumor necrosis factor-alpha induction of endothelial ephrin A1 expression is mediated by a p38. J Biol Chem 276:13771–13777
8. Cheng N, Brantley DM, Chen J (2002) The ephrins and Eph receptors in angiogenesis. Cytokine Growth Factor Rev 13:75–85
9. Claxton S, Fruttiger M (2005) Oxygen modifies artery differentiation and network morphogenesis in the retinal vasculature. Dev Dyn 233:822–828
10. Eichmann A, Makinen T, Alitalo K (2005) Neural guidance molecules regulate vascular remodeling and vessel navigation. Genes Dev 19:1013–1021
11. Feldheim DA, Nakamoto M, Osterfield M, Gale NW, DeChiara TM, Rohatgi R, Yancopoulos GD, Flanagan JG (2004) Loss-of-function analysis of EphA receptors in retinotectal mapping. J Neurosci 24:2542–2550
12. Frisen J, Yates PA, McLaughlin T, Friedman GC, O'Leary DD, Barbacid M (1998) Ephrin-A5 (AL-1/RAGS) is essential for proper retinal axon guidance and topographic mapping in the mammalian visual system. Neuron 20:235–43
13. He S, Ding Y, Zhou J, Krasnoperov V, Zozulya S, Kumar SR, Ryan SJ, Gill PS, Hinton DR (2005) Soluble EphB4 regulates choroidal endothelial cell function and inhibits laser-induced choroidal neovascularization. Invest Ophthalmol Vis Sci 46:4772–4779
14. Hindges R, McLaughlin T, Genoud N, Henkemeyer M, O'Leary DD (2002) EphB forward signaling controls directional branch extension and arborization required for dorsal-ventral retinotopic mapping. Neuron 35:475–487
15. Inatani M (2005) Molecular mechanisms of optic axon guidance. Naturwissenschaften 92:549–561
16. Iso T, Maeno T, Oike Y, Yamazaki M, Doi H, Arai M, Kurabayashi M (2006) Dll4-selective Notch signaling induces ephrinB2 gene expression in endothelial cells. Biochem Biophys Res Commun 341:708–714
17. Klein R (2004) Eph/ephrin signaling in morphogenesis, neural development and plasticity. Curr Opin Cell Biol 16:580–589
18. Lambot MA, Depasse F, Noel JC, Vanderhaeghen P (2005) Mapping labels in the human developing visual system and the evolution of binocular vision. J Neurosci 25:7232–7237
19. Lemke G, Reber M (2005) Retinotectal mapping: new insights from molecular genetics. Annu Rev Cell Dev Biol 21:551–580
20. Mann F, Harris WA, Holt CE (2004) New views on retinal axon development: a navigation guide. Int J Dev Biol 48:957–964
21. Martin G, Schlunck G, Hansen LL, Agostini HT (2004) Differential expression of angioregulatory factors in normal and CNV-derived human retinal pigment epithelium. Graefes Arch Clin Exp Ophthalmol 242:321–326
22. Martin G, Ehlken C, Lange C, Gogaki EG, Fiedler U, Schaffner F, Hansen LL, Augustin HG, Agostini HT (2006) Involvement of the ephrinB2/EphB4 system in physiological retinal vascularization and in oxygen-induced angioproliferative retinopathy. (submitted)
23. McBride JL, Ruiz JC (1998) Ephrin-A1 is expressed at sites of vascular development in the mouse. Mech Dev 77:201–4

24. McLaughlin T, O'Leary DD (2005) Molecular gradients and development of retinotopic maps. Annu Rev Neurosci 28:327–355

25. Mukouyama YS, Shin D, Britsch S, Taniguchi M, Anderson DJ (2002) Sensory nerves determine the pattern of arterial differentiation and blood vessel branching in the skin. Cell 109:693–705

26. Ogawa K, Pasqualini R, Lindberg RA, Kain R, Freeman AL, Pasquale EB (2000) The ephrin-A1 ligand and its receptor, EphA2, are expressed during tumor neovascularization. Oncogene 19:6043–52

27. Oster SF, Deiner M, Birgbauer E, Sretavan DW (2004) Ganglion cell axon pathfinding in the retina and optic nerve. Semin Cell Dev Biol 15:125–136

28. Pasquale EB (2005) Eph receptor signalling casts a wide net on cell behaviour. Nat Rev Mol Cell Biol 6:462–475

29. Shawber CJ, Das I, Francisco E, Kitajewski J (2003) Notch signaling in primary endothelial cells. Ann N Y Acad Sci 995:162–170

30. Steinle JJ, Meininger CJ, Forough R, Wu G, Wu MH, Granger HJ (2002) Eph B4 receptor signaling mediates endothelial cell migration and proliferation via the phosphatidylinositol 3-kinase pathway. J Biol Chem 277:43830–43835

31. Steinle JJ, Meininger CJ, Chowdhury U, Wu G, Granger HJ (2003) Role of ephrin B2 in human retinal endothelial cell proliferation and migration. Cell Signal 15:1011–1017

32. Umeda N, Ozaki H, Hayashi H, Oshima K (2004) Expression of ephrinB2 and its receptors on fibroproliferative membranes in ocular angiogenic diseases. Am J Ophthalmol 138:270–279

33. Wang HU, Chen ZF, Anderson DJ (1998) Molecular distinction and angiogenic interaction between embryonic arteries and veins revealed by ephrin-B2 and its receptor Eph-B4 [see comments]. Comment in: Cell 93:661–4; Cell 93:741–53

34. Weinstein BM (2005) Vessels and nerves: marching to the same tune. Cell 120:299–302

35. Williams SE, Mann F, Erskine L, Sakurai T, Wei S, Rossi DJ, Gale NW, Holt CE, Mason CA, Henkemeyer M (2003) Ephrin-B2 and EphB1 mediate retinal axon divergence at the optic chiasm. Neuron 39:919–935

36. Zamora DO, Davies MH, Planck SR, Rosenbaum JT, Powers MR (2005) Soluble forms of EphrinB2 and EphB4 reduce retinal neovascularization in a model of proliferative retinopathy. Invest Ophthalmol Vis Sci 46:2175–2182

4

4 Hematopoietic Stem Cells in Vascular Development and Ocular Neovascularization

N. Sengupta, M.B. Grant, S. Caballero, M.E. Boulton

Core Messages

- Hematopoietic stem cells (HSCs) have defined developmental origins
- HSCs reside in the bone marrow and repopulate the blood and vascular components
- HSCs and their differentiated progeny can be identified and isolated by specific surface markers
- Adult neovascularization is now recognized to occur by recruitment and incorporation of pre-

cursor and stem cells as well as by the growth of mature resident cells
- Endothelial progenitor cells (EPCs) are major contributors to aberrant ocular neovascularization associated with proliferative diabetic retinopathy and age-related macular degeneration
- EPC recruitment, migration and differentiation are dependent on systemic and local mediators
- EPCs can be manipulated in culture
- HSCs are involved in ocular neovascularization

4.1 Background

Neovascular diseases of the eye include retinopathy of prematurity (ROP), proliferative diabetic retinopathy (PDR), and the exudative or "wet" form of age-related macular degeneration (ARMD). Together these diseases affect all age groups and are the leading causes of vision impairment in developed nations [77].

Pathologies of the vascular system, including ocular neovascularization, involve the endothelium. Endothelial cells comprise the innermost layer of all blood vessels, cardiac valves, and several other body cavities. They are responsible for maintaining vessel patency, preventing thrombosis, and relaxation and contraction of vessels [80]. The migration, proliferation, and organization of endothelial cells and their precursors into tubules defines the process of neovascularization, whether it be during development or during adult reparative or pathological processes.

Vasculogenesis has been used to characterize *de novo* formation of blood vessels from less differentiated precursor cells, and is considered to occur solely during embryonic development. During embryogenesis the so-called hemangioblast gives rise to the vasculature as well as to erythrocytes and blood leukocytes. Postnatal neovascularization is classically thought to involve angiogenesis, i.e., the creation of new vessels by the proliferation, migration and tubule formation of resident, mature endothelial

cells. With the knowledge that endothelial progenitor cells (EPCs) and their undifferentiated antecedents, hematopoietic stem cells (HSCs), participate in postnatal neovascularization, it is now recognized that both pathological and physiological postnatal neovascularization consists of a combination of both angiogenesis and vasculogenesis.

4.2 Developmental Origins of HSCs

During mammalian development, the primitive hematopoietic cells consist of precursors with erythroid and/or myeloid potential and, in the mouse, appear in the blood islands of the yolk sac beginning 7 days after fertilization [10]. Progenitors endowed with long-term reconstitution (LTR) activity are generated solely in the intraembryonic para-aortic splanchnopleura and not in the yolk sac [27]. After circulation is established, multipotent precursors with LTR activity can then be found in the blood [31, 92] and these cells then go to the yolk sac [45]. The para-aortic splanchnopleura becomes the aorta, gonads, and mesonephros (AGM). The hematopoietic precursors present in the AGM at 10.5 days postfertilization have LTR capacity when transferred to adult normal recipient mice [92].

The AGM, also called the splanchnic mesoderm, is the site of origin of intraembryonic HSCs [29, 46]. These HSCs exist as clusters of basophilic cells located within the floor of the aorta from 10 to 10.5 days after fertilization. These hematopoietic intra-aortic clusters have been described in all vertebrate embryos, including humans [44]. Other structures potentially involved in intraembryonic HSC generation are the subaortic patches (SAPs), located below the aortic floor, which express the transcription factor GATA-3 [84] and the AA4.1 antigen [106]. SAPs disappear concomitantly with the cessation of AGM-HSC production at 12 days postfertilization [43, 84].

AGM-hematopoietic activity has been shown to derive from cells harboring markers shared by endothelial and hematopoietic lineages. These cells also lack the panhematopoietic marker CD45 [28]. Thus the endothelium of the aortic floor is believed to display "hemogenic" activity and is able to give rise to HSCs. Recently, Bertrand showed that CD45$^+$ cells contain only macrophage precursors while CD45$^{-/lo}$c-kit$^+$AA4.1$^+$CD41$^+$ AGM fraction contains multipotent cells which account for the bulk of multipotent precursors at this stage and contain the LTR activity [16]. When HSCs are released into the circulation they seed the fetal liver and eventually the adult bone marrow [97].

4.2.1 HSCs Lack Regional Patterning

Most tissues are patterned so that progenitors in different locations are programmed to have different properties. For example, stem cells from different regions of the nervous system acquire intrinsic differences in their properties as they migrate through distinct environments [47].

HSCs by virtue of their function also migrate through diverse environments throughout life. Keil and coworkers asked whether HSCs change as they are exposed to diverse environments. This ability would affect their plasticity. They observed significant differences in hematopoiesis between the fetal liver and fetal spleen. However, they were not able to detect phenotypic, functional, or gene expression differences between the HSCs in these organs. Their work suggests no regional differences between HSCs. They were also unable to detect phenotypic, functional, or gene expression differences between HSCs in different adult bone marrow compartments. Their failure to detect differences among stem cells from different regions of the hematopoietic system unlike the nervous system at the same time during development suggests that the hematopoietic system has evolved mechanisms to prevent the spatial reprogramming of HSC properties as they migrate between distinct environments [66].

4.3 Defining the Adult HSCs

Adult HSCs are mostly quiescent cells that comprise ~0.05–0.1% of the bone marrow [60, 91, 124]. Upon activation, HSCs proliferate and differentiate into progenitor cells which can be found in the peripheral blood [59]. These progenitor cells may then differentiate further as required. At the same time, HSCs may divide to yield undifferentiated progeny with identical HSC characteristics, fulfilling the two criteria for a stem cell: self-renewal and the ability to give rise to differentiated progeny.

HSCs in the adult are found in the bone marrow and continue to repopulate cells of the blood and lymph throughout life [85]. In addition, HSCs are now known to have hemangioblast activity, i.e., they can differentiate into all components of the vascular system (vessels as well as blood components, Fig. 4.1). HSCs have been reported to have the potential to differentiate into a variety of other tissues, including liver [74], muscle [39], and neuronal cells [20, 35, 87], thus demonstrating a role for stem cell plasticity in tissue maintenance and repair.

4.3.1 HSC Self-Renewal

Stem cells are heralded as a limitless source for tissue or organ regeneration because of their self-renewal capacity. However, the self-renewal capacity has not been quantified and remains an issue of debate. Most researchers would consider the most reliable indicator of self-renewal capacity to be the long-term multilineage repopulating activity, detectable by transplantation experiments.

Evidence for self-renewal of HSCs has been provided by retroviral marking studies in which HSC clones tagged with retroviral integration sites were transplanted into secondary recipients [32, 65]. A high degree of HSC purification enabled successful long-term reconstitution with single HSCs [36, 102]. Since then transplantation of single CD34$^{/low}$ and c-kit$^+$ Sca-1$^+$ lineage (KSL) cells has been successfully accomplished by many investigators [37, 51]. We observed donor-derived CD34KSL cells in the recipients' bone marrow indicating self-renewal and expansion of the originally transplanted single CD34KSL cells [51]. Studies by Ema, however, demonstrated that when single CD34KSL cells were sorted and transplanted into secondary recipients, the reconstitution capacity of the CD34KSL cells appeared significantly diminished. These data imply that while HSCs do self-renew in bone marrow of primary recipients, their capacity to self-renew declines [37].

In attempting to better characterize and thus quantify self-renewal, additional endpoints have

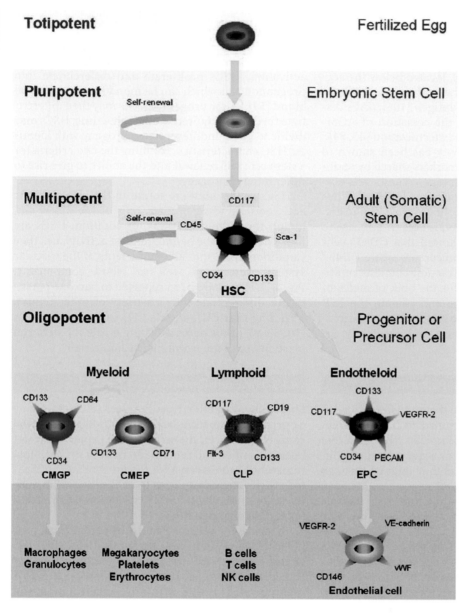

Fig. 4.1. Plasticity and differentiation potential of stem and precursor cells. The differentiation potential, also called plasticity, varies with the age and lineage commitment. The most plastic cell is the totipotent fertilized egg. This cell differentiates into pluripotent embryonic stem cells. The adult multipotent HSC is derived from these embryonic stem cells and its differentiation potential is debated. However, it is known that the HSC is at least multipotent, giving rise to numerous oligopotent progenitor/precursor cells. These cells include the common macrophage/granulocyte precursor (CMGP), the common megakaryocyte/erythrocyte precursor (CMEP) and the common lymphocyte precursor (CLP). We are coining the word "endotheloid" to describe the endothelial precursor which has recently been shown to derive from HSCs. The plasticity of a particular cell type can be partially determined by the expression of different cell surface markers. These cell surface markers can be used to determine lineage potential

been developed. Such terms as competitive repopulation [88], repopulating units (RUs) [55], and competitive repopulating units (CRUs) [126] have been used to express stem cell activity quantitatively. RUs indicate the amount of repopulating activity and CRUs the actual number of stem cells. Thus these two measurements complement one another. When both values are determined for an experiment the mean activity of stem cells (MAS) can be calculated (MAS = RU/CRU). Ema and coworkers demonstrated a great diversity of repopulating activity in HSCs using these tools [37]. While all myeloid, B-lymphoid, and T-lymphoid lineages were reconstituted, the degrees of reconstitution in each lineage varied. They determined that the proliferation capacity of HSCs is dissociated from the multilineage differentiation capacity and that this heterogeneity of HSCs is likely to result from their different levels of self-renewal capacity. Measurement of CRUs given by single HSCs suggested that the greater the self-renewal capacity the higher the repopulating activity. Similarly, HSCs with low RU regenerate themselves less than those with high RU. The heterogeneity in self-renewal capacity that these investigators observed implied that HSC self-renewal is not an unlimited capability [37].

These data may explain why bone marrow cells cannot be serially transplanted in mice more than four to six times, and support the generation-age model [111]. Allsopp and colleagues postulated that a molecular mechanism responsible for the decline in repopulating ability associated with aging could include shortening of telomeres in HSCs after transplantation [2]. However, when they overexpressed telomerase to prevent telomere shortening in HSCs they were unable to prevent the decline in stem cell activity. They concluded that unless telomeres are critically short, telomere shortening itself does not seem to be a limiting factor for the repopulating ability of HSCs [2].

Homing is another factor that could potentially influence the repopulating ability of HSCs. Enhanced homing ability may explain higher repopulating ability in Lnk-deficient HSCs [37]. Lnk is a family of adaptor proteins implicated in integration and regulation of multiple signaling events. Only Lnk-deficient, and not wild-type, HSCs homed to "incomplete niches" that express very low levels of cognate ligands. The authors postulated that Lnk-deficient HSCs stay longer in niches allowing the maintenance of self-renewal [37]. Nevertheless, all these studies were performed in HSCs that express foreign genes or proteins to facilitate tracking. Thus, these genetically manipulated HSCs may behave differently than normal ones and may have different self-renewal characteristics as well as altered transdifferentiation.

4.3.2 HSC Pluripotency/Plasticity

Recent studies support the possibility that HSCs may have broad potential to differentiate into various cell types. Our studies using HSC for bone transplantation support that HSCs significantly contribute to new retinal, choroidal and iris endothelium following mechanical injury [51, 52, 119]. Sahara and coworker tested the pluripotency of HSCs by comparing vascular lesions induced by mechanical injury after bone marrow reconstitution with total bone marrow cells, KSL cells, or a single HSC cell (Tip-SP CD34⁻KSL cell with the strongest dye-efflux activity) harboring gfp [114]. The lesions contained a significant number of gfp⁺ cells in the total bone marrow and KSL groups, whereas gfp⁺ cells were rarely detected in the HSC group. Sahara's studies suggest that it is rare for a highly purified HSCs to transdifferentiate into vascular cells, whereas the KSL fraction of bone marrow cells contained a distinct population that could substantially contribute to vascular lesion formation. The authors concluded that although the KSL fraction is considered to be enriched in HSCs, mesenchymal stem cells or multipotent cells, cells more primitive than HSCs were present in this fraction and may be responsible for the vascular lesions observed. They also suggest that non-hematopoietic cells in the KSL fraction might be responsible for the KSL-derived endothelial-like cells

or smooth muscle-like cells that some investigators observed in the vascular lesion [114]. In contrast, we showed using single cell transplants that HSCs form endothelial cells but not smooth muscle-type cells in a model of retinal neovascularization [51]. Clearly more studies need to be performed to fully understand HSC potential.

Numerous investigators have demonstrated plasticity or trans/dedifferentiation of HSCs. As pointed out by Kucia and Ratajczak many of these studies did not contain proper controls to exclude such issues as cell fusion. They describe that in addition to HSCs, bone marrow also harbors versatile subpopulations of tissue-committed stem cells (TCSCs) and perhaps even more primitive pluripotent stem cells (PSCs) [72]. These rare cells accumulate in bone marrow during ontogenesis and are released from the bone marrow into peripheral blood after tissue injury to regenerate damaged organs. The presence of TCSCs/PSCs in bone marrow should be considered before experimental evidence is interpreted simply as trans/dedifferentiation or plasticity of HSCs and that bone marrow-derived stem cells are heterogeneous.

Despite numerous reports showing a contribution of HSCs to regenerate different tissues, the results have not always been reproduced by other investigators [12, 94, 96, 136] and an attempt to explain this discrepancy needs to be made. The outcome of this discrepancy is that researchers are polarized in their view of stem cell plasticity. The disparity in the literature could be explained by differences in the tissue injury models employed, purity of stem cell populations used for regeneration, or inability to detect tissue chimerism due to technical limitations [75, 79].

In addition, alternative explanations for the observed HSC transdifferentiation have been proposed. For example, the chemokine stromal derived factor-1 (SDF-1) and its receptor (CXCR4) are crucial to stem cell homing and recruitment [17]. It has been shown that CXCR4⁺ TCSCs circulate in the body and compete for occupancy of SDF-1-positive niches in various tissue/organs [71]. Hematopoietic, skeletal muscle satellite, liver oval, neural and other tissue-specific stem/progenitor cells express CXCR4 and circulate at low levels in the peripheral blood to maintain stem cell pools in distant parts of the body. Bone marrow is a source of SDF-1, stem cell chemoattractants and survival factors. In the bone marrow not only HSCs but also differentiated circulating TCSCs can compete with HSCs for stem cell niches.

Stem cell plasticity has been explained in some systems by cell fusion [4, 22, 23, 132, 146]. Donor-derived HSCs or monocytes were observed to fuse with differentiated cells in recipient tissues which led to the creation of daughter cells that have a double number of chromosomes in their nuclei and express

4

cell surface and cytoplasmic markers that are derived from both parental cells. However, fusion as a major contributor to the observed donor-derived chimerism has been excluded in several recently published studies in various organs [3, 54, 64, 120]. An even more controversial interpretation of this is the possibility of so-called therapeutic cell fusion [22, 23].

While our studies and others support the hypothesis that HSCs can adopt a tissue-specific phenotype by transdifferentiation, this view remains controversial. Others believe that the process really involves fusion and not transdifferentiation [132, 135]. Campbell and coworkers documented polyploidization of vascular smooth muscle cells in response to mechanical and humoral stimuli [24]. Thus, in some settings cell fusion can account for part of the accumulation of bone marrow-derived cells in the repair of injured tissue including vascular lesions. However, we have been unable to detect polyploidy in HSC-derived cells within vascular lesions of the retina, choroid or iris following injury. While fusion may represent a significant component of the HSC repair observed in the liver and in vascular smooth muscle in some vascular beds, we consider it a rare event in the vasculature of the eye.

4.4 The HSC Niche

Stem cells are thought to occupy discrete spaces consisting of cellular elements, matrix, signaling molecules and pathways, collectively called a niche (Fig. 4.2) [90, 118]. The precise nature of these cellular and molecular components (the microenvironment) is thought to maintain the typical features of stem cells, i.e., quiescence, maintenance or expansion. Interaction of HSCs with their niche is critical for adult hematopoiesis in the bone marrow. HSCs must maintain the balance between quiescence and self-renewal in the stem cell niche as well as maintain long-term hematopoiesis.

The HSC niche comprises the endosteal surface of the bone marrow cavity, with osteoblasts being the most important cellular elements in maintaining HSCs [48, 81, 147]. Various signaling pathways and soluble molecules, including Wnt, Notch, transforming growth factor β/bone morphogenic protein and Hedgehog, are known to be involved in all stem cell niches [42]. Of these, Wnt signaling appears to be most important for HSC self-renewal [109].

HSCs can be removed from the bone marrow of a donor and successfully engrafted by infusion into peripheral blood. Engraftment requires the transmigration of circulating HSCs from peripheral blood into the recipient's niche. This engraftment depends upon the physical availability of niche space, as well

as homing signals. The spatial requirement is usually met by either chemical or radiological bone marrow ablation. The homing signals that induce this transmigration are poorly understood and only now beginning to be studied. It is thought that some of the same signals that induce HSC mobilization from the bone marrow are also involved in repopulating the HSC niche [1, 142].

4.4.1 Molecular Mechanisms for Maintenance in the Niche

The Tie2-angiopoietin system appears to play a critical role in regulating proliferation, adhesion and survival of HSCs. Tie2$^+$ HSCs have a higher long-term reconstitution activity than do Tie2 HSCs. Arai and coworkers put forth the idea that localization of Tie2$^+$ HSCs on the bone surface is regulated by stem-cell-specific adhesion molecules such as N-cadherin [6]. Once the HSCs localize to osteoblasts, angiopoietin-1 (Ang-1) produced by osteoblasts may activate its receptor, Tie2, on HSCs and promote tight adhesion of HSCs in the niche, resulting in quiescence and enhanced survival of HSCs.

Tie2/Ang-1 signaling induces adhesion of HSCs to fibronectin and collagen [116, 128]. Following Ang-1 binding, phosphorylation of Tie2 results in activation of the phosphatidylinositol 3-kinase (PI3-K)/Akt signaling pathway to promote endothelial cell survival [68]. Ang-1 leads to morphologic changes in HSCs in a β1-integrin-dependent manner and increases adherence of HSCs to the bone surface in vivo [7, 68]. The tight cell attachment of HSCs to stromal cells likely affects the cell-cycle status of HSCs. Since integrins trigger signaling that promotes cell survival, cell adhesion enhanced by Tie2/Ang-1 signaling may protect HSCs from stress in combination with PI3-kinase/Akt signaling. Ang-1 administration provides clinical benefits that protect HSCs from anticancer therapy [25]. Thus Ang-1 maintains the in vivo repopulating ability of HSCs by inhibiting cell division and promotes quiescence of HSCs in vivo and Tie2 signaling keeps HSCs in the niche [7]. While Tie2 is required for postnatal bone marrow hematopoiesis it is not required for embryonic hematopoiesis. Tie2 is critical for the maintenance and survival of HSCs in the adult bone marrow and Tie2-deficient or kinase-deficient Tie2 cells are unable to occupy the adult bone marrow niche when competing with wild-type cells [7].

4.5 HSC Mobilization

Adult HSCs expand and differentiate exclusively in the bone marrow, from which they can be mobilized into the bloodstream. Cell-cell and cell-matrix inter-

Fig. 4.2. Hematopoietic stem cells can be induced to leave the stem cell niche, differentiate, and participate in neovascularization. This simplified schematic depicts some of the more well-known molecules and events involved in HSC mobilization. The lower half of the figure shows a damaged retinal microvessel, which releases several key molecules including SDF-1, VEGF and IL-8. These molecules act as mobilization signals, reaching the bone marrow stroma via the circulation, where they interact with numerous components of the HSC niche (*top half of the figure*). SDF-1 and IL-8 are both involved in G-CSF-mediated release and differentiation of HSC. SDF-1 and VEGF both act on osteoclasts, inducing their release of membrane-bound SCF. VEGF also acts directly to increase HSC proliferation, and may possibly influence their differentiation. SDF-1 mediates the release of MMP-9 from osteoclasts. Free MMP-9 can then cleave kit ligand, which influences HSCs, and MMP-9 can induce the release of HSCs from the stroma by impacting cell adhesion. Together these events result in the proliferation of HSCs and their differentiation into first multipotent progenitors and then into endothelial progenitors. The precise temporal sequence of differentiation and concomitant changes in

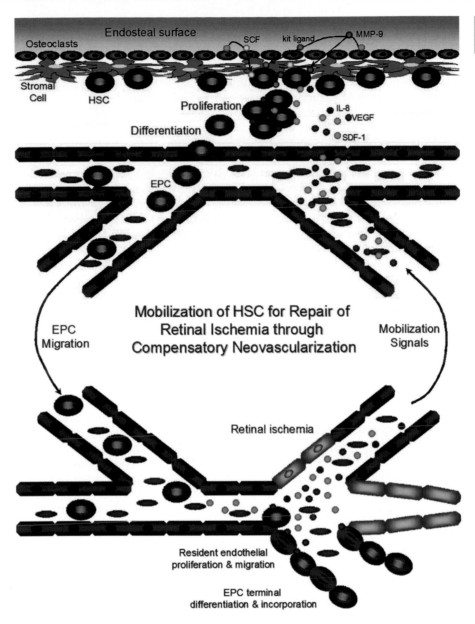

surface marker expression remain to be fully described, but it is likely that by the time mobilized cells translocate from the niche into the circulation they are fully committed to a specific terminal differentiation pathway. These circulating EPCs are then induced to leave the circulation by locally high concentrations of mobilization factors (among other, less clear signals) where they then differentiate fully and participate in reparative or regenerative processes. Here they are shown integrating with resident endothelium to form a compensatory collateral vessel

actions are crucial to the proliferation and differentiation of HSCs in the bone marrow niche. Depending on the circumstances, HSCs can traffic from the marrow into the circulation in large numbers, a process termed mobilization [26]. Mobilization of HSCs is mediated by a complex interplay of changes in integrin-mediated adhesion as well as chemokine and growth factor receptor signaling.

Exogenous administration of granulocyte colony stimulating factor (G-CSF) is the standard method of inducing stem and progenitor cell release into the circulation in a clinical setting [26]. The chemokines interleukin 8 (IL-8) [112] and stromal-derived factor 1 (SDF-1) [56] both participate in G-CSF-mediated stem cell mobilization. SDF-1, as well as VEGF and PlGF (vascular endothelial growth factor and placental growth factor, respectively), induces osteoclasts to secrete the metalloprotease MMP-9, whose action results in the shedding of the membrane-bound cytokine stem cell factor (SCF) from the bone marrow, releasing it into the circulation. Inhibiting SDF-1 has been shown to reduce the degree of stem cell involvement in induced neovascularization [119].

4.5.1 The SDF-1/CXCR4 Axis

Endothelial cells in developing vascular beds express the SDF-1 receptor CXCR4, and animals deficient in either SDF-1 or CXCR4 do not form large vessels within the gastrointestinal tract, clearly indicating a role for this chemokine ligand/receptor in vascular development [82, 95, 127, 148]. CD34+ cells express functional CXCR4 and migrate in response to SDF-1 [104]. CXCR4 expression by endothelial cells of various origins is also well documented [93]. In endothelial cells, CXCR4 expression is increased after treatment with VEGF or basic fibroblast growth factor (FGF2) [115]. SDF-1 has also been shown to stimulate VEGF expression in a number of cells [67]. Any tissue production of SDF-1 results in mobilization of hematopoietic progenitors from the marrow to that tissue [105].

4.6 Surface Marker Expression – HSC Identification

In the adult, HSCs can be distinguished from mature blood cells by their lack of lineage-specific markers and the presence of certain other cell-surface antigens [141]. Human cells express CD133 (prominin-1, a pentaspan membrane protein) [86] while murine HSCs express the cell surface molecules c-kit (CD117) and stem cell antigen-1 (Sca-1; Ly-6 A/E), and lack markers of differentiated peripheral blood cells (lineage markers, or Lin).

Most investigative studies have been performed on mice. CD117 is a receptor tyrosine kinase that is also expressed on mast cells, germ cells, and melanocytes. CD117 mutant mice have defects in HSC activity, lack pigmentation, and are sterile [15]. Mice deficient in c-kit have macrocytic anemia, decreased megakaryocyte numbers, decreased mast cell numbers, and defects in maintenance of B- and T-cell compartments [34, 137, 138]. CD117 appears to play a key role in the establishment of HSCs in the bone marrow "niche." Blocking c-kit function in progenitor cells prior to transplant allowed normal homing to the marrow but defects occurred in lodging in the niche [33].

Sca-1 is a cell surface protein found not only on hematopoietic but also on mammary gland, cardiac, and mesenchymal stem cells in the mouse [11, 78,

124, 139]. Sca-1 is necessary for normal HSC activity as $Sca-1^{-/-}$ mice have defects in short-term competitive transplantation and serial transplantation [63]. In addition, Sca-1 knockout cells form fewer CFUs upon transplantation [63]. Because of these transplantation defects, it is believed that Sca-1 plays a role in HSC self-renewal.

Sca-1 belongs to a family of proteins bearing a UPAR (urokinase plasminogen activator receptor) domain. The UPAR domain is about 90 amino acids long and is important in cell adhesion and migration, modulates integrin function and regulates degradation of the extracellular matrix [18, 108]. Furthermore, UPAR directly regulates integrin expression and function by signaling and/or binding to integrins [18]. Therefore, Sca-1 is believed to modulate integrin function in HSC.

In addition to modulating HSC behavior, Sca-1 suppresses T-cell proliferation and alters signaling through the T-cell receptor (TCR) [125]. Sca-1 downregulation is required for T-cell differentiation [13]. $Sca-1^{-/-}$ mice have decreased megakaryocyte and platelet formation [63]. Interestingly, Bradfute and coworkers recently showed that Sca-1 affects CD117 expression and lineage fate of peripheral blood cells after transplantation and human CD34+ cell activity, but is not critical for self-renewal [19].

Although Sca-1 is routinely used to isolate many stem cells in the mouse, its biological role in HSC function is still only partially characterized. For example, such questions as how Sca-1 affects HSC engraftment and proliferation and how overexpression of Sca-1 affects HSC function remain unanswered.

4.7 Surface Marker Expression – HSC Isolation

Isolation of mouse HSCs is a relatively straightforward, although time-consuming, process. The basic principle involves identifying specific cell populations with antibodies conjugated to either magnetic particles or fluorochromes, and then sorting the cells thus identified using magnetic or fluorescence activated cell sorting (MACS or FACS), respectively. HSCs and their progeny can be identified immunologically through their expression of distinct surface markers, the composition of which changes during differentiation. Typically some surface markers appear while others disappear (Fig. 4.1). Sca-1 and CD117 are most often used to distinguish HSCs from other cell types. These two markers yield a population of cells that is 95% HSCs. Additional purification may be accomplished with considerable effort by first depleting lineage cells using antibodies to B220 (B-lymphocytes), CD3, CD4 and CD8 (T-lym-

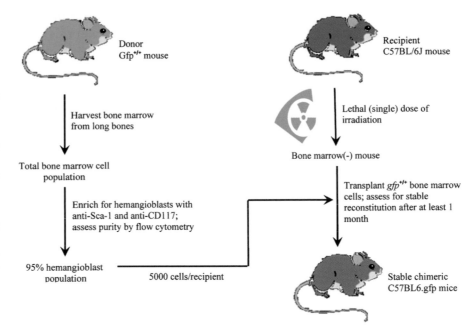

Fig. 4.3. The generation of chimeric animals is depicted in schematic form. The bone marrow of recipient C57 BL/6J mice is depleted with lethal irradiation. At the same time, sorted HSCs from *gfp*^+/+ donor mice are transplanted into these recipients. After 1 month, stable reconstitution is established through flow cytometry. These chimeric mice are then used for subsequent experiments. With the transplantation technique, the end functionality of the donor cells can be determined based on the presence of the gfp label in the cells

phocytes), CD11b (macrophages), Gr-1 (granulocytes), and TER119 (erythroblasts). This approach first uses magnetic bead-conjugated antibodies to deplete the lineage cells, followed by FACS using fluorescent-conjugated antibodies to CD117 and Sca-1 of the remaining cell fraction. For general reconstitution of bone marrow-ablated recipient animals, the use of FACS (CD177 and Sca-1) is usually sufficient. The following protocol is used routinely in our laboratory.

4.8 Methodologies

4.8.1 Extraction of HSCs from Donor Mice

1. Use 10- to 20-week-old donor mice. TgN(GFPU)5Nagy (Jackson Laboratories, Bar Harbor, ME), bred in-house for homozygosity at the *gfp* locus, are used if tagged HSCs are needed. If HSCs are being used to reconstitute irradiated recipient mice, one donor mouse is needed for every ten recipients (Fig. 4.3).
2. Put to death donor mice; dissect away both hind legs and place on ice.
3. Working within a sterile field such as a laminar flow hood, dislocate the knee joint and cut through the joint to separate the thigh from the calf.
4. Dissect soft tissue away from the femur and tibia and slice off the joints, exposing the bone marrow.
5. Expel the marrow using a tuberculin syringe with ~0.5–1 ml sterile PBS (phosphate buffered saline) for each bone into a sterile centrifuge tube.
6. After collecting all of the marrow, homogenize by repeated pipetting.

7. Centrifuge 300×*g*/10 min/4°C; decant and resuspend in 2 ml of sterile PBS.
8. Aliquot into four tubes as follows:
   a) 500 μl sterile PBS, 30 μl BM cells.
   b) 500 μl sterile PBS, 30 μl BM cells, 1 μl phycoerythrin (PE)-conjugated rat anti-mouse Sca-1 (BD Pharmingen, San Jose, CA).
   c) 500 μl sterile PBS, 30 μl BM cells, 1 μl allophycocyanin (APC)-conjugated rat anti-mouse CD117 (BD Pharmingen).
   d) To the remaining BM cells add 120 μl of PE-anti-Sca-1 and 120 μl APC-anti-CD117.
9. Incubate 15 min/4°C.
10. Centrifuge 300×*g*/10 min/4°C; decant and resuspend pellets in PBS (1 ml for tubes a, b and c, 10 ml for tube d).
11. Strain through filter tubes (cell-strainer cap, e.g., Falcon 2235) to remove large cell clumps and other debris.
12. Sort HSCs in tube d using a FACS Vantage SE Turbosort (BD Biosciences, San Jose, CA) or similar device to collect the fraction of cells positive for both CD117 and Sca-1; tubes a, b and c are used for calibration. Collect HSCs into sterile PBS with 30% (v/v) serum during sort.
13. Combine sorted HSC fractions and wash by centrifugation 300×*g*/10 min/4°C.

4.8.2 Reconstitution of Bone Marrow-Ablated Recipient Mice

1. Recipient animals: 10–16 week old C57BL6/J.
2. Irradiate mice (850–900 rad) to ablate bone marrow.
3. Introduce ~5,000 HSCs into the systemic circulation of each recipient animal, either by tail vein or retro-orbital sinus injection.

4

4. Maintain animals on antibacterial water (Bactrim or Baytril) for at least 2 weeks.
5. Assess bone marrow reconstitution after at least 1 month. A successful reconstitution will result in a high percentage of gfp$^+$ leukocytes. Using flow cytometry analysis after a tail bleed, animals with at least 25% gfp$^+$ peripheral blood cells are considered reconstituted.

4.8.3 EPC Culture

Cell lines have been utilized to study many aspects of stem cell behavior. Stromal cell lines have been generated from the AGM region to study the molecular events involved in hematopoiesis and HSC differentiation within the AGM microenvironment [97]. These lines successfully maintained cultures of murine fetal liver-derived [97], marrow-derived [144], and human cord blood-derived cells [73]. Much focus has been given to characterizing growth factors believed responsible for the maintenance of hematopoiesis and differentiation in vitro [99].

Soluble factors are used for the preparation of medium required for the maintenance or expansion of stem cells in vitro. Commonly added factors are TPO, Flk2-ligand (FL), SCF, IL-6, G-CSF, and IL-3. Although some media have met with a degree of success, most investigators reported only a moderate expansion of stem cells in culture. It is difficult to maintain stem cells in culture without the presence of a supportive feeder layer of cells. Whether stem cells need to be in contact with these stromal cells is still a matter of debate. Both the AGM-S3 and DAS104–4 cell lines were incapable of maintaining early progenitors from fetal liver [97] or cord blood CD34$^+$ cells [144] when they were not in direct contact with the stromal cells during culture. Thus stromal cells may be necessary for the maintenance of bone marrow HSCs because they provide anchorage as well as growth factors [98, 101]. As discussed below secreted molecules are equally important for the maintenance of stem cells in non-contact culture.

In our laboratory, EPCs are cultured using the endocult liquid medium kit (Stem Cell Technologies, Vancouver, CA) per manufacturer's protocol. Briefly, 16 ml of human peripheral blood is collected into a CPT tube (BD Biosciences, San Jose, CA) and spun to obtain the mononuclear cell fraction (MNC). The MNCs are washed twice with PBS/2% FBS and the pellet resuspended in 3 ml of endocult medium. The cells are plated onto a six-well fibronectin-coated dish (BD Biosciences, San Jose, CA) and kept at 37 °C, 5% CO_2 for 2 days. After 2 days, the non-adherent EPCs are harvested and further cultured for an additional 3 days to allow formation of endothelial colonies. Colonies are defined as a central core of "round" cells with elongated "sprouting" cells at the periphery and are classified as early outgrowth colony forming unit-endothelial cells or CFU-ECs.

CD34$^+$ cells are isolated from peripheral human blood using EasySep magnet kit (Stem Cell Technologies) per manufacturer's protocol. Briefly 2×10^8 MNCs/ml are isolated from peripheral blood using CPT tubes. EasySep positive selection cocktail is added (100 µl/ml cells) to the MNCs and incubated at room temperature for 15 min. Following the incubation, 50 µl/ml of EasySep magnetic nanoparticles are added to the MNCs and incubated at room temperature for 10 min. The cells are washed five times within the magnet. The positively selected cells are then removed from the magnet and kept in HPGM media containing stem cell factor (25 ng/ml), thrombopoietin (50 ng/ml) and flk/flt ligand (50 ng/ml, R & D Systems, Minneapolis, MN). These factors help to keep the cells in the undifferentiated state.

CD34$^+$ cells were obtained from the peripheral blood of healthy subjects using flow cytometry, labeled via PKH67 dye (Sigma, St. Louis, MO) and injected into postnatal day 1 mouse pups intravitreally. The pups were then subjected to the oxygen-induced retinopathy model as described by Smith [122]. The eyes were then removed, dissected, and examined for the presence of the green cells colocalizing with the vessels (Fig. 4.4).

4.8.3.1 Supportive Cells and Soluble Factors

Several different stromal cell lines have been reported to support HSC maintenance in culture. Oostendorp and colleagues compared gene expression profiles of two HSC-supportive and four HSC-non-supportive cell lines and focused on genes that might be involved in the common mechanism of HSC maintenance [100]. Thirty-one genes were found to be differentially expressed more than twofold in the HSC-supportive stromal cell lines compared with non-supportive lines. One-third of the genes expressed at a higher level in HSC-supporting stromal cells are secretory proteins. The authors favored the view that maintenance of HSCs is not supported by alternatives to VCAM-1-dependent adhesion pathways but by mechanisms dependent on soluble molecules. This favors the view that HSC maintenance on embryonic AGM-derived stromal cells is supported by contact-independent mechanisms [100].

The HSC-supportive cell line AFT024 expresses high levels of pleiotrophin (PTN), thrombospondin-2 (TSP2) and insulin-like growth factor binding protein-3 (IGFBP-3) [53]. PTN is a heparin-binding cytokine that in the embryo is expressed in metanephric and developing liver mesenchymal cells [9].

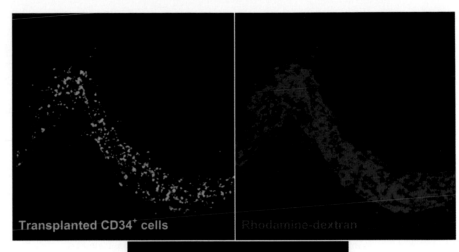

Transplanted CD34⁺ cells

Rhodamine-dextran

Merge

Fig. 4.4. Large numbers of transplanted PKH67⁺ CD34⁺ cells were seen along the entire length of blood vessels after vaso-obliteration. The animals were perfused with rhodamine-dextran. The colocalization of the gfp⁺ cells and the blood vessel can be seen in the *bottom panel*, indicating that not only stem cells, but progenitor cells can be involved in differentiation into vascular cell types. These cells can then incorporate into and increase the patency of functional vessels

PTN is known as a guidance molecule for neurites and osteoblasts, probably by using surface-expressed syndecan 3 [70]. Whether PTN deficiency also affects HSCs has not been determined. PTN binds to several cell-surface molecules besides syndecan 3, including anaplastic lymphoma kinase [5], protein tyrosine phosphatase receptor Z (also known as RPTP-β) [83], and cytoplasmic nucleolin [129]. Expression of the multifunctional cytoplasm-nucleolus shuttling protein nucleolin is observed in HSCs and is able to import endocytosed PTN into the nucleus [107].

IGFBP-3 was also highly expressed on AFT024 and the long-term culture-supportive cell line HS27a [49]. Our laboratory has extensively studied IGFBP-3 function in cultured endothelial cells. However, the function of IGFBP-3 in HSCs is much less clear. IGFBP-3 is found at high levels in serum, where it forms a heterotrimeric complex with IGF and the acid-labile subunit. IGFBP-3 was differentially expressed between HSC-supportive and non-supportive cell lines. These findings suggest that IGFBP-3 could be one of the molecules commonly involved in the regulation of HSC behavior. Although IGFBP-3 binds IGF and modulates its availability, IGFBP-3 generates IGF-independent signaling. IGFBP-3 signals through an IGF receptor-independent pathway to phosphorylate Smad-2 and -3 and downregulate Smad-4 [38], which are involved in signaling through TGFβ family members. The responsible IGFBP-3 receptor, however, has not yet been identified. IGFBP-3 contains a nuclear localization sequence and IGFBP-3 binds intranuclear target genes, including p53, and retinoic acid receptor α [76]. Thus, IGFBP-3 may affect HSC maintenance not only in its secreted form but also through its nuclear counterpart.

Other factors overexpressed in HSC-supportive cell lines include cathepsin K, FGF-7 (fibroblast growth factor), and pentaxin-related gene. Cathepsin K is a cysteine protease that is normally expressed by osteoclasts and that has been shown to play a role in bone resorption. One could postulate that the proteolytic activity of cathepsin K might be beneficial in the formation of the marrow niche.

After 7 days of culture on fibronectin, CD34⁺ mononuclear cells display an endothelial cell pheno-

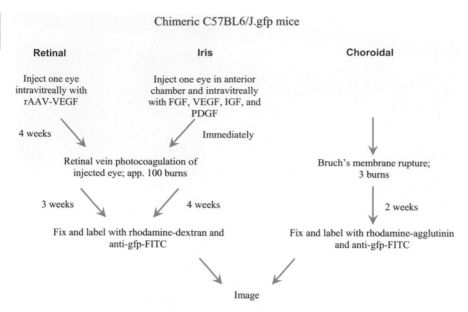

Fig. 4.5. The time course depiction of the three models of ocular neovascularization we have developed: retinal, iris, and choroidal neovascularization

type, incorporate acetylated LDL, produce nitric oxide (NO) when stimulated with VEGF, and express PECAM and Tie-2 receptor [8, 62].

Postnatal neovascularization has previously been considered synonymous with proliferation and migration of preexisting endothelial cells resident within parent vessels, i.e., angiogenesis [40]. However, the finding that bone marrow-derived cells may home to sites of neovascularization and differentiate in situ is consistent with vasculogenesis [110], a critical paradigm for the establishment of vascular networks in the embryo. Asahara identified cells derived from bone marrow capable of differentiating into neovasculature as CD34+ expressing cells [8]. CD34+ cells represent a population of EPCs. Injection of healthy CD34+ cells accelerates revascularization of ischemic limbs [117] and revascularization during wound healing in diabetic mice [121]. Thus, these cells presumably are key mediators of endothelial repair. Individuals with type I diabetes, ischemic heart disease, hypertension, and risk factors for atherosclerosis have reduced numbers of (and dysfunctional) circulating EPCs [131, 134].

We have demonstrated that HSCs provide functional hemangioblast activity during retinal neovascularization [50, 51] and that HSCs give rise to a variety of lineages including the CD34+ EPCs. This is a newly described mechanism of endothelial repair and represents a significant source of cells for neoangiogenic vessels. We postulated that defects in EPC function may be responsible for the development and persistence of acellular capillaries in diabetic retinopathy since lack of repair of acellular capillaries contributes to the development of retinal ischemia and represents an irreversible step in the pro-

gression of this disease [21, 41]. EPCs isolated from patients with Type 1 diabetes have a decreased rate of migration, and incubation with an NO donor alters the EPC cytoskeleton, normalizing their rate of migration. EPC migration can be stimulated by activation of growth factor receptors [57, 61, 69, 130, 133, 143] such as vascular endothelial growth factor 1 and 2 (VEGFR1 and VEGFR2) and cytokine receptors such as CXCR4 [30, 58, 133, 145]. These same receptors can influence NO production and are themselves regulated by NO [14, 89, 103, 123, 140]. The balance of receptor expression and activation has profound effects on the detrimental or beneficial action of the NO generated. In the diabetic state, CXCR4 and VEGFR1/R2 activation results in dysregulated and diminished NO bioavailability.

4.9 Mouse Models of HSC Involvement in Ocular Neovascularization

Since one of the hallmarks of proliferative diabetic retinopathy and age-related macular degeneration is retinal and choroidal neovascularization, respectively, we sought to mimic this by laser injury. In both models chimeric mice were used to provide a method to distinguish resident vasculature from the donor-derived contribution to the neovascular complex (Fig. 4.5).

4.9.1 Preretinal Neovascularization

After stable reconstitution is established, chimeric mice are injected intravitreally with a recombinant AAV vector expressing the full-length VEGF$_{165}$. Four weeks later, the mice are anesthetized, their eyes

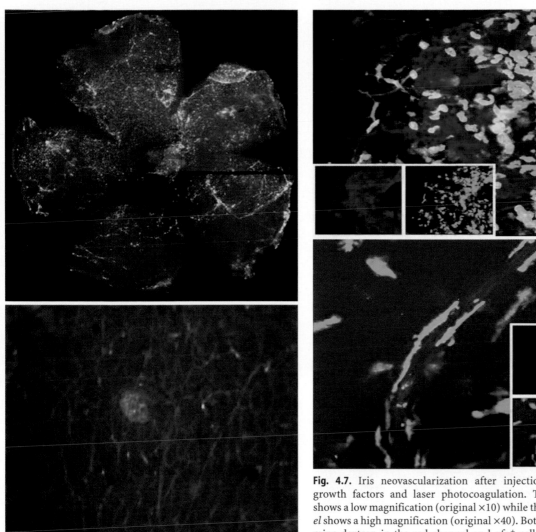

Fig. 4.7. Iris neovascularization after injection of various growth factors and laser photocoagulation. The *top panel* shows a low magnification (original ×10) while the *bottom panel* shows a high magnification (original ×40). Both show rhodamine-dextran in the red channel and gfp⁺ cells in the green channel. Note the large number of gfp⁺ cells in both images, especially in the *bottom panel*, where they can be seen incorporating with the resident vasculature

Fig. 4.6. Retinal neovascularization. The *top panel* (original image ×4) shows a composite image of a neural retina from a chimeric mouse that had branch retinal vein occlusion. The *bottom image* (original image ×100) shows a close-up from the same tissue. The red fluorescence shows blood vessels while the green fluorescence shows HSC-derived cells. The colocalization between the two channels is especially obvious in the *lower panel*, indicating that HSCs do differentiate into blood vessels

dilated and branch retinal vessel occlusion is performed by laser photocoagulation. An argon green laser system (HGM Corporation, Salt Lake City, UT) is used for photocoagulation with the aid of a 78-diopter lens; the laser is applied to selected venous sites next to the optic nerve. Venous occlusion (approximately 50–100 burns) is accomplished using laser parameters of 1-s duration, 50 µm spot size, and 50–100 mW intensity. Three weeks after laser injury, mice are killed and their eyes are enucleated. The eyes may then be examined immunohistochemically (flat-mounts and cryosections) using epifluorescence and confocal microscopy for the presence of gfp⁺ cells and their colocalization with endothelial cells in the neural retina (Fig. 4.6) [51].

4.9.2 Iris Neovascularization

Chimeric mice are also used to induce iris neovascularization. A cocktail of growth factors including VEGF, IGF-1, and FGF2 are injected intravitreally at the time of laser. An argon green laser is used with a 78-diopter lens at an intensity of 150 mW. Multiple burns are applied to completely occlude the branch

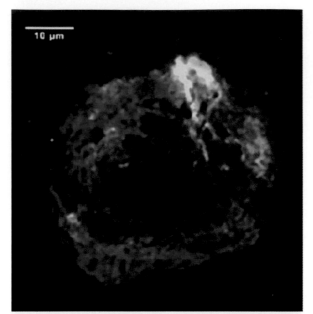

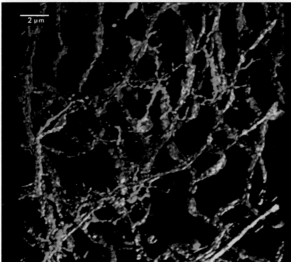

Fig. 4.8. Choroidal neovascularization. The *top panel* shows a CNV lesion from a chimeric mouse in which Bruch's membrane was ruptured. The *bottom image* shows a close-up from the same tissue. The red fluorescence shows blood vessels while the green fluorescence shows HSC-derived cells. The colocalization between the two channels is especially obvious in the *lower panel*

vessels to stimulate compensatory neovascularization in the iris. Peak neovascularization is seen 4 weeks after the laser injury (Fig. 4.7).

4.9.3 Choroidal Neovascularization

To induce choroidal neovascularization (CNV), stably reconstituted mice are anesthetized and their eyes dilated and subjected to laser injury in a manner similar to Ryan's [113]. A 532 nm laser (Iridex, Mountain View, CA) coupled to a slit lamp set to a 100 µm spot focused on the retinal pigment epithelium (RPE) is used to deliver a 250 mW pulse for 0.1 s. Three burns are applied approximately one disk diameter from the optic nerve in three quadrants. This laser application results in a bubble, indicative of Bruch's membrane rupture, more than 95 % of the time. Neovascularization of the choriocapillaris occurs within this lesion and peaks at 2–3 weeks after injury, at which time the animals are put to death and the eyes removed and fixed. For flat-mount examination the anterior segment is dissected and discarded, and the neural retina is removed. Figure 4.8 depicts a typical flat-mounted posterior cup that was reacted with rhodamine-conjugated *Ricinus communis* agglutinin I (Vector Laboratories, Burlingame, VT) to delineate vasculature, and shows colocalization of gfp+ HSC-derived cells with the CNV [119].

4.10 Conclusion

Ocular neovascular diseases such as ROP, PDR, and ARMD affect all age groups and cause vision impairment for millions in developed nations. It was believed that aberrant neovascularization as in these diseases was from the migration, proliferation, and organization of resident endothelial cells. However, it is now known that HSCs participate in postnatal neovascularization along with resident cells, signifying that neovascularization consists of both angiogenesis and vasculogenesis.

Soon after fertilization, stem/precursor cells are in the blood islands of the yolk sac. Upon further development, the cells then reside in the AGM, the blood, the fetal liver, and eventually the bone marrow. After activation, bone marrow HSCs proliferate and differentiate into progenitor cells that are found in the peripheral blood [59]. The progenitor cells can then differentiate further if needed. Also, in order to be considered a true stem cell, the HSCs must yield undifferentiated progeny with identical characteristics. HSCs are now known to have hemangioblast activity, differentiating into all of the components of the vascular system. It has been shown that HSCs also have the potential to differentiate into several other types of tissue, such as liver, muscle and neurons, demonstrating their plasticity.

In addition to plasticity, stem cells are also characterized by their self-renewal capacity. Although stem cells are considered an endless source for tissue or organ regeneration, the extent has not yet been quantified and is thus debatable. Most researchers would consider the most reliable indicator of self-renewal capacity to be the long-term multilineage repopulating activity, detectable by transplantation experiments. The capacity of HSCs to self-renew is quite heterogeneous. This translates to differences in proliferation capacity and multilineage differentiation capacity. This may provide an explanation as to why bone marrow cells cannot be serially transplanted in mice indefinitely. Homing could also influence the repopulating ability of HSCs. In some systems, stem cell plasticity has been explained by cell fusion. Donor-derived cells fused with differentiated cells in resident tissue, leading to cells with double the normal number of chromosomes in their nuclei. However, fusion is not thought to account for all issues of (trans)differentiation, especially in the eye.

As well as being plastic and self-renewing, stem cells are also thought to occupy niches. The qualities of the niche are thought to promote stem cell quiescence, maintenance or expansion. Interaction of HSCs with their niche is thought to be necessary for adult hematopoiesis in the bone marrow. HSCs must maintain a balance between quiescence and self-renewal in the stem cell niche as well as maintain long-term hematopoiesis. Cell engraftment after transplantation depends upon the physical availability of niche space, as well as homing signals. The spatial requirement is usually met through either chemical or radiological bone marrow ablation. Some of the same signals that induce HSC mobilization from the bone marrow are also involved in repopulating the bone marrow niche.

Stem cells must leave their bone marrow niche to respond to injury/ischemia. Based upon stimuli (such as SDF-1) outside of the bone marrow, the HSCs can be mobilized and released into the circulation. They can then home to where they are needed, often an ischemic area, and then differentiate at some point between their origin and their destination. This homing often involves the binding of cell surface ligands and receptors, either of which can be on the HSCs.

In order to examine intricacies of HSCs, they often must be removed and purified from the entire bone marrow population. Characterizing HSCs involves identifying specific cell populations with antibodies conjugated to either magnetic particles or fluorochromes, and then sorting the cells thus identified using MACS or FACS. HSCs and their progeny can be identified immunologically through the presence or lack of certain cell surface markers, the composition of which changes during differentiation. Sca-1 and CD117 are most often used to distinguish HSCs from other cell types. The presence of these two markers yields a population of cells that is 95 % HSCs. Isolated cell lines have been used to study many aspects of stem cell behavior, especially characterizing the growth factors responsible for cell maintenance and

differentiation in vitro. In addition stromal cell lines have been reported to support the maintenance of HSCs in culture.

We used murine HSCs as a tool to study stem cell maintenance, migration and differentiation. Since a hallmark of PDR and wet ARMD is ocular neovascularization, we mimicked this by laser injury. In both models, gfp chimeric mice provided a technique to distinguish resident vasculature from the donor-derived contribution to the neovascularization. Our studies that use HSCs for transplantation support that these stem cells significantly contribute to neovasculature; these can then be used for examining the mechanisms underlying stem cell homing, differentiation, and involvement in ocular neovascularization.

References

1. Abkowitz JL, Robinson AE, Kale S, Long MW, Chen J (2003) Mobilization of hematopoietic stem cells during homeostasis and after cytokine exposure. Blood 102:1249–1253
2. Allsopp RC, Morin GB, DePinho R, Harley CB, Weissman IL (2003) Telomerase is required to slow telomere shortening and extend replicative lifespan of HSCs during serial transplantation. Blood 102:517–520
3. Almeida-Porada G, Porada C, Zanjani ED (2004) Plasticity of human stem cells in the fetal sheep model of human stem cell transplantation. Int J Hematol 79:1–6
4. Alvarez-Dolado M, Pardal R, Garcia-Verdugo JM, Fike JR, Lee HO, Pfeffer K, Lois C, Morrison SJ, Alvarez-Buylla A (2003) Fusion of bone-marrow-derived cells with Purkinje neurons, cardiomyocytes and hepatocytes. Nature 425:968–973
5. Amet LE, Lauri SE, Hienola A, Croll SD, Lu Y, Levorse JM, Prabhakaran B, Taira T, Rauvala H, Vogt TF (2001) Enhanced hippocampal long-term potentiation in mice lacking heparin-binding growth-associated molecule. Mol Cell Neurosci 17:1014–1024
6. Arai F, Hirao A, Ohmura M, Sato H, Matsuoka S, Takubo K, Ito K, Koh GY, Suda T (2004) Tie2/angiopoietin-1 signaling regulates hematopoietic stem cell quiescence in the bone marrow niche. Cell 118:149–161
7. Arai F, Hirao A, Suda T (2005) Regulation of hematopoietic stem cells by the niche. Trends Cardiovasc Med 15:75–79
8. Asahara T, Murohara T, Sullivan A, Silver M, van der Zee R, Li T, Witzenbichler B, Schatteman G, Isner JM (1997) Isolation of putative progenitor endothelial cells for angiogenesis. Science 275:964–967
9. Asahina K, Sato H, Yamasaki C, Kataoka M, Shiokawa M, Katayama S, Tateno C, Yoshizato K (2002) Pleiotrophin/heparin-binding growth-associated molecule as a mitogen of rat hepatocytes and its role in regeneration and development of liver. Am J Pathol 160:2191–2205
10. Auerbach R, Huang H, Lu L (1996) Hematopoietic stem cells in the mouse embryonic yolk sac. Stem Cells 14:269–280
11. Baddoo M, Hill K, Wilkinson R, Gaupp D, Hughes C, Kopen GC, Phinney DG (2003) Characterization of mesenchymal stem cells isolated from murine bone marrow by negative selection. J Cell Biochem 89:1235–1249

4

12. Balsam LB, Wagers AJ, Christensen JL, Kofidis T, Weissman IL, Robbins RC (2004) Haematopoietic stem cells adopt mature haematopoietic fates in ischaemic myocardium. Nature 428:668–673
13. Bamezai A, Palliser D, Berezovskaya A, McGrew J, Higgins K, Lacy E, Rock KL (1995) Regulated expression of Ly-6A.2 is important for T cell development. J Immunol 154:4233–4239
14. Beauchamp MH, Sennlaub F, Speranza G, Gobeil F, Jr., Checchin D, Kermorvant-Duchemin E, Abran D, Hardy P, Lachapelle P, Varma DR, Chemtob S (2004) Redox-dependent effects of nitric oxide on microvascular integrity in oxygen-induced retinopathy. Free Radic Biol Med 37:1885–1894
15. Bernstein A, Forrester L, Reith AD, Dubreuil P, Rottapel R (1991) The murine W/c-kit and Steel loci and the control of hematopoiesis. Semin Hematol 28:138–142
16. Bertrand JY, Giroux S, Golub R, Klaine M, Jalil A, Boucontet L, Godin I, Cumano A (2005) Characterization of purified intraembryonic hematopoietic stem cells as a tool to define their site of origin. Proc Natl Acad Sci U S A 102:134–139
17. Blades MC, Ingegnoli F, Wheller SK, Manzo A, Wahid S, Panayi GS, Perretti M, Pitzalis C (2002) Stromal cell-derived factor 1 (CXCL12) induces monocyte migration into human synovium transplanted onto SCID Mice. Arthritis Rheum 46:824–836
18. Blasi F, Carmeliet P (2002) uPAR: a versatile signalling orchestrator. Nat Rev Mol Cell Biol 3:932–943
19. Bradfute SB, Graubert TA, Goodell MA (2005) Roles of Sca-1 in hematopoietic stem/progenitor cell function. Exp Hematol 33:836–843
20. Brazelton TR, Rossi FMV, Keshet GI, Blau HM (2000) From Marrow to Brain: Expression of Neuronal Phenotypes in Adult Mice. Science 290:1775–1779
21. Bresnick GH, Engerman R, Davis MD, de Venecia G, Myers FL (1976) Patterns of ischemia in diabetic retinopathy. Trans Sect Ophthalmol Am Acad Ophthalmol Otolaryngol 81:OP694–709
22. Camargo FD, Chambers SM, Goodell MA (2004) Stem cell plasticity: from transdifferentiation to macrophage fusion. Cell Prolif 37:55–65
23. Camargo FD, Finegold M, Goodell MA (2004) Hematopoietic myelomonocytic cells are the major source of hepatocyte fusion partners. J Clin Invest 113:1266–1270
24. Campbell JH, Tachas G, Black MJ, Cockerill G, Campbell GR (1991) Molecular biology of vascular hypertrophy. Basic Res Cardiol 86 Suppl 1:3–11
25. Cho CH, Kammerer RA, Lee HJ, Yasunaga K, Kim KT, Choi HH, Kim W, Kim SH, Park SK, Lee GM, Koh GY (2004) Designed angiopoietin-1 variant, COMP-Ang1, protects against radiation-induced endothelial cell apoptosis. Proc Natl Acad Sci U S A 101:5553–5558
26. Cottler-Fox MH, Lapidot T, Petit I, Kollet O, DiPersio JF, Link D, Devine S (2003) Stem cell mobilization. Hematology (Am Soc Hematol Educ Program) 419–437
27. Cumano A, Ferraz JC, Klaine M, Di Santo JP, Godin I (2001) Intraembryonic, but not yolk sac hematopoietic precursors, isolated before circulation, provide long-term multilineage reconstitution. Immunity 15:477–485
28. de Bruijn MF, Ma X, Robin C, Ottersbach K, Sanchez MJ, Dzierzak E (2002) Hematopoietic stem cells localize to the endothelial cell layer in the midgestation mouse aorta. Immunity 16:673–683
29. de Bruijn MF, Speck NA, Peeters MC, Dzierzak E (2000) Definitive hematopoietic stem cells first develop within the major arterial regions of the mouse embryo. EMBO J 19:2465–2474
30. De Falco E, Porcelli D, Torella AR, Straino S, Iachininoto MG, Orlandi A, Truffa S, Biglioli P, Napolitano M, Capogrossi MC, Pesce M (2004) SDF-1 involvement in endothelial phenotype and ischemia-induced recruitment of bone marrow progenitor cells. Blood 104:3472–3482
31. Delassus S, Cumano A (1996) Circulation of hematopoietic progenitors in the mouse embryo. Immunity 4:97–106
32. Dick JE, Magli MC, Huszar D, Phillips RA, Bernstein A (1985) Introduction of a selectable gene into primitive stem cells capable of long-term reconstitution of the hemopoietic system of W/Wv mice. Cell 42:71–79
33. Driessen RL, Johnston HM, Nilsson SK (2003) Membrane-bound stem cell factor is a key regulator in the initial lodgment of stem cells within the endosteal marrow region. Exp Hematol 31:1284–1291
34. Ebbe S, Phalen E, Stohlman F, Jr. (1973) Abnormalities of megakaryocytes in W-WV mice. Blood 42:857–864
35. Eglitis MA, Mezey E (1997) Hematopoietic cells differentiate into both microglia and macroglia in the brains of adult mice. PNAS 94:4080–4085
36. Ema H, Nakauchi H (2000) Expansion of hematopoietic stem cells in the developing liver of a mouse embryo. Blood 95:2284–2288
37. Ema H, Sudo K, Seita J, Matsubara A, Morita Y, Osawa M, Takatsu K, Takaki S, Nakauchi H (2005) Quantification of self-renewal capacity in single hematopoietic stem cells from normal and Lnk-deficient mice. Dev Cell 8:907–914
38. Fanayan S, Firth SM, Baxter RC (2002) Signaling through the Smad pathway by insulin-like growth factor-binding protein-3 in breast cancer cells. Relationship to transforming growth factor-beta 1 signaling. J Biol Chem 277:7255–7261
39. Ferrari G, Cusella-De Angelis G, Coletta M, Paolucci E, Stornaiuolo A, Cossu G, Mavilio F (1998) Muscle regeneration by bone marrow-derived myogenic progenitors. Science 279:1528–1530
40. Folkman J (1971) Tumor angiogenesis: therapeutic implications. N Engl J Med 285:1182–1186
41. Fong DS, Aiello L, Gardner TW, King GL, Blankenship G, Cavallerano JD, Ferris FL, 3rd, Klein R (2004) Retinopathy in diabetes. Diabetes Care 27 Suppl 1:S84–87
42. Fuchs E, Tumbar T, Guasch G (2004) Socializing with the neighbors: stem cells and their niche. Cell 116:769–778
43. Garcia-Porrero JA, Godin IE, Dieterlen-Lievre F (1995) Potential intraembryonic hemogenic sites at pre-liver stages in the mouse. Anat Embryol (Berl) 192:425–435
44. Godin I, Cumano A (2002) The hare and the tortoise: an embryonic haematopoietic race. Nat Rev Immunol 2:593–604
45. Godin I, Dieterlen-Lievre F, Cumano A (1995) Emergence of multipotent hemopoietic cells in the yolk sac and para-aortic splanchnopleura in mouse embryos, beginning at 8.5 days postcoitus. Proc Natl Acad Sci U S A 92:773–777
46. Godin I, Garcia-Porrero JA, Dieterlen-Lievre F, Cumano A (1999) Stem cell emergence and hemopoietic activity are incompatible in mouse intraembryonic sites. J Exp Med 190:43–52
47. Goh EL, Ma D, Ming GL, Song H (2003) Adult neural stem cells and repair of the adult central nervous system. J Hematother Stem Cell Res 12:671–679
48. Gong JK (1978) Endosteal marrow: a rich source of hematopoietic stem cells. Science 199:1443–1445

49. Graf L, Iwata M, Torok-Storb B (2002) Gene expression profiling of the functionally distinct human bone marrow stromal cell lines HS-5 and HS-27a. Blood 100:1509–1511

50. Grant MB, Caballero S, Brown GA, Guthrie SM, Mames RN, Vaught T, Scott EW (2003) The contribution of adult hematopoietic stem cells to retinal neovascularization. Adv Exp Med Biol 522:37–45

51. Grant MB, May WS, Caballero S, Brown GA, Guthrie SM, Mames RN, Byrne BJ, Vaught T, Spoerri PE, Peck AB, Scott EW (2002) Adult hematopoietic stem cells provide functional hemangioblast activity during retinal neovascularization. Nat Med 8:607–612

52. Guthrie SM, Caballero S, Mames RN, Grant MB, Scott EW (2005) Analysis of the vascular potential of hematopoietic stem cells. Methods Mol Med 105:369–380

53. Hackney JA, Charbord P, Brunk BP, Stoeckert CJ, Lemischka IR, Moore KA (2002) A molecular profile of a hematopoietic stem cell niche. Proc Natl Acad Sci USA 99: 13061–13066

54. Harris RG, Herzog EL, Bruscia EM, Grove JE, Van Arnam JS, Krause DS (2004) Lack of a fusion requirement for development of bone marrow-derived epithelia. Science 305:90–93

55. Harrison DE (1993) Competitive repopulation in unirradiated normal recipients. Blood 81:2473–2474

56. Hattori K, Heissig B, Tashiro K, Honjo T, Tateno M, Shieh JH, Hackett NR, Quitoriano MS, Crystal RG, Rafii S, Moore MA (2001) Plasma elevation of stromal cell-derived factor-1 induces mobilization of mature and immature hematopoietic progenitor and stem cells. Blood 97:3354–3360

57. Herbrig K, Pistrosch F, Oelschlaegel U, Wichmann G, Wagner A, Foerster S, Richter S, Gross P, Passauer J (2004) Increased total number but impaired migratory activity and adhesion of endothelial progenitor cells in patients on long-term hemodialysis. Am J Kidney Dis 44: 840–849

58. Hiasa K, Ishibashi M, Ohtani K, Inoue S, Zhao Q, Kitamoto S, Sata M, Ichiki T, Takeshita A, Egashira K (2004) Gene transfer of stromal cell-derived factor-1alpha enhances ischemic vasculogenesis and angiogenesis via vascular endothelial growth factor/endothelial nitric oxide synthase-related pathway: next-generation chemokine therapy for therapeutic neovascularization. Circulation 109:2454–2461

59. Ikuta K, Uchida N, Friedman J, Weissman IL (1992) Lymphocyte development from stem cells. Annu Rev Immunol 10:759–783

60. Ikuta K, Weissman IL (1992) Evidence that hematopoietic stem cells express mouse c-kit but do not depend on steel factor for their generation. Proc Natl Acad Sci USA 89: 1502–1506

61. Imanishi T, Hano T, Nishio I (2004) Angiotensin II potentiates vascular endothelial growth factor-induced proliferation and network formation of endothelial progenitor cells. Hypertens Res 27:101–108

62. Ishikawa F, Livingston AG, Minamiguchi H, Wingard JR, Ogawa M (2003) Human cord blood long-term engrafting cells are CD34+ CD38. Leukemia 17:960–964

63. Ito CY, Li CY, Bernstein A, Dick JE, Stanford WL (2003) Hematopoietic stem cell and progenitor defects in Sca-1/ Ly-6A-null mice. Blood 101:517–523

64. Jang YY, Collector MI, Baylin SB, Diehl AM, Sharkis SJ (2004) Hematopoietic stem cells convert into liver cells within days without fusion. Nat Cell Biol 6:532–539

65. Keller G, Paige C, Gilboa E, Wagner EF (1985) Expression of a foreign gene in myeloid and lymphoid cells derived from multipotent haematopoietic precursors. Nature 318:149–154

66. Kiel MJ, Iwashita T, Yilmaz OH, Morrison SJ (2005) Spatial differences in hematopoiesis but not in stem cells indicate a lack of regional patterning in definitive hematopoietic stem cells. Dev Biol 283:29–39

67. Kijowski J, Baj-Krzyworzeka M, Majka M, Reca R, Marquez LA, Christofidou-Solomidou M, Janowska-Wieczorek A, Ratajczak MZ (2001) The SDF-1-CXCR4 axis stimulates VEGF secretion and activates integrins but does not affect proliferation and survival in lymphohematopoietic cells. Stem Cells 19:453–466

68. Kim I, Kim JH, Moon SO, Kwak HJ, Kim NG, Koh GY (2000) Angiopoietin-2 at high concentration can enhance endothelial cell survival through the phosphatidylinositol 3'-kinase/Akt signal transduction pathway. Oncogene 19: 4549–4552

69. Kim SY, Park SY, Kim JM, Kim JW, Kim MY, Yang JH, Kim JO, Choi KH, Kim SB, Ryu HM (2005) Differentiation of endothelial cells from human umbilical cord blood AC133-CD14+ cells. Ann Hematol 84:417–422

70. Kinnunen T, Raulo E, Nolo R, Maccarana M, Lindahl U, Rauvala H (1996) Neurite outgrowth in brain neurons induced by heparin-binding growth-associated molecule (HB-GAM) depends on the specific interaction of HB-GAM with heparan sulfate at the cell surface. J Biol Chem 271:2243–2248

71. Kucia M, Ratajczak J, Ratajczak MZ (2005) Bone marrow as a source of circulating CXCR4+ tissue-committed stem cells. Biol Cell 97:133–146

72. Kucia M, Reca R, Jala VR, Dawn B, Ratajczak J, Ratajczak MZ (2005) Bone marrow as a home of heterogenous populations of nonhematopoietic stem cells. Leukemia 19: 1118–1127

73. Kusadasi N, Oostendorp RA, Koevoet WJ, Dzierzak EA, Ploemacher RE (2002) Stromal cells from murine embryonic aorta-gonad-mesonephros region, liver and gut mesentery expand human umbilical cord blood-derived CAF-C(week6) in extended long-term cultures. Leukemia 16: 1782–1790

74. Lagasse E, Connors H, Al-Dhalimy M, Reitsma M, Dohse M, Osborne L, Wang X, Finegold M, Weissman IL, Grompe M (2000) Purified hematopoietic stem cells can differentiate into hepatocytes in vivo. Nat Med 6:1229–1234.

75. Lakshmipathy U, Verfaillie C (2005) Stem cell plasticity. Blood Rev 19:29–38

76. Lee KW, Cohen P (2002) Nuclear effects: unexpected intracellular actions of insulin-like growth factor binding protein-3. J Endocrinol 175:33–40

77. Lee P, Wang CC, Adamis AP (1998) Ocular neovascularization: an epidemiologic review. Surv Ophthalmol 43: 245–269.

78. Lee WY, Jin DK, Oh MR, Lee JE, Song SM, Lee EA, Kim GM, Chung JS, Lee KH (2003) Frequency analysis and clinical characterization of spinocerebellar ataxia types 1, 2, 3, 6, and 7 in Korean patients. Arch Neurol 60:858–863

79. Lemischka I (2002) Rethinking somatic stem cell plasticity. Nat Biotechnol 20:425

80. Libby P, Aikawa M, Kinlay S, Selwyn A, Ganz P (2000) Lipid lowering improves endothelial functions. Int J Cardiol 74 Suppl 1:S3-S10

81. Lord BI, Testa NG, Hendry JH (1975) The relative spatial

4

distributions of CFUs and CFUc in the normal mouse femur. Blood 46:65–72

82. Ma Q, Jones D, Borghesani PR, Segal RA, Nagasawa T, Kishimoto T, Bronson RT, Springer TA (1998) Impaired B-lymphopoiesis, myelopoiesis, and derailed cerebellar neuron migration in CXCR4- and SDF-1-deficient mice. Proc Natl Acad Sci U S A 95:9448–9453

83. Maeda N, Noda M (1998) Involvement of receptor-like protein tyrosine phosphatase zeta/RPTPbeta and its ligand pleiotrophin/heparin-binding growth-associated molecule (HB-GAM) in neuronal migration. J Cell Biol 142:203–216

84. Manaia A, Lemarchandel V, Klaine M, Max-Audit I, Romeo P, Dieterlen-Lievre F, Godin I (2000) Lmo2 and GATA-3 associated expression in intraembryonic hemogenic sites. Development 127:643–653

85. Martin-Rendon E, Watt SM (2003) Exploitation of stem cell plasticity. Transfus Med 13:325–349

86. Marzesco AM, Janich P, Wilsch-Brauninger M, Dubreuil V, Langenfeld K, Corbeil D, Huttner WB (2005) Release of extracellular membrane particles carrying the stem cell marker prominin-1 (CD133) from neural progenitors and other epithelial cells. J Cell Sci 118:2849–2858

87. Mezey E, Chandross KJ, Harta G, Maki RA, McKercher SR (2000) Turning Blood into Brain: Cells Bearing Neuronal Antigens Generated in Vivo from Bone Marrow. Science 290:1779–1782

88. Micklem HS, Ford CE, Evans EP, Ogden DA, Papworth DS (1972) Competitive in vivo proliferation of foetal and adult haematopoietic cells in lethally irradiated mice. J Cell Physiol 79:293–298

89. Milkiewicz M, Hudlicka O, Brown MD, Silgram H (2005) Nitric oxide, VEGF, and VEGFR-2: interactions in activity-induced angiogenesis in rat skeletal muscle. Am J Physiol Heart Circ Physiol 289:H336–343

90. Moore KA (2004) Recent advances in defining the hematopoietic stem cell niche. Curr Opin Hematol 11:107–111

91. Morrison SJ, Uchida N, Weissman IL (1995) The biology of hematopoietic stem cells. Annu Rev Cell Dev Biol 11:35–71

92. Muller A, Medvinsky A, Strouboulis J, Grosveld F, Dzierzak E (1994) Development of hematopoietic stem cell activity in the mouse embryo. Immunity 1:291–301

93. Murdoch C (2000) CXCR4: chemokine receptor extraordinaire. Immunol Rev 177:175–184

94. Murry CE, Soonpaa MH, Reinecke H, Nakajima H, Nakajima HO, Rubart M, Pasumarthi KB, Virag JI, Bartelmez SH, Poppa V, Bradford G, Dowell JD, Williams DA, Field LJ (2004) Haematopoietic stem cells do not transdifferentiate into cardiac myocytes in myocardial infarcts. Nature 428:664–668

95. Nagasawa T, Hirota S, Tachibana K, Takakura N, Nishikawa S, Kitamura Y, Yoshida N, Kikutani H, Kishimoto T (1996) Defects of B-cell lymphopoiesis and bone-marrow myelopoiesis in mice lacking the CXC chemokine PBSF/SDF-1. Nature 382:635–638

96. Nygren JM, Jovinge S, Breitbach M, Sawen P, Roll W, Hescheler J, Taneera J, Fleischmann BK, Jacobsen SE (2004) Bone marrow-derived hematopoietic cells generate cardiomyocytes at a low frequency through cell fusion, but not transdifferentiation. Nat Med 10:494–501

97. Ohneda O, Fennie C, Zheng Z, Donahue C, La H, Villacorta R, Cairns B, Lasky LA (1998) Hematopoietic stem cell maintenance and differentiation are supported by embryonic aorta-gonad-mesonephros region-derived endothelium. Blood 92:908–919

98. Oostendorp RA, Dormer P (1997) VLA-4-mediated interactions between normal human hematopoietic progenitors and stromal cells. Leuk Lymphoma 24:423–435

99. Oostendorp RA, Harvey KN, Kusadasi N, de Bruijn MF, Saris C, Ploemacher RE, Medvinsky AL, Dzierzak EA (2002) Stromal cell lines from mouse aorta-gonads-mesonephros subregions are potent supporters of hematopoietic stem cell activity. Blood 99:1183–1189

100. Oostendorp RA, Robin C, Steinhoff C, Marz S, Brauer R, Nuber UA, Dzierzak EA, Peschel C (2005) Long-term maintenance of hematopoietic stem cells does not require contact with embryo-derived stromal cells in cocultures. Stem Cells 23:842–851

101. Oostendorp RA, Spitzer E, Reisbach G, Dormer P (1997) Antibodies to the beta 1-integrin chain, CD44, or ICAM-3 stimulate adhesion of blast colony-forming cells and may inhibit their growth. Exp Hematol 25:345–349

102. Osawa M, Hanada K, Hamada H, Nakauchi H (1996) Long-term lymphohematopoietic reconstitution by a single CD34-low/negative hematopoietic stem cell. Science 273:242–245

103. Ostendorf T, Van Roeyen C, Westenfeld R, Gawlik A, Kitahara M, De Heer E, Kerjaschki D, Floege J, Ketteler M (2004) Inducible nitric oxide synthase-derived nitric oxide promotes glomerular angiogenesis via upregulation of vascular endothelial growth factor receptors. J Am Soc Nephrol 15:2307–2319

104. Peichev M, Naiyer AJ, Pereira D, Zhu Z, Lane WJ, Williams M, Oz MC, Hicklin DJ, Witte L, Moore MA, Rafii S (2000) Expression of VEGFR-2 and AC133 by circulating human CD34(+) cells identifies a population of functional endothelial precursors. Blood 95:952–958

105. Pelus LM, Bian H, Fukuda S, Wong D, Merzouk A, Salari H (2005) The CXCR4 agonist peptide, CTCE-0021, rapidly mobilizes polymorphonuclear neutrophils and hematopoietic progenitor cells into peripheral blood and synergizes with granulocyte colony-stimulating factor. Exp Hematol 33:295–307

106. Petrenko O, Beavis A, Klaine M, Kittappa R, Godin I, Lemischka IR (1999) The molecular characterization of the fetal stem cell marker AA4. Immunity 10:691–700

107. Phillips RL, Ernst RE, Brunk B, Ivanova N, Mahan MA, Deanehan JK, Moore KA, Overton GC, Lemischka IR (2000) The genetic program of hematopoietic stem cells. Science 288:1635–1640

108. Ploug M, Ellis V (1994) Structure-function relationships in the receptor for urokinase-type plasminogen activator. Comparison to other members of the Ly-6 family and snake venom alpha-neurotoxins. FEBS Lett 349:163–168

109. Reya T, Duncan AW, Ailles L, Domen J, Scherer DC, Willert K, Hintz L, Nusse R, Weissman IL (2003) A role for Wnt signalling in self-renewal of haematopoietic stem cells. Nature 423:409–414

110. Risau W, Lemmon V (1988) Changes in the vascular extracellular matrix during embryonic vasculogenesis and angiogenesis. Dev Biol 125:441–450

111. Rosendaal M, Hodgson GS, Bradley TR (1979) Organization of haemopoietic stem cells: the generation-age hypothesis. Cell Tissue Kinet 12:17–29

112. Rothe L, Collin-Osdoby P, Chen Y, Sunyer T, Chaudhary L, Tsay A, Goldring S, Avioli L, Osdoby P (1998) Human osteoclasts and osteoclast-like cells synthesize and release high basal and inflammatory stimulated levels of

the potent chemokine interleukin-8. Endocrinology 139: 4353–4363

113. Ryan SJ (1979) The development of an experimental model of subretinal neovascularization in disciform macular degeneration. Trans Am Ophthalmol Soc 77:707–745

114. Sahara M, Sata M, Matsuzaki Y, Tanaka K, Morita T, Hirata Y, Okano H, Nagai R (2005) Comparison of various bone marrow fractions in the ability to participate in vascular remodeling after mechanical injury. Stem Cells 23:874–878

115. Salcedo R, Wasserman K, Young HA, Grimm MC, Howard OM, Anver MR, Kleinman HK, Murphy WJ, Oppenheim JJ (1999) Vascular endothelial growth factor and basic fibroblast growth factor induce expression of CXCR4 on human endothelial cells: In vivo neovascularization induced by stromal-derived factor-1alpha. Am J Pathol 154:1125–1135

116. Sato A, Iwama A, Takakura N, Nishio H, Yancopoulos GD, Suda T (1998) Characterization of TEK receptor tyrosine kinase and its ligands, Angiopoietins, in human hematopoietic progenitor cells. Int Immunol 10:1217–1227

117. Schatteman GC, Hanlon HD, Jiao C, Dodds SG, Christy BA (2000) Blood-derived angioblasts accelerate blood-flow restoration in diabetic mice. J Clin Invest 106:571–578

118. Schofield R (1978) The relationship between the spleen colony-forming cell and the haemopoietic stem cell. Blood Cells 4:7–25

119. Sengupta N, Caballero S, Mames RN, Butler JM, Scott EW, Grant MB (2003) The role of adult bone marrow-derived stem cells in choroidal neovascularization. Invest Ophthalmol Vis Sci 44:4908–4913

120. Shefer G, Wleklinski-Lee M, Yablonka-Reuveni Z (2004) Skeletal muscle satellite cells can spontaneously enter an alternative mesenchymal pathway. J Cell Sci 117:5393–5404

121. Sivan-Loukianova E, Awad OA, Stepanovic V, Bickenbach J, Schatteman GC (2003) CD34+ blood cells accelerate vascularization and healing of diabetic mouse skin wounds. J Vasc Res 40:368–377

122. Smith LE, Wesolowski E, McLellan A, Kostyk SK, D'Amato R, Sullivan R, D'Amore PA (1994) Oxygen-induced retinopathy in the mouse. Invest Ophthalmol Vis Sci 35:101–111

123. Sonveaux P, Martinive P, DeWever J, Batova Z, Daneau G, Pelat M, Ghisdal P, Gregoire V, Dessy C, Balligand JL, Feron O (2004) Caveolin-1 expression is critical for vascular endothelial growth factor-induced ischemic hindlimb collateralization and nitric oxide-mediated angiogenesis. Circ Res 95:154–161

124. Spangrude GJ, Aihara Y, Weissman IL, Klein J (1988) The stem cell antigens Sca-1 and Sca-2 subdivide thymic and peripheral T lymphocytes into unique subsets. J Immunol 141:3697–3707

125. Stanford WL, Haque S, Alexander R, Liu X, Latour AM, Snodgrass HR, Koller BH, Flood PM (1997) Altered proliferative response by T lymphocytes of Ly-6A (Sca-1) null mice. J Exp Med 186:705–717

126. Szilvassy SJ, Humphries RK, Lansdorp PM, Eaves AC, Eaves CJ (1990) Quantitative assay for totipotent reconstituting hematopoietic stem cells by a competitive repopulation strategy. Proc Natl Acad Sci U S A 87:8736–8740

127. Tachibana K, Hirota S, Iizasa H, Yoshida H, Kawabata K, Kataoka Y, Kitamura Y, Matsushima K, Yoshida N, Nishikawa S, Kishimoto T, Nagasawa T (1998) The chemokine receptor CXCR4 is essential for vascularization of the gastrointestinal tract. Nature 393:591–594

128. Takakura N, Watanabe T, Suenobu S, Yamada Y, Noda T, Ito Y, Satake M, Suda T (2000) A role for hematopoietic stem cells in promoting angiogenesis. Cell 102:199–209.

129. Take M, Tsutsui J, Obama H, Ozawa M, Nakayama T, Maruyama I, Arima T, Muramatsu T (1994) Identification of nucleolin as a binding protein for midkine (MK) and heparin-binding growth associated molecule (HB-GAM). J Biochem (Tokyo) 116:1063–1068

130. Tepper OM, Capla JM, Galiano RD, Ceradini DJ, Callaghan MJ, Kleinman ME, Gurtner GC (2005) Adult vasculogenesis occurs through in situ recruitment, proliferation, and tubulization of circulating bone marrow-derived cells. Blood 105:1068–1077

131. Tepper OM, Galiano RD, Capla JM, Kalka C, Gagne PJ, Jacobowitz GR, Levine JP, Gurtner GC (2002) Human endothelial progenitor cells from type II diabetics exhibit impaired proliferation, adhesion, and incorporation into vascular structures. Circulation 106:2781–2786

132. Terada N, Hamazaki T, Oka M, Hoki M, Mastalerz DM, Nakano Y, Meyer EM, Morel L, Petersen BE, Scott EW (2002) Bone marrow cells adopt the phenotype of other cells by spontaneous cell fusion. Nature 416:542–545

133. Valgimigli M, Rigolin GM, Fucili A, Porta MD, Soukhomovskaia O, Malagutti P, Bugli AM, Bragotti LZ, Francolini G, Mauro E, Castoldi G, Ferrari R (2004) CD34+ and endothelial progenitor cells in patients with various degrees of congestive heart failure. Circulation 110: 1209–1212

134. Vasa M, Fichtlscherer S, Aicher A, Adler K, Urbich C, Martin H, Zeiher AM, Dimmeler S (2001) Number and migratory activity of circulating endothelial progenitor cells inversely correlate with risk factors for coronary artery disease. Circ Res 89:E1–7

135. Vassilopoulos G, Wang PR, Russell DW (2003) Transplanted bone marrow regenerates liver by cell fusion. Nature 422:901–904

136. Wagers AJ, Sherwood RI, Christensen JL, Weissman IL (2002) Little evidence for developmental plasticity of adult hematopoietic stem cells. Science 297:2256–2259

137. Waskow C, Paul S, Haller C, Gassmann M, Rodewald HR (2002) Viable c-Kit(W/W) mutants reveal pivotal role for c-kit in the maintenance of lymphopoiesis. Immunity 17:277–288

138. Waskow C, Rodewald HR (2002) Lymphocyte development in neonatal and adult c-Kit-deficient (c-KitW/W) mice. Adv Exp Med Biol 512:1–10

139. Welm BE, Tepera SB, Venezia T, Graubert TA, Rosen JM, Goodell MA (2002) Sca-1(pos) cells in the mouse mammary gland represent an enriched progenitor cell population. Dev Biol 245:42–56

140. Wilkinson-Berka JL (2004) Vasoactive factors and diabetic retinopathy: vascular endothelial growth factor, cycoloxygenase-2 and nitric oxide. Curr Pharm Des 10:3331–3348

141. Wognum AW, Eaves AC, Thomas TE (2003) Identification and isolation of hematopoietic stem cells. Arch Med Res 34:461–475

142. Wright DE, Wagers AJ, Gulati AP, Johnson FL, Weissman IL (2001) Physiological migration of hematopoietic stem and progenitor cells. Science 294:1933–1936

143. Wu X, Rabkin-Aikawa E, Guleserian KJ, Perry TE, Masuda Y, Sutherland FW, Schoen FJ, Mayer JE, Jr., Bischoff J (2004) Tissue-engineered microvessels on three-dimen-

4

sional biodegradable scaffolds using human endothelial progenitor cells. Am J Physiol Heart Circ Physiol 287: H480–487

144. Xu MJ, Tsuji K, Ueda T, Mukouyama YS, Hara T, Yang FC, Ebihara Y, Matsuoka S, Manabe A, Kikuchi A, Ito M, Miyajima A, Nakahata T (1998) Stimulation of mouse and human primitive hematopoiesis by murine embryonic aorta-gonad-mesonephros-derived stromal cell lines. Blood 92:2032–2040

145. Yamaguchi J, Kusano KF, Masuo O, Kawamoto A, Silver M, Murasawa S, Bosch-Marce M, Masuda H, Losordo DW, Isner JM, Asahara T (2003) Stromal cell-derived factor-1 effects on ex vivo expanded endothelial progenitor cell recruitment for ischemic neovascularization. Circulation 107:1322–1328

146. Ying QL, Nichols J, Evans EP, Smith AG (2002) Changing potency by spontaneous fusion. Nature 416:545–548

147. Zhang J, Niu C, Ye L, Huang H, He X, Tong WG, Ross J, Haug J, Johnson T, Feng JQ, Harris S, Wiedemann LM, Mishina Y, Li L (2003) Identification of the haematopoietic stem cell niche and control of the niche size. Nature 425:836–841

148. Zou YR, Kottmann AH, Kuroda M, Taniuchi I, Littman DR (1998) Function of the chemokine receptor CXCR4 in haematopoiesis and in cerebellar development. Nature 393:595–599

5 Inflammation as a Stimulus for Vascular Leakage and Proliferation

A.M. Joussen, A.P. Adamis

Core Messages

- Diabetic retinopathy shows many of the characteristics of an inflammatory disease
- Diabetic retinal vascular leakage, capillary non-perfusion, and endothelial cell damage are caused, in part, by retinal leukocyte stasis in early experimental diabetes
- Vascular endothelial growth factor (VEGF) plays key roles in mediating both ischemia-related neovascularization as well as retinal leukocyte stasis
- Leukocytes adhere to the retinal vasculature via intercellular adhesion molecule-1 (ICAM-1) and CD18
- FasL mediated apoptosis is involved in vascular remodeling upon ischemia and diabetes
- These pathological processes are similar to those underlying the leukocyte-mediated pruning of the retinal vasculature during normal development

In the past few decades, our knowledge of the mechanisms underlying retinal vasoproliferation has increased greatly (see Chapters 2, 3.1, 3.2 and 3.3). While vasoproliferation was once considered to be mainly a consequence of ischemia, current evidence also supports a contribution of inflammatory mechanisms. Inflammation is also highly related to vascular leakage in diseases that are known to result in retinal and macular edema. Recently, inflammatory mechanisms have gained interest with respect to the retinal pathology following ischemia, as well as in diseases such as diabetic retinopathy (DR) and sickle cell retinopathy (see Chapter 27.1). In this chapter, the discussion will focus on the published data relating to the inflammatory mechanisms in ischemic retinal diseases such as DR. The definition of inflammation in this setting is the involvement of any leukocyte-mediated pathology in the course of the disease.

We examined several lines of evidence, including correlative studies of elevated levels of inflammatory mediators in the presence of DR and the impact of anti-inflammatory agents on the disease. As a central focus, we will discuss in detail a series of preclinical studies that support a causal linkage between inflammation and two principal characteristics of the pathology associated with DR, ischemia-linked neovascularization and the breakdown of the blood-retinal barrier (BRB), together with the role of VEGF in mediating these events. These studies have provided good evidence that the vascular damage that is seen in DR is mediated by processes that are very similar to those that regulate retinal vascularization during normal development.

5.1 Evidence for Inflammation in the Pathogenesis of Diabetic Retinopathy

Essentials

- Inflammatory mediators are upregulated in DR
- Diabetic retinal pathology can be inhibited by anti-inflammatory agents
- VEGF is a key mediator of inflammatory changes in the diabetic retina
- Angiopoietin-1 regulates vascular permeability and expression of inflammatory mediators in diabetic retinopathy

5.1.1 Upregulation of Inflammatory Mediators in Diabetic Retinopathy

Both clinical and preclinical studies have associated the development of DR with elevated ocular levels of inflammatory mediators. McLeod et al. [70] reported that levels of intercellular adhesion molecule-1 (ICAM-1), an important adhesive molecule for circulating leukocytes, were elevated throughout the vasculature of diabetic patients, whereas the distribution was much more restricted in nondiabetic subjects; moreover, this elevation was accompanied by a significantly higher number of neutrophils in both the choroid and retina. Limb et al. [65] reported that levels of ICAM-1 as well as other adhesion molecules such as vascular cell adhesion molecule (VCAM-1) and E-selectin were elevated in patients with proliferative DR, while Funatsu et al. [33] have also reported elevated levels of ICAM-1 in patients with diabetic macular edema (DME). These patients also showed elevated vitreous levels of vascular endothelial growth factor (VEGF) [33], which upregulates ICAM-1 expression [107]. As discussed below, VEGF may in fact be a key factor mediating inflammatory events in the diabetic eye, and DR-correlated elevation of VEGF levels was first reported over a decade ago [1, 4]. Since then diabetes-associated elevations of VEGF in the vitreous have been reported by a number of groups, together with increases in a variety of other factors, including interleukin-6 [31], angiotensin II [32], angiopoietin 2 [109], erythropoietin [110] and stroma-derived factor-1 (SDF-1). SDF-1 is itself a stimulator of VEGF expression [14] and an important mediator of cell migration and adhesion [60].

Tumor necrosis factor-α (TNF-α) is a proinflammatory cytokine that has also been implicated in the pathogenesis of diabetic retinopathy [64, 66, 92]; moreover, susceptibility to diabetic retinopathy has been associated with TNF-α gene polymorphism [39]. TNF-α is found in the extracellular matrix, endothelium, and vessel walls of fibrovascular tissue in proliferative diabetic retinopathy [64], and in the vitreous from eyes with this complication [29, 101]. TNF-α can stimulate VEGF expression by the retinal pigment epithelium [84] and in choroidal neovascular membranes [38], and has been implicated as an inducer of pathological angiogenesis in the retina [34].

The evidence provided by these correlative measurements of inflammatory mediators has been supplemented by other approaches. The advent of high-density microarray technology [15, 21, 48, 100], with its capacity for simultaneous monitoring of thousands of genes, provides a unique opportunity for a high-throughput analysis of diabetic retinopathy at the molecular level. In an analysis of retinal gene expression in streptozotocin-induced diabetes in rats, numerous genes operative in inflammatory reactions were found to be upregulated [51]. Prominent among these were the genes for macrophage migration inhibitory factor (MIF), a proinflammatory lymphokine that is believed to be involved in maintaining neutrophils in the vasculature and in

facilitating their adhesion and local release of cytokines [82, 93], as well as a number of genes for adhesion molecules and apoptosis. While the findings from these approaches are purely correlative, and are not able to differentiate between potential molecular mechanisms, they nonetheless can provide important clues as to the nature of processes that may be involved in the pathogenesis of DR.

Finally, correlative studies have also been carried out examining the levels of serum factors in patients with DR [73]. These workers reported that the levels of the chemokines RANTES (Regulated on Activation, Normal T-cell Expressed and Secreted) and SDF-1α were significantly elevated in patients with at least severe nonproliferative diabetic retinopathy compared to patients with less severe diabetic retinopathy. Positive immunostaining was observed in the inner retina for RANTES and monocyte chemoattractant protein-1 (MCP-1) in patients with diabetes. In keeping with earlier findings, staining was also strongly positive throughout the diabetic retina for ICAM-1, while normal retinal tissues showed little reactivity.

5.1.2 Diabetic Retinal Pathology Can Be Inhibited by Anti-inflammatory Agents

Induction of adhesion molecules on endothelial cells by proinflammatory cytokines such as TNF-α [65] and VEGF is mediated at the molecular level by the activation of a redox-sensitive transcription factor, nuclear factor (NF)-κB [108]. NF-κB upregulates ICAM-1 and various inflammatory genes such as cyclooxygenase (COX)-2 [63]. The cyclooxygenases COX-1 and COX-2 are key enzymes in the conversion of arachidonic acid to prostaglandin H_2, the common precursor for all other eicosanoids. COX-1 is expressed ubiquitously and generates eicosanoids with cytoprotective function whereas COX-2 is an immediate-early gene expressed at sites of acute inflammation and generates eicosanoids with a proinflammatory role that create a positive feedback loop by further activating NF-κB and inflammatory cytokines [69, 85].

The observation that arthritic diabetic patients receiving high daily doses of aspirin exhibit reduced symptoms of diabetic retinopathy led to the hypothesis that anti-inflammatory treatment could prove beneficial [88]. Aspirin, in low doses (8 mg/day), inhibits platelet aggregation, predominantly via acetylation of COX-1 and reduction of thromboxane A_2 production. In intermediate doses (2–4 g/day), aspirin inhibits both COX-1 and COX-2, blocking prostaglandin production, and is antipyretic [86]. In high doses (6–8 g/day), it is a potent anti-inflammatory drug suitable for the treatment of rheumatic disorders. The mechanism of this action is unclear, although it seems to be COX and prostaglandin independent [111]. It should be noted, however, that while aspirin was originally found to inhibit platelet aggregation in vitro and to retard the development of microaneurysms in patients with diabetic retinopathy [59], the results of the Early Treatment of Diabetic Retinopathy Study (ETDRS) and the Wisconsin Epidemiologic Study of Diabetic Retinopathy did not identify any beneficial effect of low and intermediate amounts of aspirin use in diabetic retinopathy [25].

Recently, interest has arisen in the use of specific COX-2 inhibitors that do not bear the unwanted gastrointestinal side effects of general anti-inflammatory drugs such as aspirin. In a rat model of diabetic retinopathy, aspirin and the selective COX-2 inhibitor meloxicam reduce leukostasis and vascular leakage, in part, through inhibition of TNF-α and NF-κB activation; similar results were seen with the anti-TNF-α agent etanercept, a soluble TNF-α-receptor/Fc fusion protein [54]. All three agents prevented the upregulation of ICAM-1 and eNOS that is observed early in the course of diabetic retinopathy [54]. Finally, in a clinical study, Sfikakis et al. [98] have reported that intravenous administration of infliximab, a monoclonal antibody to TNF-α, led to alleviation reduction of macular thickness, as well as alleviation of symptoms in patients with DME. Taken together, these various findings suggest that inhibition of inflammation holds promise as a method for alleviating the retinal vascular symptoms that accompany diabetes, and simultaneously provides support for the idea that inflammation is a contributing factor to the overall retinal pathology.

5.1.3 VEGF Is a Key Mediator of Inflammatory Changes in the Diabetic Retina: General Considerations

VEGF is a pluripotent growth factor which exhibits a number of properties that are important in promoting both the ischemia-promoted neovascularization and the vascular leakage that are characteristic of DR. It is the most potent known promoter of angiogenesis, acting in a variety of roles, including endothelial cell mitogen, chemoattractant and survival factor ([7, 10, 62]; see [27] for a review), and an extensive body of research has established its importance for ocular vascularizing diseases (reviewed in [81]). Two other properties are of direct relevance for diabetes-related retinal pathology: VEGF expression is upregulated by hypoxia, so that its levels increase in response to local ischemia [87, 99], and it is also the most potent known enhancer of endothelial cell permeability, some 50,000 times greater than histamine [97].

The injection of VEGF into normal nondiabetic eyes recapitulates many of the retinal vascular changes triggered by diabetes, including ICAM-1 upregulation, leukocyte adhesion, vascular permeability and capillary nonperfusion [20, 68, 77, 105]. In brain endothelium, VEGF-induced ICAM-1 upregulation is mediated by nitric oxide (NO) [90]. NO, a molecule with both cytotoxic and signaling capabilities, can be generated by various NO synthases (NOS). The endothelial isoform, eNOS, is upregulated in diabetic neural ganglia [115].

Evidence is developing that causally links VEGF to inflammation in a variety of experimental systems. VEGF has been shown to trigger leukocyte adhesion in experimental models of tumors [72] and skin [20]. In the eye, Flt-1-based VEGF inhibition was demonstrated to inhibit inflammation-related choroidal neovascularization [42]; a detailed discussion of related experiments in the diabetic eye is provided in Sect. 5.2 below. Similar anti-inflammatory effects following VEGF inhibition have also been described in a model of rheumatoid arthritis [74]. Taken together, the evidence suggests that VEGF-induced inflammation might not be limited to the eye but rather represents a common regulatory mechanism in disease.

Finally, it should be noted that the enhanced neutrophil integrin expression that characterizes diabetes is not affected by VEGF inhibition [51]. This finding is in general agreement with data showing that VEGF inhibition does not alter neutrophil CD11b and CD18 expression induced by conditioned medium from a tumor cell line [113]. Nevertheless, retinal VEGF may still serve to attract leukocytes to the retinal vasculature. It is well known that monocytes, a leukocyte class operative in diabetic retinopathy [70, 96], migrate in response to VEGF [11, 17]. Moreover, via their own VEGF, leukocytes may serve to amplify the direct effects of VEGF when they adhere to endothelium. VEGF is present in neutrophils [35], monocytes [44], eosinophils [43], lymphocytes [30], and platelets [79]. Thus, although VEGF may not directly alter leukocyte surface integrin expression in diabetes, it may still alter the biology of leukocytes.

5.1.4 Angiopoietin-1 Regulates Vascular Permeability and Expression of Inflammatory Mediators in Diabetic Retinopathy

Angiopoietin-1 and angiopoietin-2 are ligands for Tie-2, an endothelial-specific receptor tyrosine kinase which regulates embryonic vasculogenesis and angiogenesis [95]. Tie-2 receptors also have been shown to be widely expressed in the adult quiescent vasculature [112]. Whereas angiopoietin-1 is a receptor agonist [102], angiopoietin-2 represents a

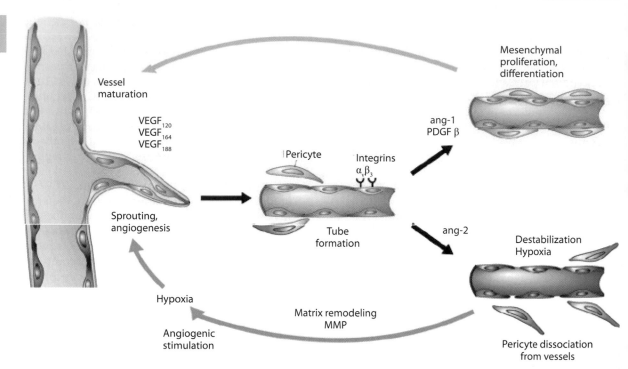

Fig. 5.1. Determinants of angiogenesis and vascular permeability. The VEGF family factors and their receptors mediate angiogenesis in response to hypoxia. They affect the expression of integrins and mediate cell-cell interactions, e.g., through cell adhesion molecules. The angiopoietins determine the fate of the vessel toward maturation or dissociation. Whereas angiopoietin-1 promotes mesenchymal proliferation, differentiation and stabilization of the vessel, angiopoietin-2 leads to destabilization of the vessel and dissociation of pericytes. The recent findings concerning angiopoietin are beginning to elucidate the effect of the angiopoietins in the pathogenesis of disease

natural angiopoietin-1/Tie-2 inhibitor [41, 83]. In contrast to VEGF, which can initiate vessel growth in vitro and in vivo [2, 5], neither angiopoietin-1 nor angiopoietin-2 alone can promote in vivo neovascularization [9]. There is, however, evidence that the VEGF functions are coordinated and complementary to angiopoietins expressed on endothelial cells, as VEGF stimulates anigopoietin-2 but not angiopoietin-1 or Tie-2 expression in vitro [83].

Angiopoietins act later in vascular development, being involved in remodeling, maturation and stabilization [37, 40, 41]. Angiopoietin-1 is thought to help maintain and stabilize the mature vessels by promoting interactions between the endothelial cells and the surrounding support cells [41, 71, 102], while angiopoietin-2 is expressed at sites of vascular remodeling [40, 41, 102]. It has been proposed that pathological overexpression of angiopoietin-2 can destabilize the constitutive linkages between angiopoietin-1 and Tie-2 that help to maintain vessel stability; in the absence of VEGF this destabilization leads to vessel regression, while in the presence of VEGF it facilitates the angiogenic response [41, 102]. Figure 5.1 depicts the involvement of angiopoietins in new vessel growth and vessel maturation.

Vascular leakage is the hallmark of diseases associated with neovascularization and plays a major role in the pathogenesis of diabetic retinopathy, with retinal edema being the single most important cause of vision loss in diabetes. Angiopoietin-1 acts to stabilize vessels, acting as an anti-permeability factor that protects against leakage, including leakage induced by VEGF and other inflammatory agents (103, 104; for a review see [49]). Recently, Joussen et al. [52] reported that angiopoietin-1 could prevent increases in vascular permeability caused by experimental diabetes in rats. Seven days following the induction of experimental diabetes by intraperitoneal injection of streptozotocin, adult rat eyes received a single, unilateral injection of vehicle, 50 ng, 200 ng or 800 ng angiopoietin-1; contralateral control eyes received an injection of phosphate-buffered saline (PBS). Twenty-four hours later, vascular permeability was quantified by measuring extravasation of Evans blue dye [114]. Retinal vascular permeability was increased approximately twofold in the 1-week diabetic eyes and was not altered by the PBS injections. In marked contrast, the diabetes-related vascular permeability in the angiopoietin-1 treated eyes was dose dependently suppressed. Moreover, angiopoietin-1 also potently suppressed the expression of endogenous retinal VEGF and ICAM-1. Given that VEGF and ICAM-1 are important mediators of leukocyte adhesion (see Sect. 5.2), it is not surprising that intravitreal angiopoietin-1 also led to a dose-dependent inhibition of diabetes-associated leukocyte adhesion [52].

5.2 Cellular and Molecular Mechanisms Mediating Diabetes-Associated Vascular Damage and the Neovascularizing Response to Ischemia

Even though there is no animal model system that accurately reproduces the entire course of clinical diabetic retinopathy, rodent models do exist which resemble the clinical condition in several respects, including the importance of retinal leukostasis. These models have demonstrated that some of the processes which contribute to vascular damage in ischemic retinal conditions are very similar to those which underlie the physiological remodeling of the nascent vasculature during normal retinal development.

- Leukocytes mediate retinal vascular remodeling during development and vaso-obliteration in disease
- VEGF and leukocyte invasion are important factors in regulating both ischemia-mediated ocular neovascularization and vascular damage in DR
- T-lymphocytes are negative regulators of pathological retinal neovascularization
- Monocytes are positive regulators of pathological retinal neovascularization
- Retinal cell injury and death are Fas/FasL-dependent

5.2.1 Leukocytes Mediate Retinal Vascular Remodeling During Development and Vaso-obliteration in Disease

In studies of vascularization in the rat retina, Ishida et al. [45] have shown that pruning of the vasculature occurs in normal development and that similar processes are involved in the pathological obliteration of retinal vasculature during conditions of hyperoxia. During normal development, the retinal vasculature extends to fill the peripheral avascular retina; beginning on postnatal (P) day 5, these investigators reported a considerable increase in the number of adherent cells which they identified as leukocytes using immunochemical staining for CD45 (leukocyte common antigen). The number of leukocytes subsequently decreased, reaching a plateau from P6 to P11 as the vascular pruning moved peripherally.

As the vasculature grew toward the periphery, the zone containing the leukocytes also migrated toward the periphery, staying two to three disk diameters behind the leading edge of vascular growth (p. 783). In its wake was a pruned and remodeled secondary vasculature. When the leukocyte and vascular changes were compared within a zone five disc diameters from the disc, it was apparent that the decrease in both vascular density and branch-surrounded spaces was preceded by an increase in leukocyte density, indicating a cause-effect relationship. By P12, all pruning was completed and the leukocytes had all but disappeared [45].

Studies of the mechanism of local adhesion of these leukocytes revealed key roles for leukocyte integrin CD18 and its corresponding ligand, ICAM-1 [45]. Western blotting showed that ICAM-1 expression was elevated in P4P6 retinas actively undergoing remodeling as compared to normal adult rat retinas, in which vascular remodeling was absent. Moreover, immunochemical studies in P8 retinas demonstrated that the upregulation of ICAM-1 expression was localized to the region of the vasculature containing the adherent leukocytes. In addition, there was very little ICAM-1 expression at P1 or P12, times at which there were very few leukocytes present. During the P4P6 period, when the biggest changes in leukocyte numbers and vascular density took place, CD18 inhibition by a neutralizing antibody confirmed the causal involvement of leukocytes in vascular pruning. Both the number of adherent leukocytes and the spaces in the vasculature were significantly reduced. Similarly, genetically transformed mice lacking CD18 showed much less vascular pruning at P9 than wild-type mice.

Additional immunocytochemical experiments [45], using an antibody to CD8 and CD25, revealed that many of the adherent leukocytes were T-lymphocytes (p. 784); moreover, administration of a neutralizing antibody to CD2, an antigen that is essential for T-lymphocyte adhesion, significantly suppressed vascular pruning ([45], p. 785). T-lymphocytes are the principal locus of expression of Fas, while its ligand FasL is much more widely expressed [80]. Apoptosis, mediated through the interaction of Fas-FasL, was the principal means of effecting vascular pruning; intraperitoneal administration of a FasL-neutralizing antibody led to a 75% decrease of vascular pruning at P6 [45].

These basic molecular mechanisms underlying vascular pruning during normal development were also found to be operative in a pathological model of hyperoxia-induced ischemia [45]. P2 mice, together with their nursing mothers, were placed in an 80% oxygen environment for up to 12 h, leading to dramatic vaso-obliteration. This decrease in vascularization occurred shortly after an increase in the number of adherent leukocytes. As had occurred during physiological pruning, ICAM-1 expression was elevated in parallel with leukocyte accumulation, and vaso-obliteration was significantly diminished by administration of an antibody to CD18, the ligand for ICAM-1. Similarly, CD18-deficient mice showed much less vaso-obliteration of retinal vasculature than did wild-type controls when similarly exposed to oxygen. The importance of FasL-mediated apoptosis was also confirmed, as vaso-obliteration was significantly reduced in response to administration of a FasL-neutralizing antibody ([45], p. 785).

5.2.2 VEGF and Leukocyte Invasion Are Important Factors in Regulating Both Ischemia-Mediated Ocular Neovascularization and Vascular Damage in Diabetic Retinopathy

The pathophysiology of DR is complex, with vascular changes including both neovascularization and damage to the existing retinal vasculature. A number of biochemical pathways are believed to be important in linking hyperglycemia to vascular abnormalities in the retina ([16]; [28], pp. 2,540–1). Avascular capillaries and microaneurysms in the retinal vasculature, characteristic histopathologic alterations in diabetic retinopathy, have been demonstrated in a variety of long-term animal models of diabetes [13, 23, 57, 58]. Alterations in the retinal vasculature include death of pericytes, thickening of the basement membrane, and adhesion of leukocytes to the endothelium, leading to blockages and capillary dropout, with resultant local ischemia ([16], pp. 442–3; [75]). These phenomena reflect an accelerated death rate in both pericytes and endothelial cells, which is evident before the capillary dropouts [78].

As discussed earlier, one aspect of these biochemical changes in the diabetic retina is the upregulation in levels of VEGF and other growth factors. It should be noted, moreover, VEGF is upregulated by hypoxia [27], a direct consequence of the local nonperfusion and ischemia [18]. Retinal cell types that express VEGF and which respond to hypoxia by its upregulation include all classes of neurons, glia, endothelial cells, pericytes, and the retinal pigment epithelium [6, 12, 26]. As discussed above, VEGF is a key mediator of ocular neovascularization; the following sections discuss recent studies in rodent model systems that have provided further clues as to the mechanisms that mediate its actions in promoting both neovascularization in DR and the breakdown of the BRB.

5.2.3 Regulation of Ischemia-Mediated Retinal Vascularization

Ishida et al. [47] have provided evidence that VEGF plays a key role in ischemia-mediated retinal vascularization, and that this process is both positively and negatively regulated by inflammatory cells. Using a mouse model system approximating retinopathy of prematurity, they induced the formation of an ischemic retina through exposure to a high oxygen environment; when oxygen levels were returned to normal, an aberrant neovascularization ensued, with proliferation into the vitreous. This pathological revascularization was compared to the physiological vascularization that occurs during retinal development [47].

During physiological vascularization, retinal VEGF levels were upregulated, with the $VEGF_{164}$ isoform approximately twice as abundant as $VEGF_{120}$. VEGF levels were further increased during pathological vascularization, with most of the increase occurring with $VEGF_{164}$, which reached levels more than tenfold greater than $VEGF_{120}$. The importance of VEGF for both forms of vascularization was demonstrated by their inhibition following intravitreous injection of a VEGF-receptor-FC fusion protein, which inactivates all VEGF isoforms. Interestingly, injection of pegaptanib, an aptamer which binds $VEGF_{164}$ but not $VEGF_{120}$, was just as effective as the fusion protein in inhibiting pathological vascularization, while having little effect on physiological vascularization [47].

The inflammatory nature of the ischemia-induced pathological vascularization was further evidenced by the accumulation of adherent leukocytes on the leading edge of the pathological vascularization very shortly after induction of growth; these leukocytes were not present at very early stages of physiological vascularization. Moreover, as had been seen during both physiological and pathological vascular pruning (see above), immunohistochemical studies revealed the presence of T-lymphocytes on the pathological vascular fronds. Inactivation of T-lymphocytes with a neutralizing antibody to CD2 led to a significant increase in pathological vascularization, suggesting that they were acting to control its extent [47].

Finally, an additional feature of the pathological vascularization was the presence of monocytes, identified immunohistochemically. Inactivation of monocytes by clodronate liposomes significantly reduced the pathological vascularization, but not physiological vascularization [47]. Similar findings have been reported by several other groups [24, 94, 106]. As circulating monocytes are both attracted by VEGF [11, 17, 107] and also express it [36], especially in conditions of hypoxia [47], this component of the inflammatory response may serve as an amplificatory mechanism, while T-lymphocyte-mediated apoptosis serves as a negative control.

5.2.4 Diabetes Associated Vascular Damage Is Accelerated by Leukostasis and Fas-FasL-Mediated Apoptosis

A series of investigations into the mechanisms underlying vascular damage in diabetes have suggested that an inflammatory process, similar to that which mediates vascular pruning, contributes to breakdown of the BRB, with both VEGF and inflammatory leukocytes again exerting important influences. In the specific case of DR, intravitreal injection of VEGF in monkeys led to many of the features characteristic of the diabetic pathology, including intraretinal and preretinal neovascularization, with the induced blood vessels characterized by endothelial cell hyperplasia, microaneurysm formation, hemorrhage and edema, capillary occlusion and ischemia [105].

Several mechanisms are believed to mediate VEGF's actions in promoting vascular leakage, including increases in endothelial cell fenestrations [91] and damage to tight junctions [8]. Of particular interest in the context of inflammatory mechanisms is VEGF's upregulation of the expression of ICAM-1. Miyamoto et al. [77], using a rodent model of diabetes, reported that the early stages of the disease were characterized by leukocyte entrapment in capillaries, followed by local nonperfusion and leakage. This leukostasis was correlated with increased expression of ICAM-1. Blockade of this ICAM-1 with a monoclonal antibody led to significant reductions in both nonperfusion and leakage [77], with concomitant reductions in diabetes-associated leukostasis and injury or death of endothelial cells [50]. Similar elevations of ICAM-1 and leukocyte numbers have also been observed in human eyes of patients with DR [70], suggesting that the animal models do reflect the human disease in this regard.

These phenomena, together with increases in vascular permeability and breakdown of the blood-retinal barrier, were found to be directly correlated with increases in retinal levels of VEGF in diabetic animals [89]; in addition, these effects were blocked by suppression of VEGF action through administration of a soluble VEGF-receptor construct [53]. In a parallel experiment, intravitreally injected VEGF led to increased retinal expression of ICAM-1, together with increased leukocyte adhesion and increased vascular permeability; all of these responses were significantly reduced by intravenous administration of an antibody to ICAM-1 [77]. Finally, diabetic mice

in which ICAM-1 or its leukocyte-bearing ligand CD18 have been genetically ablated show significant reductions in leukostasis and endothelial cell injury, with reductions in the number of pericyte ghosts and acellular capillaries after 11 months [56].

Taken together, these experiments support a mechanism whereby the diabetes-associated elevation of VEGF leads to increased ICAM-1 expression, followed by increased leukocyte adhesion, endothelial cell injury and increased vascular permeability. As in the case of physiological vascular pruning, Fas-FasL-mediated apoptosis appears to be the final step in the inflammatory damage. Joussen et al. [55] reported that in streptozotocin-induced diabetes in rats, FasL expression was increased in neutrophils, while Fas expression was simultaneously upregulated in the retinal vasculature. In vitro assays revealed that leukocytes from the diabetic rats, but not from controls, could induce endothelial cell apoptosis; moreover, inhibition of FasL-mediated apoptosis in vivo by systemic administration of an anti-FasL antibody significantly inhibited endothelial cell apoptosis and BRB breakdown.

Finally, these phenomena further emphasize the specifically inflammatory nature of the $VEGF_{165}$ isoform. In nondiabetic rats, injection of $VEGF_{164}$ (the rodent equivalent of human $VEGF_{165}$) was approximately twice as potent as the rodent $VEGF_{120}$ isoform in mediating upregulation of ICAM-1, leukocyte adhesion and the induction of BRB breakdown that is characteristic of diabetes [46]. These findings were accompanied by experiments in diabetic rats, where intravitreous injection of pegaptanib, which specifically targets $VEGF_{165/164}$, significantly inhibited leukostasis and BRB breakdown, both in early and in late diabetes [46]. Moreover, this work has proved to have direct clinical application, as intravitreous pegaptanib was found to provide significant clinical benefit in treating both the leakage characteristic of DME [19] as well as retinal revascularization in diabetic eyes ([3]; see also Chapter 19.3.2, this volume).

5.3 Conclusions

While there remain many questions to be answered as to the underlying events that ultimately lead to vision loss in diabetes, there is evidence for a major contribution from inflammatory phenomena. This is true both for the vascular damage occurring in the diabetic retina and for the neovascularization which is induced in response to the ischemia that attends the loss of capillary blood flow. These findings have already been the basis for the demonstrated clinical benefit of pegaptanib, an agent which inactivates the inflammatory isoform of VEGF. There is every reason to expect that further benefit may ensue when other components which regulate these inflammatory events, such as ICAM-1, CD18, TNF-α and the Fas-FasL-mediated apoptosis pathway, are similarly targeted.

References

1. Adamis AP, Miller JW, Bernal MT, D'Amico DJ, Folkman J, Yeo TK, Yeo KT (1994) Increased vascular endothelial growth factor levels in the vitreous of eyes with proliferative diabetic retinopathy. Am J Ophthalmol 118:445–450

2. Adamis AP, Shima DT, Tolentino MJ, Gragoudas ES, Ferrara N, Folkman J, D'Amore PA, Miller JW (1996) Inhibition of VEGF prevents retinal ischemia-associated iris neovascularization in a primate. Arch Ophthalmol 114:66–71

3. Adamis AP, Altaweel M, Bressler NM, Cunningham ET Jr, Davis MD, Goldbaum M, Gonzales C, Guyer DR, Barrett K, Patel M, Macugen Diabetic Retinopathy Study Group (2006) Changes in retinal neovascularization after pegaptanib (Macugen) therapy in diabetic individuals. Ophthalmology 113:23–28

4. Aiello LP, Avery RL, Arrigg PG, Keyt BA, Jampel HD, et al. (1994) Vascular endothelial growth factor in ocular fluid of patients with diabetic retinopathy and other retinal disorders. N Engl J Med 331:1480–1487

5. Aiello LP, Pierce EA, Foley ED, Takagi H, Chen H, Riddle L, Ferrara N, King G, Smith LEH (1995a) Suppression of retinal neovascularization in vivo by inhibition of vascular endothelial growth factor (VEGF) using soluble VEGF-receptor chimeric proteins. Proc Natl Acad Sci USA 92:10457–10461

6. Aiello LP, Northrup JM, Keyt BA, Takagi H, Iwamoto MA (1995b) Hypoxic regulation of vascular endothelial growth factor in retinal cells. Arch Ophthalmol 113:1538–1544

7. Alon T, Hemo I, Itin A, Pe'er J, Stone J, Keshet E (1995) Vascular endothelial growth factor acts as a survival factor for newly formed retinal vessels and has implications for retinopathy of prematurity. Nat Med 1:1024–1028

8. Antonetti D, Barber AJ, Hollinger LA, Wolpert EB, Gardner TW (1999) Vascular endothelial growth factor induces rapid phosphorylation of tight junction proteins occludin and zonula occluden 1. J Biol Chem 274:23463–23467

9. Asahara T, Chen D, Takahashi T, Fujikawa K, Kearney M, Magner M, Yancopoulos GD, Isner JM (1998) Tie2 receptor ligands, angiopoietin-1 and angiopoietin-2, modulate VEGF-induced postnatal neovascularization. Circ Res 83:233–240

10. Asahara T, Takahashi T, Masuda H, Kalka C, Chen D, Iwaguro H, Inai Y, Silver M, Isner JM (1999) VEGF contributes to postnatal neovascularization by mobilizing bone marrow-derived endothelial progenitor cells. EMBO J 18:3964–3972

11. Barleon B, Sozzani S, Zhou D, Weich HA, Mantovani A, Marme D (1996) Migration of human monocytes in response to vascular endothelial growth factor (VEGF) is mediated via the VEGF receptor flt-1. Blood 87:3336–3343

12. Blaauwgeers HG, Holtkamp GM, Rutten H, Witmer AN, Koolwijk P, Partanen TA, et al. (1999) Polarized vascular endothelial growth factor secretion by human retinal pigment epithelium and localization of vascular endothelial growth factor receptors on the inner choriocapillaris. Evidence for a trophic paracrine relation. Am J Pathol 155:421–428

13. Boeri D, Cagliero E, Podesta F, Lorenzi M (1994) Vascular wall von Willebrand factor in human diabetic retinopathy. Invest Ophthalmol Vis Sci 35:600–607

14. Brooks HL Jr, Caballero S Jr, Newell CK, Steinmetz RL, Watson D, Segal MS, Harrison JK, Scott EW, Grant MB (2004) Vitreous levels of vascular endothelial growth factor and stromal-derived factor 1 in patients with diabetic retinopathy and cystoid macular edema before and after intraocular injection of triamcinolone. Arch Ophthalmol 122:1801–1807

15. Brown PO, Botstein D (1999) Exploring the new world of the genome with DNA microarrays. Nat Genetics 21:S33–S37

16. Caldwell RB, Bartoli M, Behzadian MA, El-Remessy AE, Al-Shabrawey M, Platt DH, Caldwell RW (2003) Vascular endothelial growth factor and diabetic retinopathy: pathophysiological mechanisms and treatment perspectives. Diabetes Metab Res Rev 19:442–455

17. Clauss M, Gerlach M, Gerlach H, Brett J, Wang F, Familletti PC, Pan YC, Olander JV, Connolly DT, Stern D (1990) Vascular permeability factor: a tumor-derived polypeptide that induces endothelial cell and monocyte procoagulant activity, and promotes monocyte migration. J Exp Med 172:1535–1545

18. Comer GM, Ciulla TA (2004) Pharmacotherapy for diabetic retinopathy. Curr Opin Ophthalmol 15:508–518

19. Cunningham ET Jr, Adamis AP, Altaweel M, Aiello LP, Bressler NM, D'Amico DJ, Goldbaum M, Guyer DR, Katz B, Patel M, Schwartz SD; Macugen Diabetic Retinopathy Study Group (2005) A phase II randomized double-masked trial of pegaptanib, an anti-vascular endothelial growth factor aptamer, for diabetic macular edema. Ophthalmology 112:1747–1757

20. Detmar M, Brown LF, Schon MP, Elicker BM, Velasco P, Richard L, Fukamura D, Monsky D, Claffey KP, Jain RK (1998) Increased microvascular density and enhanced leukocyte rolling and adhesion in the skin of VEGF transgenic mice. J Invest Dermatol 111:1–6

21. Duggan DJ, Bittner M, Chen Y, Meltzer P, Trent JM (1999) Expression profiling using cDNA microarrays. Nat Genet 21:S10–S14

22. Early Treatment Diabetic Retinopathy Study Research Group (1991) Early Treatment Diabetic Retinopathy Study design and baseline patient characteristics. Ophthalmology 98:741–756

23. Engerman RL, Kern TS (1995) Retinopathy in galactosemic dogs continues to progress after cessation of galactosemia. Arch Ophthalmol 113:355–358

24. Espinosa-Heidmann DG, Suner IJ, Hernandez EP, Monroy D, Csaky KG, Cousins SW (2003) Macrophage depletion diminishes lesion size and severity in experimental choroidal neovascularization. Invest Ophthalmol Vis Sci 44:3586–3592

25. ETDRS Investigators (1992) Aspirin effects on mortality and morbidity in patients with diabetes mellitus. ETDRS Report No. 14. JAMA 268:1292–1300

26. Famiglietti EV, Stopa EG, McGookin ED, Song P, LeBlanc V, Streeten BW (2003) Immunocytochemical localization of vascular endothelial growth factor in neurons and glial cells of human retina. Brain Res 969:195–204

27. Ferrara N (2004) Vascular endothelial growth factor: basic science and clinical progress. Endocr Rev 25:581–611

28. Fong DS, Aiello LP, Ferris FL 3rd, Klein R (2004) Diabetic retinopathy. Diabetes Care 27:2540–2553

29. Franks WA, Limb GA, Stanford MR, Ogilvie J, Wolstencroft

RA, Chignell AH, Dumonde DC (1992) Cytokines in human intraocular inflammation. Curr Eye Res 11:187–191

30. Freeman MR, Schneck FX, Gagnon ML, Corless C, Soker S, Niknejad K, Peoples GE, Klagsbrun M (1995) Peripheral blood T lymphocytes and lymphocytes infiltrating human cancers express vascular endothelial growth factor: a potential role for T cells in angiogenesis. Canc Res 55:4140–4145

31. Funatsu H, Yamashita H, Shimizu E, Kojima R, Hori S (2001) Relationship between vascular endothelial growth factor and interleukin-6 in diabetic retinopathy. Relationship between vascular endothelial growth factor and interleukin-6 in diabetic retinopathy. Retina 21:469–477

32. Funatsu H, Yamashita H, Ikeda T, Nakanishi Y, Kitano S, Hori S (2002) Angiotensin II and vascular endothelial growth factor in the vitreous fluid of patients with diabetic macular edema and other retinal disorders. Am J Ophthalmol 133:537–543

33. Funatsu H, Yamashita H, Sakata K, Noma H, Mimura T, Suzuki M, Eguchi S, Hori S (2005) Vitreous levels of vascular endothelial growth factor and intercellular adhesion molecule 1 are related to diabetic macular edema. Ophthalmology 112:806–816

34. Gardiner TA, Gibson DS, de Gooyer TE, de la Cruz VF, McDonald DM, Stitt AW (2005) Inhibition of tumor necrosis factor-alpha improves physiological angiogenesis and reduces pathological neovascularization in ischemic retinopathy. Am J Pathol 166:637–644

35. Gaudry M, Bregerie O, Andrieu V, El Benna J, Pocidalo MA, Hakim J (1997) Intracellular pool of vascular endothelial growth factor in human neutrophils. Blood 41:4153–4161

36. Grossniklaus HE, Ling JX, Wallace TM, Dithmar S, Lawson DH, Cohen C, et al. (2002) Macrophage and retinal pigment epithelium expression of angiogenic cytokines in choroidal neovascularization. Mol Vis 8:119–126

37. Hanahan D (1997) Signaling vascular morphogenesis and maintenance. Science 277:48–50

38. Hangai M, He S, Hoffmann S, Lim JI, Ryan SJ, Hinton DR (2006) Sequential induction of angiogenic growth factors by TNF-alpha in choroidal endothelial cells. J Neuroimmunol 171:45–56

39. Hawrami K, Hitman GA, Rema M, Snehalatha C, Viswanathan M, Ramachandran A, Mohan V (1996) An association in non-insulin-dependent diabetes mellitus subjects between susceptibility to retinopathy and tumor necrosis factor polymorphism. Hum Immunol 46:49–54

40. Holash J, Maisonpierre PC, Compton D, Boland P, Alexander CR, Zagzag D, Yancopoulos GD, Wiegand SJ (1999a) Vessel cooption, regression, and growth in tumors mediated by angiopoietins and VEGF. Science 284:1994–1998

41. Holash J, Wiegand SJ, Yancopoulos GD (1999b) New model of tumor angiogenesis: dynamic balance between vessel regression and growth mediated by angiopoietins and VEGF. Oncogene 18:5356–5362

42. Honda M, Sakamoto T, Ishibashi T, Inomata H, Ueno H (2000) Experimental subretinal neovascularization is inhibited by adenovirus-mediated soluble VEGF/flt-1 receptor gene transfection: a role of VEGF and possible treatment for SRN in age-related macular degeneration. Gene Ther 7:978–985

43. Horiuchi T, Weller PF (1997) Expression of vascular endothelial growth factor by human eosinophils: upregulation by granulocyte macrophage colony-stimulating factor and interleukin-5. Am J Respir Cell Mol Biol 17:70–77

44. Iijima K, Yoshikawa N, Connolly DT, Nakamura H (1993) Human mesangial cells and peripheral blood mononuclear cells produce vascular permeability factor. Kidney Int 44:959–966

45. Ishida S, Yamashiro K, Usui T, Kaji Y, Ogura Y, Hida T, Honda Y, Oguchi Y, Adamis AP (2003a) Leukocytes mediate retinal vascular remodeling during development and vaso-obliteration in disease. Nat Med 9:781–788

46. Ishida S, Usui T, Yamashiro K, Kaji Y, Ahmed E, Carrasquillo KG, et al. (2003b) Vegf164 is proinflammatory in the diabetic retina. Invest Ophthalmol Vis Sci 44:2155–2162

47. Ishida S, Usui T, Yamashiro K, Kaji Y, Amano S, Ogura Y, Hida T, Oguchi Y, Ambati J, Miller JW, Gragoudas ES, Ng YS, D'Amore PA, Shima DT, Adamis AP (2003c) VEGF164-mediated inflammation is required for pathological, but not physiological, ischemia-induced retinal neovascularization. J Exp Med 198:483–489

48. Iyer VR, Eisen MB, Ross DT, Schuler G, Moore T, Lee JCF, Trent M, Staudt LM, Hudson JJ, Boguski MS, Lashkari DL, Shalon D, Botstein D, Brown PO (1999) The transcriptional program in response to human fibroblasts to serum. Science 293:83–87

49. Jain RK, Munn LL (2000) Leaky vessels? Call Ang1! [news]. Nat Med 6:131–132

50. Joussen AM, Murata T, Tsujikawa A, Kirchhof B, Bursell SE, Adamis AP (2001a) Leukocyte-mediated endothelial cell injury and death in the diabetic retina. Am J Pathol 158:147–152

51. Joussen AM, Huang S, Poulaki V, Camphausen K, Beecken WD, Kirchhof B, Adamis AP (2001b) In vivo retinal gene expression in early diabetes. Invest Ophthalmol Vis Sci 42:3047–3057

52. Joussen AM, Poulaki V, Tsujikawa A, Qin W, Qaum T, Xu Q, Moromizato Y, Bursell SE, Wiegand SJ, Rudge J, Ioffe E, Yancopoulos GD, Adamis AP (2002a) Suppression of diabetic retinopathy with angiopoietin-1. Am J Pathol 160:1683–1693

53. Joussen AM, Poulaki V, Qin W, Kirchhof B, Mitsiades N, Wiegand SJ, Rudge J, Yancopoulos GD, Adamis AP (2002b) Retinal vascular endothelial growth factor induces intercellular adhesion molecule-1 and endothelial nitric oxide synthase expression and initiates early diabetic retinal leukocyte adhesion in vivo. Am J Pathol 160:501–509

54. Joussen AM, Poulaki V, Mitsiades N, Kirchhof B, Koizumi K, Dohmen S, Adamis AP (2002c) Nonsteroidal anti-inflammatory drugs prevent early diabetic retinopathy via TNF-alpha suppression. FASEB J 16:438–440

55. Joussen AM, Poulaki V, Mitsiades N, Cai WY, Suzuma I, Pak J, Ju ST, Rook SL, Esser P, Mitsiades CS, Kirchhof B, Adamis AP, Aiello LP (2003) Suppression of Fas-FasL-induced endothelial cell apoptosis prevents diabetic blood-retinal barrier breakdown in a model of streptozotocin-induced diabetes. FASEB J 17:76–78

56. Joussen AM, Poulaki V, Le ML, Koizumi K, Esser C, Janicki H, Schraermeyer U, Kociok N, Fauser S, Kirchhof B, Kern TS, Adamis AP (2004) A central role for inflammation in the pathogenesis of diabetic retinopathy. FASEB J 18:1450–1452

57. Kern TS, Engerman RL (1995) Galactose-induced retinal microangiopathy in rats. Invest Ophthalmol Vis Sci 36:490–496

58. Kern TS, Engerman RL (1996) A mouse model of diabetic retinopathy. Arch Ophthalmol 114:986–990

59. Khosla PK, Seth V, Tiwari HK, Saraya AK (1982) Effect of aspirin on platelet aggregation in diabetes mellitus. Diabetologia 23(2):104–7

60. Kucia M, Jankowski K, Reca R, Wysoczynski M, Bandura L, Allendorf DJ, Zhang J, Ratajczak J, Ratajczak MZ (2004) CXCR4-SDF-1 signalling, locomotion, chemotaxis and adhesion. J Mol Histol 35:233–245

61. Kuroki M, Voest EE, Amano S, Beerepoot LV, Takashima S, Tolentino M, Kim RY, Rohan RM, Colby KA, Yeo KT, Adamis AP (1996) Reactive oxygen intermediates increase vascular endothelial growth factor expression in vitro and in vivo. J Clin Invest 98:1667–1675

62. Leung DW, Cachianes G, Kuang WJ, Goeddel DV, Ferrara N (1989) Vascular endothelial growth factor is a secreted angiogenic mitogen. Science 246:1306–1309

63. Lim JW, Kim H, Kim KH (2001) Nuclear factor-kappaB regulates cyclooxygenase-2 expression and cell proliferation in human gastric cancer cells. Lab Invest 81:349–360

64. Limb GA, Chignell AH, Green W, LeRoy F, Dumonde DC (1996) Distribution of TNF alpha and its reactive vascular adhesion molecules in fibrovascular membranes of proliferative diabetic retinopathy. Br J Ophthalmol 80:168–173

65. Limb GA, Hickman-Casey J, Hollifield RD, Chignell AH (1999a) Vascular adhesion molecules in vitreous from eyes with proliferative diabetic retinopathy. Invest Ophthalmol Vis Sci 40:2453–2457

66. Limb GA, Webster L, Soomro H, Janikoun S, Shilling J (1999b) Platelet expression of tumour necrosis factor-alpha (TNF-alpha), TNF receptors and intercellular adhesion molecule-1 (ICAM-1) in patients with proliferative diabetic retinopathy. Clin Exp Immunol 118:213–218

67. Lu M, Kuroki M, Amano S, Tolentino M, Keough K, Kim I, Bucala R, Adamis AP (1998) Advanced glycation end products increase retinal vascular endothelial growth factor expression. J Clin Invest 101:1219–1224

68. Lu M, Perez V, Ma N, Miyamoto K, Peng HB, Liao JK, Adamis AP (1999) VEGF increases retinal vascular ICAM-1 expression in vivo. Invest Ophthalmol Vis Sci 40:1808–1812

69. McCormack K (1998) Roles of COX-1 and COX-2. J Rheumatol 25:2279–2281

70. McLeod DS, Lefer DJ, Merges C, Lutty GA (1995) Enhanced expression of intracellular adhesion molecule-1 and P-selectin in the diabetic human retina and choroid. Am J Pathol 147:642–653

71. Maisonpierre PC, Suri C, Jones PF, Bartunkova S, Wiegand SJ, Radziejewski C, Compton D, McClain J, Aldrich TH, Papadopoulos N, Daly TJ, Davis S, Sato TN, Yancopoulos GD (1997) Angiopoietin-2, a natural antagonist for Tie2 that disrupts in vivo angiogenesis. Science 277:55–60

72. Melder RJ, Koenig GC, Witwer BP, Safabakhsh N, Munn LL, Jain RK (1996) During angiogenesis, vascular endothelial growth factor and basic fibroblast growth factor regulate natural killer cell adhesion to tumor endothelium. Nat Med 2:992–997

73. Meleth AD, Agron E, Chan CC, Reed GF, Arora K, Byrnes G, Csaky KG, Ferris FL 3rd, Chew EY (2005) Serum inflammatory markers in diabetic retinopathy. Invest Ophthalmol Vis Sci 46:4295–4301

74. Miolata J, Maciewiez R, Kendrew J, Fledmann M, Paleolog E (2000) Treatment with soluble VEGF receptor reduces disease severity in murine collagen-induced arthritis. Lab Invest 80:1195–1205

75. Miyamoto K, Hiroshiba N, Tsujikawa A, Ogura Y (1998) In vivo demonstration of increased leukocyte entrapment in retinal microcirculation of diabetic rats Invest Ophthalmol Vis Sci 39:2190–2194

76. Miyamoto K, Khosrof S, Bursell SE, Rohan R, Murata T, Clermont A, Aiello LP, Ogura Y, Adamis AP (1999) Prevention of leukostasis and vascular leakage in streptozotocin-induced diabetic retinopathy via intercellular adhesion molecule-1 inhibition. Proc Natl Acad Sci USA 96:10836–10841

77. Miyamoto K, Khosrof S, Bursell SE, Moromizato Y, Aiello LP, Ogura Y, Adamis AP (2000) Vascular endothelial growth factor-induced retinal vascular permeability is mediated by intercellular adhesion molecule-1 (ICAM-1). Am J Pathol 156:1733–1739

78. Mizutani M, Kern TS, Lorenzi M (1996) Accelerated death of retinal microvascular cells in human and experimental diabetic retinopathy. J Clin Invest 97:2883–2890

79. Mohle R, Green D, Moore MA, Nachman RL, Rafii S (1997) Constitutive production and thrombin-induced release of vascular endothelial growth factor by human megakaryocytes and platelets. Proc Natl Acad Sci USA 94:663–668

80. Nagata S, Golstein P (1995) The Fas death factor. Science 267:1449–1456

81. Ng EW, Adamis AP (2005) Targeting angiogenesis, the underlying disorder in neovascular age-related macular degeneration. Can J Ophthalmol 40:352–368

82. Nishihira J (1998) Novel pathophysiological aspects of macrophage migration inhibitory factor (review). Int J Mol Med 2:17–28

83. Oh H, Takagi H, Suzuma K, Otani A, Matsumura M, Honda Y (1999a) Hypoxia and vascular endothelial growth factor selectively up-regulate angiopoietin-2 in bovine microvascular endothelial cells. J Biol Chem 274:15732–15739

84. Oh H, Takagi H, Takagi C, Suzuma K, Otani A, Ishida K, Matsumura M, Ogura Y, Honda Y (1999b) The potential angiogenic role of macrophages in the formation of choroidal neovascular membranes. Invest Ophthalmol Vis Sci 40:1891–1898

85. Pairet M, Engelhardt G (1996) Distinct isoforms (COX-1 and COX-2) of cyclooxygenase: possible physiological and therapeutic implications. Fundam Clin Pharmacol 10:1–11

86. Pillinger MH, Capodici C, Rosenthal P, Kheterpal N, Hanft S, Philips MR, Weissmann G (1998) Modes of action of aspirin-like drugs: salycylates inhibit erk activation and integrin-dependent neutrophil adhesion. Proc Natl Acad Sci USA 95:14540–14545

87. Plate KH, Breier G, Weich HA, Risau W (1992) Vascular endothelial growth factor is a potential tumour angiogenesis factor in human gliomas in vivo. Nature. 359: 845–888

88. Powell EDU, Field RA (1964) Diabetic retinopathy in rheumatoid arthritis. Lancet 2:17–18

89. Qaum T, Xu Q, Joussen AM, Clemens MW, Qin W, Miyamoto K, Hassessian H, Wiegand SJ, Rudge J, Yancopoulos GD, Adamis AP (2001) VEGF-initiated blood-retinal barrier breakdown in early diabetes. Invest Ophthalmol Vis Sci 42:2408–2413

90. Radisavljevic Z, Avraham H, Avraham S (2000) Vascular endothelial growth factor upregulates ICAM-1 expression via the phosphtidylinositol3 OH-kinase/AKT/nitric oxide pathway and modulates migration of brain microvascular endothelial cells. J Biol Chem 275:20770–20774

91. Roberts WG, Palade GE (1995) Increased microvascular permeability and endothelial fenestration induced by vascular endothelial growth factor. J Cell Sci 108:2369–2379

92. Safieh-Garabedian B, Dardenne M, Kanaan SA, Atweh SF, Jabbur SJ, Saade NE (2000) The role of cytokines and pros-

taglandin-E(2) in thymulin induced hyperalgesia. Neuro-pharmacology 39:1653–1661

93. Sakane S, Nishihira J, Hirokawa J, Yoshimura H, Honda T, Aoki K, Tagami S, Kawakami Y (1999) Regulation of macrophage migration inhibitory factor (MIF) expression by glucose and insulin in adipocytes in vitro. Mol Med 5:361–371

94. Sakurai E, Anand A, Ambati BK, van Rooijen N, Ambati J (2003) Macrophage depletion inhibits experimental choroidal neovascularization. Invest Ophthalmol Vis Sci 44:3578–3585

95. Sato A, Iwama A, Takakura N, Nishio H, Yancopoulos GD, Suda T (1998) Characterization of TEK receptor tyrosine kinase and its ligands, Angiopoietins, in human hematopoietic progenitor cells. Int Immunol 10:1217–1227

96. Schröder S, Palinski W, Schmid-Schönbein GW (1991) Activated monocytes and granulocytes, capillary nonperfusion, and neovascularization in diabetic retinopathy. Am J Pathol 139:81–100

97. Senger DR, Connolly DT, Van de Water L, Feder J, Dvorak HF (1990) Purification and NH2-terminal amino acid sequence of guinea pig tumor-secreted vascular permeability factor. Cancer Res 50:1774–1778

98. Sfikakis PP, Markomichelakis N, Theodossiadis GP, Grigoropoulos V, Katsilambros N, Theodossiadis PG (2005) Regression of sight-threatening macular edema in type 2 diabetes following treatment with the anti-tumor necrosis factor monoclonal antibody infliximab. Diabetes Care 28:445–447

99. Shweiki D, Itin A, Soffer D, Keshet E (1992) Vascular endothelial growth factor induced by hypoxia may mediate hypoxia-initiated angiogenesis. Nature 359:843–845

100. Southern E, Mir K, Schepinov M (1999) Molecular interactions on microarrays. Nat Genet 21:S5–S9

101. Spranger J, Meyer-Schwickerath R, Klein M, Schatz H, Pfeiffer A (1995) TNF-alpha Konzentration in Glaskörper. Anstief bei neovaskulären Erkrankungen und proliferativer diabetischer Retinopathie. Med Klin 90:134–137

102. Suri C, Jones PF, Patan S, Bartunkova S, Maisonpierre PC, Davis S, Sato TN, Yancopoulos GD (1996) Requisite role of angiopoietin-1, a ligand for the TIE2 receptor, during embryonic angiogenesis. Cell 87:1171–1180

103. Thurston G, Suri C, Smith K, McClain J, Sato TN, Yancopoulos GD, McDonald DM (1999) Leakage-resistant blood vessels in mice transgenically overexpressing angiopoietin-1. Science 286:2511–2514

104. Thurston G, Rudge JS, Ioffe E, Zhou H, Ross L, Croll SD, Glazer N, Holash J, McDonald DM, Yancopoulos GD (2000) Angiopoietin-1 protects the adult vasculature against plasma leakage. Nat Med 6:460–463

105. Tolentino MJ, Miller JW, Gragoudas ES, Jakobiec FA, Flynn E, Chatzistefanou K, Ferrara N, Adamis AP (1996) Intravitreous injections of vascular endothelial growth factor produce retinal ischemia and microangiopathy in an adult primate. Ophthalmology 103:1820–1828

106. Tsutsumi C, Sonoda KH, Egashira K, Qiao H, Hisatomi T, Nakao S, Ishibashi M, Charo IF, Sakamoto T, Murata T, Ishibashi T (2003) The critical role of ocular-infiltrating macrophages in the development of choroidal neovascularization. J Leukoc Biol 74:25–32

107. Usui T, Ishida S, Yamashiro K, Kaji Y, Poulaki V, Moore J, et al. (2004) Vegf 164(165) as the pathological isoform: Differential leukocyte and endothelial responses through vegfr1 and vegfr2. Invest Ophthalmol Vis Sci 45:368–374

108. Wallach D (1997) Cell death induction by TNF: a matter of self control. Trends Biochem Sci 22:107–109

109. Watanabe D, Suzuma K, Suzuma I, Ohashi H, Ojima T, Kurimoto M, Murakami T, Kimura T, Takagi H (2005a) Vitreous levels of angiopoietin 2 and vascular endothelial growth factor in patients with proliferative diabetic retinopathy. Am J Ophthalmol 139:476–481

110. Watanabe D, Suzuma K, Matsui S, Kurimoto M, Kiryu J, Kita M, Suzuma I, Ohashi H, Ojima T, Murakami T, Kobayashi T, Masuda S, Nagao M, Yoshimura N, Takagi H (2005b) Erythropoietin as a retinal angiogenic factor in proliferative diabetic retinopathy. N Engl J Med 353:782–792

111. Weissmann G (1991) Aspirin. Sci Am 264:84–96

112. Wong MP, Chan SY, Fu KH, Leung SY, Cheung N, Yuen ST, Chung LP (2000) The angiopoietins, tie2 and vascular endothelial growth factor are differentially expressed in the transformation of normal lung to non-small cell lung carcinomas. Lung Cancer 29:11–22

113. Wu QD, Wang JH, Bouchier-Hayes D, Redmond HP (2000) Neutrophil-induced transmigration of monocytes is attenuated in patients with diabetes mellitus: a potential predictor for the individual capacity to develop collaterals. Circulation 102:185–190

114. Xu Q, Qaum T, Adamis AP (2001) Sensitive blood-retinal barrier breakdown quantitation using Evans blue. Invest Ophthalmol Vis Sci 42:789–794

115. Zodochne DW, Verge VM, Cheng C, Hoke A, Jolley C, Thomsen K, Rubin I, Laurtzen M (2000) Nitric oxide synthase activity and expression in experimental diabetic neuropathy. J Neuropathol Exp Neurol 59:798–807

6 The Neuronal Influence on Retinal Vascular Pathology

A.J. BARBER, H.D. VAN GUILDER, M.J. GASTINGER

Core Messages

- Diabetic retinopathy is clinically characterized as a vascular disease
- Increased vascular permeability is a critical pathology that leads to macular edema
- Vascular permeability indicates a loss of the blood-retina barrier phenotype
- The blood-retina barrier is induced by signals from neural tissue
- Diabetes increases apoptosis and degeneration of retinal neurons
- Neuroglial cells of the retina exhibit a stress reaction in response to diabetes
- Vascular endothelial growth factor is produced by the neural retina
- Vascular endothelial growth factor increases retinal vascular permeability
- Diabetic retinopathy should be considered a neurovascular degeneration, involving both neural and vascular components of the retina

6.1 Introduction

Much of the effort to understand diabetic retinopathy has focused on vascular pathology. Due to the influence of diabetes on systemic physiology, it is thought that many changes in retinal vascular cells likely stem from direct biochemical disruptions such as hyperglycemia. Other contributing factors may include circulating cytokines, advanced glycation end products, cholesterol, albumin, and electrolytes, and more complex functional changes such as reduced elasticity in erythrocytes and other blood cells. These and other changes could be responsible for generating the vascular pathologies that have been well established in diabetic retinopathy. There is increasing evidence, however, for involvement of the neural elements of the retina, the neurons and glial cells, and it is no longer clear if the pathological changes in these cells result from vascular dysfunction, such as the reduction in effectiveness of the blood-retinal barrier, or if diabetic physiology induces neural pathology, which in turn gives rise to vascular changes. A third possibility is that early vascular and neural responses to diabetes are independent phenomena that are triggered by different factors.

This chapter will review evidence for early neural changes in the retinas of humans with diabetes as well as diabetic animal models. We suggest the hypothesis that chronic neurodegeneration of the retina begins soon after the onset of diabetes, and that these changes may be responsible for changes in visual function, as well as the loss of integrity of the blood-retina barrier. Neurodegeneration and vascular pathology, particularly increases in vascular permeability, are two characteristics of diabetic retinopathy that are established in several models of diabetes. Chronic neurodegeneration appears to have an early onset in animal models of diabetes and may explain some of the functional deficits in both humans and animals [12]. Loss of integrity of the blood-retinal barrier, indicated by increased permeability to water, solutes and proteins, has been recognized as a hallmark of diabetes for many years [28, 111]. The interaction between these two consequences of diabetes has not been investigated, however, and their causal relationship is not well understood. We explore some of the evidence that retinal neurodegeneration and vascular permeability in diabetes are linked. We also briefly outline some of the published methods used by our laboratory to assess vascular permeability and apoptosis in the retina.

6.2 The Phenotype of the Retinal Vasculature Is Determined by the Tissue It Serves

It is generally accepted that the development and maintenance of blood vessels are directed by local signals from the tissues they perfuse. This is particularly the case of the central nervous system, in which the vascular endothelial cells express proteins that form specialized tight junctions, which render the blood vessels impermeable to water and solutes in the lumen [93]. This is achieved by the expression of tight junction proteins such as occludin, claudins, zonula occludins, and several others, which make a tight seal between vascular endothelial cells [10, 25, 103]. (For more information on tight junctions see Chapter 8.1, this volume.) The vascular endothelial cells also express many unique transporters that regulate the entry of other substances, such as amino acids, into the sensitive neural parenchyma, and express unique basement membrane characteristics [14, 75] (Fig. 6.1). Together these properties establish physiological boundaries, termed the blood-brain and blood-retina barrier [26, 27].

The unique phenotype of the vasculature in the brain and retina is induced by the neural tissue. In a classic study on blood-brain barrier development, Stewart and Wiley demonstrated that transplanting avascular tissue from neonatal quail brain into the coelomic cavity of chick embryos causes the endothelial cells that grow into the quail neural tissue and adopt blood-brain barrier properties, becoming less

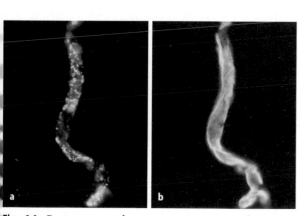

Fig. 6.1. Basement membrane components contribute to a unique phenotype of vascular barriers. Sections from rat brain and retina were labeled by immunohistochemistry for von Willebrand factor and agrin. **a** von Willebrand factor is abundant in all blood vessels and appears as a punctuate cytoplasmic antigen. **b** The same vessel was also labeled for agrin, a basement membrane glycoprotein that is only expressed on vessels in tissue that have a blood-tissue barrier. The unique properties of vessels that form blood-tissue barriers are induced by signals from the tissue that they serve. (From Fig. 1 in [14]. Reproduced with permission of *Developmental Dynamics*)

permeable to circulating dye. Conversely, chick brain vessels lose their barrier properties after growing into quail coelomic tissue grafts [94]. It is now widely accepted that the signals required to induce the barrier phenotype in vascular endothelial cells of the central nervous system are derived from the glial cells, although the specific identity of these signals is still unclear [49, 89, 101].

The retinal vasculature has a similar phenotype to that of the brain [9]. The components comprising the vascular endothelial basement membrane in the retina and brain are also similar [14], as is the neural source of the barrier phenotype. For example, retinal Müller cells injected into the iris of rat eyes causes the vasculature in that region to adopt barrier properties [100]. Also, culture media from astrocytes increase the expression of the tight junction protein, ZO-1, in primary bovine retinal endothelial cells [35, 36]. These data suggest that glia manufacture a soluble factor that induces the expression of the features of the blood-retina barrier in vascular endothelial cells.

6.3 Diabetes Causes a Loss of the Blood-Retina Barrier Phenotype

It is well established that diabetes increases the permeability of the vasculature in the retina [28]. The most direct method that has been used to measure retinal vascular permeability is by imaging fluorescein leaking from the blood vessels into the vitreous and retinal tissue [31, 51, 60]. Other approaches that have been used to study this phenomenon in animal models include measuring the leak of radiolabeled tracers, Evans blue dye, and albumin [83, 106, 116]. Immunohistochemical approaches in specimens from both humans and rats with diabetes have also identified regions around major blood vessels that contain more albumin, suggesting increased permeability to protein as well as water and small molecules [107, 108, 109]. This increase occurs soon after the onset of diabetes, beginning in the larger superficial vessels and spreading to the capillary bed [13, 80]. This suggests that diabetes induces a progressive loss of barrier properties in the retinal vasculature.

6.4 The Increase in Vascular Permeability Is Likely Due to an Increase in the Expression of VEGF

The evidence currently suggests that the primary factor increasing blood-retina barrier permeability in diabetes is the potent growth and proliferation protein, vascular endothelial growth factor (VEGF). VEGF-induced increases in retinal vascular permeability occur through regulation of vascular endothelial tight junction proteins [9]. VEGF content is increased in the vitreous of patients with proliferative diabetic retinopathy [4, 6]. Similarly, diabetes increases the expression of both VEGF mRNA and protein in STZ-diabetic rat retina within as little as 1 week of the onset of diabetes, correlating with increases in vascular permeability that can be prevented by treatment with VEGF Trap, a high-affinity VEGF receptor blocker [67]. This suggests that VEGF is a primary inducer of retina vascular permeability in diabetes.

6.5 VEGF Originates from the Neural Tissue of the Retina in Diabetes

The primary source of retinal VEGF in diabetes remains to be identified, but recent research utilizing immunofluorescent microscopy indicates that this growth factor originates from specific cell types within the neural retina. The ganglion cells and amacrine cells appear to be potent sources of VEGF, which has also been detected in the nerve fiber layer, the retinal pigment epithelial cell layer, the outer plexiform layer, the photoreceptor layer and the inner retina [34]. VEGF content is known to be increased in diabetic retinopathy and in many co-morbid complications including ischemia and oxidative stress [32, 72]. Immunohistochemical studies indicate that neurons and glial cells are the main source of VEGF in diabetes in both humans and rodent models [7, 110]. VEGF protein detected by immunohistochemistry appears to accumulate at the vascular endothelial cell walls [61]. mRNA transcript encoding VEGF protein, however, is localized to discrete neuronal populations, but not endothelial cells, suggesting that VEGF protein is synthesized in neurons and migrates to the blood vessel walls [40, 67]. These data indicate that the neural tissue of the retina is the most likely source of VEGF, and that diabetes-induced increases in VEGF expression may be caused by the response of neurons and glia to physiological stresses imposed by diabetes.

6.6 Injury and Neurodegeneration in the Brain Are Accompanied by Increased VEGF Expression and Vascular Abnormalities

Neurodegenerative diseases often involve both neural and vascular pathology. Growing evidence suggests that changes within the neural tissue and vasculature are intimately linked. While VEGF is identified as playing a central role in diabetic retinopathy, it is also well recognized in studies of acute neurodegeneration stemming from cerebral ischemia and stroke [53]. VEGF mRNA expression is increased in

an animal model of brain ischemia induced by experimental heart attack, increasing between 24 and 48 h after the ischemic event and returning to normal about 7 days later [78]. VEGF expression is also known to increase in humans following transient stroke, and similar results have been obtained after transient forebrain ischemia in rats, as well as in retinal ischemia [57, 88]. Cerebral ischemia-reperfusion is followed by an acute increase in vascular permeability with a distinctive time course [48, 77, 112]. It is likely that VEGF plays a role in this transient loss of blood-brain barrier integrity [57, 78], but may also be responsible for some degree of neuroprotection, because blocking VEGF increases the infarct volume in a rat model of brain ischemia [117]. Topical application of VEGF after cerebral ischemia also reduces infarct volume, suggesting that the growth factor may have dual roles, including beneficial effects on neuronal cell survival [44].

VEGF also appears to be involved in chronic neurodegenerative diseases. Increases in VEGF expression have been noted in brain tissue from patients with Alzheimer's disease, with the immunoreactivity being most prominent within brain astrocytes and the walls of large blood vessels [53]. VEGF and the inflammatory cytokine, TGF-beta, are also found to be increased in the cerebrospinal fluid of patients with Alzheimer's disease as well as those with vascular dementia [98]. Immunoreactivity for VEGF colocalizes with plaques containing beta-amyloid in Alzheimer's brains, again suggesting that it may have an important role in the neuropathogenesis of the disease [118]. One possible outcome of elevated VEGF in these brains is the reduction of neuronal apoptosis, because it can be neuroprotective in some circumstances [50, 73, 91, 95–97, 113, 117]. A genetic study demonstrated that two single-nucleotide polymorphisms within the promoter region of the VEGF gene have a significantly higher incidence in a population with Alzheimer's disease, compared to a healthy control group, suggesting that some VEGF promoter polymorphisms are either more toxic to the brain, or less able to prevent neurodegeneration [29]. These data support a role for VEGF in other neurodegenerative diseases.

It is also becoming increasingly clear that vascular pathology occurs in Alzheimer's disease. Vascular lesions and breakdown of the blood-brain barrier may contribute to cerebral degeneration [52]. The vascular lesions include loss of endothelial cell markers, increased vascular inflammatory markers, basement membrane thickening and intracerebral hemorrhages and microinfarcts [79]. It is possible that these changes are accompanied by transient increases in cerebrovascular permeability, because serum albumin and IgG immunoreactivity are also evident in amyloid plaques [114]. The vascular lesions of Alzheimer's disease are less well established than the neuronal pathology, presumably because of the difficulties in examining the three-dimensional structure of the vascular structure. These abnormalities are, however, reminiscent of the vascular lesions well established in the retinas as a result of diabetes, suggesting that similar mechanisms may be at work in the two diseases.

In central nervous system degenerations like hypoxic ischemia and Alzheimer's disease, the loss of neurons by apoptosis is well established. Data suggest that important vascular complications may contribute to these diseases, some of which may be mediated by VEGF. A similar relationship between the neural tissue and the blood vessels of the retina likely exists in diabetic retinopathy.

6.7 Diabetes Increases Neural Apoptosis, Leading to a Cumulative Reduction in the Thickness of the Inner Layers of the Retina

Neurodegeneration of the retina can provide a direct explanation for vision loss in diabetic retinopathy. Numerous studies now show that diabetes increases retinal apoptosis, reduces the number of retinal ganglion cells, and causes atrophy of the inner plexiform layer, which is composed entirely of neuronal and glial projections. Diabetes also reduces survival of retinal neurons [15, 43], induces reactive changes typical of pathological insults in retinal glial cells [16, 64, 83], and causes the appearance of abnormal swellings on centrifugal axons [37], giving rise to the concept that neurodegeneration occurs in diabetic retinopathy [12].

Recent studies of retinal cell loss in diabetes have quantified apoptosis by the number of TUNEL-positive cells in whole and sectioned retinas from STZ-diabetic rats, compared to age-matched controls. Increased numbers of apoptosis-positive cells are consistently counted in whole rat retinas after 1, 3, 6 and 12 months of STZ diabetes (Fig. 6.2). Similar data have been obtained from two postmortem retinas of humans with diabetes [15]. Despite several reports of increased apoptosis in the diabetic retina, the specific identities of apoptotic cells are less well established. Studies have reported TUNEL-positive labeling in trypsin-digested retinas, in which only the vasculature remains, implying that the cells are of vascular origin [54, 63]. Interestingly, many apoptotic cells detected by TUNEL staining and active caspase-3 immunoreactivity in STZ-diabetic rat retina are distinct from the vasculature, indicating a neural identity [33, 39] (Fig. 6.3). Although TUNEL-positive cells have been detected in the outer nuclear

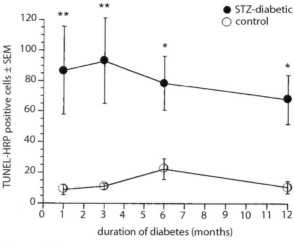

Fig. 6.2. Diabetes increases apoptosis in the retina. Apoptosis was quantified in whole retinas of STZ rats after 1, 3, 6 and 12 months of diabetes. The number of apoptotic cells, identified by TUNEL, was standardized to the surface area of each whole-mount retina, calculated by image analysis. The retinas from STZ-diabetic rats contained significantly more TUNEL-positive cells at all time points studied, compared to the age-matched control rats ($*p < 0.01$, $**p < 0.001$). (From Fig. 4 in [15]. Reproduced with permission of the American Society for Clinical Investigation)

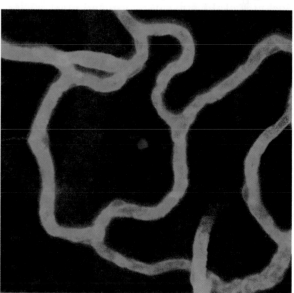

Fig. 6.3. Apoptosis in the neural retina. A whole-mounted rat retina was labeled for the vascular basement membrane antigen, agrin (*green*) and the active form of the apoptosis enzyme, caspase-3 (*red*). The majority of cells with positive immunoreactivity for active caspase-3 were spatially separated from the vasculature, indicating that they were not vascular cells, and more likely to be part of the neural tissue

layer, the most prevalent diabetes-induced cell loss appears to occur in the inner retina, where ganglion, bipolar and amacrine cells are depleted [33, 38, 76, 86]. For example, in rats after 7.5 months of STZ diabetes there is a 10% reduction in large cell bodies in

6

the retinal ganglion cell layer [15]. Furthermore, diabetes significantly reduces the thickness of the inner plexiform and inner nuclear layers. The thickness of the inner plexiform and inner nuclear layers was diminished by 22% and 14% respectively. Together these data suggest a loss of neurons and their processes from the inner part of the retina (Fig. 6.4). Additional studies demonstrate that 1 month of STZ diabetes decreases the thickness of the inner plexiform layer by 10% in Sprague-Dawley rats and by nearly 16% in Brown Norway rats [5]. Apoptosis indicated by TUNEL-positive and active caspase-3-immunoreactive cells in the ganglion cell layer has been further confirmed by the appearance of fragmented DNA in electron micrographs. These data are consistent with the observation that the number of axons in the optic nerve decreases in STZ-diabetic rats, implying a loss of retinal ganglion cells [85]. Collectively, these data suggest that there is a predominant loss of retinal ganglion, bipolar and amacrine neurons, and the projections needed to communicate between these cells in the retinas of diabetic rats.

Diabetes also increases apoptosis and reduces the thickness of the inner plexiform and inner nuclear layers in the retinas of Ins2^{Akita} diabetic mice [18]. This mouse is spontaneously diabetic due to a point mutation on the second insulin gene, causing degeneration of pancreatic beta cells. The mice develop significant hyperglycemia 4–5 weeks after birth [119]. After 4 weeks of diabetes in these mice, the number of cells detected and quantified by counting cells labeled with an antibody to the active form of caspase-3 is significantly higher in retinas from diabetic mice compared to control littermates, suggesting an increase in apoptosis [18].

Morphological changes in the Ins2^{Akita} mouse retinas are also similar to those observed in STZ-diabetic rats, including reduced inner plexiform layer thickness. In mice that had been diabetic for 22 weeks there was a 16.7% reduction in the thickness of the central part of the inner plexiform layer and a 27% reduction in the peripheral part of this layer compared to non-diabetic littermates in both the central and peripheral retina [18] (Fig. 6.4B, C). A similar study on STZ-diabetic mice also reported a loss of 20–25% of the cell bodies in the ganglion cell layer after 14 weeks of diabetes [62].

These studies demonstrate that retinal degeneration is an effect of diabetic pathology rather than the method of diabetes induction, as they occur in multiple species and in both drug-induced and transgenic models. Collectively, these data suggest that diabetes leads to loss of neurons in the inner retina, as well as the projections necessary for communication between these cells. Importantly, the increase in apo-

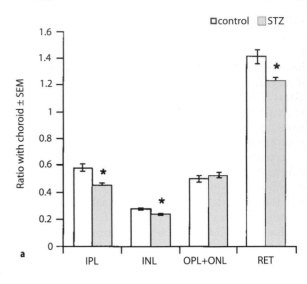

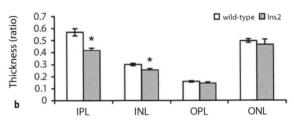

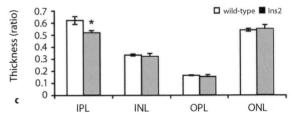

Fig. 6.4. Degeneration of the inner retina. The layers of the inner retina become less thick over long periods of diabetes in both rats and mice. **a** Retinal morphology was measured in sections from streptozotocin-diabetic rats after 7.5 months. The thickness of the inner plexiform and inner nuclear layers was significantly less in STZ rats compared to controls (*$p < 0.001$), while there was no difference in the outer layers (OPL+ONL). **b** In Ins2^{Akita} mice, after 5.5 months of hyperglycemia, the thickness of IPL and INL was significantly less in the peripheral retina compared to littermate controls (*$p < 0.05$). **c** The thickness of the IPL in the central part of the retina was also significantly less in the Ins2^{Akita} diabetic mice, compared to controls (*$p < 0.05$), but the thickness of the central INL was unchanged by diabetes. *IPL* inner plexiform layer, *INL* inner nuclear layer, *OPL* outer plexiform layer, *ONL* outer nuclear layer, *RET* total retinal thickness, *STZ* streptozotocin-diabetic rats, *Ins2* Ins2^{Akita} diabetic mice. (**a** from Fig. 7 in [15]. Reproduced with permission of the American Society for Clinical Investigation. **b, c** From Fig. 8 in [18]. Reproduced with permission of *Investigative Ophthalmology and Visual Science*)

ptosis occurs soon after the onset of experimental diabetes, corresponding to the earliest changes in vascular permeability [9, 13].

6.8 Specific Neurons Are Lost by Apoptosis in Diabetes

Several studies have reported that apoptosis in the retina increases with diabetes, but the identity of the dying cells is less well established. Some studies have reported TUNEL positive labeling in trypsine digest retinas, implying that the cells were vascular [54, 63]. Other work, however, has focused on identifying specific subsets of retinal neurons undergoing apoptosis in diabetes. Gastinger et al. [38] recently demonstrated that active caspase-3 immunoreactivity colocalizes with NeuN, a nuclear antigen found in the nuclei of most types of mature neurons (Fig. 6.5). This study also measured a significant depletion of amacrine cell numbers in whole retinas from Ins2^{Akita} mice after 6 months of diabetes, including both dopaminergic and cholinergic cells, identified by immunoreactivity to tyrosine hydroxylase and choline acetyltransferase, respectively. Diabetes has been shown to cause degeneration of a third type of amacrine cells that produce nitric oxide (NO), decreasing the number of NO-immunoreactive cells and attenuating the neuroprotective effect of NO [41a]. This is particularly interesting in light of the regulatory role of NO in vasodilation and ocular blood flow. Data from other histology studies demonstrate that several subsets of neurons are affected adversely by diabetes, including NO-producing bipolar cells, photoreceptors, horizontal cells and retinal ganglion cells [1, 62, 76]. Therefore, diabetes increases the attrition of many types of neuron in the retina, leading to chronic neurodegeneration.

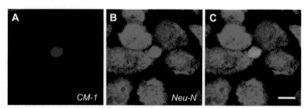

Fig. 6.5. Retinal neurons undergo apoptosis in diabetes. Whole retinas from STZ rats after 1 month of diabetes were labeled by immunohistochemistry for the active form of the apoptosis enzyme, caspase-3, and nuclear antigen unique to neurons, Neu-N. **A** Cells undergoing apoptosis were positive for caspase-3 immunoreactivity (*red*). **B** The same cell also contained immunoreactivity for Neu-N (*green*), identifying it as a neuron. **C** The two antigens colocalized (*yellow*) and the apoptotic cell was surrounded by Neu-N positive cells that were not undergoing apoptosis. (From Fig. 2 in [39]. Reproduced with permission of *Investigative Ophthalmology and Visual Science*)

6.9 Neurons in the Retina Have Morphological Characteristics of Neurodegeneration

Retinas from diabetic rats may exhibit axonal degeneration similar to that observed in other neurodegenerative diseases. A histological study of cross-sections of optic nerve from STZ-diabetic rats found that there was a reduction in the number of axons, while the density of glial cells was increased [85]. Functional deficits in retrograde transport may also occur within the optic nerve, indicated by a decreased accumulation of Fluorogold in large and medium sized retinal ganglion cells when injected into the dorsal lateral geniculate nucleus of STZ-diabetic rats [121, 122]. In a similar study, inhibition of glucose disposal through the polyol pathway with an aldose reductase inhibitor normalized the rate of retrograde transport of Fluorogold through the optic nerve, suggesting that the reduction of retrograde transport is due to metabolic effects of hyperglycemia [47].

Abnormalities such as axoplasmic swellings have also been noted on centrifugal axons labeled by immunohistochemistry for phosphorylated heavy neurofilament protein, which forms part of the cytoskeleton structure in many axons [37]. A preliminary study reported the presence of abnormal swellings and constrictions on retinal ganglion cell axons in Ins2^{Akita} mice crossbred with Thy1.YFP mice, in which yellow fluorescent protein was expressed under the control of the Thy1 promoter [38]. In the retina, Thy1 is expressed exclusively by ganglion cells, and this cross results in a subpopulation of endogenously fluorescent retinal ganglion cells expressing YFP throughout all their neuronal projections, including the axon. These histological characteristics are reminiscent of those noted in degeneration of the sciatic nerve induced by transection and diabetes [41, 99], and suggest that diabetes causes neurodegeneration and abnormalities in the retinal ganglion cell axons.

Another preliminary report suggests that a 1-month duration of diabetes leads to a reduction in the expression of synaptic vesicle-associated proteins such as VAMP2, SNAP-25, synaptophysin, and synapsin 1 [105]. These data are in concordance with other studies demonstrating that synaptic proteins, such as synaptophysin, are depleted from the brain during diabetes and in Alzheimer's disease, and suggest that the mechanisms of functional communication between neurons in the retina are compromised soon after the onset of diabetes [23, 70]. Taken together these observations suggest that diabetes induces neurodegenerative characteristics in the retina, including loss of synaptic integrity and axonal transport, which may accompany or even precede neuronal apoptosis.

6.10 The Glial Cells of the Retina React to Diabetes As if an Injury Has Occurred

The glial cells of the retina and brain provide metabolic support for neurons. The highly specialized Müller cells span the retina from inner to outer limiting membranes while the astrocytes form a monolayer close to the inner limiting membrane. The very high metabolic demand and abundant neurotransmitter activity of the retina makes the glial cells extremely important to retinal function, because they regulate potassium, calcium and proton currents, as well as neurotransmitters such as glutamate [68, 81].

In diabetes the expression pattern of the glial cell intermediate filament, glial fibrillary acidic protein (GFAP), is altered in both the astrocytes and Müller cells of the rat retina. Diabetes reduces GFAP content in astrocytes, which normally exhibit high levels of GFAP in the retina, while markedly upregulating GFAP expression in Müller cells, which do not express this protein under normal conditions (Fig. 6.6). These observations are significant because GFAP expression is an established response to several types of stress and injury in the brain and retina [45, 71, 74]. Other studies have also shown that diabetes alters GFAP expression in both human and rat retina [11, 43, 58, 64, 83].

Other metabolic functions of glial cells are also affected by diabetes. The rate of conversion of glutamate into glutamine is reduced by diabetes [58], as are glutamate oxidation and glutamine synthesis [59]. Together these data show that major metabolic processes supporting glutamate neurotransmission are compromised by diabetes, which could lead to inappropriate neuronal communication and possibly to excitotoxicity due to excess stimulation of neuronal ionotropic glutamate receptors.

Like the brain, the retina contains microglial cells, which are normally quiescent but adopt a reactive state during infection and injury. The microglial cells also become reactive in the retinas of mice and rats with diabetes [18, 83, 120] (Fig. 6.7). When quiescent these cells have long fine processes that appear to interact with other cells including neurons. When the retina is stressed by neuroinflammation, ischemia, or some other types of damage, however, microglia adopt a reactive morphology in which their processes become contracted and swollen. The activation of microglia may in part be due to increased expression of cytokines, but this process may be blocked by anti-inflammatory drugs such as minocycline [56]. Taken together these data suggest that the microglia and macroglia of the retina respond to diabetes in a manner that suggests an underlying stress or metabolic insult.

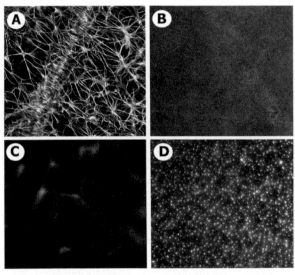

Fig. 6.6. Retinal macroglia express a stress reaction in diabetes. The stress reaction of astrocytes and Müller cells was visualized by immunohistochemistry for glial fibrillary acidic protein (GFAP) in whole-mount retinas from control and STZ-injected rats after 4 months of diabetes. **A** In control rats the astrocytes, which form a monolayer at the superficial surface of the retina, express high levels of GFAP. **B** In STZ rats GFAP expression in the astrocytes is reduced. **C** Focusing deeper into the retina reveals that the Müller cells do not express GFAP in the control rats. **D** Focusing at a similar level in the retinas of STZ diabetic rats reveals that GFAP expression is high in the Müller cell processes. (From Fig. 2 in [16]. Reproduced with permission of *Investigative Ophthalmology and Visual Science*)

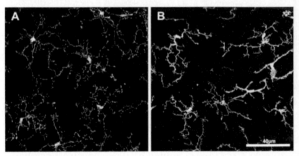

Fig. 6.7. Retinal microglia become reactive in diabetes. Whole retinas from Ins2[Akita] diabetic mice, after 2 months of hyperglycemia, were labeled by immunohistochemistry for the calcium binding protein called Iba1, which is unique to microglia in the retina. **A** In non-diabetic littermates the microglia had a quiescent morphology, with many long, fine processes. **B** In diabetic mice the microglia of some regions had a reactive morphology with short swollen processes. (From Fig. 6 in [18]. Reproduced with permission of *Investigative Ophthalmology and Visual Science*)

6.11 There is a Loss of Function in Diabetic Retinopathy That Begins Soon After the Onset of Diabetes

The onset of vision loss often precedes the clinically measurable changes in the vasculature, and likely stems from disruptions of the neural retina. Diabetes-induced neuronal degeneration has been observed in postmortem human retinas as well as in rodent models of diabetes [3, 15, 18, 20, 62, 65, 76, 87, 115]. Such retinal neurodegeneration is a fundamental pathological feature of other high-incidence retinal diseases, including glaucoma and age-related macular degeneration, that involve functional vision loss [2, 55]. Normal vision is dependent on intact signal transduction pathways through the network of neurons in the retina, which is further dependent on undisrupted cell-cell interactions between neuronal, glial, vascular and epithelial cells. Due to the interdependent nature of these cells, the degeneration of one subpopulation has the potential to impact the function and overall health of the entire retina in a feed-forward manner. This is very likely one contributing mechanism to the progressive nature of diabetic retinopathy, emphasizing the need for early intervention.

Deficits in vision occurring among people with diabetes include decreases in night vision, contour and contrast sensitivity, detail discrimination and visual acuity [46, 66, 90, 102]. Although normal daytime vision is the product of complex neuronal processing, various aspects of vision can be attributed to inputs from distinct neuronal types in the retina. Clinical studies have detected a loss of neuronal function with basic vision examinations and psychophysical testing, correlated with underlying neurophysiological disruptions using electroretinography. The electroretinogram (ERG) provides an assessment of neuronal electrical responses to visual stimuli and can be used in humans as well as animal models of diabetes. Studies of humans with diabetes using this technology have demonstrated retinal dysfunction that precedes visible vascular changes characteristic of diabetic retinopathy [24, 30, 42]. Data from these studies suggest that the function of the inner retina is most affected by diabetes, with reduced amplitude and frequency of amacrine cell responses to visual stimuli, indicated by aberrant oscillatory potentials [19, 104]. In conclusion, data showing changes in the electrophysiological response of the retina suggest that diabetes rapidly induces a loss of function that may precede the gross vascular pathology [21].

6.12 Conclusions

Neurodegeneration in the retina is accompanied by loss of neural function, which may begin soon after the onset of diabetes. These early changes may be reflected in the increase in expression of VEGF, which occurs within the neurons and glial cells, presumably as a response to stresses induced by diabetes. VEGF, and possibly other vasoactive factors, causes the increases in vascular permeability and proliferation that are more widely recognized as diabetic retinopathy. VEGF is currently being studied as a potential target for a variety of novel therapies [8, 69, 84, 92]. Since retinal function may not be reversible, however, other current approaches should involve direct targeting of survival of the neural retina with therapeutic agents such as ACE inhibitors [22], cannabidiol [33], and growth factors such as brain derived neurotrophic factor [87], insulin and IGF-I [17, 82, 86], which may exert a neuroprotective effect, reducing the numbers of cells lost to diabetes and perhaps limiting the increased expression of VEGF that leads to angiogenesis and vascular permeability.

6.13 Methodology

Here we summarize two of the methodologies that were developed to collect the apoptosis and immunohistochemical data discussed earlier.

A. Terminal dUTP nick end labeling (TUNEL) on whole retina

1. Retinas are dissected on ice immediately and fixed for 20 min in 10% normal buffered formalin and rinsed before mounting flat onto coated microscope slides, by four radial cuts. Samples are placed at 4°C overnight and allowed to dehydrate.
2. At the time of staining, retinas are rehydrates in PBS, fixed further in formalin or 4% paraformaldehyde, 10 min @r.t., rinsed in PBS (twice for 5 min each) and dehydrated in 50%, 75, 95%, 100% ethanol, 20 min each.
3. The retinas are cleared in xylene (2 times for 10 min) and then stored in xylene overnight at 4°C. This step removes much of the lipid in the tissue.
4. The next day the samples are allowed to come to room temperature, placed in fresh xylene, and incubated at 60°C for 30 min, then rehydrated in ethanol, 100%, 95%, 75%, 10 min each, then in 50% ethanol for 5 min, before two 5 min rinses in water.
5. The retinas are permeabilized in 0.3% Triton X-100 (15 min), washed in TRIS/NaCl buffer (100 mM TRIS, 150 mM NaCl, pH 8.0) and then digested with 20 µg/ml proteinase K in TRIS/NaCl buffer, for 30 min. This step is stopped by rinsing in water.
6. Endogenous peroxidase activity is quenched with 3% H_2O_2 for 10 min.
7. After further rinsing in saline the TUNEL protocol is carried out using an in situ terminal transferase labeling kit. Terminal transferase is stored at –20°C and prepared fresh in incubation buffer containing digoxigenin labeled nucleotides. The tissue is covered with plastic coverslips and incubated at 37°C for 1 h. Then the enzyme reaction is washed off.

6

8. The tissue is incubated with an HRP-conjugated anti-digoxigenin antibody, washed three times in PBS and then exposed to HRP substrate to develop a visible precipitate.

9. The positive cells are counted by sequentially scanning across the entire retina with a light microscope and ×40 objective. The total number of positive cells is expressed per area of each retina to standardize for variation in the samples due to dissection.

B. Immunohistochemistry for confocal microscopy on whole retina

1. Dissect whole retinas carefully using a dissection microscope and fine scissors and forceps. Keep the retina as intact as possible with no cuts or tears. Handle retinas gently by edges using very fine pointed forceps.

2. Fix each whole retina in paraformaldehyde, and rinse in PBS. The optimum concentration of fixative and duration of this step must be empirically determined for every antigen and primary antibody. Optimum fixation can vary between 10 min and several hours in 2% or 4% paraformaldehyde.

3. Retinas are transferred to 200 µl of 10% serum containing 0.3% Triton X-100, in 96-well microtiter plate wells, for 1–2 h.

4. Transfer to separate wells containing 200 µl of primary antibody and incubate at 4°C for 3–4 days. The optimal concentration for each antibody must be empirically determined.

5. Transfer retinas to six-well culture plates and wash repeatedly in 10 ml volumes of wash buffer (PBS with 0.3% Triton X-100). Each wash should last 1 h, and then incubate in wash buffer overnight.

6. Transfer retinas back to a 96-well plate for the secondary antibody incubation. Fluorescent conjugates for each secondary antibody should be chosen carefully to be compatible with the available microscopes. The concentrations of secondary antibodies must also be optimized empirically.

7. Repeat the washing step as described in step 5. Then mount the tissue on microscope slides by making four radial cuts to flatten each retina, using fine scissors, pointed forceps and a soft brush. This step is best done with a dissection microscope. Tissue should be mounted in a non-fluorescent mounting media. The volume of media should be adjusted to account for the thickness of the retinal tissue.

References

1. Abu-El-Asrar AM, Dralands L, Missotten L, Al-Jadaan IA, Geboes K (2004) Expression of apoptosis markers in the retinas of human subjects with diabetes. Invest Ophthalmol Vis Sci 45(8):2760–2766

2. Adler R, Curcio C, Hicks D, Price D, Wong F (1999) Cell death in age-related macular degeneration. Mol Vision 5:31

3. Agardh E, Bruun A, Agardh CD (2001) Retinal glial cell immunoreactivity and neuronal cell changes in rats with STZ-induced diabetes. Curr Eye Res 23(4):276–84

4. Aiello LP, Avery RL, Arrigg PG, Keyt BA, Jampel HD, Shah ST, Pasquale LR, Thieme H, Iwamoto MA, Park JE (1994) Vascular endothelial growth factor in ocular fluid of patients with diabetic retinopathy and other retinal disorders. N Engl J Med 331(22):1480–7

5. Aizu Y, Oyanagi K, Hu J, Nakagawa H (2002) Degeneration of retinal neuronal processes and pigment epithelium in the early stage of the streptozotocin-diabetic rats. Neuropathology 22(3):161–70

6. Ambati J, Chalam KV, Chawla DK, D'Angio CT, Guillet EG, Rose SJ, Vanderlinde RE, Ambati BK (1997) Elevated gamma-aminobutyric acid, glutamate, and vascular endothelial growth factor levels in the vitreous of patients with proliferative diabetic retinopathy. Arch Ophthalmol 115(9):1161–6

7. Amin RH, Frank RN, Kennedy A, Eliott D, Puklin JE, Abrams GW (1997) Vascular endothelial growth factor is present in glial cells of the retina and optic nerve of human subjects with nonproliferative diabetic retinopathy. Invest Ophthalmol Vis Sci 38(1):36–47

8. Amrite AC, Ayalasomayajula SP, Cheruvu NP, Kompella UB (2006) Single periocular injection of celecoxib-PLGA microparticles inhibits diabetes-induced elevations in retinal PGE2, VEGF, and vascular leakage. Invest Ophthalmol Vis Sci 47(3):1149–60

9. Antonetti DA, Barber AJ, Khin S, Lieth E, Tarbell JM, Gardner TW (1998) Vascular permeability in experimental diabetes is associated with reduced endothelial occludin content: vascular endothelial growth factor decreases occludin in retinal endothelial cells. Diabetes 47(12):1953–9

10. Antonetti DA, Barber AJ, Hollinger LA, Wolpert EB, Gardner TW (1999) Vascular endothelial growth factor induces rapid phosphorylation of tight junction proteins occludin and zonula occluden-1. A potential mechanism for vascular permeability in diabetic retinopathy and tumors. J Biol Chem 274(33):23463–7

11. Asnaghi V, Gerhardinger C, Hoehn T, Adeboje A, Lorenzi M (2003) A role for the polyol pathway in the early neuroretinal apoptosis and glial changes induced by diabetes in the rat. Diabetes 52(2):506–511

12. Barber AJ (2003) A new view of diabetic retinopathy: a neurodegenerative disease of the eye. Progr Neuropsychopharmacol Biol Psychiatry 27(2):283–290

13. Barber AJ, Antonetti DA (2003) Mapping the blood vessels with paracellular permeability in the retinas of diabetic rats. Invest Ophthalmol Vis Sci 44(12):5410–5416

14. Barber AJ, Lieth E (1997) Agrin accumulates in the brain microvascular basal lamina during development of the blood-brain barrier. Dev Dyn 208(1):62–74

15. Barber AJ, Lieth E, Khin SA, Antonetti DA, Buchanan AG, Gardner TW (1998) Neural apoptosis in the retina during experimental and human diabetes. Early onset and effect of insulin. J Clin Invest 102(4):783–791

16. Barber AJ, Antonetti DA, Gardner TW (2000) Altered expression of retinal occludin and glial fibrillary acidic protein in experimental diabetes. Invest Ophthalmol Vis Sci 41(11):3561–8

17. Barber AJ, Nakamura M, Wolpert EB, Reiter CEN, Seigel GM, Antonetti DA, Gardner TW (2001) Insulin rescues retinal neurons from apoptosis by a phosphatidylinositol 3-kinase/Akt-mediated mechanism that reduces the activation of caspase-3. J Biol Chem 276(35):32814–32821

18. Barber AJ, Antonetti DA, Kern TS, Reiter CE, Soans RS, Krady JK, Levison SW, Gardner TW, Bronson SK (2005) The Ins2Akita mouse as a model of early retinal complications in diabetes. Invest Ophthalmol Vis Sci 46(6):2210–8

19. Bearse MA, Jr, Han Y, Schneck ME, Barez S, Jacobsen C, Adams AJ (2004) Local multifocal oscillatory potential

abnormalities in diabetes and early diabetic retinopathy. Invest Ophthalmol Vis Sci 45(9):3259–65

20. Bloodworth JM, Jr (1962) Diabetic retinopathy. Diabetes 11:1–22

21. Bresnick GH (1986) Diabetic retinopathy viewed as a neurosensory disorder. Arch Ophthalmol 104:989–990

22. Bui BV, Armitage JA, Tolcos M, Cooper ME, Vingrys AJ (2003) ACE inhibition salvages the visual loss caused by diabetes. Diabetologia 46(3):401–8

23. Callahan LM, Vaules WA, Coleman PD (2002) Progressive reduction of synaptophysin message in single neurons in Alzheimer disease. J Neuropathol Exp Neurol 61(5): 384–95

24. Caputo S, Di Leo MA, Falsini B, Ghirlanda G, Porciatti V, Minella A, Greco AV (1990) Evidence for early impairment of macular function with pattern ERG in type I diabetic patients. Diabetes Care 13(4):412–8

25. Chehade JM, Haas MJ, Mooradian AD (2002) Diabetes-related changes in rat cerebral occludin and zonula occludens-1 (ZO-1) expression. Neurochem Res 27(3):249–52

26. Cunha-Vaz JG (1983) Studies on the pathophysiology of diabetic retinopathy. The blood-retinal barrier in diabetes. Diabetes 32 Suppl 2:20–7

27. Cunha-Vaz JG (2004) The blood-retinal barriers system. Basic concepts and clinical evaluation. Exp Eye Res 78(3): 715–21

28. Cunha-Vaz J, Faria de Abreu JR, Campos AJ (1975) Early breakdown of the blood-retinal barrier in diabetes. Br J Ophthalmol 59(11):649–56

29. Del Bo R, Scarlato M, Ghezzi S, Martinelli Boneschi F, Fenoglio C, Galbiati S, Virgilio R, Galimberti D, Galimberti G, Crimi M, Ferrarese C, Scarpini E, Bresolin N, Comi GP (2005) Vascular endothelial growth factor gene variability is associated with increased risk for AD. Ann Neurol 57(3):373–80

30. Di Leo MA, Falsini B, Caputo S, Ghirlanda G, Porciatti V, Greco AV (1990) Spatial frequency-selective losses with pattern electroretinogram in type 1 (insulin-dependent) diabetic patients without retinopathy. Diabetologia 33(12): 726–30

31. Do Carmo A, Ramos P, Reis A, Proenca R, Cunha-Vaz JG (1998) Breakdown of the inner and outer blood retinal barrier in streptozotocin-induced diabetes. Exp Eye Res 67(5):569–575

32. Duh E, Aiello LP (1999) Vascular endothelial growth factor and diabetes: the agonist versus antagonist paradox. Diabetes 48(10):1899–906

33. El-Remessy AB, Al-Shabrawey M, Khalifa Y, Tsai NT, Caldwell RB, Liou GI (2006) Neuroprotective and blood-retinal barrier-preserving effects of cannabidiol in experimental diabetes. Am J Pathol 168(1):235–44

34. Famiglietti EV, Stopa EG, McGookin ED, Song P, LeBlanc V, Streeten BW (2003) Immunocytochemical localization of vascular endothelial growth factor in neurons and glial cells of human retina. Brain Res 969(1–2):195–204

35. Gardner TW (1995) Histamine, ZO-1 and increased blood-retinal barrier permeability in diabetic retinopathy. Trans Am Ophthalmol Soc 93:583–621

36. Gardner TW, Lieth E, Khin SA, Barber AJ, Bonsall DJ, Lesher T, Rice K, Brennan WA, Jr (1997) Astrocytes increase barrier properties and ZO-1 expression in retinal vascular endothelial cells. Invest Ophthalmol Vis Sci 38(11):2423–7

37. Gastinger MJ, Barber AJ, Khin SA, McRill CS, Gardner TW, Marshak DW (2001) Abnormal centrifugal axons in strep-

tozotocin-diabetic rat retinas. Invest Ophthalmol Vis Sci 42(11):2679–2685

38. Gastinger MJ, Conboy E, Bronson SK, Barber AJ (2006) Retinal ganglion cells undergo pathological changes in Ins2Akita diabetic mice. ARVO 2006 Abstract #2059

39. Gastinger MJ, Singh RS, Barber AJ (2006) Loss of cholinergic and dopaminergic amacrine cells in streptozotocin-diabetic rat and Ins2Akita-diabetic mouse retinas. Invest Ophthalmol Vis Sci 47(7):3143–50

40. Gerhardinger C, Brown LF, Roy S, Mizutani M, Zucker CL, Lorenzi M (1998) Expression of vascular endothelial growth factor in the human retina and in nonproliferative diabetic retinopathy. Am J Pathol 152(6):1453–62

41. Glass JD, Griffin JW (1991) Neurofilament redistribution in transected nerves: evidence for bidirectional transport of neurofilaments. J Neurosci 11(10):3146–54

41a. Goto R, Doi M, Ma N, Semba R, et al. (2005) Contribution of nitric oxide-producing cells in normal and diabetic rat retina. Jpn J Ophthalmol 49:363–70

42. Greco AV, Di Leo MA, Caputo S, Falsini B, Porciatti V, Marietti G, Ghirlanda G (1994) Early selective neuroretinal disorder in prepubertal type 1 (insulin-dependent) diabetic children without microvascular abnormalities. Acta Diabetologica 31(2):98–102

43. Hammes HP, Federoff HJ, Brownlee M (1995) Nerve growth factor prevents both neuroretinal programmed cell death and capillary pathology in experimental diabetes. Mol Med 1(5):527–34

44. Hayashi T, Abe K, Itoyama Y (1998) Reduction of ischemic damage by application of vascular endothelial growth factor in rat brain after transient ischemia. J Cereb Blood Flow Metabol 18(8):887–95

45. Huxlin KR, Dreher Z, Schulz M, Dreher B (1995) Glial reactivity in the retina of adult rats. Glia 15(2):105–18

46. Hyvarinen L, Laurinen P, Rovamo J (1983) Contrast sensitivity in evaluation of visual impairment due to macular degeneration and optic nerve lesions. Acta Ophthalmol 61(2):161–70

47. Ino-Ue M, Zhang L, Naka H, Kuriyama H, Yamamoto M (2000) Polyol metabolism of retrograde axonal transport in diabetic rat large optic nerve fiber. Invest Ophthalmol Vis Sci 41(13):4055–8

48. Ito U, Go KG, Walker JT, Jr, Spatz M, Klatzo I (1976) Experimental cerebral ischemia in Mongolian gerbils III. Behaviour of the blood-brain barrier. Acta Neuropathol 34(1): 1–6

49. Janzer RC, Raff MC (1987) Astrocytes induce blood-brain barrier properties in endothelial cells. Nature 325(6101): 253–7

50. Jin KL, Mao XO, Greenberg DA (2000) Vascular endothelial growth factor: Direct neuroprotective effect in in vitro ischemia. PNAS 97(18):10242–10247

51. Jones CW, Cunha-Vaz JG, Rusin MM (1982) Vitreous fluorophotometry in the alloxan- and streptozocin-treated rat. Arch Ophthalmol 100(7):1141–5

52. Kalaria RN (1999) The blood-brain barrier and cerebrovascular pathology in Alzheimer's disease. Ann N Y Acad Sci 893:113–25

53. Kalaria RN, Cohen DL, Premkumar DR, Nag S, LaManna JC, Lust WD (1998) Vascular endothelial growth factor in Alzheimer's disease and experimental cerebral ischemia. Brain Res Mol Brain Res 62(1):101–5

54. Kern TS, Tang J, Mizutani M, Kowluru RA, Nagaraj RH, Romeo G, Podesta F, Lorenzi M (2000) Response of capil-

6

lary cell death to aminoguanidine predicts the development of retinopathy: comparison of diabetes and galactosemia. Invest Ophthalmol Vis Sci 41(12):3972–8

55. Kerrigan LA, Zack DJ, Quigley HA, Smith SD, Pease ME (1997) Tunel-positive ganglion cells in human primary open-angle glaucoma. Arch Ophthalmol 115(8):1031–1035

56. Krady JK, Basu A, Allen CM, Xu Y, Lanoue KF, Gardner TW, Levison SW (2005) Minocycline reduces proinflammatory cytokine expression, microglial activation, and caspase-3 activation in a rodent model of diabetic retinopathy. Diabetes 54(5):1559–65

57. Lee MY, Ju WK, Cha JH, Son BC, Chun MH, Kang JK, Park CK (1999) Expression of vascular endothelial growth factor mRNA following transient forebrain ischemia in rats. Neurosci Lett 265(2):107–10

58. Lieth E, Barber AJ, Xu B, Dice C, Ratz MJ, Tanase D, Strother JM (1998) Glial reactivity and impaired glutamate metabolism in short-term experimental diabetic retinopathy. Diabetes 47(5):815–20

59. Lieth E, LaNoue KF, Antonetti DA, Ratz M (2000) Diabetes reduces glutamate oxidation and glutamine synthesis in the retina. Exp Eye Res 70(6):723–730

60. Lobo CL, Bernardes RC, Santos FJ, Cunha-Vaz JG (1999) Mapping retinal fluorescein leakage with confocal scanning laser fluorometry of the human vitreous. Arch Ophthalmol 117(5):631–637

61. Lutty GA, McLeod DS, Merges C, Diggs A, Plouet J (1996) Localization of vascular endothelial growth factor in human retina and choroid. Arch Ophthalmol 114(8):971–7

62. Martin PM, Roon P, Van Ells TK, Ganapathy V, Smith SB (2004) Death of retinal neurons in streptozotocin-induced diabetic mice. Invest Ophthalmol Vis Sci 45(9):3330–6

63. Mizutani M, Kern TS, Lorenzi M (1996) Accelerated death of retinal microvascular cells in human and experimental diabetic retinopathy. J Clin Invest 97(12):2883–90

64. Mizutani M, Gerhardinger C, Lorenzi M (1998) Muller cell changes in human diabetic retinopathy. Diabetes 47(3):445–9

65. Mohr S, Xi X, Tang J, Kern TS (2002) Caspase activation in retinas of diabetic and galactosemic mice and diabetic patients. Diabetes 51(4):1172–1179

66. Mortlock KE, Chiti Z, Drasdo N, Owens DR, North RV (2005) Silent substitution S-cone electroretinogram in subjects with diabetes mellitus. Ophthal Physiol Optics 25(5):392–9

67. Murata T, Nakagawa K, Khalil A, Ishibashi T, Inomata H, Sueishi K (1996) The relation between expression of vascular endothelial growth factor and breakdown of the blood-retinal barrier in diabetic rat retinas. Lab Invest 74(4):819–25

68. Newman E, Reichenbach A (1996) The Muller cell: a functional element of the retina. Trends Neurosci 19(8):307–12

69. Ng EW, Shima DT, Calias P, Cunningham ET, Jr, Guyer DR, Adamis AP (2006) Pegaptanib, a targeted anti-VEGF aptamer for ocular vascular disease. Nature Rev Drug Discovery 5(2):123–32

70. Nitta A, Murai R, Suzuki N, Ito H, Nomoto H, Katoh G, Furukawa Y, Furukawa S (2002) Diabetic neuropathies in brain are induced by deficiency of BDNF. Neurotoxicol Teratol 24(5):695–701

71. O'Callaghan JP (1991) Assessment of neurotoxicity: use of glial fibrillary acidic protein as a biomarker. Biomed Env Sci 4(1–2):197–206

72. Ogata N, Yamanaka R, Yamamoto C, Miyashiro M, Kimoto T, Takahashi K, Maruyama K, Uyama M (1998) Expression of vascular endothelial growth factor and its receptor, KDR, following retinal ischemia-reperfusion injury in the rat. Curr Eye Res 17(11):1087–96

73. Oosthuyse B, Moons L, Storkebaum E, Beck H, Nuyens D, Brusselmans K, Van Dorpe J, Hellings P, Gorselink M, Heymans S, Theilmeier G, Dewerchin M, Laudenbach V, Vermylen P, Raat H, Acker T, Vleminckx V, Van Den Bosch L, Cashman N, Fujisawa H, Drost MR, Sciot R, Bruyninckx F, Hicklin DJ, Ince C, Gressens P, Lupu F, Plate KH, Robbrecht W, Herbert JM, Collen D, Carmeliet P (2001) Deletion of the hypoxia-response element in the vascular endothelial growth factor promoter causes motor neuron degeneration. Nature Genet 28(2):131–8

74. Osborne NN, Larsen AK (1996) Antigens associated with specific retinal cells are affected by ischaemia caused by raised intraocular pressure: effect of glutamate antagonists. Neurochem Int 29(3):263–70

75. Pardridge WM, Connor JD, Crawford IL (1975) Permeability changes in the blood-brain barrier: causes and consequences. CRC Crit Rev Toxicol 3(2):159–99

76. Park SH, Park JW, Park SJ, Kim KY, Chung JW, Chun MH, Oh SJ (2003) Apoptotic death of photoreceptors in the streptozotocin-induced diabetic rat retina. Diabetologia 46(9):1260–8

77. Petito CK (1979) Early and late mechanisms of increased vascular permeability following experimental cerebral infarction. J Neuropathol Exp Neurol 38(3):222–34

78. Pichiule P, Chavez JC, Xu K, LaManna JC (1999) Vascular endothelial growth factor upregulation in transient global ischemia induced by cardiac arrest and resuscitation in rat brain. Brain Res Mol Brain Res 74(1–2):83–90

79. Premkumar DR, Cohen DL, Hedera P, Friedland RP, Kalaria RN (1996) Apolipoprotein E-epsilon4 alleles in cerebral amyloid angiopathy and cerebrovascular pathology associated with Alzheimer's disease. Am J Pathol 148(6):2083–95

80. Qaum T, Xu Q, Joussen AM, Clemens MW, Qin W, Miyamoto K, H Hassessian H, Wiegand SJ, Rudge J, Yancopoulos GD, Adamis AP (2001) VEGF-initiated blood-retinal barrier breakdown in early diabetes. Invest Ophthalmol Vis Sci 42(10):2408–2413

81. Reichenbach A, Stolzenburg JU, Eberhardt W, Chao TI, Dettmer D, Hertz L (1993) What do retinal Muller (glial) cells do for their neuronal 'small siblings'? J Chem Neuroanat 6(4):201–13

82. Reiter CE, Gardner TW (2003) Functions of insulin and insulin receptor signaling in retina: possible implications for diabetic retinopathy. Progr Retin Eye Res 22(4):545–62

83. Rungger-Brandle E, Dosso AA, Leuenberger PM (2000) Glial reactivity, an early feature of diabetic retinopathy. Invest Ophthalmol Vis Sci 41(7):1971–1980

84. Saishin Y, Saishin Y, Takahashi K, Lima e Silva R, Hylton D, Rudge JS, Wiegand SJ, Campochiaro PA (2003) VEGF-TRAP(R1R2) suppresses choroidal neovascularization and VEGF-induced breakdown of the blood-retinal barrier. J Cell Physiol 195(2):241–8

85. Scott TM, Foote J, Peat B, Galway G (1986) Vascular and neural changes in the rat optic nerve following induction of diabetes with streptozotocin. J Anat 144:145–152

86. Seigel GM, Lupien SB, Campbell LM, Ishii DN (2006) Systemic IGF-I treatment inhibits cell death in diabetic rat retina. J Diabetes Complications 20(3):196–204

87. Seki M, Tanaka T, Nawa H, Usui T, Fukuchi T, Ikeda K, Abe H, Takei N (2004) Involvement of brain-derived neurotro-

phic factor in early retinal neuropathy of streptozotocin-induced diabetes in rats: therapeutic potential of brain-derived neurotrophic factor for dopaminergic amacrine cells. Diabetes 53(9):2412–9

88. Slevin M, Krupinski J, Slowik A, Rubio F, Szczudlik A, Gaffney J (2000) Activation of MAP kinase (ERK-1/ERK-2), tyrosine kinase and VEGF in the human brain following acute ischaemic stroke. Neuroreport 11(12):2759–64

89. Sobue K, Yamamoto N, Yoneda K, Hodgson ME, Yamashiro K, Tsuruoka N, Tsuda T, Katsuya H, Miura Y, Asai K, Kato T (1999) Induction of blood-brain barrier properties in immortalized bovine brain endothelial cells by astrocytic factors. Neurosci Res 35(2):155–64

90. Sokol S, Moskowitz A, Skarf B, Evans R, Molitch M, Senior B (1985) Contrast sensitivity in diabetics with and without background retinopathy. Arch Ophthalmol 103(1):51–4

91. Sondell M, Lundborg G, Kanje M (1999) Vascular endothelial growth factor has neurotrophic activity and stimulates axonal outgrowth, enhancing cell survival and Schwann cell proliferation in the peripheral nervous system. J Neurosci 19(14):5731–40

92. Spaide RF, Fisher YL (2006) Intravitreal bevacizumab (Avastin) treatment of proliferative diabetic retinopathy complicated by vitreous hemorrhage. Retina 26(3):275–8

93. Staddon JM, Rubin LL (1996) Cell adhesion, cell junctions and the blood-brain barrier. Curr Opin Neurobiol 6(5):622–7

94. Stewart PA, Wiley MJ (1981) Developing nervous tissue induces formation of blood-brain barrier characteristics in invading endothelial cells: a study using quail-chick transplantation chimeras. Dev Biol (Orlando) 84(1):183–92

95. Sun FY, Guo X (2005) Molecular and cellular mechanisms of neuroprotection by vascular endothelial growth factor. J Neurosci Res 79(1–2):180–4

96. Sun Y, Jin K, Xie L, Childs J, Mao XO, Logvinova A, Greenberg DA (2003) VEGF-induced neuroprotection, neurogenesis, and angiogenesis after focal cerebral ischemia. J Clin Invest 111(12):1843–51

97. Svensson B, Peters M, Konig HG, Poppe M, Levkau B, Rothermundt M, Arolt V, Kogel D, Prehn JH (2002) Vascular endothelial growth factor protects cultured rat hippocampal neurons against hypoxic injury via an antiexcitotoxic, caspase-independent mechanism. J Cereb Blood Flow Metabol 22(10):1170–5

98. Tarkowski E, Issa R, Sjogren M, Wallin A, Blennow K, Tarkowski A, Kumar P (2002) Increased intrathecal levels of the angiogenic factors VEGF and TGF-beta in Alzheimer's disease and vascular dementia. Neurobiol Aging 23(2):237–43

99. Terada M, Yasuda H, Kikkawa R (1998) Delayed Wallerian degeneration and increased neurofilament phosphorylation in sciatic nerves of rats with streptozocin-induced diabetes. J Neurol Sci 155(1):23–30

100. Tout S, Chan-Ling T, Hollander H, Stone J (1993) The role of Muller cells in the formation of the blood-retinal barrier. Neuroscience 55(1):291–301

101. Tran ND, Schreiber SS, Fisher M (1998) Astrocyte regulation of endothelial tissue plasminogen activator in a blood-brain barrier model. J Cereb Blood Flow Metabol 18(12):1316–24

102. Trick GL, Burde RM, Gordon MO, Kilo C, Santiago JV (1988) Retinocortical conduction time in diabetics with abnormal pattern reversal electroretinograms and visual evoked potentials. Doc Ophthalmol 70(1):19–28

103. Tserentsoodol N, Shin BC, Suzuki T, Takata K (1998) Colocalization of tight junction proteins, occludin and ZO-1, and glucose transporter GLUT1 in cells of the blood-ocular barrier in the mouse eye. Histochem Cell Biol 110(6):543–51

104. Tzekov R, Arden GB (1999) The electroretinogram in diabetic retinopathy. Surv Ophthalmol 44(1):53–60

105. VanGuilder HD, Ellis RW, Freeman WM, Barber AJ (2006) Streptozotocin-diabetes decreases synaptic protein expression in rat retina. ARVO 2006 Abstract #1733

106. Vinores SA (1995) Assessment of blood-retinal barrier integrity. Histol Histopathol 10(1):141–54

107. Vinores SA, Gadegbeku C, Campochiaro PA, Green WR (1989) Immunohistochemical localization of blood-retinal barrier breakdown in human diabetics. Am J Pathol 134(2):231–5

108. Vinores SA, Campochiaro PA, Lee A, McGehee R, Gadegbeku C, Green WR (1990) Localization of blood-retinal barrier breakdown in human pathologic specimens by immunohistochemical staining for albumin. Lab Invest 62(6):742–50

109. Vinores SA, McGehee R, Lee A, Gadegbeku C, Campochiaro PA (1990) Ultrastructural localization of blood-retinal barrier breakdown in diabetic and galactosemic rats. J Histochem Cytochem 38(9):1341–52

110. Vinores SA, Youssri AI, Luna JD, Chen YS, Bhargave S, Vinores MA, Schoenfeld CL, Peng B, Chan CC, LaRochelle W, Green WR, Campochiaro PA (1997) Upregulation of vascular endothelial growth factor in ischemic and non-ischemic human and experimental retinal disease. Histol Histopathol 12(1):99–109

111. Wallow IH, Engerman RL (1977) Permeability and patency of retinal blood vessels in experimental diabetes. Invest Ophthalmol Vis Sci 16(5):447–61

112. Westergaard E, Go G, Klatzo I, Spatz M (1976) Increased permeability of cerebral vessels to horseradish peroxidase induced by ischemia in Mongolian gerbils. Acta Neuropathol 35(4):307–25

113. Wick A, Wick W, Waltenberger J, Weller M, Dichgans J, Schulz JB (2002) Neuroprotection by hypoxic preconditioning requires sequential activation of vascular endothelial growth factor receptor and Akt. J Neurosci 22(15):6401–6407

114. Wisniewski HM, Kozlowski PB (1982) Evidence for blood-brain barrier changes in senile dementia of the Alzheimer type (SDAT). Ann N Y Acad Sci 396:119–29

115. Wolter JR (1961) Diabetic retinopathy. Am J Ophthalmol 51:1123–1139

116. Xu Q, Qaum T, Adamis AP (2001) Sensitive blood-retinal barrier breakdown quantitation using Evans blue. Invest Ophthalmol Vis Sci 42(3):789–94

117. Yang ZJ, Bao WL, Qiu MH, Zhang LM, Lu SD, Huang YL, Sun FY (2002) Role of vascular endothelial growth factor in neuronal DNA damage and repair in rat brain following a transient cerebral ischemia. J Neurosci Res 70(2):140–9

118. Yang SP, Bae DG, Kang HJ, Gwag BJ, Gho YS, Chae CB (2004) Co-accumulation of vascular endothelial growth factor with beta-amyloid in the brain of patients with Alzheimer's disease. Neurobiol Aging 25(3):283–90

119. Yoshioka M, Kayo T, Ikeda T, Koizumi A (1997) A novel locus, Mody4, distal to D7Mit189 on chromosome 7 determines early-onset NIDDM in nonobese C57BL/6 (Akita) mutant mice. Diabetes 46(5):887–94

120. Zeng XX, Ng YK, Ling EA (2000) Neuronal and microglial

6

response in the retina of streptozotocin-induced diabetic rats. Vis Neurosci 17(3):463–471

121. Zhang L, Inoue M, Dong K, Yamamoto M (1998) Alterations in retrograde axonal transport in optic nerve of type I and type II diabetic rats. Kobe J Med Sci 44(5–6): 205–15

122. Zhang LX, Ino-ue M, Dong K, Yamamoto M (2000) Retrograde axonal transport impairment of large- and medium-sized retinal ganglion cells in diabetic rat. Curr Eye Res 20(2):131–136

7 Hypoxia in the Pathogenesis of Retinal Disease

V. Poulaki

Core Messages

- Hypoxic signaling is mediated via the hypoxia-inducible factor (HIF) pathway including HIF-1 subunits that contain binding sites to hypoxia-response elements
- HIF-1 regulates target genes including angiogenic mediators, glucose transporters, cytokines, apoptotic genes, regulators of erythropoiesis and genes involved in DNA repair
- von Hippel-Lindau disease (VHL) regulates vascular endothelial growth factor (VEGF) through HIF-1α dependent and independent processes
- Hypoxia, hyperglycemia and reactive oxygen intermediates, growth factors and inflammatory mediators upregulate VEGF

7.1 Introduction

Oxygen cannot passively diffuse for more than a radius of 100 µm around capillaries. As a result, adequate O_2 supply to each cell depends on effective regulation of the integrity and function of the vascular network. On the other hand, high O_2 tissue levels would result in reactive oxygen species (ROS) generation and cellular damage. Therefore, an optimal O_2 concentration is needed to avoid hypoxia or ROS-mediated cellular injury. The retinal tissue is very active metabolically and, therefore, exquisitely dependent on adequate O_2 supply for its function [4]. The delivery of oxygen to the retina is dependent not only on systemic blood pressure, hemoglobin content and integrity of local vasculature, but on the level of intraocular pressure and local autoregulatory mechanisms as well. Hypoxia and its sequelae are implicated in the pathogenesis of most retinal diseases, especially those that involve pathologic neovascularization. This is due to the potent stimulation of production of vascular endothelial growth factor (VEGF), mediated by the hypoxia-inducible factor (HIF)-1 pathway, in response to hypoxia.

7.2 The HIF Pathway and Its Role in Hypoxia Signaling

Essentials

- The HIF family includes HIF-1 and less well characterized proteins such as HIF-2α, HIF-3α, ARNT2, and ARNT3
- The HIF-1 complex consists of two subunits: HIF-1α and HIF-1β
- HIF-1 subunits contain binding sites to hypoxia-response elements through which they regulate their target genes and coactivators such as CREB, CBP and p300
- HIF-1β is constitutively expressed in normoxic cells whereas HIF-1α is tightly regulated by hypoxia both at the transcriptional level and the post-transcriptional level
- Under normoxic conditions HIF-1α is hydroxylated, bound to von Hippel-Lindau disease (VHL) protein, which targets it for destruction by the proteasome. Hydroxylation is regulated by factors such as FIH-1-p300 and CBP and Fe chelating agents such as CoCl. During hypoxia hydroxylation is inhibited secondary to the inhibition of hydroxylases
- HIF-2α may play a more important role in chronic hypoxia because it is not degraded
- HIF-1α can be induced in an oxygen-independent manner by cytokines, growth fac-

tors, stress or increased temperature through the PI3K-AKT-mTOR pathway
- HIF-1 regulates multiple target genes such as angiogenic mediators, glucose transporters, cytokines, apoptotic genes, regulators of erythropoiesis and proteins involved in DNA repair
- VHL regulates VEGF through HIF-1α dependent and independent processes

HIF-1 is a transcription factor that plays a pivotal role in cellular and systemic homeostatic responses to hypoxia. The HIF complex consists of two subunits: the 120-kDa HIF-1α and HIF-1β or ARNT, which has two isoforms (774 and 789 kDa, 92 and 94 kDa, respectively) that differ by the presence of the sequence encoded by a 45-bp alternative exon [131, 132]. HIF-1 binds to consensus and ancillary hypoxia-response elements (HREs) present in the promoter or enhancer regions of target genes [43, 70, 131, 143]. The core sequence of the consensus HIF-1 binding site is 5'-(A/G)CGTG-3'. HIF-1 has been identified in all metazoan species that have been analyzed from *C. elegans* to *H. sapiens*, underscoring the importance of HIF-1 for the response of multicellular organisms to cellular O_2 changes [110]. The HIF family also includes structurally related proteins (HIF-2α, HIF-3α, ARNT2, ARNT3) with similar but less well characterized function [12].

HIF-1α and HIF-1β expression are required for cardiovascular, skeletal and CNS development and embryonic survival in mice [49, 78, 105]. HIF-1 regu-

lates, in response to hypoxia or other stimuli, the transcription of a broad range of genes that mediate responses to the hypoxic environment, including regulation of energy metabolism, angiogenesis, erythropoiesis, cell cycle, and apoptosis. HIF-1α protein accumulates in cells exposed to 1% oxygen and decays rapidly upon return of the cells to 20% oxygen [131, 140]. In normoxic cells, the HIF-1α protein is degraded rapidly by the ubiquitin/proteasomal pathway. In addition, HIF-1α can be induced in an oxygen-independent manner by various cytokines through the PI3K-AKT-mTOR pathway.

7.2.1 Structure of the HIF Transcription Factor Complex and Regulation of Its Activity

HIF-1 is a heterodimeric protein that is composed of HIF-1α and HIF-1β subunits. The HIF-1α gene contains 15 exons [50] and is located on chromosome 14q21-q24 [111]. The HIF-1β (ARNT) gene contains 22 exons, varying in size from 25 to 214 bp, and spans 65 kb on chromosome 1q21. Both alpha and beta subunits contain basic helix-loop-helix and Per-ARNT-Sim (PAS) domains at their amino-terminal part, which are important for heterodimerization and binding to hypoxia-response elements (HREs). HIF-1α also contains, in its carboxy-terminal part, binding sites for coactivators such as CREB binding protein (CBP) and p300, which allow interaction with RNA polymerase II and the transcription complex [110].

HIF-1β mRNA and protein are expressed constitutively, even in normoxic cells. It is HIF-1α protein expression that is tightly regulated by O_2 levels and is responsible for the inducibility of the complex. HIF-

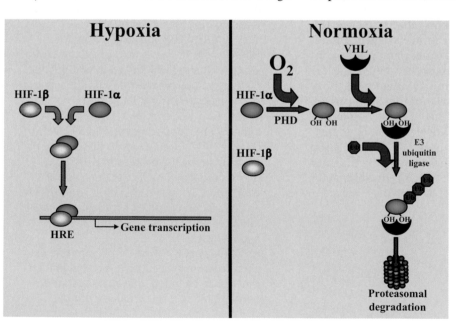

Fig. 7.1. Transcriptional control of HIF-1α regulated genes in hypoxia and normoxia. In normoxia, HIF-1α is heavily hydroxylated and bound to the VHL gene product that targets it to proteosomal degradation. Therefore it cannot bind to HIF-1α responsive genes. In hypoxia, HIF-1α is not hydroxylated and is free to form heterodimeric complexes with HIF-1β. The HIF-1α and HIF-1β complexes regulate gene transcription through binding to HIF responsive elements (HRE)

1α protein expression can be regulated by hypoxia in at least two ways:

- **Transcriptional level:** The mRNA for HIF-1α is upregulated in hypoxic/ischemic tissues [10]. This effect requires hours to occur.
- **Post-transcriptional level:** The HIF-1α protein is stabilized in hypoxic cells. This effect occurs within minutes. Jewell et al. [51] demonstrated that the HIF-1α protein accumulates in the nucleus within 2 min of anoxic/hypoxic exposure. HIF-1α has a half-life of 1–5 min in normoxic conditions [131, 140]. HIF-1α protein expression is negatively regulated in normoxic cells by hydroxylation of two conserved proline residues (Pro-402 and Pro-564). This reaction is O_2-dependent and the O_2 concentration is rate limiting for enzymatic activity, thus providing a mechanism for the direct regulation of HIF-1α protein levels by O_2. This hydroxylation promotes HIF-1 binding to the von Hippel Lindau protein (VLH) and, through that, to the E3 ubiquitin-protein ligase complex, which ubiquitinates HIF-1 and targets it for destruction by the proteasome [46, 58, 107]. This is the main mechanism of HIF-1 inactivation in normoxia and it is the inhibition of this pathway in O_2-deprived cells that leads to the rapid accumulation of HIF-1. The HIF-1α prolyl hydroxylases contain Fe(II) at the active site, which can be chelated by desferrioxamine (DFX) or replaced by Co(II), thus inactivating the enzyme. This explains the observation that cell treatment with $CoCl_2$ or desferrioxamine mimics hypoxia in activating the HIF-1α pathway and stimulating VEGF expression. Missense mutations and/or deletions of the HIF-1α regions involved in this degradation pathway result in stabilization of the protein and overexpression even under normoxic conditions [123].

Moreover, HIF-1α transcriptional activity can be inhibited by:

- Inactivation of the transactivating capacity of HIF-1 via hydroxylation of its carboxy-terminal domain at Asn-803 by factor inhibiting HIF-1 (FIH-1). This enzymatic reaction is also O_2-dependent and prevents the interaction of HIF-1α with the coactivators CBP and p300 [33, 52, 57, 101]. This provides an additional O_2-dependent, VHL-independent, mechanism for inhibition of HIF-1 activity.
- Degradation of HIF-1α mRNA, which occurs in prolonged hypoxia. This is an autoregulatory mechanism that prevents prolonged activation of the HIF-1 pathway. HIF-2α mRNA is resistant to this mechanism, suggesting that in chronic hypoxia, HIF-2α remains active. This provides a mechanism for differential gene expression in acute vs. chronic hypoxia [128].

7.2.2 Downstream Targets of HIF

In hypoxic cells, overall protein synthesis is suppressed. However, a specific set of transcripts is upregulated. HIF-1 binds to promoter/enhancer elements and plays a pivotal role in stimulating expression of genes involved in energy metabolism, angiogenesis, cell proliferation and apoptosis (for a detailed list, see [110] and [109]). Prominent targets are genes for:

- **Mediators of angiogenesis and vascular physiology,** such as vascular endothelial growth factor (VEGF), VEGF receptor FLT-1, adrenomedullin, endothelin-1, heme oxygenase 1, nitric oxide synthase 2, and plasminogen activator inhibitor 1. HIF-1 increases VEGF mRNA levels via stimulation of transcription by binding to a hypoxia response element located 1 kb 5-prime to the transcriptional start site of VEGF. Also, hypoxia, via HIF-1, may increase the stability of VEGF mRNA [71, 72, 75].
- **Glucose transporters** (glucose transporter 1, glucose transporter 3) and *glycolytic enzymes* [adenylate kinase 3, aldolase A, aldolase C, enolase 1 (ENO1), glucose transporter 1, glucose transporter 3, glyceraldehyde-3-phosphate dehydrogenase, hexokinase 1, hexokinase 2, insulin-like growth factor 2 (IGF-2), lactate dehydrogenase A, phosphoglycerate kinase 1, pyruvate kinase M].
- **Cytokines and intracellular mediators of cell proliferation and apoptosis:** Bcl2/adenovirus EIB 19-kDa-interacting protein 3 (BNIP3) is a cell death factor that is a member of the Bcl-2 proapoptotic family recently shown to induce necrosis rather than apoptosis. The BNIP3 promoter contains a functional HIF-1-responsive element (HRE) and is potently activated by both hypoxia and forced expression of HIF-1α [20]. BNIP3 and its homologue, Nip3-like protein X, are molecular effectors of hypoxia-mediated apoptosis [66, 117]. Moreover, HIF-1 induces CXCR4 expression [119].
- **Regulators of erythropoiesis:** HIF-1 upregulates erythropoietin, ceruloplasmin, transferrin, and transferrin receptor mRNAs [109]. Erythropoietin is produced by the kidney in response to hypoxia and has a pivotal role in regulating erythropoiesis [37, 38]. An additional role of erythropoietin has been discovered and appears to be of particular importance for CNS and retina diseases. Preconditioning with erythropoietin protects neurons in models of ischemic and degen-

erative damage [28, 29]. Similarly, acute hypoxia in the adult mouse retina dose-dependently stimulates expression of erythropoietin, fibroblast growth factor 2 and vascular endothelial growth factor via hypoxia-inducible factor-1α (HIF-1α) stabilization and protects retinal morphology and function against light-induced apoptosis by inhibiting caspase activation. The erythropoietin receptor required for erythropoietin signaling localizes to photoreceptor cells. The protective effect of hypoxic preconditioning is mimicked by systemically applied erythropoietin that crosses the blood-retinal barrier and prevents apoptosis even when given therapeutically after light insult [42]. Activation of the HIF-1 pathway and/or application of erythropoietin may, through the inhibition of apoptosis, be beneficial for the treatment of different forms of retinal disease [9].

- **Proteins involved in DNA repair:** HIF-1α inhibits MSH2 and MSH6 expression [64], suggesting that HIF-1α may be responsible for the genetic instability characteristic of cells undergoing hypoxic stress.

7.2.3 Activation of HIF by Non-hypoxic Stimuli

HIF-1 can also be activated independently of O_2 levels. As mentioned already, desferrioxamine and cobalt ions inhibit the prolyl hydroxylases by chelating or replacing, respectively, the Fe(II) ion in the prolyl hydroxylase catalytic site [110]. Therefore, they inhibit HIF-1α degradation, thus stabilizing its expression. This mechanism explains the stimulatory effect of desferrioxamine and cobalt ions on VEGF expression [47]. Moreover, the hydroxylation reaction requires 2-oxoglutarate as a substrate and generates succinate as a side product [110]. As a result, succinate inhibits HIF-1α alpha prolyl hydroxylases in the cytosol, leading to stabilization and activation of HIF-1α. This provides a link between the tricarboxylic acid (TCA) cycle (i.e., the energy-producing metabolic pathways of the cell) and HIF-1α activation, VEGF expression and angiogenesis [108]. Moreover, this link may explain why succinate dehydrogenase (SDH) and fumarate hydratase, both enzymes of the TCA cycle, function as tumor suppressors [109].

In addition, HIF-1α can be induced in an oxygen-independent manner by various cytokines, hormones, growth factors or other stimuli (nitric oxide, increased temperature, or mechanical stress) through the PI3K-AKT-mTOR pathway [79, 82, 97, 98]. These signaling pathways may play a key role in HIF-1 activation, VEGF expression and neovascularization in the retina. For example, in the diabetic retina, IGF-I potently stimulates VEGF expression and vitreous

IGF-I levels correlate with the presence and severity of diabetic retinal neovascularization [81]. Intravitreous IGF-I injection dose-dependently increases retinal Akt, JNK, HIF-1α, NF-κB and AP-1 activity, and VEGF levels and causes microangiopathy [26]. In vitro, IGF-I potently stimulates VEGF expression in RPE cells [103]. IGF-I stimulates VEGF promoter activity in vitro, mainly via HIF-1α, and secondarily via NF-κB and AP-1, as demonstrated by deletional mapping of the VEGF promoter and by electric mobility shift assays (EMSA) [96]. Systemic inhibition of IGF-I signaling with a receptor neutralizing antibody, or with inhibitors of PI-3 kinase (PI-3K), c-Jun kinase (JNK) or Akt, suppressed retinal Akt, JNK, HIF-1α, NF-κB and AP-1 activity, VEGF expression, as well as ICAM-1 levels, leukostasis and blood-retinal barrier breakdown, in a diabetic animal model [96].

Insulin is another factor that can activate HIF-1. Acute intensive insulin therapy transiently worsens diabetic retinopathy and is known epidemiologically as an independent risk factor for it. Acute intensive insulin therapy markedly increases VEGF mRNA and protein levels in the retina of diabetic rats, by activating HIF-1α via a pathway that involves p38 mitogen-activated protein kinase (MAPK) and phosphatidylinositol (PI) 3-kinase, but not p42/p44 MAPK or protein kinase C [96].

7.2.4 VHL and Its Role in Retinal Angiogenesis

von Hippel-Lindau disease (VHL), also known as angiomatosis retinae, is a hereditary autosomal-dominant cancer syndrome predisposing to a variety of malignant and benign neoplasms (frequently retinal, cerebellar, and spinal hemangioblastoma, renal cell carcinoma, pheochromocytoma, and pancreatic islet cell tumors, endolymphatic sac tumors, and benign cysts affecting a variety of organs) which is caused by germline mutations in the VHL gene, which is located on chromosome 3p25 [76]. Hemangioblastomas tumors of the stromal cells, usually of the CNS and retina, are seen in VHL disease but also occur as sporadic non-hereditary tumors, frequently caused by somatic VHL mutations. The stromal cells have sustained the genetic damage in these tumors, as detected by tissue microdissection, in situ hybridization, and immunohistochemical studies [67, 131]. It is the activation of the HIF pathway in the stromal cells that results in upregulation in expression of VEGF mRNA and protein, which then stimulates vascular growth in a paracrine fashion [14, 39, 44].

VHL associates with elongins B and C, cullin 2 and Rbx-1 in a multiprotein complex with ubiquitin ligase activity for specific substrates such as HIF-1α.

In normoxia, VHL protein recognizes the presense of hydroxylated prolines by oxygen dependent hydroxylases in the HIF-1α protein and targets it for proteasomal degradation by ubiquitylation. Therefore in normoxia the HIF-1α levels and the expression of the HIF-1α responsive genes are also low. In hypoxia, HIF-1α does not contain hydroxylated prolines because the activity of the hydroxylases is inhibited by oxygen. This inhibits the interaction with the VHL complex and the accumulation of HIF-1α and HIF-1α responsive genes. Additionally, pVHL interacts directly with the transcription factor Sp1 and suppresses the Sp1-mediated activation of the VEGF promoter [84].

The VHL complex regulates HIF-1α protein levels not only by targeting it to ubiquitylation but also ensuring that the ubiquitylated protein is degraded by the proteasome. That is mediated through the interaction of VHL-HIF-1α protein with Tat-binding protein TBP-1, an ATPase that facilitates the recruitment of the complex to the proteasome and the ATP-mediated degradation of HIF-1α during normoxia [24]. VHL complex ubiquitylates in an oxygen-dependent fashion a constituent of the RNA polymerase II complex, Rbp7, that regulates the transcription of VEGF in a gene and tissue specific manner. Therefore VHL is linked to VEGF in a non-HIF-dependent fashion [86].

Of course VHL's role extends beyond the regulation of hypoxia-induced genes. Another constituent of RNA polymerase II, Rbp1 is also dependent on VHL ubiquitylation. Only the active hyperphosphorylated form of RBp1 as the one induced in response to DNA damaging factors is subject to ubiquitylation, giving rise to the hypothesis that this interaction is a safety mechanism in order to maintain efficient DNA repair and prevent cell death. VHL also interacts with the active form of PKC, preventing it from causing cytoskeletal disorganization, and cytoskeletal components such as fibronectin. It was also recently suggested that the VHL complex inhibits transcription elongation/initiation of specific genes including HIF1 by association with different elongin BC ligases [25].

It was recently reported that adenoviral-mediated transfer of the VHL gene inhibits neovascularization in a murine laser-induced multiple branch retinal vein occlusion (BRVO) model, assessed by color photographs and fluorescein angiography (FA). VEGF mRNA expression was also significantly reduced in the adVHL-treated retina and iris. In accordance with the above, there was also an observed reduction of retinal edema, neovascularization elsewhere (NVD), new vessels elsewhere (NVE) and rubeosis [2].

7.3 VEGF

Essentials
- VEGF is an endothelial cell mitogen that promotes the formation of new vessels. It has five different isoforms and binds to two high affinity receptors (flt-1 and flk-1)
- Hypoxia, hyperglycemia, reactive oxygen intermediates, growth factors such as IGF-1 and inflammatory mediators upregulate VEGF
- Signal transduction intermediates in the VEGF activated pathways include Akt, MAPK, and STAT3

VEGF, also known as vascular permeability factor, is an endothelial cell mitogen [40] that promotes the formation of new vessels. It exists in five different isoforms of 121, 145, 165, 189, and 206 amino acids, which are derived from alternatively spliced mRNAs, of which $VEGF_{165}$ is the predominant molecular species. It binds two high-affinity receptors, the 180-kDa fms-like tyrosine kinase (Flt-1, also known as VEGFR1) and the 200-kDa kinase insert domain-containing receptor (KDR), also known as fetal liver kinase (flk) or VEGFR2, but KDR transduces the signals for endothelial proliferation and chemotaxis [35, 112].

Hypoxia stimulates VEGF mRNA expression mainly via binding of HIF-1α to consensus and ancillary hypoxia-response elements (HREs) in the VEGF promoter. Also, hypoxia may increase the stability of VEGF mRNA [71, 72]. Other inducers of VEGF are advanced glycation end products (AGEs), hyperglycemia, reactive oxygen intermediates, inflammatory mediators, prostaglandins, IGF-I and insulin.

VEGF participates in the pathogenesis and progression of a wide range of angiogenesis-dependent diseases, including cancer [40], certain inflammatory disorders, and diabetic retinopathy. VEGF stimulates endothelial cell proliferation and neovascularization via a MAPK-dependent pathway [85]. VEGF may also promote endothelial cell migration and vascular permeability [5]. The resulting vessel leakiness promotes interstitial edema and worsens hypoxia, further stimulating VEGF production. VEGF also provides endothelial cells with a cytoprotective, anti-apoptotic stimulus through Flk-1/KDR-mediated phosphorylation/activation of Akt [41]. Since Akt can stimulate VEGF expression, the latter may be part of an autocrine loop through which VEGF stimulates its own gene expression. Another loop may involve STAT3, as VEGF can activate STAT3 signaling in retinal microvascular endothelial cells via a

VEGFR2/STAT3 complex that induces STAT3 tyrosine phosphorylation, nuclear translocation and stimulation of VEGF expression [136].

7.4 Retinal Hypoxia and Retinopathy of Prematurity

Essentials

- Retinopathy of prematurity (ROP) is a disease of abnormal vasculogenesis in premature babies of very low weight
- The prevalent pathogenesis for ROP features an early occurring insult (hyperoxia, sepsis, etc.) that results in arrest of vasculogenesis that subsequently resumes without forward progress resulting in a ridge formation with neovascularization and sequelae such as hemorrhage and retinal detachment
- The pathophysiology of ROP involves the hyperoxia-induced vascular growth arrest (Phase I) and the hypoxia induced vascular growth (Phase II)
- Phase I involves the downregulation of angiogenic factors (VEGF, IGF-I) and the upregulation of anti-angiogenic factors [pigment epithelium-derived factor (PEDF), endothelin-1 and platelet-activating factor (PAF)]. Free radical intermediates generated from oxygen species or nitric oxide oxidants overwhelm the scavenging cellular mechanisms and result in microvascular dysfunction and apoptosis
- Phase II involves the HIF-1α- and HLF-mediated upregulation of VEGF
- IGF-1 is downregulated in ROP and the duration for which its levels are low correlates with the severity of the disease

7.4.1 Introduction

Retinopathy of prematurity is a disease of delayed or abnormal retinal vascular growth in premature babies especially of very low birth weight. Normally, the retina vasculogenesis begins in the 16th week of gestation as waves of mesenchymal spindle cells originating from the optic nerve lead shunts that are later covered by endothelium forming lumens. The retina is normally completely vascularized at full term (40 weeks of gestation) [6]. However, in the premature neonate, the retina remains incompletely vascularized at the time of birth and, consequently, very susceptible to toxic insults. Several theories exist on the pathogenesis of ROP, featuring an early occurring insult (high oxygen concentration, vita-min E deficiency, sepsis, intraventricular hemorrhage) to the premature vessels growing to the nasal and temporal retinal periphery, resulting in the temporary arrest of vasculogenesis. After that initial injury the vessel growth resumes but the vessels start growing without forward progress forming a vascular ridge that can be of sizeable proportions. This vascular ridge can regress or progress into the vitreous cavity, resulting in hemorrhage and retina detachment [87]. Unchecked neovascularization can lead to local scarring, contracture, and detachment of the retina, with or without intraocular hemorrhage. Extensive retinal detachment in premature infants typically causes severe visual impairment, even with successful surgical reattachment of the retina [93]. Although low birth weight, low gestational age, and use of supplemental oxygen are the most important predisposing factors for ROP, it has been suggested that genetic factors may confer additional susceptibility [135].

7.4.2 Pathophysiology of Retinopathy of Prematurity

The pathophysiology of ROP can be separated into two distinct phases [116]: Phase I involves the hyperoxia-induced vascular growth arrest and Phase II involves the relative hypoxia-induced vascular growth. Multiple factors and signal transduction pathways operate in each phase, whereas it seems that single exposure to hyperoxia does not result in the pathologic sequela of ROP but a combination of hyperoxia and hyperbarism is necessary. Hyperbarism can impede further the retinal and choroidal flow by inducing vasoconstriction in the choriocapillaris [21].

Phase I is characterized by the oxidative damage, the downregulation of angiogenic factors such as VEGF and IGF-I [69], the upregulation of anti-angiogenic factors such as PEDF [27] and modulating factors such as endothelin-1 and PAF [8]. Upon exposure to high oxygen concentrations (hyperoxia) the retinal tissue acts like an oxygen sink that is especially susceptible to oxidative damage. That is due to its high content of fatty acids with labile double bonds and a constant exposure to light that by photoexcitation can induce free radical formation and peroxidation reaction products acting on these double bonds. Oxygen radicals, peroxynitrite, hydroxyl radicals and lipid peroxidation products that are formed overwhelm the endogenous antioxidant mechanisms and cause cellular damage by reacting with membrane lipids, nucleic acids and metal containing compounds. That leads to microvascular dysfunction and selective death of the vascular endothelium that is more susceptible to damage than the pericytes

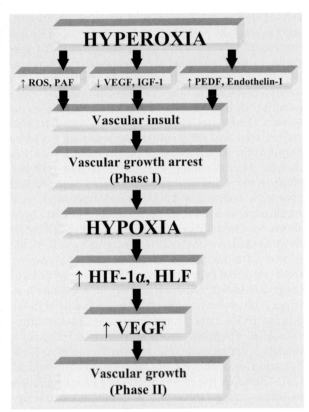

HYPEROXIA

↑ ROS, PAF   ↓ VEGF, IGF-1   ↑ PEDF, Endothelin-1

Vascular insult

**Vascular growth arrest
(Phase I)**

HYPOXIA

↑ **HIF-1α, HLF**

↑ **VEGF**

**Vascular growth
(Phase II)**

Fig. 7.2. Pathophysiology of phases I and II of retinopathy of prematurity. Phase I is characterized by vascular growth arrest that results from the hyperoxia-induced vascular insult through reactive oxygen species (ROS) and the downregulation of protective growth factors such as IGF. The resultant hyperoxia results in rebound increase of transcription factors such as HIF-1α and through them of angiogenic factors such as VEGF. This is a compensatory mechanism that promotes vascular growth in phase II

and the smooth muscle cells [19]. The importance of reactive oxygen intermediates in the pathogenesis of ROP is evident with administration of antioxidants to murine models and subsequent prevention of retinopathy. In parallel, protein levels of "free radical scavenger enzymes" such as superoxide dismutase (SOD) decrease upon exposure to constant hyperoxia whereas delivery of SOD via long-circulating liposomes reduces the oxygen-induced vaso-obliteration [90]. Among free radical donors, NO seems to play an important role in the hyperoxia-induced vasoattenuation of the early phases of ROP. NO derived oxidants such as peroxynitrite are generated in various models of ischemia and their levels are found to correlate with exposure to high concentrations of inspired oxygen in human infants who develop bronchopulmonary dysplasia. Peroxynitrite-mediated cellular injury includes the modification of important cellular proteins, the impairment of endothelial cell proliferation and migration and VEGF-

induced phosphorylation of the KDR receptor. Additionally it seems that NO and VEGF are reciprocally regulated [127]; NO donors inhibit VEGF expression in vivo whereas activation of VEGF receptors such as KDR activates endothelial NO synthetase (eNOS) that upregulates NO and inhibits VEGF expression via a negative feedback inhibition loop involving the AP-1 binding to the VEGF promoter. In agreement with the above the hyperoxia-induced VEGF downregulation in eNOS knock-out animals is less pronounced, contributing to the improved oxygen tolerance of their retinal capillaries in murine models of ROP. The same protective effect can be elicited with pharmacological inhibition of the NOS activity [19].

The oxidative stress can be augmented by the upregulation of PAF during this period. PAF seems to increase the generation of oxygen radicals and to also exert a direct and selective cytotoxic effect on the neurovascular retinal endothelial cells, inducing a form of death that seems to be intermediate between apoptosis and necrosis [68]. This cytotoxicity is independent of platelet aggregation and involves thromboxane TXA2, which along with PAF is a strong stimulant of calcium mobilization, an important element of both necrotic and apoptotic cell death processes. The PAF and TXA2 increase in cellular calcium results in the activation of specific phospholipases and proteases, the disruption of mitochondrial permeability transition pores and the arrest in ATP production that leads to energy starvation and sustains the cytotoxic cycle. The role of PAF in retinal vaso-obliteration has been demonstrated with the protective effect of PAF receptor blockers in murine models of ROP [8].

During Phase I, hyperoxia induces a downregulation in various VEGF isoforms and the VEGF receptors flk-1 and flt-1. Although the signal transduction pathway of the hyperoxia-induced VEGF downregulation is not fully elucidated, it seems to involve the kinase c-abl and to be influenced by the administration of dexamethazone and COX-2 inhibitors [92]. VEGF regression during the period of hyperoxia correlates well with the apoptosis of the vascular endothelium and the regression of newly formed capillaries. Parallel to the VEGF downregulation, PEDF, an anti-angiogenic factor, is upregulated during the period of hyperoxia [103, 114]. PEDF is likely to contribute to the regulation of blood vessel growth in the eye by creating a permissive environment for angiogenesis when oxygen is limiting (as it is in tumors and in retinopathies) and an inhibitory environment when oxygen concentrations are normal or high [114]. Given its high potency and the broad range of angiogenic inducers against which it can act, PEDF may prove to be a useful therapeutic for pathologic ocular neovascularization as well as for retinoblastomas, where its dual activities of inducing cell differentiation and inhibiting angiogenesis may be particularly effective [7].

Vasoconstriction during the hyperoxic phase can also be caused by the increased expression of endothelin-1 that is supposed to be secreted by the endothelial cells. The role of endothelin-1 is controversial and the signal transduction pathway involved is only partially known and involves the angiotensin-converting enzyme (ACE) [137]. ACE inhibition blocks hyperoxic-induced ET-1 secretion from retinal capillary endothelial cells and exerts a protective role in animal models of ROP [113].

Phase II of ROP is characterized by the relative hypoxia-induced proliferation of the retinal blood vessels when the infant is moved from the hyperoxic environment to normal room air. Hypoxia induces the upregulation of VEGF that leads to the retinal neovascularization and the invasion of the vitreous that characterizes clinical and experimental ROP [100, 121].The expression of VEGF in the innermost layers of retina falls during hyperoxia and increases on return to room air. Regression of retinal capillaries in neonatal rats exposed to high oxygen is preceded by a shut-off of VEGF production by nearby neuroglial cells. Intraocular injection of VEGF at the onset of experimental hyperoxia prevents apoptotic death of endothelial cells and rescues the retinal vasculature [3, 121].

The hypoxia-induced upregulation of VEGF is well studied and occurs through the activation of the transcriptional factor HIF-1α [131] and HLF (HIF-1α-like factor otherwise known as EPAS) [33, 75] and its binding to multiple response elements on the VEGF promoter. Knock-out mice for HLF do not exhibit the proliferative neovascular response in an ROP murine model, highlighting the role of that transcription factor in ROP [83]. Interestingly enough these mice show reduced expression levels of erythropoietin, an HIF-regulated growth factor, and restoration of the erythropoietin levels renders these mice susceptible to retinopathy at a significant level. It is conceivable that VEGF and erythropoietin collaborate in the development of neovascularization in response to hypoxia. Recently, a dominant negative regulator of HIFs, IPAS, was isolated by Makino et al. [77] and was shown to be expressed in response to hypoxia in the retinal ganglion cell layer (GCL) and inner nuclear layer (INL), suggesting that a more complicated mechanism occurs during the neovascular response in ROP. Although knock-out mice for HIF-1 are not viable, mice deficient for HIF-1α transcriptional targets such as RTP-801 demonstrate a significant attenuation in the development of major pathologic features of this model such as retinal vaso-obliteration, retinal neovascularization and apoptosis of the INL

cells [17]. The mechanism with which RTP-801 protects against the pathological manifestation of ROP is not completely understood. Although RTP-801 loss could protect endothelial cells against hyperoxia-induced apoptosis and subsequent hypoxia-induced neovascularization, the mechanism seems more complex. In murine ROP models there always seems to be a clear correlation between the non-perfused retinal area and the neovascular response, and this relationship is perturbed in the RTP-801 knock-out mouse. An interesting alternative is that RTP-801 absence prevents the hypoxia-induced neuroretinal apoptosis exerting a direct effect rather than as a result of reduced vascular disease. RTP-801 has been shown to modulate cellular ROS levels and cell sensitivity to oxidative stress [117]. Interestingly, RTP-801 is also a transcriptional target of p53 [32], another transcriptional factor involved in the pathogenesis of oxygen-induced retinopathy [104]. p53 is known to trigger the expression of several genes involved in the cellular redox control and contribute to the mitochondrial apoptotic cascade initiated in various models of retinal ischemia. HIF-1α regulates p53 stabilization and activity via interaction with mdm2 [23]. Therefore the p53 and HIF-1α redox regulation pathways merge through the combined transcriptional regulation of RTP-801 in ROP.

On the other hand, vasculogenesis in ROP is also regulated by non-oxygen regulated factors such as IGF-I, which inhibits the retinal neovascularization when at low levels and stimulates it at high levels. IGF-I knock-out mice exhibit abnormal retinal vasculogenesis despite the presence of normal VEGF levels [116]. Low IGF-I levels in vitro prevent VEGF-induced phosphorylation of Akt, a kinase that is critical for the vascular endothelial cell survival. IGF-I levels are low in premature infants that develop ROP in comparison to the age-matched infants without the disease. The duration for which the retinal IGF levels are low correlates with the severity of the disease, making IGF-I an important player in ROP.

ROP was very prevalent in developed countries in the 1940s and 1950s, but the realization that high O_2 was responsible for it led to more prudent use of O_2 and a decrease in ROP. Two factors that are currently leading to an increase in ROP incidence are the improved availability of intensive care facilities for premature infants in developing countries and the increased survival of even more premature infants (very low birth weight infants) in developed countries [135]. The current therapy for prevention of the disease is generally to titrate the infant's blood to an adequate PaO_2 but to prevent systemic hyperoxia. Vitamin E supplementation may also be useful in decreasing the severity of ROP, possibly by functioning as an antioxidant [18, 60].

7.5 Retinal Hypoxia and Diabetic Retinopathy

Essentials

- Diabetic retinopathy (DR) is classified into two groups based on the presence of abnormal neovessels: non-proliferative and proliferative
- Hypoxia plays a significant role in the pathogenesis of DR since early hyperoxia reverses aspects of vision loss
- Growth factors such as VEGF, TNF-α, MIF and endothelin and interleukins upregulate adhesion molecules in the endothelium and leukocytes, resulting in leukostasis, reduction in blood flow and direct leukocyte-mediated endothelial apoptosis involving the Fas/FasL pathway
- AGEs and protein glycation downregulate free radical scavengers and upregulate nitric oxide and angiogenic growth factors contributing to cellular demise
- Hypoxia and hyperglycemia activate the PCK Akt and MAPK pathways that in turn upregulate transcription factors such as NF-kB and HIF-1α that are central in the upregulation of growth factors and adhesion molecules that are central in the pathogenesis of DR

7.5.1 Introduction

Diabetes mellitus (DM) is the leading cause of blindness in people between the ages of 20 and 74 in the United States, and DR will eventually affect most Type I diabetics. Blindness is 25 times more common in patients with DM than in controls. DR is classified into two main groups: non-proliferative (mild, moderate, moderately severe and severe) and proliferative (mild, moderate and high risk) diabetic retinopathy. Non-proliferative diabetic retinopathy (NPDR) is characterized by increased vascular permeability, dilation and tortuosity of the retinal veins, abnormal vascular communications between arterioles and venules, microaneurysms, intraretinal hemorrhages and cotton wool spots (areas of infarction in the nerve layer). Microvascular angiopathy results in exudation of plasma from breakdown of the blood-retinal barrier. The reabsorption of the exuded fluid results in the deposition of protein and lipid exudates ("hard exudates"). Proliferative diabetic retinopathy (PDR) is marked by the formation of neovessels in the area of the optic disk (NVD) or elsewhere (NVE). Fifty percent of Type I and 10% of Type II diabetics who had DR for 15 years will have PDR, whereas the prevalence is higher in Type II patients who require insulin [61, 62].

7.5.2 Pathophysiology of Diabetic Retinopathy

Although tissue hypoxia has been for many years suggested to play a critical role in the progression of diabetic retinopathy, the exact time when hypoxia begins has so far remained unknown. Many animal models fail to demonstrate the existence of retinal hypoxia early in the course of diabetic retinopathy, whereas we know that by the time capillary non-perfusion is evident tissue hypoxia has already occurred [133]. The diabetic retina is hypoxic even when few microaneurysms are present and before capillary dropout is present [74]. It was also demonstrated that early hyperoxia reverses the contrast sensitivity deficits and oscillatory potential reductions, stressing the importance of hypoxia in the pathogenesis of the cardinal manifestations of DR [133].

It is now known that reduction in the retinal blood flow is one of the earlier signs of diabetic retinopathy. Leukostasis, the adhesion of leukocytes in the vascular endothelium, also occurs early and is at least partially responsible for the diminution of the blood flow. Leukocytes are found in higher numbers in diabetic individuals, they possess higher amounts of adhesion molecules (ICAM-1, VCAM-1) and they are less deformable than in normal individuals. The hyperglycemic and hyperosmotic environment in combination with the upregulated expression of multiple growth factors and cytokines such as VEGF, TNF-α and interleukins contributes to the enhanced expression of the adhesion molecules in the vascular endothelium and the leukocytes [54, 63]. It is also suggested that large periods of dark adaptation aggravate tissue hypoxia by depriving the inner retina of the small amount of oxygen diffusing from the choroid during light adaptation [133].

Leukocyte adhesion to the vascular endothelium results in the endothelial cell injury and death that leads to vascular leakage, acellular capillaries and ultimately capillary dropout and retinal hypoxia [54]. This is a continuing vicious cycle since retinal hypoxia enhances even further the adhesion of the leukocytes to the endothelium and the endothelial damage. The endothelial cell death involves a Fas/FasL mediated apoptotic cell death and is prevented by the administration of neutralizing FasL antibodies in a rat model of diabetic retinopathy [56]. Both Fas and FasL are upregulated in the endothelial cells and leukocytes respectively, during the course of diabetes from various cytokines such as interferon-γ, IL-4, TNF-α and IL-1β [54]. The crosslinking

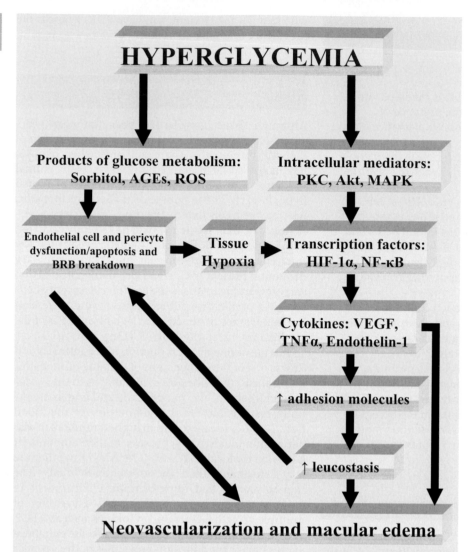

HYPERGLYCEMIA

**Products of glucose metabolism:
Sorbitol, AGEs, ROS**

**Intracellular mediators:
PKC, Akt, MAPK**

**Endothelial cell and pericyte
dysfunction/apoptosis and
BRB breakdown**

**Tissue
Hypoxia**

**Transcription factors:
HIF-1α, NF-κB**

**Cytokines: VEGF,
TNFα, Endothelin-1**

↑ adhesion molecules

↑ leucostasis

Neovascularization and macular edema

Fig. 7.3. Pathophysiology of hyperglycemia-induced macular edema and neovascularization. High glucose levels upregulate growth factors such as VEGF through the activation of kinases such as Akt and transcription factors such as HIF-1α. Upregulated VEGF promotes neovascularization directly and macular edema indirectly through the endothelial and pericyte dysfunction, and the breakdown of the blood-retina barrier

of Fas from FasL initiates an apoptotic process involving the activation of apical caspases such as caspase-8, the activation of mitochondria that funnels into the upregulation and activation of executional caspases such as caspase-3 that cleaves nucleic acids and important cellular proteins resulting in the cellular demise.

Chronic changes in the retinal metabolic pathways are central in DR and add to the hypoxic damage. Hyperglycemia results in increased intracellular glucose, accumulation of sorbitol (an intermediate metabolite), an increased lactate/pyruvate ratio and disturbance of the redox balance that results in cell damage. Protein glycation and AGEs as well as the downregulation of genes that code for free-radical scavengers contribute to the oxidative stress that results in the cellular demise, and micro- and macroangiopathy, further aggravating the cycle of hypoxia

and the upregulation of growth factors such as VEGF and TNF-α [138]. Hypoxia also activates local vascular loops that attempt to reverse the decreased vascular flow such as the stimulation of the adenosine receptor. This promotes the accumulation of superoxide radicals from the adenosine catabolism, which upregulates the nitric oxide synthase activity and ultimately affects gene expression of transcripts such as VEGF that reduce the retinal blood flow and contribute to the cellular damage [129].

Hypoxia and hyperglycemia activate the main players of multiple transduction pathways such as PKC and MAPK that contribute to the endothelial and pericyte dysfunction, the breakdown of the blood-retinal barrier and cellular death [139]. We have found that multiple activators of PKC such as diacylglycerol and PKC-binding proteins are upregulated early in diabetes whereas PKC inhibitors such as 14–3-3 are downregulated, contributing to the upregulation of VEGF [53]. In parallel, members of the ras family are activated, mediating the MAPK-induced VEGF upregulation. Activation of hsp proteins suchas hsp27 and hsp60 is also a consequence of the oxidative stress induced by hypoxia, although hsp70 is downregulated, likely contributing to the leukocyte-mediated endothelial damage observed in the diabetic retina [53]. Hypoxia and the subsequent generation of free radicals also results in the upregulation of eNOS and the generation of NO that mediates through akt the enhanced expression of adhesion molecules like ICAM-1 and growth factors such as VEGF [55]. It is possible that growth factors, cytokines, and reactive oxygen intermediates produced by inflammatory cells upregulate the proapoptotic genes as poorly in the endothelial cells and pericytes and collaborate with ischemia and oxidative stress to induce apoptosis.

Central to the pathogenesis of diabetic retinopathy is the upregulation of growth factors such as VEGF that play a major role in all its cardinal manifestations. The transcription of the VEGF gene is very sensitive to oxygen changes and it is strongly responsive to the hypoxic environment through the upregulation of the transcription factor HIF-1α. Insulin and IGF-I also upregulate VEGF during the course of diabetes via the orchestrated activation of HIF-1α through akt and NF-kB through MAPK that bind to adjacent responsive elements at the VEGF promoter [96]. Multiple proinflammatory cytokines are upregulated in the hypoxic environment of the diabetic retina. A few that will be mentioned are MIF, a lymphokine that enhances the adhesion of leukocytes in the vasculature, and endothelin B receptor [53]. Endothelin (ET), its ligand, is a potent vasoconstrictor and a permeability factor that works via a PKC-mediated mechanism and regulates extracellular matrix protein gene expression in target organs [22].

Hyperglycemia and hypoxia during diabetes result in the formation of vascular microaneurysms, venular dilatation, thickening of the retinal basement membrane, and microvascular contractile cell (pericyte) death, leading to acellular capillaries, which tend to undergo occlusion, causing retinal ischemia. "Cotton-wool" spots and soft exudates represent ischemic areas of the nerve-fiber retina layer. Platelet microthrombi can form, leading to capillary occlusion [13]. Hemorrhages and/or extravasation of fluid and retinal edema promote more hypoxia. Growth of new blood vessels in the retina in response to retinal hypoxia is the hallmark of proliferative DR.

7.6 Retinal Hypoxia and Vascular Occlusive Disease

Central and branch retinal vein occlusion (CRVO and BRVO, respectively) result in retinal ischemia, hypoxia and neovascularization. Scatter photocoagulation can be performed therapeutically to suppress fluid accumulation and limit preretinal and iris neovascularization [36]. Vitrectomy and retinal detachment procedures are occasionally required in patients with uncontrolled vitreous hemorrhage and retinal membrane formation, which threatens the integrity of the macula [99]. Scatter photocoagulation of these ischemic hypoxic areas restores the local retinal pO_2 to normal values within 2 weeks [99]. It has been proposed that photocoagulation decreases oxygen consumption in the outer retina by destroying photoreceptors. The glial scar that develops in their place allows oxygen to diffuse from the choriocapillaris into the retina without being consumed in the mitochondria of the photoreceptors, thus increasing inner retina O_2 availability. As a consequence of reversing hypoxia, VEGF production and neovascularization are suppressed. Vasoconstriction of existing vessels occurs, as well, due to elevated pO_2, thus increasing arteriolar resistance, decreasing hydrostatic pressure in capillaries and venules and suppressing fluid extravasation and edema formation, which in turn facilitates O_2 diffusion in the extracellular fluid and promotes O_2 delivery to cells, further downregulating VEGF production [120]. Hence, the reported inhibitory effect of photocoagulation on the development of retinal neovascularization could be due to a reversal effect on tissue hypoxia [99]. Vitrectomy also improves retinal oxygenation by allowing oxygen and other nutrients to be transported in water currents in the vitreous cavity from well-perfused, well-oxygenated to ischemic areas of the retina [120].

VEGF is upregulated in the eyes of animal models of CRVO, as well as in humans with CRVO [94]. Aqueous VEGF concentrations correlate with the onset, persistence, and regression of neovascularization; extent of retinal capillary non-perfusion; and vascular permeability. Following treatment with laser ablation of the hypoxic retina, neovascularization regressed only if VEGF concentrations were successfully reduced. Thus, a high aqueous VEGF level may suggest the need for initial or repeat treatment [16]. Retinal vein occlusion in the rat retina induced upregulation of VEGF and basic fibroblast growth factor (bFGF) expression, overall protein tyrosine-phosphorylation, as well as specific tyrosine-phosphorylation of PLCγ, PI3K, and MAPK within the distribution of the occluded vein [45]. BRVO increases the concentration of another HIF-depen-

7

dent factor, erythropoietin, in the vitreous fluid, which may act as an endogenous neuroprotective factor against ischemic retinal disorders [48]. On the other hand, pigment epithelium-derived factor (PEDF), a major antiangiogenic growth factor in the eye, is downregulated by hypoxia [27]. The PEDF concentrations in ocular fluids from patients with extensive non-diabetic retinal neovascularization caused by retinal vein occlusion were lower than in control patients. Levels of PEDF were restored in patients with retinal scatter photocoagulation compared with patients without previous photocoagulation [118]. These findings suggest that substitution of angiogenesis inhibitors may be an effective approach in the treatment of PDR.

Inhibition of the VEGF pathway, by either an antisense oligodeoxynucleotide against VEGF or an anti-VEGF antibody, delivered to the retina via intravitreal injection, has been demonstrated to reduce ocular angiogenesis in a non-human primate model of retinal vein occlusion [11, 1], suggesting that anti-VEGF therapy at an early stage of ischemic retinal vein occlusion may be therapeutically beneficial. Similarly, adenovirus-mediated VHL intraocular gene transfer inhibited the development of angiogenesis in a monkey model of multiple BRVO [2], confirming that the HIF/VHL pathway is crucial in the upregulation of VEGF and induction of ocular neovascularization in this condition and suggesting that gene therapy based on VHL gene delivery has potential in the treatment of human ocular neovascularization.

7.7 Other Diseases

7.7.1 Cystoid Macular Edema

Essentials
- Oxygen supplementation reduces foveal thickness in macular edema, likely reducing hypoxic damage

Oxygen plays a pathogenetic role in other diseases also. It was recently shown that supplemental oxygen with a nasal cannula reduced foveal thickness by an average of 42.5% in patients with cystoid macular edema [89]. Although the pathophysiology behind this finding is unclear, it is known that oxygen can exert its beneficial actions independently of the atmospheric pressure. It is possible that improved oxygenation breaks the cycle of hypoxia and edema in the diabetic retina, a hypothesis that is supported by the worsening of the edema after the discontinuation of the oxygen laser in all the eyes except the ones that received additional focal laser therapy. A combi-

nation of laser treatment with supplemental oxygenation can be argued for the refractory cases of cystoid macular edema. In these cases supplemental oxygenation can reduce macular thickness below a critical point so that the benefits of focal laser photocoagulation may be realized.

7.7.2 Retinal Degeneration

Essentials
- Rod loss that characterizes retinal degeneration increases oxygen concentration in the outer retina likely resulting in caspase independent oxidative cone loss that activates the mitochondria in later stages and is not associated with DNA fragmentation
- Increased inner retina oxygen concentration results in increased oxygen diffusion in the outer retina with subsequent capillary loss and cellular damage
- Reduction of oxygen flow in animal models of retinal degeneration prevents hypoxic damage

Oxygen is also central to the pathogenesis of the retinal vasculature obliteration that accompanies photoreceptor loss in retinal degenerative diseases. It is believed that when the photoreceptors die oxygen diffuses freely from the choroid to the inner retina causing vasoconstriction of the major vessels and capillary loss. Reduction of the oxygen influx in animal models of retinal degenerative diseases when they are maintained under hypoxic conditions prevents this capillary loss [95].

Because choroidal levels do not have autoregulatory mechanisms controlled by oxygen levels, oxygen levels in the outer retina vary according to oxygen consumption. Therefore one could expect that during the course of retinitis pigmentosa (RP) as the rods die and the oxygen consumption in a particular retinal area decreases, the oxygen levels increase. It was indeed shown in transgenic rat models of RP that the oxygen levels in the outer retina increase significantly as the rods degenerate and decrease in number [141].

Since it is known that high levels of oxygen are damaging to the photoreceptors, it was hypothesized that cone death in retinal degenerations could represent a form of oxidative damage [122, 126]. It was recently shown in a transgenic rhodopsin pig model of RP that after the rods degenerate the cones show evidence of oxidative damage by expressing high amounts of biomarkers for oxidative damage to lipids proteins and DNA. From the above we can naturally assume that retinal areas that have the highest concentration of rod photoreceptors such as the periphery would be affected first in retinal degenerations, whereas areas with high cone concentrations such as the macula would be spared; that is indeed the case.

Oxidative stress results in the generation of reactive oxygen species (ROS) that leads to the oxidative inactivation of the caspases [59, 134]. It seems that the oxidative damage has the key elements of an apoptotic cell death but is not dependent on caspase activation. In vitro and in animal models of RP this form of oxidative retinal cell apoptosis is not accompanied by caspase activation and is not inhibited by caspase inhibitors [30, 142]. One of the earliest events during ROS-induced photoreceptor death is the exposure of phosphatidylserine (PS) in the outer surface of the cellular membrane. Although PS exposure is traditionally a caspase dependent process, in this case it seems that it follows alternative regulatory mechanisms mainly controlled by ROS. Although in traditional forms of apoptosis the mitochondria are activated early and act as a positive loop in the activation of caspases, in this caspase independent oxidative death it seems that mitochondrial depolarization is a late step. This apoptotic death is not associated with low molecular weight DNA fragmentation at least in vitro that is a caspase dependent process and could involve other endonucleases. It is interesting to hypothesize that wide variability in the clinical phenotype of these diseases can be attributed to polymorphisms in key antioxidant enzymes such as superoxide dismutase, glutathione peroxidase, and metallothionines.

7.7.3 Glaucoma

Essentials

- Hyperbaric oxygen treatment improves visual field outcomes in glaucoma without affecting the intraocular pressure (IOP)
- Glaucoma is associated with vascular abnormalities such as vasospasm and hypotension that could be associated with altered microcirculation in the central retinal artery and hypoxic injury to the ganglion or supporting cells
- Glutamate, reactive oxygen species, nitric oxide, calcium and TNF-α are among the proposed mediators of the hypoxic damage
- HIF-1α is upregulated in animal models of glaucoma with differential localization and expression among the retinal cell subtypes that could be translated to differential sensitivity to glaucomatous damage

Growing evidence supports a vascular hypothesis in the pathogenesis of glaucoma. Patients with glaucoma treated with hyperbaric oxygen, a treatment that does not interfere with IOP regulation, demonstrated improved visual field outcomes in relation to controls [15]. Retinal photocoagulation in an animal model of glaucoma with subsequent increase in the inner retinal oxygen distribution results in increased ganglion cell survival [91]. Vascular abnormalities that occur in glaucoma patients such as vasospasm, hypotension, and various vascular perfusion defects proven by angiographic studies lead investigators to believe that reduced vascular perfusion in the optic nerve can be one of the pathogenetic mechanisms in glaucoma. The altered microcirculation in the central retinal artery can be secondary to the IOP or secondary to abnormal blood pressure, vasospasm, hemorrhage, autoregulation, increased viscosity or altered rheological characteristics of the blood. The hypoxic/oligemic insult results then in decreased oxygen supply that can contribute to the ganglion cell death in patients with glaucoma. The hypoxic insult can also result in the death of astrocytes and supporting microglia and ganglion cell axons that can account for the cupping associated with glaucoma before the ganglion cell loss. It was hypothesized that the oligemic/hypoxic insult results in astrocytic injury that is propagated among various cell types (Müller, microglia) in the form of depolarization that is called spreading depression where various ions such as calcium and potassium and glutamate play a role [73]. The dysfunctioning astrocytes can then secrete various substances such as nitric oxide, prostaglandins, and D-serine that could exacerbate the injury to the ganglion cells and the supporting glial cells. Additionally the lamina cribrosa of the optic nerve is considered a transitional zone between an efficient (myelinated) and inefficient (non-myelinated) system of action potential transmission and therefore is more susceptible to hypoxic injuries. That would lead to glutamate "leakage" from the ganglion cells and eventually increase its extracellular levels when the Müller cells' capacity to remove it from the extracellular space is saturated. The increased levels of glutamate would differentially injure the ganglion cells according to their excitatory receptor profile [31].

Many mediators and molecular pathways have been proposed to mediate the hypoxic damage such as reactive oxygen species, glutamate excitotoxicity, nitric oxide mediated oxidative stress [88], calcium toxicity [65], and tumor necrosis factor-mediated apoptosis [124]. In murine models of glaucoma, HIF-1α was shown to be expressed in higher levels compared to normal controls and exhibit a spatial relationship with the functional damage that is found in these eyes. HIF-1α immunoreactivity is higher in the inner retinal layers, giving rise to the hypothesis that ischemic axons in the optic nerve head initiate a retrograde signal to their cell bodies, upregulating this transcriptional factor [125]. Although the most prominent increase occurs in the inner retinal layer, one cannot exclude the possibility that all layers are hypoxic and this response is the result of a differential regulation of HIF-1α. It is also intriguing that HIF-1α immunolocalization is cytoplasmic in the retinal ganglion cells and nuclear in the glial cells probably due to a differential activity of the regulatory signaling cascades in the different subtypes of the retinal cells. Differential HIF-1α expression could explain the differential sensitivity to the glaucomatous damage. It is still unclear what are the molecular pathways leading to the HIF-1α upregulation although nitric oxide, cytokines such as TNF-α and reactive oxygen species have been implicated. One of the apoptotic genes that HIF-1α activates and is implicated in the glaucomatous and excitotoxic damage is p53 [34]. It is possible that HIF-1α initiates a cell death program in glaucomatous eyes through p53, which could be a transcriptional activator of neuronal apoptosis.

7.7.4 Retinal Detachment

Essentials
- Hyperoxia has a protective effect in animal models of retinal detachment by restoring the influx of oxygen from the choroids that is separated from the outer retina

Hypoxia may also be a contributing factor in vision loss in retinal detachment. It has been recently hypothesized that hyperoxia may be beneficial in preventing photoreceptor damage in retinal detachment [80, 106]. Detachment separates the inner segments from their O_2 supply, and although no consuming tissue is found under the retina, the increased distance reduces the flux of O_2 from the choroid to the photoreceptor inner segments. Hyperoxia should restore this flux, at least for detachments of moderate height, because increased choroidal pO_2 can compensate for the increased distance. For large detachments, the photoreceptors may benefit more from increased amounts of O_2 in the retinal circulation than in the choroidal circulation. The protective effect of hyperoxia on detached photoreceptors has been shown experimentally in cats, in which hyperoxia was able to save photoreceptors and prevent the activation of retinal glia normally caused by detachment.

7.8 Conclusions

Retinal hypoxia occurs in a diverse spectrum of ophthalmological disorders and is a potent stimulus for neovascularization. As visual acuity depends on the transparency of the eye structures, abnormal neovascularization is vision threatening. The VHL/HIF pathway is a pivotal regulator of the response to hypoxia and VEGF is its main effector cytokine. Pharmacological suppression of the VHL/HIF/VEGF could be an important advance for the treatment of such neovascularization and the preservation of vision.

References

1. Adamis AP, et al. (1996) Inhibition of vascular endothelial growth factor prevents retinal ischemia-associated iris neovascularization in a nonhuman primate. Arch Ophthalmol 114(1):66–71
2. Akiyama H., et al. (2004) Inhibition of ocular angiogenesis by an adenovirus carrying the human von Hippel-Lindau tumor-suppressor gene in vivo. Invest Ophthalmol Vis Sci 45(5):1289–96
3. Alon T, et al. (1995) Vascular endothelial growth factor acts as a survival factor for newly formed retinal vessels and has implications for retinopathy of prematurity. Nat Med 1(10):1024–8
4. Ames A, 3rd, et al. (1992) Energy metabolism of rabbit retina as related to function: high cost of Na⁺ transport. J Neurosci 12(3):840–53
5. Antonetti DA, et al. (1998) Vascular permeability in experimental diabetes is associated with reduced endothelial occludin content: vascular endothelial growth factor decreases occludin in retinal endothelial cells. Penn State Retina Research Group. Diabetes 47(12):1953–9
6. Azad R, Chandra P (2005) Retinopathy of prematurity. J Indian Med Assoc 103(7):370–2
7. Barnstable CJ, Tombran-Tink J (2004) Neuroprotective and antiangiogenic actions of PEDF in the eye: molecular targets and therapeutic potential. Prog Retin Eye Res 23(5):561–77
8. Beauchamp MH, et al. (2002) Platelet-activating factor in vasoobliteration of oxygen-induced retinopathy. Invest Ophthalmol Vis Sci 43(10):3327–37
9. Becerra SP, Amaral J (2002) Erythropoietin – an endogenous retinal survival factor. N Engl J Med 347(24):1968–70
10. Bergeron M, et al. (1999) Induction of hypoxia-inducible factor-1 (HIF-1) and its target genes following focal ischaemia in rat brain. Eur J Neurosci 11(12):4159–70
11. Bhisitkul RB, et al. (2005) An antisense oligodeoxynucleotide against vascular endothelial growth factor in a nonhuman primate model of iris neovascularization. Arch Ophthalmol 123(2):214–9
12. Blancher C, et al. (2000) Relationship of hypoxia-inducible factor (HIF)-1alpha and HIF-2alpha expression to vascular endothelial growth factor induction and hypoxia survival in human breast cancer cell lines. Cancer Res 60(24):7106–13
13. Boeri D, Maiello M, Lorenzi M (2001) Increased prevalence of microthromboses in retinal capillaries of diabetic individuals. Diabetes 50(6):1432–9

14. Bohling T, et al. (1996) Expression of growth factors and growth factor receptors in capillary hemangioblastoma. J Neuropathol Exp Neurol 55(5):522–7
15. Bojic L, et al. (1993) The effect of hyperbaric oxygen breathing on the visual field in glaucoma. Acta Ophthalmol (Copenh) 71(3):315–9
16. Boyd SR, et al. (2002) Correlation of increased vascular endothelial growth factor with neovascularization and permeability in ischemic central vein occlusion. Arch Ophthalmol 120(12):1644–50
17. Brafman A, et al. (2004) Inhibition of oxygen-induced retinopathy in RTP801-deficient mice. Invest Ophthalmol Vis Sci 45(10):3796–805
18. Brion LP, Bell EF, Raghuveer TS (2003) Vitamin E supplementation for prevention of morbidity and mortality in preterm infants. Cochrane Database Syst Rev 4:CD003665
19. Brooks SE, et al. (2001) Reduced severity of oxygen-induced retinopathy in eNOS-deficient mice. Invest Ophthalmol Vis Sci 42(1):222–8
20. Bruick RK (2000) Expression of the gene encoding the pro-apoptotic Nip3 protein is induced by hypoxia. Proc Natl Acad Sci U S A 97(16):9082–7
21. Calvert JW, Zhou C, Zhang JH (2004) Transient exposure of rat pups to hyperoxia at normobaric and hyperbaric pressures does not cause retinopathy of prematurity. Exp Neurol 189(1):150–61
22. Chakrabarti S, et al. (1998) Augmented retinal endothelin-1, endothelin-3, endothelinA and endothelinB gene expression in chronic diabetes. Curr Eye Res 17(3):301–7
23. Chen D, et al. (2003) Direct interactions between HIF-1alpha and Mdm2 modulate p53 function. J Biol Chem 278(16):13595–8
24. Corn PG, et al. (2003) Tat-binding protein-1, a component of the 26S proteasome, contributes to the E3 ubiquitin ligase function of the von Hippel-Lindau protein. Nat Genet 35(3):229–37
25. Czyzyk-Krzeska MF, Meller J (2004) von Hippel-Lindau tumor suppressor: not only HIF's executioner. Trends Mol Med 10(4):146–9
26. Danis RP, Bingaman DP (1997) Insulin-like growth factor-1 retinal microangiopathy in the pig eye. Ophthalmology 104(10):1661–9
27. Dawson DW, et al. (1999) Pigment epithelium-derived factor: a potent inhibitor of angiogenesis. Science 285(5425):245–8
28. Dawson TM (2002) Preconditioning-mediated neuroprotection through erythropoietin? Lancet 359(9301):96–7
29. Digicaylioglu M, Lipton SA (2001) Erythropoietin-mediated neuroprotection involves cross-talk between Jak2 and NF-kappaB signalling cascades. Nature 412(6847):641–7
30. Doonan F, Donovan M, Cotter TG (2003) Caspase-independent photoreceptor apoptosis in mouse models of retinal degeneration. J Neurosci 23(13):5723–31
31. Dreyer EB, et al. (1996) Elevated glutamate levels in the vitreous body of humans and monkeys with glaucoma. Arch Ophthalmol 114(3):299–305
32. Ellisen LW, et al. (2002) REDD1, a developmentally regulated transcriptional target of p63 and p53, links p63 to regulation of reactive oxygen species. Mol Cell 10(5):995–1005
33. Ema M, et al. (1999) Molecular mechanisms of transcription activation by HLF and HIF1alpha in response to hypoxia: their stabilization and redox signal-induced interaction with CBP/p300. EMBO J 18(7):1905–14
34. Fels DR, Koumenis C (2005) HIF-1alpha and p53: the ODD couple? Trends Biochem Sci 30(8):426–9
35. Ferrara N, et al. (1992) Molecular and biological properties of the vascular endothelial growth factor family of proteins. Endocr Rev 13(1):18–32
36. Finkelstein D (1996) Laser therapy for central retinal vein obstruction. Curr Opin Ophthalmol 7(3):80–3
37. Fisher JW (1997) Erythropoietin: physiologic and pharmacologic aspects. Proc Soc Exp Biol Med 216(3):358–69
38. Fisher JW (2003) Erythropoietin: physiology and pharmacology update. Exp Biol Med (Maywood) 228(1):1–14
39. Flamme I, Krieg M, Plate KH (1998) Up-regulation of vascular endothelial growth factor in stromal cells of hemangioblastomas is correlated with up-regulation of the transcription factor HRF/HIF-2alpha. Am J Pathol 153(1):25–9
40. Folkman J (2000) Incipient angiogenesis. J Natl Cancer Inst 92(2):94–5
41. Gerber HP, et al. (1998) Vascular endothelial growth factor regulates endothelial cell survival through the phosphatidylinositol 3'-kinase/Akt signal transduction pathway. Requirement for Flk-1/KDR activation. J Biol Chem 273(46):30336–43
42. Grimm C, et al. (2002) HIF-1-induced erythropoietin in the hypoxic retina protects against light-induced retinal degeneration. Nat Med 8(7):718–24
43. Guillemin K, Krasnow MA (1997) The hypoxic response: huffing and HIFing. Cell 89(1):9–12
44. Hatva E, et al. (1996) Vascular growth factors and receptors in capillary hemangioblastomas and hemangiopericytomas. Am J Pathol 148(3):763–75
45. Hayashi A, Kim HC, de Juan E, Jr (1999) Alterations in protein tyrosine kinase pathways following retinal vein occlusion in the rat. Curr Eye Res 18(3):231–9
46. Huang LE, et al. (1998) Regulation of hypoxia-inducible factor 1alpha is mediated by an O_2-dependent degradation domain via the ubiquitin-proteasome pathway. Proc Natl Acad Sci U S A 95(14):7987–92
47. Ijichi A, Sakuma S, Tofilon PJ (1995) Hypoxia-induced vascular endothelial growth factor expression in normal rat astrocyte cultures. Glia 14(2):87–93
48. Inomata Y, et al. (2004) Elevated erythropoietin in vitreous with ischemic retinal diseases. Neuroreport 15(5): 877–9
49. Iyer NV, et al. (1998) Cellular and developmental control of O_2 homeostasis by hypoxia-inducible factor 1 alpha. Genes Dev 12(2):149–62
50. Iyer NV, Leung SW, Semenza GL (1998) The human hypoxia-inducible factor 1alpha gene: HIF1A structure and evolutionary conservation. Genomics 52(2):159–65
51. Jewell UR, et al. (2001) Induction of HIF-1alpha in response to hypoxia is instantaneous. FASEB J 15(7):1312–4
52. Jiang BH, et al. (1997) Transactivation and inhibitory domains of hypoxia-inducible factor 1alpha. Modulation of transcriptional activity by oxygen tension. J Biol Chem 272(31):19253–60
53. Joussen AM, et al. (2001) In vivo retinal gene expression in early diabetes. Invest Ophthalmol Vis Sci 42(12):3047–57
54. Joussen AM, et al. (2002) Nonsteroidal anti-inflammatory drugs prevent early diabetic retinopathy via TNF-alpha suppression. FASEB J 16(3):438–40
55. Joussen AM, et al. (2002) Retinal vascular endothelial growth factor induces intercellular adhesion molecule-1 and endothelial nitric oxide synthase expression and initiates early diabetic retinal leukocyte adhesion in vivo. Am J Pathol 160(2):501–9
56. Joussen AM, et al. (2003) Suppression of Fas-FasL-induced endothelial cell apoptosis prevents diabetic blood-retinal

barrier breakdown in a model of streptozotocin-induced diabetes. FASEB J 17(1):76–8

57. Kallio PJ, et al. (1998) Signal transduction in hypoxic cells: inducible nuclear translocation and recruitment of the CBP/p300 coactivator by the hypoxia-inducible factor-1alpha. EMBO J 17(22):6573–86

58. Kallio PJ, et al. (1999) Regulation of the hypoxia-inducible transcription factor 1alpha by the ubiquitin-proteasome pathway. J Biol Chem 274(10):6519–25

59. Kasahara E, et al. (2005) SOD2 protects against oxidation-induced apoptosis in mouse retinal pigment epithelium: implications for age-related macular degeneration. Invest Ophthalmol Vis Sci 46(9):3426–34

60. Katz ML, Robison WG, Jr (1988) Autoxidative damage to the retina: potential role in retinopathy of prematurity. Birth Defects Orig Artic Ser 24(1):237–48

61. Klein R, et al. (1994) The Wisconsin Epidemiologic Study of diabetic retinopathy. XIV. Ten-year incidence and progression of diabetic retinopathy. Arch Ophthalmol 112(9):1217–28

62. Klein R, Klein BE, Moss SE (1989) The Wisconsin epidemiological study of diabetic retinopathy: a review. Diabetes Metab Rev 5(7):559–70

63. Koizumi K, et al. (2003) Contribution of TNF-alpha to leukocyte adhesion, vascular leakage, and apoptotic cell death in endotoxin-induced uveitis in vivo. Invest Ophthalmol Vis Sci 44(5):2184–91

64. Koshiji M, et al. (2005) HIF-1alpha induces genetic instability by transcriptionally downregulating MutSalpha expression. Mol Cell 17(6):793–803

65. Kristian T, Siesjo BK (1996) Calcium-related damage in ischemia. Life Sci 59(5–6):357–67

66. Kubasiak LA, et al. (2002) Hypoxia and acidosis activate cardiac myocyte death through the Bcl-2 family protein BNIP3. Proc Natl Acad Sci U S A 99(20):12825–30

67. Lee JY, et al. (1998) Loss of heterozygosity and somatic mutations of the VHL tumor suppressor gene in sporadic cerebellar hemangioblastomas. Cancer Res 58(3):504–8

68. Lemasters JJ (1999) Mechanisms of hepatic toxicity. V. Necrapoptosis and the mitochondrial permeability transition: shared pathways to necrosis and apoptosis. Am J Physiol 276(1):G1–6

69. Leske DA, et al. (2004) The role of VEGF and IGF-1 in a hypercarbic oxygen-induced retinopathy rat model of ROP. Mol Vis 10:43–50

70. Levy AP, et al. (1995) Transcriptional regulation of the rat vascular endothelial growth factor gene by hypoxia. J Biol Chem 270(22):13333–40

71. Levy AP, Levy NS, Goldberg MA (1996) Hypoxia-inducible protein binding to vascular endothelial growth factor mRNA and its modulation by the von Hippel-Lindau protein. J Biol Chem 271(41):25492–7

72. Levy AP, Levy NS, Goldberg MA (1996) Post-transcriptional regulation of vascular endothelial growth factor by hypoxia. J Biol Chem 271(5):2746–53

73. Lian XY, Stringer JL (2004) Energy failure in astrocytes increases the vulnerability of neurons to spreading depression. Eur J Neurosci 19(9):2446–54

74. Linsenmeier RA, et al. (1998) Retinal hypoxia in long-term diabetic cats. Invest Ophthalmol Vis Sci 39(9): 1647–57

75. Liu LX, et al. (2002) Stabilization of vascular endothelial growth factor mRNA by hypoxia-inducible factor 1. Biochem Biophys Res Commun 291(4):908–14

76. Maher ER, Kaelin WG, Jr (1997) von Hippel-Lindau disease. Medicine (Baltimore) 76(6):381–91

77. Makino Y, et al. (2001) Inhibitory PAS domain protein is a negative regulator of hypoxia-inducible gene expression. Nature 414(6863):550–4

78. Maltepe E, et al. (1997) Abnormal angiogenesis and responses to glucose and oxygen deprivation in mice lacking the protein ARNT. Nature 386(6623):403–7

79. Mazure NM, et al. (1997) Induction of vascular endothelial growth factor by hypoxia is modulated by a phosphatidyl-inositol 3-kinase/Akt signaling pathway in Ha-ras-transformed cells through a hypoxia inducible factor-1 transcriptional element. Blood 90(9):3322–31

80. Mervin K, et al. (1999) Limiting photoreceptor death and deconstruction during experimental retinal detachment: the value of oxygen supplementation. Am J Ophthalmol 128(2):155–64

81. Meyer-Schwickerath R, et al. (1993) Vitreous levels of the insulin-like growth factors I and II, and the insulin-like growth factor binding proteins 2 and 3, increase in neovascular eye disease. Studies in nondiabetic and diabetic subjects. J Clin Invest 92(6):2620–5

82. Mitsiades CS, Mitsiades N, Koutsilieris M (2004) The Akt pathway: molecular targets for anti-cancer drug development. Curr Cancer Drug Targets 4(3):235–56

83. Morita M, et al. (2003) HLF/HIF-2alpha is a key factor in retinopathy of prematurity in association with erythropoietin. EMBO J 22(5):1134–46

84. Mukhopadhyay D, et al. (1997) The von Hippel-Lindau tumor suppressor gene product interacts with Sp1 to repress vascular endothelial growth factor promoter activity. Mol Cell Biol 17(9):5629–39

85. Murata M, Kador PF, Sato S (2000) Vascular endothelial growth factor (VEGF) enhances the expression of receptors and activates mitogen-activated protein (MAP) kinase of dog retinal capillary endothelial cells. J Ocul Pharmacol Ther 16(4):383–91

86. Na X, et al. (2003) Identification of the RNA polymerase II subunit hsRPB7 as a novel target of the von Hippel-Lindau protein. EMBO J22(16):4249–59

87. Neely KA, Gardner TW (1998) Ocular neovascularization: clarifying complex interactions. Am J Pathol 153(3): 665–70

88. Neufeld AH, Sawada A, Becker B (1999) Inhibition of nitric-oxide synthase 2 by aminoguanidine provides neuroprotection of retinal ganglion cells in a rat model of chronic glaucoma. Proc Natl Acad Sci USA 96(17):9944–8

89. Nguyen QD, et al. (2004) Supplemental oxygen improves diabetic macular edema: a pilot study. Invest Ophthalmol Vis Sci 45(2):617–24

90. Niesman MR, Johnson KA, Penn JS (1997) Therapeutic effect of liposomal superoxide dismutase in an animal model of retinopathy of prematurity. Neurochem Res 22(5): 597–605

91. Nork TM, et al. (2000) Protection of ganglion cells in experimental glaucoma by retinal laser photocoagulation. Arch Ophthalmol 118(9):1242–50

92. Nunes I, et al. (2001) c-abl is required for the development of hyperoxia-induced retinopathy. J Exp Med 193(12):1383–91

93. Palmer EA (2003) Implications of the natural course of retinopathy of prematurity. Pediatrics 111(4):885–6

94. Pe'er J, et al. (1998) Vascular endothelial growth factor upregulation in human central retinal vein occlusion. Ophthalmology 105(3):412–6

95. Penn JS, Li S, Naash MI (2000) Ambient hypoxia reverses retinal vascular attenuation in a transgenic mouse model of autosomal dominant retinitis pigmentosa. Invest Ophthalmol Vis Sci 41(12):4007–13

96. Poulaki V, et al. (2002) Acute intensive insulin therapy exacerbates diabetic blood-retinal barrier breakdown via hypoxia-inducible factor-1alpha and VEGF. J Clin Invest 109(6):805–15

97. Poulaki V, et al. (2003) Regulation of vascular endothelial growth factor expression by insulin-like growth factor I in thyroid carcinomas. J Clin Endocrinol Metab 88(11): 5392–8

98. Poulaki V, et al. (2004) Insulin-like growth factor-I plays a pathogenetic role in diabetic retinopathy. Am J Pathol 165(2):457–69

99. Pournaras CJ, et al. (1990) Scatter photocoagulation restores tissue hypoxia in experimental vasoproliferative microangiopathy in miniature pigs. Ophthalmology 97(10):1329–33

100. Provis JM, et al. (1997) Development of the human retinal vasculature: cellular relations and VEGF expression. Exp Eye Res 65(4):555–68

101. Pugh CW, et al. (1997) Activation of hypoxia-inducible factor-1; definition of regulatory domains within the alpha subunit. J Biol Chem 272(17):11205–14

102. Punglia RS, et al. (1997) Regulation of vascular endothelial growth factor expression by insulin-like growth factor I. Diabetes 46(10):1619–26

103. Reynolds JD (2001) The management of retinopathy of prematurity. Paediatr Drugs 3(4):263–72

104. Rosenbaum DM, et al. (1998) The role of the p53 protein in the selective vulnerability of the inner retina to transient ischemia. Invest Ophthalmol Vis Sci 39(11):2132–9

105. Ryan HE, Lo J, Johnson RS (1998) HIF-1 alpha is required for solid tumor formation and embryonic vascularization. EMBO J 17(11):3005–15

106. Sakai T, et al. (2001) The ability of hyperoxia to limit the effects of experimental detachment in cone-dominated retina. Invest Ophthalmol Vis Sci 42(13):3264–73

107. Salceda S, Caro J (1997) Hypoxia-inducible factor 1alpha (HIF-1alpha) protein is rapidly degraded by the ubiquitin-proteasome system under normoxic conditions. Its stabilization by hypoxia depends on redox-induced changes. J Biol Chem 272(36):22642–7

108. Selak MA, et al. (2005) Succinate links TCA cycle dysfunction to oncogenesis by inhibiting HIF-alpha prolyl hydroxylase. Cancer Cell 7(1):77–85

109. Semenza GL (2000) HIF-1 and human disease: one highly involved factor. Genes Dev 14(16):1983–91

110. Semenza GL (2004) Hydroxylation of HIF-1: oxygen sensing at the molecular level. Physiology (Bethesda) 19: 176–82

111. Semenza GL, et al. (1996) Assignment of the hypoxia-inducible factor 1alpha gene to a region of conserved synteny on mouse chromosome 12 and human chromosome 14q. Genomics 34(3):437–9

112. Senger DR, et al. (1983) Tumor cells secrete a vascular permeability factor that promotes accumulation of ascites fluid. Science 219(4587):983–5

113. Sharma J, et al. (2003) Ibuprofen improves oxygen-induced retinopathy in a mouse model. Curr Eye Res 27(5):309–14

114. Shih SC, et al. (2003) Selective stimulation of VEGFR-1 prevents oxygen-induced retinal vascular degeneration in retinopathy of prematurity. J Clin Invest 112(1):50–7

115. Shoshani T, et al. (2002) Identification of a novel hypoxia-inducible factor 1-responsive gene, RTP801, involved in apoptosis. Mol Cell Biol 22(7):2283–93

116. Smith LE (2003) Pathogenesis of retinopathy of prematurity. Semin Neonatol 8(6):469–73

117. Sowter HM, et al. (2001) HIF-1-dependent regulation of hypoxic induction of the cell death factors BNIP3 and NIX in human tumors. Cancer Res 61(18):6669–73

118. Spranger J, et al. (2001) Loss of the antiangiogenic pigment epithelium-derived factor in patients with angiogenic eye disease. Diabetes 50(12):2641–5

119. Staller P, et al. (2003) Chemokine receptor CXCR4 down-regulated by von Hippel-Lindau tumour suppressor pVHL. Nature 425(6955):307–11

120. Stefansson E (2001) The therapeutic effects of retinal laser treatment and vitrectomy. A theory based on oxygen and vascular physiology. Acta Ophthalmol Scand 79(5):435–40

121. Stone J, et al. (1996) Roles of vascular endothelial growth factor and astrocyte degeneration in the genesis of retinopathy of prematurity. Invest Ophthalmol Vis Sci 37(2):290–9

122. Stone J, et al. (1999) Mechanisms of photoreceptor death and survival in mammalian retina. Prog Retin Eye Res 18(6):689–735

123. Sutter CH, Laughner E, Semenza GL (2000) Hypoxia-inducible factor 1alpha protein expression is controlled by oxygen-regulated ubiquitination that is disrupted by deletions and missense mutations. Proc Natl Acad Sci USA 97(9): 4748–53

124. Tezel G, Wax MB (2000) Increased production of tumor necrosis factor-alpha by glial cells exposed to simulated ischemia or elevated hydrostatic pressure induces apoptosis in cocultured retinal ganglion cells. J Neurosci 20(23): 8693–700

125. Tezel G, Wax MB (2004) Hypoxia-inducible factor 1alpha in the glaucomatous retina and optic nerve head. Arch Ophthalmol 122(9):1348–56

126. Travis GH (1991) Molecular characterization of the retinal degeneration slow (rds) mutation in mouse. Prog Clin Biol Res 362:87–114

127. Tsurumi Y, et al. (1997) Reciprocal relation between VEGF and NO in the regulation of endothelial integrity. Nat Med 3(8):879–86

128. Uchida T, et al. (2004) Prolonged hypoxia differentially regulates hypoxia-inducible factor (HIF)-1alpha and HIF-2alpha expression in lung epithelial cells: implication of natural antisense HIF-1alpha. J Biol Chem 279(15): 14871–8

129. Vasquez G, et al. (2004) Role of adenosine transport in gestational diabetes-induced L-arginine transport and nitric oxide synthesis in human umbilical vein endothelium. J Physiol 560(1):111–22

130. Vortmeyer AO, et al. (1997) von Hippel-Lindau gene deletion detected in the stromal cell component of a cerebellar hemangioblastoma associated with von Hippel-Lindau disease. Hum Pathol 28(5):540–3

131. Wang GL, et al. (1995) Hypoxia-inducible factor 1 is a basic-helix-loop-helix-PAS heterodimer regulated by cellular O_2 tension. Proc Natl Acad Sci USA 92(12): 5510–4

132. Wang GL, Semenza GL (1995) Purification and characterization of hypoxia-inducible factor 1. J Biol Chem 270(3): 1230–7

133. Wangsa-Wirawan ND, Linsenmeier RA (2003) Retinal oxygen: fundamental and clinical aspects. Arch Ophthalmol 121(4):547–57

134. Wenzel A, et al. (2005) Molecular mechanisms of light-

induced photoreceptor apoptosis and neuroprotection for retinal degeneration. Prog Retin Eye Res 24(2):275–306

135. Wheatley CM, et al. (2002) Retinopathy of prematurity: recent advances in our understanding. Br J Ophthalmol 86(6):696–700

136. Yahata Y, et al. (2003) Nuclear translocation of phosphorylated STAT3 is essential for vascular endothelial growth factor-induced human dermal microvascular endothelial cell migration and tube formation. J Biol Chem 278(41):40026–31

137. Yamanouchi I, Igarashi I (1991) Arterial catheters, endothelin, and ROP. Pediatrics 88(4):874–5

138. Yokoi M, et al. (2005) Elevations of AGE and vascular endothelial growth factor with decreased total antioxidant status in the vitreous fluid of diabetic patients with retinopathy. Br J Ophthalmol 89(6):673–5

139. Yokota T, et al. (2003) Role of protein kinase C on the expression of platelet-derived growth factor and endothelin-1 in the retina of diabetic rats and cultured retinal capillary pericytes. Diabetes 52(3):838–45

140. Yu AY, et al. (1998) Temporal, spatial, and oxygen-regulated expression of hypoxia-inducible factor-1 in the lung. Am J Physiol 275(4):L818–26

141. Yu DY, et al. (2000) Intraretinal oxygen levels before and after photoreceptor loss in the RCS rat. Invest Ophthalmol Vis Sci 41(12):3999–4006

142. Zeiss CJ, Neal J, Johnson EA (2004) Caspase-3 in postnatal retinal development and degeneration. Invest Ophthalmol Vis Sci 45(3):964–70

143. Zelzer E, et al. (1998) Insulin induces transcription of target genes through the hypoxia-inducible factor HIF-1alpha/ARNT. EMBO J 17(17):5085–94

8 Blood Retinal Barrier

8.1 Blood-Retinal Barrier, Retinal Vascular Leakage and Macular Edema

B.E. PHILLIPS, D.A. ANTONETTI

Core Messages
- The blood-retinal barrier forms a selective barrier, restricting ion, substrate, and water permeability, allowing for proper neuronal function
- The blood-retinal barrier has a number of specialized characteristics that allow for barrier formation
- Vascular endothelial cells and retinal pigmented epithelial cells physically form the blood-retinal barrier
- The tight junction protein complex is essential in establishing the blood-retinal barrier
- Diabetic retinopathy is characterized by increases in vascular permeability
- Diabetic retinopathy causes changes in growth factors and cytokines which modify tight junction protein content

8.1.1 Introduction

Proper retinal function requires the presence of a well-defined blood-retinal barrier (BRB). In many of the leading causes of medical blindness this BRB is compromised. Indeed, retinopathy of prematurity and age related macular degeneration both include production of aberrant vessels with poor barrier properties. Further, diabetic retinopathy, the leading cause of blindness in working age adults, involves progressive vision loss and is closely associated with macular edema [117]. Increased fluid accumulation, as well as lipid and albumin deposits, is believed to be the result of the breakdown of the BRB that normally controls the neuronal environment. Barrier dysfunction results in increased permeability which diagnostically indicates progressive retinopathy [32]. This chapter will review the normal physiology of the BRB, the changes that occur through the course of diabetic retinopathy, and the known underlying molecular mechanisms that may lead to barrier dysfunction. Elucidating the mechanisms of barrier dysfunction in diabetic retinopathy will further our understanding of its pathogenesis and provide future therapeutic targets.

8.1.2 Characteristics of the Blood-Retinal Barrier

Essentials
- The BRB comprises vascular endothelial and retinal pigmented epithelial cells
- Mechanisms of permeability across endothelial and epithelial barriers
- Facilitated water and substrate transport
- Tight junction formation

8.1.2.1 Blood-Retinal Barrier Physiology

The BRB, similar to the blood-brain barrier (BBB), is a physiologic barrier that regulates ion, protein, and water flux into the retina. This barrier allows the neural retina to establish and maintain specific substrate and ion concentrations allowing for proper neuronal function. Additionally, it regulates infiltration of immune competent cells, blood-born toxins, and various hormones that could negatively affect neuronal function and survival. The barrier is physically established by two cell types: vascular endothelial cells that form capillary beds in the ganglion cell layer and outer plexiform layer of the retina and retinal pigmented epithelial cells (RPE) that form a barrier between the choroid capillary plexus and the retina [102, 86]. Combined, these cells regulate the flux of nutrients into the retina from the blood supply.

8

8.1.2.2 Fenestrations

Cellular transport of material across the barrier can occur by two pathways: transcellular flux that is transport across the cell, or paracellular flux, transport between the cells. Transcellular flux is transport through the cell that may be passive diffusion, facilitated, or by active transport mechanisms. Fenestrations are locations in capillaries where the endothelial cell depth is reduced, and the thinning of the capillary wall facilitates transcellular transport of materials and cells out of the capillaries. Fenestrations are seen in highly permeable tissues such as the glomerular capillary wall and choroid capillary plexus [37, 102]. Endothelial cells in the BBB lack fenestrations contributing to their ability to maintain a barrier. This loss of endothelial cell fenestrations is one of the initial developmental steps of BBB formation [11, 49]. Although fenestrations have not been fully explored in the BRB, the analogous function and conservation of properties of the BRB and BBB suggest a similar paucity of fenestrations.

8.1.2.3 Endocytosis and Facilitated Diffusion

Cells continuously take up and sample extracellular material in a process called pinocytosis (Fig. 8.1.1). Membrane invaginations across the cell surface pinch off forming vesicles that move to the cell interior. These vesicles can move internally to be degraded or through the cell to be released at the basal-lateral membrane leading to non-specific transport of materials across the cell. Pinocytotic vesicles are selectively decreased in endothelial cells located in the BBB [16, 27, 149]. In order to maintain transport of necessary substrates for neuronal function, specific receptor-facilitated transport mechanisms are used to move materials across the BBB. Receptor mediated endocytosis may occur through clathrin or caveolin mediated endocytosis, which is ATP dependent [189]. A receptor binds to its specific substrate concentrating the substrate within the invagination to reduce transport of non-specific materials (Fig. 8.1.1). The transferrin receptor and the transport of iron across the BBB is an important example of receptor mediated transport across vascular endothelium [20]. Vascular endothelial cells of the BBB have increased mitochondria content, which generates a cell's energy supply of ATP [126]. Elevated mitochondria levels may be necessary to supply numerous receptor-mediated transport mechanisms.

Channel-facilitated transport is a mechanism for the diffusion of specific substrates across the BBB. This is mediated by transmembrane proteins located in the cell surface that allow for the flux of specific materials across the cell membrane. The GLUT-1 glucose transporter is highly and selectively expressed in vascular endothelial cells of the BBB and supplies the neuronal tissue with necessary glucose [128]. Together these mechanisms provide the necessary substrates for neuronal function.

8.1.2.4 Water Transport

Facilitated transport of water out of the neuronal extracellular space is essential to maintain retinal function. RPE cells regulate water content and lactic acid removal generated by high metabolic rates in the retina. RPE cells transport water out of the retina and into the choroid capillary plexus. The force gen-

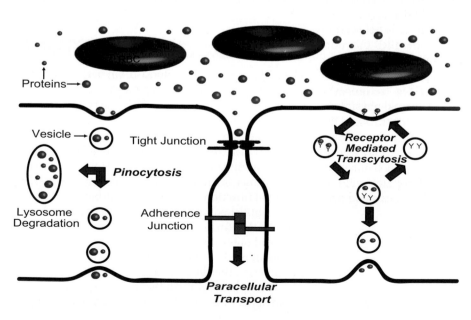

Fig. 8.1.1. A side view of two adjacent vascular endothelial cells is depicted. Paracellular flux of blood-born proteins between the cells is inhibited by the tight junction complex. Non-specific transport of materials through the cell is pinocytosis. Pinocytotic vesicles are greatly reduced in barrier forming cells. To compensate for reduced permeability, receptor mediated transcytosis allows for specific substrate transport across the barrier

erated by water flux helps to maintain retinal attachment [153]. Lactic acid and water transport across the RPE are closely linked. The movement of water out of the retina is accomplished through the use of a number of ion channels and a specific water channel, aquaporin-1 [153, 158]. Aquaporin-1 is a transmembrane protein found in water permeable tissues that forms a transcellular channel for water transport [1, 123, 153]. Water transport is dependent on the pumping of Cl^- and K^+ across the RPE basolateral membrane into the choroid capillary plexus [18, 107 – 109, 140]. The flow of these ions is driven by the apical $Na^+/K^+ATPase$ pump that elevates internal K^+ levels allowing the subsequent flux of Cl^-. In addition, lactic acid is taken up by the RPE from the neuronal extracellular space by the monocarboxylate transporter-1 and then transported from the RPE cell interior to the choroid capillary plexus by monocarboxylate transporter-3 [15, 130, 194]. Increased acidification of the RPE cell due to lactic acid buildup activates the Cl^- and K^+ pumps on the apical membrane, with subsequent diffusion across the basolateral membrane that ultimately drives water flux through aquaporin channels. The reuptake of Cl^- from the choroid capillary plexus by HCO_3^-/Cl^- exchange pumps is activated with extracellular acidification [158]. In summary, elevated retinal metabolic rates increase lactic acid production. The RPE responds to this elevated metabolism by increasing lactate transport into RPE cells, which increases water transport from the retina to the choroid capillary plexus allowing for compensatory removal of by-products from the retina. This regulation of transcellular transport requires precise control of paracellular flux between the two compartments, which is achieved by well-developed tight junctions.

8.1.2.5 Tight Junctions

The tight junctions are a protein complex that forms a continuous belt around the circumference of the cell in order to regulate paracellular flux of solutes and water (Fig. 8.1.1). The tight junction can be observed with electron microscopy as points of increased protein density and close apposition between adjacent cell membranes (Fig. 8.1.2) [48, 19]. The distance between the cell membranes is spanned by strands comprising transmembrane proteins [143], which will be explored later in this chapter. Tight junctions can be found in epithelial cell layers such as the RPE cells, and in the vascular endothelial cells of the BBB and BRB, both of which have well-developed tight junction complexes uncharacteristic of peripheral vasculature [116, 120, 177]. Rat models demonstrate occludin protein expression, a transmembrane tight junction protein, can be ini-

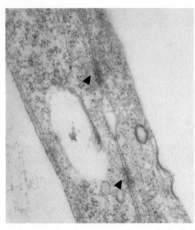

Fig. 8.1.2. An electron micrograph of two vascular endothelial cells. The *black arrows* point out the tight junction complex between the two cell membranes

tially observed in BBB vascular endothelial cells 1 week postnatally, which correlates with the maturation of the BBB [73]. Developmentally the BBB vascular endothelial cells and the RPE show increased expression of TJ proteins over time that correlates with decreases in permeability and the establishment of the BBB and BRB [40, 73, 134, 139, 155, 183, 185].

8.1.2.6 P-Glycoprotein

The BBB and BRB selectively transport soluble substrates and water across the barrier, but there are a number of lipophilic molecules that can freely diffuse across the cell membrane. The entry and accumulation of these potentially harmful compounds is restricted by multidrug-resistance proteins, including P-glycoproteins, by actively pumping specific lipophilic molecules from the cell back into the blood stream in an ATP dependent manner [96, 138]. A P-glycoprotein homozygous knock-out mouse possesses greatly elevated levels of lipophilic tracers in the brain [145, 146]. P-glycoproteins are located in BBB endothelial cells as well as a large number of different epithelial tissues including the liver, kidney, intestine, pancreas, and brain choroid plexus [14, 133]. Recently, P-glycoprotein has been shown to be expressed in retinal vascular endothelial cells, suggesting a role in the BRB [100]. Together, the tight junctions and multi-drug resistance proteins form an effective balance that reduces permeability but allows passage of specific substrates required for proper neuronal function.

8.1.3 Diabetic Retinopathy

Essentials
- Increased permeability and macular edema
- Elevated vascular endothelial growth factor
- Pericyte and endothelial cell apoptosis

8.1.3.1 Permeability and Macular Edema

Diabetic retinopathy remains the leading cause of blindness among working age adults in Western society. A recent study of patients with type I diabetes demonstrates that after 14 years of diabetes over 70% demonstrate some level of diabetic retinopathy and 10% have moderate to severe diabetic retinopathy [93]. Diabetic retinopathy causes progressive vision loss associated with breakdown of the BRB [44]. Vascular endothelial cells that line the capillaries in the inner retina have been shown by microscopic techniques as a major site of vascular dysfunction [60, 118, 175]. Increased vascular permeability and lipoprotein plaque formation occur early in diabetic retinopathy (Fig. 8.1.3) [31, 60]. Vascular permeability has been further demonstrated in early diabetic animal models. Type I diabetes can be modeled by administration of streptozocin, which kills beta cells in the pancreas [162]. The model demonstrates increased vessel permeability as early as 1 week by measuring flux of the albumin binding dye Evan's blue [190], or after 1 [43] and 3 [6] months by measuring flux of injected fluorescein isothiocyanate conjugated albumin. Increased flux, as shown by vit-

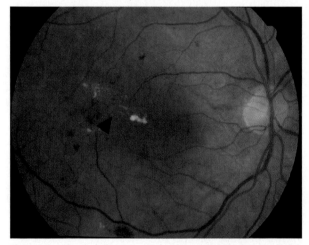

Fig. 8.1.3. A fundus photograph from a patient with non-proliferative diabetic retinopathy displaying vascular leakage. The *red spots* are vascular hemorrhages. The *black arrow* points out a lipoprotein plaque located in the macula of the eye. The build-up of lipoprotein exudates leads to macula edema, negatively affecting vision

reous fluorometry measurements, occurs in diabetic patients and diagnostically indicates progressive retinopathy [32].

While solute flux is most often measured as indicator of permeability, an increase in water permeability also occurs in diabetic retinopathy causing macular edema [166, 172, 176]. Macular edema is an increase in the thickness of the central macular region due to water accumulation and is associated with vision loss [31]. The net movement of water and solutes out of capillaries is governed by Strling's law and subsequent adaptations reviewed by Salathe [35, 142]. Starling's law states the net movement of fluid out of capillaries is determined by the sum of hydrostatic and oncotic pressures, predominantly intralumenal hydrostatic and oncotic pressure. The flux of solutes is controlled by both diffusive flux and convective forces generated as the fluid travels across the endothelial barrier. Permeability of the endothelial wall to paracellular flux of water and solutes is governed by the junctional complex. Therefore, changes in accumulation of water and solutes in the retina may be due to changes in wall permeability or may be caused by changes in hydrostatic or oncotic pressure and these parameters should be considered in the interpretation of experimental results.

8.1.3.2 Vascular Endothelial Growth Factor

Vascular endothelial growth factor (VEGF) contributes to a number of phenotypic changes seen with diabetic retinopathy. Diabetic animal models and humans with diabetes have elevated levels of VEGF in the vitreous of the retina [3, 4, 99, 119]. VEGF was originally identified as a potent permeabilizing agent on endothelial cells [150] and increases both solute and water flux in endothelial cell culture [38] and solute flux in the retina [9]. VEGF also plays a role in angiogenesis and is associated with neovascularization in diabetic retinopathy [83, 99, 147].

The cause of elevated VEGF in diabetic retinopathy has not been rigorously demonstrated but may be due to retinal ischemia after capillary non-perfusion and occlusion [2, 51, 147] or may be due to activation of an inflammatory response possibly through activation of the receptor for advanced glycation end products (AGEs) [156]. The exact mechanism that leads to capillary occlusion is not clear, but a number of observations have been made. Intercellular adhesion molecule-1 (ICAM-1) expression is elevated in diabetes [98], and allows for increased leukostasis [113]. Leukostasis is the first step of diapedesis but while no evidence exists for leukocyte infiltration in diabetic retinopathy, adhesion of leukocytes to endothelial cells may prevent normal blood flow. Leukostasis increases with diabetic retinopathy [82, 112] and inhibition of ICAM-1 has been shown to block leukostasis and improve retinopathy prognosis [79, 113]. Vasoconstriction of the vessels may further facilitate the ability of leukostasis to reduce capillary blood flow. A VEGF-induced model of retinopathy showed that VEGF alone leads to vessel narrowing [75]. Additionally, the vasoconstrictor endothelin-1 is upregulated in diabetic retinopathy and is associated with decreased blood flow [23, 163, 193]. Recently, it has been demonstrated that the inhibition of the endothelin-1 receptor blocks increases in VEGF, ICAM, and leukostasis [103]. The involvement of VEGF is further complicated by the number of VEGF and VEGF receptor isoforms and their different interactions [83]. Understanding VEGF's multiple roles will allow for better pharmacological techniques of treatment.

8.1.3.3 Apoptosis and Cell Loss

One of the early changes of diabetic retinopathy is the loss of pericytes and endothelial cells. Pericytes are smooth muscle cells that form a sheath around vascular endothelial cells. Pericytes direct vascular development by prohibiting endothelial growth, allowing for vessel stabilization, and providing necessary factors for angiogenesis [29, 53, 74, 121]. Genetic and pharmacological studies that prevent proper pericyte interaction have demonstrated increased endothelial cell growth and abnormal vascular development under these conditions [95, 168]. These pathological changes may provide a strong

stimulus for the progression into proliferative diabetic retinopathy, although the mechanisms of pericyte loss have not been fully elucidated [68]. Angiopoietin-2 (AP2), an angiopoietin-1 (AP1) antagonist, is upregulated 30-fold in the diabetic retina and has been shown to increase pericyte death [69]. AP1 promotes endothelial cell survival and is involved with vascular remodeling [36, 85, 144, 159]. The increase in AP2 affects the vasculature by contributing to pericyte cell death and reducing AP1 mediated endothelial cell survival. Brownlee [94, 174], Kowluru [88–90] and others have examined the role of elevated oxidative stress in diabetic retinopathy. Mitochondrial production of superoxides, hypothesized as a result of excess glucose metabolism, is proposed to contribute to endothelial apoptosis [87]. AGEs produced by hyperglycemia have also been shown to elevate levels of pericyte and endothelial capillary cell death [59, 97, 115, 157]. Recent studies have linked AGEs to increased oxidative stress [88]. Ongoing research will continue to uncover additional mechanisms of cell loss in this complicated system.

8.1.3.4 Routes of Paracellular Flux

Routes of paracellular flux likely vary depending on the solute examined. Routes of flux may include at least three different pores as depicted in Fig. 8.1.4 and modeled by Guo et al. [65]. Small solutes may migrate through a small pore where tight junctions are assembled. A second pore may be a larger pore formed by disassembled tight junctions with a medi-

Fig. 8.1.4. A paracellular permeability model depicting the different pores involved in paracellular flux. *Top* A capillary comprising vascular endothelial cells depicting endothelial cell loss due to diabetic retinopathy resulting in hemorrhages and increased protein flux. *Bottom* A close-up view of tight junctions between two adjacent cells near the site of cell loss. The small pore to paracellular flux is at locations of intact tight junctions. Tight junction disassembly leads to the formation of medium sized pores with increased permeability to proteins. Sites of cell death and division comprise large

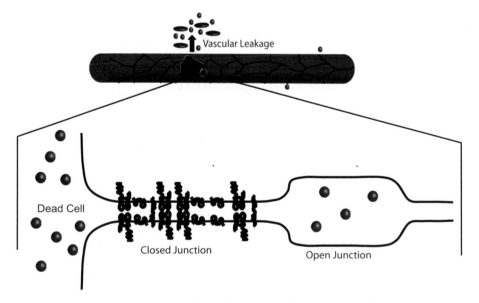

pores that allow for greatly increased permeability to proteins and possibly red blood cells. Diabetes leads to tight junction internalization and protein content reduction, which could increase medium pore formation. Vascular endothelial cell apoptosis elevation in diabetes also leads to an increased number of large pores um size. Finally, large pores formed by cell division or cell death may also contribute to paracellular flux [65]. Application of this model to diabetic retinopathy would implicate VEGF increasing both the number of medium and large pores. The medium pores are increased due to tight junction disassembly, and the large pores are increased due to the number of dividing endothelial cells in proliferative diabetic retinopathy. The large pores are further increased with endothelial cell drop out due to apoptosis. The existence of multiple changing pores that lead to increased permeability may require alternate approaches for therapy.

8.1.4 The Structure and Role of the Tight Junctions

Essentials
- Cell polarity is necessary for assembly
- Occludin and claudins create the barrier
- Changes in junctional proteins alter barrier properties

8.1.4.1 Tight Junction

The tight junction protein complex allows for the establishment of the BRB, restricting paracellular diffusion of blood-born compounds into the neuronal tissues. Understanding the normal function of the tight junction and the pathological changes that lead to increased permeability is necessary to understand disease progression. This section will cover in detail the establishment and composition of the tight junction complex and the alterations that occur in diabetes.

8.1.4.2 Polarity and Tight Junction Assembly

Establishment of cell polarity is the first step in development of the tissue barrier. The endothelial cells of the vascular retina and RPE develop distinct apical versus basal lateral surfaces. This development of cell polarity orients the cell with its surroundings and is an essential component of development, asymmetrical cell division, cell migration, and the formation of asymmetrical cell morphology [45, 61, 84]. Polarity initiation is an essential process for the formation of the cell barrier. It organizes the cytoskeleton, junctional complexes, and apical/basal cell membrane proteins in a cell tissue layer [22, 76, 122, 151]. Epithelial cell polarity has been more widely studied than endothelial polarity. The initiating step in cell polarization is contact with a basement membrane or an adjacent cell [122]. The cell

then begins to establish junctional complexes. The first junction to form is a primitive spot like junction that consists of adherene junction proteins [54, 55, 114, 160]. The protein zonula occludens-1 (ZO-1) in the spot like junctions then recruits junction adhesion molecule (JAM) and additional tight junction proteins [54, 55]. These proteins are recruited apically above the adherene junction, and go on to form the tight junction, which is the most apical junction formed between mammalian cells [114]. The transition from a spot like junction to a continuous adherens and tight junction requires the Par6/Par3/atypical PKC (aPKC) polarity complex [160]. Par6 (Partioning defect protein 6) was first identified through mutational analysis in *C. elegans* as being required for cell polarity [66]. Genetic studies of *Drosophila* and mammalian epithelial cells show a loss of cell polarization and tight junction formation with Par6 mutations [129, 192]. Similarly, a non-functional kinase dead aPKC caused Par-6, TJ, and apical/basal protein mislocalization preventing barrier formation [161]. The role of the Par6/Par3/aPKC complex in tight junction formation and maintenance has only begun to be studied and still requires additional investigation, particularly as it relates to endothelial junction formation.

8.1.4.3 Tight Junction Structure

The tight junction complex contains at least 40 proteins comprising trans-membrane and internal adapter proteins that regulate paracellular flux [62]. Transmembrane proteins that make up the tight junction are occludin, claudins, and JAMs. Occludin was the first discovered transmembrane protein of the tight junction [57]. Occludin is a 65-kDa transmembrane protein that spans the membrane four times providing two extracellular loops that are able to bind to the occludin molecule on the adjacent cell. Changes in occludin content correlate with vascular permeability making it a likely candidate in regulating the opening and closing of the tight junction [6, 73, 111, 165]. Claudins consist of at least 24 genes that are differentially expressed in various tissues. Similar to occludin, claudins span the membrane 4 times and are able to bind to other claudins and occludin on the adjacent cell. Claudins regulate small charged molecule and ion permeability [124, 167, 171, 181]. The expression of specific claudins confers the charge permeability properties displayed by that tissue. This principle has been elegantly demonstrated by point mutants of claudin 15 that reverse its charge selectivity [26].

JAM is part of a family of proteins, with JAM-A specifically localized to the tight junction [101]. JAM has a single extracellular domain and associates with a number of adhesion molecules. JAM is specifically involved in tight junction assembly and leukostasis, and may play a number of other roles involving the tight junction [12, 33, 42, 127].

Numerous adapter proteins localize just below the membrane and act as tight junction organizers and cytoskeleton anchors. The first tight junction protein identified was zonula occludens-1 (ZO-1), later joined by ZO-2 and ZO-3 [64, 71, 154]. These proteins are part of a protein family called membrane-associated guanylate kinases (MAGUK), which consist of an SH3 domain, PDZ domains, and a guanylate kinase GuK binding domain [46]. ZO-1 acts as a molecular scaffold that organizes, assembles, and links the tight junction to the cytoskeleton through a number of protein-protein interactions (Fig. 8.1.5) [47, 54, 55, 184]. ZO-1 binds to ZO-3, which then interacts with the polarity complex protein termed proteins associated with Lin7 (Pals-1), Pals1-associated tight junction protein (PATJ), and crumbs-3 (CRB-3) [136]. Pals-1, also containing PDZ domains, interacts with the Par6/Par3/aPKC complex [76], thus connecting these polarity complexes. Changes in the Pals-1 and PATJ expression interfere with establishment of polarity similarly to reduced Par6 content and aPKC activity [151, 135]. ZO-1 also interacts with AF-6, which is negatively regulated by Ras activation [91, 191]. Constitutive Ras activation has been shown to downregulate tight junction protein expression [24]. In conclusion, formation of the barrier first requires establishment of cell polarity. Cell culture models and genetic studies have so far uncovered two complexes apparently necessary for recruitment of the tight junction complex. The first, Par3, Par6, aPKC, and the second, Pals-1, PATJ, and CRB-3, are connected to the ZO-1 tight junction scaffolding proteins. Ultimately, recruitment of the transmembrane tight junction proteins occludin, claudins and JAM establishes the interaction with adjacent cells and confers barrier properties. Elucidating this complex interaction in the retina and brain will likely unveil unique mechanisms specific to the establishment of the blood-neural barrier.

8.1.4.4 Occludin

The majority of known diabetic changes in the tight junction complex involve the protein occludin. Occludin is of particular interest due to correlation between changes in content and permeability. Diabetic animal models have a 35 % decrease in occludin content of the retina with a 62 % increase in vascular permeability [6]. Changes in occludin content may be due to elevated levels of VEGF and hepatic growth factor (HGF) in the vitreous of diabetic patients [3, 80, 110, 152]. VEGF and HGF administration in vitro increased permeability, reduced occludin protein content, and led to tight junction complex internali-

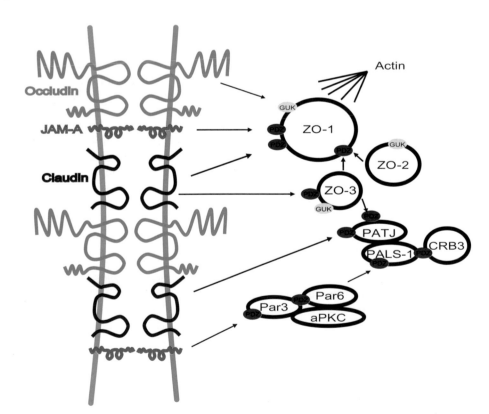

Fig. 8.1.5. A top down view of two adjacent cells displaying tight junction protein interactions. Transmembrane proteins occludin, claudin, and JAM on adjacent cells bind to either partner on the adjacent cell forming the tight junctional strands. Internal scaffolding proteins containing multiple PDZ domains organize transmembrane and polarity complex proteins allowing for junction formation. The tight junction complex is connected to the cytoskeleton through ZO-1 bridging actin and occludin. Occludin binds the guanylate kinase (GuK) binding domain on ZO-1

zation in RPE, BBB, and BRB vascular endothelial cells [25, 77, 78, 179]. Similar increases in retinal vascular permeability and occludin internalization were observed in rat models injected with VEGF [5, 6, 9, 10], and blocked by VEGF receptor inhibitors [132]. Changes in tight junction organization may be regulated by occludin phosphorylation, which is stimulated by VEGF, shear stress, and histamine [5, 39, 72] and closely correlates with regulation of barrier properties. This phosphorylation of occludin is regulated by PKC, which may explain the mechanism of action of PKC inhibitors in regulating vascular permeability [70]. Diabetes has also been characterized as a chronic low grade inflammatory state [30, 41, 50, 52, 58, 60, 67, 131], and cytokines increase permeability and disrupt tight junctions in both endothelial and epithelial cells [178].

Techniques to modify occludin expression and binding properties have been used to better understand its role in the tight junction. Glucocorticoid treatment increases the tight junction gene expression of occludin in primary retinal vascular endothelial cells leading to decreased solute and water permeability [7, 137]. Glucocorticoids also increase transelectrical epithelial resistance (TER) by de novo protein synthesis [196]. Similarly, direct overexpression of full length and COOH truncated occludin led to an increase in TER [8, 105]. Paradoxically, small molecule paracellular flux increased over time [105]. It was hypothesized that occludin acts as a seal when interacting with the tight junction protein ZO-1, but as occludin's expression outpaced ZO-1, the lack of protein-protein interaction produced more pores in the tight junction [105]. Conversely, overexpression of occludin in fibroblasts, in low calcium media to prevent adherens junction formation, conferred adhesion properties [170]. The role of occludin in regulating barrier properties was further demonstrated in *raf1* transformed epithelial cells, which express reduced occludin content and possess poor cell adhesion properties. Loss of tight junction, cell polarity, and control of cell growth are common in epithelial cell cancers [180]. Reintroduction of wild-type occludin or the second extracellular loop of occludin alone in *raf-1* transformed cells reestablished wild-type cell adhesion and prevented tumor formation [180]. These data suggest occludin not only confers adhesive properties but also contributes to organization of the tight junction complex.

Studies have also used reduction or ablation of occludin expression to study the role of occludin. A number of studies used small peptides to bind regions of the first or second extracellular loop of occludin in an attempt to impair its function. The studies did not reach a consensus on the function of individual loops, but binding peptides to either loop decreases TER, increases permeability, and disrupts localization of other tight junction proteins [92, 125, 164, 173, 188]. The discrepancies between peptide experiments are most likely due to the difference in the occludin protein sequence used from varying species. Transgenic mice that lack occludin were generated and the offspring were viable. Electron microscopy revealed tight junction formation occurring in the absence of occludin, and TER of intestinal epithelium were normal. However, the mice displayed a number of aberrant phenotypes including sterility in males, testicular atrophy, females unable to suckle, chronic inflammation of gastric epithelium, calcification of the brain, and reduced bone densities [141]. Acid secretion in the gastric corpus mucosa was later shown to be reduced due to the loss of parietal cells and hyperplasia of mucus cells [148]. These are all regions possessing endothelial or epithelial tight junction barriers suggesting a dysfunction in tight junctions despite their normal appearance by electron microscopy.

While these data point to a role for occludin in establishing barrier properties, recent data highlights the complexity of the tight junction. Tissue culture models reveal no change in epithelial TER or junctional assembly after occludin ablation [106, 195]. Stable MDCK cell lines lacking occludin expression had normal diffusive permeability to 3- and 10-kDa dextran, but demonstrated an increase in small organic cation and mannitol permeability [195]. Interpretation of these permeability changes is complicated by increased expression of claudins 3 and 4 and a decrease in claudin 1 and 7 [195]. Interestingly, occludin ablated cells failed to stimulate exchange of GTP on Rho in response to cholesterol depletion [195]. Cholesterol depletion has been shown to cause a rearrangement of the cytoskeleton, and Rho-GTP is involved in cytoskeleton remodeling [17]. Inhibition of cytoskeleton rearrangement may explain why occludin depleted cells were unable to release apoptotic cells from the cell monolayer [195]. This body of research on occludin demonstrates a role for occludin in barrier function but highlights the broad nature of occludin function and the need for continuing research to fully understand its complex role in cell biology.

8.1.4.5 Claudins

The claudin tight junction protein family regulate small charged molecule and ion permeability [124, 167, 171, 181]. Different complements of claudins are expressed in various tissues, establishing their cation or anion selectivity. It has been shown that by changing the expression profile of claudins within a tissue, its ion charge permeability preferences can be changed [171]. Claudin expression changes have been linked to a number of pathologies that display a loss of barrier function [104, 186]. Mutations in claudin-16 are the cause of familial hypomagnesemia with hypocalciuria and nephrocalcinosis. Loss of claudin-16 prevents reabsorption of Mg^{2+} and Ca^{2+} in the kidney's thick ascending loop of Henle resulting in eventual kidney failure [81]. Claudin-3 is lost with autoimmune encephalomyelitis in the brain and spinal cord [187]. Claudin-4 expression is reduced in intestinal epithelium with collagenous colitis, preventing intestinal water absorption resulting in diarrhea [21]. Loss of claudin-14 has been linked to deafness in human and mouse genetic analysis [13, 182]. Transgenic mice lacking claudin-1 expression die after birth due to transepidermal water loss [56]. Claudin-5 is an endothelial specific form of claudin and is found in the BBB [181] and BRB [9]. Transgenic mice that lack claudin-5 were only able to survive for up to 10 h after birth [124]. Like the occludin transgenic mice, electron microscopy of these mice showed normal appearing tight junctions. However, unlike the occludin ablated mice, the claudin-5 mice had increased permeability of molecular markers smaller than 800 Da. These markers freely diffused across the endothelial blood vessels in the brain [124]. Thus, claudins have been established as providing a clear role for barrier properties through transgenic animals. However, their role in BRB and the changes that may occur in leading to macular edema are not well described.

8.1.4.6 Zonula Occludens

The ZO-1, ZO-2, and ZO-3 proteins are part of a protein family called membrane-associated guanylate kinases (MAGUK) which consist of an SH3 domain, PDZ domains, and a guanylate kinase (GuK) binding domain [46]. ZO-1 acts as a molecular scaffold that organizes and assembles the tight junction [47, 54, 55, 184]. Tissue culture models with depleted ZO-1 protein have tight junctions indistinguishable from wild-type cells expressing ZO-1, similarly to occludin depletion models [106, 141, 169]. A Ca^{2+} switch assay reveals a delay in tight junction formation in cells lacking ZO-1 with no effect on the adherens junction assembly [106, 169]. Tight junction assembly delay was rescued when the SH3 domain or full length of ZO-1 was reintroduced into the cell [106]. Ablation of ZO-1 caused an increase in ZO-2 and a decrease in cingulin expression, both of which are binding partners of ZO-1 [169]. Cingulin is a scaffolding protein that is believed to mediate tight junction protein recruitment through interactions with ZO-1, ZO-2, JAM, and the cytoskeleton [12, 28, 34]. Direct ablation of either of these proteins had no

effect on tight junction assembly [63, 106]. The TER and inulin transport of the ZO-1 ablated cells were identical to wild-type levels after tight junction formation [106].

8.1.5 Summary

This chapter has reviewed the physiologic components of the BRB that establish its ability to selectively restrict substrate permeability into the neuronal tissue. Tight junction formation and maintenance in the vascular endothelial and RPE cells is a required element in barrier formation. Specific tight junction protein components are changed in a large number of pathologies with disrupted barriers, including diabetic retinopathy. The elevation of VEGF in diabetes leads to a reduction in tight junction protein content, increased permeability, and angiogenesis. The ability of glucorticoids to impede VEGF induced changes makes this class of drugs an attractive possible therapeutic, along with other methods of VEGF inhibition. Continued research and understanding of the mechanisms of BRB formation and breakdown in disease pathology will enable development of treatments for diabetic retinopathy and other eye diseases that involve macular edema.

Acknowledgements. I would like to acknowledge Timothy Bennett for the contribution of the picture seen in Fig. 8.1.3 (unpublished data).

References

1. Agre P, Preston GM, Smith BL, Jung JS, Raina S, Moon C, Guggino WB, Nielsen S (1993) Aquaporin CHIP: the archetypal molecular water channel. Am J Physiol 265:F463–76
2. Aiello LP (1997) Clinical implications of vascular growth factors in proliferative retinopathies. Curr Opin Ophthalmol 8:19–31
3. Aiello LP, Avery RL, Arrigg PG, Keyt BA, Jampel HD, Shah ST, Pasquale LR, Thieme H, Iwamoto MA, Park JE, et al. (1994) Vascular endothelial growth factor in ocular fluid of patients with diabetic retinopathy and other retinal disorders. N Engl J Med 331:1480–7
4. Amin RH, Frank RN, Kennedy A, Eliott D, Puklin JE, Abrams GW (1997) Vascular endothelial growth factor is present in glial cells of the retina and optic nerve of human subjects with nonproliferative diabetic retinopathy. Invest Ophthalmol Vis Sci 38:36–47
5. Antonetti DA, Barber AJ, Hollinger LA, Wolpert EB, Gardner TW (1999) Vascular endothelial growth factor induces rapid phosphorylation of tight junction proteins occludin and zonula occluden 1. A potential mechanism for vascular permeability in diabetic retinopathy and tumors. J Biol Chem 274:23463–7
6. Antonetti DA, Barber AJ, Khin S, Lieth E, Tarbell JM, Gardner TW (1998) Vascular permeability in experimental diabetes is associated with reduced endothelial occludin content: vascular endothelial growth factor decreases occludin in ret-

inal endothelial cells. Penn State Retina Research Group. Diabetes 47:1953–9

7. Antonetti DA, Wolpert EB, DeMaio L, Harhaj NS, Scaduto RC, Jr (2002) Hydrocortisone decreases retinal endothelial cell water and solute flux coincident with increased content and decreased phosphorylation of occludin. J Neurochem 80:667–77

8. Balda MS, Whitney JA, Flores C, Gonzalez S, Cereijido M, Matter K (1996) Functional dissociation of paracellular permeability and transepithelial electrical resistance and disruption of the apical-basolateral intramembrane diffusion barrier by expression of a mutant tight junction membrane protein. J Cell Biol 134:1031–49

9. Barber AJ, Antonetti DA (2003) Mapping the blood vessels with paracellular permeability in the retinas of diabetic rats. Invest Ophthalmol Vis Sci 44:5410–6

10. Barber AJ, Antonetti DA, Gardner TW (2000) Altered expression of retinal occludin and glial fibrillary acidic protein in experimental diabetes. The Penn State Retina Research Group. Invest Ophthalmol Vis Sci 41:3561–8

11. Bauer HC, Bauer H, Lametschwandtner A, Amberger A, Ruiz P, Steiner M (1993) Neovascularization and the appearance of morphological characteristics of the blood-brain barrier in the embryonic mouse central nervous system. Brain Res Dev Brain Res 75:269–78

12. Bazzoni G, Martinez-Estrada OM, Mueller F, Nelboeck P, Schmid G, Bartfai T, Dejana E, Brockhaus M (2000) Homophilic interaction of junctional adhesion molecule. J Biol Chem 275:30970–6

13. Ben-Yosef T, Belyantseva IA, Saunders TL, Hughes ED, Kawamoto K, Van Itallie CM, Beyer LA, Halsey K, Gardner DJ, Wilcox ER, Rasmussen J, Anderson JM, Dolan DF, Forge A, Raphael Y, Camper SA, Friedman TB (2003) Claudin 14 knockout mice, a model for autosomal recessive deafness DFNB29, are deaf due to cochlear hair cell degeneration. Hum Mol Genet 12:2049–61

14. Bendayan R, Lee G, Bendayan M (2002) Functional expression and localization of P-glycoprotein at the blood brain barrier. Microsc Res Tech 57:365–80

15. Bergersen L, Johannsson E, Veruki ML, Nagelhus EA, Halestrap A, Sejersted OM, Ottersen OP (1999) Cellular and subcellular expression of monocarboxylate transporters in the pigment epithelium and retina of the rat. Neuroscience 90:319–31

16. Bertossi M, Virgintino D, Maiorano E, Occhiogrosso M, Roncali L (1997) Ultrastructural and morphometric investigation of human brain capillaries in normal and peritumoral tissues. Ultrastruct Pathol 21:41–9

17. Boivin D, Bilodeau D, Beliveau R (1996) Regulation of cytoskeletal functions by Rho small GTP-binding proteins in normal and cancer cells. Can J Physiol Pharmacol 74:801–10

18. Botchkin LM, Matthews G (1993) Chloride current activated by swelling in retinal pigment epithelium cells. Am J Physiol 265:C1037–45

19. Brightman MW, Reese TS (1969) Junctions between intimately apposed cell membranes in the vertebrate brain. J Cell Biol 40:648–77

20. Burdo JR, Antonetti DA, Wolpert EB, Connor JR (2003) Mechanisms and regulation of transferrin and iron transport in a model blood-brain barrier system. Neuroscience 121:883–90

21. Burgel N, Bojarski C, Mankertz J, Zeitz M, Fromm M, Schulzke JD (2002) Mechanisms of diarrhea in collagenous colitis. Gastroenterology 123:433–43

22. Cau J, Hall A (2005) Cdc42 controls the polarity of the actin and microtubule cytoskeletons through two distinct signal transduction pathways. J Cell Sci 118:2579–87

23. Chakravarthy U, Hayes RG, Stitt AW, Douglas A (1997) Endothelin expression in ocular tissues of diabetic and insulin-treated rats. Invest Ophthalmol Vis Sci 38:2144–51

24. Chen Y, Lu Q, Schneeberger EE, Goodenough DA (2000) Restoration of tight junction structure and barrier function by down-regulation of the mitogen-activated protein kinase pathway in ras-transformed Madin-Darby canine kidney cells. Mol Biol Cell 11:849–62

25. Clermont AC, Cahill M, Salti H, Rook SL, Rask-Madsen C, Goddard L, Wong JS, Bursell D, Bursell SE, Aiello LP (2006) Hepatocyte growth factor induces retinal vascular permeability via MAP-kinase and PI-3 kinase without altering retinal hemodynamics. Invest Ophthalmol Vis Sci 47:2701–8

26. Colegio OR, Van Itallie CM, McCrea HJ, Rahner C, Anderson JM (2002) Claudins create charge-selective channels in the paracellular pathway between epithelial cells. Am J Physiol Cell Physiol 283:C142–7

27. Coomber BL, Stewart PA (1986) Three-dimensional reconstruction of vesicles in endothelium of blood-brain barrier versus highly permeable microvessels. Anat Rec 215:256–61

28. Cordenonsi M, D'Atri F, Hammar E, Parry DA, Kendrick-Jones J, Shore D, Citi S (1999) Cingulin contains globular and coiled-coil domains and interacts with ZO-1, ZO-2, ZO-3, and myosin. J Cell Biol 147:1569–82

29. Crocker DJ, Murad TM, Geer JC (1970) Role of the pericyte in wound healing. An ultrastructural study. Exp Mol Pathol 13:51–65

30. Crook M (2004) Type 2 diabetes mellitus: a disease of the innate immune system? An update. Diabet Med 21:203–7

31. Cunha-Vaz J, Bernardes R (2005) Nonproliferative retinopathy in diabetes type 2. Initial stages and characterization of phenotypes. Prog Retin Eye Res 24:355–77

32. Cunha-Vaz J, Lobo C, Sousa JC, Oliveiros B, Leite E, de Abreu JR (1998) Progression of retinopathy and alteration of the blood-retinal barrier in patients with type 2 diabetes: a 7-year prospective follow-up study. Graefes Arch Clin Exp Ophthalmol 236:264–8

33. Cunningham SA, Rodriguez JM, Arrate MP, Tran TM, Brock TA (2002) JAM2 interacts with alpha4beta1. Facilitation by JAM3. J Biol Chem 277:27589–92

34. D'Atri F, Citi S (2001) Cingulin interacts with F-actin in vitro. FEBS Lett 507:21–4

35. Damas J (1998) [Starling's law in 1998]. Rev Med Liege 53:425–30

36. Davis S, Aldrich TH, Jones PF, Acheson A, Compton DL, Jain V, Ryan TE, Bruno J, Radziejewski C, Maisonpierre PC, Yancopoulos GD (1996) Isolation of angiopoietin-1, a ligand for the TIE2 receptor, by secretion-trap expression cloning. Cell 87:1161–9

37. Deen WM, Lazzara MJ, Myers BD (2001) Structural determinants of glomerular permeability. Am J Physiol Renal Physiol 281:F579–96

38. DeMaio L, Antonetti DA, Scaduto RC, Jr, Gardner TW, Tarbell JM (2004) VEGF increases paracellular transport without altering the solvent-drag reflection coefficient. Microvasc Res 68:295–302

39. DeMaio L, Chang YS, Gardner TW, Tarbell JM, Antonetti DA (2001) Shear stress regulates occludin content and phosphorylation. Am J Physiol Heart Circ Physiol 281:H105–13

40. Dermietzel R, Krause D (1991) Molecular anatomy of the

blood-brain barrier as defined by immunocytochemistry. Int Rev Cytol 127:57–109

41. Duncan BB, Schmidt MI (2006) The epidemiology of low-grade chronic systemic inflammation and type 2 diabetes. Diabetes Technol Ther 8:7–17

42. Ebnet K, Suzuki A, Ohno S, Vestweber D (2004) Junctional adhesion molecules (JAMs): more molecules with dual functions? J Cell Sci 117:19–29

43. Enea NA, Hollis TM, Kern JA, Gardner TW (1989) Histamine H1 receptors mediate increased blood-retinal barrier permeability in experimental diabetes. Arch Ophthalmol 107:270–4

44. Engler C, Krogsaa B, Lund-Andersen H (1991) Blood-retina barrier permeability and its relation to the progression of diabetic retinopathy in type 1 diabetics. An 8-year follow-up study. Graefes Arch Clin Exp Ophthalmol 229:442–6

45. Etienne-Manneville S, Hall A (2003) Cdc42 regulates GSK-3beta and adenomatous polyposis coli to control cell polarity. Nature 421:753–6

46. Fanning AS, Anderson JM (1999) Protein modules as organizers of membrane structure. Curr Opin Cell Biol 11:432–9

47. Fanning AS, Ma TY, Anderson JM (2002) Isolation and functional characterization of the actin binding region in the tight junction protein ZO-1. FASEB J 16:1835–7

48. Farquhar MG, Palade GE (1963) Junctional complexes in various epithelia. J Cell Biol 17:375–412

49. Fenstermacher J, Gross P, Sposito N, Acuff V, Pettersen S, Gruber K (1988) Structural and functional variations in capillary systems within the brain. Ann N Y Acad Sci 529:21–30

50. Festa A, D'Agostino R, Jr, Howard G, Mykkanen L, Tracy RP, Haffner SM (2000) Chronic subclinical inflammation as part of the insulin resistance syndrome: the Insulin Resistance Atherosclerosis Study (IRAS). Circulation 102: 42–7

51. Frank RN (1991) On the pathogenesis of diabetic retinopathy. A 1990 update. Ophthalmology 98:586–93

52. Frohlich M, Imhof A, Berg G, Hutchinson WL, Pepys MB, Boeing H, Muche R, Brenner H, Koenig W (2000) Association between C-reactive protein and features of the metabolic syndrome: a population-based study. Diabetes Care 23:1835–9

53. Fujimoto K (1995) Pericyte-endothelial gap junctions in developing rat cerebral capillaries: a fine structural study. Anat Rec 242:562–5

54. Fukuhara A, Irie K, Nakanishi H, Takekuni K, Kawakatsu T, Ikeda W, Yamada A, Katata T, Honda T, Sato T, Shimizu K, Ozaki H, Horiuchi H, Kita T, Takai Y (2002) Involvement of nectin in the localization of junctional adhesion molecule at tight junctions. Oncogene 21:7642–55

55. Fukuhara A, Irie K, Yamada A, Katata T, Honda T, Shimizu K, Nakanishi H, Takai Y (2002) Role of nectin in organization of tight junctions in epithelial cells. Genes Cells 7: 1059–72

56. Furuse M, Hata M, Furuse K, Yoshida Y, Haratake A, Sugitani Y, Noda T, Kubo A, Tsukita S (2002) Claudin-based tight junctions are crucial for the mammalian epidermal barrier: a lesson from claudin-1-deficient mice. J Cell Biol 156: 1099–111

57. Furuse M, Hirase T, Itoh M, Nagafuchi A, Yonemura S, Tsukita S, Tsukita S (1993) Occludin: a novel integral membrane protein localizing at tight junctions. J Cell Biol 123:1777–88

58. Garcia PJ, Spellman CW (2006) Should all diabetic patients receive statins? Curr Atheroscler Rep 8:13–8

59. Gardiner TA, Anderson HR, Stitt AW (2003) Inhibition of advanced glycation end-products protects against retinal capillary basement membrane expansion during long-term diabetes. J Pathol 201:328–33

60. Gardner TW, Antonetti DA, Barber AJ, LaNoue KF, Nakamura M (2000) New insights into the pathophysiology of diabetic retinopathy: potential cell-specific therapeutic targets. Diabetes Technol Ther 2:601–8

61. Goldstein B (2000) Embryonic polarity: a role for microtubules. Curr Biol 10:R820–2

62. Gonzalez-Mariscal L, Betanzos A, Nava P, Jaramillo BE (2003) Tight junction proteins. Prog Biophys Mol Biol 81:1–44

63. Guillemot L, Hammar E, Kaister C, Ritz J, Caille D, Jond L, Bauer C, Meda P, Citi S (2004) Disruption of the cingulin gene does not prevent tight junction formation but alters gene expression. J Cell Sci 117:5245–56

64. Gumbiner B, Lowenkopf T, Apatira D (1991) Identification of a 160-kDa polypeptide that binds to the tight junction protein ZO-1. Proc Natl Acad Sci U S A 88:3460–4

65. Guo P, Weinstein AM, Weinbaum S (2003) A dual-pathway ultrastructural model for the tight junction of rat proximal tubule epithelium. Am J Physiol Renal Physiol 285:F241–57

66. Guo S, Kemphues KJ (1996) Molecular genetics of asymmetric cleavage in the early *Caenorhabditis elegans* embryo. Curr Opin Genet Dev 6:408–15

67. Haffner SM (2006) The metabolic syndrome: inflammation, diabetes mellitus, and cardiovascular disease. Am J Cardiol 97:3A–11A

68. Hammes HP (2005) Pericytes and the pathogenesis of diabetic retinopathy. Horm Metab Res 37 Suppl 1:39–43

69. Hammes HP, Lin J, Wagner P, Feng Y, Vom Hagen F, Krzizok T, Renner O, Breier G, Brownlee M, Deutsch U (2004) Angiopoietin-2 causes pericyte dropout in the normal retina: evidence for involvement in diabetic retinopathy. Diabetes 53:1104–10

70. Harhaj NS, Felinski EA, Wolpert EB, Sundstrom JM, Gardner TW, Antonetti DA (2006) VEGF activation of protein kinase C stimulates occludin phosphorylation and contributes to endothelial permeability. IOVS (in press)

71. Haskins J, Gu L, Wittchen ES, Hibbard J, Stevenson BR (1998) ZO-3, a novel member of the MAGUK protein family found at the tight junction, interacts with ZO-1 and occludin. J Cell Biol 141:199–208

72. Hirase T, Kawashima S, Wong EY, Ueyama T, Rikitake Y, Tsukita S, Yokoyama M, Staddon JM (2001) Regulation of tight junction permeability and occludin phosphorylation by Rhoa-p160ROCK-dependent and -independent mechanisms. J Biol Chem 276:10423–31

73. Hirase T, Staddon JM, Saitou M, Ando-Akatsuka Y, Itoh M, Furuse M, Fujimoto K, Tsukita S, Rubin LL (1997) Occludin as a possible determinant of tight junction permeability in endothelial cells. J Cell Sci 110(14):1603–13

74. Hirschi KK, D'Amore PA (1997) Control of angiogenesis by the pericyte: molecular mechanisms and significance. EXS 79:419–28

75. Hofman P, van Blijswijk BC, Gaillard PJ, Vrensen GF, Schlingemann RO (2001) Endothelial cell hypertrophy induced by vascular endothelial growth factor in the retina: new insights into the pathogenesis of capillary nonperfusion. Arch Ophthalmol 119:861–6

76. Hurd TW, Gao L, Roh MH, Macara IG, Margolis B (2003) Direct interaction of two polarity complexes implicated in epithelial tight junction assembly. Nat Cell Biol 5:137–42

77. Jiang WG, Martin TA, Matsumoto K, Nakamura T, Mansel

RE (1999) Hepatocyte growth factor/scatter factor decreases the expression of occludin and transendothelial resistance (TER) and increases paracellular permeability in human vascular endothelial cells. J Cell Physiol 181: 319–29

78. Jin M, Barron E, He S, Ryan SJ, Hinton DR (2002) Regulation of RPE intercellular junction integrity and function by hepatocyte growth factor. Invest Ophthalmol Vis Sci 43:2782–90

79. Joussen AM, Murata T, Tsujikawa A, Kirchhof B, Bursell SE, Adamis AP (2001) Leukocyte-mediated endothelial cell injury and death in the diabetic retina. Am J Pathol 158:147–52

80. Katsura Y, Okano T, Noritake M, Kosano H, Nishigori H, Kado S, Matsuoka T (1998) Hepatocyte growth factor in vitreous fluid of patients with proliferative diabetic retinopathy and other retinal disorders. Diabetes Care 21: 1759–63

81. Kausalya PJ, Amasheh S, Gunzel D, Wurps H, Muller D, Fromm M, Hunziker W (2006) Disease-associated mutations affect intracellular traffic and paracellular Mg^{2+} transport function of Claudin-16. J Clin Invest 116:878–91

82. Kinukawa Y, Shimura M, Tamai M (1999) Quantifying leukocyte dynamics and plugging in retinal microcirculation of streptozotosin-induced diabetic rats. Curr Eye Res 18:49–55

83. Kliche S, Waltenberger J (2001) VEGF receptor signaling and endothelial function. IUBMB Life 52:61–6

84. Knoblich JA (2001) Asymmetric cell division during animal development. Nat Rev Mol Cell Biol 2:11–20

85. Koblizek TI, Weiss C, Yancopoulos GD, Deutsch U, Risau W (1998) Angiopoietin-1 induces sprouting angiogenesis in vitro. Curr Biol 8:529–32

86. Korte GE, Burns MS, Bellhorn RW (1989) Epithelium-capillary interactions in the eye: the retinal pigment epithelium and the choriocapillaris. Int Rev Cytol 114:221–48

87. Kowluru RA (2005) Diabetic retinopathy: mitochondrial dysfunction and retinal capillary cell death. Antioxid Redox Signal 7:1581–87

88. Kowluru RA (2005) Effect of advanced glycation end products on accelerated apoptosis of retinal capillary cells under in vitro conditions. Life Sci 76:1051–60

89. Kowluru RA, Atasi L, Ho YS (2006) Role of mitochondrial superoxide dismutase in the development of diabetic retinopathy. Invest Ophthalmol Vis Sci 47:1594–9

90. Kowluru RA, Kennedy A (2001) Therapeutic potential of anti-oxidants and diabetic retinopathy. Expert Opin Investig Drugs 10:1665–76

91. Kuriyama M, Harada N, Kuroda S, Yamamoto T, Nakafuku M, Iwamatsu A, Yamamoto D, Prasad R, Croce C, Canaani E, Kaibuchi K (1996) Identification of AF-6 and canoe as putative targets for Ras. J Biol Chem 271:607–10

92. Lacaz-Vieira F, Jaeger MM, Farshori P, Kachar B (1999) Small synthetic peptides homologous to segments of the first external loop of occludin impair tight junction resealing. J Membr Biol 168:289–97

93. Lecaire T, Palta M, Zhang H, Allen C, Klein R, D'Alessio D (2006) Lower-than-expected prevalence and severity of retinopathy in an incident cohort followed during the first 4–14 years of type 1 diabetes: The Wisconsin Diabetes Registry Study. Am J Epidemiol 164:143–50

94. Lin J, Bierhaus A, Bugert P, Dietrich N, Feng Y, Vom Hagen F, Nawroth P, Brownlee M, Hammes HP (2006) Effect of R-(+)-alpha-lipoic acid on experimental diabetic retinopathy. Diabetologia 49:1089–96

95. Lindblom P, Gerhardt H, Liebner S, Abramsson A, Enge M, Hellstrom M, Backstrom G, Fredriksson S, Landegren U,

96. Ling V (1997) Multidrug resistance: molecular mechanisms and clinical relevance. Cancer Chemother Pharmacol 40 Suppl:S3–8

97. Lu M, Kuroki M, Amano S, Tolentino M, Keough K, Kim I, Bucala R, Adamis AP (1998) Advanced glycation end products increase retinal vascular endothelial growth factor expression. J Clin Invest 101:1219–24

98. Lu M, Perez VL, Ma N, Miyamoto K, Peng HB, Liao JK, Adamis AP (1999) VEGF increases retinal vascular ICAM-1 expression in vivo. Invest Ophthalmol Vis Sci 40:1808–12

99. Lutty GA, McLeod DS, Merges C, Diggs A, Plouet J (1996) Localization of vascular endothelial growth factor in human retina and choroid. Arch Ophthalmol 114:971–7

100. Maines LW, Antonetti DA, Wolpert EB, Smith CD (2005) Evaluation of the role of P-glycoprotein in the uptake of paroxetine, clozapine, phenytoin and carbamazepine by bovine retinal endothelial cells. Neuropharmacology 49: 610–7

101. Mandell KJ, Parkos CA (2005) The JAM family of proteins. Adv Drug Deliv Rev 57:857–67

102. Marmor MF, Wolfensberger TJ (1998) The retinal pigment epithelium: function and disease. Oxford University Press, New York

103. Masuzawa K, Goto K, Jesmin S, Maeda S, Miyauchi T, Kaji Y, Oshika T, Hori S (2006) An endothelin type A receptor antagonist reverses upregulated VEGF and ICAM-1 levels in streptozotocin-induced diabetic rat retina. Curr Eye Res 31:79–89

104. Mazzon E, Puzzolo D, Caputi AP, Cuzzocrea S (2002) Role of IL-10 in hepatocyte tight junction alteration in mouse model of experimental colitis. Mol Med 8:353–66

105. McCarthy KM, Skare IB, Stankewich MC, Furuse M, Tsukita S, Rogers RA, Lynch RD, Schneeberger EE (1996) Occludin is a functional component of the tight junction. J Cell Sci 109(9):2287–98

106. McNeil E, Capaldo CT, Macara IG (2006) Zonula occludens-1 function in the assembly of tight junctions in Madin-Darby canine kidney epithelial cells. Mol Biol Cell 17:1922–32

107. Miller SS, Edelman JL (1990) Active ion transport pathways in the bovine retinal pigment epithelium. J Physiol 424: 283–300

108. Miller SS, Hughes BA, Machen TE (1982) Fluid transport across retinal pigment epithelium is inhibited by cyclic AMP. Proc Natl Acad Sci U S A 79:2111–5

109. Miller SS, Steinberg RH (1977) Active transport of ions across frog retinal pigment epithelium. Exp Eye Res 25: 235–48

110. Mitamura Y, Takeuchi S, Matsuda A, Tagawa Y, Mizue Y, Nishihira J (2000) Hepatocyte growth factor levels in the vitreous of patients with proliferative vitreoretinopathy. Am J Ophthalmol 129:678–80

111. Mitic LL, Anderson JM (1998) Molecular architecture of tight junctions. Annu Rev Physiol 60:121–42

112. Miyamoto K, Hiroshiba N, Tsujikawa A, Ogura Y (1998) In vivo demonstration of increased leukocyte entrapment in retinal microcirculation of diabetic rats. Invest Ophthalmol Vis Sci 39:2190–4

113. Miyamoto K, Khosrof S, Bursell SE, Moromizato Y, Aiello LP, Ogura Y, Adamis AP (2000) Vascular endothelial

growth factor (VEGF)-induced retinal vascular permeability is mediated by intercellular adhesion molecule-1 (ICAM-1). Am J Pathol 156:1733–9

114. Miyoshi J, Takai Y (2005) Molecular perspective on tight-junction assembly and epithelial polarity. Adv Drug Deliv Rev 57:815–55

115. Moore TC, Moore JE, Kaji Y, Frizzell N, Usui T, Poulaki V, Campbell IL, Stitt AW, Gardiner TA, Archer DB, Adamis AP (2003) The role of advanced glycation end products in retinal microvascular leukostasis. Invest Ophthalmol Vis Sci 44:4457–64

116. Morcos Y, Hosie MJ, Bauer HC, Chan-Ling T (2001) Immunolocalization of occludin and claudin-1 to tight junctions in intact CNS vessels of mammalian retina. J Neurocytol 30:107–23

117. Moss SE, Klein R, Klein BE (1998) The 14-year incidence of visual loss in a diabetic population. Ophthalmology 105: 998–1003

118. Murata T, Ishibashi T, Inomata H (1993) Immunohistochemical detection of blood-retinal barrier breakdown in streptozotocin-diabetic rats. Graefes Arch Clin Exp Ophthalmol 231:175–7

119. Murata T, Nakagawa K, Khalil A, Ishibashi T, Inomata H, Sueishi K (1996) The relation between expression of vascular endothelial growth factor and breakdown of the blood-retinal barrier in diabetic rat retinas. Lab Invest 74:819–25

120. Nagy Z, Peters H, Huttner I (1984) Fracture faces of cell junctions in cerebral endothelium during normal and hyperosmotic conditions. Lab Invest 50:313–22

121. Nehls V, Denzer K, Drenckhahn D (1992) Pericyte involvement in capillary sprouting during angiogenesis in situ. Cell Tissue Res 270:469–74

122. Nelson WJ (2003) Adaptation of core mechanisms to generate cell polarity. Nature 422:766–74

123. Nielsen S, Smith BL, Christensen EI, Agre P (1993) Distribution of the aquaporin CHIP in secretory and resorptive epithelia and capillary endothelia. Proc Natl Acad Sci U S A 90:7275–9

124. Nitta T, Hata M, Gotoh S, Seo Y, Sasaki H, Hashimoto N, Furuse M, Tsukita S (2003) Size-selective loosening of the blood-brain barrier in claudin-5-deficient mice. J Cell Biol 161:653–60

125. Nusrat A, Brown GT, Tom J, Drake A, Bui TT, Quan C, Mrsny RJ (2005) Multiple protein interactions involving proposed extracellular loop domains of the tight junction protein occludin. Mol Biol Cell 16:1725–34

126. Oldendorf WH, Cornford ME, Brown WJ (1977) The large apparent work capability of the blood-brain barrier: a study of the mitochondrial content of capillary endothelial cells in brain and other tissues of the rat. Ann Neurol 1:409–17

127. Ostermann G, Weber KS, Zernecke A, Schroder A, Weber C (2002) JAM-1 is a ligand of the beta(2) integrin LFA-1 involved in transendothelial migration of leukocytes. Nat Immunol 3:151–8

128. Pardridge WM, Boado RJ, Farrell CR (1990) Brain-type glucose transporter (GLUT-1) is selectively localized to the blood-brain barrier. Studies with quantitative western blotting and in situ hybridization. J Biol Chem 265:18035–40

129. Petronczki M, Knoblich JA (2001) DmPAR-6 directs epithelial polarity and asymmetric cell division of neuroblasts in *Drosophila*. Nat Cell Biol 3:43–9

130. Philp NJ, Yoon H, Grollman EF (1998) Monocarboxylate transporter MCT1 is located in the apical membrane and MCT3 in the basal membrane of rat RPE. Am J Physiol 274:R1824–8

131. Pickup JC, Crook MA (1998) Is type II diabetes mellitus a disease of the innate immune system? Diabetologia 41: 1241–8

132. Qaum T, Xu Q, Joussen AM, Clemens MW, Qin W, Miyamoto K, Hassessian H, Wiegand SJ, Rudge J, Yancopoulos GD, Adamis AP (2001) VEGF-initiated blood-retinal barrier breakdown in early diabetes. Invest Ophthalmol Vis Sci 42:2408–13

133. Rao VV, Dahlheimer JL, Bardgett ME, Snyder AZ, Finch RA, Sartorelli AC, Piwnica-Worms D (1999) Choroid plexus epithelial expression of MDR1 P glycoprotein and multidrug resistance-associated protein contribute to the blood-cerebrospinal-fluid drug-permeability barrier. Proc Natl Acad Sci U S A 96:3900–5

134. Rizzolo LJ (1997) Polarity and the development of the outer blood-retinal barrier. Histol Histopathol 12:1057–67

135. Roh MH, Fan S, Liu CJ, Margolis B (2003) The Crumbs3-Pals1 complex participates in the establishment of polarity in mammalian epithelial cells. J Cell Sci 116:2895–906

136. Roh MH, Liu CJ, Laurinec S, Margolis B (2002) The carboxyl terminus of zona occludens-3 binds and recruits a mammalian homologue of discs lost to tight junctions. J Biol Chem 277:27501–9

137. Romero IA, Radewicz K, Jubin E, Michel CC, Greenwood J, Couraud PO, Adamson P (2003) Changes in cytoskeletal and tight junctional proteins correlate with decreased permeability induced by dexamethasone in cultured rat brain endothelial cells. Neurosci Lett 344:112–6

138. Rothenberg M, Ling V (1989) Multidrug resistance: molecular biology and clinical relevance. J Natl Cancer Inst 81:907–10

139. Rubin LL, Staddon JM (1999) The cell biology of the blood-brain barrier. Annu Rev Neurosci 22:11–28

140. Rymer J, Miller SS, Edelman JL (2001) Epinephrine-induced increases in $[Ca^{2+}]$(in) and KCl-coupled fluid absorption in bovine RPE. Invest Ophthalmol Vis Sci 42: 1921–9

141. Saitou M, Furuse M, Sasaki H, Schulzke JD, Fromm M, Takano H, Noda T, Tsukita S (2000) Complex phenotype of mice lacking occludin, a component of tight junction strands. Mol Biol Cell 11:4131–42

142. Salathe EP, Venkataraman R (1982) Interaction of fluid movement and particle diffusion across capillary walls. J Biomech Eng 104:57–62

143. Sasaki H, Matsui C, Furuse K, Mimori-Kiyosue Y, Furuse M, Tsukita S (2003) Dynamic behavior of paired claudin strands within apposing plasma membranes. Proc Natl Acad Sci U S A 100:3971–6

144. Sato TN, Tozawa Y, Deutsch U, Wolburg-Buchholz K, Fujiwara Y, Gendron-Maguire M, Gridley T, Wolburg H, Risau W, Qin Y (1995) Distinct roles of the receptor tyrosine kinases Tie-1 and Tie-2 in blood vessel formation. Nature 376:70–4

145. Schinkel AH, Smit JJ, van Tellingen O, Beijnen JH, Wagenaar E, van Deemter L, Mol CA, van der Valk MA, Robanus-Maandag EC, te Riele HP, et al. (1994) Disruption of the mouse mdr1a P-glycoprotein gene leads to a deficiency in the blood-brain barrier and to increased sensitivity to drugs. Cell 77:491–502

146. Schinkel AH, Wagenaar E, Mol CA, van Deemter L (1996) P-glycoprotein in the blood-brain barrier of mice influ-

ences the brain penetration and pharmacological activity of many drugs. J Clin Invest 97:2517–24

147. Schlingemann RO, van Hinsbergh VW (1997) Role of vascular permeability factor/vascular endothelial growth factor in eye disease. Br J Ophthalmol 81:501–12

148. Schulzke JD, Gitter AH, Mankertz J, Spiegel S, Seidler U, Amasheh S, Saitou M, Tsukita S, Fromm M (2005) Epithelial transport and barrier function in occludin-deficient mice. Biochim Biophys Acta 1669:34–42

149. Sedlakova R, Shivers RR, Del Maestro RF (1999) Ultrastructure of the blood-brain barrier in the rabbit. J Submicrosc Cytol Pathol 31:149–61

150. Senger DR, Perruzzi CA, Feder J, Dvorak HF (1986) A highly conserved vascular permeability factor secreted by a variety of human and rodent tumor cell lines. Cancer Res 46:5629–32

151. Shin K, Straight S, Margolis B (2005) PATJ regulates tight junction formation and polarity in mammalian epithelial cells. J Cell Biol 168:705–11

152. Shinoda K, Ishida S, Kawashima S, Wakabayashi T, Matsuzaki T, Takayama M, Shinmura K, Yamada M (1999) Comparison of the levels of hepatocyte growth factor and vascular endothelial growth factor in aqueous fluid and serum with grades of retinopathy in patients with diabetes mellitus. Br J Ophthalmol 83:834–7

153. Stamer WD, Bok D, Hu J, Jaffe GJ, McKay BS (2003) Aquaporin-1 channels in human retinal pigment epithelium: role in transepithelial water movement. Invest Ophthalmol Vis Sci 44:2803–8

154. Stevenson BR, Siliciano JD, Mooseker MS, Goodenough DA (1986) Identification of ZO-1: a high molecular weight polypeptide associated with the tight junction (zonula occludens) in a variety of epithelia. J Cell Biol 103:755–66

155. Stewart PA, Hayakawa K (1994) Early ultrastructural changes in blood-brain barrier vessels of the rat embryo. Brain Res Dev Brain Res 78:25–34

156. Stitt A, Gardiner TA, Alderson NL, Canning P, Frizzell N, Duffy N, Boyle C, Januszewski AS, Chachich M, Baynes JW, Thorpe SR (2002) The AGE inhibitor pyridoxamine inhibits development of retinopathy in experimental diabetes. Diabetes 51:2826–32

157. Stitt AW (2003) The role of advanced glycation in the pathogenesis of diabetic retinopathy. Exp Mol Pathol 75: 95–108

158. Strauss O (2005) The retinal pigment epithelium in visual function. Physiol Rev 85:845–81

159. Suri C, Jones PF, Patan S, Bartunkova S, Maisonpierre PC, Davis S, Sato TN, Yancopoulos GD (1996) Requisite role of angiopoietin-1, a ligand for the TIE2 receptor, during embryonic angiogenesis. Cell 87:1171–80

160. Suzuki A, Ishiyama C, Hashiba K, Shimizu M, Ebnet K, Ohno S (2002) aPKC kinase activity is required for the asymmetric differentiation of the premature junctional complex during epithelial cell polarization. J Cell Sci 115:3565–73

161. Suzuki A, Yamanaka T, Hirose T, Manabe N, Mizuno K, Shimizu M, Akimoto K, Izumi Y, Ohnishi T, Ohno S (2001) Atypical protein kinase C is involved in the evolutionarily conserved par protein complex and plays a critical role in establishing epithelia-specific junctional structures. J Cell Biol 152:1183–96

162. Szkudelski T (2001) The mechanism of alloxan and streptozotocin action in B cells of the rat pancreas. Physiol Res 50:537–46

163. Takagi C, Bursell SE, Lin YW, Takagi H, Duh E, Jiang Z, Clermont AC, King GL (1996) Regulation of retinal hemodynamics in diabetic rats by increased expression and action of endothelin-1. Invest Ophthalmol Vis Sci 37:2504–18

164. Tavelin S, Hashimoto K, Malkinson J, Lazorova L, Toth I, Artursson P (2003) A new principle for tight junction modulation based on occludin peptides. Mol Pharmacol 64:1530–40

165. The Diabetes Control and Complications Trial Research Group (1993) The effect of intensive treatment of diabetes on the development and progression of long-term complications in insulin-dependent diabetes mellitus. N Engl J Med 329:977–86

166. The Diabetes Control and Complications Trial Research Group (1998) Early worsening of diabetic retinopathy in the Diabetes Control and Complications Trial. Arch Ophthalmol 116:874–86.

167. Turksen K, Troy TC (2004) Barriers built on claudins. J Cell Sci 117:2435–47

168. Uemura A, Ogawa M, Hirashima M, Fujiwara T, Koyama S, Takagi H, Honda Y, Wiegand SJ, Yancopoulos GD, Nishikawa S (2002) Recombinant angiopoietin-1 restores higher-order architecture of growing blood vessels in mice in the absence of mural cells. J Clin Invest 110:1619–28

169. Umeda K, Matsui T, Nakayama M, Furuse K, Sasaki H, Furuse M, Tsukita S (2004) Establishment and characterization of cultured epithelial cells lacking expression of ZO-1. J Biol Chem 279:44785–94

170. Van Itallie CM, Anderson JM (1997) Occludin confers adhesiveness when expressed in fibroblasts. J Cell Sci 110(9):1113–21

171. Van Itallie CM, Fanning AS, Anderson JM (2003) Reversal of charge selectivity in cation or anion-selective epithelial lines by expression of different claudins. Am J Physiol Renal Physiol 285:F1078–84

172. Vialettes B, Silvestre-Aillaud P, Atlan-Gepner C (1994) [Outlook for the future in the treatment of diabetic retinopathy]. Diabetes Metab 20:229–34

173. Vietor I, Bader T, Paiha K, Huber LA (2001) Perturbation of the tight junction permeability barrier by occludin loop peptides activates beta-catenin/TCF/LEF-mediated transcription. EMBO Rep 2:306–12

174. Vincent AM, Brownlee M, Russell JW (2002) Oxidative stress and programmed cell death in diabetic neuropathy. Ann N Y Acad Sci 959:368–83

175. Vinores SA, McGehee R, Lee A, Gadegbeku C, Campochiaro PA (1990) Ultrastructural localization of blood-retinal barrier breakdown in diabetic and galactosemic rats. J Histochem Cytochem 38:1341–52

176. Vitale S, Maguire MG, Murphy RP, Hiner CJ, Rourke L, Sackett C, Patz A (1995) Clinically significant macular edema in type I diabetes. Incidence and risk factors. Ophthalmology 102:1170–6

177. Vorbrodt AW, Dobrogowska DH (2003) Molecular anatomy of intercellular junctions in brain endothelial and epithelial barriers: electron microscopist's view. Brain Res Brain Res Rev 42:221–42

178. Walsh SV, Hopkins AM, Nusrat A (2000) Modulation of tight junction structure and function by cytokines. Adv Drug Deliv Rev 41:303–13

179. Wang W, Dentler WL, Borchardt RT (2001) VEGF increases BMEC monolayer permeability by affecting occludin expression and tight junction assembly. Am J Physiol Heart Circ Physiol 280:H434–40

180. Wang Z, Mandell KJ, Parkos CA, Mrsny RJ, Nusrat A (2005) The second loop of occludin is required for suppression of Raf1-induced tumor growth. Oncogene 24: 4412–20

181. Wen H, Watry DD, Marcondes MC, Fox HS (2004) Selective decrease in paracellular conductance of tight junctions: role of the first extracellular domain of claudin-5. Mol Cell Biol 24:8408–17

182. Wilcox ER, Burton QL, Naz S, Riazuddin S, Smith TN, Ploplis B, Belyantseva I, Ben-Yosef T, Liburd NA, Morell RJ, Kachar B, Wu DK, Griffith AJ, Riazuddin S, Friedman TB (2001) Mutations in the gene encoding tight junction claudin-14 cause autosomal recessive deafness DFNB29. Cell 104:165–72

183. Williams CD, Rizzolo LJ (1997) Remodeling of junctional complexes during the development of the outer blood-retinal barrier. Anat Rec 249:380–8

184. Wittchen ES, Haskins J, Stevenson BR (1999) Protein interactions at the tight junction. Actin has multiple binding partners, and ZO-1 forms independent complexes with ZO-2 and ZO-3. J Biol Chem 274:35179–85

185. Wolburg H, Lippoldt A (2002) Tight junctions of the blood-brain barrier: development, composition and regulation. Vascul Pharmacol 38:323–37

186. Wolburg H, Wolburg-Buchholz K, Kraus J, Rascher-Eggstein G, Liebner S, Hamm S, Duffner F, Grote EH, Risau W, Engelhardt B (2003) Localization of claudin-3 in tight junctions of the blood-brain barrier is selectively lost during experimental autoimmune encephalomyelitis and human glioblastoma multiforme. Acta Neuropathol (Berl) 105:586–92

187. Wolburg H, Wolburg-Buchholz K, Liebner S, Engelhardt B (2001) Claudin-1, claudin-2 and claudin-11 are present in tight junctions of choroid plexus epithelium of the mouse. Neurosci Lett 307:77–80

188. Wong V, Gumbiner BM (1997) A synthetic peptide corresponding to the extracellular domain of occludin perturbs the tight junction permeability barrier. J Cell Biol 136: 399–409

189. Wu X, Zhao X, Baylor L, Kaushal S, Eisenberg E, Greene LE (2001) Clathrin exchange during clathrin-mediated endocytosis. J Cell Biol 155:291–300

190. Xu Q, Qaum T, Adamis AP (2001) Sensitive blood-retinal barrier breakdown quantitation using Evans blue. Invest Ophthalmol Vis Sci 42:789–94

191. Yamamoto T, Harada N, Kano K, Taya S, Canaani E, Matsuura Y, Mizoguchi A, Ide C, Kaibuchi K (1997) The Ras target AF-6 interacts with ZO-1 and serves as a peripheral component of tight junctions in epithelial cells. J Cell Biol 139:785–95

192. Yamanaka T, Horikoshi Y, Suzuki A, Sugiyama Y, Kitamura K, Maniwa R, Nagai Y, Yamashita A, Hirose T, Ishikawa H, Ohno S (2001) PAR-6 regulates aPKC activity in a novel way and mediates cell-cell contact-induced formation of the epithelial junctional complex. Genes Cells 6: 721–31

193. Yokota T, Ma RC, Park JY, Isshiki K, Sotiropoulos KB, Rauniyar RK, Bornfeldt KE, King GL (2003) Role of protein kinase C on the expression of platelet-derived growth factor and endothelin-1 in the retina of diabetic rats and cultured retinal capillary pericytes. Diabetes 52:838–45

194. Yoon H, Fanelli A, Grollman EF, Philp NJ (1997) Identification of a unique monocarboxylate transporter (MCT3) in retinal pigment epithelium. Biochem Biophys Res Commun 234:90–4

195. Yu AS, McCarthy KM, Francis SA, McCormack JM, Lai J, Rogers RA, Lynch RD, Schneeberger EE (2005) Knockdown of occludin expression leads to diverse phenotypic alterations in epithelial cells. Am J Physiol Cell Physiol 288:C1231–41

196. Zettl KS, Sjaastad MD, Riskin PM, Parry G, Machen TE, Firestone GL (1992) Glucocorticoid-induced formation of tight junctions in mouse mammary epithelial cells in vitro. Proc Natl Acad Sci U S A 89:9069–73

8

8.2 MRI Studies of Blood-Retinal Barrier: New Potential for Translation of Animal Results to Human Application

B.A. Berkowitz

Core Messages

- Blood-retinal barrier breakdown can contribute to abnormal accumulation of water in the retina
- Traditional methods for clinical evaluation of macular edema include fluorescein angiography, optical coherence tomography, and vitreous fluorophotometry
- Dynamic contrast enhanced MRI (DCE-MRI) with gadolinium diethylenetriaminepentaace-

tic acid (Gd-DTPA) allows for in vivo quantitative measurement of blood-retinal barrier breakdown
- DCE-MRI was assessed in a variety of experimental and clinical conditions (e.g., focal retinal photocoagulation, endophthalmitis, diabetic retinopathy)
- DCE-MRI allows for an exact determination of blood-retinal breakdown

8.2.1 Introduction

Abnormal accumulation of water in the retina (i.e., retinal edema) is commonly associated with retinal vessel diseases such as diabetes. Edema that develops in the macula secondary to diabetes is often linked with major visual loss [21]. However, despite decades of study, current treatment options for retinal edema remain limited to, primarily, laser photocoagulation. However, photocoagulation is destructive, does not restore lost vision, and is helpful in only about 50% of the patients [1]. Better appreciation of the underlying pathophysiology associated with edema formation is likely to improve diagnosis and medical care for retinal edema. Currently it is understood that fluid build-up in the retina can develop in intracellular and/or extracellular compartments. The blood-retinal barrier (BRB) refers to tight cell-cell junctions of the retinal vascular endothelium (also known as inner BRB) as well as tight junctions of the retinal pigment epithelium (or outer BRB). Intracellular edema (or cytotoxic edema) is defined as cellular swelling that occurs without opening of the BRB. Extracellular (or vasogenic) edema is characterized by retinal thickening in association with loss of BRB integrity [35]. Consequently, the development of optimal strategies for treating retinal edema may depend on determining the ratio of the contribution of intra- and extracellular mechanisms to edema and measuring how this ratio changes between patients, between different retinopathies, and during disease progression [32].

Traditional methods for evaluating the efficacy of medical therapy for retinal edema or deciding on endpoints for therapeutic intervention (e.g., visual acuity and seven-field stereoscopic fundus photography) detect only gross changes that occur late in the course of the disease. With these endpoints, studies require large numbers of patients with many years of follow-up before evidence-based conclusions can be drawn [35]. Fluorescein angiography (FA), an essential tool in the current clinical diagnosis and management of retinal diseases, is a rapid and straightforward photographic technique that allows localization of fluorescein leakage due to BRB breakdown, but cannot quantify BRB damage in physiologic terms, such as the permeability surface area product (PS). In other words, only subjective determination of the magnitude of BRB damage is possible with FA and this prevents analytic staging of treatment efficacies (using, for example, BRB PS) within and between patients over time. Furthermore, as a tracer, fluorescein is far from ideal because, among other variables, it is bound to serum protein, has a complex pharmacokinetic profile that complicates evaluating the fluorescein concentration plasma integral, is metabolized to different fluorescein-based metabolites which can confound interpretation, and enters the vitreous through a passive mechanism but is actively removed from vitreous to choroid [49, 50, 51]. Importantly, the location of BRB damage, as measured by FA, is not always linked with regions of edema [37]. For these reasons, FA remains a suboptimal approach for quantitatively evaluating new or existing treatment responses. Alternatively, retinal fluid buildup in either intra- or extracellular compartments can produce an increase in retinal thickness that is measurable using analytic techniques such as ocular coherence tomography (OCT). Because it only measures retinal thickness, OCT is unable to determine whether the type of edema is intra- or extracellular [32]. In addition, interrogating the exact same retinal region during repeated exams using clinically available OCT machines can be difficult. This registration problem may limit the accuracy and applicability of OCT measurement of longitudinal changes in retinal thickness. Therefore, while the methods mentioned have a proven usefulness in the clinic, they are also somewhat limited in their diagnostic and prognostic applicability.

Vitreous fluorophotometry (VFP), a fluorescein-based method, was developed to address some of these concerns [12]. Unlike FA, which presents a two-dimensional map of fluorescence across the retina, but not fluorescein concentrations, VFP measures a one-dimensional spatial profile of fluorescein concentration in the vitreous that can be used to quantitate BRB damage. However, VFP is not widely avail-able, it cannot be performed in the presence of optical opacities (such as cataract), its interpretation can be confounded by changes in vitreous fluidity, and it suffers from the problems associated with fluorescein that are listed above [35, 50, 51]. Additionally, data must be carefully collected to take into account the influence of the potentially confounding variables associated with the use of fluorescein that can confound physiologic interpretation of the VFP data.

8.2.2 Magnetic Resonance Imaging

Magnetic resonance imaging (MRI) provides a pan-retinal measure of ocular anatomy from inferior to superior (or nasal to temporal) ora serrata and is not dependent upon the clarity of the ocular media. Therefore, optical distortions such as diabetic cataracts do not degrade MRI images. MRI has been successfully adapted to laboratory animal, as well as human, subjects, making it possible to conduct translational research between humans and a wide range of species including mice and rats (Fig. 8.2.1). In the mid to late 1980s, proof-of-concept principle with this technique was achieved in studies of mon-

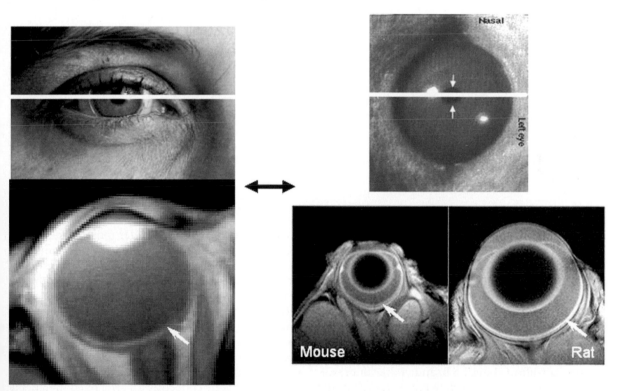

Fig. 8.2.1. Representative high resolution images [in-plane resolution is 0.39×0.39 mm^2 (human) and 23.4 µm^2 (rat and mouse)] measured from a single slice (*white line, upper images*) in humans (*left panels*) and rodents (*right panels*). Note that the rodent images are scaled differently than the human images and that the dissimilar appearance of the lens in the human and rat images is due to the different pulse sequences used. The *white line* in the posterior region of the eye represents the retina/choroid complex (*white arrow*). Acquisition of artifact-free high resolution images in both clinical and preclinical settings highlights the ability of MRI to be used to conduct translational research between humans and other species including rats and mice, as well as other species

Fig. 8.2.2. Schematic illustrating how the Gd-DTPA contrast agent is injected intravenously in a rodent. Gd-DTPA is distributed throughout the body, including the retina. In tissues with non-fenestrated blood vessels, such as the retina, Gd-DTPA is maintained inside the vascular space. The *dotted arrows* indicate that Gd-DTPA only leaks out of the circulation in the presence of damaged BRB. In this case, the presence of Gd-DTPA is readily detected by the vitreous water (Fig. 8.2.3)

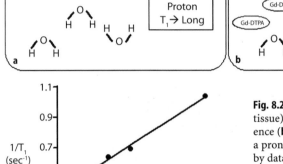

Fig. 8.2.3. Diagrams in *top row* (**a, b**) outline what happens to vitreous (and tissue) water spin lattice relaxation time (T_1) in the absence (**a**) and presence (**b**) of Gd-DTPA. Due to its paramagnetic nature, Gd-DTPA produces a pronounced shortening of vitreous water T_1 time. Importantly, as shown by data collected in a water phantom with different amounts of Gd-DTPA, the T_1 relaxation rate ($1/T_1$) is linearly related to Gd-DTPA concentration (**c**). Using this calibration curve, and the well established relationship between T_1 and signal intensity on a T_1-weighted image (see text), the concentration of Gd-DTPA in the vitreous can be determined. Collection of sequential images prior to and post Gd-DTPA administration allows the influx of Gd-DTPA to be measured and converted to a BRB PS (see text)

keys and rabbits where damage to the blood ocular barriers was quantified using a non-fluorescein based method: dynamic contrast enhanced MRI (DCE-MRI) with gadolinium diethylenetriamine-pentaacetic acid (Gd-DTPA, 590 Da) [22, 48]. In this experiment, Gd-DTPA, also known as Gad, was injected intravenously and delivered by the blood supply to the retina (Fig. 8.2.2). Gd-DTPA is a relatively safe tracer to administer intravenously, has a plasma half-life of about 1.5 h, and, unlike fluorescein, Gad does not cause yellowing of the skin and urine [53, 54]. In addition, Gd-DTPA is freely diffusible (i.e., is a non-specific extracellular marker) and follows a well defined biexponential plasma decay (i.e., it is not bound to plasma protein and is excreted mainly by the kidneys) [53, 54]. Note that gadolinium can also be bound to larger molecules, such as albumin, if necessary [59] and that gadolinium-based contrast agents are paramagnetic. Normally, Gd-DTPA does not penetrate non-fenestrated blood vessels or barriers, such as in the inner and outer BRB; thus the vitreous space remains unenhanced (Fig. 8.2.2) [8, 72]. However, when BRB becomes damaged, Gd-DTPA enters into the vitreous space via a passive diffusion mechanism. Gd-DPTA shortens the vitreous water spin-lattice (T_1) relaxation time such that by collecting a T_1 weighted image, the presence of Gd-DTPA can be readily detected as an increase (i.e., an enhancement) in signal intensity (Fig. 8.2.3). Alternatively, vitreous T_1 can be measured but this is usually a relatively longer procedure than simply collecting a T_1 weighted image. Another

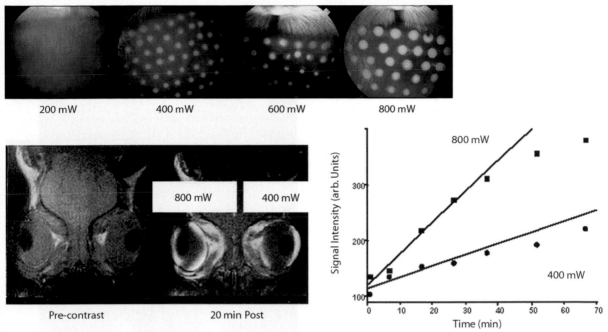

Fig. 8.2.4. *Top row:* Photographs of pigment rabbit fundus 2 days after argon laser panretinal photocoagulation procedure. Rabbit eyes were treated with the laser power set as indicated below each fundus photograph [8]. These photographs were obtained just prior to MRI examination. *Left panel:* Representative real-time, T_1-weighted, coronal proton images acquired during before (*left image*) and post (*right image*) contrast injection. In this animal, both eyes were treated with a similar panretinal photocoagulation procedure, except the laser powers were different between OS and OD, as indicated [8]. Note that the signal intensity change in the anterior chamber is similar for each eye. This is used as an internal check of the goodness of the intravenous injection. *Right panel:* Representative time course of the mean vitreous signal intensity from OS (*solid squares*) and OD (*solid circles*) of the animal in the left panel. Linear analysis of the first 20 min post Gd-DTPA injection is presented as *solid lines* [8]. The resulting slope was called the leakiness parameter since it reflects the amount of Gd-DTPA entering the vitreous space through the disrupted retina

approach, the quantification of increased intraretinal Gd-DTPA concentration, is considerably more difficult because at low spatial resolution plasma based-GD-DTPA cannot be easily distinguished from retinal tissue-only Gd-DTPA. Thus, most efforts have investigated changes in vitreous signal intensity following Gd-DTPA injection. Berkowitz et al. validated the use of DCE-MRI to sensitively and accurately measure BRB PS in rabbits and humans [8–10, 61, 62, 71]. In rabbits, grid photocoagulation was applied to inferior retina at different power levels (200, 400, 600, and 800 mW) producing progressively larger disruption of BRB that could be readily detected using DCE-MRI (Fig. 8.2.4) [8]. Vitreous signal intensities were measured at each time point and the initial change in signal intensity with time analyzed to produce a slope (Fig. 8.2.4) that could be interpreted as a measure of BRB "leakiness" and was correlated with laser power (Fig. 8.2.5) [8]. In humans, retinas of patients with proliferative diabetic retinopathy were examined by both FA and DCE-MRI [71]. As shown in Fig. 8.2.6, the location and severity of enhancement, judged by visual inspection

of the images, corresponded to the fluorescein angiographic and/or clinical appearance of preretinal neovascularization.

Studies in the early 1990s demonstrated that the increase in vitreous signal intensity was linearly related to Gd-DTPA concentration (Fig. 8.2.3) [8, 9]. Knowing the Gd-DTPA vitreous level and the Gd-DTPA plasma concentration time course allows one to determine the extent of the leak in the retina, or, in other words, BRB PS [8, 9]. To calculate PS, the Simplified Early Enhancement method of Tofts and Berkowitz was used [9, 61]. This method requires estimates of the vitreous T_1 in the absence of Gd-DTPA (T_{10}), the relaxivity of Gd-DTPA (R_1, s^{-1} mM^{-1}), and the Gd-DTPA concentration plasma time course parameters [9, 61]. Vitreous T_{10} was previously reported to be approximately 3.5 s. This value has been verified in control rodents using a homogenous excitation and surface coil reception while collecting gradient recalled echo images at different flip angles. As expected for a T_{10} of approximately 3.5 s, a maximum vitreous signal intensity was found at a flip angle of 12°–13° (based on the

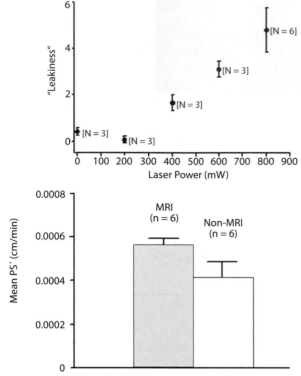

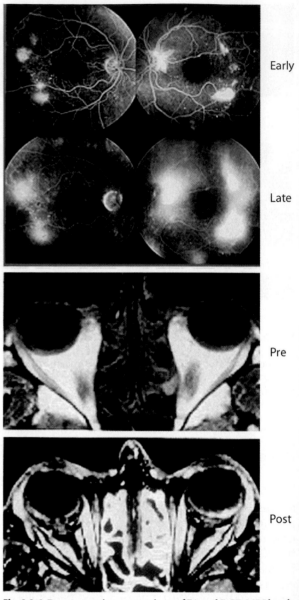

Early

Late

Pre

Post

Fig. 8.2.5. *Top graph:* The relationship between the leakiness parameter (defined in Fig. 8.2.4) and the laser power setting described in Fig. 8.2.4 for the vitreous. Note that a similar analysis of signal intensity changes in the anterior chamber was also performed (not shown) and no discernible correlation was found [8]. *Bottom graph:* In a separate experiment we developed and applied the theory of converting the leakiness parameter into a measure of BRB PS. The accuracy and precision of this approach was investigated in rabbits pretreated with sodium iodate (30 mg/kg intravenously) and summarized here [9]. The MRI-derived PS normalized to the area of leaky retina (*PS'*) was compared to a similarly normalized PS calculated using a classical physiologic method and agreement was found between the two methods ($P > 0.05$). Furthermore, PS' values in eyes with normal vitreous were not different ($P > 0.05$) compared with that from contralateral gas-compressed vitrectomized eyes (data not shown) [10]

Fig. 8.2.6. Representative comparison of FA and DCE-MRI localization of BRB damage. *Top panel:* Early and late FA of a 41-year-old man with a 22-year history of insulin dependent diabetes (20/20 OD, 20/20 OS) and no history of photocoagulation. *Bottom panel:* Pre- and postcontrast administration of same patient in *top panel*. Note agreement in location and severity of enhancement on FA and DCE-MRI

Ernst angle formula, data not shown) [47]. Gd-DTPA relaxivity is constant at a set temperature and field strength and so the previously reported value of 4.5 s/mM was used [9]. To determine the pharmacokinetic parameters after a bolus of Gd-DTPA in rats, blood samples were obtained in separate experiments in heparinized tubes during the precontrast period, and 1, 3, 7, 15, 30, and 60 min postinjection. Following centrifugation, the plasma fraction was obtained for NMR analysis. Inversion recovery T_1 experiments were performed on the water signal of the plasma fraction at room temperature. From the T_1 value, the amount and thus concentration of Gd-DTPA was determined from a cali-

bration curve obtained at room temperature in a separate phantom study. The unidirectional rate constant, k, is given by:

$$K = E / R_1 T_k D \{(\beta_1[1 - \exp(-m_1 T)]/m_1) + (\alpha_2[1 - \exp(-m_2 t)] / m_2)\} \quad (8.2.1)$$

where D is the Gd-DTPA dose, $T_k = T_R \exp(-T_R/T_{10})/(1 - \exp(-T_R/T_{10}))$, T_R is the repetition time, $a_{1,2}$ are the Gd-DTPA plasma amplitudes, and $m_{1,2}$ are the rate constants of each plasma component. Thus, k can be found from a single measurement of enhancement, provided R_1, T_k, and the plasma parameters are known. Setting the vitreous volume in the slice $Vv = A_{\text{region-of-interest}}(\text{slice thickness})$, PS then is:

$$PS = kA_{\text{region-of-interest}} \text{ (slice thickness)} \qquad (8.2.2)$$

This equation assumes that all tracer in the slice originated from the portion of retina in that slice. This assumption has been verified experimentally [7].

To determine the accuracy of this approach, the outer BRB was first destroyed via an intravenous injection of sodium iodate [9]. Sodium iodate causes extensive disruption of the tight cell-cell junctions of the retinal pigment epithelium (outer BRB) within 24 h post-treatment [20]. At this time, BRB PS was determined using DCE-MRI, and eyes were then enucleated and frozen prior to measuring vitreous levels of Gd-DTPA [9]. BRB PS was then calculated using standard physiologic modeling [9]. The plasma time course was separately determined and used in both MRI and physiologic assessments. As shown in Fig. 8.2.5, agreement was found between the DCE-MRI and "gold standard" PS measures. Furthermore, following sodium iodate exposure, vitreous Gd-DTPA distribution in gas-compressed vitrectomized eyes was different from that of eyes with normal vitreous (Fig. 8.2.7). The BRB PS calculated from control eyes and those exposed to sodium iodate were not different ($P > 0.05$) [10]. In other words, changes in vitreous fluidity are not a confounding factor for DCE-MRI BRB PS measurements. More recent work has confirmed the linearity and accuracy of the technique (Fig. 8.2.8) [7]. A linear relationship between vitreous enhancement (E) and Gd-DTPA dose was demonstrated when the injected Gd-DTPA dose was compared to the detected signal in sodium iodate treated rats (Fig. 8.2.8) [7]. Further, DCE-MRI BRB PS measurements were in agreement with PS measured using a more conventional approach with radiolabeled sucrose [7]. The concurrence between DCE-MRI BRB PS and "gold standard" values, given the methodological differences between the studies [e.g., destructive versus non-destructive tissue analysis and different tracers (Gd-DTPA and sucrose)] provides strong support for the underlying assumptions, linearity, and accuracy of the DCE-MRI approach.

Quantifying BRB damage using DCE-MRI is relatively new to ophthalmology. To help understand its weaknesses and strengths, comparisons with known clinical methods have been performed. One disadvantage is that DCE-MRI BRB measurements cannot be performed in a doctor's office. However, DCE-MRI using Gd-DTPA is widely available in most medical centers and we envision its use, not as a pri-

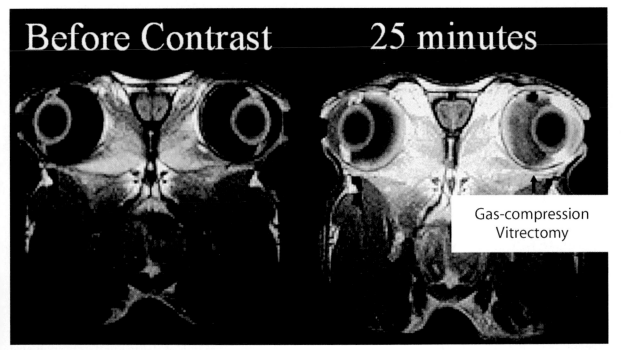

Fig. 8.2.7. Representative T1-weighted MRI of a sodium iodate treated rabbit with a control eye (*left*) and a gas-compressed vitrectomized eye (*right*) before and after Gd-DTPA injection. Note the heterogeneous Gd-DTPA distribution in the eye with normal vitreous (consistent with diffusion) and the more homogeneous distribution in the vitrectomized eye (consistent with bulk fluid flow and mixing). The *dark circular region* in the gas-compressed vitrectomized eye represents a residual gas bubble

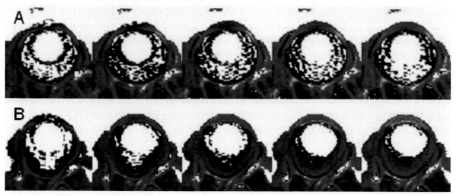

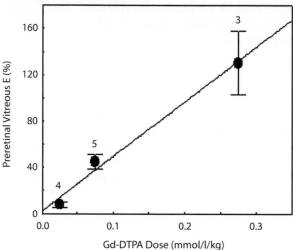

Fig. 8.2.8. *Top images:* Parameter maps of the percentage of change in signal intensity (enhancements, *E*) from precontrast levels at 3, 7, 9, 15, 19 min postcontrast injection for representative (**A**) control and (**B**) sodium iodate-treated rats. Brighter colors (e.g., *yellow*) represent greater Gd-DTPA levels. *White* is background and represents no change. The same color scale was used for all parameter maps. The rapid anterior chamber filling with contrast agent seen in all animals was expected and was used in each animal to confirm that the injection was good. Note only the sodium iodate-treated rats demonstrated coherent increases in vitreous *E* over time. *Bottom graph:* Plot of the percentage of change in vitreous signal intensity (*E*) vs. Gd-DTPA dose in sodium iodate-treated rats. A significant linear relationship was found ($r = 0.91$, $P < 0.005$). The numbers of animals used to generate each datum are listed *above each symbol.* *Error bars* represent the SEM

mary screening tool, but as a useful method when quantitative measures of BRB damage need to be assessed over time, such as in a drug clinical trial, or when fluorescein methods are not applicable or produce equivocal results. A potential weakness of DCE-MRI was postulated by Plehwe et al., who measured T_1 in selected volumes of vitreous after administration of Gd-DTPA and suggested that, at the lower limit of detection, MRI has a 1,200-fold lower sensitivity for detection of Gd-DTPA by reduction of T_1 than that for the detection of fluorescent material by VFP [48]. However, Plehwe et al. did not present data to support this theoretical hypothesis and subsequent studies have clearly demonstrated that their prediction was incorrect [42, 70, 71]. To further demonstrate this, direct comparison of the detection sensitivity of VFP and DCE-MRI was performed using higher doses of Gd-DTPA, acquisition of a thicker slice, and use of T_1-weighted images emphasizing relative differences in T_1 between pixels rather than measurement of absolute T_1. These studies were performed under a variety of experimental conditions, including focal retinal photocoagulation and cryopexy, grid photocoagulation obscured by vitreous hemorrhage, albino fundus, and endophthalmitis

(Table 8.2.1, personal communication with Harsha Sen, Noburo Ando, Brian Conway, and Eugene de Juan). DCE-MRI was found to reliably detect Gd-DTPA leakage from three diode laser burns, the smallest photocoagulation lesion distinguishable from untreated eyes by VFP (Table 8.2.1). These results strongly suggest that the sensitivity of MRI is at least comparable to that of VFP when based on detectable lesion size (Table 8.2.1). Conditions such as vitreous hemorrhage, albino fundus, and endophthalmitis can confound VFP data but had no effect on the analysis of the DCE-MRI data (some of this data is presented in Table 8.2.1, personal communication with Harsha Sen, Noburo Ando, Brian Conway, and Eugene de Juan) [10]. In addition, we also confirmed that increased vitreous fluidity, which can occur with age and disease and confound the accuracy of BRB damage assessments, as measured by the VFP method, does not affect the BRB PS as assessed by DCE-MRI (Fig. 8.2.7) [10].

The spatial sensitivity of DCE-MRI BRB, which is considered a strength of the method, was evaluated by taking advantage of the retinal anatomy of the rabbit. The rabbit retina is largely avascular except for a band of blood vessels that extends temporally

Table 8.2.1. Comparison of VFP and DCE-MRI BRB assessment (mean ± SEM)

Treatment	VFP (ng/ml)	N	P	PS' ($\times 10^{-4}$ cm/s)	N	P
Control	10.4+0.5	19	–	−0.14+0.13	6	–
3 laser burns	18.4+1.4	6	<0.001	0.46+0.12	6	0.007
6 laser burns	33.9+1.6	3	<0.001	0.60+0.1	5	0.002
Cryopexy	43.5+8.1	6	<0.001	1.26+0.4	4	<0.001
Albino	191+40.1	6	<0.001	−0.01+0.12	5	0.53

VFP was measured from vitreous fluorescence 2 mm anterior to retina 1 h after intravenous fluorescein injection.

N = number of eyes, P = p value for difference from control using Student's unpaired t-test, PS' = permeability surface area product normalized per unit area determined from the MR images

and nasally. This anatomy greatly simplifies the separate identification and quantitation of inner and outer BRB damage on an MRI slice perpendicular to the band of blood vessels. To demonstrate this, pigmented rabbits were treated with intravenous sodium iodate 30 mg/kg (a specific RPE cell poison), intravitreal *N*-ethylcarboxamidoadenosine (NECA, which specifically disrupts the vascular BRB), or retinal diode laser photocoagulation [60]. The pattern of enhancement observed in eyes treated with NECA clearly showed only inner BRB damage in contrast to the panretinal leakage that appeared following sodium iodate exposure. Simultaneous leakage from the outer and inner BRB in eyes treated with dense retinal laser photocoagulation could be localized and measured independently. Taken together, these data demonstrate that the detection and spatial sensitivity of DCE-MRI is at least as good as fluorescein-based methods for monitoring BRB breakdown.

Nearly 15 years of research has also shown that DCE-MRI has wide applicability with regard to the study of BRB breakdown in various conditions and retinopathy models. It has been established that DCE-MRI BRB PS measurements are capable of detecting subtle changes in vascular permeability following ischemia/reperfusion, proliferative vitreoretinopathy, endophthalmitis, argon vs. diode photocoagulation, and ocular surgery, in addition to the diabetic applications discussed above [4, 7, 23, 24, 42, 67, 69–71]. Further, quantifiable changes in BRB permeability have proven useful in the assessment of various therapeutic interventions, including the effects of corticosteroid (triamcinolone acetonide) treatment [4, 24, 58, 70]. DCE-MRI using Gd-DTPA is routinely available in most medical centers and has many advantages over FA and VFP including the absence of optical limitations, the ability to produce images in any of three dimensions, and use of a tracer that is non-metabolized, non-toxic, non-actively transported, and has a well defined plasma concentration curve. Collecting DCE-MRI is thus relatively robust (i.e., can be done under a variety of adverse conditions) and uses a tracer that allows considerable simplification of, and confidence in, the data analysis. Importantly, these results lay the foundation for future translational studies in which the exact same biomarker and technique, BRB integrity measured by DCE-MRI, can be used to evaluate drug treatments in animal models of retinopathy and in human clinical studies.

To further explore the usefulness of DCE-MRI, additional comparisons have been made with non-clinical methods for assessing BRB damage as applied to the streptozocin-induced hyperglycemic rat, a model of diabetic retinopathy. The diabetic rat develops retinal pathological damage similar to the early stages that are characteristic of the non-proliferative stage of the human disease. Many investigators have attempted to examine early BRB in diabetic rat models as indirect evidence for increased risk of extracellular (or vasogenic) edema formation. The results of these studies, however, have been quite controversial [35]. Using a variety of non-MRI detection methods and tracers in diabetic or galactose-fed rats, some investigators have reported little or no change in BRB permeability in early diabetes [11, 19, 20, 25, 26, 29–31, 36, 63, 68]. In contrast, using a histologic detection method with an endogenous tracer (extravasation of albumin) only a subset of diabetic rats are reported to develop some leakage [44, 66, 67]. In further contrast, there are many reports, using either fluorescein, albumin, or Evans blue dye as the tracer, of BRB damage occurring as early as 2 weeks after inducing diabetes in rats [18, 27, 43, 66, 73].

To the best of our knowledge, these fundamental discrepancies regarding if and when BRB opening occurs in diabetic rats have not been resolved. It is noteworthy that a number of confounding variables have been shown to substantially affect detection of barrier integrity using fluorescein, albumin, and Evans blue dye, such as the size and properties of the probe, interpretation of increased albumin-based tracers, rapid binding of Evans blue dye to tissue protein, possible contamination from the aqueous humor, as well as the issues listed above that unambiguous interpretation of increased passive BRB

8

damage following fluorescein injection requires careful attention to a number of potential confounding factors [5, 11, 15, 17, 41]. We have now identified yet another potential factor that might help explain the contradictory conclusions of studies of diabetes-induced alteration in retinal permeability. Most of the experimental methods used to assess BRB integrity require animal death and/or enucleation before determining the extent of BRB damage. Using DCE-MRI, we have shown that BRB becomes leaky immediately before and after death, possibly causing an artificial increase in retinal permeability in methods that require enucleation or retinal isolation to assess permeability (Fig. 8.2.9) [7]. In control rats, signal intensities change were minimal while the animal was alive ($P > 0.05$), indicating an intact BRB. However, clear evidence for BRB damage ($P < 0.05$) was found before loss of respiratory movement (i.e., death) (Fig. 8.2.9). In separate studies, BRB opening was also evident within minutes after sacrifice in animals with intravenous KCl injection (data not shown). Thus, near death conditions were associated with a rapid breakdown in BRB permeability. We speculate that control and diabetic rats might demonstrate this near death artifact to varying degrees producing different extents of BRB opening. In addition, extensive manipulation of the rats prior to BRB assessment may exacerbate the contribution of the death artifact. For example, Xu et al. allowed Evans blue dye to circulate for 90 min in ketamine/xylazine anesthetized rats before opening the chest cavity and perfusing via the left ventricle [72]. Note that the pneumothorax produced by opening of the chest

cavity will significantly compromise the animals' physiology and soon lead to the animals' death. In addition, the perfusion fluid maintained at pH 3.5 was used because it allowed optimal binding of Evans blue dye to albumin. It is possible that a pH 3.5 solution could produce a change in cell shape and BRB disruption. Nagy, Szabo, and Huttner found that cerebral perfusion with acidic pH buffer induced substantial blood-brain barrier leakage [45]. We have preliminary DCE-MRI data (not shown) suggesting that progressively lower arterial pHs will increase vitreous signal intensity enhancements (E) by Gd-DTPA in control rats. Although more work is needed, it appears possible that the use of low pH wash solutions alone can induce BRB damage. In addition, the rat was then perfused for 2 min at a physiological pressure of 120 mm Hg. However, ketamine/xylazine anesthetic will lower mean arterial blood pressure [73]. The act of perfusing the animal at a physiological pressure of 120 mm Hg may produce a relative acute hypertensive event. Such events are associated with damage to the blood-brain barrier [40]. It is also possible that some dilution of extravascular dye by the perfusate also occurred. Taken together, these considerations raise the possibility that some combination of death, pneumothorax manipulation, acidic flush solution, and in situ perfusion could artificially induce BRB damage so that some of the Evans blue dye leaked into the retina.

To avoid these potentially confounding factors, including the death artifact, DCE-MRI was applied to rats that were diabetic for either 2, 4, 6, or 8 months [7]. No evidence for passive BRB leakage in rats was found until after 6 months of diabetes (Fig. 8.2.10) [7]. In addition, we did not see increased vitreous enhancement (E) following injection of gadolinium bound to human serum albumin (Fig. 8.2.11). It is possible that in vivo and ex vivo methods measure different aspects of BRB integrity. The MRI approach only measures tracer levels in the vitreous, while ex vivo approaches only monitor tracer concentrations intraretinally. Nonetheless, it is noteworthy that, given the extensive validation of the DCE-MRI technique discussed above, there does not appear to be evidence for early BRB damage in diabetic rats.

We next asked the question: what if DCE-MRI is not sensitive enough to detect a subtle diabetes induced lesion that might occur early in the time course of the disease process? If previous studies are correct and BRB damage occurred as early as 2 weeks of hyperglycemia, then, we reasoned, retinal edema might have developed by 3 months of diabetes. Interestingly, Park et al. claimed to have detected an increase in inner nuclear layer thickness at 1 week after inducing diabetes in rats [46]. However, no sta-

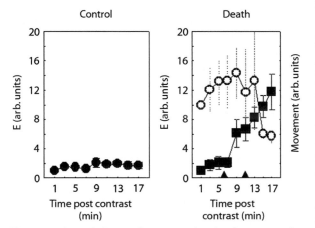

Fig. 8.2.9. Plots of vitreous humor % signal enhancement (*E*, *filled symbols*), normalized to the 1 min time point value, as a function of time postcontrast injection in control rats (*n* = 6, *left panel*), and control rats receiving additional urethane boluses (2×0.2 ml) i.v. (*arrowheads; n* = 3, *right panel*). In the *right panel*, respiratory movement (*open circle*), evident as low-intensity image smearing in the phase-encode direction, is also plotted and used to assess time of animal death

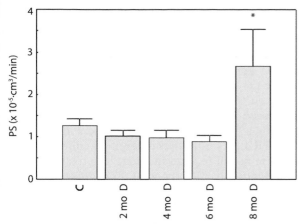

Fig. 8.2.10. Summary of BRB PS from control rats ($n=10$), and 2 months ($n=5$), 4 months ($n=5$), 6 months ($n=5$), and 8 months ($n=4$) diabetic rats. A significant difference (*$P<0.05$) was only found between the control and 8 month diabetic groups. These results are similar to other studies that find no evidence for early BRB damage [11, 19, 20, 25, 26, 29–31, 36, 63, 68]. We reasoned that perhaps an albumin-based contrast agent, as used in other studies of BRB damage, might reveal BRB damage. However, as shown in Fig. 11, no leakage was detected with a larger contrast agent

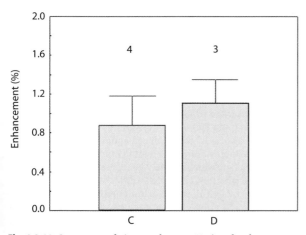

Fig. 8.2.11. Summary of vitreous humor % signal enhancement (E) in control and 3 month diabetic rats measured using a higher MW contrast agent [human serum albumin-(Gd-DTPA)30 (MW 93 kDa)] than Gd-DTPA (MW 590 Da). No between-group differences ($P>0.05$) in vitreous signal intensity were found with albumin-(Gd-DTPA)30. These data provide additional evidence of lack of early BRB damage (Fig. 8.2.10)

tistical analysis of the data was presented to confirm that retina thickness actually increased over control values [46]. If edema was present in the study of Park et al., then it was brief since retinal thickness at later time points was not increased. Consequently, the relevance of the findings of Park et al. to more chronic edema typically found clinically is unclear [46]. In contrast, we, and others, have reported a decrease in retinal thickness relative to controls in regions about

the optic nerve in rat after 1 month of diabetes [6, 38, 46]. The degree of retinal thinning appears to remain relatively constant (10–15%) after the induction of diabetes [6, 38, 46]. This lack of evidence for chronic edema is not consistent with the reported BRB damage starting at 2 weeks of diabetes and continuing, nor is it consistent with the formation of intracellular edema. In other words, if there is BRB damage starting at 2 weeks of diabetes, and edema does not develop, how important is that BRB damage? Our data, and that of others, suggests the alternative: that edema does not form in diabetic rat retinas by 3 months (which is consistent with an intact BRB) [11, 19, 20, 25, 26, 29–31, 36, 63, 68].

Although diabetic rats do not appear to develop an early passive leakage through damaged BRB, it remains of value to measure how well treatments, such as receptor antagonists, correct increases in BRB PS induced by exogenous administered soluble mediators, such as vascular endothelial growth factor (VEGF), also known as vascular permeability factor (VPF) [28]. This approach could help optimize treatment dosing, route, and scheduling. For example, a single intravitreous injection of VEGF/VPF is reported to induce a three- to fourfold increase in BRB permeability without disruption of retinal architecture [2, 3, 16, 39, 52]. Derevjanik et al. reported evidence of a roughly fourfold increased BRB damage to mannitol (182 Da) 6 h after intravitreous injection of 10^{-6} M VEGF in mice [16]. These findings are supported by DCE-MRI studies that found a threefold increase in BRB PS in the rat under similar conditions [7]. Since the dose of VEGF used in this latter study was higher than that used in other studies, it may appear that the sensitivity of DCE-MRI is lower. For example, Xu et al. utilized a tenfold lower dose (50 ng) of VEGF and reported a fourfold increase in BRB leakage [72]. However, based on our calculations (shown below), the VEGF dose used in the present study (and by Derevjanik et al. [16]) and in the work of Xu et al. are both higher than needed to achieve substantial equilibrium binding. Assuming a vitreous volume is 50 µl in the rat [34], our VEGF dose of 0.5 µg corresponds to a vitreous concentration of 0.24×10^{-6} M and the dose used by Xu et al. corresponds to a vitreous level of 0.24×10^{-7} M. Since the K_d of VEGF for its two receptors is $\sim 10^{-10}$–10^{-11} M [13, 14], both VEGF doses are clearly higher than needed to achieve substantial equilibrium binding. DCE-MRI BRB PS measurements appear to be a sensitive and quantitative approach for monitoring treatment efficacy on VEGF/VPF-induced BRB damage in rats.

Finally, we have demonstrated another advantage of DCE-MRI: providing a useful surrogate marker of BRB breakdown in patients with diabetic macular edema (DME) [64]. We found that the mean slope of the signal enhancement function was significantly greater in diabetic patients than in controls, indicating a greater time-dependent accumulation of contrast in the pre-macular vitreous of diabetic patients [64]. These data are consistent with results from previous VFP studies that demonstrate significantly greater contrast influx in diabetic patients with edema than in control groups [55, 56, 57]. Previously reported differences between patients with non-clinically significant edema and clinically significant edema subgroups [55, 56, 57] were not found in this study and may be due to the different sizes of the subgroups studied by VFP and DCE-MRI. We think that DCE-MRI will be most beneficially applied together with other quantitative methods for evaluating edema, such as OCT, since DCE-MRI provides a snapshot in time of BRB dysfunction that may not be obtained from potentially more slowly changing structural measures but may be independent of retinal thickening. Future studies comparing DCE-MRI with OCT are now warranted.

8.2.3 Conclusions

In summary, it would appear that clinical management of retinal edema based only on measurement of either retinal thickness or BRB integrity is not adequate for evaluating medical treatment of edema because the ratio of the contribution of intra- and extracellular mechanisms to edema or how this ratio changes between patients, between different retinopathies, and over time is not well defined [32]. In principle, quantitatively measuring both retinal thickness and BRB integrity can provide enough information to substantially improve diagnosis and prognosis of retinal edema. For example, finding that an increase in retinal thickness occurs with an intact BRB would imply the presence of intracellular edema. As another example, if treatments corrected a BRB defect but have no effect on elevated retinal thickness, it would imply a strong intracellular component. Currently, such distinctions cannot be made. One reason is that present methods for detecting (e.g., FA) and/or measuring (e.g., VFP) BRB damage clinically are suboptimal, as described above. Simpler, more available, robust, and accurate measures of BRB PS are needed, especially to quantify the impact of medical therapy on edema. In this chapter, we highlight DCE-MRI as addressing this need, since clinical MRI machines are widely available, in vivo measures of BRB PS are obtained without any manipulation of the eye, and the method is sensitive and well validated against a range of other methods. While DCE-MRI is not expected to be screening tool, we anticipate its increasing use in the clinic when fluorescein-based

methods are not applicable or produce equivocal results. As a research tool for use in clinical trials or in experimental settings, DCE-MRI BRB PS studies hold even more promise. Further, DCE-MRI BRB PS studies are applicable experimentally and have helped define the conditions under which other techniques reported to measure the extent of BRB breakdown should be validated (e.g., they should be linear and provide a physiologic measure of PS). Finally, DCE-MRI BRB PS studies have identified animal models that are useful (e.g., evaluating drug treatments for VEGF-induced BRB opening) or not (e.g., early changes in the diabetic rat).

References

1. Aiello LP (2002) The potential role of PKC beta in diabetic retinopathy and macular edema. Surv Ophthalmol 47 Suppl 2:S263–S269
2. Aiello LP, Bursell SE, Clermont A, Duh E, Ishii H, Takagi C, et al. (1997) Vascular endothelial growth factor-induced retinal permeability is mediated by protein kinase C in vivo and suppressed by an orally effective beta-isoform-selective inhibitor. Diabetes 46(9):1473–1480
3. Alikacem N, Yoshizawa T, Nelson KD, Wilson CA (2000) Quantitative MR imaging study of intravitreal sustained release of VEGF in rabbits. Invest Ophthalmol Vis Sci 41(6):1561–1569
4. Ando N, Sen HA, Berkowitz BA, Wilson CA, de Juan E, Jr (1994) Localization and quantitation of blood-retinal barrier breakdown in experimental proliferative vitreoretinopathy. Arch Ophthalmol 112(1):117–122
5. Antonetti DA, Barber AJ, Khin S, Lieth E, Tarbell JM, Gardner TW (1998) Vascular permeability in experimental diabetes is associated with reduced endothelial occludin content: vascular endothelial growth factor decreases occludin in retinal endothelial cells. Penn State Retina Research Group. Diabetes 47(12):1953–1959
6. Barber AJ, Lieth E, Khin SA, Antonetti DA, Buchanan AG, Gardner TW (1998) Neural apoptosis in the retina during experimental and human diabetes. Early onset and effect of insulin. J Clin Invest 102(4):783–791
7. Berkowitz BA, Roberts R, Luan H, Peysakhov J, Mao X, Thomas KA (2004) Dynamic contrast-enhanced MRI measurements of passive permeability through blood retinal barrier in diabetic rats. Invest Ophthalmol Vis Sci 45(7):2391–2398
8. Berkowitz BA, Sato Y, Wilson CA, de Juan E (1991) Blood-retinal barrier breakdown investigated by real-time magnetic resonance imaging after gadolinium-diethylenetriaminepentaacetic acid injection. Invest Ophthalmol Vis Sci 32(11):2854–2860
9. Berkowitz BA, Tofts PS, Sen HA, Ando N, de Juan E, Jr (1992) Accurate and precise measurement of blood-retinal barrier breakdown using dynamic Gd-DTPA MRI. Invest Ophthalmol Vis Sci 33(13):3500–3506
10. Berkowitz BA, Wilson CA, Tofts PS, Peshock RM (1994) Effect of vitreous fluidity on the measurement of blood-retinal barrier permeability using contrast-enhanced MRI. Magn Reson Med 31(1):61–66
11. Caspers-Velu LE, Wadhwani KC, Rapoport SI, Kador PF

(1995) Permeability of the blood-retinal and blood-aqueous barriers in galactose-fed rats. J Ocul Pharmacol Ther 11(3):469–487

12. Cunha-Vaz JG (2004) The blood-retinal barriers system. Basic concepts and clinical evaluation. Exp Eye Res 78(3): 715–721

13. Cunningham SA, Stephan CC, Arrate MP, Ayer KG, Brock TA (1997) Identification of the extracellular domains of Flt-1 that mediate ligand interactions. Biochem Biophys Res Commun 231(3):596–599

14. Cunningham SA, Tran TM, Arrate MP, Brock TA (1999) Characterization of vascular endothelial cell growth factor interactions with the kinase insert domain-containing receptor tyrosine kinase. A real time kinetic study. J Biol Chem 274(26):18421–18427

15. Dallal MM, Chang SW (1994) Evans blue dye in the assessment of permeability-surface area product in perfused rat lungs. J Appl Physiol 77(2):1030–1035

16. Derevjanik NL, Vinores SA, Xiao WH, Mori K, Turon T, Hudish T, et al. (2002) Quantitative assessment of the integrity of the blood-retinal barrier in mice. Invest Ophthalmol Vis Sci 43(7):2462–2467

17. DiMattio J (1991) In vivo use of neutral radiolabelled molecular probes to evaluate blood-ocular barrier integrity in normal and streptozotocin-diabetic rats. Diabetologia 34(12):862–867

18. Enea NA, Hollis TM, Kern JA, Gardner TW (1989) Histamine H1 receptors mediate increased blood-retinal barrier permeability in experimental diabetes. Arch Ophthalmol 107(2):270–274

19. Ennis SR (1990) Permeability of the blood-ocular barrier to mannitol and PAH during experimental diabetes. Curr Eye Res 9(9):827–838

20. Ennis SR, Betz AL (1986) Sucrose permeability of the blood-retinal and blood-brain barriers. Effects of diabetes, hypertonicity, and iodate. Invest Ophthalmol Vis Sci 27(7): 1095–1102

21. Ferris FL, III, Patz A (1984) Macular edema. A complication of diabetic retinopathy. Surv Ophthalmol 28 Suppl: 452–461

22. Frank JA, Dwyer AJ, Girton M, Knop RH, Sank VJ, Gansow OA, et al. (1986) Opening of blood-ocular barrier demonstrated by contrast-enhanced MR imaging. J Comput Assist Tomogr 10(6):912–916

23. Funatsu H, Wilson CA, Berkowitz BA, Sonkin PL (1997) A comparative study of the effects of argon and diode laser photocoagulation on retinal oxygenation. Graefes Arch Clin Exp Ophthalmol 235(3):168–175

24. Garner WH, Scheib S, Berkowitz BA, Suzuki M, Wilson CA, Graff G (2001) The effect of partial vitrectomy on blood-ocular barrier function in the rabbit. Curr Eye Res 23(5):372–381

25. Grimes PA (1985) Fluorescein distribution in retinas of normal and diabetic rats. Exp Eye Res 41(2):227–238

26. Grimes PA (1988) Carboxyfluorescein distribution in ocular tissues of normal and diabetic rats. Curr Eye Res 7(10):981–988

27. Jones CW, Cunha-Vaz JG, Rusin MM (1982) Vitreous fluorophotometry in the alloxan- and streptozocin-treated rat. Arch Ophthalmol 100(7):1141–1145

28. Kent D, Vinores SA, Campochiaro PA (2000) Macular oedema: the role of soluble mediators. Br J Ophthalmol 84(5): 542–545

29. Kirber WM, Nichols CW, Grimes PA, Winegrad AI, Laties AM (1980) A permeability defect of the retinal pigment epithelium. Occurrence in early streptozocin diabetes. Arch Ophthalmol 98(4):725–728

30. Lightman S, Pinter G, Yuen L, Bradbury M (1990) Permeability changes at blood-retinal barrier in diabetes and effect of aldose reductase inhibition. Am J Physiol 259: R601–R605

31. Lightman S, Rechthand E, Terubayashi H, Palestine A, Rapaport S, Kador P (1987) Permeability changes in blood-retinal barrier of galactosemic rats are prevented by aldose reductase inhibitors. Diabetes 36:1271–1275

32. Lobo C, Bernardes R, Faria dA, Jr, Cunha-Vaz JG (1999) Novel imaging techniques for diabetic macular edema. Doc Ophthalmol 97(3–4):341–347

33. Luan H, Roberts R, Sniegowski M, Goebel DJ, Berkowitz BA (2006) Retinal thickness and subnormal retinal oxygenation response in experimental diabetic retinopathy. Invest Ophthalmol Vis Sci 47(1):320–8

34. Lukaszew RA, Mullins CM, Penn JS, Berkowitz BA (1997) Noninvasive and quantitative staging of hyaloidopathy in experimental retinopathy of prematurity. Invest Ophthalmol Vis Sci 38(4):S747

35. Lund-Andersen H (2002) Mechanisms for monitoring changes in retinal status following therapeutic intervention in diabetic retinopathy. Surv Ophthalmol 47 Suppl 2:S270–S277

36. Maepea O, Karlsson C, Alm A (1984) Blood-ocular and blood-brain barrier function in streptozocin-induced diabetes in rats. Arch Ophthalmol 102(9):1366–1369

37. Marmor MF (1999) Mechanisms of fluid accumulation in retinal edema. Doc Ophthalmol 97(3–4):239–249

38. Martin PM, Roon P, Van Ells TK, Ganapathy V, Smith SB (2004) Death of retinal neurons in streptozotocin-induced diabetic mice. Invest Ophthalmol Vis Sci 45(9):3330–3336

39. Mathews MK, Merges C, McLeod DS, Lutty GA (1997) Vascular endothelial growth factor and vascular permeability changes in human diabetic retinopathy. Invest Ophthalmol Vis Sci 38(13):2729–2741

40. Mayhan WG (2001) Regulation of blood-brain barrier permeability. Microcirculation 8(2):89–104

41. Menzies SA, Hoff JT, Betz AL (1990) Extravasation of albumin in ischaemic brain oedema. Acta Neurochir Suppl (Wien) 51:220–222

42. Metrikin DC, Wilson CA, Berkowitz BA, Lam MK, Wood GK, Peshock RM (1995) Measurement of blood-retinal barrier breakdown in endotoxin-induced endophthalmitis. Invest Ophthalmol Vis Sci 36(7):1361–1370

43. Murata T, Ishibashi T, Khalil A, Hata Y, Yoshikawa H, Inomata H (1995) Vascular endothelial growth factor plays a role in hyperpermeability of diabetic retinal vessels. Ophthalmic Res 27(1):48–52

44. Murata T, Nakagawa K, Khalil A, Ishibashi T, Inomata H, Sueishi K (1996) The relation between expression of vascular endothelial growth factor and breakdown of the blood-retinal barrier in diabetic rat retinas. Lab Invest 74(4):819–825

45. Nagy Z, Szabo M, Huttner I (1985) Blood-brain barrier impairment by low pH buffer perfusion via the internal carotid artery in rat. Acta Neuropathol (Berl) 68(2): 160–163

46. Park SH, Park JW, Park SJ, Kim KY, Chung JW, Chun MH, et al. (2003) Apoptotic death of photoreceptors in the streptozotocin-induced diabetic rat retina. Diabetologia 46(9): 1260–1268

47. Pelc NJ (1993) Optimization of flip angle for T1 dependent contrast in MRI. Magn Reson Med 29(5):695–699

48. Plehwe WE, McRobbie DW, Lerski RA, Kohner EM (1988) Quantitative magnetic resonance imaging in assessment of the blood-retinal barrier. Invest Ophthalmol Vis Sci 29(5):663–670

49. Prager TC, Chu HH, Garcia CA, Anderson RE (1982) The influence of vitreous change on vitreous fluorophotometry. Arch Ophthalmol 100(4):594–596

50. Prager TC, Chu HH, Garcia CA, Anderson RE, Field JB, Orzeck EA, et al. (1983) The use of vitreous fluorophotometry to distinguish between diabetics with and without observable retinopathy: effect of vitreous abnormalities on the measurement. Invest Ophthalmol Vis Sci 24(1):57–65

51. Prager TC, Wilson DJ, Avery GD, Merritt JH, Garcia CA, Hopen G, et al. (1981) Vitreous fluorophotometry: identification of sources of variability. Invest Ophthalmol Vis Sci 21(6):854–864

52. Qaum T, Xu Q, Joussen AM, Clemens MW, Qin W, Miyamoto K, et al. (2001) VEGF-initiated blood retinal barrier breakdown in early diabetes. Invest Ophthalmol Vis Sci 42(10):2408–2413

53. Runge VM, Clanton JA, Lukehart CM, Partain CL, James AE, Jr (1983) Paramagnetic agents for contrast-enhanced NMR imaging: a review. AJR Am J Roentgenol 141(6):1209–1215

54. Runge VM, Clanton JA, Price AC, Wehr CJ, Herzer WA, Partain CL, et al. (1985) The use of Gd DTPA as a perfusion agent and marker of blood-brain barrier disruption. Magn Reson Imaging 3(1):43–55

55. Sander B, Larsen M, Engler C, Moldow B, Lund-Andersen H (2002) Diabetic macular oedema: the effect of photocoagulation on fluorescein transport across the blood-retinal barrier. Br J Ophthalmol 86(10):1139–1142

56. Sander B, Larsen M, Engler C, Strom C, Moldow B, Larsen N, et al. (2002) Diabetic macular oedema: a comparison of vitreous fluorometry, angiography, and retinopathy. Br J Ophthalmol 86(3):316–320

57. Sander B, Larsen M, Moldow B, Lund-Andersen H (2001) Diabetic macular edema: passive and active transport of fluorescein through the blood-retina barrier. Invest Ophthalmol Vis Sci 42(2):433–438

58. Sato Y, Berkowitz BA, Wilson CA, de Juan E, Jr (1992) Blood-retinal barrier breakdown caused by diode vs argon laser endophotocoagulation. Arch Ophthalmol 110(2):277–281

59. Schmiedl U, Ogan MD, Moseley ME, Brasch RC (1986) Comparison of the contrast-enhancing properties of albumin-(Gd-DTPA) and Gd-DTPA at 2.0 T: and experimental study in rats. AJR Am J Roentgenol 147(6):1263–1270

60. Sen HA, Berkowitz BA, Ando N, de Juan E, Jr (1992) In vivo imaging of breakdown of the inner and outer blood-retinal barriers. Invest Ophthalmol Vis Sci 33(13):3507–3512

61. Tofts PS, Berkowitz BA (1993) Rapid measurement of capillary permeability using the early part of the dynamic Gd-DTPA MRI enhancement curve. J Magn Reson Series B 102:129–136

62. Tofts PS, Berkowitz BA (1994) Measurement of capillary permeability from the Gd enhancement curve: a comparison of bolus and constant infusion injection methods. Magn Reson Imaging 12(1):81–91

63. Tornquist P, Alm A, Bill A (1990) Permeability of ocular vessels and transport across the blood-retinal-barrier. Eye 4(2):303–309

64. Trick GL, Liggett J, Levy J, Adamsons I, Edwards P, Desai U, et al. (2005) Dynamic contrast enhanced MRI in patients with diabetic macular edema: initial results. Exp Eye Res 81(1):97–102

65. Vinores SA, Derevjanik NL, Mahlow J, Berkowitz BA, Wilson CA (1998) Electron microscopic evidence for the mechanism of blood-retinal barrier breakdown in diabetic rabbits: comparison with magnetic resonance imaging. Pathol Res Pract 194(7):497–505

66. Vinores SA, McGehee R, Lee A, Gadegbeku C, Campochiaro PA (1990) Ultrastructural localization of blood-retinal barrier breakdown in diabetic and galactosemic rats. J Histochem Cytochem 38(9):1341–1352

67. Vinores SA, Van Niel E, Swerdloff JL, Campochiaro PA (1993) Electron microscopic immunocytochemical evidence for the mechanism of blood-retinal barrier breakdown in galactosemic rats and its association with aldose reductase expression and inhibition. Exp Eye Res 57(6):723–735

68. Wallow IH (1983) Posterior and anterior permeability defects? Morphologic observations on streptozotocin-treated rats. Invest Ophthalmol Vis Sci 24(9):1259–1268

69. Wilson CA, Berkowitz BA, Funatsu H, Metrikin DC, Harrison DW, Lam MK, et al. (1995) Blood-retinal barrier breakdown following experimental retinal ischemia and reperfusion. Exp Eye Res 61(5):547–557

70. Wilson CA, Berkowitz BA, Sato Y, Ando N, Handa JT, de Juan E, Jr (1992) Treatment with intravitreal steroid reduces blood-retinal barrier breakdown due to retinal photocoagulation. Arch Ophthalmol 110(8):1155–1159

71. Wilson CA, Fleckenstein JL, Berkowitz BA, Green ME (1992) Preretinal neovascularization in diabetic retinopathy: a preliminary investigation using contrast-enhanced magnetic resonance imaging. J Diabetes Complications 6(4):223–229

72. Xu Q, Qaum T, Adamis AP (2001) Sensitive blood-retinal barrier breakdown quantitation using Evans blue. Invest Ophthalmol Vis Sci 42(3):789–794

73. Yi-Ming W, Shu H, Miao CY, Shen FM, Jiang YY, Su DF (2004) Asynchronism of the recovery of baroreflex sensitivity, blood pressure, and consciousness from anesthesia in rats. J Cardiovasc Pharmacol 43(1):1–7

9 Retinal Blood Flow

L. Schmetterer, G. Garhöfer

Core Messages

- Two blood systems support the human eye: the retinal and uveal vasculatures
- Retinal blood flow is characterized by a low perfusion rate, a high vascular resistance, and a high oxygen extraction
- Choroidal blood flow is characterized by a high perfusion rate, a low vascular resistance and a low oxygen extraction
- Several methods exist for the assesment of retinal perfusion parameters including Laser Doppler methods, angiographic techniques or Doppler ultrasound based systems
- Retinal circulation is highly autoregulated to keep blood flow constant despited changes of perfusion pressure
- Retinal blood flow can adapt to changing metabolic requirents
- Blood flow is strongly dependent on retinal oxygen tension, and vasoactive substances produced by the endothelial cells most importantly nitric oxide, prostacyclin (PCI$_2$) and endothelin-1

9.1 Anatomy

The human eye is supplied by two vascular systems: the retinal and the uveal vessels. The uveal vascular system consists of the iris, the ciliary body and the choroid. The outer layers of the retina including the photoreceptors are nourished by the choroid, whereas the inner layers of the retina including the retinal ganglion cells are supplied by the retina. Approximately 65 % of the oxygen consumed by the retina is delivered from the choroid.

In humans both the choroidal and the retinal vessels are supplied from branches of the ophthalmic artery (Fig. 9.1). The central retinal artery provides the blood to the retina; the central retinal vein drains the retina. The anterior choroid is supplied from the long posterior ciliary arteries, whereas the posterior choroid is supplied from the short posterior ciliary arteries. The choroid drains into the vortex veins.

The retinal blood flow is characterized by a low perfusion rate (approximately 40 – 80 µl/min; Fig. 9.2), a high vascular resistance, a high oxygen extraction and accordingly a high arteriovenous oxygen extraction (35 – 40 %). By contrast, the choroid shows the highest perfusion rate per gram tissue of all vascular beds within the human body (approximately 1,200 – 2,000 µl/min), a low vascular resistance, a low oxygen extraction and a low arteriovenous oxygen

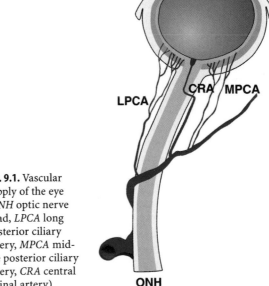

Fig. 9.1. Vascular supply of the eye (*ONH* optic nerve head, *LPCA* long posterior ciliary artery, *MPCA* middle posterior ciliary artery, *CRA* central retinal artery)

extraction (3 – 5 %). The driving force of blood flow in the eye is the ocular perfusion pressure (OPP). As in other vascular beds the perfusion pressure is the difference between the arterial and the venous pressures. The situation in the eye is, howev-

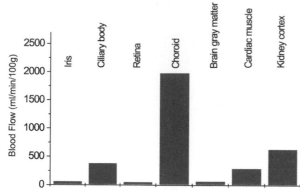

Fig. 9.2. Blood flow rate in several tissues as determined by radioactive microsphere technique in cynomolgus monkeys. Data are expressed as volumetric flow/100 g tissue. (Adapted from [1])

er, specific, because the venous pressure almost equals the intraocular pressure (IOP).

Retinal vessels cover the inner two-thirds of the retina, whereas the photoreceptors are avascular. The larger retinal arteries and veins are located in the nerve fiber layer. The density of capillaries is highest in the central part of the retina, where three to four layers are formed, and becomes lower in the periphery, where only a single layer exists. The very center of the retina is, however, avascular to allow as much as light as possible to reach the central photoreceptors. The extreme periphery of the retina is also avascular.

Retinal capillaries have a small caliber of 5–7 µm, whereas the caliber of choroidal capillaries is considerably wider (>10 µm). In the retinal capillaries adjacent endothelial cells are connected by continuous tight junctions forming the blood-retinal barrier. The blood-brain barrier prevents diffusion of even small molecules like glucose and lactate and is only permeable for lipid soluble substances such as oxygen and carbon dioxide. Accordingly, passage of a larger molecule through the blood-retinal barrier requires an active transport system. Compared to capillaries in other vascular beds, the high density of pericytes is striking, with a ratio of endothelial cells to pericytes of 1:1. Since pericytes change their tone in response to several vasoactive agents, retinal vascular resistance and blood flow may well be regulated by intravasal factors including circulating hormones. Retinal blood vessels are not innervated distal to the lamina cribrosa. This has important functional consequences, because the retinal blood flow, unlike the choroidal blood flow, is not under neural control.

9.2 Quantification of Retinal Blood Flow

Measurement of retinal blood flow in humans is not easy and requires sophisticated and expensive equip-

ment. One approach is to measure blood flow velocities in retinal arteries and/or veins using laser Doppler velocimetry. This approach is based on the fact that the Doppler shift of light is directly proportional to the blood velocity when the vessel is illuminated with laser light. To gain information on volumetric blood flow, measurements of retinal vessel diameters are required. Such measurements have been performed with fundus-camera-based techniques using either photography or continuous recordings with a video system. Calculation of cross sectional area of the vessel and multiplication with mean flow velocity allows for an estimation of volumetric blood flow. This approach is, however, limited to larger vessels and calculation of total retinal blood flow requires measurements on all visible vessels entering the optic nerve head, which is a time consuming approach.

Laser Doppler flowmetry (LDF) is a related approach, but the laser light is not directed onto a retinal vessel, but onto vascularized tissue with no visible larger vessels. Based on a scattering theory of light in tissue, which assumes complete randomization of light directions impinging on the erythrocytes, relative measures of the mean velocity of the erythrocytes and the blood volume can be obtained. Relative values of blood flow are calculated as the product of velocity and volume. Interindividual comparison of LDF data is hampered by the fact that flow values are not solely influenced by the perfusion rate of the tissue, but also by the scattering properties of tissue, which may vary considerably among subjects and may generally be altered in retinal vascular disease. The method has, however, considerable potential in investigating reactivity of the ocular vasculature to stimuli such as flicker light, changes in perfusion pressure and changes in oxygen and carbon dioxide levels. Scanning laser Doppler flowmetry combines the principles of scanning laser tomography and LDF. The commercially available Heidelberg retina flowmeter (HRF) provides a two-dimensional flow map of the retina and the optic nerve head (Fig. 9.3). Measurements in the retina are, however, hampered by the fact that a contribution from the underlying choriocapillaris cannot be excluded.

Other approaches are based on fluorescein angiography as used in clinical practice to obtain information on the vascular anatomy and integrity of the retinal vasculature. Several techniques have been proposed to extract quantitative data from these angiograms. Densitometric measurements allow for measurement of relative concentrations of fluorescein in retinal vessels for each consecutive image. The mean retinal circulation time (MCT) is defined as the difference between venous and arterial times and is proportional to retinal blood velocity. This approach does, however, require that the complete dye dilution

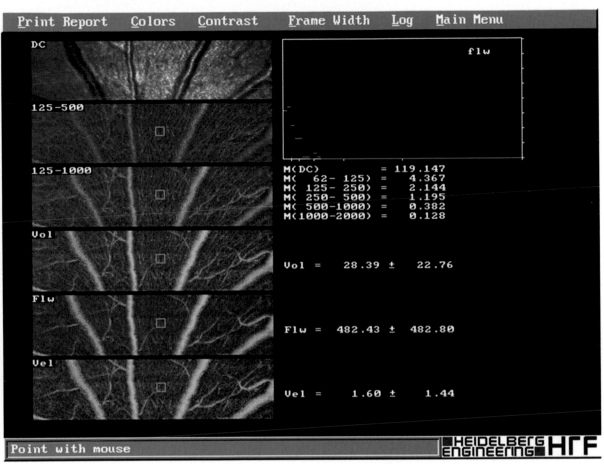

Fig. 9.3. Two dimensional flow mapping using the Heidelberg retina flowmeter. The parameters of volume (*Vol*), flow (*Flw*) and velocity (*Vel*) are indicated. (Courtesy of Georg Michelson)

curves have been recorded and that the dye bolus reaching the eye after intravenous injection can clearly be identified. Measurements of arteriovenous passage time (AVP) overcome part of these problems, because it does not depend on the shape of the dye dilution curve. The AVP is defined as the time between the first appearance of the dye in a retinal artery and the corresponding vein (Fig. 9.4). The method, however, requires that the blood of the area supplied by this artery is drained via the adjacent vein. Whether this is true remains to be elucidated. In addition, the appearance of the dye in retinal vessels may be hampered by leakage of fluorescein from retinal vessels in retinal vascular disease or the formation of arteriovenous shunt vessels. The mean dye velocity (MDV) is obtained by measuring fluorescence intensity along two points of a vessel (Fig. 9.5) and provides an estimate of blood flow velocity along the segment of the vessel. Finally, macular blood flow velocities have been quantified by tracking hyperfluorescent and hypofluorescent dots as they pass through perifoveal capillaries assumed to represent leukocytes and erythrocytes,

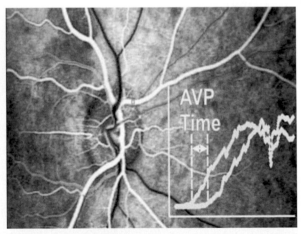

Fig. 9.4. The arteriovenous passage time (*AVP*) is calculated from the first appearance of the dye in the artery and the corresponding vein. (From [9])

respectively. This approach, however, does require excellent image quality, because otherwise these hyperfluorescent and hypofluorescent dots cannot be identified unequivocally in consecutive images.

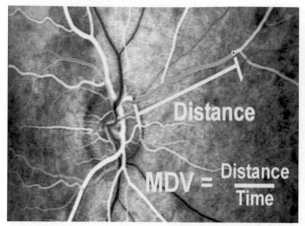

Fig. 9.5. The time between arrival of the dye at two positions along a vessel can be used to calculate the mean dye velocity (*MDV*). (From [9])

Based on the acoustical Doppler effect the blood flow velocity in the retrobulbar central retinal artery can be measured using color Doppler imaging (Fig. 9.6). Information on volumetric retinal blood flow cannot be obtained, because the resolution of ultrasound does not allow for measurement of the diameter of the vessel. A variety of parameters is extracted from these measurements including peak systolic flow velocity (PSV), end diastolic flow velocity (EDV) and mean flow velocity (MFV). In addition, a resistive index is calculated as RI = (PSV-EDV)/PSV. It is, however, unclear whether the RI represents an adequate measure of retinal vascular resistance, and factors other than resistance may influence RI as well.

The blue field entoptic technique is based on the blue field entoptic phenomenon, which can be seen best when looking into blue light. Under these conditions many tiny corpuscles can be seen flying around an area of the center of the fovea. This phenomenon is based on the fact that the red, but not white, blood cells absorb short wavelength light. Accordingly, the passage of a white blood cell is perceived as a flying corpuscle. To extract quantitative data from this technique, a simulated particle field is shown to the subject under study. By adjusting the number and the mean velocity of the particles in the simulated particle field with their own perception, an estimate of perimacular white blood cell flux can be extracted as the product of number and mean velocity of the leukocytes. Accordingly, this method is subjective in nature and requires sufficient cooperation from the subject. In addition, leukocyte flux may not be proportional to retinal blood flow in all clinical conditions.

9.3 Blood Flow Regulation

Autoregulation is defined as the intrinsic ability of a vascular bed to change its vascular resistance in response to changes in perfusion pressure. In this very strict sense autoregulation cannot be investigated in humans, because ocular perfusion pressure cannot be modified without affecting other factors such as neural input, circulating hormones and the

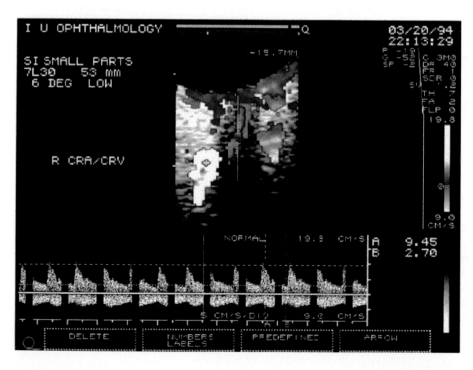

Fig. 9.6. Measurement of flow velocities in central retinal artery and vein using color Doppler imaging (CDI). (From [5])

metabolic environment. In the retina autoregulation plays an important functional role, because the retina has a constant oxygen demand. Accordingly, intact autoregulation ensures adequate oxygenation of the inner retina during physiological changes in ocular perfusion pressure. Abnormal autoregulation may therefore be detrimental for the eye, resulting in ischemia and hypoxia.

Autoregulatory mechanisms may be more easily investigated in the retina than in many other vascular beds, because the retinal vessels lack autonomic innervation. A decrease in ocular perfusion pressure is normally induced by an experimental increase in intraocular pressure as achieved for instance with a suction cup. This leads to a parallel decrease in ocular perfusion pressure, because the intraocular pressure almost equals the pressure in retinal veins. Using this technique effective autoregulation was evidenced by an adaptation of retinal vascular resistance in face of the decrease in perfusion pressure (Fig. 9.7). Autoregulatory phenomena during an increase in perfusion pressure are investigated either during a decrease in intraocular perfusion pressure, as achieved for instance after releasing a suction cup, or during isometric exercise and the concomitant increase in arterial blood pressure. Alternatively, systemic administration of vasoconstrictor substances has been used to study autoregulatory mechanisms in the retina, but this approach has the disadvantage that effects of increased arterial blood pressure cannot necessarily be separated from direct vasoconstrictor effects of the drug. During isometric exercise the retina shows some autoregulatory capacity as evidenced from the plateau in the pressure/flow relationship, with an upper limit of autoregulation of 100–110 mmHg (Fig. 9.8).

Another important aspect of blood flow regulation in the retina is the strong dependence on arterial oxygen tension, with hyperoxia inducing vasodilatation and hypoxia inducing vasoconstriction (Fig. 9.9). This again appears to be related to the constant oxygen demand of the retina. Carbon dioxide is also an important regulator of blood flow, with high-

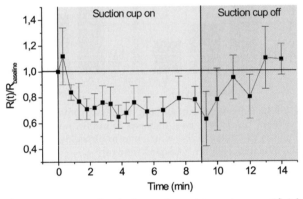

Fig. 9.7. Response of retinal vascular resistance to an artificial increase in IOP to 27 mmHg in healthy subjects. (Adapted from [10])

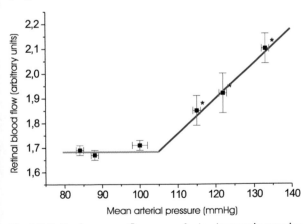

Fig. 9.8. Retinal pressure/flow curve during isometric exercise. The retinal blood flow is autoregulated up to mean arterial pressures of 100–110 mmHg. (Adapted from [12])

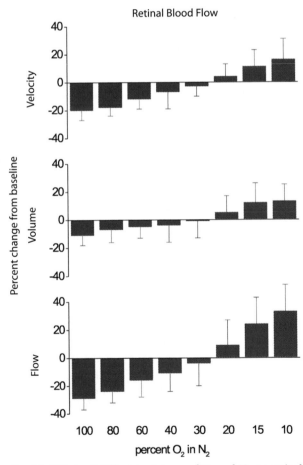

Fig. 9.9. Effects of different mixtures of O_2 and N_2 on retinal blood flow parameters in healthy subjects as assessed with the HRF. (Adapted from [15])

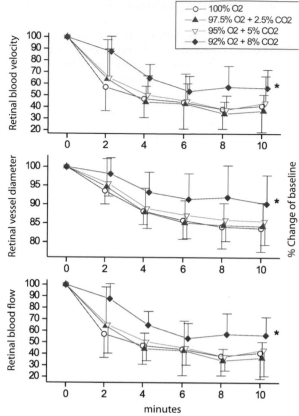

Fig. 9.10. Effects of different mixtures of O_2 and CO_2 on retinal blood flow velocity, retinal vessel diameters and retinal blood flow in healthy subjects. (Adapted from [7])

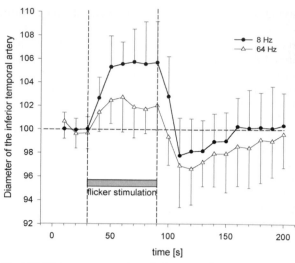

Fig. 9.11. Flicker responses in retinal arteries during stimulation with 8 Hz (*solid circles*) and 64 Hz (*open triangles*) in healthy subjects

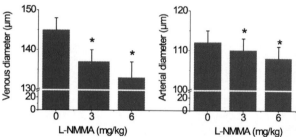

Fig. 9.12. Dose dependent effect of the nitric synthase inhibitor L-NMMA on retinal vessel diameter in healthy humans. Data indicate that nitric oxide contributes to physiological tone in the retinal vasculature. (Adapted from [6])

er arterial CO_2 levels leading to vasodilatation and lower levels producing vasoconstriction as in the brain. Accordingly, administration of carbogen (a mixture of CO_2 with O_2) has been proposed as a therapeutic approach in retinal ischemic and hypoxic disease. It appears, however, that the addition of CO_2 to high concentrations of O_2 in the inhalate does not sufficiently counteract the pronounced vasoconstrictor effects of hyperoxia (Fig. 9.10).

The retina has the capability to change its metabolic turnover and blood flow in response to neural stimulations. As in the brain this phenomenon is called neurovascular coupling. Flickering light of various frequencies has been used to investigate this phenomenon in some detail. The increase in retinal blood flow following flicker stimulation is in the order of 30–40%, with most of the effect occurring in the smaller vessels. Using high-resolution imaging of retinal vessels, flicker-induced vasodilatation may, however, also be seen in larger vessels (Fig. 9.11). Combining electroretinography with blood flow measurements revealed that the increase in retinal and optic nerve head blood flow after flicker stimulation is closely related to retinal ganglion cell activity.

In recent years it has been discovered that the endothelium plays a key role in regulation of retinal vascular tone as it does in other vascular beds. The vascular endothelium produces a number of vasoactive substances including prostacyclin (PGI_2) from arachidonic acid, nitric oxide from L-arginine and the potent vasoconstrictor endothelin-1. Nitric oxide plays a key role in the maintenance of vascular tone in the retina, because inhibition of NO synthase with L-arginine analogues causes dose-dependent vasoconstriction (Fig. 9.12). In addition, nitric oxide appears to play a role in several agonist-induced vasodilator effects including histamine, insulin and hypercapnia. Endothelin-1 induces potent vasoconstriction in the human retina with more pronounced effects on the smaller vessels than on the larger vessels. This potent vasoconstrictor effect may be reversed with a specific endothelin$_A$ receptor antagonist. Under physiological conditions endothelin-1 appears, however, to contribute little to retinal vascular tone, because administration of an endothelin$_A$

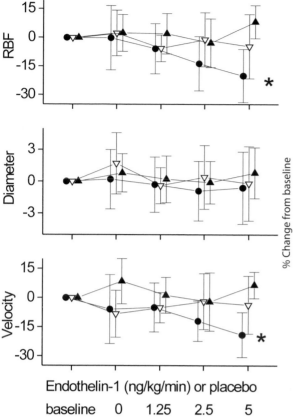

% Change from baseline

Fig. 9.13. Effect of endothelin-1 and the endothelin$_a$ receptor antagonist BQ-123 on retinal blood flow parameters. *Solid circles* endothelin-1 + placebo, *solid up triangle* placebo + BQ-123, *open down triangle* endothelin-1 + BQ-123. (Adapted from [8])

receptor antagonist alone has little effect on retinal blood flow (Fig. 9.13). Endothelin-1 has also been shown to play a role in hyperoxia-induced vasoconstriction in the retina.

9.4 Blood Flow in Retinal Vascular Disease

Retinal blood flow abnormalities have been investigated in some detail in patients with diabetes. Vascular abnormalities appear to be an early event in diabetes including venous vasodilatation. There is, however, a number of a contradicting results regarding retinal blood flow in diabetes, which may be related to the different techniques used for the assessment of retinal perfusion parameters. Moreover, it is difficult to investigate retinal blood flow in diabetes under controlled conditions, because both insulin and glucose induce vasodilatation in the eye. There is, however, a variety of studies indicating retinal vascular dysregulation in diabetes. This includes abnormal retinal autoregulation in response to changes in perfusion pressure, altered retinal oxygen reactivity and altered retinal flicker-induced vasodilatation. The alteration in retinal autoregulation in diabetes is more severe in the presence of systemic hypertension, making the retina extremely sensitive to retinal hyper- and hypoperfusion. The mechanism underlying changes in retinal oxygen-reactivity in the diabetic retina is largely unclear and may be related to retinal hypoxia, low grade inflammation, endothelial dysfunction, or metabolic changes in the diabetic milieu. Flicker-induced vasodilatation in the diabetic retina is largely reduced, but it is not entirely clear whether this is related to neuronal dysfunction or to alterations in neurovascular coupling. Taken together, alterations of retinal blood flow in diabetes are complex and influenced by a variety of factors (Fig. 9.14). However, due to the lack of longitudinal studies linking these perfusion abnormalities to the progression of diabetic retinopathy, the exact role of retinal vascular dysregulation in the pathophysiology of diabetes-induced alterations in the retina remains obscure.

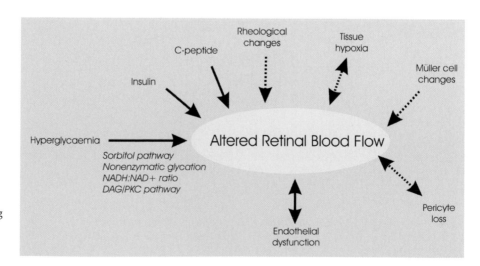

Fig. 9.14. Factors contributing to abnormal retinal blood flow in diabetes. (Adapted from [14])

9

In clinical routine quantification of retinal blood flow does not yet play a role in patients with retinal vascular disease. This is related to the fact that all currently available systems for the measurement of retinal blood flow are expensive and time-consuming and require significant technical expertise. In addition, large scale studies clearly linking alterations in retinal blood flow to the development and progression of retinal vascular disease are lacking. Accordingly, new techniques are required for the quantification of retinal ischemia and hypoxia in humans to gain more insight into the physiopathology of retinal vascular disease.

References

1. Alm A, Bill A (1973) Ocular and optic nerve blood flow at normal and increased intraocular pressures in monkeys (*Macaca irus*): a study with radioactively labelled microspheres including flow determinations in brain and some other tissues. Exp Eye Res 15(1):15–29
2. Cioffi GA, Granstam E, Alm A (2003) Ocular circulation. In: Kaufman PL, Alm A (eds) Adler's physiology of the eye, 10th edn. Mosby, St. Louis, pp 747–784
3. Delaey C, van de Voorde J (2000) Regulatory mechanisms in the retinal and choroidal circulation. Ophthalmic Res 32:249–56
4. Haefliger IO, Flammer J, Beny JL, Luscher TF (2001) Endothelium-dependent vasoactive modulation in the ophthalmic circulation. Prog Retin Eye Res 20:209–225
5. Harris A, Chung HS, Ciulla TA, et al. (1999) Progress in measurement of ocular blood flow and relevance to our understanding of glaucoma and age-related macular degeneration. Prog Retin Eye Res 18:669–687
6. Huemer K-H, Garhöfer G, Zawinka C, Golestani E, Litschauer B, Schmetterer, Dorner GT (2003) Effects of dopamine on human retinal vessel diameter and its modulation during flicker stimulation. Am J Physiol Heart Circ Physiol 284:358–363
7. Luksch A, Garhöfer G, Imhof A, Polak K, Polska E, Dorner GT, Anzenhofer S, Wolzt M, Schmetterer L (2002) Effect of inhalation of different mixtures of O_2 and CO_2 on retinal blood flow. Br J Ophthalmol 86:1143–1147
8. Polak K, Luksch A, Frank B, Jandrasits K, Polska E, Schmetterer L (2003) Regulation of human retinal blood flow by endothelin-1. Exp Eye Res 76:633–640
9. Rechtman E, Harris A, Kumar R, Cantor LB, Ventrapragada S, Desai M, Friedman S, Kageman L, Garozzi HJ (2003) An update on retinal circulation assessment technologies. Curr Eye Res 27:329–343
10. Riva CE, Grunwald JE, Petrig BL (1986) Autoregulation of human retinal blood flow. Invest Ophthalmol Vis Sci 27:1706–1712
11. Riva CE, Logean E, Falsini B (2005) Visually evoked hemodynamical response and assessment of neurovascular coupling in the optic nerve head and retina. Prog Retin Eye Res 24:183–215
12. Robinson F, Riva CE, Grunwald JE, Petrig Bl, Sinclair SH (1986) Retinal blood flow autoregulation in response to an acute increase in blood pressure. Invest Ophthalmol Vis Sci 27:722–726
13. Schmetterer L, Polak K (2001) Role of nitric oxide in the control of ocular blood flow. Prog Ret Eye Res 20:823–847
14. Schmetterer L, Wolzt M (1999) Ocular blood flow and associated functional deviations in diabetic retinopathy. Diabetologia 42:387–405
15. Strenn K, Menapace R, Rainer G, Findl O, Wolzt M, Schmetterer L (1997) Reproducibility and sensitivity of scanning laser Doppler flowmetry during graded changes in PO_2. Br J Ophthalmol 81:360–364

10 Genetic Approach to Retinal Vascular Disease
10.1 Gene Therapy for Proliferative Ocular Disease

T.J. McFarland, J.T. Stout

10

Core Messages

- Gene therapy is a process whereby specific genes are delivered into cells to restore a missing function, alter an aberrant function or give cells a new function
- There are two approaches to gene delivery: ex vivo (the introduction of the genetic material occurs in cells that have been removed from the body) and in vivo (genes are delivered to cells inside the body)
- Gene therapy can target both somatic and germ line cells, although somatic cells have been the primary focus of most therapeutic protocols
- There are a variety of gene delivery methods, ranging from the simple incubation of target cells with naked DNA to complex biological systems such as genetically engineered viruses
- A myriad of retinal disease and pathologies may benefit from these approaches
- In a limited number of clinical trials, gene therapy for ocular disease has proven safe. Compartmentalization of the eye, relative immune isolation and the characteristics of the terminally differentiated intraocular cells may be important factors in the relative safety of these approaches

10.1.1 Introduction

Development of novel molecular biology techniques in the 1970s and 1980s furnished scientists with new tools to advance the study and treatment of human disease. Progress in the understanding of bacterial and viral biology led to innovations in molecular cloning and chimeric plasmid construction. Advances in nucleic acid sequencing allowed researchers to gain a better understanding of genes and the ability to study gene mutations. Unraveling the minutiae of molecular events involved in gene transcription and translation furthered the analysis of cellular pathways and their complex interrelationships. Production of proteins ex vivo allowed physicians to treat diseases such as diabetes with synthetic human insulin, ending the dependency on animal sources.

These advances enabled researchers to propose a new question: can an inherited disorder and/or pathological event be treated at the genetic level? This question introduced the science of gene therapy. Gene therapy was initially an attempt to replace defective genes with functional ones. Investigators believed this was the best (and perhaps only) way to treat rare, genetically well-defined, loss-of-function diseases. The first gene therapy trial commenced in 1990, was

directed by W. French Anderson and Michael Blaese and was designed to treat a severe combined immunodeficiency syndrome (SCID) caused by a deficiency of the enzyme adenosine deaminase (ADA) [1]. Since then more than 350 somatic cell gene therapy trials have been conducted or are underway. Although gene therapy holds considerable promise, ethical and safety issues are important to consider. Sadly, despite extensive pre-clinical studies, severe complications have been observed in at least two different human clinical trials [6, 7]. While currently there are no commercially available gene therapy-based medications, as safe and effective drugs are developed a variety of disease processes are likely to be amenable to this form of therapy.

10.1.2 Delivery Systems

Essentials

- Viral, non-viral and physical methods exist for gene transfer
- Expression efficiency can vary greatly depending on the chosen vehicle
- All delivery systems have certain advantages and disadvantages (see Table 10.1.1)

Table 10.1.1. Vehicle selection criteria

Delivery vehicle	Pros	Cons
Retroviral	Integrating vectors capable of transducing both mitotic and non-mitotic cells Long-term stable expression has been shown in a wide variety of cell types including various layers of the retina Lentiviral components do not elicit a strong immune response when delivered systemically Lentiviral vectors can be produced transiently with high reproducibility in a relatively short amount of time (2 weeks)	Integration is random, a potential problem if the insertion of the transgene interrupts a crucial gene requiring di-allelic expression, a tumor suppressor for example Recombination with native HIV theoretically creating hybrid virions capable of infecting numerous cell types and/or regaining the ability to replicate is a concern. This is being addressed with vector manipulation and the use of lentiviruses that do not infect human cells (FIV, EIAV and BIV)
Adenoviral	Large transgene carrying capacity Capable of infecting both proliferating and stationary stage cells. Important for the use in ocular gene transfer applications High transduction efficiencies with strong short-term transgene expression	Adenoviral vectors do not integrate into host's cell genome, resulting in transient expression Multiple administrations of the vector would be required for some disease modalities Systemic delivery has been shown to elicit strong immune responses. A potential issue when delivered to the eye
AAV	AAV is not associated with human disease, does not illicit inflammatory responses, and approximately 80% of the population is sero-positive for the virus AAV has the potential to integrate into the host chromosomes, however most evidence suggests that recombinant AAV vectors reside episomally in the nucleus AAV is capable of infecting both dividing and non-dividing cells; transduce at high frequencies with a large receptor tropism	AAV has a relatively small transgene carrying capacity (approx. 5 kb). The use of mini-promoters and hetero-dimer vectors could increase gene payload Some cells appear to be resistant to infection by AAV, although many cells of the retina have been shown permissive to infection Production of high titer preparations is often difficult and requires a helper virus for full assembly
Chemical	Packaging the DNA decreases the chance of degradation during delivery The major reason for the use of non-viral delivery methods is that they pose no real safety or long-term health associated risks Plasmid DNA is very easy to produce at high concentrations with great purity and is cost effective Plasmid DNA does not mount any kind of immune response There are generally no size constraints associated with non-viral delivery methods	Delivery is not very efficient with low transient expression common Often display cytotoxic effects by disruption and damage to the cell's membrane Targeted gene delivery to specific sites is unreliable and transfection usually occurs in the first contacted cells. This is most common during systemic administration
Physical	The benefits of physical delivery are similar to that of non-viral methods	Transient expression usually requires repeat administrations Naked DNA is often cleared by the lymphatic system quite rapidly and degradation can occur limiting transfection considerably Some physical delivery methods may cause significant cellular damage, in some cases resulting in cell degradation and/or death Targeting specific cell types is often difficult thus requiring a bystander effect in order to achieve a desired outcome

10.1.2.1 Retroviral (Fig. 10.1.1)

Retroviruses were the first virus-mediated means of gene transfer. The original vectors were developed from oncoretroviruses such as Molony murine leukemia virus, Rous sarcoma virus, spleen necrosis virus, and avian leukosis virus [3]. These offered suitable integrative gene expression but are unable to infect non-dividing cells. Retroviruses that were capable of infecting both proliferating and stationary cells, like those of the retina, became a desirable alternative. Identification and study of the lentivirus HIV-1 led to research for its use in gene transfer methods. Lentiviral vectors have many advantages when compared to other retroviral vectors, such as large carrying capacity, ease of production/purification of high titer particles and less likelihood of integrating into a spot that will cause deleterious mutations [27]. In 2003, the first human clinical trial using a HIV-1 based lentiviral vector (VRX496) to treat HIV-positive patients was initiated [19].

- Lentiviruses are single-stranded RNA viruses 80–130 nm in size with an approximate 9.2 kb genome [14]. The RNA is packaged within a viral core surrounded by an envelope, derived from the host's cell membrane. The envelope contains viral glycoproteins used for binding cellular receptors.
- Lentiviral vectors have undergone several generations of construction and manipulations in the laboratory. Each version has increased the vectors' safety of use and transgene expression in vivo and in vitro.

- Current vectors are considered self-inactivating (SIN) due to deletions in the 3' long terminal repeat (LTR), rendering native viral promoters inactive [13]. Implementation of several enhancing factors such as the central polypurine tract (cPPT) and the woodchuck hepatitis virus posttranscriptional regulatory element (WPRE) have increased transgene uptake and expression stability, respectively [22].

10.1.2.2 Adenovirus (Fig. 10.1.2)

Adenoviruses were first used for gene transfer in 1985 [2]. Since then, they have been used in hundreds of gene therapy experiments.

- Adenoviruses are non-enveloped icosahedral viral particles approximately 80 nm in size. The genome consists of double stranded DNA, 36 kb in length [17].
- The wild type genome is transcribed in two non-exclusive phases labeled early and late. Early gene products are involved in cell cycle hijacking, immune response evasion, and replication. Late genes are responsible for viral assembly and packaging [21].
- Recombinant adenoviral vectors are produced replication deficient. This is accomplished by removing the "early" expressing genes. Current vectors are considered gutless or lacking all viral genes, retaining only *cis*-acting elements such as the inverted terminal repeats (ITRs). This strategy introduces the necessity to produce viral particles in a helper dependent manner. Genes

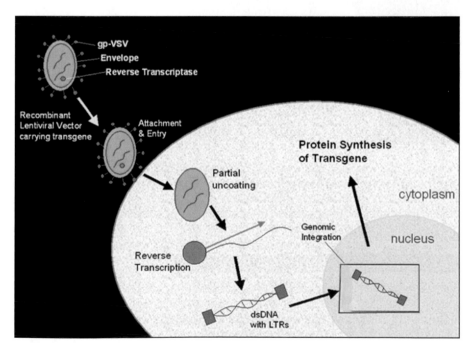

Fig. 10.1.1. Lentiviral vector

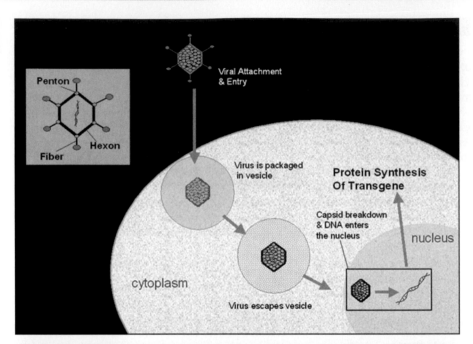

Fig. 10.1.2. Adenoviral vector

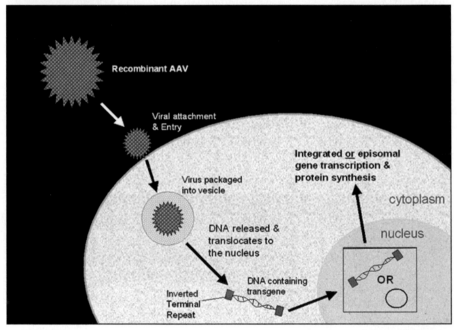

Fig. 10.1.3. Adeno associated virus

required for structural components and replication are provided in *trans* via co-transfection with plasmids containing necessary genes or with packaging cell lines that stably produce required components.

10.1.2.3 AAV (Fig. 10.1.3)

The use of adeno-associated viral (AAV) vectors in gene transfer has gained ever-increasing popularity in the last 10 years. AAV is now a well-studied vehicle for viral mediated gene transfer into ocular tissue.

- AAV is a small icosahedral single stranded DNA virus, ranging 20–25 nm in size. Its genome is small, composed of 4.7 kb of DNA with two open reading frames (ORF) (coding for the Rep and Cap genes) [18].
- Recombinant AAV is considered gutless retaining only the inverted tandem repeats (ITR). First-generation AAV vectors retained the Rep gene (involved with integration events) but it was later removed to increase vector carrying capacity and is usually sup-

plied in *trans* along with the structural capsid gene (Cap) [4].

10.1.2.4 Non-viral (Fig. 10.1.4)

Non-viral gene transfer methods traditionally employ chemicals (cationic liposomes and polymers) to deliver plasmid DNA. Basically, the DNA is encapsulated inside a vehicle by physical and chemical interactions and delivered to cells either systemically or locally. The DNA enters the cell via an endosome-mediated process, from which the plasmid escapes, traverses the nucleus and undergoes transcription [20].

- **Cationic lipids** are fluid-filled compartments composed of phospholipids similar to the lipid bilayer of a cell's membrane. Aqueous plasmid mixtures when combined with the lipids form micelle-like structures due to hydrophobic interactions.
- **Cationic polymers** are made up of repeating units of small molecular weight natural (peptides) or synthetic (polyethylenimine) molecules that form covalent high molecular weight chains or complex branches, as seen with dendrimers. Functional groups convey charge and physical properties, which allow the polymer complex to envelope the DNA, creating suitable transport vehicles [8].

Development of these carriers to improve the above processes, and thus enhance gene delivery in vivo, is ongoing.

10.1.2.5 Physical (Fig. 10.1.5)

Delivery of plasmid DNA without the aid of associated molecules is dependent on physical incorporation. A wide variety of methods have been developed to achieve this objective. The efficiency of plasmid uptake and gene expression depends on the method employed. To date, there are half a dozen or so technologies available to researchers, some of which are as efficient as viral based methods [23, 28].

- **Direct injection** is by far the most basic delivery method. It simply involves direct or systemic administration of naked plasmid DNA. The plasmid backbone is normally bacterial derived with a "therapeutic" gene of choice under the transcriptional control of a eukaryotic promoter.
- **Electroporation** is the delivery of plasmid DNA through small channels in the cell membrane created via microelectric pulses. This method has shown increasing success with the advent of new cell-specific solutions and modifications to the electric field permitting rapid DNA uptake to the cell's nucleus.
- **Hydroporation** administration involves the injection of extremely large volumes of DNA. The pressure created from the rapid influx of fluid is equivalent to that of blood pressure. This sudden rise in pressure gives cells a transient hyperpermeability. This technique has been shown to work well in organs such as the liver, but would most likely not be useful for ocular therapeutics due to the sheer volume and pressure required.

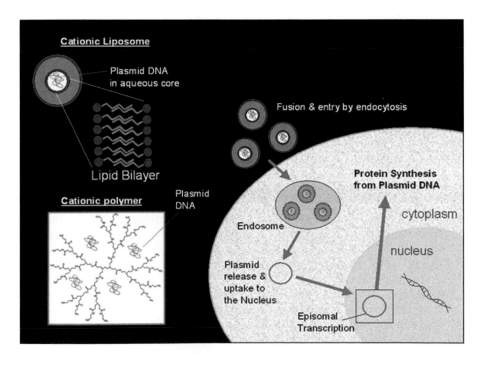

Fig. 10.1.4. Chemical mediated gene transfer

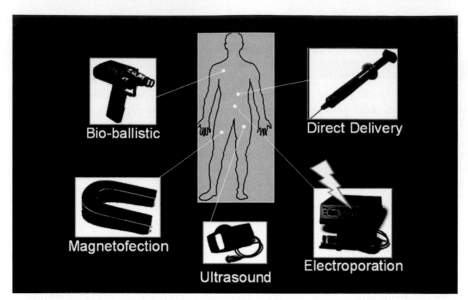

Fig. 10.1.5. Physical methods of gene delivery

- **Ultrasound** has been shown to increase transient permeability of cell membranes. By using contrasting agents (Optison), delivery of the plasmid DNA can be greatly enhanced compared to without. Ultrasound is a commonly used imaging aid in ophthalmology, so its use may have practical applications.
- **Magnetofection** relies on magnetic nanoparticles that surround the DNA. These complexes are then guided by magnetic fields to place them in close proximity to the targeted cells. There has been preliminary success with this application both in vitro and in vivo with DNA uptake several hundred times higher than naked DNA alone. This technique is fairly novel and long-term risks associated with using these nanoparticles have yet to be determined.
- **Bioballistic** delivery involves coating the DNA with gold particles, which can then be delivered by a "gene-gun." The DNA is basically shot into the desired tissue via high-pressure air. This method has been used for DNA vaccination but normally results in very low transient expression of the transgene.

10.1.3 Target Diseases

Essentials
- A small number of retinal vascular diseases account for most blindness in developed countries (Fig. 10.1.6.)
- Molecular advances are continually elucidating the pertinent biologic pathways and identifying the genes involved, thus offering possible targets for gene therapy (see Table 10.1.2)

Table 10.1.2. Genetic targets to modulate disease

Target gene class	ROP	AMD	PDR	CRO
Angiogenic	VEGF	PEDF	VEGF	Tie-2
	IGF	VEGF	RAGE	VEGF
	NRP-1	Tie-2	NRP-1	Timps
	FLK-1	Timps	FLK-1	MMPs
		MMPs	IGF	
			FGF	
Immunostatic	EphrinB	CFH	IL-6	
			IL-10	
Transcriptional	VHL		VHL	
	Hif-1alpha		Hif-1alpha	
Apoptosis			Bcl-2	BclXL
				Bcl-2

10.1.3.1 Retinopathy of Prematurity

Retinopathy of prematurity (ROP) accounts for a significant proportion of childhood blindness in developed countries. The disease is classified as having two independent stages [25].

Stage 1: Normal vascularization occurs in utero at month 4 of gestation. Premature birth and standard oxygen therapy halts ocular vasculogenesis, leaving a considerable portion of the eyecup avascular.

Stage 2: Return to environmental normoxia stimulates a hypoxic/ischemic response in the eye. The avascular zone of the retina becomes metabolically active, initiating several factors including VEGF to be upregulated and inducing a neovascular state. This in turn results in aberrant blood vessel formation and pathology similar to that seen in proliferative vitreoretinopathy.

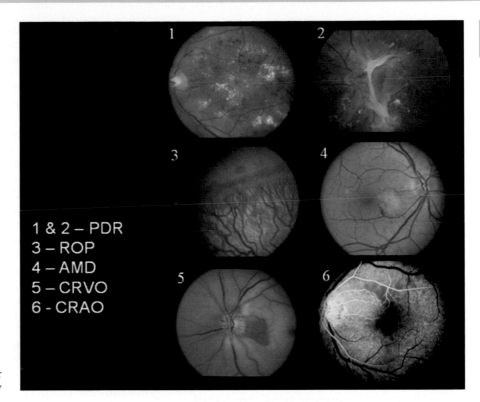

1 & 2 – PDR
3 – ROP
4 – AMD
5 – CRVO
6 - CRAO

Fig. 10.1.6. Target diseases for retinal vascular gene therapy

Current treatments involve either cryo- or laser ablation (the latter is standard) of the avascular zones in order to prevent further neovascularization. Treatment success is dependent on the severity of retinopathy and the number of zones involved [11].

10.1.3.2 Age Related Macular Degeneration

Age related macular degeneration (AMD) is the leading cause of blindness in patients over the age of 60 in the developed world [29]. Its exact etiology is not fully understood and it is considered to be a multifactorial disease, associated with both environmental and genetic stimuli. Patients with vascular AMD present with drusen, geographic atrophy, subretinal neovascularization, and central vision loss [16]. Extensive clinical trials have led to the development of preventive measures such as dietary supplements [24] and life-style risk assessments [12]. Despite this, AMD continues to be a major cause of morbidity in the elderly, a steadily growing population. Currently, treatment options include surgery, photocoagulation, and photodynamic therapy and the injection of a variety of anti-angiogenic substances [5].

10.1.3.3 Proliferative Diabetic Retinopathy

Proliferative diabetic retinopathy (PDR) is associated with uncontrolled hyperglycemia as a result of diabetes mellitus and is a leading cause of blinding

disease in patients under age 30. The proliferative process is intimately related to the vascular structure of the retina, specifically the interactions between the endothelial cells and the pericytes [9]. Pathology is believed to involve the breakdown of the blood-retinal barrier, which then leads to vascular permeability, macular edema and neovascularization. The exact cause of this breakdown is not entirely understood but may involve endothelial tight gap junction collapse with subsequent introduction of blood sugar by-products (advanced glycosylation end products) that stimulate pericyte death and the upregulation of growth factors [26].

10.1.3.4 Retinal Vaso-occlusive Disorders

A leading cause of blindness in patients over age 60 is retinal vaso-occlusive disease. Occlusions can affect either the veins or arteries that supply the outer retina. Central retinal artery occlusions (CRAO) and central retinal vein occlusions (CRVO) often create the most significant disease states compared to peripheral branch obstructions.

CRAO: Results in abrupt and massive visual loss, with visual outcomes of 20/200 or lower [30]. Occlusion of the central retinal artery or branch arteries can have a variety of etiologies ranging from transient nighttime hypotension to cardiovascular disease and

10

diabetes. CRAO appears to be more common in males than females and is usually diagnosed too late for effective therapeutic intervention. The time required for irreversible loss of vision in an experimental model of CRAO was less than 250 min [10]. The ensuing pathology is believed to be associated with ischemic initiation of the apoptotic cascade [30] with loss of retinal function, namely a significant reduction in ganglion cell layer response during the electroretinogram (ERG) [15]. The ischemia also results in neovascular complications.

CRVO or BRVO: Considered the most prevalent vaso-occlusive disorder in the US, it results from blockage in the central retinal vein or branch veins. It is classified as having two distinct forms (ischemic and non-ischemic) that initiate cell death much more slowly than CRAO. Non-ischemic CRVO is normally of acute onset and has a far better visual outcome than the ischemic form. Ischemic CRVO is associated with retinal hemorrhage, tortuous veins, cotton wool spots, and markedly decreased visual acuity with poor prognosis. The underlying derivation of CRVO is multifactorial (high cholesterol, emboli, and inflammatory disease may all play a role) with patients over age 50 most susceptible. Current treatments for CRVO involve anticoagulants and steroids.

10.1.4. Fundamentals of Gene Therapy

Essentials
- In vitro experimentation is paramount to gene therapy advancement
- Angiogenic models can emulate in vivo conditions
- There are two approaches for gene delivery to the body
- Cell derivation...does this matter?

10.1.4.1 In Vitro

In vitro technologies are the foundation of basic science research and have led to the most detailed understanding of retinal pathology. In vitro experiments usually involve specific cell cultures or tissues masses. Cells germane to the study of retinal vascular disease have for the most part been established and

are commercially available. Several endothelial cell properties such as cell migration, tube formation, and proliferation can be assayed to study angiogenesis. There are a number of beneficial tissue-based assays that have been developed to bridge the gap between cellular and animal research, such as the aortic ring and chick embryo assay. The advantage to these types of experiments is the ability to control one has over many key parameters. For example, environmental parameters such as hypoxia are possible via hypoxia chambers and incubators offering the ability to mimic pathological conditions using premixed gas preparations.

At one time, vectors for experimental use had to be engineered and prepared using laborious methods. Today, commercially available kits including detailed protocols and reagents make vector production practical and affordable for most laboratories. Kits are currently available for producing adenoviral, lentiviral, and AAV-based vectors.

10.1.4.2 Ex Vivo Versus In Vivo

There are two approaches to introduce therapeutic genes into the body. The ex vivo approach entails removing a cell or tissue type, delivering a gene, and reintroducing those cells to the affected site. This approach allows cells to be transduced in culture and then transplanted to a target site, where they can mimic normal tissue and deliver therapy to the affected area. This was shown to be possible in the eye when retroviral-transduced RPE cells were transplanted under the retina and shown to inhibit VEGF-induced choroidal neovascularization in a rabbit model. This approach offers several potential advantages. The gene transfer can be monitored and tested for function, insertional events (dependent on delivery approach), and expression level. This provides the assurance that what will be introduced into the patient is what was initially desired. There are cases, however, in which this approach may not be feasible as diagnostically testing every adverse possibility is time-consuming and extreme situations may not afford this option. In such cases, in vivo delivery is the only option. Direct or local administration is considered far safer then systemic delivery and offers fewer possible complications. The advent of better targeting vectors will also improve in vivo administration in the future. This could be accomplished with the use of advanced viral pseudo-typing methods that would allow specific cells to be targeted based on outer membrane receptor profiles. Another solution relies on tight control of gene insertion. Placing a gene precisely where it is desired in the genome could afford greater safety and stability.

10.1.4.3 Somatic Cells Versus Stem Cells

Somatic cells are differentiated cells that comprise the majority of tissues in the body. To date, most gene therapy research has focused on gene transfer into somatic cells. These cells offer the advantage of being easy to acquire, biologically static, and potentially easier to control.

Gene therapy using somatic cells:

- Can act as factories to produce a secretable form of the transgene-product, supplying surrounding cells with said "therapeutic" protein.
- Can be targeted themselves, resulting in single cell effects such as with the use of suicide genes in cancer research.
- Can be targeted to initiate cellular cascade events that stimulate surrounding cells via a bystander effect or activation of downstream products.

Stem cells are undifferentiated cells that become specialized given specific stimuli. Stem cells fall into three main categories, totipotent, pluripotent or multipotent.

- Totipotent cells have the ability differentiate into any cell type and form complete organisms or tissues/organs.
- Pluripotent cells are true stem cells but lack the ability to form extraembryonic membranes (i.e., placenta cells). Embryonic stem cells and germ line cells fall into this class.
- Multipotent cells are true stem cells but can only differentiate into a subclass of specialized cells, for example hematopoietic stem cells that give rise to blood cells.

Totipotent and pluripotent gene transfer still remains a novel therapeutic strategy but offers the possibility of organ regeneration and gene replacement therapy during embryonic development. The use of multipotent cells, on the other hand, has much promise for use in retinal vascular gene therapy protocols. Multipotent cells possess specific cellular markers (i.e., CD34); therefore, it is possible to isolate the stem cells based on these characteristics with established methods. The isolated cells, such as hematopoietic stem cells (HSCs), can be transduced so as to harbor therapeutic genes and can be subsequently delivered to the retina by intravitreal injection. Recently this method resulted in direct targeting of endothelial progenitor cells (EPCs), a subset of HSCs, to retinal astrocytes. A marked inhibition of angiogenesis was observed after employing this technique in lineage negative EPCs derived from bone marrow using the T2-tryptophanyl-tRNA synthetase gene. The direct targeting and incorporation of these cells into newly developed blood vessels carrying therapeutic agents make this an attractive therapeutic approach.

HSCs have been isolated from peripheral blood and bone marrow with comparable cell quality. The main differences between the two sources are cell yield and the procedures employed. Bone marrow contains a significantly higher proportion of CD34+ cells compared to peripheral blood. The use of cytokines has been shown to increase peripheral blood levels to that of bone marrow but long term risks associated with cytokine use are unknown. Bone marrow isolation is also an invasive procedure requiring anesthesia, wound recovery, and pain management. For these reasons peripheral blood may offer a more convenient means of harvest. Regardless of source, ex vivo gene transfer into hematopoietic stem cells is an excellent therapeutic prospect without the pitfalls associated with allograft transplantation.

10.1.5 Genes as Drugs/Clinical Trials

Essentials
- Human gene therapy for ocular disease
- Anti-proliferative genes enter the limelight

Human ocular gene therapy trials using adenoviral mediated gene transfer have shown efficacy for the treatment of retinoblastoma and ocular melanoma. Both treatments have employed the herpes simplex thymidine kinase gene for its cell suicide properties. Although these are not retinal vascular diseases per se, they do demonstrate the use of gene therapy in the eye. Recently, the first clinical trial use of an anti-angiogenic gene for the treatment of a proliferative ocular disease has been demonstrated.

In 2004–2005, the use of adenoviral-mediated delivery of the *PEDF* gene for the treatment of neovascular AMD was reviewed in a Phase I clinical trial. This multicenter study sponsored by GenVec was performed to evaluate dose response in 28 patients diagnosed with neovascular AMD. The dose escalation study assessed viral load ranging from 10^6 to $10^{9.5}$ viral particles. Side effects appeared to be minor, including mild inflammation and moderate transient intraocular pressure spikes. Patients were clinically evaluated 13 times over a 12-month period. Tests performed included a full ophthalmic exam, physical exam, blood chemistry, viral cultures, and viral protein ELISA. None of the treated patients showed evidence of viral replication from urine or sputum samples. There was also no consistent or robust pattern of anti-adenoviral reactive antibodies detected. Other observations suggested a dose dependent reduction in lesion size over the study period. These observations, while not a specific trial endpoint, suggest two points: that a single treatment may provide sustained therapeutic results for several months and that the use of *PEDF* may benefit patients with neovascular AMD. Phase II clinical assessment for this drug is underway.

10.1.6 Ethical/Safety Concerns

The past 20 years of scientific advances have led to the development of human gene therapy research. This same research has given rise to many bioethical and safety questions concerning the use of gene therapy and future implications. Gene therapy has both somatic and germ line implications. Currently, somatic based research abounds and although safety is an issue, it is less of an ethical quandary. On the other hand, germ line manipulation using gene transfer is a subject that raises many ethical questions.

- What is normal variation?
- What is health?
- Can genetic prevention lead to genetic "enhancement" or "selection"?
- Should embryonic stem cells be used in research? In treatment?

To date, the use of germ line stem cells for ocular gene therapy has not been explored significantly; this discussion will focus on safety issues concerning gene transfer into somatic and multipotent stem cells.

Integrative vectors such as retroviruses pose the risk of transgene insertion into areas of the genome that would disrupt native gene expression. This can be detrimental in cases where downstream oncogenes are upregulated or when tumor suppressor genes can no longer function adequately. This fear was realized after a human trial for the treatment of X linked SCID resulted in three of the patients developing leukemia after T cell transduction.

Another potential adverse effect of gene therapy is related to immune response. The two main issues involve immune attack on vector components such as viral proteins or on the exogenous "therapeutic" gene product itself. First, viruses can encourage a significant immune mobilization; this was made tragically apparent after the death of a patient who was systemically administered a lethal dose of adenovirus. Since then, strict guidelines have been developed to prevent this type of complication. Second, gene replacement therapy in which a functional copy of a gene is introduced into the body could elicit immune responses because this form of the protein is immunologically novel to the body.

Ocular gene therapy may have an advantage over gene therapy at other sites in the body, as the eye is a compartmentalized structure that is essentially immunoprivileged. Experiments using ocular gene therapy do not generally give rise to immune complications, as discussed above in the description of the AdPEDF.11 clinical trial. For these reasons, the eye may be the most suitable organ in which to test and evaluate gene therapy. Advances in vector construction have and will continue to evolve addressing integrative issues and concerns.

References

1. Anderson WF (1990) September 14, 1990: the beginning. Hum Gene Ther 1:371–372
2. Breyer B, Jiang W, Cheng H, Zhou L, Paul R, Feng T, He TC (2001) Adenoviral vector-mediated gene transfer for human gene therapy. Curr Gene Ther 1:149–162
3. Buchschacher GL, Jr (2001) Introduction to retroviruses and retroviral vectors. Somat Cell Mol Genet 26:1–11
4. Buning H, Braun-Falco M, Hallek M (2004) Progress in the use of adeno-associated viral vectors for gene therapy. Cells Tissues Organs 177:139–150
5. Byrne S, Beatty S (2003) Current concepts and recent advances in the management of age-related macular degeneration. Ir J Med Sci 172:185–190
6. Carmen IH (2001) A death in the laboratory: the politics of the Gelsinger aftermath. Mol Ther 3:425–428
7. Chinen J, Puck JM (2004) Successes and risks of gene therapy in primary immunodeficiencies. J Allergy Clin Immunol 113:595–603; quiz 604
8. De Laporte L, Cruz Rea J, Shea LD (2006) Design of modular non-viral gene therapy vectors. Biomaterials 27:947–954
9. Gardner TW, Antonetti DA, Barber AJ, LaNoue KF, Levison SW (2002) Diabetic retinopathy: more than meets the eye. Surv Ophthalmol 47 Suppl 2:S253–262
10. Hayreh SS, Zimmerman MB, Kimura A, Sanon A (2004) Central retinal artery occlusion. Retinal survival time. Exp Eye Res 78:723–736
11. Hutcheson KA (2003) Retinopathy of prematurity. Curr Opin Ophthalmol 14:286–290
12. Hyman L, Neborsky R (2002) Risk factors for age-related macular degeneration: an update. Curr Opin Ophthalmol 13:171–175
13. Lever AM, Strappe PM, Zhao J (2004) Lentiviral vectors. J Biomed Sci 11:439–449
14. Levy JA, Conrat HF, Owens RA (1994) Viruses using reverse transcription during replication. In: Virology. Prentice Hall, New York, pp 125–141
15. Machida S, Gotoh Y, Tanaka M, Tazawa Y (2004) Predominant loss of the photopic negative response in central retinal artery occlusion. Am J Ophthalmol 137:938–940
16. Maguire MG (1999) Natural history. In: Berger JW, Fine SL, Maguire MG (eds) Age related macular degeneration. Mosby, St. Louis, pp 17–31
17. McConnell MJ, Imperiale MJ (2004) Biology of adenovirus and its use as a vector for gene therapy. Hum Gene Ther 15:1022–1033
18. Merten OW, Geny-Fiamma C, Douar AM (2005) Current issues in adeno-associated viral vector production. Gene Ther 12 Suppl 1:51–61

19. Morris KV (2005) VRX-496 (VIRxSYS). Curr Opin Invest Drugs 6:209–215
20. Montier T, Delepine P, Pichon C, Ferec C, Porteous DJ, Midoux P (2004) Non-viral vectors in cystic fibrosis gene therapy: progress and challenges. Trends Biotechnol 22:86–592
21. Parks RJ (2000) Improvements in adenoviral vector technology: overcoming barriers for gene therapy. Clin Genet 58:1–11
22. Quinonez R, Sutton RE (2002) Lentiviral vectors for gene delivery into cells. DNA Cell Biol 21:937–951
23. Schmidt-Wolf GD, Schmidt-Wolf IG (2003) Non-viral and hybrid vectors in human gene therapy: an update. Trends Mol Med 9:67–72
24. Seddon JM, Ajani UA, Sperduto RD, Hiller R, Blair N, Burton TC, Farber MD, Gragoudas ES, Haller J, Miller DT, et al. (1994) Dietary carotenoids, vitamins A, C, and E, and advanced age-related macular degeneration. Eye Disease Case-Control Study Group. JAMA 272:1413–1420
25. Smith LE (2003) Pathogenesis of retinopathy of prematurity. Semin Neonatol 8:469–473
26. Stitt AW (2003) The role of advanced glycation in the pathogenesis of diabetic retinopathy. Exp Mol Pathol 75:95–108
27. Trono D (2003) Virology: picking the right spot. Science 300:1670–1671
28. Wells DJ (2004) Gene therapy progress and prospects: electroporation and other physical methods. Gene Ther 11:1363–1369
29. Zarbin MA (2004) Current concepts in the pathogenesis of age-related macular degeneration. Arch Ophthalmol 122:598–614
30. Zhang Y, Cho CH, Atchaneeyasakul LO, McFarland T, Appukuttan B, Stout JT (2005) Activation of the mitochondrial apoptotic pathway in a rat model of central retinal artery occlusion. Invest Ophthalmol Vis Sci 46:2133–2139

10 I

10.2 Norrin and Its Role During Angiogenesis of the Retina

M. Scholz, E.R. Tamm

Core Messages

- Norrie disease (ND) is an X-linked retinal dysplasia
- The ND (*NDP*) gene encodes for norrin, a protein causing Norrie disease
- Norrin is critically required for the formation of the inner retinal plexus and is involved in the completion of the final secondary capillary layer, but does not include formation of the initial angiogenic sprouting
- Norrie disease mutant mice demonstrate structural abnormalities in the retinal vasculature

10.2.1 Introduction

Norrin or Norrie disease protein is a 131-amino-acid-long, secreted protein that forms disulfide-bonded oligomers, which associate with the extracellular matrix upon secretion [11]. Sequence analysis indicates a cysteine knot motif similar to that seen in growth factors of the cysteine knot family [8]. Norrin is predominantly expressed in brain, retina and the olfactory bulb [1, 3]. Mutations in the ND gene (*NDP*) that encodes norrin cause Norrie disease, an X-linked retinal dysplasia [2, 7]. The biochemical function(s) of norrin have been unclear for several years. Recent studies strongly indicate that norrin is critically required to induce the formation of the deep retinal capillary layers and the capillaries of the inner ear [10, 12, 19], a function that is mediated by high affinity binding to *frizzled-4* (*Fz4*) and activation of the classical *Wnt* pathway [19]. In contrast to the function of other growth factors that modulate angiogenesis in the developing retina, norrin does not induce initial angiogenic sprouting, but is rather involved in the completion of a final secondary capillary layer.

10.2.2 Norrie Disease

Norrie disease (ND) is an X-linked, rare inherited disorder that presents with congential or early childhood blindness because of severe retinal dysplasia [16, 17, 18]. About one-third of individuals with Norrie disease develop progressive hearing loss and some degree of mental retardation. The classical finding has been reported as a grayish-yellow, glistening, elevated mass that replaces the retina and is visible through a clear lens. These masses have been referred to as pseudoglioma. Vascular changes appear to be associated, including vascularization of the mass from the vitreous, often resulting in vitreous hemorrhage. The affected gene (*NDP*) is localized on chromosome Xp 11.3 and encodes for Norrie disease protein or norrin [2, 7]. Various mutations in *NDP* have been described (nucleotide deletions, nonsense, frameshift, splice site and missense mutations), which commonly appear to result in loss of function of norrin [1].

10.2.3 Norrie Disease Mutant Mice

To understand the molecular pathogenesis of Norrie disease, mutant mice with a targeted disruption of *Ndp* ($Ndp^{y/-}$) were generated [3]. $Ndp^{y/-}$ mutant mice show fibrous masses in the vitreous body and an overall disorganization of the retinal ganglion cell layer [3]. The outer plexiform layer may disappear resulting in a juxtaposed inner and outer nuclear layer. In regions of juxtaposition, the outer segments of photoreceptors are no longer present. Starting from postnatal day 9, the number of retinal neurons in the inner retinal layers of $Ndp^{y/-}$ mutant mice substantially decreases as compared to wild-type littermates [13]. There are also pronounced structural abnormalities in the retinal vasculature, which include the complete absence of the two intraretinal capillary beds and the formation of vascular membranes at the vitreal surface of the retina [6, 12, 13]. In addition, the hyaloid vasculature persists [6, 9, 13]. The changes in retinal structure correlate with a loss of function as dark-adapted ERGs in $Ndp^{y/-}$ mice show a dramatic loss of the positive b-wave, which is mostly shaped by the neurons of the inner retina, particularly the bipolar cells [10, 14]. In addition, no substantial oscillatory potentials, a set of higher frequency wavelets usually superimposed on the b-wave, are observed. Overall, ERG waveforms in $Ndp^{y/-}$ mice closely match those obtained in normal eyes in a state of retinal hypoxia [15]. An abnormal

vasculature has also been observed in the inner ear of $Ndp^{y/-}$ mice, a finding that is consistent with the observation that the mice develop progressive hearing loss up to profound deafness [12].

10.2.4 In Vivo Overexpression of Norrin

To learn more directly about the function of norrin, a mouse model with transgenic ocular overexpression of ectopic norrin by means of a strong lens-specific promoter has been developed [10]. The pheno-

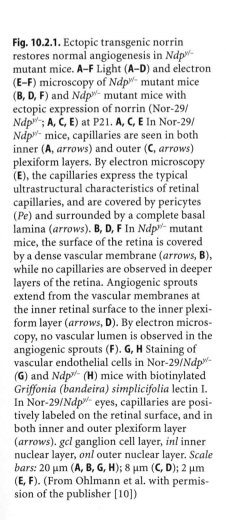
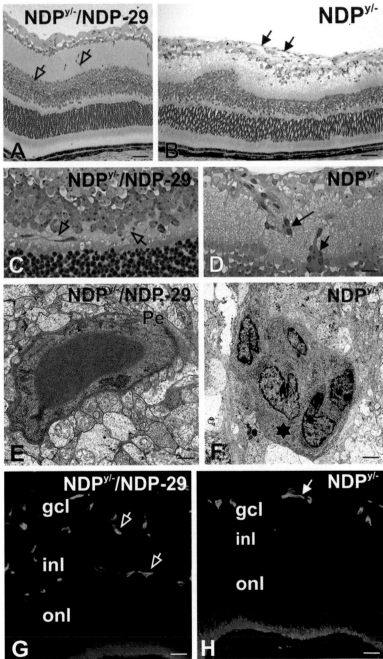

Fig. 10.2.1. Ectopic transgenic norrin restores normal angiogenesis in $Ndp^{y/-}$ mutant mice. **A–F** Light (**A–D**) and electron (**E–F**) microscopy of $Ndp^{y/-}$ mutant mice (**B, D, F**) and $Ndp^{y/-}$ mutant mice with ectopic expression of norrin (Nor-29/$Ndp^{y/-}$; **A, C, E**) at P21. **A, C, E** In Nor-29/$Ndp^{y/-}$ mice, capillaries are seen in both inner (**A**, *arrows*) and outer (**C**, *arrows*) plexiform layers. By electron microscopy (**E**), the capillaries express the typical ultrastructural characteristics of retinal capillaries, and are covered by pericytes (*Pe*) and surrounded by a complete basal lamina (*arrows*). **B, D, F** In $Ndp^{y/-}$ mutant mice, the surface of the retina is covered by a dense vascular membrane (*arrows*, **B**), while no capillaries are observed in deeper layers of the retina. Angiogenic sprouts extend from the vascular membranes at the inner retinal surface to the inner plexiform layer (*arrows*, **D**). By electron microscopy, no vascular lumen is observed in the angiogenic sprouts (**F**). **G, H** Staining of vascular endothelial cells in Nor-29/$Ndp^{y/-}$ (**G**) and $Ndp^{y/-}$ (**H**) mice with biotinylated *Griffonia (bandeira) simplicifolia* lectin I. In Nor-29/$Ndp^{y/-}$ eyes, capillaries are positively labeled on the retinal surface, and in both inner and outer plexiform layer (*arrows*). *gcl* ganglion cell layer, *inl* inner nuclear layer, *onl* outer nuclear layer. *Scale bars:* 20 μm (**A, B, G, H**); 8 μm (**C, D**); 2 μm (**E, F**). (From Ohlmann et al. with permission of the publisher [10])

10

type of these animals indicates that norrin induces growth of ocular capillaries, as lenses of transgenic mice with ectopic expression of norrin show significantly more capillaries in the hyaloid vasculature that surrounds the lens during development. In vitro, lenses of transgenic mice in coculture with microvascular endothelial cells induce proliferation of the cells. These effects may be mediated through the additional action of other factors, as mRNA for vascular endothelial growth factor (VEGF) and placental growth factor (PlGF) is found in higher amounts in the eyes of transgenic animals than in wild-type littermates. To learn if the ectopic expression of norrin could restore normal retinal vascular development and remodeling in $Ndp^{y/-}$ mice, the animals were crossbred with norrin overexpressing mice. In the resulting animals, both retinal capillary beds are present and can be visualized by conventional light microscopy, electron microscopy and labeling of vascular endothelial cells with biotinylated *Griffonia (Bandeiraea) simplicifolia* lectin I (Fig. 10.2.1). The capillaries show the typical ultrastructural characteristics of retinal capillaries, as they are covered by pericytes and surrounded by a complete basal lamina. In summary, the ectopic transgenic expression of lens-derived norrin completely restores normal vascular development of the retina in $Ndp^{y/-}$ mice. In addition, no abnormal vascular sprouting or signs of retinal neovascularization are observed, and the formation of abnormal dense vascular membranes at the inner retinal surface of $Ndp^{y/-}$ mice is completely prevented.

The improvement in structure correlates with restoration of neuronal function in the retina, as both the positive b-wave and the oscillatory potentials are completely restored.

It is of interest to note that ectopic norrin not only restores normal capillarization of the retina, but also prevents the progressive loss of retinal ganglion cells (RGCs) seen in $Ndp^{y/-}$ mutant mice. A possible explanation is that hypoxia induces RGC death in $Ndp^{y/-}$ mice, and that the restoration of retinal capillaries caused by the presence of ectopic norrin maintains normal oxygenation of the retina. Still, a more direct neurotrophic effect of norrin appears to be possible, as transgenic mice with ectopic expression of norrin show more BrdU-labeled retinal progenitor cells at embryonic day 14.5 and thicker retinas in postnatal life.

10.2.5 Conclusion

Ectopic norrin induces growth of retinal capillaries, an effect that completely respects the normal vascular architecture of the retina and does not involve abnormal vascular sprouting or signs of retinal neo-

vascularization. An intriguing aspect of this observation is that pharmacologic modulation of norrin might be used not only for treatment of the vascular abnormalities associated with Norrie disease, but also for other vascular disorders of the retina [5, 10]. The expression of norrin's receptor, Fz4, has been found in cultured human dermal microvascular endothelial cells and in endothelial cells from adult lung capillaries isolated ex vivo [4], indicating that ectopic norrin and Fz4 signaling could also be used for treatment of vascular disorders outside the eye.

References

1. Berger W (1998) Molecular dissection of Norrie disease. Acta Anat (Basel) 162:95–100
2. Berger W, van de PD, Warburg M, Gal A, Bleeker-Wagemakers L, de Silva H, Meindl A, Meitinger T, Cremers F, Ropers HH (1992) Mutations in the candidate gene for Norrie disease. Hum Mol Genet 1:461–465
3. Berger W, van de PD, Bachner D, Oerlemans F, Winkens H, Hameister H, Wieringa B, Hendriks W, Ropers HH (1996) An animal model for Norrie disease (ND): gene targeting of the mouse ND gene. Hum Mol Genet 5:51–59
4. Favre CJ, Mancuso M, Maas K, McLean JW, Baluk P, McDonald DM (2003) Expression of genes involved in vascular development and angiogenesis in endothelial cells of adult lung. Am J Physiol Heart Circ Physiol 285:H1917–1938
5. Gariano RF, Gardner TW (2005) Retinal angiogenesis in development and disease. Nature 438:960–966
6. Luhmann UF, Lin J, Acar N, Lammel S, Feil S, Grimm C, Seeliger MW, Hammes HP, Berger W (2005) Role of the Norrie disease pseudoglioma gene in sprouting angiogenesis during development of the retinal vasculature. Invest Ophthalmol Vis Sci 46:3372–3382
7. Meindl A, Berger W, Meitinger T, van de PD, Achatz H, Dorner C, Haasemann M, Hellebrand H, Gal A, Cremers F (1992) Norrie disease is caused by mutations in an extracellular protein resembling C-terminal globular domain of mucins. Nat Genet 2:139–143
8. Meitinger T, Meindl A, Bork P, Rost B, Sander C, Haasemann M, Murken J (1993) Molecular modelling of the Norrie disease protein predicts a cystine knot growth factor tertiary structure. Nat Genet 5:376–380
9. Ohlmann AV, Adamek E, Ohlmann A, Lutjen-Drecoll E (2004) Norrie gene product is necessary for regression of hyaloid vessels. Invest Ophthalmol Vis Sci 45:2384–2390
10. Ohlmann A, Scholz M, Goldwich A, Chauhan BK, Hudl K, Ohlmann AV, Zrenner E, Berger W, Cvekl A, Seeliger MW, Tamm ER (2005) Ectopic norrin induces growth of ocular capillaries and restores normal retinal angiogenesis in Norrie disease mutant mice. J Neurosci 25:1701–1710
11. Perez-Vilar J, Hill RL (1997) Norrie disease protein (norrin) forms disulfide-linked oligomers associated with the extracellular matrix. J Biol Chem 272:33410–33415
12. Rehm HL, Zhang DS, Brown MC, Burgess B, Halpin C, Berger W, Morton CC, Corey DP, Chen ZY (2002) Vascular defects and sensorineural deafness in a mouse model of Norrie disease. J Neurosci 22:4286–4292
13. Richter M, Gottanka J, May CA, Welge-Lussen U, Berger W, Lutjen-Drecoll E (1998) Retinal vasculature changes in Norrie disease mice. Invest Ophthalmol Vis Sci 39:2450–2457

14. Ruether K, van de PD, Jaissle G, Berger W, Tornow RP, Zrenner E (1997) Retinoschisislike alterations in the mouse eye caused by gene targeting of the Norrie disease gene. Invest Ophthalmol Vis Sci 38:710–718

15. Tazawa Y, Seaman AJ (1972) The electroretinogram of the living extracorporeal bovine eye. The influence of anoxia and hypothermia. Invest Ophthalmol 11:691–698

16. Warburg M (1961) Norrie disease: a new hereditary bilateral pseudotumor of the retina. Acta Ophthalmol 39:757–772

17. Warburg M (1963) Norrie disease: atrofia bulborum hereditarum. Acta Ophthalmol 41:134–146

18. Warburg M (1966) Norrie disease: a congenital oculo-acoustico-cerebral degeneration. Acta Ophthalmol 89(Suppl): 1–147

19. Xu Q, Wang Y, Dabdoub A, Smallwood PM, Williams J, Woods C, Kelley MW, Jiang L, Tasman W, Zhang K, Nathans J (2004) Vascular development in the retina and inner ear: control by Norrin and Frizzled-4, a high-affinity ligand-receptor pair. Cell 116:883–895

Section II
General Concepts in the Diagnosis and Treatment of Retinal Vascular Disease

11 A Practical Guide to Fluorescein Angiography

H. Heimann, S. Wolf

Core Messages

- Fluorescein angiography remains the gold standard for diagnosis and treatment decisions of the most commonly treated retinal disorders, e.g., exudative age-related macular degeneration, diabetic retinopathy and retinal venous occlusions. Despite the recent advancements and growing popularity of optical coherence tomography and indocyanine green (ICG) angiography, fluorescein angiography is still indispensable for the identification of choroidal neovascularizations in age-related macular degeneration and macular ischemia in retinal vascular diseases, and, therefore, is likely to remain a key examination method for patients with retinal and choroidal diseases
- The outstanding value of fluorescein angiography derives from its representation of the integrity of the inner and outer blood-retina barriers. The detection of extravasated unbound fluorescein with fluorescein angiography helps to identify even subtle defects of the blood-retina barriers and thereby helps to establish the diagnosis and to set up treatment strategies
- There is no standard protocol for the interpretation of fluorescein angiographies. Usually, between 30 and 40 pictures are taken from 10 s to 10 min after the intravenous injection of 5 ml 10% sodium fluorescein. The interpretation is usually divided into arterial phase (12–15 s), arteriovenous phase (15–35 s) and late phase (5 or 10 min). Ideally, pseudostereo-angiographies with the option of stereo-viewing on review of the pictures should be performed
- Fluorescein angiography is considered to be a safe diagnostic procedure. Mild reactions, e.g., nausea, occur in about 1:20 angiographies. Moderate complications, e.g., urticaria or dyspnea, can be seen in about 1:60 cases. Severe complications, e.g., anaphylactic shock, have been described in 1:2,000 angiographies. The risk for lethal complications of an angiography has been estimated to be 1:220,000. Fluorescein angiography should not be performed in cases with known allergic reactions to fluorescein

11.1 History

Fluorescein is one of the most potent artificial dyes. It was first synthesized in 1871 by Adolf von Baeyer, a German chemist who in 1905 received the Nobel Prize in chemistry for his work on organic dyes.

The occurrence of fluorescein in the anterior chamber following intravenous injection was first described by Paul Ehrlich in 1882. Further landmarks in the use of fluorescein for posterior segment diseases were the detection of increased fluorescein leakage in chorioretinitis (Burk 1910) and its role in the differential diagnosis of choroidal tumors after intravenous administration (MacLean and Maumenee 1955) [7].

The basic concept of fluorescein angiography (FA) was first described in 1959 by Flocks, Miller and Chao following initial in vivo experiments in cats. The first FA on a human was performed by Novotny and Alvis soon afterwards [1]. As a landmark in the interpretation of FA and a reference guide to date the *Stereoscopic Atlas of Macular Diseases* was first published by Donald Gass in 1969 [3].

Fluorescein angiography has been in routine use for the diagnosis and treatment of posterior segment diseases for about 40 years now. It accelerated the process of understanding many retinal and choroidal diseases. In parallel with the introduction of laser photocoagulation, it built the foundation of a whole new subspecialty branch of ophthalmology, the "medical retina." Large multicenter randomized trials, basing their initial classification and subsequent course of the disease on FA, set evidence-based standards for the treatment of diabetic retinopathy,

11 II

venous occlusive diseases and age related macular degeneration (AMD) [2, 3, 5, 6].

Although electrophysiology and more recently optical coherence tomography and indocyanine green (ICG) angiography have been vital additional methods for the examination of retinal and choroidal diseases, FA still remains the most important investigation next to ophthalmoscopy for the majority of retinal and choroidal diseases. It continues to be the "gold standard" in the diagnosis and treatment of the three most prevalent diseases seen in the field of "medical retina" – age related macular degeneration, diabetic retinopathy and retinal venous occlusions. Clinical and research studies, as well as the decision regarding funding of certain therapeutic options, are therefore in large part based on the results of FA.

A recent spur to the use of FA has been the new treatment methods available for AMD, e.g., photodynamic therapy and intravitreal injection of anti-vascular endothelial growth factor (VEGF) substances; as a consequence, the number of angiographies performed in tertiary centers has more than doubled over the past 5 years.

Digital imaging systems have overtaken classic b/w film techniques as the standard method of recording FA in most institutions. Comparing both techniques, the advantages of digital systems are the immediate availability of the angiography for diagnosis and treatment with a resolution sufficient for most cases; the possibility of image enhancement and image analysis; the easy storage of images; and the speedy distribution of FA to computers within a department or elsewhere. Disadvantages are that digital systems are more expensive, achieve a lower resolution of single images and imply a technically more complicated method of viewing pseudo-stereo images compared to classic film recordings.

11.2 Concept

Fluorescence is defined as the emission of absorbed radiated energy in the form of electromagnetic radiation with similar or longer wavelength. In the context of fluorescence angiography, this means that an exciting light reaching the fluorescing molecule causes emission of light of a different wavelength that then again can selectively be detected. In practice, the dye molecules not bound to proteins function as manifold sources of light within the tissue. This enables identification of the localization of the dye and results in a much higher contrast than that achievable by the dying properties alone.

Fluorescein sodium ($C_{20}H_{10}O_5Na_2$, MW 376) is the most commonly used dye worldwide for FA. Indocyanine green is another dye used for FA.

Fluorescein angiography is used for diagnosis of posterior segment diseases, as a treatment guideline and as a tool for the evaluation of the development and course of the disorder during subsequent treatment or observation.

It is most often employed for diseases of the retina or at the retinal-choroidal junction and, to a lesser extent, for disease of the choroid and the optic nerve head.

The value of the fluorescein sodium for examination of posterior segment diseases in contrast to other dyes derives from the relation of its intraocular distribution to the integrity of the blood-retinal barrier. Unbound fluorescein does not permeate intact inner and outer blood-retinal barriers in significant quantity. If, however, these structures are injured, the permeation and visualization of fluorescein by FA allows identification and localization of the damaged structures as well as that of newly formed vessels or vascular irregularities.

Both the absorption and emission spectrums of sodium fluorescein are dependent on various factors; in intravenous administration they are 465 nm (absorption) and 525 nm (emission).

Approximately 70–80% of the sodium fluorescein binds to plasma proteins; 20–30% does not bind. Injected fluorescein is diluted in the bloodstream to a factor of approximately 600 and disperses throughout the whole body. With the exception of the central nervous system vessels, the retina, and to a limited degree the iris, unbound sodium fluorescein can freely permeate all blood vessels of the body.

The intravenous administration of fluorescein sodium during FA is well tolerated in most cases. Mild reactions, e.g., nausea, occur in about 1:20 angiographies. Moderate complications, e.g., urticaria or dyspnea, can be seen in about 1:60 cases. Severe complications, e.g., anaphylactic shock, have been described in 1:2,000 angiographies. The risk of lethal complications with an angiography has been estimated to be 1:220,000.

11.3 Performing Fluorescein Angiography

The reasons for performing FA, the procedure and its possible side effects should be explained to the patient. As FA is an invasive procedure with potential severe side effects, written consent is required in most countries and institutions.

Equipment: fundus camera with automatic film transport, or a digital camera or scanning laser ophthalmoscope; an electronic flash unit with an exciter filter of 465–490 nm (blue green spectrum); and a recording filter of 520–530 nm (green-yellow spectrum). Pseudo-stereo viewing usually requires particular software and viewing systems (digital angiography) or magnifying glasses (film angiography).

Parallel fluorescein and ICG angiographies are currently obtainable with specialized equipment only and are usually not employed in daily practice.

The pupil should be dilated and intravenous access secured.

The staff on duty must be trained in resuscitation and emergency procedures in the rare event of a severe allergic reaction to the dye. Drugs and equipment needed to treat anaphylactic reactions must be readily available.

The angiography is usually performed by a specialized photographer or an ophthalmologist. The experience and skill of the photographer are of vital importance for achieving high quality angiograms. In contrast, angiographies of poor quality are more often than not useless and require a repeat procedure many hours or days afterwards.

A standard image of 30° is used for most diseases; alternatively 20° (high magnification) and 50° or 60° (wide angle) exposures are used.

Color fundus photograph and b/w exposures of both eyes should be taken before angiography.

The angiography is started by an intravenous bolus administration of 5 ml 10% sodium fluorescein (2.5 ml for patients with renal insufficiency). The quicker the injection, the higher the contrast in the early phases of the angiography. A stopwatch is set off at the start of the injection to monitor the chronology of the exposures.

First exposures are obtained after 10–15 s at first appearance of the dye within the choroidal and retinal circulations (pre-arterial and early arteriovenous phase). Rapid sequential exposures (approximately one image per second) are then taken (arteriovenous phase) until the maximum fluorescence is reached after 25–35 s.

After completion of the arteriovenous phase, additional exposures of the fundus periphery (for example, in diabetic retinopathy) or the fellow eye (for example, in macular degeneration) can be made.

Pseudo-stereo photographs can be obtained by lateral shifts of the camera on sequential frames. A true stereoangiography requires specialized equipment where two pictures can be taken with a fixed angle at the same time.

Angiography is concluded with the last exposures being taken 5, 10 or 15 min after the initial injection (late phase).

There is no universal protocol for the timing of exposures. Protocols used in different institutes and studies vary, depending on the profile of the disease and the institution. For most clinical studies, certification of the equipment, the photographer and the examiner is required.

Usually, about 30 pictures are taken per angiography session. Typical shots resembling the early arteriovenous phase, arteriovenous phases and late phases can be selected and printed for documentation. Storing all images for analysis is recommended for all cases.

In situations where the angiography could not be performed in a suitable way, it can be repeated after 2–4 h in most cases. However, in larger studies, a 48-h interval is generally recommended.

11.4 Interpretation of Fluorescein Angiography

For the assessment of angiographies, the appropriate conditions should be in place. For example, digital angiographies should be viewed on a large high resolution screen with stereo-viewing capabilities wherever possible. FA captured on film should either be viewed with magnifying glasses on a light box or enlarged to high quality prints. In some reading centers, angiographies on b/w film are still the gold standard due to the higher resolution and easier stereo-viewing of these images.

Assessment and interpretation of angiography are biased by the quality and timing of the angiography and the expertise of the examiner. Intra- and interobserver differences similar to those found in the evaluations of chest X-rays exist even among expert examiners. In case of doubt, additional advice should be sought, which nowadays is facilitated by the possibility of mailing digital angiographies via the internet.

As there is no standardized procedure for the recording of the angiography, neither is there a standardized protocol for its interpretation. Most investigators would base their interpretation on the evaluation of color and red-free pictures, the early arteriovenous phase, the arteriovenous phase and the late phase. Pseudo-stereo interpretation should be standard for interpretation of exudative AMD lesions as most treatment guidelines are based on stereo-images.

During angiography, the fluorescein dye fills the retinal and choroidal vessels in a reproducible manner (Figs. 11.1–11.4). Exceptions from the standard angiography patterns are identified and classified in relation to the anatomical structures and the timepoint at which they can be detected during the course of the angiography. The most common diseases examined with fluorescein angiography and their typical findings are illustrated in Table 11.1.

11 II

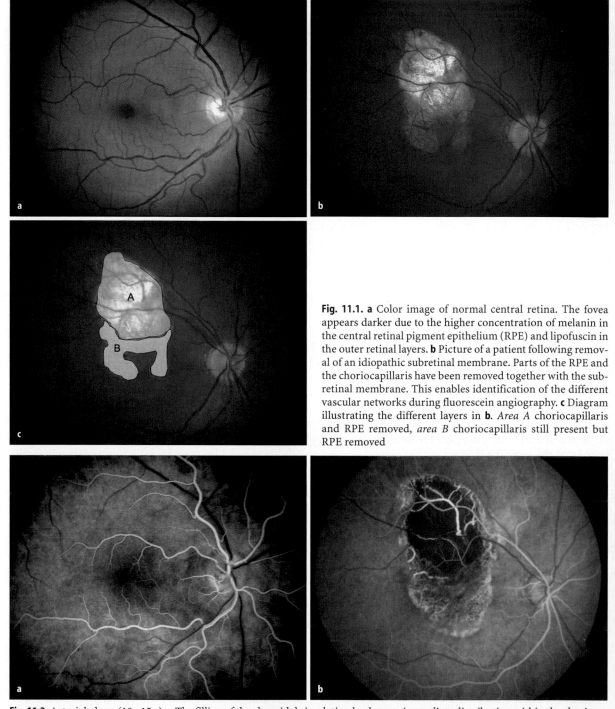

Fig. 11.1. a Color image of normal central retina. The fovea appears darker due to the higher concentration of melanin in the central retinal pigment epithelium (RPE) and lipofuscin in the outer retinal layers. **b** Picture of a patient following removal of an idiopathic subretinal membrane. Parts of the RPE and the choriocapillaris have been removed together with the subretinal membrane. This enables identification of the different vascular networks during fluorescein angiography. **c** Diagram illustrating the different layers in **b**. *Area A* choriocapillaris and RPE removed, *area B* choriocapillaris still present but RPE removed

Fig. 11.2. Arterial phase (10–15 s). **a** The filling of the choroidal circulation leads to an immediate distribution within the choriocapillaris. Fluorescein leaks out of the choriocapillaris and the emitting light in addition is scattered by the RPE, resulting in a background hyperfluorescence. This background fluorescence is blocked by the higher pigment content within the fovea. The retinal arteries are filled with fluorescein. **b** Filling of a posterior ciliary artery can be seen in area A. These vessels do not leak fluorescein. They quickly fill the choriocapillaris (area B) where a fine meshwork of channels is filled with fluorescein that immediately leaks out of the vessel lumen. Compare the "scattering" effect of the RPE in areas still containing this layer in front of the choriocapillaris

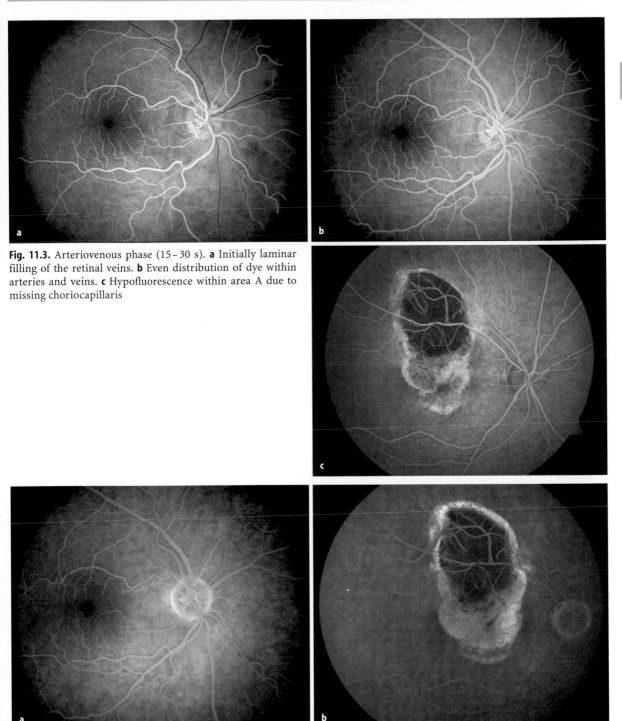

Fig. 11.3. Arteriovenous phase (15–30 s). **a** Initially laminar filling of the retinal veins. **b** Even distribution of dye within arteries and veins. **c** Hypofluorescence within area A due to missing choriocapillaris

Fig. 11.4. Late phase (1–10 min). **a** Washout of fluorescein out of the retinal vessels. The dye has been washed out of the vessels of the choriocapillaris. The choroidal vessels are seen as dark shadows against this background and that of the sclera. **b** Demonstrating persistent "background fluorescence"

Table 11.1. Synopsis of typical fluorescein angiography findings in most commonly examined entities

Disease		Localization	Clinical picture	Early arteriovenous phase	Arteriovenous phase	Late phase	Essentials
AMD							
Classic CNV		Central macula Retina/RPE/ choriocapillaris	Grayish protrusion in central macula ± intraretinal blood ± exudates ± RPE changes ± surrounded by pigmented ring	Early, well demarcated hyperfluorescence	Increasing hyperfluorescence leaking from the edges of the CNV	Persisting hyperfluorescence	Early, well demarcated hyperfluorescence with increasing and persisting leakage. Vessel structure of CNV sometimes visible
Occult CNV		Central macula Retina/RPE/cho-riocapillaris	Grayish tumor in central macula ± intraretinal blood ± exudates ± RPE changes	Hypofluorescence from retinal edema	Pinpoint or larger hyperfluorescences starting in the middle to late AV phase	Increasing leakage of unknown or fibrovascular origin	No bright and early hyperfluorescence. Increasing leakage of unknown origin. Stereo viewing essential
Pigment epithelial detachment		Macula Retina/RPE	Orange/yellowish dome-shaped elevation with smooth borders	Hypofluorescence from retinal edema	Increasing, smooth hyperfluorescence	Increasing and persistent leakage with smooth borders	Notching of RPE detachment points at possible occult CNV. Consider ICG angiography
RAP (Retinal angiomatous proliferation)		Central macula Retina/RPE/cho-riocapillaris	More exudates, localized intraretinal hemorrhages, retinal anastomosis can sometimes be seen clinically as retinal vessel of constant caliber „dipping" into the CNV	Filling of the CNV from communicating retinal vessel. Early bright hyperfluorescence	Like classic or predominantly classic CNV	Like classic or predominantly classic CNV	Consider high speed ICG. Clinical picture points at diagnosis
Pigment epithelial tear		Central macula RPE	Well demarcated, often L-shaped area brighter than surrounding macular tissue bordered by hyperpigmented patch to one side	Early, well demarcated hyperfluorescence in the area of missing RPE (window defect), bordered by hypofluorescent patch (retracted RPE)	Increasing hyperfluorescence in the area of the tear	Persistent hyperfluorescence in the area of the tear	Clinical picture points at diagnosis. Tear often follows treatment of occult CNV (or spontaneously)

11

	Location	Ophthalmoscopy	Early	Late	Comments
Drusen	Macula, mostly around the fovea — Retina/RPE	Yellowish round protusions	Mostly slightly hypofluorescent but some can have early, well demarcated hyperfluorescence	Increasing hyperfluorescence — No leakage	Persisting or diminishing hyperfluorescence — No leakage	No leakage from edges of drusen
Geographic atrophy	Central macula — Initially slightly eccentric — RPE	Well demarcated area void of RPE	Early, sharply demarcated hyperfluorescence (window defect)	Hyperfluorescence parallels choroidal filling — No leakage from the edges	Decreasing hyperfluorescence parallels regression of choroidal filling	No leakage — No thickness on stereo — Consider autofluorescence for follow-up
Diabetic retinopathy **Focal edema**	Slightly eccentric — Retina	Clustered groups of microaneurysm associated with retinal thickening and exudates	Hypofluorescence from retinal thickening	Filling of microaneurysm parallels filling of retinal vessels — Slow, increasing leakage around aneurysm	Increasing and persisting leakage around aneurysm	Consider focal laser for areas of leakage
Diffuse edema	Central macula — Retina	Diffuse thickening ± exudates ± microaneurysm	Hypofluorescence from retinal thickening	Diffuse leakage over parafoveal vessels	Increasing and persisting leakage ± cystoid configuration	Consider grid treatment — Consider OCT for follow-up
Ischemic retinopathy	Central macula — Retina	Diffuse thickening ± visible ghost vessels ± IRMA or neovascularizations bordering ischemic areas	Hypofluorescence over affected areas	Persisting hypofluorescence due to absent retinal vessels	Persisting hypofluorescence or leakage from bordering neos or IRMA	No benefit from focal laser
Proliferative retinopathy	On the optic nerve head or along the major vessel arcades — Epiretinal	Visible epiretinal neovascularizations ± sheets of fibrous tissue ± traction on retina	Early filling of neovascularizations parallel filling of retinal vasculature	Leakage over neovascularizations	Persisting diffuse epiretinal leakage over neovascularizations	Clinical picture — Minute neovascularizations can sometimes be documented by FA only

11 II

Table 11.1. (*Cont.*)

Disease	Localization	Clinical picture	Early arteriovenous phase	Arteriovenous phase	Late phase	Essentials
Tumors						
Choroidal melanoma	Anywhere on the fundus Choroid	Pigmented or non pigmented choroidal tumor Occasionally breakthrough through Bruchs membrane or retinal infiltration ± subretinal hemorrhages ± drusen on surface	Hypofluorescence though pigment and tumor volume Early filling of tumor vessels characteristic but not always present	Increasing hyperfluorescence within tumor ± pinpoint leakages ± increasing leakage from tumor vessels	Persisting hyperfluorescence within tumor	„Second circulation" within tumor supports diagnosis but is not always present
Choroidal hemangioma	Posterior pole Choroid	Reddish or grayish tumor with surrounding fluid	Very early filling of tumor vessels parallel to choroidal filling before retinal vessels	Increasing hyperfluorescence within tumor	Persisting leakage within tumor and surrounding tissues	Very early frames (before retinal filling) essential
Miscellaneous						
Central serous retinopathy	Posterior pole RPE	Yellowish lesion in the macula	Hypofluorescence caused by subretinal fluid	Pinpoint leakage, slowly expanding to more widespread leakage	Persisting leakage within the space of detached retina	In questionable cases look for possible sources of leakage outside the major vessel arcades
Chronic central serous retinopathy	Posterior pole RPE	Pigmentary changes in the macula	Early window defects in areas of depigmentation of RPE	Either window defects or leakages in cases of active lesions	Persisting window defects or leakage in active lesions	Important differential diagnosis to occult CNV, appears less prominent than CNV Clinical course and stereos important for diagnosis

	Location	Clinical findings	Early phase	Mid phase	Late phase	Comments
Retinal teleangiectasis	Posterior pole Retina	Red dot-like lesions around fovea ±pigmentary changes ±exudates	Early filling of teleangiectasis parallel to retinal vessels	Increasing moderate leakage from teleangiectasis	Persisting leakage	Stereos and clinical pictures essential for identifying lesion within the retina OCT often shows thinning or outer retinal layers	
Cystoid macular edema	Fovea and central macula Retina	Cystoid spaces or bleb- or hole like lesions in the fovea	Hypofluorescence due to fluid within cystoid spaces	Slowly accumulating hyperfluorescence within cystoid spaces	Increasing leakage filling the cystoid spaces	Clinical picture important Consider OCT as initial examination technique or in questionable cases	
Macular hole	Central macula	Yellowish lesion or round retinal defect around the fovea with raised edges	Hypofluorescence in the area of the hole	Occasionally some hyperfluorescence within yellowish deposits in the area of the hole or in cystoid spaces around the hole	Occasionally persisting hyperfluorescence within or around the hole or cystoid edema at the edges	Clinical picture and OCT more important for diagnosis	
Macular pucker	Macula Retina	Distortion of macular vessels Epiretinal membrane	Distortion of retinal vessels is better visualized	Mild hyperfluorescence secondary to leakage from retinal vessels	Increasing hyperfluorescence from leaking vessels	Diagnosis can usually be made without angiography OCT more important	
Retinal vasculitis	Segmental or all quadrants Retina and choroid	Vascular sheathing, hemorrhages, vitritis	Attenuated filling of affected arteries/veins Weak early hyperfluorescence through leakage from affected vessels	Increasing hyperfluorescence over affected vessels ±filling defects	Increasing hyperfluorescence through leakage	Extensive differential diagnosis Angiography sometimes augments identification of affected vessels	

Table 11.1. (Cont.)

Disease	Localization	Clinical picture	Early arteriovenous phase	Arteriovenous phase	Late phase	Essentials
Choroidal vasculitis	Segmental or all quadrants Choroid and retina	Pigmentary changes or "white dots" ±vitritis	Hypofluorescent spots in affected choroidal areas	Transition from hypofluorescence to hyperfluorescence in affected areas	Hyperfluorescence in affected areas	Extensive differential diagnosis ICG angiography usually more helpful in identifying affected areas
Retinal vascular occlusions						
Central retinal vein occlusion (CRVO)	All retinal quadrants ±macular edema	Dot and flame-shaped intra- and preretinal hemorrhages, dilated veins, (cystoid) macular edema, cotton wool spots	Hypofluorescence through edematous retina and hemorrhages	Delayed filling of retinal veins, leakage over veins, hypofluorescence in areas of capillary dropout	Diffuse increased hyperfluorescence through leakage from dysfunctioning veins ±cystoid or diffuse leakage in the macula	Classification of ischemic versus non-ischemic central retinal vein occlusion according to fluorescein angiography is controversial
Branch vein occlusion	One/two retinal quadrants ±macular edema	Dot and flame-shaped intra- and preretinal hemorrhages, dilated veins, (cystoid) macular edema, cotton wool spots	Hypofluorescence through edematous retina and hemorrhages in the affected quadrant(s)	Delayed filling of retinal veins, leakage over veins, hypofluorescence in areas of capillary dropout in the affected quadrant(s)	Diffuse increased hyperfluorescence through leakage from dysfunctioning veins ±cystoid or diffuse leakage in the macula	Indication for laser coagulation based on fluorescein angiography according to Branch Vein Occlusion Study
Central retinal artery/arterial branch occlusion	Retina All quadrants or one to two quadrants	Retinal edema Cherry red spot in the macula ±visible plaques in the arterial stem and/or branches	Delayed and segmented or no filling of affected retinal arteries Choroidal filling visible unless occlusion of ophthalmic artery is present	Delayed and segmented filling of arteries and veins in the affected quadrants	Diffuse leakage through dysfunctioning retinal vessels in the affected quadrants	Fluorescein angiography usually unnecessary for diagnosis

Two basic patterns of abnormalities can be found during angiography: hyperfluorescence and hypofluorescence.

Hyperfluorescence is defined as an increase in fluorescein detection in a particular structure compared to standard angiography. It can be caused by an increased permeation of dye (e.g., damage to the integrity of the blood-retinal barriers), newly developed structures containing fluorescein (e.g., neovascularizations of the retina or choroid) or an increased visibility of normal fluorescein concentrations (e.g., window defects allowing better visualization of choroidal fluorescence through a window in the retinal pigment epithelium).

The term "leakage" refers to a subgroup of hyperfluorescence and is defined as passage of fluorescein visible on angiography across barriers which it normally does not cross in a significant amount (e.g., inner and outer retinal blood barrier). It can further be subdivided into "staining" (leakage into a solid structure, e.g., scar tissue) or "pooling" (leakage into preformed hollow spaces, e.g., serous pigment epithelial detachment) or a combination of both. The terms "leakage" and "staining" have recently been used to distinguish active from fibrotic choroidal neovascularizations following photodynamic therapy. Here, "leakage" is defined as increasing leakage through the late phases of angiography in contrast to "staining" as a sign of inactive CNV (usually attenuation of hyperfluorescence towards the late phase).

Hypofluorescence is defined as a decrease of fluorescein detection compared to standard angiography. It can be caused by blockage of exciting or emitting light (e.g., by haemorrhages) or a decrease of fluorescein containing structures or their fluorescein content (e.g., occlusion of retinal or choroidal vessels).

11.5 Quantitative Evaluation of Fluorescein Angiography

By means of densitometric measurements the filling times of arterioles, venules, or defined areas of the retina or choroid can be evaluated from fluorescein angiography [8]. Early, middle, and late phases are differentiated; the examiner precisely specifies the time after injection of the fluorescein or the time after the dye first appears on the optic disk.

Arm-retina time: The normal arm-retina time is about 10–20 s. A prolonged arm-retina time indicates a generalized circulatory disorder (e.g., cardiovascular insufficiencies, carotid occlusion).

Arterial filling time should be faster than 1 s. Slow arterial filling indicates reduced perfusion pressure that may be related to a stenosis of the central retinal or ophthalmic arteries, increased intraocular pressure, or severe impairment of systemic circulation.

Arteriovenous passage time is the time period between the inflow of the dye into retinal arteries and the appearance in the corresponding venules. Various studies have indicated that arteriovenous passage time is prolonged in retinal vascular disorders like diabetic retinopathy, retinal vascular occlusion, inflammatory retinal and choroidal disease, and glaucoma.

11.6 Avoiding Unnecessary Angiographies

As with all invasive diagnostic tests, fluorescein angiography should only be employed if the diagnosis cannot be established without the test, the angiography serves as a treatment guideline or is needed for interpretation of the course of a disease during follow-up examinations. FA should not be used as a substitute for a thorough clinical examination or for documentation purposes only.

Due to the increasing demand for FA following the development of new treatment strategies for the treatment of exudative AMD, the capacities for FA are nowadays often stretched to the limit within tertiary centers. In addition to the potential side effects of angiography and the health-economic aspects of performing unnecessary examinations, this is another reason to avoid FAs that are not needed for the above reasons.

Appropriate communication with the photographer is essential. Before angiography, the photographer should be sure about the structures to be examined and particular points of importance during the angiography (e.g., which phase of angiography, which region of the retina, which magnification is to be used and additional exposures of the periphery or the fellow eye). The setting up of standardized forms for angiography and standard protocols for particular diseases (e.g., for exudative AMD or diabetic macular edema) within the department is advisable.

In our experience, superfluous angiographies are frequently performed in the following situations:

- Non-exudative AMD. For documentation purposes, color photographs or autofluorescence examinations are sufficient.
- End-stage exudative AMD. There is no therapeutic option available at present and the diagnosis can usually be established without angiography.
- Clinically significant macular edema in diabetic retinopathy. The diagnosis is based on ophthalmoscopy and not on angiography. FA should only be performed if laser treatment is planned (to identify sources of leakage and areas of ischemia). Optical coherence tomography (OCT)

11 II

seems to be particularly helpful for monitoring the course of clinically significant macular edema (CSME) following treatment with laser photocoagulation or intravitreal triamcinolone.

- Retinal venous occlusions. The diagnosis can usually be established without angiography. The use of FA for the differentiation of ischemic versus non-ischemic venous occlusions is highly controversial. An evidenced-based necessity of FA is at present established only in branch vein occlusions according to the guidelines of the Branch Vein Occlusion Study (see Chapter 21.3).
- Intraocular tumors. Often, angiography is inadequately included in the standard workup of intraocular tumors. In experienced institutions, the diagnosis can be established based on clinical examination and ultrasonography in most cases.

If the diagnosis cannot be established by other methods, angiography is seldom able to achieve that essential bit of information needed for differential diagnosis.

References

1. Blacharski PA (1985) Twenty-five years of fluorescein angiography. Arch Ophthalmol 103(9):1301–2
2. Bressler NM (2002) Verteporfin therapy of subfoveal choroidal neovascularization in age-related macular degeneration: two-year results of a randomized clinical trial including lesions with occult with no classic choroidal neovascularization – verteporfin in photodynamic therapy report 2. Am J Ophthalmol 133(1):168–9
3. Early Treatment Diabetic Retinopathy Study Research Group (1995) Focal photocoagulation treatment of diabetic macular edema. Relationship of treatment effect to fluorescein angiographic and other retinal characteristics at baseline: ETDRS report no. 19. Arch Ophthalmol 113(9):1144–55
4. Gass JD (1997) Stereoscopic atlas of macular diseases: Diagnosis and treatment, 4th edn. Mosby, St. Louis
5. Macular Photocoagulation Study Group (1991) Subfoveal neovascular lesions in age-related macular degeneration. Guidelines for evaluation and treatment in the macular photocoagulation study. Arch Ophthalmol 109(9):1242–57
6. The Branch Vein Occlusion Study Group (1984) Argon laser photocoagulation for macular edema in branch vein occlusion. Am J Ophthalmol 98(3):271–82
7. Wessing A (1968) Fluoreszenzangiografie der Retina. Lehrbuch und Atlas. Georg-Thieme Verlag, Stuttgart
8. Wolf S, Arend O, Reim M (1994) Measurement of retinal hemodynamics with scanning laser ophthalmoscopy: reference values and variation. Surv Ophthalmol 38 Suppl: S95–100

12 Optical Coherence Tomography in the Diagnosis of Retinal Vascular Disease

A. WALSH, S. SADDA

Core Messages

- OCT provides both cross-sectional visualization and clinically relevant quantitative measurements of ocular tissues
- OCT is an objective and quantitative method of standardizing disease monitoring both in clinical trials and in clinical practice
- Fluorescein angiography and OCT provide unique yet complementary information that may necessitate the use of both in the evaluation of patients with retinal vascular diseases
- OCT has greatly improved the evaluation and description of the vitreomacular interface
- OCT is useful in monitoring response to retinal vascular disease therapies and in guiding re-treatmant decisions
- Conventional OCT has substantial limitations both in hardware and in software
- Future OCT systems should be faster and provide more clinically relevant measurements

12.1 Overview

Retinal vascular diseases, in particular diabetic retinopathy and retinal venous occlusive disorders, are important causes of visual loss and blindness. Other important retinal vascular diseases which can affect visual function include arterial occlusive disease, parafoveal telangiectasis, Coat's disease, vasculitides, macroaneurysms, and hypertensive retinopathy. Despite the various etiologies and underlying pathogenic processes, the mechanisms of visual loss are frequently similar among these diseases. One such common final pathway is the development of occlusions of the microcirculation (capillaries) with attendant retinal ischemia. The most frequent sequela, however, is a compromise in retinal vascular permeability leading to leakage and exudation with accumulation of fluid, lipid, and proteins within the retina [40] or in the subretinal space. Structural alterations are also a frequent outcome of retinal vascular disease. These changes include the development of cystoid spaces in the retina and vitreomacular traction.

The discovery and development of optical coherence tomography (OCT) has had a significant impact on the diagnosis and management of retinal vascular diseases, particularly in the identification of subtle structural alterations of the retina and in the detection and quantification of macular edema.

Although it was first described only 15 years ago, OCT is now positioned to play a major role in clinical trials and clinical practice for the foreseeable future. The application of OCT for the diagnosis and management of retinal vascular diseases is discussed in this chapter.

12.2 Evaluation

The diagnostic armamentarium available to retinal specialists in the early 1990s closely resembled technologies available to practitioners nearly 2 decades prior. This stagnant situation changed rapidly in the mid-1990s, however, with the near-simultaneous explosion of computer and digital imaging technologies. Digital imaging simplified fundus photography and fluorescein angiography by reducing patient wait times and the infrastructure that was needed to develop film negatives. Intranet- and internet-based integration of imaging devices located at different sites and made by different manufacturers enabled ophthalmologists to access data quickly and efficiently. At the same time, knowledge and understanding of computer technology became widespread, enabling a rapid sharing of technical clinical data.

Yet, during this explosion of computer and imaging technology, one thing remained almost unchanged – the means of interpreting and evaluating

12 II

diagnostic imaging data. Time-consuming evaluations of color images for retinal diseases such as diabetic retinopathy were still based on subjective analyses by human graders. Angiographic assessments still required training and expertise, and were plagued by inconsistencies and intergrader variability [32, 38]. Few investigators took advantage of the newly available digital information from these images to perform advanced quantitative analyses.

During that time, Carl Zeiss Meditec, Inc. (Dublin, CA) was quietly developing a fledgling technology that would soon revolutionize the field – OCT. Introduction of this disruptive technology required the development of novel hardware and software as well as a new market in which to sell the instrument. Most importantly, Zeiss chose to make a radical departure from industry standards. Instead of simply providing a fundus image to be viewed and interpreted by the clinician, Zeiss chose to automate the extraction of quantitative information from OCT images. This was a dramatic departure from the industry norm, which only required that fundus images be provided for subjective evaluations, without measurements or automated assessments.

With slit-lamp biomicroscopy firmly established as the gold-standard clinical evaluation and angiography as the principal tool for further disease clarification, many clinicians in the mid 1990s were skeptical of this new technology. Soon, however, unique benefits of noninvasive cross-sectional imaging became apparent. For example, vitreomacular traction syndrome (VMT), which was previously clinically unapparent in many patients, was able to be visualized and diagnosed with confidence using OCT. Macular holes and cysts could also be visualized and monitored. Furthermore as intravitreal therapies for retinal diseases were introduced, OCT's value in the quantification of retinal thickening and subretinal exudation became certain. The optimal integration of OCT data with other clinical findings to synthesize a clinical care plan is still an evolving subject.

12.2.1 Optical Coherence Tomography

Essentials
- Provides both cross-sectional visualization and clinically relevant quantitative measurements
- Based on low-coherence interferometry
- Measures reflectivity of tissue interfaces with axial resolution less than 10 μm
- Good reproducibility in patients with diabetes

A complete discussion of OCT is beyond the scope and purpose of this chapter. The reader is referred to several excellent texts on this topic [27, 30, 33, 34]. Briefly, OCT is based on the principle of low-coherence interferometry. Akin to B-scan ultrasonography, OCT uses differences in the reflection of light, instead of sound echoes, to render two-dimensional images (tomograms) of the retina. Various light sources, typically superluminescent diodes or lasers with very short pulses (i.e., femtosecond lasers), are used to generate broad bandwidth light. The depth or axial resolution of OCT is based on the bandwidth of these sources, and the lateral resolution is determined by the diameter of the focused probe beam.

The light is split into two different paths: a reference beam projected inside the instrument and a sample beam focused on the tissue of interest. In time domain OCT, differences in the time of flight for these two light paths are measured using a Michelson-type interferometer. In Fourier domain OCT, these differences are characterized with a spectrometer and Fourier-based mathematical calculations. Differences in the optical characteristics of ocular tissues result in the different reflectivity intensities that are measured by OCT.

A single point of light reflected off the retina forms an *A-scan*, which contains information about the axial location of these tissue interfaces. These single points of light can be laterally aligned to form *B-scan* images, which often use a false-color display to depict interface intensities: highly reflective interfaces are rendered in white and red, medium level reflections are shown as yellow and green, and features with low reflectivity are depicted in blue. OCT data can be viewed en face as a *C-scan* or in dense three-dimensional cubes (3D OCT) by capturing B-scans or C-scans in rapid succession. Whereas the axial resolution of clinical ophthalmic echography is limited to greater than 100 μm, differences less than 10 μm can be discerned with conventional OCT instruments. Newer instruments, potentially coupled with adaptive optics devices, can resolve structures that are separated by less than 2 μm in the axial direction [109].

Scan acquisition is painless for the subject, but requires cooperation and steady fixation. Due to inherent speed limitations in conventional time-domain OCT technology, a radial pattern of scan line capture is often used as a compromise between acquisition time and imaging density. Even with this compromise, time-domain OCT instruments require the subject to maintain steady fixation for many seconds at a time, which may be difficult for patients with macular diseases [15, 38]. Therefore, use of an external fixation light for the fellow eye has been advocated when the acuity of the eye being examined is less than 20/300 [38].

Software included with conventional OCT instruments attempts to measure the thickness of the retina by first identifying the anteriormost highly reflective layer (inner border of neurosensory retina) and then identifying the posterior extent of the retina just anterior to the highly reflective retinal pigment epithelium (RPE) layer [15]. Many studies have verified the reproducibility of these OCT measurements by acquiring several measurements of the same eye at the same visit [3, 6, 9, 25, 28, 29, 45, 49, 50, 56, 58, 112]. OCT measurements also appear to be reproducible in patients with diabetes. Baseline differences in retinal thickness measurements between normal subjects and diabetic patients without clinically significant macular edema (CSME) may exist and be subject to diurnal variation [9, 25, 43, 82]. Male gender, high BMI, and longer axial lengths may also be associated with increased retinal-thickness measurements and may need to be considered in clinical studies utilizing OCT [112]. Investigators have also found that high-resolution scans acquired after mydriasis provide more consistent results [75], and total macular volume measurements may be more consistent than foveal thickness values [6].

These studies provide evidence that OCT is both accurate and reproducible. Quantitative measurements of retinal thickness provide clinicians with actual numbers which can be used to treat patients with complex diseases. In addition, the images themselves are an excellent way to assess and document clinical status and help patients to understand and visualize their diseases.

12.2.2 Biomicroscopic Examination

Essentials
- OCT may be more sensitive for the detection of CSME than non-contact or contact-lens biomicroscopy
- Diffuse or subclinical (between 200 and 300 μm) retinal thickening and absence of hard exudates in the central macula both predict failure of slit-lamp biomicroscopy for the detection of CSME
- Diagnosis by OCT cannot yet replace the clinical diagnosis of CSME

Macular edema is a common final pathway for many retinal vascular diseases, and its clinical detection is critical for the prevention of vision loss. One of the best studied examples of this is CSME. The clinical definition of CSME is based on the assessment of the diabetic fundus with stereoscopic slit-lamp biomicroscopy through a pharmacologically dilated pupil [18]. Since non-contact lens biomicroscopy may fail to detect CSME in as many as 10 % of cases, many clinicians advocate the use of contact lens biomicroscopy to prevent blinking and eye movements, increase axial magnification, and facilitate the detection of subtle areas of edema [5, 10]. Regardless of the technique chosen, the results of this time-consuming examination are subjective, ephemeral, and at best semiquantitative.

OCT, on the other hand, is a high-resolution, cross-sectional, objective, and reproducible method of documenting the retinal thickness and morphology of macular diseases. Several studies suggest that OCT may be more sensitive than slit-lamp biomicroscopy for detecting small changes in retinal thickness [28, 54, 91, 116]. Lattanzio and colleagues documented a progressive increase in mean foveal thickness in patient groups advancing from no edema (228 μm) to non-clinically significant macular edema (322 μm) to clinically significant macular edema (476 μm) [44]. This suggests that OCT's graded measurements may be more sensitive to small changes than categorical assessments by clinicians. Yasukawa et al. found that all eyes with CSME by non-contact biomicroscopy had detectable retinal thickening by OCT, whereas almost one-third of the eyes graded clinically as non-edematous had retinal thickening by OCT [117]. Brown et al. discovered that even with contact lens examination, more than half of cases with retinal thickness greater than 200 μm by OCT had clinically undetectable retinal thickening and almost one-fifth of cases with retinal thickness greater than 300 μm were also undetectable [5]. In their final assessment, they found that almost one-fourth of diabetic macular edema (DME) cases discovered with OCT had undetectable thickening on clinical examination. Other investigators have also found that OCT has a higher sensitivity than clinical examination for the detection of other features of diabetic retinopathy such as intraretinal cysts and subretinal fluid [67].

Investigators using OCT as the gold-standard definition of DME have found that diffuse thickening, thickening between 200 and 300 μm, and absence of hard exudates in the central macula are all predictors for the failed detection of CSME by slit-lamp biomicroscopy [5, 10, 28, 116]. Since more than half of the errors in the diagnosis of DME may be on the conservative side [8], some clinicians advocate the use of OCT if macular edema is not detected using standard clinical methods [5]. However, Browning also found a substantial percentage of patients undergoing focal laser therapy for CSME who did not have a single OCT zone thickened beyond normal ranges [7].

12 II

Since OCT was developed after the completion of the Early Treatment Diabetic Retinopathy Study, we do not have robust evidence for its use as an adjunct or replacement for slit-lamp biomicroscopy, nor do we have guidance on treatment recommendations for subclinical or OCT-evident macular edema. Nevertheless, even in the absence of proof from robust clinical trials, OCT is still becoming a standard of clinical practice for the evaluation of macular diseases, and will likely become a permanent feature in treatment and management decisions for diabetic macular edema.

12.2.3 Stereoscopic Fundus Photography

Essentials
- Reading center protocols for fundus image assessment are not typically accessible to the clinician in practice
- OCT may be a more objective and quantitative method of standardizing retinal thickness assessments both in clinical trials and in clinical practice
- Diagnosis by OCT cannot yet replace the photographic diagnosis of CSME

Fundus photograph grading systems developed for the purposes of clinical trials provide a reproducible, semiquantitative assessment of disease stage, but are relatively complicated, time-consuming, and difficult to use in daily practice [25]. The excellent spatial resolution of conventional fundus imaging makes it unlikely to be supplanted in the foreseeable future. However, variability in the axial resolution of stereoscopic photographs, either from photographic technique or image quality, calls into question the advisability of basing future macular edema trials on this imperfect analysis. Furthermore, the continued use of reading center evaluations in clinical trials propagates a fundamental disconnect between clinical trials and clinical practice: conclusions from many trials are based on photographic examinations at centralized reading centers, while clinical treatment decisions are made by physicians in practice who have neither the time nor the specific training to implement reading center protocols. Therefore, conclusions and recommendations from those studies may be improperly applied to patient care, directing some patients to unnecessary treatments and others to improper diagnosis or monitoring [8].

OCT is a widely available method of standardizing clinical data collection, and it is a more objective and quantitative modality than fundus photography. Current OCT instrumentation does not possess the spatial resolution of conventional fundus imaging devices, but it can render cross-sectional and topographic findings in greater detail. In addition, the standardized numerical results provided by OCT examinations have the potential to facilitate both the conduct of research as well as the ultimate application of its recommendations to clinical practice. Strom and colleagues recently attempted to bridge the gap between traditional reading center assessments and modern OCT examinations by demonstrating good agreement between stereoscopic fundus photograph grading and OCT measurements for the detection of diabetic macular edema [98]. Although they suggest that stereo fundus photographs may be more sensitive than OCT for the detection of retinal thickening, it is likely that reproducible, quantitative results will ultimately establish OCT as a better objective evaluation of retinal thickness than expert subjective assessments of fundus images.

12.2.4 Fluorescein Angiography

Essentials
- Fluorescein angiography and OCT each provide unique pieces of information that may be necessary for the appropriate management of patients with retinal vascular diseases
- OCT retinal thickness measurements may correlate better with visual acuity in DME than fluorescein angiography
- OCT retinal thickness measurements may not correlate well with visual acuity in venous occlusive disease

Fluorescein angiography (FA) is a highly effective method of evaluating retinal vascular disorders because it provides important information about macular perfusion and patterns of leakage. Many morphologic changes, such as intraretinal edema and neurosensory retinal detachments, are not specifically imaged with angiography, although the dynamic process of fluorescein leakage adequately demonstrates the underlying compromise of the blood-retinal barrier and disturbances in the equilibrium between fluid extravasation and reabsorption [67]. Furthermore, current methods of FA interpretation require expert knowledge and are subjective, qualitative, and time-consuming [32, 38].

OCT provides unparalleled cross-sectional rendering of in situ retinal morphologic changes such as cystoid edema and subretinal fluid. Although existing OCT instruments do not provide dynamic information about fluid movement within or underneath the retina, they automatically extract reproducible measurements of retinal thickness and volume. In fact, many investigators have found that these parameters may have a stronger correlation with visual acuity in DME than FA findings [11, 25, 28, 29, 35, 44, 59, 66, 70]. Other groups have found close associations between FA and OCT findings [2]. One study by Kang and colleagues found good correlations between focal leakage on FA and intraretinal edema on OCT, as well as between diffuse cystoid leakage on FA and outer retinal or subretinal fluid accumulation on OCT [39]. In contrast to these observations, studies in venous occlusive disease have failed to demonstrate strong correlations between OCT measurements and FA findings [11, 45].

Therefore, it appears that FA and OCT each have a unique role in the assessment of retinal vascular diseases. Treatment decisions that are based on the presence of fluid within or under the retina may be better assessed by OCT, whereas disease entities characterized by dynamic leakage may be better assessed with angiography. Since neither modality currently provides a complete answer, concurrent FA and OCT may be required to sufficiently assess the extent of damage from retinal vascular diseases. However, further developments in OCT technology, such as Doppler OCT and OCT angiography, may provide a comprehensive solution in a single imaging modality.

12.2.5 Ultrasound

Essentials
- The axial resolution of noninvasive B-scan ultrasound is less than OCT
- Echography may be an option for the detection of macular edema in cases with poor posterior pole visibility or poor patient cooperation

B-scan ultrasonography is a useful method for imaging the vitreous, the peripheral retina, and the posterior pole in cases in which media opacities or patient cooperation prohibit adequate visualization through the pupil. The resolution of clinical echography devices, however, is limited by their operating frequency, and cannot approach the resolution of OCT when used to noninvasively image the posterior pole. In one study by Lai et al., B-scan ultrasonography was shown to have a high degree of sensitivity (91%) and specificity (96%) for detecting retinal thickening when compared to OCT measurements [41]. However, this same study demonstrated only a mod-

erate correlation ($r = 0.65$) between qualitative thickness measurements made with the ultrasound device and quantitative measurements made with OCT [41]. Therefore, echography has not been shown to be as effective at quantifying macular thickening as OCT, and should most likely be reserved for use in the select cases outlined above.

12.3 Diagnosis

OCT is a useful tool for the diagnosis and management of retinal vascular disorders because of its ability to measure retinal thickness and render intraretinal details. When coupled with other imaging modalities, such as FA, OCT may also be helpful in understanding the pathogenesis of the disorder and in optimizing treatments. OCT may also be useful for the diagnosis of macular edema in the presence of media opacities, such as asteroid hyalosis, that prevent accurate biomicroscopic assessment [7]. Furthermore, diagnoses such as the vitreomacular traction syndrome that may have been missed in past decades are readily apparent with cross-sectional imaging. And, even in cases with easily observable findings, OCT can be used to establish a baseline macular volume and monitor subsequent responses to treatment. Serial measurements can then be compared to values obtained from a normal population to enable earlier and more accurate diagnosis and treatment [28].

Retinal vascular disorders typically produce three categories of findings on OCT: retinal thickening, intraretinal cystoid changes, and/or serous detachments of the neurosensory retina. The pattern of findings on OCT can be helpful in making a diagnosis or in determining the patient's prognosis, and will be discussed in more detail below [11, 70].

12.3.1 Intraretinal Edema

Essentials
- Intraretinal edema is evident as retinal thickening associated with decreased optical backscattering
- Branch retinal vein occlusions produce a characteristic asymmetric pattern of swelling
- Retinal edema by OCT should be scrutinized for evidence of vitreomacular traction

Edema within the retina can be detected on OCT as an area of retinal thickening associated with decreased optical backscattering [11, 15, 28, 66]. When displayed in false color, these areas appear

12 **II**

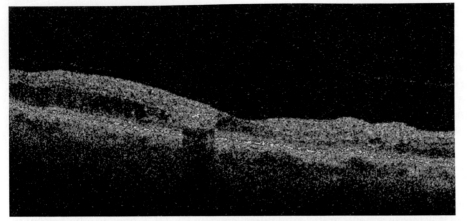

Fig. 12.1. Outer retinal edema is evident in the left half of this B-scan as decreased optical backscattering (*dark color*) in an area of retinal thickening. This can be compared to the right half of the scan, which is not as thick and has a brighter signal in the outer retina

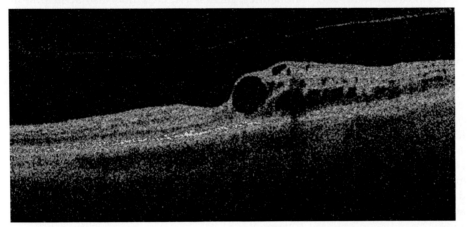

Fig. 12.2. Abrupt retinal thickening, such as this juxtafoveal cystoid macular edema, respecting a meridian is characteristic of branch retinal vein occlusions. A detached posterior hyaloid face is also evident on this image as a thin, somewhat discontinuous line above the vitreoretinal interface

darker than adjacent, nonedematous tissue (Fig. 12.1). If the edema occurs adjacent to or within the fovea, the typical foveal depression may be reduced, eliminated, or even inverted. Branch retinal vein occlusions (BRVOs) produce a characteristic, abrupt disparity in retinal thickness along the horizontal raphe that is characteristic of the asymmetric edema seen in this disorder (Fig. 12.2).

Otani's definition of outer, sponge-like swelling further specifies that intraretinal edema is typically ill-defined and widespread [66]. This is in contrast to cystoid macular edema (discussed below), which may have more focal and well-defined areas of involvement. Thickening from primary intraretinal edema should be differentiated from the thickening that may occur secondary to retinal traction [15]. A brief examination of the vitreoretinal interface for the presence of an epiretinal membrane or evidence of traction from the posterior vitreous should elucidate this cause.

12.3.2 Cystoid Macular Edema

Essentials
- Cystoid spaces are round, hyporeflective regions on OCT, typically in the outer retinal layers
- Pseudophakic cystoid changes may occur in the inner retina
- Findings on OCT correlate relatively well with fluorescein angiography
- Cystoid changes are spread evenly across the macula in CRVO, while they respect the horizontal raphe in BRVO
- Foveal cystoid changes may be a feature of idiopathic juxtafoveal telangiectasis

Cystoid macular edema (CME) may be a common final pathway for many retinal vascular diseases including diabetic macular edema, retinal vein occlusions, hypertensive retinopathy, and idiopathic juxtafoveal telangiectasis. Histopathologic reports indicate that intraretinal cystoid spaces in CME may vary in their location and size according to the

II 12

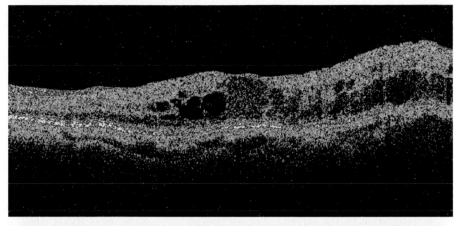

Fig. 12.3. Round, hyporeflective regions in the outer retina are indicative of cystoid macular edema. This can be compared to the more diffuse, non-cystoid outer retinal edema present to the right

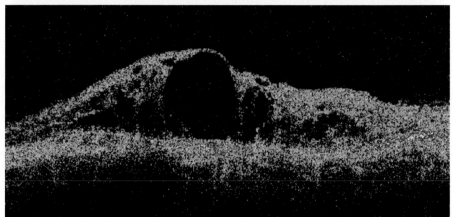

Fig. 12.4. Large cystoid spaces, such as this region in the fovea, can span the entire thickness of the retina

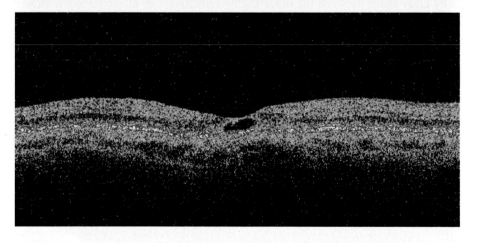

Fig. 12.5. Foveal and juxtafoveal cystoid spaces may be found in isolation in idiopathic juxtafoveal telangiectasis

underlying disease etiology [104]. This can now be corroborated in situ by OCT imaging in which cystoid spaces appear as round, hyporeflective regions within the neurosensory retina (Fig. 12.3) [15]. These features typically occur in the outer retinal layers, but when they increase in size, they can span the entire thickness of the retina and extend to the internal limiting membrane (Fig. 12.4) [28]. Pseudophakic cystoid macular edema is a notable exception because hyporeflective spaces may occur in the inner retina instead of the outer retinal layers [11]. Small cystoid spaces often have well-defined boundaries, while larger spaces may have somewhat diffuse boundaries [11, 28].

The distribution of cystoid spaces within the retina can be a clue to the underlying disease etiology. For example, cystoid spaces in CRVO tend to be spread evenly across the macula, whereas cystoid spaces in BRVO typically respect the horizontal raphe (Fig. 12.2) [55]. And although CME usually occurs in association with sponge-like retinal swelling [70], it may appear in isolation in certain diseases such as idiopathic juxtafoveal telangiectasis. In this syndrome, small, often horizontally oriented, oblong cystoid spaces are found in the fovea or temporal inner retina on OCT, yet may be associated with a normal foveal contour (Fig. 12.5) [55].

12.3.3 Serous Retinal Detachment

Essentials
- Defined as fluid separating the neurosensory retina from the RPE
- Evident on OCT as an optically clear space between the retina and RPE
- May be undetectable using biomicroscopy or FA
- May be present in up to one-third of DME cases, more than two-thirds of BRVOs, and eight out of ten CRVOs
- Subfoveal fluid may be associated with slower recovery of vision and resorption of retinal edema

Serous retinal detachment (SRD) occurs when fluid separates the neurosensory retina from the RPE. It can be recognized on OCT by the presence of an optically clear space between the inner highly reflective line believed to correspond to the photoreceptor outer segments and the outer highly reflective line corresponding to the RPE (Fig. 12.6).

Although SRDs are most commonly observed in diseases affecting the choroid and the RPE (e.g., choroidal neovascularization or central serous choroidopathy), they can also be a clinically significant feature of retinal vascular diseases. In contrast to their striking appearance on OCT, SRDs in retinal vascular diseases are frequently not apparent biomicro-

scopically or by angiography [69]. The relative insensitivity of biomicroscopy and fluorescein angiography in detecting SRDs is likely due in part to concomitant retinal pathology (e.g., retinal edema) in these patients. Nevertheless, this insensitivity suggests that routine OCT imaging of patients with these diseases may assist the clinician in developing a more complete morphological description of the retinal vascular disease features present in a given patient.

In studies of patients with diabetic macular edema, SRDs have been reported to be present on OCT scans in 15–31 % of cases [11, 39, 69, 70]. SRDs, however, appear to be more frequent in patients with venous occlusive disease, and are present in up to 82 % of CRVOs [68] and 71 % of BRVOs [95]. Although the mechanism of subretinal fluid accumulation is not entirely certain, the phenomenon has been studied in histopathologic series. Wolter and coworkers hypothesized that the deterioration of RPE function due to inflammation or associated ischemia could play a role [111]. The rapidity of fluid movement out of the intravascular space has also been postulated to play an important role, particularly in venous occlusive disease. Concomitant vitreomacular traction has also been proposed as a contributing factor in the accumulation of subretinal fluid in some cases.

Regardless of the mechanism, the recognition of an SRD appears to have clinical significance. Ohashi et al. observed a slower resorption of intraretinal edema and a slower, attenuated recovery of vision in patients with subfoveal SRD [60]. Some investigators have also suggested incorporating pars plana vitrectomy (PPV) into the management of patients with DME associated with SRD. Interestingly, the presence or thickness of subretinal fluid does not appear to correlate with the thickness of the overlying retina. Furthermore, there does not appear to be a difference in visual acuity whether or not there are cystoid changes in the retina associated with SRD [11].

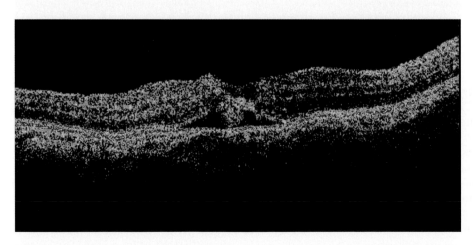

Fig. 12.6. Subretinal fluid, such as this subfoveal pocket, creates an optically clear space anterior to the highly reflective RPE band but external to the photoreceptor outer segments

12.3.4 Vitreomacular Traction Syndrome

Essentials
- OCT has greatly improved the evaluation and description of the vitreomacular interface
- Partial posterior vitreous detachments may be more common in patients with DME
- Epiretinal membranes may be depicted in better detail with OCT
- Measurement of vitreomacular traction may be useful in managing patients with retinal vascular diseases

Retinal vascular diseases are frequently associated with breakdown of the blood-retinal barrier. Fluorophotometry studies suggest that transudate from retinal vessels can migrate not only into the retina, but also into the vitreous cavity [85]. Early histologic and biochemical changes in the vitreous of patients with diabetes include elevated levels of early and advanced glycation end products (AGEs) and greater amounts of collagen cross-linking [85]. Stitt observed that AGEs may provide the substrate for the collagen cross-linking observed in these patients, and suggested that these cross-links could explain the vitreous changes observed clinically in these patients [96]. Changes in the vitreous of eyes with retinal vascular disease may also contribute to the development of macular edema, particularly cystoid macular edema, by providing a substrate for traction on the retina [83, 84]. Schepens described two types of traction: traction with a narrow vitreous strand, and traction from a broad vitreoretinal adhesion [83].

Biomicroscopically, relationships between the posterior hyaloid and the retina can be difficult to visualize, particularly in patients with broad adhe-

sions. OCT has dramatically improved the evaluation of this relationship, and increased the recognition of vitreomacular traction although it may be necessary to adjust the signal-to-noise ratio of commercially available OCT instruments to accentuate the posterior hyaloid face [23, 51, 55]. In patients with a completely detached posterior vitreous, the posterior hyaloid may be visible as a thin, hyperreflective signal anterior to and separate from the retina (Fig. 12.2) [11]. This signal will be absent in patients with a completely attached posterior vitreous. In patients with a partially attached vitreous, this thin hyperreflective membrane can be observed with broad or focal insertions to the retinal surface [105]. The resulting anteroposterior distortion of the retinal contour may produce a characteristic peaked retinal appearance (Fig. 12.7) or a focal neurosensory detachment.

The superiority of OCT in imaging the vitreoretinal interface led Gaucher to define a new staging system for vitreous separation in diabetic patients, combining OCT and clinical criteria: Stage 0 was defined as the absence of a posterior vitreous detachment (PVD), with no Weiss ring visible and the posterior hyaloid not visible on OCT; stage 1, as a perifoveolar PVD with foveolar attachment (posterior hyaloid partially attached to macula on at least one OCT scan); stage 2, as an incomplete PVD with residual attachment to the optic nerve (no Weiss ring but posterior hyaloid remains visible on all OCT scans but is detached from macula); and stage 3, as a complete PVD (Weiss ring seen on funduscopy) [23]. Applying this staging system, stage 1 PVDs were found to be more frequent in patients with DME (53%) compared to those without edema (22%), consistent with the presumed importance of vitreomacular traction in DME. Catier and colleagues looked at patients with macular edema of various etiologies and concluded that patients with DME were much more likely to have a partial PVD than in other causes of mac-

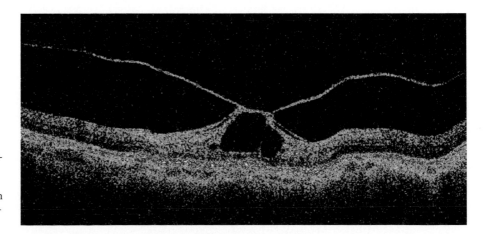

Fig. 12.7. The posterior hyaloid inserting into the fovea may produce the characteristic peaked appearance of vitreomacular traction which, as in this case, is often associated with cystoid macular edema

ular edema [11]. Of interest, however, Uchino's group found this stage of PVD in more than half of normal patients in their study [105].

Epiretinal membranes (ERMs) are another form of preretinal traction that may be present in retinal vascular diseases and can be visualized with OCT. Typically, ERMs can be distinguished from the posterior hyaloid membrane by their thicker, more consistent appearance and higher reflectivity. They may be easier to detect when separated from the inner retina, but can also be recognized by secondary retinal distortion or disruption of the normal foveal depression [55]. OCT can be used to document the opacity and thickness of an epiretinal membrane, its distance from the inner retina, its effects on the underlying retina, such as distortion, edema, or neurosensory detachment, or a patient's response to treatment.

An advanced form of epiretinal traction, fibrovascular proliferation secondary to proliferative diabetic retinopathy, can lead to two distinct anatomic configurations: traction retinal detachment and tractional retinoschisis [36]. Traction retinal detachments are diagnosed by observing optically clear subretinal spaces in the presence of epiretinal proliferation. Retinoschisis, on the other hand, can be described as a separation of the neurosensory retina into two layers connected by bridging columnar bands of tissue without accompanying subretinal fluid. Differentiating between these two conditions may be of prognostic importance in choosing the best management option [36].

Description of the vitreoretinal interface by OCT has affected the management of patients with DME. For instance, early vitrectomy may be considered in patients with evidence of significant vitreomacular traction [51, 88]. Shah and coworkers studied the prognostic significance of vitreomacular traction in 33 patients undergoing vitrectomy in their prospective study [88]. They observed that patients with evi-

dence of clinical and/or OCT macular traction showed significant improvements in postoperative visual acuity compared with patients without traction. Yamada and coworkers performed a similar study comparing the prognostic significance of two types of partial PVDs which they distinguished by OCT: (1) incomplete V-shaped detachments, and (2) partial detachments temporal to the fovea but with nasal attachment [113]. They found that anatomic outcomes appeared to be more favorable in the former group.

Beyond its ability to confirm the release of vitreomacular traction forces after vitrectomy, OCT has also been used to evaluate the resolution of neurosensory detachments and to study vitreomacular relationships following YAG capsulotomy [11, 107].

12.3.5 Miscellaneous Findings

Essentials
- Structures such as blood vessels, hemorrhages, neovascularization, and lipid often appear as hyperreflective interfaces with posterior shadowing
- Posterior shadowing may confound automated retinal thickness calculations in these areas
- Sparse sampling with conventional radial-line scanning protocols makes detection of microaneurysms difficult

Blood vessels can be identified on OCT as circular or semicircular, hyperreflective or hyporeflective regions with varying degrees of posterior shadowing (Fig. 12.8). Retinal hemorrhages, particularly if thick, may also appear as areas of hyperreflectivity with posterior shadowing (Fig. 12.9). Retinal neovascularization can present as a focal area of hyperref-

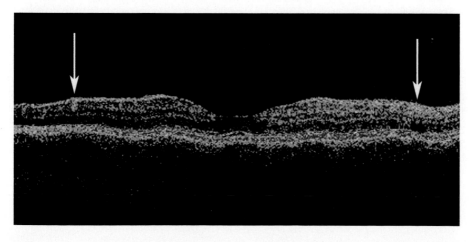

Fig. 12.8. Vessels (*white arrows*) can appear as circular or semicircular structures with marked posterior shadowing

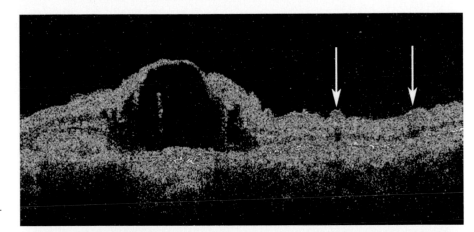

Fig. 12.9. Intraretinal hemorrhages (*white arrows*) may appear as hyperreflective regions with variable posterior shadowing

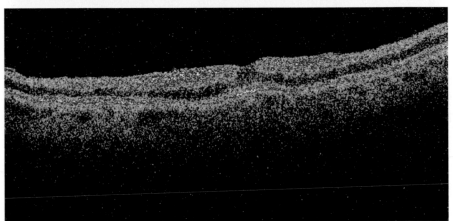

Fig. 12.10. This patient presented 2 weeks after the onset of acute, severe, painless vision loss. A cherry red spot was present on clinical examination. The patient was diagnosed with a central retinal artery occlusion. Note the hyperreflective inner retina and absence of recognizable inner retinal layers

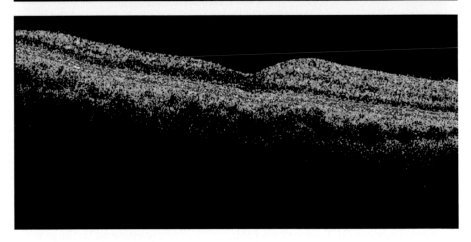

Fig. 12.11. This patient with Susac syndrome suffered repeated branch retinal artery occlusions (BRAO). Chronic retinal atrophy from an old BRAO is evident by comparing the retinal thickness and structure on the left (abnormal) versus the right

lectivity on the retinal surface, but may be difficult to distinguish from a non-vascularized epiretinal membrane. Inner retinal ischemia may be evident on OCT as increased reflectivity in the nerve fiber layer and inner retinal layers (Fig. 12.10). Although retinal artery occlusions may have this appearance in the acute phase, they often progress to longer-term retinal atrophy (Fig. 12.11).

Traditionally, other features typical of retinal vascular diseases such as microaneurysms and intrare-tinal microvascular abnormalities have been difficult to visualize on OCT [28]. This is likely due, at least in part, to the poor sampling density of the peripheral macula that is characteristic of conventional radial-line scanning protocols. Next-generation OCT devices employing high-speed, Fourier-domain technology can overcome this limitation by significantly increasing the density of the scans in the macula, while also increasing the axial resolution of the image (Fig. 12.12).

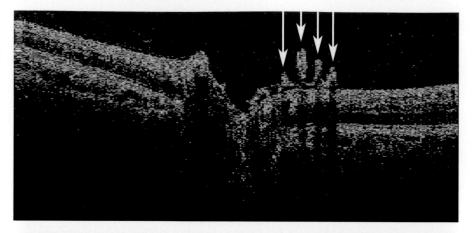

Fig. 12.12. Fourier domain OCT scan along the edge of the optic nerve demonstrating neovascularization of the disc (*white arrows*) in this patient with proliferative diabetic retinopathy

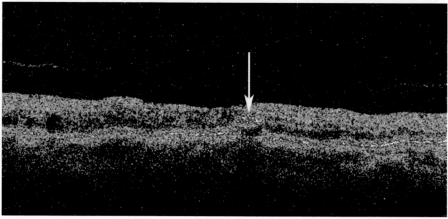

Fig. 12.13. Juxtafoveal lipid exudation (*white arrow*) in this patient with diabetic retinopathy often casts a dark or absolute shadow in the outer retina

Exudations of lipid and protein are common features of retinal vascular diseases. Due to their relative optical opacity, hard exudates, which accumulate in the retina, can be frequently identified on OCT images as focal areas of high optical reflectivity with dense shadowing that extends posteriorly (Fig. 12.13) [28, 70]. Although they occur most commonly in the outer plexiform layer, hard exudates can be observed in almost any retinal layer as well as in the subretinal space. One important consequence of the posterior shadowing caused by these and other structures is obscuration of the posterior retinal boundary which, in turn, can confound calculations of the retinal thickness by naive OCT algorithms [28].

12.4 Management

OCT has many potential applications in the management of patients with retinal vascular disease, particularly in patients with macular edema. Quantification of the extent of retinal thickening by OCT can assist in the selection of patients who may benefit from treatment, identify which treatment is indicated, guide the application of the treatment, and allow precise monitoring of the response to treatment [28,

70]. The morphologic detail available in OCT scans can also be used to define classification or staging systems for retinal disease which, in turn, can guide management. Panozo et al. devised a classification system for DME which takes into account five parameters: retinal thickness, diffusion, volume, morphology, and presence of vitreous traction [71]. Applying this classification system, in a retrospective study of 169 patients undergoing one of three treatments (laser, vitrectomy, or intravitreal triamcinolone) for diabetic macular edema, patients with earlier OCT stages (e.g., diffuse edema without subretinal fluid) were observed to have better outcomes compared to those with more advanced stages [72]. OCT findings may also have prognostic implications which can affect treatment decisions. For instance, in patients with diabetic retinopathy and good vision (20/20) undergoing panretinal photocoagulation (PRP), patients with preoperative parafoveal thickness higher than 300 μm had a higher incidence of persistent decreased vision after PRP [92]. This observation may help identify a group of patients who may benefit from a less aggressive laser strategy, with a staged delivery of treatment and a longer interval between PRP sessions.

12.4.1 Focal and Panretinal Photocoagulation

Essentials
- OCT does not yet replace the clinical definitions of CSME
- It may be possible to exclude some patients with clinical foveal edema from unnecessary laser treatment by demonstrating normal foveal retinal thickness by OCT
- Focal laser can cause a significant reduction in OCT-measured retinal thickness
- PRP may cause a transient increase in macular thickening by OCT

The Early Treatment Diabetic Retinopathy Study (ETDRS) demonstrated that focal laser photocoagulation could reduce the risk of moderate vision loss (defined as a doubling of the visual angle) in patients with CSME diagnosed by contact lens biomicroscopy and stereoscopic fundus photography [18]. As described in Sect. 12.2.2, correlating OCT measurements with existing definitions of CSME has proven difficult due to the limitations in existing scanning protocols and analysis algorithms, which are designed to provide a dense sampling of the fovea, but rely on extensive interpolation in the extrafoveal areas that may be involved by CSME. The development of grid scanning protocols for the Stratus OCT [80] and dense scanning patterns facilitated by FD-OCT may eventually address this limitation for non-foveal zones. In the interim, many clinicians have utilized OCT for assessment of foveal edema not detected by the clinician [5, 8]. It is uncertain, however, whether patients with OCT-detected "subclinical edema" are at risk for moderate vision loss or whether these patients would benefit from focal laser therapy. Brown et al. deemed that "it does not seem reasonable at this time to extrapolate the results of the Early Treatment Diabetic Retinopathy Study to cases of subclinical edema" [5]. OCT may be of some value, however, in excluding patients who in fact do not have significant foveal thickening (i.e., a false-positive clinical exam), thereby averting the risks of treatment toxicities such as paracentral scotoma or choroidal neovascularization (CNV) [8].

Although the ultimate utility of OCT in the identification of candidates for focal laser treatment is still uncertain, OCT has been used in studies evaluating the anatomic effects on the retina following photocoagulation [44]. In a series by Rivellesse et al., patients undergoing focal laser for diabetic CSME had a significant reduction in retinal thickness after therapy [71]. The anatomic effects of laser on macular thickness have also been studied in patients receiving PRP for proliferative diabetic retinopathy. Shimura et al. evaluated 72 eyes of 36 patients undergoing PRP treatment spread over 4 sessions [90]. Patients were treated weekly in one eye and biweekly in the other. Macular thickness by OCT was observed to transiently increase in both eyes, but to a greater degree in weekly treated eyes. Recovery of macular thickening to normal levels was also faster in biweekly treated eyes, but no statistically significant difference in visual outcomes was observed between the two groups. Although this was a small series, the observations are consistent with previous clinical observations in the ETDRS that showed that macular edema may worsen after PRP [19]. The OCT findings may also explain the apparent reduction in macular function on multifocal electroretinography (mfERG) in the absence of a change in visual acuity observed by Lovestam-Adrian [46]. Although Lovestam did not observe changes on OCT, OCTs were not obtained in the immediate perioperative period, but only 6 months after PRP.

12.4.2 Pharmacologic Therapy

Essentials
- OCT is useful in monitoring the response to pharmacologic therapies for retinal vascular disease
- Changes in OCT appear to correlate well with changes in angiographic leakage
- OCT retinal thickness parameters are becoming important as entry criteria and outcome variables in clinical trials for retinal vascular disease
- Longitudinal changes in OCT are useful for guiding retreatment decisions

The development of pharmacotherapies for retinal vascular disease has been a major focus of clinical research since the discovery of the role of vascular endothelial growth factor (VEGF) in disease pathophysiology. Anatomic improvements in retinal thickness on OCT have been an important variable in studies evaluating the effectiveness of these agents.

Although it likely has multiple mechanisms of action in addition to its anti-VEGF effect, intravitreal triamcinolone acetonide (IVTA) has been widely used for the treatment of various retinal vascular diseases. A number of small studies have demonstrated a reduction in central retinal thickening (foveal center thickness or average foveal subfield thickness) following a single intravitreal injection of triamcinolone in patients with macular edema associated with diabetes [14, 37, 42, 47, 52] and retinal venous occlusive disease [108]. Follow-up was variable in these studies, but a

50% reduction in central thickening on OCT was a typical response by 3 months after treatment [47]. In some series, the reduction in retinal thickness correlated with reduced leakage on angiography and an improvement in visual acuity [53]. OCT has also been used to document the time course of retinal thickness changes following triamcinolone injection, and has demonstrated that macular edema tends to recur after 3 months [99]. Recurrence of edema as identified by OCT is a useful criterion for clinicians when considering potential retreatment. Large-scale randomized clinical trials are currently in progress (SCORE, DRCR) to evaluate the efficacy of IVTA for patients with macular edema from retinal vascular disease. Retinal thickness measurements from OCT are being used as outcome variables as well as criteria for enrollment and retreatment in these studies.

Studies using OCT findings as outcome variables are currently underway for other anti-angiogenic therapies (including pegaptanib, ranibizumab, and bevacizumab) used for retinal vascular disease. Rosenfeld reported an improvement from 20/200 to 20/50 with resolution of CME in one patient with a CRVO just 1 week after an intravitreal injection of 1.0 mg of bevacizumab (Avastin) [79]. A randomized, sham-controlled study of 172 patients with DME (two-thirds of whom received treatment) revealed a mean reduction of OCT central retinal thickness of 68 μm and a modest visual benefit in patients receiving a 0.3 mg dose of pegaptanib [16].

12.4.3 Surgery

Surgical approaches for retinal vascular diseases have gained increasing popularity and OCT findings have proven to be useful not only in monitoring the response to therapy, but also in selecting suitable candidates for surgery.

12.4.3.1 Pars Plana Vitrectomy for Macular Edema

Essentials

- Detection of vitreomacular traction by OCT aids in the selection of patients for vitrectomy
- Postoperative OCT is used to confirm normalization of retinal morphology following surgery
- Diffuse diabetic macular edema on OCT, even in the absence of traction, may benefit from vitrectomy
- OCT-detected neurosensory detachment may predict a higher risk of postoperative subfoveal lipid exudates

Pars plana vitrectomy has been advocated as a potential treatment for patients with refractory DME, particularly in patients with suspected VMT. OCT has revolutionized the selection of patients for vitrectomy by providing definitive evidence of tractional effects on the macula. Giovannini et al. identified two patterns of maculopathy in patients with VMT: a thickening of the superior profile of the OCT tomogram, or the disappearance and inversion of the physiologic foveal depression [24]. Following surgery, many of the eyes in this series showed a normalization of the retinal morphology. Additional retrospective series have demonstrated similar results with surgery, showing varying but significant reductions in OCT foveal center retinal thickness following vitrectomy [64, 78, 97]. Several studies have also shown functional improvements [24, 73, 78] in visual acuity and multifocal electroretinography [101, 102] in conjunction with the apparent anatomic benefits of reduction in retinal edema evident on OCT.

Other investigators have advocated the use of PPV for diffuse refractory DME, even in the absence of OCT evidence of macular traction [73, 100]. Patel and coworkers observed a reduction in OCT retinal thickness in patients undergoing PPV for diffuse DME, including a small subgroup of patients who had diffuse low-lying elevation of the retina and no vitreous traction [73]. Despite these favorable anatomic results demonstrated by OCT, the true benefit and role of PPV in diffuse DME remains uncertain. In a randomized trial of vitrectomy versus additional focal laser in patients with persistent (following previous laser) diabetic macular edema without vitreous traction, surgery was not shown to be better than laser in terms of anatomic (macular thickness) or functional (visual acuity) outcomes [100].

In addition to the lack of convincing randomized clinical trial data to support its use, PPV for DME is not without significant risk. Although reduced OCT retinal thickness was observed, Yamamoto and colleagues observed a variety of intra- and postoperative complications including retinal tears, retinal detachment, vitreous hemorrhage, neovascular glaucoma, epiretinal membrane formation, and lamellar macular hole [115]. Another complication following surgery that was observed by Yamamoto as well as by Otani was the accumulation of hard exudates in the subretinal space, a finding typically associated with a worse visual prognosis [65]. Otani also noticed that the presence of a serous neurosensory detachment on OCT before or after surgery was associated with an increased tendency toward the formation of these subfoveal exudates.

The use of OCT to quantify reduction in macular edema following vitrectomy has not been restricted to patients with diabetic retinopathy. Sekiryu and coworkers demonstrated dramatic reductions in OCT retinal thickness in a small series of patients undergoing PPV for massive foveal cystoid macular edema associated with CRVOs [86].

12.4.3.2 Radial Optic Neurotomy for Central Retinal Vein Occlusions

Essentials
- OCT can demonstrate reduction in macular edema following radial optic neurotomy
- Dramatic reductions in edema can be observed on OCT within a few months of surgery
- Morphologic response on OCT often does not correlate with visual improvement
- Retinal ischemia may limit the therapeutic benefit

Ompremcak and coworkers have advocated the use of radial optic neurotomy (RON) for treating patients with persistent macular edema associated with central retinal venous occlusive disease [62, 63]. The procedure involves the use of a blade to make an incision at the nasal edge of the optic nerve. Although the mechanism of action is unknown, many investigators believe that this procedure may increase the development of collateral vessels that serve to decompress the venous bed [20, 22, 94, 103]. In a series of 117 consecutive patients with CRVO and severe visual loss, Ompremcak and coworkers observed anatomical resolution in 95 % and visual improvement in 71 % of patients [63].

OCT has been a useful tool in monitoring the therapeutic response to RON as well as observing postoperative complications such as a macular hole [93]. In several small series, dramatic reductions in OCT retinal thickness were observed by several months after surgery [48, 57, 74]. These beneficial anatomic results, however, often did not correlate with improvements in visual acuity, particularly in patients with ischemic occlusions [48].

12.4.3.3 Arteriovenous Sheathotomy for Branch Retinal Vein Occlusion

Essentials
- Manipulation of the arteriovenous sheath (sheathotomy) may relieve persistent macular edema in BRVO
- Sheathotomy may relieve venous compression and mobilize thrombus

Essentials
- Dramatic reductions in macular edema have been observed on OCT
- OCT edema reduction may be rapid, and visible by first postoperative day

Several investigators have also advocated the use of arteriovenous sheathotomy for treating persistent macular edema in patients with BRVO [61, 87]. Supporters of this approach believe that manipulation of the common adventitial sheath shared by the retinal artery and vein at arteriovenous crossing points can relieve the compression of the venous lumen and mobilize the presumed thrombus. Rapid reperfusion of the retinal circulation, including at the time of surgery, has been demonstrated by these investigators. Fujii et al. demonstrated a rapid, dramatic, and sustained reduction (from 450 to 228 μm) in retinal edema by OCT and an improvement in visual acuity, beginning within 1 day after the sheathotomy in a patient with a BRVO [21].

12.5 Future Directions

Despite the tremendous advances in hardware design over the last decade, conventional OCT devices acquire a relatively limited data set and use outdated software fraught with imperfections. Discoveries in the realm of OCT hardware have far outpaced advancements in OCT software. The clinical utility and success of the next generation OCT devices are dependent on careful attention to both hardware and software optimization. Future OCT devices will also likely have broad scale applicability, not only in ophthalmology, but also in many other fields of medicine including cardiology, gastroenterology, oncology, orthopedics, otolaryngology and possibly urology.

12.5.1 Current OCT Limitations

Essentials
- Scan acquisition speed and fixation instability can compromise OCT thickness measurements, especially in the foveal center
- Low sampling density of existing radial line scan patterns requires interpolation of data over 95 % of the area being mapped
- The outer retinal boundary is consistently detected incorrectly by current StratusOCT algorithms
- Moderate and severe errors in retinal boundary detection are frequent, especially in cases with subretinal disease

12 II

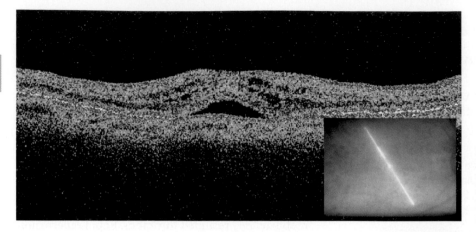

Fig. 12.14. Video image (*inset*) shows radial line passing through the center of this full-thickness macular hole, but the OCT image shows a tangential slice

Although OCT technology has spearheaded a quantitative revolution in the diagnosis and management of retinal disease, there are important hardware and software limitations of which the clinician must be aware when interpreting OCT studies. An important limitation of current StratusOCT hardware is the scan acquisition speed. Each "high-resolution" 6-mm macular radial line scan composed of 512 A-scans takes more than 1 s to be acquired. Consequently, the accuracy of the line scan may be affected by eye movements, particularly in patients with macular diseases associated with poor fixation. Fixation instability also hampers registration and correlation of OCT line scans with other imaging modalities such as color fundus photographs or fluorescein angiograms. Although an infrared image of the fundus indicating the approximate location of the line scan is obtained at the end of every line scan, the precise fundus location of the scan is not provided. This limitation is illustrated by Fig. 12.14 in which the infrared image suggests that the line scan passes through the center of an apparent full-thickness macular hole, but the corresponding OCT line scan actually passes through an adjacent area of retina that fails to include the macular hole.

The limitation presented by slow scan acquisition speed is further accentuated when multiple radial line scans are obtained to generate a retinal thickness map. Fixation instability results in misalignment of the centerpoints of the line scans and an inaccurate center retinal thickness. To address this problem, faster scanning protocols such as the Fast Macular Thickness Scan have been developed, but these sacrifice transverse resolution (only 128 A-scans per 6-mm radial line) for higher scanning speed. Slow scan speeds also limit the number of line scans that can be obtained for constructing macular thickness maps, resulting in the need for significant interpolation between sample points (particularly in the peripheral macula). In fact, typical radial-line protocols mea-

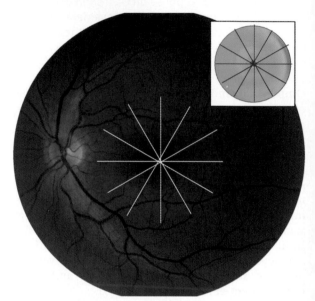

Fig. 12.15. Radial line scanning protocols (*white lines*) cover only a small portion (<5% shown as *red lines on inset map*) of the retinal area being mapped

sure less than 5% of the retinal area being mapped (Fig. 12.15). This translates into interpolated values covering 95% of the area being measured. Use of alternative scan patterns such as concentric circles rather than radial lines (Fig. 12.16) may provide more even sampling of the retina and reduce interpolation errors, but is still subject to registration errors related to the slow scanning speed of time-domain OCT.

C-scan OCT has been proposed as a potential hardware solution to the registration problems inherent in StratusOCT. The images obtained by C-scan OCT, however, present retinal anatomy in an orientation that is unfamiliar to most clinicians, thereby limiting the usefulness of this approach.

The StratusOCT analysis software also has significant shortcomings. For example, the generation of a

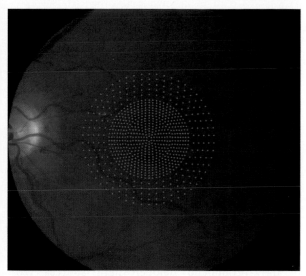

Fig. 12.16. Macular grid scanning protocol for OCT. Unlike radial line protocols, which concentrate the sampled spots along the six radial lines, grid scanning protocols spread the spots across the macula in a pattern of concentric rings, thereby permitting a more even and dense sampling of the macula

retinal thickness map, the key quantitative output of macular OCT scans, is dependent on accurate identification of the inner and outer retinal boundaries. Hee and coworkers have described the use of standard edge-detection algorithms for identification of the retinal boundaries in the first-generation OCT devices (OCT-1, Carl Zeiss Meditec), with the two strongest edges in each tomogram believed to correspond to the vitreoretinal interface and the retinal pigment epithelium [27]. Although the algorithms for retinal boundary detection utilized by the StratusOCT have not been published, it is believed that they may also rely in part on edge-detection routines. Although these algorithms were validated in normal patients as well as in some patients with DME and found to be reproducible, it has recently been suggested that even in normal patients, the interfaces detected by these conventional algorithms do not appear to correspond to the correct anatomic boundaries of the retina [76]. The root of this problem lies in the assumption that the first inner highly reflective layer (HRL) corresponds to the RPE interface. Recent studies suggest that the inner HRL is actually part of the neurosensory retina (corresponding to photoreceptor outer segments) and that the outer HRL is likely to be the true location of the RPE [76]. Thus, Stratus OCT likely underestimates the true retinal thickness in both normal subjects and patients with retinal disease.

Even if one accounts for this "expected" misidentification of the outer retinal border, boundary detection errors are still a frequent problem with conven-

tional OCT analysis algorithms [77]. Sadda and coworkers reviewed each of the 6 radial-line scans from 200 eyes of 200 consecutive patients undergoing OCT imaging for macular disease, and graded errors in retinal boundary detection in each line scan using a semiquantitative scale [81]. Errors in retinal boundary detection were observed in 92% of cases, with moderate or severe errors in 33%. Errors were more common in scans containing subretinal fluid or pigment epithelial detachments. Errors were less frequent, however, in patients with retinal vascular disease and macular edema, which was likely related to the lower frequency of subretinal pathology and disruption of the hyperreflective RPE interface in these patients. Subretinal fluid and vitreomacular traction, as described in this chapter, are still features of retinal vascular disease which can compromise the accuracy of conventional algorithms in these cases. Current StratusOCT software allows the clinician to manually place caliper marks and measure retinal thickness at selected locations, but does not allow for global correction of errors across the entire image or image set. In addition, because the RPE interface is defined as the outer retinal boundary, subretinal fluid is frequently (but variably) included in the measurements of retinal thickness. The failure to distinguish the subretinal space from the retinal thickness represents a loss of potential clinically useful information. For example, some therapies may preferentially affect retinal edema but not subretinal fluid. Current analysis software also does not allow quantification of other OCT features of retinal vascular disease such as retinal cysts and lipid exudates.

12.5.2 Clinical Applications

Essentials

- OCT definitions of macular edema may not correlate with traditional definitions of CSME
- OCT grid scanning protocols and region growing software can identify zones of thickening which may correlate with prior definitions of CSME
- Automated CSME-detection algorithms are more reproducible than the human gold standard
- OCT parameters could be used to define a new objective definition of CSME

Retinal thickness measurements have become an integral part of clinical management of patients with retinal vascular disease. Measurements have been used to screen patients for eligibility in clinical trials of macular edema and to monitor the response of patients to various interventions. An important area of uncertainty, however, is the relationship between OCT measurements of retinal thickness and the subjective standard of CSME defined in previous landmark clinical trials of diabetic retinopathy.

Brown and coworkers defined an entity of "subclinical edema" to distinguish retinal thickening apparent on OCT which was not detected by biomicroscopy or on stereoscopic fundus photographs [5]. It is uncertain, however, whether patients with subclinical edema would benefit from focal laser treatment. To more precisely identify and define CSME, Sadda et al. evaluated a concentric grid scanning protocol and region-growing software algorithm designed to automatically outline and quantify areas of CSME [80]. The performance of the algorithm for identification of CSME was compared against the clinical gold standard (biomicroscopy and/or stereoscopic photographs), and sensitivity and specificity was found to be over 85%. Secondary analysis revealed that cases of disagreement between the automated algorithm and the clinical assessment were due to the variability and subjectivity of the clinical gold standard.

These observations suggest that OCT parameters could potentially be used to redefine a new objective standard for CSME, which could be used in future clinical trials of macular edema and retinal vascular disease. An objective OCT metric will also facilitate comparative analyses among clinical trials.

12.5.3 The Future of OCT Hardware

Essentials
- Improvements in the axial resolution of future OCT devices will depend principally on advances in light source technology
- Continued speed optimizations will further enhance the three-dimensional depiction of tissues
- Doppler OCT may be able to bridge the gap with angiography by providing dynamic OCT information
- Functional imaging represents a new challenge for OCT imaging in the next decade

The hardware that enables ophthalmic OCT has undergone a tremendous evolution in the last 2 decades – even more so in the last 5 years. Significant improvements have occurred in both the axial resolution and speed of these devices.

Most of the improvement in axial resolution that has occurred in recent OCT devices has come from changes in the light source used for the OCT instrument [17]. The superluminescent diode (SLD), a technology which could be considered the workhorse of OCT because it combines high bandwidth with low cost, has improved steadily over the last 5 years. Ongoing research into the use of multiple SLD arrays and new SLD technologies, such as those incorporating quantum dots, will likely enhance the resolution of future OCT devices. OCT instruments based on laser light sources, such as doped rare-earth and titanium:sapphire (Ti:Al$_2$O$_3$) lasers [17], offer superb image quality but have limited commercial feasibility due to their prohibitive costs. This crucial problem could be mitigated, however, by the development of inexpensive laser systems or the use of alternative light sources. For example, several research groups are currently studying the potential of ultra-broadband halogen light sources [26]. Another exciting area of OCT hardware research involves swept source OCT [13]. In swept source OCT, the moving reference arm of time domain OCT is replaced with a sweeping frequency laser. Finally, several investigators are now attempting to integrate adaptive optics systems into OCT devices [31]. Adaptive optics systems utilize wavefront sensing technologies and deformable mirrors to remove optical aberrations from reflected light. Adaptive optics may be able to improve both the axial and transverse resolution of future OCT instruments so that specific populations of retinal cells can be evaluated.

Improvement in the scanning speed of OCT devices has been an important step forward in OCT hardware development, as it mitigates problems related to patient fixation instability and sparse sampling of the macula. Most of the speed improvements evident in the newest commercial OCT devices are due to Fourier or spectral-domain technology [1], which typically allows scan acquisition 50 times faster than with time domain devices. The faster scanning speeds possible with FD OCT instruments enable dense sampling of the fundus region being mapped. This enables pseudo-realistic three-dimensional reconstructions of fundus structures (3D OCT) and better rendering of invariant landmarks, such as blood vessels and the optic nerve. The latter capability facilitates alignment of OCT data between visits and with other fundus imaging modalities [110]. Full-field OCT devices have an even greater speed advantage than standard FD OCT [4]. By capturing spectral information an entire 'field' at a time with a two-dimensional detector instead of the standard linear array, full field OCT systems remove the time penalty of conventional systems in which the scanning spot mechanism captures a single A-scan at a time.

Other promising avenues of OCT hardware research include polarization-sensitive OCT (PS OCT), Doppler OCT, and functional OCT. PS OCT [12] takes advantage of the native polarization characteristics inherent in many ocular tissues (such as the nerve fiber layer) to improve signal differentiation. Doppler OCT utilizes the motion data encoded within the OCT signal and may allow detailed mapping of retinal blood vessels and blood flow, thereby providing information previously only accessible through invasive means such as retinal angiography [118]. Although it lacks the dynamic leakage information of fluorescein angiography, it may provide a bridge that expands the capabilities of OCT towards angiography. Finally, functional imaging represents an entirely new paradigm focusing on the health and function of tissues – not just on their structural representation. Current research efforts in functional OCT are taking advantage of the spectral capabilities that are built into the new OCT systems. Coupled with adaptive optics, functional OCT could act as a powerful new tool both for research and clinical purposes.

12.5.4 The Future of OCT Software

Essentials
- Current OCT software measurement techniques will be inadequate in the era of Fourier domain OCT
- Future OCT software needs greater accuracy and clinical knowledge integration
- Future OCT software should provide more clinically relevant measurements

Despite the giant leap toward automated quantification made by the first generation of OCT software, subsequent software advances have been slow and greatly outpaced by hardware developments. The shortcomings of currently available OCT software will become even more apparent in the era of Fourier-domain instrumentation where larger amounts of data are acquired in a shorter amount of time and a larger number of mathematical calculations are required to distill the data into clinically useful numbers. Clinicians who, in the past, could easily interpret findings from six *B-scans* during a busy clinic may not be able to review the hundreds of *B-scans* that are produced by 3D-OCT instruments in just a few seconds. Furthermore, in this era of quantitative medicine driven by objective measurements, subjective clinical assessments will be replaced by objective numerical data from instruments such as OCT.

Two major enhancements will be necessary in the next generation of OCT software. First, the accuracy of OCT measurements must be improved. Since clinical decisions will rely more and more on the precision of OCT quantification, the standards for the next generation of OCT software should be set higher than currently available systems which are prone to errors. This should be an achievable goal since higher resolution image data provides the opportunity for more detailed tissue analyses including delineation and measurement of ocular tissue components such as individual retinal layers and subretinal lesions [89]. One difficult and unsolved problem, however, is how to keep pace with the evolution in our understanding of the histologic and pathologic correlates to OCT measurements. For example, it is now believed that the current StratusOCT system incorrectly detects the true outer retinal boundary. This knowledge was not available to the designers of the system nor was the system intended to learn or incorporate new data into its analysis algorithm. Ideally, future OCT software systems will be able to adapt to advances in clinical knowledge by incorporating new clinical understanding into diagnostic measurements in the field.

The other major challenge facing future OCT software systems will be to improve the relevance of clinical measurements. Although retinal thickness is an important measurement in many retinal diseases, lumping all intraretinal and subretinal findings into a single measurement does not take full advantage of the cross-sectional image data acquired by OCT. In an era of intravitreal anti-VEGF therapies, for instance, it would be useful to learn the individual volumes of the retinal tissue, cystoid spaces, subretinal fluid, subretinal tissue, and sub-RPE lesions instead of just the aggregate retinal thickness or macular volume. Furthermore, with the improved image registration capabilities of FD-OCT devices, these measurements can be compared between visits to look for clinically relevant changes in disease states. Registration will also enable integration of other diagnostic modalities such as color imaging, angiography, perimetry, and multifocal electroretinography [106, 114]. Point-to-point localization of OCT findings on fundus images or angiography will also both assist the care of the patient and advance our understanding of the pathophysiology underlying the disease.

OCT has revolutionized the practice of ophthalmology. Continued hardware and software enhancements should improve the relevance and precision of OCT measurements. These advanced capabilities will enhance the clinician's ability to manage complex diseases, make critical treatment decisions, and potentially even diagnose diseases at an earlier stage.

References

1. Alam S, Zawadzki RJ, Choi S, Gerth C, Park SS, Morse L, Werner JS (2006) Clinical application of rapid serial Fourier-domain optical coherence tomography for macular imaging. Ophthalmology 113:1425–31
2. Antcliff RJ, Marshall J (1999) The pathogenesis of edema in diabetic maculopathy. Semin Ophthalmol 14:223–32
3. Baumann M, Gentile RC, Liebmann JM, Ritch R (1998) Reproducibility of retinal thickness measurements in normal eyes using optical coherence tomography. Ophthalmic Surg Lasers 29:280–5
4. Blazkiewicz P, Gourlay M, Tucker JR, Rakic AD, Zvyagin AV (2005) Signal-to-noise ratio study of full-field Fourier-domain optical coherence tomography. Appl Opt 44:7722–9
5. Brown JC, Solomon SD, Bressler SB, Schachat AP, DiBernardo C, Bressler NM (2004) Detection of diabetic foveal edema: contact lens biomicroscopy compared with optical coherence tomography. Arch Ophthalmol 122:330–5
6. Browning DJ (2004) Interobserver variability in optical coherence tomography for macular edema. Am J Ophthalmol 137:1116–7
7. Browning DJ, Fraser CM (2004) Optical coherence tomography to detect macular edema in the presence of asteroid hyalosis. Am J Ophthalmol 137:959–61
8. Browning DJ, McOwen MD, Bowen RM, Jr, O'Marah TL (2004) Comparison of the clinical diagnosis of diabetic macular edema with diagnosis by optical coherence tomography. Ophthalmology 111:712–5
9. Browning DJ, Fraser CM (2005) Regional patterns of sight-threatening diabetic macular edema. Am J Ophthalmol 140:117–24
10. Brun SC, Bressler SB, Maguire MG, Heiner C, Bressler NM, Schachat AP (1993) A comparison of fundus biomicroscopy and 90 diopter lens examination in the detection of diabetic clinically significant macular edema [ARVO abstract]. Invest Ophthalmol Vis Sci 34:718
11. Catier A, Tadayoni R, Paques M, Erginay A, Haouchine B, Gaudric A, Massin P (2005) Characterization of macular edema from various etiologies by optical coherence tomography. Am J Ophthalmol 140:200–6
12. Cense B, Chen TC, Park BH, Pierce MC, de Boer JF (2004) Thickness and birefringence of healthy retinal nerve fiber layer tissue measured with polarization-sensitive optical coherence tomography. Invest Ophthalmol Vis Sci 45:2606–12
13. Choma MA, Hsu K, Izatt JA (2005) Swept source optical coherence tomography using an all-fiber 1300-nm ring laser source. J Biomed Opt 10:44009
14. Ciardella AP, Klancnik J, Schiff W, Barile G, Langton K, Chang S (2004) Intravitreal triamcinolone for the treatment of refractory diabetic macular oedema with hard exudates: an optical coherence tomography study. Br J Ophthalmol 88:1131–6
15. Coker JG, Duker JS (1996) Macular disease and optical coherence tomography. Curr Opin Ophthalmol 7:33–8
16. Cunningham ET, Jr, Adamis AP, Altaweel M, Aiello LP, Bressler NM, D'Amico DJ, Goldbaum M, Guyer DR, Katz B, Patel M, Schwartz SD (2005) A phase II randomized double-masked trial of pegaptanib, an anti-vascular endothelial growth factor aptamer, for diabetic macular edema. Ophthalmology 112:1747–57
17. Drexler W (2004) Ultrahigh-resolution optical coherence tomography. J Biomed Opt 9:47–74
18. Early Treatment Diabetic Retinopathy Study Research Group (1985) Photocoagulation for diabetic macular edema. Early Treatment Diabetic Retinopathy Study report number 1. Arch Ophthalmol 103:1796–806
19. Early Treatment Diabetic Retinopathy Study Research Group (1991) Early photocoagulation for diabetic retinopathy. ETDRS report number 9. Ophthalmology 98:766–85
20. Friedman SM (2003) Optociliary venous anastomosis after radial optic neurotomy for central retinal vein occlusion. Ophthalmic Surg Lasers Imaging 34:315–7
21. Fujii GY, de Juan E, Jr, Humayun MS (2003) Improvements after sheathotomy for branch retinal vein occlusion documented by optical coherence tomography and scanning laser ophthalmoscope. Ophthalmic Surg Lasers Imaging 34:49–52
22. Garciia-Arumii J, Boixadera A, Martinez-Castillo V, Castillo R, Dou A, Corcostegui B (2003) Chorioretinal anastomosis after radial optic neurotomy for central retinal vein occlusion. Arch Ophthalmol 121:1385–91
23. Gaucher D, Tadayoni R, Erginay A, Haouchine B, Gaudric A, Massin P (2005) Optical coherence tomography assessment of the vitreoretinal relationship in diabetic macular edema. Am J Ophthalmol 139:807–13
24. Giovannini A, Amato GP, Mariotti C, Ripa E (1999) Diabetic maculopathy induced by vitreo-macular traction: evaluation by optical coherence tomography (OCT). Doc Ophthalmol 97:361–6
25. Goebel W, Kretzchmar-Gross T (2002) Retinal thickness in diabetic retinopathy: a study using optical coherence tomography (OCT). Retina 22:759–67
26. Grieve K, Paques M, Dubois A, Sahel J, Boccara C, Le Gargasson JF (2004) Ocular tissue imaging using ultrahigh-resolution, full-field optical coherence tomography. Invest Ophthalmol Vis Sci 45:4126–31
27. Hee MR, Izatt JA, Swanson EA, Huang D, Schuman JS, Lin CP, Puliafito CA, Fujimoto JG (1995a) Optical coherence tomography of the human retina. Arch Ophthalmol 113:325–32
28. Hee MR, Puliafito CA, Wong C, Duker JS, Reichel E, Rutledge B, Schuman JS, Swanson EA, Fujimoto JG (1995b) Quantitative assessment of macular edema with optical coherence tomography. Arch Ophthalmol 113:1019–29
29. Hee MR, Puliafito CA, Duker JS, Reichel E, Coker JG, Wilkins JR, Schuman JS, Swanson EA, Fujimoto JG (1998) Topography of diabetic macular edema with optical coherence tomography. Ophthalmology 105:360–70
30. Hee MR, Fujimoto JG, Ko TH, et al. (2004) Interpretation of the optical coherence tomography image. In: Schuman JS, Puliafito CA, Fujimoto JG (eds) Optical coherence tomography of ocular diseases, 2nd edn. Slack Inc., Thorofare, NJ
31. Hermann B, Fernandez EJ, Unterhuber A, Sattmann H, Fercher AF, Drexler W, Prieto PM, Artal P (2004) Adaptive-optics ultrahigh-resolution optical coherence tomography. Opt Lett 29:2142–4
32. Holz FG, Jorzik J, Schutt F, Flach U, Unnebrink K (2003) Agreement among ophthalmologists in evaluating fluorescein angiograms in patients with neovascular age-related macular degeneration for photodynamic therapy eligibility (FLAP-study). Ophthalmology 110:400–5
33. Huang D, Swanson EA, Lin CP, Schuman JS, Stinson WG, Chang W, Hee MR, Flotte T, Gregory K, Puliafito CA, et al. (1991) Optical coherence tomography. Science 254:1178–81
34. Huang D, Kaiser PK, Lowder CY, Traboulsi E (2006) Retinal imaging, 1st edn. Elsevier, Philadelphia, p 640

35. Hussain A, Hussain N, Nutheti R (2005) Comparison of mean macular thickness using optical coherence tomography and visual acuity in diabetic retinopathy. Clin Exp Ophthalmol 33:240–5

36. Imai M, Iijima H, Hanada N (2001) Optical coherence tomography of tractional macular elevations in eyes with proliferative diabetic retinopathy. Am J Ophthalmol 132:458–61

37. Jonas JB, Sofker A (2001) Intraocular injection of crystalline cortisone as adjunctive treatment of diabetic macular edema. Am J Ophthalmol 132:425–7

38. Kaiser RS, Berger JW, Williams GA, Tolentino MJ, Maguire AM, Alexander J, Madjarov B, Margherio RM (2002) Variability in fluorescein angiography interpretation for photodynamic therapy in age-related macular degeneration. Retina 22:683–90

39. Kang SW, Park CY, Ham DI (2004) The correlation between fluorescein angiographic and optical coherence tomographic features in clinically significant diabetic macular edema. Am J Ophthalmol 137:313–22

40. Klein R, Klein BE, Moss SE (1984) Visual impairment in diabetes. Ophthalmology 91:1–9

41. Lai JC, Stinnett SS, Jaffe GJ (2003) B-scan ultrasonography for the detection of macular thickening. Am J Ophthalmol 136:55–61

42. Lam DS, Chan CK, Tang EW, Li KK, Fan DS, Chan WM (2004) Intravitreal triamcinolone for diabetic macular oedema in Chinese patients: six-month prospective longitudinal pilot study. Clin Exp Ophthalmol 32:569–72

43. Larsen M, Wang M, Sander B (2005) Overnight thickness variation in diabetic macular edema. Invest Ophthalmol Vis Sci 46:2313–6

44. Lattanzio R, Brancato R, Pierro L, Bandello F, Iaccher B, Fiore T, Maestranzi G (2002) Macular thickness measured by optical coherence tomography (OCT) in diabetic patients. Eur J Ophthalmol 12:482–7

45. Lerche RC, Schaudig U, Scholz F, Walter A, Richard G (2001) Structural changes of the retina in retinal vein occlusion – imaging and quantification with optical coherence tomography. Ophthalmic Surg Lasers 32:272–80

46. Lovestam-Adrian M, Andreasson S, Ponjavic V (2004) Macular function assessed with mfERG before and after panretinal photocoagulation in patients with proliferative diabetic retinopathy. Doc Ophthalmol 109:115–21

47. Martidis A, Duker JS, Greenberg PB, Rogers AH, Puliafito CA, Reichel E, Baumal C (2002) Intravitreal triamcinolone for refractory diabetic macular edema. Ophthalmology 109:920–7

48. Martinez-Jardon CS, Meza-de Regil A, Dalma-Weiszhausz J, Leizaola-Fernandez C, Morales-Canton V, Guerrero-Naranjo JL, Quiroz-Mercado H (2005) Radial optic neurotomy for ischaemic central vein occlusion. Br J Ophthalmol 89:558–61

49. Massin P, Vicaut E, Haouchine B, Erginay A, Paques M, Gaudric A (2001) Reproducibility of retinal mapping using optical coherence tomography. Arch Ophthalmol 119:1135–42

50. Massin P, Erginay A, Haouchine B, Mehidi AB, Paques M, Gaudric A (2002) Retinal thickness in healthy and diabetic subjects measured using optical coherence tomography mapping software. Eur J Ophthalmol 12:102–8

51. Massin P, Duguid G, Erginay A, Haouchine B, Gaudric A (2003) Optical coherence tomography for evaluating diabetic macular edema before and after vitrectomy. Am J Ophthalmol 135:169–77

52. Massin P, Audren F, Haouchine B, Erginay A, Bergmann JF, Benosman R, Caulin C, Gaudric A (2004) Intravitreal triamcinolone acetonide for diabetic diffuse macular edema: preliminary results of a prospective controlled trial. Ophthalmology 111:218–24; discussion 224–5

53. Micelli Ferrari T, Sborgia L, Furino C, Cardascia N, Ferreri P, Besozzi G, Sborgia C (2004) Intravitreal triamcinolone acetonide: valuation of retinal thickness changes measured by optical coherence tomography in diffuse diabetic macular edema. Eur J Ophthalmol 14:321–4

54. Moreira RO, Trujillo FR, Meirelles RM, Ellinger VC, Zagury L (2001) Use of optical coherence tomography (OCT) and indirect ophthalmoscopy in the diagnosis of macular edema in diabetic patients. Int Ophthalmol 24:331–6

55. Moshfeghi AA, Mavrofrides EC, Puliafito CA (2006) Optical coherence tomography and retinal thickness assessment for diagnosis and management. In: Ryan SJ (ed) Retina, 4th edn. Vol. 2. Elsevier, Philadelphia, p 1889

56. Muscat S, Parks S, Kemp E, Keating D (2002) Repeatability and reproducibility of macular thickness measurements with the Humphrey OCT system. Invest Ophthalmol Vis Sci 43:490–5

57. Nagpal M, Nagpal K, Bhatt C, Nagpal PN (2005) Role of early radial optic neurotomy in central retinal vein occlusion. Indian J Ophthalmol 53:115–20

58. Neubauer AS, Priglinger S, Ullrich S, Bechmann M, Thiel MJ, Ulbig MW, Kampik A (2001) Comparison of foveal thickness measured with the retinal thickness analyzer and optical coherence tomography. Retina 21:596–601

59. Nussenblatt RB, Kaufman SC, Palestine AG, Davis MD, Ferris FL, 3rd (1987) Macular thickening and visual acuity. Measurement in patients with cystoid macular edema. Ophthalmology 94:1134–9

60. Ohashi H, Oh H, Nishiwaki H, Nonaka A, Takagi H (2004) Delayed absorption of macular edema accompanying serous retinal detachment after grid laser treatment in patients with branch retinal vein occlusion. Ophthalmology 111:2050–6

61. Opremcak EM, Bruce RA (1999) Surgical decompression of branch retinal vein occlusion via arteriovenous crossing sheathotomy: a prospective review of 15 cases. Retina 19:1–5

62. Opremcak EM, Bruce RA, Lomeo MD, Ridenour CD, Letson AD, Rehmar AJ (2001) Radial optic neurotomy for central retinal vein occlusion: a retrospective pilot study of 11 consecutive cases. Retina 21:408–15

63. Opremcak EM, Rehmar AJ, Ridenour CD, Kurz DE (2006) Radial optic neurotomy for central retinal vein occlusion: 117 consecutive cases. Retina 26:297–305

64. Otani T, Kishi S (2000) Tomographic assessment of vitreous surgery for diabetic macular edema. Am J Ophthalmol 129:487–94

65. Otani T, Kishi S (2001) Tomographic findings of foveal hard exudates in diabetic macular edema. Am J Ophthalmol 131:50–4

66. Otani T, Kishi S, Maruyama Y (1999) Patterns of diabetic macular edema with optical coherence tomography. Am J Ophthalmol 127:688–93

67. Ozdek SC, Erdinc MA, Gurelik G, Aydin B, Bahceci U, Hasanreisoglu B (2005) Optical coherence tomographic assessment of diabetic macular edema: comparison with fluorescein angiographic and clinical findings. Ophthalmologica 219:86–92

68. Ozdemir H, Karacorlu M, Karacorlu S (2005a) Serous mac-

12 II

ular detachment in central retinal vein occlusion. Retina 25:561–3

69. Ozdemir H, Karacorlu M, Karacorlu S (2005b) Serous macular detachment in diabetic cystoid macular oedema. Acta Ophthalmol Scand 83:63–6

70. Panozzo G, Gusson E, Parolini B, Mercanti A (2003) Role of OCT in the diagnosis and follow up of diabetic macular edema. Semin Ophthalmol 18:74–81

71. Panozzo G, Parolini B, Gusson E, Mercanti A, Pinackatt S, Bertoldo G, Pignatto S (2004) Diabetic macular edema: an OCT-based classification. Semin Ophthalmol 19:13–20

72. Parolini B, Panozzo G, Gusson E, Pinackatt S, Bertoldo G, Rottini S, Pignatto S (2004) Diode laser, vitrectomy and intravitreal triamcinolone. A comparative study for the treatment of diffuse non tractional diabetic macular edema. Semin Ophthalmol 19:1–12

73. Patel JI, Hykin PG, Schadt M, Luong V, Fitzke F, Gregor ZJ (2006) Pars plana vitrectomy for diabetic macular oedema: OCT and functional correlations. Eye 20:674–80

74. Patelli F, Radice P, Zumbo G, Fasolino G, Marchi S (2004) Optical coherence tomography evaluation of macular edema after radial optic neurotomy in patients affected by central retinal vein occlusion. Semin Ophthalmol 19:21–4

75. Paunescu LA, Schuman JS, Price LL, Stark PC, Beaton S, Ishikawa H, Wollstein G, Fujimoto JG (2004) Reproducibility of nerve fiber thickness, macular thickness, and optic nerve head measurements using StratusOCT. Invest Ophthalmol Vis Sci 45:1716–24

76. Pons ME, Garcia-Valenzuela E (2005) Redefining the limit of the outer retina in optical coherence tomography scans. Ophthalmology 112:1079–85

77. Ray R, Stinnett SS, Jaffe GJ (2005) Evaluation of image artifact produced by optical coherence tomography of retinal pathology. Am J Ophthalmol 139:18–29

78. Rosenblatt BJ, Shah GK, Sharma S, Bakal J (2005) Pars plana vitrectomy with internal limiting membranectomy for refractory diabetic macular edema without a taut posterior hyaloid. Graefes Arch Clin Exp Ophthalmol 243:20–5

79. Rosenfeld PJ, Fung AE, Puliafito CA (2005) Optical coherence tomography findings after an intravitreal injection of bevacizumab (avastin) for macular edema from central retinal vein occlusion. Ophthalmic Surg Lasers Imaging 36:336–9

80. Sadda SR, Tan O, Walsh AC, Schuman JS, Varma R, Huang D (2006a) Automated detection of clinically significant macular edema by grid scanning optical coherence tomography. Ophthalmology 113:1196 e1–2

81. Sadda SR, Wu Z, Walsh AC, Richine L, Dougall J, Cortez R, LaBree LD (2006b) Errors in retinal thickness measurements obtained by optical coherence tomography. Ophthalmology 113:285–93

82. Schaudig UH, Glaefke C, Scholz F, Richard G (2000) Optical coherence tomography for retinal thickness measurement in diabetic patients without clinically significant macular edema. Ophthalmic Surg Lasers 31:182–6

83. Schepens CL, Avila MP, Jalkh AE, Trempe CL (1984) Role of the vitreous in cystoid macular edema. Surv Ophthalmol 28 Suppl:499–504

84. Sebag J, Balazs EA (1984) Pathogenesis of cystoid macular edema: an anatomic consideration of vitreoretinal adhesions. Surv Ophthalmol 28 Suppl:493–8

85. Sebag J, Buckingham B, Charles MA, Reiser K (1992) Biochemical abnormalities in vitreous of humans with proliferative diabetic retinopathy. Arch Ophthalmol 110:1472–6

86. Sekiryu T, Yamauchi T, Enaida H, Hara Y, Furuta M (2000) Retina tomography after vitrectomy for macular edema of central retinal vein occlusion. Ophthalmic Surg Lasers 31:198–202

87. Shah GK, Sharma S, Fineman MS, Federman J, Brown MM, Brown GC (2000) Arteriovenous adventitial sheathotomy for the treatment of macular edema associated with branch retinal vein occlusion. Am J Ophthalmol 129:104–6

88. Shah SP, Patel M, Thomas D, Aldington S, Laidlaw DA (2006) Factors predicting outcome of vitrectomy for diabetic macular oedema: results of a prospective study. Br J Ophthalmol 90:33–6

89. Shahidi M, Wang Z, Zelkha R (2005) Quantitative thickness measurement of retinal layers imaged by optical coherence tomography. Am J Ophthalmol 139:1056–61

90. Shimura M, Yasuda K, Nakazawa T, Kano T, Ohta S, Tamai M (2003) Quantifying alterations of macular thickness before and after panretinal photocoagulation in patients with severe diabetic retinopathy and good vision. Ophthalmology 110:2386–94

91. Shimura M, Saito T, Yasuda K, Tamai M (2005a) Clinical course of macular edema in two cases of interferon-associated retinopathy observed by optical coherence tomography. Jpn J Ophthalmol 49:231–4

92. Shimura M, Yasuda K, Nakazawa T, Tamai M (2005b) Visual dysfunction after panretinal photocoagulation in patients with severe diabetic retinopathy and good vision. Am J Ophthalmol 140:8–15

93. Shukla D, Rajendran A, Kim R (2006) Macular hole formation and spontaneous closure after vitrectomy for central retinal vein occlusion. Graefes Arch Clin Exp Ophthalmol

94. Spaide RF, Klancnik JM, Jr (2004) Still motion animation of the retina technique. Retina 24:315–7

95. Spaide RF, Lee JK, Klancnik JK, Jr, Gross NE (2003) Optical coherence tomography of branch retinal vein occlusion. Retina 23:343–7

96. Stitt AW, Moore JE, Sharkey JA, Murphy G, Simpson DA, Bucala R, Vlassara H, Archer DB (1998) Advanced glycation end products in vitreous: Structural and functional implications for diabetic vitreopathy. Invest Ophthalmol Vis Sci 39:2517–23

97. Stolba U, Binder S, Gruber D, Krebs I, Aggermann T, Neumaier B (2005) Vitrectomy for persistent diffuse diabetic macular edema. Am J Ophthalmol 140:295–301

98. Strom C, Sander B, Larsen N, Larsen M, Lund-Andersen H (2002) Diabetic macular edema assessed with optical coherence tomography and stereo fundus photography. Invest Ophthalmol Vis Sci 43:241–5

99. Tewari HK, Sony P, Chawla R, Garg SP, Venkatesh P (2005) Prospective evaluation of intravitreal triamcinolone acetonide injection in macular edema associated with retinal vascular disorders. Eur J Ophthalmol 15:619–26

100. Thomas D, Bunce C, Moorman C, Laidlaw DA (2005) A randomised controlled feasibility trial of vitrectomy versus laser for diabetic macular oedema. Br J Ophthalmol 89:81–6

101. Toth CA, Birngruber R, Boppart SA, Hee MR, Fujimoto JG, DiCarlo CD, Swanson EA, Cain CP, Narayan DG, Noojin GD, Roach WP (1997a) Argon laser retinal lesions evaluated in vivo by optical coherence tomography. Am J Ophthalmol 123:188–98

102. Toth CA, Narayan DG, Boppart SA, Hee MR, Fujimoto JG, Birngruber R, Cain CP, DiCarlo CD, Roach WP (1997b) A

comparison of retinal morphology viewed by optical coherence tomography and by light microscopy. Arch Ophthalmol 115:1425–8

103. Traynor MP, Conway BP (2004) Collateral vessel formation after radial optic neurotomy. Retina 24:616–7

104. Tso MO (1982) Pathology of cystoid macular edema. Ophthalmology 89:902–15

105. Uchino E, Uemura A, Ohba N (2001) Initial stages of posterior vitreous detachment in healthy eyes of older persons evaluated by optical coherence tomography. Arch Ophthalmol 119:1475–9

106. van Velthoven ME, de Vos K, Verbraak FD, Pool CW, de Smet MD (2005) Overlay of conventional angiographic and en-face OCT images enhances their interpretation. BMC Ophthalmol 5:12

107. Watanabe M, Oshima Y, Emi K (2000) Optical cross-sectional observation of resolved diabetic macular edema associated with vitreomacular separation. Am J Ophthalmol 129:264–7

108. Williamson TH, O'Donnell A (2005) Intravitreal triamcinolone acetonide for cystoid macular edema in nonischemic central retinal vein occlusion. Am J Ophthalmol 139:860–6

109. Wojtkowski M, Srinivasan V, Fujimoto JG, Ko T, Schuman JS, Kowalczyk A, Duker JS (2005a) Three-dimensional retinal imaging with high-speed ultrahigh-resolution optical coherence tomography. Ophthalmology 112(10):1734–46

110. Wojtkowski M, Srinivasan V, Fujimoto JG, Ko T, Schuman JS, Kowalczyk A, Duker JS (2005b) Three-dimensional retinal imaging with high-speed ultrahigh-resolution optical coherence tomography. Ophthalmology 112:1734–46

111. Wolter JR (1981) The histopathology of cystoid macular edema. Albrecht Von Graefes Arch Klin Exp Ophthalmol 216:85–101

112. Wong AC, Chan CW, Hui SP (2005) Relationship of gender, body mass index, and axial length with central retinal thickness using optical coherence tomography. Eye 19:292–7

113. Yamada N, Kishi S (2005) Tomographic features and surgical outcomes of vitreomacular traction syndrome. Am J Ophthalmol 139:112–7

114. Yamamoto S, Yamamoto T, Hayashi M, Takeuchi S (2001) Morphological and functional analyses of diabetic macular edema by optical coherence tomography and multifocal electroretinograms. Graefes Arch Clin Exp Ophthalmol 239:96–101

115. Yamamoto T, Hitani K, Tsukahara I, Yamamoto S, Kawasaki R, Yamashita H, Takeuchi S (2003) Early postoperative retinal thickness changes and complications after vitrectomy for diabetic macular edema. Am J Ophthalmol 135:14–9

116. Yang CS, Cheng CY, Lee FL, Hsu WM, Liu JH (2001) Quantitative assessment of retinal thickness in diabetic patients with and without clinically significant macular edema using optical coherence tomography. Acta Ophthalmol Scand 79:266–70

117. Yasukawa T, Kiryu J, Tsujikawa A, Dong J, Suzuma I, Takagi H, Ogura Y (1998) Quantitative analysis of foveal retinal thickness in diabetic retinopathy with the scanning retinal thickness analyzer. Retina 18:150–5

118. Yazdanfar S, Rollins AM, Izatt JA (2003) In vivo imaging of human retinal flow dynamics by color Doppler optical coherence tomography. Arch Ophthalmol 121:235–9

13 General Concepts in Laser Treatment for Retinal Vascular Disease

13 II

F. Rüfer, J. Roider

Core Messages

- The first retinal coagulation was performed in 1949 with sunlight by Meyer-Schwickerath. It was replaced in the 1950s by xenon high pressure lamps
- The development of laser technology began in 1960
- The melanin in the pigment granules is the main absorbing structure in the retina. The first ruby lasers had undesirable side effects such as retinal hemorrhages and retinal tears
- The first commercial argon laser systems enabled the Diabetic Retinopathy Study to be performed between 1972 and 1975
- Different laser types such as krypton laser (647 nm or 530 nm), dye laser (560–680 nm), Nd:YAG laser (1,064 nm or 532 nm) or diode laser (810 nm) show only slight differences in histology. The relative absorption maximum of hemoglobin can be avoided with longer wavelengths. The coagulated area is replaced by unspecific scar tissue
- The mechanisms of retinal laser treatment are not yet fully understood. The relation between need for oxygen and oxygen supply is opti-

mized by the retinal coagulation. But various growth factors and their time dependent course must also be taken into account in complex explanatory models
- By destruction of retinal tissues the risk of severe loss of vision within 2 years can be reduced by 50% in diabetic retinopathy
- If neovascularizations of the retina or of the iris exist, the eyes benefit from full scatter panretinal laser coagulation. Treatment does not lead to improvement of visual acuity
- The risk of development of neovascularization can be reduced by a modified sector laser coagulation in branch retinal vein occlusion
- The retinal pigment epithelium (RPE) can be selectively treated using short repetitive laser pulses with minimal heat diffuse to the adjacent retinal layers. Thus treatment of the macula is possible without damage of the overlying photoreceptors. Currently the therapeutic benefit of the procedure is being evaluated in multicenter trials in diseases such as diabetic macular edema, central serous retinopathy, and drusen in age-related macular degeneration (AMD)

13.1 History of Laser Treatment

Today light coagulation is a common treatment procedure and the basis of treatment in many retinal diseases. Its origins go back to Meyer-Schwickerath 1949 [46], who initially used sunlight as an energy source. However, sunlight was unsuitable for several reasons. A system of several mirrors and a long exposure time were necessary. The dependence on the weather forecast was an obvious problem. With the development of xenon high pressure lamps (Fig. 13.1) at the beginning of the 1950s, enough power for light coagulation of the ocular fundus became available [34]. Much scientific work was published at this time which showed the therapeutic

effect of the light coagulation and which led to a wider experience [47, 48]. The main focuses of clinically based xenon coagulation were retinopexy and the treatment of proliferative diabetic retinopathy.

13.2 Laser Sources

With the invention of laser technology in 1960 by Maiman [39], a rapid development of laser photocoagulation began. The first laser available was a ruby laser. Initially the focus on experimental examination was on interactions between pulsed ruby laser and retinal tissue [41]. It turned out that the histological effects of the laser light in the retina could be explained by absorption in different structures. The

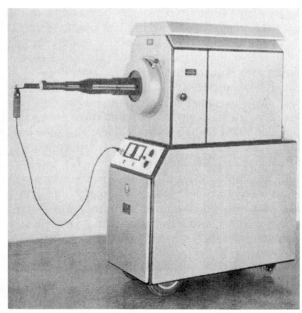

Fig. 13.1. Xenon high pressure lamp (Mayer-Schwickerath 1959)

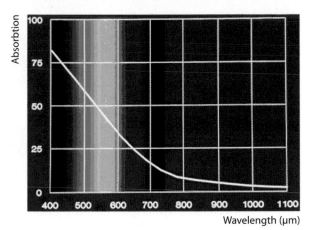

Fig. 13.2. Light absorption in the human retinal pigment epithelium

main absorber in the retina is represented by melanin in the pigment granules; other absorbers such as hemoglobin are of less importance. The short pulses of the ruby laser (694 nm) varied between 20 and 1,000 ns depending on the technical mode [42]. Short laser pulses initially prevented its clinical application becoming widespread, undesirable side-effects like retinal hemorrhages and tears occurred and no sufficient adhesion, e.g., in retinopexy, could be achieved [43, 44].

With the introduction of the argon laser technology at the beginning of the 1970s, there were systems available which enabled a broad clinical approach because of their continuous emission and powerful outcome. The first commercial argon laser systems in ophthalmology were used in 1971 [32]. Between 1972 and 1975 the Diabetic Retinopathy Study was performed, which was a large-scale prospective randomized multicenter clinical trial. It showed that treatment of proliferative diabetic retinopathy by coagulation brings great advantages and thus the risk of massive loss of vision can be reduced by 50% [66, 68, 69]. For this reason the study was stopped in favor of coagulation treatment. It also turned out that there are no differences between the therapeutic effects of xenon and argon laser coagulation; however, the technical advantages of the argon laser were obvious, since the argon laser could be coupled in mirror systems more precisely and applied to the fundus more accurately. With further development of laser techniques more types of new laser systems became available. Initially all sufficiently powerful continuous wave laser systems were tested experi-

mentally and clinical trials were carried out. The various laser types like krypton laser (647 nm or 530 nm), dye laser (560–680), Nd:YAG laser (1,064 nm or 532 nm) or diode laser (810 nm) differ mainly in wavelengths and in technical parameters such as size of apparatus, efficacy, type of current necessary for operation and necessity for cooling. As the main difference between the various laser systems is just the wavelength, only a small difference in histology could be expected if pulse durations of 100 ms or more are used. The relative absorption maximum of hemoglobin can be avoided with longer wavelengths while the absorption rate in retinal pigment epithelium is decreasing [6, 33]. Thus, using longer wavelengths leads to increasing damage of deeper structures, e.g., the choroid. A practical advantage of longer wavelengths is the better transmission through cataract lenses [61]. None of the different types of lasers showed a significant clinical advantage in coagulation of subretinal membranes [60, 65, 80] or in treatment of vasoproliferative diseases [3, 54]. All different lasers similarly led to a considerable heat conduction out of the absorbing structures into the adjacent tissue layers as exposure times exceed 100 ms or more in clinical applications [73].

13.2.1 Histological Findings After Light and Laser Coagulation

Historically, the first histological examinations were done with retinal tissues coagulated by xenon light. The retinal xenon lesions showed irreversible tissue damage. Areas of mild coagulation (Fig. 13.3) as seen on ophthalmoscopy showed destruction of the choriocapillaris, the retinal pigment epithelium and the photoreceptors including the outer nuclear layer [71, 76]. Destruction of the photoreceptors meant irre-

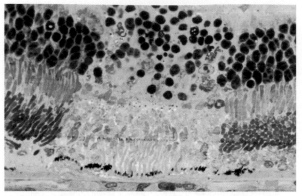

Fig. 13.3. Mild lesion after laser coagulation with a green laser (514 nm), exposure time 100 ms. The lesion was not visible clinically

Fig. 13.4. Histology of laser lesion (514 nm), which was clinically visible (exposure time 100 ms)

versible focal loss of function in the coagulated areas. More severe lesions (Fig. 13.4) using laser power as in clinical applications showed additional damage leading to pyknotic cells in the inner nuclear layer and in the ganglion cell layer [9, 24, 71]. Using laser histology, lesions do not look different as long as laser exposures between 100 and 500 ms are used. During the 1st h, probably even during the first seconds, the effect is intensified biologically by intra- or intercellular edema [35]. Besides damage of the retinal pigment epithelium (RPE), the photoreceptors were always affected. These findings are independent of the wavelength of the laser used [5, 36, 43, 76]. In stronger lesions the inner layers of the retina are also involved.

Damage of the inner retinal layers (ganglion cell layer, nerve fiber layer) is undesirable. Even if power settings for threshold lesions are used, damage of photoreceptors cannot be avoided, as there is a wide intra- and interindividual range in the variety of retinal thickness and pigmentation. Retinal vessels are usually not damaged primarily by retinal coagulation, as the power is too low and the spot size is too

large, which theoretically has to be matched to the diameter of a retinal vessel of a few micrometers [8]. Temperature elevations in the RPE using common clinical coagulation parameters do not lead to coagulation effects in retinal capillaries [40]. Because of different absorption characteristics in the RPE, theoretically different effects on the tissue are expected for different wavelengths. Actually there are no differences in the damage of the RPE and the outer retinal layers using the different wavelengths of the different lasers such as krypton laser [40], Nd:YAG laser [58] and diode laser [10]. By using longer wavelengths (e.g., krypton laser) the inner retinal layers can be protected as shown in animal experiments [2]. For applications in the macula the frequency doubled Nd:YAG laser (532 nm) has a small advantage over the argon laser because of its lower absorption in macula xanthophyll [23]. On the other hand, long wavelength light increasingly penetrates into the choroid, leading to choroidal bleedings [2, 37, 72] compared with histological effects in the retina and the choroid following treatment with Nd:YAG laser (1064 nm) and argon laser coagulation. After coagulation with a 1,064 nm Nd:YAG laser there were found severe changes in the choroid, even some occlusions of major choroidal vessels, which is in accordance with the reduced absorption.

The biological response to light or laser coagulation is similarly independent of the laser light used. It seems to depend only on the amount of tissue damaged. The repair process of a laser burn is unspecific to laser lesions and similar to the repair process of retinal detachment [38, 72]. There are only minor differences between different animal species [77] After laser coagulation the coagulated area gradually regenerates by proliferating glia cells originating from the intact retina and adjacent choroid. The retinal pigment epithelium contributes to the scarring process in several ways [75]. The blood-retina barrier begins to reform after 7 days [31]. The RPE barrier is restored anatomically and functionally [75]. After 3 months microglia can be found in the coagulated area [50]. Defects in the outer nuclear layer are replaced by Müller cells. Moderate exposure leads to a differentiated regeneration without adhesions between glia tissues and RPE cell layer [74]. Stronger lesions cause retinal defects with retinal adhesion to the RPE or, if missing, with retinal adhesion to Bruch's membrane [74]. In those adhesions proliferating cells or RPE cells layered on Bruch's membrane are involved. The findings on repair of the choriocapillary layer are different. While recapillarization was found in cats 30 days after coagulation [55], there were persisting defects in monkeys [15]. Treatment with drugs of retinal damage, e.g., in accidental laser injuries, shows a statistical effect on the amount

Fig. 13.5. Laser lesion replaced by scar tissue

of area destroyed, but cannot change the irreversible destruction of the photoreceptors itself [59, 62].

To summarize, photocoagulation is an unspecific method. Using common laser parameters, the RPE, the choriocapillary layer, the photoreceptors and to a certain degree more distant structures are damaged. The coagulated area is replaced by unspecific scar tissue, which leads to irreversible visual field loss.

13.2.2 Mechanisms of Treatment

In proliferative diabetic retinopathy one of the first explanations of the effect of laser coagulation was the destruction of the oxygen consuming photoreceptors, while the relation between need for oxygen and

oxygen supply was optimized by the retinal coagulation [79]. Other animal experiments showed early changes in the RPE, leading to the assumption that the therapeutic effect follows a destruction and replacement of these specifically damaged RPE cells [70]. A different explanatory model has been proposed by Marshall et al. [40], who showed after coagulation of porcine eyes proliferative dividing cells of venous capillaries. These cell activities were seen even in the inner retinal layers, suggesting a long distance effect of a biological substance. This substance likely is a vascular endothelial growth factor (VEGF), which is released oxygen dependently and which can be detected both after retinal ischemia and in ocular neovascularization. VEGF is a strong angiogenic factor [25]. To explain the beneficial effects of retinal laser photocoagulation, Wilson et al. examined gene expressions of retina, RPE and choroid in mice eyes 3 days after retinal laser coagulation [78]. Among 265 differentially classified genes, an increased expression of angiotensin II type 2 receptor was found, which is involved in inhibition of VEGF expression and VEGF-induced angiogenesis. The same study also found a decreased expression of calcitonin receptor-like receptor (CRLR) precursor, interleukin-1 (IL)-1βm, the fibroblast growth factors (FGF-14 and FGF-16), and plasminogen activator inhibitor II (PAI2), which may also contribute to the anti-angiogenetic effects of laser therapy. Ogata et al. showed an upregulation of pigment epithelium-derived factor (PEDF) in cultured human retinal pigment epithelial cells after photocoagulation [53].

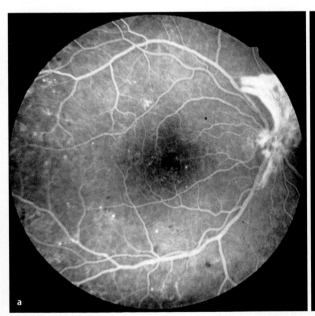
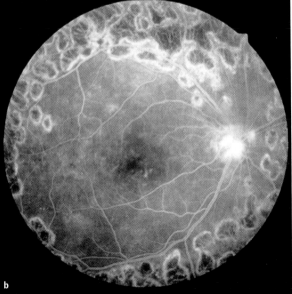

Fig. 13.6. a Image of fluorescein angiogram of a patient with a severe proliferation of the disk (NVD). **b** Image of fluorescein angiogram after panretinal laser coagulation. The NVD has significantly regressed

PEDF was recently shown to be a potent inhibitor of ocular angiogenesis [14]. Ogata et al. found not only that anti-angiogenic factors but also angiogenic factors were upregulated transiently after laser photocoagulation in cultured RPE cells. Via reverse transcription polymerase chain reaction a significant increase in the expression level of basic fibroblast growth factor (bFGF) was detected, with a maximum 6 h after photocoagulation and a decreasing level after 72 h to less than before photocoagulation. bFGF is a strong angiogenic factor [52] which also accelerates wound healing [45]. Therefore, its use has been suggested to promote the proliferation of laser injured RPE cells in culture. Among the angiogenetic factors, Ogata et al. also found increased expression after 6 h of kinase insert domain-containing receptor-1 (KDR/flk-1), Ets-1, which is a transcription factor expressed in endothelial cells during angiogenesis, and nuclear factor kappa B (NF-κB), which may regulate the initiation of angiogenesis [63, 64] as well as VEGF [52]. Interestingly, after 72 h KDR/flk-1, Ets-1, NF-κB and VEGF gradually decreased to a level lower than that before photocoagulation. These experimental findings in cultured RPE cells indicate that not only the detection of various growth factors, but also their time dependent course must be taken into account in complex explanation models, which are not yet fully understood.

13.3 Standards and Indications for Panretinal Laser Coagulation

13.3.1 Full Scatter Panretinal Laser Coagulation

13.3.1.1 Diabetes

The present guidelines for panretinal laser coagulation are mainly based on experience in diabetic retinopathy. The treatment of proliferative diabetic retinopathy is based on the Diabetic Retinopathy Study (DRS) [67, 68]. The DRS Study was a prospective, randomized and multicenter trial. A total of 1,700 patients with proliferative and non-proliferative diabetic retinal changes were examined between 1972 and 1975, looking into whether a retinal xenon or argon laser coagulation of the retina has a beneficial effect in comparison to the natural cause. The treatment parameters were exposure times of 100 ms and spot diameters of between 500 and 1,000 μm. In one session 500–1,000 laser spots were applied. With such intense destruction of retinal tissues the risk of severe loss of vision within 2 years could be reduced by 50%. Therefore the study was stopped in favor of treatment for all patients. The systematic analysis of the retinal changes showed four high risk factors for

acute loss of sight, which are generally accepted as indications for a dense (full scatter) panretinal laser coagulation. While eyes with only one risk factor are at a relative risk of 4.2–6.8% of severe loss of vision, the risk in eyes with four risk factors rises to 37%. The risk factors/indications are:

- Presence of vitreous or preretinal hemorrhage
- Location of new vessels on or near the optic disk (NVD)
- Presence of new vessels "elsewhere" (NVE)
- Severity of new vessels (proliferation area greater than one-fourth of the optic disk size)

13.3.1.2 Central Retinal Vein Occlusion

Another indication for a full scatter panretinal laser coagulation arises from central retinal vein occlusion (CRVO). The main complications of a central vein occlusion apart from macular edema are neovascularizations of the retina and of the iris. The incidence of retinal neovascularizations is correlated highly with the degree of ischemia [27]. The guidelines for treatment of CRVO are based on the central retinal vein occlusion study, which was performed from 1988 until 1992 [13]. It showed that macular edema could be reduced after grid laser coagulation, and also that there was no improvement in visual acuity. There was also no effect of prophylactic panretinal laser coagulation to prevent neovascularizations of the iris. But if neovascularizations of the retina or of the iris exist, the treated eyes clearly benefit from full scatter panretinal laser coagulation.

Guidelines for the Clinic
Suitable exposure times are 100–200 ms and a spot size of 500 μm. The laser application should lead to a mild white retinal lesion. The distance between the laser spots should be 0.5–1 laser spot. In patients with four diabetic risk factors the range of laser spots varies between 1,000 and 2,000 depending on the spot size. It is recommended to apply laser lesions in two to four sessions, e.g., 2 weeks apart, to avoid serous choroidal detachment and patient discomfort [16]. It should be remembered not to laser onto intraretinal bleedings, as absorption of the laser beam by the hemoglobin in the inner retinal layers may lead to overcoagulation and damage of the retinal layer. In central retinal vein occlusion with proliferation and neovascularization, often even more laser spots are necessary. Regression of neovascularization can be expected after 4–6 weeks.

13.3.1.3 Branch Retinal Vein Occlusion

The natural course of branch retinal vein occlusions is characterized by macular edema and vitreous hemorrhage from retinal neovascularizations. About 30–50% of patients with branch retinal vein occlusion (BRVO) recover to visual acuities of 0.5 or even better without therapy [26, 49]. Cases with poor visual acuity are caused by ischemia. In two-thirds of patients a macular edema leads to loss of vision. Neovascularization can develop if there are large areas of ischemia. The treatment guidelines for BRVO are based on the Branch Vein Occlusion Study [11, 51]. It showed that visual acuity was better in the treatment group after 3 years and also proved that the risk of development of neovascularization could be reduced by a modified sector laser coagulation.

Guidelines for the Clinic
Retinal laser coagulation should be done not earlier than 3–6 months after the appearance of a branch retinal vein occlusion. Laser photocoagulation should be done only if retinal hemorrhage has significantly cleared. Otherwise the inner layer of the retina will be destroyed. For the treatment of macular edema, exposure times of 100 ms and a spot size of 100 μm are recommended. The distance between spots should be 2–3 spot diameters. The area of the edema should be treated in a dense grid. After occurrence of neovascularizations a sector retinal laser coagulation as described in Sect. 13.3.1 is indicated.

13.3.2 Mild Scatter Panretinal Laser Coagulation

The treatment guidelines for non-proliferative diabetic retinopathy are to be regarded as the results of the Early Treatment Diabetic Retinopathy Study (ETDRS) [17–19]. Similarly to the DRS it was a multicenter, prospective and randomized trial including 3,711 patients with non-proliferative or early proliferative retinal changes. For classification, mainly retinal changes, which were seen on fundoscopy, were taken into account. Analyzing these risk factors led to the current valid indication of treating severe non-proliferative diabetic retinopathies. For classification of a severe non-proliferative diabetic retinopathy the 4:2:1 rule has been proven [29]:

A severe non-proliferative diabetic retinopathy is present:

- If either intraretinal bleeding occurs in 4 quadrants
- Or if venous beading occurs in at least 2 quadrants
- Or if intraretinal microvascular abnormalities (IRMA) occur in at least one quadrant

Guidelines for the Clinic
A mild scatter panretinal laser coagulation is carried out similarly to the full scatter panretinal laser coagulation except for the amount of laser spots: In total 600 laser spots of 500 μm at greater distances are applied.

Complications of Panretinal Laser Coagulation

Depending on the retinal area destroyed, visual field loss and disturbances of dim vision are the most frequent complications. If the area outside the arcades is destroyed, visual field is only about 20 degrees. In rare cases macular edema can lead to a loss of visual acuity after coagulation. These possible complications indicate overtreatment should be avoided. A reason for the failure of the laser therapy is often an

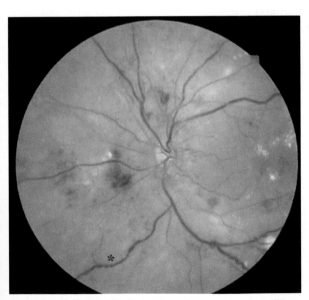

Fig. 13.7. Fundus image of a patient with severe non-proliferative diabetic retinopathy: Severe venous beading (see *asterisk*) is conceivable

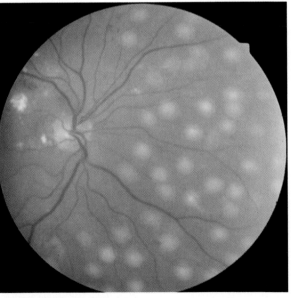

Fig. 13.8. Acute lesions after mild scatter panretinal laser coagulation 1 h after treatment

13 II

insufficient coagulation with too small and too few spots.

A therapeutic effect after panretinal laser coagulation can be seen usually not earlier than 6 weeks after coagulation. If there is no regression of the risk factors, an additional laser coagulation is to be carried out. However, the DRS studies showed that in proliferative diabetic retinopathy the risk of severe loss of vision can be avoided only in 50 %, even if laser treatment is extensive and appropriately performed.

13.4 Focal Laser Application

The ETDRS demonstrated that focal laser coagulation in diabetic macular edema reduces the risk of moderate loss of vision. There was a significant loss of vision in 5 % of the treated eyes, compared to 8 % in not treated eyes after 1 year. After 2 years the loss of vision averaged out at 7 % in treated vs. 16 % in untreated eyes, and after 3 years it was 12 % vs. 24 % [17]. The diagnosis of clinically significant macular edema (CSM) is made by fundoscopy. It is present and should be treated by focal laser coagulation if:

- There is a clinical retinal thickening within 500 μm distance from the center of the macula
- There is hard exudation within 500 μm distance from the center of the macula with retinal thickening in the bordering retina
- There is a retinal thickened area by the size of at least one papilla diameter within the distance of one papilla diameter from the center of the macula

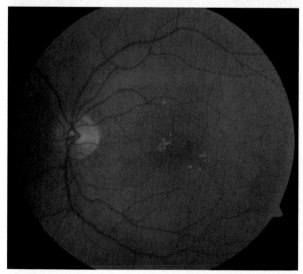

Fig. 13.9. Diabetic patient with clinically significant macular edema

Guidelines for the Clinic
Focal laser coagulation and its complications:
The placement of the laser coagulation spots has to be decided by fluorescein angiography. Areas with edema should be treated close to the leakage. Suitable exposure times are 100 ms and a spot size of 100 μm. The beginning power should not exceed 70–80 mW. The laser application should lead to a mild gray retinal lesion. The first spots should be placed distant from the fovea to adjust the power. The distance between the laser spots should be one laser spot apart. Clinical results can be expected not before 3 months after treatment. The most frequent complication is a disturbance of reading by irreversible destruction of photoreceptors. To reduce the risk of complications a modified grid laser coagulation can be performed alternatively in widespread diffuse macular edema. The modified grid laser coagulation is performed similarly to the focal laser coagulation, but the distance between spots should be 2–3 spot diameters, treating the whole area of the edema. A rare complication can be secondary CNV development

13.5 Subthreshold Laser Coagulation for Retinal Disease

The benefit of retinal laser treatment has traditionally been attributed to the destruction of retinal tissue as described above. Heat conduction from the irradiated RPE into the retina in a typical laser lesion leads to irreversible thermal denaturation of the outer and inner segments [7, 76, 77]. For a variety of retinal diseases, which are probably associated with a destruction of the RPE, selective treatment of the RPE might be sufficient. Thus the overlying photoreceptors can be spared in order to avoid visual field loss, which is especially useful in the macula. If the damaged RPE was regenerating in the healing process, due to migration and proliferation of the adjacent RPE, minimal, destructive, selective RPE treatment might be optimal.

Heat diffuses out of the absorbing RPE layer at a speed of roughly 1 μm/μs. Hence, traditional laser exposures of 100 ms and more result in considerable heat conduction. The spatial and temporal temperature distribution can be calculated by mathematical models, and can be verified experimentally [4, 56]. Only a small temperature difference exists between the RPE and the neural retina after a 100-ms exposure. This difference is about 18 % from the RPE to 5 μm into the retina. The pulse duration needed to spare the neural retina can be estimated by the thermal relaxation time or the time interval required for the heat to diffuse out of an absorbing tissue [1, 4, 57]. Considering a size of an RPE cell of about 10 μm, high temperatures can be limited to the RPE cell itself, if the exposure time is of the order of microseconds rather than the customary millisecond settings. Since no more laser energy is delivered at the end of a laser pulse, the temperature quickly smooths out. If the tissue between repetitive laser pulses has suffi-

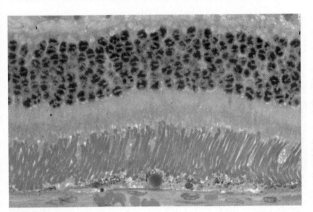

Fig. 13.10. Histology of the retina 2 h after selective RPE treatment. The RPE is significantly destroyed

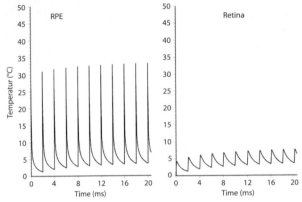

Fig. 13.11. Temperature-time course within the RPE and the retina during application of repetitive ms-laser pulses. Despite significant temperature elevations within the RPE the average temperature in the retina is low, sparing the photoreceptors

cal pilot study, the focus of treatment was on three pathological conditions: diabetic macular edema, central serous retinopathy, and drusen in age-related macular degeneration (AMD). Treatment was performed using a chain of repetitive laser pulses with a frequency-doubled Nd:YLF laser. In a pilot study, the selectivity of retinal pigment epithelium treatment was investigated. Microperimetry was performed directly on top of laser lesions during a follow-up period of up to 1 year. A repetitive pulsed Nd:YLF laser was applied in 17 patients, using pulse energies of 20–130 μJ. To find the necessary energy, test exposures were performed in the inferior macular region. Seventy-three of 179 test lesions were followed at various times by performing microperimetry directly on top of the laser lesions. All test lesions were at the threshold of RPE disruption and none of the laser effects was visible by means of ophthalmoscopy during photocoagulation. After exposure with 500 pulses, retinal defects could be detected in up to 73% of patients (100 μJ) after the first day. Most defects were no longer detectable after 3 months. After exposure with 100 pulses, no defects could be detected with 70 and 100 μJ after 1 day, and the neural retina remained undamaged during the follow-up period. Thus selective retinal pigment epithelium damage could be achieved. In a multicenter clinical trial visual and morphologic outcome of patients with focal DMP treated with SRT was evaluated [20].

Sixty eyes in 60 patients were treated with SRT using a frequency doubled Q-switched Nd:YLF laser (527 nm). Each laser exposure contained a train of 30 pulses, each with a duration of 1.7 μs, at a repetition rate of 100 Hz. The SRT laser lesions were not visible ophthalmoscopically during treatment, but were detectable by fluorescein angiography. Median foveal retinal thickness, measured by optical coherence tomography (OCT), was 244 μm and 230 μm at baseline at 6 months. The maximum retinal thickness measured by radial OCT scans in the treated edematous macula area decreased from 351 μm at baseline to 330 μm after 6 months. FA leakage decreased in 31.1%, remained stable in 52.1% and increased in 15.8% of the patients after 6 months. Visual acuity results at 6 months showed that 39.6% of patients had an improvement of greater than 1 line, 49.1% demonstrated a stabilization of VA within ±1 line and 11.3% had a reduction of greater than 1 line. According to these first clinical results, SRT offers the potential for earlier treatment and the possibility of treating closer to the fovea without side effects associated with conventional argon laser treatment. Based on these findings, SRT has also shown that the destruction of the photoreceptors is not always necessary and laser treatment has to be individually adapted to the underlying diseases.

cient time to cool down completely to baseline, high temperatures can be achieved inside the RPE, keeping temperatures low in the adjoining photoreceptors at the same time. Figure 13.10 shows the histological effect after exposure to a chain of 500 repetitive 5-μs laser pulses.

In animal experiments, it has been shown that the RPE may respond in several different ways after injury. Adjacent RPE cells are able to spread out to fill defects with hypertrophy. This has been shown in rabbits after photocoagulation and after surgically induced RPE defects and in monkeys after retinal detachment [30]. A new RPE barrier is quickly restored. In treatment of diabetic macular edema, the beneficial effect of laser coagulation is thought to be mediated by restoration of a new RPE barrier [12]. Similar is the rationale of therapy in central serous retinopathy (CSR). Another possible target is the therapy of drusen. Drusen disappear after photocoagulation of the surrounding tissues. The value of prophylactic treatment of drusen is actually being studied by several groups [21, 22, 28]. In a first clini-

13

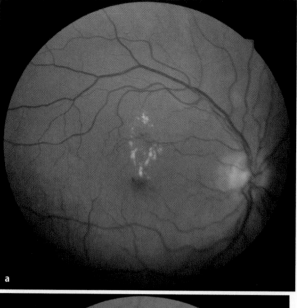

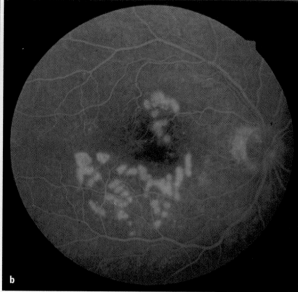

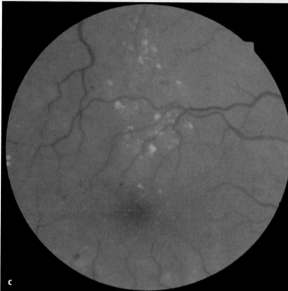

Fig. 13.12. a Focal diabetic macular edema before treatment by SRT. **b** The same fundus 2 h after SRT – the lesions are visible only in the fluorescein angiogram and show the pattern of treatment. **c** Fundus image 6 months after SRT. The hard exudates have resolved

References

1. Anderson RR, Parrish JA (1983) Selective photothermolysis: precise microsurgery by selective absorption of pulsed radiation. Science 220:524–527
2. Apple DJ, Thomas G, Swartz M, Kavka-Van Norman D (1984) Comparative histopathology and ultrastructure of human krypton and argon laser lesions. Doc Ophthalmol Proc Series 36. Junk Publ, The Hague, pp 45–60
3. Balles MW, Puliafito CA, D'Amico DJ, Jacobson JJ, Birngruber R (1990) Semiconductor diode laser photocoagulation in retinal vascular disease. Ophthalmology 97:1553–1561
4. Birngruber R (1980) Thermal modeling in biological tissues. In: Hillenkamp F, Pratesi R, Sacchi CA (eds) Lasers in biology and medicine. Plenum, New York, pp 77–97
5. Birngruber R (1984) Die Lichtbelastung unbehandelter Netzhautareale bei der Photokoagulation. Fortschr Ophthalmol 81:147–149
6. Birngruber R, Fujimoto JG, Puliafito CA, Schoenlein RW, Lin WZ, Gawande A (1988) [Retinal effects produced by femtosecond laser pulse]. Fortschr Ophthalmol 85:699–704
7. Birngruber R, Gabel VP, Hillenkamp F (1983) Experimental studies of laser thermal retinal injury. Health Phys 44:519–531
8. Boergen KP, Birngruber R, Hillenkamp F (1981) Laser-induced endovascular thrombosis as a possibility of selective vessel closure. Ophthalmic Res 13:139–150
9. Bornfeld N, Gerke E, El-Hifnawi E (1982) Licht- und rasterelektronenmikroskopische Untersuchungen an frischen und vernarbten Xenon-Lichtkoagulations-Herden der Netzhaut von "mini-pigs". Fortschr Ophthalmol 79:151–154
10. Brancato R, Pratesi G, Leoni G, Trabucchi G, Vanni U (1989) Histopathology of diode and argon laser lesions in rabbit retina. Invest Ophthalmol Vis Sci 30:1504–1510
11. Branch Vein Occlusion Study Group (1986) Argon laser scatter photocoagulation for prevention of neovascularization and vitreous hemorrhage in branch vein occlusion. A randomized clinical trial. Arch Ophthalmol 104:34–41
12. Bresnick GH (1983) Diabetic maculopathy. A critical review highlighting diffuse macular edema. Ophthalmology 90:1301–1317
13. Central Vein Occlusion Study Group (1993) Central vein occlusion study of photocoagulation therapy. Baseline findings. Online J Curr Clin Trials Doc 95:6021
14. Dawson DW, Volpert OV, Gillis P, Crawford SE, Xu H, Benedict W, Bouck NP (1999) Pigment epithelium-derived factor: a potent inhibitor of angiogenesis. Science 285:245–248
15. Diddie KR, Ernest JT (1977) The effect of photocoagulation on the choroidal vasculature and retinal oxygen tension. Am J Ophthalmol 84:62–66
16. Doft BH, Blankenship GW (1982) Single versus multiple treatment sessions of argon laser panretinal photocoagula-

tion for proliferative diabetic retinopathy. Ophthalmology 89:772–779

17. Early Treatment Diabetic Retinopathy Study Research Group (1985) Photocoagulation for diabetic macular edema. Early Treatment Diabetic Retinopathy Study Report No. 1. Arch Ophthalmol 103:1796–1806

18. Early Treatment Diabetic Retinopathy Study Research Group (1991) Early photocoagulation for diabetic retinopathy. ETDRS Report No. 9. Ophthalmology 98:766–785

19. Early Treatment Diabetic Retinopathy Study Research Group (1991) Grading diabetic retinopathy from stereoscopic color fundus photographs – an extension of the modified Airlie House classification. ETDRS Report No. 10. Ophthalmology 98:786–806

20. Elsner H, Liew SHM, Klatt C, Hamilton P, Marshall J, Pörksen E, Laqua H, Brinkmann R, Birngruber R, Roider J (2005) Selective Retina Therapy (SRT) Multicenter Clinical Trial: 6 Month Results in Patients With Diabetic Maculopathy. Poster Presentation, The ARVO Annual Meeting, Program Summary Book 1463/B232, 63

21. Figueroa MS, Regueras A, Bertrand J, Aparicio MJ, Manrique MG (1997) Laser photocoagulation for macular soft drusen. Updated results. Retina 17:378–384

22. Frennesson CI (2003) Prophylactic laser treatment in early age-related maculopathy: an 8-year follow-up in a randomized pilot study shows a reduced incidence of exudative complications. Acta Ophthalmol Scand 81:449–454

23. Gabel VP, Birngruber R (1979) Klinische Folgerungen aus der Xanthophylleinlagerung in der Netzhautmitte. Ber Dtsch Ophthalmol Ges 76:475–478

24. Gloor BP (1969) Zellproliferation, Narbenbildung und Pigmentation nach Lichtkoagulation (Kaninchen-Versuche). Klin Monatsbl Augenheilkd 154:633–648

25. Guerrin M, Moukadiri H, Chollet P, Moro F, Dutt K, Malecaze F, Plouet J (1995) Vasculotropin/vascular endothelial growth factor is an autocrine growth factor for human retinal pigment epithelial cells cultured in vitro. J Cell Physiol 164:385–394

26. Gutman FA, Zegarra H (1974) The natural course of temporal retinal branch vein occlusion. Trans Am Acad Ophthalmol Otolaryngol 78:OP178–OP192

27. Hayreh SS, Rojas P, Podhajsky P, Montague P, Woolson RF (1983) Ocular neovascularization with retinal vascular occlusion – III. Incidence of ocular neovascularization with retinal vein occlusion. Ophthalmology 90:488–506

28. Ho AC, Maguire MG, Yoken J, Lee MS, Shin DS, Javornik NB, Fine SL (1999) Laser-induced drusen reduction improves visual function at 1 year. Choroidal Neovascularization Prevention Trial Research Group. Ophthalmology 106:1367–1373

29. Initiativgruppe Früherkennung diabetischer Augenerkrankungen. Stadieneinteilung und Lasertherapie der diabetischen Retinopathie und Makulopathie. Contact address: Prof. Dr. P. Kroll, Universitäts-Augenklinik, Robert-Koch-Strasse 4, D-35037 Marburg. Report No. 1

30. Inomata H (1975) Wound healing after xenon arc photocoagulation in the rabbit retina. Identification of the proliferating cells in the lesion by light and electron microscopic autoradiography using 3H-thymidine. Ophthalmologica 170:462–474

31. Johnson RN, McNaught EI, Foulds WS (1977) Effect of photocoagulation on the barrier function of the pigment epithelium. II. A study by electron microscopy. Trans Ophthalmol Soc UK 97:640–651

32. L'Esperance FA, Jr (1969) The treatment of ophthalmic vascular disease by argon laser photocoagulation. Trans Am Acad Ophthalmol Otolaryngol 73:1077–1096

33. L'Esperance FA, Jr (1989) Ophthalmic lasers, vol. I. Mosby, St. Louis

34. Littmann H (1957) Der Zeiss-Lichtkoagulator nach Meyer-Schwickerath mit Xenonhochdrucklampe. Ber Dtsch Ophthalmol Ges 61:311–316

35. Lorenz B (1988) Morphologic changes of chorioretinal argon laser burns during the first hour post exposure. Lasers Life Sci 2:207–226

36. Lorenz B, Birngruber R, Vogel A (1989) Quantifizierung der Wellenlängenabhängigkeit laserinduzierter Aderhauteffekte. Fortschr Ophthalmol 86:644–654

37. Lorenz B, Ganson N, Stein HP, Birngruber R (1986) CW-Neodym:YAG-Laserphotokoagulation am Kaninchenfundus-Effekt auf Aderhaut und Sklera im Vergleich mit Argonlaserphotokoagulation. Fortschr Ophthalmol 83:436–440

38. Machemer R, Laqua H (1975) Pigment epithelium proliferation in retinal detachment (massive periretinal proliferation). Am J Ophthalmol 80:1–23

39. Maiman TH (1960) Stimulated optical radiation in ruby. Nature 187:493–497

40. Marshall J, Clover G, Rothery S (1984) Some new findings of retinal irradiation by krypton and argon lasers. In: Birngruber R, Gabel V-P (eds) Laser treatment and photocoagulation of the eye. Junk Publ, The Hague, pp 21–37

41. Marshall J, Mellerio HJ (1968) Histology of retinal lesions produced with Q-switched lasers. Exp Eye Res 7:225–230

42. Marshall J, Mellerio HJ (1976) Laser irradiation of retinal tissue. Br Med Bull 26:156–160

43. Marshall J, Mellerio J (1967) Histology of the formation of retinal laser lesions. Exp Eye Res 6:4–9

44. Marshall J, Mellerio J (1971) Disappearance of retino-epithelial scar tissue from ruby laser photocoagulations. Exp Eye Res 12:173–174

45. McGee GS, Davidson JM, Buckley A, Sommer A, Woodward SC, Aquino AM, Barbour R, Demetriou AA (1988) Recombinant basic fibroblast growth factor accelerates wound healing. J Surg Res 45:145–153

46. Meyer-Schwickerath G (1949) Koagulation der Netzhaut mit Sonnenlicht. Ber Dtsch Ophthalmol Ges 55:256–259

47. Meyer-Schwickerath G (1954) [Light coagulation; a method for treatment and prevention of the retinal detachment.] Albrecht Von Graefes Arch Ophthalmol 156:2–34

48. Meyer-Schwickerath G (1956) [Photo-coagulation of the ocular fundus and the retina.] Ann Ocul (Paris) 189:533–548

49. Michels RG, Gass JD (1974) The natural course of retinal branch vein obstruction. Trans Am Acad Ophthalmol Otolaryngol 78:OP166–OP177

50. Miki T, Sato T, Hiromori T, Ohsawa E, Mii T (1984) Effect of xenon arc photocoagulation in normal and pathologic conditions on the rabbit eye. In: Birngruber R, Gabel VP (eds) Laser treatment and photocoagulation of the eye. Junk Publ, The Hague, pp 81–93

51. Miller SD (1985) Argon laser photocoagulation for macular edema in branch vein occlusion. Am J Ophthalmol 99:218–219

52. Ogata N, Ando A, Uyama M, Matsumura M (2001) Expression of cytokines and transcription factors in photocoagulated human retinal pigment epithelial cells. Graefes Arch Clin Exp Ophthalmol 239:87–95

53. Ogata N, Tombran-Tink J, Jo N, Mrazek D, Matsumura M (2001) Upregulation of pigment epithelium-derived factor after laser photocoagulation. Am J Ophthalmol 132:427–429

54. Olk RJ (1990) Argon green (514 nm) versus krypton red (647 nm) modified grid laser photocoagulation for diffuse diabetic macular edema. Ophthalmology 97:1101–1112

55. Perry DD, Risco JM (1982) Choroidal microvascular repair after argon laser photocoagulation. Am J Ophthalmol 93:787–793

56. Roider J, Birngruber R (1995) Solution of the heat conduction equation. In: Welch AJ, Van Gemert M (eds) Optical-thermal response of laser irradiated tissue. Plenum Press, New York, pp 385–409

57. Roider J, Hillenkamp F, Flotte T, Birngruber R (1993) Microphotocoagulation: selective effects of repetitive short laser pulses. Proc Natl Acad Sci U S A 90:8643–8647

58. Roider J, Schiller M, El-Hifnawi E, Birngruber R (1994) Retinale Photokoagulation mit einem gepulsten, frequenzverdoppelten Nd:YAG-Laser (532 nm). Ophthalmologe 91:777–782

59. Rosner M, Tchirkov M, Dubinski G, Naveh N, Tso MOM (1996) Methylprednisolone ameliorates laser induced retinal injury in rats. Invest Ophthalmol Vis Sci 37(Suppl):696

60. Schachat AP (1991) Management of subfoveal choroidal neovascularization. Arch Ophthalmol 109:1217–1218

61. Schmidt-Erfurth U, Vogel A, Birngruber R (1992) The influence of wavelength on the laser power required for retinal photocoagulation in cataractous human eyes. Laser Light Ophthalmol 5:69–78

62. Schuscherba ST, Cross ME, Pizarro JM, Ujimori V, Nemeth TJ, Bowman PD, Stuck BE, Marshall J (1996) Pretreatment with hydroxyethyl starch-deferoxamine but not glucocorticoids limits neutrophil infiltration in acute retinal laser injury. Invest Ophthalmol Vis Sci 37(Suppl):695

63. Shono T, Ono M, Izumi H, Jimi SI, Matsushima K, Okamoto T, Kohno K, Kuwano M (1996) Involvement of the transcription factor NF-kappaB in tubular morphogenesis of human microvascular endothelial cells by oxidative stress. Mol Cell Biol 16:4231–4239

64. Stoltz RA, Abraham NG, Laniado-Schwartzman M (1996) The role of NF-kappaB in the angiogenetic response of coronary microvessel endothelial cells. Proc Natl Acad Sci U S A 93:2832–2837

65. The Canadian Ophthalmology Study Group (1993) Argon green vs krypton red laser photocoagulation of extrafoveal choroidal neovascular lesions: one year results in age-related macular degeneration. Arch Ophthalmol 111:181–185

66. The Diabetic Retinopathy Study Research Group (1976) Preliminary report on effects of photocoagulation therapy. Am J Ophthalmol 81:383–396

67. The Diabetic Retinopathy Study Research Group (1978) Photocoagulation treatment of proliferative diabetic retinopathy: the second report of diabetic retinopathy study findings. Ophthalmology 85:82–106

68. The Diabetic Retinopathy Study Research Group (1979) Four risk factors for severe visual loss in diabetic retinopathy. The third report from the Diabetic Retinopathy Study. Arch Ophthalmol 97:654–655

69. The Diabetic Retinopathy Study Research Group (1981) Photocoagulation treatment of proliferative diabetic retinopathy. Clinical application of Diabetic Retinopathy Study (DRS) findings, DRS Report No. 8. Ophthalmology 88:583–600

70. Tso MOM, Cunha-Vaz JG, Shih C, et al. (1980) Clinicopathologic study of blood-retinal barrier in experimental diabetes mellitus. Arch Ophthalmol 98:2032–2040

71. Tso MOM, Wallow IH, Elgin S (1977) Experimental photocoagulation of the human retina. I. Correlation of physical, clinical, and pathologic data. Arch Ophthalmol 95:1035–1040

72. Van der Zypen E, Fankhauser F, Loertsher HP (1984) Retinal and choroidal repair following low power argon and Nd-YAG laser irradiation. In: Birngruber R, Gabel VP (eds) Doc Ophthalmol Proc Series 36. Junk Publ, The Hague, pp 61–70

73. Vogel A, Birngruber R (1992) Temperature profiles in human retina and choroid during laser coagulation with different wavelengths ranging from 514 to 810 nm. Lasers Light Ophthalmol 5:9–16

74. Wallow IH (1981) Long-term changes in photocoagulation burns. Dev Ophthalmol 2:318–327

75. Wallow IH (1984) Repair of the pigment epithelial barrier following photocoagulation. Arch Ophthalmol 102:126–135

76. Wallow IH, Birngruber R, Gabel VP, Hillenkamp F, Lund OI (1975) [Retinal reactions to intense light. I. Threshold lesions. Experimental, morphological and clinical studies of pathological and therapeutic effects of laser and white light]. Adv Ophthalmol 31:159–232

77. Wallow IH, Tso MOM, Fine BSF (1973) Retinal repair after experimental xenon arc photocoagulation. 1. A comparison between rhesus monkey and rabbit. Am J Ophthalmol 75:32–52

78. Wilson AS, Hobbs BG, Shen WY, Speed TP, Schmidt U, Begley CG, Rakoczy PE (2003) Argon laser photocoagulation-induced modification of gene expression in the retina. Invest Ophthalmol Vis Sci 44:1426–1434

79. Wolbarsht ML, Landers MB, Stefansson E (1981) Vasodilatation and the etiology of diabetic retinopathy: a new model. Ophthalmic Surg 12:104–107

80. Yannuzzi LA (1982) Krypton red laser photocoagulation for subretinal neovascularisation. Retina 2:29–46

14 The Role of Photodynamic Therapy in Retinal Vascular Disease

B. Jurklies, N. Bornfeld

Core Messages

- Photodynamic therapy (PDT) has improved functional outcome in the treatment of choroidal neovascularization (CNV) due to various clinical disorders. While PDT is used as a first line therapy for the treatment of choroidal hemangiomas, a small number of case series and studies have investigated the role of PDT in the treatment of retinal disorders, such as retinal angioma, vasoproliferative tumors, and parafoveal telangiectasis
- PDT is a non-thermal, photochemical, two-step modality with preferential selectivity for the vascularized target tissue of the choroids compared to the vessels of the retina. It requires:
  - Photoactivable compound (photosensitizer) preferentially accumulated in the target tissue
  - Oxygen
  - Light of a wavelength meeting the absorption maximum of the photosensitizer and activating the photosensitizer
- While photocoagulation induces immediately visible thermal effects with light pulses of high intensity and short duration inducing a non-selective coagulation necrosis, PDT needs different parameters compared to photocoagulation, with a longer duration of light exposure, a smaller number of treatment spots, larger spot size, and a lower light intensity, resulting in a photothrombosis which is invisible on ophthalmoscopy during the treatment procedure.
- Light exposure after accumulation of the photosensitizer in the target tissue may result in activation and transformation of the photosensitizer to the excited triplet state. The highly reactive triplet state molecule may react with:
  - Tissue substrates (Type I reaction)
  - Ground state oxygen (Type II reaction), resulting in production of singlet oxygen
  - The Type II reaction may represent the major part of the photodynamic reaction

- PDT may induce vascular, cellular and immuno-modulatory effects, resulting in deformation of cell organelles, peroxidation of lipid membranes, increased permeability of cell membranes, damage to the endothelium, thrombus formation, vessel occlusion, upregulation of various interleukin factors, activation of lymphocyte subtypes and immunosuppressive effects
- Verteporfin (benzoporphyrin derivative monoacid A) has been shown to be an effective photosensitizer in animal experiments and clinical multicenter studies and is the only photosensitizer for the clinical use and the therapy of ocular disorders
  - The liposomal delivered formulation may facilitate the binding to plasma lipoproteins and to LDL receptors in the target tissue, to enhance the phototoxic effects in the target tissue
  - The short half-life reduces the risk of toxic side effects and photosensitivity
  - The absorption peak at 692 nm is used for clinical application
- The effects of PDT on vascular lesions during the follow-up consist of:
  - Damage to the vascular endothelium
  - Exudation due to vascular leakage (hours after PDT session)
  - Photothrombosis within the vessels (hours to days after treatment)
  - Recanalization and reproliferation due to activation of angiogenic factors, e.g., vascular endothelial growth factor (VEGF) (weeks after PDT session)
  - Fibrosis and deactivation following retreatments (weeks to months after PDT)
- Treatment of choroidal lesions using the recommended parameters may be associated with a selectivity of PDT expressed more for the choroidal than for the retinal layers. However, effects of PDT on the choroidal layer have been observed for both the choroidal lesion and the normal choroid beneath the vascular lesion

- Toxic effects and side effects of PDT with verteporfin may consist of:
  - Visual disturbances (abnormal vision, decreased vision, visual feeled defects)
  - Acute severe visual loss and visual loss of 6 lines
  - Hemorrhages
  - Alterations of the retinal pigment epithelium (RPE), retinal pigment epithelial tear of the RPE
  - Injection side adverse effects
  - Allergic reactions
  - Photosensitivity
- The current treatment modality is based on multicenter studies treating CNVs using verteporfin (dose of 6 mg/m^2 body surface area), and exposing a light dose of 50 J/cm^2 within 83 s (irradiance of 600 mW/m^2). Sev-eral authors suggested that this light dose may be not sufficiently effective for the treatment of angiomas and hemangiomas and recommended a light dose of 100 J/cm^2
- The effects of PDT on retinal lesions are at least in part limited by the small number of cases and case series. However, the incidence of most of these diseases may be low
- PDT may represent an alternative treatment modality for retinal capillary hemangiomas (RCAs) located in the periphery, with significant exudation, which may not be evaluable for laser coagulation
- In addition, treatment effects due to PDT have been observed on vasoproliferative tumors resolving the exudation
- However, PDT may not be useful for treating parafoveal teleangiectasis without any CNV

14.1 Introduction

Photodynamic therapy (PDT) with verteporfin has significantly improved functional outcome in the treatment of choroidal neovascularization (CNV) in various clinical disorders [86]. While CNV due to age-related macular degeneration [4, 53, 76–80, 84] and pathologic myopia [83, 85] was the primary focus, several studies showed that the treatment effects of PDT were not only restricted to these two underlying disorders. Recently, several clinical studies observed PDT as a treatment modality of CNV due to ocular histoplasmosis syndrome [52, 55, 68], choroiditis [71, 89], angioid streaks [31, 37, 68], Stargardt's disease [81], symptomatic choroidal hemangioma [30, 61, 82] and other causes [15, 72]. While PDT has become an established treatment modality for various choroidal disorders, its role in the treatment of retinal diseases remains to be determined.

This chapter reports on the molecular and biophysical mechanisms of photodynamic therapy, the characteristics of the photosensitizer verteporfin, the steps required to complete the PDT, and current treatment results for various retinal disorders.

14.2 Photodynamic Therapy

Photodynamic therapy is a non-thermal, photochemical, two-step treatment modality; this allows the treatment of a vascularized target tissue with preferential selectivity. It requires the application of a non-toxic photoactivable compound (photosensitizer), the preferential accumulation of the photosensitizer in the target tissue, oxygen and the activation of the photosensitizer by non-thermal light exposure of a specific wavelength. This procedure results in a sequence of photochemical and photobiological effects in the target tissue. The selectivity of this treatment modality is achieved by a preferential accumulation of the photosensitizer in the target area and the light exposure, which meets the absorption maximum of the photosensitizer and is confined to the target area [14].

14.2.1 Effects of Light on Biological Tissue

The therapeutic effects of light on biological tissues in ophthalmology may be transmitted by photomechanical, photothermal, and photochemical means. Photodisruption uses light of high power applied in very short-term pulses and very small light spots to irradiate a small tissue volume [73]. These high-intensity electric fields to a small focal laser volume are able to generate a plasma of positively and negatively charged ions which induce a shock wave when applied with durations of nanoseconds or shorter [73]. These shock waves mechanically disrupt tissue in the vicinity of the target volume [8]. Photodisruption with nanosecond Nd:YAG laser pulses is applied for laser iridotomies and capsulotomies of the posterior capsule. Photocoagulation induces thermal effects with light pulses of high energy and short duration. A non-selective coagulation necrosis is achieved in biological tissues with light absorbing fundus chromophores (melanin of the retinal pigment epithelium, hemoglobin). It is associated with irreversible protein denaturation and cellular injury and involves the structure and function of adjacent tissues (e.g., retina). Thermal effects are applied by, e.g., laser photocoagulation for the treatment of

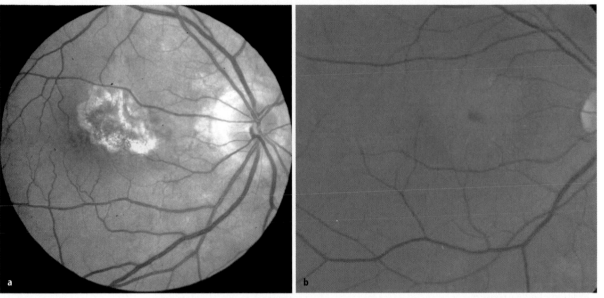

Fig. 14.1. a Scar after several sessions of laser photocoagulation due to an extrafoveal CNV and several extrafoveal and juxtafoveal recurrences of CNV. **b** In contrast, no effects are detectable on the fundus after a PDT session

Fig. 14.2. Photodynamic therapy consisting of intravenous administration of the photosensitizing agent (Ia), the distribution and accumulation of the photosensitizing agent (Ib), particularly within the target tissue, and the light application to the target tissue (II). Both steps are necessary to induce effects of photodynamic therapy within the target tissue

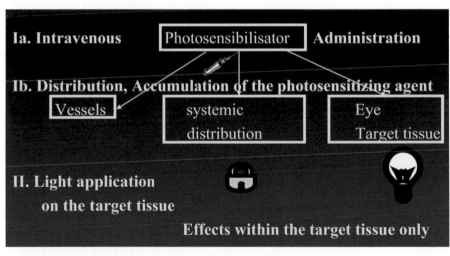

extrafoveal CNVs to achieve an obliteration of the neovascular tissue [10]. Their restrictions in the treatment of subfoveal lesions are well known and have emphasized the search for modalities resulting in less impairment of retinal function (Fig. 14.1a). In contrast, photodynamic therapy represents a photochemical procedure [14]: It involves first the (intravenous) administration of a target tissue localizing a photosensitizing agent, followed by activation of the photosensitizer in the target tissue with a light of a wavelength that is specific for absorption by the photosensitizing agent. Both steps are a precondition for the photosensitized reactions in the target tissue and are depicted in Fig. 14.2. While these photosensitizing reactions do not require any light doses with

thermal or mechanical effects, the involvement of oxygen in the target tissue is essential. Therefore, compared to the findings after laser coagulation of a CNV, PDT does not reveal any signs that would be visible within the area of the CNV immediately after the treatment procedure, when performed with low light doses in a therapeutic range (Fig. 14.1b). The preconditions necessary for PDT comprise the following:

- (Intravenously applied) photosensitizer
- Light of a wavelength specific for the absorption by the photosensitizer
- Oxygen!
- Non-thermal and photochemical reaction

14.2.2 Differences Between PDT and Laser Coagulation

The differences between laser coagulation and photodynamic therapy are presented in Table 14.1, which shows that both techniques require different parameters regarding duration of light exposure, number of exposures, spot size, intensity of light and visibility immediately after the performance of the procedure.

Table 14.1. Differences between laser coagulation and photodynamic therapy according to the duration of light exposure, number of exposures, spot size, intensity of light and visibility after the treatment session. The parameters of laser coagulation and PDT are compared ($\uparrow$ high, long, $\downarrow$ low, short)

Parameters	Laser coagulation	PDT
Duration of light exposure	$\downarrow$	$\uparrow$
Number of treatment spots	$\uparrow$	$\downarrow$
Treatment spot size	$\downarrow$	$\uparrow$
Intensity of light	$\uparrow$	$\downarrow$
Visibility immediately after treatment	$\uparrow$	$\downarrow$

14.2.3 Mechanisms of Photodynamic Therapy

The photochemical processes in the target tissue after this basic two-step procedure are shown in Fig. 14.3. The term photodynamic reaction is used to distinguish photosensitized reactions from other photochemical processes suggesting that oxygen is consumed [14]. Briefly, after administration of the photosensitizing agent and accumulation within the target area, light energy is absorbed ($h\nu$) by the pho-

tosensitizer. The photosensitizer is transformed from the ground singlet to the excited singlet state. This state of activity may decay non-radiatively by internal conversion, emit photons deriving fluorescence, or undergo intersystem crossing at the electron level to the excited triplet state. It represents a highly reactive, short-living molecule, which may emit phosphorescence with conversion back to the ground state. Alternatively, it initiates photochemical reactions by transferring energy via two pathways: It may induce Type I and Type II reactions by interacting with neighboring tissue substrates (Type I) or ground state ($^3O_2$) oxygen (Type II), respectively (Fig. 14.3). In the Type I reaction, the excited triplet state molecule reacts directly with tissue substrates, and an electron or hydrogen atom transfer takes place leading to superoxide-anion (O_2^-). In Type II reaction, the excited triplet state molecule reacts with ground state oxygen leading to excited state singlet oxygen ($^1O_2$) formation, while the photosensitizer regenerates to the ground state (Fig. 14.3). This Type II reaction with singlet oxygen production is suggested to mediate the major part of the photodynamic reaction, although it has not been detected in vivo. Singlet oxygen has a short life time ($< 0.04 \, \mu s$) and a short radius of action ($< 0.02 \, \mu m$) [13, 14]. Therefore the photodynamic reaction mediated tissue damage is closely related to the localization of the sensitizing agent [14]. In experimental setups one photon is able to induce one singlet oxygen molecule. Nevertheless, the reactive oxygen species, singlet oxygen, superoxide-anion, hydroxyl radicals and the triplet ground state photosensitizer are involved in mediating tissue injury at the cellular, subcellular, nucleic acid, enzyme and cell membrane

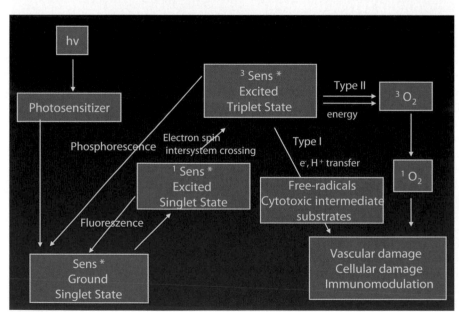

Fig. 14.3. Mechanisms of the photodynamic reaction

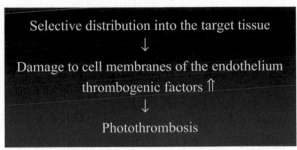

Fig. 14.4. The effects of PDT inducing a photothrombotic reaction within the vessels

levels [14]: In addition, PDT has effects on vessels and immunomodulating factors.

Destructive effects on the cell membranes can be observed shortly after light exposure [14, 41], resulting in swelling, reduction in active membrane transport systems, increased permeability of cell membranes, inhibition of plasma membrane enzymes, lipid peroxidation, deformation of cell organelles, mitochondria and membrane bound enzymes [14, 20, 40–42, 75, 88].

At the vascular level photodynamic vascular thrombosis is the result of cell damage of the endothelium with platelet aggregation, thrombus formation, vascular occlusion, and hemorrhage mediated by the photodynamic reaction [17, 25, 26, 56]. A summary of vascular actions resulting in photothrombotic effects is shown in Fig. 14.4. It has been suggested that the effects of photothrombosis are directly correlated with the amount of circulating photosensitizing agent at the time of irradiation [14, 17].

In addition, immunologic effects may be mediated, at least in part as a result of the inflammatory reaction after PDT: An upregulation of interleukins, IL-1β, IL-2, tumor necrosis factor-α, IL-10, and granulocyte stimulating factors; migration of macrophages and neutrophils; and activation of B-lymphocytes and subtypes of T-lymphocytes has been reported [12, 16, 32, 33]. In addition, transient immunosuppressive effects after PDT have been observed: The survival of skin allografts has been prolonged at low concentrations of verteporfin, which might be mediated by reduced amounts of antigen presenting cells [44].

In summary, the effects of PDT in the target tissue may be expressed by cellular, vascular and immunologic mechanisms. While the vascular effects of PDT play an important role in mechanisms of action, the relative contribution of these mechanisms depends on the type of photosensitizer [14, 21].

14.2.3.1 Light

Depending on the irradiation the PDT is a linear process. The effect of PDT is at least in part determined by the total amount of light energy delivered to the target tissue. This fluence, expressed in J/cm^2, is a product of the time of illumination in seconds and the power density, expressed in mW/cm^2. To achieve a fluence of 50 J/cm^2, light with a power density of 600 mW/cm^2 has to be applied for 83 s. The light source should match the absorption spectrum of the photosensitizing agent. In ophthalmology, lasers are the source of choice because of their several advantages: They represent a light source with monochromatic and high energy light, which allows the application of a constant fluence rate homogeneously within the light spot. The illumination is transmitted via an optic fiber system and focused on the fundus via a contact lens. Diode lasers are the source of choice at present.

The penetration depth of light illumination in PDT depends on the wavelength used. The depth of penetration is about 2–3 mm for light at a wavelength of 630 nm and increases up to 5–6 mm at a wavelength of 700–800 nm [24]. However, the effects of PDT are influenced by the specific characteristics of the photosensitizer [14]. The loss of light intensity due to the absorption in the layers of the fundus has been suggested to be negligible, particularly at a power density of 600 mW/cm^2 and at wavelengths used at present in the clinic, because the effects of PDT could be confirmed in the treatment of occult CNV with hemorrhages and in the treatment of other choroidal lesions [61, 84].

It has been shown that the relative selectivity between the retina and the choroids was reduced when the time between intravenous administration and illumination was reduced. Shortening this time enhanced the occlusive vascular effects on the retina and choroid with a loss of retinochoroidal selectivity [34].

14.2.3.2 Photosensitizer

Several factors influence at least in part the effects and selectivity: the intravenous concentration and the formulation of the photosensitizing agent [57, 58], and duration of the infusion [34]. Therefore, at the time of illumination a sufficient concentration of photosensitizer has to be present in the target tissue in order to generate sufficient photodynamic activity [14]. In addition, phototoxic effects within the tissue correlate with the applied dose of the light and the drug, respectively [34, 56]. To achieve identical effects, the concentration of the photosensitizer is inversely correlated with the light dose [38].

14 II

Various cellular and subcellular sites (cell nuclei, lysosomes, mitochondria, cell membranes, endothelial cells of the vasculatures) have been reported to be influenced by photosensitizers [14]. Photosensitizing agents are more likely to penetrate into the interstitium, when they are hydrophilic, and have a higher chance of being confined to the vessel wall when they are lipophilic [14]. In addition, the probability of cell inactivation per quantum absorbed light is higher for lipophilic than for hydrophilic sensitizers [42]. While in animal experiments a bolus injection has been used, it has been suggested that this mode of application enhanced the relative selectivity for effects of PDT on the neovasculature [25, 57].

The selectivity of a photosensitizer could be increased by coupling it to specific carriers, e.g., markers, antibodies, lipoproteins and markers associated with cellular proliferation [57–59]. An increased expression of LDL receptors in malignant and neovascular endothelium has been suggested to be responsible for the selective affinity of photosensitizers, particularly of the porphyrin derivatives, to the proliferating endothelium and tumor vessels [57]. Higher concentrations of photosensitizers have been observed in the tumor tissue compared to normal tissues (e.g., skin, muscles, brain, lung) [21]. In addition, coupling to the LDL receptors and internalizing the LDL-receptor-photosensitizer complex into the cell may increase the intracellular effects of the PDT after illumination [29, 59].

14.3 Verteporfin in PDT

There are sufficient numbers of photosensitizers under investigation. Nevertheless, verteporfin is the only photosensitizer currently approved for clinical use in clinical ophthalmology.

14.3.1 Characteristics

Verteporfin, benzoporphyrine derivative monoacid ring A (BPD-MA), is a second generation porphyrin derivative and consists of two regioisomers, which differ only in the location of the carboxylic acd and methyl ester on the C and D rings of the chlorine-type molecule. BPD-MA consists of a reduced porphyrin cycle with a cyclohexadiene ring fused at ring A. This cyclohexadiene ring may be responsible for the high photosensitizing potency. There are two absorption peaks, at 400 nm and at 692 nm, respectively. However, in order to avoid possible light toxicity, only the absorption peak at 692 nm has relevance for therapeutic use. Light at specific wavelengths can be efficiently absorbed by verteporfin 4–10 times more than hematoporphyrin [50]. In addition, this peak at 692 nm is free from competition for light by

hemoglobin, which absorbs light below 600 nm [50]. It has a half-life of 5–6 h, a fast clearance rate and is metabolized in the liver to inactive metabolites, where the monoacid regioisomers are metabolized to a diacid. Only 4% are cleared via kidneys. Verteporfin does not cause any significant photosensitivity of the skin 24 h after intravenous injection [49, 50].

A 10- to 70-fold increased cytotoxicity toward non-adherent human cell lines (leukemia cells, human lymphocytes, mouse mastocytoma cells) was found to develop compared with hematoporphyrin [48, 49].

It is lipophil which facilitates the accumulation in cell membranes, tumor cells improving the efficacy of the photosensitizer BPD-MA. To increase the selective localization and photodynamic action in the target tissue, BPD-MA was coupled to human low-density lipoprotein, suggesting that LDL metabolism is increased in neovascular tissue and tumor cells [2, 57]. Therefore, for clinical use, verteporfin has a liposomal delivered formulation, to facilitate binding to plasma lipoproteins, enhance the phototoxic effects in the target tissue and internalize verteporfin into the target cells via LDL-receptor mediated binding [2, 57–59].

14.3.2 Effects of PDT in Animal Experiments

Fluorescein microscopy detected BPD-MA in the RPE and choroids 5 min after the injection, with increasing concentrations within 20 min and staining including the photoreceptor outer segments. Therefore, it is detected rapidly in the vessels of the retina and choroids, and the RPE. However, in the retina and choroid, it is detected for no longer than 2 h, confirming the fast elimination time of BPD-MA [22]. Kramer et al. [34] observed BPD-MA in the CNV within 1–30 min after injection of 2 mg/kg, persisting for up to 2.5 h. Fluorescence in the normal choroidal and retinal vessels occurred and faded earlier: 5 min for choroidal vessels and 20 min for retinal vessels. However, traces of fluorescence were observed in the RPE up to 24 h after infusion [34].

Experimental CNV was induced in cynomolgus monkeys following argon laser coagulation and using a laser injury model by Ryan [54]. Effective closure of the CNV demonstrated by fundus photography, fluorescein angiography and electron microscopy could be observed [38] following intravenous injection of 1–2 mg/kg BPD-MA for 5 min. BPD-MA was activated 1–81 min after completion of dye injection using a light dose of 50, 75, 100, and 150 J/cm^2 at an irradiance of 150, 300, and 600 mW/cm^2, respectively. The endothelial cells of the CNV were necrotic or missing. The vessels were filled with platelets, neutrophils, erythrocytes and fibrin. The pericytes were vacuolized. Associated damage and loss of photoreceptors was also noted. Damage to the RPE, pyknotic nuclei in the outer nuclear layer, and loss of photoreceptors was noted, while the inner retina appeared almost unchanged.

Husain et al. [25] used BPD-MA at a dose of 0.375 mg/kg, infused for 10 min (fast infusion rate) and 32 min (slow infusion rate). Irradiation was followed 32–105 min after beginning infusion with a fluence of 150 J/cm^2 at 600 mW/cm^2. CNV was closed as demonstrated by fluorescein angiography 24 h after PDT, when irradiation was performed 32 min (fast infusion rate) and 32–55 min (slow infusion rate) after the start of the infusion. Histological examination revealed that the choriocapillaris was closed under the CNV. Histopathologic examination 4 weeks after PDT showed that the underlying choriocapillaris was reopened and the overlying neurosensory retina had some separation of the outer segments, swelling of the outer plexiform layer and pyknosis of cells from the outer nuclear layer. Selectivity of the PDT effects to normal retina and choriocapillaris was investigated. It demonstrated some damage to the RPE and choriocapillars and some damage to the photoreceptors in all of the illuminated eyes. The neurosensory retina showed up to 40% of pyknosis in the outer nuclear layer, when treated 30–40 min after start of infusion at a fast infusion rate. At a slow infusion rate and illumination within 65 min after start of infusion damage to the choriocapillaris, the RPE and disarray of the photoreceptors and up to 20% pyknosis of the outer nuclear layer was observed, while larger choroidal vessels remained intact.

Kramer et al. [34] further investigated the dye dosimetry and optimal treatment parameters, including time of laser irradiation after dye injection, to achieve selective closure of an experimentally induced CNV in the cynomolgus monkey. Dye doses of 0.25, 0.375, 0.5 and 1 mg/kg were studied. The light parameters were kept constant (irradiance 600 mW/cm^2, fluence 150 J/cm^2). Closure of CNV was observed at all dye doses: The lower the dose, the shorter the time interval after the injection in which the illumination induced an occlusion of the CNV. The experiments revealed optimal treatment parameters at a dye dose of 0.375 mg/kg, which correlates approximately to 6 mg/kg body surface area, with an illumination applied at 20–50 min after start of infusion. Eighty-five percent of the CNVs were closed using these treatment parameters. Investigating the effects on normal structures revealed damage of the RPE and some misalignments of the photoreceptor outer segments and nuclei at all dye doses. Damage of the larger choroidal vessels or retinal vessels or significant pyknosis of the outer nuclear layer was

observed at a dose of 1 mg/kg, 0.5, and 0.375 mg/kg, when illuminated within 50 min, 20 min, and 10 min after start of bolus injection, respectively. Therefore, the selectivity of the PDT with BPD-MA to the CNV may be reduced, when the time between illumination and start of infusion is short: Damage to retinal and larger choroidal vessels was observed, when illumination was performed within 5 min of the infusion, because the dye concentration may be equal in the normal choroid, retinal vessels and the CNV, respectively.

The effects of PDT (0.375 mg/kg BPD-MA, fluence 150 J/cm^2, irradiance 600 mW/cm^2) on experimentally induced CNVs in cynomolgus monkeys were observed in the long term, when follow-up was performed for 4 and 7 weeks: CNV was closed in 71% 4 weeks after PDT, while the damage to the normal RPE, choriocapillaris and photoreceptors histologically showed some recovery with preservation of the neurosensory retina 7 weeks after PDT [26]. These effects on CNV and normal structures have been observed with bolus injection, and infusion over 10 and 32 min. The effects of repeated treatments at 2 weekly intervals on cynomolgus monkeys have been observed using different doses of BPD-MA (6, 12, 18 mg/m^2 body surface area) at a fluence of 100 J/cm^2 and an irradiance of 600 mW/cm^2 20 min after administration of the drug. Damage to the normal retina and choroids was only minimal at a dose of 0.47 mg/kg (corresponding to 6 mg/m^2 body surface area), while higher doses increased the risk of significant damage to normal structures [47].

14.3.3 Effects of PDT in (Normal) Human Tissue

Following the findings in the animal experiments, the effects of PDT on CNVs due to age-related macular degeneration, pathologic myopia, angioid streaks and presumed ocular histoplasmosis syndrome have been investigated in a phase I/II trial as a proof of principle [39, 60, 68]. Fluorescein leakage was completely absent 1 week after PDT, and only minor leakage was found 4 weeks after PDT. Compared to the findings prior to PDT, leakage was present in the majority of patients and at least in part reduced in most patients 12 weeks after PDT, when illumination was performed 15–20 min after start of the infusion. Nevertheless an increase in size of the CNV was observed in some patients, particularly when light was applied 30 min after start of the infusion. Therefore, the interval between start of the infusion and illumination was reduced. Multiple treatments did not appear to have significant side effects on visual function. In this study, significant effects of PDT on the CNV were observed using light doses within a

Table 14.2. Effects of PDT on vascular lesions during the follow-up

Effect of PDT	Development during the follow-up time
Selective distribution of the photo-sensitizer	
Take-up on the endothelium of the vessels	
Damage of the vascular endothelium	
Exudation due to vascular leakage	Hours
Photothrombosis within the vessels	Hours to days
Recanalization	Weeks
Reproliferation	Weeks to months
Fibrosis and deactivation of the CNV following retreatment	Weeks to months

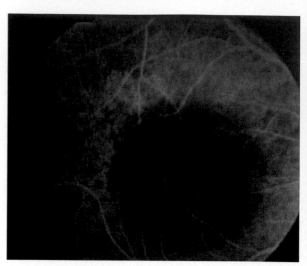

Fig. 14.5. Hypofluorescence detected by fluorescein angiography 1 week after PDT. The area of hypofluorescence correlates with the size of the laser spot and involves at least in part "normal" fundus areas not directly involved in the lesion components

range of 25–150 J/cm². Treatment parameters with a verteporfin concentration of 6 mg/m² body surface area administered intravenously over 10 min and activated with a light dose of 50 J/cm² applied over 83 s at a light intensity of 600 mW/cm² 15 min after start of infusion have been shown be to optimal for the treatment of CNVs [39, 60, 68].

The effects of PDT on CNV and vascular tumors during the follow-up are summarized in Table 14.2. They may at least in part be similar in both CNV and vascular tumors, respectively. Within the first hours after the PDT damage to the vascular endothelium, a breakdown is induced of the vascular tight junctions leading to an increase in leakage and exudation into the tissue. Following photothrombosis an occlusion of the CNV and at least in part the choroidal tissues adjacent to the CNV occurs within the first days after PDT. While PDT may at least in part enhance angiogenic stimuli (VEGF, PEDF) within the treated area, recanalization and reproliferation of the CNV may take place during the following weeks. This corresponds to a reappearance of fluorescein leakage and results in retreatment of the lesion. Deactivation and involution of the lesion may occur in the majority of patients when a sufficient number of retreatments has been performed.

Ultrastructural examination by electron and light microscopy and immunohistochemistry performed on a CNV 4 weeks after PDT revealed evidence of endothelial cell degeneration with platelet aggregation and thrombus formation of the peripheral vessels of the CNV. Some vessels showed nuclear swelling with peripheral chromatin clumping and cytoplasm vacuolization of the endothelial cells, and other vessels had an attenuation of the cytoplasm with retracted processes. The latter was suggested to be a change preceding a break in the endothelium. Findings of erythrocytes in a vessel cast with degenerated endothelial cells and pericytes, and macrophages wrapped around the occluded vessel and around the basal lamina of the vessel, may at least in part represent stages of vessel resorption. In addition, degenerated pericytes and sloughing of degenerated endothelial cells into the lumen were present. Clot fragmentation and reendothelialization observed in some parts may have been responsible for the restoration and recanalization of the occluded vessels. The vessels in the center of the membrane were unremarkable and may at least in part correlate with a reperfusion after occlusion due to the PDT [19].

In addition, the effects of PDT observed during the first weeks after the treatments are not limited to the lesion component [56, 62, 63]: The hypofluorescence detected by fluorescein angiography correlates particularly with the size of the treatment spot and may at least in part involve (normal) areas of the choroid beneath the lesion component (Fig. 14.5). Again, histopathologic studies confirmed the suggestion that hypofluorescence of the "normal" choroid represents a transient non-perfusion of the choroid within the treated area: PDT with light doses of 50 J/cm² induced a homogenous occlusion of the choriocapillaries with damage of the endothelial cells, represented by swelling, destruction and detachment from basement membrane. Extravasation of inflammatory cells and erythrocytes to the extracellular space was found. However, the distribution of some regular endothelial cells suggested a reorganization of new vessels to recanalize the area of occluded vessels. Sometimes areas of vacuolization of the RPE were observed. Exposure of PDT with light doses of 100 J/cm² induced additional effects on deeper choroidal vessels with a larger lumen and at least in part on the RPE with local cellular vacuolization and separation from Bruch's membrane. How-ever, no effects of the PDT on photoreceptors, capillaries of the retina and the optic nerve and the ganglion cell layer were observed with light doses of 50 J/cm^2 and 100 J/cm^2, respectively (Table 14.2).

14.3.4 Toxic Effects and Adverse Effects of PDT with Verteporfin

One of the major side effects is the increased photosensitivity particularly of the skin as long as the photosensitizing agent is circulating in its active form in the organism [14]. The clearance and the retention

Table 14.3. Adverse effects and visual loss of 6 lines and more judged as clinically relevant observed within the TAP, VIP and VIM multicenter studies

| | % of eyes | | | |
	TAP, 2001 AMD	VIP, 2001 AMD	VIP, 2001 Myopia	VIM, 2004[j] AMD
Visual disturbances[a]	22.1[b]	42[c]	23[d]	5/13
Acute severe visual loss[e]	0.7[h]	4.4[f]		0/3
Injection site adverse effects[i]	15.9	8	10	8/15
Infusion-related back pain	2.5	2.2	1	3/5
Allergic reactions	2	1	4	0/0
Photosensitivity	3.5	0.4	4	0/0
Visual loss ≥ 6 lines	18.2	29[g]	11	18/13

The numbers represent the percentage of eyes with adverse effects observed in the verteporfin group

[a] Visual disturbances included reports of abnormal vision, decreased vision and visual field defects of the verteporfin group

[b] TAP study group after 24 months including abnormal vision (10.2 %), decreased vision (14.4 %), and visual field defects (6 %) of the verteporfin subgroup

[c] VIP study group after 24 months including abnormal vision (20 %), decreased vision (30 %), and visual field defects (15 %) of the verteporfin subgroup

[d] VIP study group, CNV in pathologic myopia, after 24 months including abnormal vision (9 %), decreased vision (16 %), and visual field defects (4 %) of the verteporfin subgroup

[e] Visual loss of at least 4 lines within 7 days after the treatment

[f] Percentage of the verteporfin subgroup with occult CNV in AMD including extensive subretinal exudation (0.4 %), subretinal RPE hemorrhage (1.3 %), and no obvious cause (2.6 %)

[g] Percentage of the verteporfin subgroup with occult CNV in AMD

[h] TAP study 3 of 402 patients of the verteporfin subgroup

[i] Injection site adverse effects include pain, edema, extravasation, inflammation, hemorrhage, hypersensitivity, discoloration, and fibrosis

[j] Numbers are presented for the two subgroups with reduced and standard light doses of 25 J/cm^2 and 50 J/cm^2, respectively

time of a photosensitizer may at least in part determine the risk of phototoxic side effects on the skin [49, 50]. Using a dose of 6 mg/m^2 verteporfin, no phototoxic effects on the skin were observed 24 h and more after infusion [76, 77].

The spectrum of possible side effects after PDT with verteporfin has been observed in several prospective, double-masked, randomized, multicenter, placebo-controlled studies [76, 77, 83, 84, 87]. Adverse effects evaluated to be clinically relevant and observed for the verteporfin subgroup of the multicenter studies are listed in Table 14.3.

Visual disturbances included reports of abnormal vision, decreased vision and visual field defects observed in 10.2 %, 14.4 % and 6 % for the verteporfin group of the TAP study, respectively [77]. They were detected more for lesions with occult CNV [83, 84] than for lesions with classic CNV [77]. An acute severe loss of visual acuity within 7 days after PDT was observed more for patients with occult CNV compared to patients with lesions consisting of predominantly classic CNV.

Allergic reactions, infusion related pain, reactions of the injection site due to infusion of verteporfin and photosensitivity of the skin may be additional systemic adverse effects observed in the multicenter studies. A severe visual loss of 6 lines and more during the complete follow-up time was observed for 18 %, 29 %, 11 % and 13 – 18 % of the verteporfin subgroup patients in the TAP study, VIP study (occult CNV), VIP study (CNV due to pathologic myopia) and VIM study, respectively.

The results of the phase I/II human trial showed an occlusion of retinal vessels using a light dose of 150 J/cm^2 applied 15 min after infusion of verteporfin (6 mg/m^2 body surface area) [68]. However, no effects on retinal vessels, the ganglion cell layer and photoreceptors have been observed by histopathologic studies after PDT with light doses of 100 J/cm^2 and 50 J/cm^2 [56, 62].

In addition, alterations of the RPE following a treatment session of PDT have been reported in several case series and prospective studies (Fig. 14.6). They have been observed particularly in young patients (females) with classic CNV and in the presence of choroidal hemangiomas [30, 61], respectively. While the pigment mottling and focal atrophy of the RPE correlated with the size of the treatment spot, the retinal function of these areas was not significantly affected. Enhanced photochemical effects on the RPE, an increased sensitivity of the RPE, focal atrophy of the RPE following occlusion of the choroid, inherent effects of the RPE and hormonal status of these patients have been discussed as possible causes of these reactions [45, 90].

14 ▌▌

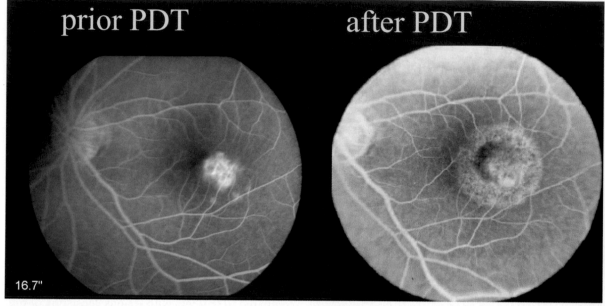

Fig. 14.6. Mottling of the RPE and window defects of the RPE. The area of the RPE atrophy corresponds with the lesion size of the laser spot

14.4 Current Treatment Recommendations

The current treatment recommendations of CNV are based on experimental animal studies and on prospective multicenter studies for the treatment of lesions due to age-related macular degeneration [76, 77, 85, 86] and pathologic myopia [83–85]: Verteporfin is administered intravenously at a dose of 6 mg/m² body surface area over a period of 10 min. Five minutes after the cessation of infusion, light exposure (laser emitting light of 692 nm) with an irradiance of 600 mW/m² is started, delivering 50 J/cm² within 83 s. The size of the light spot should completely cover the lesion consisting of CNV, and features that can occlude the boundaries of the CNV like hemorrhages, blocked hypofluorescence due to RPE, blood or fibrosis, and serous detachments of the RPE [86].

Based on the reports about treatment of choroidal hemangioma, light exposure with a light dose of 100 J/cm² over an interval of 166 s has been applied by the majority of authors for the treatment of angiomatous retinal lesions. Some authors performed the first PDT session with a light dose of 50 J/cm² and observed recurrent exudation during the follow-up [5–7]. However, treatment sessions with a light dose of 100 J/cm² have been reported to be more effective according to the resolution of exudation, regression of angiomatous lesions, stabilization of visual function and number of recurrences, respectively [5–7].

14.5 Verteporfin in Retinal Vascular Disease

The positive effects of PDT with verteporfin in the treatment of CNV of various causes have stimulated trials for the treatment of retinal disorders. However, most of these trials represent at least in part prospective case series, while the incidence of most of these diseases is low compared to the incidence of exudative AMD.

- Retinal capillary hemangioma
- Vasoproliferative tumor
- Parafoveal teleangiectasis

14.5.1 Retinal Capillary Hemangioma

14.5.1.1 Characteristics

Retinal capillary hemangiomas (RCH) may occur as a solitary tumor or as the most frequent and earliest manifestation of the systemic von Hippel-Lindau (VHL) syndrome [43, 67]. Clinically, the majority of RCH are localized in the (temporal) periphery of the retina as a reddish-pink tumor (Fig. 14.7). However, a minority of the VHL gene carriers have juxtapapillary presentations of RCH. Intraretinal and subretinal exudation may result in exudative retinal detachment involving the macula, the juxtapapillary area and loss of visual acuity. In addition, tractional detachment of the retina, epiretinal membranes, hemorrhages of the vitreous

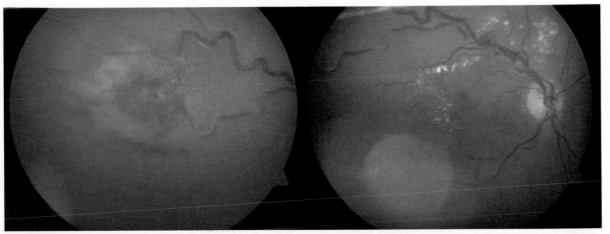

Fig. 14.7. Retinal capillary hemangioma with exudation involving the posterior pole and the macula, visual acuity 20/60 prior to PDT

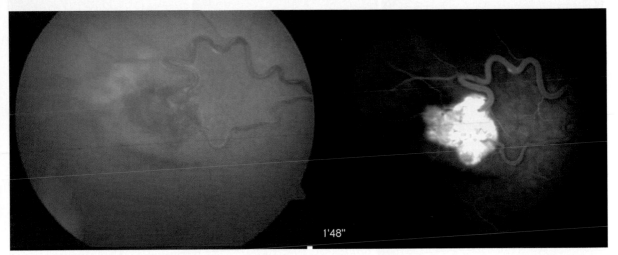

Fig. 14.8. RCH with subretinal exudation prior to the first PDT session

and the retina may be observed in RCH [67]. The typical dilated arterial feeder vessel and the draining retinal venous vessels may not be present and difficult to detect in smaller RCH and RCH of the optic disk, respectively. Fluorescein angiography gives rapid filling via a retinal feeder vessel in the early phases and staining with fluorescein leakage of the RCH in the late phases, respectively (Fig. 14.8).

14.5.1.2 Treatment Options

Treatment depends on the size, location and secondary complications due to the RC. It includes laser photocoagulation only of small RC with a lesion size of less than 1.5 mm in diameter [69], cryotherapy [66, 69] of small anterior localized RCA (external beam, plaque), radiotherapy [35], and vitreoretinal surgery [28]. Juxtapapillary RC

may be a therapeutic challenge, because of the high risk of causing irreversible injuries to the optic nerve, the nerve fiber layer and retinal vessels. Therefore, treatment of juxtapapillary lesions is recommended only when central retinal function may be reduced [36].

14.5.1.3 Effects of PDT on Retinal Angiomas

Treatment of RCA with PDT has been reported in small case series with RC [1, 5, 6, 51, 74]. PDT may be a treatment option particularly for RCA with a peripheral localization. The majority of authors have used a light dose of 100 J/cm^2 5 min after intravenous administration of verteporfin (6 mg/kg body surface area).

14 II

Peripheral RCA

The effects of PDT on RCA of the periphery during the follow-up are shown in Figs. 14.8–14.12. The subretinal exudation clearly increased after the first treatment due to a transient decompensation of the endothelium and the vessels of the angiomatous lesion (Fig. 14.9). However, 1 week after PDT (Fig. 14.10), occlusion of the RCA documented by fluorescein angiography and resolution of the subretinal exudation was observed at least in part, which resulted in an increase in the visual acuity. Partial recanalization of the RCA may occur during the 3-month

follow-up examinations and may lead to a repetition of the PDT session. Occlusion of the RCA without significant exudation was observed 3 months after the final PDT. As shown in Fig. 14.11, some exudation was present 6 months after the first PDT, while the diameter of the RCH decreased compared to the baseline. The effects of the second PDT are depicted in Fig. 14.12, consisting of an area of hypofluorescence around the RCH and some exudation detected by fluorescein angiography.

Laser photocoagulation of small RCA has been reported to be very effective. PDT may be an alternative and effective treatment option and at least in

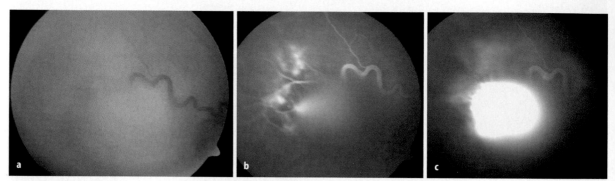

Fig. 14.9. Fundus photography (**a**) and fluorescein angiography (**b, c**) of an RCH 24 h after the first PDT. Extensive leakage compared to the baseline is detected by fluorescein angiography

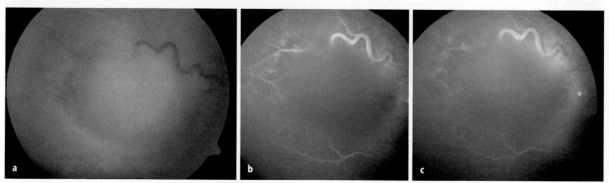

Fig. 14.10. Fundus photography (**a**) and fluorescein angiography (**b, c**) of an RCH 1 week after the first PDT. Extensive leakage compared to the baseline is detected by fluorescein angiography

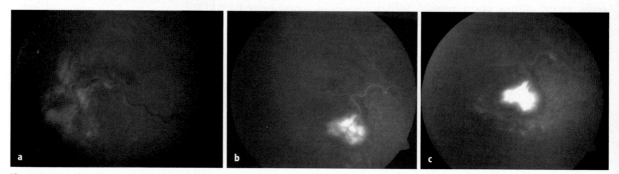

Fig. 14.11. Fundus photography (**a**) and fluorescein angiography (**b, c**) of an RCH 6 months after the first PDT and prior to the second PDT session. The RCH is smaller in diameter and the extension of exudation is decreased compared to the baseline, respectively. However, exudation is not completely resolved

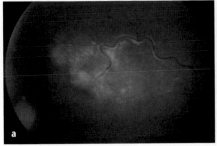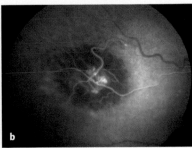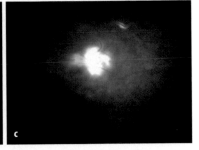

Fig. 14.12. Fundus photography (**a**) and fluorescein angiography (**b, c**) of an RCH 24 h after the second PDT. Hypofluorescence around the RCH corresponding to the diameter of the laser spot confirms the retreatment performed hours previously. In addition, a small leakage is detected by fluorescein angiography

part useful for RCA with significant exudation, particularly for RCA with a size which may not be evaluable for laser treatment. However, the observations of the literature are based on case reports and small case series.

Juxtapapillary RCA

A prospective case series of five patients [64] with symptomatic juxtapapillary RCA was treated with PDT using 6 mg/kg (body surface area) verteporfin and a light dose of 100 J/cm². Twelve months after beginning the PDT treatment sessions subretinal exudation with involvement of the macula was resolved, and the size of the RCA was regressed in all out of five patients. The final mean visual acuity decreased from 0.22 (range 0.5 – 0.025) to 0.1 (range 0.32 – 0.01). A decrease of 1, 3 and 10 lines was observed in three patients. The visual acuity of the remaining two patients was stable (0.05) and increased slightly (from 0.063 to 0.1). However, occlusion of retinal vessels and ischemia of the optic nerve occurred in three out of five patients. Therefore the positive effects of PDT on the subretinal exudation and the size of RCA were limited by the adverse effects on visual function of patients with juxtapapillary RCA.

14.5.2 Vasoproliferative Tumor

14.5.2.1 Characteristics

This vascularized retinal tumor lesion is located particularly inferotemporally and anterior to the equator of the fundus periphery. Usually the artery feeder vessel and the draining vein are dilated to a lesser extent compared to those of the RCA. Fluorescein angiography reveals features similar to those of the RCA. However, in addition to subretinal exudation, hemorrhages, preretinal fibrosis and gliosis of the macula with a decrease in visual function may be possible findings. They may be observed idiopathi-

cally and in combination with retinitis pigmentosa, intermediate uveitis, inflammatory diseases, ocular toxocariasis, sickle cell retinopathy, Coats' disease, and chronic retinal detachment, respectively [23, 67]. Therapeutic interventions may be necessary for vasoproliferative tumor (VTR) with progressive exudation, symptoms and decrease in visual function.

14.5.2.2 Treatment Recommendations

While laser photocoagulation, cryocoagulation and brachytherapy with ruthenium plaques are usually the treatment forms of VTR [3, 23, 67], PDT has been shown to be an effective alternative as reported in case reports and small series of prospective case series [7, 9, 51]. Effective treatment sessions have been performed with a light dose of 100 J/cm² and 5 min after infusion of verteporfin (6 mg/kg body surface area).

14.5.2.3 Effects of PDT

Similar to the observations found with RCA after PDT occlusion of the VTR, vessels can be detected after PDT. Resolution of the subretinal exudation, a decrease in tumor size and an increase in visual function have been reported after up to two treatment sessions, while recurrences were not detected after a follow-up time of 7 – 12 months [7, 9, 51].

14.5.3 Parafoveal Teleangiectasis

14.5.3.1 Characteristics

Parafoveal retinal teleangiectasis is characterized by dilated parafoveal capillaries, which may lead to exudation and macular edema. Other findings may be right angle venules, capillary occlusion, atrophy and hyperplasia of the RPE, retinal crystals and lipids, and CNV due to parafoveal teleangiectasis [18]. The findings, which may be unilateral, bilateral, focal, diffuse around the parafoveal region, depend on the subtype of parafoveal teleangiectasis. Fluorescein angiography reveals irregular parafoveal capillaris, which may be combined with leakage into the retina and macula, defects of the RPE [18].

14.5.3.2 Treatment Recommendations

Laser photocoagulation should be considered in the presence of visual loss due to macular edema and exudation, particularly for Type I teleangiectasis [18]. However, other treatment modalities are currently under investigation.

14.5.3.3 Effects of PDT on Parafoveal Teleangiectasis

A few case reports investigated the treatment of juxtafoveal telangiectasis with PDT. No beneficial effects on the visual acuity and on the macular edema were observed after final PDT [11]. In contrast, case reports and case series [27, 46, 70] reported at least in part positive effects of PDT on CNV due to idiopathic juxtafoveal retinal teleangiectasis: Compared to the baseline examinations prior to PDT and after a mean follow-up of 21 months, visual acuity was stabilized in two eyes (decrease and increase of less than 1 line), improved in three eyes (1 lines) and decreased in two eyes (decrease of 1 line), while no leakages were detected after the final examination of the seven study eyes [46]. However, an atrophy of the RPE corresponding with the size of the laser spot followingPDT has been documented in patients with CNV due to parafoveal teleangiectasis [65]. Therefore, prospective studies are required to confirm the beneficial effects of PDT on patients with CNV due to parafoveal teleangiectasis.

14.6 Conclusions

Photodynamic therapy with verteporfin has significantly improved the prognosis of several diseases of the macula involving the choroidea, choriocapillaris and retina. There is at least in part a selectivity of the PDT treatment which may contribute to stabilization of the visual function. PDT has been established for the treatment of CNV (depending on the combination of the lesion type) and choroidal hemangiomas, respectively. However, experience of PDT treatment of retinal diseases is limited. This may at least in part be caused by the low incidence of specific retinal diseases suitable for the treatment option of PDT. However, the results of several case series suggest that PDT represents an effective treatment option for retinal capillary hemangiomas and vasoproliferative tumors. This may depend on the clinical findings and the localization of the lesion. Further investiga-

tions will define the role of PDT in the treatment of retinal diseases as single therapy and as a combination with pharmacologic and anti-angiogenic agents, respectively.

References

1. Aaberg TM Jr, Aaberg TM Sr, Martin DF, Gilman JP, Myles R (2006) Three cases of large retinal capillary hemangioma treated with verteporfin and photodynamic therapy. Arch Ophthalmol 123:328–332
2. Allison BA, Waterfield E, Richter AM, Levy JG (1991) The effects of plasma lipoproteins on in vitro tumor cell killing and in vivo tumor photosensitization with benzoporphyrin derivative. Photochem Photobiol 54:709–715
3. Anastassiou G, Bornfeld N, Schüler AO, Schilling H, Weber S, Flühs D, Jurklies B, Sauerwein W (2006) Ruthenium-106 plaque brachytherapy for symptomatic vasoproliferative tumours of the retina. Br J Ophthalmol 90:447–450
4. Arnold JJ, Blinder KJ, Bressler NM, Bressler SB, Burdan A, Haynes L, Lim JI, Miller JW, Potter MJ, Reaves A, Rosenfeld PJ, Sickenberg M, Slakter JS, Soubrane G, Strong HA, Stur M, Treatment of Age-Related Macular Degeneration with Photodynamic Therapy Study Group, Verteporfin in Photodynamic Therapy Study Group (2004) Acute severe visual acuity decrease after photodynamic therapy with verteporfin: case reports from randomized clinical trials – TAP and VIP report No. 3. Am J Ophthalmol 137:683–696
5. Atebara NH (2002) Retinal capillary hemangioma treated with verteporfin photodynamic therapy. Am J Ophthalmol 134:788–790
6. Bakri SJ, Sears JE, Singh AD (2005) Transient closure of a retinal capillary hemangioma with verteporfin. Retina 25:1103–1104
7. Barbazetto IA, Smith RT (2003) Vasoproliferative tumor of the retina treated with PDT. Retina 23:565–567
8. Berger JW (1999) Macular laser tissue interactions: implications for treatment and prophylaxis, Chapter 14. In: Berger JW, Fine SL, Maguire MG (eds) Age-related macular degeneration. Mosby, St. Louis, pp 249–261
9. Blasi MA, Scupola A, Tiberti AC, Sasso P, Balestrazzi E (2006) Photodynamic therapy for vasoproliferative retinal tumors. Retina 26:404–409
10. Bloom SM, Brucker AJ (1997) Principles of photocoagulation. In: Laser surgery of the posterior eye segment, Chapter 1. Lippincott-Raven, Philadelphia, pp 3–36
11. De Lahitte GD, Cohen SY, Gaudric A (2004) Lack of apparent short-term benefit of photodynamic therapy in bilateral, acquired, parafoveal teleangiectasis without subretinal neovascularization. Am J Ophthalmol 138:892–894
12. De Vree WJ, Essers MC, Koster JF, Sluiter W (1997) Role of interleukin 1 and granulocyte colony stimulating factor in photofrin-based photodynamic therapy of rat rhabdomyosarcoma tumors. Cancer Res 57:2555–2558
13. Dougherty TJ (1984) Photodynamic therapy of malignant tumors. Crit Rev Oncol Hematol 2:83–116
14. Dougherty TJ, Gomer CJ, Henderson B, Jori G, Kessel D, Korbelik M, Moan J, Peng Q (1998) Photodynamic therapy. J Natl Cancer Inst 90:889–905
15. Ergun E, Tittl M, Stur M (2004) Photodynamic therapy with Verteporfin in subfoveal choroidal neovascularization to central serous chorioretinopathy. Arch Ophthalmol 122:37–41

16. Evans S, Matthews W, Perry R, Fraker D, Norton J, Pass HI (1990) Effect of photodynamic therapy on tumor necrosis production by murine macrophages. J Natl Cancer Inst 82:34–39
17. Fingar VH (1996) Vascular effects of photodynamic therapy. J Clin Laser Med Surg 14:323–328
18. Gass JD, Blodi B (1993) Idiopathic juxtafoveolar retinal teleangiectasis. Update of classification and follow-up study. Ophthalmology 100:1536–1546
19. Ghazi NA, Jabbour NM, De La Cruz ZC, Green WR (2001) Clinicopathologic studies of age-related macular degeneration with classic subfoveal choroidal neovascularization treated with photodynamic therapy. Retina 21:478–486
20. Gibson SL, Murant RS, Hilf R (1988) Photosensitizing effects of hematoporphyrin derivative and Photofrin II on the plasma membrane enzymes 5'nucleotidase, Na$^+$ K$^+$ ATPase, and Mg^{2+} ATPase in R3230AC rat mammary adenocarcinomas. Cancer Res 48:3360–3366
21. Gomer CJ (1989) Photodynamic therapy in the treatment of malignancies. Semin Hematol 26:24–34
22. Haimovici R, Kramer M, Miller JW, et al. (1997) Localization of the lipoprotein delivered benzoporphyrin derivative in the rabbit. Curr Eye Res 16:83–90
23. Heimann H, Bornfeld N, Vij O, Coupland SE, Bechrakis NE, Kellner U, Foerster MH (2000) Vasoproliferative tumors of the retina. Br J Ophthalmol 84:1162–1169
24. Henderson BW, Dougherty TJ (1992) How does photodynamic therapy work? Photochem Photobiol 55:145–157
25. Husain D, Miller JW, Michaud N, Connolly E, Flotte T, Gragoudas E (1996) Intravenous infusion of liposomal benzoporphyrin derivative for photodynamic therapy of experimental choroidal neovascularization. Arch Ophthalmol 114:978–986
26. Husain D, Kramer M, Kenny AG, Michaud N, Flotte T, Gragoudas E, Miller JW (1999) Effects of photodynamic therapy on experimental choroidal neovascularization and normal retina and choroid up to 7 weeks after treatment. Invest Ophthalmol Vis Sci 40:2322–2331
27. Hussain N, Das T, Sumasri K, Ram LS (2005) Bilateral sequential photodynamic therapy for subretinal neovascularization with type 2A parafoveal teleangiectasis. Am J Ophthalmol 140:333–335
28. Johnson MW, Flynn HW, Gass JDM (1992) Pars plana vitrectomy and direct diathermy for complications of multiple retinal angiomas. Ophthalmic Surg 23:47–50
29. Jori G (1990) Factors controlling the selectivity and efficiency of tumor damage in photodynamic therapy. Lasers Med Sci 5:115–120
30. Jurklies B, Anastassiou G, Ortmans S, Schüler A, Schilling H, Schmidt-Erfurth U, Bornfeld N (2003) Photodynamic therapy using Verteporfin in circumscribed choroidal hemangioma. Br J Ophthalmol 87:84–89
31. Jurklies B, Bornfeld N, Schilling H (2006) Photodynamic therapy using Verteporfin for choroidal neovascularization associated with angioid streaks – long-term effects. Ophthalmic Res 38:209–217
32. Korbelik M (1996) Induction of tumor immunity by photodynamic therapy. J Clin Laser Med Surg 14:329–334
33. Korbelik M, Krosl G (1994) Enhanced macrophage cytotoxicity against tumor cells treated with photodynamic therapy. Photochem Photobiol 60:497–502
34. Kramer M, Miller JW, Michaud N, Moulton RS, Hasan T, Flotte T, Gragoudas E (1996) Liposomal benzoporphyrin derivative Verteporfin photodynamic therapy. Selective treatment of choroidal neovascularization in monkeys. Ophthalmology 103:427–438
35. Kreusel KM, Bornfeld N, Lommatzsch A, et al. (1998) Ruthenium-106 brachytherapy for peripheral retinal capillary hemangioma. Ophthalmology 105:1386–1392
36. McDonald HR (2003) Diagnostic and therapeutic challenges. Juxtapapillary retinal capillary hemangioma. Retina 23:86–91
37. Menchini U, Visgili G, Introini U, Bandello F, Ambesi-Impiombato M, Pece A, Parodi MB, Giacomelli G, Capobianco B, Varano M, Brancato R (2004) Outcome of choroidal neovascularization in angioid streaks after photodynamic therapy. Retina 24:763–771
38. Miller JW, Walsh AW, Kramer M, Hasan T, Michaud N, Flotte T, Haimovici R, Gragoudas E (1995) Photodynamic therapy of experimental choroidal neovascularization using lipoprotein-delivered benzoporphyrin. Arch Ophthalmol 113:810–818
39. Miller JW, Schmidt-Erfurth U, Sickenberg M, Pournaras C, Laqua H, Barbazetto I, Zografos L, Piguet B, Donati G, Lane A-M, Birngruber R, van den Berg H, Strong A, Manjuris U, Todd G, Fsadni M, Bressler NM, Gragoudas E (1999) Photodynamic therapy with verteporfin for choroidal neovascularization caused by age-related macular degeneration: results of a single treatment in a phase 1 and 2 study. Arch Ophthalmol 117:1161–1173
40. Moan J, Christensen T (1981) Cellular uptake and photodynamic effects of hematoporphyrin. Photochem Photobiol 2:291–299
41. Moan J, McGhie J, Jacobsen PB (1983) Photodynamic effects of cells in vitro exposed to hematoporphyrin derivative in light. Photochem Photobiol 37:599–604
42. Moan J, Peng Q, Evensen JF, Berg K, Western A, Rimington C (1987) Photosensitizing efficiencies, tumor and cellular uptake of different photosensitizing drugs relevant for photodynamic therapy of cancer. Photochem Photobiol 46: 713–721
43. Moore AT, Maher ER, Rosen P, et al. (1991) Ophthalmological screening for von Hippel Lindau disease. Eye 5:723–728
44. Obochi MO, Ratkay LG, Levy JG (1997) Prolonged skin allograft survival after photodynamic therapy associated with modification of donor skin antigenicity. Transplantation 63:810–817
45. Postelmans L, Pasteels B, Coquelet P, El Ouardighi H, Verougstraete C, Schmidt-Erfurth U (2004) Severe pigment epithelial alterations in the treatment area following photodynamic therapy for classic choroidal neovascularization in young females (2004). Am J Ophthalmol 138:803–808
46. Potter MJ, Szabo SM, Sarraf D, Michels R, Schmidt-Erfurth U (2006) Photodynamic therapy for subretinal neovascularization in type 2A idiopathic juxtafoveolar teleangiectasis. Can J Ophthalmol 41:34–37
47. Reinke MH, Canakis C, Husain D, et al. (1999) Verteporfin photodynamic therapy retreatment of normal retina and choroids in the cynomolgus monkey. Ophthalmology 106: 1915–1923
48. Richter AM, Kelly B, Chow J, et al. (1987) Preliminary studies on a more effective phototoxic agent than hematoporphyrin. J Natl Cancer Inst 79:1327–1332
49. Richter AM, Cerruti-Sola S, Sternberg ED, Dolphin D, Levy JG (1990) Biodistribution of tritiated benzoporphyrine derivative (3H-BPD-MA) a new potent photosensitizer in normal and tumor bearing mice. J Photochem Photobiol 53:231–244

50. Richter AM, Waterfield E, Jain AK, Allison B, Sternberg ED, Dolphin D, Levy JG (1991) Photosensitising potency of structural analogues of benzoporphyrin derivative (BPD) in a mouse tumor model. Br J Cancer 63:87–93

51. Rodriguez-Coleman H, Spaide R, Yanuzzi L (2002) Treatment of angiomatous lesions of the retina with photodynamic therapy. Retina 22:228–232

52. Rosenfeld PJ, Saperstein DA, Bressler NM, Reaves TA, Sickenberg M, Rosa RH, Sternberg P, Aaberg TM Sr, Aaberg TM Jr, for the Verteporfin in Ocular Histoplasmosis Study Group (2004) Photodynamic therapy with Verteporfin in ocular histoplasmosis: Uncontrolled, open-label 2-year study. Ophthalmology 111:1725–1733

53. Rubin GS, Bressler N, Treatment of Age-Related Macular Degeneration with Photodynamic Therapy (TAP) Study Group (2002) Results from the treatment of age-related macular degeneration with photodynamic therapy (TAP) investigation – TAP report No.4. Retina 22:536–544

54. Ryan SJ (1982) Subretinal neovascularization: natural history of an experimental model. Arch Ophthalmol 100:1804–1809

55. Saperstein DA, Rosenfeld PJ, Bressler NM, Rosa RH, Sickenberg M, Sternberg P, Aaberg TM Sr, Aaberg TM Jr, Reaves TA, for the Verteporfin in Ocular Histoplasmosis (VOH) Study group (2002) Ophthalmology 109:1499–1505

56. Schlötzer-Schrehardt U, Viestenz A, Naumann GOH, Laqua H, Michels S, Schmidt-Erfurth U (2002) Dose-related structural effects of photodynamic therapy on choroidal and retinal structures of human eyes. Graefes Arch Clin Exp Ophthalmol 240:748–757

57. Schmidt-Erfurth U, Bauman W, Gragoudas E (1994a) Photodynamic therapy of experimental choroidal melanoma using lipoprotein-delivered benzoporphyrin. Ophthalmology 101:89–99

58. Schmidt-Erfurth U, Hasan T, Gragoudas E, Michaud N, Flotte TJ, Birngruber R (1994b) Vascular targeting in photodynamic occlusion of subretinal vessels. Ophthalmology 101:1953–1961

59. Schmidt-Erfurth U, Diddens H, Birngruber R, Hasan T (1997) Photodynamic targeting of human retinoblastoma cells using covalent low-density lipoprotein conjugates. Br J Ophthalmol 75:54–61

60. Schmidt-Erfurth U, Miller JW, Sickenberg M, et al. (1999) Photodynamic therapy with verteporfin for choroidal neovascularization caused by age-related macular degeneration: results of retreatments in a phase 1 and 2 study. Arch Ophthalmol 117:1177–1187

61. Schmidt-Erfurth U, Michels S, Kusserow C, Jurklies B, Augustin A (2002a) Photodynamic therapy for symptomatic choroidal hemangioma: Visual and anatomic results. Ophthalmology 109:2284–2294

62. Schmidt-Erfurth U, Laqua H, Schlötzer-Schrehardt U, Viestenz A, Naumann GOH (2002b) Histopathological changes following photodynamic therapy in human eyes. Arch Ophthalmol 120:835–844

63. Schmidt-Erfurth U, Michels S, Barbazetto I, Laqua H (2002c) Photodynamic effects on choroidal neovascularization and physiological choroids. Invest Ophthalmol Vis Sci 43:830–841

64. Schmidt-Erfurth U, Kusserow C, Barabazetto IA, Laqua H (2002d) Benefits and complications of photodynamic therapy of papillary capillary hemangiomas. Ophthalmology 109:1256–1266

65. Shanmugam MP, Agarwal M (2005) RPE atrophy following photodynamic therapy in type 2A idiopathic parafoveal teleangiectasis. Indian J Ophthalmol 53:61–63

66. Shields JA (1993) Response of retinal capillary hemangioma to cryotherapy. Arch Ophthalmol 111:551

67. Shields JA, Shields CL (1999) Vascular tumors of the retina and optic disc. In: Shields JA, Shields CL (eds) Atlas of intraocular tumors, Chapter 17. Lippincott Williams and Wilkins, Philadelphia, pp 243–267

68. Sickenberg M, Schmidt-Erfurth U, Miller JW Pournaras CJ, Zografos L, Piguet B, Donati G, Laqua H, Barbazetto I, Gragoudas ES, Lane AM, Birngruber R, van den Bergh H, Strong HA, Manjuris U, Gray T, Fsadni M, Bressler NM (2000) A preliminary study of photodynamic therapy using verteporfin for choroidal neovascularization in pathologic myopia, ocular histoplasmosis syndrome, angioid streaks, and idiopathic causes. Arch Ophthalmol 118:327–336

69. Singh AD, Nouri M, Shields CL, Shields JA, Perez N (2002) Treatment of retinal capillary hemangioma. Ophthalmology 109:1799–1806

70. Snyers B, Verougstraete C, Postelmans L, Leys A, Hykin P (2004) Photodynamic therapy of subfoveal neovascular membrane in type 2A idiopathic juxtafoveolar retinal teleangiectasis. Am J Ophthalmol 137:812–819

71. Spaide RF, Freund KB, Slakter J, Sorenson J, Yanuzzi LA, Fisher Y (2002a) Treatment of subfoveal choroidal neovascularization associated with multifocal choroiditis and panuveitis with photodynamic therapy. Retina 22:545–549

72. Spaide RE, Donsoff I, Lam DL, Yanuzzi LA, Jampol LM, Slakter J, Sorenson J, Freund KB (2002b) Treatment of polypoidal choroidal vasculopathy with photodynamic therapy. Retina 22:529–535

73. Steinert RF, Puliafito CA (1985) The neodymium:YAG laser in ophthalmology: principles and clinical practice of photodisruption. Saunders, Philadelphia

74. Szabo A, Gehl Z, Seres A (2005) Photodynamic (verteporfin) therapy for retinal capillary hemangioma, with monitoring of feeder and draining blood vessel diameters. Acta Ophthalmol Scan 83:512–513

75. Thomas JP, Girotti AW (1989) Role of lipid peroxidation in hematoporphyrin derivative-sensitized photokilling of tumor cells: protective effects of glutathione peroxidase. Cancer Res 49:1682–1686

76. Treatment of Age-Related Macular Degeneration with Photodynamic Therapy (TAP) Study Group (1999) Photodynamic therapy of subfoveal choroidal neovascularization in age-related macular degeneration with verteporfin: one-year results of 2 randomized clinical trials – TAP report 1. Arch Ophthalmol 117:1329–1345

77. Treatment of Age-Related Macular Degeneration with Photodynamic Therapy (TAP) Study Group (2001) Photodynamic therapy of subfoveal choroidal neovascularization in age-related macular degeneration with verteporfin: two-year results of 2 randomized clinical trials – TAP report 2. Arch Ophthalmol 119:198–207

78. Treatment of Age-Related Macular Degeneration with Photodynamic Therapy (TAP) Study Group (2002a) Photodynamic therapy of subfoveal choroidal neovascularization in age-related macular degeneration with verteporfin: additional information regarding baseline lesion composition's impact on vision outcomes – TAP report 3. Arch Ophthalmol 120:1443–1454

79. Treatment of Age-Related Macular Degeneration with Photodynamic Therapy (TAP) Study Group (2002b) Photodynamic therapy of subfoveal choroidal neovascularization in

age-related macular degeneration with verteporfin: three-year results of an open-label extension of 2 randomized clinical trials – TAP report 5. Arch Ophthalmol 120: 1307–1314

80. Treatment of Age-Related Macular Degeneration with Photodynamic Therapy (TAP) Study Group and Verteporfin in Photodynamic Therapy (VIP) Study Groups (2003) Effect of lesion size, visual acuity, and lesion composition on visual acuity change with and without verteporfin therapy for choroidal neovascularization secondary to age-related macular degeneration: TAP and VIP report 1 (2003). Am J Ophthalmol 136:407–418

81. Valmaggia C, Niederberger H, Helbig H (2002) Photodynamic therapy for choroidal neovascularization in fundus flavimaculatus. Retina 22:111–113

82. Verbraak FD, Schlingemann RO, Keunen JE, de Smet MD (2003) Longstanding symptomatic choroidal hemangioma managed with limited PDT as initial or salvage therapy. Graefes Arch Clin Exp Ophthalmol 241:891–898

83. Verteporfin in Photodynamic Therapy Study Group (2001a) Photodynamic therapy of subfoveal choroidal neovascularization in pathologic myopia with Verteporfin. 1-year results of a randomized clinical trial – VIP report No. 1. Ophthalmology 108:841–852

84. Verteporfin in Photodynamic Therapy Study Group (2001b) Verteporfin therapy of subfoveal choroidal neovascularization in age-related macular degeneration: Two year results of a randomized clinical trial including lesions with occult with no classic choroidal neovascularization – Verteporfin in photodynamic therapy report 2. Am J Ophthalmol 131:541–560

85. Verteporfin in Photodynamic Therapy Study Group (2003) Verteporfin of subfoveal choroidal neovascularization in pathologic myopia. 2 year results of a randomized clinical trial – VIP report No. 3. Ophthalmology 110:667–673

86. Verteporfin Roundtable Participants (2005) Guidelines for using Verteporfin (Visudyne) in photodynamic therapy for choroidal neovascularization due to age-related macular degeneration and other causes: Update. Retina 25:119–134

87. Visudyne in Minimally Classic Choroidal Neovascularization Study Group (2005) Verteporfin therapy of subfoveally minimally classic choroidal neovascularization in age-related macular degeneration. Arch Ophthalmol 123:448–457

88. Volden G, Christensen T, Moan J (1981) Photodynamic membrane damage of hematoporphyrin derivative treated NHIK 3025 cells in vitro. Photobiochem Photobiophy 3: 105–111

89. Wachtlin J, Heimann H, Behme T, Foerster MH (2003a) Long-term results after photodynamic therapy with choroidal neovascularization secondary to inflammatory chorioretinal diseases. Graefes Arch Clin Exp Ophthalmol 241:899–906

90. Wachtlin J, Behme T, Heimann H, Kellner U, Foerster MH (2003b) Concentric retinal pigment epithelium atrophy after a single photodynamic therapy. Graefes Arch Clin Exp Ophthalmol 241:518–521

15 Cryosurgery in Retinal Vascular Disease

B. KIRCHHOF, A.M. JOUSSEN

15 ▌▌

Core Messages

- Cryotherapy is applied to generate chorioretinal scars, to reduce retinal ischemia or to occlude abnormal retinal vessels
- Repetitive treatments are necessary to achieve permanent obliteration of vascular abnormalities
- Cryotherapy is superior to laser coagulation in the case of media opacity and in shallow (exudative or tractional) retinal detachments
- The freeze zone of cryoburns has a better depth than the zone of destruction after laser treatment. Therefore cryotherapy is especially useful for smaller tumors
- The freeze zone of cryoburns is wider, about 3–8 mm, than laser burns. Transition and demarcation of the intact retina are less abrupt. Therefore cryotherapy is especially useful when a strong vitreoretinal adhesion is attempted. The shallow transition to the intact retina weakens the retina less than laser burns and thereby largely avoids tears at the edge of the chorioretinal adhesions

---

Cryotherapy has a long history in the treatment of retinal vascular disease. Freezing the retina to create inflammation in the area of application began as early as 1918, when Schöler applied solid carbon dioxide to the sclera and described choroidal inflammation [9]. In the treatment of proliferative retinopathy, cryotherapy today has largely been replaced by laser photocoagulation. However, the transscleral mode of application renders cryotherapy particularly useful in the case of hazy media or cataracts. It is also useful in treating more peripheral lesions that cannot be easily visualized at a slit lamp.

Cryotherapy induces a greater breakdown of the blood-ocular barrier, which has been implicated in the risk of cystoid macular edema, choroidal detachment and exudative retinal detachment. Laser flare photometry showed a greater increase in aqueous flare and a slower recovery of visual acuity after limited external retinal cryotherapy compared to laser coagulation. However, this difference did not affect visual acuity 10 weeks after treatment [13].

Cryotherapy in retinal detachment surgery has long been associated with a number of postoperative events, including macular pucker and proliferative vitreoretinopathy (PVR) due to dispersion of viable pigment epithelial cells and breakdown of the blood-ocular barrier. However, blood-ocular barrier breakdown is clinically irrelevant if excessive cryoapplication is avoided, and if applied to eyes with uncomplicated retinal detachment, and without significant preoperative PVR. In comparison, transscleral diode laser did not show better results in retinal detachment surgery [12].

In the treatment of ischemic retinal diseases, mobilization of retinal pigment epithelium (RPE) cells and risk of PVR does not apply as for the lack of retinal holes. Blood-retinal barrier breakdown, however, will be temporarily aggravated.

The pathology of cryotherapy lesions has been well investigated. The formation of retinal scars includes desmosomal connection between the Müller cells and the basement membrane of RPE. Processes of Müller cells infiltrate the collagen lamellae of Bruch's membrane [3–5]. Laboratory studies have shown that there is little difference in the strength of the chorioretinal scar created by laser photocoagulation and by cryotherapy [1, 15].

It usually takes about 5–7 days for a chorioretinal scar to complete. The repair and remodeling after cryotherapy vary depending on the intensity of the application. For treatment of vascular abnormalities repetitive freezing of the choroid and the outer retinal layers (3 times) is mandatory.

Although laser photocoagulation in many indications has replaced cryocoagulation, cryotherapy is still used as adjunctive treatment in the therapy of vascular disease.

15.1 Technique of Cryotherapy and Equipment

Essentials
Equipment needed:
- Cryoprobe
- Indirect ophthalmoscope
- Condensing 20- and 28-dpt lenses for indirect ophthalmoscopy
- Lid speculum
- Topical anesthetic
- Local anesthetic for injection (e.g., 4% Xylocaine without epinephrine)
- Mydriatic eye drops

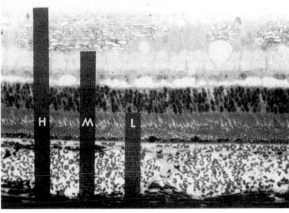

Fig. 15.1. A section of normal retina demonstrating penetration of cryogenic necrosis (represented by *columns*) to the outer limiting membrane after a light application (*L*); to the nerve fiber layer after a medium application (*M*); and to the internal limiting membrane after heavy application (*H*). (With permission from Ingrid Kreissig (ed): *Minimal surgery for retinal detachment*. Thieme, Stuttgart, 2000, p. 104)

A variety of probe tips and cryotherapy machines are available; most are gas operated. Most cryosurgical units use either carbon dioxide or nitrous oxide gas and are based on the Joule-Thompson principle that a sudden drop in temperature occurs when pressurized gas is allowed to expand through a narrow aperture. The tube is defrosted by a passage of another warm gas, or the same gas under low pressure. A silicone sleeve limits the cooling effect to the tip of the probe. When nitrous oxide gas is used, a pressure of 600 psi is sufficient to produce cooling up to −89°C.

The surgeon should be certain that the shaft of the probe is insulated, and test the freezing before usage. Treatment should be controlled via indirect ophthalmoscopy. The freezing should be terminated soon after a distinct whitening of the neurosensory retina is observed.

In general subconjunctival anesthesia is sufficient for cryotherapy.

A quantity of 0.2–0.3 ml of 4% Xylocaine is injected subconjunctivally in the quadrant requiring therapy. After 10–15 min, sufficient anesthesia is achieved and treatment may be started. For treatment of children general anesthesia should be given priority.

For applications close to the posterior pole it is necessary to open the conjunctiva.

Usually a few seconds are sufficient to achieve a sustained freezing until the retina first turns white. After thawing, a faint gray area representing intraretinal edema is all that remains of the cryoapplication. Prolonged freezing is necessary when subretinal exudation increases the distance between the retina and cryoprobe. The probe should not be removed until thawing is completed to avoid retinal breaks.

Postoperative medications are rarely necessary, but antibiotic ointment can be applied.

Complications of treatment itself, such as corneal abrasion, subconjunctival hemorrhage and intraocular hemorrhage, are rare. If an intraocular hemorrhage under treatment should occur, pressure on the eye helps to stop the bleeding. It is possible to perforate the sclera (thin sclera in myopia or reoperation) with the cryoprobe when using as a scleral depressor. In that case the wound needs to be treated like a rupture or penetration of other origin.

Depending on the duration of freezing, chorioretinal scars develop that vary in scleral penetration. Cryogenic necrosis reaches the outer limiting membrane after light application, but can also include the full retinal thickness including the superficial retinal vasculature after heavy application (Fig. 15.1).

Endocryocoagulation is indicated during vitrectomy when endophotocoagulation is insufficient to completely obliterate pathologic retinal vessels, e.g., in Coats' disease or familial exudative vitreoretinopathy (FEVR). The endocryoprobe is inserted through the sclerotomy and the tip held without pressure in close apposition to the retina. While freezing, the tip is tightly connected to the retinal tissue. While the frozen part of the retina is rigid, the surrounding tissue can easily tear. Thus the tip of the endoprobe must not be moved while freezing. Complete thawing should be awaited before removing the probe from the retina. Unlike laser burns, cryotherapy does not show an immediate treatment effect on the retina.

15 II

15.2 Indications of Cryosurgery in Retinal Vascular Disease

15.2.1 Cryotherapy for Retinopathy of Prematurity

Jandeck et al. discuss in detail elsewhere in this book the value of cryotherapy in retinopathy of prematurity (ROP) (Chapter 20.2). The benefit of cryotherapy for treatment of threshold ROP for both structure and visual function that was demonstrated in the CRYO-ROP study was recently shown to persist over 15 years of follow-up [8]. However, new retinal detachments may occur as late as 10 years after treatment and suggest value in long-term, regular follow-up of eyes that experience threshold ROP.

Comparing laser photocoagulation with cryotherapy in patients with threshold ROP, laser-treated eyes had better structural and functional outcome [7]. Similarly, in a report by Jandeck [2], the results of laser therapy were superior to those of cryotherapy, indicating that laser treatment is the therapy of choice: an "unfavorable outcome," as described in the CRYO-ROP study, occurred in 1 of 91 (1%) eyes with laser treatment and in 3 of 46 (6.5%) eyes with cryotherapy. Temporal dragging of vessels was noticed in 6 of 91 eyes (6.6%) with laser treatment vs 7 of 46 eyes (15.2%) with cryotherapy, respectively. Visual acuity ≥ 20/25 was achieved in 39.2% of eyes with laser therapy and in 17.6% with cryotherapy ($p < 0.05$).

Nevertheless, while the primary choice for treatment of the peripheral avascular zone on ROP is laser photocoagulation, additional cryotherapy can be valuable in cases with media opacities such as cataracts or a tunica vasculosa lentis that absorb the laser light and prevent sufficient treatment. Cryotherapy is also to be considered in addition to laser therapy in patients with very central disease when the larger spots of the cryoprobe are advantageous to filling the avascular area. However, cryotherapy should only be used to fill the gaps when treatment with photocoagulation is incomplete.

15.2.2 Cryotherapy for Diabetic Retinopathy and Retinal Ischemic Disease

The gold standard for the treatment of neovascularization associated with diabetic retinopathy or central vein occlusion is laser photocoagulation.

Today, transscleral peripheral retina cryotherapy should not be applied as first line treatment, but is often feasible in situations (such as media opacity) that preclude the use of transpupillary photocoagulation and may help to reduce peripheral ischemia in cases of rubeosis iridis if vitrectomy is not indicated for any reason (see Chapter 17).

Cryotherapy of the anterior retina can further be valuable to prevent fibrovascular ingrowth at sclerotomy sites in patients undergoing pars plana vitrectomy (PPV) for proliferative diabetic retinopathy [14]. Peripheral retinal cryotherapy was also suggested in phakic patients with postoperative diabetic vitreous hemorrhage [6]. In these cases, peripheral cryotherapy (often augmented, when possible, by additional posterior pole endolaser photocoagulation) may be used to supplement previous retinal ablative therapy.

15.2.3 Cryotherapy for Vascular Abnormalities and Exudative Retinopathies: Coats' Disease, FEVR and Small Peripheral Hemangiomas

Since the 1970s, photocoagulation has been the first line therapy for vessel abnormalities [11]. Still, in cases with severe exudative detachment, cryotherapy can be advantageous in occluding the pathological vessels. A combination of photocoagulation and cryotherapy should be considered in these patients.

In a series of 117 patients (124 eyes) with Coats' disease, the primary management was observation in 22 eyes (18%), cryotherapy in 52 (42%), laser photocoagulation in 16 (13%), various methods of retinal detachment surgery in 20 (17%), and enucleation in 14 (11%) [10].

Similarly, additional cryotherapy should be used in cases with familial exudative retinopathy when photocoagulation remains insufficient to occlude the pathological vessels (Chapter 22.5).

It is of note that in cases with large subretinal deposits and exudation, repetitive freeze-thaw cycles can be required for effective treatment (Figs. 15.2,

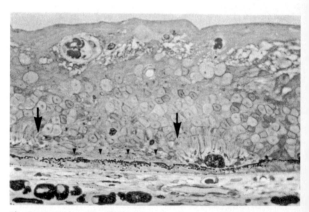

Fig. 15.2. Pigment epithelium and retina 4 weeks after cryocoagulation of the RPE and retina. The outer limiting membrane (*large arrows*) is ruptured and retinal glial cells interconnect with RPE cells (*small arrows*). Of note, there is a hyalinization of the retinal vessel walls. (With permission from H. Laqua: *Retinale Adhäsionen nach Netzhautoperation.* Graefes Archiv klin Exp Ophthalmol 1977; 203:119–131)

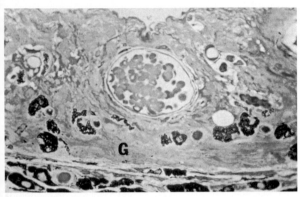

Fig. 15.3. Pigment epithelium and retina 4 weeks after cryocoagulation of the RPE and retina. While large retinal vessels are still perfused, smaller vessels are obliterated and hyalinized within the scar tissue. (With permission from H. Laqua: *Retinale Adhäsionen nach Netzhautoperation.* Graefes Archiv klin Exp Ophthalmologie 1977; 203:119–131)

15.3), although treatment is limited by its associated inflammation. Cryotherapy may cause an increase in exudation early after treatment. This will usually regress within 2 or 3 weeks after therapy. Only patients with persistent bullous exudation should be watched carefully for retinal breaks and rhegmatogenous detachment.

If the vascular pathology does not regress 4–6 weeks after treatment, re-treatment with either cryotherapy or photocoagulation should be considered until the pathological vessels become completely occluded. This will cause secondary regression of exudates, which can take several months.

Cryotherapy is further effective in the treatment of vascular tumors; however, side effects such as increased exudation, hemorrhages and incomplete regression of large tumors should be considered when deciding on cryotherapy as a combined or sole treatment. Small peripheral tumors are most likely to benefit from cryotherapy (see Chapters 28.2–28.4).

References

1. Han DP, Nash RW, Blair JR, O'Brien WJ, Medina RR (1995) Comparison of scleral tensile strength after transscleral retinal cryopexy, diathermy, and diode laser photocoagulation. Arch Ophthalmol 113:1195–1199

2. Jandeck C, Kellner U, Heimann H, Foerster MH (2005) Anatomische und funktionelle Ergebnisse nach Laserkoagulation und Cryotherapy bei ROP. Ophthalmologe 102:33–38

3. Kreissig I, Lincoff H (1974) [Animal experiments about retinal scarring (author's transl.)]. Albrecht Von Graefes Arch Klin Exp Ophthalmol 190:165–182

4. Kreissig I, Lincoff H (1974) [Bruch's membrane and its structural changes after cryopexis (author's transl.)]. Klin Monatsbl Augenheilkd 164:71–89

5. Lincoff H, O'Connor P, Kreissig I (1970) [Retinal adhesion after cryopexy]. Klin Monatsbl Augenheilkd 156:771–783

6. Neely KA, Scroggs MW, McCuen BW 2nd (1999) Peripheral retinal cryotherapy for postvitrectomy diabetic vitreous hemorrhage in phakic eyes. Am J Ophthalmol 127:740–741

7. Ng EY, Connolly BP, McNamara JA, Regillo CD, Vander JF, Tasman W (2002) A comparison of laser photocoagulation with cryotherapy for threshold retinopathy of prematurity at 10 years: part 1. Visual function and structural outcome. Ophthalmology 109:928–934

8. Palmer EA, Hardy RJ, Dobson V, Phelps DL, Quinn GE, Summers CG, Krom CP, Tung B; Cryotherapy for Retinopathy of Prematurity Cooperative Group (2005) 15-year outcomes following threshold retinopathy of prematurity: final results from the multicenter trial of cryotherapy for retinopathy of prematurity. Arch Ophthalmol 123:311–318

9. Schöler F (1918) Experimentelle Erzeugung von Aderhaut-Netzhautentzündung durch Kohlensäureschnee. Klin Monatsbl Augenheilkd 60:1

10. Shields JA, Shields CL, Honavar SG, Demirci H, Cater J (2001) Classification and management of Coats disease: the 2000 Proctor Lecture. Am J Ophthalmol 131:572–583

11. Spitznas M, Joussen F, Wessing A (1976) Treatment of Coats' disease with photocoagulation. Albrecht Von Graefes Arch Klin Exp Ophthalmol 199:31–37

12. Steel DH, West J, Campbell WG (2000) A randomized controlled study of the use of transscleral diode laser and cryotherapy in the management of rhegmatogenous retinal detachment. Retina 20:346–357

13. Veckeneer M, Van Overdam K, Bouwens D, Feron E, Mertens D, Peperkamp E, Ringens P, Mulder P, Van Meurs J (2001) Randomized clinical trial of cryotherapy versus laser photocoagulation for retinopexy in conventional retinal detachment surgery. Am J Ophthalmol 132:343–347

14. Yeh PT, Yang CM, Yang CH, Huang JS (2005) Cryotherapy of the anterior retina and sclerotomy sites in diabetic vitrectomy to prevent recurrent vitreous hemorrhage: an ultrasound biomicroscopy study. Ophthalmology 112:2095–2102

15. Zauberman H (1969) Tensile strength of chorioretinal lesions produced by photocoagulation, diathermy, and cryopexia. Br J Ophthalmol 53:749–752

16 Vitrectomy in Retinal Vascular Disease: Surgical Principles

A.M. JOUSSEN, AND B. KIRCHHOF

16 II

Core Messages

- Vitreoretinal surgery allows the treatment of even advanced proliferative diseases of the retina. Main indications for vitrectomy include the destruction of ischemic retina and pathological vessels, the removal of dense vitreous opacities, the release of retinal traction and retinal detachment repair
- Technical prerequisites include vitrectomy in a closed system. Wide-angle viewing systems allow visualization of the peripheral retina during surgery. Endo-photocoagulation facilitates destruction of ischemic retina and abnormal vessels

- Silicone oil as an inert tamponade agent may help to prevent growth factor accumulation in the vitreous cavity. Retinal fibrovascular re-proliferations are generally based on the remaining posterior vitreous. Chromovitrectomy by triamcinolone acetonide can help to detect vitreous remnants
- The iris-lens diaphragm physiologically prevents growth-factor accumulation in the anterior segment and should be maintained if possible. Cataract surgery has certain peculiarities, e.g., a large rhexis is required that facilitates later inspection of the peripheral retina. Silicone lenses should be avoided if future silicone oil surgery is likely

---

16.1 Introduction

No advance in the treatment of vitreoretinal diseases has been as significant as the introduction of pars plana vitrectomy by Machemer in 1971 [44]. The closed system allowed for a safe intraocular manipulation and constant viewing of the retina. During the past few decades the instrumentation has advanced, but the same principles still apply.

Vitrectomy not only removes opacities but releases vitreoretinal traction. In proliferative disease additional laser photocoagulation of ischemic retina is required. This chapter addresses the indications and surgical techniques.

16.2 Prerequisites for Vitreoretinal Surgery in Retinal Vascular Disease: Microscope Requirements and Wide-Angle Viewing Systems

Standard equipment for pars plana vitrectomy includes a vitreous cutter and aspiration device, an infusion of BSS or similar, an air pump, a contact or non-contact viewing system, and a microscope.

The stereo operating microscope should allow a magnification of up to 30-fold. It should be equipped with coaxial illumination, a power zoom, power focusing, and X-Y positioning. The microscope needs to be fitted with laser filters for both the surgeon and the assistant. An integrated video system serves the purpose of documentation and co-visualization for the personnel or other observers.

The initial visualization in vitrectomy was performed using a hand-held, plano-concave contact lens (e.g., Hoffman or Landers lens), which allowed control of the central fundus. For a better view to the peripheral retina, a biconcave lens is available; however, the viewing angle is limited to 20–35 degrees.

Currently, 130-degree wide-angle viewing systems are commonly used.

The inverted image from the wide-angle lens system is corrected through a stereoscopic diagonal inverter. Unless the contact lens is self-stabilizing, a skilled assistant is needed. Non-contact wide-angle systems (e.g. Biom, Oculus, Germany) can be managed by the surgeon alone. A non-contact system is atraumatic to the cornea, yet sufficient visualization of the peripheral retina can still be achieved when rotating the eye.

Table 16.1 lists the currently available wide-angle viewing systems.

Table 16.1. Wide-angle viewing systems

Non-contact systems	• BIOM/SDI Inverted image; different miniature, indirect viewing lenses cover a range of field of views from 70 to 110 degrees) • EIBOS Upright image; a 90- and a 60-diopter lens allow a field of view of 125 and of 100 degrees
Contact systems	• For all contact systems, miniaturization is mandatory to allow sufficient freedom of movement of vitreoretinal instruments. That is why conventional diagnostic panfunduscopic lenses (as they are used, e.g., for retinal laser coagulation) are inappropriate for vitrectomy • AVI contact lens systems Inverted image Miniature contact lenses correspond to a viewing angle of 68 and 130 degrees • Volk modified the AVI System Reinverting operating lens system (ROLS), Volk miniature indirect contact lenses with both standard lenses and the self-stabilizing lenses (58, 85, 156 diopters)

16.3 Surgical Techniques

16.3.1 Three-Port Vitrectomy

For vitreoretinal surgery in vascular disease 20-gauge tools have become standard and are strongly recommended. Recently, 23-gauge instruments have been introduced to avoid sutures of sclera and conjunctiva. The use of 25-gauge instruments is controversial, since the vitrectomy part is considerably more time consuming. The authors question progress through surgical instruments smaller than 23 gauge in vitrectomy (see Chapter 20.3).

A 1.4-mm linear incision rounds out to a required 0.89-mm-diameter hole for 20-gauge instrumentation. We first place the sclerotomies for the infusion at the 3 or 9 o'clock position, using a 20-gauge beaver blade. The distance to the limbus is 3.5 mm in phakic eyes and 3 mm in aphakic and pseudophakic eyes.

Before the infusion, which is stabilized by a 7-0 suture, is opened we try to recognize the tip of the infusion canula in the vitreous cavity and free of choroid. Especially a hypotonic eye is at risk of subchoroidal or subretinal infusion. The second and third sclerotomy are placed superotemporally and superonasally.

Usually the core vitreous is removed first. As long as the posterior hyaloid is attached there is hardly any risk of peripheral iatrogenic holes. In the absence of preretinal neovascularization we attempt to induce a posterior vitreous detachment by short phases of high suction (up to 360 mm Hg) with the vitrectomy probe. Once the posterior vitreous is detached, we trim towards the base and try to avoid pulling on the peripheral retina as much as possible.

In case preretinal neovascularizations lock the hyaloid to the retina posterior vitreous, separation and trimming is attempted elsewhere. The then isolated areas of adherent vitreous cortex, possibly complemented by fibrous tissue, are separately addressed using horizontally angled cutting scissors. Tight and wide adhesions between vitreous and retina should be separated using (bimanual) delamination techniques (see Chapter 19.2.2). Panretinal photocoagulation concludes the procedure. Even though preretinal neovascularizations cannot reform in a vitrectomized eye, panretinal laser coagulation is still obligatory to address or protect from rubeosis of the iris and from rubeosis of the ciliary body.

The necessity of destroying large parts of the peripheral retina will probably not be overcome by new anti-vascular endothelial growth factor (VEGF) agents, unless we are willing to trade the permanent laser induced reduction of ischemia for endlessly repeated intravitreal injections. Whenever possible, we attempt to laser prior to vitrectomy. Laser coagulation eases vitreous separation. Possible obliteration of preretinal neovascularizations reduces the risk of bleeding during vitrectomy. The latter aspect may currently also be achieved by intravitreal injection of a VEGF antagonist. Nevertheless, the effectiveness of anti-VEGF therapy as a pre-treatment or an adjunct to surgery has to be investigated in large scale studies prior to a broad clinical application in retinal vascular disease.

Vitrectomy is concluded by filling the vitreous cavity either by BSS, gas, or silicone as needed (see below). The intraocular pressure is adjusted to normal or to a slightly elevated level. The sclerotomies are sutured tightly by 7-0 Vicryl.

So far, the authors believe that there is no advantage to 25-gauge systems as there are no medical advantages in favor of the 25- or 23-gauge systems, but only patient comfort related to the smaller conjunctival lesions. The overall operating time is no shorter [20, 30].

Surgical Technique
- Pars plana is today commonly performed with the help of a wide-angle viewing system, attached to an operating microscope. The sequence of surgical steps is as follows and requires modification according to the patient's requirements:
- Access to the vitreous cavity via three ports in the pars plana
- Removal of the posterior vitreous, including the posterior hyaloid
- Removal of tractional components by dissection techniques (e.g., delamination, en-block resection)
- Treatment of avascular areas and abnormal vessels with endo-photocoagulation or intraoperative exo-cryopexy

16 II

If tamponade is required:
- Fluid-air exchange with subsequent exchange of air to SF6 20% or C3F8
- Silicone oil can be exchanged directly from BSS, heavy liquids or air
- Adjustment of the intraocular pressure

16.3.2 „Chromovitrectomy"

A near complete removal of vitreous helps to prevent proliferative vitreoretinopathy (PVR). Remnants of the hyaloid serve as a scaffold for fibrovascular proliferation, e.g., in proliferative diabetic retinopathy. Due to the transparency of the vitreous, it is difficult to visualize isolated patches of vitreous on the retina. Corticosteroid crystals entangle on rough surfaces and thereby mark otherwise invisible remnants of vitreous on the retina [32, 59, 73]. No retinal toxicity has been noted in doses of 2–4 mg in vitrectomized and non-vitrectomized eyes [46]. In surgery for retinal vascular disease, it may help to prevent fibrin exudation because of its inherent anti-inflammatory and proliferative characteristics.

A complete removal of epiretinal fibrous membranes might help to prevent recurrence of PVR. Dyes such as trypan blue are helpful in staining epiretinal membranes. Indocyanine green selectively stains the inner limiting membrane (ILM). The epiretinal membrane is depicted in negative contrast. Remnants of ILM are easily distinguished from the unstained underlying nerve fiber layer [4].

Surgical Technique for Chromovitrectomy
- After core vitrectomy one to two drops of triamcinolone acetonide in aqueous suspension (40 mg/ml) are injected via a flute needle into the vitreous cavity. Remnants of the vehicle are quickly washed out. However, to reduce the concentration of the stabilizers, the aqueous suspension can alternatively be allowed to stand for 30 min, when the crystals sediment and separate from the vehicle. The pellet of concentrated steroid is then re-suspended to the original concentration in saline
- After dispersion of the crystals in the vitreous cavity, active aspiration with a vitrectomy probe or soft cannulated extrusion needle is applied
- Posterior vitreous detachment, if not already present, can be created by maximal suction (200–300 mm Hg) just nasal to the optic disk. Once posterior vitreous separation occurs, some of the triamcinolone suspended in the midvitreous cavity settles on the dependent retinal surface

16.3.3 Heavy Liquids

Perfluorocarbon liquids (PFCLs) have become an indispensable tool to vitreoretinal surgery. The high specific gravity of PFCLs allows the intraoperative hydrokinetic manipulation of the retina. PFCL are addressed as the "third hand" in protecting the retina from inadvertent aspiration into the cutter during "shaving" of the vitreous base. PFCLs stabilize the retina during peeling and delamination of epiretinal membranes. PFCLs displace liquefied submacular and epiretinal blood. Thereby PFCLs facilitate endolaser photocoagulation and intraoperative retinal reattachment. PFCLs can be exchanged against air or silicone oil.

Although widely used, retinal damage from PFCLs is reported after short-term use as well as from extended tamponades [7–9, 16, 19, 48, 64, 65]. Impurities of perfluorocarbon preparation [47, 65] may be among the limiting factors for long-term tolerance as well as dispersion of the liquids, foam cell reaction, photoreceptor toxicity and preretinal membrane formation and tractional retinal detachment. The rate of dispersion is related to factors such as turbulence, viscosity, and solubility. Miyamoto et al. described a retinal gliosis after a 1-month tamponade with perfluoroether [48]. Similarly, hypertrophy of Müller cells with bump-like protrusions into the photoreceptor interspaces was observed after only 6 days tamponade by perfluoro-octane and perfluoropolyether [16]. Whitish precipitates in the vitreous cavity and behind the lens have been reported with admixtures of vitreous and liquid perfluorocarbon bubbles or F6H8. Such precipitates represent condensed collagen without inflammatory alterations [15, 74].

In contrast to these observations, Nahib et al. observed an unchanged retinal anatomy after 6 weeks tamponade with perfluorophenanthrene [50]. Similarly, Flores described no adverse reaction in the retina after perfluoro-octylbromide tamponade of up to 6 months [19].

Although the changes described with perfluoro-octane, perfluorodecalin and semi-fluorinated alkanes are similar, the influence of the physical properties of these substances on retinal damage remains unknown. Perfluorodecalin differs from F6H8 with respect to its molecular weight and its specific gravity (Table 16.2). Recently, Wong and coworkers demonstrated in a model eye chamber that a complete vitreous fill of F6H8 increased the pressure on the retina by only 0.52 mm Hg [72]. Such a pressure rise is small compared to the diurnal pressure changes of the normal eye [13, 72]. Although the influence of gravity was not responsible for retinal toxicity in experimental studies (Mackiewicz, Joussen, unpublished), up until now the use of PFCLs as a long term substitute cannot be recommended. While the long term effects of PFCLs are still poorly understood, they undoubtedly are an indispensable intraoperative tool.

Table 16.2. Properties of perfluorodecalin, perfluoro-octane and F6H8

	Molecular weight (g/mol)	Specific gravity (g/cm³) (25°C)	Boiling point (°C)	Refractive index (20°C)	Surface tension against air (mN/m)	Interface tension against water (mN/m)	Viscosity (mPas) (25°C)
Perfluoro-decalin	462	1.93	142	1.313	19.0	57.8	5.68
Perfluoro-octane	438	1.76	105	1.270	14.0	55.0	1.20
F6H8	432	1.33	223	1.343	19.7	45.3	3.44

Tips and tricks for the use of liquid perfluorocarbons during surgery:

• Liquid perfluorocarbon and blood: Liquid submacular blood or epiretinal blood can be displaced to the periphery by liquid perfluorocarbon. Inject PFCL slowly, to avoid a stronger jet of PFCL. Such a jet stream can create retinal holes and PFCL ends up subretinally. Filling the eye with PFCL facilitates endophotocoagulation in two ways: firstly by displacement of blood, and secondly by appositioning the retina against the pigment epithelium, where the heat evolves

• Direct exchange of silicone oil against liquid perfluorocarbon: Filling the eye with PFCL until the level of the infusion is indicated by bubble formation. Bubbles of PFCL may end up subretinally, either through peripheral retinal holes or via retinotomies/retinectomies. Subretinal "pearls" of PFCL tend to slide towards the macula during PFCL-silicone oil exchange. Subretinal PFCL is often perceived only postoperatively. Submacular PFCL must be removed. It causes slowly progressive visual loss. Bubbles of PFCL may be become entangled in the vitreous base and retained in the eye. Therefore, one should avoid a complete or near complete PFCL coat of the retina. Therefore, allow some time for the PFCL to flow to the posterior pole

• PFCL can be exchanged against air or silicone oil. For the latter, the silicone oil should be layered over the PFCL. The adhesion of PFCL to silicone oil evacuates most of the remaining water through the sclerotomies. The less water that remains in the eye, the more complete the silicone oil fill will be

• Liquid perfluorocarbon is helpful in the removal of "sticky silicone oil": Seldomly removal of heavy silicone oil is complicated by a rather tight adhesion to the retina (see below). Adhesion of heavy silicone to PFCL is even greater than adhesion of heavy silicone to the retina. Therefore PFCL can be used to collect heavy silicone oil from the surface of the retina like a "sticky tape." Finally, the conglomerate of PFCL and heavy silicone can be removed from the eye via the flute needle

16.3.4 Vitreous Tamponades

16.3.4.1 Gas Tamponades

The use of intraocular gases was first reported for the treatment of retinal detachment in 1911 by Ohm, who treated two patients by injecting air into the vitreous cavity after drainage of subretinal fluid [55]. Gases gained popularity in the treatment of retinal detachment due to their unsurpassed interfacial tension against water. The greater the interfacial tension the larger the retinal hole to be blocked can be. Norton [54], Machemer [42], and Lincoff [40, 41] developed the application to the eye and today a family of straight chain perfluorocarbon gases is available with varying ability of expansion and longevity.

For advanced cases of neovascular retinal disease a long-term tamponade is usually required. Nevertheless, there are patients in whom silicone oil for a variety of reasons is avoided and in whom a gas tamponade, usually with sulfur hexafluoride (SF6), is beneficial.

Sulfur hexafluoride (SF6) is a colorless non-toxic gas, approximately 5 times heavier than air [45]. SF6 is chemically inert and hydrolysis requires extremely high temperatures.

Intraoperatively, gases facilitate the view to the retinal periphery. Gases do not mix with blood and do not convey (tumor) cells. Thus, endolaser coagulation of the peripheral retina or endoresection of choroidal melanomas is best performed under gas. On the other hand, localization of retinal holes is hampered by a gas fill.

Tips and tricks for surgery with gas tamponades:

• *Reproducible gas filling*
Before the infusion line is removed from the air filled eye, the vitreous cavity is flushed by 50 cc of 20% SF6. The air-SF6 admixture is injected via a 30-gauge needle through a separate pars plana puncture. Surplus gas is pushed back into the air infusion line and air pump

• *Avoid drying of retina*
Welch et al. have shown that the air jet of the infusion line causes large temporal field defects, from dried retina typically nasal to the disk [67]. It is advisable to use humidified air or plug the sclerotomies when instruments are removed from the eye for extended periods

• *Improving vision to the retina*
During fluid gas exchange, fogging of an artificial intraocular lens (IOL) is often encountered on the posterior surface within the area of a capsulotomy. Healon, applied to the back of the IOL, eliminates the fogging and reestablishes full visibility

16.3.4.2 Extended Vitreous Tamponade with Silicone Oil

Silicone oil was introduced to vitreoretinal surgery in 1962 by Cibis [11]. With the advent of vitrectomy

16

[43] and through Ando's 6 o'clock iridectomy [5], silicone oil has widened our indications to, e.g., severely injured eyes so far beyond remedy.

Silicone oils are linear synthetic organic polymers with a common macromolecular backbone made of siloxane (-Si-O) repeating units. The major differences among silicone fluids reside in the chemical structures of radical side groups, in radical termination of the chains, and in the size distribution of chains. In ophthalmology, the term silicone oil has been used to designate any of the viscous hydrophobic polymeric compounds based on siloxane chemistry.

Physical Properties and Clinical Consequences of Silicone Oils

In the eye, the dynamics of the silicone oil tamponade involves buoyancy, interfacial surface tension, and viscosity.

In a liquid, the cohesive forces generated by the molecular attraction between closely packed molecules generate friction, a resistance to fluid flow. The amplitude of this resistance is represented by the fluid coefficient of kinematic viscosity [17, 18], usually expressed in centistokes (cSt). Increasing a silicone oil's molecular weight results in an increased polymer chain length and, consequently, an increase in viscosity.

Clinically used silicone oils are highly purified polydimethylsiloxanes with a viscosity of 1,000–5,000 cSt. Low-viscosity silicone oils are preferred by some surgeons because of their quicker filling into the vitreous cavity and removal from the vitreous cavity. While initially silicone oils of 1,000 cSt were used, there is currently a trend to use oils of higher viscosity.

One of the main problems in the clinical use of silicone oil is emulsification. Fine silicone oil droplets can cause a secondary glaucoma by blocking aqueous outflow. Heidenkummer and coworkers investigated the effect of viscosity of highly purified polydimethylsiloxanes in the presence of biological detergents (albumin, acidic alpha-1-glycoprotein, fibrin, fibrinogen, gamma-globulins, and very-low-density lipoprotein). Silicone oil at 5,000 cSt was in all cases distinctly more stable than the silicone oils with a viscosity up to 4,000 cSt independent of the detergent [23].

Interestingly, viscosity did not affect functional and anatomical outcome after retinal detachment. There was no significant difference in the rate of retinal re-detachment at each of the follow-up intervals [60].

Patients with proliferative retinal disease and likewise patients with PVR are more prone to emulsification due to their high amount of biological detergents from blood-retina barrier breakdown. Thus

silicone oil of 5,000 cSt seems to be superior to 1,000 cSt silicone oil from the authors' point of view. There are, however, at this point no randomized clinical trials available to support this hypothesis.

Heavier than Water Silicone Oils

Perfluoroalcanes are heavier-than-water transparent liquids which are tolerated by the retina [74]. Perfluorhexyloctane (F6H8) has recently gained attention as a long term vitreous substitute [33]. However, surgeons are concerned about inflammation in the eye [35, 62, 66]. Although F6H8 is chemically and biologically inert [74], its low viscosity promotes emulsification. Minute bubbles subsequently trigger chemotaxis of inflammatory cells and phagocytosis [36].

Heavy silicone oil theoretically offers the chance to reduce the rate of tractional late retinal re-detachments [56] through displacement of the PVR stimulating modulators from the inferior retina. Preliminary consecutive observations of 40 patients treated by a heavy silicone oil tamponade [69] and further series [63, 74] do not disprove this hypothesis.

The heavy silicone oil Densiron 68 (HSO 68-1500) is a transparent homogeneous liquid which is slightly heavier (1.06 g/cm^3) than water. Densiron 68 is a mixture of 5,000 mPas silicone oil (specific gravity of 0.97 g/cm^3) and of 3.5 mPas F6H8 (specific gravity 1.33 g/cm^3). Densiron 68 has a low viscosity (1,480 mPas).

While the use of silicone oil in vitreoretinal surgery for vascular disease is common, heavy silicone oils are currently gaining importance and not only in the treatment of inferior PVR. The stickiness of the heavy silicone oil is sometimes reported as a problem during removal. On the other hand, the stickiness adds to the effect of gravity, namely by blocking access of inflammatory mediators within the vitreous from the retina. The stickiness brings about a "sealing" effect, preventing spread of hemorrhage from retinal or chorioretinal wounds in the inferior retinal periphery, e.g., after chorioretinal biopsy or translocation. Finally, heavy silicone oil exempts the patient from positioning (macular hole surgery). It even achieves closure of macular holes refractive to conventional gas tamponade (Rizzo, unpublished).

Silicone Oil as a Drug Carrier

Current treatment approaches in retinal vascular disease focus on prevention in earlier stages of the disease (see, e.g., Chapter 19). Surgery will probably be reserved for late stage disease and rare entities with otherwise untreatable vascular abnormalities.

With inert tamponades such as silicone oil, there are two possible ways of delivering the drugs: either the agents concentrate in the remaining watery phase, or they dissolve within the silicone oil. In the best case scenario, the silicone oil can act as a slow release system.

Nevertheless these slow release systems using silicone oil as a drug carrier are currently at an experimental level and are not clinically applicable. When used for retinal tamponade in experimental proliferative vitreoretinopathy, silicone oil was investigated as a vehicle for delivery of a lipophilic antiproliferative agent [3]. However, histopathologic examination of the eyes injected with this antiproliferative agent and silicone oil indicated some retinal disorganization even at the lower therapeutic levels. Similarly, retinoic acid was evaluated in silicone and silicone-fluorosilicone copolymer oils in a rabbit model of proliferative vitreoretinopathy. Retinoic acid, in this model, was found to be useful in PVR treated with silicone oil or for eyes treated intraoperatively with heavier-than-water silicone oil when it is used as a short-term retinal tamponade [51]. To improve release profiles, biodegradable microspheres of poly (DL-lactide-co-glycolide) (PLGA) for intraocular sustained release of ganciclovir were prepared using a dispersion of ganciclovir in fluorosilicone oil (FSiO) that was further dispersed in an acetone solution of PLGA (50/50 and inherent viscosity 0.41 dl/g), and emulsified in silicone oil with a surfactant [27]. So far, there is no broad clinical application.

For retinal vascular disease the admixture of triamcinolone is of great interest, given the effect on vascular permeability, e.g., in macular edema (see Chapters 19.3.1 – 19.3.3).

The clinical outcome and complications associated with intravitreal injection of unaltered triamcinolone acetonide in conjunction with pars plana vitrectomy and silicone oil injection for the treatment of complicated proliferative diabetic retinopathy with tractional retinal detachment and severe proliferative vitreoretinopathy were recently presented in retrospective evaluations. Up to 4 mg of triamcinolone acetonide can be safely injected in silicone-filled, vitrectomized eyes without any significant retinal toxicity [34, 49]. It should, however, be expected that the triamcinolone crystals cumulate in the watery phase and form a sticky tablet-like deposit that can remain for several months.

When to Remove a Silicone Oil Tamponade?

Although a long-term tamponade is possible with silicone oil, secondary glaucoma due to emulsification is a considerable threat.

Chronic intraocular pressure elevation occurs in a minority (11%) of patients who are treated with silicone oil. Most of these eyes are effectively treated by anti-glaucoma medications. Eyes that do not respond to medical therapy may be effectively managed by glaucoma drainage implant placement in an inferior quadrant [2].

A retrospective analysis of 87 consecutive cases investigated the outcome after silicone oil removal in patients with severe proliferative diabetic retinopathy (48 eyes) and 39 eyes with complex proliferative vitreoretinopathy or giant retinal tears after trauma. In 75% of proliferative diabetic retinopathy patients the retina remained attached; this is in contrast to patients with proliferative vitreoretinopathy, of whom only 48.5% remained stable [31]. Interestingly, the success was independent of the duration of intraocular silicone oil tamponade in proliferative diabetic retinopathy.

There is no general rule as to the required duration of silicone oil tamponade. It is assumed that the silicone oil exerts some decompartmentalization effect and thus contributes to the settlement of active proliferation by evacuating the growth factor reservoir of the vitreous. Usually active proliferative disease after vitrectomy, endolaser coagulation and silicone oil tamponade will require a minimum of 8–10 weeks to settle. The authors favor silicone oil removal at the earliest possible time point (mostly after about 3 months) to avoid secondary cataract formation, optic atrophy, or glaucoma.

Silicone oil in the anterior chamber is associated with a specific complication profile and results in secondary glaucoma, decreased visual acuity by corneal decompensation and band keratopathy.

To avoid silicone oil passage into the anterior chamber in phakic and pseudophakic eyes, intraoperative overfill should be avoided. Prompt surgical removal of the silicone oil from the anterior chamber is usually the preferred method. We favor a technique with two paracenteses. A superior paracentesis in the 12 o'clock position allows outflow of the oil. With pressure of a spatula on the posterior lip, this paracentesis can be opened. The eye is rotated inferiorly. If all the silicone oil can be removed with this technique, no further intervention is required. In some cases, complete removal of the silicone oil requires viscoelastics (e.g., hyaluronic acid) to be injected via a second paracentesis at 6 o'clock and express the silicone oil via the superior paracentesis. Usually the viscoelastic is subsequently rinsed out; however, if left behind in the anterior chamber a close monitoring of the intraocular pressure is required in the postoperative phase. The superior paracentesis is finally closed with a 10-0 nylon suture.

In aphakia, an inferior peripheral iridectomy (Ando iridectomy) prevents postoperative migration of silicone oil into the anterior chamber [5]. The peripheral iridectomy must be kept open. Subtenon steroids may help to resolve fibrinogenic reactions. A "plugged" iridectomy may be reopened by YAG laser pulses. If unsuccessful, a surgical reopening of the Ando iridectomy should be performed.

Tips and tricks for surgery with silicone oil tamponades:

- "Sub-silicone oil" proliferation is probably not a chemical effect of the oil but the consequence of compartmentalization, that is concentration of wound healing factors (blood ocular barrier breakdown) under the oil bubble

- Direct exchange of heavy liquids (perfluoroctane, perfluorodecaline) with silicone is possible just the same as via an intermediate step of air

- Silicone oil removal should be aimed at and is usually possible after about 3 months

- Heavy silicone deserves further investigation as the physical means to inhibit PVR of the inferior retina via displacement of growth factors, such as hemostypicum, and as tamponade medium for persistent macular holes allowing supine positioning

---

16.3.5 Endolaser Coagulation

The endolaser has significantly altered the indication profile of vitreous surgery for retinal vascular disease. When initially introduced to vitreoretinal surgery, endo-photocoagulation was used to treat retinal tears, stop bleeding, apply scatter photocoagulation, coagulate ciliary processes and enlarge the pupil in eyes with miotic pupils and iris neovascularization [10, 57]. These indications are so far still valid and in use.

Today endolaser probes are in use for argon and diode lasers. The intensity of the laser burn depends on the duration and power settings of the laser and the working distance of the endolaser tip from the retinal surface and the angle thereof. Furthermore, the degree of underlying pigmentation is important and blood on the retinal surface should be avoided (see heavy liquids) in order to prevent excessive retinal scarring.

In panretinal endolaser photocoagulation the laser pattern is similar to that obtained using the slit lamp or indirect laser delivery systems. Scleral depression in conjunction with an illuminated probe eases the application to the retinal periphery without lens touch.

In the gas or air filled eye the view to the peripheral retina is facilitated. However, retinal holes are more difficult to discern under air. It may be useful to coagulate and mark the central rim of retinal breaks under liquid perfluorocarbon, and complete retinopexy of the peripheral rim after the fluid-air exchange.

Laser coagulation of abnormal retinal vessels aims at the destruction of these vessels. Long expo-

sure times are important and repetitive treatment may be required. Nevertheless, caution is necessary, e.g., in coagulation of aneurysms, in applying endolaser to areas of vascularized retina in order to prevent inadvertent vascular occlusions.

Complications of endolaser photocoagulation include retinal necrosis, choroidal neovascularization and retinal tears that occur after very intensive spots. "Popping" effects should be avoided. These can be avoided by careful adaptation of the laser energy and distance of the endoprobe.

16.3.6 Lens Management and Compartmentalization

In eyes with dense cataracts not only the patient's visual acuity, but also the fundus view is obscured and the necessary panretinal photocoagulation is not possible to an adequate extent. In some cases the use of a krypton laser with a wavelength in the range of 600 nm is advantageous in transcending a nuclear cataract.

In cases of proliferative retinopathy, panretinal photocoagulation prior to cataract surgery is advised. Otherwise, small incision surgery allows for photocoagulation within a short time after cataract extraction.

Phaco-emulsification with implantation of a posterior chamber lens is today generally accepted in eyes with proliferative retinopathy. Previous reports on the stimulation of rubeosis following IOL implantation showed no or insufficient laser treatment. In order to facilitate later panretinal photocoagulation or vitrectomy, a large capsulorrhexis is required as well as an IOL with a large optic. The combination of cataract removal, vitrectomy and endophotocoagulation was only in small case series reported to be associated with a higher risk for neovascularization of the iris [6, 37]. Acryl is the recommended lens material, as these IOLs can be folded and implanted in small incision surgery. Silicone oil tends to stick like glue to silicone lenses.

Cataract surgery without IOL implantation is best performed by endophaco-emulsification. Complete removal of the capsule prevents secondary fibrosis and closure of an eventual (Ando) iridectomy.

Tips and tricks for lens management during vitreous surgery for proliferative retinal diseases:

- It is worthwhile discussing keeping the iris-lens diaphragm in order to avoid growth-factor diffusion to the anterior segment via the vitreous cavity

- Silicone IOLs should be avoided in all patients with vascular disease as they might require silicone oil tamponade at some point, which then may stick to the lens material

- Large capsulorrhexis and large optics facilitate future inspection of the peripheral retina
- Removal of IOLs in association with vitreous surgery can easily be performed via pars plana
- If no lens implantation is planned, a pars plana lentectomy can easily be performed in conjunction with vitrectomy

16.4 Indications for Vitreoretinal Surgery in Retinal Vascular Disease

With the current techniques, the three major indications for vitreoretinal surgery in retinal vascular disease are:

- Destruction of ischemic retina and pathological vessels
- Removal of dense vitreous opacities
- Release of traction and retinal detachment repair

16.4.1 Destruction of Ischemic Retina and Pathological Vessels

It has been shown previously that anti-VEGF therapies are able to reduce rubeosis and limit macular edema by enhancing vascular stability. Recent data has identified the molecular mechanisms which are involved in the ischemic reaction (see Chapter 7). Nevertheless, anti-VEGF therapy alone will not amend large avascular areas and in the retina, but photocoagulation will remain the treatment of choice to reduce the stimulus for neovascularization sustainably. Panretinal photocoagulation reduces the release of VEGF, causes the RPE to release inhibitory substances (TGF-β) and increases choroidal oxygen transport to the retina [1, 58]. Argon, krypton, or preferably diode-pumped, frequency-doubled YAG (532 nm) lasers can be used. A detailed rationale of photocoagulation is given in Chapter 13 and in Sect. 16.3.5 of this chapter. In rare instances neovascularization does not regress despite supposedly complete panretinal coagulation: If there is any doubt about a sufficient photocoagulation, it is advisable to further intensify the treatment and increase the number of laser spots per area. If it is certain that panretinal photocoagulation is complete, then vitrectomy plus silicone oil tamponade aims at regression/prevention of rubeosis and ciliary body neovascularization, and at remedy of preretinal neovascularization. Preretinal neovascularization requires the presence of hyaloid [70].

Vitreoretinal surgery and vitrectomy itself cannot reduce ischemia as well, but posterior vitreous detachment may help to remove the scaffold for cellular vitreoretinal traction. Removal of the vitreous gel is also considered to limit accumulation of growth factors. Still, vitreoretinal surgery alone without destruction of the ischemic sites is insufficient and thus vitrectomy is mostly performed to allow for sufficient endophotocoagulation.

16.4.2 Removal of Dense Vitreous Hemorrhages

One of the most frequent reasons for vitrectomy is persistent vitreous and preretinal hemorrhage. Vitreous hemorrhage is most likely a consequence of ruptured neovascularization at the vitreoretinal interface secondary to a (usually partial) posterior vitreous detachment. Vitreous hemorrhage can occur in any proliferative disease.

„Early vitrectomy" should be considered in eyes with vitreous hemorrhage, precluding laser application, not resolving within 4–8 weeks. The general aim of the treatment is an early adequate panretinal photocoagulation (including the outer retinal periphery). As described above, it is advantageous to perform the panretinal photocoagulation intraoperatively using an illuminated laser probe and scleral indentation to complete the treatment to the peripheral retina.

Timing of the vitrectomy depends on the underlying disease, the age of the patient, and visual acuity of the contralateral eye. In cases of dense hemorrhage, a preoperative ultrasound examination (B-scan) is advisable to assess the macula. If retinal detachment is close to extending into the macula, or if the macula has detached only recently, then vitrectomy should be performed in due course. Early vitrectomy is further advisable in eyes lacking previous panretinal photocoagulation. In diabetics, a severe progressive proliferation of the fellow eye is another reason to perform vitrectomy instantly [12, 61]. In these conditions surgical treatment of vitreous hemorrhage in fellow eyes may help to prevent progression to a tractional retinal detachment. In any case of associated anterior segment neovascularization (either rubeosis iridis or manifest neovascular glaucoma) vitreous hemorrhage is an indication for early surgical intervention. Only instant vitrectomy and complete panretinal photocoagulation is able to inhibit progression of the neovascular process and to prevent occlusion of the chamber angle [29].

16.4.3 Release of Traction

In proliferative vitreoretinal diseases (diabetes, PVR), vitrectomy may not only remove growth factors from the vitreous cavity but also the scaffold to which contractile cells adhere and by which tractional forces are transmitted to the retina. Removal of the vitreous or spontaneous posterior vitreous separa-

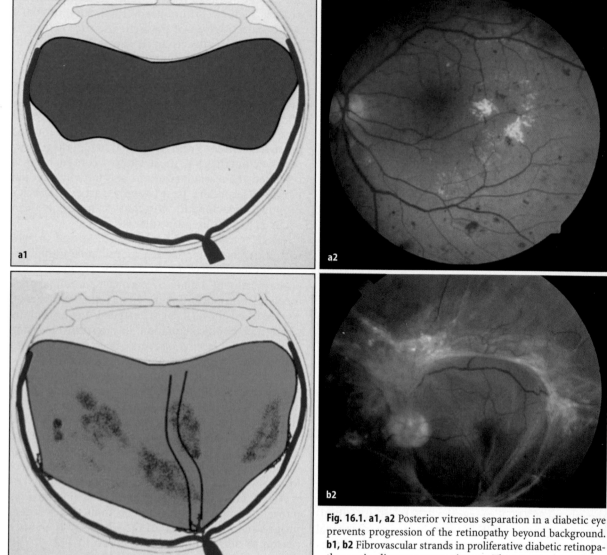

Fig. 16.1. a1, a2 Posterior vitreous separation in a diabetic eye prevents progression of the retinopathy beyond background. **b1, b2** Fibrovascular strands in proliferative diabetic retinopathy require direct contact to vitreous. The preretinal vessels peg the hyaloid to the retina, allowing only partial posterior vitreous separation and vitreoretinal traction in advanced stages

tion prevents vitreoretinal traction in diabetes, since the contractile cells in proliferative diabetic retinopathy selectively attach to vitreous fibers. Although vitrectomy obviates vitreoretinal traction, vitrectomy cannot prevent neovascularization in the vitreous base, and on the ciliary body, and on the iris (rubeosis). Therefore vitrectomy plus panretinal photocoagulation are indispensable partners in the treatment of proliferative diabetic retinopathy (Fig. 16.1).

Despite the fact that pars plana vitrectomy for vitreous hemorrhage, e.g., in diabetics without retinal detachment, provides considerable visual improvement for the patient, the functional results in cases of complicated tractional detachment are disappoint-

ing although a good anatomical result is achieved. Helbig and coworkers report an overall intraoperative reattachment rate of 86% with persistent reattachment of 82% within 6 months postoperatively [25, 26] (for details see Chapter 19.2.2). We advise an encircling band in patients with combined tractional-rhegmatogenous detachments.

There are rare exceptions which should be treated with caution. Young children are especially prone to PVR development. In these patients iatrogenic retinal breaks should be avoided. Similarly patients with rare vascular abnormalities and secondary vitreous alterations such as FEVR (see Chapter 22.5) require adjusted treatment. Here, a similar massive PVR reaction has to be expected after retinotomies.

In severe ocular trauma, vitrectomy eliminates a considerable part of the risk of vitreoretinal traction. However, contractile cells attach not only to vitreous collagen but also to retinal surfaces. Trauma derived strands are able to detach the retina without vitreous interposition. Thus, in severe ocular trauma in addition to vitrectomy additional regimens are recommended (pharmacological adjuncts: 5-FU, daunomycin).

In PVR as a complication of rhegmatogenous retinal detachment, contractile cells adhere to vitreous collagen and to the retinal surface. Since floating cells settle down, star folds develop typically on the inferior retina. Usually the posterior hyaloid is already detached at the time of retinal detachment. Thus vitreoretinal traction (anterior-posterior, circumferential) is most likely to be found at the peripheral retina, in the area of the inferior vitreous base (Figs. 16.1, 16.2).

PVR is considered an undesired wound healing response. It requires both contractile cells and inflammatory mediators deriving from blood ocular barrier breakdown. If one of the two "partners" is missing, PVR is unlikely to occur. PVR is not a typical complication of uveitis, where there is prominent blood ocular barrier breakdown but no contractile cells to respond to. PVR is not a typical complication of retinal holes when the retina is attached. There are potentially contractile cells (RPE cells), but inflammatory modulators are absent. Rhegmatogenous retinal detachment, however, features both blood ocular barrier breakdown plus contractile cells. The RPE is exposed to the vitreous milieu via the retinal hole. In untreated or longstanding rhegmatogenous retinal detachments PVR is a common manifestation (Fig. 16.3).

Physical separation of cells and growth factors is thought to reduce recurrence of PVR. Since PVR is

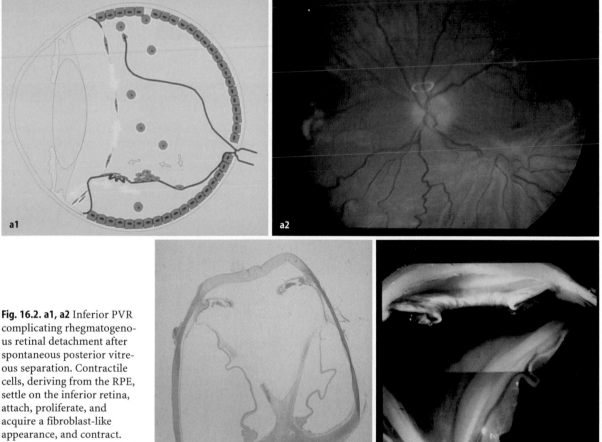

Fig. 16.2. a1, a2 Inferior PVR complicating rhegmatogenous retinal detachment after spontaneous posterior vitreous separation. Contractile cells, deriving from the RPE, settle on the inferior retina, attach, proliferate, and acquire a fibroblast-like appearance, and contract. **b1, b2** Ocular trauma related PVR (perforating injury). Fibroblasts use incarcerated vitreous as scaffold to grow contractile vitreal and vitreoretinal strands

16 II

Inflammation   +   Cells

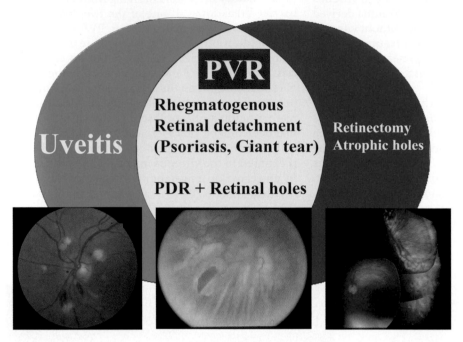

PVR

Rhegmatogenous
Retinal detachment
(Psoriasis, Giant tear)

Uveitis

Retinectomy
Atrophic holes

PDR + Retinal holes

Fig. 16.3. Uveitis (*left*) shows blood ocular barrier breakdown, and no retinal holes. Atrophic holes in attached retina (*right*) are usually not seen along with blood ocular barrier damage. Rhegmatogenous retinal detachment (*middle*) has an overlap of both responsive fibroblastic cells and inflammatory environment. The retina is detached and wrinkled indicating indirectly epiretinal membrane formation

predominantly observed at the lower retina, a heavier than water tamponade, either semifluorinated fluorocarbon, or heavier than water silicone oil, is presently being tested in a prospective randomized study (HSO, Cologne, Germany) (Fig. 16.4).

A relatively new indication for vitrectomy is persistent macular edema. Diabetic macular edema is a consequence of a blood-retinal barrier breakdown, but also tractional forces in the macular area are involved in its pathogenesis.

Since 1996, surgical intervention for macular edema has been more frequently reported. The observation that a posterior vitreous detachment is less frequently found in patients with diffuse diabetic macular edema, led to the assumption that a posterior vitreous detachment could be therapeutically efficient [52, 53]. Hikichi and coworkers reported a resorption of macular edema in 55% of patients after posterior vitreous detachment compared to 25% with attached posterior vitreous [28].

Since then, multiple authors have demonstrated that vitrectomy including removal of the posterior vitreous results in a reduction of macular edema and potential improvement of visual acuity [21, 22, 39]. So far, there is no randomized clinical controlled trial available proving the advantage of ILM peeling in macular edema (see Chapters 19.3.1–19.3.3).

References

1. Alder VA, Cringle SJ, Brown M (1974) The effect of regional retinal photocoagulation on vitreal oxygen tension. Invest Ophthalmol Vis Sci 13:863–868
2. Al-Jazzaf AM, Netland PA, Charles S (2005) Incidence and management of elevated intraocular pressure after silicone oil injection. J Glaucoma 14:40–46
3. Arroyo MH, Refojo MF, Araiz JJ, Tolentino FI, Cajita VN, Elner VM (1993) Silicone oil as a delivery vehicle for BCNU in rabbit proliferative vitreoretinopathy. Retina 13:245–250
4. Bardak Y, Cekic O, Tig SU (2005) Comparison of ICG-assisted ILM peeling and triamcinolone-assisted posterior vitreous removal in diffuse diabetic macular oedema. Eye 20(12):1357–9
5. Beekhuis WH, Ando F, Zivojnovic R, Mertens DAE, Peperkamp E (1987) Basal iridectomy at 6 o'clock in the aphakic eye treated with silicone oil: prevention of keratopathy and secondary glaucoma. Br J Ophthalmol 71:197–200
6. Blankenship GW, Flynn HW Jr, Kokame G (1989) Posterior chamber intraocular lens insertion during pars plana lensectomy and vitrectomy for complications of proliferative diabetic retinopathy. Am J Ophthalmol 108:1–4
7. Bryan JS, Friedman SM, Mames RN, Margo CE (1994) Experimental vitreous replacement with perfluorotri-*n*-propylamine. Arch Ophthalmol 112:1098–1102
8. Chang S, Lincoff H, Zimmerman NJ, Fuchs W (1981) Giant retinal tears. Surgical techniques and results using perfluorocarbon liquids. Arch Ophthalmol 197:761–766
9. Chang A, Zimmerman NJ, Iwamoto T, Ortiz R, Faris D (1987) Experimental vitreous replacement with perfluorotributylamine. Am J Ophthalmol 103:29–37

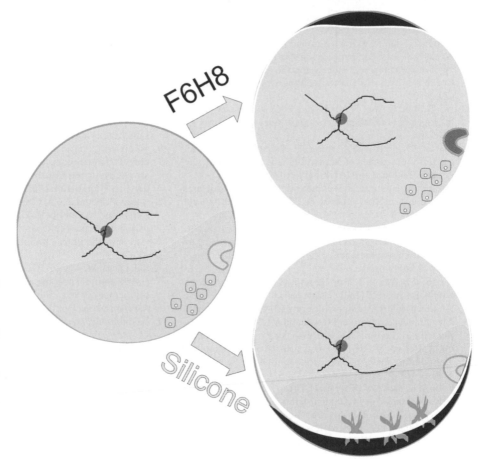

Fig. 16.4. Schematic representation of an eye with rhegmatogenous retinal detachment at risk of PVR, with retinal hole and epiretinal cells (*left side*). Conventional lighter than water silicone oil fill (*right bottom*) allows cells and inflammatory environment together. Contractile retinal membranes form. A heavier than water tamponade (*right top*) separates the stimulating milieu (*top*) from the cells (*bottom*)

10. Charles S (1981) Endophotocoagulation. Retina 1:117–120

11. Cibis PA, Becker B, Okun E, Canaan S (1962) The use of liquid silicone in retinal detachment surgery. Arch Ophthalmol 68:590–9

12. Diabetic Retinopathy Vitrectomy Study Research Group (1990) Early vitrectomy for severe vitreous hemorrhage in diabetic retinopathy. Four-year results of a randomized trial: Diabetic Retinopathy Vitrectomy Study Report 5. Arch Ophthalmol 108:958–964

13. Doi M, Refojo MF (1994) Histopathology of rabbit eyes with intravitreous silicone-fluorosilicone copolymer oil. Exp Eye Res 59:737–746

14. Eckardt C (2005) Transconjunctival sutureless 23-gauge vitrectomy. Retina 25:208–211

15. Eckard C, Nicolai U (1993) Klinische und histologische Befunde nach mehrwöchiger intraokularer Tamponade mit Perfluorodecalin. Ophthalmologe 90:443–447

16. Eckard C, Nicolai U, Winter M, Knop E (1991) Experimental intraocular tolerance to liquid perfluorooctane and perfluoropolyether. Retina 11:375–384

17. Ference M, Lemon HB, Stephenson RJ (1956) Analytical experimental physics, 2nd edn. University of Chicago Press, Chicago

18. Feynman RP, Leighton RB, Sands M (1972) The Feynman lectures on physics, 7th edn. Addison-Wesley, Reading, MA

19. Flores Aquilar M, Munguia D, Loeb E, Crapotta JA, Vuong C, Shakiba S (1995) Intraocular tolerance of perfluorooctylbromide (Perflubrom). Retina 15:3–13

20. Fujii GY, De Juan E Jr, Humayun MS, Chang TS, Pieramici DJ, Barnes A, Kent D (2002) Initial experience using the transconjunctival sutureless vitrectomy system for vitreoretinal surgery. Ophthalmology 109:1814–1820

21. Gandorfer A, Messmer EM, Ulbig MW, Kampik A (2000) Resolution of diabetic macular edema after surgical removal of the posterior hyaloid and the inner limiting membrane. Retina 20:126–133

22. Harbour JW, Smiddy WE, Flynn HW Jr, Rubsamen PE (1996) Vitrectomy for diabetic macular edema associated with a thickened and taut posterior hyaloid membrane. Am J Ophthalmol 121:405–413

23. Heidenkummer HP, Kampik A, Thierfelder S (1992) Experimental evaluation of in vitro stability of purified polydimethylsiloxanes (silicone oil) in viscosity ranges from 1000 to 5000 centistokes. Retina 12(3 Suppl):S28–32

24. Helbig H (2002) Diabetische Traktionsablatio. Klin Monatsbl Augenheilkd 219(4):186–190

25. Helbig H, Kellner U, Bornfeld N, Foerster MH (1996) Grenzen und Möglichkeiten der Glaskörperchirurgie bei diabetischer Retinopathie. Ophthalmologe 93(6):647–654

26. Helbig H, Kellner U, Bornfeld N, Foerster MH (1998) Vitrektomie bei diabetischer Retinopathie: Ergebnisse, Risikofaktoren, Komplikationen. Klin Monatsbl Augenheilkd 212(5):339–342

27. Herrero-Vanrell R, Ramirez L, Fernandez-Carballido A, Refojo MF (2000) Biodegradable PLGA microspheres loaded with ganciclovir for intraocular administration. Encap-

sulation technique, in vitro release profiles, and sterilization process. Pharm Res 17:1323–1328

28. HikichiT, Fujio N, Akiba J, Azuma Y, Takahashi M, Yoshida A (1997) Association between the short-term natural history of diabetic macular edema and the vitreomacular relationship in type II diabetes mellitus. Ophthalmology 104: 473–477

29. Ho T, Smiddy WE, Flynn HWJ (1992) Vitrectomy in the management of diabetic eye disease. Surv Ophthalmol 37: 190–202

30. Ibarra MS, Hermel M, Prenner JL, Hassan TS (2005) Longer-term outcomes of transconjunctival sutureless 25-gauge vitrectomy. Am J Ophthalmol 139:831–836

31. Kampik A, Hoing C, Heidenkummer HP (1992) Problems and timing in the removal of silicone oil. Retina 12(3 Suppl):S11–6

32. Kampougeris G, Cheema R, McPherson R, Gorman C (2006) Safety of triamcinolone acetonide (TA)-assisted pars plana vitrectomy in macular hole surgery. Eye 2006 Feb 3; Epub ahead of print

33. Kirchhof B, Wong D, Van Meurs J, Hilgers RD, Macek M, Lois N, Schrage NF (2002) Use of perfluorohexyloctane as a long-term internal tamponade agent in complicated retinal detachment surgery. Am J Ophthalmol 133:95–101

34. Kivilcim M, Peyman GA, El-Dessouky ES, Kazi AA, Cheema R, Hegazy H (2000) Retinal toxicity of triamcinolone acetonide in silicone-filled eyes. Ophthalmic Surg Lasers 31:474–478

35. Kobuch K, Menz IH, Hoerauf H, Dresp JH, Gabel VP (2001) New substances for intraocular tamponades: perfluorocarbon liquids, hydrofluorocarbon liquids and hydrofluorocarbon-oligomers in vitreoretinal surgery. Graefes Arch Clin Exp Ophthalmol 239:635–642

36. Kociok N, Gavranic C, Kirchhof B, Joussen AM (2005) Influence on membrane-mediated cell activation by vesicles of silicone oil or perfluorohexyloctane. Graefes Arch Clin Exp Ophthalmol 243:345–358

37. Kokame GT, Flynn HW Jr, Blankenship GW (1989) Posterior chamber intraocular lens implantation during diabetic pars plana vitrectomy. Ophthalmology 96:603–610

38. Lewis H (2001) The role of vitrectomy in the treatment of diabetic macular edema. Am J Ophthalmol 131:123–125

39. Lewis H, Abrams GW, Blumenkranz MS, Campo RV (1992) Vitrectomy for diabetic macular traction and edema associated with posterior hyaloid traction. Ophthalmology 99: 753–759

40. Lincoff A, Haft D, Ligget P, Reifer C (1980) Intravitreal expansion of perfluorocarbon bubbles. Arch Ophthalmol 98:1646

41. Lincoff H, Mardirossian J, Lincoff A, Liggett P, Iwamoto T, Jakobiec F (1980) Intravitreal longevity of three perfluorocarbon gases. Arch Ophthalmol 98(9):1610–1611

42. Machemer R (1977) Intravitreous injection in sulphur hexafluoride gas (SF6). In: Freeman HM, Hirose T, Schepens CL (eds) Vitreous surgery and advances in fundus diagnosis and treatment. Appleton Century Crofts, New York

43. Machemer R, Aaberg TM (1979) Vitrectomy, 2nd edn. Grune & Stratton, New York

44. Machemer R, Buettner H, Norton EW, Parel JM (1971) Vitrectomy: a pars plana approach. Trans Am Acad Ophthalmol Otolaryngol 75:813–820

45. Matheson Gas Products (1971) Matheson gas book, 5th edn. Matheson Gas Products, Milwaukee, WI

46. McCuen BW 2nd, Bessler M, Tano Y, Chandler D, Machemer R (1981) The lack of toxicity of intravitreally administered triamcinolone acetonide. Am J Ophthalmol 91(6): 785–8

47. Meinert H, Knoblich A (1993) The use of semifluorinated alkanes in blood-substitutes. Biomater Artif Cells Immobilization Biotechnol 21(5):583–595

48. Miyamoto K, Refolo MF, Tolentino FI, Fournier GA, Albert DM (1984) Perfluoroether liquid as a long-term vitreous substitute. An experimental study. Retina 4:264–268

49. Munir WM, Pulido JS, Sharma MC, Buerk BM (2005) Intravitreal triamcinolone for treatment of complicated proliferative diabetic retinopathy and proliferative vitreoretinopathy. Can J Ophthalmol 40:598–604

50. Nahib M, Peyman GA, Clark LC Jr, Hoffman RE, Miceli M, Abou-Steit M, Tawalol M, Liu KR (1989) Experimental evaluation of perfluorophenanthrene as a high specific gravity vitreous substitute: a preliminary report. Ophthalmic Surg 20:286–293

51. Nakagawa M, Refojo MF, Marin JF, Doi M, Tolentino FI (1995) Retinoic acid in silicone and silicone-fluorosilicone copolymer oils in a rabbit model of proliferative vitreoretinopathy. Invest Ophthalmol Vis Sci 36:2388–2395

52. Nasrallah FP, Jalkh AE, Van Coppenolle F, Kado M, Trempe CL, McMeel JW, Schepens CL (1988) The role of the vitreous in diabetic macular edema. Ophthalmology 95:1335–1339

53. Nasrallah FP, van de Velde F, Jalkh AE, Trempe CL, McMeel JW, Schepens CL (1989) Importance of the vitreous in young diabetics with macular edema. Ophthalmology 96:1511–1516

54. Norton EW, Aaberg T, Fung W, Curtin VT (1969) Giant retinal tears. I. Clinical management with intravitreal air. Am J Ophthalmol 68:1011–1021

55. Ohm J (1911) Über die Behandlung der Netzhautablösung durch operative Entfernung der subretinalen Flüssigkeit und Einspritzung von Luft in den Glaskörper. Graefes Arch Klin Ophthalmol 79:442–450

56. Petersen J, Ritzau-Tondrow U, Vogel M (1986) Fluorosilicone oil heavier than water: a new aid in vitreoretinal surgery. Klin Monatsbl Augenheilkd 189:228–232

57. Peyman GA, Grisolano JM, Palacia MN (1980) Intraocular photocoagulation with the argon-krypton laser. Arch Ophthalmol 98:2062–2064

58. Puliafito CA, Deutsch TF, Boll J, To K (1987) Semiconductor endophotocoagulation of the retina. Arch Ophthalmol 105: 424–427

59. Rodrigues EB, Meyer CH, Kroll P (2005) Chromovitrectomy: a new field in vitreoretinal surgery. Graefes Arch Clin Exp Ophthalmol 243:291–293

60. Scott IU, Flynn HW Jr, Murray TG, Smiddy WE, Davis JL, Feuer WJ (2005) Outcomes of complex retinal detachment repair using 1000- vs 5000-centistoke silicone oil. Arch Ophthalmol 123:473–478

61. Smiddy WE, Feuer W, Irvine WD, Flynn HW Jr, Blankenship GW (1995) Vitrectomy for complications of proliferative diabetic retinopathy. Functional outcomes. Ophthalmology 102(11):1688–1695

62. Stefaniotou MI, Aspiotis MV, Kitsos GD, Kalogeropoulos CD, Asproudis IC, Psilas KG (2002) Our experience with perfluorohexyloctane (F6H8) as a temporary endotamponade in vitreoretinal surgery. Eur J Ophthalmol 12(6): 518–522

63. Tognetto D, Minutola D, Sanguinetti G, Ravalico G (2005) Anatomical and functional outcomes after heavy silicone

oil tamponade in vitreoretinal surgery for complicated retinal detachment: a pilot study. Ophthalmology 112:1574

64. Velikay M, Wedrich A, Stolba U, Datlinger P, Li Y, Binder S (1993) Experimental long-term vitreous replacement with purified and non purified perfluorodecaline. Am J Ophthalmol 116:565–570

65. Velikay M, Stolba U, Wedrich A, Li Y, Datlinger P, Binder S (1995) The effect of the chemical stability and purification of perfluorocarbon liquids in experimental extended-term vitreous substitution. Graefes Arch Clin Exp Ophthalmol 223:26–30

66. Vote B, Wheen L, Cluroe A, Teoh H, McGeorge A (2003) Further evidence for proinflammatory nature of perfluorohexyloctane in the eye. Clin Exp Ophthalmol 31:408–414

67. Welch JC (1997) Dehydration injury as a possible cause of visual field defect after pars plana vitrectomy for macular hole. Am J Ophthalmol 124:698–699

68. Winter M, Eberhardt W, Scholz C, Reichenbach A (2000) Failure of potassium siphoning by Muller cells: a new hypothesis of perfluorocarbon liquid-induced retinopathy. Invest Ophthalmol Vis Sci 41:256–261

69. Wolf S, Schon V, Meier P, Wiedemann P (2003) Silicone oil-RMN3 mixture ("heavy silicone oil") as internal tamponade for complicated retinal detachment. Retina 23:335–342

70. Wong HC, Sehmi KS, McLeod D (1989) Abortive neovascular outgrowths discovered during vitrectomy for diabetic vitreous haemorrhage. Graefes Arch Clin Exp Ophthalmol 227(3):237–240

71. Wong D, Van Meurs JC, Stappler T, Groenewald C, Pearce IA, McGalliard JN, Manousakis E, Herbert EN (2005) A pilot study on the use of a perfluorohexyloctane/silicone oil solution as a heavier than water internal tamponade agent. Br J Ophthalmol 89:662–665

72. Wong D, Williams R, Stappler T, Groenewald C (2005) What pressure is exerted on the retina by heavy tamponade agents? Graefes Arch Clin Exp Ophthalmol 243:474–477

73. Yamaguchi T, Inoue M, Ishida S, Shinoda K (2006) Detecting vitreomacular adhesions in eyes with asteroid hyalosis with triamcinolone acetonide. Graefes Arch Clin Exp Ophthalmol 13:1–4

74. Zeana D, Becker J, Kuckelkorn R, Kirchhof B (1999) Perfluorohexyloctane as a long-term vitreous tamponade in the experimental animal. Int Ophthalmol 23:17–24

17 Treatment of Rubeotic Secondary Glaucoma

T. Schlote, K.U. Bartz-Schmidt

17 ‖

Core Messages

- Rubeotic glaucoma is a late manifestation of ischemic retinal disease
- In the early form the chamber angle is open; later stages present with angle closure
- Panretinal photocoagulation and cryotherapy are indicated to reduce retinal ischemia
- Pars plana vitrectomy and endolaser coagulation in combination with silicone oil endotam-

ponade may have an additional effect on ischemia
- Transscleral cyclophotocoagulation is indicated for advanced cases or drainage implants
- Retinectomy can create a posterior outflow for aqueous
- Anti-VEGF and steroid injections are under investigation

17.1 Definition and Classification

In most cases (97%) neovascular glaucoma is caused by ischemia [5]. The most common underlying diseases are proliferative diabetic retinopathy and proliferative retinopathy after central retinal vein occlusion (CRVO). Rare types of neovascular glaucoma are of non-ischemic nature (e.g., tumor-induced, inflammatory disease) but should be separated because of different treatment strategies (Table 17.1).

Neovascular glaucoma due to retinal ischemia can be separated into three different forms:
- Early neovascular glaucoma with open angle
- Late neovascular glaucoma with closed angle
- Absolute glaucoma with or without painful blind eye (Fig. 17.1a–c)

Table 17.1. Underlying diseases causing neovascular glaucoma

Retinal ischemia	Proliferative diabetic retinopathy
	Central retinal vein occlusion
	Central retinal artery occlusion
	Ischemic ophthalmopathy
	Long-standing retinal detachment
	Sickle cell disease
	Retinopathia prematurorum
	Radiation retinopathy
	X-linked retinoschisis
Inflammatory diseases	Intermediate/posterior uveitis
	Fuchs' heterochromic cyclitis
	Behçet's disease
	Eales' disease
	Vogt-Koyanagi-Harada syndrome
	Systemic lupus erythematosus
	Sympathetic ophthalmia
Intraocular tumors	Retinoblastoma
	Melanoma of the iris/choroidea
	Lymphoma
	Metastasis

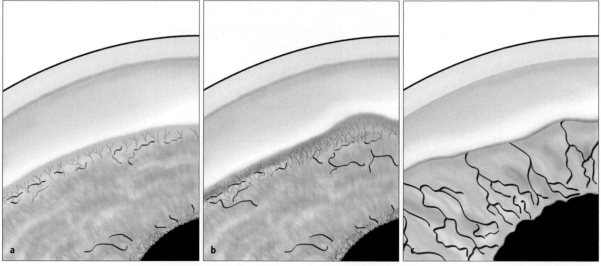

Fig. 17.1. a Open angle without peripheral anterior synechiae and the beginning of angle neovascularization. **b** Open angle with strong neovascularization and focal peripheral anterior synechiae. **c** Complete secondary angle closure glaucoma due to peripheral anterior synechiae. No angle structures are visible

17.2 Prognosis

Rubeotic secondary glaucoma has to be considered as a late manifestation of a progressive underlying ischemic retinal disease. In most cases visual acuity is markedly reduced caused by retinal and optic disk changes (e.g., macular edema, optic atrophy) and only secondarily by glaucomatous optic neuropathy.

Clinical-pathological examination of enucleated eyes demonstrated that secondary angle closure glaucoma accounts for 40–60% of all enucleated glaucoma eyes [12, 32]. The most common type of glaucoma was neovascular glaucoma. Rubeosis iridis was seen in nearly half of all enucleated glaucoma eyes.

17.3 Treatment

The treatment strategy depends on the stage of neovascular glaucoma (Fig. 17.2).

The aims of treatment are:

- Treatment of retinal ischemia
- Regression of rubeosis iridis and angle neovascularization
- Avoidance of irreversible secondary angle closure by peripheral anterior synechiae (PAS)
- Control of intraocular pressure
- Maintenance of visual acuity

17.3.1 Early Rubeotic Secondary Glaucoma (Open Angle Type)

Early rubeotic secondary glaucoma is characterized by neovascularization of the angle but the absence or only the beginning of peripheral anterior synechiae (Figs. 17.1a, b, 17.3a).

It is important to recognize that new vessels are able to grow in the angle without slit-lamp evidence of iris neovascularization (Fig. 17.4). Browning et al. [7] found angle neovascularization without iris neovascularization in 12% of eyes with CRVO. The CRVO study reported about 10% of eyes with nonischemic CRVO and 6% of eyes with ischemic CRVO with angle neovascularization without iris neovascularization [4]. Similar observations have been described in patients with diabetes mellitus [6]. Gonioscopically fine irregular vessels that may overgrow the trabecular meshwork can be seen. There is no general recommendation of how often gonioscopy should be performed.

Treatment of retinal ischemia is the most important step to resolve iris and angle neovascularization and remains the most important factor for intraocular pressure (IOP) regulation in early (open-angle) rubeotic glaucoma. In patients with clear optic media, **panretinal photocoagulation** (PRP) is performed. Ohnishi et al. [34] documented regression of rubeosis in 68% of patients and normalization of IOP in 42% of patients treated with PRP.

Transient IOP increase can be treated with topical drugs reducing aqueous humor production (beta-blockers, carboanhydrase inhibitors, α_2-analogues). Prostaglandin analogues probably are less effective

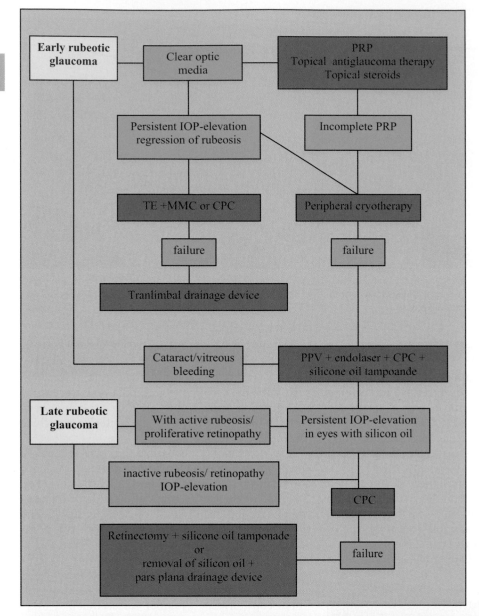

Fig. 17.2. Therapeutic guidelines for the treatment of neovascular glaucoma (*PRP* panretinal photocoagulation, *PPV* pars plana vitrectomy, *CPC* cyclophotocoagulation, *TE+MMC* trabeculectomy with mitomycin C)

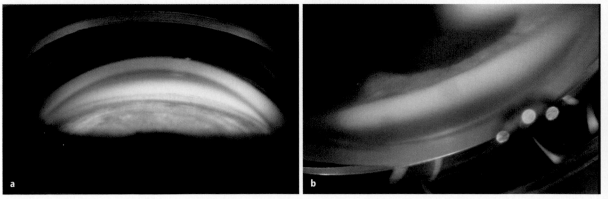

Fig. 17.3. Gonioscopic findings in neovascular glaucoma: **a** early neovascularization of the angle without anterior synechiae; **b** secondary angle closure

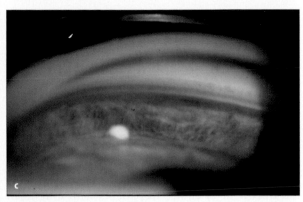

Fig. 17.3c. Siderosis of the angle

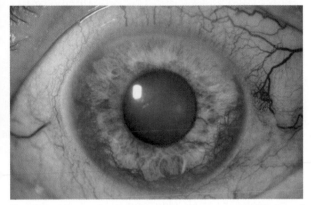

Fig. 17.4. Advanced rubeosis iridis with neovacularization of the angle in an eye with proliferative diabetic retinopathy

in glaucoma eyes associated with intraocular inflammation and should be avoided.

Rubeosis is always associated with an incomplete blood-aqueous humor-barrier leading to a persistent high level of serum proteins and cells within the anterior segment of the eye. This inflammatory trigger is one important mechanism in the induction and progression of peripheral anterior synechiae. Rimexolone, a high-potent steroid derivate with a reduced IOP – increasing risk compared to other topical steroids, can be used to reduce this inflammatory reaction.

When adequate PRP (1,200–1,600 laser spots) is not completely achievable, other modalities should be considered and added, including panretinal cryotherapy or transscleral diode laser retinopexy. **Panretinal cryotherapy** is effective in causing regression of rubeosis. Sihota et al. [40] achieved IOP control in 82% of patients at 1 year after panretinal cryotherapy.

In recent years, **pars plana vitrectomy** and lensectomy (if necessary) with **endolaser photocoagulation** have developed into the most appropriate therapeutic option in patients with proliferative ischemic retinopathy and progression of rubeosis and reti-

nopathy after inadequate panretinal photocoagulation. It is the primary therapeutic decision in patients with rubeosis and proliferative ischemic retinopathy with vitreous haze due to intravitreal bleeding.

It has to be regarded that a high level of vascular endothelial growth factor (VEGF) is maintained within the vitreous cavity after vitrectomy in proliferative diabetic retinopathy, and probably in other ischemic retinopathies as well [19]. Therefore vitrectomy without endotamponade is probably insufficient for the control of progressive ischemic retinopathy with manifest rubeotic glaucoma. Antiproliferative surgery including pars plana vitrectomy, endolaser coagulation of the retina and ciliary body combined with **silicone oil endotamponade** is necessary. Bartz-Schmidt et al. [3] reported normalization of IOP in 72% of patients with uncontrolled rubeotic glaucoma associated with proliferative diabetic retinopathy or central retinal vein occlusion [3]. Silicon oil tamponade prevents postoperative complications and supports regression of rubeosis by separating the anterior from the posterior segment.

Persistent IOP elevation with open angle or limited anterior synechiae after regression of rubeosis can be treated with all antiglaucomatous drugs except pilocarpine. Pilocarpine may enhance the disruption of the blood-aqueous humor barrier and the formation of posterior synechiae.

In refractory cases, **trabeculectomy** may be performed in patients without previous vitrectomy. However, trabeculectomy without antimetabolites has shown a worse outcome in neovascular glaucoma. Mietz et al. [28] reported a failure rate of 80% in this group of patients. Therefore trabeculectomy is now recommended in association with the primary use of **antimetabolites**. Despite this modification, the failure rate of filtration surgery is still high. Hyang and Kim [18] reported a success rate of 29% in patients with neovascular glaucoma 12 months after trabeculectomy with mitomycin C. Mandal et al. [25] observed a surgical success rate of 67% in 15 eyes with neovascular glaucoma and a follow-up between 2 and 82 months.

After failure of trabeculectomy, **translimbal implantation of drainage devices** (Molteno, Baerveldt, Ahmed) or transscleral cyclophotocoagulation can be performed. The classical translimbal implantation is an appropriate method in eyes with only a partially occluded angle. Mermoud and colleagues [27] reported the success of single-plate Molteno implantation in 60 eyes with neovascular glaucoma. The success rate was 62% at 1 year, but decreased to 10% at 5 years. Loss of light perception was seen in 48% and progression to phthisis bulbi occurred in 18%. The authors concluded that Molteno implanta-

17

tion may be used for pain relief and avoidance of enucleation.

Alternatively, **transscleral cyclophotocoagulation** can be successfully used, especially in older patients (see Sect. 17.3.2). One study from Taiwan reported comparable results of diode laser cyclophotocoagulation and trabeculectomy in patients with neovascular glaucoma [23]. Another study compared the results of tube-shunt surgery versus noncontact Nd:YAG cyclophotocoagulation [13]. In this study satisfactory control of IOP was achieved in 37% of eyes treated with noncontact cyclophotocoagulation compared to 67% receiving a tube shunt surgery. Using contact cyclophotocoagulation more patients can be controlled satisfactorily (see Sect. 17.3.2).

17.3.2 Advanced Rubeotic Secondary Glaucoma (Angle Closure Type)

Late rubeotic secondary glaucoma is characterized by complete closure of the angle due to excessive peripheral anterior synechiae (Figs. 17.1c, 17.3b). Typical biomicroscopic visible changes of the anterior segment of the eye include corneal changes (epithelial edema, band keratopathy, peripheral neovascularization), strong rubeosis and covering of the iris by a fibrovascular membrane leading to pupil distortion, ectopium uveae, heterochromia due to siderosis after repeated intraocular bleeding (Fig. 17.5), posterior synechiae and often cataract.

In the presence of secondary angle closure, IOP often remains increased despite control of the ischemic retina and regression of iris neovascularization. Topical antiglaucomatous therapy is mostly insufficient. Only aqueous humor-reducing drugs may reduce the IOP, because the angle is closed by fibrovascular membranes.

In the presence of an active ischemic retina, cyclophotocoagulation (endo- or transscleral) is often necessary in combination with pars plana vitrectomy, endolaser photocoagulation and silicone oil tamponade.

Using contact diode laser **transscleral cyclophotocoagulation** different success rates have been reported in the literature ranging from 56% to 87% within 1–3 years of follow-up [33, 36]. There may be an increased risk of hypotonia and phthisis compared to other types of glaucoma [29].

Transscleral diode laser cyclophotocoagulation is the treatment of choice in patients with long-term intravitreal silicone oil tamponade. Sivagnavel et al. [41] reported an overall success rate of 50%. Some patients developed hypotonia despite intravitreal silicone oil. Han et al. [17] achieved complete success in 55% of their patients with silicone oil tamponade after retinal detachment. Another aspect is whether repeated vitrectomies or later removal of the silicone oil may increase the risk of hypotonia after cyclophotocoagulation. Nabili and Kirkness [31] reported five out of ten eyes with diabetic neovascular glaucoma and previous repeated vitrectomies developed hypotonia after diode laser cyclophotocoagulation.

Even after repeated cyclophotocoagulation, an elevated IOP is seen in a substantial number of eyes with secondary angle glaucoma. Therefore, alternative strategies are needed. Relatively new treatment options are **pars-plana-modified drainage implants** (Baerveldt implant, Ahmed glaucoma valve) (Fig. 17.6). In 2000 and 2002 the first clinical reports were published concerning the efficacy of a pars plana modified Baerveldt implant with a Hofmann elbow [8, 24]. The studies reported a high success rate (>90%) in a relatively small group of eyes. One study directly compared the efficacy of Nd:YAG cyclopho-

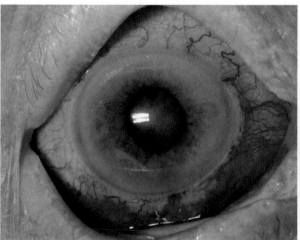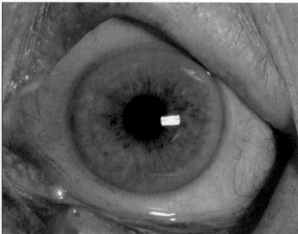

Fig. 17.5. Heterochromia due to siderosis of the right eye in a patient with rubeotic glaucoma and repeated intraocular bleeding due to proliferative diabetic retinopathy

tocoagulation and pars plana modified Baerveldt implant in neovascular glaucoma [8]. After 6 months of follow-up, the success rate for cyclophotocoagulation was 77% and for Baerveldt implantation it was 95%. The modified Ahmed glaucoma valve with pars plana clip is another drainage device with resisted flow. IOP regulation without further antiglaucomatous therapy (complete success) was reported in 64% of patients with secondary angle closure glaucoma 1 year after surgery [37]. So far little is known about the long-term efficacy of these implants.

Another option is **retinectomy,** which opens a new posterior outflow path for the aqueous to the absorbing choroid (Fig. 17.7). In 2003, long term results 5 years after surgery (retinectomy with intraocular gas tamponade) were reported [22]. Nearly 53% of all treated eyes showed a long term regulated IOP. On the other hand, 48% of eyes developed retinal complications. The high risk of complications may be reduced by primarily used silicone oil tamponade.

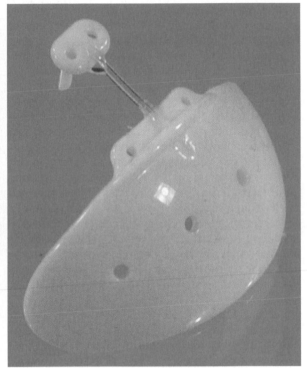

Fig. 17.6. Pars-plana-modified Baerfeld implant with Hofmann elbow

17.3.3 Blind Painful Eye with Rubeotic Secondary Glaucoma

The primary aim of treatment of a painful blind eye is an eye free of complaints accompanied by a cosmetically acceptable situation (Fig. 17.8). There is no clear therapeutic guideline concerning the treatment of a painful blind eye with (neovascular) glaucoma. Three different causes of pain should be considered:

- Pain due to an increased intraocular pressure
- Pain due to ciliary body spasm in persistent intraocular inflammation
- Pain due to ocular surface diseases

Concerning the *high intraocular pressure* in a blind painful eye, medical treatment is rarely successful. Intraocular surgery should be avoided because of the (low) risk of sympathetic ophthalmia [14]. **Transscleral cyclophotocoagulation** may be effective in the reduction of intraocular pressure and pain resolution [36]. Martin and Broadway [25] reported a success rate (resolution of pain) of 95% after diode laser cyclophotocoagulation [26]. Successful treat-

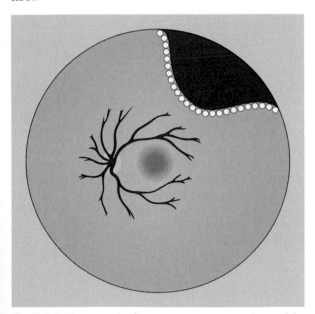

Fig. 17.7. Retinectomy in the temperosuperior quadrant of the retina

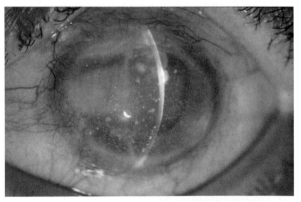

Fig. 17.8. Blind painful eye with advanced rubeotic glaucoma after multiple anterior and posterior segment surgery. Band keratopathy, peripheral corneal neovascularization; endothelial precipitates are visible

ment was associated with a reduction of IOP of more than 30% of the initial value.

Alternatively, retrobulbar injection of alcohol or **chlorpromazine** can be performed or may be added in patients after failure of cyclophotocoagulation. Chen et al. [9] reported a successful treatment using 1–2 ml chlorpromazine (25 mg/ml) in 80% of 20 patients. More than 50% of these patients were free of pain for more than 1 year.

Enucleation remains the ultima ratio, but will resolve the pain in the overwhelming majority of patients (90%) [10, 39]. Especially in cases of low intraocular pressure and beginning phthisis associated with pain, enucleation resolves the pain and may contribute to a better cosmetic situation.

Inflammation-induced pain should be treated with anti-inflammatory agents. Topical corticosteroids of high potency (dexamethasone, prednisolone) in combination with cycloplegics may resolve pain in eyes with active non-infectious intraocular inflammation. Seldomly non-infectious scleritis may occur and should be treated with oral corticosteroids for several weeks.

Pain due to disturbances of the ocular surface including dry eye, decompensation of the cornea, band keratopathy or trophic corneal ulceration may complicate the situation in a blind eye. In some cases pain can be resolved by artificial tears, therapeutic contact lens, or removal of band keratopathy. Sometimes treatment of trophic corneal changes by amnion membrane transplantation or covering with conjunctiva may be necessary [38].

17.3.4 Treatment Options Under Investigation

There is no doubt that new treatment strategies are needed to improve the long term prognosis of rubeotic secondary glaucoma and retain a useful vision over many years.

The first major aim of new therapeutic procedures is the *regression and control of neovascularization/rubeosis.* Several antiangiogenic substances are under evaluation. Intravitreal injection of **triamcinolone** acetonide may result in a regression of rubeosis and was also reported to reduce the intraocular pressure in patients with neovascular glaucoma due to diabetic retinopathy and ischemic central retinal vein occlusion [20]. The question is whether triamcinolone may be used as an additional tool in the treatment of neovascular glaucoma. There is a substantial risk of a further increase of intraocular pressure in glaucoma eyes, so the risk-benefit ratio in a single patient cannot be calculated [21].

The high angioproliferative potency of vascular endothelial growth factor (VEGF) resulting in iris neovascularization and neovascular glaucoma has been demonstrated by a range of experimental and clinical studies [42, 43]. Now, **anti-VEGF therapy** (pegaptanib) seems to be effective in the treatment of neovascular age-related macular degeneration [16]. The potency of pegaptanib on iris neovascularization should be evaluated in future studies.

A further area of research is **gene therapy**. Experimental studies demonstrated that ocular angiogenesis can be inhibited by an adenovirus carrying the human von Hippel-Lindau tumor-suppressor gene [1]. VEGF expression was significantly reduced after intraocular gene transfer in a monkey model.

Squalamine, an antiangiogenic aminosterol, inhibited iris neovascularization after systemic injection in a primate model [15]. A further potentially useful agent may be anecortave, a synthetic derivate of cortisol. Specific chemical modifications to the cortisol structure have resulted in the creation of a potent inhibitor of blood vessel growth. First comparative studies using **anecortave** as a posterior juxtascleral depot reported a regression of subfoveal choroidal neovascularization in age-related macular degeneration within 12 months [2, 11]. At present no data are known about the possible additive effect of anecortave in the treatment of proliferative retinopathies and iris neovascularization/neovascular glaucoma.

Finally, Muller et al. [30] reported the possible treatment of iris neovascularization with **photodynamic therapy** with verteporfin [30].

The *second major aspect* includes procedures that may allow an early effective control of the intraocular pressure in manifest rubeotic (secondary angle closure) glaucoma decreasing the risk of overtreatment and phthisis. At present different treatments (cyclophotocoagulation, drainage devices, retinectomy) are used for the same situation. One important step is to compare these different procedures by prospective, randomized, multicenter trials. In addition, other new surgical ideas are being introduced. **Suprachoroidal seton implantation** (a modified Krupin eye valve) has been reported to allow the drainage of aqueous humor from the anterior chamber to the suprachoroidal space [35]. Several patients with painful blind eyes due to neovascular glaucoma in proliferative diabetic retinopathy were successfully treated. This new treatment option should be proven in a larger patient group under well designed study conditions.

References

1. Akiyama H, Tanaka T, Itakura H, Kanai H, Maeno T, Doi H, Yamazaki M, Takahashi K, Kimura Y, Kishi S, Kurabayashi M (2004) Inhibition of ocular angiogenesis by an adenovirus carrying the human von Hippel-Lindau tumor-suppressor gene in vivo. Invest Ophthalmol Vis Sci 45:1289–96
2. Augustin AJ, D'Amico DJ, Mieler WF, Schneebaum C, Beasley C (2005) Safety of posterior juxtascleral depot administration of the angiostatic cortisone anecortave acetate for treatment of subfoveal choroidal neovascularisation in patients with age-related macular degeneration. Graefes Arch Clin Exp Ophthalmol 243:9–12
3. Bartz-Schmidt KU, Thumann G, Psichias A, Krieglstein GK, Heimann H (1999) Pars plana vitrectomy, endolaser coagulation of the retina and the ciliary body combined with silicone oil endotamponade in the treatment of uncontrolled neovascular glaucoma. Graefes Arch Clin Exp Ophthalmol 237:969–75
4. Central Retinal Vein Occlusion Study Group (1993) Baseline and early natural history report. The Central Retinal Vein Occlusion Study. Arch Ophthalmol 11:1087–95
5. Brown GC, Magargal LE, Schachat A, Shah H (1984) Neovascular glaucoma. Etiologic considerations. Ophthalmology 91:315–20
6. Browning DJ (1991) Risk of missing angle neovascularisation by omitting screening gonioscopy in patients with diabetes mellitus. Am J Ophthalmol 112:212
7. Browning DJ, Scott AQ, Peterson CB, Warnock J, Zhang Z (1998) The risk of missing angle neovascularisation by omitting screening gonioscopy in acute retinal vein occlusion. Ophthalmology 105:776–84
8. Chalam KV, Gandham S, Gupta S, Tripathi BJ, Tripathi RC (2002) Pars plana modified Baerveldt implant versus neodymium:YAG cyclophotocoagulation in the management of neovascular glaucoma. Ophthalmic Surg Lasers 33:383–393
9. Chen TC, Ahn Yuen SJ, Sangalang MA, Fernando RE, Leuenberger EU (2002) Retrobulbar chlorpromazine injections for the management of blind and seeing painful eyes. J Glaucoma 11:209–13
10. Custer PL, Reistad CE (2000) Enucleation of blind, painful eyes. Ophthalmic Plast Reconstr Surg 16:326–9
11. D'Amico DJ, Goldberg MF, Hudson H, Jerdan JA, Krueger DS, Luna SP, Robertson SM, Russel S, Singerman L, Slakter JS, Yannuzzi L, Zilliox P, Anecortave Acetate Clinical Study Group (2003) Anecortave acetate as monotherapy for treatment of subfoveal neovascularisation in age-related macular degeneration: twelve-month clinical outcomes. Ophthalmology 110:2372–83
12. De Gottrau P, Holbach LM, Naumann GOH (1994) clinicopathological review of 1146 enucleations (1980–1990). Br J Ophthalmol 78:260–5
13. Eid TE, Katz LJ, Spaeth GL, Augsburger JJ (1997) Tube-shunt surgery versus neodymium:YAG cyclophotocoagulation in the management of neovascular glaucoma. Ophthalmology 104:1692–700
14. Gasch AT, Foster CS, Grosskreutz CL, Paquale LR (2000) Postoperative sympathetic ophthalmia. Int Ophthalmol Clin 40:69–84
15. Genaidy M, Kazi AA, Peyman GA, Paissos-Machado E, Farahat HG, Williams JI, Holroyd KJ, Blake DA (2002) Effect of squalamine on iris neovascularisation in monkeys. Retina 22:772–8
16. Gragoudas ES, Adamis AP, Cunningham ET Jr, Feinsod M, Guyer DR, VEGF Inhibition Study in Ocular Neovascularisation Clinical Trial Group (2004) Pegaptanib for neovascular age-related macular degeneration. N Engl J Med 351:2805–16
17. Han SK, Park Kh, Kim DM, Chang BL (1999) Effect of diode laser trans-scleral cyclophotocoagulation in the management of glaucoma after intravitreal silicone oil injection for complicated retinal detachment. Br J Ophthalmol 83:713–17
18. Hyang SM, Kim SK (2001) Mid-term effects of trabeculectomy with mitomycin C in neovascular glaucoma patients. Korean J Ophthalmol 15:98–106
19. Itakura H, Kishim S, Kotajima N, Murakami M (2004) Persistent secretion of vascular endothelial growth factor into the vitreous cavity in proliferative diabetic retinopathy after vitrectomy. Ophthalmology 111:1880–4
20. Jonas JB, Hayler JK, Sofker A, Panda-Jonas S (2001) Regression of neovascular iris vessels by intravitreal injection of crystalline cortisone. J Glaucoma 10:284–7
21. Jonas JB, Degenring RF, Kreissig I, Akkoynun I, Kamppeter BA (2005) Intraocular pressure elevation after intravitreal triamcinolone acetonide injection. Ophthalmology 112:593–8
22. Joussen AM, Walter P, Jonescu-Cypers CP, Koizumi K, Poulaki V, Bartz-Schmidt KU, Krieglstein GK, Kirchhof B (2003) Retinectomy for treatment of intractable glaucoma: long term results. Br J Ophthalmol 87:1094–1103
23. Kuang TM, Liu CJ, Chou CK, Hsu WM (2004) Clinical experience in the management of neovascular glaucoma. J Chin Med Assoc 67:131–5
24. Luttrull JK, Avery RL, Baerveldt G, Easley KA (2000) Initial experience with pneumatically stented Baerveldt implant modified for pars plana insertion for complicated glaucoma. Ophthalmology 107:143–9
25. Mandal AK, Majji AB, Mandal SP, Das T, Jalali S, Gothwal VK, Jain SS, Nutheti R (2002) Mitomycin-C augmented trabeculectomy for neovascular glaucoma. A preliminary report. Indian J Ophthalmol 50:287–93
26. Martin KR, Broadway DC (2001) Cyclodiode laser therapy for painful, blind glaucomatous eyes. Br J Ophthalmol 85:474–6
27. Mermoud A, Salmon JF, Alexander P, Straker C, Murray AD (1993) Molteno tube implantation for neovascular glaucoma. Long-term results and factors influencing the outcome. Ophthalmology 100:897–902
28. Mietz H, Raschka B, Krieglstein GK (1999) Risk factors for failures of trabeculecotomies performed without antimetabolites. Br J Ophthalmol 83:814–21
29. Mistlberger A, Liebmann JM, Tschiderer H, Ritch R, Ruckhofer J, Grabner G (2001) Diode laser transscleral cyclophotocoagulation for refractory glaucoma. J Glaucoma 10:288–93
30. Muller VA, Ruokonen P, Schellenbeck M, Hartmann C, Tetz M (2003) Treatment of rubeosis iridis with photodynamic therapy with verteprofin – A new therapeutic and prophylactic option for patients with risk of neovascular glaucoma. Ophthalmic Res 35:60–4
31. Nabili S, Kirkness CM (2004) Trans-scleral diode laser cyclophoto-coagulation in the treatment of diabetic neovascular glaucoma. Eye 18:352–6
32. Naumann GD, Portwich E (1976) Ätiologie und Hauptursache für 1000 Enukleationen (Eine klinikopathologische Studie). Klin Monatsbl Augenheilkd 168:622–30

33. Oguri A, Takahashi E, Tomita G, Yamamoto T, Jikihara S, Kitazawa Y (1998) Transscleral cyclophotocoagulation with the diode laser for neovascular glaucoma. Ophthalmic Surg Lasers 29:722–7

34. Ohnishi Y, Ishibashi T, Sagawa T (1994) Fluorescein gonio-angiography in diabetic neovascularisation. Graefes Arch Clin Exp Ophthalmol 232:199–204

35. Ozdamar A, Aras C, Karacorlu M (2003) Suprachoroidal seton implantation in refractory glaucoma: a novel surgical technique. J Glaucoma 12:354–9

36. Schlote T, Derse M, Rassmann K, Nicaeus T, Dietz K, Thiel H-J (2001) Efficacy and safety of contact transscleral diode laser cyclophotocoagulation for advanced glaucoma. J Glaucoma 10:294–301

37. Schlote T, Ziemssen F, Bartz-Schmidt KU (2006) Pars plana modified Ahmed Glaucoma Valve for treatment of refractory glaucoma: A pilot study. Graefes Arch Exp Clin Ophthalmol 244(3):336–41

38. Seitz B, Gruterich M, Cursiefen C, Kruse FE (2005) Konservative und chirurgische Behandlung der neurotrophen Keratopathie. Ophthalmologe 102:15–26

39. Sha-Desai SD, Tyers AG, Manners RM (2000) Painful blind eye: efficacy of enucleation and evisceration in resolving ocular pain. Br J Ophthalmol 84:437–8

40. Sihota R, Sandramouli S, Sood NN (1991) A prospective evaluation of anterior retinal cryoablation in neovascular glaucoma. Ophthalmic Surg 22:256–9

41. Sivagnavel V, Ortiz-Hurtado A, Williamson TH (2005) Diode laser trans-scleral cyclophotocoagulation in the management of glaucoma in patients with long-term intra-vitreal silicone oil. Eye 19:253–7

42. Tolentino MJ, Miller JW, Gragoudas ES, Chatzistefanou K, Ferrara N, Adamis AP (1996) Vascular endothelial growth factor is sufficient to produce iris neovascularisation and neovascular glaucoma in a nonhuman primate. Arch Ophthalmol 114:964–70

43. Tripathi RC, Li J, Tripathi BJ, Chalam KV, Adamis AP (1998) Increased level of vascular endothelial growth factor in aqueous humor of patients with neovascular glaucoma. Ophthalmology 105:232–7

18 Intravitreal Injections: Guidelines to Minimize the Risk of Endophthalmitis

I.U. Scott, H.W. Flynn, Jr

Core Messages

- Although the intravitreal injection procedure dates back to 1911, the recent development of pharmacotherapies for various posterior segment diseases has led to a rapid increase in the use of this technique
- Despite the lack of studies assessing the impact of various technical factors on the risk of endophthalmitis associated with this procedure, clinical experience has provided important lessons for safe administration
- Injection procedure guidelines include consideration of preexisting conditions, such as active external infection, eyelid abnormalities, povidone-iodine, lid scrubs, preinjection topical antibiotics, lid speculum, drape, gloves, anesthesia, single-use medication bottles for pupil dilation and anesthesia, and postinjection topical antibiotics
- Guidelines for follow-up include consideration of procedure-related risks and the specific needs of the patient
- Both infectious and non-infectious endophthalmitis have been reported following intravitreal injections. Typical clinical features of non-infectious endophthalmitis include absence of pain and eyelid edema and no increased conjunctival injection compared with that present immediately after injection

---

18.1 Introduction

Intravitreal injection was reported by Ohm in 1911 as a technique to introduce air for retinal tamponade and repair of retinal detachment [28]. Intravitreal administration of pharmacotherapies dates to the mid-1940s with the use of penicillin to treat endophthalmitis [34, 35]. Since that time, use of the intravitreal injection technique has steadily increased, with its usage being focused primarily on the treatment of retinal detachment [7, 32], endophthalmitis [8, 31], and cytomegalovirus (CMV) retinitis [13, 43]. The increasing confidence in the efficacy and safety of intravitreal injections, in conjunction with the development of additional pharmacotherapies, has led to a recent rapid increase in the use of this technique for the administration of various pharmacotherapies (e.g., ranibizumab [6], pegaptanib sodium [9, 41, 42]) for age-related macular degeneration (AMD) and intravitreal triamcinolone for macular edema associated with a variety of etiologies, such as diabetic retinopathy [21], central retinal vein occlusion [10, 36], branch retinal vein occlusion [5, 17, 30, 37], uveitis [2, 44], and birdshot retinochoroidopathy [22].

18.2 Risk of Endophthalmitis Associated with the Intravitreal Injection Procedure

With the recent widespread use of the intravitreal injection technique, there has been increased concern regarding the risk of endophthalmitis following intravitreal injection. The risk of endophthalmitis in eyes with CMV retinitis treated with intravitreal injection has been estimated to be 1.3% per eye and 0.1% per injection [16]. The risk of endophthalmitis following intravitreal triamcinolone acetonide injection has been estimated to be approximately 0.8% for non-infectious endophthalmitis [16, 24, 25] and 0.6–0.16% for infectious endophthalmitis [16, 26]. In general, endophthalmitis has been reported very rarely after intravitreal injection of gas or agents other than antiviral agents and triamcinolone acetonide, with a combined prevalence of approximately 0.1% for all other compounds [16]. Combined analysis of two concurrent prospective, randomized, double-masked, multicenter, controlled clinical trials of intravitreal pegaptanib therapy for subfoveal choroidal neovascularization secondary to age-related macular degeneration demonstrated that the rate of endophthalmitis was 12/890 (1.3%) patients over a

54-week period (the protocol included an intravitreal injection of pegaptanib into one eye of each patient every 6 weeks for 48 weeks) [9]. Eight of the 12 cases of endophthalmitis occurred during the first year of the study; after protocol changes were made relating to the injection procedure (e.g., use of a sterile eyelid speculum, a sterile drape, and gloves), no further cases of endophthalmitis occurred. This suggests that the injection technique is important in reducing the risk of endophthalmitis following intravitreal injection. Strategies that may be important in reducing the risk of endophthalmitis include attention to issues before, during, and after the injection. Guidelines developed as a result of round-table deliberations conducted after a review of published and unpublished studies and case series are summarized in Tables 18.1 and 18.2 [1].

Table 18.1. Guideline areas with strong agreement

- Povidone-iodine for ocular surface, eyelids, and eyelashes
- Use speculum and avoid contamination of the needle with eyelashes or eyelid margin
- Avoid extensive massage of eyelids either pre- or postinjection (to avoid expressing meibomian glands)
- Avoid injecting patients who have active eyelid or ocular adnexal infection
- Dilate pupil
- Use adequate anesthetic for a given patient (topical drops and/or subconjunctival injection)
- Avoid prophylactic or postinjection paracentesis

Table 18.2. Guideline areas with no clear consensus

- Most did not want to use a povidone-iodine flush, and preferred drops; no benefits attributed to waiting for the povidone-iodine to dry
- Most did not use a sterile drape
- Most advocated use of gloves
- Use of pre- or postinjection topical antibiotics – little published scientific data to support reduction in endophthalmitis
- Intraocular pressure (IOP) may be checked following injection – no agreement on IOP level at which physicians are comfortable to discharge patient
- Dilated funduscopic examination can be performed following injection to confirm central retinal artery perfusion and intraocular location of drug.
  No consensus on whether patients are competent to self-report signs and symptoms of endophthalmitis or other adverse events; no consensus regarding the need for clinical follow-up examination versus telephone interchange with physician or nurse

18.3 Injection Procedure Guidelines (Figs. 18.1 – 18.9)

Active external infection, including significant blepharitis, should be treated prior to injection. In addition, eyelid abnormalities such as ectropion are

reported risk factors for endophthalmitis and should be considered. Ocular surface bacteria may be the most common sources of bacteria causing postoperative endophthalmitis [3, 12, 20, 20, 38]. Thus, one strategy to reduce the risk of endophthalmitis is to reduce or eliminate the bacteria on the patient's ocular surface and eyelids. While this may be achieved in various ways (povidone-iodine, topical antibiotics, eyelid hygiene and sterile isolation of the surgical site), povidone-iodine is the only agent that has been demonstrated to reduce the risk of postoperative endophthalmitis in a prospective study of cataract surgery [37]. It is unknown whether the application of povidone-iodine in the form of drops versus a flush impacts on the ability of this agent to prevent endophthalmitis. Lid scrubs have been reported to be associated with a significant increase in bacterial flora [4]; thus, excessive eyelid manipulation should be avoided (although the efficacy of lid scrubs in

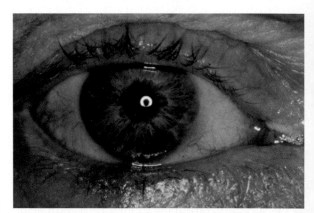

Fig. 18.1. Dilated and telangiectatic vessels along lower eyelid margin consistent with significant meibomian gland disease. This patient's eyelid disease was treated prior to administering an intravitreal injection of medication

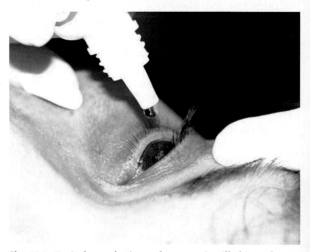

Fig. 18.2. Topical anesthetic eyedrops are instilled into the eye before the povidone-iodine eyelid scrubs are performed since povidone-iodine can cause ocular discomfort

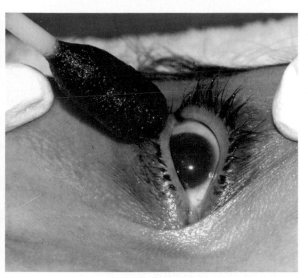

Fig. 18.3. The eyelid margins are scrubbed with povidone-iodine

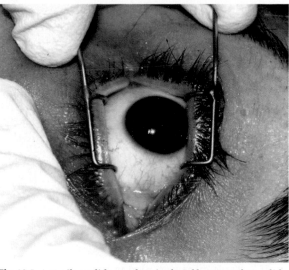

Fig. 18.5. A sterile eyelid speculum is placed between the eyelids. We typically instill additional povidone-iodine eyedrops after speculum placement

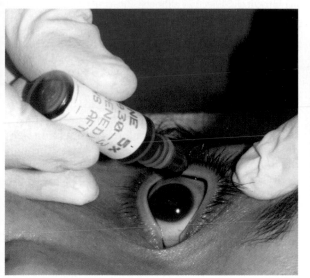

Fig. 18.4. Povidone-iodine eyedrops are instilled into the eye

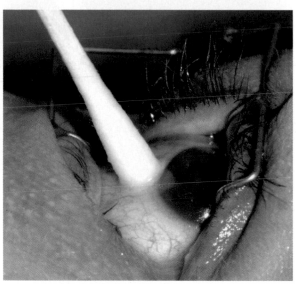

Fig. 18.6. A cotton-tipped swab soaked in sterile xylocaine 2% is held against the injection site to achieve local anesthesia

combination with povidone-iodine has not been reported). Since true contact allergy to povidone-iodine is rare, a reported history of such an allergy should be verified with a skin patch test.

Topical antibiotics have been demonstrated to reduce ocular surface bacteria significantly, but have not been proven to have a significant impact on reducing the risk of endophthalmitis [11, 15, 29, 40]. In a study of 35 patients who underwent intraocular surgery [15], a combination of antibiotics and povidone-iodine resulted in 83% of eyes having sterile cultures, compared to 40% of eyes treated with povidone-iodine alone and 31% of eyes treated with antibiotic alone (a solution consisting of polymyxin B sulfate, neomycin sulfate, and gramicidin); these results provide evidence of a synergistic effect between antibiotics and povidone-iodine.

During the injection procedure, the use of a sterile lid speculum is recommended in order to avoid needle contact with lids and lashes. The use of a sterile drape is optional but gloves, part of universal precautions, are appropriate. Sterile topical anesthetic is administered as the first step in the procedure. Ophthalmologists may consider subconjunctival anesthesia, but this requires additional instrumentation and manipulation which may be associated with increased surface flora. If subconjunctival anesthesia is used, keep in mind that the needle used for intravitreal injection passes through the subconjunctival space filled with anesthetic and that surface bacteria may have been introduced beneath the conjunctiva.

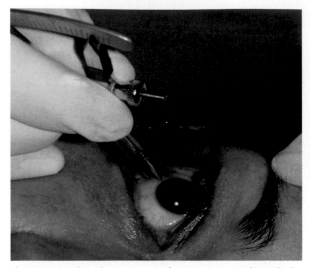

Fig. 18.7. Sterile calipers are used to measure and mark the desired distance posterior to the limbus

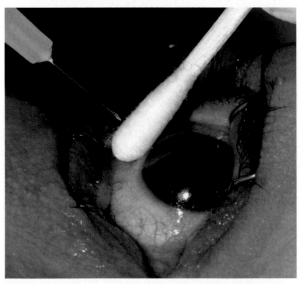

Fig. 18.8. The eye is stabilized with a cotton-tipped swab during the injection procedure. The cotton-tipped swab can then be rolled over the injection site following the injection to minimize reflux of drug and/or liquid vitreous

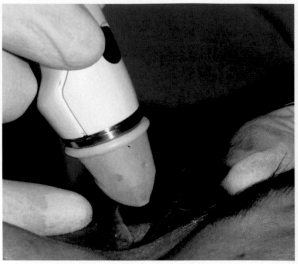

Fig. 18.9. We generally measure the intraocular pressure following intravitreal injection although this is not mandatory; it is important to perform a dilated funduscopic examination following the injection procedure to confirm central retinal artery perfusion and intraocular location of drug (when possible), as well as to investigate for such injection-related complications as retinal tear, retinal detachment, or vitreous hemorrhage

Although lidocaine gel has been used with increased frequency for anterior segment surgery cases in recent years, and has been reported as providing satisfactory patient comfort during intravitreal injection procedures while causing less chemosis and hemorrhage than subconjunctival anesthesia [19], another study identified lidocaine gel as a potential risk factor for endophthalmitis following cataract surgery [23]. Lidocaine gel may serve as a barrier, reducing the ability of povidone-iodine to contact the ocular surface and reduce the risk of endophthalmitis. Even if povidone-iodine is administered prior to lidocaine gel (in an attempt to bypass the potential barrier effect), it should be recognized that the commercially available lidocaine gel is not prepared as a sterile formulation and, therefore, the injection needle may become contaminated as it passes through the lidocaine gel and before it enters the intravitreal cavity. Care should be taken to avoid pressure to the eyelids, eyelid margins, and the adnexa due to the potential for release of resident bacteria. According to one study [14], 13% of the ophthalmic drugs obtained from multiple-use medication bottles tested positive for bacteria and 21% of the bottle tips were culture-positive. Single-use medication bottles should be considered for pupil dilation and anesthesia, although this is not critical if povidone-iodine is applied after the instillation of anesthetic and dilating agents.

After the injection procedure, topical antibiotics have been shown to lower the number of bacteria on the ocular surface and may enter the anterior chamber; however, their efficacy in reducing the risk of endophthalmitis has not been proven. Prophylactic antibiotics, if administered, should provide appropriate coverage for likely pathogens and be relatively low cost.

18.4 Guidelines for Follow-up

Because of the relatively low risk of endophthalmitis associated with intravitreal injection, no study has compared different strategies for follow-up. As clinicians' comfort level with this procedure has increased, most clinicians no longer evaluate patients one day following intravitreal injection. However, patients should be instructed to contact the ophthalmologist immediately with signs and symptoms of complications (e.g., increased ocular redness or discomfort compared to right after the injection, decreased vision) and it is prudent to contact the patient within 1 week of the procedure to inquire about such symptoms.

18.5 Non-infectious Endophthalmitis (Figs. 18.10, 18.11)

Both infectious and non-infectious endophthalmitis have been reported following intravitreal injections. Non-infectious endophthalmitis may represent dispersion of drug crystals (such as has been reported in association with triamcinolone acetonide) [24] in the anterior chamber and vitreous cavity or an acute inflammatory reaction to a component in the drug formulation. The incidence of non-infectious endophthalmitis following intravitreal triamcinolone acetonide injection has been reported to range from 0.2% to 1.6% [18, 27, 33, 39]. Typical clinical features of non-infectious endophthalmitis include absence of pain and eyelid edema and no increased conjunctival injection compared with that present immediately after injection. Such patients should be monitored carefully to rule out progressive inflammation due to early endophthalmitis. In addition, patients should be instructed to contact the ophthalmologist immediately if they note any change in their ocular symptoms, such as pain, decreased vision, or increased ocular redness compared to that present right after the injection procedure.

18.6 Summary

Intravitreal injection is becoming an increasingly widespread technique in the management of a variety of posterior segment diseases and is an important part of the retina specialist's armamentarium. In order to optimize the outcomes associated with intravitreal injection, careful attention should be paid to reducing the risk of postinjection endophthalmitis. Ultimately, the outcomes of treatment depend not only on the safety and efficacy of the pharmacotherapy being delivered, but also on the safety and potential adverse events associated with the procedure itself.

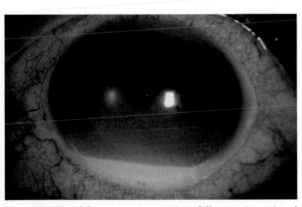

Fig. 18.10. Pseudohypopyon in a patient following intravitreal triamcinolone acetonide injection

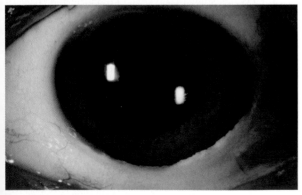

Fig. 18.11. Same eye as in Fig. 18.10 following spontaneous improvement of the pseudohypopyon (no treatment was administered)

References

1. Aiello LP, Brucker AJ, Chang S, et al. (2004) Evolving guidelines for intravitreous injections. Retina 24:S3–S19
2. Antcliff RJ, Spalton DJ, Stanford MR, et al. (2001) Intravitreal triamcinolone for uveitic cystoid macular edema: an optical coherence tomography study. Ophthalmology 108: 765–772
3. Bannerman TL, Rhoden DL, McAllister SK, et al. (1997) The source of coagulase-negative staphylococci in the Endophthalmitis Vitrectomy Study. A comparison of eyelid and intraocular isolates using pulsed-field gel electrophoresis. Arch Ophthalmol 115:357–361
4. Bucci FA, Amico LM, Guerino A, Evans R (2003) The effect of preop lid scrubs and levofloxacin on eyelid and conjunctival cultures prior to cataract surgery. Association for Research in Vision and Ophthalmology, Fort Lauderdale, FL
5. Cekic O, Change S, Tseng JJ, et al. (2005) Intravitreal triamcinolone injection for treatment of macular edema secondary to branch retinal vein occlusion. Retina 25:851–855
6. Chang TS, Tonnu IQ, Globe DR, Fine J (2004) Longitudinal changes in self-reported visual functioning in AMD patients in a randomized controlled phase I/II trial of Lucentis (ranibizumab; rHuFAV v2). Invest Ophthal Vis Sci 45:E-Abstract 3098
7. Cibis PA, Becker B, Okun E, Canaan S (1962) The use of liquid silicone in retinal detachment surgery. Arch Ophthalmol 68:590–599
8. Forster RK, Zachary IG, Cottingham AJ Jr, Norton EWD (1976) Further observations on the diagnosis, cause, and treatment of endophthalmitis. Am J Ophthalm 81:52–56
9. Gragoudas ES, et al., for the VEGF Inhibition Study in Ocular Neovascularization Clinical Trial Group (2004) Pegaptanib

for neovascular age-related macular degeneration. N Engl J Med 351:2805–2816

10. Greenberg PB, Martidis A, Rogers AH, et al. (2002) Intravitreal triamcinolone acetonide for macular oedema due to central retinal vein occlusion. Br J Ophthalmol 86:247–248

11. Grimes S, Mein C, Trevino S (1991) Preoperative antibiotic and povidone-iodine preparation of the eye. Ann Ophthalmol 23:263–266

12. Han DP, Wisniewski SR, Wilson LA, et al. (1996) Spectrum and susceptibilities of microbiologic isolates in the Endophthalmitis Vitrectomy Study. Am J Ophthalmol 122:1–17

13. Henry K, Cantrill H, Fletcher C, et al. (1987) Use of intravitreal ganciclovir (dihydroxy propoxymethyl guanine) for cytomegalovirus retinitis in a patient with AIDS. Am J Ophthalmol 103:17–23

14. Hovding G, Sjursen H (1982) Bacterial contamination of drops and dropper tips of in-use multidose eye drop bottles. Acta Ophthalmol (Copenh) 60:213–222

15. Isenberg S, Apt L, Yoshimori R, Khwarg S (1985) Chemical preparation of the eye in ophthalmic surgery IV: comparison of povidone-iodine on the conjunctiva with a prophylactic antibiotic. Arch Ophthalmol 103:1340-1342

16. Jager RD, Aiello LP, Patel SC, et al. (2004) Risks of intravitreous injection: a comprehensive review. Retina 24:676–698

17. Jonas JB, Akkoyun I, Kamppeter B, et al. (2005) Branch retinal vein occlusion treated by intravitreal triamcinolone acetonide. Eye 19:65–71

18. Jonas JB, Kreissig I, Degenring RF (2003) Endophthalmitis after intravitreal injection of triamcinolone acetonide. Arch Ophthalmol 121:1663–1664

19. Kozak I, Cheng L, Freeman WR (2005) Lidocaine gel anesthesia for intravitreal drug administration. Retina 25:994–998

20. Leong JK, Shah R, McCluskey PJ, et al. (2002) Bacterial contamination of the anterior chamber during phacoemulsification cataract surgery. J Cataract Refract Surg 28:826–833

21. Martidis A, Duker JS, Greenberg PB, et al. (2002) Intravitreal triamcinolone for refractory diabetic macular edema. Ophthalmology 109:920–927

22. Martidis A, Duker JS, Puliafito CA (2001) Intravitreal triamcinolone for refractory cystoid macular edema secondary to birdshot retinochoroidopathy. Arch Ophthalmol 119:1380–1383

23. Miller JJ, Scott IU, Flynn HW Jr, et al. (2005) Acute-onset endophthalmitis after cataract surgery (2000–2004): incidence, clinical settings, and visual acuity outcomes after treatment. Am J Ophthalmol 139:983–987

24. Moshfeghi AA, Scott IU, Flynn HW Jr, Puliafito CA (2004) Pseudohypopyon after intravitreal triamcinolone acetonide injection for cystoid macular edema. Am J Ophthalmol 138:489–492

25. Moshfeghi DM, Kaiser PK, Bakri SJ, Kaiser RS, Maturi RK, Sears JE, Scott IU, Belmont J, Beer PM, Quiroz-Mercado H, Mieler WF (2005) Presumed sterile endophthalmitis following intravitreal triamcinolone acetonide injection. Ophthalmic Surg Lasers Imaging 36:24–29

26. Moshfeghi DM, Kaiser PK, Scott IU, Sears JE, Benz M, Sinesterra JP, Kaiser RS, Bakri SJ, Maturi RK, Belmont J, Beer PM, Murray TG, Quiroz-Mercado H, Mieler WF (2003) Acute endophthalmitis following intravitreal triamcinolone acetonide injection. Am J Ophthalmol 136:791–796

27. Nelson ML, Tennant MT, Sivalingam A, et al. (2003) Infectious and presumed noninfectious endophthalmitis after

intravitreal triamcinolone acetonide injection. Retina 23: 686–691

28. Ohm J (1911) Über die Behandlung der Netzhautablösung durch operative Entleerung der Subretinalen Flössigkeit und Einspritzung von Luft in den Glaskörper. Albrecht von Graefes Arch Ophthalmol 79:442–450

29. Osher RH, Amdahl LD, Cheetham JK (1994) Antimicrobial efficacy and aqueous humor concentration of preoperative and postoperative topical trimethoprim/polymyxin B sulfate versus tobramycin. J Cataract Refract Surg 20:3–8

30. Ozkiris A, Evereklioglu C, Erkilic K, Ilhan O (2005) The efficacy of intravitreal triamcinolone acetonide on macular edema in branch retinal vein occlusion. Eur J Ophthalmol 15:96–101

31. Peyman GA, Vastine DW, Raichand M (1978) Experimental aspects and their clinical application. Ophthalmology 85:374–385

32. Rosengren B (1952) 300 cases operated upon for retinal detachment; method and results. Acta Ophthalmol (Copenh) 30:117–122

33. Roth DB, Chieh J, Spirn MJ, et al. (2003) Non-infectious endophthalmitis associated with intravitreal triamcinolone injection. Arch Ophthalmol 121:1279–1282

34. Rycroft B (1945) Penicillin and the control of deep intraocular infection. Br J Ophthalmol 29:57–87

35. Schneider J, Frankel SS (1947) Treatment of late postoperative intraocular infections with intraocular injection of penicillin. Arch Ophthalmol 37:304–307

36. Scott IU, Ip MS (2005) It's time for a clinical trial to investigate intravitreal triamcinolone for macular edema associated with retinal vein occlusion: the SCORE study. Arch Ophthalmol 123:581–582

37. Speaker MG, Menikoff JA (1991) Prophylaxis of endophthalmitis with topical povidone-iodine. Ophthalmology 98:1769–1775

38. Speaker MG, Milch FA, Shah MK, et al. (1991) Role of external bacterial flora in the pathogenesis of acute postoperative endophthalmitis. Ophthalmology 98:639–649; discussion 650

39. Sutter FK, Gillies MC (2003) Pseudo-endophthalmitis after intravitreal injection of triamcinolone. Br J Ophthalmol 87:972–974

40. Ta CN, Egbert PR, Singh K, et al. (2002) Prospective randomized comparison of 3-day versus 1-hour preoperative ofloxacin prophylaxis for cataract surgery. Ophthalmology 109:2036–2041

41. The Eyetech Study Group (2002) Preclinical and phase 1A clinical evaluation of an anti-VEGF pegylated aptamer (EYE001) for the treatment of exudative age-related macular degeneration. Retina 22:143–152

42. The Eyetech Study Group (2003) Anti-vascular endothelial growth factor therapy for subfoveal choroidal neovascularization secondary to age-related macular degeneration: phase II study results. Ophthalmology 110:979–986

43. Vitravene injection (fomivirsen sodium intravitreal injectable). Approval letter, pages 1–4. Vol 2004. U.S. Food and Drug Adminstration website. Available at: www.fda.gov/cder/foi/nda/98/20961_Vitravene_Approv.pdf. Accessed April 29, 2004

44. Young S, Larkin G, Branley M, Lightman S (2001) Safety and efficacy of intravitreal triamcinolone for cystoid macular oedema in uveitis. Clin Exp Ophthalmol 29:2–6

Section III
Pathology, Clinical Course and Treatment of Retinal Vascular Diseases

19 Grading of Diabetic Retinopathy

M. Larsen, W. Soliman

Core Messages

- The purposes of clinical retinopathy grading are:
  - Identification of eyes that have reached a threshold for treatment of retinopathy
  - Definition of an appropriate follow-up interval, to minimize the risk of progression to levels of retinopathy above a treatment threshold before follow-up, estimated on the basis of the current level of retinopathy and its likely rate of progression
  - Staging of microvascular complications for use in the management of systemic conditions
- The purposes of research retinopathy grading are:
  - To assess retinopathy levels and retinopathy progression with high sensitivity and reproducibility using a grading scale with a documented relation to long term visual outcome

The grading of diabetic retinopathy is based on the concept that a hierarchy of stages can be defined, where the higher the grade, the higher the risk of suffering visual loss. The grading scale reflects the natural course of the disease in its unrelenting and most devastating form. In reality, diabetic retinopathy can regress both spontaneously and after therapeutic intervention. To construct or validate a grading scale, rates of progression to visual loss or another endpoint that is meaningful to the patient must be measured. Grading scales constructed for scientific and regulatory purposes, such as the study of new interventions, are made with the purpose of achieving maximum sensitivity for change in retinopathy. Interventional trials may then result in selected grades being validated and found useful for guiding interventions such as photocoagulation for proliferative diabetic retinopathy or diabetic macular edema. The timely identification of patients who reach interventional thresholds is the purpose subserved by simpler grading scales used in clinical screening practice.

In the scientific and regulatory environment, diabetic retinopathy grading is an accepted surrogate measure of outcome, but only because it has a documented relation to visual outcome. Interventions that are potent enough to improve visual outcome can obviously be assessed on a functional measure rather than a surrogate morphological endpoint.

In the clinical management of diabetic retinopathy, intervention should be made or at least considered before visual loss has occurred. Thus, in principle, measures of visual function are not sensitive enough to detect a need for treatment, unless the retinopathy has progressed beyond threshold. Clinical retinopathy assessment is also used to estimate the safe follow-up interval (time to next screening visit) and to decide when to stop treatment.

Retinopathy grading for scientific purposes is done based on fundus photography because of the need for lasting documentation. Stereoscopic fundus photography is the reference standard. In clinical practice, stereoscopic slit-lamp biomicroscopy using a corneal contact lens is the best standard of care. Direct ophthalmoscopy is of little use other than determining the absence of retinopathy because the field of view is restricted and because the lack of stereoscopic viewing makes the direct observation of preretinal proliferations and macular edema difficult or impossible.

Systematic evaluation of the diabetic fundus begins with the identification of the number and location of fundus lesions by class (hemorrhage, hard exudate, cotton wool spot, intraretinal microvascular abnormality, neovascular proliferation, fibrosis, edema, other lesions). This information is then used to determine the severity of retinopathy, which summarizes the information about individual lesion types in a given eye (per-eye grading) or in a given patient (per-patient grading) where usually the severity score for the most severely affected eye is used to represent that patient's severity score. Severity levels are represented, ideally, as a one-dimensional hierarchic scale ranging from absence of retinopathy to the most advanced stages associated with severe visual loss. The levels of the scale reflect the natural progression of the disease in that the higher the retinopathy level, the greater the risk of visual loss or, in other words, the closer the patient is to losing sight.

Diabetic retinopathy grading has a practical application in fundus photographic screening systems, where photography substitutes for examination by an ophthalmologist. It can be argued that a comprehensive clinical examination by a fully trained ophthalmologist, repeated at regular intervals for the entire duration of diabetes, with fundus photography used for documentation only, is the best standard of care. The practical experience is, however, that diabetic retinopathy remains the leading cause of blindness, most of it probably preventable, in the working age population in industrialized nations. The major problems appear to be poor compliance, uneven access to examination by ophthalmologist, and uncoordinated management of the many specialist services that combine to make good diabetes care [13]. In consequence, photographic screening programs have been instituted in settings that range from remote rural primary health care centers to highly specialized diabetology centers. The screening protocols are much simpler than the research protocols, consisting for instance of only two-field non-stereoscopic digital fundus photography, visual acuity with refractometer correction, a simplified grading system, and decision algorithms mandating referral to an ophthalmologist at varying levels of retinopathy, ranging from any retinopathy to retinopathy that closely approaches treatment threshold.

Digital image analysis has been developed in an attempt to supplement or substitute the visual evaluation of fundus photographs [9, 10]. Despite good results in clinical tests, their utility in a practical setting remains to be fully evaluated.

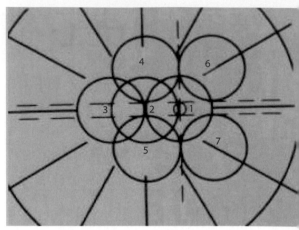

Fig. 19.1. ETDRS fundus photography: location of the seven 30-degree fields. Field 1 is centered on the optic disk. Field 2 is centered on the center of the macula. Field 3 is centered just temporal to the macula. Fields 4–7 are tangential to horizontal lines passing through the upper and lower poles of the disk and to a vertical line passing through its center

vitreous hemorrhage [6]. The scale can be used to describe overall retinopathy severity and change in severity over time [7].

ETDRS grading is based on the classification of various types of lesions on a multi-step scale (absent, questionable, definitely present but less than standard photograph 1 or written definition, equal to or worse than the new standard photograph but less severe than standard photograph 2, equal to or worse than standard photograph 2, etc.). For some lesions a standard photograph provided the dividing lines between grades. For others a written definition was used. Five standard photographic fields were defined to include most of the fundus within 30 degrees of the center of the macula.

ETDRS fundus photography includes seven stereoscopic photographic fields, each covering 30 degrees (Fig. 19.1). The photographs are recorded on photographic film as diapositives.

ETDRS fundus photography grading is made by visual inspection of each stereoscopic pair of diapositives on a retroilluminated white glass plate using a stereo viewer (+15 diopters) and comparison, when needed, with reference diapositives. The grade for each lesion is recorded separately for each standard field. The lesions graded in the revised modified Airlie classification comprise:

19.01 Early Treatment Diabetic Retinopathy Study Grading – An Extension of the Modified Airlie House Classification

The best validated grading system for diabetic retinopathy, in terms of its outcome-related evidence base, is the extension of the Airlie House classification that was developed for the Early Treatment Diabetic Retinopathy Study (ETDRS) [5]. On the basis of a prospective evaluation of the relative risk of retinopathy progression and visual loss, a scale was developed that divides diabetic retinopathy into 13 levels ranging from absence of retinopathy to severe

- Lesions graded in Fields 3–7 (extramacular area)
  - Microaneurysms
  - Hemorrhages and/or microaneurysms (H/Ma)
  - Drusen
  - Hard exudates (HE)
  - Soft exudates (cotton-wool patches) (SE)

- Intraretinal microvascular abnormalities (IRMA)
- Venous beading (VB)
- Venous narrowing
- Venous loops and/or reduplication (VLR)
- Venous sheathing
- Perivenous exudates (PVEX)
- Arteriolar narrowing (AN)
- Arteriolar sheathing
- Arteriovenous nicking (AVN)
- New vessels elsewhere (more than one disk diameter from the disk) (NVE) (in fields 3 – 7 and in field 1 outside the area defined for NVD)
- Dilated tips of new vessels elsewhere (in fields 3 – 7 and in field 1 outside the area defined for NVD)
- Fibrous proliferation elsewhere (more than one disk diameter from the disk) (FPE) (in fields 3 – 7 and in field 1 outside the area defined for NVD)
- Plane of proliferation elsewhere (PP) (in fields 3 – 7 and in field 1 outside the area defined for NVD)
- Lesions graded in all fields
  - Preretinal hemorrhages (PRH)
  - Vitreous hemorrhage (VH)
  - Scars of prior photocoagulation
- Lesions graded in Fields 2 – 7
  - Retinal elevation (traction, detachment)
- Lesions graded in Field 1
  - Neovascularization on or within one disk diameter of the disk (NVD)
  - Dilated tips of the new vessels of the disk
  - Fibrous proliferation on or within one disk diameter of the disk (FPD)
  - Plane of proliferation on or within one disk diameter of the disk
  - Papillary swelling
- Lesions graded in Field 2
  - Hard exudate rings
  - Posterior vitreous detachment
  - Retinal thickening (edema)
  - Hard exudates within one disk diameter of the center of the macula and at the center of the macula
  - Cystoid spaces
  - Other lesions within one disk diameter of the center of macula

The lesion grades are summarized for each eye into an ETDRS level that defines the diabetic retinopathy severity (Table 19.1).

The scale of ETDRS levels (Table 19.1) does not cover macular edema, which is graded separately [5]. Using a grid overlay on one photograph in the Field 2 stereo pair, the grader evaluates the size and location of thickened retina, the maximum thickness of the thickened retina, the degree of cystoid edema, and the location of the hard exudate that is often found with it. The ETDRS defined diabetic macular edema as *thickening and/or hard exudate within 1 disk diameter of the center of the macula* (Table 19.2), given that the total amount of hard exudate within the Field 1 photograph exceeded that in a certain standard photograph. The study of diabetic macular edema and the effect of photocoagulation for diabetic macular edema led to the definition of the term clinically significant macular edema (CSME) for edema that involves or threatens the center of the macula (even if visual acuity is not yet reduced) as assessed by stereo contact lens biomicroscopy or stereo photography. CSME currently serves as the reference threshold for intervention by thermal laser photocoagulation. CSME is defined as:

- Retinal thickening at or within 500 µm of the center of the macula and/or
- hard exudates at or within 500 µm of the center of the macula, if associated with thickening of the adjacent retina and/or
- a zone or zones of retinal thickening 1 disk area in size at least part of which is within 1 disk diameter of the center of the macula.

Additional abnormalities evaluated within 1 disk diameter of the foveal center include epiretinal and preretinal new vessels, epiretinal and preretinal fibrosis, abnormal pigmentation, retinal tension lines, and macular hole.

Optical coherence tomography (OCT) is a valuable tool for assessment of retinal thickening. Its inclusion in ongoing trials on the treatment of diabetic macular edema will likely result in the development of methods for translation between visual grading and OCT thickness mapping and in a better understanding of the relation between thickening of the macula and visual function as well as visual prognosis and potential for recovery if the edema is resolved.

ETDRS fundus photography and grading is the scientific and regulatory standard. It places considerable demand on the patient and the resources for photography, and the grading is complex, for which reason simplified protocols have been developed, two of which will be described below.

19.02 EURODIAB Grading

The EURODIAB grading system is a simplified system, based on experience with the seven-field 30° ETDRS grading system, but adapted to enable large epidemiological studies, in which participating cen-

19

ETDRS level	ETDRS severity	ETDRS definition	EURODIAB level
10	No retinopathy	Diabetic retinopathy absent	0
20	Very mild NPDR	Microaneurysms only	1
35	Mild NPDR	Hard exudates, cotton-wool spots, and/or mild retinal hemorrhages	1
43	Moderate NPDR	43A: retinal hemorrhages moderate (>photograph 1[a]) in four quadrants or severe (≥ photograph 2A) in one quadrant 43B: mild IRMA (<photograph 8A) in one to three quadrants	2
47	Moderate NPDR	47A: both level 43 characteristics 47B: mild IRMA in four quadrants 47C: severe retinal hemorrhage in two to three quadrants 47D: venous beading in one quadrant	2
53A–D	Severe NPDR	53A: ≥ 2 level 47 characteristics 53B: severe retinal hemorrhages in four quadrants 53C: moderate to severe IRMA (≥ photograph 8A) in at least one quadrant 53D: venous beading in at least 2 quadrants	3
53E	Very severe NPDR	≥ 2 level 53A–D characteristics	3
61	Mild PDR	NVE <0.5 disk area in one or more quadrants	5
65	Moderate PDR	65A: NVE ≥ 0.5 disk area in one or more quadrants 65B: NVD <photograph 10A (<0.25–0.33 disk area)	5
71, 75	High-risk PDR	NVD ≥ photograph 10A, or NVD <photograph 10A or NVE ≥ 0.5 disk area plus VH or PRH, or VH or PRH obscuring ≥ 1 disk area	5
81, 85	Advanced PDR	Fundus partially obscured by VH and either new vessels ungradable or retina detached at the center of the macula	5
(not classified)			4 (photocoagulated)

Table 19.1. Revised modified Airlie House diabetic retinopathy classification (ETDRS)

NPDR non-proliferative diabetic retinopathy, *PDR* proliferative diabetic retinopathy, *IRMA* intraretinal microvascular abnormalities, *NVE* new vessels elsewhere, *NVD* new vessels on or within 1 disk diameter of the optic disk, *PRH* pre-retinal hemorrhage, *VH* vitreous hemorrhage. The definition for each level assumes that the definition for any higher level is not met. NPDR levels 35 and above all require presence of microaneurysms. The corresponding EURODIAB classification scale is shown in the left column
[a] See [5] for definition of standard photographs

ters have a limited experience in fundus photography [1]. The photographic protocol comprises two 45° color diapositive photographs per eye, one macular field photograph with the optic disk just inside the edge of the frame on its horizontal meridian and one disk/nasal field centered nasal of the disk with the disk one disk diameter in from the temporal edge of the frame on its horizontal meridian (Figs. 19.2–19.13). The diabetic retinopathy severity scale comprises no retinopathy, non-proliferative retinopathy, and proliferative retinopathy. Non-proliferative retinopathy is defined as the presence of one or more microaneurysms, hemorrhages, and/or hard exudates. Proliferative retinopathy is defined as any new vessels, fibrous proliferations, preretinal hemorrhage, vitreous hemorrhage, or photocoagulation scars. Per-patient grading is determined by the eye with the most advanced retinopathy [11]. Because fundus photography was not stereoscopic, macular edema (thickening) was not gradable. Retinopathy was assessed after grading of individual lesion types, from top down. This means that the first step is to determine whether proliferative diabetic retinopathy is present. In the event that it is, the retinopathy level of the eye is 5; if not, the level is not 5 and one proceeds to evaluate if the eye is level 4, and so on.

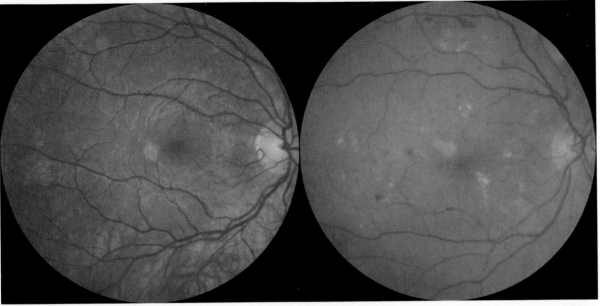

Fig. 19.2. Hemorrhages and microaneurysm (HMA) standard macular field fundus photographs 1 (*left*) and 2 (*right*) according to the EURODIAB diabetic retinopathy grading system

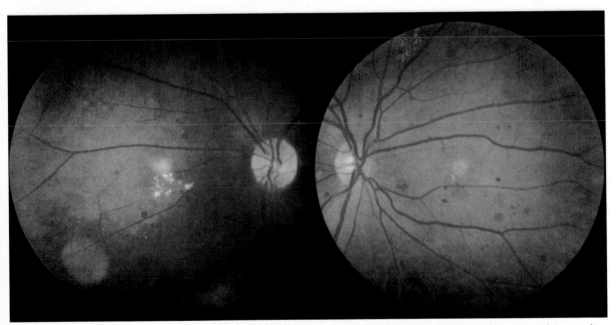

Fig. 19.3. Hemorrhages and microaneurysm (HMA) standard disk/nasal field fundus photographs 1 (*left*) and 2 (*right*) according to the EURODIAB diabetic retinopathy grading system

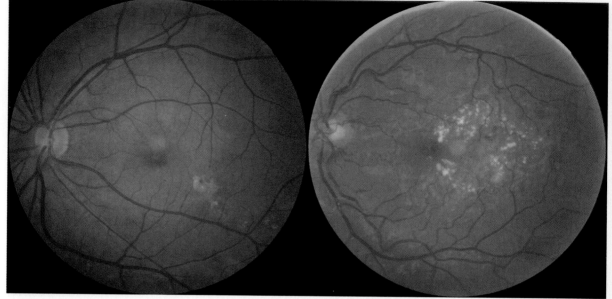

Fig. 19.4. Hard exudate (HE) standard macular field fundus photographs 1 (*left*) and 2 (*right*) according to the EURODIAB diabetic retinopathy grading system

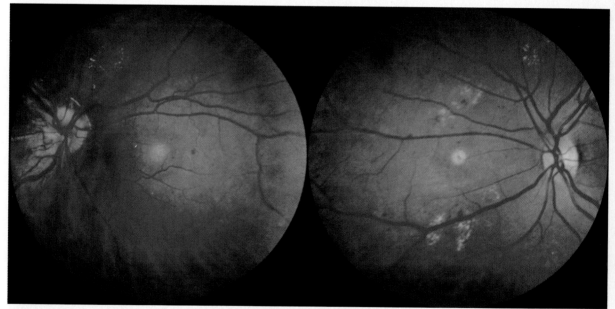

Fig. 19.5. Hard exudate (HE) standard disk/nasal field fundus photographs 1 (*left*) and 2 (*right*) according to the EURODIAB diabetic retinopathy grading system

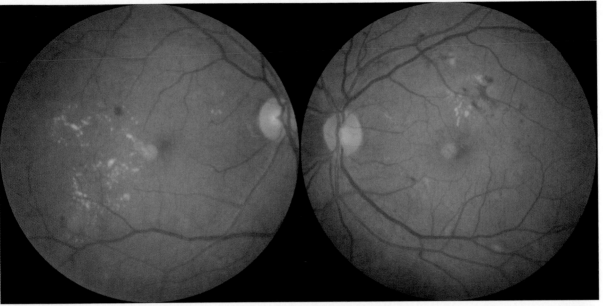

Fig. 19.6. Cotton wool spots (CWS) standard macular field fundus photographs 1 (*left*) and 2 (*right*) according to the EURODIAB diabetic retinopathy grading system

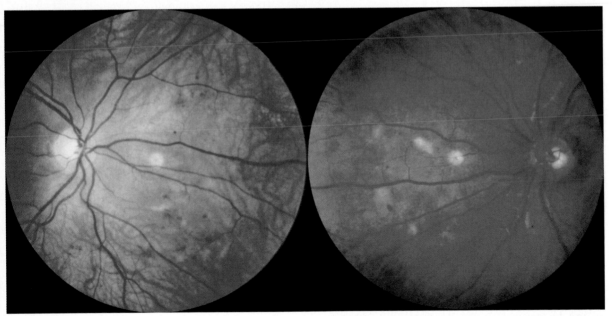

Fig. 19.7. Cotton wool spots (CWS) standard disk/nasal field fundus photographs 1 (*left*) and 2 (*right*) according to the EURODIAB diabetic retinopathy grading system

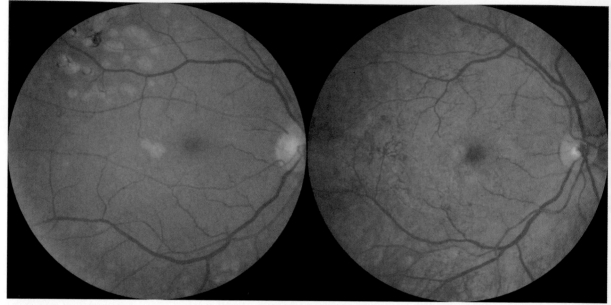

Fig. 19.8. Intraretinal microvascular abnormalities (IRMA) standard macular field fundus photographs 1 (*left*) and 2 (*right*) according to the EURODIAB diabetic retinopathy grading system

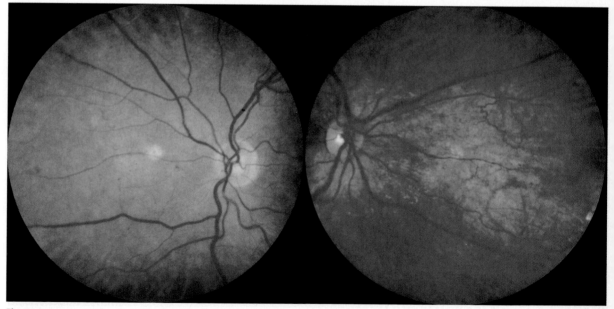

Fig. 19.9. Intraretinal microvascular abnormalities (IRMA) standard disk/nasal field fundus photographs 1 (*left*) and 2 (*right*) according to the EURODIAB diabetic retinopathy grading system

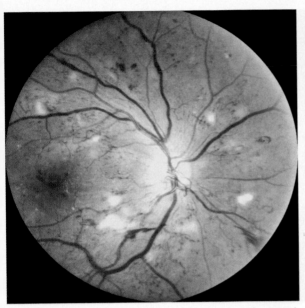

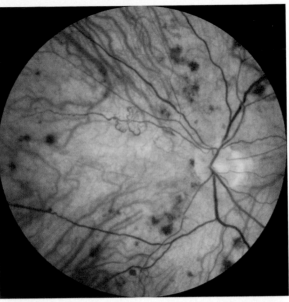

Fig. 19.10. Venous beading (VB) standard fundus photograph according to the EURODIAB diabetic retinopathy grading system

Fig. 19.11. New vessels elsewhere (NVE) and new vessels on the disk (NVD) standard fundus photograph according to the EURODIAB diabetic retinopathy grading system

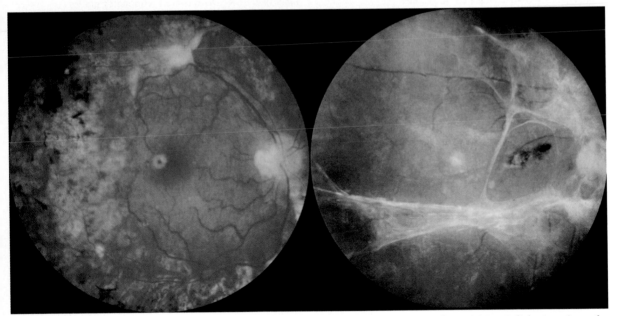

Fig. 19.12. Fibrous proliferation elsewhere (FPE) standard fundus photograph according to the EURODIAB diabetic retinopathy grading system

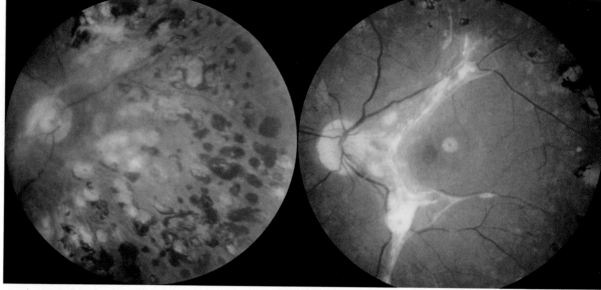

Fig. 19.13. Fibrous proliferation on the disk (FPD) standard fundus photographs 1 (*left*) and 2 (*right*) according to the EURODIAB diabetic retinopathy grading system

Table 19.2. EURODIAB single lesion grading

Single standard photograph
0 Lesion absent
1 Lesion questionable (grader >50% but less than 90% sure)
2 Definitely present but <SP1
3 Lesion present ≥ SP1
Two standard photographs
0 Lesion absent
1 Lesion questionable (grader >50% but less than 90% sure)
2 Definitely present but <SP1
3 Lesion present ≥ SP1 but < SP2
4 Lesion present ≥ SP2

Table 19.3. EURODIAB lesion scale definition (estimated number relative to standard photographs)

Lesion	Abbrev.	Field(s)	Standard(s)
Hemorrhages and	HMA	M	SP1 + SP2
Microaneurysms		D/N	SP1 + SP2
Hard exudates	HE	M	SP1 + SP2
		D/N	SP1 + SP2
Cotton wool spots	CWS	M	SP1 + SP2
		D/N	SP1 + SP2
Intraretinal microvascular	IRMA	M	SP1 + SP2
abnormalities		D/N	SP1 + SP2
Venous beading	VB	M + D/N	SP1
New vessels elsewhere	NVE	M + D/N	SP1
Fibrous proliferation	FPE	M + D/N	SP1 + SP2
elsewhere			
New vessels disk	NVD	D/N	SP1
Fibrous proliferation disk	FPD	D/N	SP1 + SP2
Pre-retinal hemorrhage	PRH	M + D/N	SM1
Vitreous hemorrhage	VH	M + D/N	SM1
Scars of photocoagulation	PC	M + D/N	SM1

M macular field, *D/N* disk/nasal field, *SP* standard photograph, *SM* standard measure (50% or more of field involved)

Table 19.4. EURODIAB overall retinopathy grades

Level	Severity
0	No retinopathy
1	Minimal non-proliferative retinopathy: HMA = grade 2 to 3 in 1 or 2 fields and/or HE = grade 2 to 4 in 1 or 2 fields
2	Moderate non-proliferative retinopathy: HMA = grade 4 in only 1 field *or* HMA = grade 2 to 3 in 1 or 2 fields plus: CWS = grade 2 to 3 in 1 or 2 fields and/or IRMA = grade 2 in 1 or 2 fields and/or VB = grade 2 in 1 to 2 fields
3	Severe non-proliferative (pre-proliferative) retinopathy: HMA = grade 4 in both fields *or* HMA =grade 2 to 4 in 1 or 2 fields plus: CWS = grade 4 in 1 or 2 fields and/or IRMA = grade 3 in 1 or 2 fields and/or VB = grade 3 in 1 or 2 fields
4	Photocoagulated: Scars of photocoagulation in any field
5	Proliferative: Any of the following: New vessels (disk or elsewhere) Fibrous proliferations (disk of elsewhere) Pre-retinal hemorrhage Vitreous hemorrhage

NPDR non-proliferative diabetic retinopathy, *HMA* hemorrhage and microaneurysm, *HE* hard exudates, *CWS* cotton wool spots, *IRMA* intraretinal microvascular abnormalities, *VB* venous beading. All appearances are in comparison with the EURODIAB standard photographs [1]

19.03 International Clinical Diabetic Retinopathy Severity Scale

The proposed International Clinical Diabetic Retinopathy and Diabetic Macular Edema Disease Severity Scale was developed by a consensus panel of experts to provide a severity scale for common usage that is targeted at solving practical clinical issues, for instance in retinopathy screening and to improve communication and coordination of care among physicians who care for patients with diabetes [12].

The five-stage disease severity classification for diabetic retinopathy includes three stages of low risk, a fourth stage of severe non-proliferative retinopathy, and a fifth stage of proliferative retinopathy. Diabetic macular edema is classified as apparently present or apparently absent. If training and equipment allow the screener to make a valid decision, macular edema is further categorized as a function of its distance from the central macula.

The system was developed with reference to the ETDRS grading system. It may help standardize clinical retinopathy grading where many systems are in use that are nominally different but vary little in their fundamental concept.

References

1. Aldington SJ, Kohner EM, Meuer S, Klein R, Sjolie AK (1995) Methodology for retinal photography and assessment of diabetic retinopathy: the EURODIAB IDDM Complications Study. Diabetologia 38:437–444
2. Davis MD, Fisher MR, Gangnon RE, Barton F, Aiello LM, Chew EY, Ferris FL 3rd, Knatterud GL (1998) Risk factors for high-risk proliferative diabetic retinopathy and severe visual loss: Early Treatment Diabetic Retinopathy Study Report #18. Invest Ophthalmol Vis Sci 39(2):233–52
3. Diabetic Retinopathy Study Research Group (1981) Diabetic Retinopathy Study. Report Number 6. Design, methods, and baseline results. Report Number 7. A modification of the Airlie House classification of diabetic retinopathy. Invest Ophthalmol Vis Sci 21(1):1–226
4. Early Treatment Diabetic Retinopathy Study Research Group (1985) Photocoagulation for diabetic macular edema. Early Treatment Diabetic Retinopathy Study Report Number 1. Arch Ophthalmol 103(12):1796–806
5. Early Treatment Diabetic Retinopathy Study Research Group (1991) Grading diabetic retinopathy from stereoscopic color fundus photographs – an extension of the modified Airlie House classification. ETDRS Report Number 10. Ophthalmology 98:786–806
6. Early Treatment Diabetic Retinopathy Study Research Group (1991) Fundus photographic risk factors for progression of diabetic retinopathy. ETDRS Report Number 12. Ophthalmology 98:823–833

Table 19.5. Diabetic Retinopathy Disease Severity Scale

Proposed disease severity level	Findings observable on dilated ophthalmoscopy
• No apparent retinopathy	• No abnormalities
• Mild non-proliferative diabetic retinopathy	• Microaneurysms only
• Moderate non-proliferative diabetic retinopathy	• More than just microaneurysms but less than severe non-proliferative diabetic retinopathy
• Severe non-proliferative diabetic retinopathy	• Any of the following: more than 20 intraretinal hemorrhages in each of 4 quadrants; definite venous beading in 2+ quadrants; prominent intraretinal microvascular abnormalities in 1+ quadrant *and* no signs of proliferative retinopathy
• Proliferative diabetic retinopathy	• One or more of the following: neovascularization, vitreous/preretinal hemorrhage

Proposed disease severity level findings observable on dilated ophthalmoscopy
• Diabetic macular edema apparently absent. No apparent retinal thickening or hard exudates in posterior pole
• Diabetic macular edema apparently present. Some apparent retinal thickening or hard exudates in posterior pole

If diabetic macular edema is present, it can be categorized as follows:

Proposed disease severity Level	Findings observable on dilated ophthalmoscopy[a]
• Diabetic macular edema present	• Mild diabetic macular edema: some retinal thickening or hard exudates in posterior pole but distant from the center of the macula
	• Moderate diabetic macular edema: retinal thickening or hard exudates approaching the center of the macula but not involving the center
	• Severe diabetic macular edema: retinal thickening or hard exudates involving the center of the macula

[a] Hard exudates are a sign of current or previous macular edema. Diabetic macular edema is defined as retinal thickening, and this requires a three-dimensional assessment that is best performed by a dilated examination using slit-lamp biomicroscopy and/or stereo fundus photography

7. Early Treatment Diabetic Retinopathy Study Research Group (1998) Risk factors for high-risk proliferative diabetic retinopathy and severe visual loss. ETDRS Report No. 18. Invest Ophthalmol Vis Sci 39:233–52

8. Gardner TW, Sander B, Larsen M, Kunselman A, Have TT, Lund-Andersen H, Reimers J, Hubbard L, Blankenship GW, Quillen DA, Brod RD, Wilmarth MH, Hansen HP, Parving H-H, Davis MD (2007) An Extension of the Early Treatment Diabetic Retinopathy Study (ETDRS) System for Grading of Diabetic Macular Edema in the Astemizole Retinopathy Trial. Curr Eye Res (accepted for publication)

9. Larsen M, Godt J, Larsen N, Lund-Andersen H, Sjolie AK, Agardh E, Kalm H, Grunkin M, Owens DR (2003) Automated detection of fundus photographic red lesions in diabetic retinopathy. Invest Ophthalmol Vis Sci 44(2):761–766

10. Larsen N, Godt J, Grunkin M, Lund-Andersen H, Larsen M (2003) Automated detection of diabetic retinopathy in a fundus photographic screening population. Invest Ophthalmol Vis Sci 44:767–771

11. van Hecke MV, Dekker JM, Stehouwer CD, Polak BC, Fuller JH, Sjolie AK, Kofinis A, Rottiers R, Porta M, Chaturvedi N (2005) EURODIAB prospective complications study: Diabetic retinopathy is associated with mortality and cardiovascular disease incidence: the EURODIAB prospective complications study. Diabetes Care 28(6):1383–1389

12. Wilkinson CP, Ferris FL 3rd, Klein RE, Lee PP, Agardh CD, Davis M, Dills D, Kampik A, Pararajasegaram R, Verdaguer JT (2003) Global Diabetic Retinopathy Project Group: Proposed international clinical diabetic retinopathy and diabetic macular edema disease severity scales. Ophthalmology 110(9):1677–82

13. Zoega GM, Gunnarsdottir T, Bjornsdottir S, Hreietharsson AB, Viggosson G, Stefansson E (2005) Screening compliance and visual outcome in diabetes. Acta Ophthalmol Scand 83(6):687–90

19.1 Nonproliferative Diabetic Retinopathy

19.1.1 Nonproliferative Stages of Diabetic Retinopathy: Animal Models and Pathogenesis

T.S. KERN, S. MOHR

Core Messages

- Diabetes-induced degeneration of retinal capillaries during the early (nonproliferative) stages of diabetic retinopathy appears to contribute to the later progression of the retinopathy. Thus, inhibiting the degeneration of retinal capillaries is likely to be a meaningful therapeutic target to inhibit visual loss in diabetes

- All laboratory animal species tested to date have been found to develop at least the early microvascular stages of diabetic retinopathy, but progression to the more advanced lesions has been less common
- Multiple diverse therapies have been found to inhibit the early stages of diabetic retinopathy in animals, and anti-inflammatory effects seem to be common to a number of the therapies

Diabetic retinopathy is today the leading cause of acquired blindness among young adults throughout the developed world. It classically has been regarded as a disease of the microvasculature of the retina, and the natural history of the disease has been divided into an early, nonproliferative (or background) stage, and a later, proliferative stage. Nonproliferative diabetic retinopathy currently is diagnosed ophthalmoscopically based on the presence of retinal vascular abnormalities, including dilation of retinal veins, retinal microaneurysms, intraretinal microvascular abnormalities (which include intraretinal new vessels), areas of capillary nonperfusion, retinal hemorrhages, cotton wool spots (infarctions within the nerve fiber layer), edema, and exudates. Proliferative diabetic retinopathy is diagnosed based on the presence of new vessels on the surface of the retina. The resulting preretinal new vessels are the major cause of vitreous hemorrhage and consequent visual loss. Retinal edema is the other major contributor to visual impairment in diabetes.

Available evidence suggests that abnormalities that occur in the early stages of the retinopathy underlie the progression of the ocular disease, ultimately leading to neovascularization. Molecular mechanisms involved in the neovascular response, and development of therapies to inhibit this process, are major areas of research interest at present, but are beyond the scope of this chapter. In this chapter, we will focus on the use of animal models to investigate the pathogenesis of the early, nonproliferative stages of diabetic retinopathy and therapeutic approaches

to the inhibition of the retinopathy that have come out of such research. This review will be restricted to in vivo studies.

19.1.1.1 Early Stages of Diabetic Retinopathy: Histopathology

Histologically, the early stages of diabetic retinopathy in man and animals are characterized by the presence of saccular capillary microaneurysms, pericyte-deficient capillaries, and obliterated and acellular capillaries [41]. Pericyte loss is evident as an excessive number of pericyte "ghosts" on viable capillaries, the "ghost" referring to a pocket in basement membrane that formally was occupied by the pericyte. Acellular capillaries apparently were functional capillaries that degenerated until only a basement membrane tube remains. Acellular, degenerate capillaries are not perfused, and are regarded as histologic markers of nonperfused capillaries. Although devoid of nuclei, these degenerate vessels sometimes are not truly acellular, and may be filled with cytoplasmic processes of glial cells [48]. Whether the invasion of retinal capillaries by glia in diabetes is secondary to capillary degeneration, or whether it initiates vessel occlusion and degeneration, is not known.

Capillary occlusion initially occurs in single, isolated capillaries, and in the early stages has no clinical consequences. As more and more capillaries become occluded, however, the retina presumably becomes ischemic, causing elaboration of one or

19 III

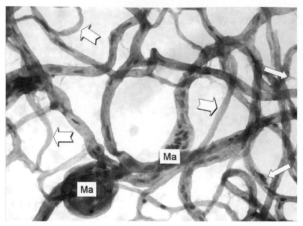

Fig. 19.1.1.1. Retinal microvasculature of diabetic patient (isolated by the trypsin digest technique). *Ma* microaneurysm, *narrow arrow* pericyte ghost, *wide arrow* acellular capillary

more growth factors, such as vascular endothelial growth factor (VEGF). Thus, capillary vaso-obliteration and the resulting increase in acellular capillaries appears to represent a discrete event that progressively contributes to the development of retinal ischemia, and ultimately neovascularization. Mechanisms believed to contribute to the capillary occlusion and/or obliteration in diabetes include: (1) occlusion of the vascular lumen by white blood cells or platelets [18, 75, 78], and (2) death of cells of the vessel wall as a result of exogenous (cytokines, etc.) or endogenous abnormalities within the vascular cells themselves.

Retinal capillary endothelial cells and pericytes have been found to die by an apoptotic-like (TUNEL-positive) process in diabetic humans, and in diabetic or experimentally galactosemic rats [122]. The rate of death of these cells in hyperglycemic rats becomes significantly greater than normal at about 6–8 months of disease, but is not demonstrable sooner. Although statistically greater than normal, the number of TUNEL-positive capillary cells detected in the retina of diabetic rats is quite small at any given duration of diabetes. For example, diabetes of 8 months duration resulted in only 9±6 TUNEL-positive capillary cells in diabetic rats [122]. This is not unexpected, since apoptotic cells are known to be rapidly cleared by phagocytosis [37]. Although the duration that dying cells are able to be stained using the TUNEL technique is known to be short (less than 1 day), it is still reasonable to ask whether or not the cumulative effect of this small number of cells undergoing apoptosis over a period of months (in rodents) or years (in humans) is sufficient to account for the capillary degeneration that is characteristic of diabetic retinopathy. The ability of a therapy to inhibit diabetes-induced apoptosis of capillary cells has been found to pre-

dict the ability of the therapy to inhibit the degeneration of retinal capillaries [95], thus suggesting that the apoptotic process is an important component in the pathogenesis of the retinopathy.

Damage to nonvascular cells of the retina (including ganglion cells) in diabetic humans also has been detected histologically [17], and functionally [22], and a possible role of the neural disease in the pathogenesis of diabetic retinopathy has been postulated [20]. Consistent with a possible role of apoptosis in the death of retinal neurons, numerous initiator and effector caspases have been found to become activated in the retina of diabetic animals or in retinal cells incubated in elevated glucose concentration [123]. Compared to capillary cells, more neurons become TUNEL-positive or caspase 3-positive in retinas of diabetic rats, and do so at a faster rate [14]. As with capillary cells, however, the rate of neuronal death also is very small compared to the total number of neurons.

Apoptosis might not be the only form of cell death occurring in diabetic retinopathy. Joussen and coworkers [75] have reported that only 9 days of diabetes caused an increase in intracellular accumulation of propidium iodide (commonly used as a marker of necrosis) in the retinal vasculature. Thus, focusing solely on apoptotic cell death might underestimate the total number of cells dying at any time. Nevertheless, propidium iodide likely overestimates necrosis, since it has been found to enter even viable cells having impaired membrane integrity [151].

The basement membrane that surrounds retinal capillaries thickens in diabetes, and had been postulated to play a role in development of the retinopathy in previous decades. This view had become less popular in recent years, but that now needs to be reexamined again in light of recent findings that the degeneration of retinal endothelial cells and pericytes in galactose-fed rats can be significantly inhibited by inhibiting synthesis of fibronectin, a component of the basement membrane [145].

Although all vascular cells should be exposed to the same concentration of blood glucose in a given individual, there is unexplained regional variability in susceptibility to diabetes-induced microvascular disease between embryologically similar tissues, and even within the same retina. Microvessels isolated from cerebral cortical of dogs with either diabetes or experimental galactosemia of 5 years duration possessed none of the microaneurysms, acellular capillaries, and pericyte ghosts that occurred in retinal vessels of the same animals [91]. Similar conclusions have been reached in diabetic humans [42]. In addition, microaneurysms and acellular capillaries have been found to develop in a nonuniform distribution even within the same retina in diabetic patients [33] and in experimentally diabetic or galactosemic dogs, lesions being more common in the superior and temporal portions of the retina [89]. Likewise, neovascularization in diabetic patients has been noted to be more common in the superior and temporal portions of the retina than in other regions [155]. Currently available hypotheses regarding the pathogenesis of diabetic retinopathy do not account adequately for the geographically unequal distribution of the microvascular disease, and suggest that local factors are important. The only biochemical abnormality identified to date that shows a similar nonuniform distribution across the retina in patients (and thus might contribute to the development of the histopathology) is a diabetes-induced increase in activity of the proinflammatory caspase, caspase 1 [154].

19.1.1.2 Animal Models of the Early Stages of Diabetic Retinopathy

Many animal species have been studied as possible models of diabetic retinopathy, and lesions that have been found to develop in various models are summarized in Table 19.1.1.1. The early stages of diabetic retinopathy have been found to develop in all mammalian species studied who have had diabetes for relatively long durations. Each of the different species have been found to have their own advantages and disadvantages, so their use will depend on desired endpoints and experimental design. Most of the studies of diabetic retinopathy in animals to date have focused on insulin-deficient models (type 1 diabetes), as opposed to models of insulin resistance (type 2 diabetes).

Studies in humans have shown considerable similarities in the retinopathy that develops in the different types of diabetes [99]; and in animals, retinal lesions have been found to be the same whether due to experimental induction of diabetes (alloxan, streptozotocin, growth hormone, pancreatectomy) or spontaneous diabetes (unpublished). Likewise, experimental elevation of blood hexose level by feeding galactose has resulted in development of a diabetic-like retinopathy in dogs [52, 53, 80, 81, 152], rats [59, 87, 142, 143, 164], and mice [90].

The morphologic similarity of retinal lesions in diabetes and the experimental galactosemia has suggested that the two retinopathies share a common final pathway leading from elevated blood hexose level to the retinal vascular disease [53]. Consistent with this, both diabetes- and galactose-induced retinopathies have been inhibited in rodents by antioxidants [108] or deletion of ICAM [76], but the retina in galactosemia differed from diabetes with respect

Table 19.1.1.1. Diabetes-induced retinal histopathology in various species

	Human	Dog (3 – 5 yrs)	Rat (½ – 1½ yrs)	Mouse (½ – 1 yrs)
Nonproliferative stages				
Microaneurysms	+	+	±	0
Acellular capillaries	+	+	+	+
Pericyte loss	+	+	+	+
IRMA	+	+	0	0
Hemorrhages	+	+	0	0
BM thickening	+	+	+	+
Neurodegeneration	+	+	+	±
Proliferative stage				
Intraretinal neovascularization	+	+	0	0
Preretinal neovascularization	+	0	0	0

to activation of proapoptotic caspases [123] and in the ability of aminoguanidine to inhibit capillary cell death and degeneration [95].

19.1.1.2.1 Dogs and Cats

Diabetic dogs have been shown repeatedly to develop morphologic lesions of the retinopathy that are indistinguishable from those of background retinopathy seen in diabetic patients, including capillary microaneurysms, acellular (and nonperfused) capillaries, pericyte ghosts, varicose and dilated capillaries [or intraretinal microvascular abnormalities (IRMAs)] [48]. The lesions in diabetic dogs can be inhibited by strict regulation of glycemia with exogenous insulin [50]. Microaneurysms, leukocyte and platelet plugging of vessels, and degenerating endothelial cells likewise were observed in cats after several years of diabetes [71]. Capillary aneurysms, degeneration of capillary cells, and vaso-obliteration usually do not begin to appear in these animals until about 2–3 years after induction of diabetes. The cost, slow development of lesions, and lack of availability of antibodies or molecular biology techniques have made dog and cat models less suitable for current studies to understand molecular mechanisms involved in the development of diabetic retinopathy. Neovascularization has been observed to develop *within the retina* in long-term diabetic dogs, but pre-retinal neovascularization has not been detected.

19.1.1.2.2 Rats

The streptozotocin-diabetic rat has been the primary model for research into the pathogenesis of the vascular lesions of diabetic retinopathy [87]. A significant increase in the number of degenerated capillaries and TUNEL-positive capillary cells has been detected after about 8 months of diabetes [122]. Others have reported death of retinal capillary cells after as little as 1 week of diabetes [75], but possible adverse consequences of the recent administration of the cytotoxin, streptozotocin, were not ruled out. Spontaneously diabetic BB rats exhibit retinal lesions similar to those observed in alloxan diabetic rats, including pericyte loss, basement membrane thickening, and an absence of microaneurysms after about 14 months of hyperglycemia [27]. In contrast, a model of type 2 diabetes, the Zucker diabetic fatty rat, reportedly developed no degenerative lesions of the retinal microvasculature, and in fact, capillary cell nuclear density was found to be greater than normal in these animals [38].

Loss of neural cells of the ganglion cell layer has been reported in diabetic rats [14]. Other retinal cell types also are being reported as lost in diabetes [5]. The rat offers practical advantages in terms of costs, housing requirements, and available reagents, although the lesions that characterize the advanced stages of diabetic retinopathy in humans have not been observed to develop in most strains of diabetic rats. The nonobese, Spontaneously Diabetic Torii rat has been reported to show evidence of retinal capillary nonperfusion and neovascularization in the retina at the extraordinarily short duration of 5–10 weeks of diabetes [119]. In light of the early reports, a recent report is surprising since it describes that many of the animals showed proliferative retinopathy without evidence of vascular nonperfusion [167].

19.1.1.2.3 Mice

Recent studies have begun to characterize the development of retinopathy in several mouse models. Spontaneously diabetic Akita mice (Ins2^{Akita}), *db/db* mice, and streptozocin-diabetic C57Bl/6J mice develop the early vascular pathology characteristic of diabetic retinopathy – acellular capillaries, pericyte loss, and capillary cell apoptosis – beginning at about 6 months of diabetes [13, 15, 58]. The acellular capillaries and pericyte ghosts become more numerous with increasing duration of diabetes (through 18 months of diabetes) [58]. Diabetes also has been reported to cause loss of retinal ganglion cells in Ins2^{Akita} and C57Bl/6J mice [13]. In contrast to the reported loss of retinal ganglion cells from C57Bl/6 mice after only 14 weeks of diabetes [116], however, diabetic C57Bl/6 mice in our hands did not show detectable loss of ganglion cells by any of three independent methods (number of cells in ganglion cell layer of retinal cross-sections, retrograde labeling of

retinal ganglion cells with fluorescent dye, or TUNEL staining) during up to 1 year duration of diabetes [58].

Genetically modified mice now are being used to explore the pathogenesis of the retinopathy in diabetic mice. Mice deficient in the genes encoding for the leukocyte adhesion molecules CD18, ICAM-1, and lipid regulation have been used to study the pathogenesis of diabetic retinopathy [76], and studies involving numerous other genes (including iNOS, PARP, COX2) are ongoing (Kern, unpublished). The principal advantages of mice are cost, availability of reagents, and ability to generate genetically modified animals for study. The principal disadvantages of mice for studies of the retinopathy are the small size of the retina (and consequently the quantitatively small number of lesions that can be detected per retina), and the difficulty in unambiguously identifying and quantifying pericytes and pericyte "ghosts." Diabetes-induced retinal neovascularization has not been detected in any purely diabetic mouse model to date.

19.1.1.2.4 Primates

Microaneurysms and retinal hemorrhages were detected in primates after 10–15 years of chronic hyperglycemia [19, 47, 110], as were retinal ischemia, other lesions, alterations in the blood-retinal barrier, and maculopathy [73, 74, 97, 98, 160]. The main advantages of primates in the study of diabetic retinopathy are: (1) the presence of a macula, thus allowing investigation of macular edema, a major cause of visual impairment in diabetes, and (2) the likelihood that many antibodies and molecular probes generated against human samples will work also on the primates. The major disadvantages of this model are the cost and slow development of lesions. Pre-retinal neovascularization has not been detected in diabetic primates.

19.1.1.3 Insulin Therapy to Inhibit Development of Retinopathy

Intensive insulin therapy has been shown to inhibit development of vascular lesions of diabetic retinopathy in patients [43], dogs [50], and rats (transplanted with exogenous islets [64]). Therapeutically, improved glycemic control is less effective at inhibiting the development of retinopathy if initiation of the therapy is delayed; lesions of the retinopathy were observed to continue to develop for at least some interval after elimination of elevated blood hexose level in diabetic patients [43], diabetic dogs [55] and rats [150], and in some, but not all, studies of galactosemic animals [57, 82, 138]. A possible biochemical basis of this continued progression of retinopathy even after elimination of hyperglycemia has been suggested in studies of diabetic rats and galactosemic rats where oxidative stress and nitric oxide production remained significantly greater than normal even months after elimination of hyperglycemia [104].

19.1.1.4 Is Diabetic Retinopathy a Chronic Inflammatory Disease?

Retinas from diabetic animals exhibit biochemical and physiological abnormalities which, in composite, have features that include inflammatory processes. In retinas of diabetic animals, induction of iNOS and COX-2, as well as increase in nitric oxide and prostaglandins [24, 25, 44, 45, 125], have been reported, presumably as a consequence of a diabetes-induced translocation of NF-κB to the nucleus of retinal cells [176]. Inhibition of COX2 has been reported to inhibit the diabetes-induced upregulation of retinal VEGF [12] and increase in retinal vessel permeability [78], and to inhibit death of retinal endothelial cells cultured in diabetic-like concentrations of glucose [44]. Less selective cyclo-oxygenase inhibitors have inhibited development of the retinopathy in diabetic dogs and rodents [92], as well as the increase in vascular permeability in diabetic rodents [78]. Blocking FasL in vivo has been shown to prevent endothelial cell damage, vascular leakage, and platelet accumulation in diabetes, indicating that the Fas/FasL system can contribute to the diabetes-induced damage that leads to development of the retinopathy [77]. Increases in TNF-α and IL-1β levels have been shown in the vitreous of diabetic patients and in retinas of STZ-induced diabetic rodents [78]. Diabetic mice genetically deficient in TNF-α have been reported in an abstract to be protected from galactose-induced retinopathy [111]. Activity of caspase-1, the enzyme responsible for interleukin-1b production, is increased in retinas of diabetic mice and galactose-fed mice, and diabetic humans, and retinal Müller cells incubated in elevated glucose concentration [123].

Leukostasis is an important component of the inflammatory process, and adhesion of leukocytes to the vascular wall has been found to be significantly increased in retinas of diabetic animals [75]. Investigators have reported that diabetes increased expression of ICAM-1 (intracellular adhesion molecule-1) in retina [78] and interaction of this protein with the CD18 adhesion molecule on monocytes and neutrophils contributed to the diabetes-induced increase in leukostasis within retinal vessels [120]. Leukostasis has been postulated to be an important factor in death of retinal endothelial cells in diabetes. Consistent with this postulate, mice that are genetically deficient in either ICAM or CD18 developed significantly less microvascular pathology and capillary leakage than did wild-type controls [76]. Leukostasis, however, seems not sufficient to account for diabetes-induced changes in retinal blood flow [1].

Platelets also accumulate in the retinal vasculature in the diabetes [18]. Platelet microthrombi are present in the retinas of diabetic rats and humans and have been spatially associated with apoptotic endothelial cells [18].

Deposition of C5b-9, the terminal product of complement activation, has been observed within retinal blood vessels of diabetic rats and humans [172]. Endogenous inhibitors of complement activation, including CD55, CD59, and DAF, have been observed to have subnormal expression or impaired function as a result of nonenzymatic glycation [2, 40, 136]. Whether or not inhibition of the complement system can inhibit development of lesions characteristic of the retinopathy remains to be learned.

Breakdown of the blood-retinal barrier breakdown, another early event in the development of diabetic retinopathy, has been attributed to increases in leukostasis, cytokines, and growth factors. Increased permeability of the blood retinal barrier contributes to retinal edema and visual impairment in diabetic patients, and there has been considerable effort directed towards developing means to assess increased vascular permeability within the retina, and to identify therapies to inhibit this defect. Therapies that have been found to inhibit the diabetes-induced increase in vascular permeability within the retina include aldose reductase inhibitors [34, 134, 157–159], protein kinase C inhibitors [6], tyrosine kinase inhibitors [124], aspirin [78], a cyclooxygenase 2 inhibitor [78], steroids [153], VEGF receptor antagonist, and TNF-α receptor antagonists [78].

Taken together, these numerous defects in diabetes are consistent with a diabetes-induced inflammatory response in the retina. Interestingly, the stimulus for these inflammatory-like changes appears to be glucose itself, since many of these changes can be detected in retinal cells incubated in vitro in diabetic-like concentrations of glucose. Thus, as long as the blood sugar remains elevated, this inflammatory response is inappropriately maintained, and apparently contributes to the development of tissue damage.

19.1.1.5 Therapies

The concept that diabetic retinopathy includes aspects of an inflammatory disease suggests new therapeutic strategies and targets at which to inhibit the retinopathy. A variety of therapies reported are listed below, listing their effects on inflammatory changes in the retina where possible. Their postulated targets are summarized in Fig. 19.1.1.2 and Table 19.1.1.2.

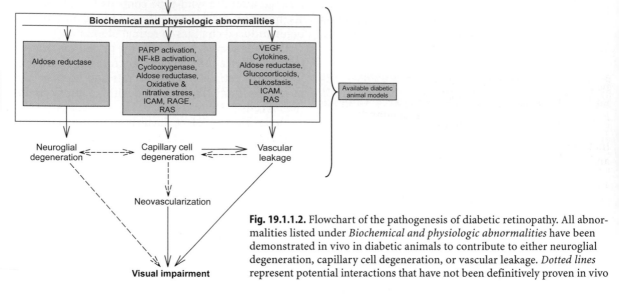

Fig. 19.1.1.2. Flowchart of the pathogenesis of diabetic retinopathy. All abnormalities listed under *Biochemical and physiologic abnormalities* have been demonstrated in vivo in diabetic animals to contribute to either neuroglial degeneration, capillary cell degeneration, or vascular leakage. *Dotted lines* represent potential interactions that have not been definitively proven in vivo

Table 19.1.1.2. Inhibition of diabetes-induced retinal injury in animals

Suspected biochemical or physiological defect	Pharmacologic inhibitor/gene modification
Aldose reductase[a]	Sorbinil [35]
Growth factor regulation	Angiopoietin-1 [79]
Cyclooxygenase	Aspirin [78], meloxicam [78], celecoxib [12], nepafenac [94]
Permeability, inflammation	Glucocorticoids [153]
TNF-α	Etanercept [78]
Inhibition of glycolysis	Benfotiamine [68]
Leukostasis	ICAM knockout [76], CD18 knockout [76]
NF-κB	Salicylates [175]
Oxidative stress	Antioxidants [108], nicanartine [67], lipoic acid [107]
Nitrative stress Protein cross-linking Dicarbonyl stress	Aminoguanidine [95]
Poly(ADP-ribose) polymerase (PARP)	PJ34 [176]
Protein kinases	Genistein [124], GF109203X [166]
RAGE	Soluble RAGE [15]
Renin-angiotensin system	Captopril [173]
VEGF	VEGF trapA [135], VEGF receptor signal inhibitor SU5416 [32]

[a] Controversial [56]

19.1.1.5.1 Insulin and Other Glucose-Lowering Drugs

The best evaluated therapeutic agent that inhibits the development of diabetic retinopathy is insulin or other glucose-lowering drugs. This group of agents have been found to inhibit development and pro-gression of retinopathy in type 1 diabetes, in type 2 diabetes, and in diabetic dogs [50]. Recent studies have demonstrated a pro-survival action of insulin on retinal cells independent of its effect on glucose concentrations [137].

19.1.1.5.2 Aspirin and Salicylates

In 5-year studies of diabetic dogs, aspirin significantly inhibited the formation of acellular capillaries and retinal hemorrhages but was less effective on microaneurysms and pericyte ghosts in those animals [92]. Prospective clinical trials to assess the possible effect of aspirin on diabetic retinopathy in patients have yielded contradictory results. Aspirin treatment resulted in a statistically significant (although weak) inhibition of the mean yearly increase in the number of microaneurysms in the DAMAD trial [36], whereas no beneficial effect was observed on any aspect of retinopathy in the ETDRS trial [46]. The lack of effect of aspirin in the ETDRS likely is attributable, in part, to the greater severity of retinopathy at the onset than in the DAMAD trial, especially since animal and clinical studies have shown that retinopathy tends to resist arrest once it is initiated [55]. Perhaps more importantly, both patient studies utilized doses of aspirin that were less than those used in the dog studies. The concentration of aspirin used in the dog study was moderate (22 mg/kg BW), consistent with doses used to treat arthritis in patients (equivalent to approximately 1.5 g of aspirin per day in a 70 kg patient), whereas doses used in the DAMAD trial and ETDRS were lower.

The ability of three different nonsteroidal anti-inflammatory drugs (aspirin, sodium salicylate and sulfasalazine) to inhibit the development of the retinopathy recently was examined in experimentally diabetic rats [175]. Moderate doses of each of these salicylate derivatives significantly inhibited the diabetes-induced increase in retinal capillary cell death and formation of acellular capillaries. Aspirin is known to inhibit production of prostaglandins, but salicylate has much less of this activity, suggesting that the common action of aspirin, sodium salicylate and sulfasalazine to inhibit retinopathy was not primarily mediated by inhibition of prostaglandins. Each of these salicylates can inhibit the NF-κB pathway (which is involved in the pathogenesis of the inflammatory response), and evidence indicates that the drugs did indeed inhibit activation of the NF-κB, as well as retinal expression of ICAM, iNOS, COX2 (gene products regulated by NF-κB dependent transcription). These data demonstrate that moderate doses of salicylates inhibit early lesions of diabetic retinopathy, seemingly via inhibition of NF-κB, and are consistent with the premise that diabetic retinopathy is a chronic inflammatory disease.

19.1.1.5.3 Cyclooxygenase Inhibitor

Nepafenac is a potent inhibitor of cyclooxygenases that can be applied in eye drops, and meloxicam is a selective inhibitor of COX-2. Both compounds have been found to inhibit diabetes-induced leukocyte adhesion in retinal vessels of diabetic rats [78]. Meloxicam also reduced eNOS levels, inhibited NF-κB activation in the diabetic retina, and modestly, but significantly, reduced TNF-α levels in the retina; but its effect on histologic lesions of diabetic retinopathy was not studied. Long-term topical administration of nepafenac significantly inhibited the diabetes-induced increase in the number of TUNEL-positive capillary cells, acellular capillaries, and pericyte ghosts [94].

19.1.1.5.4 Eternacept

Eternacept is a soluble TNF-α receptor that acts as competitive inhibitor to block effects of TNF-α binding to cells. Eternacept reduced leukocyte adherence in all retinal blood vessel types of rats diabetic for 1 week compared to control [78]. Eternacept did not reduce retinal VEGF levels, but it inhibited blood-retinal barrier breakdown and NF-κB activation in the diabetic retina. No effects of the therapy on histologic lesions of the retinopathy were reported in diabetes, but mice genetically deficient in TNF-α have been reported in an abstract to be protected from galactose-induced retinopathy [111].

19.1.1.5.5 Aminoguanidine

Aminoguanidine has been found to dramatically inhibit the development of retinopathy in 5-year studies of diabetic dogs [92], and in shorter studies of diabetic rats [95]. Moreover, aminoguanidine also has been found to inhibit the diabetes-induced increase in apoptosis of retinal capillary cells in vivo [95]. Aminoguanidine reacts with the highly reactive dicarbonyls (such as glyoxal and methylglyoxal) and as a result blocks late sequelae of nonenzymatic glycation including formation of AGEs and protein cross-links [165, 169, 178]. Aminoguanidine also inhibits oxidative stress [61], activation of protein kinase C [102], and induction of VEGF [59], iNOS [45], and COX2 [44].

19.1.1.5.6 Antioxidants

Pharmacologic doses of mixtures of antioxidants have been found to partially, but significantly, inhibit the development of acellular capillaries and pericyte ghosts in diabetic rats. To date, mixtures of α-tocopherol and ascorbate [108], Trolox, acetylcysteine and selenium [108], α-tocopherol alone (Kern, unpublished), and lipoic acid [107] have been found to significantly inhibit the development of acellular capillaries in retinas of diabetic rodents. The antioxidant and lipid-lowering agent, nicanartine, significantly inhibited diabetes-induced alterations in the number of retinal capillary endothelial cells and pericytes in rats but had no effect on the formation of acellular capillaries [67]. Antioxidants have been found to inhibit development of inflammatory changes in retinas of diabetic animals, including NF-κB and leukostasis [105].

19.1.1.5.7 Aldose Reductase Inhibitors

Activation of the polyol pathway in retinas from diabetic rats has been shown by numerous investigators to contribute to the development of a wide variety of biochemical and physiological abnormalities. Excessive activity of the polyol pathway in the retina has been implicated in hyperglycemia-induced alterations in redox state, oxidative stress, nitric oxide generation, and VEGF production [59]. Nevertheless, the effect of aldose reductase inhibitors on the development of vascular lesions of the retinopathy has been controversial. Studies in diabetic rats have demonstrated that inhibition of the polyol pathway inhibited thickening of retinal capillary basement membrane and neurodegeneration [11] and capillary degeneration [35], whereas 5-year studies in diabetic dogs showed no inhibition of diabetes-induced microvascular lesions despite total inhibition of the diabetes-induced polyol accumulation in retina [56]. Effects of aldose reductase inhibitors have been controversial in galactose-fed rats [142] and dogs [81]. A deficiency in studies of the polyol pathway to date is that extent to which the pathway was inhibited in the retina has been assessed only based on steady-state levels of polyol, not on in vivo measurement of flux through the pathway. Moreover, many studies have not made any effort to demonstrate the extent to which the pathway was inhibited in the retina in vivo, so that data from different studies are difficult to compare.

19.1.1.5.8 Antisense Oligonucleotides Against Fibronectin

Thickening of retinal capillary basement membrane and overexpression of basement membrane components are closely associated with the development of diabetes-induced vascular disease. Intravitreal injection of antisense oligonucleotides targeted against fibronectin was found to significantly reduce fibronectin mRNA and protein level in retina of galactose-fed rats, to partially inhibit thickening of the retinal capillary basement membrane, and to inhibit formation of acellular capillaries in this galactosemic model [145].

19.1.1.5.9 Benfotiamine

Administration of benfotiamine to the diabetic rats for 36 weeks significantly inhibited the development of acellular capillaries in retinas of diabetic rats [68]. This lipid-soluble thiamine derivative was used since it is known to activate transketolase. The authors postulated that benfotiamine should convert the accumulated fructose-6-phosphate and glyceraldehyde-3-phosphate into pentose-5-phosphates, thereby diverting sugar metabolism away from the impairment in glycolysis. Consistent with this hypothesis, they demonstrated that benfotiamine significantly inhibited several hyperglycemia-induced abnormalities including activities of protein kinase C and the hexosamine pathway and formation of advanced glycation endproducts. Which of these (or other) pathways is the ultimate cause of the microvascular pathology in the retina of these diabetic rats remains to be clarified.

19.1.1.5.10 Blood Pressure Medications

Clinical studies have detected a beneficial effect of ACE inhibitors on diabetic retinopathy. The EUCLID Study Group investigated the effect of ACE inhibitor, lisinopril, on retinopathy in normotensive type 1 diabetic patients. In that 24-month study, retinopathy was found to have progressed by at least one level in 23.4% of control patients and in only 13.2% of patients treated with lisinopril [30]. Thus, ACE inhibitor therapy resulted in a 50% reduction in the progression of the retinopathy. The UKPDS clinical trial likewise revealed that the ACE inhibitor, captopril, inhibited a two-step progression of retinopathy by 34% over a 9-year period [161]. The observed inhibition of retinopathy by ACE inhibitors remains unexplained. Corrosion cast and SEM studies of Otsuka Long-Evans Tokushima Fatty (OLETF) diabetic rats showed that cilazapril therapy inhibited diabetes-induced changes in capillary tortuosity, caliper and density in long-term studies [16]; and a recent abstract reported that captopril dramatically inhibited the early stages of diabetic retinopathy (capillary degeneration) [173].

19.1.1.5.11 Nerve Growth Factor

Administration of nerve growth factor significantly inhibited the diabetes-induced obliteration of retinal capillaries in rats [63] and mice (Kern, unpublished). The mechanism of this action remains under investigation, but one possibility under investigation is the known induction of antioxidant enzymes by nerve growth factor [62, 129, 130, 146, 177].

19.1.1.5.12 Protein Kinase Inhibitors

Inhibitors of protein kinase C have been demonstrated to inhibit a variety of diabetes-induced biochemical and physiological abnormalities in the retina [6], but they have not yet been reported to inhibit diabetes-induced histopathology of the retinal vasculature in animal models. LY333531 inhibited diabetes-induced increases in leukostasis and decreases in retinal blood flow in rats [1]. Clinical trials using this inhibitor have been evaluated in advanced stages of diabetic retinopathy (where the results did not achieve statistical significance) [133]. Less selective inhibitors of protein kinases have been found to inhibit diabetes-induced increases in retinal permeability [124].

19.1.1.5.13 Pyridoxamine

Pyridoxamine was found to significantly inhibit the diabetes-induced dropout of retinal capillaries at 29 weeks of diabetes in rats [149]. Simultaneously, it inhibited upregulation of laminin protein and mRNA for extracellular matrix proteins in the retina. Pyridoxamine is an inhibitor of the formation of advanced glycation endproducts and lipoxidation endproducts.

19.1.1.5.14 PARP Inhibitor

Administration of a potent PARP inhibitor (PJ34) for 9 months to diabetic rats significantly inhibited the diabetes-induced death of retinal microvascular cells and the development of early lesions of diabetic retinopathy, including acellular capillaries and pericyte ghosts [176]. Evidence suggested that it exerted this beneficial effect by regulating activation of the transcription factor, NF-κB. In bovine retinal endothelial cells, PARP interacted directly with subunits of NF-κB, and inhibition of PARP blocked the hyperglycemia-induced increase in NF-κB and proinflammatory gene products regulated by it.

19.1.1.5.15 Soluble RAGE

Soluble RAGE (receptor for Advanced Glycation Endproducts), a competitor of cellular RAGE for its ligands, was administered to assess the role of RAGE in diabetes-induced alterations in retinal function and capillary degeneration [15]. Diabetic (*db/db*) and hyperlipidemic (apoE$^{-/-}$) mice were crossed, and capillary degeneration was accelerated in animals showing both defects. Attenuation of the RAGE axis with soluble RAGE ameliorated neuronal dysfunction in the retina and reduced the development of capillary lesions in retinas from these animals.

19.1.1.6 Summary

Animal models of diabetic retinopathy have provided a wealth of information pertaining to biochemical, physiological, and histopathologic abnormalities that contribute to the development of the early stages of diabetic retinopathy. The lack of absolute specificity of most available pharmacologic therapies adds a confounding factor in investigations to assess the role of specific biochemical abnormalities in the pathogenesis of retinopathy. Genetically modified rodents will likely overcome this difficulty in the future and thus provide an important tool to help further dissect the pathogenesis of the retinopathy. At present, our understanding of the sequence of events that ultimately lead to diabetic retinopathy in visual loss is incomplete. Directing therapy at broader abnormalities, such as illustrated by inflammation, may have therapeutic advantages over highly specific inhibitors in terms of prevention of treatment of diabetic complications.

Acknowledgements. This work was funded by PHS grants EY00300, DK57733, the Medical Research Service of the Department of Veteran Affairs, the Kristin C. Dietrich Diabetes Research Award (TK), a career Development Award from the Juvenile Diabetes Research Foundation (CDA-2–2000–390), and PHS grant (EY014380) to S.M.

References

1. Abiko T, Abiko A, Clermont AC, Shoelson B, Horio N, Takahashi J, Adamis AP, King GL, Bursell SE (2003) Characterization of retinal leukostasis and hemodynamics in insulin resistance and diabetes: role of oxidants and protein kinase-C activation. Diabetes 52:829–837
2. Acosta J, Hettinga J, Fluckiger R, Krumrei N, Goldfine A, Angarita L, Halperin J (2000) Molecular basis for a link between complement and the vascular complications of diabetes. Proc Natl Acad Sci U S A 97:5450–5455
3. Addison DJ, Garner A, Ashton N (1970) Degeneration of intramural pericytes in diabetic retinopathy. Br Med J 1: 264–266
4. Agardh CD, Agardh E, Zhang H, Ostenson CG (1997) Altered endothelial/pericyte ratio in Goto-Kakizaki rat retina. J Diabetes Complications 11:158–162
5. Agardh E, Bruun A, Agardh CD (2001) Retinal glial cell immunoreactivity and neuronal cell changes in rats with STZ-induced diabetes. Curr Eye Res 23:276–284
6. Aiello LP, Bursell SE, Clermont A, Duh E, Ishii H, Takagi C, Mori F, Ciulla TA, Ways K, Jirousek M, et al. (1997) Vascular endothelial growth factor-induced retinal permeability is mediated by protein kinase C in vivo and suppressed by an orally effective beta-isoform-selective inhibitor. Diabetes 46:1473–1480
7. Aizu Y, Katayama H, Takahama S, Hu J, Nakagawa H, Oyanagi K (2003) Topical instillation of ciliary neurotrophic factor inhibits retinal degeneration in streptozotocin-induced diabetic rats. Neuroreport 14:2067–2071
8. Aizu Y, Oyanagi K, Hu J, Nakagawa H (2002) Degeneration of retinal neuronal processes and pigment epithelium in the early stage of the streptozotocin-diabetic rats. Neuropathology 22:161–170
9. Akagi Y, Kador PF (1990) Effect of aldose reductase inhibitors on the progression of retinopathy in galactose-fed dogs. Exp Eye Res 50:635–639
10. Antonetti DA, Wolpert EB, DeMaio L, Harhaj NS, Scaduto RC, Jr (2002) Hydrocortisone decreases retinal endothelial cell water and solute flux coincident with increased content and decreased phosphorylation of occludin. J Neurochem 80:667–677
11. Asnaghi V, Gerhardinger C, Hoehn T, Adeboje A, Lorenzi M (2003) A role for the polyol pathway in the early neuroretinal apoptosis and glial changes induced by diabetes in the rat. Diabetes 52:506–511
12. Ayalasomayajula SP, Kompella UB (2003) Celecoxib, a selective cyclooxygenase-2 inhibitor, inhibits retinal vascular endothelial growth factor expression and vascular leakage in a streptozotocin-induced diabetic rat model. Eur J Pharmacol 458:283–289
13. Barber AJ, Antonetti DA, Kern TS, Reiter CE, Soans RS, Krady JK, Levison SW, Gardner TW, Bronson SK (2005) The Ins2Akita mouse as a model of early retinal complications in diabetes. Invest Ophthalmol Vis Sci 46:2210–2218
14. Barber AJ, Lieth E, Khin SA, Antonetti DA, Buchanan AG, Gardner TW (1998) Neural apoptosis in the retina during experimental and human diabetes. Early onset and effect of insulin. J Clin Invest 102:783–791
15. Barile GR, Pachydaki SI, Tari SR, Lee SE, Donmoyer CM, Ma W, Rong LL, Buciarelli LG, Wendt T, Horig H, et al. (2005) The RAGE axis in early diabetic retinopathy. Invest Ophthalmol Vis Sci 46:2916–2924
16. Bhutto IA, Lu ZY, Takami Y, Amemiya T (2002) Retinal and

choroidal vasculature in rats with spontaneous diabetes type 2 treated with the angiotensin-converting enzyme inhibitor cilazapril: corrosion cast and electron-microscopic study. Ophthalmic Res 34:220–231

17. Bloodworth JM, Jr, Molitor DL (1965) Ultrastructural aspects of human and canine diabetic retinopathy. Invest Ophthalmol 4:1037–1048

18. Boeri D, Maiello M, Lorenzi M (2001) Increased prevalence of microthromboses in retinal capillaries of diabetic individuals. Diabetes 50:1432–1439

19. Bresnick G, Engerman R, Davis MD, de Venecia G, Myers FL (1976) Patterns of ischemia in diabetic retinopathy. Trans Am Acad Ophthalmol Otolaryngol 81:694–709

20. Bresnick GH (1986) Diabetic retinopathy viewed as a neurosensory disorder. Arch Ophthalmol 104:989–990

21. Bresnick GH, Palta M (1987) Oscillatory potential amplitudes: Relation to severity of diabetic retinopathy. Arch Ophthal 105:929–933

22. Bresnick GM, Palta M (1987) Predicting progression of severe proliferative diabetic retinopathy. Arch Ophthalmol 105:810–814

23. Bursell SE, King GL (1999) Can protein kinase C inhibition and vitamin E prevent the development of diabetic vascular complications? Diabetes Res Clin Pract 45:169–182

24. Carmo A, Cunha-Vaz JG, Carvalho AP, Lopes MC (1999) L-arginine transport in retinas from streptozotocin diabetic rats: correlation with the level of IL-1 beta and NO synthase activity. Vision Res 39:3817–3823

25. Carmo A, Cunha-Vaz JG, Carvalho AP, Lopes MC (2000) Nitric oxide synthase activity in retinas from non-insulin-dependent diabetic Goto-Kakizaki rats: correlation with blood-retinal barrier permeability. Nitric Oxide 4:590–596

26. Chakrabarti S, Sima AAF (1989) Effect of aldose reductase inhibition and insulin treatment on retinal capillary basement membrane thickening in BB rats. Diabetes 38: 1181–1186

27. Chakrabarti S, Sima AAF, Tze WJ, Tai J (1987) Prevention of diabetic retinal capillary pericyte degeneration and loss by pancreatic islet allograft. Curr Eye Res 6:649–658

28. Chakrabarti S, Zhang WX, Sima AA (1991) Optic neuropathy in the diabetic BB-rat. Adv Exp Med Biol 291:257–264

29. Chang WP, Dimitriadis E, Allen T, Dunlop M, Cooper M, Larkins RG (1991) The effect of aldose reductase inhibitors on glomerular prostaglandin production and urinary albumin excretion in experimental diabetes mellitus. Diabetologia 34:225–231

30. Chaturvedi N, Sjolie AK, Stephenson JM, Abrahamian H, Keipes M, Castellarin A, Rogulja-Pepeonik Z, Fuller JH (1998) Effect of lisinopril on progression of retinopathy in normotensive people with type 1 diabetes. The EUCLID Study Group. EURODIAB Controlled Trial of Lisinopril in Insulin-Dependent Diabetes Mellitus. Lancet 351:28–31

31. Craven PA, DeRubertis FR (1989) Sorbinil suppresses glomerular prostaglandin production in the streptozotocin diabetic rat. Metabolism 38:649–654

32. Cukiernik M, Hileeto D, Evans T, Mukherjee S, Downey D, Chakrabarti S (2004) Vascular endothelial growth factor in diabetes induced early retinal abnormalities. Diabetes Res Clin Pract 65:197–208

33. Cunha-Vaz JG (1972) Diabetic retinopathy. Human and experimental studies. Trans Ophthal Soc UK 92:111–124

34. Cunha-Vaz JG, Mota MC, Leite EL, Abreu JR, Ruas MA (1986) Effect of sorbinil on the blood-retinal barrier in early diabetic retinopathy. Diabetes 35:575–578

35. Dagher Z, Park YS, Asnaghi V, Hoehn T, Gerhardinger C, Lorenzi M (2004) Studies of rat and human retinas predict a role for the polyol pathway in human diabetic retinopathy. Diabetes 53:2404–2411

36. DAMAD Study Group (1989) Effect of aspirin alone and aspirin plus dipyridamole in early diabetic retinopathy: a multicenter randomized controlled clinical trial. Diabetes 38:491–498

37. Danial NN, Korsmeyer SJ (2004) Cell death: critical control points. Cell 116:205–219

38. Danis RP, Yang Y (1993) Microvascular retinopathy in the Zucker diabetic fatty rat. Invest Ophthalmol Vis Sci 34: 2367–2371

39. Das A, Frank RN, Zhang NL, Samadini E (1990) Increases in collagen type IV and laminin in galactose-induced retinal capillary basement membrane thickening – prevention by an aldose reductase inhibitor. Exp Eye Res 50: 269–280

40. Davies CS, Harris CL, Morgan BP (2005) Glycation of CD59 impairs complement regulation on erythrocytes from diabetic subjects. Immunology 114:280–286

41. Davis MD, Kern TS, Rand LI (1997) Diabetic retinopathy. In: Alberti KGMM, Zimmet P, DeFronzo RA (eds) International textbook of diabetes mellitus. John Wiley & Sons, New York, pp 1413–1446

42. de Oliveira F (1966) Pericytes in diabetic retinopathy. Br J Ophthalmol 50:134–143

43. Diabetes Control and Complications Trial Research Group (1993) The effect of intensive treatment of diabetes on the development of long-term complications in insulin-dependent diabetes mellitus. N Engl J Med 329:977–986

44. Du Y, Sarthy V, Kern T (2004) Interaction between NO and COX pathways in retinal cells exposed to elevated glucose and retina of diabetic rats. Am J Physiol 287:R735–741

45. Du Y, Smith MA, Miller CM, Kern TS (2002) Diabetes-induced nitrative stress in the retina, and correction by aminoguanidine. J Neurochem 80:771–779

46. Early Treatment Diabetic Retinopathy Research Group (1991) Effects of aspirin treatment on diabetic retinopathy. Ophthalmology 98:757–765

47. Engerman RL (1979) Development of retinopathy in diabetic monkeys: Excerpta Medican International Congress Series No. 481: Tenth Congress of the International Diabetes Federation, 58

48. Engerman RL (1989) Pathogenesis of diabetic retinopathy. Diabetes 38:1203–1206

49. Engerman RL, Bloodworth JMB, Jr (1965) Experimental diabetic retinopathy in dogs. Arch Ophthalmol 73:205–210

50. Engerman RL, Bloodworth JMB, Jr, Nelson S (1977) Relationship of microvascular disease in diabetes to metabolic control. Diabetes 26:760–769

51. Engerman RL, Finkelstein D, Aguirre G, Diddie K, Fox R, Frank R, Varma S (1982) Appropriate animal models for research on human diabetes mellitus and its complications. Ocular complications. Diabetes 31(Suppl 1):82–88

52. Engerman RL, Kern TS (1982) Experimental galactosemia produces diabetic-like retinopathy. Diabetes 31(Suppl):26A

53. Engerman RL, Kern TS (1984) Experimental galactosemia produces diabetic-like retinopathy. Diabetes 33:97–100

54. Engerman RL, Kern TS (1985) Diabetic retinopathy: Is it a consequence of hyperglycemia? Diabetic Med 2:200–203

55. Engerman RL, Kern TS (1987) Progression of incipient diabetic retinopathy during good glycemic control. Diabetes 36:808–812

56. Engerman RL, Kern TS (1993) Aldose reductase inhibition

fails to prevent retinopathy in diabetic and galactosemic dogs. Diabetes 42:820–825

57. Engerman RL, Kern TS (1995) Retinopathy in galactosemic dogs continues to progress after cessation of galactosemia. Arch Ophthalmol 113:355–358

58. Feit-Leichman RA, Kinouchi R, Takeda M, Fan Z, Mohr S, Kern TS, Chen DF (2005) Vascular damage in a mouse model of diabetic retinopathy: relation to neuronal and glial changes. Invest Ophthalmol Vis Sci:46:4281–4287

59. Frank RN, Amin R, Kennedy A, Hohman TC (1997) An aldose reductase inhibitor and aminoguanidine prevent vascular endothelial growth factor expression in rats with long-term galactosemia. Arch Ophthalmol 115:1036–1047

60. Gardiner TA, Anderson HR, Stitt AW (2003) Inhibition of advanced glycation end-products protects against retinal capillary basement membrane expansion during long-term diabetes. J Pathol 201:328–333

61. Giardino I, Fard AK, Hatchell DL, Brownlee M (1998) Aminoguanidine inhibits reactive oxygen species formation, lipid peroxidation, and oxidant-induced apoptosis. Diabetes 47:1114–1120

62. Guegan C, Ceballos-Picot I, Chevalier E, Nicole A, Onteniente B, Sola B (1999) Reduction of ischemic damage in NGF-transgenic mice: correlation with enhancement of antioxidant enzyme activities. Neurobiol Dis 6:180–189

63. Hammes H-P, Federoff HJ, Brownlee M (1995) Nerve growth factor prevents both neuroretinal programmed cell death and capillary pathology in experimental diabetes. Mol Med 1:527–534

64. Hammes H-P, Klinzing I, Wiegand S, Bretzel RG, Cohen AM, Federlin K (1993) Islet transplantation inhibits diabetic retinopathy in the sucrose-fed diabetic Cohen diabetic rat. Invest Ophthalmol Vis Sci 34:2092–2096

65. Hammes H-P, Martin S, Federlin K, Geisen K, Brownlee M (1991) Aminoguanidine treatment inhibits the development of experimental diabetic retinopathy. Proc Natl Acad Sci U S A 88:11555–11558

66. Hammes H-P, Syed S, Uhlmann M, Weiss A, Federlin K, Geisub K, Brownlee M (1995) Aminoguanidine does not inhibit the initial phase of experimental diabetic retinopathy in rats. Diabetologia 38:269–273

67. Hammes HP, Bartmann A, Engel L, Wulfroth P (1997) Antioxidant treatment of experimental diabetic retinopathy in rats with nicanartine. Diabetologia 40:629–634

68. Hammes HP, Du X, Edelstein D, Taguchi T, Matsumura T, Ju Q, Lin J, Bierhaus A, Nawroth P, Hannak D, et al. (2003) Benfotiamine blocks three major pathways of hyperglycemic damage and prevents experimental diabetic retinopathy. Nat Med 9:294–299

69. Hammes HP, Strodter D, Weiss A, Bretzel RG, Federlin K, Brownlee M (1995) Secondary intervention with aminoguanidine retards the progression of diabetic retinopathy in rat model. Diabetologia 38:656–660

70. Hammes HP, Weiss A, Fuhrer D, Kramer HJ, Papavassilis C, Grimminger F (1996) Acceleration of experimental diabetic retinopathy in the rat by omega-3 fatty acids. Diabetologia 39:251–255

71. Hatchell DL, Braun RD, Lutty GA, McLeod DS, Toth CA (1995) Progression of diabetic retinopathy in a cat. Invest Ophthalmol Vis Sci 36:S1067

72. Ishii H, Jirousek MR, Koya D, Takagi C, Xia P, Clermont A, Bursell S-E, Kern TS, Ballas LM, Heath WF, et al. (1996) Amelioration of vascular dysfunctions in diabetic rats by an oral PKC β inhibitor. Science 272:728–731

73. Johnson MA, Lutty GA, McLeod DS, Otsuji T, Flower RW, Sandagar G, Alexander T, Steidl SM, Hansen BC (2005) Ocular structure and function in an aged monkey with spontaneous diabetes mellitus. Exp Eye Res 80:37–42

74. Jonasson O, Jones CW, Bauman A, John E, Manaligod J, Tso MOM (1985) The pathophysiology of experimental insulin-deficient diabetes in the monkey. Ann Surg 201:27–39

75. Joussen AM, Murata T, Tsujikawa A, Kirchhof B, Bursell SE, Adamis AP (2001) Leukocyte-mediated endothelial cell injury and death in the diabetic retina. Am J Pathol 158: 147–152

76. Joussen AM, Poulaki V, Le ML, Koizumi K, Esser C, Janicki H, Schraermeyer U, Kociok N, Fauser S, Kirchhof B, et al. (2004) A central role for inflammation in the pathogenesis of diabetic retinopathy. FASEB J 18:1450–1452

77. Joussen AM, Poulaki V, Mitsiades N, Cai WY, Suzuma I, Pak J, Ju ST, Rook SL, Esser P, Mitsiades CS, et al. (2003) Suppression of Fas-FasL-induced endothelial cell apoptosis prevents diabetic blood-retinal barrier breakdown in a model of streptozotocin-induced diabetes. FASEB J 17:76–78

78. Joussen AM, Poulaki V, Mitsiades N, Kirchhof B, Koizumi K, Dohmen S, Adamis AP (2002) Nonsteroidal anti-inflammatory drugs prevent early diabetic retinopathy via TNF-alpha suppression. FASEB J 16:438–440

79. Joussen AM, Poulaki V, Tsujikawa A, Qin W, Qaum T, Xu Q, Moromizato Y, Bursell SE, Wiegand SJ, Rudge J, et al. (2002) Suppression of diabetic retinopathy with angiopoietin-1. Am J Pathol 160:1683–1693

80. Kador PF, Akagi Y, Takahashi Y, Ikebe H, Wyman M, Kinoshita JH (1990) Prevention of retinal vessel changes associated with diabetic retinopathy in galactose-fed dogs by aldose reductase inhibitors. Arch Ophthalmol 108:1301–1309

81. Kador PF, Akagi Y, Terubayashi H, Wyman M, Kinoshita JH (1988) Prevention of pericyte ghost formation in retinal capillaries of galactose-fed dogs by aldose reductase inhibitors. Arch Ophthalmol 106:1099–1102

82. Kador PF, Takahashi Y, Akagi Y, Neuenschwander H, Greentree W, Lackner P, Blessing K, Wyman M (2002) Effect of galactose diet removal on the progression of retinal vessel changes in galactose-fed dogs. Invest Ophthalmol Vis Sci 43:1916–1921

83. Kador PF, Takahashi Y, Sato S, Wyman M (1994) Amelioration of diabetes-like retinal changes in galactose-fed dogs. Prev Med 23:717–721

84. Kador PF, Takahashi Y, Wyman M, Ferris F, III (1995) Diabeteslike proliferative retinal changes in galactose-fed dogs. Arch Ophthalmol 113:352–354

85. Kato N, Yashima S, Suzuki T, Nakayama Y, Jomori T (2003) Long-term treatment with fidarestat suppresses the development of diabetic retinopathy in STZ-induced diabetic rats. J Diabetes Complications 17:374–379

86. Keogh RJ, Dunlop ME, Larkins RG (1997) Effect of inhibition of aldose reductase on glucose flux, diacylglycerol formation, protein kinase C, and phospholipase A2 activation. Metabolism 46:41–47

87. Kern TS, Engerman RL (1994) Comparison of retinal lesions in alloxan-diabetic rats and galactose-fed rats. Curr Eye Res 13:863–867

88. Kern TS, Engerman RL (1995) Galactose-induced retinal microangiopathy in rats. Invest Ophthalmol Vis Sci 36: 490–496

89. Kern TS, Engerman RL (1995) Vascular lesions in diabetes are distributed non-uniformly within the retina. Exp Eye Res 60:545–549

90. Kern TS, Engerman RL (1996) A mouse model of diabetic retinopathy. Arch Ophthalmol 114:986–990

91. Kern TS, Engerman RL (1996) Capillary lesions develop in retina rather than cerebral cortex in diabetes and experimental galactosemia. Arch Ophthalmol 114:306–310

92. Kern TS, Engerman RL (2001) Pharmacologic inhibition of diabetic retinopathy: Aminoguanidine and aspirin. Diabetes 50:1636–1642

93. Kern TS, Kowluru R, Engerman RL (1996) Dog and rat models of diabetic retinopathy. In: Shafrir E (ed) Lessons from animal diabetes. Smith-Gordon, London, pp 395–408

94. Kern TS, Miller C, Du YP, Zheng L, Mohr S, Ball S, Bingaman DP (2005) Topical administration of Nepafenac inhibits diabetes-induced retinal microvascular disease and underlying abnormalities of retinal metabolism and physiology. (submitted)

95. Kern TS, Tang J, Mizutani M, Kowluru RA, Nagaraj RH, Romeo G, Podesta F, Lorenzi M (2000) Response of capillary cell death to aminoguanidine predicts the development of retinopathy: comparison of diabetes and galactosemia. Invest Ophthalmol Vis Sci 41:3972–3978

96. Kim SH, Chu YK, Kwon OW, McCune SA, Davidorf FH (1998) Morphologic studies of the retina in a new diabetic model; SHR/N:Mcc-cp rat. Yonsei Med J 39:453–462

97. Kim SY, Johnson MA, McLeod DS, Alexander T, Hansen BC, Lutty GA (2005) Neutrophils are associated with capillary closure in spontaneously diabetic monkey retinas. Diabetes 54:1534–1542

98. Kim SY, Johnson MA, McLeod DS, Alexander T, Otsuji T, Steidl SM, Hansen BC, Lutty GA (2004) Retinopathy in monkeys with spontaneous type 2 diabetes. Invest Ophthalmol Vis Sci 45:4543–4553

99. Klein R, Davis MD, Moss SE, Klein BE, DeMets DL (1985) The Wisconsin Epidemiologic Study of Diabetic Retinopathy. A comparison of retinopathy in younger and older onset diabetic persons. Adv Exp Med Biol 189:321–335

100. Kowluru RA (2003) Effect of reinstitution of good glycemic control on retinal oxidative stress and nitrative stress in diabetic rats. Diabetes 52:818–823

101. Kowluru RA, Engerman RL, Kern TS (1999) Effects of aminoguanidine on hyperglycemia-induced retinal metabolic abnormalities. Diabetes 48 (Suppl 1):A19

102. Kowluru RA, Engerman RL, Kern TS (2000) Abnormalities of retinal metabolism in diabetes or experimental galactosemia VIII. Prevention by aminoguanidine. Curr Eye Res 21:814–819

103. Kowluru RA, Koppolu P (2002) Diabetes-induced activation of caspase-3 in retina: effect of antioxidant therapy. Free Radic Res 36:993–999

104. Kowluru RA, Koppolu P (2002) Termination of experimental galactosemia in rats, and progression of retinal metabolic abnormalities. Invest Ophthalmol Vis Sci 43:3287–3291

105. Kowluru RA, Koppolu P, Chakrabarti S, Chen S (2003) Diabetes-induced activation of nuclear transcriptional factor in the retina, and its inhibition by antioxidants. Free Radic Res 37:1169–1180

106. Kowluru RA, Kowluru A, Chakrabarti S, Khan Z (2004) Potential contributory role of H-Ras, a small G-protein, in the development of retinopathy in diabetic rats. Diabetes 53:775–783

107. Kowluru RA, Odenbach S (2004) Effect of long-term administration of alpha-lipoic acid on retinal capillary cell death and the development of retinopathy in diabetic rats. Diabetes 53:3233–3238

108. Kowluru RA, Tang J, Kern TS (2001) Abnormalities of retinal metabolism in diabetes and experimental galactosemia. VII. Effect of long-term administration of antioxidants on the development of retinopathy. Diabetes 50:1938–1942

109. Kusner LL, Sarthy VP, Mohr S (2004) Nuclear translocation of glyceraldehyde-3-phosphate dehydrogenase: a role in high glucose-induced apoptosis in retinal Muller cells. Invest Ophthalmol Vis Sci 45:1553–1561

110. Laver N, Robison WG, Jr, Hansen BC (1994) Spontaneously diabetic monkeys as a model for diabetic retinopathy (ARVO abstract). Invest Ophthalmol Vis Sci 35(Suppl):1733

111. Le LM, Poulaki V, Koizumi K, Fauser S, Kirchhof B, Joussen AM (2003) Reduced histopathological alterations in long-term diabetic TNF-R deficient mice (ARVO abstract). Invest Ophthalmol Vis Sci 44:3894

112. Lefer DJ, McLeod DS, Merges C, Lutty GA (1993) Immunolocalization of ICAM-1 (CD54) in the posterior eye of sickle cell and diabetic patients. Invest Ophthalmol Vis Sci 34:1206

113. Lieth E, Gardner TW, Barber AJ, Antonetti DA (2000) Retinal neurodegeneration: early pathology in diabetes. Clin Exp Ophthalmol 28:3–8

114. Linsenmeier RA, Braun RD, McRipley MA, Padnick LB, Ahmed J, Hatchell DL, McLeod DS, Lutty GA (1998) Retinal hypoxia in long-term diabetic cats. Invest Ophthalmol Vis Sci 39:1647–1657

115. Lu ZY, Bhutto IA, Amemiya T (2003) Retinal changes in Otsuka Long-Evans Tokushima Fatty rats (spontaneously diabetic rat) – possibility of a new experimental model for diabetic retinopathy. Jpn J Ophthalmol 47:28–35

116. Martin PM, Roon P, Van Ells TK, Ganapathy V, Smith SB (2004) Death of retinal neurons in streptozotocin-induced diabetic mice. Invest Ophthalmol Vis Sci 45:3330–3336

117. McCaleb ML, McKean ML, Hohman TC, Laver N, Robison WG, Jr (1991) Intervention with the aldose reductase inhibitor, tolrestat, in renal and retinal lesions of streptozotocin-diabetic rats. Diabetologia 34:695–701

118. McLeod DS, Lefer DJ, Merges C, Lutty GA (1995) Enhanced expression of intercellular adhesion molecule-1 and P-selectin in the diabetic human retina and choroid. Am J Pathol 147:642–653

119. Miao G, Ito T, Uchikoshi F, Kamei M, Akamaru Y, Kiyomoto T, Komoda H, Nozawa M, Matsuda H (2004) Stage-dependent effect of pancreatic transplantation on diabetic ocular complications in the Spontaneously Diabetic Torii rat. Transplantation 77:658–663

120. Miyamoto K, Khosrof S, Bursell SE, Rohan R, Murata T, Clermont AC, Aiello LP, Ogura Y, Adamis AP (1999) Prevention of leukostasis and vascular leakage in streptozotocin-induced diabetic retinopathy via intercellular adhesion molecule-1 inhibition. Proc Natl Acad Sci U S A 96:10836–10841

121. Miyamura N, Bhutto IA, Amemiya T (1999) Retinal capillary changes in Otsuka Long-Evans Tokushima fatty rats (spontaneously diabetic strain). Electron-microscopic study. Ophthalmic Res 31:358–366

122. Mizutani M, Kern TS, Lorenzi M (1996) Accelerated death of retinal microvascular cells in human and experimental diabetic retinopathy. J Clin Invest 97:2883–2890

123. Mohr S, Tang J, Kern TS (2002) Caspase activation in retinas of diabetic and galactosemic mice and diabetic patients. Diabetes 51:1172–1179

III 19

124. Nakajima M, Cooney MJ, Tu AH, Chang KY, Cao J, Ando A, An GJ, Melia M, de Juan E, Jr (2001) Normalization of retinal vascular permeability in experimental diabetes with genistein. Invest Ophthalmol Vis Sci 42:2110–2114

125. Naveh-Floman N, Weissman C, Belkin M (1984) Arachidonic acid metabolism by retinas of rats with streptozotocin-induced diabetes. Curr Eye Res 3:1135–1139

126. Nonaka A, Kiryu J, Tsujikawa A, Yamashiro K, Miyamoto K, Nishiwaki H, Honda Y, Ogura Y (2000) PKC-beta inhibitor (LY333531) attenuates leukocyte entrapment in retinal microcirculation of diabetic rats. Invest Ophthalmol Vis Sci 41:2702–2706

127. Obrosova IG, Minchenko AG, Vasupuram R, White L, Abatan OI, Kumagai AK, Frank RN, Stevens MJ (2003) Aldose reductase inhibitor fidarestat prevents retinal oxidative stress and vascular endothelial growth factor overexpression in streptozotocin-diabetic rats. Diabetes 52:864–871

128. Ou P, Wolff SP (1993) Aminoguanidine: A drug proposed for prophylaxis in diabetes inhibits catalase and generates hydrogen peroxide in vitro. Biochemical Pharmacol 46:1139–1144

129. Pan Z, Perez-Polo R (1993) Role of nerve growth factor in oxidant homeostasis: glutathione metabolism. J Neurochem 61:1713–1721

130. Pan Z, Sampath D, Jackson G, Werrbach-Perez K, Perez-Polo R (1997) Nerve growth factor and oxidative stress in the nervous system. Adv Exp Med Biol 429:173–193

131. Park SH, Park JW, Park SJ, Kim KY, Chung JW, Chun MH, Oh SJ (2003) Apoptotic death of photoreceptors in the streptozotocin-induced diabetic rat retina. Diabetologia 46:1260–1268

132. Picard S, Parthasarathy S, Fruebis J, Wiztum JL (1991) Aminoguanidine inhibits oxidative modification of low density lipoprotein and the subsequent increase in uptake by macrophage scavenger receptors. Proc Soc Natl Acad Sci USA 89:6876–6880

133. PKC-DRS Study Group (2005) The effect of ruboxistaurin on visual loss in patients with moderately severe to very severe nonproliferative diabetic retinopathy: initial results of the Protein Kinase C beta Inhibitor Diabetic Retinopathy Study (PKC-DRS) multicenter randomized clinical trial. Diabetes 54:2188–2197

134. Pugliese G, Tilton RG, Speedy A, Chang K, Province MA, Kilo C, Williamson JR (1990) Vascular filtration function in galactose-fed versus diabetic rats: The role of polyol pathway activity. Metabolism 39:690–697

135. Qaum T, Xu Q, Joussen AM, Clemens MW, Qin W, Miyamoto K, Hassessian H, Wiegand SJ, Rudge J, Yancopoulos GD, et al. (2001) VEGF-initiated blood-retinal barrier breakdown in early diabetes. Invest Ophthalmol Vis Sci 42:2408–2413

136. Qin X, Goldfine A, Krumrei N, Grubissich L, Acosta J, Chorev M, Hays AP, Halperin JA (2004) Glycation inactivation of the complement regulatory protein CD59: a possible role in the pathogenesis of the vascular complications of human diabetes. Diabetes 53:2653–2661

137. Reiter CE, Gardner TW (2003) Functions of insulin and insulin receptor signaling in retina: possible implications for diabetic retinopathy. Prog Retin Eye Res 22:545–562

138. Robison WG, Jr, Jacot JL, Glover JP, Basso MD, Hohman TC (1998) Diabetic-like retinopathy: early and late intervention therapies in galactose-fed rats. Invest Ophthalmol Vis Sci 39:1933–1941

139. Robison WG, Jr, Kador PF, Akagi Y, Kinoshita JH, Gonzalez R, Dvornik D (1986) Prevention of basement membrane thickening in retinal capillaries by a novel inhibitor of aldose reductase, tolrestat. Diabetes 35:295–299

140. Robison WG, Jr, Kador PF, Kinoshita JH (1983) Retinal capillaries: basement membrane thickening by galactosemia prevented with aldose reductase inhibitor. Science 221:1177–1179

141. Robison WG, Jr, Laver NM, Jacot JL, Glover JP (1995) Sorbinil prevention of diabetic-like retinopathy in the galactose-fed rat model. Invest Ophthalmol Vis Sci 36:2368–2380

142. Robison WG, Jr, Nagata M, Laver N, Hohman TC, Kinoshita JH (1989) Diabetic-like retinopathy in rats prevented with an aldose reductase inhibitor. Invest Ophthalmol Vis Sci 30:2285–2292

143. Robison WGJ, Laver NM, Jacot JL, Chandler ML, York BM, Glover JP (1997) Efficacy of treatment after measurable diabeticlike retinopathy in galactose-fed rats. Invest Ophthalmol Vis Sci 38:1066–1073

144. Romeo G, Liu WH, Asnaghi V, Kern TS, Lorenzi M (2002) Activation of nuclear factor-kappaB induced by diabetes and high glucose regulates a proapoptotic program in retinal pericytes. Diabetes 51:2241–2248

145. Roy S, Sato T, Paryani G, Kao R (2003) Downregulation of fibronectin overexpression reduces basement membrane thickening and vascular lesions in retinas of galactose-fed rats. Diabetes 52:1229–1234

146. Sampath D, Perez-Polo R (1997) Regulation of antioxidant enzyme expression by NGF. Neurochem Res 22:351–362

147. Seki M, Tanaka T, Nawa H, Usui T, Fukuchi T, Ikeda K, Abe H, Takei N (2004) Involvement of brain-derived neurotrophic factor in early retinal neuropathy of streptozotocin-induced diabetes in rats: therapeutic potential of brain-derived neurotrophic factor for dopaminergic amacrine cells. Diabetes 53:2412–2419

148. Soulis-Liparota T, Cooper M, Dunlop M, Jerums G (1995) The relative roles of advanced glycation, oxidation and aldose reductase inhibition in the development of experimental diabetic nephropathy in the Sprague-Dawley rat. Diabetologia 38:387–394

149. Stitt A, Gardiner TA, Alderson NL, Canning P, Frizzell N, Duffy N, Boyle C, Januszewski AS, Chachich M, Baynes JW, et al. (2002) The AGE inhibitor pyridoxamine inhibits development of retinopathy in experimental diabetes. Diabetes 51:2826–2832

150. Su EN, Alder VA, Yu DY, Yu PK, Cringle SJ, Yogesan K (2000) Continued progression of retinopathy despite spontaneous recovery to normoglycemia in a long-term study of streptozotocin-induced diabetes in rats. Graefes Arch Clin Exp Ophthalmol 238:163–173

151. Sugiyama T, Kobayashi M, Kawamura H, Li Q, Puro DG (2004) Enhancement of P2X(7)-induced pore formation and apoptosis: an early effect of diabetes on the retinal microvasculature. Invest Ophthalmol Vis Sci 45:1026–1032

152. Takahashi Y, Augustin W, Wyman M, Kador PF (1993) Quantitative analysis of retinal vessel changes in galactose-fed dogs. J Ocular Pharmacol 9:257–269

153. Tamura H, Miyamoto K, Kiryu J, Miyahara S, Katsuta H, Hirose F, Musashi K, Yoshimura N (2005) Intravitreal injection of corticosteroid attenuates leukostasis and vascular leakage in experimental diabetic retina. Invest Ophthalmol Vis Sci 46:1440–1444

154. Tang J, Mohr S, Du Y, Kern TS (2003) Non-uniform distribution of lesions and biochemical abnormalities within the retina of diabetic humans. Curr Eye Res 27:7–13

155. Taylor E, Dobree JH (1970) Proliferative diabetic retinopathy. Site and size of initial lesions. Br J Ophthal 54:11–18

156. Tilton RG, Chang K, Hasan KS, Smith SR, Petrash JM, Misko TP, Moore WM, Currie MG, Corbett JA, McDaniel ML, et al. (1993) Prevention of diabetic vascular dysfunction by guanidines. Inhibition of nitric oxide synthase versus advanced glycation end-product formation. Diabetes 42: 221–232

157. Tilton RG, Chang K, Pugliese G, Eades DM, Province MA, Sherman WR, Kilo C, Williamson JR (1989) Prevention of hemodynamic and vascular albumin filtration changes in diabetic rats by aldose reductase inhibitors. Diabetes 37:1258–1270

158. Tilton RG, Chang K, Weigel C, Eades D, Sherman WR, Kilo C, Williamson JR (1988) Increased ocular blood flow and 125I-albumin permeation in galactose-fed rats: inhibition with sorbinil. Invest Ophthalmol Vis Sci 29:861–868

159. Tilton RG, Pugliese G, LaRose LS, Faller AM, Chang K, Province MA, Williamson JR (1991) Discordant effects of the aldose reductase inhibitor sorbinil on vascular structure and function in chronically diabetic and galactosemic rats. J Diab Compl 5:230–237

160. Tso MOM, Kurosawa A, Benhamou E, Bauman A, Jeffrey J, Jonasson O (1988) Microangiopathic retinopathy in experimental diabetic monkeys. Tr Am Ophth Soc 86:390–418

161. UK Prospective Diabetes Study Group (1998) Tight blood pressure control and risk of macrovascular and microvascular complications in type 2 diabetes: UKPDS 38. UK Prospective Diabetes Study Group. BMJ 317:703–713

162. United Kingdom Prospective Diabetes Study (1998) Intensive blood-glucose control with sulphonylureas or insulin compared with conventional treatment and risk of complications in patients with type 2 diabetes. Lancet 352: 837–853

163. Van den Enden MK, Nyengaard JR, Ostrow E, Burgan JH, Williamson JR (1995) Elevated glucose levels increase retinal glycolysis and sorbitol pathway metabolism. Implications for diabetic retinopathy. Invest Ophthalmol Vis Sci 36:1675–1685

164. Vinores SA, Campochiaro PA (1989) Prevention or moderation of some ultrastructural changes in the RPE and retina of galactosemic rats by aldose reductase inhibition. Exp Eye Res 49:494–510

165. Vlassara H (1994) Recent progress on the biologic and clinical significance of advanced glycation end products. J Lab Clin Med 124:19–30

166. Xu X, Zhu Q, Xia X, Zhang S, Gu Q, Luo D (2004) Blood-retinal barrier breakdown induced by activation of protein kinase C via vascular endothelial growth factor in streptozotocin-induced diabetic rats. Curr Eye Res 28: 251–256

167. Yamada H, Yamada E, Higuchi A, Matsumura M (2005) Retinal neovascularisation without ischaemia in the spontaneously diabetic Torii rat. Diabetologia 48:1663–1668

168. Yamashiro K, Tsujikawa A, Ishida S, Usui T, Kaji Y, Honda Y, Ogura Y, Adamis AP (2003) Platelets accumulate in the diabetic retinal vasculature following endothelial death and suppress blood-retinal barrier breakdown. Am J Pathol 163:253–259

169. Yang S-W, Vlassara H, Peten EP, He C-J, Striker GE, Striker LJ (1994) Advanced glycation end products up-regulate gene expression found in diabetic glomerular disease. Proc Natl Acad Sci U S A 91:9436–9440

170. Yuuki T, Kanda T, Kimura Y, Kotajima N, Tamura J, Kobayashi I, Kishi S (2001) Inflammatory cytokines in vitreous fluid and serum of patients with diabetic vitreoretinopathy. J Diabetes Complications 15:257–259

171. Zeng XX, Ng YK, Ling EA (2000) Neuronal and microglial response in the retina of streptozotocin-induced diabetic rats. Vis Neurosci 17:463–471

172. Zhang J, Gerhardinger C, Lorenzi M (2002) Early complement activation and decreased levels of glycosylphosphatidylinositol-anchored complement inhibitors in human and experimental diabetic retinopathy. Diabetes 51:3499–3504

173. Zhang J, Xi X, Gao L (2005) Captopril inhibits acellular capillary formation in diabetic retinopathy (ARVO abstract). Invest Ophthalmol Vis Sci 46:B405

174. Zheng L, Du Y, Miller C, Bingaman D, Kern T (2003) Topical nepafenac: effects on diabetes-induced increases in retinal leukostasis and superoxide production (ARVO abstract). Invest Ophthalmol Vis Sci

175. Zheng L, Kern TS (2005) Non-steroidal anti-inflammatory drugs (NSAIDs) inhibit development of early stages of diabetic retinopathy (abstract). Diabetes 54 (Suppl1): A227

176. Zheng L, Szabo C, Kern TS (2004) Poly(ADP-ribose) polymerase is involved in the development of diabetic retinopathy via regulation of nuclear factor-kappaB. Diabetes 53: 2960–2967

177. Zhou Z, Chen H, Zhang K, Yang H, Liu J, Huang Q (2003) Protective effect of nerve growth factor on neurons after traumatic brain injury. J Basic Clin Physiol Pharmacol 14:217–224

178. Zimmerman GA, Meistrell M III, Bloom O, Cockroft KM, Bianchi M, Risucci D, Broome J, Farmer J, Cerami A, Vlassara H, et al. (1995) Neurotoxicity of advanced glycation end-products during focal stroke and neuroprotective effects of aminoguanidine. Proc Natl Acad Sci U S A 92:3744–3748

19.1.2 Pharmacological Approach and Current Clinical Studies

19.1.2.1 Protein Kinase C Inhibitors

A. Girach, D.S. Fong

Core Messages

- Protein kinase C (PKC) is a family of intracellular serine/threonine kinases that are involved in cellular signaling processes
- Elevated glucose levels lead to the generation of advanced glycation end products via the polyop pathway and finally to the de novo synthesis of diacylglycerol (DAG), resulting in upregulation of tissue-specific isoforms of DAG
- A tissue specific PKC-β inhibitor is ruboxistaurin (LY333531)
- Preclinical studies have demonstrated an inhibition of vascular endothelial growth factor (VEGF)-induced permeability
- Several phase III studies have been performed demonstrating that there is a dose dependent statistically significant reduction in the rate of moderate visual loss
- Ruboxistaurin had a statistically significant impact on diabetic macular edema (DME) progression. Ruboxistaurin as given orally has been well tolerated in clinical trials and has shown benefit in prevention of vision loss in patients with non-proliferative diabetic retinopathy

19.1.2.1.1 Introduction

The World Health Organization (WHO) estimates that 171 million persons are currently affected and 366 million will be affected with diabetes mellitus by the year 2030 [20]. Diabetic retinopathy is present in about 40% of diabetic patients 40 years and older [10]. After 20 years of diabetes, almost every patient has some form of retinopathy. After 30 years of diabetes, proliferative diabetic retinopathy is present in 70% of people with Type 1 diabetes [14]. One mechanism in the pathophysiology of diabetic retinopathy is the activation of protein kinase C (PKC).

19.1.2.1.2 Protein Kinase C Family of Isoenzymes

Intracellular kinases are enzymes that add phosphate groups to cellular proteins. There are two main types of intracellular kinases: Tyrosine kinases are enzymes that phosphorylate proteins at tyrosine sites. Serine/threonine kinases are enzymes that phosphorylate proteins at serine/threonine sites. PKC is a serine/threonine kinase that is distributed throughout the body and is known to play an important part in a variety of cellular signaling processes (Table 19.1.2.1.1).

Table 19.1.2.1.1. Protein kinase C involvement in multiple cellular/vascular processes [6]

- Ion channel gating
- Permeability
- Receptor function
- Cytoskeletal structure
- Proliferation
- Apoptosis
- Cell division
- Transcription

The PKCs are a family of structurally and functionally related proteins derived from alternative splicing of single mRNA transcripts. There are currently at least 13 known isoenzymes of the PKC family. These can be divided into conventional, novel and atypical types, based on their degree of calcium and phospholipid dependency (Table 19.1.2.1.2). Certain PKC isoforms are expressed preferentially in different tissues of the body. The vascular, retinal and renal tissues express PKC-βI and PKC-βII more so than other isoenzymes. In particular, the retina is known to preferentially express the PKC-βI, PKC-βII and PKC-δ [15]. One other characteristic of PKC is that their activity has been shown to correlate with increasing plasma glucose concentration (Fig. 19.1.2.1.1).

Table 19.1.2.1.2. PKC isoforms

Conventional	Novel	Atypical
α	δ	ι
βI	ε	ξ
βII	η	λ
γ	θ	
	μ	
Calcium dependent	Calcium independent	Calcium independent
Phospholipid dependent		Phospholipid independent

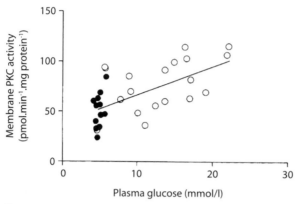

Fig. 19.1.2.1.1. Linear regression between membrane monocyte PKC activity and fasting plasma glucose. Data points represent the pooled experimental observations in 19 patients with diabetes (*white circles*) and 14 control subjects (*black circles*) ($r^2 = 0.4008$, $p = 0.0001$). [5]

19.1.2.1.3 PKC Activation and Diabetic Retinopathy

Hyperglycemia from diabetes is associated with increased incidence and progression of diabetic retinopathy [11]. These elevated levels of glucose lead to an increased flux through the polyol pathway, the generation of advanced glycation endproducts and the generation of reactive oxygen species. Hyperglycemia stimulates a de novo synthesis of diacylglycerol (DAG), which cause an upregulation of tissue-specific isoforms of PKC, leading to translocation of these isoforms from the cytosol to the membrane. This translocation to the membrane form of PKC then causes a cascade of events (Table 19.1.2.1.1), which ultimately lead to the development of diabetic complications (Fig. 19.1.2.1.2). In addition, upregulation of PKC also leads to stimulation of vascular endothelial growth factor (VEGF) expression. Increased VEGF can in turn increase activation of PKC [1, 21].

In animal models, Suzuma and colleagues [16] looked at transgenic mice which overexpressed PKC-β in the endothelium of their vasculature, and found an increased level of neovascularization in response to ischemia. Conversely, in mice expressing dominant negative PKC-β, they found a markedly diminished neovascularization response to the same ischemic insult. In addition to its role in neovascularization, PKC-β overexpression also leads to increased endothelial permeability, a factor in diabetic macular edema (DME) [13].

19.1.2.1.4 PKC-β Inhibitor – Ruboxistaurin

The role of PKC in the pathophysiology of diabetic retinopathy suggests that inhibition of the PKC-β enzyme, by a specific PKC-β inhibitor, might prevent or reduce the risk of diabetic retinopathy. After extensive screening, ruboxistaurin (LY333531) was discovered in 1994. Ruboxistaurin is a bisindolylmaleimide compound that is an orally administered PKC-β specific inhibitor (Fig. 19.1.2.1.3), that has high levels of βI and βII isoform-specific inhibitor activity (Table 19.1.2.1.3). Preclinical studies of ruboxistaurin have shown activity in diabetic retinopathy (DR).

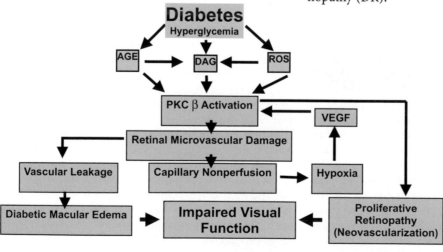

Fig. 19.1.2.1.2. Metabolic pathway of diabetes induced impaired visual function. *AGE* advanced glycation endproducts, *DAG* diacylglycerol, *ROS* reactive oxygen species, *PKC* protein kinase C, *VEGF* vascular endothelial growth factor

Fig. 19.1.2.1.3. Chemical structure of ruboxistaurin (LY333531)

Table 19.1.2.1.3. Kinase selectivity of ruboxistaurin (LY333531). (Adapted from [9])

Kinase	IC$_{50}$ (nM)
PKC-α	360
PKC-βI	4.7
PKC-βII	5.9
PKC-γ	300
PKC-δ	250
PKC-ε	600
PKC-ξ	>100,000
PKC-η	52
PKA	>100,000
Ca calmodulin	6,200
Casein kinase	>100,000
Src-tyrosine kinase	>100,000
Rat brain PKC	3,200

PKC protein kinase C

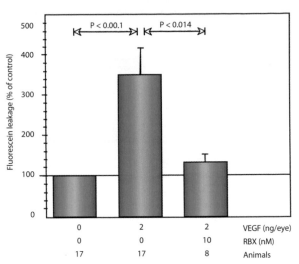

Fig. 19.1.2.1.4. Ruboxistaurin blockade of VEGF-induced increased retinal vascular permeability. *VEGF* vascular endothelial growth factor, *RBX* ruboxistaurin. [1]

19.1.2.1.4.1 Pre-clinical

Experimental studies carried out by Ishii et al. confirmed elevated levels of PKC activity in streptozotocin induced diabetic rats when compared to normal rats. This elevation of retinal PKC activity was reduced to near normal levels by oral administration of ruboxistaurin. In addition, Ishii et al. monitored mean retinal circulation time (MRCT) and found an abnormally high MRCT in diabetic rats that could be normalized with treatment of oral ruboxistaurin. Aiello et al. showed that ruboxistaurin could block VEGF-induced increased vascular permeability in rats [1]. In another model, histological evaluation of the retinal distribution of intravenously injected 70 kDa lysine-conjugated fluorescein dextran was performed. Rats received intravitreal injection of vehicle alone in one eye and ruboxistaurin, a selective PKC β inhibitor (LY333531), (10 nmol/l) in the contralateral eye, followed by intravitreal injection of VEGF (0.5 nmol/l final) in both eyes. VEGF-treated eyes demonstrated diffuse fluorescein staining throughout the retinal tissue, while combined treatment with VEGF and ruboxistaurin demonstrated significantly less fluorescein leakage to near normal levels (Fig. 19.1.2.1.4). In a porcine model of neovascularization of the retina, following retinal vascular occlusion, treatment with oral ruboxistaurin resulted in a significant reduction in neovascularization over 3 months [7].

19.1.2.1.4.2 Clinical

Ruboxistaurin has been studied in clinical trials; the key ones relevant for ophthalmology are summarized below:

- **Endothelial dysfunction and ruboxistaurin**
  Hyperglycemia is associated with endothelial dysfunction, and alters the endothelial-dependent vasodilatory response to acetylcholine. To study this effect, forearm blood flow was measured in response to increasing arterial infusions of endothelial-dependent vasodilator methacholine under hyperglycemic conditions. In a placebo controlled, double masked crossover study in healthy volunteers, inhibition by ruboxistaurin led to an increase in forearm blood flow and prevented the reduction in the nitric oxide mediated vasodilation induced by a hyperglycemic state [3].

- **Phase 1b study**
  Ocular, systemic safety and pharmacodynamic effects of ruboxistaurin were studied in a placebo controlled, double-masked, randomized 28 day study of ruboxistaurin (8 mg twice per day, 16 mg per day, or 16 mg twice per day) or placebo in 29 patients with diabetes. These patients had either no or mild non-proliferative diabetic retinopathy. The results showed that oral ruboxistaurin was well tolerated and there were no clinically relevant safety concerns. Interestingly, there was a dose-dependent reduction in mean retinal circulation time with ruboxistaurin, when compared to placebo, $p = 0.046$ (Fig. 19.1.2.1.5) [2]. Similar results were obtained with retinal blood flow. This pivotal study demonstrated, for

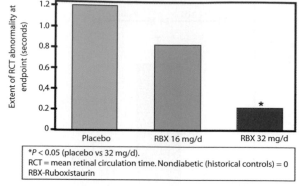

*P < 0.05 (placebo vs 32 mg/d).
RCT = mean retinal circulation time. Nondiabetic (historical controls) = 0
RBX-Ruboxistaurin

Fig. 19.1.2.1.5. Impact of ruboxistaurin on mean retinal circulation time. *p < 0.05 (placebo vs 32 mg/day). *RCT* mean retinal circulation time, non-diabetic (historical controls) = 0. *RBX* ruboxistaurin. (Adapted from [2])

the first time, that a pharmacodynamic impact of ruboxistaurin could be seen as early as 28 days in these patients with diabetes, that the 32 mg dose seemed to be the most efficacious dose and confirmed the animal findings on mean retinal circulation time found by Ishii et al. [8].

19.1.2.1.5 PKC-DRS Trial

The PKC-Diabetic Retinopathy Study was a phase 3 multicenter, double-masked, randomized, placebo controlled trial of 252 patients, randomized to either placebo, 8 mg, 16 mg or 32 mg ruboxistaurin given orally once per day. The trial duration was 3 years and used the primary endpoint of either a 3-step progression of diabetic retinopathy (DR) on 7-field color stereo fundus photography or occurrence of panretinal photocoagulation (PRP).

Eligible patients had one eye with at least moderately severe non-proliferative DR (NPDR), equivalent to Early Treatment Diabetic Retinopathy Study (ETDRS) retinopathy grading scale 47b–53e, without prior PRP, and a best-corrected visual acuity (VA) of at least 45 letters on the ETDRS visual acuity chart (Snellen equivalent = 20/125). Any level of DME was allowed at baseline, as was prior focal/grid photocoagulation.

At baseline, the four groups were well matched and there were no statistically or clinically significant differences, except that the 16 mg group had a higher mean body mass index. There was no statistically or clinically significant difference between the groups in their ophthalmic characteristics. The baseline mean best-corrected VA was 80 letters (Snellen equivalent = 20/25) for the placebo group and this was well matched compared to the ruboxistaurin groups.

After 3 years of follow-up, the primary endpoint of DR progression or PRP did not reveal any benefit

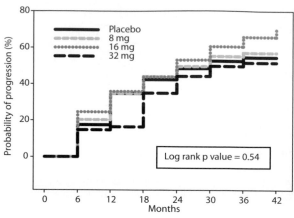

Fig. 19.1.2.1.6. Primary endpoint: time to progression of retinopathy or PRP [18]

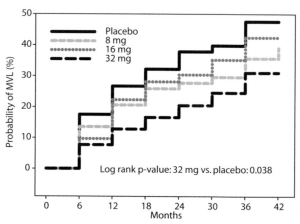

Fig. 19.1.2.1.7. PKC-DRS Trial. Secondary endpoint: occurrence of moderate visual loss. [18]

for any of the ruboxistaurin treated groups (Fig. 19.1.2.1.6).

However, an interesting finding from the PKC-DRS trial was noted in the pre-determined secondary endpoint of time to occurrence to moderate visual loss (15 or more letter loss on ETDRS VA chart). There was a statistically significant reduction in the event rate to moderate visual loss in the 32 mg ruboxistaurin arm when compared to placebo, in a pairwise comparison (Fig. 19.1.2.1.7).

19.1.2.1.6 PKC-DMES (MBBK) Trial

The PKC-Diabetic Macular Edema Study was a phase 3 multicenter, double-masked, randomized, parallel, placebo controlled trial involving 686 patients, randomized to either placebo or 4 mg, 16 mg, or 32 mg ruboxistaurin given orally once per day. The trial duration was 3 years. The primary endpoint was a composite endpoint: progression of DME to within 100 µm from the center of the macula on 7-

field color stereo fundus photography (as graded by a reading center) or occurrence of focal/grid photocoagulation.

Patients' study eyes had to have: DME between 300 and 3,000 μm from the center of macula at baseline, mild or moderately severe NPDR, equivalent to ETDRS retinopathy grading scale 20–47a, without prior PRP or focal/grid laser photocoagulation, and a best-corrected VA of at least 75 letters on the ETDRS VA chart (Snellen equivalent = 20/32). At baseline, the four groups were well matched for their medical and ophthalmic characteristics with no clinically relevant differences between them.

Surprisingly, there were no statistically significant differences between any of the ruboxistaurin treated group and placebo for the primary endpoint of progression of DME to a sight-threatening stage or application of focal/grid laser photocoagulation (Fig. 19.1.2.1.8). However, there seemed to be an imbalance in the site-to-site application and reasons for application of focal/grid laser photocoagulation. When the secondary endpoint of progression of DME alone (without the laser photocoagulation

component) was examined, there was a reduction in the progression of DME, towards a sight-threatening stage (within 100 μm from center of macula), by the 32 mg ruboxistaurin group, when compared to placebo (Fig. 19.1.2.1.9).

19.1.2.1.7 PKC-DRS2 (MBCM) Trial

The PKC-Diabetic Retinopathy Study 2 trial was started while the PKC-DRS and PKC-DMES studies were ongoing, and was initially set up to mimic the PKC-DRS trial. The original primary endpoint of the PKC-DRS2 trial was progression of DR or PRP treatment, but when the moderate vision loss results from PKC-DRS trial were known, and given that a vision loss endpoint would be a better outcome measure, the primary endpoint of the PKC-DRS2 trial was changed to sustained moderate vision loss, without any unmasking of this trial.

The PKC-DRS2 trial was a phase 3 multicenter, double-masked, randomized, parallel, placebo controlled trial involving 685 patients, randomized to either placebo or 32 mg ruboxistaurin given orally once per day. The trial was 3 years in duration. The new primary endpoint was sustained moderate vision loss (SMVL) at 3 years. SMVL was defined as moderate vision loss (≥ 15 letter loss on ETDRS VA chart) sustained for 6 months at the end of the trial, or the last 6 months of study participation if a patient discontinued early.

Patients needed to have at least one eye eligible with: moderately severe-very severe NPDR, equivalent to ETDRS retinopathy grading scale 47a–53e, without prior PRP, and a best-corrected VA of at least 45 letters on the ETDRS VA chart (Snellen equivalent = 20/125). Any level of DME was allowed at baseline, as was prior focal/grid photocoagulation.

At baseline, there were no statistically or clinically significant differences noted between the placebo and 32 mg ruboxistaurin groups, with the mean baseline best-corrected VA of 77 ETDRS chart letters in both groups (Snellen equivalent 20/32).

The primary endpoint of SMVL occurred in 9.1% of placebo patients, as compared to 5.5% of 32 mg ruboxistaurin patients. This equated to a 40% risk reduction in vision loss, in addition to standard of care, by ruboxistaurin, and this was statistically significant ($p = 0.034$) [19].

Other secondary endpoint results indicate:

- Mean VA at 3 years was statistically significantly lower in the ruboxistaurin group compared to placebo eyes ($p = 0.012$).
- In a categorical analysis of baseline-to-endpoint change, 2.4% of placebo eyes gained ≥ 15 letters as compared to 4.9% of ruboxistaurin eyes ($p = 0.027$).

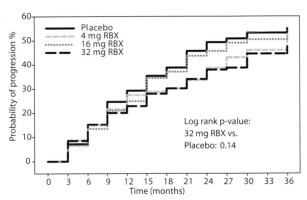

Fig. 19.1.2.1.8. PKC-DMES Trial. Primary endpoint: progression of DME to sight-threatening stage or focal/grid laser photocoagulation. [17]

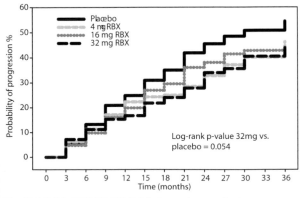

Fig. 19.1.2.1.9. PKC-DMES Trial. Secondary endpoint: progression of DME to sight-threatening stage. [17]

19 **III**

- No statistically significant difference in the outcome of progression of NPDR to proliferative DR was observed.
- The favorable impact of ruboxistaurin on vision loss, compared to placebo, was observed irrespective of baseline DME level (Fig. 19.1.2.1.10) or baseline DR level (Fig. 19.1.2.1.11) or prior focal/grid laser photocoagulation at baseline (Fig. 19.1.2.1.12).
- Ruboxistaurin had a statistically significant impact on DME progression, to within 100 μm from the center of macula, in eyes with clinically significant macular edema at baseline ($p = 0.003$).
- Ruboxistaurin treatment resulted in a statistically significant reduction in the application of initial focal/grid laser photocoagulation, in eyes of patients which were laser-naïve at baseline ($p = 0.008$).

19.1.2.1.8 Safety of Ruboxistaurin

To date, 2,113 patients with at least one diabetic microvascular complication have been treated with ruboxistaurin for up to 3 years in placebo-controlled clinical trials, representing 3,326 patient-years of exposure. Safety data from 1,408 placebo-treated and 1,396 ruboxistaurin 32 mg/day-treated patients were combined from 11 phase 2 or 3 clinical trials. Of the 51 deaths reported, none was considered related to study drug by the investigators. In the placebo treat-

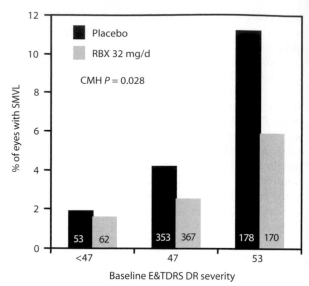

Fig. 19.1.2.1.11. PKC-DRS2 Trial. Sustained moderate vision loss by baseline diabetic retinopathy level. *SMVL* sustained moderate vision loss, *DR* diabetic retinopathy, *RBX* ruboxistaurin, *CMH* Cochran-Mantel-Haenszel test. [19]

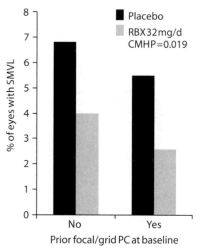

Fig. 19.1.2.1.12. PKC-DRS2 Trial. Sustained moderate vision loss by prior focal/grid photocoagulation at baseline. *SMVL* sustained moderate vision loss, *RBX* ruboxistaurin, *PC* photocoagulation, *CMH* Cochran-Mantel-Haenszel test. [19]

ed group, 2.1% ($n = 30$) died, while 1.5% ($n = 21$) of ruboxistaurin 32 mg/day-treated patients died. A total of 23.2% of placebo-treated patients experienced one serious adverse event, compared with 20.8% of ruboxistaurin 32 mg/day-treated patients.

Treatment-emergent adverse events (TEAEs) with a >10% incidence in the ruboxistaurin 32 mg/day group were nasopharyngitis, influenza, and cough. Incidence of these TEAEs was not significantly different between treatment groups.

Adverse drug reactions occurring significantly more often in the ruboxistaurin 32 mg/day-treated

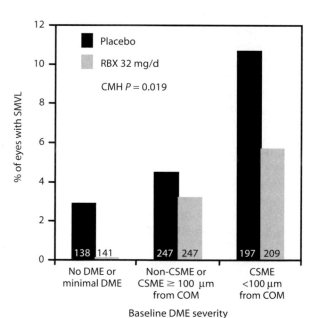

Fig. 19.1.2.1.10. PKC-DRS2 Trial. Sustained moderate vision loss by baseline macular edema level. *SMVL* sustained moderate vision loss, *DME* diabetic macular edema, *CSME* clinically significant macular edema, *COM* center of macula, *RBX* ruboxistaurin, *CMH* Cochran-Mantel-Haenszel test [19]

patients, as compared to placebo-treated patients, include: chalazion, posterior capsular opacification, dyspepsia, increased blood creatine phosphokinase levels, urgency of micturition and superficial thrombosis [12].

19.1.2.1.9 Conclusion

Protein kinase C enzyme upregulation appears to be a critical step in the pathogenesis of diabetic retinopathy. Selective inhibition of the PKC-β enzyme, by ruboxistaurin, has been shown to be of benefit not only in animal models, but also in clinical trials. Although not aproved yet, ruboxistaurin, given once per day orally, appears to be well tolerated in the clinical trials so far and has shown benefit in the prevention of vision loss in patients with non-proliferative diabetic retinopathy, even when added to standard of care.

References

1. Aiello LP, Bursell S-E, Clermont A, et al. (1997) Vascular endothelial growth factor-induced retinal permeability is mediated by protein kinase C in vivo and suppressed by an orally effective β-isoform-selective inhibitor. Diabetes 46: 1473–1480
2. Aiello LP, Clermont A, Arora V, et al. (2006) Inhibition of PKC-β by oral administration of Ruboxistaurin is well tolerated and ameliorates diabetes-induced retinal hemodynamic abnormalities in patients. Invest Ophthalmol Vis Sci 47: 86–92
3. Beckman J, Goldfine A, Gordon M, et al. (2002) Inhibition of protein kinase C-β prevents impaired endothelium-dependent vasodilation caused by hyperglycemia in humans. Circ Res 90:107–111
4. Campochiaro PA; C99-PKC412–003 Study Group (2004) Reduction of diabetic macular edema by oral administration of the kinase inhibitor PKC412. Invest Ophthalmol Vis Sci 45(3):922–31
5. Ceolotto G, Gallo A, Miola M, et al. (1999) Protein kinase C activity is acutely regulated by plasma glucose concentration in human monocytes in vivo. Diabetes 48(6): 1316–1322
6. Curtis T, Scholfield C (2004) The role of lipids and protein kinase Cs in the pathogenesis of diabetic retinopathy. Diabetes Metab Res Rev 20:28–43
7. Danis R, Bingaman D, Jirousek M, et al. (1998) Inhibition of intraocular neovascularization due to retinal ischemia in pigs by PKC-β inhibition with LY333531. Invest Ophthalmol Vis Sci 39:171–9

8. Ishii H, Jirousek MR, Koya D, et al. (1996) Amelioration of vascular dysfunctions in diabetic rats by an oral PKC β inhibitor. Science 272:728–731
9. Jirousek M, Gillig J, Gonzalez C, et al. (1996) (S)-13-[(dimethylamino)methyl]-10,11,14,15-tetrohydro-4,9:16,21-dimetheno-1H,13H-dibenzo[e,k]pyrrolo[3,4-h][1,4,13]-oxadiazacyclohexadecene-1,3(2H)-d ione (LY333531) and related analogues: isozyme selective inhibitors of protein kinase C-β. J Med Chem 39:2664–2671
10. Kempen J, O'Colmain B, Leske M, et al. for the Eye Diseases Prevalence Research Group (2004) The prevalence of diabetic retinopathy among adults in the United States. Arch Ophthalmol 122:552–563
11. Klein R, Klein BE, Moss SE, et al. (1988) Glycosylated haemoglobin predicts the incidence and progression of diabetic retinopathy. JAMA 260:2864–71
12. McGill J, King G, Berg P, et al. (2006) Clinical safety of the selective PKC-β inhibitor, Ruboxistaurin. Expert Opin Drug Saf 5(6):835–845
13. Nagpala PG, Malik A, Vuong P, et al. (1996) PKC-β overexpression augments phorbol ester-induced increase in endothelial permeability. J Cell Physiol 166:249–255
14. Orchard T, Dorman J, Maser R, et al. (1990) Prevalence of complications in IDDM by sex and duration. Pittsburgh Epidemiology of Diabetes Complications Study II. Diabetes 39:1116–1124
15. Park J-Y, Takahara N, Gabriele A, et al. (2000) Induction of endothelin-I expression by glucose: an effect of protein kinase C activation. Diabetes 49:1239–1248
16. Suzuma K, Takahara N, Suzuma I, et al. (2002) Characterisation of protein kinase-β isoforms action on retinoblastoma protein phosphorylation, vascular endothelial growth factor-induced endothelial cell proliferation, and retinal neovascularisation. Proc Natl Acad Sci USA 99:721–726
17. The PKC-DMES Study Group (2006) Effect of ruboxistaurin in patients with diabetic macular edema: 30-month results of the randomized PKC-DMES clinical trial. Arch Ophthalmol (in press)
18. The PKC-DRS Study Group (2005) The effect of ruboxistaurin on visual loss in patients with moderately severe to very severe nonproliferative diabetic retinopathy: initial results of the PKC-DRS multicenter randomized clinical trial. Diabetes 54:2188–97
19. The PKC-DRS2 Study Group (2006) The effect of ruboxistaurin on visual loss in patients with diabetic retinopathy. Ophthalmology (in press)
20. Wild S, Roglic G, Green A, et al. (2004) Global prevalence of diabetes: estimates for the year 2000 and projections for 2030. Diabetes Care 27(5):1047–1053
21. Xia P, Aiello LP, Ishii H, et al. (1996) Characterization of vascular endothelial growth factor's effect on the activation of protein kinase C, its isoforms, and endothelial cell growth. J Clin Invest 98:2018–2026

19.1.2.2 Somatostatin Analogues

G.E. LANG

Core Messages

- Synthetic analogues of the naturally occurring growth hormone inhibitor, somatostatin, are good candidates for pharmaceutical therapy of diabetic retinopathy (DR). They block production of growth hormone and insulin like growth factor 1

- We have evidence so far that somatostatin analogues can reduce progression of DR and preserve visual acuity
- Somatostatin analogues are safe and effective in the treatment of cystoid macular edema and proliferative diabetic retinopathy

---

Laser treatment and vitreoretinal surgery have markedly improved the prognosis for visual problems due to diabetic retinopathy (DR), but nevertheless DR is still the leading cause of blindness in industrialized countries in adults of working age. Therefore we need new therapeutic approaches in the treatment of DR.

19.1.2.2.1 Pathogenesis of Diabetic Retinopathy

The pathophysiology of DR is complex. The principal causes of DR are biochemical, hemodynamic, and endocrine. Hyperglycemia results in the production of advanced glycation endproducts, activation of the polyol pathway and changes in the cellular signal transduction. The results are hyperviscosity, proinflammatory milieu, reduced flexibility of white and red blood cells, increased adhesion of leukocytes and increased oxidative stress. All stages of DR are characterized by the overexpression of a number of different growth factors, the most important being vascular endothelial growth factor (VEGF) and insulin-like growth factor 1 (IGF-1), leading to manifestation and progression of the disease [7].

Clinically DR is subdivided into nonproliferative and proliferative stages. Nonproliferative DR is characterized by the occurrence of microaneurysms, intraretinal hemorrhages, cotton wool spots, hard exudates, intraretinal microvascular abnormalities (IRMA), and venous beading. Proliferative DR exhibits preretinal neovascularization, vitreous hemorrhage and iris neovascularization leading to angle closure glaucoma. Macular edema can occur in any stage of DR.

Histological findings are early thickening of basement membranes and loss of intramural pericytes and vascular endothelial cells. Due to this cellular loss the vascular reactivity is reduced. Damage of the tight junctions of the retinal vascular endothelial cells results in a breakdown of the inner blood-retinal barrier, leading to the manifestation of a diabetic macular edema. The more advanced stages of DR show progressive occlusion of capillaries resulting in retinal hypoxia. The hypoxic retina produces angiogenic growth factors like VEGF and IGF-1. The result is the development of preretinal and iris neovascularization [14, 17, 18].

Diabetic retinopathy occurs after 20 years in 95% of all type 1 and 50–80% of all type 2 diabetic patients. Proliferative DR is found after 20 years in 50% of type 1 and 10–30% of type 2 diabetic patients. Clinically significant macular edema occurs after 15 years in 15% of type 1 and 25% of type 2 patients.

19.1.2.2.2 Somatostatin and Somatostatin Analogues

Somatostatin (somatotropin-release inhibitory factor) is a cyclic tetradecapeptide which was isolated from hypothalamic extracts and regulates growth hormone secretion. This neuropeptide is also produced in different other human organs like hypophysis, gut, exocrine and endocrine glands and also in the retina. Somatostatin is distributed throughout diverse cells and its action affects several biological processes like neurotransmission, hormone secretion, cell proliferation, membrane stabilization and

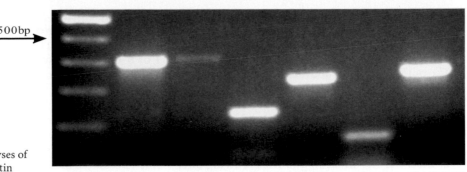

Fig. 19.1.2.2.1. RT-PCR analyses of the expression of somatostatin receptors 1–5 in the human retina

MWM SSTR1 SSTR2 SSTR3 SSTR4 SSTR5 GADPH

III 19

inflammation. Native somatostatin has a short half-life of about 3 min. The synthetic somatostatin analogue Sandostatin® LAR® has been developed for patients requiring long-term therapy and consists of octreotide acetate microencapsulated by a biodegradable polymer.

Somatostatin acts via five G-protein coupled somatostatin receptor subtypes (SSTR1–5) (Fig. 19.1.2.2.1) that were identified by molecular cloning studies. Somatostatin inhibits the release of different hormones and enzymes [8]. The classical action of somatostatin to inhibit growth hormone (GH) release is mediated by SSTR2 and 5. After binding somatostatin, the somatostatin receptors generate a transmembrane signal. This results in a reduction of the calcium concentration and activation of tyrosine phosphatases. Somatostatin is a postreceptor antagonist of growth factors acting by inhibition of signal transduction [17]. Octreotide has high activity at SSTR2, good activity at SSTR5 and moderate activity at SSTR3. It is inactive at SSTR1 and 4.

Growth hormone is produced in the anterior pituitary, resulting in the synthesis of IGF-1. IGF-1 increases the cellular uptake of glucose in the tissue and acts as surviving, growth and progression factor. It presumably acts as a key signal for cells going into the mitotic cycle. IGF-1 stimulates somatostatin secretion and inhibits GH secretion [11].

Somatostatin and SSTRstr have been identified in human retina and are produced in the retina.

Somatostatin analogues have been shown to exhibit pharmacological properties similar to the natural hormone somatostatin. Sandostatin (octreotide acetate) is a synthetic octapeptide analogue of somatostatin. Sandostatin is a potent and specific somatostatin receptor type 2 agonist. It is 45 times more potent in terms of inhibition of growth factor secretion relative to native somatostatin. The original injectable form of octreotide required a dosing regimen of three subcutaneous injections per day. Sandostatin LAR is a long-acting delivery system, administered i.m. once a month. After i.m. injection of Sandostatin LAR, the serum octreotide concentration reaches an initial peak within 1 h and reaches a plateau concentration at around day 14. Steady-state octreotide serum concentrations are achieved after the third injection.

Somatostatin has an antiangigenic effect. Somatostatin is found in the vitreous and in diabetic patients reduced levels of somatostatin were reported [13]. Somatostatin does appear to also have an effect on fluid transport from the retinal pigment epithelium to the choroid, a process that is important in the development of macular edema.

19.1.2.2.2.1 Role of Somatostatin, GH and IGF-1 in Diabetic Retinopathy

There is evidence that GH/IGF-1 excess and interactions between IGF-1 and VEGF play key roles in the development and progression of DR. The anterior pituitary secretes GH, which results in the synthesis of IGF-1 in multiple tissues. GH mediates both systemic and local IGF-1 production. IGF-1 acts as growth and progression factor.

Poulsen [20] found a relation of pituitary hormone and DR in a patient with proliferative DR, who suffered a postpartal pituitary insufficiency (Sheehan syndrome). Five years after the event retinopathy had regressed, indicating the important role of GH and IGF-1 in DR. In the following 20 years after the first description in 1953, diabetic patients underwent pituitary ablation for the treatment of proliferative diabetic retinopathy. These patients showed a regression of neovascularization and a significant reduction in the number of patients progressing to blindness. It is assumed that the treatment effect was related to postsurgical GH deficiency. Interestingly in diabetic patients with dwarfism due to GH deficiency, diabetic retinopathy is absent.

There is also strong evidence to indicate a hypersecretion of GH in diabetics. In patients with DR a substantial hypersecretion of GH was found. Further evidence for the important role of GH in the pathogenesis of DR includes the case report of the development of retinal changes strongly mimicking diabetic retinopathy in two nondiabetic patients treated with GH [15].

The mitogenic effects of GH are mediated by IGF-1. Interestingly in patients with DR elevated serum levels of IGF-1 were found. Somatostatin reduces systemic IGF-1 release via suppression of GH. Additional evidence for the role of IGF-1 in DR comes from studies of the normoglycemic reentry phenomenon. This term describes the worsening of DR occurring after rapid lowering of longstanding severe hyperglycemia in patients with DR. A rapid rise of serum IGF-1 levels has been found to accompany worsening of DR when insulin therapy is intensified and results in improved diabetic control [6]. It is further known that somatostatin, GH and IGF-1 play a role in the manifestation and progression of DR. Retinal hypoxia leads to an increased intraocular synthesis of both IGF-1 and VEGF. In eyes of diabetic patients increased vitreal IGF-1 levels were found. The highest levels of IGF-1 were present in patients with proliferative DR. In patients who had undergone vitrectomy after laser treatment the intravitreal VEGF levels were reduced but not the IGF levels [22], indicating the important role of IGF-1 in late stages of DR especially in cases where laser treatment alone is not effective.

19.1.2.2.2.2 Experimental Studies

Several laboratory studies support the hypothesis of the modes of action of octreotide. Somatostatin and IGF-1 receptors have been demonstrated on retinal vascular endothelial cells [1]. In preclinical studies it was shown that GH can stimulate the proliferation of human retinal endothelial cells. Sall et al. [21] found in cultured human retinal pigment epithelial cells that IGF-1 induced a dose dependent increase in IGF-1R phosphorylation and in VEGF mRNA levels. Somatostatin and octreotide inhibited IGF-1 receptor phosphorylation and decreased VEGF production. Both IGF-1R phosphorylation and accumulation of VEGF mRNA were inhibited by physiological levels of somatostatin and octreotide (1 nM). These results demonstrate a somatostatin and octreotide mediated attenuation of IGF-1R signal transduction and VEGF mRNA accumulation via somatostatin receptor type 2. The role of the GH-IGF-1 axis has also been studied in transgenic and MK678, an octreotide-like somatostatin analogue – treated mice models showing a significant reduction of retinal neovascularization. These studies indicate that the interactions between IGF-1 and its receptors are necessary for maximal induction of retinal neovascularization by VEGF. These findings also indicate the essential role for IGF-1 in the late stages of DR, being

a permissive factor in angiogenesis and indicating the relationship between VEGF and IGF-1 receptors.

It has been shown [9] that octreotide inhibits IGF-1 or b-FGF stimulated endothelial cell proliferation at a low concentration of 10 nM of octreotide. In preclinical studies octreotide also directly inhibits endothelial cell proliferation, indicating additional mechanisms of antiangiogenic action, probably by direct SSTR mediated inhibition [1, 23].

These studies emphasize the importance of paracrine and autocrine effects of octreotide. These data from preclinical studies strongly suggest a role for the use of octreotide as a therapeutic option in diabetic retinopathy.

19.1.2.2.2.3 Mechanism of Action of Octreotide in Diabetic Retinopathy

Because growth hormone and IGF-1 are important mediators of angiogenesis in the retina, the use of synthetic somatostatin analogues in the treatment of DR is at present a promising area of pharmacological research. DR develops as a result of imbalance of pro- and antiangiogenic factors. VEGF and IGF-1 are major players in the pathogenesis. But DR only develops if at the same time there is a lack of natural angiogenesis inhibitors like transforming growth factor β (TGF-β) and pigment epithelium derived factor (PEDF) [2, 4].

Octreotide acts via paracrine and autocrine effects on retinal endothelial cells [9]. It binds to the SSTR and inhibits endothelial cell growth stimulated by growth factors like VEGF and IGF-1.

Somatostatin analogues have two different clinical mechanisms of action. First they can stabilize the blood-retinal barrier in patients with diabetic macular edema. Second they inhibit the neoangiogenesis via IGF-1 inhibition in patients with advanced stages of DR, resulting in regression of preretinal neovascularization or iris neovascularization (Fig. 19.1.2.2.2). The synthetic somatostatin analogue octreotide has been proven to be effective in small series of patients and two phase III trials have been completed recently [18].

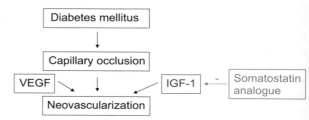

Fig. 19.1.2.2.2. Somatostatin analogues inhibit the IGF-1 induced neovascularization in diabetic retinopathy

19.1.2.2.3 Current Clinical Use of Octreotide

Octreotide is in clinical use for the treatment of tumors like GH producing pituitary adenoma. It is approved for the treatment of acromegaly, malignant carcinoma syndrome, VIPoma, and gastropancreatic neuroendocrine tumors. Octreotide inhibits the pituitary release of GH from the tumor and lowers IGF-1 plasma levels. As overproduction of GH and IGF-1 plays an important role in the pathogenesis of DR, octreotide is under investigation for the treatment of DR.

19.1.2.2.3.1 Treatment of Diabetic Retinopathy with Somatostatin Analogues

According to the clinical experience described above, the idea of treating DR with somatostatin analogues was born more than 20 years ago. Somatostatin analogues were used up to now in prospective studies or on a compassionate basis.

Sandostatin and Sandostatin LAR are more selective and have enhanced activities in comparison to somatostatin. In vitro and in vivo studies have confirmed that somatostatin analogues are specific and potent inhibitors of GH and IGF-1. Octreotide reduces elevated levels of GH and IGF-1. Preliminary evidence suggests that octreotide provides clinical benefits in DR in terms of reduced progression of DR, improved visual acuity and regression of macular edema.

Several clinical studies with somatostatin analogues in patients with DR have been reported in the literature. Most of the studies have been small or open trials, case reports, and studies of short duration.

Mallet et al. [19] reported significant regression of retinal neovascularization in patients with severe proliferative diabetic retinopathy showing progression of DR despite panretinal photocoagulation. The patients were treated with SMS 201-995 (octreotide) by subcutaneous infusions for up to 20 months.

Böhm et al. [3] treated 9 of 18 patients with persistent proliferative DR with vitreous hemorrhage despite having received full scatter laser photocoagulation with octreotide as subcutaneous injections in a dose of 100 µg tid three times daily for up to 3 years; the other nine patients served as control. A significantly reduced incidence of vitreous hemorrhages and number of vitrectomies in the octreotide-treated group was found. In the octreotide group visual acuity remained stable; in the control group visual acuity significantly decreased. Neovascularizations regressed in 85 % of the patients in the treated group and were stable in 15 %; in the control group neovas-

cularizations increased in 42 % and were unchanged in 58 %. Octreotide was well tolerated throughout the 3 years. Insulin dose had to be reduced up to 50 % in insulin-dependent patients. No patient had a severe hypoglycemic episode.

Grant et al. [10] studied the effect of octreotide in 23 type 1 and 2 diabetics with preproliferative and early proliferative DR. The patients were treated with maximally tolerated doses of octreotide (200–5,000 µg/day subcutaneously). Patients were followed for 15 months or until laser photocoagulation was required for both eyes. In the treated group, significantly fewer patients developed high risk characteristics. In the octreotide treated group only one of 22 eyes required laser treatment due to high risk proliferative DR whereas 9 of 24 eyes in the control group had to be treated with laser. Octreotide treatment showed a significant reduction of progression of DR.

Kuijpers et al. [16] described a nondiabetic patient with idiopathic cystoid macular edema who was treated with octreotide 100 µg tid s.c. The cystoid macular edema improved during treatment period, but recurred after treatment cessation and responded again when octreotide treatment was recommenced.

In another case report, Sandostatin LAR showed a beneficial effect in the treatment of diabetic cystoid macular edema (CME). The CME in both eyes was refractive to vitrectomy and periocular steroids. The patient was treated with Sandostatin LAR 20 mg every 4 weeks. One year treatment resulted in complete resolution of CME in the right eye and marked improvement in the fellow eye. Corrected visual acuity was 20/40 in the right eye and 20/100 in the left eye [12].

Chantelau and Frystyk [5] reported patients in whom the progression of diabetic retinopathy during intensified metabolic control was treated with manipulation by administration of octreotide. Octreotide lowered total IGF-1 levels. Macular edema resolved partly and visual acuity improved. Intensified insulin therapy in poorly controlled type 1 patients is able to cause florid diabetic retinopathy with acute macular edema. These changes may improve by administration of octreotide by downregulation of IGF-1.

Octreotide was under investigation in two phase III (802 and 804) multicenter, double-masked, randomized, placebo-controlled trials initiated by Novartis in 1999. Included were type 1 and 2 diabetic patients with moderate to severe nonproliferative DR to early proliferative DR. The patients were treated with the long acting octreotide (Sandostatin LAR, Novartis), which was injected intramuscularly once a month for up to 6 years. In the treatment arms the patients were randomized to either 20 or 30 mg. Primary outcome was DR progression; secondary outcomes include change in visual acuity and macular edema progression. The studies were completed and the results are expected to be published shortly.

19.1.2.2.3.2 Side Effects

Because of its inhibitory action on growth hormone, glucagon and insulin release, Sandostatin LAR affects glucose regulation. Octreotide results in a reduction of the blood glucose level in patients treated with insulin, requiring fewer insulin doses. Therefore insulin dosis has to be reduced by 25–50% in most insulin-treated diabetics with octreotide treatment. Close daily monitoring of blood glucose levels is mandatory under octreotide treatment because of the risk of hypoglycemia [18]. Hypoglycemic episodes occur most often 10–14 days after the first Sandostatin LAR injection if the insulin dose is not reduced. In non-insulin dependent diabetics, Sandostatin LAR may cause either a decrease or increase in blood glucose, depending on the effects of glucagon secretion.

Gastrointestinal side effects are the most common adverse events. Diarrhea and abdominal pain occur in one-third of patients at the beginning of octreotide treatment. They begin with the first injection and can increase during the following two injections but improve in most patients during steady state. Nausea and vomiting are less common side effects. Hypothyroidism and gallstones are rare in prolonged use of Sandostatin LAR.

Local injection site reactions to Sandostatin LAR are rare and of short duration, including pain, swelling or rash.

19.1.2.2.3.3 Treatment Recommendations

The results of clinical use of octreotide have been variable. The most favorable results have been described in advanced stages of DR and high dosage regimens.

Personal observations with compassionate use have shown that especially patients with persistent retinal neovascularization or vitreous hemorrhage

after panretinal laser photocoagulation or postvitrectomy respond well to octreotide treatment. Patients with iris neovascularization also benefit from octreotide treatment (Fig. 19.1.2.2.3). In patients pretreated with octreotide, surgical procedures seem to be easier to perform concerning removal of preretinal membranes that can be peeled off more easily and seem to be more fragile. This experience suggests that especially the neovascular stages respond well to octreotide treatment. The reason might be that the late stages of DR are most likely IGF-1 driven.

About 30% of patients have mild to moderate diarrhea and gastrointestinal side effects at the beginning of treatment. The side effects usually resolve within 3 months. Pancreatic enzymes can be administered to reduce side effects. Rarely patients withdraw from treatment due to severe diarrhea. Gallbladder stones and thyroid problems are rare. Sandostatin LAR 20 mg or preferentially 30 mg once a month can be used for the treatment of DR. To reduce gastrointestinal side effects lower doses can be used at the beginning of treatment.

Sufficient evidence for GH and IGF-1 involvement in the progression of DR supports the potential utility of octreotide in the treatment of DR. However, not

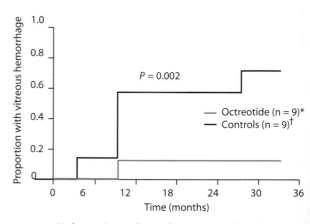

*1 dense vitreous hemorrhage not requiring vitrectomy
†5 dense vitreous hemorrhages, 3 requiring surgery

Fig. 19.1.2.2.3. Risk of vitreous hemorrhage and vitreoretinal surgery was significantly reduced in patients with octreotide treatment [3]

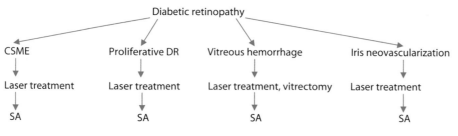

Fig. 19.1.2.2.4. Decision making in the treatment with somatostatin analogues. *CSME* clinically significant macular edema, *DR* proliferative diabetic retinopathy, *SA* somatostatin analogues (off label)

all patients respond to octreotide treatment. There are unfortunately no hard criteria to identify who benefits from octreotide treatment. In patients with treatment failure somatostatin analogues may have inadequate penetration of the blood-retina barrier after intramuscular administration. Intravitreal administration might be more effective; however, Sandostatin LAR formulation is not suitable for intravitreal injection.

19.1.2.2.3.4 Follow-up and Monitoring

Sandostatin LAR is not approved for treatment of diabetic retinopathy. However, off-label treatment of diabetic retinopathy can be considered after failure of the gold standard treatment regimen, i.e., laser treatment and vitrectomy. This is in either patients with proliferative diabetic retinopathy or iris neovascularization or those with diabetic macular edema. Ophthalmological follow-up is recommended at 3-month intervals. Sandostatin LAR administration should be monitored by a diabetologist or endocrinologist because especially in insulin-dependent patients severe hypoglycemic episodes are possible. Therefore insulin dosis has to be adjusted appropriately. Other possible side effects like gallstones and hypothyroidism also have to be monitored by regular ultrasound and thyroid status checks. It is mandatory that patients are compliant and closely self-control their blood glucose levels.

References

1. Baldysiak-Figiel A, Lang GK, Kampmeier J, Lang GE (2004) Octreotide prevents growth factor-induced proliferation of bovine retinal endothelial cells under hypoxia. J Endocrinol 180:417–424

2. Böhm BO, Feldmann B, Lang GK, Lang GE (1999) Treatment of diabetic retinopathy with long-acting somatostatin analogues. In: Lamberts SWJ (ed) Octreotide: The next decade. BioScientifica, Bristol, pp 241–257

3. Böhm BO, Lang GK, Jehle PM, Feldmann B, Lang GE (2001) Octreotide reduces vitreous hemorrhage and loss of visual acuity risk in patients with high-risk proliferative diabetic retinopathy. Horm Metab Res 33:300–306

4. Böhm BO, Lang GE, Volpert O, Jehle PM, Kurkhaus A, Rosinger S, Lang GK, Bouck N (2003) Low content of the natural ocular anti-angiogenic agent pigment epithelium-derived factor (PEDF) in aqueous humor predicts progression of diabetic retinopathy. Diabetologica 46:394–400

5. Chantelau E, Frystyk J (2005) Progression of diabetic retinopathy during improved metabolic control may be treated with reduced insulin dosage and/or somatostatin analogue administration – a case report. Growth Horm IGF Res 15:130–135

6. Chantelau E, Kohner EM (1997) Why some cases of retinopathy worsen when diabetic control improves. Br Med J 315:1105–1106

7. Fauser S, Krohne T, Kirchhof B, Joussen AM (2003) Die diabetische Makulopathie-Klinik und Therapie. Klin Monatsbl Augenheilkd 220:526–531

8. Grant MB, Caballerso S Jr (2005) The potential role of octreotide in the treatment of diabetic retinopathy. Treat Endocrinol 4:199–203

9. Grant MB, Caballero S, Millard W (1993) Inhibition of IGF-1 and b-FGF stimulated growth of human retinal endothelial cells by the somatostatin analogue, octreotide: a potential treatment for ocular neovascularization. Regulatory Peptides 48:267–278

10. Grant MB, Mames RN, Fitzgerald C, Hahariwala KM, Cooper-DeHoff R, Caballero S, Estes KS (2000) The efficacy of octreotide in the therapy of severe nonproliferative and early proliferative diabetic retinopathy. Diabetes Care 23:504–509

11. Grant MB, Caballero S, Smith LEH (2002) Somatostatin in diabetic eye diseases. In: Lamberts SWJ, Ghigo E (eds) The expanding role of octreotide II: Advances in endocrinology and eye diseases. BioScientifica, Bristol, pp 165–184

12. Hernaez-Ortega MC, Soto-Pedre E, Martin JJ (2004) Sandostatin LAR for cystoid diabetic macular edema: a 1-year experience. Diabetes Res Clin Pract 64:71–72

13. Hernandez C, Carrasco E, Casamitjana R, Deulofeu R, Garcia-Arumi J, Simo R (2005) Somatostatin molecular variants in the vitreous fluid: a comparative study between diabetic patients with proliferative diabetic retinopathy and nondiabetic control subjects. Diabetes Care 28:1941–1947

14. Joussen AM, Fauser S, Krohne TU, Lemmen K-D, Lang GE, Kirchhof B (2003) Diabetische Retinopathie: Pathophysiologie und Therapie einer hypoxieinduzierten Entzündung. Ophthalmologe 100:363–370

15. Koller EA, Green I, Gertner JM, Bost M, Malozowski SN (1998) Retinal changes mimicking diabetic retinopathy in two nondiabetic, growth hormone-treated patients. J Clin Endocrinol Metabol 83:2380–2383

16. Kuijpers RWAM (1998) Treatment of cystoid macular edema with octreotide. N Engl J Med 338:624–626

17. Lang GE (2004) Therapie der diabetischen Retinopathie mit Somatostatinanaloga. Ophthalmologe 101:290–293

18. Lang GE (2007) Pharmacological treatment of diabetic retinopathy. Ophthalmologica 222 (in press)

19. Mallet B, Vialettes B, Haroche S, Escoffier P, Gastaut P, Taubert JP, Vague P (1992) Stabilization of severe proliferative diabetic retinopathy by long-term treatment with SMS 201–995. Diabete et Metabolisme 18:438–444

20. Poulsen JE (1996) Diabetes and anterior pituitary insufficiency: final course and postpartum study of a diabetic patient with Sheehan's syndrome. Diabetes 15:73–77

21. Sall JW, Klisovic DD, O'Sorisio MS, Katz SE (2004) Somatostatin inhibits IGF-1 mediated induction of VEGF in human retinal pigment epithelial cells. Exp Eye Res 79:465–476

22. Spranger J, Mohlig M, Osterhoff M, Bühnen J, Blum WF, Pfeiffer AFH (2001) Retinal photocoagulation does not influence intraocular levels of IGF-1, IGF- and IGF-BP3 in proliferative diabetic retinopathy – evidence for combined treatment of PDR with somatosatin analogues and retinal photocoagulation. Horm Met Res 33:312–316

23. Spraul CW, Baldysiak-Figiel A, Lang GK, Lang GE (2002) Octreotide inhibits growth factor-induced bovine choriocapillary endothelial cells in vitro. Graefes Arch Clin Exp Ophthalmol 240:227–231

19.2   Proliferative Diabetic Retinopathy

19.2.1 A Surgical Approach to Proliferative Diabetic Retinopathy

H. Helbig

Core Messages

- Vitreous surgery for proliferative diabetic retinopathy can clear media opacities and relieve traction on the retina
- Vitrectomy can posititvely influence the development of the retinopathy, since the schaffold for fibrovascular proliferations is removed and possible by improving oxygen supply to the inner retina
- Laser treatment of the retina is essential to reduce the risk for complications of surgery such as iris rubeosis, fibrovascular reproliferations and rebleeding
- Functional outcome of surgery depends on the primary microvascular disease and the degree of retinal ischemia
- Cataract surgery can have a negative influence on diabetic macular edema
- Neovascular glaucoma requires aggressive surgical intervention with intense coagulation treatment, nevertheless the prognosis is guarded

19.2.1.1 Rationale for Surgery in Diabetic Retinopathy

Diabetic retinopathy is a disease of the retinal microvasculature. Surgery does not treat the primary disease and is not a causative therapeutic approach; it can only address secondary complications of the primary microvascular problem. Complications of diabetic retinopathy which can be treated surgically and which are targets for surgery include [13, 15, 21, 30, 55]:

- Removing media opacities, especially vitreous hemorrhage
- Relieving traction on the retina by active or atrophic fibrovascular membranes
- Reattachment of detached retina, tractional or rhegmatogenic
- Enabling necessary coagulation treatment of the ischemic retina
- Removing vitreous as a scaffold for neovascularizations
- Improving retinal metabolism by facilitating diffusion of oxygen, nutrients and growth factors from the vitreous cavity to the retina and vice versa

Since the primary microvascular disease is not addressed by surgery, functional results of surgery in diabetic retinopathy may be limited due to ischemic retinal damage.

19.2.1.2 Indications for Surgery in Diabetic Retinopathy: A Practical Approach

19.2.1.2.1 Does the Literature Help?

Many studies have been published on the results of vitreous surgery for complications of diabetic retinopathy. Evidence based medicine differentiates various levels of evidence for a therapeutic procedure. The highest level of evidence is provided by randomized prospective trials. The Vitrectomy for Diabetic Retinopathy Study (VDRS) was such a randomized prospective study comparing "early" vitreous surgery with observation. The DRVS studied two groups: vitreous hemorrhage and severe proliferative diabetic retinopathy with useful vision. For eyes with vitreous hemorrhage this study only showed a benefit for surgery in younger diabetics [10, 12]. For eyes with progressive proliferative diabetic retinopathy a positive effect of surgery was detected for eyes with large active neovascularizations [11]. These results, however, have to be interpreted with caution. Randomized trials usually follow a strict protocol and run a long time before results become available. They are usually not adapted to changing technical developments and improved understanding of the pathophysiology. They may therefore give an answer to a question that is no longer relevant.

Many other non-randomized studies describe retrospectively the outcome of surgery for diabetic retinopathy. Most studies have subdivided the surgical indications into different groups [9, 58–60]:

- Vitreous hemorrhage
- Tractional detachment of the macula
- Tractional rhegmatogenic retinal detachment
- Severe, progressive proliferative retinopathy

The individual patient often does not fit so easily into a single category. Patients with a vitreous hemorrhage commonly present with circumscribed tractional detachment or active neovascularization. Definition of macular detachment is not uniform. Some authors include eyes with detachment of the peripheral macula close to the major vascular arcades and a full vision in this group [49, 59], others only eyes with completely detached foveas [19].

Moreover, surgical techniques, instrumentation and understanding of the pathophysiology have considerably improved in recent years. These changes include:

- Availability of endolaser [48, 54]
- Extracapsular instead of intracapsular cataract removal [3, 5]
- Wide-angle viewing systems [61]
- Perfluorocarbon liquids [27]
- Peeling of the inner limiting membrane [14]
- Supplementary drug treatment (e.g., intravitreal triamcinolone) [28]

The results of surgery have therefore improved, complications have become less common and indications for surgery have been extended. Older studies are therefore of limited value for the quantitative estimation of the chances and risks of surgery, but they have helped very much in advancing our indications for surgery, improving our understanding of the pathophysiology, and developing intraoperative strategies.

Many factors have to be considered before an individual decision for or against surgery can be made, including status of the fellow eye, general health status, duration of visual loss, past and expected clinical course of the disease, lens status, extent of macular ischemia and last but not least patient preferences. All these factors are difficult to control in studies of a disease with such a wide variation of clinical features. Thus for a patient with diabetic retinopathy, recommendations for surgery are commonly based more on individual weighing of risks and benefits than on high level evidence based knowledge.

19.2.1.2.2 Vitreous Hemorrhage

The first historic pars plana vitrectomy was performed in an eye with diabetic vitreous hemorrhage [34, 35]. Vitreous hemorrhage is a typical indication

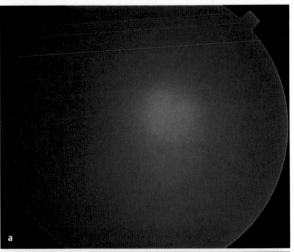

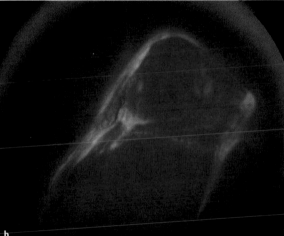

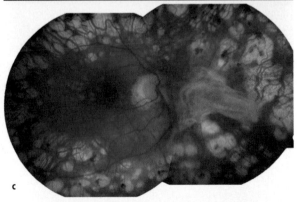

Fig. 19.2.1.1. a Diffuse vitreous hemorrhage; the position of the optic disk can just barely be suspected. Visual acuity was HM. **b** Vitreous and subhyaloidal hemorrhage. Extensive fibrovascular membranes can be seen and tractional detachment of the retina is suspected. Vision is HM. **c** Tractional retinal detachment nasally to the fovea. The membranes are atrophic after scatter laser treatment and the traction does not threaten the fovea. Vision was 20/25 and no further therapy was necessary

19 III

for surgery. However, the individual decision for surgery is commonly not as easy. On the one hand, the surgical risk in a "simple" vitreous hemorrhage is low. On the other hand, the blood may clear spontaneously. As long as there is no retinal detachment, iris rubeosis or macular edema, no irreversible damage will occur when surgery is postponed. Therefore it is not possible to give a clear general recommendation of how long to wait for surgery. In a very dense or recurrent hemorrhage it may be prudent to operate within a few days. Other patients may benefit from waiting many months for the hemorrhage to clear. It is left to the surgeon's and patient's combined judgement to estimate which hemorrhage may clear rapidly enough to wait.

Vitreous hemorrhage occurs if neovascularizations tear. This commonly happens in eyes which undergo partial vitreous detachment or contraction of the fibrovascular membranes. These vitreous changes may occur spontaneously but may be accelerated after scatter laser treatment of the retina. Retinal photocoagulation induces fibrotic transformation, contraction of neovascular membranes, and partial vitreous detachment. Thus, scatter photocoagulation may eventually trigger vitreous hemorrhage shortly after treatment before neovascularizations become atrophic.

In eyes with known traction on the central retina or evidence of traction in ultrasound echography [6], delaying surgery may cause irreversible damage to the macula. In this situation surgery should not be delayed. Figure 19.2.1.1a shows an example of a vitreous hemorrhage which can be observed. The hemorrhage is not very dense and no traction on the retina is present. In contrast, in Figure 19.2.1.1b tractional membranes are visible and surgery is recommended. Iris neovascularizations are an indirect sign of severe retinal ischemia requiring retinal coagulation treatment. Eyes with iris rubeosis and vitreous hemorrhage therefore need vitreous surgery to allow adequate photocoagulation, before irreversible obstruction of the chamber angle occurs.

19.2.1.2.3 Extrafoveal Tractional Detachment

Not all eyes with extrafoveolar tractional retinal detachment need surgery. In many cases the situation may remain stable for many years and a detached retina may even spontaneously reattach. Risk factors for visual loss have been the presence of active neovascularizations, progression of the detachment and the occurrence of vitreous hemorrhage [8].

Before advanced vitreous surgery was available, retinal photocoagulation was considered to be dan-

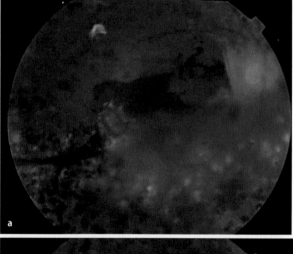

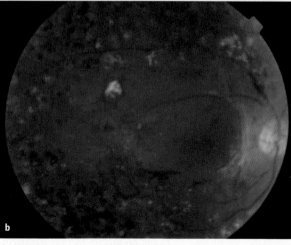

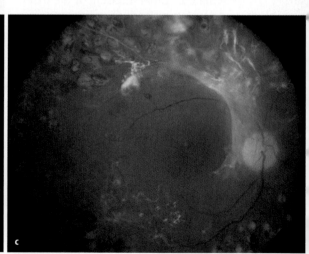

Fig. 19.2.1.2. a Severe proliferative diabetic retinopathy with subhyaloidal hemorrhage and fibrovascular membranes. Visual acuity was 20/200. **b** Three months after supplementary scatter laser treatment hemorrhage has resolved, but active neovascularizations are still present, and visual acuity is 20/60. **c** One year later, after additional scatter laser treatment, the fibrovascular membranes became atrophic and vision remained stable at 20/60

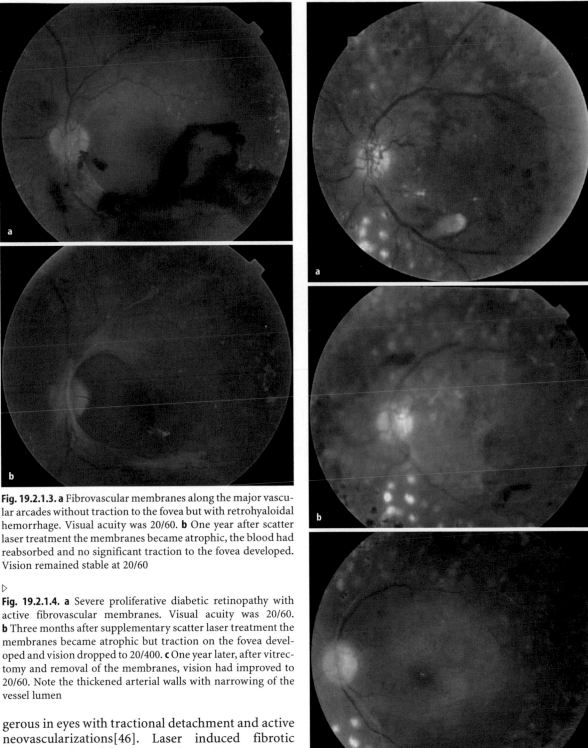

Fig. 19.2.1.3. a Fibrovascular membranes along the major vascular arcades without traction to the fovea but with retrohyaloidal hemorrhage. Visual acuity was 20/60. **b** One year after scatter laser treatment the membranes became atrophic, the blood had reabsorbed and no significant traction to the fovea developed. Vision remained stable at 20/60

▷

Fig. 19.2.1.4. a Severe proliferative diabetic retinopathy with active fibrovascular membranes. Visual acuity was 20/60. **b** Three months after supplementary scatter laser treatment the membranes became atrophic but traction on the fovea developed and vision dropped to 20/400. **c** One year later, after vitrectomy and removal of the membranes, vision had improved to 20/60. Note the thickened arterial walls with narrowing of the vessel lumen

gerous in eyes with tractional detachment and active neovascularizations[46]. Laser induced fibrotic transformation of the neovascularizations may possibly increase traction on the retina and the tractional detachment may progress. Without surgery the visual prognosis of these eyes was poor. Today we recommend performing scatter treatment to induce atrophic regression of the new vessels. It is important, however, in this situation that eyes with extrafo-

veal tractional detachment must be carefully watched after laser treatment. If the detachment progresses and the fovea is threatened, immediate surgery should be performed. With a detached fovea the

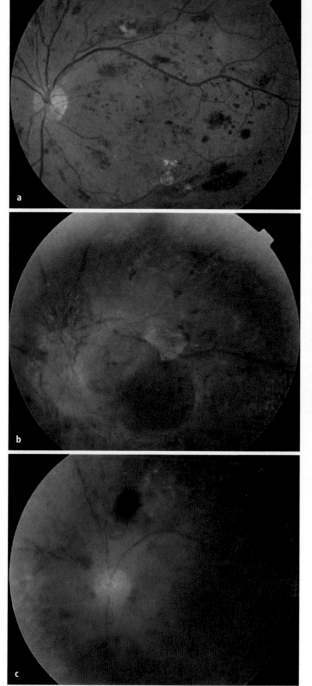

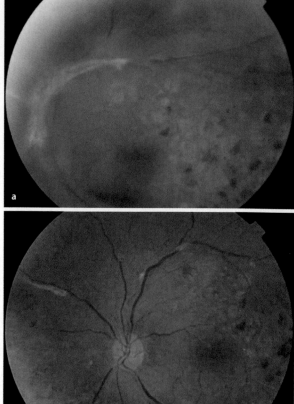

Fig. 19.2.1.6. a Fibrovascular membranes along the major vascular arcades with traction to the fovea and with vitreous hemorrhage despite dense scatter laser treatment. Visual acuity was 20/400. **b** One year after vitrectomy and removal of the membranes the eye became stable and vision improved to 20/40

Fig. 19.2.1.5. a Severe non-proliferative diabetic retinopathy, vision 20/30. **b** One year later severe progressive proliferative diabetic retinopathy with traction to the macula and vitreous hemorrhage had developed despite scatter laser treatment. Vision had dropped to 20/400. **c** Three weeks after vitrectomy, endolaser and removal of the membranes, the retina is flat. Vision later improved to 20/60. **c** One year later, slow resolution of exudates and edema was observed, and vision had improved to 20/80.

visual acuity drops and even after anatomically successful surgery the visual prognosis is guarded in eyes with diabetic retinopathy [17]. On the other hand, many eyes may be stabilized by laser treatment alone, not requiring surgery. Even if surgery may later become necessary, laser induces partial transformation of active new vessels into atrophic fibrotic membranes and partial vitreous detachment. This will make surgery technically easier and may reduce the risk of postoperative hemorrhages.

Figure 19.2.1.1c shows an example of a tractional retinal detachment nasally to the fovea, not threatening the fovea. Scatter laser treatment has already been performed, the membrane is atrophic and the situation appears to be stable without further therapy. Figures 19.2.1.2 and 19.2.1.3 show the course of the disease in eyes where active fibrovascular membranes without foveal traction could be stabilized

III 19

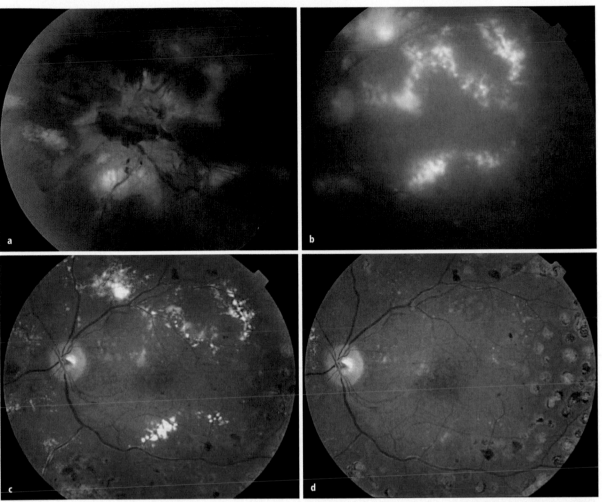

Fig. 19.2.1.7. a Vitreous and subhyaloidal hemorrhage in severe proliferative diabetic retinopathy with active fibrovascular membranes. Visual acuity was HM. **b** Two weeks after vitreous surgery and scatter laser treatment extensive hard exudates in the macula had become visible, and vision had improved to 20/200. **c** One year later, slow resolution of exudates and edema was observed, and vision had improved to 20/80. **d** Two years later, exudates had completely reabsorbed and vision was 20/40

with scatter laser treatment. In the eyes shown in Figs. 19.2.1.4 – 19.2.1.6, laser treatment did not stabilize the retinopathy and traction on the fovea developed. In these cases vitreous surgery was necessary to stabilize vision.

In some cases traction on the fovea may cause tractive macular edema, even if the fovea itself is not detached. Focal laser treatment is not helpful and not recommended in this situation. The best therapy for tractional edema is to remove the traction surgically to allow resolution of the macula edema. Figure 19.2.1.7 shows the slow reabsorption of tractional macular edema after surgical removal of the membranes.

19.2.1.2.4 Tractional Detachment of the Fovea

If the fovea itself is tractively elevated, vision is usually very poor. For this situation surgery is the only

therapeutic option. However, even after anatomically successful surgery, functional results are usually disappointing and much worse than after surgery for a non-diabetic rhegmatogenous retinal detachment. Often, eyes with tractional detachment of the macula have advanced macular ischemia, which is limiting visual recovery. Risk factors for poor visual outcome are extension of the retinal detachment, duration of macular detachment and iris rubeosis (indicating severe retinal ischemia) [17]. In eyes with this combination of risk factors, we have to consider not performing any surgery, especially if the fellow eye is good and general health is poor. Patients with advanced diabetic eye disease usually suffer from other severe manifestations of the microvascular and macrovascular diabetic complications. Extended and repeated surgery may be a serious stress factor for patients with impaired general health. Life expec-

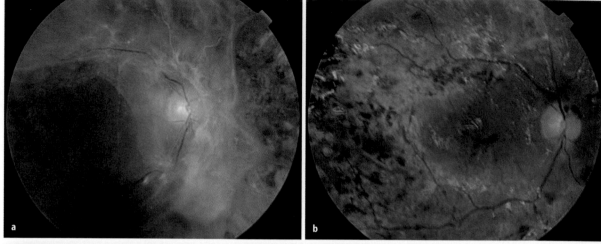

Fig. 19.2.1.8. a Long-standing diabetic tractional retinal detachment of the macula by atrophic fibrotic membranes. Visual acuity was LP. **b** After vitrectomy and silicon tamponade the retina is attached but vision is still LP

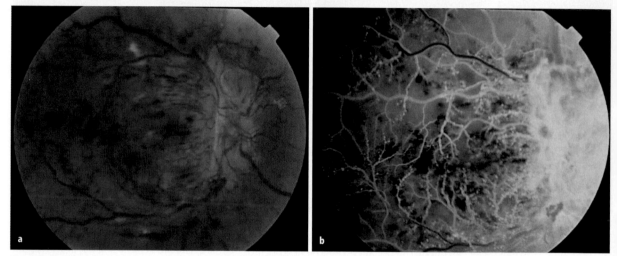

Fig. 19.2.1.9. a Diabetic tractional retinal detachment, visual acuity CF. **b** Fluorescein angiography showing severe diffuse retinal ischemia with largely occluded capillary bed

tancy is also limited with advanced diabetic complications, and is not much better than in patients with malignomas [18]. These non-ocular factors have to be considered if a decision to perform surgery is made in a situation with a reduced prognosis *quo ad visum*. If both eyes have reduced vision, surgical attempts will nevertheless have to be performed despite the poor prognosis. Figure 19.2.1.8 shows a patient with long-standing tractional detachment of the macula in his only eye. After vision had dropped to LP, surgery was performed but vision did not improve despite anatomically successful surgery. Figure 19.2.1.9 shows the fundus of an eye with recent tractional detachment of the fovea and a vision of HM. The fellow eye had good vision. Fluorescein angiography showed severe retinal ischemia. Therefore no surgery was recommended.

19.2.1.2.5 Tractional Rhegmatogenous Retinal Detachment

Diabetic fibrovascular membranes may not only elevate the retina, but in rare cases also create tears in the retina. A rhegmatogenous retinal detachment may develop. Clinically these tears may be difficult to identify, because they can be small and hidden behind membranes or blood. On the other hand, circumscribed thinning of the retina, e.g., in areas of former photocoagulation, can mimic retinal tears. For the diagnosis of a rhegmatogenous detachment one should look for a bullous and mobile retinal detachment, in contrast to a concave, immobile, tent-like shape of a tractive detachment. Diabetic tractional-rhegmatogenous detachment in most cases progresses rapidly and therefore has to be operated

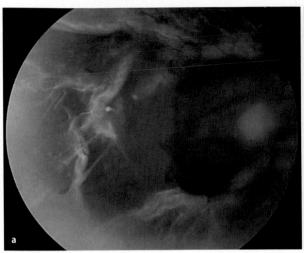

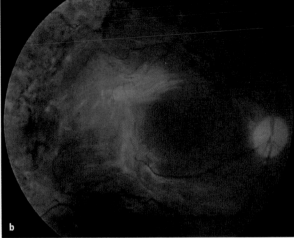

III 19

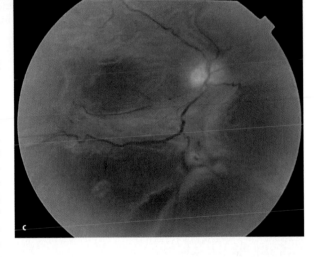

Fig. 19.2.1.10. a Tractional rhegmatogenous retinal detachment, visual acuity HM. **b** After vitrectomy with gas tamponade and endolaser, the retina is attached and vision improved to 20/100. Eight weeks after surgery the patient noted metamorphopsia. Proliferative vitreoretinopathy with epiretinal membranes and wrinkling of the retina was observed. **c** Two weeks later the retina was completely detached

on immediately like other rhegmatogenous detachments. Technically it may be difficult to remove all tractive tissue tightly adhering to the mobile detached retina. Therefore the rate of intraoperatively created holes was relatively high [58]. Reproliferations after surgery are unfortunately not too rare and constitute a combination of diabetic fibrovascular membranes and the typical proliferative vitreoretinopathy (PVR) after rhegmatogenous retinal detachment. Figure 19.2.1.10 shows such an eye with tractional rhegmatogenous retinal detachment which developed severe reproliferations 2 months after primarily successful surgery.

19.2.1.2.6 Macular Edema

Treatment of macular edema is described in detail in the next chapter. Here only tractional macular edema is mentioned. Traction by atrophic fibrovascular membranes may present with distortion of the central retina, creating a picture similar to macular pucker. A typical example is shown in Fig. 19.2.1.11. Traction of fibrovascular membranes may also cause a type of macular edema with typical hard exudates. Tractional edema has a good chance to resolve after vitrectomy (see Fig. 19.2.1.7). Visual recovery is dependent on the degree of macular ischemia.

19.2.1.2.7 Surgery for Neovascular Glaucoma

Neovascular glaucoma is a severe complication of advanced proliferative diabetic retinopathy. Patho-

physiologically it is assumed that the ischemic retina is the source of production of vasoproliferative growth factors, which can diffuse to the anterior segment of the eye, especially in aphakic and vitrectomized eyes. Growth of fibrovascular membranes in the chamber angle obstructs aqueous outflow and intraocular pressure rises, often to very high levels. Therapy should be directed first to the cause of the neovascular stimulus. Extensive photocoagulation of the retina or cryotherapy to the peripheral retina should be applied [47]. This treatment can induce regression of neovascularizations; however, intraocular pressure commonly remains elevated because the vascular membranes in the chamber angle do not completely disappear. They may undergo fibrotic transformation and still obstruct the aqueous outflow. Therefore treatment to lower intraocular pressure is usually necessary. Conventional glaucoma surgery has poor success rates in these eyes. They have a disrupted blood aqueous barrier and intraocular fluids contain high concentrations of cytokines stimulating fibrosis. Rapid obstruction of the trabe-

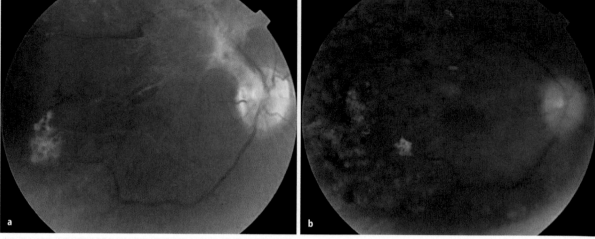

Fig. 19.2.1.11. a Atrophic diabetic fibrovascular membranes with traction on the fovea creating a lamellar hole. Visual acuity was 20/400. **b** Three months after peeling of the membrane, vision had improved to 20/60

culectomy outflow is therefore common even with the use of antimetabolites. Glaucoma drainage devices have been successfully implanted [24], but the rate of complications is not low with this procedure. Many clinicians therefore use cyclodestructive measures to lower intraocular pressure [45]. Either transscleral cryotherapy or transscleral laser treatment to the ciliary body is used. During vitreoretinal or cataract surgery direct endolaser treatment to the ciliary processes can be applied to reduce aqueous humor production [2].

19.2.1.2.8 Cataract Surgery

Cataract surgery in eyes with proliferative diabetic retinopathy has two goals. On the one hand, removing the opaque lens will improve vision for the patient. On the other hand, it enables the ophthalmologist to adequately view the retina and to perform photocoagulation to the retina if necessary. Unfortunately cataract surgery may have a negative effect on the development of the retinopathy. Preretinal and iris neovascularizations may be stimulated to develop after cataract surgery [5]. The most common problem, however, is worsening of macular edema, which is more common in diabetics even without visible retinopathy [41]. The mechanism is not quite clear yet, but the surgically induced production of inflammatory mediators and the facilitated diffusion of cytokines between the anterior and the posterior segment after removal of the lens may play a role.

19.2.1.3 Surgical Principles

19.2.1.3.1 Complete Removal of the Vitreous

One of the primary goals of surgery is to relieve all traction on the retina, to reattach the retina, and to improve retinal function. Complete removal of vitreous and fibrovascular tissue is also important for the subsequent course of the disease. Eyes with a completely detached vitreous rarely develop posterior neovascularizations. New vessels do not develop if they have no substrate to grow on. Especially a partially attached vitreous provides such a scaffold for the outgrowth of fibrovascular membranes. Vitrectomy removes this substrate for fibrovascular tissue. Remnants of attached vitreous or epiretinal fibrovascular membranes after vitrectomy form an excellent substrate for reproliferations. It is therefore necessary to remove all epiretinal tissue as completely as possible.

Diabetic fibrovascular membranes behave differently from epiretinal membranes in macular pucker requiring a different surgical approach. Pucker membranes grow flat on the surface of the retina without invasive or infiltrative ingrowth into the retina. These membranes can be peeled off in a single piece, once a free edge of the membrane can be grasped with an intraocular forceps. It is not necessary to cut adhesions to the retina with sharp instruments. Diabetic fibrovascular membranes grow out from retinal vessels forming tight connections to the retina. Attempts to peel these membranes without cutting the adhesions with sharp instruments may create retinal tears. Various techniques have been described for this purpose. Either the membrane is cut into pieces with vertically cutting scissors between the sites of adhesion to the retina ("segmentation" technique [43]). The remaining stumps are trimmed with the vitreous cutter. Other approaches involve severing the membrane from the retina in a single piece by cutting the adhesions with horizontally cutting scissors ("delamination" [42] or "en bloc" [40] technique).

Diabetic neovascular membranes primarily grow from retinal vessels. To remove them, a vessel has to be cut or torn off. This creates an opening of the vessel lumen and a potential source for a vitreous hemorrhage. Visible bleeding stumps may be coagulated intraoperatively using endodiathermia depending on the localization. If the source of hemorrhage is a large vessel or the optic disk, this may not be possible. The bleeding often stops spontaneously or after increasing the infusion pressure. These tiny tears in retinal vessels, however, may cause postoperative hemorrhages in the early postoperative phase.

19.2.1.3.2 Application of Endolaser Whenever Possible

One of the most important steps in diabetic vitreous surgery is coagulation treatment of the retina [13, 15, 30, 55]. Complications requiring vitreous surgery are almost always an indicator for an active and unstable retinopathy. Therefore almost all eyes undergoing vitreous surgery require endolaser treatment of the ischemic retina. Moreover, the surgery itself may even worsen some aspects of the retinopathy. Preretinal neovascularizations participate in oxygen and nutrient supply to the retina; this is the biologic function of the new vessels. If active neovascularizations are removed from the retinal surface, this may further reduce the vascular supply to the inner retina. Thus, vitrectomy may even worsen retinal ischemia and stimulate the production of vasoproliferative growth factors. Moreover, after removal of the vitreous, these growth factors may more easily reach the anterior segment [56]. Progression of iris rubeosis is a severe complication of vitreous surgery [51]. This complication, however, may be significantly reduced by intraoperative endolaser treatment to the retina [20].

19.2.1.3.3 Check the Periphery with Wide-Angle Viewing System for Retinal Tears

Preparation of tightly adhering fibrovascular membranes from the retina is often time consuming and requires repeated exchange of instruments in the eye. Moreover, the vitreous is commonly partly attached and may be sticky in diabetic eyes, possibly because it contains blood and fibrin due to a breakdown of the blood-retina barrier. In diabetic vitrec-

tomies iatrogenic retinal holes in the vitreous base may therefore occur more often than with other indications for vitrectomy, leading to retinal detachment if not treated properly [51]. Careful inspection of the retinal periphery, preferably with a wide-angle viewing system [61] under indentation, is therefore mandatory.

19.2.1.3.4 Cataract Surgery

If the lens has to be operated on, intracapsular surgery should be avoided, to leave a barrier between the anterior and the posterior segment intact. Intracapsular surgery is associated with an increased risk for neovascular glaucoma [1]. If the capsule of the lens is left in place in extracapsular cataract surgery, diabetic eyes have an increased risk of postoperative fibrin exudation and formation of synechia between the lens capsule and the iris due to an impaired blood-retinal barrier [16, 23, 53]. There is no absolute contraindication for the implantation of an intraocular lens (IOL) in these eyes [3, 4]. Even in eyes with iris rubeosis, an IOL may be implanted [31]; extensive coagulation treatment of the retina, however, is mandatory in these cases [25, 26]. The IOL in diabetic eyes should be implanted safely into the capsular bag and a large opening of the anterior lens capsule should be provided to improve visualization of the fundus. Silicone lenses should be avoided. If later in the course of the disease liquid silicone has to be installed in the vitreous cavity, the liquid silicone may form tight adhesions to the silicone IOL, forming droplets on the IOL which are optically very disturbing, commonly requiring IOL exchange [62]. Combined vitreoretinal and cataract surgery can be successfully performed [32, 52]. The inflammatory response with fibrin in the anterior chamber and the formation of synechia, however, is probably lower in a two step procedure [53]. We therefore prefer to perform cataract surgery in the same procedure as the vitreoretinal procedure only if lens opacities are disturbing posterior segment visualization. Since progression of macular edema is a common complication after cataract surgery, the injection of intravitreal triamcinolone at the end of the surgery may be considered [28, 29, 50]. This supplementary treatment may avoid worsening and even improve macular edema [57] after cataract surgery.

19.2.1.3.5 Silicone

Introduction of liquid silicone has expanded our armamentarium for vitreous surgery in diabetic retinopathy [33, 37–39]. Retinal holes may be created during preparation of diabetic fibrovascular membranes, especially if the tractional detachment is long-standing and the retina itself has become thin and atrophic. In these eyes instillation of liquid silicone may be used as a tamponade for the retinal defects avoiding rhegmatogenous retinal detachment after vitreous surgery. Silicone must not be used in eyes with tractional retinal detachment without completely removing the tractional membranes [44]. Since water soluble growth factors concentrate in high levels in the thin interface between retina and silicone, a strong stimulus for proliferation of fibrovascular membranes is present leading to reproliferations and retinal redetachment under the silicone. This can be avoided in many cases by complete removal of all fibrovascular tissues before instillation of silicone.

Another indication for liquid silicone is recurrent vitreous hemorrhage after vitrectomy. Since silicone fills the vitreous cavity, a clear optic axis is provided. Visual rehabilitation is improved in these eyes by liquid silicone [22].

A third indication for liquid silicone can be severe anterior segment neovascularization [7, 20, 36]. Filling the vitreous cavity with liquid silicone provides a barrier for diffusion from the posterior to the anterior segment of the eye. In eyes with severe neovascularization of the anterior segment, silicone may reduce the diffusion of vasoproliferative growth factors from the ischemic retina to the iris. Together with retinal coagulation therapy, silicone is a treatment option for severe anterior segment neovascularization.

References

1. Aiello LM, Wand M, Liang G (1983) Neovascular glaucoma and vitreous hemorrhage following cataract surgery in patients with diabetes mellitus. Ophthalmology 90:814–820
2. Bartz-Schmidt KU, Thumann G, Psichias A, Krieglstein GK, Heimann K (1999) Pars plana vitrectomy, endolaser coagulation of the retina and the ciliary body combined with silicone oil endotamponade in the treatment of uncontrolled neovascular glaucoma. Graefes Arch Clin Exp Ophthalmol 237:969–975
3. Benson GT, Flynn HW, Blankenship GW (1989) Posterior chamber intraocular lens implantation during diabetic pars plana vitrectomy. Ophthalmology 96:603–610
4. Benson WE, Brown GC, Tasman W, McNamara JA, Vander JF (1993) Extracapsular cataract extraction with placement of a posterior chamber lens in patients with diabetic retinopathy. Ophthalmology 100:730–738
5. Blankenship GW (1980) The lens influence on diabetic vitrectomy results. Report of a prospective randomized study. Arch Ophthalmol 98:2196–2198
6. Capeans C, Santos L, Tourino R, Otero JL, Gomez-Ulla F, Sanchez-Solario M (1997) Ocular echography in the prognosis of vitreous hemorrhage in type II diabetes mellitus. Int Ophthalmol 21:269–275
7. Castellarin A, Grigorian R, Bhagat N, Del Priore L, Zarbin MA (2003) Vitrectomy with silicone oil infusion in severe diabetic retinopathy. Br J Ophthalmol 87:318–321
8. Charles S, Flinn CE (1981) The natural history of diabetic extramacular traction retinal detachment. Arch Ophthalmol 99:66–68
9. de Bustros S, Thompson JT, Michels RG, Rice TA (1987) Vitrectomy for progressive proliferative diabetic retinopathy. Arch Ophthalmol 105:196–199
10. Diabetic Retinopathy Vitrectomy Study Group (1985) Early vitrectomy for severe vitreous hemorrhage in diabetic retinopathy. Two-year results of a randomized trial. Diabetic Retinopathy Vitrectomy Study report 2. The Diabetic Retinopathy Vitrectomy Study Research Group. Arch Ophthalmol 103:1644–1652
11. Diabetic Retinopathy Vitrectomy Study Group (1988) Early vitrectomy for severe proliferative diabetic retinopathy in eyes with useful vision. Clinical application of results of a randomized trial – Diabetic Retinopathy Vitrectomy Study Report 4. The Diabetic Retinopathy Vitrectomy Study Research Group. Ophthalmology 95:1321–1334
12. Diabetic Retinopathy Vitrectomy Study Group (1990) Early vitrectomy for severe vitreous hemorrhage in diabetic retinopathy. Four-year results of a randomized trial: Diabetic Retinopathy Vitrectomy Study Report 5. Arch Ophthalmol 108:958–964
13. Gandorfer A, Kampik A (2000) Pars plana vitrectomy in diabetic retinopathy. From pathogenetic principle to surgical strategy. Ophthalmologe 97:325–330
14. Gandorfer A, Messmer EM, Ulbig MW, Kampik A (2000) Resolution of diabetic macular edema after surgical removal of the posterior hyaloid and the inner limiting membrane. Retina 20:126–133
15. Helbig H, Sutter FK (2004) Surgical treatment of diabetic retinopathy. Graefes Arch Clin Exp Ophthalmol 242:704–709
16. Helbig H, Kellner U, Bornfeld N, Foerster MH (1996) Cataract surgery and YAG-laser capsulotomy following vitrectomy for diabetic retinopathy. Ger J Ophthalmol 5:408–414
17. Helbig H, Kellner U, Bornfeld N, Foerster MH (1996) Grenzen und Möglichkeiten der Glaskörperchirurgie bei diabetischer Retinopathie. Ophthalmologe 93:647–654
18. Helbig H, Kellner U, Bornfeld N, Foerster MH (1996) Life expectancy of diabetic patients undergoing vitreous surgery. Br J Ophthalmol 80:640–643
19. Helbig H, Kellner U, Bornfeld N, Foerster MH (1998) Vitrektomie bei diabetischer Retinopathie: Ergebnisse, Risikofaktoren, Komplikationen. Klin Monatsbl Augenheilkd 212:339–342
20. Helbig H, Kellner U, Bornfeld N, Foerster MH (1998) Rubeosis iridis after vitrectomy for diabetic retinopathy. Graefes Arch Clin Exp Ophthalmol 236:730–733
21. Ho T, Smiddy WE, Flynn HW, Jr (1992) Vitrectomy in the management of diabetic eye disease. Surv Ophthalmol 37:190–202
22. Hoerauf H, Roider J, Bopp S, Lucke K, Laqua H (1995) Endotamponade with silicon oil in severe proliferative retinopathy with attached retina. Ophthalmologe 92:657–662
23. Honjo M, Ogura Y (1998) Surgical results of pars plana vitrectomy combined with phacoemulsification and intraocular lens implantation for complications of proliferative diabetic retinopathy. Ophthalmic Surg Lasers 29:99–105
24. Honrubia FM, Gomez ML, Hernandez A, Grijalbo MP (1984) Long-term results of silicone tube in filtering sur-

gery for eyes with neovascular glaucoma. Am J Ophthalmol 97:501–504

25. Hykin PG, Gregson RM, Hamilton AM (1992) Extracapsular cataract extraction in diabetics with rubeosis iridis. Eye 6:296–299

26. Hykin PG, Gregson RM, Stevens JD, Hamilton PA (1993) Extracapsular cataract extraction in proliferative diabetic retinopathy. Ophthalmology 100:394–399

27. Imamura Y, Minami M, Ueko M, Satoh B, Ikeda T (2003) Use of perfluorocarbon liquid during vitrectomy for severe proliferative diabetic retinopathy. Br J Ophthalmol 87:563–566

28. Jonas JB, Sofker A, Degenring R (2003) Intravitreal triamcinolone acetonide as an additional tool in pars plana vitrectomy for proliferative diabetic retinopathy. Eur J Ophthalmol 13:468–473

29. Jonas JB, Hayler JK, Sofker A, Panda-Jonas S (2001) Intravitreal injection of crystalline cortisone as adjunctive treatment of proliferative diabetic retinopathy. Am J Ophthalmol 131:468–471

30. Joussen A, Llacer H, Mazciewicz J, Kirchhof B (2004) Chirurgische Therapie der diabetischen Retinopathie und Makulopathie. Ophthalmologe 101:1138–1146

31. Kuchle M, Handel A, Naumann GO (1998) Cataract extraction in eyes with diabetic iris neovascularization. Ophthalmic Surg Lasers 29:28–32

32. Lahey JM, Francis RR, Kearney JJ (2003) Combining phacoemulsification with pars plana vitrectomy in patients with proliferative diabetic retinopathy: a series of 223 cases. Ophthalmology 110:1335–1339

33. Lucke K (1993) Silicone oil in surgery of complicated retinal detachment. Ophthalmologe 90:215–238

34. Machemer R (1995) Reminiscences after 25 years of pars plana vitrectomy. Am J Ophthalmol 119:505–510

35. Machemer R, Buettner H, Norton RWD, Parel JM (1971) Vitrectomy: A pars plana approach. Trans Am Acad Ophthalmol Otolaryngol 75:813–820

36. McCuen BW, 2nd, Rinkoff JS (1989) Silicone oil for progressive anterior ocular neovascularization after failed diabetic vitrectomy. Arch Ophthalmol 107:677–682

37. McCuen BW, 2nd, de Juan E, Jr, Landers MB, 3rd, Machemer R (1985) Silicone oil in vitreoretinal surgery, part 2: Results and complications. Retina 5:198–205

38. McCuen BW, 3rd, de Juan E, Jr, Machemer R (1985) Silicone oil in vitreoretinal surgery, part 1: Surgical techniques. Retina 5:189–197

39. McLeod D (1986) Silicone-oil injection during closed microsurgery for diabetic retinal detachment. Graefes Arch Clin Exp Ophthalmol 224:55–59

40. Meier P, Wiedemann P (1997) Vitrectomy for traction macular detachment in diabetic retinopathy. Graefes Arch Clin Exp Ophthalmol 235:569–574

41. Menchini U, Bandello F, Brancato R, Camesasca FL, Galdini M (1993) Cystoid macular oedema after extracapsular cataract extraction and intraocular lens implantation in diabetic patients without retinopathy. Br J Ophthalmol 77:208–211

42. Meredith TA (1997) Epiretinal membrane delamination with a diamond knife. Arch Ophthalmol 115:1598–1599

43. Meredith TA, Kaplan HJ, Aaberg TM (1980) Pars plana vitrectomy techniques for relief of epiretinal traction by membrane segmentation. Am J Ophthalmol 89:408–413

44. Messmer E, Bornfeld N, Oehlschlager U, Heinrich T, Foerster MH, Wessing A (1992) Epiretinal membrane formation after pars plana vitrectomy in proliferative diabetic retinopathy. Klin Monatsbl Augenheilkd 200:267–272

45. Nabili S, Kirkness CM (2004) Trans-scleral diode laser cyclophoto-coagulation in the treatment of diabetic neovascular glaucoma. Eye 18:352–356

46. Packer AJ (1987) Vitrectomy for progressive macular traction associated with proliferative diabetic retinopathy. Arch Ophthalmol 105:1679–1682

47. Pauleikhoff D, Gerke E (1987) Photocoagulation in diabetic rubeosis iridis and neovascular glaucoma. Klin Monatsbl Augenheilkd 190:11–16

48. Peyman GA, Salzano TC, Green JL (1981) Argon endolaser. Arch Ophthalmol 99:2037–2038

49. Rice TA, Michels RG, Rice EF (1983) Vitrectomy for diabetic traction retinal detachment involving the macula. Am J Ophthalmol 95:22–33

50. Sakamoto T, Miyazaki M, Hisatomi T, et al. (2002) Triamcinolone-assisted pars plana vitrectomy improves the surgical procedures and decreases the postoperative blood-ocular barrier breakdown. Graefes Arch Clin Exp Ophthalmol 240:423–429. Epub 2002 Mar 2015

51. Schachat AP, Oyakawa RT, Michels RG, Rice TA (1983) Complications of vitreous surgery for diabetic retinopathy. II. Postoperative complications. Ophthalmology 90:522–530

52. Senn P, Schipper I, Perren B (1995) Combined pars plana vitrectomy, phacoemulsification, and intraocular lens implantation in the capsular bag: a comparison to vitrectomy and subsequent cataract surgery as a two-step procedure. Ophthalmic Surg Lasers 26:420–428

53. Shinoda K, O'Hira A, Ishida S, et al. (2001) Posterior synechia of the iris after combined pars plana vitrectomy, phacoemulsification, and intraocular lens implantation. Jpn J Ophthalmol 45:276–280

54. Smiddy WE (1992) Diode endolaser photocoagulation. Arch Ophthalmol 110:1172–1174

55. Smiddy WE, Flynn HW, Jr (1999) Vitrectomy in the management of diabetic retinopathy. Surv Ophthalmol 43:491–507

56. Stefansson E, Landers MB, 3rd, Wolbarsht ML (1982) Vitrectomy, lensectomy, and ocular oxygenation. Retina 2:159–166

57. Sutter FK, Simpson JM, Gillies MC (2004) Intravitreal triamcinolone for diabetic macular edema that persists after laser treatment: three-month efficacy and safety results of a prospective, randomized, double-masked, placebo-controlled clinical trial. Ophthalmology 111:2044–2049

58. Thompson JT, de Bustros S, Michels RG, Rice TA (1987) Results and prognostic factors in vitrectomy for diabetic traction-rhegmatogenous retinal detachment. Arch Ophthalmol 105:503–507

59. Thompson JT, de Bustros S, Michels RG, Rice TA (1987) Results and prognostic factors in vitrectomy for diabetic traction retinal detachment of the macula. Arch Ophthalmol 105:497–502

60. Thompson JT, de Bustros S, Michels RG, Rice TA, Glaser BM (1986) Results of vitrectomy for proliferative diabetic retinopathy. Ophthalmology 93:1571–1574

61. Virata SR, Kylstra JA (2001) Postoperative complications following vitrectomy for proliferative diabetic retinopathy with sew-on and noncontact wide-angle viewing lenses. Ophthalmic Surg Lasers 32:193–197

62. Weber U, Bullerkotte J (1998) Cataract operation after silicone oil surgery. Oil drop adhesion to silicone lenses. Ophthalmologe 95:219–224

III 19

19.2.2 Laser Coagulation of Proliferative Diabetic Retinopathy

W. SOLIMAN, M. LARSEN

19 III

Core Messages

- Photocoagulation remains the most powerful treatment against proliferative diabetic retinopathy (PDR) and several types of exudative maculopathy
- The objective of photocoagulation treatment in PDR is to arrest and induce regression of neovascularization, and hence to prevent vitre-

ous hemorrhage, traction retinal detachment, and visual loss
- The fundamental mechanism of action of photocoagulation on retinal neovascularization involves tissue destruction through non-invasive application of light that is absorbed in natural chromophores and subsequently converted to heat

19.2.2.1 History of Photocoagulation

It is difficult to determine who first proposed that retinal degeneration or post-traumatic atrophy protects against diabetic retinopathy, but the first person on record who intentionally applied photothermal retinal injury in the treatment of diabetic retinopathy was Gerd Meyer-Schwickerath in 1960 [16]. His first experiments were made using a heliostat, an optical system for the collection of sunlight. A commercial instrument from Carl Zeiss, Inc. with a xenon arc lamp light source replaced this instrument, its beam delivery system being based on the

direct ophthalmoscope and a moving mirror blocking the physician's view of the fundus while the thermal pulse was applied. This bulky instrument with its excessive heat loss was replaced by the laser, which found one of its first successful commercial applications in fundus photocoagulation instruments. Today, the best instruments are integrated into slit-lamp biomicroscopes, allowing fundus viewing during the application of thermal energy while using monochromatic light with well-defined absorption characteristics, the light source producing negligible heat loss and not needing noisy ventilation. Nevertheless, the fundamental characteris-

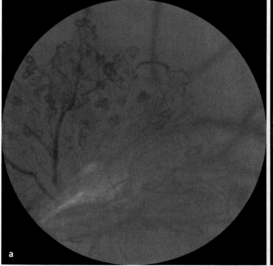
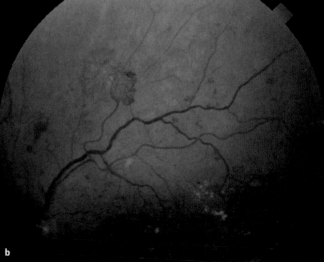

Fig. 19.2.2.1. Preretinal proliferations of new vessels in proliferative diabetic retinopathy

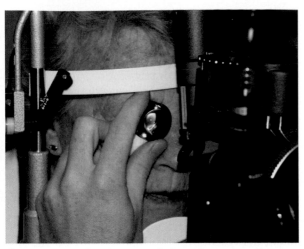

Fig. 19.2.2.2. Fundus photocoagulation using a corneal contact lens

tics of the retinal injury remain essentially the same.

Initially, the concept of photocoagulation treatment for neovascularization included direct thermal coagulation of the new vessels. This approach was abandoned when it was shown that partial ablation of the outer retina was sufficient to achieve the full therapeutic potential and because the energy needed to coagulate preretinal new vessels is high enough to cause damage to blood vessels and nerve fibers in the inner layers of the underlying neurosensory retina. Thus, it emerged that a good photocoagulation lesion for the treatment of proliferative diabetic retinopathy is one that leads to damage of only the outer retina, that is the retinal pigment epithelium and the photoreceptor layer. A good distribution of photocoagulation lesions is one that spares central vision and leaves a contiguous network of intact peripheral retina while ablating a sufficiently large part of the outer retina to eliminate the formation of the hypoxia-induced messenger molecules that drive neovascularization [4]. While it is doubtful that retinal pigment epithelium (RPE) damage is necessary to achieve therapeutic benefit, it is only thanks to its content of pigment that visible light can be absorbed and converted into the thermal energy that leads to secondary coagulation of the photoreceptors.

19.2.2.2 Mechanisms of Action

Absorption of light in the RPE leads to heat production and, if the flux of energy is sufficiently high, this will result in the coagulation of the RPE and the photoreceptor outer segments, the latter being in so close contact with the melanin-containing elements of the RPE that collateral damage is unavoidable in the attached retina. Primary damage may also occur in

the choriocapillaris. To induce cell loss, the heat should be sufficient to denaturize the tissue proteins, i.e., to fry them, without inducing evaporation, i.e., without boiling the tissue, because the consequent explosive effect may cause undesirable damage to the tissue, such as a rupture of Bruch's membrane that predisposes to secondary development of a subretinal neovascular membrane.

Considerable effort has been made to determine the optimum wavelength of light for photocoagulation treatment. The choice of wavelengths was initially limited to what lasers were available. The first one, the ruby laser, emits in the red, at 694 nm. In this part of the spectrum, the transparency of the retinal pigment epithelium is high and the energy required to obtain the desired response – bleaching of the outer retina – is so high that the undesired deposition of heat in the choroid and sclera causes considerable pain and a need for retrobulbar anesthesia. This was one of several reasons why the argon laser became an attractive alternative, its 488 nm and 514.5 nm lines of blue and green light being highly absorbed in the RPE. Two arguments against blue and green light are that the scatter of light in the aging or cataractous lens is relatively high and that the yellow xanthophyll of the neurosensory retina makes light of these colors, especially blue, unattractive for photocoagulation of subfoveal neovascular membranes. The red krypton 647 nm laser was proposed as an alternative, with advantages also in terms of less lens scatter and better penetration in the presence of vitreous hemorrhages [19], but the results of the treatment of subfoveal neovascularization remained pitiful anyway and the krypton laser has only found limited clinical use. Currently, diode lasers at the yellow end of the green spectrum, 532 nm, are an attractive option because of their compact design and relatively low lens scatter.

The mechanism of action of laser photocoagulation in proliferative diabetic retinopathy may be viewed as a simple restitution of the balance between oxygen demand and oxygen supply in the face of widespread capillary perfusion loss. Tissue oxygen measurements indeed support that photocoagulation improves oxygenation of the inner layers of the retina by destroying parts of the highly metabolically active photoreceptor layer, thus allowing more oxygen to perfuse to the inner layers of the retina [17, 21].

19.2.2.3 Clinical Trials and Indications for Retinal Photocoagulation Treatment

The Diabetic Retinopathy Study (DRS), the first report of which was published in 1976, was designed to assess the effect of argon or xenon arc lamp photocoagulation on proliferative diabetic retinopathy

III 19

(PDR). The DRS demonstrated that the risk of severe vision loss was reduced by approximately 50% following photocoagulation in eyes with high-risk PDR.

High-risk characteristics:
- Neovascularization on the disk (NVD) ≥ 1/3 disk area with or without vitreous hemorrhage
- Vitreous hemorrhage or preretinal hemorrhage with any NVD
- Vitreous hemorrhage plus neovascularization elsewhere (NVE) > 1/2 disk area

Severe visual loss was defined as visual acuity < 5/200 at each of two consecutive visits scheduled at 4 month intervals. The photocoagulation technique consisted of scattered burns, i.e., photocoagulation lesions, spaced about one burn width apart, from the posterior pole, sparing the macula, to the equator. For argon laser photocoagulation, 800–1,600 burns of 500 µm diameter and an exposure time of 0.1 s were given, at an energy setting sufficient to produce moderately intense burns.

For eyes with severe non-proliferative diabetic retinopathy (NPDR) or PDR without high-risk characteristics, the DRS recommended regular follow-up and prompt treatment if high-risk characteristics develop [1]. To assess the potential benefit of earlier treatment, the Early Treatment Diabetic Retinopathy Study (ETDRS) was designed. The ETDRS defined severe NPDR as the presence of any of the three following characteristics (known as the 4-2-1 rule):

- Dot/blot hemorrhages and microaneurysms in 4 quadrants
- Venous beading in 2 quadrants
- Intraretinal microvascular abnormality in 1 quadrant

The ETDRS compared early scatter photocoagulation versus deferral of photocoagulation in patients with mild to severe NPDR or early PDR with or without macular edema. The results of the study led to a revised recommendation, lowering the threshold for scatter photocoagulation in older patients with type II diabetes to very severe NPDR (at least two criteria of the 4-2-1 rule) or early PDR without DRS high-risk characteristics. In younger patients with type I diabetes, the ETDRS recommendation was that photocoagulation be deferred until DRS high-risk characteristics develop [9, 10]. It should be noted that in general, proliferative diabetic retinopathy is characterized by preretinal new vessels in an eye with partial posterior vitreous detachment, neovascularization that remains on the surface of the eye being difficult to distinguish from what is called intraretinal microvascular abnormalities. Consequently, the gold standard for fundus examination is *stereoscopic* fundus biomicroscopy.

19.2.2.4 Clinical Practice

In clinical practice, a number of factors should be considered before the decision to recommend scatter photocoagulation is taken:

- Systemic conditions such as poor metabolic control, recent improvement of metabolic control, arterial hypertension, pregnancy and renal failure are associated with a poorer visual prognosis and may justify earlier scatter photocoagulation.
- An aggressive course of PDR in the first eye to be affected suggests that early and widespread photocoagulation treatment in fellow eye may be needed.

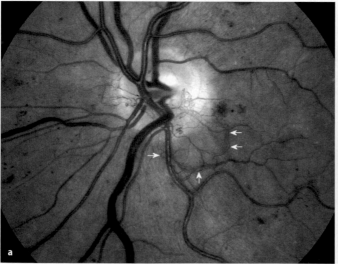 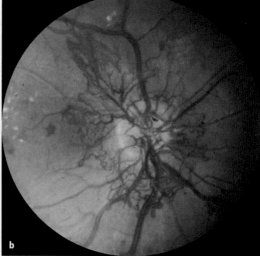

Fig. 19.2.2.3. Optic nerve head neovascularization in PDR. *Arrows* indicate rim of neovascular tuft (*left*)

- Early treatment beyond standard guidelines may be warranted in patients who have demonstrated poor compliance with therapy or retinopathy screening. Reservations against retrobulbar anesthesia coupled with a fear of pain may warrant more sessions with smaller spot diameters but higher numbers of less painful burns.
- Optic media opacities such as cataract or vitreous hemorrhage should prompt consideration of early photocoagulation treatment.
- Impending cataract surgery in a patient with active PDR should lead to photocoagulation treatment before cataract surgery because this procedure may accelerate the progression of retinopathy [12].
- Coexisting clinically significant macular edema should be treated before PDR because photocoagulation may lead to aggravation of macular edema and transient or permanent visual acuity reduction [15, 14].
- The presence of anterior segment neovascularization should lead to prompt and aggressive treatment because progression to neovascular glaucoma is associated with a particularly poor visual outcome.

19.2.2.5 Preparations for Photocoagulation: Information and Consent

Before administering photocoagulation treatment, it is wise to ensure that the patient is fully informed about a list of relevant issues. The following section contains background information that will give the physician a background for answering the patient's questions.

- **What is the aim of the treatment and the criteria of effectiveness?**
  The intention is to improve visual outcome above the outcome of the spontaneous course of the disease. We try to stabilize the condition and prevent further deterioration of the case, which is blindness in the worst-case scenario. Visual loss cannot always be avoided, but with access to modern, comprehensive diabetes care, very few patients go blind and only a minor fraction of patients lose reading vision. The majority of patients with PDR only, i.e., without macular edema, will have unchanged visual acuity after photocoagulation treatment. The short-term criteria of effectiveness are the elimination, reduction, or at least the stabilization of retinal neovascularization and the prevention of transformation of neovascular proliferations into preretinal traction fibrosis. Long-term success is defined by the absence of vitreous hemorrhage, retinal detachment, and major visual loss. Clinical data suggest that these goals are attainable in nearly every patient who has access to high-quality diabetes care and makes use of it, including retinopathy screening and timely photocoagulation treatment. Only the combined approach to systemic and ocular prevention of visual loss from diabetic retinopathy can achieve such results. In many settings, a large proportion of patients present to the ophthalmologist only when they have developed advanced proliferative diabetic retinopathy with visual loss. In such patients, the prognosis is guarded, and the outcome may vary from excellent to poor. The information to be presented to the patient may range from assuring a nervous patient that the visual prognosis is good, to warning the patient that additional procedures may be necessary to control the retinopathy, not because photocoagulation does not help and not because it has deleterious effects, but because it cannot stand alone.

- **Is photocoagulation for PDR a safe procedure?**
  Photocoagulation for diabetic retinopathy is generally a very safe procedure. The single – most dramatic adverse event that can happen during fundus photocoagulation therapy is the inadvertent placement of a photocoagulation lesion in the fovea. Poor patient compliance is rarely the cause, because it will be noted during the initiation of the procedure and this will lead to the adoption of appropriate measures, e.g., the use of the first session as a training session with only few lesions being placed and anti-anxiety medication being prescribed for use before the following session, or to the use of retrobulbar anesthesia or, in very rare cases, full anesthesia and sedation.

- **What safety measures can be adopted?**
  To avoid an accidental lesion in the fovea, keep meticulous track of where your aiming beam is pointing at all times. If you shift your gaze to the laser control panel or any other object outside the fundus field of view, you must retrace your position in relation to the fovea. The optic nerve head is usually the safest landmark from which to move into the periphery. Select an interstitium between two vessel branches and proceed in the peripheral direction, as far as a contact lens without internal mirrors permits. Then proceed to the next interstitium, always staying on the outside of the temporal vascular arcades and at least three disk diameters away from the center of the fovea. If using a three-mirror lens, the risk of accidentally being in the central opening while thinking you are in one of the peripheral mirrors can be minimized by using low enough magnification to view all mirrors simultaneously. Photocoagulation

should be applied between vessels. Accidental application of photothermal energy to a retinal vein may cause vitreous hemorrhage. This is usually self-limiting and small and does not require treatment. Application of short bursts of high energy over a small spot can lead to choroidal hemorrhage into the subretinal space. This may lead to subretinal neovascularization, but outside the retinal vascular arcades it is usually of little consequence. In the macula, there is increasing propensity for progression to subfoveal neovascularization the closer the lesion is to the fovea.

- **Is photocoagulation painful?**
  Photocoagulation in the posterior pole is essentially painless, whereas some pain is noted as the equator is approached, and the pain can be intense from the equator and forward. The pain often varies from one application to the next, but it tends to be most intense at the horizontal meridians where the major choroidal nerves are found. The pain is often referred to the neck. Topical anesthesia enables the use of a contact lens, but it does not reduce the pain associated with fundus photocoagulation. Most patients can undergo full treatment using topical anesthesia only, depending on the spot size, energy settings and number of applications given. To relieve pain, retrobulbar, peribulbar, or subtenon anesthesia may be used. Some ophthalmologists use a mild sedative, e.g., diazepam 5 mg, and an analgesic, e.g., 1 g paracetamol given 1 h before the treatment, while others suggest that the pain is best counteracted by establishing a good rapport with the patient, thus minimizing pain through the relief of anxiety.

- **How long does it take to do retinal photocoagulation and how many sessions are needed?**
  Full treatment for PDR should be divided into at least two sessions per eye. If no retrobulbar anesthesia is applied, a higher number of sessions are often used, but it should probably not exceed six per eye. The duration of the photocoagulation procedure is typically between 5 and 15 min.

- **What is the earliest effect of laser?**
  Immediately after treatment, the treated eye will see very little because of the diffuse photobleaching from the scattered laser light. The eye will recover from the dazzling within half an hour. Some patients report seeing the pattern of laser photocoagulation and occasionally also photopsia for some months after the treatment.

- **Will vision improve after laser coagulation?**
  Retinal photocoagulation for PDR is intended to improve long-term central vision outcome. Photocoagulation does not restore vision, but it may

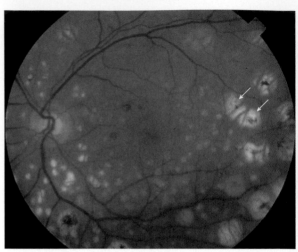

Fig. 19.2.2.4. Photocoagulation lesions of varying diameters as seen years after treatment using 500 μm as the largest spot size. The expansion of lesions to a diameter of about 1,000 μm and occasional confluence of scars (*arrows*) is attributable to creeping atrophy at the rim of lesions

stabilize vision where vision would otherwise have been lost. This is one reason why photocoagulation should be applied before visual loss has occurred.
Photocoagulation for PDR induces multiple small scotomata in the peripheral visual fields and with time the scotomata can reach confluence by the process of creeping atrophy (see below). Patients rarely complain of this peripheral field visual loss. There are several explanations for this. Thus, peripheral vision is often impaired in PDR, even before photocoagulation is performed. Additionally, peripheral field scotomata are invisible, as is the physiological blind spot, because of the psychophysical filling-in phenomenon.

- **What is the meaning of the term "panretinal"?**
  The term panretinal ("all-retinal") photocoagulation is a misnomer, because it may be misinterpreted to mean that the entire retina should be photocoagulated, which would be a disaster. A photocoagulation scar is a blind spot. Consequently, it is only by leaving enough untreated retina that symptomatic visual loss can be avoided. The term panretinal photocoagulation is used to describe a treatment where photocoagulation lesions are scattered evenly over the fundus *except* in the macula (within 2 disk diameters of the foveal center). The intention is to save central vision, at the expense of some peripheral vision. By distributing the peripheral lesions in a non-confluent manner, the absolute scotomata in the patient's peripheral visual field remain small enough to be of no practical consequence.

Because photocoagulation spares the inner retina, a photocoagulation lesion is not associated with an arcuate scotoma and the field defect is confined to the photocoagulated area.

- **Should fluorescein angiography be done before treatment of PDR and during follow-up?**
  The diagnosis and monitoring of PDR should be based primarily on stereoscopic biomicroscopy. Angiography is rarely needed. Cases with persistent preretinal proliferations that are difficult to detect by biomicroscopy do not necessarily need supplementary treatment, but obviously they should be watched carefully for signs of progression (Fig. 19.2.2.5).

- **What are the guidelines for follow-up?**
  Timely and sufficient photocoagulation of preretinal new vessels in PDR will cause the proliferations to be eliminated, reduced in size, or transformed into fibrous tissue without traction effects on the retina. Therefore, photocoagulation for PDR should always be followed up, to determine if supplementary treatment or vitrectomy is advisable. Photocoagulation for PDR is given in a titrated manner, the initial sessions preferably sparing most or all of the area within the temporal vascular arcades. After the initial series of sessions has been completed, the patient should be seen again about 3 months later, at which time regression of the vessels can be seen in favorable cases. If continued activity is seen, additional therapy may be warranted. Often, it will be possible to extend the treatment more anteriorly,

and this should be done before adding photocoagulation within the temporal vascular arcades.

- **Is there a role for intravitreal angiostatic therapy in PDR?**
  Angiostatic agents are becoming available for the treatment of retinal disease. It is possible that they will find a role in the treatment of PDR, beginning as adjuvant therapy in cases where photocoagulation is likely to be insufficient to prevent visual loss.

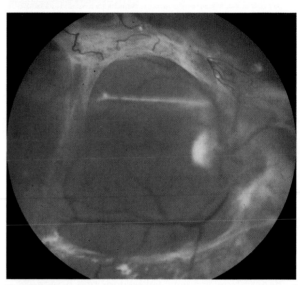

Fig. 19.2.2.6. Proliferative diabetic retinopathy in late untreated fibrotic stage. Further contraction of the macula-encircling preretinal fibrosis may result in sudden loss of central vision because of foveal detachment

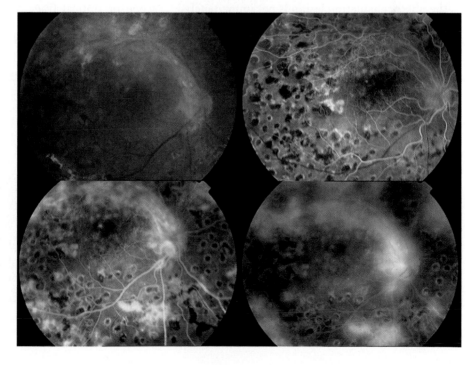

Fig. 19.2.2.5. Residual preretinal fibrovascular proliferations after full photocoagulation treatment for proliferative diabetic retinopathy. Photocoagulation scars with a dark or ash-gray appearance are typical of people with a pigmented sclera

19 **III**

- **Is late vitreous hemorrhage a sign of failed therapy?**
  Vitreous hemorrhage may occur long after the completion of otherwise successful photocoagulation treatment for PDR. The likely source of hemorrhage is small residual new vessels that suffer traction from a shrinking vitreous. The hemorrhage tends to be minor and to resolve spontaneously. Few cases require vitrectomy.

19.2.2.6 Consent to Treatment

Formal requirements for obtaining the patient's consent to treatment vary between countries, but good practice should always include giving careful information to the patient about available options, the objectives of treatment, potential complications, postoperative precautions, and guidelines for follow-up.

19.2.2.7 Photocoagulation Protocol

19.2.2.7.1 Wavelength/Color

- Green argon 514.5 nm light or green diode laser 532 nm light are currently the most favored retinal photocoagulation in PDR. Red krypton 647 nm has also been found to be effective, its main potential advantage in PDR being better penetration of a very yellow lens or a brown cataract than green light [24]. The infrared diode laser at 810 nm causes more intense pain than the green lasers and has the disadvantage that the photocoagulation lesion is not immediately visible [3, 2].

19.2.2.7.2 Anesthesia

Retrobulbar anesthesia, peribulbar anesthesia, and subtenonal anesthesia are all effective pain-relieving procedures [11, 15, 18, 22, 25]. The use of smaller spot sizes and multiple treatment sessions are an alternative method of reducing pain.

19.2.2.7.3 Biomicroscopic Contact or Pre-corneal Lenses Used for Photocoagulation

A number of excellent lenses are available for use in fundus photocoagulation together with the laser-equipped slit-lamp biomicroscope. They differ in magnification, field of view, image quality, ease of use, and in whether the image is inverted or erect. Wide-field lenses provide the best orientation with respect to fundus landmarks, whereas the Goldmann 3-mirror lens provides the best access to the most peripheral fundus. The laser spot magnification is defined as the linear magnification relative to the laser spot as projected through the Goldmann 3-mirror lens. The laser spot magnification is inversely proportional to the magnification of the fundus image. If the laser spot is magnified by a factor of 2, then the energy should increase, in principle, by the square, i.e., a factor of 4. This rule is only a crude guideline, however, and it is necessary to perform a renewed titration of the energy setting after a change in spot size or magnification.

Retinal photocoagulation can be made in the fully sedated patient using a special indirect ophthalmoscope or optical fiber delivery during vitrectomy. The desired distribution and tissue effect are essentially the same as when using the laser-equipped slit-

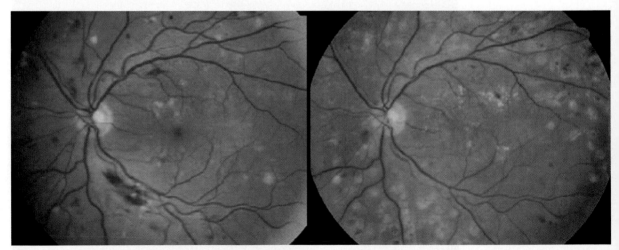

Fig. 19.2.2.7. Severe non-proliferative diabetic retinopathy before (*left*) and after photocoagulation treatment (*right*) outside the temporal vascular arcades. Retinopathy lesions in the treated areas have undergone marked regression, whereas within the arcades a foveal center-involving diabetic macular edema has developed

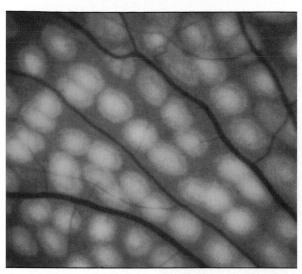

Fig. 19.2.2.8. Fundus photograph recorded years after scatter treatment for proliferative diabetic retinopathy. The initial spacing of photocoagulation lesions ("burns") is indicated by the bright center of the lesions, the intended distance between spots being 1 burn width. The darker rim around the white center is attributable to the phenomenon of creeping atrophy. Note that the lesions are placed between the major retinal vessels

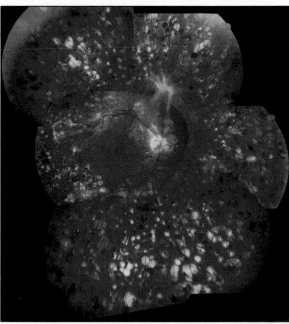

Fig. 19.2.2.9. Fundus photographic montage demonstrating extent of retinal photocoagulation treatment for proliferative diabetic retinopathy, as seen 2 years after its completion. Note the complete fibrotic involution of preretinal neovascularization 3 disk diameters above the optic disk and partial fibrotic involution of a large neovascularization at the end of the superior temporal vascular arcade. Note also the variable appearance of the photocoagulation scars, ranging from dark-brown hyperpigmentation to white unpigmented (or depigmented) sclera. The variation in lesion diameter is partly attributable to the increase in magnification with increasing eccentricity

lamp biomicroscope. Treatment at the biomicroscope should generally be done using a contact lens, because it gives better control of eye movements than a non-contact pre-corneal lens.

19.2.2.7.4 Parameters and End-points

The DRS prescribed the use of 800–1,600 moderately intense argon laser applications placed 1 burn width apart, using a 500-µm-diameter spot size and an exposure time of 0.1 s [7]. Posterior treatment borders should be about one-half to one disk diameter nasal to the disk, and no closer than 2 disk diameters above, temporal to, and below the center of the macula and extending peripherally to the quator of the eye. The same parameters were described by the ETDRS, except that the range of burns was from 1,200 to 1,600 [23].

The light intensity is titrated by finding a peripheral fundus location of average pigmentation. The aiming beam is set to the desired spot size, e.g., 200 µm, and pulse length, usually 0.1 s, and a relatively low intensity, e.g., 200 mW for a 532 nm laser. A coagulation pulse is delivered and the tissue reaction is observed. The desired end-point is a pale lesion that matches the diameter of the aiming beam. White lesions that are larger than the aiming beam diameter should be avoided.

19.2.2.7.5 Placement and Extent of Treatment?

The strategy of treatment is to eliminate the stimulus for neovascularization by ablating a large number of small areas of the outer retina [23]. Consequently, neovascularizations, fibrous tissue, and detached retina should not be treated directly.

19.2.2.7.6 Sessions

Intense photocoagulation is associated with the risk of inducing progression of diabetic macular edema or even serous detachment of the macula (Fig. 19.2.2.11) or serous detachment of the ciliary body and secondary angle closure [8]. To avoid such complications, full photocoagulation treatment or PDR should be divided into no less than two sessions per eye, separated by at least 2 weeks when two sessions are used per eye by and at least 4 days if three or more sessions are used [23]. Deviations from these rules may be warranted in non-compliant patients. In patients with significant vitreous hemorrhage the

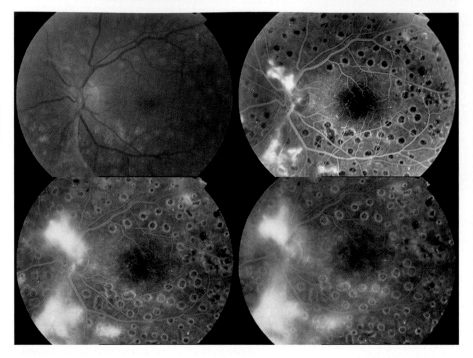

Fig. 19.2.2.10. Persistent PDR after an initial photocoagulation session. In several regions of the fundus, the distance between lesions is larger than recommended and supplementary treatment should be done with variable spot size lesions. The dark angiographic appearance of a photocoagulation scar is attributable to loss of choriocapillaris perfusion, whereas the bright rim of the scar is caused by loss of retinal pigment epithelium baring the underlying intact choriocapillaris. (Photographs courtesy of Khaled Abdelazeem, Assiut University)

approach is one of treating as much as possible per session while waiting for the hemorrhage to clear or to perform vitrectomy.

19.2.2.7.7 Postoperative Treatment

A short acting cycloplegic may be given after extensive treatment, to prevent synechiae and to decrease ciliary body swelling. In patients with a history of iritis, a short course of corticosteroids may be given to prevent the induction of a relapse.

19.2.2.7.8 Follow-up and Retreatment

- The objective of follow-up is to assess retinopathy activity and to complete treatment as needed. Successful treatment reduces the extent and the fullness of the new vessels or at least it blocks the growth of new vessels. If administered early enough in the course of retinopathy, the new vessels may disappear completely. Late treatment of advanced retinopathy may lead to extensive preretinal fibrosis, the subsequent shrinking of which does not appear to be influenced by photocoagulation treatment. The assessment of progression can be made on:
  - the appearance of the rim of neovascularizations, freshly formed new vessels being dilated, without any visible fibrosis, and forming a dense network;
  - comparison with a baseline written description of the fundus; or

  - (3) comparison with a baseline fundus photograph.
- The appearance of neovascularizations that were not present at baseline indicates that the extent and/or intensity of treatment have been insufficient to eliminate the stimulus for neovascularization. Notably physicians who are beginners can be seen to use too low power settings because they rely on standard values while neglecting the need to titrate settings to give the intended tissue response in the individual eye.
- Increased vitreous hemorrhage need not signify failure of treatment because it can occur as the consequence of vitreal detachment, etc.
- Treatment is not complete until photocoagulation lesions have been evenly distributed over the entire peripheral retina, the distance between lesions being one lesion diameter. Termination of treatment before the stipulated standard of 1,200 to 1,600 lesions of 500 µm diameter or an equivalent area covered using smaller spot diameters can occasionally be seen, suggesting that such treatment may be warranted. Such treatment has not been validated and apparent successes can sometimes be shown to be based on cases treated before they had reached conventional treatment thresholds.
- Every examination should include assessment of the iris for rubeosis and gonioscopy should be done if neovascularization of the anterior segment is suspected.

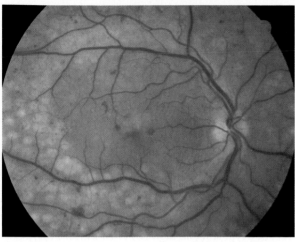

Fig. 19.2.2.11. Serous detachment of the entire macula following extensive photocoagulation treatment for proliferative diabetic retinopathy

According to the ETDRS protocol, additional treatment burns should be placed in between the previous scars, anterior, or posterior to them but should spare the central macula within 500 µm of its center and the size of the burn should not exceed 200 µm at the area between 500 µm and 1,500 µm from the macula [23].

19.2.2.8 Complications of Photocoagulation for PDR

Intraoperative and postoperative complications of fundus photocoagulation include:

- Corneal epithelial abrasion secondary to the use of a contact lens.
- Iridocyclitis.
- Angle narrowing because of forward movement and rotation of the ciliary body which may lead to increased intraocular pressure and severe pain [5].
- Transient accommodative paralysis and mydriasis may result from the injury of nerves in the choroid innervating the anterior segment.
- Confluent lesions may lead to symptomatic peripheral visual field loss. Decreased dark adaptation is commonly found [20]. Extensive treatment may lead to choroidal detachment, exudative retinal detachment, choroidal hemorrhage, retinal tears and detachment as well as progression of tractional retinal detachment, possibly caused by transient exudation. Choroidal neovascular membrane formation may occur even from peripheral lesions, especially if the burn resulted in choroidal hemorrhage.
- Macular edema may develop or worsen after photocoagulation, especially in patients with perifoveal capillary non-perfusion. Recovery may occur within weeks, but sometimes visual loss may be permanent [13]. In consequence, it is recommended that macular edema, if present, be treated at least 1 week before photocoagulation for PDR is initiated.
- Accidental photocoagulation of the fovea may occur if the treating physician loses track of the fundus landmarks. Sudden eye movements are rarely a problem in the treatment of PDR. Although some recovery may occur within the first months, a lesion in the center of the fovea will cause permanent symptomatic visual loss.
- Vitreous hemorrhage can result from rupture of neovascular vessels during treatment. It is usually of limited extent and resolves spontaneously over a few months.

Most of these complications can be minimized or avoided by dividing the treatment into multiple sessions and by avoiding large spot sizes and high light intensity.

References

1. Aiello L, Berrocal J, Davis M, Ederer F, Goldberg MF, Harris JE, Klimt CR, Knatterud GL, Margherio RR, McLean EN, et al. (1973) The diabetic retinopathy study. Arch Ophthalmol 90(5):347–8
2. Bandello F, Brancato R, Lattanzio R, Trabucchi G, Azzolini C, Malegori A (1996) Double-frequency Nd:YAG laser vs. argon-green laser in the treatment of proliferative diabetic retinopathy: randomized study with long-term follow-up. Lasers Surg Med 19(2):173–6
3. Bandello F, Brancato R, Trabucchi G, Lattanzio R, Malegori A (1993) Diode versus argon-green laser panretinal photocoagulation in proliferative diabetic retinopathy: a randomized study in 44 eyes with a long follow-up time. Graefes Arch Clin Exp Ophthalmol 231(9):491–4
4. Beetham WP, Aiello LM, Balodimos MC, Koncz L (1970) Ruby laser photocoagulation of early diabetic neovascular retinopathy. Preliminary report of a long-term controlled study. Arch Ophthalmol 83(3):261–72
5. Blondeau P, Pavan P, Phelps C (1981) Acute pressure elevation following panretinal photocoagulation. Arch Ophthalmol 99:1239–1241
6. Cook HL, Newsom RS, Mensah E, Saeed M, James D, Ffytche TJ (2002) Etonox as an analgesic agent during panretinal photocoagulation. Br J Ophthalmol 86:1107–1108
7. Diabetic Retinopathy Study Research Group (1981) Photocoagulation treatment of proliferative diabetic retinopathy: clinical application of diabetic retinopathy study (DRS) finding. DRS Report Number 8. Ophthalmology 88:583–600
8. Doft BH, Blankenship GW (1982) Single versus multiple treatment sessions of argon laser panretinal photocoagulation for proliferative diabetic retinopathy. Ophthalmology 89(7):772–9
9. Early Treatment Diabetic Retinopathy Study Research Group. Early photocoagulation for diabetic retinopathy

III 19

ETDRS Report Number 9 (1991) Ophthalmology 98(Suppl): 766–785

10. Ferris F (1996) Early photocoagulation in patients with type II diabetes or I. Trans Am Ophthalmol Soc 94:504–537

11. Guise PA (2003) Sub-tenon anesthesia: a prospective study of 6,000 blocks. Anesthesiology 98:964–968

12. Jaffe GJ, Burton TC (1988) Progression of nonproliferative diabetic retinopathy following cataract extraction. Arch Ophthalmol 106:745–749

13. Kleiner R, Elman M, Murphy R, et al. (1988) Transient severe visual loss after panretinal photocoagulation. Am J Ophthalmol 106: 298–306

14. McDonald HR, Schatz H (1985) Macular edema following panretinal photocoagulation. Retina 5(1):5–10

15. McDonald HR, Schatz H (1985) Visual loss following panretinal photocoagulation for proliferative diabetic retinopathy. Ophthalmology 92(3):388–93

16. Meyer-Schwickerath G (1960) Light coagulation (translated by Drance SM). Mosby, St Louis, MO

17. Molnar I, Poitry S, Tsacopoulos M, Gilodi N, Leuenberger PM (1985) Effect of laser photocoagulation on oxygenation of the retina in miniature pigs. Invest Ophthalmol Vis Sci 26(10):1410–4

18. Ripart J, Lefrant JY, de La Coussaye JE, Prat-Pradal D, Vivien B, Eledjam JJ (2001) Peribulbar versus retrobulbar anesthesia for ophthalmic surgery: an anatomical comparison of extraconal and intraconal injections. Anesthesiology 94:56–62

19. Ryan SJ (ed) (2001) Retina, vol 2, chap 61. Mosby, St. Louis, MO

20. Seiberth V, Alexandridis E, Feng W (1987) Function of the diabetic retina after panretinal argon laser coagulation. Graefes Arch Clin Exp Ophthalmol 225(6):385–90

21. Stefansson E, Hatchell DL, Fisher BL, Sutherland FS, Machemer R (1986) Panretinal photocoagulation and retinal oxygenation in normal and diabetic cats. Am J Ophthalmol 101(6):657–64

22. Stevens JD, Foss AJ, Hamilton AM (1993) No-needle one-quadrant sub-tenon anaesthesia for panretinal photocoagulation. Eye 7:768–771

23. The Early Treatment Diabetic Retinopathy Study Research Group (1987) Techniques for scatter and local photocoagulation treatment of diabetic retinopathy: Early Treatment Diabetic Retinopathy Study Report no. 3. Int Ophthalmol Clin 27(4):254–64

24. The Krypton Argon Regression Neovascularization Study (1993) Randomized comparison of krypton versus argon scatter photocoagulation for diabetic disc neovascularization. The Krypton Argon Regression Neovascularization Study report number 1. Ophthalmology 100(11):1655–64

25. Weinberger D, Ron Y, Lichter H, Rosenblat I, Axer-Siegel R, Yassur Y (2000) Analgesic effect of topical sodium, diclofenac 0.1% drops during retinal laser photocoagulation. Br J Ophthalmol 84:135–137

19.3 Diabetic Macular Edema

19.3.1 Therapeutic Approaches to (Diabetic) Macular Edema

A.M. JOUSSEN

Core Messages

- Breakdown of the blood-retinal barrier occurs as a consequence of a variety of conditions such as metabolic alterations, ischemia, hydrostatic or mechanical forces or inflammation
- Laser treatment of macular edema is controversial for diffuse edema and is not indicated in ischemic forms
- Surgical inner limiting membrane (ILM) peeling is thought to lower the tractional forces and the diffusion of substances through the vitreous

- Intravitreal triamcinolone is successfully used to reduce macular edema despite ongoing discussion about the formulation and dosage. However, controlled prospective clinical trials are required to investigate its efficacy in restoring or maintaining visual function
- The search for a specific pharmacological treatment is ongoing on the basis of new findings regarding the involvement of cytokines and growth factors in the formation of macular edema. Vascular endothelial growth factor (VEGF) inhibitors are currently under investigation in clinical studies

This chapter will review the current knowledge of the pathogenesis of diabetic macula edema (DME), and will discuss the rationale and efficacy of surgical approaches.

Macula edema, however, is not solely related to diabetes, but general principles resulting in macular edema can be found in a variety of diseases.

We discuss laser treatment and its applicability to different forms of macular edema as well as other surgical and pharmacological options.

- Among the growth factors involved, vascular endothelial growth factor plays a dominant role as a mediator of vascular leakage
- Inflammatory phenomena are causally linked to vascular cell death and leakage
- Mechanical factors involved include a strong vitreoretinal adhesion and an altered posterior vitreous structu

19.3.1.1 Macular Edema as a Result of Various Disease Mechanisms

Essentials

- Macular edema is a common phenomenon in different diseases, resulting from either metabolic alterations, ischemia, hydrostatic and mechanical forces, inflammation, or pharmacotoxic effects, or a combination thereof
- Blood-retinal barrier breakdown may occur to a variable extent via dysfunction of intercellular junction, increased transcellular transport, or increased endothelial cell destruction

19.3.1.1.1 Causes of Macular Edema

Macular edema is a common phenomenon in various diseases where fluid accumulates in between the retinal cells. Both the focal and diffuse as well as the more cystic form are characterized by extracellular accumulation of fluid, specifically in Henle's layer and the inner nuclear layer of the retina. The compartmentalization of the accumulated fluid is likely due in part to the relative barrier properties of the inner and outer plexiform layers. The fluid originates from the intravascular compartment.

The classic pattern of cystoid macular edema with its petaloid appearance originating from the fluorescein leakage of perifoveal capillaries may be seen in cases of advanced edema of various origins. These include postsurgical cystoid macular edema

19 III

Table 19.3.1.1. Causes of macular edema in relation to the underlying disorders

Disease group	Disorder	Pathogenesis
Metabolic alterations	Diabetes Retinitis pigmentosa Inherited CME (aut. dom.)	• Abnormal glucose metabolism • Aldose reductase • CME: leakage at the level of RPE • Müller cell disease: leakage from perifoveolar capillaries
Ischemia	• Vein occlusion • Diabetic retinopathy • Severe hypertensive retinopathy • HELLP syndrome • Vasculitis, collagenosis	• Inner blood-retinal barrier (retinal capillary hypoperfusion) • Outer blood-retinal barrier (ischemic hypoperfusion of the choroid: serous detachment)
Hydrostatic forces	Retinal vascular occlusions: • Venous occlusion • Arterial hypertension • Low IOP	• Increased intravascular pressure • Failure of the BRB
Mechanical forces	• Vitreous traction on the macula	• Epiretinal membranes with tangential traction • Vitreomacular traction syndrome
Inflammation	• Intermediate uveitis • Postoperative CME • Diabetic macular edema • Choroidal inflammatory diseases	• Mediated by prostaglandins • *CME is indication for treatment* • Perivascular leukocytic infiltrates • Diabetic leukostasis mediates vascular leakage by endothelial cell apoptosis • Vogt-Koyanagi-Harada Syndrome • Birdshot Retinochoroidopathy
Pharmacotoxic effects	e.g. • Epinephrine (in aphakia) • Betaxolol • Latanoprost	mostly via prostaglandins

as well as cystoid edema associated with one of the following conditions: diabetes, vascular occlusion, hypertensive retinopathy, epiretinal membranes, intraocular tumors (e.g., melanoma, choroidal hemangioma), intraocular inflammation (e.g., pars planitis), macroaneurysm, retinitis pigmentosa, choroidal neovascularization and radiation retinopathy.

Given the heterogeneous etiology of macular edema its effective treatment depends upon a better understanding of its pathogenesis. In general, formation of macular edema is related to metabolic changes, ischemia, hydrostatic forces, and inflammatory and toxic mechanisms that influence the formation of macular edema to various degrees in the different conditions (Table 19.3.1.1).

Metabolic alterations have a causal role in diabetic maculopathy, but also in inherited diseases such as the autosomal dominant form of macular edema or macula edema in retinitis pigmentosa. Furthermore, ischemia of the inner or outer blood-retinal barrier leads to macular edema. Decreased perfusion of the retinal capillaries is seen, e.g., in vein occlusion and diabetic retinopathy, whereas ischemia plus decreased perfusion of the choroid with associated serous retinal detachment occurs in severe hyperten-

sive retinopathy, in eclampsia and in rheumatoid disorders. Following retinal vascular occlusion the intravascular pressure increases and leads to dysfunction of the blood-retinal barrier. Similarly, hydrostatic forces are effective in arterial hypertension or in eyes with low intraocular pressure and may cause fluid accumulation in the macula. Mechanical traction such as in epiretinal membranes or in vitreomacular traction syndrome promotes macular edema by physical forces.

The importance of inflammation in macular edema is discussed in more detail below. Inflammation apparently plays a role in intermediate uveitis, postoperative cystoid macular edema (Irvine-Gass syndrome), diabetic macular edema and various forms of choroidal inflammatory diseases including Vogt-Koyanagi-Harada syndrome and bird shot retinochoroidopathy. All prostaglandin-like pharmacological agents, even if applied topically, can induce macular edema via a cytokine response similar to inflammatory conditions.

The current therapy for macular edema targets conditions where mechanical traction, hydrostatic force, or inflammation play a pathogenetic role in the formation of macular edema. Unfortunately, even the currently available surgical and pharmacological treatments have suboptimal results in many cases. Therefore, there is an obvious need for the development of a more effective and targeted treatment that can be satisfied only by the better understanding of the pathophysiology of the macular edema formation, which differs according to the underlying disease.

19.3.1.1.1.1 Specificities of Diabetic Macular Edema

Diabetic macular edema is the most common cause of visual impairment in patients with diabetes mellitus and affects approximately 75,000 new patients in the United States every year [17, 94]. The incidence of macular edema significantly increases with increasing severity of diabetes in both early-onset and late-onset diabetic patients [94–96] (Tab 19.3.1.2).

Diabetic macular edema tends to be a chronic disease. Although spontaneous recovery is not uncommon, 24% of eyes with clinically significant macular edema (CSME) and 33% of eyes with center-involving CSME will have a moderate visual loss (15 or more letters on the ETDRS chart) within 3 years if untreated [30, 40].

The Early Treatment Diabetic Retinopathy Study (ETDRS) defined DME as retinal thickening or presence of hard exudates within one disk diameter of the center of the macula. To characterize the severity of macular edema the term "clinically significant macular edema" (CSME) is used.

CSME as defined by ETDRS includes any one of the following lesions [30]:
- Retinal thickening at or within 500 μm from the center of the macula
- Hard exudates at or within 500 μm from the center of the macula, if there is thickening of the adjacent retina
- And an area or areas of retinal thickening at least 1 disk area in size, at least part of which is within 1 disk diameter of the center of the macula

---

19.3.1.1.2 Molecular and Cellular Alterations Leading to Macular Edema

The breakdown of the blood-retinal barrier seems to be the most important mechanism that explains the extravasation of fluid although changes to the retinal blood flow may play a role [5] (see Chapters 8.1, 9). The blood-retinal barrier consists of the retinal pigment epithelium layer (outer blood-retinal barrier) and the vascular endothelium (inner blood-retinal barrier) that prohibit the passage of macromolecules and circulating cells from the vascular compartment to the extracellular and therefore intraretinal space.

In general, an increase in passive permeability through the endothelium can occur via three general mechanisms (Fig. 19.3.1.1):

- Dysfunction of the intercellular junctions
- Increased transcellular transport
- Increased endothelial cell destruction

The initial site of damage that results in the increased vascular permeability is controversial to date. Although the impairment of the perivascular supporting cells such as pericytes and glial cells might play a role, the endothelial cell dysfunction and injury seem more likely to be the first pathogenetic step towards the breakdown of the blood-retinal barrier early in the course of the disease. In order to dissect the molecular and pathophysiologic mechanisms that lead to the accumulation of fluid in the macular area, we have chosen diabetic macular edema as a model.

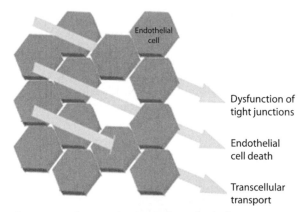

Fig. 19.3.1.1. Three mechanisms of vascular leakage

Table 19.3.1.2. Prevalence of diabetic macular edema. (Adapted from [94])

Early onset diabetes		Late onset diabetes (type II diabetes)			
Duration less than 5 years	Duration of 20 years	Duration less than 5 years		Duration of 20 years	
Insulin-dependent	Insulin-dependent	Non-insulin-dependent	Insulin-dependent	Non-insulin-dependent	Insulin-dependent
0%	32%	3%	5–8%	18%	38%

19.3.1.1.2.1 Cell-to-Cell Junctions and Vascular Permeability

Fluid homeostasis and endothelial permeability is mostly regulated by intercellular junctions in the non-diseased retina. Intercellular junctions are complex structures formed by the assembly of a transmembranous and cytoplasmic/cytoskeletal protein component. At least four different types of endothelial junctions have been described: tight junctions, gap junctions, adherence junctions and syndesmos. Tight junctions are the most apical component of the intercellular cleft (Fig. 19.3.1.2).

Although the molecular structure of tight junctions generally appears to be similar in all barrier systems, there are some differences between epithelial and endothelial tight junctions, and between tight junctions of peripheral and retinal endothelial cells [154]. In contrast to tight junctions in epithelial systems, structural and functional characteristics of tight junctions in endothelial cells respond promptly to ambient factors. It is likely that inflammatory agents increase permeability by binding to specific receptors that transduce intercellular signals, which in turn cause cytoskeletal reorganization and widening of the interendothelial clefts. Endothelial junctions also regulate leukocyte extravasation. Once leukocytes have adhered to the endothelium, a coordinated opening of interendothelial cell junctions occurs.

19.3.1.1.2.2 Cellular Interaction and Vascular Permeability

Leukocytic infiltration of the retinal tissue characterizes many inflammatory diseases such as diabetes, pars planitis, and choroidal inflammatory diseases.

In diabetes, activated leukocytes adhere to the retinal vascular endothelium [119, 153]. Increased leukostasis is one of the first histologic changes in dia-

betic retinopathy and occurs prior to any apparent clinical pathology.

Adherent leukocytes play a crucial role in diabetic retinopathy by directly inducing endothelial cell death in capillaries [82] causing vascular obstruction and vascular leakage. Endothelial cell death precedes the formation of acellular capillaries [153]. With time, however, acellular capillaries prevail and become widespread. Although the mechanism of this destructive process remains elusive, it is clear that the interaction between the altered leukocytes and the endothelial cells and the subsequent endothelial damage represents a crucial pathogenic step [82, 86, 119].

19.3.1.1.2.3 Growth Factors, Vasoactive Factors, and Vascular Permeability

The disruption of endothelial integrity leads to retinal ischemia and vascular endothelial growth factor (VEGF)-mediated iris and retinal neovascularization [84, 119, 120]. VEGF is 50,000 times more potent than histamine in causing vascular permeability [13, 22, 23, 39, 92, 159]. Previous work has shown that retinal VEGF levels correlate with diabetic blood-retinal barrier breakdown in rodents [36] and humans [4]. Flt-1(1 – 3Ig)F$_c$, a soluble VEGF receptor, reverses early diabetic blood-retinal barrier breakdown and diabetic leukostasis in a dose-dependent manner [84]. Early blood-retinal barrier breakdown localizes, in part, to retinal venules and capillaries of the superficial inner retinal circulation [139] and can be sufficiently reduced by VEGF inhibition. Although VEGF is only one of the cytokines involved in the pathogenesis of the vascular leakage, it is likely to be one of the most effective therapeutic targets (see Chapters 3.2, 19.1.1, 19.1.2.1, 19.1.2.2).

There are several other vasoactive factors and biochemical pathways affected by sustained hyperglycemia and known to be involved in diabetic macular edema, which are discussed in more detail in other chapters.

High glucose concentration leads to increased diacylglycerol (DAG) by two pathways: de novo synthesis and through dehydrogenation of phosphatidylcholine (PC). Increased levels of DAG-mediated protein kinase C (PKC) occur. Several studies have shown that a decrease in retinal blood flow occurs with PKC activation. Conversely, inhibition of PKC with LY333531 (Eli Lilly, Indianapolis, IN) normalized decreased retinal blood flow in diabetic rats [72, 182] (see Chapters 19.1.1, 19.1.2.1, 19.1.2.2).

PKC activation causes vasoconstriction by increasing the expression of endothelins (especially endothelin-1, ET-1). The expression of endothelins can be induced by a variety of growth factors and

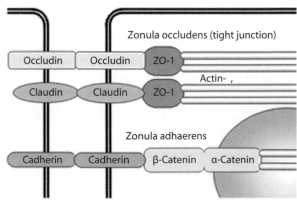

Fig. 19.3.1.2. Tight junctions in endothelial cells

cytokines including thrombin, tumor necrosis factor-α (TNF-α), transforming growth factor-β (TGF-β), insulin, and vasoactive substances including: angiotensin II, vasopressin, and bradykinin. Heparin inhibits endothelins, most likely via inhibition of PKC [102, 142, 170].

Furthermore, retinal vascular endothelial cells are very sensitive to histamine. Several studies have documented increased vascular histamine synthesis in diabetic rats and humans [50, 51, 129]. The administration of histamine reduces ZO-1 protein expression and thus correlates with vascular permeability. The H_1 receptor stimulates PKC that has been implicated in increased retinal vascular permeability [48]. Interestingly, Aiello and coworkers showed that administration of LY333531, a PKC-β isoform-selective inhibitor, does not significantly decrease histamine-induced permeability but VEGF-induced permeability. In contrast, administration of non-isoform-selective PKC inhibitors did significantly suppress histamine-induced permeability [3].

Furthermore, in vascular endothelial cells, advanced glycation end products (AGE) may affect the gene expression of endothelins (ET-1) and modify VEGF expression. The AGE-stimulated increased VEGF expression is dose and time dependent and additive to hypoxia [111, 176].

19.3.1.1.2.4 Endothelial Cell Death and Vascular Permeability

Blood-retinal barrier breakdown is at least in part due to endothelial cell damage and apoptosis. The pro-apoptotic molecule Fas-ligand (FasL) induces apoptosis in cells that carry its receptor Fas (CD 95) [20]. There is evidence that FasL is expressed on vascular endothelium, where it functions to inhibit leukocyte extravasation. The expression of FasL on vascular endothelial cells might thus prevent detrimental inflammation by inducing apoptosis in leukocytes as they attempt to enter the vessel. In fact, during inflammation and the ensuing TNF-α release, the retinal endothelium upregulates several adhesion molecules [178] that mediate the adherence of the leukocytes, but also downregulates FasL, thus allowing the leukocyte survival and migration to active sites of inflammation and infection. In experimental diabetic retinopathy, inhibition of Fas-mediated apoptotic cell death reduces vascular leakage [85]. The cumulative endothelial cell death during the course of diabetes plays a causal role in the pathogenesis of the diabetic vascular leakage and maculopathy.

19.3.1.1.2.5 Extracellular Matrix Alterations and Vascular Permeability

Degradation of the extracellular matrix affects endothelial cell function at many levels causing endothelial cell lability which is required for cellular invasion and proliferation, or influencing the cellular resistance and therefore the vascular permeability. The degradation and modulation of the extracellular matrix is exerted by matrix metalloproteinases, a family of zinc binding, calcium dependent enzymes [29, 114]. Elevation of MMP-9 and MMP-2 expression has been shown in diabetic neovascular membranes [25, 149], although a direct effect of glucose on MMP-9 expression in vascular endothelial cells could not be shown [54]. It is probable that MMPs participate at various stages during the course of the blood-retinal barrier dysfunction and breakdown. Their actions include early changes of the endothelial cell resistance with influence on the intercellular junction formation and function [38] to active participation in the endothelial and pericyte cell death [12] that occurs late in the course of the disease.

19.3.1.1.2.6 Transcellular Transport and Vascular Permeability

In addition to all the above, an important factor that is involved in the regulation of fluid homeostasis is the active cellular transport of nutrients and fluid via pinocytosis. Despite the fact that pinocytic transport is critically involved in the transepithelial fluid exchange, its role in the pathogenesis of increased vascular leakage in diabetes is just emerging [1, 43]. However, the molecular factors that are involved in the pinocytic fluid transport, how they are influenced from disease stages and how they contribute to the increased vascular permeability are unclear.

It is currently known that one of the factors involved in the regulation of pinocytic transport is VEGF. Vascular endothelial growth factor increases vascular permeability not only by disrupting the intercellular tight junctions between the retinal endothelial cells but also by inducing the formation of fenestrations and vesiculovacuolar organelles. The role of VEGF in the disruption of the pinocytic transport that is translated into increased vascular permeability in disease states is still controversial [64]. Whereas, in highly permeable blood vessels the number of pinocytotic vesicles at the endothelial luminal membrane transporting plasma IgG is significantly increased, no fenestrations or vesicles were found in the endothelial cells of the VEGF affected eyes when examined by electron microscopy.

Besides the factors that act in concert with VEGF and are discussed above, sustained hyperglycemia

19 III

increases generation of free radicals and nitric oxide, which can further affect several metabolic pathways and cause oxidative damage and subsequent disorganization of the blood-retinal barrier [93].

The knowledge of the basic mechanisms involved in vascular leakage is essential for the development of an effective clinical treatment. With the growing understanding of the pathophysiology of the macular edema, the therapeutic thinking is likely to change from a merely symptomatic treatment (either surgical or medical) to a treatment that targets specifically the causal factors involved in its formation (e.g., cytokine or growth factor inhibition).

19.3.1.1.3 Mechanical Factors Involved in the Formation of Macula Edema

Clinical and anatomic evidence indicates that abnormalities in the structure of the vitreoretinal interface may play an important role in the pathogenesis of DME [60, 109, 168]. It was suggested that vitreoretinal adhesions in diabetic eyes are stronger than the shear forces of traction from vitreous shrinkage and this in turn may lead to the development of vitreomacular traction and subsequently to macular edema [169]. Nevertheless, the risk of developing diffuse macular edema was 3.4-fold lower in a group of eyes with complete posterior vitreous attachment or complete vitreoretinal separation compared to eyes with vitreomacular adhesion [110].

The vitreous humor is a gel-like structure composed mostly of water (99%), hyaluronic acid, and collagen. A structural barrier between the vitreous cavity and the retina is formed by the inner limiting membrane (ILM), which is localized between the innermost layer of the retina and the outer boundary of the vitreous. The ILM shows typical ultrastructural characteristics of a basal lamina, is found in close contact with the foot processes of Müller cells, and contains proteins that are typically found in basal laminae such as collagen type IV and laminin [18]. Striated collagen fibrils of the vitreous cortex insert into the inner portion of the ILM [65], which is also known as the hyaloid membrane of the vitreous. Detachment of the posterior hyaloid membrane with ageing or pathology results in a condensation of the posterior vitreous surface (membrana hyaloidea posterior). In youth, there is adhesion between the vitreous cortex and the ILM that is stronger than the Müller cells themselves and Müller cell foot processes become separated from their main cell body and remain connected to the posterior aspect of the ILM when this is separated from the retinal surface [156].

There has been a controversial discussion regarding the embryonic origin of the ILM, which can be demonstrated as early as 4 weeks after gestation in the human eye [144, 163]. Traditionally, the ILM has been considered to be synthesized by Müller cells. This concept has been challenged by data presented by Sarthy and coworkers, who investigated the expression of collagen type IV during development of the mouse eye [150]. Because collagen IV is an integral component of all basal laminae, the detection of its mRNA can be used to identify cellular sources of basal lamina production. By in situ hybridization at embryonic day 12, no or sparse mRNA for collagen type IV was found in the retina, while strong labeling was seen in and around the lens, especially in hyaloid vessels in the tunica vasculosa lentis. In contrast, collagen type IV itself could be readily detected in the ILM by immunohistochemistry. Thus ILM collagen type IV is very likely not produced in the retina itself, but rather in the lens and tunica vasculosa lentis. From there it is apparently deposited on the inner retinal surface to form the ILM.

In support of this are data by Halfter et al., which show that also other ILM proteins such as perlecan, laminin-1, nidogen and collagen XVIII are expressed predominantly in lens and ciliary body, but are not detected in the retina [58]. Taken together, ILM proteins appear to originate largely from lens and ciliary body, although a contribution of retinal glial cells in ILM synthesis cannot be excluded.

Gandorfer et al. describe a continuous layer of native vitreous collagen covering the ILM and a thickened premacular vitreous in diabetic patients [47]. In these patients, peeling of the ILM is difficult, but may be rewarding, as peeling is thought to release macular edema by improving fluid movement between retina and vitreous.

While the ILM is probably not directly involved in the formation of diabetic macular edema (but removed in surgical attempts to treat persistent macular edema), the posterior vitreous itself may play the dominant role in the pathogenesis of DME [123, 124, 158]. Sustained hyperglycemia can affect several biochemical pathways that can lead to liquefaction and destabilization of the vitreous gel [155, 157, 158]. Such destabilization of the central vitreous with persistent attachment of the vitreous cortex to the retina can also induce traction on the macula and contribute to the development of macular edema [96, 152, 155]. Sebag and coworkers found that although the collagen content in controls and diabetics is the same, the major crosslink in vitreous collagen is over twofold greater in diabetics than controls. Similarly, the levels of early glycation products were threefold higher, and advanced glycation end products were 10–20 times more abundant in vitreous of diabetics than of controls [157].

In a study by Nasarallah and coworkers, diabetic patients who are 60 years of age or older with macular edema had a significantly higher prevalence of attached posterior vitreous than diabetics without macular edema. Furthermore, naturally occurring posterior vitreous detachment due to loosening of the vitreoretinal adhesion, liquefaction of vitreous gel, and/or gel shrinkage [44, 90] is rare in diabetic patients 50 years of age or younger. In diabetic patients, the vitreoretinal adhesion often remains strong despite gel liquefaction and shrinkage that produces a posterior vitreous detachment [171]. Posterior vitreous detachment in younger diabetics results from diabetes-related vitreous contraction that simultaneously causes traction on the macula that may lead to macular edema [123, 124, 158].

Diffuse diabetic macula edema has been found in association with an attached, thickened, and taut posterior hyaloid [87]. As immunocytochemical staining for cytokeratin (found in retinal pigment epithelial cells) and glial fibrillary acidic protein (found in astrocytes and Müller cells) demonstrated the existence of cells in the premacular posterior hyaloid, there is a possible role for cell infiltration in the development or maintenance of macular edema. It remains to be elucidated whether these cells in the posterior vitreous cause macular edema physiologically rather than mechanically through the production of cytokines.

19.3.1.2 Diagnosis and Current Imaging Modalities

The traditional methods of evaluating macular diseases such as slit-lamp biomicroscopy and stereo fundus photography are relatively insensitive at determining small changes in retinal thickness [19, 160, 189].

Fluorescein angiography is a standard method used to evaluate patients with diabetic macula edema and is sensitive for detection of fluid leakage [91, 162]. Fluorescein angiographic findings in diabetic macula edema can be categorized into three different types of leakage: (1) focal leakage: well-defined focal area of leakage from microaneurysms or dilated capillaries; (2) diffuse leakage: presence of widespread leakage from ill-defined sources; and (3) diffuse cystoid leakage: diffuse leakage and pooling of dye in the cystic spaces of the macula in the late phase of the angiogram [91]. The most important information gained from fluorescein angiography is, however, whether there is macular ischemia or not and thus whether any treatment approach would be effective or not. Nevertheless, retinal thickness cannot be quantified by angiographic means.

The problem of most other available diagnostic techniques is the lack of sufficient resolution to provide structural images of the central retina: The resolution of standard clinical ultrasound is limited by the sound wave in ocular tissue to approximately 150 μm, whereas high-frequency ultrasound biomicroscopy offers a resolution of approximately 20–40 μm; however, its penetration is limited to the first 4 mm of the anterior segment [133]. The resolution of confocal imaging techniques including scanning laser ophthalmoscope (SLO) and scanning laser tomography (SLT) may be limited by ocular aberrations [179].

More recently diagnostic techniques have been developed for high-resolution imaging of the retina and detection of the retinal thickness (see Chapter 12). Optical coherence tomography (OCT) is anon-invasive technique developed for cross-sectional imaging in biological systems. It is based on determination of the time-of-flight delay of light from different depths of tissue using low-coherence interferometry [62, 66, 138]. OCT uses infrared illumination of the fundus to take images and thus is more comfortable and well tolerated by patients than more invasive techniques such as fluorescein angiography.

Up to now, several studies have shown that OCT is a useful technique for quantitative measurement of retinal thickness in patients with diabetes [132, 166]. Optical coherence tomography of DME has revealed three basic structural changes in the neurosensory retina: retinal swelling, cystoid macular edema, and serous retinal detachment. Otani and coworkers found that retinal swelling was the most common change in the structure of the retina (88%) [132]. The retinal thickness at the central fovea with DME was 250–1,000 μm (mean 470±180 μm), whereas normally the retinal thickness is approximately 130 μm [132]. OCT can be used to calculate the standardized change in macular thickness (CSMT). As shown by Chan and Duker, CSMT is a highly useful method for evaluation and comparison of the different therapeutic modalities for DME [21]. Moreover, OCT is a more sensitive and specific method for objective evaluation of the attachment of vitreous strands to the edges of fovea and macula and for measurement of macular thickness [123, 130]. Unfortunately, OCT image quality can be affected by media opacities, and the reliability of the data depends on the skill of the OCT operator.

Another accurate and very sensitive imaging technique for diagnosing and monitoring a large spectrum of macular diseases is the retinal thickness analyzer (RTA). RTA imaging is based on projecting a thin laser slit beam (green helium-neon laser, 40 μW) obliquely onto the retina. The backscattered light is analyzed by a fundus camera. Depth precision and depth resolution of the RTA are 5–10 μm and 50 μm, respectively [97, 106]. Because of the small variation in the measures of the normal retinal thickness, the RTA can be very useful for early detection of macular edema [9]. Similar to OCT imaging, RTA measurements are very sensitive to media opacities.

The retinal thickness measurements obtained by both instruments, OCT and RTA, are very similar for both normal subjects as well as patients with macular edema [97, 137].

Retinal thickening may correlate better with areas of retinal dysfunction than does the amount of fluorescein leakage [126].

The early detection of retinal thickness alterations is important for diagnosing and monitoring diabetic macula edema and for offering treatment in the early stages of the disease. OCT and the RTA imaging may play an important role in identifying subclinical retinal thickening and quantitative monitoring of macular edema.

19.3.1.3 Treatment of Macular Edema

Essentials
- In an effort to reduce macular edema, at least with some rationale, different approaches have been used and found effective in certain conditions. Laser coagulation, pharmacological approaches, and surgical measures are most frequently used
- While focal laser treatment is recommended in patients with focal macular edema, there is no confirmed evidence that grid laser treatment improves diffuse diabetic macular edema
- In contrast to the routine clinical treatment, the use of carbonic anhydrase inhibition for macular edema is not based upon scientific evidence
- Intravitreal triamcinolone acetonide is gaining attention for the treatment of macular edema not only as an additive during surgery or in the treatment of persistent macular edema, but also as primary treatment in diffuse macular edema. Randomized multicenter studies are ongoing
- VEGF inhibitors are promising for the reduction of vascular leakage. Clinical studies using VEGF inhibitors are currently being performed with two specific molecules; however, neither substance is currently available for clinical use outside clinical trials

19.3.1.3.1 Laser Treatment

Many studies have demonstrated a beneficial effect of photocoagulation therapy for DME (Tables 19.3.1.3, 19.3.1.4, 19.3.1.5) [14, 30, 70].

The exact mechanism of action of laser photocoagulation-induced resolution of DME is unknown. A detailed discussion can be found in Chapter 13. In short, a laser-induced destruction of oxygen-consuming photoreceptors is discussed as well as cell death and scarring (involving gliosis and RPE hyperplasia) induced by the temporary rise in tissue temperature after laser photocoagulation. Oxygen that normally diffuses from the choriocapillaries into the outer retina can now diffuse through the laser scar to the inner retina, thus relieving inner retinal hypoxia [16, 181]. There is contrasting data as to whether an increased preretinal oxygen partial pressure is involved and allows for microvascular repair in the treated areas [121, 135]. When studying the diameter of retinal arterioles, venules, and their macular branches before and after macular laser photocoagulation in eyes with DME, the macular arteriolar branches were found to be constricted by 20.2 % and the venular branches 13.8 %. This was attributed to an improved retinal oxygenation caused by the laser treatment which leads to autoregulatory vasoconstriction, improving the diabetic macula edema [53].

According to another theory, the beneficial effect of laser photocoagulation is due to an enhanced proliferation of retinal pigment epithelial and endotheli-

Table 19.3.1.3. Functional outcome of the eyes with clinically significant macular edema (CSME) with less/more diabetic retinopathy (DR) treated with different techniques of laser photocoagulation (LP) in the ETDRS. Follow-up was 5 years

Type of treatment	Severe visual loss (VA< 5/200)		Moderate visual loss (15 letters or more)	
	CSME with mild DR	CSME with advanced DR	CSME with mild DR	CSME with advanced DR
Immediate focal LP/delayed full scatter LP	1.0 %	4.7 %	22.4 %	26.2 %
Immediate full scat- ter LP/ delayed focal LP	0.9 %	3.8 %	29.8 %	24.1 %
Immediate focal LP/delayed mild scatter LP	2.2 %	4.0 %	19.5 %	24.1 %
Immediate mild scatter/delayed focal LP	1.2 %	4.1 %	21.8 %	25.7 %
Deferral	2.9 %	6.5 %	30.2 %	32.1 %

Table 19.3.1.4. Recently published studies investigating the efficacy of panretinal photocoagulation on diabetic diffuse macular edema (table constructed by Dr. A. Lux)

Study	Eyes	Diagnosis	Therapy	Observation period	Initial visual acuity	Final visual acuity	Improvement/stabilization of VA	Macula edema
Zein [193]	35	PDR	Triam + PRP	9 months	20/286	20/80		84% resorbed
Zein [193]	35	PDR	Grid + PRP	9 months	20/282	20/256		46% resorbed
Yi [190]	238	DR	Focal PC	1 year			137 = 57%	91.6% resorbed
Yi [190]	84	DR	PRP + focal PC	1 year			52 = 61.9%	
Shimura [161]	72	Severe NPDR, mild PDR	PRP				89–92%	
Rema [143]	261	PDR	PRP	1 year			73%, 77%, 100%	
Lee [108]	52	PDR	Grid + PRP	1 year			87%	93% resorbed
Kyto [103]	1	PDR	Triam + PRP	9–15 weeks	0.05	0.25		From 691 to 239 μm
Gardner [49]	18	PDR	PRP					In 13 eyes reduced
Dogru [28]	39	PDR	PRP	5 year			28%	
Bandello [11]	65	PDR	Light PRP/PRP	22 months	0.12/0.14	0.18/0.27		

Triam intravitreal injection of triamcinolone, *PRP* panretinal photocoagulation, *focal LC* targeted photocoagulation of microaneurysms in the macular area, *grid LC* scatter photocoagulation close to the macula, *PDR* proliferative diabetic retinopathy, *DR* diabetic retinopathy, not further specified, *NPDR* non-proliferative diabetic retinopathy

Table 19.3.1.5. Published studies investigating the efficacy of grid laser treatment on diabetic diffuse macular edema

Studies		Patients	Eyes treated	Eyes control	Observation period	V/A after grid laser			Untreated			
						Better	Unchanged	Worse	Better	Unchanged	Worse	
ETDRS	Prosp. randomized	No separate evaluation for focal and diffuse macular edema										
Olk [127]	Prosp. randomized	92	42	37	24 months	45.2	45.2	9.5	8.1	48.5	43.2	Sign.
Lee et al. [107]	Retrospective		302		36 months	14.5	60.9	9.5				
Ladas et al. [105]	Prosp. randomized		26	23	24 months	15.4	65.4	19.2	0	56.5	43.5	p=0.049
			24	22	36 months	8.3	54.2	37.5	0	40.9	59.1	ns
Wiznia [186]	Prosp. randomized		35	35	17 months	22.9	0.6	17.1	5.7	54.3	0.4	ns
McNaught et al. [116]	Prosp. randomized		37	35	12 months (distant and near V/A)	72	14	14	17	66	17	p<0.001
		29	18	17	24 months (only near V/A)	61	11	28	18	47	35	ns
Tachi et al. [168]	Retrospective	32	41		12 months	12	39	49				
McDonald et al. [115]	Retrospective		89	0	3–33 months	17	77	6				
Bailey et al. [10]	Prospective	33			9 months			28.9				

Worse ≥ deterioration >2 lines; Better ≥ 2 lines better (given in percentages)

al cells leading to a repair and restoration of the blood-retinal barrier [177]. The RPE cells may respond to the injury in several ways: if the lesion is relatively small, the RPE defect can be filled by cell spreading; if the defect is relatively large, the cells can proliferate to resurface the area, and the RPE can produce cytokines (e.g., TGF-β) that antagonize the permeabilizing effects of VEGF [15, 52].

Laser therapy is well established in diabetic retinopathy as well as in diseases with peripheral retinal ischemia. The Early Treatment Diabetic Retinopathy Study (ETDRS) was designed to evaluate the effects of argon laser photocoagulation for macular edema in a prospective, randomized, multicenter clinical trial. Among the subgroup of eyes with mild to moderate non-proliferative diabetic retinopathy with macular edema, visual acuity improved in 16 %, remained unchanged in 77 %, and worsened in 7 % of treated eyes, whereas visual acuity improved in 11 %, remained unchanged in 73 %, and worsened in 16 % of untreated eyes after 2 years of follow-up. After 3 years of follow-up, vision worsened in 12 % of treated eyes compared to 24 % of untreated eyes. Furthermore, the ETDRS Reading Center identified and analyzed 350 eyes with CSME, which had retinal thickening involving the center of the macula. After 1 year of follow-up, retinal thickening in the center of the macula was present in only 35 % of eyes assigned to immediate photocoagulation compared to 63 % of eyes assigned to deferred photocoagulation. There was a statistically significant benefit of laser photocoagulation in this group of eyes (Table 19.3.1.3) [30]. Nevertheless, a current literature review indicates that at least in selected groups, a beneficial effect of PRP with respect to macula edema can be identified (Table 19.3.1.4). The exact relationship between peripheral ischemia and macula edema and thus the relevance of peripheral panretinal photocoagulation in these patients as a treatment for macula edema still remains to be determined.

The standard guidelines for "focal" laser photocoagulation for DME have been provided by the ETDRS [35]. Direct treatment of leaking microaneurysms and "grid" treatment of diffuse macular edema or non-perfused thickened retina have been suggested for mild and moderate non-proliferative diabetic retinopathy (NPDR), and combination scatter laser photocoagulation and focal laser photocoagulation has been suggested for DME in selected cases of severe NPDR and in eyes with PDR (Tables 19.3.1.3, 19.3.1.5). The risk of severe visual loss (30 letters with final VA < 5/200) was not significantly different between the treatment and control groups in each category of retinopathy (mild and advanced retinopathy). The risk of moderate visual loss (15 letters), however, was significantly lower after the first year of follow-up in eyes assigned to the laser treatment in each category (less/more severe retinopathy) compared to the group of eyes in the control group. At early time points, the risk of moderate visual loss in the group of eyes with DME and more severe diabetic retinopathy was higher (statistically significant, $p < 0.01$) in the treatment group compared with the eyes in the control group [34]. In spite of the fact that

the 5-year risk of severe visual loss in each category (DME with less/more severe retinopathy) is lower in the treated group compared with the control group, the rates are very low in all groups.

There was no indication that the development of severe visual loss might be influenced by the timing of focal photocoagulation in the eyes with macular edema assigned to early scatter photocoagulation [34]. The most effective strategy for reducing the risk of moderate visual loss in eyes with macular edema and less severe retinopathy was immediate focal photocoagulation with delayed scatter (added only if a more severe retinopathy developed), whereas eyes assigned to immediate full scatter and delayed focal photocoagulation had an increased risk of moderate visual loss during the first 16 months of follow-up and thereafter were similar to the eyes assigned to deferral.

The ETDRS investigators suggested that the reduced rate of moderate visual loss is mostly due to the effects of early focal photocoagulation, which should be considered for all eyes with CSME. Focal photocoagulation has been associated with a lower risk of moderate visual loss, an increased chance of visual improvement, less loss of color vision, and minor visual field changes [17, 30–34].

Focal burns to microaneurysms surrounding the center of the macula should not be larger than 50–100 μm, with a duration of 0.05–0.1 s. The preferred endpoint is a whitening or darkening of microaneurysm.

Grid pattern (if at all indicated!) burns are located above, below, and temporal to the center of the macula (spot size: 50–100 μm; duration: 0.05–0.1 s, preferred endpoint: mild RPE whitening). Importantly, grid treatment is not placed within 500 μm of the center of the macula or within 500 μm from the disk margin, but can extend up to 2 disk diameters to the macular area.

Central laser coagulation is definitely not recommended in eyes with ischemic maculopathy (Table 19.3.1.6). Although laser photocoagulation has been proven beneficial for the treatment of DME, it can be associated with several complications such as full-thickness retinal break, choroidal neovascularization (Fig. 19.3.1.3), subretinal fibrosis, or symptomatic scotomata [56, 59]. These complications can cause symptomatic visual loss.

Table 19.3.1.6. Grading of the foveal avascular zone (on fluorescein angiograms). (Adapted from [35])

- Grade 0: < 300 μm
- Grade 1: = 300 μm
- Grade 2: > 300 μm and < 500 μm
- Grade 3: > or = 500 μm
- Grade 4: cannot grade

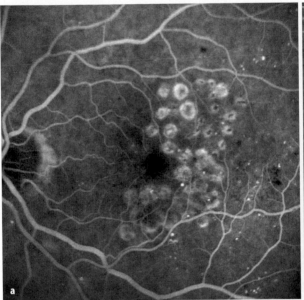
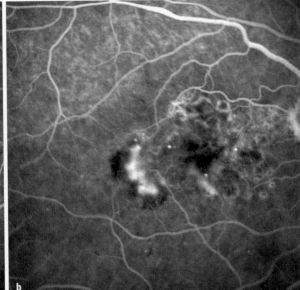
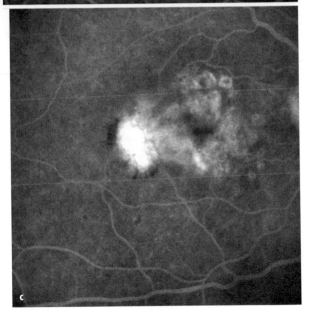

Fig. 19.3.1.3. Complications after grid laser coagulation: **a** enlarged central RPE scars of the left eye, no obvious macula edema; **b** early phase fluorescein angiography demonstrates CNV lesion growing from earlier photocoagulation scars of the right eye; **c** late phase with increased leakage of fluorescein

While focal laser coagulation reduces hypoxic areas and directly occludes leaky microaneurysms, the rationale for grid laser treatment in diffuse macular edema is not yet well established. Potentially, grid laser may produce its effect by thinning the retina, bringing retinal vessels closer to choroidal vessels, and permitting the retinal vessels to constrict by autoregulation, thereby decreasing retinal blood flow and consequently decreasing edema formation [184].

Currently there is no confirmed evidence that grid laser treatment improves diffuse diabetic macular edema (Table 19.3.1.5). Olk and coworkers demonstrated in a randomized study including 303 eyes that 3 years after grid laser treatment 14.5% of the treated eyes improved by 2 lines or more, 60.9% did not change significantly and 24.6% deteriorated by more than 2 lines. The results, however, were not compared to a non-treated control group [127, 128]. Data from other studies showed similar results. The comparability of the different studies is limited due to the different criteria of inclusion, exclusion, monitoring and treatment. Despite the lack of functional improvement (visual acuity) there is a reduction of retinal thickness (anatomical edema) after grid laser treatment as shown in several studies [107, 128].

In conclusion, the 3-year risk of massive visual loss from macular edema without focal laser treatment is about 30%, compared to 15% after focal laser treatment [30, 32]. Interestingly, scatter (panretinal) laser coagulation was not effective, but may be even deleterious in patients with mild diabetic retinopathy.

Prophylactic treatment of a non-significant macular edema is not advantageous over no treatment. Prophylactic laser coagulation is therefore not justified [2, 30, 71]. Laser coagulation of diabetic macular edema should only be considered when the edema is clinically significant (CSME) (see above).

19.3.1.3.2 Medical Treatment

Medical treatment of macular edema so far is best established in postsurgical and predominantly inflammatory edema, e.g., in uveitis, but is gaining importance in the treatment of diabetic macular edema. The majority of therapeutic strategies inhibit the release of inflammatory mediators and therefore target the pathogenetic factors responsible for the altered vascular permeability. The remaining treatments are mostly symptomatic and include carbonic anhydrase inhibitors and methods that increase blood flow and oxygenation (e.g., hyperbaric oxygenation, diuresis and dialysis [17, 98, 99, 112, 118, 136, 167, 172]), which are so far not applicable for diabetic macula edema.

Medical treatment for diabetic macular edema consists of therapeutic agents that are collectively categorized into three groups: corticosteroids, cyclooxygenase inhibitors and carboanhydrase inhibitors.

19.3.1.3.2.1 Carbonic Anhydrase Inhibitors and Non-steroidal Anti-inflammatory Drugs

The rationale of carbonic anhydrase inhibitors as a therapeutic agent in the treatment of macular edema is to improve the ability of the retinal pigment epithelial cells (RPE cells) to pump fluid out of the retina [112]. Currently, there are no available randomized studies that confirm a beneficial effect of carbonic anhydrase inhibitors in the treatment of macular edema. Non-randomized observations demonstrated improved visual function in patients with postsurgical macular edema, e.g., after cataract surgery or buckling procedures [24, 180]. The effect lasts only as long as the patient takes the drug (on-off effect, tachyphylaxis) [180]. The favorable reports that were described at first regarding the application of carbonic anhydrase inhibition in patients with macular edema secondary to retinitis pigmentosa are not supported by long-term observations. With the continuous use of methazolamide a rebound phenomenon is observed [41, 42].

In contrast to the routine clinical treatment, the use of carbonic anhydrase inhibition for macular edema is not based upon scientific evidence to date. The use of carbonic anhydrase inhibitors for diabetic macular edema is not recommended.

As cyclooxygenase inhibitors (NSAIDs) block the synthesis and release of prostaglandins, non-steroidal drugs have been investigated in the prophylaxis and therapy of postsurgical cystoid macular edema.

It is clear that NSAIDs target the inflammatory mediators that are responsible for the edema formation and although they may not be an optimal standalone treatment they can be used as steroid-sparing agents. There may be several explanations why NSAIDs cannot improve vision in diabetic macular edema, such as chronic edema, inflammation and ischemia that induce permanent structural alterations. Although effects on diabetic vascular leakage were achieved in preclinical studies [83, 84], there is so far no clinical evidence for an effect of NSAIDs in diabetic macular edema.

19.3.1.3.2.2 Corticosteroids

Steroids are currently regaining attention with the growing use of intravitreal triamcinolone (for details see Chapter 18). Corticosteroids block the release of arachidonic acid from cell membranes and thus reduce the synthesis of prostaglandins. Furthermore, they inhibit the migration of leukocytes and the release of proinflammatory mediators such as TNF-α and VEGF. Steroids specifically stabilize endothelial tight junctions and increase their numbers [8, 146]. As discussed previously, this is especially important to the development of macular edema.

Routes of administration are manifold, including topical, periocular, oral, and intravenous routes. Subtenon injections of corticosteroids are widely used in patients with asymmetric or unilateral uveitis [191]. The advantages of the periocular injections are high concentrations of corticosteroids in the posterior eye, and reduction of the adverse effects compared to systemic administration. Intraocular levels of corticosteroids are identical between subtenon and retrobulbar administration [172]. For oral administration, the initial high dose (1 – 1.5 mg/kg) is subsequently decreased according to clinical effect [26, 45, 145].

Recent publications suggest that the intravitreal application of triamcinolone seems to be a promising therapeutic method for macular edema that fails to respond to conventional treatment [75, 113]. Martidis et al. published a prospective, non-comparative, interventional case series to determine if intravitreal injection of triamcinolone acetonide is safe and effective in treating diabetic macular edema unresponsive to prior laser photocoagulation [113]. Sixteen eyes with clinically significant diabetic macular edema (CSME) that failed to respond to at least two previous sessions of laser photocoagulation were included in the study. The response of the laser treatment was measured by clinical examination and optical coherence tomography (OCT) at least 6 months after initial laser therapy. Eyes with a residual central macular thickness of more than 300 μm (normal: 200 μm) and visual loss from baseline were offered intravitreal injection of 4 mg triamcinolone acetonide. In this study, the mean improvement in visual acuity measured 2.4, 2.4, and 1.3 Snellen lines at the 1, 3, and 6 month follow-up intervals, respectively. The central macular thickness as measured by OCT decreased by 55%, 57.5%, and 38%, respectively, over these same intervals from an initial pretreatment mean of 540.3 μm (±96.3 μm). Intraocular pressure exceeded 21 mm Hg in five, three, and one eye(s), respectively, during these intervals. One eye exhibited cataract progression at 6 months. No other complications were noted over a mean follow-up of 6.2 months. Reinjection was performed in three of eight eyes after 6 months because of recurrence of macular edema.

Similar pilot studies were performed in patients with uveitis, central vein occlusion, and cystoid macular edema after cataract surgery [6, 55, 75, 192]. In most published reports, complications do not appear to be prohibitive; however, all reports demonstrate a limited number of selected cases.

Further randomized studies are therefore warranted to assess long-term efficacy and need for retreatment. Preliminary data of Jonas et al. suggests that there is no tachyphylaxis in visual acuity or intraocular pressure outcomes after repeated intravitreal injections of triamcinolone acetonide [81].

Reviewing the published data on intravitreal injections of triamcinolone acetonide, the therapeutic window seems very wide. The dose range of intravitreally injected triamcinolone acetonide varies from 2 mg [6] to 4 mg [113, 192] and even 25 mg in a single report [76]. Interestingly, reaccumulation of fluid in cystoid spaces occurs between 6 weeks and 3 months after injection, and this does not seem to be dose dependent. Repeated injections at intervals ranging from 10 weeks [77–80] to more than 6 months [6] show a variable treatment response.

There are currently no data on the pharmacokinetic profile of intravitreal triamcinolone, which might be altered from a previous vitrectomy. Physiological intravitreal cortisol levels are reported to be 5.1 ng/ml, and vitreous levels after peribulbar injections are in the range of 13 mg/ml. The effective dose of the triamcinolone acetonide is further influenced by the mandatory washes of the widely used stabilizing agent benzylethanol during the preparation of the injection that even if standardized alter the remaining amounts of the drug in the solution. Additionally, an inhibitory effect of the stabilizing agent on the drug cannot be excluded.

Jaffe and coworkers [73, 74] constructed a fluocinolone acetonide drug delivery device that releases fluocinolone acetonide in a linear manner over an extended period. A clinical phase III study by Bausch & Lomb investigated the efficacy of 0.5 mg (slow release) fluocinolone acetonice in 80 patients with diffuse diabetic macular edema. Patients receiving the implant showed a statistically significant regression of retinal thickness after 6 months in comparison to the control group. Furthermore, 80% of the eyes in the treatment group demonstrated a stable or improved visual acuity compared to only 50% of the eyes of the control group.

Complications of intravitreal triamcinolone acetonide delivery systems comprise retinal detachment, vitreous hemorrhage, increased intraocular pressure, cataract formation, and pseudohypopyon. Elevation of the intraocular pressure after triamcinolone acetonide of more than 5 mm Hg has been reported in up to 30% of eyes [185]. It is therefore prudent that patients are asked about any history of a previous steroid response. The incidence of culture-positive endophthalmitis following intravitreal triamcinolone is as rare as 0.87% in a large, multicenter, retrospective case series [122]. It occurs rapidly (median 7.5 days) and can result in severe loss of vision and the eye. The risk of endophthalmitis is considerably higher compared to other intravitreally injected drugs. Engstrom and Holland reported the rate of endophthalmitis following intravitreal ganciclovir injection as 0.29% (4 cases in 1,372 injections) [37]. The greater risk of endophthalmitis following intravitreal triamcinolone injection may be partly due to a small-size bias or an increased susceptibility to infections in diabetic individuals. Roth and coworkers reported on seven patients who developed a clinical picture simulating endophthalmitis following intravitreal triamcinolone injection [148]. Extensive signs of inflammation developed 1–2 days after injection, at an earlier time point than in bacterial endophthalmitis. Vitreous taps were sterile and inflammation resolved spontaneously with recovery to preinjection visual acuity or better. This inflammatory response might be a response to the stabilizing additive benzylethanol (see above).

In any case, it is recommended to follow a sterile protocol for intravitreal injections of triamcinolone acetonide (see Table 19.3.1.7).

Table 19.3.1.7. Recommendations when performing intravitreal triamcinolone acetonide injections. (Modified after [122])

- Strict adherence to a sterile technique with the use of a lid speculum
- If the patient is immunocompromised, consideration should be given to delaying the procedure
- Any external ocular inflammation/infection should be controlled prior to administering intravitreal triamcinolone acetonide injections
- Intravitreal triamcinolone acetonide injections should only be performed in the presence of a filtering bleb after careful consideration of the risk of endophthalmitis vs. the possibility of visual improvement
- Patients should be seen by an ophthalmologist at least once in the immediate postoperative period (1–7 days) after intravitreal triamcinolone acetonide injections
- Patients should be instructed to return to their ophthalmologist immediately at the first sign of any visual disturbance or pain
- Prompt intervention should be instituted, including obtaining a vitreous specimen for culture and administering intravitreal antibiotics

19 III

19.3.1.3.2.3 Antiangiogenic Treatment

The rationale and treatment are discussed in detail in Chapters 19.3.2 and 19.3.3.

VEGF is one of the most potent angiogenic growth factors to date. It is secreted by a variety of normal and cancer cells, acting as an endothelial cell mitogen and permeability factor, which is why it was originally named vascular permeability factor (VPF).

VEGF antagonists might be effective in diabetic macular edema through inhibition of vascular leakage. There are currently two products that underwent phase III clinical trials in the treatment of neovascular age-related macular degeneration: the anti-VEGF pegylated aptamer (pegaptanib, Macugen, Eye-Tech Pharmaceuticals – Pfizer) and an anti-VEGF humanized neutralizing antibody fragment (ranibizumab, Lucentis, Genentech – Novartis). Bevacizumab (Avastin) is the full-length antibody that ranibizumab was derived from and is in use for cancer treatment. Currently the off-label use of bevacizumab is gaining attention. Nevertheless, unless large scale reports with a considerable follow-up are published, such off-label treatment cannot be recommended for broad clinical use. So far, all anti-VEGF drugs need to be injected intravitreally repetitively. The results of clinical trials for diabetic macular edema are discussed in Chapters 19.3.1, 19.3.2 and 19.3.3.

19.3.1.3.3 Surgical Approaches

Essentials
- Vitrectomy with or without peeling of posterior hyaloid membrane may be beneficial for the treatment of DME in eyes that are resistant to laser photocoagulation and/or steroid injection
- Vitrectomy without ILM peeling might be beneficial in patients with a taut posterior hyaloid lacking a posterior vitreous detachment
- The ILM originates from the lens and ciliary body and is not pathogenically involved in diabetic macular edema. Nevertheless, ILM peeling is performed to reduce the diffusion barrier towards the vitreous cavity and to prevent epiretinal membrane formation
- The complications encountered with PPV for diabetic macula edema should be considered: cataract formation, epiretinal membrane formation, fibrinoid syndrome, development of hard exudates, macular ischemia, neovascular glaucoma, retinal detachment, retinal tear, tractional rhegmatogenous retinal detachment, vitreous hemorrhage

- There are no studies which have proved that hyperbaric oxygenation and plasma membrane filtration can exert a beneficial effect on diabetic macula edema

There is clinical evidence that traction forces at the vitreoretinal interface may play an important role in the pathogenesis of macular edema. Several authors have studied vitrectomy for persistent macular edema and have suggested that release of the tractional forces at the vitreomacular interface may improve resolution of the macular edema and restore visual acuity (Tables 19.3.1.8, 19.3.1.9). The mechanism by which the release of traction resolves macular edema is only partially understood (see Sect. 19.3.1.1.3 above). Some investigators hypothesize that vitrectomy may improve perifoveal retinal microcirculation [88]. Some hypothesize that the posterior hyaloid exerts tangential tractional forces that lead to shallow macular detachment and that are released with surgery [89]. The observation that patients with a diffuse diabetic macula edema have a reduced incidence of posterior vitreous detachment generated the idea that posterior vitreous detachment during vitrectomy could be used as a therapeutic measure in patients with macula edema [123, 124]. Hikichi and coworkers report a 55 % resolution of diabetic macular edema following posterior vitreous detachment. In contrast only a 25 % resolution of diabetic macular edema was observed in patients with attached hyaloid [63]. Peeling of the inner limiting membrane of the retina ensures complete release of tractional forces, removes a potential diffusional barrier, and inhibits reproliferation of fibrous astrocytes [46].

It is, however, still a matter of investigation whether such mechanical interventions effectively resolve macular edema and allow for a long-term functional benefit for the patient.

19.3.1.3.3.1 Pars Plana Vitrectomy with Posterior Vitreous Detachment

It was demonstrated that a surgically induced posterior vitreous detachment in patients with a diffuse diabetic macular edema leads to a reduction of macular edema with a subsequent increase in visual acuity [46, 60, 109].

In the study by Pendergast and coworkers, 27 (49.1 %) of 55 eyes demonstrated improvement in best-corrected visual acuity of 2 or more lines. Fifty-two (94.5 %) of the 55 vitrectomized eyes showed improvement in clinically significant macular edema and in 45 eyes (81.8 %) the macular edema resolved completely during a mean period of 4.5 months

Table 19.3.1.8. Non-randomized case control studies investigating PPV without ILM peeling for diabetic macular edema. Review of the literature and clinical results

Authors	Number of patients	Eyes	Anatomical starting situation	Observation period (months)	Visual acuity			Resolution of edema		
					Improved	Unchanged	Worse	Complete	Partial	Unchanged
Lewis et al. [109]	10	10	Diffuse	16	90% (after cataract surgery)	None	10%	None	80%	20%
Van Effenterre et al. [174]	18	22	Diffuse	14	86%	13.6%	None	54%	45%	None
Harbour et al. [60]	7	7	Diffuse	14.5	60%	30%	10%	50%	20%	30%
Tachi et al. [168]	41	58	Diffuse	12	53.4%	31.1%	15.6%	98%	None	2%
Micelli-Ferrari et al. [117]	18	18	9 diffuse 9 cystoid	12	Only intermediate improvement			Only intermediate improvement		
Pedergast et al. [134]	50	56	Diffuse	23.2	49.1%	41.8%	9.1%	81.8%	12.7%	5.5%
La Heij et al. [104]	19	21	11 diffuse 2 cystoid 8 unknown	3	47.6% (71%)	19%	10%	100%	None	None
Yamamoto et al. [187]	29	30	30 diffuse 12 cystoid	10.8	43%	50%	7%	20%	57%	23%
Sato et al. [151]	40	45	Cystoid	6	51%	47%	2%	Unknown	Unknown	Unknown
Otani and Kishi [130, 131]	7	7 (7 controls)	Diffuse	5	57%	43%	None	None	100%	None
Ikeda et al. [68]	3	3	Diffuse	5	100%	None	None	100%	None	None
Ikeda et al. [69]	5	5	Diffuse		100%	None	None	60%	20%	20%
Yang et al. [188]	11	13	Diffuse	14.8	100%	None	None	85%	15%	None

Table 19.3.1.9. Non-randomized case control studies investigating PPV with ILM peeling for diabetic macular edema. Review of the literature and clinical results

Authors	Number of patients	Eyes	Anatomical starting situation	Observation period (months)	Visual acuity			Resolution of edema		
					Improved	Unchanged	Worse	Complete	Partial	Unchanged
Gandorfer et al. [46]	11	12	Diffuse	16	92%	8%	None	100%	None	
Kumagai et al. [101]	103	135	Diffuse	20	?	None	10%	None	80%	20%
Ndoye Roth et al. [125]	15	19	Cystoid	9.5	57.8%	86%	86%	86%	86%	86%
Dillinger et al. [27]	55	60	60 diffuse 25 cystoid	3	43%	30%	10%	50%	20%	30%
Kuhn et al. [100]	27	30	16 diffuse 14 cystoid	12	66%	31.1%	15.6%	98%	None	2%
Radetzky et al. [140]	5	5	Diffuse	9.4	60%	Intermediate	Intermediate	Intermediate	Intermediate	Intermediate
Rosenblatt et al. [147]	20	26	Diffuse and cystoid	8	50%	38.5%	11.5%	81.8%	12.7%	5.5%
Stefaniotou et al. [164]	52	73	Diffuse	?	63%	19%	10%	100%	None	None

(range 1–13 months). Eyes with macular ischemia and preoperative best-corrected visual acuity of 20/200 or less tended to respond less favorably to vitrectomy than eyes lacking these characteristics. All eyes had at least 6 months of follow-up after surgery, with a mean follow-up of 23.2 months [134].

Lewis et al. and similarly Harbour et al. demonstrated in eyes with diabetic macula edema and associated macular traction that pars plana vitrectomy (PPV) with removal of the posterior hyaloid face resulted in reduced macular edema and in improved visual acuity [60, 109]. Ikeda and coworkers reported the results of PPV with the creation of a posterior hyaloid detachment in three eyes with cystoid diabetic macula edema [68]. Macular edema resolved in 100% of eyes and was associated with an improvement of visual acuity. La Heij et al. reported that macular edema resolved in 100% of eyes that underwent PPV with induction of a posterior vitreous detachment [104]. Postoperatively, visual acuity significantly improved. Interestingly, eyes without preoperative macular laser photocoagulation had a significantly higher percentage of visual improvement than eyes with preoperative macular laser treatment. Similarly, the report by Otani and Kishi, who studied 13 eyes with diffuse or cystoid diabetic macula edema and no posterior vitreous detachment, emphasizes the superiority of posterior vitreous detachment in these eyes over laser photocoagulation. Visual acuity improved in 38% of eyes by more than two lines after PPV and removal of the posterior hyaloid, although previous laser treatment did not exert any functional benefit [131].

Yamamoto and coworkers retrospectively reviewed records of 30 eyes with or without posterior vitreous detachment in combination with or without epimacular membranes [187]. They found an improvement of postoperative function that was independent of the preoperative anatomical situation.

Nevertheless, there are few studies demonstrating a long-term follow-up. Micelli-Ferrari and coworkers prospectively studied 18 eyes with diffuse and cystoid DME that underwent PPV. Visual acuity improved significantly 1 month after surgery in the group with diffuse diabetic macula edema, but decreased to preoperative values 10 months after surgery [117].

Overall most studies demonstrated an improvement of function after vitrectomy; however, there was a strong correlation between the preoperative and postoperative visual acuity.

If judging the final benefit, the complications encountered with pars plana vitrectomy for diabetic macula edema should be considered; these comprise cataract formation (up to 63%), retinal detachment (10%), retinal tear (14–21%), tractional rhegmatogenous retinal detachment (5–8%), and vitreous hemorrhage (12–16%), epiretinal membrane formation (8–12%), fibrinoid syndrome (8%), glaucoma (2–8%), development of hard exudates (4%), macular ischemia (11%), and neovascular glaucoma (4–8%).

19.3.1.3.3.2 Vitrectomy and Peeling of the Inner Limiting Membrane

The rationale for vitrectomy (removal of the hyaloid) plus peeling of the internal limiting membrane is the postulated improvement of fluid diffusion from the retina to the vitreous cavity. ILM peeling is the technique of choice for macular holes, macular edema, and improved pucker surgery. After pars plana vitrectomy, ILM is dissected from the retinal surface. Staining with indocyanine green (ICG) or other dyes can be used to facilitate visualization with a subsequent peeling of the ILM with a microforceps. The ILM can usually be torn like the anterior lens capsule in patients with macular hole, but may be more fragile and adhesive in patients with diabetic edema. The risks of retinal breaks and consecutive retinal detachment (about 5–8%), cataract (up to 63%), and endophthalmitis in any vitreous surgery have to be considered (see above).

However, the foot processes of Müller cells, which adhere to the ILM, may be damaged by traumatic peeling of the ILM off the retinal surface. Histological studies demonstrated cellular elements resembling the plasma membrane of Müller cells and other undetermined retinal structures adherent to the retinal side of the ILM [61]. There is an ongoing discussion as to whether the use of ICG and the toxicity thereof leads to a more traumatic removal of the membrane compared to ILM peeling without staining. Arguments for the use of ICG are the better control of manipulation and the lower risk of iatrogenic damage. In general, with the use of ICG, surgery tends to be performed at an earlier stage and with better initial visual acuity.

Vitrectomy including removal of the ILM aids the resolution of diffuse diabetic macular edema and improvement of visual acuity and prevents epiretinal membrane formation.

Gandorfer et al. described 12 eyes with CSME, 11 of which had previous laser photocoagulation treatment without improvement and that underwent PPV with removal of the posterior hyaloid and ILM peeling. Intraoperatively, 10 patients had a thickened posterior hyaloid attached to the macula. Postoperatively, macular edema completely or partially resolved in 10 eyes, and visual acuity improved in 11 eyes [46].

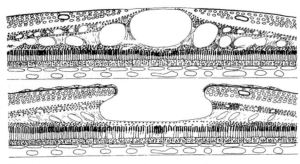

Fig. 19.3.1.4. Cystic macular edema – pseudo-hole

The advantage of ILM peeling over vitrectomy alone is the complete release of tractional forces and inhibition of reproliferation of fibrous astrocytes, which seems to be prudent in the eyes of patients with diabetes and advanced vitreoretinal interface disease of the macula. Looking at the histological appearance of a cystic macular edema as demonstrated schematically in Fig. 19.3.1.4, large bullous edema can potentially be worsened after ILM peeling and can proceed to a pseudo-macular hole. In this respect careful examination and the use of optical coherence tomography might help to exclude patients with large bullae.

Our own data suggest that ILM peeling is ineffective in resolving the macular edema in CRVO and proliferative diabetic retinopathy. Improvements were apparent only in non-proliferative diabetic retinopathy: A retrospective review analyzed a series of 23 eyes from 23 patients with persistent macular edema after PPV with ICG assisted peeling of the ILM. The main diagnoses were uveitis (anterior, intermedia, posterior and panuveitis) ($n=9$), central retinal vein occlusion (CRVO) ($n=4$), diabetic retinopathy (DR) ($n=5$), vitreoretinal traction syndrome ($n=2$), and Irvine-Gass syndrome ($n=3$). Visual acuity improved in 9 out of 23 patients after 3 months and in 6 out of 21 and in 7 out of 21 patients after 6 months. This improvement was predominantly seen in patients with uveitis (5/9), or diabetic maculopathy (3/5); one patient with Irvine-Gass syndrome showed significant reduction, whereas one with vitreoretinal traction showed improvement in visual acuity. The group of CRVO patients showed no significant change during the follow-up. The use of endotamponade did not influence the outcome. Patients with uveitis and non-proliferative diabetic maculopathy demonstrated a transient benefit. The lack of long-term improvement is in accordance with the hypothesis that ILM peeling cannot interfere with the mechanism of macular edema [140].

However, it is not clear that the ILM peeling is necessary for PPV to be an effective treatment for DME. Even among patients in whom DME is not associated with clinically evident posterior hyaloidal thickening or traction, PPV can lead to resolution of macular edema and improved vision.

Tachi and Ogino performed vitrectomy on 58 eyes with DME without posterior vitreous detachment and without a thick posterior hyaloid [168]. Eleven eyes had undergone laser photocoagulation for DME previously. At 12 months follow-up, macular edema resolved in 98% of eyes, and visual acuity improved in 53% of eyes.

Ikeda and coworkers reported their experience with five eyes with cystoid DME and a detached posterior hyaloid membrane that underwent PPV [69]. Macular edema resolved in 60% of eyes, and visual acuity improved by two lines in 80% of eyes. In two eyes macular edema persisted. In one of these two eyes macular edema resolved after grid photocoagulation. In the second eye, macular edema did not resolve even after photocoagulation.

Yang reported that PPV with posterior hyaloid removal could be beneficial in eyes with DME with massive hard exudates that have responded poorly to conventional laser photocoagulation [188]. Macular edema and hard exudates significantly decreased in all 13 eyes (100%) and visual acuity improved in 11/13 eyes (85%).

In conclusion, long-term data on the effectiveness of ILM peeling for resolution of macular edema are not available. Additional randomized studies have to be performed that should include data on reading performance and retinal thickness to better correlate anatomical alterations to clinically relevant functional parameters.

19.3.1.3.4 Modification of Systemic Blood Flow

Hyperbaric oxygen is thought to alter the blood flow via vasoconstriction, and to facilitate the reformation of damaged junctional complexes in the vessel wall. Thus, hyperbaric oxygen works predominantly at the level of the inner blood retinal barrier. Intermittent hyperbaric oxygen for 21 days showed an improvement in visual acuity in patients with chronic cystoid macula edema (CME) after cataract extraction [136]. The improvement in visual acuity, however, does not correlate to a reduction of macular edema. It is possible that hyperbaric oxygen alters macular ischemia or affects the anterior segment of the eye. In uveitis-associated CME, hyperbaric oxygen had no significant effect [99, 118]. Other rheological treatments such as plasma membrane filtration demonstrated good effects in initial studies, but did not enter large-scale prospective studies [183]. Still there is no evidence that rheological measures have any effect on diabetic macula edema.

19 III

19.3.1.4 Discussion: Open Questions and Technical Aspects

Most of the presented data for the surgical methods and pharmacological treatments represent small case series. In order to further evaluate the discussed treatment approaches randomized prospective studies in a large population are needed. However, not only the endpoint criteria and the measurement approaches should be evaluated, but also the best time point of treatment or the question of whether to treat ischemic forms of macular edema and how.

Many studies presenting positive treatment results use the anatomical reduction of macular edema as their endpoint. It is well known that the anatomical endpoint (decrease in retinal thickness, or a "dry macula" on angiography) in many cases differs considerably from the functional endpoint (visual acuity and reading ability). Grid laser has been shown to be efficacious in reducing vascular leakage; however, it does not improve visual acuity (see Table 19.3.1.2). Evaluation of trials with different protocols remains difficult. Thus, to evaluate the efficacy of different treatment approaches a prospective randomized design is necessary, which should emphasize functional rather than anatomical endpoints. Reading ability is an excellent measure of macular function. The measurement should include reading acuity, as well as maximum reading speed. For this purpose, the standardized Radner Reading Charts provide clinically reliable and reproducible results for individuals with normal eyesight and for patients with visual impairment. These reading test systems, which take into account the current international standards for visual acuity measurements (EN ISO 8596, NAS-NRC) and the psychophysical requirements for controlling optical item interactions, can provide reliable measures for clinical and scientific analyses of reading performance [141, 165].

Retinal thickness measurement with optical coherence tomography (OCT) is an established method for quantitative assessment of macular edema. Retinal thickness is determined as micrometers maximal thickness within 500 μm around the fovea. The normal thickness values for patients without edema are as follows: central part of central subfield 155 μm; central subfield as a whole 180 μm; inner superior, nasal and inferior subfields 260 μm; inner temporal subfield 250 μm; outer superior, inferior, and temporal subfields 230 μm; outer nasal subfield 250 μm. A desired effect for any treatment is decreasing retina thickness to within normal (175–200 μm). A reduction of 75–150 μm from baseline may be shown to be clinically significant. It is important to consider the high variability in fixation in patients with macular edema. For quantitative comparison of retinal thickness in repeated measurements during treatment, care should be taken that the system software chooses the point of measurement independently of the patient's fixation.

The OCT is currently unable to distinguish the ILM from the outer retinal layers. However, it might be useful to exclude any kind of retinal traction. As discussed above, a large edematous cyst might be clinically difficult to distinguish from a full thickness macular hole associated with a macular pucker. OCT can easily answer these questions, but does not answer the question of ischemia, which in most cases determines whether to treat or not. Almost the only indication for fluorescein angiography in diabetic patients and patients with macula edema is to exclude macular ischemia (Table 19.3.1.6).

Ischemic maculopathy remains untreatable in the area of laser coagulation. A foveal avascular zone of more than 500 μm should be considered ischemic and thus according to the current recommendations cannot be treated. However, there is still hope that ischemic maculopathy can be treated pharmacologically. For ischemic ophthalmopathy this has been demonstrated in a preliminary report using intravitreal triamcinolone, which was demonstrated to reduce iris neovascularization and increase visual function [77]. In contrast it is important to note that antiangiogenic drugs such as VEGF antagonists potentially increase ischemia and they need to be carefully investigated in this respect.

Early intervention in macular edema is undoubtedly advantageous, as the risk of ultrastructural alterations induced by a persistent macular edema increases with time. It is well known that with time the central avascular zone and the areas of ischemia are likely to increase. Late treatment risks transition to untreatable ischemic forms of macular edema. To date, however, most surgical approaches will only be considered for persistent macular edema unresponsive to laser treatment or pharmacological approaches. As a general rule surgical and medical treatment could be considered for eyes with a BCVA 20/50 + 3 (68 ETDRS letters) and 20/320 (25 ETDRS letters).

The different treatment approaches are likely to affect the clinical course of macular edema at variable time points and for different time periods. While the effect of intravitreal steroids is known to deteriorate with time, similar fluctuations are likely for other treatments. Especially for the surgical options long-term data is currently unavailable. Thus, any treatment option should be evaluated for the duration of the anticipated beneficial effect and beyond. As macular edema requiring treatment appears to be mostly chronic, a follow-up of only 6–8 weeks is inefficient.

19.3.1.5 Summary

Diabetic macular disease is considered a structural alteration of the macula in any of the following conditions:

- Collection of intraretinal fluid in the macula with or without exudates (lipids) and with or without cystoid changes
- Non-perfusion of parafoveal capillaries with or without intraretinal fluid
- Traction in the macula by fibrous tissue proliferation that drags the retinal tissue, causing surface wrinkling or detachment of the macula
- Intraretinal or preretinal hemorrhage in the macula
- Lamellar or full-thickness retinal hole formation
- Combination of the above

Treatment by laser coagulation is limited to focal edema, but is controversial in diffuse edema and has proven to be ineffective in ischemic diabetic maculopathy. Currently, pharmacological approaches which inhibit growth factor activity (anti-VEGF therapies) as well as anti-inflammatory strategies are being established. Furthermore, ongoing clinical trials are analyzing the attempt to surgically reduce the diffusion barrier by ILM peeling in comparison to triamcinolone acetonide.

References

1. Abrass CK (1998) Measurement of the rates of basal pinocytosis of horseradish peroxidase and internalization of heat-aggregated IgG by macrophages from normal and streptozotocin-induced diabetic rats. Immunology 65:411–415
2. Aiello LM, Ferris FL, 3rd (1987) Photocoagulation for diabetic macular edema. Arch Ophthalmol 105:1163
3. Aiello LP, Bursell S-E, Clermont A, et al. (1997) Vascular endothelial growth factor – induced retinal permeability is mediated by protein kinase C in vivo and suppressed by an orally effective beta isoform selective inhibitor. Diabetes 46:1473–1480
4. Amin RH, Frank RN, Kennedy A, Eliott D, Puklin JE, Abrams GW (1999) Vascular endothelial growth factor is present in glial cells of the retina and optic nerve of human subjects with nonproliferative diabetic retinopathy. Invest Ophthalmol Vis Sci 38:36–47
5. Antcliff RJ, Marshall J (1999) The pathogenesis of edema in diabetic maculopathy. Semin Ophthalmol 14:223–232
6. Antcliff RJ, Spalton DJ, Stanford MR, Graham EM, Ffytche TJ, Marshall J (2001) Intravitreal triamcinolone for uveitic cystoid macular edema: an optical coherence tomography study. Ophthalmology 108:765–772
7. Antonetti DA, Barber AJ, Khin S, Lieth E, Tarbell JM, Gardner TW (1998) Vascular permeability in experimental diabetes is associated with reduced endothelial occludin content: vascular endothelial growth factor decreases occludin in retinal endothelial cells. Penn State Retina Research Group. Diabetes 47:1953–1959
8. Antonetti DA, Wolpert EB, DeMaio L, Harhaj NS, Scaduto RC, Jr (2002) Hydrocortisone decreases retinal endothelial cell water and solute flux coincident with increased content and decreased phosphorylation of occludin. J Neurochem 80:667–677
9. Asrani S, Zou S, d'Anna S, Vitale S, Zeimer R (1999) Noninvasive mapping of the normal retinal thickness at the posterior pole. Ophthalmology 106:269–273
10. Bailey CC, Sparrow JM, Grey RH, Cheng H (1999) The National Diabetic Retinopathy Laser Treatment Audit. III. Clinical outcomes. Eye 13:151–159
11. Bandello F, Brancato R, Meschini U, Virgili G, Lanzetta P, Ferrari E, Incorvaia C (2001) Light panretinal photocoagulation (LPRP) versus classic panretinal photocoagulation (CPRP) in proliferative diabetic retinopathy. Semin Ophthalmol 16:12–18
12. Behzadian MA, Wang XL, Windsor LJ, Ghaly N, Caldwell RB (2001) TGF-beta increases retinal endothelial cell permeability by increasing MMP-9: possible role of glial cells in endothelial barrier function. Invest Ophthalmol Vis Sci 42:853–859
13. Berse B, Brown LF, Van de Water L, Dvorak HF, Senger DR (1992) Vascular permeability factor (vascular endothelial growth factor) gene is expressed differentially in normal tissues, macrophages, and tumors. Mol Biol Cell 3:211–220
14. Blankenship GW (1979) Diabetic macular edema and argon laser photocoagulation: a prospective randomized study. Ophthalmology 86:69–76
15. Boulton ME, Xiao M, Khaki A (1995) Changes in growth factor expression in pig eyes following scatter laser photocoagulation. Invest Ophthalmol Vis Sci 36(Suppl):95
16. Bresnick GH (1983) Diabetic maculopathy. A critical review highlighting diffuse macular edema. Ophthalmology 90:1301–1317
17. Bresnick GH (1986) Diabetic macular edema. A review. Ophthalmology 93:989–997
18. Bron AJ, Tripathi RC, Tripathi BJ (1997) The inner limiting membrane, Chapter 14. In: Wolff's anatomy of the eye. Chapman & Hall, London, p 488
19. Brown JC, Solomon SD, Bressler SB, Schachat AP, DiBernardo C, Bressler NM (2004) Detection of diabetic foveal edema. Contact lens biomicroscopy compared with optical coherence tomography. Arch Ophthalmol 122:330–335
20. Cardier JE, Schulte T, Kammer H, Kwark J, Cardier M (1999) Fas (CD95-Apo-1) antigen expression and function in murine liver endothelial cells: implications for the regulation of apoptosis in liver endothelial cells. FASEB J 13:1950–1960
21. Chan A, Duker JS (2005) A standardized method for reporting changes in macular thickening using optical coherence tomography. Arch Ophthalmol 123:939–943
22. Chomczynski P, Sacchi N (1987) Single-step method of RNA isolation by acid guanidinium thiocyanate-phenol-chloroform extraction. Anal Biochem 162:156–159
23. Connolly DT, Olander JV, Heuvelman D, et al. (1989) Human vascular permeability factor: Isolation from U937 cells. J Biol Chem 264:20017–20024
24. Cox SN, Hay E, Bird AC (1988) Treatment of chronic macular edema with acetazolamide. Arch Ophthalmol 106:1190–1195
25. Das A, McGuire PG, Eriqat C, Ober RR, DeJuan E Jr, Williams GA, McLamore A, Biswas J, Johnson DW (1999) Human diabetic neovascular membranes contain high levels of urokinase and metalloproteinase enzymes. Invest Ophthalmol Vis Sci 40:809–813

26. Dick AD (1994) The treatment of chronic uveitic macular edema. Br J Ophthalmol 78:1–2

27. Dillinger P, Mester U (2004) Vitrectomy with removal of the internal limiting membrane in chronic diabetic macular oedema. Graefes Arch Clin Exp Ophthalmol 242:630–637

28. Dogru M, Nakamura M, Inoue M, Yamamoto M (1999) Long-term visual outcome in proliferative diabetic retinopathy patients after panretinal photocoagulation. Jpn J Ophthalmol 43(3):217–24

29. Dollery CM, McEwan JR, Henney AM (1995) Matrix metalloproteinases and cardiovascular disease. Circ Res 77:863–868

30. Early Treatment Diabetic Retinopathy Study Research Group (1985) Photocoagulation for diabetic macular edema. Early Treatment Diabetic Retinopathy Study Report Number 1. Arch Ophthalmol 103:1796–1806

31. Early Treatment Diabetic Retinopathy Study Research Group (1987) Treatment techniques and clinical guidelines for photocoagulation of diabetic macular edema. Early Treatment Diabetic Retinopathy Study Report Number 2. Ophthalmology 94:761–774

32. Early Treatment Diabetic Retinopathy Study Research Group (1987) Techniques for scatter and local photocoagulation treatment of diabetic retinopathy: Early Treatment Diabetic Retinopathy Study Report No. 3. Int Ophthalmol Clin 27:254–264

33. Early Treatment Diabetic Retinopathy Study Research Group (1987) Photocoagulation for diabetic macular edema: Early Treatment Diabetic Retinopathy Study Report No. 4. Int Ophthalmol Clin 27:265–272

34. Early Treatment Diabetic Retinopathy Study Research Group (1991) Early photocoagulation for diabetic retinopathy. ETDRS Report Number 9. Ophthalmology 98:766–785

35. Early Treatment Diabetic Retinopathy Study Research Group (1991) Classification of diabetic retinopathy from fluorescein angiography. ETDRS Report No. 11. Ophthalmology 98:807–822

36. Engerman RL, Kern TS (1998) Retinopathy and tissue hexose in drug-treated animals. Arch Ophthalmol 116:543–544

37. Engstrom RE Jr, Holland GN (1995) Local therapy for cytomegalovirus retinopathy. Am J Ophthalmol 120:376–85

38. Fernandez-Patron C, Zouki C, Whittal R, Chan JSD, Davidge ST, Filep JG (2001) Matrix metalloproteinases regulate neutrophil-endothelial cell adhesion through generation of endothelin-1. FASEB J 15:2230–2240

39. Ferrara N, Houck K, Jakeman L, Leung DW (1992) Molecular and biological properties of the vascular endothelial growth factor family of proteins. Endocr Rev 13:18–32

40. Ferris FL 3rd, Patz A (1984) Macular edema. A complication of diabetic retinopathy. Surv Ophthalmol 28 (Suppl): 452–461

41. Fishman GA, Glenn AM, Gilbert LD (1993) Rebound of macular edema with continued use of methazolamidine in patients with retinitis pigmentosa. Arch Ophthalmol 111: 1640–1646

42. Fishman GA, Gilbert LD, Anderson RJ, Marmor MF, Weleber RG, Viana MA (1994) Effects of methazolamidine on chronic macular edema in patients with retinitis pigmentosa. Ophthalmology 101:687–693

43. Fitzgerald ME, Caldwell RB (1990) The retinal microvasculature of spontaneously diabetic BB rats: structure and luminal surface properties. Microvasc Res 39:15–27

44. Foos RY, Wheeler NC (1982) Vitreoretinal juncture. Synchysis senilis and posterior vitreous detachment. Ophthalmology 89:1502–1512

45. Freeman G (2001) Cystoid macular oedema in uveitis: an unsolved problem. Eye 15:12–17

46. Gandorfer A, Messmer EM, Ulbig MW, Kampik A (2000) Resolution of diabetic macular edema after surgical removal of the posterior hyaloid and the inner limiting membrane. Retina 20:126–133

47. Gandorfer A, Rohleder M, Grosselfinger S, Haritoglou C, Ulbig M, Kampik A (2005) Epiretinal pathology of diffuse diabetic macular edema associated with vitreomacular traction. Am J Ophthalmol 139:638–652

48. Gardner T (1995) Histamine, ZO-1 and increased blood retinal permeability in diabetic retinopathy. Trans Am Ophthalmol Soc 93:583–621

49. Gardner TW, Eller AW, Friberg TR (1991) Reduction of severe macular edema in eyes with poor vision after panretinal photocoagulation for proliferative diabetic retinopathy. Graefes Arch Clin Exp Ophthalmol 229(4):323–8

50. Gilbert RE, Kelly DJ, Cox AJ, Wilkinson-Berka JL, Rumble JR, Osicka T, Panagiotopoulos S, Lee V, Hendrich EC, Jerums G, Cooper ME (2000) Angiotensin converting enzyme inhibition reduces retinal overexpression of vascular endothelial growth factor and hyperpermeability in experimental diabetes. Diabetologia 43:1360–1367

51. Gill DS, Barradas MA, Fonseca VA, Dandona P (1989) Plasma histamine concentrations are elevated in patients with diabetes mellitus and peripheral vascular disease. Metabolism 38:243–247

52. Glaser BM, Campochiaro PA, Davis JL Jr, Jerdan JA (1987) Retinal pigment epithelial cells release inhibitors of neovascularization. Ophthalmology 94:780–784

53. Gottfredsdottir MS, Stefansson E, Jonasson F, Gislason I (1993) Retinal vasoconstriction after laser treatment for diabetic macular edema. Am J Ophthalmol 115:64–67

54. Grant MB, Caballero S, Tarnuzzer RT, Bass KE, Ljubimov AV, Spoerri PE, Galardy RE (1998) Matrix metalloproteinases expression in human retinal microvascular cells. Diabetes 47:1311–1317

55. Greenberg PB, Martidis A, Rogers AH, Duker JS, Reichel E (2002) Intravitreal triamcinolone acetonide for macular oedema due to central retinal vein occlusion. Br J Ophthalmol 86:247–248

56. Guyer DR, D'Amico DJ, Smith CW (1992) Subretinal fibrosis after laser photocoagulation for diabetic macular edema. Am J Ophthalmol 113:652–656

57. Halfter W, Dong S, Schurer B, Ring C, Cole GJ, Eller A (2005) Embryonic synthesis of the inner limiting membrane and vitreous body. Invest Ophthalmol Vis Sci 46: 2202–2209

58. Halfter W, Willem M, Mayer U (2005) Basement membrane-dependent survival of retinal ganglion cells. Invest Ophthalmol Vis Sci 46:1000–1009

59. Han DP, Mieler WF, Burton TC (1992) Submacular fibrosis after photocoagulation for diabetic macular edema. Am J Ophthalmol 113:513–521

60. Harbour JW, Smiddy WE, Flynn HW Jr, Rubsamen PE (1996) Vitrectomy for diabetic macular edema associated with a thickened and taut posterior hyaloid membrane. Am J Ophthalmol 121:405–413

61. Haritoglou C, Gandorfer A, Gass CA, Schaumberger M, Ulbig MW, Kampik A (2002) Indocyanine green-assisted peeling of the internal limiting membrane in macular hole surgery affects visual outcome: a clinicopathologic correlation. Am J Ophthalmol 134:836–841

62. Hee MR, Izatt JA, Swanson EA, Huang D, Schuman JS, Lin

CP, Puliafito CA, Fujimoto JG (1995) Optical coherence tomography of the human retina. Arch Ophthalmol 113:325 – 332

63. Hikichi T, Fuijio N, Akiba J, Azuma Y, Takahashi M, Yoshida A (1997) Association between the short-term natural history of macular edema and the vitreomacular relationship in type II diabetes mellitus. Ophthalmology 104:473 – 478

64. Hofman P, Blauwegers HG, Tolentino MJ, Adamis AP, Nunes Cardozo BJ, Vrensen GF, Schlingemann RO (2000) VEGF-A induced hyperpermeability of blood-retinal barrier endothelium in vivo is predominantly associated with pinocytotic vesicular transport and not with formation of fenestrations. Vascular endothelial growth factor-A. Curr Eye Res 21:637 – 645

65. Hogan MJ, Alvaroda JA, Weddell JE (1971) Retina. In: Histology of the human eye. WB Saunders, Philadelphia, pp 393 – 522

66. Huang D, Swanson EA, Lin CP, Schuman JS, Stinson WG, Chang W, Hee MR, Flotte T, Gregory K, Puliafito CA, et al. (1991) Optical coherence tomography. Science 254:1178 – 1181

67. Ikeda T, Sato K, Katano T, Hayashi Y (1999) Attached posterior hyaloid membrane and the pathogenesis of honeycombed cystoid macular edema in patients with diabetes. Am J Ophthalmol 127:478 – 479

68. Ikeda T, Sato K, Katano T, Hayashi Y (1999) Vitrectomy for cystoid macular oedema with attached posterior hyaloid membrane in patients with diabetes. Br J Ophthalmol 83:12 – 14

69. Ikeda T, Sato K, Katano T, Hayashi Y (2000) Improved visual acuity following pars plana vitrectomy for diabetic cystoid macular edema and detached posterior hyaloid. Retina 20:220 – 222

70. Interim Report of Multicentered Controlled Study (1975) Photocoagulation in treatment of diabetic maculopathy. Lancet 2:1110 – 1113

71. Isenberg SJ, Apt L, Yoshimori R, Khwarg S (1985) Chemical preparation of the eye in ophthalmic surgery. IV. Comparison of povidone-iodine on the conjunctiva with a prophylactic antibiotic. Arch Ophthalmol 103:1340 – 1342

72. Ishii H, Jirousek MR, Koya D, et al. (1996) Amelioration of vascular dysfunction in diabetic rats by an oral PKC beta inhibitor. Science 272:728 – 731

73. Jaffe GJ, Ben-Nun J, Guo H, Dunn JP, Ashton P (2000) Fluocinolone acetonide sustained drug delivery device to treat severe uveitis. Ophthalmology 107:2024 – 2033

74. Jaffe GJ, Yang CH, Guo H, Denny JP, Lima C, Ashton P (2000) Safety and pharmacokinetics of an intraocular fluocinolone acetonide sustained delivery device. Invest Ophthalmol Vis Sci 41:3569 – 3575

75. Jonas JB, Sofker A (2001) Intraocular injection of crystalline cortisone as adjunctive treatment of diabetic macular edema. Am J Ophthalmol 132:425 – 427

76. Jonas JB, Kreissig I, Degenring RF (2002) Intravitreal triamcinolone acetonide as treatment of macular edema in central retinal vein occlusion. Graefes Arch Clin Exp Ophthalmol 240:782 – 783

77. Jonas JB, Kreissig I, Degenring RF (2003a) Intravitreal triamcinolone acetonide as treatment of ischemic ophthalmopathy. Eur J Ophthalmol 13:575 – 576

78. Jonas JB, Kreissig I, Degenring RF (2003b) Neovascular glaucoma treated by intravitreal triamcinolone acetonide. Acta Ophthalmol Scand 81:540 – 541

79. Jonas JB, Kreissig I, Hugger P, Sauder G, Panda-Jonas S, Degenring R (2003c) Intravitreal triamcinolone acetonide for exudative age related macular degeneration. Br J Ophthalmol 87:462 – 468

80. Jonas JB, Kreissig I, Sofker A, Degenring RF (2003d) Intravitreal injection of triamcinolone for diffuse diabetic macular edema. Arch Ophthalmol 121:57 – 61

81. Jonas JB, Spandau UH, Kamppeter BA, Vossmerbaeumer U, Harder B, Sauder G (2006) Repeated intravitreal high-dosage injections of triamcinolone acetonide for diffuse diabetic macular edema. Ophthalmology 113:800 – 804

82. Joussen AM, Murata T, Tsujikawa A, Kirchhof B, Bursell SE, Adamis AP (2001) Leukocyte-mediated endothelial cell injury and death in the diabetic retina. Am J Pathol 158(1):147 – 152

83. Joussen AM, Poulaki V, Mitsiades N, Kirchhof B, Koizumi K, Dohmen S, Adamis AP (2002a) Potential use of non-steroidal anti-inflammatory drugs for prevention of diabetic vascular changes: Aspirin prevents diabetic leakage and leukocyte adhesion through inhibition of TNF-a. FASEB J 16:438 – 440

84. Joussen AM, Qin W, Poulaki V, Wiegand S, Yancopoulos GD, Adamis AP (2002b) Endogenous VEGF induces retinal ICAM-1 and eNOS expression and initiates early diabetic retinal leukostasis. Am J Pathol 160:501 – 509

85. Joussen AM, Poulaki V, Mitsiades N, Cai WY, Suzuma I, Pak J, Ju ST, Rook SL, Esser P, Mitsiades CS, Kirchhof B, Adamis AP, Aiello LP (2003) Suppression of Fas-FasL-induced endothelial cell apoptosis prevents diabetic blood-retinal barrier breakdown in a model of streptozotocin-induced diabetes. FASEB J 17:76 – 78

86. Joussen AM, Poulaki V, Le ML, Koizumi K, Esser C, Janicki H, Schraermeyer U, Kociok N, Fauser S, Kirchhof B, Kern TS, Adamis AP (2004) A central role for inflammation in the pathogenesis of diabetic retinopathy. FASEB J 18: 1450 – 1452

87. Jumper JM, Embabi SN, Toth CA, McCuen BW II, Hatchell DL (2000) Electron immunocytochemical analysis of posterior hyaloid associated with diabetic macular edema. Retina 20:63 – 68

88. Kadonosono K, Itoh N, Ohno S (2000) Perifoveal microcirculation before and after vitrectomy for diabetic cystoid macular edema. Am J Ophthalmol 130:740 – 744

89. Kaiser PK, Riemann CD, Sears JE, Lewis H (2001) Macular traction detachment and diabetic macular edema associated with posterior hyaloidal traction. Am J Ophthalmol 131:44 – 49

90. Kakehashi A, Schepens CL, Trempe CL (1994) Vitreomacular observations: I. Vitreomacular adhesion and hole in the premacular hyaloid. Ophthalmology 101:1515 – 1521

91. Kang SW, Park CY, Ham DI (2004) The correlation between fluorescein angiographic and optical coherence tomographic features in clinically significant diabetic macular edema. Am J Ophthalmol 137:313 – 322

92. Keck PJ, Hauser SD, Krivi G, Sanzo K, Warren T, Feder J, Connolly DT (1989) Vascular permeability factor, an endothelial cell mitogen related to PDGF. Science 246:1309 – 1312

93. Kent D, Vinores SA, Campochiaro PA (2000) Macular oedema: the role of soluble mediators. Br J Ophthalmol 84: 542 – 545

94. Klein R, Klein BE, Moss SE, Davis MD, DeMets DL (1984) The Wisconsin Epidemiologic Study of Diabetic Retinopathy, IV: diabetic macular edema. Ophthalmology 91:1464 – 1474

95. Klein R, Klein BEK, Moss SE, Cruickshanks KJ (1995) The

Wisconsin epidemiologic study of diabetic retinopathy. XV. The long term incidence of macular edema. Ophthalmology 102:7–16

96. Klein R, Klein BE, Moss SE, Cruickshanks KJ (1998) The Wisconsin Epidemiologic Study of Diabetic Retinopathy: XVII. The 14-year incidence and progression of diabetic retinopathy and associated risk factors in type 1 diabetes. Ophthalmology 105:1801–1815

97. Konno S, Akiba J, Yoshida A (2001) Retinal thickness measurements with optical coherence tomography and the scanning retinal thickness analyzer. Retina 21:57–61

98. Krott R, Heller R, Heimann K (1999) Adjuvante hyperbare Sauerstofftherapie (HBO) bei cystoidem Makulaödem – erste Ergebnisse. Klin Monatsbl Augenheilkd 215:144

99. Krott R, Heller R, Aisenbrey S, Bartz-Schmidt KU (2000) Adjunctive hyperbaric oxygenation in macular edema of vascular origin. Undersea Hyper Med 27:195–204

100. Kuhn F, Kiss G, Mester V, Szijarto Z, Kovacs B (2004) Vitrectomy with internal limiting membrane removal for clinically significant macular oedema. Graefes Arch Clin Exp Ophthalmol 242:402–408

101. Kumagai K, Ogino N, Furukawa M, Demizu S, Atsumi K, Kurihara H, Iwaki M, Ishigooka H, Tachi N (2002) [Internal limiting membrane peeling in vitreous surgery for diabetic macular edema]. Nippon Ganka Gakkai Zasshi 106:590–594

102. Kuwabara T, Cogan DG (1963) Retinal vascular patterns. VI. Mural cells of the retinal capillaries. Arch Ophthalmol 69:492–502

103. Kyto JP, Angerman S, Lumiste E, Paloheimo M, Summanen PA (2005) Intravitreal triamcinolone acetonide as an adjuvant therapy to panretinal photocoagulation for proliferative retinopathy with high risk characteristics in type 1 diabetes: case report with 22 weeks follow-up. Acta Ophthalmol Scand 83(5):605–8

104. La Heij EC, Hendrikse F, Kessels AG, Derhaag PJ (2001) Vitrectomy results in diabetic macular oedema without evident vitreomacular traction. Graefes Arch Clin Exp Ophthalmol 239:264–270

105. Ladas ID, Theodossiadis GP (1993) Long-term effectiveness of modified grid laser photocoagulation for diffuse diabetic macular edema. Acta Ophthalmol (Copenh) 71:393–397

106. Landau D, Schneidman EM, Jacobovitz T, Rozenman Y (1997) Quantitative in vivo retinal thickness measurements in healthy subjects. Ophthalmology 104(4):639–42

107. Lee CM, Olk RJ (1991) Modified grid laser photocoagulation for diffuse diabetic macular edema. Long-term visual results. Ophthalmology 98:1594–1602

108. Lee CM, Olk RJ, Akduman L (2000) Combined modified grid and panretinal photocoagulation for diffuse diabetic macular edema and proliferative diabetic retinopathy. Ophthalmic Surg Lasers 31(4):292–300

109. Lewis H, Abrams GW, Blumenkranz MS, Campo RV (1992) Vitrectomy for diabetic macular traction and edema associated with posterior hyaloidal traction. Ophthalmology 99:753–759

110. Lopes de Faria JM, Jalkh AE, Trempe CL, McMeel JW (1999) Diabetic macular edema: risk factors and concomitants. Acta Ophthalmol Scand 77:170–175

111. Lu M, Kuroki M, Amano S, Tolentino M, Keough K, Kim I, Bucala R, Adamis AP (1998) Advanced glycation end products increase retinal vascular endothelial growth factor expression. J Clin Invest 101:1219–1224

112. Marmor MF, Maak T (1982) Enhancement of retinal adhesion and subretinal fluid absorption by acetazolamide. Invest Ophthalmol Vis Sci 23:121–124

113. Martidis A, Duker JS, Greenberg PB, Rogers AH, Puliafito CA, Reichel E, Baumal C (2002) Intravitreal triamcinolone for refractory diabetic macular edema. Ophthalmology 109:920–927

114. Matrisian LM (1992) The matrix-degrading metalloproteinases. Bioassays 14:455–463

115. McDonald HR, Schatz H (1985) Grid photocoagulation for diffuse macular edema. Retina 5:65–72

116. McNaught EI, Foulds WS, Allan D (1988) Grid photocoagulation improves reading ability in diffuse diabetic macular oedema. Eye 2:288–296

117. Micelli Ferrari T, Cardascia N, Durante G, Vetrugno M, Cardia L (1999) Pars plana vitrectomy in diabetic macular edema. Doc Ophthalmol 97:471–474

118. Miyamoto H, Ogura Y, Honda Y (1995) Hyperbaric oxygen treatment for macular edema after retinal vein occlusion-fluorescein angiographic findings and visual prognosis. Nippon Ganka Akkai Zasshi 99:220–225

119. Miyamoto K, Khosrof S, Bursell S-E, Rohan R, Murata T, Clermont A, Aiello LP, Ogura Y, Adamis AP (1999) Prevention of leukostasis and vascular leakage in streptozotocin-induced diabetic retinopathy via intercellular adhesion molecule-1 inhibition. Proc Natl Acad Sci USA 96:10836–10841

120. Mizutani M, Kern TS, Lorenzi M (1996) Accelerated death of retinal microvascular cells in human and experimental diabetic retinopathy. J Clin Invest 97:2883–2890

121. Molnar I, Poitry S, Tsacopoulos M, Gilodi N, Leuenberger PM (1985) Effect of laser photocoagulation on oxygenation of the retina in miniature pigs. Invest Ophthalmol Vis Sci 26(10):1410–4

122. Moshfeghi DM, Kaiser PK, Scott IU, Sears JE, Benz M, Sinesterra JP, Kaiser RS, Bakri SJ, Maturi RK, Belmont J, Beer PM, Murray TG, Quiroz-Mercado H, Meiler WF (2003) Acute endophthalmitis following intravitreal triamcinolone acetonide injection. Am J Ophthalmol 136:793–796

123. Nasarallah FP, Jalkh AE, Van Coppenolle F, Kado M, Trempe CL, McMeel JM, Schepens CL (1988) The role of vitreous in diabetic macular edema. Ophthalmology 95:1335–1339

124. Nasarallah FP, Van Coppenolle F, Jalkh AE, Trempe CL, McMeel JM, Schepens CL (1989) Importance of the vitreous in young diabetics with macular edema. Ophthalmology 96:1511–1516

125. Ndoye Roth PA, Grange JD, Hajji Z (2003) [Diabetic cystoid macular edema and vitrectomy. Preliminary results on 19 cases]. J Fr Ophthalmol 26:38–46

126. Nussenblatt RB, Kaufman SC, Palestine AG, et al. (1987) Macular thickening and visual acuity. Measurement in patients with cystoid macular edema. Ophthalmology 94:1134–1139

127. Olk RJ (1986) Modified grid argon (blue-green) laser photocoagulation for diffuse diabetic macular edema. Ophthalmology 93:938–950

128. Olk RJ (1990) Argon green (514 nm) versus krypton red (647 nm) modified grid laser photocoagulation for diffuse diabetic macular edema. Ophthalmology 97:1101–1112

129. Orlidge A, Hollis TM (1982) Aortic endothelial and smooth muscle histamine metabolism in experimental diabetes. Atherosclerosis 2:142–150

III 19

130. Otani T, Kishi S (2000) Tomographic assessment of vitreous surgery for diabetic macular edema. Am J Ophthalmol 129:487–494

131. Otani T, Kishi S (2002) A controlled study of vitrectomy for diabetic macular edema. Am J Ophthalmol 134:214–219

132. Otani T, Kishi S, Maruyama Y (1999) Patterns of diabetic macular edema with optical coherence tomography. Am J Ophthalmol 127:688–693

133. Pavlin CJ, Harasiewicz K, Sherar MD, Foster FS (1991) Clinical use of ultrasound biomicroscopy. Ophthalmology 98:287–295

134. Pendergast SD, Hassan TS, Williams GA, Cox MS, Margherio RR, Ferrone PJ, Garretson BR, Trese MT (2000) Vitrectomy for diffuse diabetic macular edema associated with a taut premacular posterior hyaloid. Am J Ophthalmol 130:178–186

135. Perry DD, Risco JM (1982) Choroidal microvascular repair after argon photocoagulation. Am J Ophthalmol 93:787–793

136. Pfoff DS, Thom SR (1987) Preliminary report on the effect of hyperbaric oxygen on cystoid macular edema. J Cat Refr Surg 13:136–140

137. Polito A, Del Borrello M, Isola M, Zemella N, Bandello F (2005) Repeatability and reproducibility of fast macular thickness mapping with stratus optical coherence tomography. Arch Ophthalmol 123(10):1330–7

138. Puliafito CA, Hee MR, Lin CP, et al. (1995) Imaging of macular diseases with optical coherence tomography. Ophthalmology 102:217–229

139. Qaum T, Xu Q, Joussen AM, Qin W, Clemens ME, Yancopoulos GD, Adamis AP (2001) Early diabetic blood-retinal barrier breakdown is VEGF-dependent. Invest Ophthalmol Vis Sci 42:2408–2413

140. Radetzky S, Walter P, Koizumi K, Kirchhof B, Joussen AM (2004) Visual outcome of patients with macular edema after pars plana vitrectomy (PPV) and indocyanine green (ICG) assisted internal limiting membrane (ILM) peeling. Graefes Arch Ophthalmol 242:273–278

141. Radner W, Obermayer W, Richter-Mueksch S, Willinger U, Velikay-Parel M, Eisenwort B (2002) The validity and reliability of short German sentences for measuring reading speed. Graefes Arch Clin Exp Ophthalmol 240:461–467

142. Ramachandran E, Frank RN, Kennedy A (1993) Effects of endothelin on cultured bovine retinal microvascular pericytes. Invest Ophthalmol Vis Sci 34:586–595

143. Rema M, Sujatha P, Pradeepa R (2005) Visual outcomes of pan-retinal photocoagulation in diabetic retinopathy at one-year follow-up and associated risk factors. Indian J Ophthalmol 53(2):93–9

144. Rhodes RH (1979) A light microscopic study of the developing human neural retina. Am J Anat 154:195–209

145. Rojas B, Zafirakis P, Christen W, Markomichelakis NN, Foster CS (1999) Medical treatment of macular edema in patients with uveitis. Doc Ophthalmol 97:399–407

146. Romero IA, Radewicz K, Jubin E, Michel CC, Greenwood J, Couraud PO, Adamson P (2003) Changes in cytoskeletal and tight junctional proteins correlate with decreased permeability induced by dexamethasone in cultured rat brain endothelial cells. Neurosci Lett 334:112–116

147. Rosenblatt BJ, Shah GK, Sharma S, Bakal J (2005) Pars plana vitrectomy with internal limiting membranectomy for refractory diabetic macular edema without a taut posterior hyaloid. Graefes Arch Clin Exp Ophthalmol 243:20–25

148. Roth DB, Chieh J, Spirn MJ, Green SN, Yarian DL, Chaudhry NA (2003) Noninfectious endophthalmitis associated with intravitreal triamcinolone injection. Arch Ophthalmol 121:1279–1282

149. Salzmann J, Limb GA, Khaw PT, Gregor ZJ, Webster L, Chingnell AH, Charteris DG (2000) Matrix metalloproteinases and their natural inhibitors in fibrovascular membranes of proliferative diabetic retinopathy. Br J Ophthalmol 84:1091–1096

150. Sarthy V (1993) Collagen IV mRNA expression during development of the mouse retina: an in situ hybridization study. Invest Ophthalmol Vis Sci 34:145–152

151. Sato Y, Lee Z, Shimada H (2002) Vitrectomy for diabetic cystoid macular edema. Jpn J Ophthalmol 46:315–322

152. Schneeberger SA, Hjelmeland LM, Tucker RP, Morse LS (1997) Vascular endothelial growth factor and fibroblast growth factor 5 are colocalized in vascular and avascular epiretinal membranes. Am J Ophthalmol 124:447–454

153. Schröder S, Palinski W, Schmidt-Schönbein GW (1991) Activated monocytes and granulocytes, capillary nonperfusion, and neovascularization in diabetic retinopathy. Am J Pathol 139:81–100

154. Schulze C, Firth JA (1993) Immunohistochemical localization of adherens junction components in blood-brain barrier microvessels of the rat. J Cell Sci 104:773–782

155. Sebag J (1987) Aging of the vitreous. Eye 2:254–262

156. Sebag J (1991) Age-related differences in the human vitreoretinal interface. Arch Ophthalmol 109:966–971

157. Sebag J (1992) Anatomy and pathology of the vitreo-retinal interface. Eye 6:541–552

158. Sebag J, Buckingham B, Charles MA, Reiser K (1992) Biochemical abnormalities in vitreous of humans with proliferative diabetic retinopathy. Arch Ophthalmol 110:1472–1476

159. Senger DR, Van De Water L, Brown LF, et al. (1993) Vascular permeability factor (VPF, VEGF) in tumor biology. Cancer Metastas Rev 12:303–324

160. Shahidi M, Ogura Y, Blair NP, et al. (1991) Retinal thickness analysis for quantitative assessment of diabetic macular edema. Arch Ophthalmol 109:1115–1119

161. Shimura M, Yasuda K, Nakazawa T, Kano T, Ohta S, Tamai M (2003) Quantifying alterations of macular thickness before and after panretinal photocoagulation in patients with severe diabetic retinopathy and good vision. Ophthalmology 110(12):2386–94

162. Smith RT, Lee CM, Charles HC, Farber M, Cunha-Vaz JG (1987) Quantification of diabetic macular edema. Arch Ophthalmol 105:218–222

163. Spira AW, Hollenberg MJ (1973) Human retinal development: ultrastructure of the inner retinal layers. Dev Biol 31:1–21

164. Stefaniotou M, Aspiotis M, Kalogeropoulos C, Christodoulou A, Psylla M, Ioachim E, Alamanos I, Psilas K (2004) Vitrectomy results for diffuse diabetic macular edema with and without inner limiting membrane removal. Eur J Ophthalmol 14:137–143

165. Stifter E, Konig F, Lang T, Bauer P, Richter-Muksch S, Velikay-Parel M, Radner W (2004) Reliability of a standardized reading chart system: variance component analysis, test-retest and inter-chart reliability. Graefes Arch Clin Exp Ophthalmol 242:31–39

166. Strom C, Sander B, Larsen N, Larsen M, Lund-Andersen H (2002) Diabetic macular edema assessed with optical coherence tomography and stereo fundus photography. Invest Ophthalmol Vis Sci 43(1):241–5

167. Suttorp-Schulten MS, Riemslag FC, Rothova A, van der Kley AJ, Riemslag FC (1996) Long-term effect of repeated hyperbaric oxygen therapy on visual acuity in inflammatory cystoid macular oedema. Br J Ophthalmol 81:329

168. Tachi N, Ogino N (1996) Vitrectomy for diffuse macular edema in cases of diabetic retinopathy. Am J Ophthalmol 8:258–260

169. Tagawa H, McMeel JW, Furukawa H, et al. (1986) Role of the vitreous in diabetic retinopathy. I. Vitreous changes in diabetic retinopathy and in physiologic aging. Ophthalmology 93:596–601

170. Takagi C, Bursell SE, Lin YW, Takagi H, Duh E, Jiang Z, Clermont AC, King GL (1996) Regulation of retinal hemodynamics in diabetic rats by increased expression and action of endothelin-1. Invest Ophthalmol Vis Sci 37: 2504–2518

171. Takahashi M, Trempe CL, Maguire K, McMeel JW (1981) Vitreoretinal relationship in diabetic retinopathy: a biomicroscopic evaluation. Arch Ophthalmol 99:241–245

172. Thach AB, Dugel PU, Flindall RJ, Sipperley JO, Sneed SR (1997) A comparison of retrobulbar versus subtenon's corticosteroid therapy for cystoid macular edema refractory to topical medications. Ophthalmology 104:2003–2008

173. Tokuyama T, Ikeda T, Sato K (2000) Effects of haemodialysis on diabetic macular leakage. Br J Ophthalmol 84: 1397–1400

174. van Effenterre G, Guyot-Argenton C, Guiberteau B, Hany I, Lacotte JL (1993) [Macular edema caused by contraction of the posterior hyaloid in diabetic retinopathy. Surgical treatment of a series of 22 cases]. J Fr Ophtalmol 16: 602–610

175. Verbraeken H (1996) Therapeutic pars plana vitrectomy for uveitis: a retrospective study of the long-term results. Graefes Arch Clin Exp Ophthalmol 234:288–293

176. Vlassara H (1997) Recent progress in advanced glycation end products and diabetic complications. Diabetes 46: S19–25

177. Wallow IH (1984) Repair of the pigment epithelial barrier following photocoagulation. Arch Ophthalmol 102(1): 126–35

178. Walsh K, Sata M (1999) Is extravasation a Fas-regulated process? Mol Med Today 5:61–67

179. Webb RH, Hughes GW, Delori FC (1987) Confocal scanning laser ophthalmoscope. Applied Optics 26:1492–1499

180. Weene LE (1992) Cystoid macular edema after scleral buckling responsive to acetazolamide. Ann Ophthalmol 24:423–424

181. Weiter JJ, Zuckerman R (1980) The influence of the photoreceptor-RPE complex on the inner retina. An explanation for the beneficial effects of photocoagulation. Ophthalmology 87:1133–1139

182. Whiteside C, Dlugosz J (2002) Mesangial cell protein kinase C isozyme activation in diabetic milieu. Am J Physiol Renal Physiol 282:F975–980

183. Widder RA, Brunner R, Walter P, Luke C, Bartz-Schmidt KU, Heimann K, Borberg H (1999) Improvement of visual acuity in patients suffering from diabetic retinopathy after membrane differential filtration: a pilot study. Transfus Sci 21:201–206

184. Wilson D, Finkelstein D, Quingley H, Green W (1988) Macular grid photocoagulation: an experimental animal study on the primate retina. Arch Ophthalmol 106:100–105

185. Wingate RJ, Beaumont PE (1999) Intravitreal triamcinolone and elevated intraocular pressure. Aust NZ J Ophthalmol 27:431–432

186. Wiznia RA (1979) Photocoagulation of nonproliferative exudative diabetic retinopathy. Am J Ophthalmol 88:22–27

187. Yamamoto T, Akabane N, Takeuchi S (2001) Vitrectomy for diabetic macular edema: The role of posterior vitreous detachment and epimacular membrane. Am J Ophthalmol 132:369–377

188. Yang CM (2000) Surgical treatment for severe diabetic macular edema with massive hard exudates. Retina 20: 121–125

189. Yasukawa T, Kiryu J, Tsujikawa A, et al. (1998) Quantitative analysis of foveal retinal thickness in diabetic retinopathy with the scanning retinal thickness analyzer. Retina 18:150–155

190. Yi Q, Bamroongsuk P, McCarty DJ, Mukesh BN, Harper CA (2003) Clinical outcomes following laser photocoagulation treatment for diabetic retinopathy at a large Australian ophthalmic hospital. Clin Exp Ophthalmol 31(4): 305–9

191. Yoshikawa K, Kotake S, Ichiishi A, Sasamoto Y, Kosaka S, Matsuda J (1995) Posterior sub-tenon injections of repository corticosteroids in uveitis patients with cystoid macular edema. Jpn J Ophthalmol 39:71–76

192. Young S, Larkin G, Branley M, Lightman S (2001) Safety and efficacy of intravitreal triamcinolone for cystoid macular oedema in uveitis. Clin Exp Ophthalmol 29:2–6

193. Zein WM, Noureddin BN, Jurdi FA, Schakal A, Bashshur ZF (2006) Panretinal photocoagulation and intravitreal triamcinolone acetonide for the management of proliferative diabetic retinopathy with macular edema. Retina 26:137–142

19.3.2 Pegaptanib for Diabetic Macular Edema

A.P. Adamis, B. Katz

Core Messages

- Vascular endothelial growth factor (VEGF), a potent inducer of angiogenesis and vascular permeability, has been strongly implicated in the etiology of ocular vascular diseases, including diabetic retinopathy (DR) and diabetic macular edema (DME)
- Studies in animal models of diabetes have shown that the $VEGF_{164/165}$ isoform acts as an especially potent inflammatory cytokine in mediating breakdown of the blood-retinal barrier (BRB)
- Intravitreous injection of pegaptanib, an aptamer specific for $VEGF_{165}$, has reversed BRB breakdown in diabetic rodents

- In a phase 2 trial involving 172 subjects with DME, intravitreously injected pegaptanib produced clinical benefits, compared to sham injection, in all three principal outcomes: visual acuity, retinal thickness and need for photocoagulation
- In this same trial, 8 of 13 patients showing retinal revascularization at study entry experienced regression or decreased fluorescein leakage in response to treatment with pegaptanib; such regression was not seen in three sham-injected eyes or in four untreated fellow eyes showing revascularization
- These findings warrant confirmation in pivotal phase 3 trials

19.3.2.1 Background

19.3.2.1.1 New Therapeutic Strategies: Targeting Vascular Endothelial Growth Factor

Vision loss associated with diabetes is caused both by retinal neovascularization and by damage to the retinal vasculature, leading to breakdown in the blood-retinal barrier (BRB) and/or ischemia. Diabetic macular edema (DME), a direct reflection of this vascular damage, causes a significant component of vision loss associated with diabetic retinopathy (DR), especially in patients suffering from type 2 diabetes [18]. Currently, laser photocoagulation is the standard of care in treating retinal complications of diabetes, and while it has contributed significantly to reducing the incidence of severe vision loss, it is basically a destructive intervention that does not address the underlying pathophysiology. Indeed, it is accompanied by frank destruction of neural tissue and can lead to perceptions of nyctalopia, visual field constriction, and dyschromatopsia. A progression in the severity of retinopathy after treatment is not uncommon [18]. There is thus a need for newer therapies with fewer side effects, especially approaches that counter retinopathic change through targeting the underlying pathophysiology of DR, rather than relying on *ex post facto* ablation.

One major area of investigation is the use of angiogenesis inhibitors, with vascular endothelial growth factor (VEGF) a principal target for inhibition. Over the past 15 years, an extensive body of research has established that VEGF is a key regulator of both physiological and pathological angiogenesis, playing a variety of roles in promoting blood vessel growth and vascular permeability (**see Callout 1**). Alternative splicing of the human gene yields at least six biologically active isoforms, composed of 121, 145, 165, 183, 189, and 206 amino acids [25], with $VEGF_{165}$, the most abundant isoform, being principally responsible for diabetes-associated ocular pathology [35, 39]. It is important to note, however, that the family of VEGF isoforms is much more than a promoter of angiogenesis as it acts in a wide variety of cellular processes (**see Callout 2**). Accordingly, strategies targeting VEGF in the clinical arena must pay particular heed to the potential for adverse events when inhibiting all isoforms of VEGF. In this regard, systemically administered VEGF inhibitors have been associated with an increased incidence of hypertension, proteinuria, bleeding and thromboembolic events [34, 37, 40, 41, 54, 70].

19 **III**

Callout 1 – Actions of VEGF in promoting angiogenesis

- Endothelial cell mitogen [48]
- Endothelial cell survival factor [7]
- Chemoattractant for bone marrow-derived endothelial cells [12, 20]
- Chemoattractant for monocyte lineage cells [13, 19]
- Inducer of synthesis of endothelial nitric oxide synthase and consequent elevation of nitric oxide, itself a promoter of angiogenesis [61, 76]
- Inducer of synthesis of enzymes promoting blood vessel extravasation:
  - Matrix metalloproteinases [33, 45]
  - Plasminogen activator [62]

Callout 2 – Additional physiological processes involving VEGF

- Bone growth [31, 66]
- Wound healing [22, 58]
- Female reproductive cycling [26, 66]
- Vasorelaxation [49]
- Glomerulogenesis [42]
- Protection of hepatic cells [47]
- Skeletal muscle regeneration [11]
- Neural survival factor [57, 60, 70]
- Trophic support of choriocapillaris [15, 52]

19.3.2.1.2 VEGF and Diabetic Retinopathy

The pathophysiology of DR is complex, with the products of several biochemical pathways being potential mediators in the relationship between hyperglycemia and retinal vascular damage. These include polyols, advanced glycation end products and reactive oxygen intermediates [18]. Anatomical correlates of the progression of DR include death of capillary pericytes, basement membrane thickening, and entrapment of leukocytes, leading to capillary blockages and local hypoxia [18, 55]. Upregulation of VEGF is likely to occur either directly, through stimulation by metabolites such as advanced glycation end products [44] and reactive oxygen intermediates [50], or indirectly, through the local hypoxia induced by capillary dropout. VEGF is synthesized by a wide range of retinal cell types [1, 5, 24] and this synthesis is significantly increased in hypoxic conditions [5, 15]. Clinical findings have confirmed that VEGF levels are elevated in both DR [2, 4, 16, 27, 51] and DME [17, 28–30].

It is now well established that increases in ocular concentrations of VEGF are closely linked both to the aberrant growth of new vessels and to increased exudation and tissue edema. This edema further exacerbates the vision loss associated with DR (see **Callout 3**) and other ocular neovascularizing syndromes such as age-related macular degeneration (AMD)

[2–4, 6, 9, 32, 43, 46, 51, 53, 59, 67, 72–75]. The neovascularization and edema signal two of VEGF's salient properties: (1) as a central promoter of angiogenesis [25] and (2) as the most potent known enhancer of vascular permeability [68]. The increase in permeability reflects several VEGF-mediated processes, including induction of fenestrations in the endothelium [64], dissolution of tight junctions [10], and of promotion adherence to the retinal vasculature by leukocytes which then act to damage the endothelium [38, 39].

Callout 3 – Key findings linking VEGF to the pathophysiology of DR

- VEGF levels are elevated in eyes of patients suffering from DR [2, 4, 16, 51] and DME [17, 28–30]
- Experimental elevation of VEGF in normal primate eyes induces many changes typical of DR, including formation of microaneurysms and exudation following BRB breakdown [73, 74]
- Studies with rodent models of diabetes have revealed that DR is associated with increases in retinal VEGF levels (Fig. 19.3.2.1), which underlie a local inflammation and consequent vascular damage and leakage [35, 39, 63]
- One VEGF isoform ($VEGF_{165}$ in humans, $VEGF_{164}$ in rodents) is especially important in mediating this inflammation [35, 39, 77]
- Elevation of retinal levels of $VEGF_{164}$ occur in parallel with BRB breakdown in diabetic rodents [63]
- Intravitreous injection of $VEGF_{164}$ induces BRB breakdown in normal animals more potently than $VEGF_{120}$ [35]

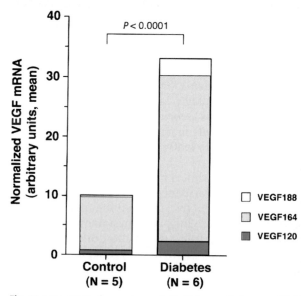

Fig. 19.3.2.1. Retinal vascular endothelial growth factor mRNA levels are increased in early diabetes. (Adapted from [63])

19.3.2.1.3 Pegaptanib

Pegaptanib is a nuclease-resistant, pegylated 28-nucleotide RNA aptamer (Fig. 19.3.2.2) that binds to the $VEGF_{164/165}$ isoform at high affinity (200 pM) while showing little activity toward the $VEGF_{120/121}$ isoform [56] (**see Callout 4**). Pegaptanib inhibits $VEGF_{164/165}$ from binding to its cellular receptors, preventing the initiation of downstream signaling events. From the perspective of DR and DME, two of the most important cellular processes which are inhibited are VEGF's actions in promoting angiogenesis and in enhancing vascular permeability. In experiments with cultured endothelial cells, pegaptanib inhibited the induction of mitogenesis by $VEGF_{165}$, but not by $VEGF_{121}$, consistent with pegaptanib's specificity for $VEGF_{165}$ [14]. In addition, in the Miles assay for vascular permeability, the VEGF-induced increase in vascular leakage was inhibited by 83% when VEGF was pre-incubated with pegaptanib [65]. In subsequent studies, using a rodent model of retinopathy of prematurity, intravitreous injection of pegaptanib was shown to inhibit pathological revascularization, but not the physiological vascularization of the retina [36], suggesting that pegaptanib treatment might be relatively harmless to the normal retinal vasculature. Further evidence of pegaptanib's sparing of physiological tissues came from studies of retinal ischemia in which $VEGF_{120}$ was shown to be sufficient to exert a neuroprotective effect [51]. Most importantly for DR, in experiments with diabetic rodents, intravitreous injection of pegaptanib was shown to cause restoration of the BRB (Fig. 19.3.2.3 [35]).

Pegaptanib includes a 40-kDa polyethylene glycol moiety at the 5' terminal [65], a change that prolongs intravitreal half-life [23]. Pivotal phase 3 trials [32] have already demonstrated that intravitreously ad-

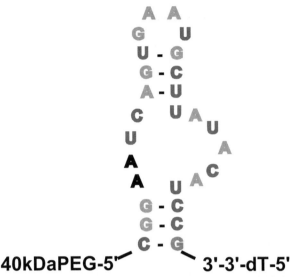

Fig. 19.3.2.2. Pegaptanib – secondary structure. *PEG* polyethylene glycol (With permission from [56])

ministered pegaptanib is effective in treating neovascular AMD and is associated with a low risk of adverse events such as endophthalmitis, traumatic lens injury or retinal detachment; where these occurred they were related to the injection procedure rather than to the study drug itself. The success of the trials led to pegaptanib's approval for neovascular AMD by regulatory authorities in the United States, Canada and Brazil, and its recommendation for market authorization by the Committee for Human Medicinal Products of the European Union.

Taken together with the data implicating $VEGF_{165}$ in the pathophysiology of DR/DME, and the positive safety record of pegaptanib in clinical trials [71], these findings led to a phase 2 trial to examine pegaptanib's utility as a treatment for DME.

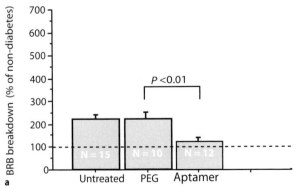

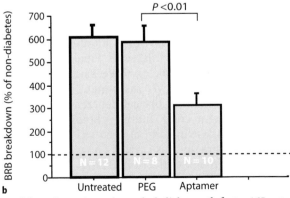

Fig. 19.3.2.3. Suppression of diabetic blood-retinal barrier (*BRB*) breakdown by antivascular endothelial growth factor 165 aptamer. Compared with polyethylene glycol (*PEG*) alone, treatment with pegaptanib resulted in **a** 82.6% blockade of early diabetic BRB breakdown ($P < 0.01$) and **b** 55.0% blockade of established diabetic BRB breakdown ($P < 0.01$). (Adapted from [35], with permission)

19 III

Callout 4 – Pegaptanib

- Nuclease-resistant RNA aptamer, specific for the $VEGF_{165}$ isoform [65]
- Inhibits $VEGF_{165}$-mediated cellular actions, including increases in vascular permeability and endothelial cell proliferation [14, 65]
- Intravitreous pegaptanib is already approved as a treatment for neovascular AMD
- Intravitreous pegaptanib can reverse BRB breakdown in diabetic rodents [35]
- Intravitreous pegaptanib inhibits pathological but not physiological vascularization [36]

19.3.2.2 Phase 2 Trial – Intravitreous Pegaptanib as a Treatment for DME

19.3.2.2.1 Design

The study was a randomized, sham-controlled, double-masked, dose-finding phase 2 trial which enrolled 172 patients 18 years and older with type I or type II diabetes, visual acuity (VA) between 20/50 and 20/320 and clinically significant DME affecting the center of the macula (see Callout 5). Only patients judged not likely to need photocoagulation therapy for 16 weeks were enrolled. Principal exclusion criteria included photocoagulation or other retinal treatments within the previous 6 months, abnormalities preventing VA or photographic measurements, severe cardiac disease, clinically significant peripheral vascular disease, uncontrolled hypertension, and glycosylated hemoglobin levels $\geq 13\%$ [21].

Patients were randomized to four treatment arms (0.3 mg, 1 mg or 3 mg pegaptanib or sham injection), with stratification by study site, size of the thickened retina area (≤ 2.5 disk areas versus > 2.5 disk areas), and baseline VA (letter score ≥ 58 versus < 58). Injections were given at baseline and every 6 weeks thereafter for a minimum of three and a maximum of six injections. Final assessments were made at week 36, or 6 weeks after the last injection. Refraction, VA assessment, an ophthalmologic examination, optical coherence tomography (OCT), and color fundus photography were conducted at baseline and at each visit, while fluorescein angiography was carried out at baseline and 6 weeks after the last injection. Overall, 169 patients received at least one injection, and more than 90% of patients in each treatment group completed the study. Prespecified efficacy criteria included VA, retinal thickness as measured by OCT, and the need for rescue photocoagulation therapy. In addi-

tion, patients found to show diabetic neovascularization at baseline were evaluated for the impact of pegaptanib treatment upon its advance or regression [21].

19.3.2.2.2 Results

19.3.2.2.2.1 Principal Endpoints – Visual Acuity, Retinal Thickness, Retinal Volume, and Need for Photocoagulation

Pegaptanib treatment was superior to sham injection according to all prespecified endpoints. Mean change in VA in the 0.3 mg pegaptanib-treated group was +4.7 letters compared to –0.4 letters for sham ($P = 0.04$; Table 19.3.2.1). Pegaptanib treatment also resulted in more patients gaining ≥ 0, ≥ 5, ≥ 10, and ≥ 15 letters of VA (Fig. 19.3.2.4). Mean change in center point retinal thickness was –68 µm in the 0.3 mg pegaptanib arm compared to +3.7 µm in the sham group ($P = 0.02$), and pegaptanib treatment resulted in significantly more patients experiencing decreases in thickness of 75 and 100 µm (Table 19.3.2.2). As well, macular volume decreased 58 mm³ in the 0.3 mg pegaptanib arm but increased 12 mm³ with sham ($P = 0.009$) [data on file: (OSI) Eyetech, Inc. and Pfizer Inc. 2005]. OCT center point thickness at baseline and change in thickness from baseline to week 36 had a modest correlation with VA at baseline or change in VA from baseline to week 36 ($R^2 = 0.18$). Lastly, in the 0.3 mg pegaptanib arm, only 25% of patients required further treatment with photocoagulation, compared to 48% in the sham group ($P = 0.042$; Table 19.3.2.3) [21].

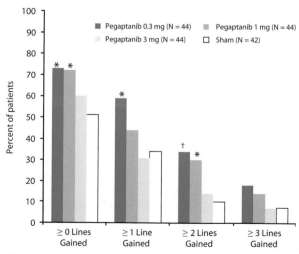

Fig. 19.3.2.4. Percentage of patients treated with pegaptanib sodium maintaining or gaining visual acuity from baseline to week 36 (intention-to-treat population, $N = 172$). *$P < 0.05$, †$P < 0.01$. (With permission from [21])

Table 19.3.2.1. Changes from baseline to week 36 in visual acuity (intention-to-treat population, $N=172^a$) (ANCOVA model). (With permission from [21])

Visual acuity	Pegaptanib 0.3 mg ($N=44$)	1 mg ($N=44$)	3 mg ($N=42$)	Sham ($N=42$)
Mean change (letters) from baseline to:				
Week 0	+0.4	−0.0	+0.2	+0.9
Week 6	+1.8	+2.9	+3.6	+1.4
Week 12	+3.5	+4.3	+2.5	+1.3
Week 30	+5.4	+4.1	+2.3	+0.6
Week 36	+4.7	+4.7	+1.1	−0.4
P value vs. sham at week 36[b]	0.04	0.05	0.55	

ANCOVA analysis of covariance
[a] For missing baseline data, day 0 data were used for the analysis. For missing data at subsequent time points, the last observation was carried forward. Missing relevant data for one patient each in the 1 mg and sham groups
[b] ANCOVA model adjusted for baseline retinal thickening area and baseline vision (*P* values of pairwise comparisons unadjusted for multiplicity)

Table 19.3.2.2. Changes from baseline to week 36 in retinal thickness of the center point of the central subfield (intention-to-treat population, $N=172^a$). (With permission from [21])

Retinal thickness	Pegaptanib 0.3 mg ($N=44$)	1 mg ($N=44$)	3 mg ($N=42$)	Sham ($N=42$)
Mean at baseline (µm)	476.0	451.7	424.7	423.2
Mean change at week 36 (µm)	−68.0	−22.7	−5.3	+3.7
(95% CI)	(−118.9, −9.88)	(−76.9, 33.8)	(−63.0, 49.5)	
P value vs. sham[b]	0.02	0.44	0.81	
≥ 75 µm decrease from baseline:				
Number (%) at week 36	21 (49)	11 (28)	9 (25)	7 (19)
Odds ratio (95% CI)	4.1 (1.5, 11.3)	1.7 (0.6, 5.0)	1.4 (0.5, 4.4)	
P value vs. sham[c]	0.008	0.283	0.596	
≥ 100 µm decrease from baseline:				
Number (%) at week 36	18 (42)	10 (26)	7 (19)	6 (16)
Odds ratio (95% CI)	3.7 (1.3, 10.8)	1.8 (0.6, 5.5)	1.3 (0.4, 4.2)	
P value vs. sham[c]	0.021	0.303	0.829	
≥ 200 µm decrease from baseline:				
Number (%) at week 36	5 (12)	3 (8)	2 (6)	1 (3)
Odds ratio (95% CI)	4.7 (0.5, 42.5)	3.0 (0.3, 30.2)	2.1 (0.2, 24.4)	
P value vs. sham[c]	0.126	0.304	0.678	

CI confidence interval
[a] For missing baseline data, day 0 data were used for the analysis. For missing data at week 36, the last observation was carried forward. Missing relevant data for one patient in the 0.3 mg, five patients in the 1 mg, six patients in the 3 mg, and five patients in the sham groups
[b] Analysis of covariance model adjusted for baseline retinal thickening area, baseline vision, and baseline retinal thickness. *P* value for difference in least square means between each dose group and sham
[c] Cochran-Mantel-Haenszel test adjusted for baseline retinal thickening area and baseline vision. *P* value for difference in odds ratios between each dose group and sham

Table 19.3.2.3. Patients receiving focal/grid laser at week 12 or later in study eye (intention-to-treat population, $N=172^a$). (With permission from [21])

Focal photocoagulation	Pegaptanib 0.3 mg ($N=44$)	1 mg ($N=44$)	3 mg ($N=42$)	Sham ($N=42$)
Yes	11 (25)	13 (30)	17 (40)	20 (48)
No	33 (75)	31 (70)	25 (60)	22 (52)

Comparisons	Odds ratio	95% CI	P value
0.3 mg vs. sham	0.37	0.15, 0.91	0.042
1 mg vs. sham	0.46	0.19, 1.12	0.090
3 mg vs. sham	0.75	0.32, 1.77	0.537

19.3.2.2.2.2 Retinal Revascularization – Retrospective Analysis

Fundus photographs and fluorescein angiograms were analyzed for changes between baseline, week 36 and week 52. Nineteen patients in all were found to have retinal revascularization in the study eye, 16 of whom were available for full analysis. Four of the 16 patients also had neovascularization in the fellow eye. Thirteen patients had received pegaptanib while the other three received sham injections. At 36 weeks, 8 of the 13 patients in the pegaptanib groups (61%) showed regression of neovasculariza-

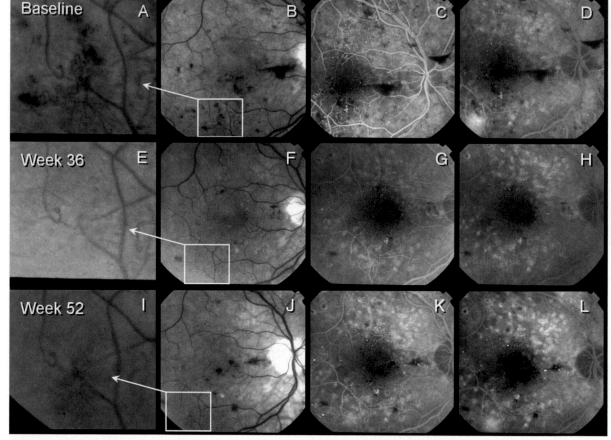

Fig. 19.3.2.5. First row: baseline visit shows **A** magnification of retinal neovascularization elsewhere (NVE); **B** red-free photograph showing the location of the neovascularization along the inferotemporal arcade, and fluorescein angiograms with areas of capillary nonperfusion in the early phase frame (**C**) and leakage from the NVE in the late phase frame (**D**). Second row: 36 weeks after six periodic pegaptanib injections (and 6 weeks since most recent injection) shows regression of NVE on red-free photographs (**E, F**) with less apparent microaneurysms in early phase (**G**) and regression of leakage from NVE in late phase (**H**). Third row: 52 weeks after study entry and 22 weeks since last pegaptanib injection, shows reappearance of NVE on red-free photographs (**I, J**) with reappearance of leakage from NVE in early and late phase frames (**K, L**). (With permission from [8])

tion, as assessed by fundus photography or angiography or both, while no regression was seen in the three sham-treated patients or in the four fellow eyes. Three of the eight patients with regressed neovascularization experienced a recurrence between weeks 36 and 52 after pegaptanib therapy was discontinued (Fig. 19.3.2.5) [8].

19.3.2.2.2.3 Safety

Pegaptanib was well tolerated at all administered doses. Adverse events were transient and associated with the injection procedure, rather than with the study drug. One case of endophthalmitis occurred among 652 injections (0.15% per injection); it was successfully treated and resolved without severe loss of vision [21].

Callout 5 – Phase 2 Trial
- Phase 2 trial, randomizing 172 patients with DME, to pegaptanib (0.3 mg, 1 mg, or 3 mg) or sham injection [21]
- Pegaptanib treatment proved superior to sham for all prespecified endpoints, including measurements of VA, retinal thickness and need for further additional photocoagulation therapy for DME
- Mean change in VA in the 0.3 mg pegaptanib treatment group was +4.7 letters compared to −0.4 letters for sham
- Eight of 13 pegaptanib-treated patients with revascularization in the study eye experienced its regression during the trial; regression of revascularization was not seen in any of three sham-treated patients or in the four fellow eyes [8]
- Pegaptanib was well tolerated, with only one case of endophthalmitis occurring among 652 injections (0.15%)

19.3.2.3 Conclusions

The encouraging results of this trial support the strategy of targeting the underlying pathophysiology of DME and indicate that pegaptanib treatment offers significant clinical benefits. The possibility of treating DR safely by selective VEGF blockade, be the loss of vision caused by increased permeability or generation of neovascularization, validates the singular role VEGF plays in the clinical manifestations of diabetes within the eye. Pivotal phase 3 trials are justified to confirm the aforementioned observations and are currently under way.

Acknowledgements. The authors would like to acknowledge the editorial assistance of Lauren Swenarchuk, PhD.

References

1. Adamis AP, Shima DT, Yeo KT, Yeo TK, Brown LF, Berse B, D'Amore PA, Folkman J (1993) Synthesis and secretion of vascular permeability factor/vascular endothelial growth factor by human retinal pigment epithelial cells. Biochem Biophys Res Commun 193:631–638
2. Adamis AP, Miller JW, Bernal MT, D'Amico DJ, Folkman J, Yeo TK, Yeo KT (1994) Increased vascular endothelial growth factor levels in the vitreous of eyes with proliferative diabetic retinopathy. Am J Ophthalmol 118:445–450
3. Adamis AP, Shima DT, Tolentino MJ, Gragoudas ES, Ferrara N, Folkman J, D'Amore PA, Miller JW (1996) Inhibition of vascular endothelial growth factor prevents retinal ischemia-associated iris neovascularization in a nonhuman primate. Arch Ophthalmol 114:66–71
4. Aiello LP, Avery RL, Arrigg PG, Keyt BA, Jampel HD, Shah ST, Pasquale LR, Thieme H, Iwamoto MA, Park JE, Nguyen HV, Aiello LM, Ferrara N, King GL (1994) Vascular endothelial growth factor in ocular fluid of patients with diabetic retinopathy and other retinal disorders. N Engl J Med 331:1480–1487
5. Aiello LP, Northrup JM, Keyt BA, Takagi H, Iwamoto MA (1995) Hypoxic regulation of vascular endothelial growth factor in retinal cells. Arch Ophthalmol 113:1538–1544
6. Aiello LP, Pierce EA, Foley ED, Takagi H, Chen H, Riddle L, Ferrara N, King GL, Smith LEH (1995) Suppression of retinal neovascularization *in vivo* by inhibition of vascular endothelial growth factor (VEGF) using soluble VEGF-receptor chimeric proteins. Proc Natl Acad Sci USA 92:10457–10461
7. Alon T, Hemo I, Itin A, Pe'er J, Stone J, Keshet E (1995) Vascular endothelial growth factor acts as a survival factor for newly formed retinal vessels and has implications for retinopathy of prematurity. Nat Med 1:1024–1028
8. Altaweel M, Adamis AP, Bressler NM, Cunningham ET, Davis MD, Goldbaum M, Gonzales C, Guyer DR, Katz B, Patel M, Macugen Diabetic Retinopathy Study Group (2006) Changes in retinal neovascularization following pegaptanib (Macugen®) therapy in diabetic individuals. Ophthalmology 113:23–28
9. Amano S, Rohan R, Kuroki M, Tolentino M, Adamis AP (1998) Requirement for vascular endothelial growth factor in wound- and inflammation-related corneal neovascularization. Invest Ophthalmol Vis Sci 39:18–22
10. Antonetti D, Barber AJ, Hollinger LA, Wolpert EB, Gardner TW (1999) Vascular endothelial growth factor induces rapid phosphorylation of tight junction proteins occludin and zonula occluden 1. J Biol Chem 274:23463–23467
11. Arsic N, Zacchigna S, Zentilin L, Ramirez-Correa G, Pattarini L, Salvi A, Sinagra G, Giacca M (2004) Vascular endothelial growth factor stimulates skeletal muscle regeneration in vivo. Mol Ther 10:844–854
12. Asahara T, Takahashi T, Masuda H, Kalka C, Chen D, Iwaguro H, Inai Y, Silver M, Isner JM (1999) VEGF contributes to postnatal neovascularization by mobilizing bone marrow-derived endothelial progenitor cells. EMBO J 18:3964–3972
13. Barleon B, Sozzani S, Zhou D, Weich HA, Mantovani A, Marme D (1996) Migration of human monocytes in response to vascular endothelial growth factor (VEGF) is mediated via the VEGF receptor flt-1. Blood 87:3336–3343
14. Bell C, Lynam E, Landfair DJ, Janjic N, Wiles ME (1999) Oligonucleotide NX1838 inhibits VEGF165-mediated cellular responses in vitro. In Vitro Cell Dev Biol Anim 35:533–542
15. Blaauwgeers HG, Holtkamp GM, Rutten H, Witmer AN, Koolwijk P, Partanen TA, Alitalo K, Kroon ME, Kijlstra A, van Hinsbergh VW, Schlingemann RO (1999) Polarized vascular endothelial growth factor secretion by human retinal pigment epithelium and localization of vascular endothelial growth factor receptors on the inner choriocapillaris. Evidence for a trophic paracrine relation. Am J Pathol 155:421–428
16. Boulton M, Foreman D, Williams G, McLeod D (1998) VEGF localisation in diabetic retinopathy. Br J Ophthalmol 82:561–568
17. Brooks HL Jr, Caballero S Jr, Newell CK, Steinmetz RL, Watson D, Segal MS, Harrison JK, Scott EW, Grant MB (2004) Vitreous levels of vascular endothelial growth factor and stromal-derived factor 1 in patients with diabetic retinopathy and cystoid macular edema before and after intraocular injection of triamcinolone. Arch Ophthalmol 122:1801–1807
18. Caldwell RB, Bartoli M, Behzadian MA, El-Remessy AE, Al-Shabrawey M, Platt DH, Caldwell RW (2003) Vascular endothelial growth factor and diabetic retinopathy: pathophysiological mechanisms and treatment perspectives. Diabetes Metab Res Rev 19:442–455
19. Clauss M, Gerlach M, Gerlach H, Brett J, Wang F, Familletti PC, Pan YC, Olander JV, Connolly DT, Stern D (1990) Vascular permeability factor: a tumor-derived polypeptide that induces endothelial cell and monocyte procoagulant activity, and promotes monocyte migration. J Exp Med 172:1535–1545
20. Csaky KG, Baffi JZ, Byrnes GA, Wolfe JD, Hilmer SC, Flippin J, Cousins SW (2004) Recruitment of marrow-derived endothelial cells to experimental choroidal neovascularization by local expression of vascular endothelial growth factor. Exp Eye Res 78:1107–1116
21. Cunningham ET Jr, Adamis AP, Altaweel M, Aiello LP, Bressler NM, D'Amico DJ, Goldbaum M, Guyer DR, Katz B, Patel M, Schwartz SD, Macugen Diabetic Retinopathy Study Group (2005) A phase II randomized double-masked trial of pegaptanib, an anti-vascular endothelial growth factor aptamer, for diabetic macular edema. Ophthalmology 112:1747–1757
22. Deodato B, Arsic N, Zentilin L, Galeano M, Santoro D, Torre V, Altavilla D, Valdembri D, Bussolino F, Squadrito F, Giacca M (2002) Recombinant AAV vector encoding human VEGF165 enhances wound healing. Gene Ther 9:777–785
23. Drolet DW, Nelson J, Tucker CE, Zack PM, Nixon K, Bolin R, Judkins MB, Farmer JA, Wolf JL, Gill SC, Bendele RA (2000) Pharmacokinetics and safety of an anti-vascular endothelial growth factor aptamer (NX1838) following injection into the vitreous humor of rhesus monkeys. Pharm Res 17: 1503–1510

24. Famiglietti EV, Stopa EG, McGookin ED, Song P, LeBlanc V, Streeten BW (2003) Immunocytochemical localization of vascular endothelial growth factor in neurons and glial cells of human retina. Brain Res 969:195–204

25. Ferrara N (2004) Vascular endothelial growth factor: basic science and clinical progress. Endocr Rev 25:581–611

26. Fraser HM, Wilson H, Rudge JS, Wiegand SJ (2005) Single injections of vascular endothelial growth factor trap block ovulation in the macaque and produce a prolonged, dose-related suppression of ovarian function. J Clin Endocrinol Metab 90:1114–1122

27. Funatsu H, Yamashita H, Shimizu E, Kojima R, Hori S (2001) Relationship between vascular endothelial growth factor and interleukin-6 in diabetic retinopathy. Retina 21:469–477

28. Funatsu H, Yamashita H, Ikeda T, Nakanishi Y, Kitano S, Hori S (2002) Angiotensin II and vascular endothelial growth factor in the vitreous fluid of patients with diabetic macular edema and other retinal disorders. Am J Ophthalmol 133:537–543

29. Funatsu H, Yamashita H, Ikeda T, Mimura T, Eguchi S, Hori S (2003) Vitreous levels of interleukin-6 and vascular endothelial growth factor are related to diabetic macular edema. Ophthalmology 110:1690–1696

30. Funatsu H, Yamashita H, Sakata K, Noma H, Mimura T, Suzuki M, Eguchi S, Hori S (2005) Vitreous levels of vascular endothelial growth factor and intercellular adhesion molecule 1 are related to diabetic macular edema. Ophthalmology 112:806–816

31. Gerber HP, Vu TH, Ryan AM, Kowalski J, Werb Z, Ferrara N (1999) VEGF couples hypertrophic cartilage remodeling, ossification and angiogenesis during endochondral bone formation. Nat Med 5:623–628

32. Gragoudas ES, Adamis AP, Cunningham ET Jr, Feinsod M, Guyer DR; VEGF Inhibition Study in Ocular Neovascularization Clinical Trial Group (2004) Pegaptanib for neovascular age-related macular degeneration. N Engl J Med 351:2805–2816

33. Hiratsuka S, Nakamura K, Iwai S, Murakami M, Itoh T, Kijima H, Shipley JM, Senior RM, Shibuya M (2002) MMP9 induction by vascular endothelial growth factor receptor-1 is involved in lung-specific metastasis. Cancer Cell 2:289–300

34. Hurwitz H, Fehrenbacher L, Novotny W, Cartwright T, Hainsworth J, Heim W, Berlin J, Baron A, Griffing S, Holmgren E, Ferrara N, Fyfe G, Rogers B, Ross R, Kabbinavar F (2004) Bevacizumab plus irinotecan, fluorouracil, and leucovorin for metastatic colorectal cancer. N Engl J Med 350:2335–2342

35. Ishida S, Usui T, Yamashiro K, Kaji Y, Ahmed E, Carrasquillo KG, Amano S, Hida T, Oguchi Y, Adamis AP (2003a) VEGF164 is proinflammatory in the diabetic retina. Invest Ophthalmol Vis Sci 44:2155–2162

36. Ishida S, Usui T, Yamashiro K, Kaji Y, Amano S, Ogura Y, Hida T, Oguchi Y, Ambati J, Miller JW, Gragoudas ES, Ng YS, D'Amore PA, Shima DT, Adamis AP (2003) VEGF164-mediated inflammation is required for pathological, but not physiological, ischemia-induced retinal neovascularization. J Exp Med 198:483–489

37. Johnson DH, Fehrenbacher L, Novotny WF, Herbst RS, Nemunaitis JJ, Jablons DM, Langer CJ, DeVore RF 3rd, Gaudreault J, Damico LA, Holmgren E, Kabbinavar F (2004) Randomized phase II trial comparing bevacizumab plus carboplatin and paclitaxel with carboplatin and paclitaxel alone in previously untreated locally advanced or metastatic non-small-cell lung cancer. J Clin Oncol 22:2184–2191

38. Joussen A, Murata T, Tsujikawa A, Kirchhof B, Bursell S-E, Adamis AP (2001) Leukocyte-mediated endothelial cell injury and death in the diabetic retina. Am J Pathol 158:147–152

39. Joussen AM, Adamis AP (2006) Inflammation as a stimulus for vascular leakage and proliferation. In: Joussen AM, Gardner TW, Kirchhof B, Ryan SJ (eds) Retinal vascular disease. Springer, Berlin Heidelberg New York

40. Kabbinavar F, Hurwitz HI, Fehrenbacher L, Meropol NJ, Novotny WF, Lieberman G, Griffing S, Bergsland E (2003) Phase II, randomized trial comparing bevacizumab plus fluorouracil (FU)/leucovorin (LV) with FU/LV alone in patients with metastatic colorectal cancer. J Clin Oncol 21:60–65

41. Kabbinavar FF, Schulz J, McCleod M, Patel T, Hamm JT, Hecht JR, Mass R, Perrou B, Nelson B, Novotny WF (2005) Addition of bevacizumab to bolus fluorouracil and leucovorin in first-line metastatic colorectal cancer: results of a randomized phase II trial. J Clin Oncol 23:3697–3705

42. Kitamoto Y, Tokunaga H, Tomita K (1997) Vascular endothelial growth factor is an essential molecule for mouse kidney development: glomerulogenesis and nephrogenesis. J Clin Invest 99:2351–2357

43. Krzystolik MG, Afshari MA, Adamis AP, Gaudreault J, Gragoudas ES, Michaud NA, Li W, Connolly E, O'Neill CA, Miller JW (2002) Prevention of experimental choroidal neovascularization with intravitreal anti-vascular endothelial growth factor antibody fragment. Arch Ophthalmol 120:338–346

44. Kuroki M, Voest EE, Amano S, Beerepoot LV, Takashima S, Tolentino M, Kim RY, Rohan RM, Colby KA, Yeo KT, Adamis AP (1996) Reactive oxygen intermediates increase vascular endothelial growth factor expression in vitro and in vivo. J Clin Invest 98:1667–1675

45. Lamoreaux WJ, Fitzgerald ME, Reiner A, Hasty KA, Charles ST (1998) Vascular endothelial growth factor increases release of gelatinase A and decreases release of tissue inhibitor of metalloproteinases by microvascular endothelial cells in vitro. Microvasc Res 55:29–42

46. Lashkari K, Hirose T, Yazdany J, McMeel JW, Kazlauskas A, Rahimi N (2000) Vascular endothelial growth factor and hepatocyte growth factor levels are differentially elevated in patients with advanced retinopathy of prematurity. Am J Pathol 156:1337–1344

47. LeCouter J, Moritz DR, Li B, Phillips GL, Liang XH, Gerber HP, Hillan KJ, Ferrara N (2003) Angiogenesis-independent endothelial protection of liver: role of VEGFR-1. Science 299:890–893

48. Leung DW, Cachianes G, Kuang WJ, Goeddel DV, Ferrara N (1989) Vascular endothelial growth factor is a secreted angiogenic mitogen. Science 246:1306–1309

49. Liu MH, Jin H, Floten HS, Ren Z, Yim AP, He GW (2002) Vascular endothelial growth factor-mediated, endothelium-dependent relaxation in human internal mammary artery. Ann Thorac Surg 73:819–824

50. Lu M, Kuroki M, Amano S, Tolentino M, Keough K, Kim I, Bucala R, Adamis AP (1998) Advanced glycation end products increase retinal vascular endothelial growth factor expression. J Clin Invest 101:1219–1224

51. Malecaze F, Clamens S, Simorre-Pinatel V, Mathis A, Chollet P, Favard C, Bayard F, Plouet J (1994) Detection of vascular endothelial growth factor messenger RNA and vascular

endothelial growth factor-like activity in proliferative diabetic retinopathy. Arch Ophthalmol 112:1476–1482

52. Marneros AG, Fan J, Yokoyama Y, Gerber HP, Ferrara N, Crouch RK, Olsen BR (2005) Vascular endothelial growth factor expression in the retinal pigment epithelium is essential for choriocapillaris development and visual function. Am J Pathol 167:1451–1459

53. Miller JW, Adamis AP, Shima DT, D'Amore PA, Moulton RS, O'Reilly MS, Folkman J, Dvorak HF, Brown LF, Berse B, Yeo T-K, Yeo K-T (1994) Vascular endothelial growth factor/vascular permeability factor is temporally and spatially correlated with ocular angiogenesis in a primate model. Am J Pathol 145:574–584

54. Miller KD, Chap LI, Holmes FA, Cobleigh MA, Marcom PK, Fehrenbacher L, Dickler M, Overmoyer BA, Reimann JD, Sing AP, Langmuir V, Rugo HS (2005) Randomized phase III trial of capecitabine compared with bevacizumab plus capecitabine in patients with previously treated metastatic breast cancer. J Clin Oncol 23:792–799

55. Miyamoto K, Hiroshiba N, Tsujikawa A, Ogura Y (1998) In vivo demonstration of increased leukocyte entrapment in retinal microcirculation of diabetic rats. Invest Ophthalmol Vis Sci 39:2190–2194

56. Ng EWM, Shima DT, Calias P, Cunningham ET Jr, Guyer DR, Adamis AP (2006) Pegaptanib, a targeted anti-VEGF aptamer for ocular vascular disease. Nat Rev Drug Discov 5:123–132

57. Nishijima K, Ng YS, Zhong L, Bradley J, Schubert W, Jo N, Akito J, Samuelsson SJ, Robinson GS, Adamis AP, Shima DT (2007) Vascular endothelial growth factor-A is a survival factor for retinal neurons and a critical neuroprotectant during the adoptive response to ischemic injury. Am J Pathol 171:53–67

58. Nissen NN, Polverini PJ, Koch AE, Volin MV, Gamelli RL, DiPietro LA (1998) Vascular endothelial growth factor mediates angiogenic activity during the proliferative phase of wound healing. Am J Pathol 152:1445–1452

59. Ohno-Matsui K, Hirose A, Yamamoto S, Saikia J, Okamoto N, Gehlbach P, Duh EJ, Hackett S, Chang M, Bok D, Zack DJ, Campochiaro PA (2002) Inducible expression of vascular endothelial growth factor in adult mice causes severe proliferative retinopathy and retinal detachment. Am J Pathol 160:711–719

60. Oosthuyse B, Moons L, Storkebaum E, Beck H, Nuyens D, Brusselmans K, Van Dorpe J, Hellings P, Gorselink M, Heymans S, Theilmeier G, Dewerchin M, Laudenbach V, Vermylen P, Raat H, Acker T, Vleminckx V, Van Den Bosch L, Cashman N, Fujisawa H, Drost MR, Sciot R, Bruyninckx F, Hicklin DJ, Ince C, Gressens P, Lupu F, Plate KH, Robberecht W, Herbert JM, Collen D, Carmeliet P (2001) Deletion of the hypoxia-response element in the vascular endothelial growth factor promoter causes motor neuron degeneration. Nat Genet 28:131–138

61. Papapetropoulos A, Garcia-Cardena G, Madri JA, Sessa WC (1997) Nitric oxide production contributes to the angiogenic properties of vascular endothelial growth factor in human endothelial cells. J Clin Invest 100:3131–3139

62. Pepper MS, Ferrara N, Orci L, Montesano R (1991) Vascular endothelial growth factor (VEGF) induces plasminogen activators and plasminogen activator inhibitor-1 in microvascular endothelial cells. Biochem Biophys Res Commun 181:902–906

63. Qaum T, Xu Q, Joussen AM, Clemens MW, Qin W, Miyamoto K, Hassessian H, Wiegand SJ, Rudge J, Yancopoulos GD, Adamis AP (2001) VEGF-initiated blood-retinal barrier breakdown in early diabetes. Invest Ophthalmol Vis Sci 42:2408–2413

64. Roberts WG, Palade GE (1997) Neovasculature induced by vascular endothelial growth factor is fenestrated. Cancer Res 57:765–772

65. Ruckman J, Green LS, Beeson J, Waugh S, Gillette WL, Henninger DD, Claesson-Welsh L, Janjic N (1998) 2'-Fluoropyrimidine RNA-based aptamers to the 165-amino acid form of vascular endothelial growth factor (VEGF165). Inhibition of receptor binding and VEGF-induced vascular permeability through interactions requiring the exon 7-encoded domain. J Biol Chem 273:20556–20567

66. Ryan AM, Eppler DB, Hagler KE, Bruner RH, Thomford PJ, Hall RL, Shopp GM, O'Neill CA (1999) Preclinical safety evaluation of rhuMAbVEGF, an antiangiogenic humanized monoclonal antibody. Toxicol Pathol 27:78–86

67. Schwesinger C, Yee C, Rohan RM, Joussen AM, Fernandez A, Meyer TN, Poulaki V, Ma JJ, Redmond TM, Liu S, Adamis AP, D'Amato RJ (2001) Intrachoroidal neovascularization in transgenic mice overexpressing vascular endothelial growth factor in the retinal pigment epithelium. Am J Pathol 158:1161–1172

68. Senger DR, Connolly DT, Van de Water L, Feder J, Dvorak HF (1990) Purification and NH2-terminal amino acid sequence of guinea pig tumor-secreted vascular permeability factor. Cancer Res 50:1774–1778

69. Skillings JR, Johnson DH, Miller K, Kabbinavar F, Bergsland E, Holmgren E, Holden SN, Hurwitz H, Scappaticci F (2005) Arterial thromboembolic events (ATEs) in a pooled analysis of 5 randomized, controlled trials (RCTs) of bevacizumab (BV) with chemotherapy [abstract]. J Clin Oncol 23:3019

70. Storkebaum E, Lambrechts D, Carmeliet P (2004) VEGF: once regarded as a specific angiogenic factor, now implicated in neuroprotection. Bioessays 26:943–954

71. The Eyetech Study Group (2003) Anti-vascular endothelial growth factor therapy for subfoveal choroidal neovascularization secondary to age-related macular degeneration: phase II study results. Ophthalmology 110:979–986

72. Tolentino MJ, Miller JW, Gragoudas ES, Chatzistefanou K, Ferrara N, Adamis AP (1996) Vascular endothelial growth factor is sufficient to produce iris neovascularization and neovascular glaucoma in a nonhuman primate. Arch Ophthalmol 114:964–970

73. Tolentino MJ, Miller JW, Gragoudas ES, Jakobiec FA, Flynn E, Chatzistefanou K, Ferrara N, Adamis AP (1996) Intravitreous injections of vascular endothelial growth factor produce retinal ischemia and microangiopathy in an adult primate. Ophthalmology 103:820–828

74. Tolentino MJ, McLeod DS, Taomoto M, Otsuji T, Adamis AP, Lutty GA (2002) Pathologic features of vascular endothelial growth factor-induced retinopathy in the nonhuman primate. Am J Ophthalmol 133:373–385

75. Tripathi RC, Li J, Tripathi BJ, Chalam KV, Adamis AP (1998) Increased level of vascular endothelial growth factor in aqueous humor of patients with neovascular glaucoma. Ophthalmology 105:232–237

76. Uhlmann S, Friedrichs U, Eichler W, Hoffmann S, Wiedemann P (2001) Direct measurement of VEGF-induced nitric oxide production by choroidal endothelial cells. Microvasc Res 62:179–189

77. Usui T, Ishida S, Yamashiro K, Kaji Y, Poulaki V, Moore J, Moore T, Amano S, Horikawa Y, Dartt D, Golding M, Shima DT, Adamis AP (2004) VEGF164(165) as the pathological isoform: differential leukocyte and endothelial responses through VEGFR1 and VEGFR2. Invest Ophthalmol Vis Sci 45:368–374

19.3.3 Ranibizumab for the Treatment of Diabetic Macular Edema

D.V. Do, Q.D. Nguyen, S.M. Shah, J.A. Haller

Core Messages
- Vascular endothelial growth factor (VEGF) is a critical stimulus for diabetic macular edema (DME)
- Ranibizumab is a humanized antigen-binding fragment (Fab) that inhibits all VEGF isoforms
- Phase II/III clinical trials for age-related macular degeneration (AMD) have demonstrated the safety and tolerability of ranibizumab
- The READ-1 study is an open-label phase I study investigating the effect of intravitreal injections of ranibizumab in 20 patients with DME
- The primary outcome of the READ-1 study was foveal thickness measured by optical coherence tomography at 7 months compared to baseline
- The READ-2 study, a phase II multicenter trial, is underway

19.3.3.1 Introduction

Diabetic retinopathy is the most prevalent cause of vision loss among working age adults, and diabetic macular edema (DME) is the most common cause of moderate vision loss in individuals with diabetes mellitus [9]. Although vision loss with DME is frequent, effective treatment with laser photocoagulation, metabolic control, intraocular steroids, and novel pharmacological therapy such as vascular endothelial growth (VEGF) factor blockers may help to decrease DME and may reverse vision loss associated with this disease. This chapter reviews the pathogenesis of DME, the rationale for the role of anti-VEGF agents, and preliminary clinical studies on the use of ranibizumab (Lucentis, Genentech Inc., South San Francisco, CA) for DME.

19.3.3.2 Pathogenesis

Diabetic macular edema occurs from leakage of plasma into the central retina, resulting in thickening of the retina because of excess interstitial fluid. The excess interstitial fluid within the macula results in stretching and distortion of photoreceptors, which eventually leads to decreased vision. Histopathologic studies have demonstrated that microaneurysms are likely responsible for focal leakage that may be seen in eyes with DME. Microaneurysms are thought to form because of hyperglycemia-induced pericyte death, which weakens the walls of retinal vessels resulting in the formation of small aneurysms which lose their barrier qualities and leak [12]. In addition to focal leakage caused by microaneurysms, eyes with DME may also demonstrate diffuse leakage from retinal capillaries that do not show visible structural changes such as microaneurysms. This pattern of diffuse leakage may be due to microscopic damage to retinal vessels that are not visible in images obtained during fluorescein angiography. It is possible that diffuse leakage results from the presence of excessive amounts of permeability factors.

19.3.3.3 Vascular Endothelial Growth Factor

Retinal hypoxia has been implicated in the pathogenesis of DME [13]. Hypoxia causes increased expression of VEGF, a potent inducer of vascular permeability that has been shown to cause leakage from retinal vessels [4, 16].

The VEGF family, which includes VEGF-A, VEGF-B, VEGF-C, VEGF-D, and placental growth factor, plays an important role in angiogenesis and vascular permeability [11, 15, 19]. Studies have demonstrated that VEGF-A is a primary activator of angiogenesis and vascular permeability, whereas other VEGF family members play a lesser role in angiogenesis. Nine VEGF-A isoforms are produced through alternate splicing of the mRNA of the human VEGF-A gene: $VEGF_{121}$, $VEGF_{145}$, $VEGF_{148}$, $VEGF_{162}$, $VEGF_{165}$, $VEGF_{165b}$, $VEGF_{183}$, $VEGF_{189}$, and $VEGF_{206}$ [20]. Among the nine isoforms, $VEGF_{165}$ is the most abundantly expressed VEGF-A isoform, and it plays a critical role in angiogenesis. However, other isoforms, such as $VEGF_{121}$, $VEGF_{183}$, and $VEGF_{189}$, are also commonly expressed in various tissues [8].

Animal studies have demonstrated that VEGF-A plays a vital role in the pathogenesis of ocular diseases in which neovascularization and increased vascular permeability occur, such as proliferative diabetic retinopathies, macular edema, and neovascular age-related macular degeneration (AMD) [1, 5, 6]. Overexpression of VEGF-A has been reported to cause ocular neovascularization and macular edema in monkeys [10]. In addition, elevated levels of VEGF-A have been identified in the vitreous of patients with diabetic retinopathy [2], and higher VEGF-A levels are found in the vitreous and aqueous humor of patients with diabetic macular edema [5, 6]. Evidence from animal and clinical studies has clearly demonstrated a pathologic role for overexpression of VEGF; therefore anti-VEGF therapies are likely to have an important role in the treatment of DME and proliferative diabetic retinopathy.

19.3.3.4 Ranibizumab

19.3.3.4.1 Biology

Ranibizumab is an FDA-approved humanized antigen-binding fragment (Fab) designed to bind and inhibit all VEGF-A isoforms and their biologically active degradation products. The vitreous half-life of ranibizumab after intravitreal administration in monkeys is 3 days with very low systemic exposure following intravitreal administration in both humans and monkeys [7, 17].

19.3.3.4.2 Clinical Experience in Age-Related Macular Degeneration

Phase I/II/III clinical trials in patients with choroidal neovascularization (CNV) secondary to AMD have shown intravitreal injection of ranibizumab to be safe and well tolerated. In addition, two pivotal phase III clinical trials [Minimally Classic/Occult Trial of the Anti-VEGF Antibody Ranibizumab In the Treatment of Neovascular AMD (MARINA) Trial and the Anti-VEGF Antibody for the Treatment of Predominantly Classic Choroidal Neovascularization in AMD (ANCHOR) Trial] demonstrated the efficacy of ranibizumab in patients with neovascular AMD; these studies were the first to show visual improvement, not just a stabilization of visual acuity, in individuals with neovascular AMD.

The MARINA Trial was a randomized, double-masked, sham-controlled clinical trial of patients with minimally classic or occult with no classic CNV secondary to AMD who were treated with monthly intravitreal ranibizumab (0.3 or 0.5 mg) or sham injections for 24 months. In the MARINA trial, approximately 90–92% of ranibizumab-treated patients lost fewer than 15 letters of VA compared with 53% of sham-injected patients after 24 months [16]. In addition, at 24 months of follow-up, approximately 33% of ranibizumab-treated patients experienced visual improvement of 15 or more letters compared with 4% of the sham-injected patients.

The ANCHOR Trial was a randomized, double-masked, sham-controlled clinical trial of patients with predominantly classic CNV secondary to AMD treated with intravitreal ranibizumab (0.3 or 0.5 mg) and sham photodynamic therapy (PDT) or sham injection and PDT (monthly administration for ranibizumab and every 3 months for PDT) for 24 months. After 12 months of follow-up, 94% and 96% of ranibizumab-treated patients (0.3 and 0.5 mg, respectively) lost fewer than 15 letters of VA compared with 64% of PDT-treated patients [3]. In addition, 36% and 40% of ranibizumab-treated patients experienced visual improvement of 15 or more letters (0.3 and 0.5 mg, respectively) compared with 6% of PDT-treated patients.

19.3.3.5 Ranibizumab in Diabetic Macular Edema

Nguyen and colleagues conducted an open-label study (Ranibizumab for Edema of the mAcula in Diabetes: A Phase I Study – the READ-1 Study) to investigate the effect of intravitreal injections of ranibizumab in 20 patients with DME [14]. Intraocular injections of 0.5 mg of ranibizumab were administered at study entry and at 1, 2, 4, and 6 months after entry. The injection regimen was selected to assess the effect of 3 monthly injections and then determine the impact of increasing the time interval between injections to 2 months for the last two injections. The primary outcome measure was foveal thickness measured by optical coherence tomography (OCT) at 7 months compared to baseline. Secondary outcome measures were macular volume measured by OCT and visual acuity measured by the protocol of the Early Treatment Diabetic Retinopathy Study (ETDRS) [21] at 7 months compared to baseline.

Among the first ten subjects (five men and five women) enrolled, pertinent baseline characteristics included: eight eyes that had received at least two sessions of focal/grid laser photocoagulation not less than 5 months prior to study entry (range 5–120 months), three eyes that had received intraocular steroids not less than 10 months prior to entry (range 10–20 months), and a mean foveal thickness of 503 ± 115 μm (range 326–729 μm) at baseline, indicating the presence of severe, chronic DME that was poorly responsive to standard therapies.

19 III

19.3.3.5.1 Effect on Foveal Thickness and Macular Volume

OCT scans from two subjects whose DME showed response to ranibizumab are shown in Fig. 19.3.3.1. Compared to baseline, mean foveal thickness was reduced by 246 μm at the primary endpoint of the study (7 months after the first ranibizumab injection), representing an elimination of 85% of the excess foveal thickness that had been present at baseline. OCT scans from two subjects are shown in Fig. 19.3.3.2. In addition, mean macular volume was reduced from 9.22 mm³ at baseline to 7.47 mm³ at 7 months, a reduction of 1.75 mm³ which was statistically significant ($P=0.009$). This reduction constituted 77% of the excess macular volume that was present at baseline.

19.3.3.5.2 Effect on Visual Acuity

Throughout each study time point, mean and median visual acuities were better than those at baseline. At the primary endpoint (7 months after the initial ranibizumab injection), mean and median visual acuity improved by 12.3 and 11 letters, which represents an improvement of a little more than 2 lines (Fig. 19.3.3.3).

In this study cohort, there was a strong correlation (R^2 value of 0.78) between visual acuity and foveal thickness as measured by OCT (Fig. 19.3.3.4). However, the rate of change of these two outcome measures was different, and rapid changes in foveal thickness were associated with more gradual improvements in visual acuity. Further studies are

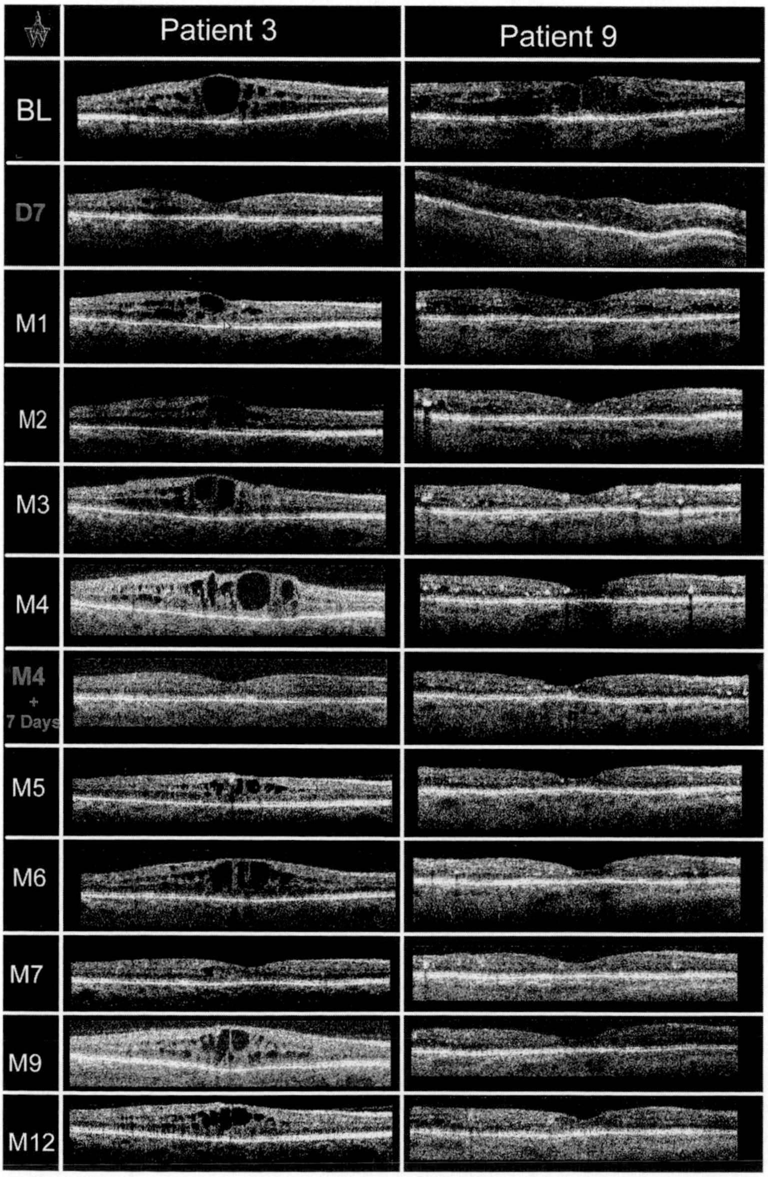

Fig. 19.3.3.1. Horizontal cross sectional optical coherence tomography (OCT) scans at all time points for patients 3 and 9 to illustrate two patterns of response over time. Seven days after the first intraocular injection of 0.5 mg of ranibizumab (*D7*), patient 3 showed a marked improvement in the appearance of the OCT scan with elimination of several large cysts and return of a normal foveal contour. At month 1 (*M1*), 1 month after the first injection, and *M2* and *M3*, 1 month after the second and third injections, respectively, the scans for patient 3 were worse than the scan at *D7*, suggesting loss of effect of ranibizumab or transient effects that are lost by 1 month after injection. An OCT scan done 7 days after the third injection (*M3+7 days*) substantiates the latter of these two possibilities. At *M4*, 2 months after the third injection, the scan showed substantial deterioration, but at *M5*, 1 month after the fourth injection, there was improvement. This was followed by deterioration at *M6*, 2 months after the fourth injection, but then at *M7*, the primary endpoint and 1 month after the fifth injection, there was improvement to the point that the scan looked more like the two previous scans that had been done 7 days after an injection than like those that had been done 1 month after an injection. Like patient 3, patient 9 also showed substantial improvement at *D7* compared to baseline with resolution of several large cysts. However, unlike patient 3, patient 9 showed continued improvement and then stability at subsequent time points regardless of the time after injection that the scan was done. This suggests that the beneficial effects of ranibizumab were more sustained in patient 9 than in patient 3. (Reprinted from [15] with permission)

Fig. 19.3.3.2. The mean excess foveal thickness at each study visit. *Each bar* represents the mean value for excess foveal thickness for all patients at the designated study visit (data for eight of ten patients at month 9). The *arrows* show when intraocular injections of 0.5 mg of ranibizumab were administered. The *bars* at baseline and the primary endpoint, month 7, are *shaded*. Compared to baseline, foveal thickness was reduced by 246 µm at the primary endpoint of the study, constituting elimination of 85 % of the excess foveal thickness that had been present at baseline. (Reprinted from [15] with permission)

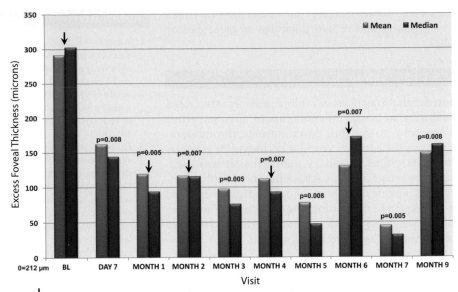

Fig. 19.3.3.3. Mean and median change in visual acuity from baseline at each study visit. The *shaded line* shows the mean change in visual acuity measured in number of letters read on an ETDRS visual acuity chart and the *unshaded line* shows the median change in visual acuity. The *arrows* show times of intraocular injection of 0.5 mg of ranibizumab. At the primary endpoint, 7 months, there was an improvement of 12.3 letters in mean visual acuity and 11 letters in median visual acuity. (Reprinted from [15] with permission)

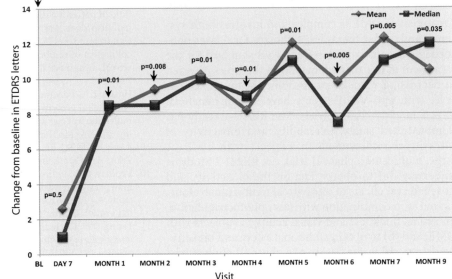

Fig. 19.3.3.4. Scatter plot of reduction in foveal thickness versus gain in visual acuity at the primary endpoint compared to baseline for all patients. The reduction in foveal thickness in micrometers on the *y*-axis is plotted against the improvement in visual acuity measured by numbers of letters read on an ETDRS visual acuity chart. There is a strong correlation with an R^2 value of 0.78. (Reprinted from [15] with permission)

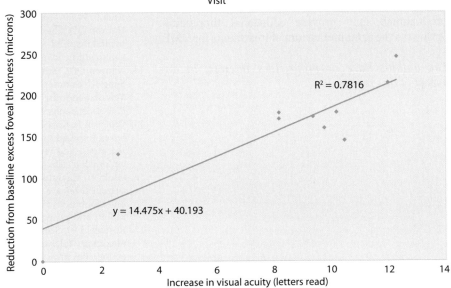

19

19 **III**

underway to investigate the correlation between visual acuity and retinal thickness as measured by OCT.

19.3.3.5.3 Safety

Intraocular injections of ranibizumab were tolerated well with no ocular inflammation or adverse events. There were no systemic adverse events, thromboembolic events, cerebral vascular accidents, or myocardial infarctions. Capillary nonperfusion was measured by image analysis at baseline and month 6 fluorescein angiograms with the investigator masked with respect to time point. The mean area of nonperfusion was 0.19812 disk areas at baseline and 0.19525 at 6 months. Therefore, no significant change in capillary nonperfusion was seen throughout the study.

19.3.3.6 Summary

Treatment of DME is complex and involves both systemic and ocular therapies. Although focal laser photocoagulation is considered the gold standard for the treatment of macular edema, novel therapies directed at decreasing vascular permeability at the molecular level with anti-VEGF agents have shown beneficial effects in early clinical trials. The READ-1 Study has demonstrated safety, tolerability, and bioactivity of intravitreal ranibizumab in subjects with DME. A large, multicenter, phase II trial, the READ-2 Study, is underway: (a) to obtain data on the bioactivity and dose interval effects of intravitreal ranibizumab alone, as well as in combination with laser photocoagulation, on retinal thickness and visual acuity in subjects with DME; and (b) to obtain additional safety and bioactivity data to aid in the design of a phase III clinical trial to evaluate ranibizumab as a therapeutic option for patients with DME. Novel anti-VEGF agents such as ranibizumab may provide additional therapeutic options to the armamentarium of treatments for DME.

The authors have no proprietary interests in any aspect of this report.

References

1. Adamis AP, Miller JW, Bernal MT, et al. (1994) Increased vascular endothelial growth factor levels in the vitreous of eyes with proliferative diabetic retinopathy. Am J Ophthalmol 118:445–450
2. Aiello LP, Avery RL, Arrigg PG, et al. (1994) Vascular endothelial growth factor in ocular fluid of patients with diabetic retinopathy and other retinal disorders. N Engl J Med 331:1480–1487
3. Brown DM, Kaiser PK, Michels M et al. (2006) Ranibizumab versus verteporfin for neovascular age-related macular degeneration. N Engl J Med 355:1432–1444
4. Derevjanik NL, Vinores SA, Xiao W-H, et al. (2002) Quantitative assessment of the integrity of the blood-retinal barrier in mice. Invest Ophthalmol Vis Sci 43:2462–2467
5. Funatsu H, Yamashita H, Noma H, et al. (2002) Increased levels of vascular endothelial growth factor and interleukin-6 in the aqueous humor of diabetics with macular edema. Am J Ophthalmol 133:70–77
6. Funatsu H, Yamashita H, Ikeda T, et al. (2002) Angiotensin II and vascular endothelial growth factor in the vitreous fluid of patients with diabetic macular edema and other retinal disorders. Am J Ophthalmol 133:537–543
7. Gaudreault J, Fei D, Rusit J, et al. (2005) Preclinical pharmacokinetics of ranibizumab (rhuFabV2) after a single intravitreal administration. Invest Ophthalmol Vis Sci 46:726–733
8. Houck KA, Ferrara N, Winer J, et al. (1991) The vascular endothelial growth factor family: identification of a fourth molecular species and characterization of alternative splicing of RNA. Mol Endocrinol 5:1806–1814
9. Klein R (1992) Retinopathy in a population-based study. Trans Am Ophthalmol Soc 90:561–594
10. Lebherz C, Maguire AM, Auricchio A, et al. (2005) Nonhuman primate models for diabetic ocular neovascularization using AAV2-mediated overexpression of vascular endothelial growth factor. Diabetes 54:1141–1149
11. Leung DW, Cachianes G, Kuang WJ, et al. (1989) Vascular endothelial growth factor is a secreted angiogenic mitogen. Science 246:1306–1309
12. Moore J, Bagby S, Ireland G, et al. (1999) Three dimensional analysis of microaneurysms in the human diabetic retina. Anat J 194:89–100
13. Nguyen QD, Shah SM, Van Anden E, et al. (2003) Supplemental inspired oxygen improves diabetic macular edema; a pilot study. Invest Ophthalmol Vis Sci 45:617–624
14. Nguyen QD, Tatlipinar S, Shah SM, Haller JA, Quinlan E, Sung JU, Zimmer-Galler I, Do DV, Campochiaro PA (2006) Vascular endothelial growth factor is a critical stimulus for diabetic macular edema. Am J Ophthalmol 142(6):961–9
15. Olofsson B, Korpelainen E, Pepper MS, et al. (1998) Vascular endothelial growth factor B (VEGF-B) binds to VEGF receptor-1 and regulates plasminogen activator activity in endothelial cells. Proc Natl Acad Sci U S A 95:11709–11714
16. Ozaki H, Hayashi H, Vinores SA, et al. (1997) Intravitreal sustained release of VEGF causes retinal neovascularization in rabbits and breakdown of the blood-retinal barrier in rabbits and primates. Exp Eye Res 64:505–517
17. Rosenfeld PJ, Schwartz SD, Blumenkranz MS, et al. (2005) Maximum tolerated dose of a humanized anti-vascular endothelial growth factor antibody fragment for treating neovascular age-related macular degeneration. Ophthalmology 112:1048–1053

18. Rosenfeld PJ, Brown DM, Heier JS, et al. (2006) Ranibizumab for neovascular age-related macular degeneration: 2-year results of the MARINA Study. N Engl J Med 355: 1419–1431

19. Senger DR, Galli SJ, Dvorak AM, et al. (1983) Tumor cells secrete a vascular permeability factor that promotes accumulation of ascites fluid. Science 219:983–985

20. Takahashi H, Shibuya M (2005) The vascular endothelial growth factor (VEGF)/VEGF receptor system and its role under physiological and pathological conditions. Clin Sci (Lond) 109(3):227–241

21. The Early Treatment Diabetic Retinopathy Study Research Group (1985) Photocoagulation for diabetic macular edema, ETDRS Report No. 1. Arch Ophthalmol 103:1644–1652

20   Retinopathy of Prematurity

20.1   Retinopathy of Prematurity: Pathophysiology of Disease

L.E.H. SMITH

Core Messages

- Despite current treatment, retinopathy of prematurity (ROP) continues to be a blinding disease. Understanding the molecular basis of the disease is necessary for prevention and treatment
- The less developed the retina at birth the worse ROP is likely to be. ROP occurs in two opposite phases. Phase I consists of delayed retinal vascular growth and vessel loss after premature birth resulting in hypoxia. Phase II consists of hypoxia-induced vascular proliferation

- Both oxygen-regulated and non-oxygen-regulated factors contribute to normal vascular development and retinal neovascularization. Vascular endothelial growth factor (VEGF) is an important oxygen-regulated factor. A critical non-oxygen-regulated growth factor is insulin-like growth factor-I (IGF-I)
- Lack of IGF-I prevents normal retinal vascular growth. Premature infants who develop ROP have low levels of serum IGF-I compared to age-matched infants without disease. Low IGF-I predicts ROP in premature infants. Restoration of IGF-I to normal levels might help prevent ROP

20.1.1   History of Retinopathy of Prematurity

Retinopathy of prematurity (ROP) was first noted in the late 1940s in preterm infants and described as retrolental fibroplasia, a total retinal detachment seen as white mass behind the lens. The disease was subsequently associated with excessive oxygen use [12, 14, 52]. Oxygen supplementation was curtailed with a decrease in ROP but with an increase in cerebral palsy and death. Supplemental oxygen is now delivered to premature infants to maintain adequate blood levels, but it is monitored carefully [37].

The incidence of ROP has increased further due most likely to factors related to prematurity itself as ever more immature infants are saved after preterm birth. Low gestational age at birth and low birth weight are stronger risk factors than controlled oxygen delivery [22]. ROP is still a major cause of blindness in children in the developed and developing world [67] despite current treatment. Although laser photocoagulation or cryotherapy of the retina reduces the incidence of blindness by about 25%, the visual outcomes after treatment are frequently poor. Prevention and/or medical treatment are urgently required.

To develop such treatments we need to understand the pathogenesis of the disease and develop medical interventions based on this understanding to prevent or treat ROP medically.

20.1.2   ROP: Disruption of Normal Vascular Development

It is necessary to understand normal retinal vascular development to understand the pathology of retinal vascular development in ROP. Retinal blood vessel development in the human fetus begins during the 4th month of gestation [26, 62] and vessels reach the most peripheral temporal aspect of the retina just before term.

Therefore, the retinas of infants born prematurely are incompletely vascularized, with a peripheral avascular zone, the area of which depends on the gestational age at birth. The more premature the infant the less the peripheral retinal vascularization. In the most premature infants to survive (postmenstrual age, PMA, 22–23 weeks) the retinal vessels at birth are found only in the posterior pole.

After premature birth into the relative hyperoxia of the extrauterine environment the vessels cease growing centripetally from the optic nerve to the periphery and some formed vessels are lost (phase I). In phase II of ROP there is vascular proliferation. It is important to understand these opposite phases of vessel loss and vessel proliferation in ROP since the same treatment depending on phase will have opposite effects. Timing of treatment is important.

20.1.3 Pathogenesis: Two Phases of ROP

Essentials
- Retinal vascularization is incomplete after premature birth and the degree of vascularization depends on the gestational age. The more immature the infant, the less the retina is vascularized
- In phase I of ROP vessel growth slows or ceases and some retinal vessels are lost. The retina becomes hypoxic
- In phase II of ROP vessels proliferate in part in response to hypoxia of non-vascularized retina, which can result in vascular leakage and retinal detachment

20.1.4 ROP: Phase I

Phase I of ROP is characterized by vessel loss. The normal retinal vascular growth that would occur in utero slows or ceases, and there is loss of some of the developed vessels. Immature vessels are particularly susceptible to oxygen [10, 11, 45, 54, 68], so this phenomenon is thought to be due in part to the influence of supplemental oxygen given to premature infants to overcome poor oxygenation secondary to lung immaturity. However, it may be due also to the relative hyperoxia of the extrauterine environment. With maturation of the premature infant, the resulting non-vascularized retina becomes increasingly metabolically active and without a blood supply, increasingly hypoxic [11, 46]. The first phase of vessel loss occurs from birth to PMA about 30 weeks.

20.1.5 ROP: Phase II

Phase II of ROP is characterized by hypoxia-induced vascular proliferation [11, 46] and starts between about 32 and 34 weeks PMA. The neovascularization phase of ROP is similar to other proliferative retinopathies such as diabetic retinopathy. The new blood vessel formation occurs at the junction between the non-vascularized retina and vascularized retina. These new vessels are leaky, and can cause tractional retinal detachments leading to blindness. If the growth of retinal blood vessels after preterm birth were normalized, the second destructive phase would not occur. Alternatively if we could attenuate the rapid proliferation of abnormal blood vessels in the second phase and allow controlled vascularization of the retina, retinal detachments could be prevented.

To accomplish these goals it is necessary to understand the growth factors involved in all aspects of ROP – both in normal retinal vascular development and in the development of neovascularization. The two phases of ROP are mirror images. The first involves growth inhibition of neural retina and the retinal vasculature and the second involves uncontrolled proliferative growth of retinal blood vessels. The controlling growth factors are likely to be deficient in phase I and in excess in phase II. Therefore control of the disease is likely to be complex and will likely require careful timing of any intervention.

20.1.6 Mouse Model of ROP

A disease model is required to study ROP. To take advantage of the genetic manipulations possible in the murine system to study the molecular pathways in retinal vascular development and in the development of ROP, we developed a mouse model of both phases of the disease [68]. The eyes of animals such as mice, rats and cats – though born full term – are incompletely vascularized at birth and are similar to the retinal vascular development of premature infants. When these neonatal animals are exposed to hyperoxia there is induced loss of some vessels and cessation of normal retinal blood vessel development, which mimics phase I of ROP [10, 11, 45, 54, 68].

When mice return to room air, the non-perfused portions of the retina become hypoxic, similar to phase II of ROP and of other retinopathies. The ischemic portions of the retina produce angiogenic factors that result in neovascularization [11, 46]. Hypoxia-inducible factors appear to be common to the proliferative phase of many eye diseases [25, 38] such as retinopathy of prematurity and diabetic retinopathy, as well as in tumor growth and wound healing. This ROP model has been useful to delineate the growth factor changes in both phases of neovascular eye diseases (Fig. 20.1.1).

20.1.7 Oxygen Regulated Factors: Vascular Endothelial Growth Factor in ROP

Essentials
- VEGF is an important factor for the development of retinal vascular proliferation in ROP. It is suppressed in phase I of ROP with hyperoxia. VEGF is markedly increased in phase II of ROP and stimulates retinal neovascularization

20 III

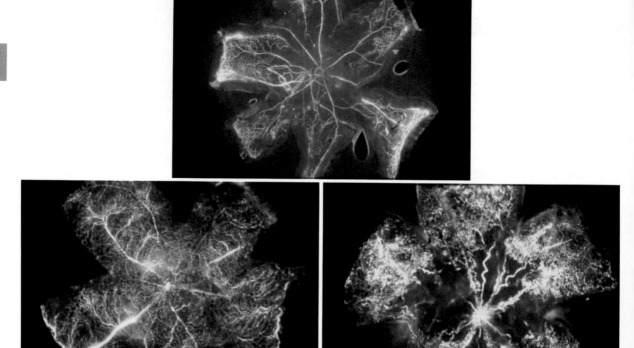

Fig. 20.1.1. The mouse model of ROP illustrates two phases of the disease. The flat mounted retinas which are perfused with FITC dextran which fills vessels illustrate the phases of ROP. The photomicrograph *on the left* is from a postnatal day 17 mouse with normal retinal vasculature. The photomicrograph *above* illustrates vessel loss after oxygen exposure, creating a hypoxic retina (phase I). The photomicrograph *on the right* illustrates the development of retinal neovascularization (*bright green areas*) as a response to vessel loss and hypoxia

- Inhibition of VEGF with anti-VEGF treatment (anti-VEGF aptamer or anti-VEGF antibody) has been successfully used clinically in other proliferative retinal vascular diseases such as age-related macular degeneration and diabetic retinopathy
- Anti-VEGF therapy for ROP in phase II to prevent retinal detachment and blindness may prove beneficial

The major perinatal risk factors described to date for ROP are oxygen and prematurity itself (postmenstrual age at birth and birth weight). We first studied oxygen-regulated factors.

In the 1940s and 1950s Michaelson and Ashton [11, 46] postulated that retinal neovascularization was caused by release of a "vasoformative factor" from the retina in response to hypoxia. It has now become widely accepted that retinal hypoxia results in the release of factors which influence new blood vessel growth [51]. Not only is hypoxia a driving force for proliferative retinopathy, or phase II of ROP, but excess oxygen is also associated with phase I with loss of vessels and cessation of normal retinal vascular development. Therefore it is likely that a growth factor or factors regulated by hypoxia and hyperoxia is important in the development of ROP.

Vascular endothelial growth factor (VEGF) is such a hypoxia/oxygen-inducible cytokine [36, 57, 66]. It was first described as a vascular permeability factor (VPF) and later described as a vascular proliferative factor, vascular endothelial growth factor or VEGF [21, 63]. VEGF is a vascular endothelial cell mitogen, which is required for tumor-associated angiogenesis [36]. Several different types of cultured retinal cells have been found to secrete VEGF under hypoxic conditions [1, 4, 5]. These characteristics make VEGF an ideal candidate for at least one of Michaelson's retinal vasoformative factors.

20.1.8 VEGF is Critical to Phase II of ROP

VEGF was shown to be required for retinal neovascularization (phase II of ROP) in studies of the mouse model of proliferative retinopathy [68]. After oxygen induction of vessel loss and subsequent hypoxia there is an increase in the expression of VEGF mRNA in the retina within 12 h. The increased expression of VEGF mRNA is sustained until the development of neovascularization [55, 68]. This occurs in the ganglion cell layer and in the inner nuclear layer consistent with expression in astrocytes and Müller cells.

These initial studies in the mouse model of ROP established the location and time course of VEGF expression in association with retinal neovascularization and a correlation with disease. To establish that a growth factor causes neovascularization, inhibition of the factor must inhibit the proliferation of blood vessels. Inhibition of VEGF with intravitreal injections of either an anti-VEGF antisense oligonucleotide or with a molecule to adsorb VEGF (VEGF receptor/IgG chimera) significantly decreased the neovascular response in the mouse model of ROP [6, 61], indicating that VEGF is a critical factor in retinal neovascularization. VEGF also has been associated with ocular neovascularization by other investigators in other animal models, confirming the central role of this cytokine in neovascular eye disease [3, 18, 48, 73, 77]. These results correspond to what is seen clinically. VEGF is elevated in the vitreous of patients with retinal neovascularization [2, 4]. VEGF was found in the retina of a patient with ROP in a pattern consistent with mouse results [77]. Based on these and other studies an anti-VEGF aptamer is now available to treat neovascularization associated with age-related macular degeneration and is in phase III clinical trials for diabetic retinopathy. An anti-VEGF antibody is in phase III of clinical trials for age-related macular degeneration and diabetic retinopathy. Clinical trials are planned for evaluation of treatment of the proliferative phase of ROP with anti-VEGF treatment.

20.1.9 VEGF in Phase I of ROP

In animal models of oxygen-induced retinopathy, there is a clear association between exposure to hyperoxia and vaso-obliteration [11, 13, 53, 68]. In the human the first phase of ROP is also likely to be triggered by hyperoxia. Further study of this association is important because the extent of non-perfusion in the initial phase of retinopathy of prematurity appears to determine the subsequent degree of neovascularization.

Premature infants normally experiencing low levels of oxygen in the intrauterine environment suffer cessation of normal retinal vessel growth and vaso-obliteration of some immature retinal vasculature when exposed to the relatively high levels of oxygen of the extrauterine environment. Supplemental oxygen may also contribute to this inhibition of vascular growth. The possibility that exposure to extrauterine oxygen causes cessation of vessel growth and vaso-obliteration secondary to suppression of VEGF was examined in animal models of ROP.

20.1.10 VEGF Role in Retinal Vessel Loss

Just as hyperoxia stimulates VEGF, hyperoxia almost totally suppresses VEGF mRNA expression in the mouse model during oxygen exposure. The suppression of VEGF mRNA production with hyperoxia causes loss or vaso-obliteration of immature retinal vessels. This loss can be prevented with intravitreal injections of exogenous VEGF or placental growth factor I, a specific ligand of VEGF receptor I [7, 56, 65]. Furthermore hyperoxia can reverse hypoxia-induced increases in VEGF, rationalizing the therapeutic use of oxygen in premature neonates with proliferative retinopathy (as used in the multicentered clinical STOP-ROP study) [23].

20.1.11 VEGF Role in Cessation of Normal Vascular Development

Normal blood vessel growth in the retina in animal models of retinal vascular development is also dependent on VEGF. There is an expanding ring of increased oxygen demand as the retina matures from the optic nerve to the periphery. Induced by this wave of "physiologic hypoxia" that precedes vessel growth [56, 72], VEGF is expressed in the retina, and blood vessels grow toward the VEGF stimulus. As the hypoxia is relieved by oxygen from the newly formed vessels, VEGF mRNA expression is suppressed, moving the wave forward.

This normal vascular development in the mouse and rat models of ROP is interrupted by oxygen exposure. Hyperoxia causes suppression of VEGF mRNA, causing loss of the physiological wave of VEGF anterior to the growing vascular front [7, 56] suppressing normal vascular development. VEGF is required for maintenance of the immature retinal vasculature and loss of VEGF with hyperoxia explains, at least in part, the first phase of ROP in the human.

20 **III**

20.1.12 Other Growth Factors in ROP

Other biochemical mediators also are almost certainly involved in the pathogenesis of ROP, even though oxygen and VEGF play a central role. Inhibition of VEGF does not completely inhibit hypoxia-induced retinal neovascularization in the second phase of ROP. In the first phase of ROP, although hyperoxia can clearly cause both cessation of vascular growth and vaso-obliteration in animal models, it is clear that clinical ROP is multifactorial. ROP persists as ever-lower gestational aged infants are saved despite the controlled use of oxygen, suggesting that other factors related to prematurity itself also are at work.

20.1.13 IGF-1 Deficiency in the Preterm Infant

Fetal growth and development during all stages of pregnancy are dependent on the insulin-like growth factors I and II (IGFs) [42]. In the first trimester they are found in embryological fluids [47] and there is a strong association between IGF concentrations and growth in human pregnancy [8, 9, 15, 16, 20, 24, 28–30, 40, 41, 43, 49, 50, 60, 74, 75]. IGF-I concentrations in fetal serum (from cordocentesis) increase with gestational age and correlate with fetal size [8, 43, 50, 60].

In the third trimester of pregnancy IGF-1 levels rise significantly [42]. In the earlier stages of the third trimester preterm birth is associated with a loss of maternal sources of IGF-I and lower levels of serum IGF-1 compared to in utero counterparts as preterm infants grow outside the womb [44]. Infants who are born very prematurely appear unable to produce adequate IGF-1 compared to term infants since IGF-I levels rise slowly after preterm birth [28]. IGF-I may be reduced further in preterm infants by poor nutrition [70], acidosis, hypothyroxinemia, and sepsis.

IGF-I is important for physical growth. Because the third trimester is associated with the rapid development of fetal tissue, loss of IGF-1 could be critical [28]. Although serum growth hormone (GH) levels in extremely preterm infants are significantly higher than term infants, serum IGF-I levels in extremely preterm infants are low. Physical growth is positively correlated with IGF-I concentrations for several months after birth whereas no relationship is observed between GH and physical growth [35]. IGF-1 appears to be particularly important for retinal and brain growth [34].

20.1.14 Growth Hormone and IGF-1 in Phase II of ROP

Since prematurity is the most significant risk factor for ROP, this suggests that growth factors such as GH and IGF-1 relating to development are critical to the disease process. Studies in the proliferative phase (phase II) of the mouse model of ROP were the first to show that IGF-1 is important in retinopathy. Because GH has been implicated in proliferative diabetic retinopathy [59, 64, 76], we considered GH and IGF-I, which mediates many of the mitogenic aspects of GH, as potential candidates for one of these growth factors.

In the mouse model of ROP, proliferative retinopathy, the second phase of ROP [69], is substantially reduced in transgenic mice expressing a GH-receptor antagonist or in wild type mice treated with a somatostatin analogue that decreases GH release [69]. GH inhibition of neovascularization is mediated through an inhibition of IGF-I, because systemic administration of IGF-I in transgenic mice with decreased GH action completely restores the neovascularization seen in control mice. Direct proof of the role of IGF-I in the proliferative phase of ROP in mice was established with an IGF-I receptor antagonist, which suppresses retinal neovascularization without altering the vigorous VEGF response induced in the mouse ROP model [71].

The role of both IGF-1 and insulin in the vascular endothelium in the ROP mouse model have been confirmed using mice with a vascular endothelial cell-specific knockout of the insulin receptor (VENIRKO) or IGF-1 receptor (VENIFARKO). VENIRKO mice show a 57% decrease in retinal neovascularization as compared with controls associated with a reduced rise in VEGF, eNOS, and endothelin-1. VENIFARKO mice showed a 34% reduction in neovascularization, suggesting that both insulin and IGF-1 signaling in endothelium play a role in retinal neovascularization [39]. Therefore, IGF-I is likely to be one of the non-hypoxia regulated factors critical to the development of ROP.

20.1.15 IGF-1 and VEGF Interaction

Hypoxia-induced VEGF production is unchanged during suppression of GH and IGF-I, causing inhibition of vascular growth. This suggests that IGF-I does not directly act through VEGF under these physiological conditions. These findings suggest a more complex role of IGF-I in retinal neovascularization [69]. Retinal neovascularization is regulated by IGF-1 at least in part through control of VEGF activation of p44/42 MAPK, establishing a hierarchical relationship between IGF-I and VEGF receptors [31, 71]. A minimal level of IGF-I is required to allow maximum VEGF stimulation of new vessel growth.

Thus low levels of IGF-I inhibit vessel growth despite the presence of VEGF. This work suggests that IGF-I serves a permissive function, and VEGF alone may not be sufficient for promoting vigorous retinal angiogenesis.

20.1.16 Low Levels of IGF-I and Phase I of ROP

We hypothesized that IGF-I is also critical to normal retinal vascular development. Suppression of IGF-1 can suppress neovascularization, in phase II of ROP, and a lack of IGF-I in the early neonatal period is associated with poor vascular growth and with subsequent proliferative ROP. IGF-I levels decrease from in utero levels after birth due to the loss of IGF-I provided by the placenta and the amniotic fluid.

In IGF-I knockout mice we found that IGF-I is critical in the normal development of the retinal vessels [32]. Retinal blood vessels grow more slowly in IGF-1 knockout mice than in normal mice, a pattern very similar to that seen in premature babies with ROP (Fig. 20.1.2). A minimum level of IGF-I is required for maximum VEGF activation of the Akt endothelial cell survival pathway. Thus loss of IGF-I could cause the disease by preventing the normal survival of vascular endothelial cells.

20.1.17 Clinical Studies: Low IGF-1 is Associated with the Degree of ROP

The greater the area of non-vascularized retina the greater the likelihood of developing more severe stages of ROP. Zone I disease is generally worse than zone II. Zone III is the least likely to develop into severe ROP. In other terms, the degree of phase I determines the degree of phase II. Normal vessel development in the retina precludes the development of proliferative ROP.

We hypothesized that prolonged low IGF-I in premature infants might be a risk factor for ROP because ROP is initiated by abnormal postnatal retinal vascular development. In 84 premature infants we conducted a prospective, longitudinal study measuring serum IGF-I concentrations weekly from birth (PMA 24–32 weeks) until discharge from the hospital. Infants were evaluated for ROP and other morbidity of prematurity: bronchopulmonary dysplasia (BPD), intraventricular hemorrhage (IVH), and necrotizing enterocolitis (NEC). Low serum IGF-I values correlated with later development of ROP (Fig. 20.1.3). The mean IGF-I level during PMA 30–33 weeks was lowest with severe ROP, intermediate with moderate ROP, and highest with no ROP. The duration of low IGF-I also correlated strongly with the severity of ROP. Each adjusted stepwise increase of 5 µg/l in mean IGF-I during PMA 30–33 weeks was associated with a 45% decreased risk of proliferative ROP. Other complications (NEC, BPD, IVH) were correlated with ROP and with low IGF-I levels. The relative risk for any morbidity (ROP, BPD, IVH, or NEC) was increased 2.2-fold if IGF-I was 33 µg/l at 33 weeks postmenstrual age. These results indicate that persistent low serum concentrations of IGF-I after premature birth are associated with later development of ROP and other complications of prematurity. In this study, IGF-I was at least as strong a determinant of risk for ROP as postmen-

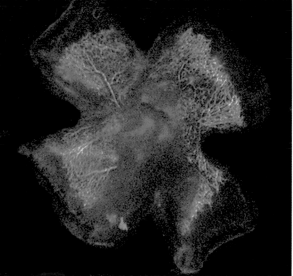

Fig. 20.1.2. Low IGF-1 inhibits normal retinal vessel growth. In IGF-1$^{-/-}$ mice retinal vessels at postnatal day 5 (*left*) are much more poorly developed than wild type controls (*right*) (with permission from [31])

strual or gestational age at birth and birth weight [32, 34] (Fig. 20.1.4). These findings suggest the possibility that increasing IGF-1 to uterine levels might prevent the disease by allowing normal retinal vascular development. If phase I is aborted the destructive second phase of vasoproliferation will not occur.

20.1.18 Low IGF-1 is Associated with Decreased Vascular Density

We have also found that very low IGF-1 directly causes decreased vascular density [33]. There was significantly less retinal vascularization in patients with genetic defects of the GH/IGF-I axis and low levels of IGF-I during and after normal retinal vessel growth as evidenced by lower number of vascular branching points compared with the reference group of normal controls. This work provides genetic evidence for a role of the GH and IGF-I system in retinal vascularization in humans. This accumulated evidence suggests that low IGF-1 is associated with vessel loss and may be detrimental by contributing to early vessel degeneration in phase I that sets the stage for hypoxia leading later to proliferative retinopathy.

20.1.19 IGF-1 and ROP

Essentials
- Postnatally low levels of IGF-1 in premature infants correlate with the severity of ROP
- Clinical trials are in the planning phase to supplement IGF-1 and IGFBP-3 to in utero levels in premature infants to evaluate if restoration of IGF-1 to normal levels can prevent or reduce the severity of ROP

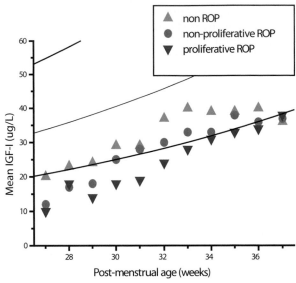

Fig. 20.1.3. Longitudinal mean IGF-I with respect to ROP severity. Mean serum IGF-I values for each postmenstrual week (weeks 29–40) and ROP stages; no ROP (stage 0, $n=37$), moderate ROP (stages 1 and 2, $n=34$), and proliferative ROP 3 (stage 3, $n=13$). The *upper, middle,* and *lower red lines* depict, respectively, the 95th, median, and 5th centiles of normal fetal IGF-I levels by using the technique of cordocentesis and an IGF-I assay similar to the one used in the present study (with permission from [34])

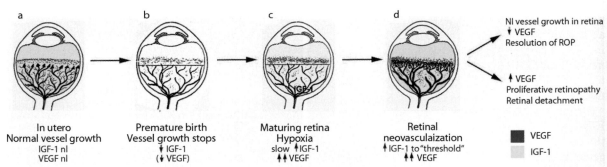

Fig. 20.1.4. Schematic representation of IGF-I and VEGF control of blood vessel development in ROP (with permission from [32]). **a** In utero, VEGF is found at the growing front of vessels. IGF-I is sufficient to allow vessel growth. **b** With premature birth, IGF-I is not maintained at in utero levels and vascular growth ceases, despite the presence of VEGF at the growing front of vessels. Both endothelial cell survival (Akt) and proliferation (mitogen-activated protein kinase) pathways are compromised. With low IGF-I and cessation of vessel growth, a demarcation line forms at the vascular front. High oxygen exposure (as occurs in animal models and in some premature infants) may also suppress VEGF, further contributing to inhibition of vessel growth. **c** As the premature infant matures, the developing but non-vascularized retina becomes hypoxic. VEGF increases in retina and vitreous. With maturation, the IGF-I level slowly increases. **d** When the IGF-I level reaches a threshold at 34 weeks gestation, with high VEGF levels in the vitreous, endothelial cell survival and proliferation driven by VEGF may proceed. Neovascularization ensues at the demarcation line, growing into the vitreous. If VEGF vitreal levels fall, normal retinal vessel growth can proceed. With normal vascular growth and blood flow, oxygen suppresses VEGF expression, so it will no longer be overproduced. If hypoxia (and elevated levels of VEGF) persists, further neovascularization and fibrosis leading to retinal detachment can occur

20.1.20 IGF-1 and Brain Development

Essentials
- Animal studies suggest that low levels of IGF-1 postnatally in preterm infants could have an effect on neural retinal development as well as on brain development and might account for abnormal neural retinal function in ROP
- Increasing postnatal IGF-1 through improved nutrition or other means might improve brain and retinal development

Perinatal low IGF-I levels may also contribute to poor neural as well as vascular retinal development and might contribute to poor neurological development in the preterm infant. IGF-1 is important for neural development in brain and retina is part of the central nervous system. ROP is associated with poor retinal function [27]. IGF-I and IGF binding proteins that modify IGF-I actions, as well as the IGF-1 receptor, are found throughout the brain during development. In cell culture IGF-I is a neural mitogen, suggesting an important role for IGF-1 in the growth and development of the central nervous system. Over- or underexpression of IGF-I in in vivo studies provide more evidence for the role of IGF-1 in central nervous system development. Transgenic mice with postnatal overexpression of IGF-1 have brains with increased numbers of neurons and increased myelination. There is inhibited brain growth in mutant mice with low IGF-1 effect (reduced IGF-I and IGF1R expression or overexpression of IGFBPs capable of inhibiting IGF actions). IGF-I has a role in recovery from neural injury [17]. IGF-I can both promote proliferation of neural cells in the embryonic central nervous system in vivo and inhibit their apoptosis during postnatal life [58].

Brain development is affected by reduction of IGF-1 levels through overexpression of IGFBP-1 in the liver, which reduces IGF-1 availability in transgenic mice [19]. The cerebral cortex is reduced in size with disorganized neuronal layers in the mice with the lowest level of IGF-1 (homozygous for IGFBP-1 overexpression). With disruption of the IGF-I gene and in a model of transgenic mice overexpressing IGFBP-1 in all tissues, including the brain, similar anomalies have been reported [19].

20.1.21 Conclusion: A Rationale for the Evolution of ROP

- Blood vessel growth is dependent on both IGF-I and VEGF. In premature infants, the absence of IGF-I (normally provided by the placenta and the amniotic fluid) inhibits blood vessel growth
- As the eye matures, it becomes oxygen starved, sending signals to increase VEGF. As the infant's organs and systems then continue to mature, IGF-I levels rise again, suddenly allowing the VEGF signal to produce blood vessels (Fig. 20.1.4)
- This neovascular proliferation of phase II of ROP can cause blindness

20.1.22 Possible Medical Intervention to Prevent ROP

- Timing is critical to any intervention. Inhibition of either VEGF or IGF-I early after birth can prevent normal blood vessel growth and precipitate the disease; whereas inhibition at the second neovascular phase might prevent destructive neovascularization
- Similarly, replacement of IGF-I early on might promote normal blood vessel growth; whereas late supplementation with IGF-I in the neovascular phase of ROP could exacerbate the disease
- In the fragile neonate, the choice of any intervention must be made very carefully to promote normal physiological development of both blood vessels and other tissue
- The finding that later development of ROP is associated with low levels of IGF-I after premature birth suggests that increasing IGF-1 to physiologic levels found in utero through better nutrition or other means might prevent the disease by allowing normal vascular development (Fig. 20.1.4)

References

1. Adamis AP, Shima DT, Yeo KT, Yeo TK, Brown LF, Berse B, D'Amore PA, Folkman J (1993) Synthesis and secretion of vascular permeability factor/vascular endothelial growth factor by human retinal pigment epithelial cells. Biochem Biophys Res Commun 193:631–638
2. Adamis AP, Miller JW, Bernal MT, D'Amico DJ, Folkman J, Yeo TK, Yeo KT (1994) Increased vascular endothelial growth factor levels in the vitreous of eyes with proliferative diabetic retinopathy. Am J Ophthalmol 118:445–450
3. Adamis AP, Shima DT, Tolentino MJ, Gragoudas ES, Ferrara N, Folkman J, D'Amore PA, Miller JW (1996) Inhibition of vascular endothelial growth factor prevents retinal ischemia-associated iris neovascularization in a nonhuman primate. Arch Ophthalmol 114:66–71
4. Aiello LP, Avery RL, Arrigg PG, Keyt BA, Jampel HD, Shah ST, Pasquale LR, Thieme H, Iwamoto MA, Park JE et al (1994) Vascular endothelial growth factor in ocular fluid of

patients with diabetic retinopathy and other retinal disorders (see comments). N Engl J Med 331:1480–1487

5. Aiello LP, Northrup JM, Keyt BA, Takagi H, Iwamoto MA (1995a) Hypoxic regulation of vascular endothelial growth factor in retinal cells. Arch Ophthalmol 113:1538–1544

6. Aiello LP, Pierce EA, Foley ED, Takagi H, Chen H, Riddle L, Ferrara N, King GL, Smith LE (1995b) Suppression of retinal neovascularization in vivo by inhibition of vascular endothelial growth factor (VEGF) using soluble VEGF-receptor chimeric proteins. Proc Natl Acad Sci USA 92:10457–10461

7. Alon T, Hemo I, Itin A, Pe'er J, Stone J, Keshet E (1995) Vascular endothelial growth factor acts as a survival factor for newly formed retinal vessels and has implications for retinopathy of prematurity. Nature Medicine 1:1024–1028

8. Arosio M, Cortelazzi D, Persani L, Palmieri E, Casati G, Baggiani AM, Gambino G, Beck-Peccoz P (1995) Circulating levels of growth hormone, insulin-like growth factor-I and prolactin in normal, growth retarded and anencephalic human fetuses. J Endocrinol Invest 18:346–353

9. Ashton IK, Zapf J, Einschenk I, MacKenzie IZ (1985) Insulin-like growth factors (IGF) 1 and 2 in human foetal plasma and relationship to gestational age and foetal size during midpregnancy. Acta Endocrinol (Copenh) 110:558–563

10. Ashton N (1966) Oxygen and the growth and development of retinal vessels. In vivo and in vitro studies. The XX Francis I. Proctor Lecture. Am J Ophthalmol 62:412–435

11. Ashton N, Ward B, Serpell G (1954) Effect of oxygen on developing retinal vessels with particular reference to the problem of retrolental fibroplasia. Br J Ophthalmol 38:397–432

12. Campbell K (1951) Intensive oxygen therapy as a possible cause of retrolental fibroplasia: a clinical approach. Med J Aust 2:48–50

13. Chan-Ling T, Tout S, Hollander H, Stone J (1992) Vascular changes and their mechanisms in the feline model of retinopathy of prematurity. Invest Ophthalmol Vis Sci 33:2128–2147

14. Crosse VM, Evans PJ (1952) Prevention of retrolental fibroplasia. Arch Ophthalmol 48:83–87

15. D'Ercole AJ, Underwood LE (1985) Somatomedin in fetal growth. Pediatr Pulmonol 1:S99–S106

16. D'Ercole AJ, Hill DJ, Strain AJ, Underwood LE (1986) Tissue and plasma somatomedin-C/insulin-like growth factor I concentrations in the human fetus during the first half of gestation. Pediatr Res 20:253–255

17. D'Ercole AJ, Ye P, O'Kusky JR (2002) Mutant mouse models of insulin-like growth factor actions in the central nervous system. Neuropeptides 36:209–220

18. Donahue ML, Phelps DL, Watkins RH, LoMonaco MB, Horowitz S (1996) Retinal vascular endothelial growth factor (VEGF) mRNA expression is altered in relation to neovascularization in oxygen induced retinopathy. Curr Eye Res 15:175–184

19. Doublier S, Duyckaerts C, Seurin D, Binoux M (2000) Impaired brain development and hydrocephalus in a line of transgenic mice with liver-specific expression of human insulin-like growth factor binding protein-1. Growth Horm IGF Res 10:267–274

20. Fant M, Salafia C, Baxter RC, Schwander J, Vogel C, Pezzullo J, Moya F (1993) Circulating levels of IGFs and IGF binding proteins in human cord serum: relationships to intrauterine growth. Regul Pept 48:29–39

21. Ferrara N, Henzel W (1989) Pituitary follicular cells secrete a novel heparin-binding growth factor specific for vascular endothelial cells. Biochem Biophys Res Commun 161:851–858

22. Flynn JT (1983) Acute proliferative retrolental fibroplasia: multivariate risk analysis. Trans Am Ophthalmol Soc 81:549–591

23. Flynn JT, Bancalari E (2000) On "supplemental therapeutic oxygen for prethreshold retinopathy of prematurity (STOP-ROP), a randomized, controlled trial. I: Primary outcomes" (editorial). J AAPOS Am Assoc Pediatr Ophthalmol Strab 4:65–66

24. Foley TP Jr, DePhilip R, Perricelli A, Miller A (1980) Low somatomedin activity in cord serum from infants with intrauterine growth retardation. J Pediatr 96:605–610

25. Folkman J, Klagsbrun M (1987) Angiogenic factors. Science 235:442–446

26. Foos R, Kopelow S (1973) Development of retinal vasculature in prenatal infants. Surv Ophthalmol 18:117–127

27. Fulton AB, Hansen RM, Petersen RA, Vanderveen DK (2001) The rod photoreceptors in retinopathy of prematurity: an electroretinographic study. Arch Ophthalmol 119:499–505

28. Giudice LC, de Zegher F, Gargosky SE, Dsupin BA, de las Fuentes L, Crystal RA, Hintz RL, Rosenfeld RG (1995) Insulin-like growth factors and their binding proteins in the term and preterm human fetus and neonate with normal and extremes of intrauterine growth. J Clin Endocrinol Metab 80:1548–1555

29. Gluckman PD, Butler JH (1983) Parturition-related changes in insulin-like growth factors-I and -II in the perinatal lamb. J Endocrinol 99:223–232

30. Gluckman PD, Johnson-Barrett JJ, Butler JH, Edgar BW, Gunn TR (1983) Studies of insulin-like growth factor -I and -II by specific radioligand assays in umbilical cord blood. Clin Endocrinol (Oxf) 19:405–413

31. Hellstrom A, Perruzzi C, Ju M, Engstrom E, Hard AL, Liu JL, Albertsson-Wikland K, Carlsson B, Niklasson A, Sjodell L et al (2001a) Low IGF-I suppresses VEGF-survival signaling in retinal endothelial cells: direct correlation with clinical retinopathy of prematurity. Proc Natl Acad Sci USA 98:5804–5808

32. Hellstrom A, Perruzzi C, Ju M, Engstrom E, Hard A-L, Liu J-L, Albertsson-Wikland K, Carlsson B, Niklasson A, Sjodell L et al (2001b) Low IGF-I suppresses VEGF-survival signaling in retinal endothelial cells: direct correlation with clinical retinopathy of prematurity. Proc Natl Acad Sci USA 98:5804–5808

33. Hellstrom A, Carlsson B, Niklasson A, Segnestam K, Boguszewski M, de Lacerda L, Savage M, Svensson E, Smith L, Weinberger D et al (2002) IGF-I is critical for normal vascularization of the human retina. J Clin Endocrinol Metab 87:3413–3416

34. Hellstrom A, Engstrom E, Hard AL, Albertsson-Wikland K, Carlsson B, Niklasson A, Lofqvist C, Svensson E, Holm S, Ewald U et al (2003) Postnatal serum insulin-like growth factor I deficiency is associated with retinopathy of prematurity and other complications of premature birth. Pediatrics 112:1016–1020

35. Hikino S, Ihara K, Yamamoto J, Takahata Y, Nakayama H, Kinukawa N, Narazaki Y, Hara T (2001) Physical growth and retinopathy in preterm infants: involvement of IGF-I and GH. Pediatr Res 50:732–736

36. Kim KJ, Li B, Winer J, Armanini M, Gillett N, Phillips HS, Ferrara N (1993) Inhibition of vascular endothelial growth

factor-induced angiogenesis suppresses tumour growth in vivo. Nature 362:841–844

37. Kinsey VE, Arnold HJ, Kalina RE, Stern L, Stahlman M, Odell G, Driscoll JM Jr, Elliott JH, Payne J, Patz A (1977) PaO2 levels and retrolental fibroplasia: a report of the cooperative study. Pediatrics 60:655–668

38. Knighton D, Hunt T, Scheuenstuhl H (1993) Oxygen tension regulates the expression of angiogenesis by macrophages. Science 221:1283–1285

39. Kondo T, Vicent D, Suzuma K, Yanagisawa M, King GL, Holzenberger M, Kahn CR (2003) Knockout of insulin and IGF-1 receptors on vascular endothelial cells protects against retinal neovascularization. J Clin Invest 111:1835–1842

40. Kubota T, Kamada S, Taguchi M, Aso T (1992) Determination of insulin-like growth factor-2 in feto-maternal circulation during human pregnancy. Acta Endocrinol (Copenh) 127:359–365

41. Langford K, Blum W, Nicolaides K, Jones J, McGregor A, Miell J (1994) The pathophysiology of the insulin-like growth factor axis in fetal growth failure: a basis for programming by undernutrition? Eur J Clin Invest 24:851–856

42. Langford K, Nicolaides K, Miell JP (1998) Maternal and fetal insulin-like growth factors and their binding proteins in the second and third trimesters of human pregnancy. Hum Reprod 13:1389–1393

43. Lassarre C, Hardouin S, Daffos F, Forestier F, Frankenne F, Binoux M (1991) Serum insulin-like growth factors and insulin-like growth factor binding proteins in the human fetus. Relationships with growth in normal subjects and in subjects with intrauterine growth retardation. Pediatr Res 29:219–225

44. Lineham JD, Smith RM, Dahlenburg GW, King RA, Haslam RR, Stuart MC, Faull L (1986) Circulating insulin-like growth factor I levels in newborn premature and full-term infants followed longitudinally. Early Hum Dev 13:37–46

45. McLeod D, Crone S, Lutty G (1996) Vasoproliferation in the neonatal dog model of oxygen-induced retinopathy. Invest Ophthalmol Vis Sci 37:1322–1333

46. Michaelson I (1948) The mode of development of the vascular system of the retina, with some observations in its significance for certain retinal diseases. Trans Ophthalmol Soc UK 68:137–180

47. Miell JP, Jauniaux E, Langford KS, Westwood M, White A, Jones JS (1997) Insulin-like growth factor binding protein concentration and post-translational modification in embryological fluid. Mol Hum Reprod 3:343–349

48. Miller JW, Adamis AP, Shima DT, D'Amore PA, Moulton RS, O'Reilly MS, Folkman J, Dvorak HF, Brown LF, Berse B et al (1994) Vascular endothelial growth factor/vascular permeability factor is temporally and spatially correlated with ocular angiogenesis in a primate model. Am J Pathol 145:574–584

49. Nieto-Diaz A, Villar J, Matorras-Weinig R, Valenzuela-Ruiz P (1996) Intrauterine growth retardation at term: association between anthropometric and endocrine parameters. Acta Obstet Gynecol Scand 75:127–131

50. Ostlund E, Bang P, Hagenas L, Fried G (1997) Insulin-like growth factor I in fetal serum obtained by cordocentesis is correlated with intrauterine growth retardation. Hum Reprod 12:840–844

51. Patz A (1982) Clinical and experimental studies on retinal neovascularization. Am J Ophthalmol 94:715–743

52. Patz A, Hoeck LE, DeLaCruz E (1952) Studies on the effect of high oxygen administration in retrolental fibroplasia. I. Nursery observations. Am J Ophthalmol 35:1248–1252

53. Penn JS, Tolman BL, Lowery LA (1993) Variable oxygen exposure causes preretinal neovascularization in the newborn rat. Invest Ophthalmol Vis Sci 34:576–585

54. Penn JS, Tolman BL, Henry MM (1994) Oxygen-induced retinopathy in the rat: relationship of retinal nonperfusion to subsequent neovascularization. Invest Ophthalmol Vis Sci 35:3429–3435

55. Pierce EA, Avery RL, Foley ED, Aiello LP, Smith LE (1995) Vascular endothelial growth factor/vascular permeability factor expression in a mouse model of retinal neovascularization. Proc Natl Acad Sci USA 92:905–909

56. Pierce EA, Foley ED, Smith LE (1996) Regulation of vascular endothelial growth factor by oxygen in a model of retinopathy of prematurity (see comments; published erratum appears in Arch Ophthalmol 1997, 115:427). Arch Ophthalmol 114:1219–1228

57. Plate KH, Breier G, Weich HA, Risau W (1992) Vascular endothelial growth factor is a potential tumour angiogenesis factor in human gliomas in vivo. Nature 359:845–848

58. Popken GJ, Hodge RD, Ye P, Zhang J, Ng W, O'Kusky JR, D'Ercole AJ (2004) In vivo effects of insulin-like growth factor-I (IGF-I) on prenatal and early postnatal development of the central nervous system. Eur J Neurosci 19:2056–2068

59. Poulsen JE (1953) Recovery from retinopathy in a case of diabetes with Simmonds' disease. Diabetes 2:7–12

60. Reece EA, Wiznitzer A, Le E, Homko CJ, Behrman H, Spencer EM (1994) The relation between human fetal growth and fetal blood levels of insulin-like growth factors I and II, their binding proteins, and receptors. Obstet Gynecol 84:88–95

61. Robinson GS, Pierce EA, Rook SL, Foley E, Webb R, Smith LE (1996) Oligodeoxynucleotides inhibit retinal neovascularization in a murine model of proliferative retinopathy. Proc Natl Acad Sci USA 93:4851–4856

62. Roth AM (1977) Retinal vascular development in premature infants. Am J Ophthalmol 84:636–640

63. Senger DR, Galli SJ, Dvorak AM, Perruzzi CA, Harvey VS, Dvorak HF (1983) Tumor cells secrete a vascular permeability factor that promotes accumulation of ascites fluid. Science 219:983–985

64. Sharp PS, Fallon TJ, Brazier OJ, Sandler L, Joplin GF, Kohner EM (1987) Long-term follow-up of patients who underwent yttrium-90 pituitary implantation for treatment of proliferative diabetic retinopathy. Diabetologia 30:199–207

65. Shih SC, Ju M, Liu N, Smith LE (2003) Selective stimulation of VEGFR-1 prevents oxygen-induced retinal vascular degeneration in retinopathy of prematurity. J Clin Invest 112:50–57

66. Shweiki D, Itin A, Soffer D, Keshet E (1992) Vascular endothelial growth factor induced by hypoxia may mediate hypoxia-initiated angiogenesis. Nature 359:843–845

67. Silverman WA (1980) Retrolental fibroplasia: a modern parable. Grune and Stratton, New York

68. Smith LE, Wesolowski E, McLellan A, Kostyk SK, D'Amato R, Sullivan R, D'Amore PA (1994) Oxygen-induced retinopathy in the mouse. Invest Ophthalmol Vis Sci 35:101–111

69. Smith LE, Kopchick JJ, Chen W, Knapp J, Kinose F, Daley D, Foley E, Smith RG, Schaeffer JM (1997a) Essential role of growth hormone in ischemia-induced retinal neovascularization. Science 276:1706–1709

70. Smith WJ, Underwood LE, Keyes L, Clemmons DR (1997b) Use of insulin-like growth factor I (IGF-I) and IGF-binding protein measurements to monitor feeding of premature infants. J Clin Endocrinol Metab 82:3982–3988

71. Smith LE, Shen W, Perruzzi C, Soker S, Kinose F, Xu X, Robinson G, Driver S, Bischoff J, Zhang B et al (1999) Regulation of vascular endothelial growth factor-dependent retinal neovascularization by insulin-like growth factor-1 receptor. Nature Medicine 5:1390–1395

72. Stone J, Itin A, Alon T, Pe'er J, Gnessin H, Chan-Ling T, Keshet E (1995) Development of retinal vasculature is mediated by hypoxia-induced vascular endothelial growth factor (VEGF) expression by neuroglia. J Neurosci 15:4738–4747

73. Stone J, Chan-Ling T, Pe'er J, Itin A, Gnessin H, Keshet E (1996) Roles of vascular endothelial growth factor and astrocyte degeneration in the genesis of retinopathy of prematurity. Invest Ophthalmol Vis Sci 37:290–299

74. Verhaeghe J, Van Bree R, Van Herck E, Laureys J, Bouillon R, Van Assche FA (1993) C-peptide, insulin-like growth factors I and II, and insulin-like growth factor binding protein-1 in umbilical cord serum: correlations with birth weight. Am J Obstet Gynecol 169:89–97

75. Wang HS, Lim J, English J, Irvine L, Chard T (1991) The concentration of insulin-like growth factor-I and insulin-like growth factor-binding protein-1 in human umbilical cord serum at delivery: relation to fetal weight. J Endocrinol 129:459–464

76. Wright AD, Kohner EM, Oakley NW, Hartog M, Joplin GF, Fraser TR (1969) Serum growth hormone levels and the response of diabetic retinopathy to pituitary ablation. Br Med J 2:346–348

77. Young TL, Anthony DC, Pierce E, Foley E, Smith LE (1997) Histopathology and vascular endothelial growth factor in untreated and diode laser-treated retinopathy of prematurity. J AAPOS 1:105–110

20.2 Clinical Course and Treatment

C. Jandeck, M.H. Foerster

Core Messages

- Retinopathy of prematurity (ROP) is a mostly bilateral vascular disease in pre-term babies, associated with lower birth weight and younger gestational age
- First presentation is an avascular retina in Zone I, II or III in combination with five different stages of retinopathy. Progression of the pathological changes may lead to a complete retinal detachment

- If fibrovascular proliferation occurs at the border of avascular and vascularized retina, treatment is indicated if a certain amount of clock hours are involved, or "plus disease" is present. Treatment of the avascular area is performed either by laser or cryotherapy. If Stage 4 or 5 occurs, retinal surgery may improve outcome
- The main goal of treatment is to prevent progression of fibrovascular proliferation and to avoid retinal detachment

Preterm babies are defined as being born before 37 weeks of gestational age (8.2% of all live births). Retinopathy of prematurity (ROP) occurs primarily in babies born before 33 weeks gestational age or in older pre-term children with severe disease and prolonged oxygen requirements. Depending on the stage of the disease, ROP can lead to blindness.

Treatment at the right time significantly reduces the risk of retinal detachment and blindness [16]. However, ROP is still the third most common cause of blindness in children [90]. This disease occurs in 27–40% of all children under 1,500 g of birth weight despite improved neonatal intensive care [7, 46, 94]. In Germany 1.3% of all live births are born at less than 32 weeks gestational age, giving a total of approximately 9,600 children per year in this category. The incidence of ROP, time of onset, rate of progression, and time of onset of "pre-threshold" disease have changed little in the last 15 years [40]. Thus, ROP continues to be a serious health concern.

20.2.1 Historical Developments and Epidemiology

Retinopathy of prematurity was first described by Terry [104]. He named this disease retrolental fibroplasia (RLF), because he recognized the cicatricial form of the disease. During the next few decades this disease became the leading cause of blindness in children in the USA and other industrial nations.

During 1951 and 1956 many studies identified uncontrolled oxygen delivery in the incubator as the main cause of the disease [61, 63]. Later oxygen treatment was reduced and the incidence rate of the disease as well as the blindness rate dramatically decreased [61, 63].

In 1950 in the USA, 50% of childhood blindness was caused by ROP. By 1965 this rate had dropped to 4% [43]. However, due to the reduction of oxygen, the mortality rate [5] and the number of neurological disturbances rose [72].

As a result, the supply of oxygen was less strictly regulated, leading again to a rise in the incidence of ROP [37, 83]. Another reason for the moderate increase in incidence rate may be due to the improved survival rate of pre-term babies. Due to better monitoring systems and advancements in neonatal intensive care units more children of even younger gestational age and lower birth weights survive.

For example, in 1950 only 8% of babies born with a birth weight of less than 1,000 g survived. The survival rate for prematurely born babies with this birth weight increased to approximately 35% in 1980 [83]. Premature babies with a birth weight of 700–799 g had a 57% chance of survival in 1983/1984 [48] and in 1995/1996 a survival rate of 84% [70].

The survival rate of pre-term babies of 24 weeks gestation in 21 German prenatal centers for the years 1995–1997 was 60–80%. The chance of survival in extremely small pre-term babies (23 completed weeks of pregnancy) rose in these 2 years from 19%

to 28 % [84]. The birth weight of these infants was approximately 400 – 500 g.

Retinopathy of prematurity occurs yearly in the USA in 14,000 – 16,000 premature infants weighing less than 1,250 g. Seven to 9 % of these children will require treatment. Despite treatment 3 – 4 % of these children will become legally blind every year (visual acuity 20/200). In Switzerland and Holland 3 – 5 % of all premature infants will develop blindness due to ROP [6, 95].

20.2.2 Explanation of Important Terms, The International Classification and the Cryo-ROP Study

Retinopathy of prematurity (ROP):
Failure of the peripheral retina to vascularize
Retinal changes at the border of vascular and avascular retina
Fibrovascular proliferation (Stage 3) at the border of vascular and avascular retina
Complete retinal detachment

The following is an attempt to describe some definitions that are required to understand this and other publications, including the International Classification, the Cryo-ROP study, and the Screening and Therapy Criteria recommended.

20.2.2.1 Terms Used to Describe Age

- **Gestational age:** Duration of the pregnancy counted in weeks from the first day of the last menstruation
- **Postnatal age:** Age of child since birth
- **Postmenstrual age:** Sum of the gestational age and the postnatal age

20.2.2.2 International Classification

The international classification of ROP was developed in 1984 and expanded in 1987 [51, 52]. This classification was accepted worldwide as a standard and made comparisons between different centers possible as well as promoting multicenter studies. The classification assists in describing the retinopathy and includes localization of the retinal involvement in zones, defining the extent of the involvement by clock hours, grading the stage or severity of retinopathy at the junction of the vascularized and avascular retina, and noting the presence or absence of dilated and tortuous posterior pole vessels (plus disease).

In July 2005 a revised international classification was published. It combined the original ICROP system and the additional changes made by a group of 15 ophthalmologists from 6 countries [53].

The location of the disease was described by

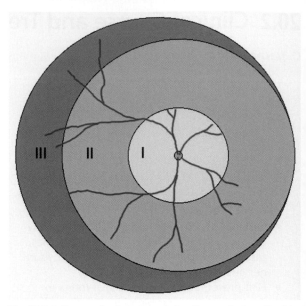

Fig. 20.2.1. Zones: the location of the disease is described by defining three concentric zones of retina, which are centered on the optic disk

defining three concentric zones of retina, which are centered on the optic disk (Fig. 20.2.1).

For the practical approach, Zone I was defined by using a 25- or 28-diopter condensing lens. The approximate temporal extent of Zone I can be determined by placing the nasal edge of the optic disk at one edge of the field of view; the limit of Zone I is at the temporal field of view.

20.2.2.2.1 Zones

- **Zone I:** Encloses a circular area centered around the optic disk with a radius equivalent to two times the distance from the optic nerve to the fovea
- **Zone II:** Extends centrifugally from the edge of Zone I to the nasal ora serrata
- **Zone III:** The residual crescent of the retina anterior to Zone II

The extent of the disease is recorded as hours of the clock (1 – 12) or as 30 degree sectors. The five stages describe the abnormal vascular response at the junction of the vascularized and avascular retina (Figs. 20.2.2 – 20.2.7) [51, 52].

20.2.2.2.2 Stages

- **Stage 1: Demarcation line**
  A flat, white line within the plana of the retina that separates the avascular retina anteriorly from the vascularized retina posteriorly (Fig. 20.2.2).

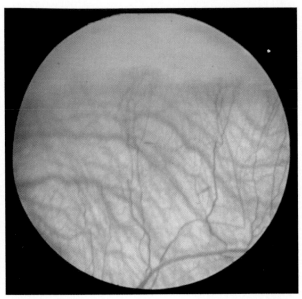

Fig. 20.2.2. Stage 1, distinct demarcation line

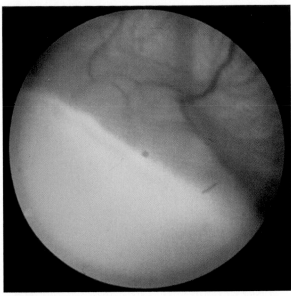

Fig. 20.2.3. Stage 2, ridge between the avascular and vascularized retina

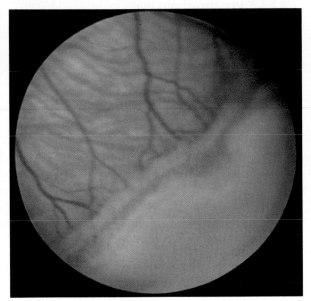

Fig. 20.2.4. Stage 3+, ridge with extraretinal proliferation

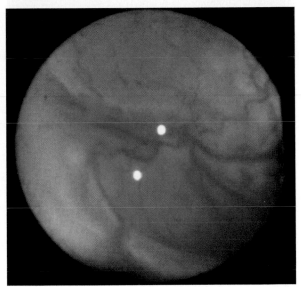

Fig. 20.2.5. Stage 4, peripheral retinal detachment in Zone I

- **Stage 2:** **Ridge**
  A prominent line that has height and width and extends above the plane of the retina. The ridge may change from white to pink and vessels may leave the plane of the retina to enter it. Small isolated tufts may be seen posterior to this ridge structure (popcorn) (Fig. 20.2.3).
- **Stage 3:** **Ridge with extraretinal fibrovascular proliferation**
  Extraretinal, fibrovascular proliferative tissue or neovascularization extends from the ridge into the vitreous. The severity

can be subdivided into mild, moderate, or severe depending on the extent of the extraretinal fibrovascular tissue infiltrating the vitreous (Fig. 20.2.4).
- **Stage 4:** **Partial retinal detachment**
  a) Extrafoveal (Fig. 20.2.5)
  b) Partial retinal detachment including the fovea
- **Stage 5:** **Total retinal detachment**
  Complete retinal detachment, mostly tractional and funnel shaped. The configuration of the funnel is described in anterior and posterior parts with an open or closed funnel (Fig. 20.2.6).

20 III

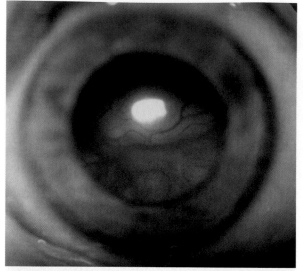

Fig. 20.2.6. Stage 5, complete retinal detachment

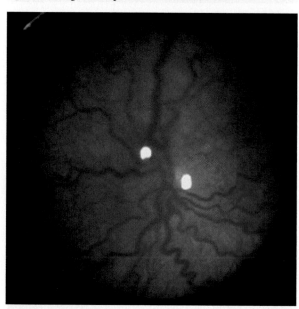

Fig. 20.2.7. Plus disease in four quadrants

20.2.2.2.3 Plus Disease

Increased venous dilatation and arteriolar tortuosity of the posterior pole in at least two quadrants. This may later increase in severity to include iris vessel engorgement, poor pupillary dilatation, and vitreous haze. A "+" symbol is added to the ROP stage number (Fig. 20.2.7).

20.2.2.2.4 Pre-Plus Disease

Vascular abnormalities of the posterior pole that are insufficient for "plus disease" but demonstrate more arterial tortuosity and more venous dilatation than normal.

20.2.2.2.5 Aggressive Posterior ROP

This is an uncommon, rapid, severe form of ROP, with a posterior location (former rush type)

20.2.2.3 Cryo-ROP Study

The multicenter Cryo-ROP study was initiated by the National Institute of Health (NIH) [16]. The following terms were defined in the Cryo-ROP Study and the ETROP Study [29] and are considered when judging the indications for therapy and the results of therapy in all subsequent studies.

20.2.2.3.1 Threshold ROP (Stage for Treatment)

Stage 3 in Zone I or Zone II with extraretinal fibrovascular proliferations of at least 5 continuous or 8 cumulative clock hours with "plus disease"

20.2.2.3.2 Pre-threshold ROP

- **Zone I:**   Any stage
- **Zone II:** Stage 2 with "plus disease"; Stage 3 less than "threshold"

20.2.2.3.3 Pre-threshold Classification Used for ETROP Study

- **Type 1 (high risk pre-threshold)**
  Zone I: Any stage with "plus disease"; Stage 3 without "plus disease"
  Zone II: Stage 2 or 3 with "plus disease"

- **Type 2 (low risk)**
  Zone I: Stage 1 or 2 without "plus disease"
  Zone II: Stage 3 without "plus disease"

20.2.2.3.4 Unfavorable Retinal Outcome

- **Stage 4B** (Partial retinal detachment), retinoschisis, or folds – all with foveal involvement
- **Stage 5** (Total retinal detachment), total retinoschisis, or retrolental membrane

20.2.2.3.5 Unfavorable Visual Outcome

- Distance visual acuity ≤20/200

20.2.3 Guidelines for Screening

The aim of all guidelines is to define a screening procedure in order to catch all premature babies whose retinopathy requires treatment [14]. Factors that must also be considered when proposing such guidelines are patient discomfort, the ophthalmologist's time, as well as financial costs, all of which need to be kept as low as possible without endangering the health of the patient or quality of care. In the UK just under 2% of babies screened require treatment and it takes 39–55 examinations to detect a single case requiring treatment [7, 42].

Different countries have different screening guidelines for the examination of pre-term infants, but the inclusion criteria for treatment and the first examination are nearly all the same. The screening protocol used at every neonatal intensive care unit should be based on the official recommendations of the country. All at-risk infants should be identified and examined at the appropriate time (Table 20.2.1).

In the United States and Canada the birth weight is the most important measurable factor. In Germany and the European countries, ultrasound is regularly performed during pregnancy to monitor the size of the baby and therefore size is used to calculate the precise gestational age. In Germany and the European countries the gestational age is the most important factor and the screening guidelines are based on this.

- **First Examination:** At 4–6 weeks of postnatal age (but not before 31 weeks postgestational age)
- **Follow-up Examinations (German Guidelines):** The time of the follow-up examination is dependent upon the respective retinal finding

---

Follow-up

After 1 week:
Vascularization in Zone I or central Zone II with or without ROP
Vascularization in Zone II with ROP Stage 2 or 3
Every ROP with plus disease
More regular check-ups should be performed in eyes with "aggressive posterior disease" and/or in eyes with very immature retina

After 2 weeks:
Vascularization in peripheral Zone II without ROP or with ROP Stage 1
Vascularization in Zone III with or without ROP

Longer intervals:
If after several examinations regression occurs
After the calculated date of birth

End of follow-up:
If the retina is completely vascularized
If an obvious regression of the peripheral retinal changes occurs, but only after the calculated date of birth (Fig. 20.2.8)

An initial examination performed in the sixth postnatal week is sufficient [18, 21, 30, 60, 86]. Retinopathies requiring treatment prior to the 6th week of life, so called "rush types," are very rare in the literature [39, 77]. Different studies [30, 46, 58, 86] have shown that the time for the development of Stage 3 ROP is prolonged for infants born very early (e.g., 23–25 weeks gestational age) compared to infants born at gestational age 28–30 weeks.

Stage 3 ROP is described only once in the literature in a child born at 26 weeks gestational age [18]. In all other instances it did not develop before 31 weeks postmenstrual age [18]. The latest occurrence of a Stage 3 ROP was recognized at 44–47 weeks postmenstrual age [30, 46, 54]. Reynolds et al. [89] combined the data of the Cryo-ROP and Light ROP Studies. In these two studies the occurrence of "prethreshold" and "threshold" was not before

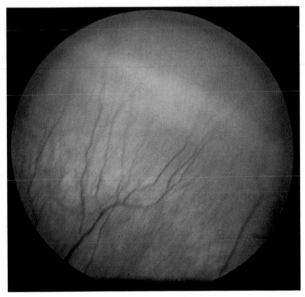

Fig. 20.2.8. Vessels growing over the ridge into the avascular retina

Table 20.2.1. Screening criteria in different countries

Country	Birth weight (g)	Gestational age (weeks)	Additional oxygen	First examination
USA 2006	≤1,500	≤32	1,500–2,000 g	4–6 wks/or 31 wks PMA
England 1996	≤1,500	≤31		6–7 wks
Canada 2000	≤1,500	≤30		4–6 wks
Sweden 1993	≤1,500	≤32		5–6 wks
Denmark	≤1,750	≤32		
Netherlands 1999	<1,500	<32		4–9 wks
Germany 2007	≤1,500	<32	>3 days	5 wks

PMA postmenstrual age

30.9 weeks or after 46.3 weeks of postmenstrual age, and not before 4.7 weeks or after 18.7 weeks of postnatal age. The time interval for the presentation of a critical stage retinopathy is cited in most of the literature as being between 32 and 42 weeks of gestation. The maximum incidence is at 37 weeks postmenstrual age. Palmer et al. [18] noticed that 5% of all children reached threshold levels indicating treatment at 31–33 weeks postmenstrual age and 5% after 42 weeks postmenstrual age. In all studies no treatment was necessary before 31 weeks postmenstrual age [16, 18, 21, 30, 60, 86]. The latest occurrence of threshold and treatment was recognized at 48 weeks postmenstrual age [18].

A further argument against examinations prior to the 6th week of life or before the 30th postmenstrual week is the clear physical stress that the child suffers during the examination. Especially in the smaller, more immature pre-term infants, the examination can lead to bradycardia, and apnea or in the worst case to an acute heart-lung failure [62, 68].

20.2.3.1 Examination Technique

The examination of the pre-term child should be performed in a darkened room. An adequate dilatation of the pupil is necessary and can be achieved under mydriasis (e.g., 2.5% phenylephrine + 0.5% tropicamide, one drop, 2–3 times, 5–10 min apart).

A second person should hold the child. After the installation of anesthetic eyedrops a lid speculum is inserted. A binocular ophthalmoscope is used and together with a localizator or muscle hook the bulbus can be rotated and indented so that the peripheral retina becomes visible (Fig. 20.2.9).

First the examiner should examine the anterior eye, noting pupil wideness, a possible tunica vasculosa lentis or a rubeosis iridis. During the retinal examination, the border of vascularization should be determined and the vessels should be checked for possible plus disease.

Documentation of the eye findings with details regarding stage and zone is important. The time of the next follow-up, a possible treatment, or the end of the screening should be recorded.

20.2.4 Indications for Coagulation Treatment

The indications for treatment vary slightly between countries, but all rely on the results of the Cryo-ROP Study and since 2003 based additionally on the result of the ETROP Study.

> **Indications for Coagulation treatment in Germany is indicated when:**
> In isolated cases an earlier treatment can be indicated (for example, in instances of swift progression and/or early distortion of the retina)

Should the findings indicate that treatment is necessary, the treatment should follow within days of the examination.

Through the screening criteria of the guidelines, it may be possible that subtle retinal changes may be diagnosed later than they actually occurred. It is, however, not important to recognize a diminishing Stage 1 or 2 that does not carry any therapeutic consequences. The goal of a screening is to detect all children who need treatment. For the later screenings (after the acute phase of ROP) for the detection of amblyopia, myopia or strabismus, it is not important to know if a lower stage of ROP occurred. There is no difference in the incidence of visual problems to premature children without ROP or to children born on term [82]. And there is no difference in the visual outcome of 10-year-old preterm children with ROP Stage 1 or without ROP [65].

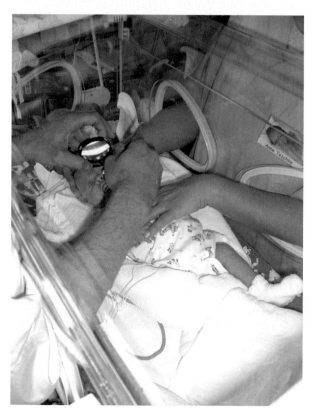

Fig. 20.2.9. Examination of a pre-term baby with muscle hook and indentator

20.2.5 Incidence of ROP and Incidence of Treatment

The incidence of any stage of ROP for children of 1,500 g in the literature is between 27% and 40% [7, 17, 30, 33, 46, 56, 88, 94]. After reaching "threshold," treatment is necessary in 1.4–10.8% [7, 17, 30, 33, 46, 57, 88, 94]. Many studies in previous years show that the incidence of ROP and the treatment incidence have diminished [9, 31, 92]. In contrast, other authors found no change in the incidence of ROP, but recognized a lower birth weight and younger gestational age in affected patients [64] (Table 20.2.2).

Table 20.2.2. Incidence of ROP and treatment in different studies

Study	Country	N	Birth weight (g)	Inci- dence	Treat- ment
Flynn [33]	USA	214	≤1,300	55.6%	4.2%
Cryo-ROP [17]	USA	4,099	<1,251	62.2%	6.1%
Light ROP [88]	USA	410	≤1,251	65.8%	5.0%
Fielder [30]	England	572	≤1,700	50.9%	4.7%
Brennan [7]	England	484	≤1,500	41.9%	5.2%
Holmström [46]	Sweden	260	<1,500	40.4%	10.8%
Schalij-Delfos [94]	Nether- lands	581	<1,500	27.4%	1.4%
Jandeck [58]	Germany	666	≤1,500	44.9%	5.7%

20.2.6 Zone I Disease

A vascularization border in Zone I during the first examination is very rare. In the Cryo-ROP study only 5% had Zone I disease, and 33% of these eyes needed treatment during follow-up. The unfavorable outcome in this group was 78%. The risk of an "unfavorable outcome" was 8.24 times higher in eyes with Zone I disease [17]. Recently performed studies indicated a better anatomical outcome after laser treatment. The unfavorable outcome was 15–36% in these studies [2, 34].

In the German guidelines and the ETROP Study, recommendations are made to treat retinopathies found with vascularization in Zone I together with blood vessel anomalies [14, 29]. In the ETROP Study Zone I disease occurred in 40%. The unfavorable outcome was reduced due to earlier treatment from 53.8% to 29.6% when compared to conventional treatment strategies.

20.2.7 Treatment Modalities

20.2.7.1 ROP Stage 3+

Nagata [74] was the first to treat retinal changes of ROP with a xenon coagulator. First he treated central to the demarcation line, but later recognized that treatment of the periphery was more successful.

Further Japanese study groups have reported successful coagulation treatments in the active stage, while Yamashita [109] was the first to apply cryo-coagulation. Studies in Israel [3] and Europe [59] have achieved good results through cryo-coagulation. The American Multicenter Cryo-ROP study from 1988 yielded such distinct proof of therapy success, so that even before the planned conclusion of the study the first results were published. Through cryo-coagulation of the avascular peripheral retina, a reduction from 43% to 21.8% in the incidence of an "unfavorable outcome" could be achieved. In addition, 15 years after the recruiting of the study children, the results show the same positive effect of the coagulation therapy. An "unfavorable outcome" with respect to the anatomical retinal presentation was found in 27.2% of treated eyes and 47.9% of untreated eyes [27].

A visual acuity of ≤20/200 (using the ETDRS Chart) was defined as an "unfavorable outcome." In the 10-year results of the Cryo-ROP Study [24], only 44.4% of treated eyes had a visual acuity of ≤20/200. In contrast, 62.1% of untreated eyes had a visual acuity of ≤20/200. With the prerequisite of retinal attachment, the percentage of eyes that had a visual acuity of 20/40 or better was the same in both groups (25.2% of treated eyes, 23.7% of untreated eyes). Contrast Sensitivity Testing [25] resulted in a significant difference between the groups, in favor of the treated eyes. The visual field of eyes with Stage 3+ was reduced by only an additional 7% after cryo-coagulation as compared to similar eyes that did not receive treatment. However, the visual field in all eyes with an advanced retinopathy (Stage 3+), regardless of treatment, was significantly reduced in comparison to eyes without ROP [26].

As a further treatment option, indirect laser coagulation (diode laser, argon laser) has been available for several years. These coagulation methods were established as a therapy for ROP [8, 49, 73] and due to the better anatomic, functional and refractive results [8, 34, 58] in comparison to cryo-coagulation, have been supported as the therapy of choice. Due to the wavelength, the diode laser ($l = 800–900$ nm) has more advantages than the argon laser ($l = 488$ and 514 nm). In cases of tunica vasculosa lentis, due to reduced absorption by the blood vessels there is less heat produced and therefore a smaller risk of induced cataract.

20.2.7.1.1 Timing of Treatment

Treatment carried out promptly after reaching the "threshold" criteria leads to a better anatomical result than a delayed coagulation [44]. It is due to this finding that the definition "within days" has been used in the German Guidelines for Ophthalmic Screening of Preterm Babies [14]. Due to the possibility of successful treatment and the very narrow time frame for the optimal therapy, it is important that all affected children be examined in timely fashion.

The large ETROP Study that appeared in December 2003, with 26 study centers and 401 randomized children, showed significantly better anatomical and functional results for treatment given at "pre-threshold" in comparison to treatment given after reaching threshold criteria. These results were further divided according to the vascularization border. Eyes with a vascularization border in Zone I and a Stage 3 benefit the most from early treatment. An "unfavorable outcome" was recorded in 30.8% of the early treatment group in contrast to 53.8% of the conventionally treated group. In view of their results the authors recommended early coagulation treatment in cases of Zone I disease and every ROP stage with "plus disease," in cases of a vascularization border in Zone I and Stage 3 with or without "plus disease" and in cases of a vascularization border in Zone II with additional presentation of Stage 2 or 3 with "plus disease."

20.2.7.1.2 Laser Treatment Versus Cryo-Coagulation

In the Cryo-ROP Study [24] an "unfavorable outcome" was described in 31% of cases after 3 months and after 10 years follow-up in 27% of cases. In the later study comparing laser and cryo-coagulation, an "unfavorable outcome" was given to 8% of laser treated eyes and to 19% of cryo-coagulated eyes. Several other authors have also compared the different results from laser and cryo-treatment [58, 75, 79, 80, 105]. In these studies, a better anatomical result could be achieved with laser treatment than with cryo-coagulation.

Some studies have shown a higher probability of cryo-coagulated eyes developing blood vessel distortion [58, 75]. The reasons for this are, however, not clear. It is postulated that the different coagulation methods cause different damage to the tissue. The cryo-coagulation changes the retina to a thin glial scar with accompanying pigment epithelial atrophy, removal of Bruch's membrane and atrophy of the choroid capillary layer [106]. Wallow et al. [107] described the histological findings after laser treatment in monkeys. In comparison to cryo-coagulation, the inner retinal layers were not damaged. In addition, a more severe breakdown of the blood-retinal barrier was described following cryo-coagulation than after laser treatment [73]. The macula-pigment epitheliopathy, which has been described by several authors following cryo-coagulation, most probably occurs due to this breakdown. Interestingly enough, this observation has not been made following laser coagulation [38, 44, 79, 93]. These changes could also account for the reduced vision [79, 93]. The macula-pigment epitheliopathy can also occur together with a macula ectopy, macular folds or retinal detachment [38].

For eyes having a visual acuity of 20/40, the publication of the Laser Study Group [58] showed a significant visual improvement for the laser treated eyes in comparison to the cryo-coagulated eyes.

In a prospective study, White and Repka [108] also proved a better visual result following laser treatment. The results from Jandeck et al. [58] showed that laser treated eyes reached a visual acuity of 20/200 significantly more often. A visual acuity of 20/25 was achieved in 39.2% of the laser treated eyes and in 17.6% of the cryo-coagulated eyes. Two other study groups [80, 93], in a comparative retrospective study, showed a better visual result after laser treatment in comparison to cryo-coagulation.

20.2.7.2 Treatment Principles

In both treatment methods, the avascular area peripheral to the border is coagulated. In order to avoid hemorrhaging, the border should not be treated. In cryo-coagulation, the entire avascular retina is confluently destroyed. In indirect laser coagulation, the treatment should be close to the ridge and the laser burns should not be separated by more than 1/4 laser beam width from one another (Fig. 20.2.10). Moving out into the periphery, the distance between burns can be increased to 1 width of the laser beam. It is recommended that a 20 diopter or 2.2 diopter condensing lens be used in conjunction with a head-mounted ophthalmoscope. The necessary number of laser burns is dependent on the size of the avascular area and the condensing lens used and can be between 600 and 2,000 burns. Consequently, the treatment of one eye can take between 20 and 40 min.

In eyes with an advanced tunica vasculosa lentis and corresponding reduced function of the laser beam, coagulation with a transscleral diode laser should be considered.

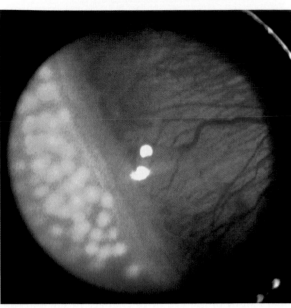

Fig. 20.2.10. Fresh laser-coagulation burns in the avascular periphery of a Stage 3 ROP

20.2.7.3 Treatment in Stages 4 and 5

The treatment of advanced stages of ROP is still controversial. In the guidelines no recommendations are given for these stages, because no controlled studies of retinal surgery in these cases are available.

The surgery part is covered in Chapter 20.3 by Quiram, Lai, and Tresee

20.2.8 Conservative Therapy

Several conservative therapies for the treatment, or rather the prophylactic treatment of, severe ROP stages have been researched. However, no method has yet been proven to be effective.

20.2.8.1 Vitamin E

The Vitamin E supplement (tocopherol) used prophylactically against the development of ROP is controversial. In a prospective, randomized, double-masked study, that compared vitamin E with a placebo, no significant difference was found with regard to the incidence of ROP between treatment and control group [15]. Other studies [87] showed that through a vitamin E supplement, a delay in the development of a Stage 3 was apparent and, consequently, the risk of blindness was reduced.

20.2.8.2 STOP-ROP Study

In the multicenter STOP-ROP (Supplemental Therapeutic Oxygen for Threshold ROP) Study [100], infants with retinal changes meeting "pre-threshold" were randomly placed in two groups. The study group received additional oxygen, in order that the oxygen saturation reached 96–99%. For the group with additional oxygen, a reduction in the advancement of the findings from "threshold" to "pre-threshold" was achieved in 48.5%. In the group without additional oxygen, the reduction rate of the advancement was 40.9%.

This difference was, however, not significant. The period of time required to reach the "threshold" criteria was simply lengthened through additional oxygen supplementation. For a subgroup from "pre-threshold" without additional "plus disease" a significant reduction in the advancement of the disease to "threshold" was illustrated through the additional oxygen supplementation (from 46% in the conventional group to 32% in the study group). The negative effects of additional oxygen were, however, a worsening of the chronic lung disease and a longer and more cost intensive hospital stay.

In earlier studies by Gaynon et al. [36] and Seiberth et al. [97], a distinct reduction in the incidence of threshold cases was found through therapy of additional oxygen. Currently no general recommendation can be given regarding additional oxygen therapy.

20.2.8.3 Light-ROP Study

An earlier theory regarding the pathogenesis of ROP involves damage to cells through free radicals. Through light exposure, an increased number of free radicals are released. The multicenter, randomized study "Light Reduction in Retinopathy of Prematurity" (LIGHT-ROP) [88] was carried out to investigate the higher incidence of ROP that occurs under circumstances of unnatural light exposure in the neonatal intensive situation. In this study, prematurely born infants were given sunglasses (reducing visible light by 97% and UV light by 100%) within a maximal time span of 24 h after birth. The sunglasses were worn up until the first fundus examination (at 4 postnatal weeks of age or 31 postmenstrual weeks). The study, with 410 patients, comprising the largest randomized study of its kind, showed that a reduction of light cannot reduce the incidence of retinopathy in high-risk children.

20.2.8.4 Surfactant

Five Cochrane Reviews evaluated trials of surfactant. In these trials prophylactic surfactant was given to high-risk pre-term babies at, or shortly after, birth to prevent RDS (respiratory distress syndrome). The

20 **III**

trials found no difference in rates of ROP but discovered a reduced mortality and a reduced rate of other respiratory problems [99].

General recommendations for laser therapy in ROP:
Diode laser. Therapy should be performed in the avascular retina
A binocular ophthalmoscope, scleral depressor, and lid speculum are essential for examination and treatment. Spot size varies according to the lens used and the distance of the head ophthalmoscope to the patient's eye; however, nearly confluent treatment is required
Exposure times differ according to the clinical appearance and depend on the laser used
Re-treatment is indicated in cases of insufficient regression or in eyes with progression
Retinal bleeding may occur during treatment in babies with high blood pressure

---

20.2.9 Late Changes

Retinopathy of prematurity (ROP) is a lifelong illness. Even after conclusion of the acute phase of ROP, further ocular problems can arise in all pre-term patients, e.g., refractive errors that appear not only following coagulation treatment, but also in cases of pre-term patients with serious retinal changes. Further ocular changes are increased incidence of strabismus and nystagmus, as well as retinal pigment changes, retinal distortion, vitreoretinal degenerations, retinal folds, retinal holes, and secondary glaucoma. Because late retinal detachments are rare, yearly check-ups are not necessary. However, risk patients should of course be informed of the possibility of a retinal detachment and its accompanying symptoms.

Cats and Tan [12] found over an observation period of 6–10 years that 55% of children with regressive ROP developed ocular changes. Other studies have shown in 59% [10], or rather 25% [35], of cases, that even without ROP during the acute phase, changes can occur that require ophthalmic examination. Each of the possible late developments are given in detail below.

20.2.9.1 Risk of Myopia

Children born on their due date have an incidence of myopia of between 6% and 9% [32]. Data found in the literature in reference to the development of myopia in prematurely born infants is not homogenous. In children with ROP, an incidence of myopia of 16–50% is given [76, 78]. The Cryo-ROP Study [22] found that the incidence of myopia rose with the increase in the stage of severity of ROP. Pre-term babies with an ROP Stage 3 under the threshold for

coagulation had a 40–62% risk of developing myopia [20, 22].

Several authors have discovered a difference in the incidence of myopia, dependent upon development of ROP in cases of prematurely born babies under the threshold criteria for treatment. In two studies [12, 76] it was found that myopia developed in 50% or rather 29% of premature babies with ROP and in 16% or rather 10% of premature babies without ROP. Other authors found only a very small [20] or no difference [11].

New studies have shown that predominantly children with ROP Stage 3 have a higher risk for myopia [82, 91]. The incidence of a refractive error in children with a retinopathy < Stage 3 was not higher than for children born on time [82].

The reason for the development of myopia in prematurely born children is the subject of frequent debate. Myopia is usually attributed to changes in the axial length of the eye or to changes in the anterior portion of the eye. In eyes with a regressive retinopathy, the axial length is longer than normal [13], normal [103] or shorter [32, 67]. Changes in the anterior portion of the eye have been accounted for through an enlarged lens [41], a flatter anterior chamber [45] or an increase in corneal curvature [32].

20.2.9.2 Visual Acuity

Several studies came to the conclusion that visual acuity in prematurely born infants is reduced until at least the 12th year of life when compared with their full-term counterparts [32, 47]. Some studies found a functional difference between eyes with and without ROP [28, 69]. Other studies however, did not find this [22]. A recently published study found that prematurely born children at 10 years of age had reduced visual acuities compared to full term children, even when children who had ROP and neurological disorders were excluded [65]. Only 53% of pre-term babies reached a visual acuity 0.7 (at 3.5 years of age) in contrast to 93% of full-term children (at 4 years of age). Thirty-four percent of prematurely born children without ROP had a visual acuity of < 0.7 in contrast to pre-term babies with ROP, whose visual acuity was under 0.7 in 61% of cases [47]. The cause for the reduced visual acuity appears to be the incomplete development of the fovea [50].

20.2.9.3 Strabismus

The incidence of strabismus is higher in premature babies than in full-term babies [30a]. In cases of full-term babies the incidence of strabismus is 5–7%. The incidence of strabismus fluctuates between 14% and 47% for premature babies with ROP and between 10% and 20% for premature babies without ROP [12, 78, 98]. In some comparative studies, a higher strabismus risk exists following the occurrence of ROP [98]; on the other hand, other studies found no difference [55, 91]. A pseudo-strabismus can occur through a macular ectopy caused by a peripheral distortion of the retina (Annette von Droste-Hülshoff syndrome) [1].

20.2.9.4 Glaucoma

A common (25–30%) late development of Stage 5 is that of secondary glaucoma, first described by Blodi [4]. The secondary glaucoma develops from retinal fibrovascular proliferation that shifts the lens anteriorly, thereby narrowing the anterior chamber. Medical therapy with miotics [4] should be tried before surgery (lensectomy or iridectomy). Angle closure glaucoma is alarmingly common among older, prematurely born patients. The enlarged lens causes a pupillary block. This form of glaucoma is, as a rule, well handled through an iridectomy or a trabeculectomy. Sometimes, however, removal of the lens is required [85].

20.2.9.5 Regressive Late Changes in the Retina

In most eyes, where the premature retinopathy did not reach threshold for treatment, the changes spontaneously resolve without consequence. However, it is possible that regressive vitreoretinal changes remain, for example, blood vessel distortion with macular shift, retinal dragging (Fig. 20.2.11), retinal folds and chorioretinal scarring (Fig. 20.2.12), peripheral blood vessel changes, peripheral avascular areas, diffuse retinal pigment-epithelial lumping

(Fig. 20.2.12), and seldomly retinal holes [51, 52]. The frequency of the distinct regressive changes varies between 2.9% and 12.3% [23, 35, 81]. In addition, changes to the vitreoretinal traction can consequently give rise to retinal foramina and retinal detachments later in life [71, 102]. This type of retinal detachment typically occurs in puberty. A successful treatment method for this type of detachment is a pars plana vitrectomy [57].

20.2.9.6 Differential Diagnosis

The differential diagnoses of ROP are dependent on the respective stage of ROP. The earlier stages are changes involving peripheral avascular retina, as in familial exudative vitreoretinopathy, incontinentia pigmenti (Bloch-Sulzberger syndrome) or Norrie disease. The later stages are associated with a leukocoria, such as congenital cataract, persistent hyperplastic primary vitreous (PHPV), retinoblastoma, ocular *Toxocara* infection, Coats' disease, uveitis and vitreal hemorrhages.

20.2.9.7 Outlook

Through the improved neonatal medicine, more and more prematurely born children are surviving at even younger ages. These pre-term children have a higher risk for retinal changes. The Cryo-ROP Study clearly showed that, through optimal screening, premature babies who received coagulation treatment at particular stages achieved a visual improvement and the rate of blindness was reduced. Because this disease is very rare, it makes sense to provide care for these babies in specialized centers.

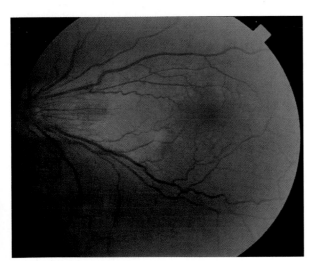

Fig. 20.2.11. Retinal dragging; visual acuity in this eye is 20/25

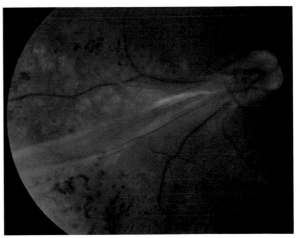

Fig. 20.2.12. Macular fold and diffuse retinal pigment epithelial lumping

20

20.2.9.8 Future Treatment Possibilities

Currently, there is no clinically established medical therapy with a localized anti-angiogenetic effect to prevent ROP. The main area interest at the moment in research is in blocking VEGF (vascular endothelial growth factor), angiopoietin or PEDF (pigment epithelial derived factor) (see Chapter 20.1).

In industrialized nations, ROP is still the cause of blindness in 6–20% of children [38, 101]. In contrast to other diseases, it is currently possible to have all children pass through an effective screening protocol, and in this way catch the advanced retinal changes early enough for appropriate treatment. The time of this screening differs for the health system of each country, and is dependent on the stage of development of the neonatal system and socioeconomic conditions of the country. Through optimal care of pre-term babies, blindness can be mostly avoided.

However, the delayed effects of the regressing retinal changes call for intensive follow-up of patients. Unfortunately, despite treatment at the optimal moment, a retinal detachment can occur in a few eyes at an acute stage. It is important in these cases to search for further possible causes such as blood transfusions, increased oxygen variability, anemia, and fetofetal transfusion syndrome. The possibility of treatment of ROP through medication, thereby avoiding surgical procedures, would be desirable for the future.

In order to make ROP an avoidable illness, more research is necessary. Knowledge of the disease mechanism would allow us to avoid the factors causing the problem or it would allow for treatment of the factors in advance of the disease.

References

1. Alfieri MC, Magli A, Chiosi E, De CG (1988) The Annette von Droste-Hulshoff syndrome. Pseudostrabismus due to macular ectopia in retinopathy of prematurity. Ophthalmic Paediatr Genet 9:13–16
2. Axer-Siegel R, Snir M, Cotlear D et al (2000) Diode laser treatment of posterior retinopathy of prematurity. Br J Ophthalmol 84:1383–1386
3. Ben Sira I, Nissenkorn I, Kremer I (1988) Retinopathy of prematurity. Surv Ophthalmol 33:1–16
4. Blodi FC (1955) Symposium: Retrolental fibroplasia (retinopathy of prematurity). Management. Trans Am Acad Ophthalmol Otolaryngol 59:35
5. Bolton DP, Cross KW (1974) Further observations on cost of preventing retrolental fibroplasia. Lancet 1:445–448
6. Bossi E, Koerner F (1995) Retinopathy of prematurity. Intensive Care Med 21:241–246
7. Brennan R, Gnanaraj L, Cottrell DG (2003) Retinopathy of prematurity in practice. I. Screening for threshold disease. Eye 17:183–188
8. Brooks SE, Johnson M, Wallace DK et al (1999) Treatment outcome in fellow eyes after laser photocoagulation for retinopathy of prematurity. Am J Ophthalmol 127:56–61
9. Bullard SR, Donahue SP, Feman SS et al (1999) The decreasing incidence and severity of retinopathy of prematurity. J AAPOS 3:46–52
10. Burgess P, Johnson A (1991) Ocular defects in infants of extremely low birth weight and low gestational age. Br J Ophthalmol 75:84–87
11. Caflisch SR, Bucher HU, Santo B et al (1998) Ophthalmologische Störungen ehemals extrem kleiner Frühgeborener im Alter von 10 Jahren. Monatsschr Kinderheilkd 146:230–234
12. Cats BP, Tan KE (1989) Prematures with and without regressed retinopathy of prematurity: comparison of long-term (6–10 years) ophthalmological morbidity. J Pediatr Ophthalmol Strabismus 26:271–275
13. Clark DI, Laws F, Wood ICJ, Laws D (1996) Axial growth of the eye and refractive error following retinopathy of prematurity (ROP) – a longitudinal study. Invest Ophthalmol Vis Sci 37:370
14. Clemens S, Eckardt C, Gerding H et al (1999) Ophthalmological screening studies in newborn infants. German Ophthalmological Society. Ophthalmologe 96:257–263
15. Committee on Fetus and Newborn (1985) Vitamin E and the prevention of retinopathy of prematurity. Pediatrics 76:315–316
16. Cryotherapy for Retinopathy of Prematurity Cooperative Group (1988) Multicenter trial of cryotherapy for retinopathy of prematurity. Preliminary results. Arch Ophthalmol 106:471–479
17. Cryotherapy for Retinopathy of Prematurity Cooperative Group (1990) Multicenter trial of cryotherapy for retinopathy of prematurity. Three-month outcome. Arch Ophthalmol 108:195–204
18. Cryotherapy for Retinopathy of Prematurity Cooperative Group (1991) Palmer EA, Flynn JT, Hardy RJ et al. Incidence and early course of retinopathy of prematurity. Ophthalmology 98:1628–1640
19. Cryotherapy for Retinopathy of Prematurity Cooperative Group (1992) Gilbert WS, Dobson V, Quinn GE, Reynolds J et al. The correlation of visual function with posterior retinal structure in severe retinopathy of prematurity. Arch Ophthalmol 110:625–631
20. Cryotherapy for Retinopathy of Prematurity Cooperative Group (1992) Quinn GE, Dobson V, Repka MX et al. Development of myopia in infants with birth weights less than 1251 grams. Ophthalmology 99:329–340
21. Cryotherapy for Retinopathy of Prematurity Cooperative Group (1994) The natural ocular outcome of premature birth and retinopathy. Status at 1 year. Arch Ophthalmol 112:903–912
22. Cryotherapy for Retinopathy of Prematurity Cooperative Group (1994) Dobson V, Quinn GE, Summers CG et al. Effect of acute-phase retinopathy of prematurity on grating acuity development in the very low birth weight infant. Invest Ophthalmol Vis Sci 35:4236–4244
23. Cryotherapy for Retinopathy of Prematurity Cooperative Group (1995) Quinn GE, Dobson V, Biglan A et al. Correlation of retinopathy of prematurity in fellow eyes in the cryotherapy for retinopathy of prematurity study. Arch Ophthalmol 113:469–473
24. Cryotherapy for Retinopathy of Prematurity Cooperative Group (2001) Multicenter Trial of Cryotherapy for Retinopathy of Prematurity: ophthalmological outcomes at 10 years. Arch Ophthalmol 119:1110–1118

25. Cryotherapy for Retinopathy of Prematurity Cooperative Group (2001) Contrast sensitivity at age 10 years in children who had threshold retinopathy of prematurity. Arch Ophthalmol 119:1129–1133

26. Cryotherapy for Retinopathy of Prematurity Cooperative Group (2001) Effect of retinal ablative therapy for threshold retinopathy of prematurity: results of Goldmann perimetry at the age of 10 years. Arch Ophthalmol 119:1120–1125

27. Cryotherapy for Retinopathy of Prematurity Cooperative Group (2005) Palmer EA, Hardy RJ, Dobson V et al. 15-year outcomes following threshold retinopathy of prematurity: final results from the multicenter trial of cryotherapy for retinopathy of prematurity. Arch Ophthalmol 123:311–318

28. Dogru M, Shirabe H, Nakamura M et al (1999) Development of grating acuity in infants with Retinopathy of Prematurity. Acta Ophthalmol Scand 77:72–75

29. Early Treatment for Retinopathy of prematurity Cooperative Group (2003) Revised indications for the treatment of Retinopathy of Prematurity. Results of the early treatment for Retinopathy of prematurity randomized trial. Arch Ophthalmol 121:1684–1696

30. Fielder AR, Shaw DE, Robinson J, Ng YK (1992) Natural history of retinopathy of prematurity: a prospective study. Eye 6:233–242

30a. Fledelius HC (1976) Prematurity and the eye. Ophthalmic 10-year follow-up of children of low and normal birth weight. Acta Ophthalmol Suppl 128:3–245

31. Fledelius HC (1996) Pre-term delivery and subsequent ocular development. A 7–10 year follow-up of children screened 1982–1984 for ROP. 1) Visual function, slit-lamp findings, and fundus appearance. Acta Ophthalmol Scand 74:288–293

32. Fledelius HC (1996) Pre-term delivery and subsequent ocular development. A 7–10 year follow-up of children screened 1982–1984 for ROP. 4) Oculometric – and other metric considerations. Acta Ophthalmol Scand 74:301–305

33. Flynn JT, Bancalari E, Bachynski BN et al (1987) Retinopathy of prematurity. Diagnosis, severity, and natural history. Ophthalmology 94:620–629

34. Foroozan R, Connolly BP, Tasman WS (2001) Outcomes after laser therapy for threshold retinopathy of prematurity. Ophthalmology 108:1644–16446

35. Gallo JE, Lennerstrand G (1991) A population-based study of ocular abnormalities in premature children aged 5 to 10 years. Am J Ophthalmol 111:539–547

36. Gaynon MW, Stevenson DK, Sunshine P et al (1997) Supplemental oxygen may decrease progression of prethreshold disease to threshold retinopathy of prematurity. J Perinatol 17:434–438

37. Gibson DL, Sheps SB, Uh SH et al (1990) Retinopathy of prematurity-induced blindness: birth weight-specific survival and the new epidemic. Pediatrics 86:405–412

38. Gilbert C, Rahi J, Eckstein M et al (1997) Retinopathy of prematurity in middle-income countries. Lancet 350:12–14

39. Goggin M, O'Keefe M (1993) Diode laser for retinopathy of prematurity–early outcome. Br J Ophthalmol 77:559–562

40. Good WV, Hardy RJ, Dobson V et al (2005) Early Treatment for Retinopathy of Prematurity Cooperative Group. The incidence and course of retinopathy of prematurity: findings from the early treatment for retinopathy of prematurity study. Pediatrics 116:15–23

41. Gordon RA, Donzis PB (1986) Myopia associated with retinopathy of prematurity. Ophthalmology 93:1593–1598

42. Haines L, Fielder AR, Scrivener R et al (2002) Royal College of Paediatrics and Child Health, the Royal College of Ophthalmologists and British Association of Perinatal Medicine. Retinopathy of prematurity in the UK I: the organisation of services for screening and treatment. Eye 16:33–38

43. Hatfield EM (1972) Blindness in infants and young children. Sight Sav Rev 42:69–89

44. Hindle NW (1982) Cryotherapy for retinopathy of prematurity: Can J Ophthalmol 17:207–212

45. Hittner HM, Rhodes LM, McPherson AR (1979) Anterior segment abnormalities in cicatricial retinopathy of prematurity. Ophthalmology 86:803–816

46. Holmstrom G, el Azazi M, Jacobson L, Lennerstrand G (1993) A population based, prospective study of the development of ROP in prematurely born children in the Stockholm area of Sweden. Br J Ophthalmol 77:417–423

47. Holmstrom G, el Azazi M, Kugelberg U (1999) Ophthalmological follow up of preterm infants: a population based, prospective study of visual acuity and strabismus. Br J Ophthalmol 83:143–150

48. Horbar JD, McAuliffe TL, Adler SM et al (1988) Variability in 28-day outcomes for very low birth weight infants: an analysis of 11 neonatal intensive care units. Pediatrics 82:554–559

49. Hunter DG, Repka MX (1993) Diode laser photocoagulation for threshold retinopathy of prematurity. A randomized study. Ophthalmology 100:238–244

50. Isenberg SJ (1986) Macular development in the premature infant. Am J Ophthalmol 101:74–80

51. International Committee for the Classification of the Late Stages of Retinopathy of Prematurity (1987) An international classification of retinopathy of prematurity. II. The classification of retinal detachment. Arch Ophthalmol 105:906–912

52. International Committee for the Classification of Retinopathy of Prematurity (1984) An international classification of retinopathy of prematurity. Arch Ophthalmol 102:1130–1134

53. International Committee for the Classification of Retinopathy of Prematurity (2005) The International Classification of Retinopathy of Prematurity revisited. Arch Ophthalmol 123:991–999

54. Jandeck C, Kellner U, Helbig H et al (1995) Natural course of retinal development in preterm infants without threshold retinopathy. Ger J Ophthalmol 4:131–136

55. Jandeck C (1999) Refraktionsfehler bei Frühgeborenen mit und ohne Koagulationsbehandlung. Orthoptik-Pleoptik 23:25–32

56. Jandeck C, Kellner U, Heimann H, Foerster MH (2004) Screening for retinopathy of prematurity: results of one centre between 1991 and 2002. Klin Monatsbl Augenheilkd 222:577–585

57. Jandeck C, Kellner U, Foerster MH (2004) Late retinal detachment in patients born prematurely: outcome of primary pars plana vitrectomy. Arch Ophthalmol 122:61–64

58. Jandeck C, Kellner U, Heimann H, Foerster MH (2005) Comparison of the anatomical and functional outcome after laser or cryotherapy for retinopathy of prematurity (ROP). Ophthalmologe 102:33–38

59. Keith CG (1982) Visual outcome and effect of treatment in stage III developing retrolental fibroplasia. Br J Ophthalmol 66:446–449

60. Kellner U, Jandeck C, Helbig H et al (1995) Evaluation of published recommendations for screening studies of retinopathy of prematurity. Ophthalmologe 92:681–684

20 III

61. Kinsey VE (1956) Retrolental fibroplasia; cooperative study of retrolental fibroplasia and the use of oxygen. AMA Arch Ophthalmol 56:481–443
62. Kumar H, Nainiwal S, Singha U et al (2002) Stress induced by screening for retinopathy of prematurity. J Pediatr Ophthalmol Strabismus 39:349–350
63. Lanman JT, Guy LP, Dancis J (1954) Retrolental fibroplasia and oxygen therapy. J Am Med Assoc 155:223–226
64. Larsson E, Carle-Petrelius B, Cernerud G et al (2002) Incidence of ROP in two consecutive Swedish population based studies. Br J Ophthalmol 86:1122–1126
65. Larsson EK, Rydberg AC, Holmstrom GE (2005) A population-based study on the visual outcome in 10-year-old preterm and full-term children. Arch Ophthalmol 123:825–832
66. Laser ROP Study Group (1994) Laser therapy for retinopathy of prematurity. Arch Ophthalmol 112:154–156
67. Laws DE, Haslett R, Ashby D et al (1994) Axial length biometry in infants with retinopathy of prematurity. Eye 8:427–430
68. Laws DE, Morton C, Weindling M et al (1996) Systemic effects of screening for retinopathy of prematurity. Br J Ophthalmol 80:425–428
69. Laws F, Laws D, Clark D (1997) Cryotherapy and laser treatment for acute retinopathy of prematurity: refractive outcomes, a longitudinal study. Br J Ophthalmol 81:12–15
70. Lemons JA, Bauer CR, Oh W et al (2001) Very low birth weight outcomes of the National Institute of Child health and human development neonatal research network, January 1995 through December 1996. NICHD Neonatal Research Network. Pediatrics 107:E1
71. Machemer R (1993) Late traction detachment in retinopathy of prematurity or ROP-like cases. Graefes Arch Clin Exp Ophthalmol 231:389–394
72. McDonald AD (1963) Cerebral Palsy in Children of Very Low Birth Weight. Arch Dis Child 38:579–588
73. McNamara JA, Tasman W, Brown GC, Federman JL (1991) Laser photocoagulation for stage 3+ retinopathy of prematurity. Ophthalmology 98:576–580
74. Nagata M, Takishima R (1968) One side of the coin of functions of public health nursing – with the establishment of home care helpers. Kango 20:130–134
75. Ng EY, Connolly BP, McNamara JA et al (2002) A comparison of laser photocoagulation with cryotherapy for threshold retinopathy of prematurity at 10 years: part 1. Visual function and structural outcome. Ophthalmology 109:928–934; discussion 935
76. Nissenkorn I, Yassur Y, Mashkowski D et al (1983) Myopia in premature babies with and without retinopathy of prematurity. Br J Ophthalmol 67:170–173
77. Nissenkorn I, Kremer I, Gilad E et al (1987) 'Rush' type retinopathy of prematurity: report of three cases. Br J Ophthalmol 71:559–562
78. Page JM, Schneeweiss S, Whyte HE, Harvey P (1993) Ocular sequelae in premature infants. Pediatrics 92:787–790
79. Paysse EA, Lindsey JL, Coats DK et al (1999) Therapeutic outcomes of cryotherapy versus transpupillary diode laser photocoagulation for threshold retinopathy of prematurity. J Aapos 3:234–240
80. Pearce IA, Pennie FC, Gannon LM et al (1998) Three year visual outcome for treated stage 3 retinopathy of prematurity: cryotherapy versus laser. Br J Ophthalmol 82:1254–1259
81. Pennefather PM, Clarke MP, Strong NP et al (1995) Ocular outcome in children born before 32 weeks gestation. Eye 9:26–30
82. Pennefather PM, Tin W, Strong NP et al (1997) Refractive errors in children born before 32 weeks gestation. Eye 11:736–743
83. Phelps DL (1981) Retinopathy of prematurity: an estimate of vision loss in the United States 1979. Pediatrics 67:924–925
84. Pohlandt F (1998) Premature delivery in borderline viability of the infant. A recommendations of the German Society of Gynecology and Obstetrics, the German Society of Pediatrics and Adolescent Medicine, the German Society of Perinatal Medicine and the Society of Neonatology and Pediatric Intensive Care Medicine. Z Geburtshilfe Neonatol 202:261–263
85. Pollard ZF (1984) Lensectomy for secondary angle-closure glaucoma in advanced cicatricial retrolental fibroplasia. Ophthalmology 91:395–398
86. Quinn GE, Johnson L, Abbasi S (1992) Onset of retinopathy of prematurity as related to postnatal and postconceptional age. Br J Ophthalmol 76:284–288
87. Raju TN, Langenberg P, Bhutani V, Quinn GE (1997) Vitamin E prophylaxis to reduce retinopathy of prematurity: a reappraisal of published trials. J Pediatr 131:844–850
88. Reynolds JD, Hardy RJ, Kennedy KA et al (1998) Lack of efficacy of light reduction in prevention of prematurity. Light reduction in Retinopathy of Prematurity (Light-ROP) Cooperative Group. N Engl J Med 338:1572–1576
89. Reynolds JD, Dobson V, Quinn GE et al (2002) CRYO-ROP and LIGHT-ROP Cooperative Study Groups. Evidence-based screening criteria for retinopathy of prematurity: natural history data from the CRYO-ROP and LIGHT-ROP studies. Arch Ophthalmol 120:1470–1476
90. Riise R (1993) Nordic registers of visually impaired children. Scand J Soc Med 21:66–68
91. Robinson R, O'Keefe M (1993) Follow-up study on premature infants with and without retinopathy of prematurity. Br J Ophthalmol 77:91–94
92. Rowlands E, Ionides AC, Chinn S et al (2001) Reduced incidence of retinopathy of prematurity. Br J Ophthalmol 85:933–935
93. Saito Y, Hatsukawa Y, Lewis J M et al (1996) Macular coloboma-like lesions and pigment abnormalities as complications of cryotherapy for retinopathy of prematurity in very low birth-weight infants. Am J Ophthalmol 122:299–308
94. Schalij-Delfos NE, Zijlmans BL, Wittebol-Post D et al (1996) Screening for retinopathy of prematurity: do former guidelines still apply? J Pediatr Ophthalmol Strabismus 33:35–38
95. Schalij-Delfos NE, Cats BP (1997) Retinopathy of prematurity: the continuing threat to vision in preterm infants. Dutch survey from 1986 to 1994. Acta Ophthalmol Scand 75:72–75
96. Schalij-Delfos NE, fort he Dutch working group on ROP (1999): Prematuren retinopathie, richtlijn voor screening. In: Van der Velde (ed): Richtlijn oogheelkunde. Deventer, Uitgava Commissie Kwaliteit van het Nederlands Oogheelkundig Gezelschap
97. Seiberth V, Linderkamp O, Akkoyun-Vardali I et al (1998) Oxygen therapy in acute retinopathy of prematurity stage 3. Invest Ophthalmol Vis Sci 39:820
98. Snir M, Nissenkorn I, Sherf I et al (1988) Visual acuity, strabismus, and amblyopia in premature babies with and without retinopathy of prematurity. Ann Ophthalmol 20:256–258

99. Soll RF (2002) Prophylactic natural surfactant extract for preventing morbidity and mortality in preterm infants (Cochrane review). Cochrane Collaboration: Cochrane Libary. Issue 2 Oxford: Update software

100. STOP-ROP (2000) Supplemental Therapeutic Oxygen for Prethreshold Retinopathy Of Prematurity, a randomized, controlled trial. I. Primary outcomes. Pediatrics 105:295–310

101. Steinkuller PG, Du L, Gilbert C et al (1999) Childhood blindness. J Aapos 3:26–32

102. Tasman W (1969) Retinal detachment in retrolental fibroplasia (RLF). Bibl Ophthalmol 79:371–376

103. Tasman W (1979) Late complications of retrolental fibroplasia. Ophthalmology 86:1724–1740

104. Terry TL (1942) Extreme prematurity and fibroblastic overgrowth of persistent vascular sheath behind each crystalline lens. I. Preliminary report. Am J Ophthalmol 25:203–204

105. Vander JF, Handa J, McNamara JA et al (1997) Early treatment of posterior retinopathy of prematurity: a controlled trial. Ophthalmology 104:1731–1735; discussion 1735–1736

106. Vrabec TR, McNamara JA, Eagle RC Jr, Tasman W (1994) Cryotherapy for retinopathy of prematurity: a histopathologic comparison of a treated and untreated eye. Ophthalmic Surg 25:38–41

107. Wallow IHL, Sponsel WE, Stevens TS (1991) Clinicopathologic correlation of diode laser burns in monkeys. Arch Ophthalmol 109:648–653

108. White JE, Repka MX (1997) Randomized comparison of diode laser photocoagulation versus cryotherapy for threshold retinopathy of prematurity: 3-year outcome. J Pediatr Ophthalmol Strabismus 34:83–87

109. Yamashita Y (1972) Studies on retinopathy of prematurity. III. Cryocautery for retinopathy of prematurity. Rinsho Ganka 26:385–393

20.3 Surgical Management of Retinopathy of Prematurity

P. QUIRAM, M. LAI, M. TRESE

Core Messages
- Eyes with previous laser ablation have improved anatomic and functional outcomes compared to eyes without previous peripheral ablation
- Stage 4A retinopathy of prematurity (ROP) often progresses to Stage 4B/5, which has a poor visual prognosis. Stage 4A detachments should be repaired before progression to Stage 4B/5
- The anatomy of an infant's eyes poses unique surgical challenges in that the pars plana is not developed and strong vitreoretinal adhesions exist
- Preoperative evaluation by visual evoked potential (VEP) is typically performed to assess visual potential in eyes with Stage 5 ROP

20.3.1 Introduction: Preoperative Evaluation and Timing of Surgical Intervention

Children who require surgical intervention for retinal detachment secondary to retinopathy of prematurity (ROP) can be divided into two groups: those who have had peripheral ablation and those who have not. The ideal situation for a child who has achieved retinal detachment is to have had previous peripheral ablation (Fig. 20.3.1). If a child needs laser intervention, they will usually reach that point between 32 and 46 weeks postmenstrual age with the peak occurring somewhere between 35 and 37 weeks postmenstrual age based on the data from the Cryo-

therapy for Retinopathy of Prematurity Study and the Early Treatment for Retinopathy of Prematurity Study [4].

Vitreous surgery intervention becomes necessary when the eye progresses to Stage 4A ROP, where a macula-on retinal detachment is present. Stage 4A retinal detachment customarily begins at the ridge tissue between the avascular and vascularized retina (Fig. 20.3.2). Natural history data from the Cryotherapy for Retinopathy of Prematurity Study suggests that Stage 4A retinal detachment presenting at a postmenstrual age of less than 46 weeks tends to be progressive in a large percentage of eyes. Data accu-

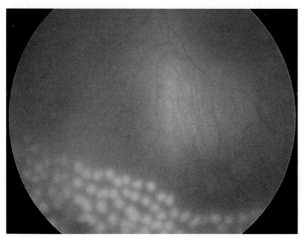

Fig. 20.3.1. Fundus photo demonstrating laser ablation of avascular retina up to, but not including, the ridge

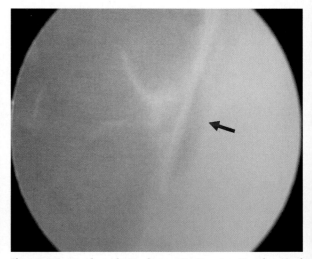

Fig. 20.3.2. Fundus photo demonstrating tractional retinal detachment at the ridge between vascular (*left*) and avascular retina (*right*)

mulated by Coats suggests that the peak incidence of retina detachment is at 41 weeks postmenstrual age [2]. There are several factors that have been identified that predict which infants will progress to Stage 4 retinopathy of prematurity. These factors include 6 clock hours of ridge elevation, 2 clock hours of plus disease, and the presence of vitreous haze. Vitreous organization can be a very significant indicator of an impending retinal detachment [6].

20.3.2 Eyes with Peripheral Ablation

Children requiring surgery for Stage 4A retinopathy of prematurity who underwent peripheral ablation are less likely to have vascularly active eyes, defined as plus disease and patent neovascular fronds. Surgery for Stage 4 detachments with previous peripheral retinal ablation is typically performed between 38 and 42 weeks postmenstrual age [1]. In eyes with active vascular proliferation and areas of untreated avascular retina, laser ablation is performed before surgery to decrease the risk of bleeding during vitreous surgery. In a long-term series with a 4-year follow-up, eyes with Stage 4B or 5 ROP with adequate peripheral ablation demonstrated a 76% reattachment rate of the posterior pole and 72% of patients had at least light perception [19]. In addition, 15% of patients achieved a visual acuity of 20/300 or better.

The assessment of vascular activity is very important and can be managed by three potential mechanisms. First, at the child's due date, 40 weeks postmenstrual age, endogenous TGFβ is produced that downregulates vascular endothelial growth factor [10, 20]. Second, if additional areas of avascular non-treated retina are present, additional laser may help return the eye to a vascularly quiet state. Finally, it may be possible to use pharmacologic agents, such as anti-VEGF agents, to control this vascular activity. At the time of preparation of this chapter, very little data is available for that mode of treatment.

20.3.3 Eyes Without Peripheral Ablation

Eyes that have not had peripheral ablation have a much lower anatomic and visual success rate. Many eyes that have not had peripheral ablation tend to progress to Stage 4B (macula-off) or Stage 5 (total) retinal detachment [3]. Furthermore, retinal detachment repair in eyes without adequate peripheral ablation results in a 50% reattachment rate [16, 17]. In addition, in eyes that reattached, only 30% had visual improvement. It is important when assessing the child preoperatively for retinal detachment to realize that the tempo of disease can be tempered by the child's race. Caucasian children tend to have a more aggressive retinopathy of prematurity and are more likely to progress to retinal detachment. Hispanic children within and outside the United States seem to have a more aggressive form of retinopathy of prematurity. In contrast, African Americans have a reduced chance of progressing to retinal detachment [14]. ROP is a bilateral disease and 85% of children will develop bilateral retinal detachment even though one eye may precede the other [11]. We have also reported that if retinal detachment does not appear by 50 weeks postmenstrual age, it most likely will not occur.

20.3.4 Treatment of Retinal Detachment

20.3.4.1 Scleral Buckling

In the past, scleral buckling has been used for the treatment of non-rhegmatogenous retinal detachment of retinopathy of prematurity [5, 18]. With the advent of lens-sparing vitrectomy for retinopathy of prematurity, we have found that scleral buckling is rarely appropriate. Scleral buckling has many disadvantages in that it requires a second operation to segment the buckle to not inhibit eye growth. It induces a very large amount of myopia (up to 12 diopters) and very rarely has good vision been reported following buckling procedures for even 4A retinopathy of prematurity. Presently, scleral buckling is not recommended for Stage 4A retinopathy of prematurity, although placement of radial elements has been employed for peripheral 4A detachments.

20.3.4.2 Lens-Sparing Vitrectomy

Because the macula is uninvolved, patients with Stage 4A ROP who undergo successful retina reattachment can potentially have good visual acuity. If the native lens can be spared during retinal reattachment, the risk of amblyopia secondary to significant anisometropia can be minimized, further increasing the visual potential. Therefore, lens-sparing vitrectomy is the procedure of choice for Stage 4A ROP. Originally described by Maguire and Trese, this procedure can be performed when there is less than 6 clock hours of retina-lens touch [9].

With the development of current vitreous surgery instrumentation and the appreciation that vitrectomy can be performed through the pars plicata in very young children where the para plana is not developed, the pathologic process causing the retinal detachment can be directly accessed (Fig. 20.3.3). In our experience, two-port vitrectomy is advantageous, allowing us to manipulate the eye in the small orbital space present in these children. The entry sites are chosen to maximize the surgical approach to the detached retina, and are verified by scleral depression to be free of retinal tissue. The sclerotomies are made in the pars plicata approximately one-half millimeter posterior to the limbus with the microvitreoretinal blade directed parallel to the visual axis to avoid lens injury (Fig. 20.3.3). We use a 19 gauge wide-angle high flow light pipe, 20 gauge vitreous cutter, and a 20 gauge infusion spatula to perform dissections. Although 25 gauge instrumentation theoretically has an advantage in that it is smaller, it is slow in removing formed vitreous and it is difficult to remove heavy proliferative tissue. In addition, manipulation of the three port 25 gauge instrumentation is more difficult with the small orbit and runs the risk of additional damage to the lens. The rigidity of the 20 gauge instrumentation allows better manipulation of the eye. The surgical goal is to free the ridge tissue from sheets of tissue that extend to the lens (Fig. 20.3.4), the ora serrata (Fig. 20.3.5), ridge to ridge (Fig. 20.3.6) and posteriorly to the area of the optic nerve stalk (Fig. 20.3.7). With all of these tractional forces removed, the reattachment rates are quite high (approximately 90 %).

Encouraging anatomic and functional outcomes have been achieved with lens-sparing vitrectomy. Three studies show an anatomic reattachment rate of

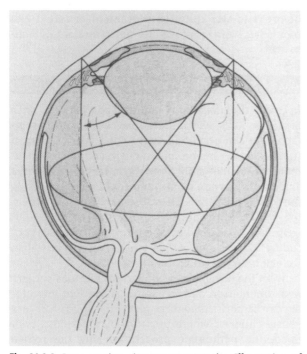

Fig. 20.3.3. Lens sparing vitrectomy entry site. Illustration of angle of entry through the pars plicata and area of influence in which lens sparing vitrectomy can be performed

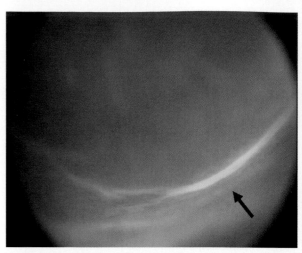

Fig. 20.3.4. Fundus photo demonstrating sheets of proliferation extending from the ridge anteriorly toward the lens

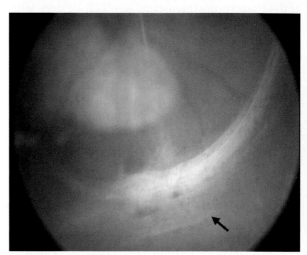

Fig. 20.3.5. Fundus photo demonstrating proliferative tissue extending from the ridge toward the ciliary process and ora serrata

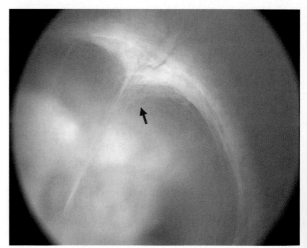

Fig. 20.3.6. Fundus photo demonstrating proliferative stalk tissue extending from the optic nerve anteriorly to intersect with ridge to ridge proliferative tissue

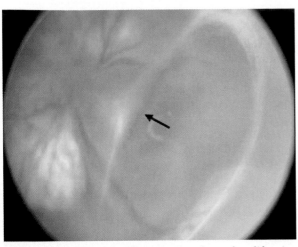

Fig. 20.3.7. Fundus photo demonstrating sheet of proliferation extending from ridge to ridge creating a funnel-shaped detachment

approximately 90% in patients with Stage 4A ROP undergoing lens-sparing vitrectomy. One study by Capone and Trese showed that 90% of eyes had retinal reattachment and fixation behavior at their last visit, with a mean follow-up period of 12 months [1]. Hubbard and colleagues reported similar anatomic success, with 84% of the eyes showing complete retinal reattachment after a median follow-up period of 13 months [8]. In another long-term study, patients with formal visual acuity testing between ages 1.8 and 6.3 years of age had a mean visual acuity of 20/58 (range 20/200 to 20/20), with 48% of the eyes achieving visual acuities of 20/40 or better [13].

20.3.4.3 Management of Stage 4B Retinopathy of Prematurity

Although peripheral ablation can be effective in stopping the progression of threshold ROP, retinal detachments can occur despite appropriate aggressive treatment. Even with close observation and treatment, some eyes will progress to Stage 4B. Compared to Stage 4A, Stage 4B retinal detachments have a much different prognosis with 72% achieving partial or complete retinal reattachment. Visually, 15% of eyes achieve 20/60 to 20/300 vision, 30% achieve 20/60 to 20/800 vision, and 48% achieve 20/60 to 20/1900 vision, which is what the authors feel is ambulatory vision. In a study by Droste and Trese, 72% achieved 20/60 to light perception vision, but 28% of eyes resulted in no light perception vision. In this study, the majority of eyes were treated with cryotherapy, which can increase effusive detachment and result in worse visual results.

Tractional 4B detachments can present in multiple conformations with asymmetric traction of posteri-

or elements, symmetrical contraction of posterior elements or contraction of anterior cellular elements. As described for 4A detachments, lens-sparing vitrectomy [9] can be performed for 4B detachments. In cases of 4B detachments where the retina touches the lens for more than 6 clock hours, lensectomy-vitrectomy-membrane peeling is necessary. We believe using two port closed vitrectomy is helpful, but occasionally three port vitrectomy can be used. Anterior infusion is provided by infusion instruments (irrigation light pipe or pick), an infusion line placed at the inferior limbus, or a bent 25-gauge needle inserted through the limbus. The iris is dissected off the lens and trimmed for visualization and to prevention of proliferation along the posterior iris. The lens is removed by the vitrector with complete removal of the lens capsule. Bimanual dissection is used to free the retina from traction. Drainage of subretinal blood is rarely performed as it is risky in the face of incomplete removal of preretinal traction. Even with the most careful vitrectomy techniques, complete removal of preretinal traction is unlikely.

20.3.4.4 Management of Stage 5 Retinopathy of Prematurity

Stage 5 retinopathy of prematurity is hopefully uncommon in hospitals that implement an appropriate screening schedule. However, Stage 5 eyes still appear, even today, particularly in developing countries. These eyes customarily have large amounts of subretinal blood and may or may not have had additional peripheral ablation. Previous studies suggest that some of the eyes may not have light perception preoperatively. To assess light perception, an awake visual evoked potential (VEP) may be used to assess the eye's light perception status; however, false negative responses are possible. Most Stage 5 retinal detachments fall into one of three configurations: an open anterior and posterior funnel, open anterior and closed posterior funnel, and closed anterior and posterior funnel [16]. Less commonly, a closed anterior and open posterior funnel may be present. These configurations are somewhat predictive in terms of anatomic results in that a closed-closed funnel has a worse prognosis than an open-open funnel.

The surgical approach in Stage 5 ROP is typically lensectomy followed by vitrectomy and membrane peeling, as described above. Techniques that take advantage of vitreous anatomy such as bimanual dissection and lamellar dissection of vitreous sheets can be very helpful in removal of layers of vitreous in an unusual retinal anatomy. Viscoelastic dissection can be used to open the funnel under low pressure and flatten retinal folds and compartmentalize bleeding while working on other areas of epiretinal tissue.

Another approach for vitreous surgery in Stage 5 ROP has been open sky vitrectomy, which requires concomitant lensectomy or an aphakic eye [7, 15]. We reserve open sky vitrectomy for eyes in which corneal opacity precludes closed vitrectomy. The technique of open sky vitrectomy allows access to more peripheral dissection of membranes with direct visualization through the open cornea.

Even if retinal reattachment is achieved, visual outcome is limited. As reported, the best vision achieved in Stage 5 eyes has been 20/200 to 20/600, and these results are quite rare [19]. The majority of eyes achieve visions of 20/800 or less. In addition, eyes can have no light perception despite having a completely attached retina. In the future, these eyes may be candidates for other types of therapy such as microelectronic or other biologic approaches.

20.3.4.5 Follow-up Care

It is important to realize that the premature children present postoperative challenges not seen in the adult population. The eye must be protected from digital damage, which can lead to bleeding. This is done with a Fox shield and elbow restraints. The child is positioned sitting up for several weeks after surgery so they cannot burrow their head, again causing the eye to bleed. In our practice, the child does not receive sedation, but the family is asked to comfort the child as much as possible to avoid extensive periods of crying with the potential for intraocular bleeding. Bleeding in the postoperative period is certainly not desirable, although in a smaller percentage of eyes, some bleeding can reabsorb, leaving the child with reasonable retinal anatomy.

In summary, the surgical management of the retinal detachment of retinopathy of prematurity has improved tremendously over the last 15 years. The child who is appropriately screened receives peripheral ablation at this point and perhaps in the future pharmacologic treatment. Early vitreous surgery will be anticipated to achieve good anatomic and visual function. In the past the association of late rhegmatogenous retinal detachment and retinopathy of prematurity has been clinically recognized. It will be interesting to follow these children who have had vitrectomy to see if their risk of rhegmatogenous retinal detachment is reduced. One of the major difficulties with vitreous surgery in children of this age is the inability to remove the posterior hyaloid despite sheets of tissue being peeled from the retina. These sheets of tissue reflect the sheet of solid and liquid vitreous that is present in the vitreous cavity of all premature children. Hopefully, the development of enzymatic products to separate the posterior hyaloid from the retina may prevent the problems of late rhegmatogenous retinal detachment. We have seen several children 3–5 years after vitrectomy who show a sheet of tissue in their vitreous cavity which may be transparent or somewhat opalescent, reflecting the detached posterior hyaloid, and thus supporting the observation that the hyaloid is generally not removed at the time of vitreous surgery. With all these considerations, however, the fate of the premature child with retinopathy of prematurity and retinal detachment is much better today than it was 2 decades ago.

References

1. Capone A, Trese MT (2001) Lens-sparing vitreous surgery for tractional Stage 4A retinopathy of prematurity retinal detachments. Ophthalmology 108:2068–2070
2. Coats D, Miller A, Brady McCreery K, Holz E, Paysse E (2004) Involution of threshold retinopathy of prematurity after diode laser photocoagulation. Ophthalmology 111:1894–1898
3. Cryotherapy for Retinopathy of Prematurity Cooperative Group (2001) Multicenter trial of cryotherapy for retinopathy of prematurity: Ophthalmological outcomes at 10 years. Arch Ophthalmol 119:1110–1118
4. Early Treatment for Retinopathy of Prematurity Cooperative Group (2003) Revised indications for the treatment of retinopathy of prematurity: results of the early treatment for retinopathy of prematurity randomized trial. Arch Ophthalmol 121:1684–1694
5. Greven C, Tasman W (1990) Scleral buckling in stages 4B and 5 retinopathy of prematurity. Ophthalmology 97:817–820
6. Hartnett MI, McColm JR (2004) Retinal features predictive of progressive stage 4 retinopathy of prematurity. Retina 24:237–241
7. Hirose T, Katsumi O, Mehta MC, Schepens CL (1993) Vision in stage 5 retinopathy of prematurity after retinal reattachment by open-sky vitrectomy. Arch Ophthalmol 111:345–349
8. Hubbard GB, Cherwick H, Burian G (2004) Lens-sparing vitrectomy for stage 4 retinopathy of prematurity. Ophthalmology 111:2274–2277
9. Maguire AM, Trese MT (1992) Lens-sparing vitreoretinal surgery in infants. Arch Ophthalmol 110:284–286
10. Mandriota SJ, Menoud PA, Pepper MS (1996) Transforming growth factor beta 1 down-regulates vascular endothelial growth factor receptor 2/flk-1 expression in vascular endothelial cells. J Biol Chem 271:11500–11505
11. Multicenter trial of cryotherapy for retinopathy of prematurity (1990) One-year outcome – structure and function. Cryotherapy for Retinopathy of Prematurity Cooperative Group (1990) Arch Ophthalmol 108:1408–1416
12. Palmer EA, Flynn JT, Hardy RJ, Phelps DL, Phillips CL, Schaffer DB, Tung B (1991) The cryotherapy for retinopathy of prematurity cooperative group: Incidence and early course of retinopathy of prematurity. Ophthalmology 98:1624–1640

13. Prenner JL, Capone A, Trese MT (2004) Visual outcomes after lens-sparing vitrectomy for stage 4A retinopathy of prematurity. Ophthalmology 111:2271–2273
14. Saunders RA, Donahue ML, Christmann LM, Pakalnis AV, Tung B, Hardy RJ, Phelps DL (1997) Racial variation in retinopathy of prematurity. The Cryotherapy for Retinopathy of Prematurity Cooperative Group. Arch Ophthalmol 115:604–608
15. Tasman W, Borrone RN, Bolling J (1987) Open sky vitrectomy for total retinal detachment in retinopathy of prematurity. Ophthalmology 94:449–452
16. Trese MT (1984) Surgical results of Stage V retrolental fibroplasias and timing of surgical repair. Ophthalmology 91:461–466
17. Trese MT (1987) Surgical therapy for stage V retinopathy of prematurity. A two-step approach. Graefes Arch Clin Exp Ophthalmol 225:266–268
18. Trese MT (1994) Scleral buckling for retinopathy of prematurity. Ophthalmology 101:23–26
19. Trese MT, Droste PJ (1998) Long-term postoperative results of a consecutive series of stages 4 and 5 retinopathy of prematurity. Ophthalmology 105:992–997
20. Zhao S, Overbeek PA (2001) Elevated TGF beta signaling inhibits ocular vascular development. Dev Biol 237:45–53

21 Vascular Occlusive Disease

21.1 Plasma Proteins – Possible Risk Factors for Retinal Vascular Occlusive Disease

E. TOURVILLE, A.P. SCHACHAT

21 III

Core Messages

- The initial history should focus on a careful review of systems to detect associated vasculitic/uveitic disease
- Middle age and older adults presenting with ocular vascular occlusive disease should have a referral to primary care physicians for evaluation of their blood pressure (acute measurement can be performed in the clinic), blood sugar and probably an assessment of serum lipids and cholesterol. Periodic visits to a family practitioner are appropriate for patients with retinal vascular disease, and patients who do not see a primary care physician should consider doing so
- In arterial ocular occlusive disease, the carotid and cardiac sources are most likely. Sedimentation rate measurement is usually appropriate
- Although no level one evidence exists, when these common risk factors have been eliminated in young patients, Hcy, APA, FVL and perhaps ATIII, protein C and S may be evaluated, especially in cases of bilateral disease, positive family history or previous thrombotic disorders

- If deficiency in one of the above parameters is identified, the management of the systemic components of the disease should be addressed by referral to a specialist familiar with this type of disorder
- The finding of an anomaly in one of the coagulation proteins could have an impact on perioperative management of thrombosis prophylaxis and birth control pill management, and lead to greater consideration being given to quit smoking
- Although no prospective randomized trial has proven the efficacy of oral folates for preventing any kind of systemic or ocular vascular event, some authorities recommend 400 μm of oral folates daily for patients with elevated plasma Hcy and low folate levels. We regard this as a CII level recommendation – and as such would not criticize doctors who do not screen for hyperhomocysteinemia or advise their patients to supplement with folates
- The ophthalmologist plays an important role in the follow-up of the eye disease itself and in managing further complications such as glaucoma, macular edema and neovascularization

21.1.1 Introduction

A complicated well-regulated balance exists between the thrombosis and fibrinolysis systems. This chapter will cover the two main categories of retinal vascular occlusive disease (RVOD): central (CVO) or branch retinal vein occlusion (BVO) and central (CAO) or branch retinal artery occlusion (BAO). This chapter will not address the overall diagnostic evaluation of patients with retinal vascular disease. Patients with CAO/BAO should be evaluated by their medical doctor for common underlying causes such as ipsilateral carotid disease and heart disease. Patients with bilateral, simultaneous CVO/BVO of course should be evaluated for common causes of hypercoagulable states such as Waldenström's or

multiple myeloma. Basic evaluation should include a medical review of systems, testing for high blood pressure and diabetes for all patients, and sedimentation rate, carotid Doppler and cardiac sonogram should always be obtained for retinal arterial disease (RAO). After a basic evaluation, many patients with RAO or retinal venous disease (RVO) are said to have idiopathic conditions. In recent years, there have been a large number of papers citing possible associations of RAO and RVO with abnormalities of plasma proteins. Are these associations real? Which should be considered and looked for in which patients? It will be some years before we can offer clear guidance on this subject, and there is a strong need for large prospective studies with contemporaneous well-matched control groups. Until these studies are available, in this chapter we summarize the pertinent literature and offer some thoughts on whether and how patients with RAO and RVO who do not have obvious causes after a "basic" evaluation should be worked up.

Many patients with RAO have an underlying cause that is known or found after a basic medical evaluation. In contrast, although retinal venous occlusive disease is common, most patients are told their condition is idiopathic. This is neither satisfying to the patient nor to the doctor. Patients want to know "Why did this happen?" and doctors want to find causes in case there are modifiable risk factors that if altered, would reduce the likelihood of a second similar event. Many consultants order a wide range of medical tests, although the classic teaching has been that for patients with BVO and CVO, unless there are simultaneous bilateral events, the only underlying medical diseases that are more common are diabetes (less true for BVO), hypertension, hypercholesterolemia, smoking, glaucoma, and atherosclerosis [15, 60–62]. Throughout this chapter those will commonly be referred to as *major risk factors*. Since patients who see their primary care practitioner will be evaluated for diabetes, hypercholesterolemia, atherosclerosis and hypertension even in the absence of venous occlusive disease, it has been our practice simply to confirm that patients do periodically see their doctor and no special testing has been advised. The ophthalmologist will consider underlying glaucoma, and both the ophthalmologist and the family physician should counsel against smoking.

In recent years, there have been a large number of papers written about possible underlying hematological risk factors which might explain why a particular patient had their retinal vascular event – the cause might not be idiopathic. Many of the papers that point out or claim an association have substantial opportunities for improvement and their conclusions may not be valid. In general, sample sizes are small, control groups are absent or not as comparable as they ought to be, the studies are retrospective, or there are other possible causes of bias or confounding that complicate the analyses and interpretations. To cite just one example of a paper that may not be correctly interpreted, consider the work of Marcucci and colleagues who write on "Thrombophilic risk factors in patients with central retinal vein occlusion [45]." They compared 100 cases and controls and conclude there is a potential role of hemostatic risk factors in the pathophysiology of CVO. After considering the article, most readers might agree. However, the methods section states "we investigated 100 consecutive, unselected patients … referred to our Thrombosis Center." The authors did not study patients with CRVO; they studied patients

with CRVO referred to thrombosis experts. This will be a select group of patients and will not likely generalize to the overall population of patients with CRVO seen by ophthalmologists.

In this chapter, we will review the literature on plasma proteins and retinal vascular disease. Levels or abnormalities of homocysteine (Hcy), methylenetetrahydrofolate reductase (MTHFR), activated protein C resistance (aPCR), antithrombin (AT) III deficiencies, protein C and S deficiencies, antiphospholipid antibodies (APA), factor II G20210A polymorphisms, lipoprotein A, factor V Leyden (FVL), factor VIII, factor XII deficiencies and von Willebrand factor will be reviewed and, where possible, we will offer guidance on which patients might be evaluated for what abnormalities and why.

We believe that the associations discussed generally remain uncertain and our comments should be viewed as guidelines and there remain patients with retinal vascular disease for whom no workup at all would be reasonable and correct. Some of the putative disease states that may increase the risk of retinal vascular disease can be managed relatively easily with a favorable risk/benefit ratio. Others may require lifelong anticoagulation with the likelihood of a severe adverse event during a patient's lifetime being a near certainty. We will offer a very brief introduction to the systemic management of these conditions since it will be necessary to weigh risks and benefits before making management decisions. In general, the medical knowledge required to counsel patients about whether to search for risk factors and if such factors are found, if they should be treated, will be beyond the expertise of most ophthalmologists. It is our prejudice that the ophthalmologist should introduce the issues to the patient, help the patient decide if further consultation with an expert is appropriate, and as needed, follow the patient along for the ophthalmic issues such as the diagnosis and management of macular edema, new vessels or glaucoma while the consultant knowledgeable in coagulation issues manages possibly associated systemic disease.

In this chapter, the relevant studies will be evaluated regarding the importance of their recommendations to clinical outcomes and the overall strength of evidence supporting these recommendations using the rating scheme of Minckler. This scheme is used as part of the manuscript submission guidelines for *Ophthalmology*, the journal of the American Academy of Ophthalmology, and is summarized in Tables 21.1.1 and 21.1.2 [31]. The system uses a "I, II, III" grading scale for the strength of the evidence and an "A, B, C" rating for the level of importance with respect to the clinical outcome.

21 III

Table 21.1.1. Grading scale for evidence based medicine [32]

Level of evidence	Description	Type of studies
I	Indicates strong evidence in support of the statement	Well-done randomized controlled clinical trials designed to address the issue in question, especially regarding the efficacy of treatment or the superiority of one treatment over another
		Well-done meta-analyses (retrospective reviews of previously published randomized controlled trials) may also constitute level "I" supporting evidence
II	Indicates that there is substantial evidence in support of the statement but the evidence lacks some qualities, thereby preventing its justifying the statement without qualification	Nonrandomized comparative trials involving sufficient subjects to demonstrate statistically significant differences between study and control groups might provide strong evidence for the efficacy of a therapy
		Cohort studies and case-control studies might provide strong evidence for or against therapy in terms of longitudinal data about disease natural history, outcome of therapy, adverse events, or specific anatomical or functional outcomes
		Well-done cross-section studies might provide strong evidence for the importance of the clinical problem
		Well-done systematic literature reviews or meta-analyses might also provide moderately strong evidence for or against a test or therapy
		Well-done randomized controlled trial dealing with the issue of interest might have been performed using too select a population and may not be clearly applicable to a broader population of interest, or it might have produced only marginally statistically significant differences between control and experimental groups
		Large consecutive case series might also fit into this category if it compares outcome only to a historical control group from the same clinical setting
III	Indicates a weak body of evidence insufficient to provide support for or against the efficacy of a test or therapy	Would generally apply to panel consensus or individual opinions, small non-comparative case series, and individual case reports
		Non-comparative studies (without controls), cohort studies with variable follow-up across the patient population studied, retrospective chart reviews with missing data, or even randomized controlled trials evaluating highly subjective outcome data

Table 21.1.2. Level of importance of recommendation to clinical outcome [32]

Level	Importance
A	Crucial to good outcome
B	Moderately important
C	Maybe, but cannot be definitely related to clinical outcome

21.1.2 Homocysteine Metabolism and the Role of MTHFR

21.1.2.1 Homocysteine

Homocysteine (Hcy) is an amino acid involved in two metabolic pathways as depicted in Fig. 21.1.1: remethylation and demethylation. Folates and vitamins B$_6$ and B$_{12}$ are required cofactors and relate

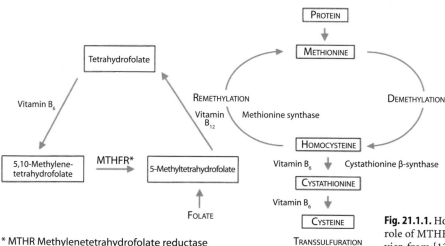

* MTHR Methylenetetrahydrofolate reductase

Fig. 21.1.1. Homocysteine metabolism and the role of MTHFR. (With permission from Elsevier, from [13])

inversely to Hcy plasma concentration. Methylenetetrahydrofolate reductase (MTHFR), a pivotal enzyme involved in Hcy metabolism, is discussed in the following section. The reader is referred elsewhere for more detailed biochemistry discussions [13, 54].

The biological explanation of how Hcy is implicated in atherogenesis is not well understood. A growing body of experimental evidence points toward endothelial cell damage either by direct exposure to high levels (1–10 mmol/L) of Hcy or by generation of reactive oxygen species. This damage leads to endothelial dysfunction creating a prothrombotic state by promoting platelet aggregation or altering the coagulation cascade [13]. To give the reader a reference frame, the generally accepted normal plasma Hcy level is below about 9–12 μmol/L; there is some variation on "normal" levels across laboratories. Hyperhomocysteinemia is divided in the literature into mild (9–15 μmol/L), moderate (15–20 μmol/L) or severe (over 20 μmol/L). As the reader delves into specific studies, different levels are used to define hyperhomocysteinemia, which makes it hard to draw uniform conclusions.

In congenital homocysteinuria, Hcy plasma levels can reach as high as 400 μmol/L. The early observation of thrombotic events and arterial lesions in these young patients led to a search for its association with various kinds of vascular occlusions. Although still debated, a large amount of data has been gathered, showing, more often than not, that even mild hyperhomocysteinemia is independently associated with heart, brain and peripheral atherosclerosis. It is noteworthy that these associations have been made through epidemiological studies and causality still remains to be demonstrated [54].

Multiple studies have established that there are higher Hcy plasma levels in male and elderly patients [54]. In the Framingham Offspring cohort, people over 65 years of age had a 23% higher Hcy plasma level compared to those younger than 45 [32]. Acquired factors can increase Hcy plasma concentration: elevated serum creatinine; low folate, vitamin B_6 or B_{12} intake; diabetes; anti-hypertensive medication; alcohol; caffeine; smoking; psoriasis; methotrexate; tricyclic antidepressant; dilantin; carbamazepine; fibrates; metformin; cyclosporin A; and trimethoprim [13, 20, 32].

21.1.2.2 Methylenetetrahydrofolate Reductase Gene 677 CT Polymorphism: the Thermolabile Form (TT MTHFR)

Mutations that result in severely reduced MTHFR activity and hyperhomocysteinemia are rare. A succession of demonstrations led to the current association of MTHFR with thrombotic events and hyper-

homocysteinemia. At first, a MTHFR with 50% specific activity in vitro and thermolability was associated with moderate elevation of Hcy and low folate levels. Then MTHRF thermolability was found to be inherited as a recessive trait in about 5% of the population compared to 17% of individuals with coronary artery disease. Subsequently, the thermolability was explained by identifying a single DNA point mutation at a polymorphic site (677 C to T transmission). This mutation causes a valine to take the place of an alanine in the MTHFR protein backbone.

The impact of thermolabile MTHFR (TT MTHFR) variant on plasma Hcy levels is unclear. A large proportion of studied populations with TT MTHFR have normal plasma Hcy concentrations. In those same samples, when people did have hyperhomocysteinemia, it was mild and certainly did not correspond with the 50% in vitro decrease in enzyme activity previously mentioned. This lack of correlation points toward other factors regulating the enzyme activity in vivo.

Clear demonstration has now been made that the hyperhomocysteinemia in patients homozygous for the 677 CT mutation related to their low folate plasma levels and could be reversed by folate supplementation. By the same token, only those with low folate levels harbored Hcy plasma elevations. This possible explanation, that the phenotypic expression of the MTHFR genotypes is dependent on the folate availability, suggests a benefit of higher folate intake of patients with the TT MTHFR.

21.1.2.3 Literature Evidence on Homocysteine

We have not identified level one evidence, that is, evidence from randomized trials that reliably confirm a relationship between elevated Hcy and retinal vascular occlusions. Table 21.1.3 describes the quality and the findings of various studies on this topic.

Cahill et al. [13] in a meta-analysis of pooled data from one prospective and 9 retrospective case-control studies compared 614 patients with any type of venous occlusive disease and 762 controls to find a significant combined difference between the two groups in Hcy plasma concentrations. A statistically significant elevation in plasma Hcy levels was found in seven of those studies comparing 465 central vein occlusions to 658 controls, in three addressing 129 branch vein occlusions with 256 controls and similarly when pooling 154 retinal artery occlusions and 358 controls. In the only prospective study [46] included in Cahill's review, they state there was no statistical significance between their subgroups age, sex and *major risk factors*. When looking closer, their pooled CVO/BRVO group had a mean age of 65.6 years, their retinal artery occlusion group was

Table 21.1.3. Studies assessing homocysteine's possible relationship with RVOD

Authors	Number pts./ctrls.	Age	% abn. pts./ctrls.	P value	Type of study	Vascular event***	Level of evidence	Asso- ciation?
M.T. Cahill	614/762	Any	N/A	<0.001	Meta-analysis	CVO/BVO RAO	II	Y
L. Di Crecchio[+]	31/31	<50	16.1/12.9	0.998	Prospective**	CVO	II	N
L. Di Crecchio[+]	31/31	<50	16.1/6.5	0.428	Prospective*	CVO	II	Y
R. Marcucci	100/100	Any	38/5	<0.005	Prospective*	CVO	III	Y
J.M. Lahey	42/59	<56	9.5/0	0.044	Prospective**	CVO	II	N
S. Boyd [11]	63/63	Any	12.4/11.6 µmol/l	0.20	Prospective*	CVO	II	N
Backhouse et al. [5]	16/	Any	6.25	NA	Retrospective*	CVO13/BVO3	III	N
C.J. Glueck	17/40	Any	Other study	NS	Prospective	CVO	III	N
Adamczuk	37/144	Any	27/5.5	<0.001	Prospective**	CVO	III	Y
L. Hansen	54/0	<70	35.2/0	NA	Retrospective	CVO23 BVO30	III	Y

[+] Same study but different control groups
Retrospective or prospective: not matched
* Prospective study matched for sex and age
** Prospective study matched for sex, age, hypertension, and hypercholesterolemia
*** Numbers following each specific RVOD represent the number of patients in each category of event

69.8 and those were compared to a 51.5-year-old control group. They did though control very well for other *major risk factors*. This is an example taken from the multiple studies used in Cahill's meta-analysis to explain why, although encompassing a large number of patients; it should be regarded as level II evidence.

In this same meta-analysis, no association was found between vitamin B_{12} and ocular vascular events but serum folate levels tended to be lower in 287 cases compared to the controls. Furthermore, the data compilation from 11 studies regarding thermolabile (TT genotype) MTHFR led to the impression Hcy levels were more likely the true risk factor than the presence of the TT genotype. Similarly, five of the studies [13, 46, 64–66] included in this meta-analysis identified Hcy as an independent risk factor for retinal vascular occlusive disease.

In a level II study, Lahey et al. evaluated 55 consecutive patients with CRVO less than 56 years old without any of the *major risk factors* and no personal or familial history of thrombosis. Four of those (9.5%) had a plasma Hcy level >12 µmol/L compared to none of their 59 healthy age-matched controls. They concluded that whether there was an association between elevated Hcy plasma levels and CVO needed further investigation.

In the level III study of Adamczuk, patients and controls were evaluated for major risk factor but hypertension and hypercholesterolemia were over five times more prevalent in their patient group. However, this study may not generalize well. Although "37 consecutive patients with CVO" were evaluated, we later read "patients were referred to our thrombosis center for ... thrombophilia screening. As discussed in the "Introduction," we need to

know better who was referred and who was not referred to eliminate the referral bias.

In 2004, Di Crecchio et al. [20] published a well-designed prospective case-control level II study on 31 patients younger than 50 and concluded that Hcy was not an independent risk factor for CVO. The main point of interest of their study was that the statistically significant difference of Hcy plasma concentration between CVO patients and age-sex matched controls (10.60 vs 9.34 µmol/L, $p = 0.003$) disappeared when the same patients were compared to controls also matched for body mass index, diabetes, hypertension and cholesterolemia (10.60 vs 10.39 µmol/L, $p = 0.674$). eaders should remember those levels are almost in the normal range of Hcy plasma concentration. Di Crecchio concluded that Hcy should be considered as a marker of atherosclerosis and the consequence of those other well-established risk factors.

Marcucci et al. investigated 100 consecutive unselected Italian patients referred to their Thrombosis Center, 6–24 months after they had a CVO. In this retrospective level III study discussed in the "Introduction," they did find elevated plasma Hcy as an independent risk factor for CVO after conducting a multivariate analysis. Although the cases and the 100 controls appeared well matched for age, sex, family history of coronary artery disease and smoking, they were nearly completely mismatched with respect to other important cardiovascular variables such as hypertension (48% of cases vs 6% of controls), hypercholesterolemia (37% of cases vs 8% of controls) and diabetes (12% vs 0%). These cardiovascular factors may relate to the prevalence of relevant plasma proteins such as lipoprotein (a) and Hcy and those imbalances call into question the certainty of some of their observations.

In his study on 54 RVOs, Hansen et al. [29] were able substantially to decrease plasma Hcy in 100 % of 14 hyperhomocysteinemic patients by prescribing 5 mg daily of folic acid over at least 2 weeks. Cahill et al. concluded estimation of plasma Hcy and folate levels should be considered for ocular vasculopathic patients. According to the actual findings in the literature discussed above, they hypothesized that reduction of plasma Hcy level by folate supplementation could decrease the occurrence of the disease in the fellow eye or other systemic vascular events. Although no prospective randomized trial has proven the efficacy of oral folates at preventing any kind of systemic or ocular vascular event, Cahill et al. recommended 400 μm of oral folates daily for patients with elevated plasma Hcy and low folate levels. We regard this as a CII level recommendation – and as such would not criticize doctors who do not screen for hyperhomocysteinemia or advise their patients to supplement with folates. Recent experience suggests that vitamin supplementation may not always be "benign." Vitamin A supplementation, for example, has been associated with increased risk of lung cancer in smokers [47, 57].

We did not find any study looking exclusively at BVO. All studies grouped them with CVO under "venous retinal disease." We have identified four studies looking more specifically for an association between BVO and Hcy levels. One from Backhouse included only three BVO patients and had no control group. The three other ones were included in Cahill's meta-analysis discussed previously. His review included all the major existing studies on Hcy and MTHFR and their association with vascular retinal diseases. He consistently found an association between hyperhomocysteinemia whether he was looking at CVO, BVO or retinal arterial occlusion. No distinction could be made in the results of each study looking at multiple types of vascular retinal occlusions, whether arterial or venous.

Despite the existence of studies pointing to mild hyperhomocysteinemia as a risk factor for systemic occlusive vascular disease, thrombosis, and stroke, the question as to whether Hcy is responsible for these events or if it is just an atherosclerosis marker remains to be answered. Most of the authors who have studied this topic have lumped together CVO, BVO, CAO and BAO and since the etiologies and pathogenesis likely differ, additional studies splitting out the various groups will be needed. The relationship, if any, with RVO remains unclear.

21.1.2.4 Literature Evidence Concerning MTHFR

In the meta-analysis by Cahill et al. discussed above in the Hcy section, no statistically significant difference was found in the proportion of the 690 patients with CVO or BVO harboring the TT MTHFR and comparing them to 2,754 controls. Nine case-control studies were reviewed and four of them did not have age-matched controls. Furthermore, all studies contained in Table 21.1.4 are also in agreement with the conclusion that MTHFR is not an independent risk factor for CVO.

Cahill's meta-analysis of the three major studies looking at CAO and BAO found no significant difference as to the prevalence of the TT MTHFR in 152 cases compared to 435 controls. Based on three level II and two level III studies evaluating 838 patients and 3,226 controls with arterial and venous retinal occlusions, the authors conclude there is a CII level of evidence and do not recommend screening for MTHFR 677CT mutation.

Table 21.1.4. Studies assessing MTHFR's possible relationship with RVOD

Authors	Number pts./ctrls.	Age	% abn. pts./ctrls.	P value	Type of study	Vascular event***	Level of evidence	Asso-ciation?
M.T. Cahill	690/2,754	Any	15.5/11.0	NA	Meta-analysis	CVO/BVO/RAO	II	N
L. Di Crecchio[+]	31/31	<50	12.9/16.1	0.998	Prospective**	CVO	II	N
L. Di Crecchio[+]	31/31	<50	12.9/12.9	0.705	Prospective*	CVO	II	N
S. Boyd	63/63	Any	35/36	0.20	Prospective*	CVO	II	N
Y.P. Adamczuk	37/144	Any	10.8/13.2	NS	Prospective**	CVO	III	N
C.J. Glueck	17/234	Any	71/49	0.22	Prospective	CVO	III	N

[+]   Same study but different control groups
Retrospective or prospective: not matched
*   Prospective study matched for sex and age
**   Prospective study matched for sex, age, hypertension, and hypercholesterolemia
***   Numbers following each specific RVOD represent the number of patients in each category of event

21 III

21.1.3 Antiphospholipid Antibodies: Lupus Anticoagulant and Anticardiolipin Antibodies

Antiphospholipid antibodies (APAs) are autoimmune immunoglobulins recognizing phospholipids in in vitro laboratory test studies that can activate the coagulation cascade and cause thrombosis. The two immunoglobulins described so far are the lupus anticoagulant (LA) and the anticardiolipin antibody (ACA). LA was originally named after plasma from lupus patients failed coagulating appropriately using in vitro phospholipid coagulation assays. For unknown reasons, activated partial thromboplastin (aPTT) time is prolonged in the presence of LA. It was later found the majority of people harboring LA did not have lupus. ACA got its name from the original radioimmunoassay test using cardiolipin as the antigen.

From 0% to 9% of the general population of healthy individuals harbor APA without any noticeable consequence [17]. The incidence of APA increases in age and may be seen in up to 14% of the healthy elderly population [62]. The ACA status is determined by identification of serum IgG or IgM. As a reference standpoint for the reader, normal values for ACA are below 10 glycopeptidolipids (GPL). Low titers are between 10 and 20 GPL and frank elevation is above 20. Usually used methods to assess the presence of LA were prolonged clotting time (mainly aPTT, Russell Viper Venom, Kaolin). Nowadays, direct LA antibody detection is possible.

It is also worth mentioning that LA tests are unreliable in individuals on anticoagulant therapy. For ACA and LA tests, sensitivity and standard values vary from one laboratory to the next and cut-off levels are relative to each study design. For this reason, sensitivity and cut-off values can explain some variability in the results from one study to the next. Furthermore, variability can be accounted for by the fact that a positive result can become negative over time and vice versa.

Primary APA syndrome is defined by the presence of ACA or LA on two occasions 6 weeks apart, associated with a triad of multiple recurrent arterial or venous thromboembolic events, thrombocytopenia and multiple fetal wastages. Livedo reticularis, renal insufficiency, pseudotumor cerebri, transverse myelitis and cardiomyopathy are also recognized features. The literature contains several case reports of retinal vascular occlusion associated with this syndrome or with established connective disease such as systemic lupus erythematosus (SLE). There seems to be general agreement to conduct blood screening for APA in patients with a positive personal history of clotting disease or systemic illness or with a positive familial history of thrombosis.

APAs can be secondary to connective tissue disease, malignancies, infection, pregnancy and AIDS. They can also be induced by medications such as steroids, phenytoin, hydralazine, quinidine, procainamide, chlorpromazine, interferon, cocaine and oral contraceptives.

Several studies have investigated the incidence of APA in patients with retinal vascular occlusive disease and found very disparate results ranging from 0% to 64%. Interesting enough though, all the studies detailed in Table 21.1.5 had a 0–8% prevalence rate for the controls.

No level I study was found in the literature. The only level II study was the one from Lahey et al. discussed earlier in the Hcy section. They found three patients positive for LA and three for ACA out of their 56 CVOs, giving them a combined prevalence of 11% of APA. None of their controls was positive. They concluded the search for APA in people < 56 years old could be helpful in counseling patients for thrombosis prophylaxis in high-risk situations.

Adamczuck's study (discussed earlier) found a statistically significant higher prevalence of antiphospholipid antibodies in CVO patients compared to their controls (5/37 compared to 3/144, $p < 0.01$) [3].

In Cobo-Soriano's study, only ACAs were found elevated and everyone in the study was negative for LA. Nine out of 40 patients (22.5%) were ACA positive. Three of those had clinical APA syndrome. This article has been the subject of criticism in the *American Journal of Ophthalmology* by Williams and Sarrafizadeh, where they disagree with their conclusion to support the routine screening of APA on a routine basis [67].

Glueck et al. (discussed later in the FVL section) found the opposite. No one was positive for ACA, but 6 of 14 patients (43%) had a Russell venom clotting time of over 39 s, therefore being positive for LA, compared to 1 of 30 controls.

We may be mistaken, but on close reading of the two papers by Abu-El-Arsrar et al., it appears obvious that the second larger paper included some or all patients from their first publication, casing some degree of duplicate data presentation. Their larger study harbored evident potential biases as 65% of their patients had *major risk factors* compared to none of the controls. Also, one patient with Behçet disease, one diagnosed with systemic lupus erythematosus (SLE) and four presenting with bilateral CVO were included in the cases group. In this same study, subgroup analysis showed no significant difference in high prevalence of APA, protein C or S and ATIII whether you were under or over 45 years of age. This may demonstrate just a general tendency of the lab results to turn out positive as they had surpris-

III 21

Table 21.1.5. Studies assessing APA's possible relationship with RVOD

Author	Number pts./ctrls.	Age	Test	% abn. pts./ctrls.	*P* value	Type of study	Vascular event***	Level of evidence	Asso-ciation?
J.M. Lahey [38]	55/28–40	<56	LA/ACA	11/0	0.00001	Prospective**	CVO	II	Y
J. Larsson [40]	37/0	<50	LA/ACA	0/	NA	Retrospective	CVO	III	N
Abu El-Asrar [1]	17/60	<45	LA/ACA	46/8	0.0002	Prospective* Not age-sex	CVO15 BVO2	III	Y
Abu El-Asrar [2]	57/74	Any	ACA	26.3/4	0.0002	Prospective*	CVO35 BVO22	III III	Y Y
Y.P. Adamczuk	37/144	Any	LA/ACA	13.5/2.1	<0.01	Prospective**	CVO	III	Y
Z.F. Bashshur [8]	24/20	Any	LA ACA	0/0 43/0	NA <0.0001	Prospective?	CVO13 BVO11	III	Y
J. Carbone	62/45/49	Any	LA/ACA	24/9/8	<0.05	Prospective	RVO47 RAO15	III	Y
Cobo-Soriano [17]	40/40	Any	LA/ACA	22.5/5	0.04	Prospective**	CVO8 CAO8 BVO18	III	Y
C.J. Glueck [24]	14/30	Any	LA/ACA	43/3 No	0.002	Retrospective	CVO	III	Y N
A. Glacet-Bernard [23]	75/38	Any	LA/ACA	5.3/2.6 (0 for RAO)	NA	Prospective**	CVO44 BVO29 RAO2	III	N
R. Marcucci	100/100	Any	LA/ACA	2/0 10/5	0.49 0.18	Retrospective*	CVO	III	N
K. Greiner [27]	76/0	Any	LA	0/	NA	Prospective?	RVO CAO BVO	III	N
K. Greiner [26]	116/0	Any	LA	0/	NA	Prospective?	CVO48 BVO33 CAO22 BAO13	III	N
L. Hansen	54/0	<70	APA	1.8/0	NA	Retrospective	CVO23 BVO30	III	N
C.D. Kalogeropoulos [35]	36/79	Any	ACA	?	NS	Prospective*	CVO13 BVO23	III	?N
C.D. Kalogeropoulos	21/79	Any	ACA			Prospective*	CAO6 BAO13	III III	N N
N. Giordano [22]	30/30	Any	ACA	6/3	NS	Retrospective**	CVO18 BVO10 CAO2	III	N
J.A. Scott [53]	45/0	<56	ACA	64	NA	Retrospective	CVO24 BVO21	III III	N N
O. Salomon [51]	21/243	Any	LA	14.3/6.6	0.19	Retrospective	CAO/BAO	III	N
S. Kadayifçilar	54/19	Any	ACA	0/0	NA	Prospective*	CVO22BVO32	III	N

Retrospective or prospective: not matched

* Prospective study matched for sex and age

** Prospective study matched for sex, age, hypertension, and hypercholesterolemia

*** Numbers following each specific RVOD represent the number of patients in each category of event

ingly high rates of thrombophilic abnormalities in both cohorts with every plasma protein they looked for.

Glacet-Bernard et al. and Carbone et al. also reported the same APA prevalence whether the patients were younger or not than 50 years old. This raises the question as to whether the elevation in APA could be secondary to the thrombosis instead of being the culprit. Carbone et al. prospectively studied 68 patients with all types of ocular vascular occlusion with no mention made about age composition of the two control groups or sex composition of the three groups. They included six anterior ischemic optic neuropathies but those will not be discussed here since it is beyond the scope of this text. The rest of the patient group was composed of 46 RVOs, 14 RAOs and 2 combined RVO-RAOs. Seven patients had retinal vasculitis, ten had thrombocytopenia, five had repeated abortions and four had livedo reticularis. Those patients were compared to a group of 45 patients with ocular inflammatory diseases and another group of 49 "randomly selected healthy volunteers." Of the 16 patients who tested positive for APA, four did so on later retesting. Two developed a stroke and four were amongst the seven "coincidental" retinal vasculitis patients. Of the three other vasculitis cases, one was ACA positive and two were ANA positive. Additionally, two of the 16 APA positives developed lupus-like disease. This cohort most likely does not represent our usual retinal vasculopathic patient.

Most of the English language papers on APA are detailed in Table 21.1.5. Lahey's level of evidence II study did find an association between CVO and APA. When we reject data duplication, six level III studies did find an association between RVO and LA or ACA as nine did not. Many of them are scrutinized throughout this chapter.

We could not find any study looking exclusively at BVO. All studies grouped them with CVO under "venous retinal disease." No distinction could be made in the results of each study looking at multiple types of vascular retinal occlusions, whether arterial or venous.

Most of the authors who have studied this topic have lumped together CVO, BVO, CAO and BAO and since the etiologies and pathogenesis likely differ, additional studies splitting out the various groups will be needed.

Based on the actual literature, mass screening for APA is definitively not recommended at this point for patients experiencing RVOD. Obvious evidence has been offered for the association of APS or SLE with RVOD [10], and APA should be looked for in patients presenting with a familial or personal history suggestive of such. Special conditions were the detection of APA, which could be beneficial for counseling the patient or altering the course of the treatment and will be discussed in greater detail at the end of this chapter.

21.1.4 Disorders of Coagulation and Anticoagulation

The human organism maintains a fine equilibrium between factors allowing the blood to circulate in its fluid form and others preventing us from bleeding to death should an accident occur. This complex balance is maintained through pro-coagulating factors closely monitored by anticoagulant proteins as described in Fig. 21.1.2. Inborn or acquired errors can happen in this well tuned cascade leading to prothrombotic states. We provide an overview of the main possible defects of the natural anticoagulants in the following section.

21.1.4.1 Factor V Leiden and Activated Protein C Resistance

Protein C is a vitamin K dependent circulating zymogen. Once activated by the complex formed by thrombin and thrombomodulin, it becomes a natural anticoagulant which degrades activated factor V and VIII. Activated protein C resistance (aPCR) is seen when factor V or V III are resistant to degradation by aPC and keep their thrombotic properties for an extended period.

Causes of acquired aPCR include pregnancy, birth control pill use and the presence of lupus anticoagulant [65]. There is general agreement in the literature that 94% of aPCR is caused by factor V Leiden (FVL) (named after the Dutch city where it was discovered). It is also called factor V R506Q and G1691A mutation of factor five for the following reasons. FVL comes from a single point mutation at nucleotide location 1691 when adenine is substituted for a guanine. The new codon created by this G1691A mutation causes the replacement of an arginine (R) by a glutamine (Q) as the 506th amino acid of the protein backbone. This is where aPC normally cleaves factor five to inactivate it. The mutant factor V R506Q is then protected from degradation by aPC. It is inherited as an autosomal dominant trait. For general understanding, a homozygous mutation usually leads to a more severe prothrombotic phenotype than a heterozygous one.

Table 21.1.6 summarizes the findings of various researchers about FVL, aPCR and the possible relationship with retinal vascular disease. We find no level I evidence study.

Perhaps the strongest study summarized in Table 21.1.6 is that of Lahey and is described in the

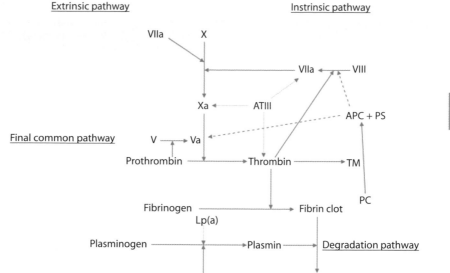

Fig. 21.1.2. The coagulation cascade. Simplified diagram of hemostatic and fibrinolytic cascade. *ATIII* antithrombin III, *PC* protein C, *PS* protein S, *TM* thrombomodulin, *APC* activated protein C, *Lp(a)* lipoprotein (a), – – → inactivating pathway, → inhibitory pathway. (Adapted from [21])

Simplified diagram of hae mostatic and fibrinolytic cascade. A TII = antithrombin III; PC = protein C; PS = protein S; TM = thrombomodulin; APC = activated protein C; Lp(a) = Lipoprotein (a);   – → = inactivating pathway;   → = inhibitory pathway.

Hcy section earlier in this chapter. He found no statistically significant difference in aPCR or FVL prevalence compared to the prevalence in well-matched controls.

The geographical prevalence of the FVL carrier rates has been shown to widely vary from 0 % in African countries, to 6 % in the USA population and up to 15 % in Greek Cypriots [48]. The FVL carrier rate is estimated at 0.6–2.9 % in southern Europe, compared to 3.4–7.9 % in northern European countries [18]. In their German population, which has a 7 % prevalence of FVL in their healthy individuals [52], Greiner et al. investigated 76 patients with any retinal vascular disease who had an *inpatient evaluation* (see next paragraph) for aPCR and other plasma proteins. Eighty-one percent of the cases had at least one, 37 % had 2 and 11 % had three of the *major risk factors*. Sixteen percent of the RVO patients had a previous history of DVT. They found 29 % of their CVO and 19 % of their BVO subjects to have aPCR. One hundred percent of their 17 cases of aPCR were due to the presence of FVL. They found the same anomaly in 19 % of their group control with deep vein thrombosis. They did have one patient with CAO and two with BAO testing positive for aPCR/FVL. This is a very small number of cases and with this type of control group, it is impossible to establish if the results could have been due to chance only.

The same authors later reported a larger study. We may be mistaken, but on close reading of the two papers, it appears that the second larger paper included some or all patients from their first publication, causing a second reporting of some of the same data. For instance, 10 of their 13 CVO patients positive for FVL in their second study were from the first one discussed above. One hundred percent of their aPCR was proven to be secondary to FVL in the second study as well. Unfortunately, in their second study also, the presence of *major risk factors* was not accounted for in the control group but 11 of 13 of their CVOs had at least one. There could easily be important imbalances in features not recorded or not reported and it is for reasons such as this that we consider this work as supplying level III evidence. They did find an association between CVO and aPCR/FVL but no such conclusions could be made with BRVO. The three RAO patients were discussed above with their first paper and the second did not offer additional cases. We do not dismiss their findings but do not offer them as conclusive.

Ciardella et al. conducted a very puzzling study in which they first evaluated the prevalence of aPCR with the old technique of standard partial thromboplastin time (PPT). This technique calculates the ratio of a PPT performed with a fixed amount of aPC added to the plasma tested, divided by a PPT performed only with the plasma. They found 45 % of their 84 RVO patients having a ratio ≤2.1 (considered positive) compared with 9 % of the control group. They then compared this old technique with a more specific and sensitive newer aPCR assay on 40 of their RVO patients. This time they found 58 % of

21 **III**

Table 21.1.6. Studies assessing FVL's and aPCR's possible relationship with RVOD

Author	Number pts./ctrls.	Age	Test	% abn. pts./ctrls.	*P* value	Type of study	Vascular event***	Level of evidence	Asso-ciation?
J.M. Lahey	55/21	<56	aPCR FVL	3.6/0	NA	Prospective**	CVO	II	N
C.J. Glueck	17/233	Any	FVL	18/3	0.02	Retrospective	CVO	III	Y
T. Williamson [68]	35/35	Any	aPCR	12.5/5	0.05	Prospective	CVO	III	Y
R. Albisinni	36/68	Any	FVL	11.1/1.5	<0.05	Retrospective**	CVO/BVO	III	Y
J. Larsson	31/0	19–48 Any	aPCR	36 26	NA	Retrospective	CVO	III	Y
J. Larsson [39]	83/101	>50	aPCR FVL	11/11	NS	Retrospective	CVO	III	N
J. Larsson	37/0	<50	aPCR	19/	NS	Retrospective	CVO	III	N
K. Greiner [26] (76)	35/209 DVT	Any	aPCR FVL	29/19	NA	Prospective	CVO	III	Y
K. Greiner [26] (76)	21/209 DVT	Any	aPCR FVL	19/19	NA	Prospective	BVO	III	N
K. Greiner [26] (76)	13/209 DVT	Any	aPCR FVL	7.7/19	NA	Prospective	CAO	III	NA
K. Greiner [26] (76)	7/209 DVT	Any	aPCR FVL	28/19	NA	Prospective	BAO	III	NA
K. Greiner [26] (116)	48/581 CAD	Any	aPCR FVL	27/7.6	<0.001	Prospective	CVO	III	Y
K. Greiner [26] (116)	33/581 CAD	Any	aPCR FVL	18.1/7.6	NA	Prospective	BVO	III	N
K. Greiner [26] (116)	21/581 CAD	Any	aPCR FVL	4.7/7.6	NA	Prospective	CAO	III	NA
K. Greiner [48](116)	14/581 CAD	Any	aPCR FVL	14.3/7.6	NA	Prospective	BAO	III	NA
A.P. Ciardella [16]	84/70 40/9 30/47	Any	aPCRo aPCRn FVL	45/9 10/11 3/2	0.00001 NS NS	Retrospective**	CVO42 BVO42	III	N
Y.P. Adamczuk	37/144	Any	FVL	0/3	NS	Prospective**	CVO	III	N
R. Marcucci	100/100	Any	aPCR FVL	19/5 12/4	<0.005 0.05	Retrospective*	CVO	III	N+
J.L. Gottlieb [25]	21/0	<50	aPCR FVL	4.7	NA	Retrospective	CVO	III	N
Backhouse	16	Any	FVL	6.25	NA	Retrospective*	CVO13/BVO3	III	N
F.Y. Demirci [19]	50/120	Any	FVL	8/9.2	<0.05	Retrospective	CVO23/BVO25	III	N
P.R. Hodgkins [30]	50/0	Any	FVL	2/	NA	Retrospective*	CVO	III	N
T. Linna [43]	46/142	<50	FVL	4.3/2.1			CVO/BVO		N
Salomon [51]	21/243	Any	FVL	14.3/6.6	0.19	Retrospective	CAO/BAO	III	N
O. Salomon [50]	102/105	Any	FVL	6.9/8.6	0.79	Retrospective	CVO/BVO	III	N
D. Kalayci [34]	52/81	Any		8/7	NS	Retrospective	CVO25 BVO27	III	N
J.A. Scott	45	<56	FVL	0	NA	Retrospective	CVO24 BVO21	III	N
L. Hansen	54/0	<70	FVL	1.8/0	NA	Retrospective	CVO23 BVO30	III	N
T. Linna [42]	46/142	<50	FVL	4.3/2.1	NS	Retrospective	CVO28 BVO18	III	N

DVT: deep vein thrombosis; CAD: coronary heart disease; Prospective: not matched
* Prospective study matched for sex and age
** Prospective study matched for sex, age hypertension, and hypercholesterolemia
*** Numbers following each specific RVOD represent the number of patients in each category of event

them to have aPCR using the old ratio compared with 10 % if the newer assay was used. The same tactic was used with nine controls and three tested ≤ 2.1 with the old technique compared to one with the new one. They concluded the first generation commercial test for aPCR is not a useful screening test. They did not find an association between aPCR/ FVL and RVO.

Several of the other studies reviewed regarding aPCR showed an obvious lack of appropriate control populations. Williamson and coworkers found a 12.5 % prevalence of aPCR using the first generation assay. This was higher than the 5 % of a historical group control from Scotland. They concluded an association between CVO and aPCR but did not, as well as in the following study from Larsson, confirm their aPCR by the almost universal FVL. As Larsson points out himself: "aPCR assays sometimes yield conflicting results when conducted on the same blood sample twice" [16]. Larsson and Williamson were the first ones to find a positive association between aPCR and CVO and are quoted frequently throughout the literature. Let us not forget they only used a functional aPCR assay, which is associated with some preanalytical and technical issues potentially affecting its accuracy [30]. Furthermore, Larrson et al. did not include a control group in their retrospective study and found 8 out of 31 patients (26 %) with aPCR in a population where the normal incidence of aPCR is around 10 – 11 %. All those patients with aPCR were less than 45 years of age, thus explaining the 36 % prevalence they calculated when getting rid of the older patients with CVO in their study. Gottlieb and Hodgkins also did not use a control group. Interestingly enough, Hodgkins who only used patients without *major risk factors* concluded there was no association of CVO with aPCR as Larsson, who did not use a sample free of *major risk factors*, concluded the opposite. In Gottlieb's study, which did not find a relationship, the only patient with FVL had a past history of thrombosis, as did 16 % of Larrson's CVO patients. Larsson et al. in a separate study on patients over 50 years old did not find an association between aPCR and CVO. Those last studies discussed are good examples to explain the lack of consensus in the literature on this topic.

Glueck et al. were one of the first groups to publish their findings and are also often quoted in support of the positive associations for FVL and lupus anticoagulant with CVO. Three of their 17 CVO patients were heterozygous for FVL. Out of their 17 patients, 6 were on antihypertensive medications, two had diabetes and one smoked more than a pack of cigarettes per day. Twenty-four percent of the patients had a past history of deep vein thrombosis, 18 % had previous avascular necrosis and 18 % had bilateral CVOs. On the other hand, the controls were 194 healthy children and 40 healthy adults that came from a separate study. The mean age of the patients was 52 compared to a mean age of 37 years old for the control group. These discrepancies and the small number of patients cast doubt on the significance of the findings.

The study by Albisinni et al. is the last of the five who did find a positive association between RVO and "genetic abnormalities." To do so, they had to pool their prothrombin and factor V mutations together as well as analyzing their CVO and BVO as a whole. Out of 36 patients with RVO, 3 had FVL, 2 had the G20210A mutation of prothrombin and one had both mutations. Of those six patients, three had a familial history of thrombosis. The retrospective design of this study and the small number of people enrolled in it, limits the ability of readers to base decision making on this report.

Salomon [51] et al. found FVL in two BAOs and one CAO out of a group of 21 RAOs but the difference in FVL did not reach statistical significance. As discussed earlier, Greiner found three RAO cases FVL positive and other case reports exist [57]. Again, this leaves us with tenuous evidence of aPCR/FVL being a putative factor for RAO.

Almost all English language studies were reviewed. Without including data duplication, five level III evidence studies found an association between CVO and aPCR and/or FVL. No such conclusions could be made with BRVO. Eighteen level III evidence studies found no such association and they are detailed in Table 21.1.6.

In general, the studies are flawed by their retrospective design, small number of patients and lack of or poor group control. Most of the studies finding an association are mismatching cases and controls or not isolating confounding factors. They are divided concerning their conclusions but most of them, including the better-designed ones, tend to deny any associations between CVO and aPCR or FVL but do not absolutely rule them out.

Since most of the 1 – 7 % of individuals with the mutant allele for FVL remain asymptomatic, other factors have to account for their venous occlusive disease, and determination of aPCR and FVL levels does not appear at this time as being a necessary part of the regular investigation for patients with CRVO or BRVO. Special conditions where the detection of aPCR or FVL could be beneficial for counseling the patient or altering the course of the treatment will be discussed in greater detail at the end of this chapter.

21.1.4.2 Natural Anticoagulant Deficiency: Protein C, S and Antithrombin III

Protein C's function was discussed in the introduction to aPCR and has a normal level of activity of 60–70% up to 140%. Its deficiency has an autosomal recessive inheritance mode. Homozygotes lacking the expression of protein C have activity levels of less than 1% and usually present with a thrombosis episode in their first few days of life. Heterozygotes have an activity level around 50% and usually remain asymptomatic for their first few decades [14]. Protein S is also a vitamin K dependent and it acts as a cofactor accelerating the effect of protein C. ATIII

Table 21.1.7. Studies assessing AT III's possible relationship with RVOD

Authors	Number pts./ctrls.	Age	% abn. pts./ctrls.	P value	Type of study	Vascular event***	Level of evidence	Association?
F. Bandello	40/40	Any	0/0	NA	Prospective**	CVO	II	N
P.R. Hodgkins [30]	50/0	Any	0/	NA	Retrospective*	CVO	III	N
Adamczuck	37/144	Any	0/0	NA	Prospective**	CVO	III	N
J.M. Lahey	55/21	<56	0/0	NA	Prospective**	CRVO	II	N
O. Teleki	45/20	Any	2/0	>0.05	Retrospective*	CVO14 BVO31	III	N
J.L. Gottlieb	21/0	<50	0	NA	Retrospective	CRVO	III	N
Backhouse et al.	16/	Any	0	NA	Retrospective*	CRVO13/BRVO3	III	N
R. Marcucci	100/100	Any	0	NA	Retrospective*	CRVO	III	N
Salomon	21	Any	0	NA	Retrospective	CAO/BAO	III	N
Abu El-Asrar	17/0	<45	5.8/	NA	Prospective* Not age-sex	CVO BVO	III	Y
Abu El-Asrar	54/0	Any	7.4/	NA	Prospective*	CVO/BVO	III	Y
L. Hansen	54/0	<70	0/0	NA	Retrospective	CVO23/BVO30	III	N
B. Bertram	167/0	Any	0.6	NA	Prospective	RVO/RAO	III	N
S. Kadayifçilar	54/19	Any	1.8/0	NA	Prospective*	CVO22/BVO32	III	N
J. Larsson	37/0	<50	2.7/	NA	Retrospective	CVO	III	N

Prospective: not matched
*   Prospective study matched for sex and age
** Prospective study matched for sex, age hypertension, and hypercholesterolemia
*** Numbers following each specific RVOD represent the number of patients in each category of event

Table 21.1.8. Studies assessing protein C's possible relationship with RVOD

Authors	Number pts./ctrls.	Age	% abn. pts./ctrls.	P value	Type of study	Vascular event***	Level of evidence	Association?
F. Bandello	40/40	Any	0/0	NA	Prospective**	CVO	II	N
J.M. Lahey	55/25	<56	0/4	NA	Prospective**	CVO	II	N
Y. Adamczuck	37/144	Any	0/0	NA	Prospective**	CVO	III	N
C.G. Glueck	17/?	Any	0	>0.1	Retrospective	RVO	III	N
O. Teleki	45/20	Any	20/0	<0.05	Retrospective*	CVO14 BVO31	III	Y
J.L. Gottlieb	21/0	<50	0	NA	Retrospective	CVO	III	N
Backhouse et al.	16/??	Any	0	NA	Retrospective*	CVO13/BVO3	III	N
R. Marcucci	100/100	Any	0	NA	Retrospective*	CRVO	III	N
K. Greiner	116/0	Any	0	NA	Prospective?	CVO48/BVO33 CAO21/BAO14	II	N
K. Greiner	76/0	Any	0/	NA	Prospective?	CVO/BVO CAO	II	N
O. Salomon	21	Any	0/	NA	Retrospective	CAO/BAO	III	N
Abu El-Asrar	17/0	<45	23.5/	NA	Prospective* Not age-sex	CVO/BVO	III	Y
Abu El-Asrar	42/0	Any	19/	NA	Prospective*	CVO/BVO	III	Y
L. Hansen	54/0	<70	0/	NA	Retrospective	CVO23/BVO30	III	N
B. Bertram [9]	167/0	Any	1.2/	NA	Prospective	RVO/RAO	III	N
S. Kadayifçilar	54/19	Any	3.7/0	NA	Prospective*	CVO22/BVO32	III	N
J. Larsson	37/0	<50	2.7/	NA	Retrospective	CVO	III	N

Prospective: not matched
*   Prospective study matched for sex and age
** Prospective study matched for sex, age hypertension, and hypercholesterolemia
*** Numbers following each specific RVOD represent the number of patients in each category of event

Table 21.1.9. Studies assessing protein S's possible relationship with RVOD

Author	Number pts./ctrls.	Age	Low protein	% abn. pts./ctrls.	P value	Type of study	Vascular event***	Level of evidence	Association?
F. Bandello	40/40	Any	S	0/0	NA	Prospective**	CVO	II	N
J.M. Lahey	55/21	<56	S	1.8·/0	NA	Prospective**	CVO	II	N
Y. Adamczuck	37/144	Any	S	0/0	NA	Prospective**	CVO	III	N
O. Teleki	45/20	Any	S	4/0	>0.05	Retrospective*	CVO14 BVO31	III	N
J.L. Gottlieb	21/0	<50	S	0	NA	Retrospective	CVO	III	N
Backhouse et al.	16/?	Any	S	0	NA	Retrospective*	CVO13/BVO3	III	N
R. Marcucci	100/100	Any	S	0	NA	Retrospective*	CRVO	III	N
K. Greiner	116/0	Any	S	0	NA	Prospective?	CVO48/BVO33 CAO22/BAO14	III	N
K. Greiner	76/0	Any	S	0/0	NA	Prospective?	CVO/BVO CAO	III	N
Abu El-Asrar	17/0	<45	S	23.5/	NA	Prospective* Not age-sex	CVO/BVO	III	Y
Abu El-Asrar	56/0	Any	S	21.4/	NA	Prospective*	CVO/BVO	III	Y
C.G. Glueck	17/	Any	S	0	>0.1	Retrospective	CVO	III	N
Salomon [57]	21	Any	S	0	NA	Retrospective	CAO/BAO	III	N
L. Hansen	54/0	<70	S	0/	NA	Retrospective	CVO23/BVO30	III	N
B. Bertram	167/0	Any	S	1.2/	NA	Prospective	RVO/RAO	III	N
S. Kadayifçilar	54/19	Any	S	1.8/0	NA	Prospective*	CVO22/BVO32	III	N
J. Larsson	37/0	<50	S	5.4/	NA	Retrospective	CVO	III	Y

Retrospective or prospective: not matched
* Prospective study matched for sex and age
** Prospective study matched for sex, age, hypertension, and hypercholesterolemia
· Bilateral CVO
*** Numbers following each specific RVOD represent the number of patients in each category of event

Table 21.1.10. Prevalence of thrombophilia and risk of thrombotic event in a normal population. (Adapted from [21])

Thrombotic condition (year discovered)	Prevalence	Relative risk
AT III (1965)	0.18%	5.5
Protein C deficiency (1981)	0.20%	6.5
Protein S deficiency (1984)	1.30%	2.4
Hyperhomocysteinemia (1984)	5–10%	2.5
Antiphospholipid syndrome (1985)	5–10%	0–8
FVL (1993)	5%	1–7
High factor VIII level (1995)	10%	4.8
Prothrombin G20210A mutation (1996)	2.3%	2.8

inactivates thrombin and other serine proteases such as factor Xa, IXa, XIIa, aPC and kallikrein.

Multiple studies detailed in Tables 21.1.7–21.1.9 did not find a single case of RVOD with deficit in any one of ATIII, protein C or S. As shown in Table 21.1.10, the prevalence of deficiencies in those anticoagulants is substantially more infrequent than the other type of thrombophilia. One might argue that the studies could have been too small to detect a possible association.

Teleki [56] did find a statistically significant association between protein C and RVO but not protein S or ATIII. Bertram et al. did find a total of 5 natural anticoagulant anomalies in their 167 patients with RVOD. Larsson found 4 of those anomalies out of 37 CVO patients and Kadayifçilar et al. found five

through their 54 patients with RVO. Lahey et al. had one patient with bilateral CVO harboring protein S deficiency. Abu El-Asrar's unusually high rates of abnormal findings have been discussed in the APA section. Other cases have been reported [28, 33]. Although we are not completely discarding the possible association between natural anticoagulant deficiency and RVOD in adults, it has not been convincingly demonstrated. As one might predict from the outcome of the several studies reviewed, extremely rarely would a test for ATIII, protein C and S turn out to be positive and their routine use is not recommended.

21.1.4.3 Prothrombin G20210A Gene Mutation

A single nucleotide G-A change in position 20210 of the prothrombin gene has been identified as a cause of hypercoagulability by an increase in the prothrombin plasma level.

Fegan [21] reviewed five studies [5, 24, 34, 41, 50] that considered a possible association of this mutation with CVO and 7 patients out of 232 had the mutation. This prevalence correlates well with the 2–5% level of carrier in the healthy Caucasian population. The study by Albisinni is the only one in the literature that found a statistically significant association of venous occlusive disease with the prothrom-

Table 21.1.11. Prothrombin G20210A gene mutation and possible relationship with RVOD

Authors	Number pts./ctrls.	Age	% abn. pts./ctrls.	P value	Type of study	Vascular event***	Level of evidence	Asso-ciation?
O. Backhouse	16/0	Any	0	NA	Retrospective*	CVO13/BVO3	III	N
C.J. Glueck	17/234	Any	0/3.8	0.41	Retrospective	RVO	III	N
Adamczuk	37/144	Any	0/2.1	NS	Prospective**	CVO	III	N
S. Boyd	63/63	Any	0/1	NA	Prospective*	CVO	II	N
R. Marcucci	100/100	Any	4/2	0.6	Retrospective*	CVO	III	N
R. Albisinni [4]	36/68	Any	8.3/0	NA	Retrospective**	CVO/BVO	III	Y
K.Greiner	116/0	Any	0	NA	Prospective	CVO48/BVO33 CAO21/BAO14	III	N
D. Kalayci [34]	52/87	Any	0/2	NA	Retrospective	CVO25/BVO27	III	N
O. Salomon [18]	102/105	Any	2.9/5.7	0.53	Retrospective	CVO/BVO	III	N
O. Salomon [51]	21/0	Any	0	NA	Retrospective	CAO/BAO	III	N
J. Larsson [41]	129/282	Any	3/1	NS	Retrospective	CVO	III	N

Retrospective or prospective: not matched
* Prospective study matched for sex and age
** Prospective study matched for sex, age, hypertension, and hypercholesterolemia
*** Numbers following each specific RVOD represent the number of patients in each category of event

Table 21.1.12. Lipoprotein A and possible relationship with RVOD

Authors	Number pts./ctrls.	Age	% abn. pts./ctrls.	P value	Type of study	Vascular event***	Level of evidence	Asso-ciation?
P.L. Lip [44]	49/36	Any	188/56 (ng/ml)	0.009	Prospective*	RVO34	III	Y
P.L. Lip	49/36	Any	203/56 (ng/ml)	0.009	Prospective*	RAO15	III	Y
F. Bandello [7]	40/40	Any	30/10	0.03/0.45	Prospective**	CVO	II	Y/N
Sagripanti [49]	14/30	Any	35.5/10 (mg/ml)	<0.01	Prospective*	CAO/BAO	III	N
C.G. Glueck	16/40	Any	50/13	0.003	Retrospective	RVO	III	Y
R. Marcucci	100/100	Any	0	NA	Prospective*	CVO	III	Y/N

Retrospective or prospective: not matched
* Prospective study matched for sex and age
** Prospective study matched for sex, age, hypertension, and hypercholesterolemia
*** Numbers following each specific RVOD represent the number of patients in each category of event

bin mutation and was discussed in the aPCR/FVL section. All the other level III studies are included in Table 21.1.11 and none of them indicates an association of prothrombin G20210A mutation with any type of RVOD.

21.1.4.4 Lipoprotein A

Lipoprotein Lp(a) is an atherogenic, hypofibrinolytic, cholesterol carrying lipoprotein that may inhibit the conversion of plasminogen to plasmin [24]. Values above 300 mg/ml are associated with an increased risk of occlusive arterial disease [7]. Table 21.1.12 details studies looking at its possible association with RVOD. Marcucci (discussed in the FVL section of this chapter) found that in the absence of hypercholesterolemia, an increased Lp(a) level over 300 mg/dl was not associated with an increased risk of CRVO, while if the patient had hypercholesterolemia, the odds ratio was 2.4. Glueck (discussed previously) also found a correlation

between low-density-lipoprotein-cholesterol levels and Lp(a).

Bandello et al. carried out a well-matched cohort on 40 CVO patients. They could only find a statistically significant difference between their 12 cases and their 4 controls with elevated Lp(a) levels when using the Rosenbaum but not the Mann-Whitney test. Two other level III evidence studies are detailed in Table 21.1.12. In these, Lip found a positive association of Lp(a) in RVO patients as Sagripanti did not find such association for RAO.

The clinical significance of finding an increase of Lp(a) as well as the management of this anomaly remains to be addressed.

21.1.4.5 Other Factors

Several other different plasma protein anomalies have been investigated. Plasminogen activator inhibitor-1 4G/5G polymorphism, anomalies in factor VII, VIII or XII or von Willebrand are reported. Some of

Table 21.1.13. Studies assessing factor VII, VIII, XII or VWF deficiencies and plasminogen activated inhibitor 4G/5G polymorphism's possible relationship with RVOD

Author	Number pts./ctrls.	Age		% abn. pts./ctrls.	P value	Type of study	Vascular event***	Level of evidence	Association?
S. Boyd	63/63	Any	VIII	115/113	0.20	Prospective*	CVO	II	N
S. Boyd	63/63	Any	VWF	115/108	0.32	Prospective*	CVO	II	N
F. Bandello	40/40	Any	VII		NA	Prospective**	CVO	II	Y
S. Kadayifçilar	54/19	Any	VII	40.9/0 43.8/0	NA	Prospective*	CVO22 BVO32	III	Y
C. Kuhli [37]	150/135	Any	XII	9.3/0.7	0.0009	Prospective**	RVO	III	Age <45: Y Age >45: N
C.J. Glueck	17/234	Any	PAI1	88/64	0.03	Retrospective	RVO	III	?
Adamczuck	37/144	Any	PAI1	21.6/23	NS	Prospective**	CVO	III	N
R. Marcucci	100/100	Any	PAI1	24/5	<0.001	Retrospective*	CVO	III	Y

Retrospective or prospective: not matched
* Prospective study matched for sex and age
** Prospective study matched for sex, age, hypertension, and hypercholesterolemia
*** Numbers following each specific RVOD represent the number of patients in each category of event

the different articles covering those topics are detailed in Table 21.1.13. Once again in our opinion, the same conclusions as with Lp(a) should apply. We need more solid data to allow the detection of those investigational plasma protein anomalies in such a way it will make an evidence based difference in the patient's management.

21.1.5 Conclusions

Periodic visits to an internist or family practitioner are appropriate for middle aged and older adults, and patients with retinal vascular disease who do not see a primary care physician should consider doing so. Patients presenting with ocular vascular occlusive disease should have an evaluation of their intraocular pressure since glaucoma is a risk factor for some of these conditions and may be seen as a complication during follow-up. The family physician will assess blood pressure and should rule out the presence of diabetes mellitus. Adults seeing primary care physicians likely will have an assessment of serum lipids and cholesterol. Although no level one evidence exists, when these common risk factors have been eliminated in young patients, Hcy, APA, FVL and perhaps ATIII, protein C and S might be looked at especially in cases of bilateral disease, positive family history or previous thrombotic disorders.

If a risk factor for retinal vascular occlusion is identified, the next question of course is what, if anything, should be done about it. The goals of treatment should be to reduce the likelihood of a future ocular occurrence as well as to reduce the chance of systemic thrombotic events. To date, there have not been any trials showing that management of a clotting-related protein alters the risk of a future retinal vascular event. We can only assume that if there is

evidence that thrombotic disease in general is reduced by such management, that the chance of future eye disease might be reduced as well.

There is good data for some of the conditions discussed in this chapter that future systemic disease can be discovered and early management instituted in selected groups of patients. It is beyond the scope of this chapter, which is written by ophthalmologists and presumably for ophthalmologists, to review the basic medicine and to teach hematology. This is why we strongly believe that the medical management, if any, needs to be recommended by appropriate specialists such as hematologists or perhaps internists. We will simply mention some examples and offer some comments about the systemic management but our review of this topic barely scratches the surface.

In this next section, we will discuss the implications of finding a positive result for each respective plasma protein anomaly. From a more general perspective, thrombosis prophylaxis could be contemplated more aggressively in surgical or prolonged immobilization settings. A stronger recommendation can be made regarding modifiable vasculopathic risk factors such as avoidance of smoking oral or contraceptive use. Vandenbroucke and coworkers [61] studied 155 women aged 15–49 years with deep vein thrombosis (DVT) and compared them to 169 controls. Women using oral contraceptives had a fourfold increase in the risk of DVT. They had a sevenfold increase if they were carriers of aPCR. The study demonstrated a 30-fold increase in the risk of thrombosis if a young woman with aPCR was using the birth control pill.

It is unknown if aPCR/FVL positive patients with RVOD or systemic thrombosis should be anticoagulated and, if so, at what dose and for how long. A calculation of the benefit to risk ratio of warfarin with a

21 **III**

target international normalized ratio (INR) of 2.5 does not support the use of long-term therapy in all patients with the factor V Leiden mutation following a first pulmonary embolism or DVT [6].

Khamashta and colleagues [36], writing in the *New England Journal of Medicine*, teach that thrombosis is the main complication of the antiphospholipid syndrome and compare high intensity anticoagulation with warfarin (to achieve an INR of 3 or more) with or without low dose aspirin to low intensity anticoagulation (INR < 3) and report that high intensity treatment is needed for optimal risk reduction. Basically, lifelong *intensive* anticoagulation would be needed. Over the years, a severe adverse event related to treatment will almost be a certainty and whether this risk is reasonable theoretically to reduce the chance of a second ocular event we believe is quite questionable. The risk benefit ratio ought to be determined by the systemic issues and we would give much less weight to the ocular issues. We believe the decision whether patients with retinal vascular disease should be anticoagulated should be left to the hematology consultant for the systemic management of their disease.

Unlike the management of antiphospholipid syndrome, which to us seems to involve significant ongoing risks, hyperhomocysteinemia may be reduced with small doses of folic acid [13, 21]. Although we doubt most ophthalmologists should be prescribing these things for their patients, we assume the long term risks of folic acid should be less than anticoagulation with coumadin and aspirin and if recommended by the consulting hematologists, should be considered based on less certain data than we would want to have than for the treatment of the antiphospholipid syndrome.

Lastly, the ophthalmologist ordering tests for AT III, protein C and S should expect an extremely low yield of positive results coming back. Whether those deficiencies are associated with RVOD in adults has yet to be established. If by any chance a patient would show such deficiency, the management of the systemic components of the disease should be addressed by referral to a specialist familiarized with this type of disorder.

References

1. Abu El-Asrar AM et al (1996) Prothrombotic states associated with retinal venous occlusion in young adults. Int Ophthalmol 20:197–204
2. Abu El-Asrar AM et al (1998) Hypercoaguable states in patients with retinal venous occlusion. Doc Ophthalmol 95:133–143
3. Adamczuk YP (2002) Central retinal vein occlusion and thrombophilia risk factors. Blood Coagul Fibrinolysis 13:623–626
4. Albisinni R et al (1998) Retinal vein occlusion and inherited conditions predisposing to thrombophilia. Thromb Haemost 80:702–703
5. Backhouse O et al (2000) Familial thrombophilia and retinal vein occlusion. Eye 14:13–17
6. Baglin C, Brown K et al (1998) Risk of recurrent venous thromboembolism in patients with the factor V Leiden mutation: effect of warfarine and prediction by precipitating factors. Br J Ophthalmol 100:764–768
7. Bandello F et al (1994) Hypercoagulability and high lipoprotein (A) levels in patients with central retinal vein occlusion. Thromb Haemost 72:39–43
8. Bashshur ZF et al (2003) Anticardiolipin antibodies in patients with retinal vain occlusion and no risk factors. Retina 23:486–490
9. Bertram B, Protein C (1995) protein S, and anti-thrombin III in acute ocular occlusive diseases. Ger J Ophthalmol 4:332–335
10. Bolling JP, Brown GC (2000) The antiphospholipid antibody syndrome. Curr Opin Ophthalmol 11:211–213
11. Boyd S et al (2001) Plasma homocysteine, methylene tetrahydrofolate reductase C677T and factor II G20210A polymorphisms, factor VIII, and VWF in central retinal vein occlusion. Br J Ophthalmol 85:1313–1315
12. Cahill M et al (2000) Raised plasma homocysteine as a risk factor for retinal occlusive disease. Br J Ophthalmol 84:154–157
13. Cahill MT et al (2003) Meta-analysis of plasma homocysteine, serum folate, serum vitamin B_{12} and thermolabile MTHFR genotype as risk factors for retinal vascular occlusive disease. Am J Ophthalmol 136:1136–1150
14. Cassels-Brown A, Minford AMB, Chatfield SL, Bradbury JA (1994) Ophthalmic manifestations of neonatal protein C deficiency. Br J Ophthalmol 78:486–487
15. Central Vein Occlusion Group (1993) Baseline and early natural history report. Arch Ophthalmol 111:1087–1095
16. Cierdella AP et al (1998) Factor V Leiden, activated protein C resistance and retinal vein occlusion. Retina 18:308–315
17. Cobo-Soriano R et al (1999) Antiphospholipid antibodies and retinal thrombosis in patients without risk factors: a prospective case-control study. Am J Ophthalmol 128:725–732
18. De Stephano V, Voso MT et al (1997), relevance of mutated factor VARG506 to GLN in Italians. Letters to the editor. Thromb Haemost 77:216–217
19. Demirci FY (1999) Prevalence of factor V Leiden in patients with retinal vein occlusion. Acta Ophthalmol Scand 77:631–633
20. Di Crecchio L et al (2004) Hyperhomocysteinemia and the MTHFR 677CT mutation in patients under 50 years of age affected by central retinal vein occlusion. Ophthalmology 111:940–945

III 21

21. Fegan CD (2002) Central retinal vin occlusion and thrombophilia. Eye 16:98–106
22. Giordano N et al (1998) Antiphospholipid antibodies in patients with retinal vascular occlusions. Acta Ophthalmol Scand 76:128–129
23. Glacet-Bernard A et al (1994) Antiphospholipid antibodies in retinal vascular occlusions. Arch Ophthalmol 112:790–795
24. Glueck CJ (1999) Heritable thrombophilia and hypofibrinolysis. Possible causes of retinal vein occlusion. Arch Ophthalmol 117:43–49
25. Gottlieb JL (1998) Activated protein C resistance, factor V Leiden and central retinal vein occlusion in young adults. Arch Ophthalmol 116:577–579
26. Greiner K (2001) Genetic thrombophilia in patients with retinal vascular occlusion. Int Ophthalmol 23:155–160
27. Greiner K et al (1999), Retinal vascular occlusion and deficiencies in the protein C pathway. Am J Ophthalmol 128:69–74
28. Greven CM, Weaver RG et al (1991) Protein S deficiency and bilateral branch retinal artery occlusion. Ophthalmology 98:33–34
29. Hansen L et al (2000) Markers of thrombophilia in retinal vein thrombosis. Acta Ophthalmol Scand 78:523–526
30. Hodgkins PR et al (1995) Factor V and antithrombin gene mutations in patients with central retinal vein occlusion. Eye 9:760–762
31. http://www.ophsource.com/periodicals/ophtha/content/infoau. guidelines (date accessed, 28 Jan 2005)
32. Jacques PF et al (2001) Determinants of plasma total homocysteine concentration in the Framingham Offspring cohort. Am J Clin Nutr 73:613–621
33. Kadayifçilar S, Özatli D, Özcebe ÖÝ, Þener EC (2001) Is activated factor VII associated with retinal vein occlusion? Br J Ophthalmol 85:1174–1178
34. Kalayci D et al (1999) Factor V Leiden and prothrombine 20210A mutation in patients with central and retinal vein occlusion. Acta Ophthalmol Scand 77:622–624
35. Kalogeropoulos CD et al (1998) Anticardiolipin antibodies and occlusive vascular disease of the eye: prospective study. Doc Ophthalmol 95:109–120
36. Khamashta MA, Cuadrado MJ et al (1995) The management of thrombosis in the antiphospholipid-antibody syndrome. N Engl J Med 332:993–997
37. Kuhli C, Scharrer I et al (2004) Factor XII deficiency: a thrombophilic risk factor for retinal vein occlusion. Am J Ophthalmol 137:459–464
38. Lahey JM et al (2002) Laboratory evaluation of hypercoagulable states in patients with central retinal vein occlusion who are less than 56 years of age. Ophthalmology 109:126–131
39. Larsson J et al (1997) Activated protein C resistance in patients with central retinal vein occlusion. Br J Ophthalmol 81:832–834
40. Larsson J et al (1999) Activated protein C resistance and anticoagulant proteins in young adults with central retinal vein occlusion. Acta Ophthalmol Scand 77:634–637
41. Larsson J, Hillarp A (1999) The prothrombin gene G20210A mutation and the platelet glycoprotein IIIa polymorphism P1^{A2} in patients with central retinal vein occlusion. Thromb Res 96:323–327
42. Linna T et al (1997) Prevalence of factor V Leiden in young adults with retinal vein occlusion. Letters to the editor. Thromb Haemost 77:214–216
43. Linna T et al (1997) Prevalence of factor V Leiden in young adults with retinal vein occlusion. Thromb Haemost 77:212–224
44. Lip PL et al (1998) Abnormalities in haemorheological factors and lipoprotein (a) in retinal vascular occlusion: implications for increased vascular risk. Eye 12:245–251
45. Marcucci R et al (2001) Thrombophilic risk factors in patients with central retinal vein occlusion. Thromb Haemost 86:772–776
46. Martin SC et al (2000) Plasma total homocysteine and retinal vascular disease. Eye 14:590–593
47. Omenn GS, Goodman GE, Thornquist MD, Balmes J, Cullen MR, Glass A, Keogh JP, Meyskens FL, Valanis B, Williams JH, Barnhart S, Hammar S (1996) Effects of a combination of beta-carotene and vitamin A on lung cancer and cardiovascular disease. N Engl J Med 334:1150–1155
48. Rees D et al (1995) World distribution of factor V Leiden. Lancet 346:1133–1134
49. Sagripanti A et al (1999) Blood coagulation parameters in retinal arterial occlusion. Graefes Arch Clin Exp Ophthalmol 237:480–483
50. Salomon O et al (1998) Analysis of genetic polymorphisms related to thrombosis and other risk factors in patients with retinal vein occlusion. Blood Coag Fibrinolysis 9:617–622
51. Salomon O et al (2001) Thrombophilia as a cause for central and branch retinal artery occlusion in patients without an apparent embolic source. Eye 15:511–514
52. Schroeder W et al (1996) Large scale screening for factor V Leiden mutation in a north-eastern German population. Haemostasis 26:233–236
53. Scott JA et al (2001) No excess of factor V :Q506 genotype but high prevalence of anticardiolipin antibodies without antiendothelial cell antibodies in retinal vein occlusion in young patients. Ophthalmologica 215:217–221
54. Selhub J (1999) Homocystine metabolism. Annu Rev Nutr 19217–246
55. Tayyanipour R et al (1998) Arterial vascular occlusion associated with factor V Leiden gene mutation. Retina 13:376–377
56. Teleki O, Protein C (1999) Protein S and antithrombin III deficiencies in retinal vein occlusion. Acta Ophthalmol Scand 77:628–630
57. The Alpha-tocopherol Beta-carotene Cancer Prevention Study Group (1994) The effect of vitamin E and beta carotene on the incidence of lung cancer and other cancers in male smokers. N Engl J Med 330:1029–1035
58. The Eye Disease Case-Control Study Group (1998) Risk factors for hemiretinal vein occlusion: comparison with risk factors for central and branch retinal vein occlusion. Ophthalmology 105:765–771
59. The Eye Disease Case Control Study Group (1993) Risk factors for branch retinal vein occlusion. Am J Ophthalmol 116:286–296
60. The Eye Disease Case Control Study Group (1996) Risk factors for central retinal vein occlusion. Arch Ophthalmol 114:545–554
61. Vandenbroucke JP, Koster T et al (1994) Increased risk of venous thrombosis in oral-contraceptive users who are carriers of factor V Leiden mutation. Lancet 344:1453–1457
62. Vila P et al (1994) Prevalence, follow-up and clinical significance of the anticardiolipin in normal subjects. Thromb Haemost 72:209–213
63. Vine AK, Samara MM (1997) Screening for resistance to activated protein C and the mutant gene for factor V:Q506

in patients with central retinal vein occlusion. Am J Ophthalmol 124:673–676

64. Werner M et al (2002) Hyperhomocysteinemia and MTHFR C677T genotypes in patients with central retinal vein occlusion. Graefes Arch Clin Exp Ophthalmol 240: 286–290

65. Werner M et al (2002) Hyperhomocysteinemia, but not MTHFR C677T mutations, as a risk factor in branch retinal vein occlusion. Ophthalmology 109:1105–1109

66. Werner M et al (2002) The role of hyperhomocysteinemia, but not MTHFR C677T mutation in patients with retinal artery occlusion. Am J Ophthalmol 134:57–61

67. Williams GA, Sarrafizadeh R (2000) Correspondence to Antiphospholipid antibodies and retinal thrombosis in patients without risk factors: a prospective case-control study. Am J Ophthalmol 130:538–539

68. Williamson TH et al (1996) Blood viscosity, coagulation and activated protein C resistance in central retinal vein occlusion: a population controlled study. Br J Ophthalmol 80:203–208

21

21.2 Central Retinal Vein Occlusion

L.L. Hansen

Core Messages
- Retinal vein occlusion is the most frequent primary vascular disorder of the retina
- Central retinal vein occlusion (CRVO) is a more subacute chronic rather than an acute entity that occurs at all ages
- Cardiovascular risk factors, some forms of thrombophilia, and hyperviscosity may trigger and facilitate the development of CRVO
- Visual outcome is variable, ranging from benign course to painful blinding

- Ischemic (non-perfused) CRVO should be differentiated from non-ischemic (perfused) CRVO, representing entities at the opposite ends of the range
- Medical treatment (hemodilution, fibrinolysis) has been tried with some success
- Newer modalities for CRVO treatment (surgical, newer steroids) are under investigation but have not yet yielded evidence-based results
- Neovascular disease cannot be prevented in all cases by panretinal photocoagulation

21.2.1 Background

Retinal vein occlusion is the most frequent primary vascular disorder of the retina and of all the retinal vascular diseases second only to the more common diabetic retinopathy. The first description of a case with central retinal vein occlusion dates from 1855 [109], but it was not until 1878 that von Michel [123] interpreted the finding as central vein thrombosis. The term central retinal vein occlusion (CRVO) is preferred nowadays as we still do not completely understand the pathogenesis of the vein blockage in the optic nerve head. In addition to CRVO, we know the less frequent occlusion of the superior or inferior retinal venous vasculature as hemiretinal vein occlusion (HRVO). The occlusion of a retinal branch vein (BRVO) is slightly more often seen than CRVO and is addressed in Chapter 21.3.

Furthermore, much confusion existed in the past regarding the varieties of CRVO, and terms such as "impending," "incipient," "partial," "incomplete CRVO," "venous stasis retinopathy," and "hemorrhagic retinopathy" [64, 66, 97] were used to describe courses ranging from very benign to disastrous. Today the most adequate and widely used designation is "non-ischemic" or "perfused" for the milder and "ischemic" or "non-perfused" for the severe form [68, 165], both representing entities at the opposite ends of the range. The adjunct "indetermi-

nate" may be employed for cases that do not fit easily into the categories mentioned.

21.2.2 Epidemiology

Essentials
- Incidence of retinal vein occlusions varies from 2 to 8 per 1,000 persons
- CRVO can occur at all ages, with a mean age between 60 and 70 years

The incidence of retinal vein occlusions in population-based studies varies from 2 to 8 per 1,000 persons and increases with age [23, 96]. CRVO can occur at all ages from 9 months to more than 90 years, but the mean age of patients is between 60 and 70 years. About 10% are younger than 50 years [10]. It affects males and females equally; yet among patients below 50 years, men prevail [38, 136, 149]. Both eyes are involved equally. Five to 11% of patients will suffer from CRVO in their other eye within 5 years [10, 75].

21.2.3 Etiology and Pathogenesis

Essentials

- Pathogenesis can be represented in three circles: (1) triggering of the venous outflow reduction by narrowing of the vein and partial thrombosis, (2) development of the retinopathy, and (3) transition into severe neovascular disease
- The extent of the blood-retinal barrier's breakdown and ischemia define the course of the retinopathy
- Triggering mostly happens at nighttime in the recumbent position probably by low blood pressure, and/or high central venous pressure
- Cardiovascular risk factors such as arterial hypertension, arteriosclerotic cardiovascular disease, diabetes mellitus, obesity, hyperlipidemia, hyperhomocysteinemia, and smoking are associated with CRVO
- Hyperviscosity and thrombophilia seem to facilitate the development of CRVO
- Local risk factors are glaucoma, trauma, and possibly inflammation

Table 21.2.1. Risk factors and associated diseases in CRVO

Cardiovascular	Atherosclerotic heart disease Arterial hypertension Diabetes mellitus Hyperlipidemia Obesity Smoking Carotid artery occlusive disease (ischemic CRVO)
Rheological abnormalities	Increased hematocrit Increased plasma viscosity Increased red cell aggregation Reduced red cell deformability
Thrombophilia	Hyperhomocysteinemia Anti-phospholipid syndrome Increased APC resistance/FV Leiden mutation (young patients) Reduced plasminogen activator inhibitor Oral contraceptives?
Local risk factors	Glaucoma Trauma? Retinal vasculitis Central artery occlusion Drusen, papilledema Arteriovenous malformation
Hyperviscosity syndromes	Polycythemia Macroglobulinemia Myeloma Leukemia

CRVO's pathogenesis certainly is multifactorial. Klien [97] speculated that three factors could lead to CRVO: (1) compression by a sclerotic central retinal artery and cribriform plate, (2) hemodynamic disturbances leading to stagnation and primary thrombus formation, and (3) degenerative and/or inflammatory disease within the vein. All these causes must still be considered, supplemented by systemic factors such as thrombophilia and rheological abnormalities [27].

When considering important factors for the triggering, development, and course of CRVO, one must diligently separate the actual cause from the facilitating risk factors and associated diseases (Table 21.2.1) that may be important in the development of the disease and, hence, its prognosis. For instance, a genuine causal relationship for cardiovascular, thrombophilic or rheological risk factors may be difficult to prove particularly as many of these factors have a high incidence in patients without CRVO as well.

21.2.3.1 Systemic Risk Factors

Cardiovascular risk factors (arterial hypertension, cardiovascular disease, diabetes mellitus, obesity, hyperlipidemia, hyperhomocysteinemia, smoking) are associated with the development and course of CRVO [10, 31, 73, 122, 156, 164]. At least one systemic risk factor is seen in two-thirds of patients over 50 years of age, often more frequently in ischemic CRVO than in its non-ischemic variant [49, 61]. The odds ratio for these factors was calculated by Wong et al. [188] and ranges between 3 and 5. Arterial hypertension (32–70%), atherosclerotic heart disease (22–50%), hyperlipidemia (30–60%) and diabetes mellitus (14–34%) are most commonly found. It is worth noting that Doppler imaging of the carotid artery did not reveal a higher percentage of occlusive disease than in age matched controls, although ischemic CRVO is more often associated with carotid artery disease [11, 61, 143, 188] than is the non-ischemic type.

Cardiovascular risk factors do not seem to be important in younger patients under 50 years of age. Only one-third suffer from arterial hypertension, and 3–9% have diabetes mellitus [10]. Nevertheless, Priluck [133], in a long-term follow-up of young adults with CRVO, observed a mortality of 12% due to vascular diseases. Migraine and mitral valve prolapse have been encountered in these patients also, although an association remains obscure [38].

Hyperviscosity of the blood caused by elevated cells and plasma proteins (e.g., leukemia, polycythemia, macroglobulinemia, myeloma) can clearly lead to bilateral venostasis with all the features of CRVO (hyperviscosity syndrome), but do minor rheological problems play a role in the development of unilateral CRVO Minor viscosity changes have been shown by several groups, indicating increased hematocrit, plasma viscosity [2, 138, 180] and red cell aggregation [47], as well as a reduced red cell deformability [132, 185]. Elevated plasma fibrogen increases the odds by a factor of 3 [188]. All these changes are more or less borderline values and far from being capable of seriously hampering blood flow as occurs in hyperviscosity syndromes; hence, they can only be regarded as risk factors in triggering and evolving CRVO.

Thrombophilia seems a logical candidate in causing or at least contributing to the development of retinal vein "thrombosis." This has prompted innumerable reports on its role during the last 20 years. A predisposing factor for thrombophilia can be diagnosed in 70–80% of patients with recurrent venous thromboembolic disease outside the eye [34]. On the other hand, these abnormalities cause venous rather than arterial thrombosis, and retinal vein thrombosis is typically associated with risk factors of arterial thromboembolism.

The existence of a thrombus in the central vein was long debated [see 179] until Green et al. [54] examined 29 eyes in which thrombi were found in nearly all of the rubeotic eyes, but we are still unsure about the situation in freshly occluded eyes. In recent years an abundant number of coagulation cascade factors have been examined in retinal vein occlusion. Thrombophilia can be caused by a lack of plasma proteins that slow down the coagulation cascade (e.g., factor XII, antithrombin III, antiphospholipid antibodies, factor V Leiden/APC resistance) and hypofibrinolysis.

Two papers have shown a decreased factor XII [1, 102] and antithrombin III [1, 185], respectively, while anti-phospholipid antibodies are claimed to be a factor by some authors [1, 3, 179] and questioned by others ([7, 45, 46]; for a complete review, see [34]).

Even more controversial is the role of an increased activated protein C (APC) resistance or protein C and S deficiency. Vossen et al. [173] reported on a large protein-C-deficient family with a history of non-ocular venous thrombosis but no signs of retinal vein occlusion. It now seems largely unequivocal that in young patients there is a frequent decrease in the APC resistance [101, 102, 144] although this is not found in all papers [21, 53]. Thus APC resistance may play a definite role in triggering a CRVO in young patients. Most authors agree that in older patients with CRVO, APC resistance is not important [7, 25, 33, 79, 107] although there are a few exceptions [56, 162, 181].

Hyperhomocysteinemia, a risk factor for systemic vascular disease, may, via its deleterious effect on vascular endothelium, induce increased platelet aggregation, lipid accumulation, and arterial thrombosis [177]. It remains controversial whether hyperhomocysteinemia, mostly dependent on a single mutation of the MTHFR gene (see [34]), is also associated with a higher risk for CRVO [8, 11, 16, 20, 26, 178], although, in a meta-analysis by Jannsen et al. [89], an overall odds ratio of 8.9 for homocysteine above the 95th percentile was calculated.

Hypofibrinolysis caused by low levels of plasminogen activator inhibitor was demonstrated by Williamson et al. [185] and confirmed by Gleuk et al. [52].

Anti-phospholipid factors (anticardiolipin antibodies, lupus anticoagulant) activate the coagulation cascade, leading to both arterial and venous thrombosis [see 34] and 6–8% of patients with anti-phospholipid syndrome present ocular manifestations. In their meta-analysis, Jannsen et al. [89] calculated an odds ratio of 3.9 for anticardiolipin based on six case-control studies.

In summary, it appears that hyperhomocysteinemia and anti-phospholipid syndrome are risk factors for CRVO and there is evidence that disorders causing hypofibrinolysis may also be important. The common hereditary thrombophilic conditions do not, however, seem to represent strong risk factors [34, 89], but one should be aware of them in young women, especially when additional risks, e.g., oral contraceptives, act synergistically [170]. This highlights the fact that atherosclerosis is an important factor in CRVO, while factors known as risk factors for venous thrombosis are not [82].

21.2.3.2 Local Risk Factors

Trauma may be associated with CRVO. A history of a preceding trauma is found in about 14% of cases [122]. The mechanism after a serious head trauma may be compression or shearing of the central vein against the lamina cribrosa and constriction by a hematoma within the optic nerve sheath. These mechanisms might induce turbulence and endothelial damage thus provoking thrombus formation (see below). However, one would then expect many cases to occur after retrobulbar anesthesia, but this has only been reported once [158]. My own clinical experience does not lend support to an important role on the part of direct or indirect trauma in triggering CRVO.

Glaucoma is the ocular disease most commonly found in association with CRVO. Glaucoma increases the risk of suffering a CRVO five- to sevenfold [23]. The incidence of glaucoma in CRVO is reported to be 23–69% when ocular hypertension is included, and vice versa 2–8% of glaucoma patients may develop CRVO [10, 38, 75, 78, 79, 111]. Hayreh pointed out that CRVO leads to a subsequent fall in the intraocular pressure (IOP) in the involved eye. It thus is important to exclude glaucoma or ocular hypertension in the fellow eye of any patient with CRVO/HRVO [75]. Furthermore, Cole et al. reported a high prevalence of arterial hypertension and hyperlipidemia in patients with both diseases [22].

Other ocular diseases and factors leading to CRVO such as arteriovenous malformation and optic nerve diseases narrowing the central vessels (drusen, papilledema) may cause a congestion in the retinal veins and precipitate CRVO [19, 43, 163]. Green suggested not only compression, but endothelial damage by the drusen as well [54]. CRVO can be observed in patients with arteriovenous fistulas in the cavernous sinus region [148]. Retinal artery occlusions slow down the flow to stasis, and in case of reperfusion after several hours, a central retinal vein thrombosis might hamper drainage. This may be the cause for combined retinal occlusions.

21.2.3.3 Pathogenesis of CRVO

One main obstacle in clarifying the pathogenesis of vein occlusion is that we have only a few cases of histopathology from fresh cases. Today we are quite certain about the thrombotic character of most cases [54], but we cannot say whether this is the starting point or the end of the story. Thus the real cause of CRVO still remains controversial and is certain to involve different factors. One should not confuse precipitating causes and risk factors, and it will be helpful if the different factors are identified that may play a role in initiating and maintaining the CRVO and that lead to either a benign course or propel the CRVO to a disastrous outcome. Pathogenesis thus can be represented in three circles: (A) triggering of the venous outflow reduction, (B) development of the retinopathy, and (C) transition into severe neovascular disease (Fig. 21.2.1).

21.2.3.3.1 Precipitating Causes of CRVO (Circle A)

Risk factors are well known, while their clear role in initiating the vicious circle is not well established. Three mechanisms seem to be involved in causing and maintaining the circle of flow turbulence, outflow reduction, and partial thrombosis: (1) a narrowing of the central vein, (2) hemodynamic disturbances, and (3) thrombophilic conditions such as thrombophilia itself or hemorheological abnormalities.

Narrowing of the central vein can be caused by an enlarged arteriosclerotic central artery following long-standing arterial hypertension and arteriosclerosis, illustrating the high incidence of cardiovascular risk factors [10, 73, 122, 156] in patients with CRVO. In this context, the passage of the vessels through a tiny, non-expandable hole in the cribrose plate seems relevant, but it has never been shown that small papillas are more prone to CRVO, as it has been demonstrated for anterior ischemic optic neuropathy. In addition to a reduced venous outflow, the ensuing turbulences in venous flow may prompt endothelial damage with partial thrombosis. It then depends on additional factors (see below) as to whether a vicious circle develops or not.

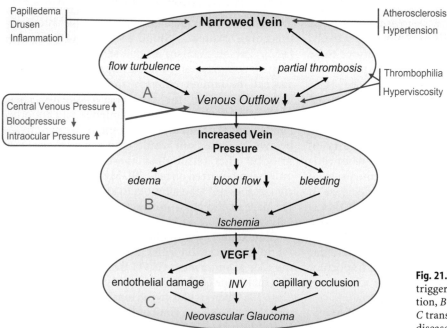

Fig. 21.2.1. Pathogenesis of CRVO: *A* triggering of the venous outflow reduction, *B* development of the retinopathy, *C* transition into severe neovascular disease

A narrowing may also occur via optic nerve swelling (papilledema, anterior ischemic optic neuropathy, drusen), but only a few cases of CRVO have been reported in such cases [19, 43, 163]. A long-standing compression of the central vein against the lamina cribrosa by increased IOP has clearly been identified as a risk factor [10, 39, 75, 77, 78, 111] while CRVO has not been reported after acute glaucoma with high IOP.

Damage to the vessel wall by trauma or inflammation with narrowing of the lumen could partly explain CRVO in young patients without cardiovascular risk factors [38, 39]. An inflammatory etiology was suggested to be a cause in young patients with CRVO [67]. More recent studies have not shown a predilection for inflammation in young patients [38, 39].

Increased thrombogenesis may certainly support the vicious circle of flow disturbance and partial thrombosis. The main risk factors for thrombophilia have been presented above (see Sect. 21.2.3.1). One cannot assume that these factors are capable of triggering CRVO as the only cause, as bilateral CRVO has never been reported.

Increased blood viscosity as caused by several minor rheological abnormalities found in patients with CRVO [2, 47, 132, 138, 180, 185] may support the development of the vein occlusion (see Sect. 21.2.3.1), as does thrombophilia. In contrast a severe hyperviscosity by highly elevated cells and/or plasma proteins usually causes a bilateral CRVO (hyperviscosity syndrome), demonstrating in those cases a causal relationship.

Hemodynamic factors seem capable of triggering CRVO by either an increase in central venous pressure (CVP) or a decrease in arterial blood pressure. The poor vision is nearly always detected by the patient the morning after a higher CVP in the recumbent position during the night and low arterial pressure when rising. In addition, the slight improvement of the visual acuity during the daytime supports the concept of better drainage when the CVP falls with the patient in an upright position. In that context, Francis et al. [40] reported that dehydration may trigger CRVO in young patients.

21.2.3.3.2 Development of the Retinopathy (Circle B)

The diminished venous outflow can easily be assumed when observing an increase in all transit times in both arterial and venous flow in fluorescein angiography. This has been demonstrated by Sinclair and Gragoudas [153] among others. They measured the time of maximum venous filling as being a good indicator for CRVO's severity. In addition, laser Doppler velocimetry revealed reduced blood flow in non-ischemic CRVO [17]. Reduced blood flow increases the viscosity of blood logarithmically, hence worsening microcirculation [146].

Increased pressure within the retinal venous vasculature usually takes some time to damage the vessels. Long-standing engorgement will cause extravasation of blood cells, affect the integrity of the endothelial blood-retinal barrier, and overflow the retina by exuding plasma. Not until a macular edema has developed will the drop in visual acuity become evident. We do not know whether increased vitreomacular adhesion might contribute to chronic edema in CRVO [116, 125].

CRVO not only leads to a decrease in capillary blood flow velocities, but also to an enlargement of perifoveal intercapillary areas [137]. Bleeding into a cyst forming in the fovea will reduce the vision more seriously. Retinal swelling, blood stasis, and occlusion of venules leads to cotton-wool spots and may contribute to ischemic development. Further progression to ischemic disease or a containment of the non-ischemic type of CRVO probably depends on finding a new balance between inflow and outflow, eventually supported by the formation of collateral disk vessels with a new drainage route [159].

21.2.3.3.3 Progression to Ischemic CRVO (Circle C)

We do not know why some cases progress to ischemic retinopathy with vast capillary dropout and others not. The risk factors mentioned may play a decisive role in such cases as is supported by the unfavorable prognosis in older people and those with diabetes mellitus [58]. The bad perfusion leads to ischemia of the inner retina with release of growth factors. Hypoxia-induced vascular endothelial growth factor VEGF is, most likely, the link between retinal ischemia and neovascular disease of the iris and retina in CRVO [131]. VEGF-induced endothelial cell hypertrophy, moreover, may cause further capillary closure in this disease [80] and may exacerbate the ischemic retinopathy. Rubeosis iridis and, if not treated, retinal neovascularization and neovascular glaucoma ensue.

21.2.4 Clinical Features

Essentials
- Symptoms: blurred vision in the involved eye after getting up, often improvement during daytime
- Fundus with hyperemic swollen papilla, engorged veins, streaky retinal bleedings, cotton-wool spots

21 **III**

- Late features: cilioretinal collaterals on the papilla, macular scar
- To classify into ischemic and non-ischemic CRVO is important, and cannot be achieved by funduscopy alone
- Diagnostic clues are visual acuity, fluorescein angiography, visual fields, pupillary light response, and electroretinography
- Laboratory testing: blood pressure, hematocrit, APC resistance in young patients
- Involvement of both eyes points to systemic disease

To adequately manage retinal vein occlusion, it is important to diligently classify the entity. Signs of arterial disease must be separated from those typical for purely venous occlusion in order to differentiate between different degrees of ischemia. The expansion of the occlusion, and, as is often realized by the patient in an advanced state only, the age of the CRVO should be estimated and corroborated with the duration of symptoms. These and additional factors (patient's age, general state of health, and cardiovascular risk) critically define the course and prognosis of CRVO and HRVO, for which unambiguous treatment guidelines only partially exist.

21.2.4.1 Symptoms and Funduscopic Features

21.2.4.1.1 Symptoms

The typical patient complaint is of blurred vision in the involved eye after getting up. This blurring may fade initially or disappear during the day and reappear the next morning (see Sect. 21.2.3.3). The vision may deteriorate over a couple of days, until it no longer improves in the course of the day. This often leads the patient to call the ophthalmologist only after 1–3 weeks [38] and differs completely from patients with retinal arterial occlusion, who suffer from a sudden loss of vision of the eye at any time during the day. Other symptoms are floaters, black spots, or seldom metamorphopsia. The visual acuity of an eye with *pure* CRVO at first presentation will never be below 0.05 (20/400). The patient usually presents with a visual acuity of 0.1–0.5 and with a loss of ability to read in the affected eye. In younger patients visual acuity often is good during the first fortnight, but then a sudden drop may occur that is independent of treatment (along with the formation of cystoid macular edema).

21.2.4.1.2 Fundus

Early features of funduscopic changes are characteristic and often provide the correct diagnosis at first glance, although the variety of signs may be considerable. The spectrum of fresh CRVO ranges from mild changes with few superficial, flame-shaped bleedings, slight engorgement of the veins, hyperemia and swelling of the optic nerve head (Fig. 21.2.2a) to dense, and more extended bleedings (Fig. 21.2.2b), strongly dilated and tortuous dark blue veins (Fig. 21.2.2c), and gigantic papilledema with vitreous bleedings on top (Fig. 21.2.2d).

Intraretinal bleedings are spread over the whole fundus and are mostly flame shaped in areas with a thick nerve fiber layer, i.e., around the optic nerve head. They can also occur as more deeply located dot and blot bleedings at the posterior pole (Fig. 21.2.3) or as spots in the periphery. There are always fewer bleedings toward the periphery. Vitreous bleedings are rare and usually occur in the papillary region of younger patients with a "papillophlebitis" type (Fig. 21.2.4).

Cotton-wool spots are signs of small capillary dropouts and hence of local ischemia (Fig. 21.2.5). They can be found in all types of CRVO (see Sect. 21.2.4.2). Macular edema occurs in all retinal vein occlusions with loss of vision. It is of the cystoid extracellular type and is usually found only after days or weeks. A blood-filled cyst is an unfavorable sign for the visual prognosis.

The swelling of the papilla is always marked, the margin often being obscured by cotton-wool spots and bleedings. Where this is not the case a hyperemic papilla helps to discern the CRVO from entities other than vein occlusion. We often find the so-called "papillophlebitis" (Figs. 21.2.2d, 21.2.4) in young patients, where affection of the optic nerve head dominates the fundus with only few bleedings in the mid and outer periphery. Some authors believe that this entity indicates an underlying inflammation ([64], see Sect. 21.2.3.3).

Late features of a long-standing CRVO can differ greatly depending on the balance between inflow and outflow that develops (see Fig. 21.2.1). The streaky bleedings become blurred or begin to slowly disappear. Engorgement of the veins becomes less, and the vessel walls lose their transparency and whiten. Cilioretinal collaterals on the papilla are a very characteristic sign of a several-weeks-to-months-old CRVO (Fig. 21.2.6a) and should not be confused with retinal neovascular tufts (see Sect. 21.2.4.3). Macular edema first increases and with the ensuing reduction after months and years, a dry, pigmented scar may be the clue indicating an old CRVO (Fig. 21.2.6b). However, the arterial inflow will not always decrease and

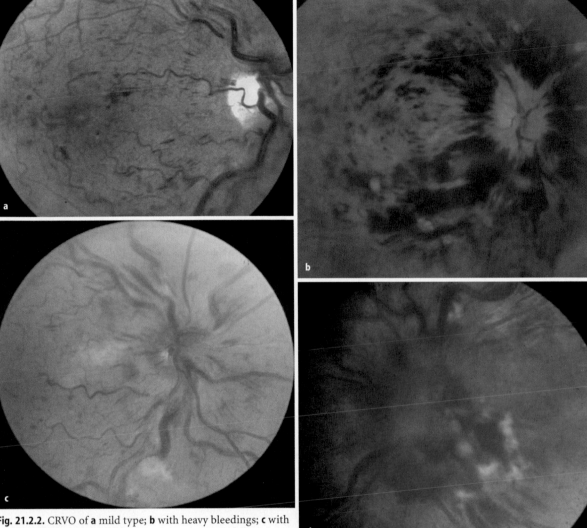

Fig. 21.2.2. CRVO of **a** mild type; **b** with heavy bleedings; **c** with few bleedings and engorgement of the veins; **d** with vitreous bleeding, all with macular edema

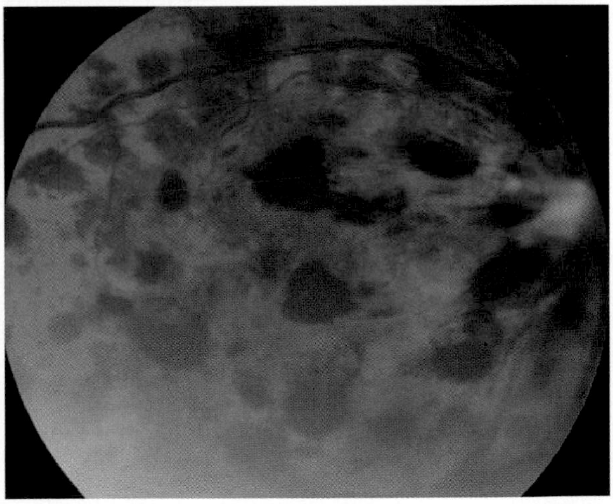

Fig. 21.2.3. CRVO with deeply located dot and blot bleedings

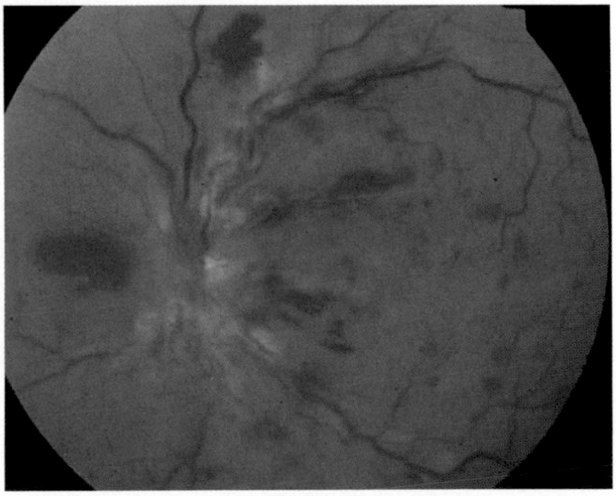

Fig. 21.2.4. CRVO with dominance of papilledema

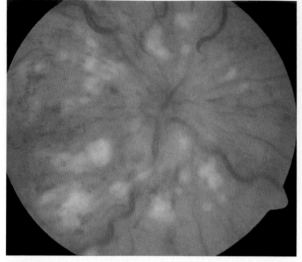

Fig. 21.2.5. CRVO with many cotton-wool spots (still not ischemic type!)

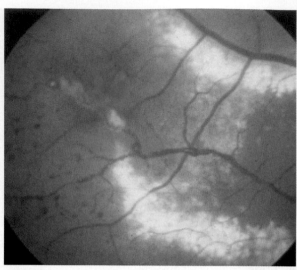

Fig. 21.2.7. CRVO of the perfused exudative type after 1 year

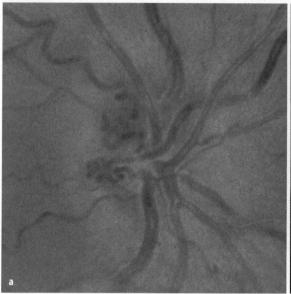

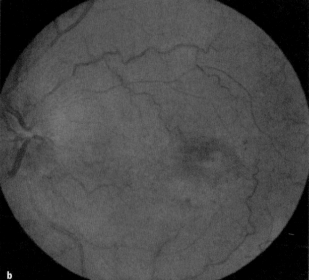

Fig. 21.2.6. CRVO after months to years with **a** collaterals, and **b** typical macular scar

adapt to the reduced venous outflow, thus permitting the reforming of an intact blood-retinal barrier. A serous retinal detachment over wide areas with hard exudates (Fig. 21.2.7) may then develop.

The first sign proving an ischemic CRVO is neovascular disease of the iris. Untreated rubeosis iridis may lead to occlusion of the chamber angle with secondary glaucoma and hyphema. Interestingly, retinal neovascularization is not encountered that often in late CRVO as in other ischemic retinopathies.

21.2.4.2 Classification of CRVO

As already mentioned, the wide range of clinical features necessitates diligent classification, as that defines the course (see Sect. 21.2.4.4) and treatment. The most important challenge is to detect patients with the ischemic form of CRVO (synonyms: non-perfused CRVO, hemorrhagic retinopathy). This cannot be done on a funduscopic basis and often not at the patient's first presentation. The ischemic and non-ischemic (venous stasis retinopathy, partial, perfused, hyperpermeable, prethrombosis) types form the two opposite sides of a continuous spectrum rather than distinct entities. Within the first

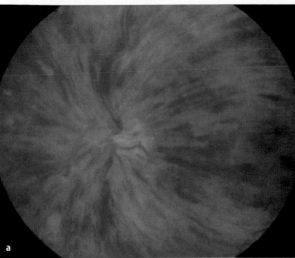
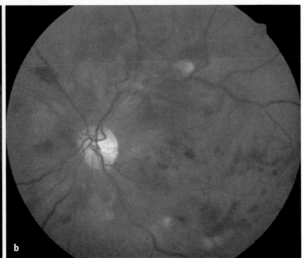

III 21

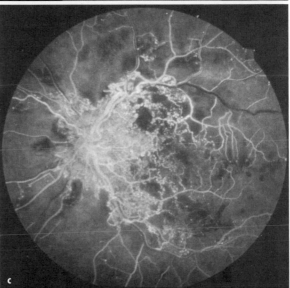

Fig. 21.2.8. CRVO: **a** intermediate type and **b** ischemic type with **c** widespread capillary occlusion

half year one should try to estimate the degree of ischemia so as not to miss the transition of non-ischemic or indeterminate forms into treatable neovascular disease (see Sect. 21.2.4.4). This applies to both CRVO and HRVO [156].

Ischemic features cannot be identified by biomicroscopy of the fundus. One weak indicator may be widespread dense bleedings, although ischemic CRVO may also occur with fewer retinal bleedings (Fig. 21.2.8a–c). All other signs that might signal serious disease, such as severe swelling of the papilla, numerous cotton-wool spots, marked dilation and tortuosity of veins and strong macular edema, cannot be correlated to the degree of ischemia, as illustrated by the example given in Fig. 21.2.8b, c. A reliable sign of ischemia is neovascularization of the iris. Fluorescein angiography (capillary dropout, prolonged transit times) is most helpful in detecting the degree of ischemia but is not always conclusive. Margagal classified CRVO based on the ischemic index, which correlated with the proportion of retinal ischemia determined by fluorescein angiography [115]. An ischemic index of 50% (corresponding to about 10 disk diameters of retinal capillary non-perfusion) was considered the threshold for a significant risk of neovascular complications. This value was later picked up by the Central Vein Occlusion Study Group as an ischemia-defining parameter [112] although it was seriously questioned by Hayreh [67, 74]. Apart from these clues, combined information of the four functional tests visual acuity, visual fields, pupillary light response, and electroretinography will enable the ophthalmologist to make a definitive diagnosis and then correctly classify the ischemic risk (Table 21.2.2).

Table 21.2.2. Significance of signs pointing to ischemic central retinal vein occlusion

Examination	Sign	Significance
Slit lamp	Iris neovascularization	+++
	Angle neovascularization	+++
Visual acuity	≤0.1 (20/200)	++
Relative afferent pupillary defect	Marked difference to unaffected eye (≥ 1.2 log units)	++
Fluorescein angiography	Capillary occlusion ≥ 10 PD	++
	Time for maximal venous filling ≥ 20 s	
Perimetry	Peripheral constriction	++
Electro-retinogram	Reduced b-wave	++
	Decreased b/a amplitude	
Funduscopy	Widespread, dense retinal bleedings	+

PD papillary diameter; the more + signs, the stronger the association to ischemic CRVO

21.2.4.3 Diagnosis

Fluorescein angiography is usually necessary in CRVO. The filling of the veins is typically delayed, producing the picture of black veins on a bright choroidal background during the early phase. Sinclair and Gragoudas [153] correlated a venous filling time of more than 20 s (between first appearance of the dye on the papilla and maximum filling of the temporal veins) to ischemic CRVO. Intensive staining of the large veins's wall occurs during the late phase, even if no generalized leakage of the small vessels is present (Fig. 21.2.9a). Strong leakage often blurs the petal form of the cystoid macular edema, which can always be found in eyes with a drop in visual acuity, but later will become obvious (Fig. 21.2.9b). Prognostically important areas with capillary occlusion/non-perfusion are difficult to discern during the first weeks because they are often concealed by heavy retinal bleedings. This seriously limits the information about capillary occlusion [69] and, hence, the prognostic value of the early angiogram (see Sect. 21.2.4.2). Lower retinal fluorescence with a smooth choroidal background combined with bud-like abortive vessels later points to massive ischemia (Fig. 21.2.8c) and allows for better classification.

Visual acuity may be quite normal initially and usually drops within a fortnight. Twenty-five to 30% are still able to read (visual acuity 0.4), while roughly 50% have fallen to 0.1 or less [62, 134]. Considering the degree of ischemia, 70–80% of patients with non-ischemic CRVO are seen with 0.1 or better while only 7–14% of ischemic eyes present with this visual acuity [71, 134]. Thus, a low visual acuity of less than 0.1 is a marker for the ischemic type with a high sensitivity [71, 168] (Fig. 21.2.2). This marker is less predictive [136] in patients younger than 50 years, with a better range in the initial visual acuity.

Perimetry can be helpful in patients with indeterminate CRVO. Almost all eyes have a central scotoma but only the peripheral fields allow for some distinction. Ischemic eyes always show considerable defects in kinetic perimetry. Hayreh et al. [71] found that 71% of non-ischemic eyes but only 8% of ischemic ones presented unremarkable Goldmann visual fields.

The **relative afferent pupillary defect** (RAPD) can be tested easily and is a reliable functional test for differentiating of ischemic from non-ischemic CRVO [152]. Hayreh et al. found a sensitivity of 80% and a specificity of 97% when using RAPD with a cutoff of ≥ 0.9 log units [71]. The prerequisite of this objective test is normal pupils and a normal fellow eye.

Electroretinography may also be useful in the differentiation of the types of CRVO. Routine ERG parameters yield a sensitivity of 80–90% and a specificity of 70–80% [70] for the b-wave amplitude under photopic and scotopic conditions. Matsui et al. [118] found that a reduced b/a-wave amplitude ratio correlated significantly with the presence of capillary dropout on fluorescein angiogram in CRVO. Williamson et al. [182] confirmed the data and simplified and shortened the protocol, thereby making the application of ERG more practical in the clinical setting. An increased photopic cone b-wave implicit time in the 30 Hz flicker ERG seems to be a good predictor for iris neovascularization. Larsson et al. [105] recommend the 30 Hz flicker ERG is best done after 3 weeks and found it superior to fluorescein angiogram.

Venous collapse pressure, as measured by an ophthalmodynamometric device in conjunction with a special Goldmann contact lens, seems to be a promising new functional test. The collapse pressure is significantly higher in eyes with ischemic CRVO than in the non-ischemic variant [90].

Laboratory testing is not necessary beyond evaluating risk factors for atherosclerotic vascular disease.

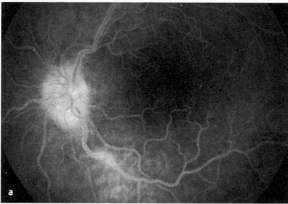

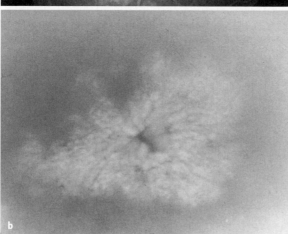

Fig. 21.2.9. Fluorescein angiography in CRVO with **a** wall staining of large veins of mild type, and **b** cystoid macular edema in a non-ischemic CRVO after 6 months

Blood pressure should always be checked and blood taken for hematocrit. In young patients, blood for APC resistance might be taken to exclude thrombophilia. History of vein thrombosis elsewhere should alert for a more extensive check-up by the specialist.

21.2.4.4 Differential Diagnosis

Differential diagnosis of CRVO is usually not a difficult problem (Table 21.2.3). Entities like diabetic and hypertensive retinopathy as well as hyperviscosity syndromes occur bilaterally. If both eyes are affected, a medical examination is mandatory! Most difficulties are encountered with early, mild non-ischemic CRVO and late forms and complications.

Table 21.2.3. Differential diagnosis of central retinal vein occlusion

	Differential diagnosis	Characteristic difference
Early CRVO	Anterior ischemic neuropathy	No peripheral bleedings
	Papilledema	No drop in visual acuity
	Hypertensive retinopathy	Both eyes involved
	Diabetic retinopathy	
	Hyperviscosity syndrome	
	Ocular ischemia	Dot and blot bleedings
		Not marked papilledema
Late CRVO	Geographic atrophy of late AMD	No collaterals on papilla
	Neovascular disease	No leakage in fluorescein
	M. Eales	angiogram

AMD age-related macular degeneration

21.2.4.4.1 Early CRVO

Anterior ischemic neuropathy (AION) with marked swelling of the papilla may simulate a mild CRVO, but can easily be discerned by the typical altitudinal visual field defect. Retinal bleedings extending to the mid-periphery strongly favor the diagnosis of CRVO, although strong swelling of the papilla may also lead to vein engorgement.

Ocular ischemia with venous stasis retinopathy [94] induced by severe carotid artery obstructive disease should not be mistaken for non-ischemic CRVO (Fig. 21.2.10a). Features that help to distinguish the two are the absence of optic disk swelling, more deeply located dot and blot bleedings, and decreased retinal artery perfusion pressure on ophthalmodynamometry. Furthermore, the bleedings in ocular ischemia are less frequent and spread over the mid-periphery. Anterior chamber flare and rubeosis iridis, combined with fair visual acuity, support ocular ischemia.

Diabetic retinopathy does not reveal papilla swelling. Furthermore, hard exudates are commonly seen with diabetic retinopathy and are rarely encountered in CRVO. When diabetic retinopathy seems more pronounced in one eye, a CRVO may be excluded by angiogram, which in turn will reveal the later vein filling in the CRVO eye.

Hypertensive retinopathy and **hyperviscosity syndrome** (Fig. 21.2.10b) can be easily ruled out by taking the blood pressure and performing laboratory tests such as sedimentation rate, blood cell count, and plasma proteins. A thorough history will reveal general symptoms not corresponding to CRVO.

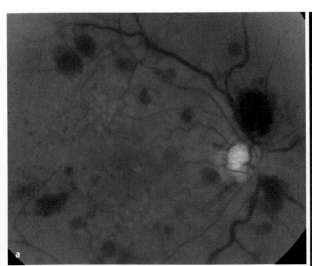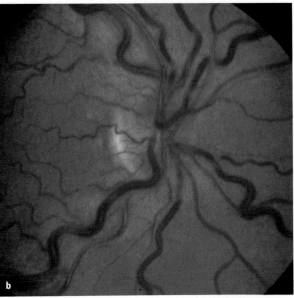

Fig. 21.2.10. Differential diagnosis of CRVO: **a** ocular ischemia in a patient with complete carotid stenosis, and **b** hyperviscosity syndrome in a patient with plasmocytoma

21.2.4.4.2 Late CRVO

Macular edema with only few or no bleedings indicates an old non-ischemic CRVO. When detected after cataract extraction it may be misleading. The diagnosis of CRVO is then supported by pigmentation beneath the fovea and proven by collaterals on the papilla (Fig. 21.2.6a, b).

Opticociliary collaterals (Fig. 21.2.6a) can easily be discerned from neovascular vessels on the disk by fluorescein angiogram, as they do not leak. When caused by a meningioma of the optic nerve the collaterals are not found in combination with a macular scar.

Macular scars after CRVO form years later and can be mistaken for the geographic atrophy of age-related macular degeneration (AMD). Hyperemia of the papilla, collaterals, and fluorescein angiogram exclude the scarred neovascular AMD (Fig. 21.2.6b).

An old CRVO is one of the most frequent causes of **neovascular glaucoma**. However, an intraocular melanoma should always be ruled out by echography if ophthalmoscopy is not possible and no history of vein occlusion exists.

21.2.4.5 Natural Course

Essentials

- Thirty-two percent of non-ischemic and 85% of ischemic CRVO eyes result in a visual acuity of 0.1 (20/400) or less after 1 year
- The better the initial visual acuity, the better the final outcome, but even non-ischemic CRVO may develop iris neovascularization
- The overall rate of neovascular disease is 16%, ranging from 10% in perfused eyes to 40% in non-perfused eyes
- The lower the initial visual acuity, the sooner iris neovascularization develops
- Patients with low vision must be observed closely during the first weeks and months so as not to miss neovascular disease

As already mentioned, retinal vein occlusion is a more subacute-chronic disease compared to the dramatic visual decline that occurs with retinal artery occlusion. Consequently, the prognosis is much more variable and difficult to predict. The natural course ranges from complete restitution to painful blindness in neovascular glaucoma. Thirty to 50% of eyes slowly progress over the first weeks and months. Unfavorable factors are given in Table 21.2.4. The final state is usually reached after 6–12 months, but can extend over years. The reasons for a poor outcome are twofold: (1) the continuous breakdown of the blood-retinal barrier leads to chronic macular edema or exudative retinal detachment, and (2) the ischemic inner retina with still functioning photoreceptor cells causes neovascular complications. The latter usually evolves between the 3rd and 6th months (see below), very seldom after a year. This makes careful observation of the non-ischemic CRVO during the 1st year mandatory. In addition, eyes with a favorable outcome have a considerable risk (2–5%) of developing a second episode within 2 years [58].

Table 21.2.4. Unfavorable factors in CRVO that may cause a poor prognosis

	Signs
Ophthalmologic factors	Ischemic CRVO (iris neovascularization within weeks)
	Bleedings in central cysts
	Dark, widespread dense bleedings
	Very low initial visual acuity ≤0.05 (20/400)
General factors	Badly treated arterial hypertension
	Diabetic retinopathy
	Patients' age over 80 years

21.2.4.5.1 Development of Visual Acuity

The visual outcome of a single case of CRVO is difficult to predict, but two large studies on its natural course have provided data for the range of the final visual acuity (FVA). The Central Vein Occlusion Study Group (CVOS) [168] found that visual outcome was largely dependent on the initial presenting visual acuity (IVA). Sixty-five percent of eyes with 20/40 (≥0.5) or better vision continued to maintain visual acuity in the same range at the end of the study. Patients in the intermediate group of 20/50 to 20/200 (0.4–0.1) had a variable outcome: 44% remained within the intermediate visual acuity group, but in 37% their FVA fell below 20/200. This means that overall only 19% ended up with a better FVA of ≥0.5.

These data fit well to the results of Quinlan et al. [134], although the patients' starting position differed slightly. In the CVOS group, 29% had an IVA >0.4, while in Quinlan's study only 17% started with this IVA. Calculated from their data shown in Figs. 21.2.7 and 21.2.8, the rates for spontaneous improvement (>2 lines) were about 15%, while 30% experienced a visual loss of more than two lines, the remaining stable group (± ≤2 lines) comprising about 55%. Interestingly, there was no large difference between the two types of CRVO, although the rate of improvement – with 20% in the ischemic group – was higher than in the non-ischemic variant, with only 13%. This reflects the latter's better IVA, which does not always allow an improvement of >2 lines.

When taking into account the final visual outcome with regard to reading ability (VA ≥ 0.4, 20/50), the CRVO's poor prognosis becomes evident. The data is not obtainable from the published CVOS results, but Quinlan et al. [134] found that 24% were initially able to read, while this rate fell to 15% after 1 year. No patient with ischemic CRVO reached this goal at the end of the study, while the rate in the non-ischemic variant fell from 35% to 23%. In the CVOS group the rate of patients with a visual acuity ≥ 0.5 fell from 29% to 26%, which corresponds somewhat to Qinlan's data. Thus one must stress that even non-ischemic CRVO cannot be regarded as a benign disease!

21.2.4.5.2 Neovascular Disease

The CVOS used the non-perfusion area (≥ 10 disk areas) in the fluorescein angiogram as a basal parameter for the risk of developing neovascular disease [168]. This has been justifiably criticized by Hayreh [69], and is supported by the CVOS data itself. It has been clearly demonstrated that all types of CRVO can progress to neovascular disease (Fig. 21.2.11a–d). The non-perfusion criteria applied to 18% of the 714 patients enrolled; 7% were indeterminate. Of the 714 eyes enrolled in the CVOS, 117 (16%) developed iris or anterior angle neovascularization (INV). These INV eyes were only partly in accordance with the non-perfusion criteria (Table 21.2.5). While 33% of the non-perfused CRVOs developed INV, the rates were 40% in the indeter-

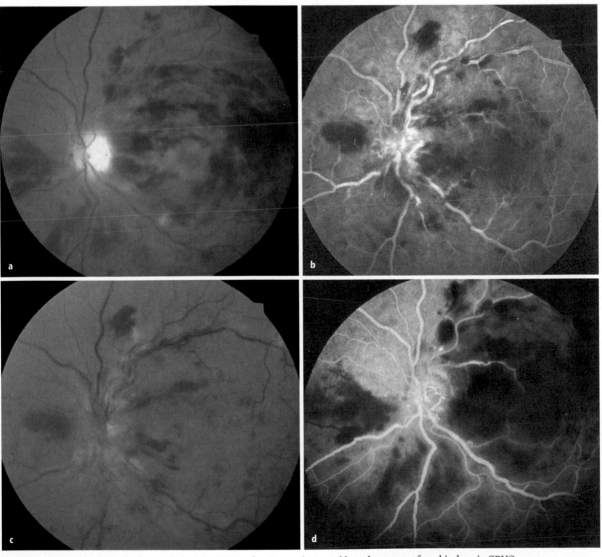

Fig. 21.2.11. Progression of **a, b** non-ischemic CRVO after 2 months to **c, d** largely non-perfused ischemic CRVO

Perfusion	Initial VA											
	≥ 0.5			0.4 – 0.1			< 0.1			All		
	N	INV	%INV	N	INV	%INV	N	INV	%INV	N	INV	%INV
Perfused	200	9	5	259	32	12	79	15	19	538	56	10
Non-perfused	8	0		32	7	22	86	34	40	126	41	33
Indeterminate	1	1	100	13	6	48	36	13	36	50	20	40
All	209	10	5	304	45	15	201	62	31	714	117	16

Table 21.2.5. Incidence of iris and/or angle neovascularization (INV) in correlation to initial visual acuity (VA) and perfusion [168]

minate and 9.5% in the perfused type. This clearly demonstrates the difficulties in providing a reliable prognosis based solely on a fluorescein angiogram.

The lower the initial visual acuity, the higher the rate of INV (Table 21.2.5). The rates were 5% (eyes with IVA > 0.4), 15% (IVA 0.1 – 0.4) and 31% (IVA < 0.1). The lower the IVA, the sooner the INV develops. Twenty percent of the INV evolves within the first 4 months, but only 7% in the group with an IVA > 0.4, while this rate is 44% in eyes with an IVA < 0.1. Though some correlation between the IVA and neovascular disease exists, development of INV cannot be excluded in patients with a good IVA! Patients with low vision must be observed closely during the first weeks and months.

Additional risk factors for neovascular disease besides low visual acuity and extensive capillary dropout seem to be venous tortuosity and extensive retinal hemorrhage. The latter may explain the high rate of INV in indeterminate CRVO because the obscured non-perfusion areas are difficult to assess. Retinochoroidal collateral veins are negatively associated with INV in late CRVO and may function in a protective manner against such an outcome [42].

Younger patients (≤ 50 years) with CRVO in general have a slightly better visual outcome [38, 149], with 42% ending up with VA ≥ 0.4 [135], and show a more variable clinical course. IVA in younger patients does not appear to be as predictive of visual or anatomic outcome as it does in older patients. On the other hand, 42% of young patients end up with a FVA ≤ 0.1 and 18% develop neovascular disease. These data do not differ much from that obtained for outcomes of older patients, thus highlighting how important careful follow-up of these patients is.

21.2.5 Treatment

Essentials

- Lowering of risk factors may inhibit CRVO's progression from a non-ischemic to the ischemic form, or reduce the frequency of recurrences or involvement of the second eye

- Randomized studies on fibrinolysis and hemodilution have demonstrated some positive effect on visual outcome but they did not prevent neovascular disease
- Isovolemic hemodilution has few contraindications, and can be used as basic treatment in most cases
- Selective or local thrombolysis continues to evolve as a potentially effective treatment
- Whether there is a definite place for intraocular steroids remains controversial and has to be evaluated in randomized studies
- The laser-induced, but often restricted, cilioretinal anastomosis formation with its wide spectrum of complications cannot be recommended
- The role of radiary optic neurotomy can only be evidenced by randomized studies and up to now cannot be recommended

As has been demonstrated above, CRVO has a wide range of outcomes and few reliable parameters for a clear-cut differentiation between a benign and an unfavorable course in the initial stage of the disease. This makes it difficult to assess the real benefit of one particular therapy and to decide on a treatment with definite risks, instead of waiting for a fortuitous course, as has always been advocated by Hayreh [69]. This also explains why there are only a few eye diseases for which such a wide range of different treatment modalities has been tried (Table 21.2.6); however, most of them were applied in preliminary pilot studies without definite success. For a review of earlier studies with additional treatment modalities including vasodilation, see Sedney [151]. Most of the studies are not only non-randomized, but have a variety of limitations which make it hard to evaluate the benefits claimed. These limitations include the lumping together of CRVO and branch vein occlusion, not differentiating between ischemic and non-ischemic CRVO or using mistaken criteria to do so, problems with study design and personal biases.

The goal of CRVO therapy should be to prevent: (1) chronic macular edema and macular scarring and (2) neovascular complications. Currently there

Table 21.2.6. Overview of recent medical therapies (controlled [CS] and prospective [PS] studies tried for CRVO). For review of earlier studies with additional treatment modalities, see Sedney [190] and for details see text [30, 99]

Rationale	Treatment	Design	Success
Cardiovascular risk factors	Lowering blood pressure	PS	(+)
	Hyperlipidemia	PS	(+)
Vasodilatation	Carbogen inhalation	PS	±
	Drug-induced	PS	±
Anticoagulation	Coumarin, heparin	PS	±
	Lowering platelet aggregation	PS	±
Thrombolysis	Systemic fibrinolysis high dose	CS	+[a]
	Systemic fibrinolysis low dose	PS	+
	Selective fibrinolysis	PS	(+)
	Retinal endovascular lysis	PS	+
	Intravitreal lysis	PS	±
Blood viscosity	Isovolemic hemodilution	CS	+
	Hypervolemic hemodilution	CS	+
	Hypovolemic hemodilution	CS	−
	Troxerutin	CS	+
Steroids	Systemic corticosteroids	PS	(+)
	Intravitreal	PS	(+)

[a] Treatment not recommended by the authors because of adverse effects

is no generally accepted treatment for the first, while the latter can be achieved by timely laser treatment, although there is debate concerning the time of intervention. The main treatments can be divided into two categories: (1) early intervention to improve the visual outcome, and (2) late treatment to avert painful blinding.

21.2.5.1 Early Treatment

As has been explained above, CRVO is a subacute or chronic disease that does not require urgent intervention within hours. Nevertheless, early treatment within the first weeks might be important to improve the chances of a reasonable visual acuity. Factors known to initiate and exacerbate the course should be dealt with aggressively within days or weeks.

21.2.5.1.1 Medical Treatment

Lowering of Risk Factors

Treatment of cardiovascular risk factors or associated medical or ocular conditions will not reverse the visual effects of CRVO [98] unless a real causal relationship such as with hyperviscosity syndromes is responsible for the occlusion ([112, 155], see below). However, lowering of risk factors may inhibit its progression from a non-ischemic to the ischemic form, or reduce the frequency of recurrences or involvement of the second eye [28]. Accordingly, arterial hypertension should be treated and cardiovascular risk factors (e.g., obesity, smoking, low physical activity, dehydration) reduced. While sleeping, the patient's upper part of the body should be slightly elevated to avoid high central venous pressure.

Glaucoma is regarded as a risk factor for CRVO, but we do not know whether a reduction in increased intraocular pressure will improve the prognosis of the disease, as supposed by Fong et al. [38]. There is little evidence that lowering intraocular pressure improves retinal blood flow in CRVO [67, 74].

Anticoagulation and Thrombolysis

Anticoagulant therapy cannot reverse thrombosis, so the rationale would, at the most, prevent the thrombus growth and stop the deterioration of CRVO. Past therapeutic regimens have included anticoagulant therapy with heparin sodium [29, 81, 171], bishydroxycoumarin [29, 140, 171] and acetylsalicylic acid [145]. Although these results were interpreted by the authors as positive, theirs and subsequent studies were not conclusive ([82], for review see [151]) and in no way fulfilled requirements of evidence-based data.

Thrombolysis seems a far more reasonable approach to dissolving a thrombus than anticoagulant therapy. Hence, there have been many attempts over the last 40 years using a variety of lytic agents [streptokinase, urokinase, recombinant tissue-type plasminogen activator (rt-PA)] and different types of access (systemic, intravitreal, retinal vessels) to restore blood flow and improve vision. Ideally, these agents should be given as soon as possible to prevent dreaded complications such as retinal ischemia and its sequelae. On the other hand, the induced lytic state places the patient at a considerable risk for systemic bleeding that can be life-threatening [30].

One of the first randomized, controlled studies for treatment of CRVO was carried out by Kohner et al. [99] in the early 1970s using *systemic streptokinase* in eyes which were predominantly ischemic. A statistically significant improvement in visual acuity was achieved but neovascular glaucoma was not averted, and breakthrough vitreous bleedings were observed. This prompted the authors not to recommend systemic lysis with streptokinase.

Oncel et al. [127] were the first to show the efficacy of rt-PA in a rabbit model with relatively fresh thrombi. Elman et al. [30] conducted a prospective clinical trial evaluating *systemic rt-PA* in 96 patients with CRVO. A visual increase of 3 lines was achieved in 42% of the eyes after 6 months, comparing well to the natural course with a 15–20% improvement. Unfortunately, although the results were very promising, one patient died of an intracranial hemorrhage.

Hattenbach et al. [63] tried to overcome the risk by using a *low-dose regimen (50 mg)* with front-loaded rt-PA. In their pilot study on ischemic CRVO, none of the 23 carefully selected patients suffered from serious bleeding. These relatively young patients (mean age 53 years) had an improvement rate (2 lines) of 44%, and 52% achieved a final visual acuity of 20/50, which is slightly better than the 42% found by Recchia et al. [136] for comparably young patients without treatment.

Lahey et al. [104] were the first to describe the use of *intravitreal rt-PA* for CRVO. Their cohort consisted of 23 eyes. After transscleral injection of about 100 μg of rt-PA, they observed a visual improvement to 0.5 or better in 34% of the eyes. In Glacet-Bernard et al.'s study [50] on 15 eyes, they concluded that intravitreal rt-PA did not significantly modify the course of the occlusion; yet Elman et al. [32] supported the positive results of Lahey.

Another attempt to circumvent this effective treatment's life-threatening complications was to avoid a lytic state by *local endovascular administration of lytic agents*. The potential advantages of this procedure are: (1) the drug is delivered to the thrombus site where rapid lysis can occur; (2) one can clearly see then the drug infusing into the retinal vein; and (3) the total dose is about 1% of the normal systemic dose. In addition, this low systemic dose is probably associated with a higher local concentration.

A French group [129, 169] used the femoral artery entry (as proposed for retinal artery occlusion by Schmidt et al. [148]) to deliver urokinase to the eye via the ostium of the ophthalmic artery. In a small pilot study, vision returned to normal within 24–48 h in four of 23 patients with CRVO, but the intervention took place in three of these patients after only 1 day of symptoms, suggesting that fibrinolysis at least is beneficial for early CRVO.

Weiss [174] chose an even more direct approach by cannulating a large retinal branch vein with a specialized cannula, and infusing a bolus of about 200 μg/ml of rt-PA toward the optic nerve. These investigators [175] reported the results of this novel surgical technique on a series of 28 patients. Of the 28 eyes with CRVO, 54% recovered >2 lines of visual acuity within 3 months, while 50% reached this goal even after a mean follow-up of 12 months. No difference was seen between ischemic and non-ischemic types. Vitreous hemorrhage was noted in 25% and new rubeosis iridis developed in 13% of the eyes. These favorable results have been corroborated by others [15] with similar results, while our group [36], despite unambiguous cannulation, was not able to confirm this success in a small pilot study.

In summary, thrombolysis continues to evolve as a potentially effective treatment. Though systemic rt-PA administration is risky and its use is limited to selected patients only, local intravitreal and endovascular delivery requires a randomized study to determine its efficacy as a promising alternative.

Viscosity Reduction

Rheological treatment does not interfere with the thrombus, but can improve the fluidity of blood and thus improve the oxygen supply in the poorly perfused retina. Though increased blood viscosity alone does not seem to play a major role in the pathogenesis of CRVO (see Sect. 2.3.1), its reduction is nevertheless useful, as blood viscosity increases logarithmically in low flow states (as in CRVO, especially in the ischemic type). Different procedures (hemodilution, plasmapheresis) and drugs (e.g., pentoxifylline, troxerutin) can affect the viscosity, but isovolemic hemodilution is the most effective [76, 125, 141, 180].

The value of rheological treatment in CRVO has been monitored in small randomized studies, one for troxerutin [48], one for pentoxifylline [24] and six for hemodilution (CRVO: [59, 62, 110, 183]; BRVO: [18, 60]).

Troxerutin improved retinal circulation times and visual acuity in a small series of 27 eyes with CRVO [48] versus placebo, and pentoxifylline [24] increased venous blood flow in a Doppler flow study. Both drugs can be helpful as supplementation when treating CRVO, however, hypertensive problems can occur with pentoxifylline.

Hemodilution. Randomized and controlled studies on hemodilution in CRVO have predominantly advocated the use of hemodilution [59, 62, 187], and only Luckie et al. [110] found no significant difference between the two groups concerning improvement rates. The data of Luckie et al. [110] are difficult to compare, as their regimens concerning hemodilution

Table 21.2.7. Course of visual acuity (with regard to initial visual acuity but not ische- mia). N = number of patients (all data taken from the scat- ter plots from the studies [59, 61, 62, 134, 168, 187])

Study type	Ref.	N	IVA ≥ 0.5	Final visual acuity ≥ 0.5	≥ 0.4	< 0.1	VA increase > 2 lines
Natural course	[168]	714	29%	28%	?	41%	?
	[134]	155	18%	19%	37%	48%	12%
Controls	[59[a], 62[a]]	30	40%	20%	40%	27%	10%
	[185[a]]	21	38%	19%	38%	43%	14%
IHD	[59[a], 62[a]]	33	30%	39%	55%	18%	46%
HHD/IHD	[185[a]]	19	42%	63%	63%	16%	37%
IHD	[61]	82	17%	35%	48%	32%	37%

VA visual acuity, *IVA* initial visual acuity, *IHD* isovolemic hemodilution, *HHD* hypervolemic hemodilution

[a] Controlled, randomized studies

Table 21.2.8. Course of visual acuity (with regard to initial visual acuity and ischemia). N = number of patients (all data taken from the scatter plots of the studies [61, 134])

Study type	Visual acuity > Ref.	Perf.	N	IVA ≥ 0.5	Final visual acuity ≥ 0.5	≥ 0.4	< 0.1	VA increase > 2 lines
Natural course	[134]	ni	107	26%	27%	54%	32%	11%
IHD	[61]	ni	47	30%	46%	63%	26%	28%
Natural course	[134]	i	48	0%	0%	0%	85%	15%
IHD	[61]	i	35	0%	20%	26%	40%	49%

VA visual acuity, *IVA* initial visual acuity, *IHD* isovolemic hemodilution, *Perf.* perfusion, *ni* non-ischemic, *i* ischemic

(hemodilution period shorter, blood replacement 0.9% saline instead of hydroxyethylstarch, thus creating a kind of hypovolemia) differed considerably. Furthermore, Luckie et al. [110] did not publish their original data, so that their outcome cannot be compared to the large studies on the natural course [134, 168]. On the other hand, the positive randomized studies [59, 62, 187] were small and not blinded, but their entrance data were comparable to those of the natural-course studies. In addition they were also supported by larger prospective non-randomized studies [51, 61].

Table 21.2.7 provides an overview over these studies' results in comparison to the natural course. The CVOS [168] has shown that initial visual acuity (IVA) is important for the final outcome (FVA). Although the percentage of control/untreated eyes with an IVA 0.5 varies from 18% to 40%, their FVA is very similar and only 10–14% improve more than two lines after 3–12 months. This rate is changed by hemodilution to 37–45%, indicating a better outcome after this rheologic treatment. The comparison of ischemic to non-ischemic types reveals that the effect is even more pronounced in ischemic types (Table 21.2.8). This adheres well to the principle that hemodilution is more effective in low flow states.

In summary, hemodilution slightly improves the final visual outcome of eyes with CRVO, but one should bear in mind at the same time that it does not prevent neovascular disease. Thus we strongly advocate isovolemic hemodilution as a basic treatment starting within the first 4–6 weeks after symptoms appeared [51]. This regimen does not interfere negatively with additional treatment modalities (intravitre-al steroids, retinal endovascular lysis, radial optic neuropathy) that are still not evidence based (see below).

Steroids

Corticosteroids can reduce the permeability of leaky vessels independent of whether this is caused by inflammation or by engorgement. Thus the macular edema can be reduced irrespective of the primary cause of CRVO and bridge the transient imbalance between inflow and outflow, which can last several months or even years (see Sect. 2.3.3). Routes of administration may be systemic, subtenon (not investigated with newer steroid generations) and intravitreal. The latter can be a crystalline triamcinolone acetonide deposit or an implanted sustained-release fluocinolone acetonide device [87].

Systemic Corticosteroids. Brückner [9] advised as early as 1955 the use of systemic corticosteroids in the treatment of CRVO. Hayreh [66, 69] claimed that among the patients with CRVO there is a small group that would respond favorably to prolonged systemic steroid treatment (up to 3 years). This group seems to be found more often in patients 50 years of age. However, the value of systemic steroids has never been evaluated within a large prospective and/or controlled study, and with the additional possibility of adverse reactions this has led to some reservations about the use of systemic steroids. We consider their use in young patients only [149].

Intravitreal crystalline triamcinolone, with its prolonged efficacy and lack of association with systemic adverse effects, was originally employed to reduce proliferative vitreoretinopathy [161, 162], but it has only been recently that ophthalmologists have taken advantage of its stabilizing effect on the blood-retinal barrier [187] to treat all kinds of macular edema [91].

Recent studies have also reported success in the treatment of cystoid macular edema in CRVO [55, 83, 91], and a series of pilot studies [6, 84, 85, 92, 100, 130, 183] ensued. In summary, 65 patients with non-ischemic CRVO and a duration of symptoms from 2 to 9 months made a significant gain of 3–5 lines in distance visual acuity. The reduction in macular edema was proven by optical coherence tomography. Both effects lasted for a maximum of 6 months and repeat injections were necessary for prolonged action, although Jonas detected triamcinolone in low, yet measurable concentrations up to 1.5 years after the intravitreal injection [91].

A rise in intraocular pressure was observed in most patients, but only one patient with intractable glaucoma was reported, in whom the triamcinolone had to be removed [93]. One must be aware of other possible complications (cataract, pseudoendophthalmitis, endophthalmitis, central artery occlusion, retinal detachment [88]).

Although the early use of triamcinolone is convincing, further study in the form of a randomized controlled study [150] is warranted to evaluate this new therapy's safety and efficacy. This NIH-funded trial (SCORE) includes patients with macular edema and a duration of symptoms of 3–8 months.

21.2.5.1.2 Surgical Treatment

Medical treatment modalities may help occasionally, but if at all, they only modestly improve visual outcome and they never prevent neovascular disease (see above). Thus early surgical treatment of CRVO based on various rationales has been attempted over the last 10 years: (1) dissolution of the thrombus as a causal treatment may circumvent the dangerous adverse reactions [30] when selectively applied (retinal endovascular lysis, REVL, intravitreal lysis, see above). This may permit the inclusion of patients who would otherwise not be recruitable for systemic low-dose lysis regimens [65]; (2) the creation of a chorioretinal anastomosis (CRA) may improve the outflow by bypassing the partly occluded central vein; (3) compression of the central vein against the lamina cribrosa fostered by an arteriosclerotic central artery may be relieved by enlarging the scleral outlet [172] and improving the outflow.

Chorioretinal Venous Anastomosis

McAllister et al. [119, 120] have advocated laser-induced iatrogenic chorioretinal anastomosis (CRA) as a treatment modality in CRVO. In eyes with non-ischemic CRVO, they created this bypass using a transretinal venipuncture technique. CRA formation via a surgical approach in ischemic CRVO [124] has also been recently tried in a pilot study.

Successful laser-induced creation of collaterals was reported in 33% [119], 37.5% [35] and 20% [13] and led to overall improvement rates >2 lines of 20–38%, which were not always associated with CRA formation [12]. Using refined techniques, investigators [108, 120, 121] later achieved CRA formation in 54–100%, and visual increase was observed in 49% of all treated patients. These results were not based on randomized studies and at best matched the visual acuity gain yielded by hemodilution (Tables 21.2.7, 21.2.8).

However, that procedure is by no means benign [12, 13, 35, 119]; it can cause immediate (intraretinal, subretinal, vitreal bleeding) and late complications (secondary neovascularization of the retina, choroid and anterior segment, subretinal fibrosis, non-clearing vitreous hemorrhage, traction retinal detachment). Because of the often restricted anastomosis formation and the wide spectrum of complications, attempts to create a CRA cannot be considered advisable.

Radial Optic Neurotomy

Opremcak et al. [128] have picked up on the idea of Vasco-Posada [172] to decompress the central vein by pushing a lancet with a sharp cutting edge on one side and an opposing blunt edge through an equal portion of the nasal optic nerve head and the adjacent sclera approximately 2 mm deep.

In their first uncontrolled retrospective pilot study [128], 11 patients with CRVO (five non-perfused) underwent radial optic neurotomy (RON) without complications. Seventy-three percent of patients demonstrated improved visual acuity after 9 months, but two developed neovascular glaucoma. This study has prompted innumerable pilot studies on RON with small patient numbers reporting positive [44, 126, 154, 176, 184] and negative results [117]. Roider et al. [139] summarized the data from five retinal centers and found a significant mean increase from logMAR 1.3 (Snellen 20/400) to 1.1 (Snellen 20/250) in 107 patients after a mean follow-up time of 6 months. Patients with angiographically identified collaterals (18/30) even showed a mean increase of 6 lines, supporting the observation of earlier studies [4, 41, 44, 154].

The procedure in no way is as benign as claimed by the initial investigators [128] as central artery injury [139, 176, 189], peripapillary retinal detachment [142], and choroidal neovascularizations [176] were described. Williamson et al. [184] were the first to report on visual field defects. This finding was

III 21

anticipated by Hayreh [72] and also reiterated by Feltgen et al. [37]. When one constantly looks for this complication, it occurs in 87% of cases [139]. The role of RON can only be evidenced by randomized studies and up to now cannot be recommended.

21.2.5.2 Late Treatment

The goal of late treatment is twofold: (1) to prevent long-standing macular edema ending up in macular scar or serous central retinal detachment with visual acuities between 0.05 (20/400) and 0.005 (hand movements) when other treatment modalities have failed, and (2) to avert painful blinding with neovascular glaucoma. The Central Vein Occlusion Study was designed to answer these questions [165].

Macular edema usually decreases spontaneously within 1–3 years, but irreversible damage of the photoreceptors often starts after as early as 3 months. The SCORE study [150] already mentioned with *intravitreal triamcinolone* will answer our question whether local steroids will be of value once other early interventions have failed.

Initial reports suggested that the grid pattern of laser photocoagulation may improve macular edema due to capillary leakage in CRVO [57, 95]. The CVOS [166] addressed this question in eyes with CRVO of at least 3 months duration and a visual acuity of 0.4 or less. Long-term results showed that the grid laser reduced angiographic leakage, but there was no visual benefit demonstrated at any point during the randomized study. Hence laser grid coagulation is not recommended in chronic macular edema of CRVO.

The intravitreal injection of inhibitors of vascular endothelial growth factor (bevacizumab, ranibizumab, pegabtanib) seems to be another upcoming, very promising treatment of macular edema [86, 157] with less adverse reactions than found with triamcinolone. However, randomized studies still lack and have to answer the question, whether it works better than other treatments and how often intravitreal have to be repeated.

Neovascular complications usually start with iris/angle neovascularization (INV/ANV) and culminate in secondary neovascular glaucoma (NVG). Several studies have evaluated whether prophylactic laser panretinal photocoagulation (PRP) in eyes with ischemic CRVO can reduce the long-term risk of INV/ANV and NVG. Some studies recommended early prophylactic PRP in eyes with ischemic CRVO [103, 114, 115], but Hayreh [71] claimed that early treatment leads to a worse outcome without prevention of neovascular disease.

In this context the CVOS [167] answered two questions: (1) does early PRP prevent anterior segment neovascularization in eyes with non-perfused CRVO (for definition see Sect. 21.2.4.5), and (2) is early PRP more effective than delaying treatment until INV/ANV is first observed? While early PRP did not prevent INV/ANV development in 20% of the eyes, the rate was only 30% in the non-treated eyes. There was an even greater resolution of INV/ANV by 1 month after PRP in non-early treatment eyes compared to early treatment eyes. Thus, PRP is useful only after INV/ANV has developed. On the other hand, this requires careful observation of eyes with ischemic or indeterminate CRVO, with monthly examinations necessary in order not to overlook INV/ANV, as well as promptly carrying out PRP in eyes with INV/ANV. If close follow-up is not possible, early PRP may be considered in high-risk patients.

When neovascular glaucoma cannot be prevented, additional cycloablations (cyclophoto- or cryocoagulation) may help to save the eye [69]. In desperate cases, peripheral retinectomy has been attempted with some success [5].

References

1. Abu el-Asrar AM, al-Momen AK, al-Amro S, Abdel Gader AG, Tabbara KF (1996–97) Prothrombotic states associated with retinal venous occlusion in young adults. Int Ophthalmol 20:197–204
2. Arend O, Remky A, Jung F, Kiesewetter H, Reim M, Wolf S (1996) Role of rheologic factors in patients with acute central retinal vein occlusion. Ophthalmology 103:80–6
3. Asherson RA, Merry P, Acheson JF, Harris EN, Hughes GR (1989) Antiphospholipid antibodies: a risk factor for occlusive ocular vascular disease in systemic lupus erythematosus and the 'primary' antiphospholipid syndrome. Ann Rheum Dis 48:358–61
4. Azad R, Verma D (2004) Does radial optic neurotomy induce surgical optociliary vessels in central retinal vein occlusion? Retina 24:182
5. Bartz-Schmidt KU, Thumann G, Psichias A, Krieglstein GK, Heimann K (1999) Pars plana vitrectomy, endolaser coagulation of the retina and the ciliary body combined with silicone oil endotamponade in the treatment of uncontrolled neovascular glaucoma. Graefes Arch Clin Exp Ophthalmol 237:969–75
6. Bashshur ZF, Ma'luf RN, Allam S, Jurdi FA, Haddad RS Noureddin BN (2004) Intravitreal triamcinolone for the management of macular edema due to nonischemic central retinal vein occlusion. Arch Ophthalmol 122:1137–40
7. Bertram B, Haase G, Remky A, Reim M (1994) Anticardiolipin antibodies in vascular occlusions of the eye. Ophthalmologe 91:768–71
8. Boyd S, Owens D, Gin T, Bunce K, Sherafat H, Perry D, Hykin PG (2001) Plasma homocysteine, methylene tetrahydrofolate reductase C677T and factor II G20210A polymorphisms, factor VIII, and VWF in central retinal vein occlusion. Br J Ophthalmol 85:1313–5
9. Brown BA, Marx JL, Ward TP, Hollifield RD, Dick JS, Brozetti JJ, Howard RS, Thach AB (2002) Homocysteine: a risk factor for retinal venous occlusive disease. Ophthalmology 109:287–90

21 **III**

10. Brown GC (1985) Central retinal vein obstruction: diagnosis and management. In: Reinicke RD (ed) Ophthalmol Annual 1:65–97

11. Brown GC, Shah HG, Magargal LE, et al. (1984) Central retinal vein obstruction and carotid artery disease. Ophthalmology 91:1627–33

12. Browning D, Rotberg M (1996) Vitreous hemorrhage complicating laser-induced chorioretinal anastomosis for central retinal vein occlusion. Am J Ophthalmol 122:588–9

13. Browning DJ, Antoszyk AN (1998) Laser chorioretinal venous anastomosis for nonischemic central retinal vein occlusion. Ophthalmology 105:670–7; discussion 677–9

14. Brückner R (1955) Zur Problematik der Zentralvenenthrombose und ihrer Therapie. Ophthalmologica 129:325–326

15. Bynoe LA, Hutchins RK, Lazarus HS, Friedberg MA (2004) Retinal endovascular surgery for central retinal vein occlusion: initial experience of four surgeons. Inv Ophthalmol Vis Sci 45S:1051

16. Cahill MT, Stinnett SS, Fekrat S (2003) Meta-analysis of plasma homocysteine, serum folate, serum vitamin B(12), and thermolabile MTHFR genotype as risk factors for retinal vascular occlusive disease. Am J Ophthalmol 136(6):1136–50

17. Chen HC, Gupta A, Wiek J, Kohner EM (1998) Retinal blood flow in nonischemic central retinal vein occlusion. Ophthalmology 105:772–5

18. Chen HC, Wiek J, Gupta A, Luckie A, Kohner EM (1998) Effect of isovolaemic haemodilution on visual outcome in branch retinal vein occlusion. Br J Ophthalmol 82:162–7

19. Chern S, Magargal LE, Annesley WH (1991) Central retinal vein occlusion associated with drusen of the optic disc. Ann Ophthalmol 23:66–9

20. Chua B, Kifley A, Wong TY, Mitchell P (2005) Homocysteine and retinal vein occlusion: a population-based study. Am J Ophthalmol 139:181–2

21. Ciardella AP, Yannuzzi LA, Freund KB, DiMichele D, Nejat M, De Rosa JT, Daly JR, Sisco L (1998) Factor V Leiden, activated protein C resistance, and retinal vein occlusion. Retina 18:308–15

22. Cole MD, Dodson PM, Hendeles S (1989) Medical conditions underlying retinal vein occlusion in patients with glaucoma or ocular hypertension. Br J Ophthalmol 73:693–698

23. David R, Zangwill L, Badarna M, et al. (1988) Epidemiology of retinal vein occlusion and its association with glaucoma and increased intraocular pressure. Ophthalmologica 197:69

24. De Sanctis MT, Cesarone MR, Belcaro G, Incandela L, Steigerwalt R, Nicolaides AN, Griffin M, Geroulakos G (2002) Treatment of retinal vein thrombosis with pentoxifylline: a controlled, randomized trial. Angiology 53 S1:S35–8

25. Demirci FY, Guney DB, Akarcay K, Kir N, Ozbek U, Sirma S, Unaltuna N, Ongor E (1999) Prevalence of factor V Leiden in patients with retinal vein occlusion. Acta Ophthalmol Scan 77:631–3

26. Di Crecchio L, Parodi MB, Sanguinetti G, Iacono P, Ravalico G (2004) Hyperhomocysteinemia and the methylenetetrahydrofolate reductase 677C-T mutation in patients under 50 years of age affected by central retinal vein occlusion. Ophthalmology 111:940–5

27. Dithmar S, Hansen LL, Holz FG (2003) Venöse retinale Verschlüsse. Ophthalmologe 7:561–77

28. Dodson PM, Kubicki AJ, Taylor KG, Kirtzinger EE (1985) Medical conditions underlying recurrence of retinal vein occlusion. Br J Ophthalmol 69:493–506

29. Duff IF, Falls HF, Linman JW (1951) Anticoagulant therapy in occlusive vascular disease of the retina. Arch Ophthalmol 46:601–13

30. Elman MJ (1996) Thrombolytic therapy for central retinal vein occlusion: results of a pilot study. Trans Am Ophthalmol Soc 94:471–504

31. Elman MJ, Bhatt AK, Quinlan PM, Enger C (1990) The risk for systemic vascular diseases and mortality in patients with central retinal vein occlusion. Ophthalmology 97:1543–48

32. Elman MJ, Raden RZ, Carrigan A (2001) Intravitreal injection of tissue plasminogen activator for central retinal vein occlusion. Trans Am Ophthalmol Soc 99:219–21

33. Faude S, Faude F, Siegemund A, Wiedemann P (1999) Activated protein C resistance in patients with central retinal vein occlusion in comparison to patients with a history of deep-vein thrombosis and a healthy control group. Ophthalmologe 96:594–9

34. Fegan CD (2002) Central retinal vein occlusion and thrombophilia. Eye 16:98–106

35. Fekrat S, Goldberg MF, Finkelstein D (1998) Laser-induced chorioretinal venous anastomosis for nonischemic central or branch retinal vein occlusion. Arch Ophthalmol 116:43–52

36. Feltgen N, Herrmann J, Agostini A, Hansen L (2005) Retinal endovascular lysis in central retinal vein occlusion: 1 year results of a pilot study. Ophthalmologe 102S

37. Feltgen N, Herrmann J, Hansen L (2005) Gesichtsfelddefekt nach radiärer Optikoneurotomie. Ophthalmologe 102: 802–4

38. Fong A, Schatz H, McDonald HR, et al. (1991) Central retinal vein occlusion in young adults (papillophlebitis). Retina 11:3–11

39. Fong ACO, Schatz H (1993) Central retinal vein occlusion in young adults. Surv Ophthalmol 37:393–417

40. Francis PJ, Stanford MR, Graham EM (2003) Dehydration is a risk factor for central retinal vein occlusion in young patients. Acta Ophthalmol Scan 81:415–6

41. Friedman SM (2003) Optociliary venous anastomosis after radial optic neurotomy for central retinal vein occlusion. Ophthal Surg Las Imag 34:315–7

42. Fuller JJ, Mason JO, White MF, McGwin G, Emond TL, Feist RM (2003) Retinochoroidal collateral veins protect against anterior segment neovascularization after central retinal vein occlusion. Arch Ophthalmol 121:332–6

43. Galvin R, Sanders MD (1980) Peripheral retinal haemorrhages with papilloedema. Br J Ophthalmol 64:262–6

44. Garcia-Arumii J, Boixadera A, Martinez-Castillo V, Castillo R, Dou A, Corcostegui B (2003) Chorioretinal anastomosis after radial optic neurotomy for central retinal vein occlusion. Arch Ophthalmol 121:1385–91

45. Girolami A, Pellati D, Lombardi AM (2005) FXII deficiency is neither a cause of thrombosis nor a protection from thrombosis. Am J Ophthalmol 139:578–9

46. Glacet-Bernard A, Bayani N, Chretien P, Cochard C, Lelong F, Coscas G (1994) Antiphospholipid antibodies in retinal vascular occlusions. A prospective study of 75 patients. Arch Ophthalmol 112:790–5

47. Glacet-Bernard A, Chabanel A, Lelong F, Samama MM, Coscas G (1994) Elevated erythrocyte aggregation in patients with central retinal vein occlusion and without conventional risk factors. Ophthalmology 101(9):483–7

48. Glacet-Bernard A, Coscas G, Chabanel A, Zourdani A, Lelong F, Samama MM (1994) A randomized, double-masked study on the treatment of retinal vein occlusion with troxerutin. Am J Ophthalmol 118:421–9

49. Glacet-Bernard A, Coscas G, Chabanel A, Zourdani A, Lelong F, Samama MM (1996) Prognostic factors for retinal vein occlusion: prospective study of 175 cases. Ophthalmology 103:551–60

50. Glacet-Bernard A, Kuhn D, Vine AK, Oubraham H, Coscas G, Soubrane G (2000) Treatment of recent onset central retinal vein occlusion with intravitreal tissue plasminogen activator: a pilot study. Br J Ophthalmol 84:609–13

51. Glacet-Bernard A, Zourdani A, Milhoub M, Maraqua N, Coscas G, Soubrane G (2001) Effect of isovolemic hemodilution in central retinal vein occlusion. Graefes Arch Clin Exp Ophthalmol 239:909–14

52. Gleuck CJ, Bell H, Vadlamani L, Gupta A, Fontaine RL, Wang P (1999) Heritable thrombophila and hypofibrinolysis. Possible causes of retinal vein occlusion. Arch Ophthalmol 117:43–9

53. Gottlieb JL, Blice JP, Mestichelli B, Konkle BA, Benson WE (1998) Activated protein C resistance, factor V Leiden, and central retinal vein occlusion in young adults. Arch Ophthalmol 116:577–9

54. Green WR, Chan CC, Hutchins GM, Terry JM (1981) Central retinal vein occlusion: A prospective histopathologic study of 29 eyes in 28 cases. Retina 1:27–55

55. Greenberg PB, Martidis A, Rogers AH, Duker JS, Reichel E (2002) Intravitreal triamcinolone acetonide for macular oedema due to central retinal vein occlusion. Br J Ophthalmol 86:247–8

56. Greiner K, Hafner G, Dick B, Peetz D, Prellwitz W, Pfeiffer N (1999) Retinal vascular occlusion and deficiencies in the protein C pathway. Am J Ophthalmol 128:69–74

57. Gutman FA, Zegarra H (1984) Macular edema secondary to occlusion of the retinal veins. Surv Ophthalmol 28:462–80

58. Hansen LL (1994) Behandlungsmöglichkeiten bei Zentralvenenverschlüssen. Ophthalmologe 91:131–145

59. Hansen LL, Danisevskis P, Arntz HR, Hovener G, Wiederholt M (1985) A randomised prospective study on treatment of central retinal vein occlusion by isovolaemic haemodilution and photocoagulation. Br J Ophthalmol 69: 108–16

60. Hansen LL, Wiek J, Arntz R (1988) Randomized study of the effect of isovolemic hemodilution in retinal branch vein occlusion. Fortschr Ophthalmol 85:514–6

61. Hansen LL, Wiek J, Schade M, Müller-Stolzenburg N, Wiederholt M (1989a) The effect and compatibility of isovolaemic haemodilution in the treatment of ischaemic and non-ischaemic central retinal vein occlusion. Ophthalmologica 199:90–99

62. Hansen LL, Wiek J, Wiederholt M (1989) A randomised prospective study of treatment of non-ischaemic central retinal vein occlusion by isovolaemic haemodilution. Br J Ophthalmol 73:895–9

63. Hattenbach LO, Wellermann G, Steinkamp GW, Scharrer I, Koch FH, Ohrloff C (1999) Visual outcome after treatment with low-dose recombinant tissue plasminogen activator or hemodilution in ischemic central retinal vein occlusion. Ophthalmologica 213:360–6

64. Hayreh SS (1976) So-called, central retinal vein occlusion. II. Venous stasis retinopathy. Ophthalmologica 172:14–37

65. Hayreh SS (1976) So-called, central retinal vein occlusion. I. Pathogenesis, terminology, clinical features. Ophthalmologica 172:1–13

66. Hayreh SS (1983) Classification of central retinal vein occlusion. Ophthalmology 90:458–474

67. Hayreh SS (2003) Management of central retinal vein occlusion. Ophthalmologica 217:167–88

68. Hayreh SS (2005) Prevalent misconceptions about acute retinal vascular occlusive disorders. Progr Ret Eye Res 24:493–519

69. Hayreh SS, Klugman MR, Beri M, Kimura AE, Podhajsky P (1990) Differentiation of ischemic from non-ischemic central retinal vein occlusion during the early acute phase. Graefes Arch Clin Exp Ophthalmol 228:201–17

70. Hayreh SS, Klugman MR, Podhajsky P, Kolder HE (1989) Electroretinography in central retinal vein occlusion: Correlation of electrographic changes with pupillary abnormalities. Graefes Arch Clin Exp Ophthalmol 227:549–61

71. Hayreh SS, Klugman MR, Podhajsky P, Servais GE, Perkins ES (1990) Argon laser panretinal photocoagulation in ischemic central retinal vein occlusion: a 10-year prospective study. Graefes Arch Clin Exp Ophthalmol 228:281–96

72. Hayreh SS, Opremcak EM, Bruce RA, Lomeo MD, Ridenour CD, Letson AD, Rehmar AJ (2002) Radial optic neurotomy for central retinal vein obstruction [comment]. Retina 22:374–7

73. Hayreh SS, Zimmerman B, McCarthy MJ, Podhajsky P (2001) Systemic diseases associated with various types of retinal vein occlusion. Am J Ophthalmol 131:61–77

74. Hayreh SS, Zimmerman MB, Beri M, Podhajsky P (2004) Intraocular pressure abnormalities associated with central and hemicentral retinal vein occlusion. Ophthalmology 111:133–41

75. Hayreh SS, Zimmerman MB, Podhajsky P (1994) Incidence of various types of retinal vein occlusion and their recurrence and demographic characteristics. Am J Ophthalmol 117:429–41

76. Heinen A, Brunner R, Hossmann V, Konen W, Roll K, Wawer T (1986) Different types of therapy having having hemorheological effects in patients with impairment of blood supply to the retina. Clin Hemorheol 6:61–79

77. Hirota A, Mishima HK, Kiuchi Y (1997) Incidence of retinal vein occlusion at the Glaucoma Clinic of Hiroshima University. Ophthalmologica 211:288–91

78. Hitchings RA, Spaeth GL (1976) Chronic RVO in glaucoma. Br J Ophthalmol 60:694–99

79. Hodgkins PR, Perry DJ, Sawcer SJ, Keast-Butler J (1995) Factor V and antithrombin gene mutations in patients with idiopathic central retinal vein occlusion. Eye 9:760–2

80. Hofman P, van Blijswijk BC, Gaillard PJ, Vrensen GF, Schlingemann RO (2001) Endothelial cell hypertrophy induced by vascular endothelial growth factor in the retina: new insights into the pathogenesis of capillary nonperfusion. Arch Ophthalmol 119:861–6

81. Holmin N, Ploman KG (1938) Thrombosis of the central vein of the retina treated with heparin. Lancet I:664–71

82. Ingerslev J (1999) Thrombophilia: a feature of importance in retinal vein thrombosis? Acta Ophthalmol Scan 77 619–21

83. Ip MS, Gottlieb JL, Kahana A, Scott IU, Altaweel MM, Blodi BA, Gangnon RE, Puliafito CA (2004) Intravitreal triamcinolone for the treatment of macular edema associated with central retinal vein occlusion. Arch Ophthalmol 122:1131–6

84. Ip MS, Kahana A, Altaweel M (2003) Treatment of central retinal vein occlusion with triamcinolone acetonide: an optical coherence tomography study. Semin Ophthalmol 18:67–73

III 21

85. Ip MS, Kumar KS (2002) Intravitreous triamcinolone acetonide as treatment for macular edema from central retinal vein occlusion. Arch Ophthalmol 120:1217–9

86. Iturralde D, Spaide RF, Meyerle CB, Klancnik JM, Yannuzzi LA, Fischer YL, Sorenson J, Slakter JS, Freund KB, Cooney M, Fine HF (2006) Intravetreal bevacizumab (AVASTIN) treatment for Macular edema in central retinal vein occlusion. Retina 26:279–84

87. Jaffe GJ, Ben-Nun J, Guo H, et al. (2000) Fluocinolone acetonide sustained drug delivery device to treat severe uveitis. Ophthalmology 107:2024–33

88. Jaissle GB, Szurman P, Bartz-Schmidt KU (2004) Nebenwirkungen und Komplikationen der intravitrealen Triamcinolonacetonid-Therapie. Ophthalmologe 101:121–8

89. Jannsen MCH, den Heijer M, Cruysberg JRM, Wollersheim H, Bredie SJH (2005) Retinal vein occlusion: a form of venous thrombosis or a complication of atherosclerosis. Thromb Haemost 93:1021–6

90. Jonas JB (2003) Ophthalmodynamometric assessment of the central retinal vein collapse pressure in eyes with retinal vein stasis or occlusion. Graefes Arch Clin Exp Ophthalmol 241:367–70

91. Jonas JB, Kreissig I, Degenring RF (2002) Intravitreal triamcinolone acetonide as treatment of macular edema in central retinal vein occlusion. Graefes Arch Clin Exp Ophthalmol 240:782–3

92. Karacorlu M, Ozdemir H, Karacorlu S (2004) Intravitreal triamcinolone acetonide for the treatment of central retinal vein occlusion in young patients. Retina 24:324–7

93. Kaushik S, Gupta V, Gupta A, Dogra MR, Singh R (2004) Intractable glaucoma following intravitreal triamcinolone in central retinal vein occlusion. Am J Ophthalmol 137:758–60

94. Kearns TP, Hollenhorst RW (1963) Venous-stasis retinopathy of occlusive disease of the artery. Proc Mayo Clin 38:304–11

95. Klein ML, Finkelstein D (1989) Macular grid photocoagulation for macular edema in retinal vein occlusion. Arch Ophthalmol 107:1297–1302

96. Klein R, Klein BE, Moss SE, et al. (2000) Epidemiology of retinal vein occlusion: the Beaver Dam Eye Study. Trans Am Ophthalmol Soc 98:133–9

97. Klien BA, Olwin JH (1956) A survey of the pathogenesis of retinal vein occlusion. Arch Ophthalmol 36:207–228

98. Kohner EM, Laatikainen L, Oughton J (1983) The management of CRVO. Ophthalmology 90:484–7

99. Kohner EM, Pettit JH, Hamilton AM, et al. (1976) Streptokinase in central retinal vein occlusion: A controlled clinical trial. Br Med J 1:550–3

100. Krepler K, Ergun E, Sacu S, Richter-Muksch S, Wagner J, Stur M, Wedrich A (2005) Intravitreal triamcinolone acetonide in patients with macular oedema due to central retinal vein occlusion. Acta Ophthalmol Scan 83:71–5

101. Kuhli C, Hattenbach LO, Scharrer I, Koch F, Ohrloff C (2002) High prevalence of resistance to APC in young patients with retinal vein occlusion. Graefes Arch Clin Exp Ophthalmol 240:163–8

102. Kuhli C, Scharrer I, Koch F, Ohrloff C, Hattenbach LO (2004) Factor XII deficiency: a thrombophilic risk factor for retinal vein occlusion. Am J Ophthalmol 137:459–64

103. Laatikainen L, Kohner EM, Khoury D, Blach RK (1977) Panretinal photocoagulation in central retinal vein occlusion: a randomised controlled clinical study. Br J Ophthalmol 61:741–53

104. Lahey JM, Fong DS, Kearney J (1999) Intravitreal tissue plasminogen activator for acute central retinal vein occlusion. Ophthalmic Surg Las 30:427–34

105. Larsson J, Andreasson S (2001) Photopic 30 Hz flicker ERG as a predictor for rubeosis in central retinal vein occlusion. Br J Ophthalmol 85:683–5

106. Larsson J, Olafsdottir E, Bauer B (1996) Activated protein C resistance in young adults with central retinal vein occlusion. Br J Ophthalmol 80:200–2

107. Larsson J, Sellman A, Bauer B (1997) Activated protein C resistance in patients with central retinal vein occlusion. Br J Ophthalmol 81:832–4

108. Leonard BC, Coupland SG, Kertes PJ, Bate R (2003) Long-term follow-up of a modified technique for laser-induced chorioretinal venous anastomosis in nonischemic central retinal vein occlusion. Ophthalmology 110:948–54

109. Liebreich R (1855) Ophthalmologische Notizen. 3. Apoplexia retinae. Graefes Arch Ophthalmol 1:346–351

110. Luckie AP, Wroblewski JJ, Hamilton P, Bird AC, Sanders M, Slater N, Green W (1996) A randomised prospective study of outpatient haemodilution for central retinal vein obstruction. Austr N Z J Ophthalmol 24:223–32

111. Luntz MH, Schenker HI (1980) Retinal vascular accidents in glaucoma and ocular hypertension. Surv Ophthalmol 25:163–167

112. Luxenberg MN, Mausolf FA (1970) Retinal circulation in the hyperviscosity syndrome. Am J Ophthalmol 70:588–598

113. Magargal LE, Brown GC, Augsburger JJ, Donoso LA (1982) Efficacy of panretinal photocoagulation in preventing neovascular glaucoma following central retinal vein occlusion. Ophthalmology 89:780–4

114. Magargal LE, Brown GC, Augsburger JJ, Parrish RK (1981) II. Neovascular glaucoma following central retinal vein occlusion. Ophthalmol 88:1095–1101

115. Magargal LE, Donoso LA, Sanborn G (1982) Retinal ischemia and risk of neovascularization following central retinal vein obstruction. Ophthalmology 89:1241–45

116. Mandelcorn MS, Nrusimhadevara RK (2004) Internal limiting membrane peeling for decompression of macular edema in retinal vein occlusion: a report of 14 cases. Retina 24:348–55

117. Martinez-Jardon CS, Meza-de Regil A, Dalma-Weiszhausz J, Leizaola-Fernandez C, Morales-Canton V, Guerrero-Naranjo JL, Quiroz-Mercado H (2005) Radial optic neurotomy for ischaemic central vein occlusion. Br J Ophthalmol 89:558–61

118. Matsui Y, Katsumi O, Mehta MC, Hirose T (1994) Correlation of electroretinographic and fluorescein angiographic findings in unilateral central retinal vein obstruction. Graefes Arch Clin Exp Ophthalmol 232:449–57

119. McAllister IL, Constable IJ (1995) Laser-induced chorioretinal venous anastomosis for treatment of nonischemic central retinal vein occlusion. Arch Ophthalmol 113:456–62

120. McAllister IL, Douglas JP, Constable IJ, Yu DY (1998a) Laser-induced chorioretinal venous anastomosis for nonischemic central retinal vein occlusion: evaluation of the complications and their risk factors. Am J Ophthalmol 126:219–29

121. McAllister IL, Vijayasekaran S, Yu DY, Constable IJ (1998b) Chorioretinal venous anastomoses: effect of different laser methods and energy in human eyes without vein occlusion. Graefes Arch Clin Exp Ophthalmol 236: 174–81

122. McGrath MA, Wechsler F, Hunyor ABL, Penny R (1978)

Systemic factors contributory to retinal vein occlusion. Arch Int Med 138:216–22

123. Michel J von (1878) Die spontane Thrombose der Vena centralis des Opticus. Graefes Arch Ophthalmol 24:37–70

124. Mirshahi A, Roohipoor R, Lashay A, Mohammadi SF, Mansouri MR (2005) Surgical induction of chorioretinal venous anastomosis in ischaemic central retinal vein occlusion: a non-randomised controlled clinical trial. Br J Ophthalmol 89:64–9

125. Neely KA, Ernest TJ, Goldstick TK, Linsenmeier RA, Moss J (1996) Isovolemic hemodilution increases retinal tissue oxygen. Graefes Arch Clin Exp Ophthalmol 234:688–94

126. Nomoto H, Shiraga F, Yamaji H, Kageyama M, Takenaka H, Baba T, Tsuchida Y (2004) Evaluation of radial optic neurotomy for central retinal vein occlusion by indocyanine green videoangiography and image analysis. Am J Ophthalmol 138:612–9

127. Oncel M, Peyman GA, Khoobehi B (1989) Tissue plasminogen activator in the treatment of experimental retinal vein occlusion. Retina 9:1–7

128. Opremcak EM, Bruce RA, Lomeo MD, Ridenour CD, Letson AD, Rehmar AJ (2001) Radial optic neurotomy for central retinal vein occlusion: a retrospective pilot study of 11 consecutive cases. Retina 21:408–15

129. Paques M, Vallee JN, Herbreteau D, Aymart A, Santiago PY, Campinchi-Tardy F, Payen D, Merlan JJ, Gaudric A, Massin P (2000) Superselective ophthalmic artery fibrinolytic therapy for the treatment of central retinal vein occlusion. Br J Ophthalmol 84:1387–91

130. Park CH, Jaffe GJ, Fekrat S (2003) Intravitreal triamcinolone acetonide in eyes with cystoid macular edema associated with central retinal vein occlusion. Am J Ophthalmol 136:419–25

131. Pe'er J, Folberg R, Itin A, Gnessin H, Hemo I, Keshet E (1998) Vascular endothelial growth factor upregulation in human central retinal vein occlusion. Ophthalmol 105:412–6

132. Peduzzi M, Codeluppi L, Poggi M, Baraldi P (1983) Abnormal blood viscosity and erythrocyte deformability in retinal vein occlusion. Am J Ophthalmol 96:399–400

133. Priluck IA, Robertson DM, Hollenhorst RW (1980) Long-term follow-up of occlusion of the central retinal vein in young adults. Am J Ophthalmol 90:190–202

134. Quinlan PM, Elman MJ, Bhatt AK, Mardesich P, Enger C (1990) The natural course of central retinal vein occlusion. Am J Ophthalmol 110:118–23

135. Radetzky S, Walter P, Fauser S, Koizumi K, Kirchhof B, Joussen AM (2004) Visual outcome of patients with macular edema after pars plana vitrectomy and indocyanine green-assisted peeling of the internal limiting membrane. Graefes Arch Clin Exp Ophthalmol 242:273–8

136. Recchia FM, Carvalho-Recchia CA, Hassan TS (2004) Clinical course of younger patients with central retinal vein occlusion. Arch Ophthalmol 122:317–21

137. Remky A, Wolf S, Knabben H, Arend O, Reim M (1997) Perifoveal capillary network in patients with acute central retinal vein occlusion. Ophthalmology 104:33–7

138. Ring CP, Pearson TC, Sanders MD, Wetherley Mein G (1976) Viscosity and retinal vein thrombosis. Br J Ophthalmol 60:397–410

139. Roider J, Hasselbach H, Rüfer F, Feltgen N, Schneider U, Bopp S, Hansen LL, Hoerauf H, Bartz-Schmidt KU (2005) Radial optic neurotomy in central vein occlusion: clinical results of 107 cases. IOVS (online) (ARVO)

140. Rosengren B (1948) The value of anti-coagulators in the treatment of retinal thrombosis. Acta Ophthalmol 26: 275–80

141. Roth S (1992) The effects of isovolumic hemodilution on ocular blood flow. Exp Eye Res 55:59–63

142. Samuel MA, Desai UR, Gandolfo CB (2003) Peripapillary retinal detachment after radial optic neurotomy for central retinal vein occlusion. Retina 23:580–3

143. Sayag D, Gotzamanis A, Brugniart C, Segal A, Ducasse A, Chambre V, Glacet-Bernard A (2002) Retinal vein occlusion and carotid Doppler imaging. J Fr Ophthalmol 25: 826–30

144. Scat Y, Morin Y, Morel C, Haut J (1995) Retinal vein occlusion and resistance to activated protein C. J Fr Ophtalmol 18:758–62

145. Schloßhardt H (1978) Beobachtungen und Erkenntnisse bei der Fotodokumentation mit Colfarit behandelter Netzhautthrombosen. Klin Mbl Augenheilk 172:895–902

146. Schmidt-Schönbein H (1978) Microrheology of erythrocytes, blood viscosity, and the distribution of blood flow in the microcirculation. Int Rev Physiol 9:1–45

147. Schmidt D, Schumacher M (1991) Occlusion of the central retinal vein due to spontaneous arteriovenous shunt of the carotid artery in the cavernous sinus. Fortschr Ophthalmol 88:683–6

148. Schmidt D, Schumacher M, Waklooh AK (1992) Microcatheter urokinase infusion in central retinal artery occlusion. Am J Ophthalmol 113: 429–34

149. Schumann M, Hansen LL, Janknecht P, Witschel H (1993) Isovolämische Hämodilution bei Zentralvenenverschlüssen von Patienten unter 50 Jahren. Klin Monatsbl Augenheilkd 203:341–46

150. Scott IU, Ip MS (2005) It's time for a clinical trial to investigate intravitreal triamcinolone for macular edema due to retinal vein occlusion: the SCORE study. Arch Ophthalmol 123:581–2

151. Sedney SC (1976) Photocoagulation in retinal vein occlusion. Doc Ophthalmol 40:1–241

152. Servais GE, Thompson HS, Hayreh SS (1986) Relative afferent pupillary defect in central retinal vein occlusion. Ophthalmology 93:301–3

153. Sinclair SH, Gragoudas ES (1979) Prognosis for rubeosis iridis following central retinal vein occlusion. Br J Ophthal 63:735–41

154. Spaide RF, Klancnik JM Jr, Gross NE (2004) Retinal choroidal collateral circulation after radial optic neurotomy correlated with the lessening of macular edema. Retina 24:356–9

155. Spalter HF (1959) Abnormal serum proteins and retinal vein thrombosis. Arch Ophthalmol 62:868–881

156. Sperduto RD, Hiller R, Chew E, Seigel D, Blair N, Burton TC, Farber MD, Gragoudas ES, Haller J, Seddon JM, Yannuzzi LA (1998) Risk factors for hemiretinal vein occlusion: comparison with risk factors for central and branch retinal vein occlusion: the eye disease case-control study. Ophthalmology 105:765–71

157. Stahl A, Agostini H, Hansen LL, Feltgen N (2007) Bevacizumab in retinal vein occlusionresults of a prospective case series. Graefes Arch Clin Exp Ophthalmol 2007 Mar 14; [Epub ahead of print]

158. Sullivan KL, Brown GC, Forman AR, et al. (1983) Retrobulbar anesthesia and retinal vascular obstruction. Ophthalmology 90:373–9

159. Takahashi K, Muraoka K, Kishi S, Shimizu K (1998) Formation of retinochoroidal collaterals in central retinal vein occlusion. Am J Ophthalmol 126:91–9

21 **III**

160. Tano Y, Chandler D, Machemer R (1980b) Treatment of intraocular proliferation with intravitreal injection of tri-amcinolone acetonide. Am J Ophthalmol 90:810–6

161. Tano Y, Sugita G, Abrams G, Machemer R (1980a) Inhibition of intraocular proliferation with intravitreal corticosteroid. Am J Ophthalmol 89:131–8

162. Tekeli O, Gursel E, Buyurgan H (1999) Protein C, protein S and antithrombin III deficiencies in retinal vein occlusion. Acta Ophthalmol Scan 77:628–30

163. Ten Doesschate MJL, Manschott WA (1985) Optic disc drusen and central retinal vein occlusion. Documenta Ophthalmol 59:27–31

164. The Central Retinal Vein Occlusion Study Group (1996) Risk factors for central retinal vein occlusion. Arch Ophthalmol 114:545–54

165. The Central Vein Occlusion Study Group (1993) Baseline and early natural history report. Arch Ophthalmol 111:1087–95

166. The Central Vein Occlusion Study Group (1995a) Evaluation of grid pattern photocoagulation for macular edema in central retinal vein occlusion. Ophthalmology 102:1425–33

167. The Central Vein Occlusion Study Group (1995b) A randomized clinical trial of early panretinal photocoagulation for ischemic central retinal vein occlusion. Ophthalmology 102:1434–44

168. The Central Vein Occlusion Study Group (1997) Natural history and clinical management of central retinal vein occlusion. Arch Ophthalmol 115:486–91

169. Vallée JN, Massin P, Aymard A, Paques M, Herbreteau D, Santiago PY, Losser MR, Gaudric A, Merland JJ (2000) Superselective ophthalmic arterial fibrinolysis with urokinase for recent severe central retinal venous occlusion: initial experience. Radiology 216:47–53

170. Vandenbroucke JP, Koster T, Briet E, Reitsma PH, Bertina RM, Rosendaal FR (1994) Increased risk of venous thrombosis in oral-contraceptive users who are carriers of factor V Leiden mutation. Lancet 344:1453–7

171. Vannas S, Raitta C (1966) Anticoagulant treatment of retinal vein occlusion. Am J Ophthalmol 62:874–79

172. Vasco-Posada J (1972) Modification of the circulation in the posterior pole of the eye. Ann Ophthalmol 4:48–59

173. Vossen CY, Naud S, Bovill EG, Weissgold DJ (2005) Normal retinal vasculature despite familial protein C deficiency. Am J Ophthalmol 139:944–5

174. Weiss JN (1998) Treatment of central retinal vein occlusion by injection of tissue plasminogen activator into a retinal vein. Am J Ophthalmol 126:142–4

175. Weiss JN, Bynoe LA (2001) Injection of tissue plasminogen activator into a branch retinal vein in eyes with central retinal vein occlusion. Ophthalmology 108:2249–57

176. Weizer JS, Stinnett SS, Fekrat S (2003) Radial optic neurotomy as treatment for central retinal vein occlusion. Am J Ophthalmol 136:814–9

177. Welch GN, Loscalzo J (1998) Homocysteine and atherothrombosis. N Engl J Med 338:1042–1050

178. Wenzler EM, Rademakers AJ, Boers GH, Cryusberg JR, Webers CA, Deutman AF (1993) Hyperhomocysteinaemia in retinal artery and retinal vein occlusion. Am J Ophthalmol 115:1162–67

179. Wiechens B, Schroder JO, Potzsch B, Rochels R (1997) Primary antiphospholipid antibody syndrome and retinal occlusive vasculopathy. Am J Ophthalmol 123:848–50

180. Wiek J, Schade M, Wiederholt M, Arntz HR, Hansen LL (1990) Haemorheological changes in patients with retinal vein occlusion after isovolaemic haemodilution. Br J Ophthalmol 74:665–9

181. Williamson TH (1997) Central retinal vein occlusion: What's the story? Br J Ophthalmol 81:698–704

182. Williamson TH, Keating D, Bradnam M (1997) Electroretinography of central retinal vein occlusion under scotopic and photopic conditions: what to measure? Acta Ophthalmologica Scand 75:48–53

183. Williamson TH, O'Donnell A (2005) Intravitreal triamcinolone acetonide for cystoid macular edema in nonischemic central retinal vein occlusion. Am J Ophthalmol 139:860–6

184. Williamson TH, Poon W, Whitefield L, Strothidis N, Jaycock P, Strothidis N (2003) A pilot study of pars plana vitrectomy, intraocular gas, and radial neurotomy in ischaemic central retinal vein occlusion. Br J Ophthalmol 87:1126–9

185. Williamson TH, Rumley A, Lowe GD (1996) Blood viscosity, coagulation, and activated protein C resistance in central retinal vein occlusion: a population controlled study. Br J Ophthalmol 80:203–8

186. Wilson CA, Berkowitz BA, Sato Y, Ando N, Handa JT, de Juan E Jr (1999) Treatment with intravitreal steroid reduces blood-retinal barrier breakdown due to retinal photocoagulation. Arch Ophthalmol 110:1155–9

187. Wolf S, Arend O, Bertram B, Remky A, Schulte K, Wald KJ, Reim M (1994) Hemodilution therapy in central retinal vein occlusion. One-year results of a prospective randomized study. Graefes Arch Clin Exp Ophthalmol 232:33–9

188. Wong TY, Larsen EK, Klein R, Mitchell P, Couper DJ, Klein BE, Hubbard LD, Siscovick DS, Sharrett AR (2005) Cardiovascular risk factors for retinal vein occlusion and arteriolar emboli: the Atherosclerosis Risk in Communities & Cardiovascular Health studies. Ophthalmology 112:540–7

189. Yamamoto S, Takatsuna Y, Sato E, Mizunoya S (2005) Central retinal artery occlusion after radial optic neurotomy in a patient with central retinal vein occlusion. Am J Ophthalmol 139:206–7

21.3 Branch Retinal Vein Occlusion

H. Hoerauf

Core Messages

- Branch retinal vein occlusion (BRVO) is a common retinal vascular disease and has public health significance
- It is important to distinguish the two different types of BRVO, major and macular BRVO, since their prognosis and management are different
- Optical coherence tomography (OCT) has become an indispensable tool for the evaluation of macular edema associated with BRVO, particularly for follow-up examinations, but fluorescein angiography (FLA) is still the gold standard at initial presentation in order to exclude macular ischemia
- While modern diagnostic techniques have greatly facilitated the establishment of the exact extent of a BRVO, only limited information is available regarding its pathophysiology, optimal therapy, the timing of therapy and secondary prevention
- The clinical course following BRVO is variable with the potential for minimal to severe permanent anatomical and functional damage
- Timely intervention can substantially reduce the incidence and severity of related complications and visual loss. A careful follow-up is recommended
- While scatter laser coagulation (LC) was shown to be very effective in causing regression of neovascularization, grid LC was less effective in reversing vision loss due to macular edema
- Patients with reduced baseline visual acuity (VA) or nonperfused macular edema have a poor visual outcome
- Pilot studies using isovolemic hemodilution, laser-shunting, thrombolytic agents, as well as surgical maneuvers such as arteriovenous sheathotomy and cannulation with tissue-type plasminogen activator administration suggest improvement over the natural history, but the evidence of currently available data is limited
- In recent years, pharmatherapeutic approaches with intravitreal injection of steroids or anti-vascular endothelial growth factor (VEGF) compounds appear promising. However, the beneficial effects on VA and macular thickness often regress as the compound is absorbed from the vitreous over several weeks to months and reinjections are necessary. Any successful therapy for BRVO would have to demonstrate proven benefit for a much longer period

21.3.1 History, Epidemiology and Classification

Essentials

- Branch retinal vein occlusion (BRVO) is three times more common than central retinal vein occlusion (CRVO)
- The prevalence of BRVO varies between 0.6% and 1.6%
- BRVO is unilateral in 90% of patients and occurs most commonly in the 7th decade
- In the majority of patients the superotemporal vein is affected
- BRVOs can be subdivided into major BRVOs and macular BRVOs with different prognosis and management
- The risk of recurrence is low

Retinal vein occlusion was first described in 1855 by Liebreich as retinal apoplexy [159] and in 1878 was recognized as thrombosis by Michel [245]. Venous occlusive diseases are among the most common reti-

21 III

nal diseases to be referred to a Retinal Center [68, 183]. Retinal vein occlusion including both central retinal vein occlusion (CRVO) and branch retinal vein occlusion (BRVO) represents the largest group of vascular retinal affections after arteriosclerotic hypertensive changes and diabetic retinopathy [49, 77]. Retinal venous occlusive disease can involve the central trunk or branches of the venous circulation. BRVO is three times more common than CRVO [51, 183]. Population-based epidemiological studies found an overall prevalence of BRVO of between 0.6% [136] and 1.6% [42, 175] increasing with age [51]. Some studies reported that patients with BRVO are older than patients with CRVO at the primary onset of the disease [92, 108] while others found no significant difference [7, 8, 198]. Based on United States (US) census figures for 2000, as many as 95,000 persons per year develop BRVO in the US alone [242]. Thus, BRVO is an ocular condition that has public health significance. Recognition of retinal vein occlusions is of particular importance because their complications are a cause of significant visual morbidity. BRVO is unilateral in 90% of patients and occurs most commonly in the 7th decade [51, 81, 122, 151], but it is not unknown in younger patients [1, 108]. At first onset of the disease 54% of patients with BRVO are 65 years or older, 41% are between 45 and 64 years old and 5% are younger than 45 years [1, 108]. In 44–60% of the patients with BRVO, the superotemporal vein is affected and most of the remainder (22–43%) occur in the inferotemporal quadrant [27, 98, 136, 151]. Occlusions in the nasal quadrants are rarely seen, but their incidence may be higher than is documented, because they are often unsymptomatic and therefore not detected. It is interesting to note that in comparisons of visual acuity (VA) between patients without BRVO and those with, only patients with BRVO affecting the superior temporal quadrant compared have a poorer VA [136].

BRVO typically occurs at an arteriovenous (AV) crossing site or at the edge of the optic disk as a hemi-central retinal vein occlusion (HCRVO). The observation that BRVO occurs at AV intersections was made over 100 years ago by Leber [153]. HCRVO pathologically is a variant of CRVO [250], which affects half of the retina due to an anatomical variation and is addressed in Chapter 24.2. BRVO can be subdivided with respect to the extent and location of

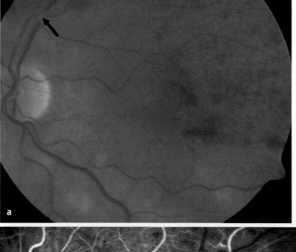

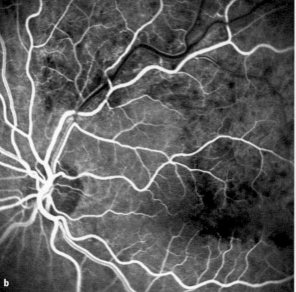

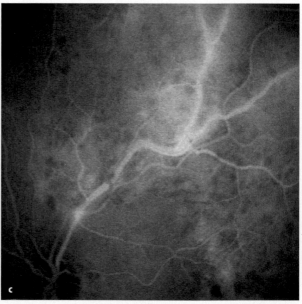

Fig. 21.3.1. a Major BRVO of the left eye affecting the superotemporal vein with superficial flame hemorrhages distal to the occlusion site (*arrow*). The associated artery appears narrowed and sclerotic. Fluorescein angiograms (FLAs) of **b** early and **c** late phase demonstrate delayed venous filling as well as venous dilation and tortuosity distal from the occlusion site, and retinal edema

the occluded area. Major BRVOs involve one of the major branch retinal veins usually near the optic disk affecting a quarter or more of the retina (Fig. 21.3.1a–c), and macular BRVOs involve one of the macular venules and a segment of the macular retina only (Fig. 21.3.2a–c) [107, 108, 109]. Clinical features, prognosis and management of the two types are different [108, 109]. Analogous to the classification of CRVO, ischemic and nonischemic types of BRVO can be differentiated.

Whereas retinal vein occlusions in general occur more frequently in males than in females, major BRVO is seen more often in females than males. Interestingly, BRVO affects the right eye more frequently than the left eye [108]. The proportion of patients who developed the same type of vein occlu-

sion in the same eye is 2.5% within 4 years. The risk of being affected by a BRVO in the fellow eye within the following 3.3 years is 4.0% for macular BRVO and 6.6% within the following 4 years for major BRVO respectively [108].

The risk of recurrence of the same branch in the same eye is very low [108] and even the risk of another major branch occlusion in the same eye is less than 1% after several years.

III 21

21.3.2 Anatomy and Histopathology

Essentials
- BRVO typically occurs at arteriovenous crossing sites
- Arterial overcrossings are at a higher risk of BRVO
- The adjacent artery is usually narrowed and sclerotic, the obstructed branch vein dilated
- At the crossing sites the outermost component of the vessel walls – the adventitia – fuse
- The common wall may be extremely thin
- Lymphocytes infiltrate the thrombus and vessel walls
- The vein recanalizes, but with persisting retinal damage around it

The crossing of retinal vessels such that the artery lies over the vein was considered the normal anatomical configuration (Fig. 21.3.3). In 1936, however,

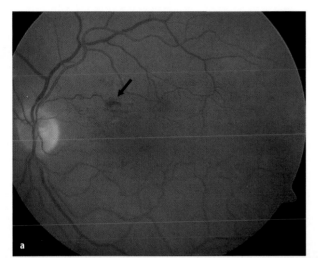

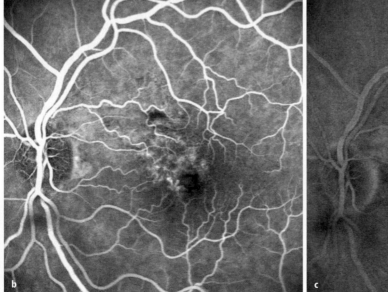

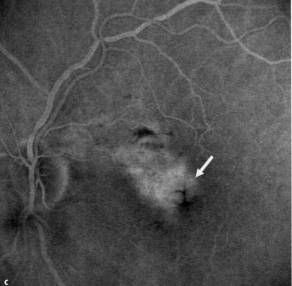

Fig. 21.3.2. Macular BRVO of the left eye with **a** fine superficial hemorrhages (*arrow*) in the area of the affected venule. **b** Early and **c** late phase FLA demonstrates leakage and cystoid macular edema (*arrow*)

Fig. 21.3.3. Schematic drawing of an arteriovenous intersection with hypertensive changes. In the majority of patients an arterial overcrossing is present at the site of occlusion and patients have a positive history of systemic hypertension

Fig. 21.3.5. Schematic drawing of an arteriovenous overcrossing. The ophthalmoscopic appearance of compression may be caused by anatomic variations where the retinal vein actually dips deep into the retina

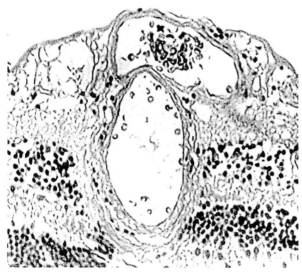

Fig. 21.3.4. Light microscopy of an arteriovenous crossing showing a common wall (from Seitz 1964) [214] predisposing the vein to degenerative changes in the arterial wall

Jensen [128] and Sallmann [207] observed that also venous overcrossings occur at 30% of all AV intersections in normal eyes. Later, the presence of both types of crossings was demonstrated histologically by Seitz [214]. These findings were confirmed clinically by Weinberg et al, who found in a large series of patients arterial overcrossings in 77.7% of eyes and venous overcrossings in 22.3% [249]. Furthermore, a significantly higher proportion of artery over vein crossings occurred in second-order veins than in first-order veins [146, 226]. But at the site of a BRVO an arterial overcrossing was present in 97.6% of eyes and a venous overcrossing only in 2.4%. Reports of a BRVO at an intersection at which the vein crosses over the artery are rare [47, 66, 128]. Thus, arterial overcrossings are at higher risk of BRVO than venous overcrossings [74, 80, 127, 248, 261]. Weinberg et al. further detected that the risk of BRVO in an eye is proportional to the number of arterial overcrossings in the eye [249] and they found that this type of overcrossing is most common in the temporal superior quadrant.

Histologic studies have demonstrated that as the branch retinal artery and vein converge on each other, the outermost components of their walls, the adventitia, fuse (Fig. 21.3.4). Histologically, the clinical impression of venous compression could not be proven [127, 214]. The vein deviates around the artery, dipping deep into the retina in arterial overcrossings (Fig. 21.3.5). Seitz attributed this prominent clinical appearance of crossing phenomena to the deeper position of the vein, rather than to true compression of the vessel. Seitz stated that the ophthalmoscopic appearance of compression is caused by thickening of the adventitial sheath and surrounding glial proliferation, both of which are less transparent tissues [214]. It has been shown that fusion of the vessel walls can continue until artery and veins share a common medium as they cross [139, 140, 219]. Recently, a histopathological examination of one patient with BRVO showed that the common wall measured 4 μm at the occlusion site [72].

Nonischemic and ischemic types of BRVO show similar histopathologic changes. They only differ in their extent of retinal destruction. The ischemic type represents a hemorrhagic infarct of the retina with an extracellular edema in the nerve fiber and ganglion cell layer. This causes dilation and congestion of vein and rhexis hemorrhages [138] from capillaries mainly in the nerve fiber layer (scattered superficial), but also in deeper layers (spherical). In both types of BRVO, retinal hemorrhages and ischemic necrotic areas reabsorb after months, resulting often in a glial scar. During recirculation the affected capillaries reopen, and with the dilation AV anastomoses develop bridging the destructed area. In later stages irreversible fibrotic alterations of the vascular structure occur [50, 196, 197].

21.3.3 Pathogenesis

Essentials

- The cause of BRVO is likely to be a multifactorial process including mechanical obstruction by degenerative changes in the arterioles, abnormal blood constituents and impedance of blood flow causing increased blood viscosity
- Endothelial damage
- Retinal veins are significantly influenced by the pathology of neighboring retinal arteries due to a common wall at the crossing site
- BRVO may lead to partial occlusion of the lumen, but only rarely to complete obstruction
- Inflammatory component in thrombus and vessel wall
- Hypoxia stimulates increase of vascular endothelial growth factor
- Vitreous attachment at the macula or at arteriovenous crossings may play an important pathogenetic role

Despite the great significance of BRVO in causing severe and sometimes permanent visual loss, its pathogenesis is not yet fully understood and has been the subject of some controversy.

Several authors have discussed the theory of mechanical obstruction [226, 248, 261] and others have suggested that BRVO is a result of arterial insufficiency [189, 197]. The pathological process at the site of the occlusion consists of degenerative changes in the vessel walls, abnormal blood constituents, and blood flow (stasis). These three classical components, known as Virchow's triad, that play a role in thrombogenesis are interrelated. The components of changes in the vessel wall and blood flow have been investigated in humans and animals [84, 138, 196, 197]. The fact that BRVOs are typically located at arterial overcrossings suggests a hemodynamic difference between arterial and venous overcrossings. Frangieh et al. found fresh or recanalized venous thrombi in histological sections of nine eyes with

BRVO at an AV crossing site where both vessels share a common adventitial sheath [80]. They concluded that a thrombus of the branch vein was probably the primary event and the other vascular changes occurred secondarily. The deviation of the retinal vein at crossing sites (Fig. 21.3.5) may lead to consecutive hemodynamic turbulence and thus predispose to thrombus formation at crossing sites, disputing the theory of mechanical compression by the artery. A difference in the distension of the vein lying beneath the artery within the retina as opposed to between the internal limiting membrane and the artery may explain the disparity of risk between arterial and venous overcrossings [248, 249]. Experimental studies have demonstrated that most of the manifestations of BRVO in humans, including nonperfusion and secondary changes in the retinal arteries, can be reproduced by experimental occlusion of the retinal vein in animals [100, 101, 114, 115, 137, 138, 201].

Although a much debated point, a complete thrombotic occlusion of the vein is generally thought not to occur. There is always evidence of some venous flow as shown by fluorescein angiography [218]. A consideration of the rheological events at the time of occlusion may add further insight into the pathophysiology of the condition. Over the last few years especially hypercoagulability has attracted particular increasing interest. It is a well documented and accepted pathological finding that the primary event of retinal vein occlusion includes endothelial cell proliferation in the vein wall associated either with degeneration of the endothelium and secondary thrombus formation or with severe phlebosclerosis [100, 101]. The exact mechanism underlying this primary event is not clear, but clinical studies indicate that it is multifactorial [59, 61, 62]. Apart from hemodynamic factors, triggering of local or systemic cardiovascular risk factors such as hypertension and hyperlipidemia have to be considered (see 3.7). The theory of arterial occlusion producing the clinical features of BRVO has been discounted [218] but arterial disease is generally agreed to play a significant role in the aetiology. Population based studies have found strong associations between BRVO and hypertension, focal arteriolar narrowing and arteriovenous nicking [136, 257]. This findings are consistent with clinical experience [74, 226, 248]. Moreover, the arterial stiffness is increased in patients with BRVO [177]. Localized arteriosclerotic processes may contribute to stasis and occlusion in adjacent retinal veins, and may explain some of the association between retinal vein occlusion and the risk factors classically associated with arterial or arteriolar disease. In BRVO impedance of blood flow is virtually

21 **III**

always at AV crossings [86] where there is either pressure on the vein or a thickened wall due to arterial disease or endothelial proliferation, or both. A triggering mechanism at the site of occlusion might set off a rheological vicious circle, the decrease in flow causing an increase in blood viscosity [200] followed by a further decrease in flow.

It was demonstrated that more eyes with BRVO develop partial vitreous separation than control eyes [212]. Thus, the vitreous may also play a role in compressing susceptible AV crossings as evidenced by studies demonstrating that eyes with decreased axial length and hyperopia are at increased risk for BRVO because of the higher likelihood of vitreomacular attachment at AV crossings [14, 165]. Interestingly, BRVO has not been reported in the literature in any eye following vitrectomy.

21.3.4 Clinical Appearance and Symptoms

21.3.4.1 Early Findings

Essentials
- Typical features include edema, and scattered superficial and deep retinal hemorrhages over a triangular retinal sector whose apex is located to the occlusion site
- The obstructed vein is characteristically dilated and tortuous distal to the occlusion, and the associated artery is usually narrowed and sclerotic

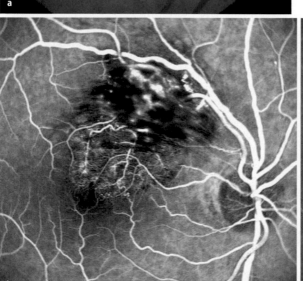

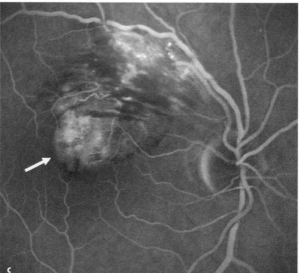

Fig. 21.3.6. a BRVO of the right eye involving the superotemporal vein with flame-shaped retinal hemorrhages radiating in a wedge. **b** Early and **c** late phase FLA show cystoid macular edema (*arrow*). **d** OCT (scan, see **a**) clearly demonstrates the associated retinal and macular edema extending to the foveal center together with cystoid spaces (*asterisk*)

- Dilated capillaries and microaneurysms are found in the area of BRVO
- Cotton wool spots may be present indicating ischemia
- The nearer the occlusion occurs to the optic disk, the greater the extent of the affected retina and the more serious the complications
- Depending on the site of occlusion cystoid macular edema may be present
- Macular BRVO may show only subtle features, such as microaneurysms in a limited sector

Due to their typical ophthalmoscopic features retinal vein occlusions are easily diagnosed (Fig. 21.3.6a–d). The first ophthalmoscopic signs are fine retinal hemorrhages at the AV crossing site. Distal from the crossing site the vein appears dilated, congested and tortuous. Retinal edema in the involved area is usually present. The involved retina demonstrates variable degrees of scattered superficial and deep retinal hemorrhages which respect the horizontal midline. Superficial hemorrhages are located in the nerve fiber layer and follow the nerve fiber layer course in an arcuate wedge of retina having its apex at the site of obstruction. The resorption of hemorrhages often takes several months up to a year. The area of retina involved depends on the size and location of the affected vein. Macular BRVO is a subgroup that may show only subtle clinical and angiographic clues, such as microaneurysms in a limited sector of the macular region [123] (Fig. 21.3.2). In contrast to its

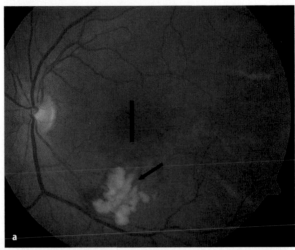

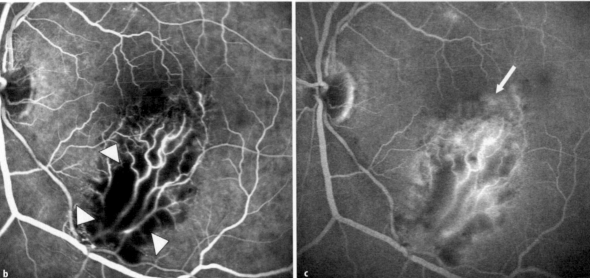

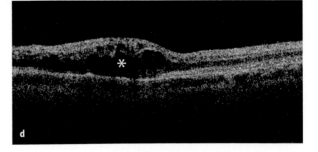

Fig. 21.3.7. Macular BRVO with **a** an intense cotton wool spot formation (*arrow*). **b** Early and **c** late phase FLA highlights the blockage at the site of cotton wool spots (*arrowheads*) and cystoid macular edema (*arrow*). **d** OCT (scan, see **a**) illustrates the cystoid changes (*asterisk*)

21 III

mild appearance visual acuity can be adversely affected due to macular edema or central ischemic damage (Fig. 21.3.7a–d). In general, the nearer the occlusion occurs to the optic disk, the greater the extent of the affected retina and the more serious the complications [27, 98]. When the ischemia results in a nerve fiber infarct, cotton wool spots develop (Fig. 21.3.7c).

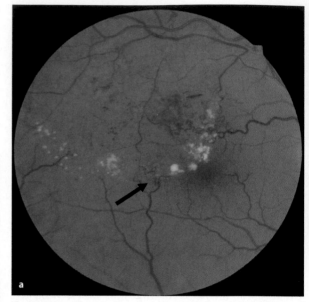

Fig. 21.3.9. a Chronic long-standing BRVO with deposition of lipid exudates and small tortuous collateral vessels crossing the horizontal raphe (*arrow*) bypassing the occluded segment is demonstrated clearly in **b** fluorescein angiography (*arrow*)

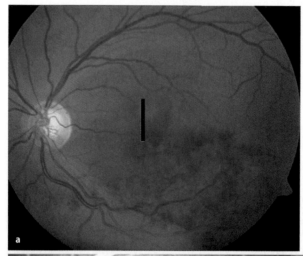

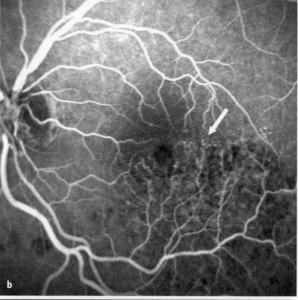

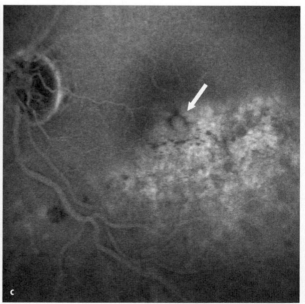

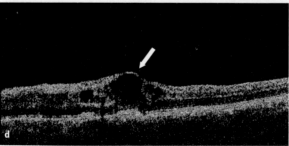

Fig. 21.3.8. a An old BRVO of the left eye affecting the inferotemporal vein with partial reabsorption of hemorrhages and tortuosity of the affected veins. **b** Early phase FLA shows numerous microaneurysms (*arrow*) in the affected area and **c** cystoid macular edema (*arrow*) in late phase FLA can be seen. **d** OCT (scan, see **a**) clearly reveals the cystoid changes (*arrow*)

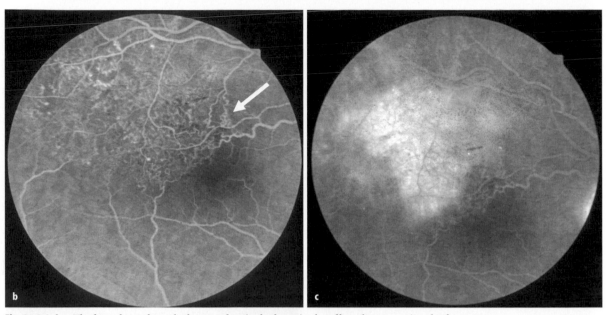

Fig. 21.3.9. b, c The late phase shows leakage and retinal edema in the affected area sparing the fovea

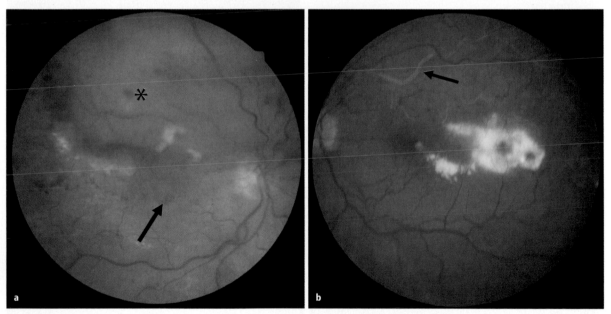

Fig. 21.3.10. a Long-standing midperipheral BRVO of the right eye with associated serous retinal detachment (*asterisk*), lipid exudates and foveal pigmentary dispersion (*arrow*). **b** Long-standing BRVO of the left eye affecting the superotemporal vein with sheathed retinal venules (*arrow*) and massive hard exudates

Retinal vein thrombosis causes increased venous pressure and may lead to retinal capillary decompensation with macular edema, which is the most frequent cause of loss of vision in BRVO. At the macula, edema is clinically recognized by a thickening of the macular retina that is often accompanied by cystoid spaces (Fig. 21.3.8a–d). Microvascular abnormalities of BRVO include dilated capillaries and microaneurysms that develop within the area of the vein occlusion. Chronic leakage from these abnormal vessels can contribute to macular edema and cause the deposition of lipid exudates in the retina (Fig. 21.3.9a–c). Large capillary or venous macroaneurysms may develop within the territory of the BRVO as well [250]. Serous retinal detachment with massive macular hard exudates has been described as a rare complication of BRVO (Fig. 21.3.10a, b) [229].

21

21.3.4.2 Late Findings and Complications

Essentials

- Occluded, sheathed retinal venules in the affected area
- Chronic leakage leading to chronic cystoid macular edema and lipid exudate deposition
- Collateral vessel formation at the edge of the affected area mostly located temporal to the fovea draining into the uninvolved quadrant
- Retinal neovascularizations (NVEs) develop in one-third of patients with major BRVO
- NVEs occur mostly at the border of perfused and nonperfused retina; neovascularization at the disk (NVD) is rare
- A significant risk for neovascularization exists when the area of capillary nonperfusion exceeds 5 disk diameters
- Neovascularization of the iris is extremely rare
- Complications of neovascularization include vitreous hemorrhage and fibrovascular membranes with consecutive tractional retinal detachment
- Visually relevant late complications are epiretinal membranes, hard exudates, chronic cystoid macular edema, retinal pigmentary dispersion, subretinal fibrosis and macular hole formation

Older BRVOs are characterized by occluded and sheathed retinal venules in the affected sector (Fig. 21.3.11). Weeks to months after the onset of BRVO, collateral vessel formation can be observed characteristically located at the edge of the involved area. Typical collaterals are usually small tortuous venous channels that cross the horizontal raphe mostly temporal to the fovea and drain into the venous circulation of the uninvolved quadrant (Fig. 21.3.12a, b). They may be difficult to distinguish

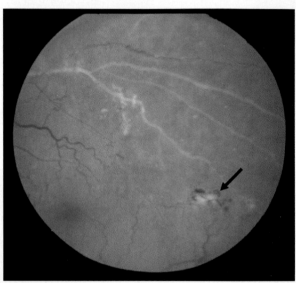

Fig. 21.3.11. Occluded and sheathed retinal venules in a long-standing BRVO affecting the superotemporal quadrant of the left eye. Neovascularization is present (*arrow*)

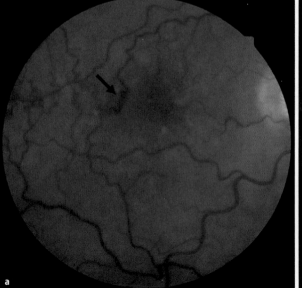

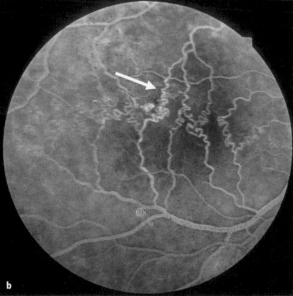

Fig. 21.3.12. a Small tortuous collateral vessels crossing the horizontal raphe between the affected and unaffected retina. **b** In FLA collateral vessels (*arrow*) show no leakage in contrast to neovascularization

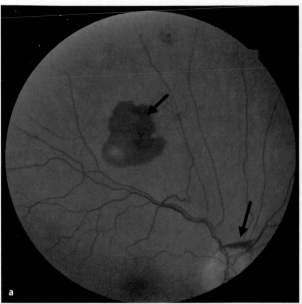

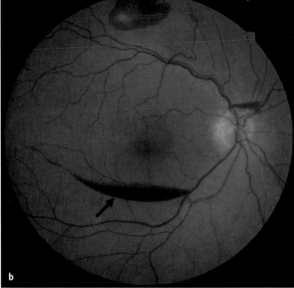

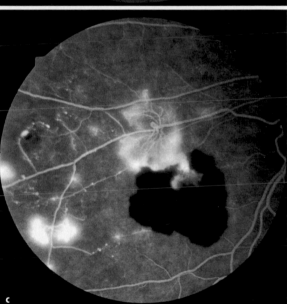

Fig. 21.3.13. a Retinal neovascularization secondary to BRVO in the superior midperiphery (*arrow*) and at the optic disk (*arrow*) of the right eye causing **b** subhyaloidal hemorrhage (*arrow*). **c** FLA reveals leakage of the new vessels

from retinal neovascularizations. These collaterals may pass from the territory of the occlusion to a point proximal to the site of occlusion or to an uninvolved vein. They may take the form of a vein to vein anastomosis, bypassing the occluded segment and then exiting through the central retinal vein. Reversal of blood flow toward the arterial system can also occur in response to the elevated venous pressure [192], although systemic hypertension can prevent this release mechanism from occurring [20]. Alternatively, arteriovenous shunts that bypass the capillary bed may occur at the AV crossing site [232], whereas in some instances unrelieved venous pressure can result in rupture of the vein wall [214].

The risk of complications can be attributed to the location of the BRVO, the extent and severity of the damage, and the adequacy of compensatory mechanisms. Retinal neovascularization may occur at the border of perfused and nonperfused retina (Fig. 21.3.13a–c) but can rarely occur away from the territory of the BRVO [76]. Neovascularization of the disk is much less common, and when it occurs, it tends to be concurrent with retinal neovascularization. A significant risk for the development of retinal neovascularization exists when the area of capillary nonperfusion exceeds 5 disk diameters (see Fig. 21.3.16) [29]. In the ischemic type of BRVO the risk for neovascularization is 36%, whereas it is 22% overall. Hayreh et al. reported a 28.8% incidence of retinal neovascularization following major BRVO.

Studies demonstrated that the majority of untreated eyes with retinal neovascularization will develop vitreous hemorrhage [27, 29, 98, 218]. In contrast to CRVO neovascularization of the iris is extremely rare. In a study on ocular neovascularization including 264 eyes with BRVO, no neovascular glaucoma was observed [109].

In advanced stages, preretinal hemorrhage (Fig. 21.3.14a–c) and vitreous hemorrhage, and more rarely fibrovascular membranes with consecutive tractional retinal detachment (Fig. 21.3.15), may develop. Rhegmatogenous retinal detachment is a rare complication of BRVO, but when breaks occur they tend to be located posterior to the equator and result from traction exerted by fibrovascular prolif-

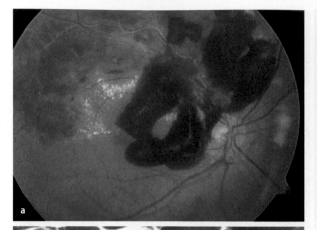

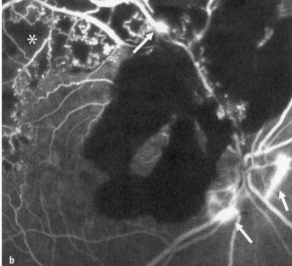

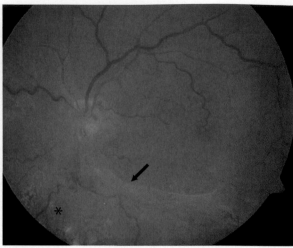

Fig. 21.3.15. A fibrovascular epiretinal membrane along the inferior vessel arcade (*arrow*) appearing inactive after scatter LC (*asterisk*) but still exerting retinal traction

eration or secondary to ischemic retinal degeneration with hole formation [210, 211]. Vitreous hemorrhages occur in approximately 7–20% of patients with BRVO [151]. Furthermore, less severe but still visually relevant complications in later stages include macular pucker, chronic cystoid macular edema, retinal pigmentary dispersion, subretinal scarring, macular hole and atrophy of inner retinal layers. Permanent and vision-limiting RPE changes can develop from long-standing edema. Rarely, exudative retinal detachment can develop within the affected area and is usually associated with ischemia [99].

21.3.5 Clinical Evaluation and Diagnostic Methods

21.3.5.1 Visual Function and Perimetry

Essentials
- Visual acuity (VA) may be unaffected to severely impaired depending on the localization of the BRVO, the severity of macular edema and the extent of capillary bed nonperfusion
- Tiny occlusions can be visually significant if the fovea is affected
- Although vision may be acutely reduced, 50–60% of patients may achieve 20/40 VA at the end of 1 year
- VA alone may be misleading in evaluating the visual function and visual field testing may be more valuable

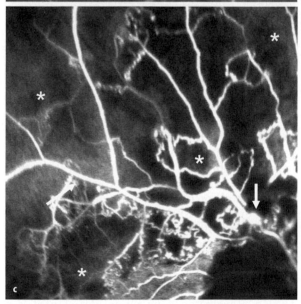

Fig. 21.3.14. a BRVO with secondary neovascularization, consecutive preretinal hemorrhage involving the macula, retinal edema, lipid exudates and sheathed venules. **b** Central and **c** peripheral FLA show extensive capillary nonperfusion (*asterisks*) and neovascularization (*arrows*)

The visual impact of a BRVO is related to the site and size of the occlusion, but even very tiny occlusions can be visually significant if the fovea is affected [123]. Visual field defects or visual loss are reasons to consult the ophthalmologist. VA is reduced in BRVO; it is because the macula has been affected by intraretinal hemorrhages, edema or ischemia. Later, other complications such as vitreous hemorrhage or retinal detachment may also impair vision. However, VA alone may be misleading in evaluating the visual function in patients with BRVO. VA increase on follow-up does not necessarily reflect a genuine visual improvement, since the patient learns by experience to fixate eccentrically [109]. Therefore, Hayreh estimates the information provided by the visual fields, plotted with the Goldman perimeter, an indispensable tool that is found to be most valuable in evaluation and management of retinal vein occlusions [109].

21.3.5.2 Angiographic Features

Essentials
- Angiographic findings in BRVO reflect changes in the permeability, caliber, and patency of retinal vessels
- Fluorescein angiography (FLA) helps to distinguish leakage without capillary nonperfusion from leakage with capillary nonperfusion
- Macular edema associated with ischemia may or may not be associated with fluorescein leakage
- An intact perifoveal capillary perfusion is the prerequisite for macular grid laser photocoagulation
- FLA maps out the extent of ischemia assisting in the detection of patients at higher risk of neovascularization and those requiring closer follow-up examinations
- A peripheral FLA may be helpful to detect the whole extent of the avascular area
- FLA helps to distinguish collateral vessel formation (which do not leak) from retinal neovascularization
- At later stages, the correct diagnosis of BRVO can often only be established with the help of FLA

Objective documentation of macular edema is most readily obtained by fluorescein angiography (FLA) since most (but not all) occurrences of macular edema are associated with a disruption in the blood-retinal barrier at the level of the retinal capillaries.

Angiographic findings in BRVO reflect changes in the permeability, caliber, and patency of retinal vessels. Venous filling in the area of the occlusion is delayed relative to the unaffected retina and often the fluorescein column is narrowed at the site of occlusion [47]. A small area of early hyperfluorescence may be observed just proximal to the occlusion site. FLA helps to distinguish leakage without capillary nonperfusion (Fig. 21.3.9a–c) from leakage with capillary nonperfusion (Fig. 21.3.16a–c). Macular edema associated with good capillary perfusion is always associated with fluorescein leakage on the fluorescein angiogram since the edema is of the vasogenic type, with leakage of the fluorescein molecule occurring through a break in the blood-retinal barrier. Macular edema associated with macular ischemia may or may not be associated with fluorescein leakage. This has therapeutic consequences: an intact perifoveal capillary perfusion is the prerequisite for macular grid laser coagulation. Retinal hemorrhages block fluorescence and sometimes make a distinct evaluation of the capillary bed impossible (Fig. 21.3.17). For the management of acute BRVO in patients with a vision of 20/40 or worse, the BRVO Study group recommends FLA evaluation for macular edema versus macular nonperfusion, and if necessary to wait for sufficient clearing of retinal hemorrhages to allow a high quality FLA for an adequate treatment decision [28, 235]. Furthermore, FLA maps out the extent of ischemia assisting in the detection of patients at higher risk of neovascularization and those requiring closer follow-up exami-

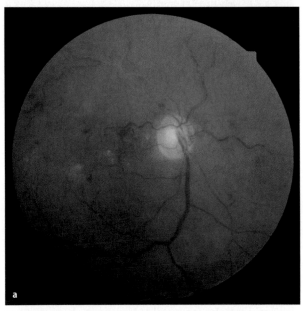

Fig. 21.3.16. a A patient with a long-standing HCRVO affecting the superior quadrants.

21 **III**

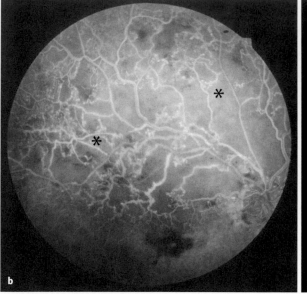
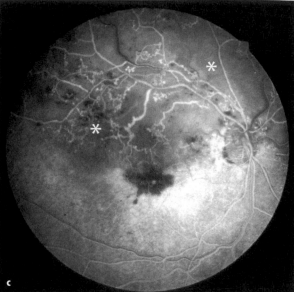

Fig. 21.3.16. b, c Illustration of how fluorescein angiography helps to map out the extent of ischemia, that is the area of marked capillary nonperfusion (*asterisks*)

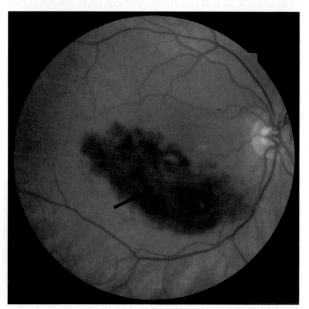

Fig. 21.3.17. Intense retinal hemorrhages (*arrow*) in acute macular BRVO prevent an angiographic evaluation of the capillary bed

Particularly, at later stages, after reabsorption of the hemorrhages, the correct diagnosis of BRVO can often only be established with the help of FLA.

nations. As such, a peripheral angiography may be helpful to detect the whole extent of the avascular area. Moreover, FLA helps to distinguish collateral vessel formation from retinal neovascularization at the edge of the affected area. In contrast to neovascularizations, shunt vessels do not leak (see Figs. 21.3.12, 21.3.13). Additional typical angiographic findings of BRVO are vascular abnormalities including capillary dilatation, microaneurysms and retinal edema.

21.3.5.3 Optical Coherence Tomography

Essentials
- OCT is helpful in determining the presence of macular edema, the foveal thickness and cystoid changes, but delivers only two-dimensional morphologic information
- OCT is able to demonstrate vitreofoveal adhesions
- OCT is fast, noninvasive and has become an indispensable tool for follow-up examinations

Optical coherence tomography (OCT) has become an extremely important and sensitive tool with which to assess the extent of macular edema in patients with BRVO and their response to treatment. A typical cross-sectional OCT image of macular edema associated with BRVO shows intraretinal cystic spaces delineating exactly the uninvolved and the affected area (see Fig. 21.3.7d). Serous detachment can be distinguished from intraretinal and subretinal fluid accumulation and is thereby influencing therapeutic strategies markedly. Moreover, OCT is advantageous in the exact evaluation of the vitreoretinal interface visualizing vitreofoveal adhesions, macular thinning, incomplete or full thickness ruptured cysts. It was demonstrated recently that foveal retinal thickness measured by OCT in patients with BRVO correlated with visual acuity and multifocal electroretinograms from the central retinal area [118]. Since OCT is less invasive and faster than FLA, it has become an indispensable tool for follow-up examinations. Often OCT demonstrates striking resolution of macular edema after treatment (see Fig. 21.3.23).

21.3.5.4 General Medical Examination and Laboratory Parameters

Essentials
- Clinical evaluation should include a detailed medical history with special emphasis on the presence of vascular risk factors
- A general medical examination should determine whether hypertension, diabetes, and hyperlipidemia might be present and should include a cardiovascular assessment
- Laboratory tests should include standard hematologic blood cell count, erythrocyte sedimentation rate, standard coagulation tests, plasma viscosity, erythrocyte aggregation, fibrinogen level, immunoglobulins including serum protein electrophoresis, TSH, serum cholesterol – and triglyceride levels, blood glucose, and creatinine
- Older patients with concurrent significant vascular diseases should not be screened for hemostatic defects
- If myeloma is suspected, Bence Jones urine proteins should be determined
- Only younger patients (50 years) or patients with recurrent CRVO/BRVO should undergo a large scale screening for the presence of thrombophilic disorders (see Sect. 3.7.3)
- An expensive complete workup in each and every BRVO patient is unwarranted

Clinical evaluation should include a detailed history with special emphasis on the presence of vascular risk factors and laboratory test results. It is important to obtain a detailed medical history covering any current drug therapy. Current opinions indicate that patients with a BRVO should be offered a laboratory risk assessment. But with our present knowledge there is no justifiable reason for a complete hemostasiological investigation like that offered to patients with spontaneous major venous thromboembolism.

The general medical examination should include a complete cardiovascular assessment. Appropriate investigations should be carried out to determine whether hypertension, diabetes, and hyperlipidemia might be present. Laboratory tests should include standard hematologic blood cell count and erythrocyte sedimentation rate, standard coagulation tests, plasma viscosity, erythrocyte aggregation, fibrinogen level, serum protein electrophoresis, serum cholesterol and triglyceride levels, blood glucose, and creatinine. Autoimmunity as a cause of thrombosis deserves particular attention especially in patients with recurrent BRVO. In this case, a test for the APC resistance phenomenon may be also considered, but this seems most relevant in patients younger than 50 years of age. Many patients with inherited heterozygous thrombophilic disorders do not develop thromboembolic events until the 2nd, 3rd or 4th decade. Scat et al. recommended the test for APC resistance only in younger patients and patients with recurrent thrombosis [209]. A further programme in these patients might include screening for hyperhomocysteinemia and such autoimmune phenomena that are known to play an important role in thrombosis, like an anticardiolipin antibody test and investigation for lupus anticoagulant. Older patients with concurrent significant vascular diseases such as diabetes, hypertension, or diffuse atherosclerosis should not be screened for hemostatic defects. Previous studies [37, 91, 171, 222] have clearly shown that evaluating elderly patients with retinal vein occlusions for the presence of antiphospholipid antibodies, and deficiencies of natural anticoagulants was not of much benefit. Therefore, it is advisable that only young patients should undergo a large scale screening for the presence of thrombophilic disorders. All tests should be standardized or samples processed centrally to avoid confusion brought about by interlaboratory variances.

An expensive complete workup in each BRVO patient is unwarranted. A routine, inexpensive hematologic evaluation is usually sufficient [109].

21.3.6 Natural Course

Essentials
- The natural history of BRVO depends on the type of occlusion, the size and location of the affected area and on whether there is associated cystoid macular edema (CME), macular nonperfusion, retinal neovascularization and vitreous hemorrhage
- Reduced baseline VA, older age and the extent of initial retinal ischemia were correlated strongly with poor visual outcome and with the development of retinal ischemia

21 **III**

- Nonischemic BRVO types may convert into ischemic types
- The impact of macular ischemia on VA and prognosis is discussed inconsistently
- Natural resolution of hemorrhages occurs over a period of between 9 and 12 months
- Fifty to 60% of untreated patients have a final VA of 20/40 or better
- Patients with VA of 20/60 or less have a small chance of spontaneous visual improvement
- The presence of collateral vessels does not correlate necessarily with better visual prognosis
- Little natural history data is available about the early and the late course of BRVO
- Regular funduscopic checks should be performed in patients especially with the ischemic type of BRVO since neovascularization and vitreous hemorrhage may also develop years after the occlusion

The prognosis of this disease is highly unpredictable because nonischemic types may convert into ischemic types within the first several weeks [108]. In each case in the initial stages, the only channel for the venous blood to exit the eye is blocked to a varying degree. Relatively little information is available on the natural history of BRVO; most of what exists is derived from clinical trials, including the Branch Retinal Vein Occlusion Study (BVOS). The natural history of BRVO depends on the size and location of the affected area and on whether there is associated CME, macular nonperfusion, retinal neovascularization and vitreous hemorrhage [46, 77, 98, 174]. Natural resolution of hemorrhages occurs between 9 and 12 months. The prognosis for visual acuity is often so variable that one can provide a patient only with a rough estimate of future outcome. Overall, the visual prognosis is good, with 50–60% of untreated patients having a final VA of 20/40 or better. About 20–25% of patients are left with 20/200 or worse, and the remainder have VA from 20/50 to 20/100. A final VA of only being able to see hand motions or worse occurs rarely [27, 29, 98, 174, 235]. The Collaborative Branch Vein Occlusion Study Group (BVOS) studied the VA prognosis for 35 untreated patients with macular edema after BRVO and loss of VA to 20/40 or worse. The study reported that only 37% of eyes showed a spontaneous resolution of the macular edema and visual improvement. After 3 years a visual acuity of less than 20/40 was found in two-thirds of untreated eyes [235], which progressed to chronic cystoid macular edema. Gutman and Zegarra reported a final visual acuity of 10/40 or less in 16 of 40 patients after 2 years [98]. Orth and Patz found that 53% of patients with BRVO achieved a final VA of 20/40 or better and 25% of the patients achieved 20/50 to 20/100 [183]. Eyes with smaller BRVO of shorter duration with less macular edema have better visual outcomes than eyes with large areas of venous obstruction, longer duration of disease, and persistent macular edema [46, 174]. As evidence for this, a study of small macular BRVO of less than 2 weeks duration demonstrated a significant spontaneous improvement in visual acuity after 3 months [187]. Typically, patients with VA of 20/60 or less have a small chance of spontaneous visual improvement. The development of collaterals across the horizontal raphe seems to help compensate for the venous occlusion in eyes without reperfusion of the retinal vein and contributes to spontaneous visual recovery [41], but the presence of collateral vessels does not correlate necessarily with better visual prognosis [236]. The collateral drainage capacity away from the affected area to areas with intact venous drainage also has a significant effect on the morphological outcome of the BRVO [41]. Collateral vessels develop usually within the first months of the occlusion and are not present at the first visit. But often by this time the potential for full visual recovery has been lost due to irreversible changes in the macular region. There is little natural history data available about the early course of BRVO. It cannot be obtained from the BVOS [235] because no patient was entered until 3 months after the occlusion due to the clinical impression that spontaneous improvement often occurred during that period. There is also inadequate long-term natural history information available. In the BVOS at 3 years there were only 35 untreated eyes with a BRVO of 3–18 months duration prior to study entry. It is unfortunately difficult to extract meaningful natural history information from such small numbers with such a variable duration of BRVO.

As in other retinal vascular disorders, exact mechanisms for the production of macular edema caused by BRVO, and for its spontaneous resolution, are poorly understood. The degree of initial macular edema was shown to be correlated with a poor final visual acuity and with persistent macular edema. Previous studies on BRVO have examined the correlation between a broken foveal capillary ring and visual prognosis [46, 77, 219]. However, the impact of macular ischemia on VA and prognosis is discussed inconsistently. In eyes with perfused macular edema, Finkelstein found a poorer prognosis [77], whereas Shilling and Jones [219] as well as Clemett [46] found a better prognosis. A retrospective small study of 23 eyes with nonperfused macular edema suggests that

90 % improve without treatment [77]. This difference was attributed to the separate mechanisms causing nonperfused and perfused edema. Initial features of the occlusion may also help to predict final prognosis. Reduced baseline VA, older age and the extent of initial retinal ischemia were correlated strongly to poor visual outcome and to the development of retinal ischemia [92].

In order to be able to judge the effectiveness of various treatment strategies, it is not only necessary to understand the mechanism through which they attain their success but it is also essential to know the natural history of the disease. Future studies analyzing the early natural course and the long-term history of the different types of BRVO are therefore imperative.

21.3.7 Associated Systemic Disorders and Risk Factors

Numerous reports about associated local, systemic and hematologic disorders have been published over the years, but only few studies are performed prospectively and differentiate between the types of vein occlusions. The studies are often contradictory, differing markedly in case selection criteria, demographics or study design. It is beyond the scope of this chapter to discuss the whole literature concerning this issue. In the following chapter possible underlying or causative diseases are summarized, but it has to be kept in mind that the presence of a particular associated systemic disease does not necessarily imply a cause-and-effect relationship [107].

21.3.7.1 Systemic Risk Factors

Essentials
- Sixty-five to 75 % of patients with BRVO have systemic hypertension, which is the main risk factor
- Further major risk factors include diabetes mellitus type 2, obesity and hyperlipidemia
- Systemic diseases with an increased risk of BRVO include hyperviscosity syndromes such as elevated erythrocyte aggregation rate, malignancy, myeloproliferative disorders and pregnancy
- Treatment of the underlying medical conditions is able to reduce the rate of BRVO recurrence

As it stands, the generally accepted systemic risk factors in retinal vein occlusion are arterial hypertension and diabetes, conditions that appear more closely correlated with arteriosclerosis and arterial thrombosis that may precipitate acute myocardial infarction and stroke, rather than being related to venous occlusion. In the case of a BRVO the close anatomical localization of arterial and venous branches have to be taken into account. Hence, the pathology of retinal veins may theoretically be significantly influenced, if not dominated, by the pathology of neighboring retinal arteries. Arteriosclerotic changes in the retinal arteries are more likely to cause BRVO than CRVO because the vessels in the optic nerve lie side by side and do not cross. Patients with BRVO are known to have another prevalence of systemic risk factors than those with CRVO or HCRVO [107], but the risk factor profile is similar [223]. The prevalence rate of arterial hypertension, peripheral vascular disease, venous disease, peptic ulcers and other gastrointestinal diseases, an increased body mass index and the number of smokers is significantly higher than in patients with CRVO [7, 8, 107]. The Eye Disease Case-Control Study Group and other similar studies have identified risk factors such as aging, diabetes, hypertension, cardiovascular risk profile, and glaucoma [236]. In the majority of reports only the association with systemic arterial hypertension and retinal vascular manifestations of hypertension have been documented [6, 7, 27, 98, 107, 236]. Sixty-five to 75 % of patients with BRVO have systemic hypertension [236]. Also patients receiving antihypertensive medication have to be considered as risk patients.

Hayreh et al. showed that the prevalence of arterial hypertension in patients with major BRVO (65.7 %) was significantly higher than in a gender-matched and age-matched control population (29.9 %). They found the same relationships in young men with macular BRVO. Patients with both types of BRVO had a higher prevalence of cerebrovascular and chronic obstructive pulmonary disease, peptic ulcers, diabetes and thyroid disorders [107]. Earlier case control studies identified hypertension, diabetes mellitus type 2, obesity and hyperlipidemia as major risk factors of BRVO [6, 63, 129, 198]. No significantly greater incidence of estrogen use or use of birth control pills in patients with BRVO compared with their control group was found [58, 129, 236]. In a recent large population based study by Wong et al., different types of retinal vein occlusion were associated with carotid artery disease, hypertension, and other cardiovascular risk factors [257]. In addition, they found an association with body mass index, current cigarette smoking and increased plasma fibrinogen levels.

Further underlying systemic diseases with an increased risk of thrombosis include hyperviscosity syndromes [200], such as elevated erythrocyte aggregation rate, malignancy, myeloproliferative disorders, pregnancy and less common conditions such as polycythemia, paraproteinemia, Behçet's disease and paroxysmal hemoglobinuria [95]. Treatment of the underlying medical conditions can reduce the severity of some of its complications [204]. It was further shown that treatment of systemic risk factors is able to reduce the rate of recurrence from approximately 10 % to 1 % [62, 64].

This underlines the importance of searching for and identifying possible systemic risk factors. A profound meta-analysis by Janssen et al., however, concluded that in investigating a new patient with retinal vein occlusion, at first, one should test for the most common systemic risk factors such as hypertension, lipid abnormalities and diabetes mellitus only [126]. An extensive and expensive workup for thrombophilic disease at the initial presentation is unwarranted in the vast majority of patients.

21.3.7.2 Local Risk Factors

Essentials
- BRVO patients show a higher incidence of hyperopia and a shorter axial length
- The association between BRVO and glaucoma is uncertain
- Glaucoma patients seem to acquire BRVO at a younger age than nonglaucomatous patients
- Testing for patients with open-angle glaucoma should be part of the evaluation of patients with BRVO

Local risk factors include ocular findings and diseases associated with the occurrence of BRVO. A higher incidence of hyperopia and a shorter axial length has been found in patients with BRVO [7, 230]. The most important local risk factor, however, is chronic open angle glaucoma, although the explanation for the association with retinal vein occlusion has yet to be elucidated. The association between CRVO and chronic open angle glaucoma has been recognized for many years with a prevalence between 10 % and 40 %, but the association between BRVO and glaucoma is uncertain [22, 92, 45, 109, 111, 112, 223, 243] and appears to be less frequent with a prevalence between 6.6 % and 15 %. A lack of correlation between BRVO and open angle glaucoma or ocular hypertension compared to that in the general population was reported by Johnston et al. [129, 163]. However, another study found similar frequency rates of BRVO and CRVO in a glaucoma population [51]. Although the exact intraocular pressure at the time of a retinal vein occlusion is not known, several authors suggest that the pressure may be raised and that the venous occlusive episode itself, subsequently, has a lowering effect on the pressure [81, 106]. It has been found that retinal vascular events tend to occur early in the course of the associated glaucoma [19, 22] and glaucoma patients seem to acquire BRVO at a younger age than nonglaucomatous patients [51]. However, the prevalence of associated medical conditions in patients with retinal vein occlusion either with or without glaucoma or ocular hypertension are markedly similar [48], implying that associated medical conditions may be of greater significance. Both conditions – glaucoma and BRVO – may not be related etiologically but rather are both manifestations of some underlying vascular abnormality [188], and there is evidence available suggesting that testing for patients with open-angle glaucoma should be part of the evaluation of patients with BRVO [223].

Whether modern antiglaucomatous drugs affecting the regulation of retinal blood flow have a positive influence following BRVO is unclear and deserves further investigation.

21.3.7.3 Hematologic Risk Factors

Essentials
- RVO has been reported to be associated with the antiphospholipid antibody syndrome, abnormalities in the physiological anticoagulant system such as protein C and protein S, factor XII deficiency, antithrombin III deficiency and defects in fibrinolytic mechanisms
- Elevated blood viscosity, variously combined with increased hematocrit, fibrinogen, lipoprotein(a), α_1-globulin or α_2-globulin levels, may be implicated in BRVO
- Patients with elevated total homocysteine are three times more likely to have BRVO
- APC resistance as a cause of BRVO is described mainly in younger patients
- Therapeutic strategies should only be drawn up together with a hematologist
- To date, there is not enough data to recommend a standard anticoagulant therapy in patients with BRVO and an underlying genetic thrombophilic abnormality

Coagulation disorders which can lead to the formation of thrombi elsewhere in the body may also contribute to the formation of thrombi in the retinal vasculature [244]. Thrombophilic disorders are characterized by a systemic dysfunction of the hemostatic pathway. However, in keeping with Virchow's triad, the phenotype must arise from local changes in blood flow, disruption of the vascular wall, or vascular-bed-restricted alterations in the balance of anticoagulant and procoagulant factors. Meanwhile many congenital, and acquired, blood protein defects are known to account for hypercoagulability and thrombosis. Among the biochemical and hemostatic disturbances that have been associated with BRVO are antiphospholipid antibody (APA) syndrome [1, 2, 10, 23, 24, 25, 33, 195, 221, 224, 238, 244, 254], abnormalities in the physiological anticoagulant system and elevated hemostatic factors such as fibrinogen and lipoprotein Lp(a) [13, 160, 162, 234]. The coagulation sequence is held in check and inhibited by specific physiological anticoagulants including antithrombin III, protein C, and protein S. Deficiencies of these natural anticoagulants are associated with recurrent systemic thromboembolic events and have also been associated with the development of a BRVO [1, 23, 24, 27, 34, 43, 44, 88, 123, 178, 183, 189, 202, 221, 222, 237, 250]. These disorders as well as factor V Leiden [44, 53, 94, 247], activated protein C (APC) resistance [96, 143, 152, 194, 209, 256] and factor XII deficiency [1, 56, 88, 144] are addressed in Chapter 21.1 in detail. The role and impact of elevated plasma total homocysteine levels [30, 31, 42, 253] and low serum folate levels [31] in retinal vascular occlusive diseases are also thoroughly discussed in Chapter 21.1.

Further rheological abnormalities include thrombophilia [120], abnormal in vivo platelet function [60], hyperviscosity [1, 36, 90, 170, 193, 200, 239] and increased inflammatory activity [61]. A raised erythrocyte sedimentation rate is an important marker of raised plasma viscosity which may also influence blood viscosity and has been described in patients with BRVO by several investigators [36, 67, 90, 190, 200, 239].

Lowe et al. reported an increased blood viscosity in patients with capillary nonperfusion or neovascularization in the chronic phase of the disease and believed that an increased viscosity may play a role in the production of ischemia [161]. Peduzzi et al. confirmed this theory and found a significantly decreased erythrocyte deformability in patients with capillary nonperfusion and retinal vein occlusion [190]. Wiek et al., however, showed that the hematological profile of patients with BRVO is not different from control subjects matched for age, gender and risk factors. They found that the determination of plasma viscosity, red cell aggregation, red cell filterability, and hematocrit could not assist in the differentiation between ischemic versus nonischemic vein occlusion [255].

The association between hematologic risk factors and BRVO is discussed inconsistently. To date, there is not enough data to recommend a standard anticoagulant therapy in patients with BRVO and an underlying genetic thrombophilic abnormality. The risks and benefits must be considered carefully in each individual case in close collaboration with a hematologist.

21.3.8 Treatment

A major problem in the clinical management of BRVO is that patients often seek the ophthalmologist weeks to months after the thrombotic incident. Possible irreversible retinal damage, if it is going to occur, has already run its course before the patient presents. At this time, the treatment requirements are more likely to be aimed at reducing the extent of the secondary complications of BRVO than using antithrombotic treatment per se. Another problem is that the rate of recurrence has not been clearly determined. Therefore, the possible requirement for secondary preventive measurements is unknown.

Treatment of BRVO has four aims:

- Improvement of hemodynamic properties and secondary prevention
- Reduction of macular edema
- Treatment of neovascular related complications
- Treatment of risk factors and associated systemic disorders

Depending on the stage of the disease, one or more of these therapeutic strategies may be indicated.

21.3.8.1 Improvement of Hemodynamic Properties and Secondary Prevention

Essentials
- No beneficial effect of anticoagulation therapy could be proven, neither for prevention nor for the therapy of BRVO
- Pentoxifyllin reduces blood viscosity by lowering the fibrinogen level and increases erythrocyte and leukocyte deformability, but no studies analyzing its effect on the outcome in BRVO patients exist
- For isovolemic hemodilution and troxerutin a treatment benefit was shown in prospective randomized trials, but these therapies are controversial and have not gained general acceptance
- Systemic thrombolysis with low dose intravenous recombinant tissue plasminogen activator (rt-PA) is a treatment option; however, the risk of life threatening hemorrhages has to be considered

The question of whether a therapeutic measure to alter the blood constituents could modify and moderate the morbidity of BRVO is very difficult to assess.

No beneficial effect of anticoagulation therapy with heparin or coumarin derivates could be proven, neither for prevention nor for the therapy of BRVO [57]. A review of the benefit of antithrombotic agents concluded that hemorrhagic complications seemed to outweigh the possible benefits [120]. Since systemic complications and progression of retinal hemorrhages have been described, the risks of secondary prevention seem inadequate [120, 250]. The only exception is the category of severe coagulation disorders (see Sect. 21.3.7.3), which might justify a prophylactic treatment.

Acetylsalicylic acid is often prescribed even after venous occlusions. No beneficial effect has been proven so far and also with this treatment is implicated the risk of new retinal hemorrhages [120] (Fig. 21.3.18). Pentoxifyllin reduces blood viscosity by lowering the fibrinogen level and increases erythrocyte and leukocyte deformability, but no studies analyzing its effect on the outcome in BRVO patients exist. Hemodilution decreases the hematocrit levels and should lower whole blood viscosity and improve retinal microcirculation. It leads to a decrease of AV passage time in the affected branch and to a significant reduction of erythrocyte aggregation [199]. Conclusions drawn from published hemodilution treatment studies [89, 103, 199], some being quite optimistic but of limited size, must include a great deal of caution. For isovolemic hemodilution [103] and troxerutin [89], a treatment benefit was shown in prospective randomized trials. Hansen et al. reported a visual acuity of 0.4 (20/50) or more in 85% of the hemodiluted patients in contrast to 33% of the nonhemodiluted patients after 1 year, and most improvements occurred in patients with ischemic

BRVO [103]. Due to the small number of patients included in these studies these therapies are controversial and have not gained general acceptance. Shalid et al. concluded in a recent review that isovolemic hemodilution is of limited benefit and should be avoided in patients with concurrent cardiovascular, renal or pulmonary morbidity [217].

The lesson learned from antithrombotic treatment in general is that there is not much chance of achieving reperfusion if antithrombotic measures are introduced weeks after precipitation of venous thrombosis.

21.3.8.2 Reduction of Macular Edema

21.3.8.2.1 Macular Grid Laser Coagulation

Essentials
- Macular grid laser coagulation (LC) is an effective treatment for persistent, perfused macular edema associated with BRVO
- In cases of persistent (>3 months) decreased vision of 20/40 or less and macular edema without capillary nonperfusion (FLA necessary!), macular grid LC in the affected area is recommended
- When foveal ischemia is present, no grid LC should be performed
- It is necessary to wait for sufficient clearing of retinal hemorrhages before LC is performed
- A pretreatment with intravitreal steroids may be helpful in very edematous retina
- In cases of persisting macular edema, grid LC can be repeated after 3 months and visual improvement can be achieved despite multiple treatments
- For macular BRVO no benefit of grid LC versus the natural course could be proven

21.3.8.2.2 Macular Grid Laser Photocoagulation in BRVO – Technique

Essentials
- Argon green laser should be used using a contact area centralis lens (e.g., Goldmann) with 100–200 µm spot size, 100–200 mW power and 0.1–0.2 s duration in topical anesthesia
- Medium white laser spots should be spaced approximately one burn width apart to the area of leakage no closer than 500 µm to the fovea (see Fig. 21.3.20)

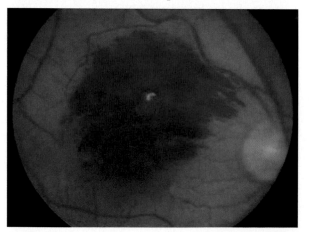

Fig. 21.3.18. Anticoagulation therapy may increase the risk of fresh and dense hemorrhages in the area of BRVO

- The power should be increased in 50 mW increments from the initial setting until the desired intensity is achieved
- Two to three rows with 100 μm spots should be applied along the edge of the foveal avascular zone (FAZ) followed by 200 μm spots to the major vascular arcades
- In very edematous retina, it is advisable to increase pulse duration rather than power
- Excessive eye movements may require retrobulbar anesthesia in rare cases
- Patients should be seen at 3 month intervals after grid LC and at 6–12 months once the macular edema is resolved

21.3.8.2.3 Corticosteroids

Essentials
- Increased levels of vascular endothelial growth factor (VEGF) and IL-6 were found in vitreous samples of patients with BRVO indicating an inflammatory component in BRVO
- Steroids may be useful in stabilizing the blood-retinal barrier, decreasing vascular permeability, and perivascular and intraluminal inflammation
- High dose oral steroid therapy was found to improve vision only over the short term, and was associated with the systemic side effects of steroids
- Intravitreal triamcinolone acetate (IVTA) can reduce macular edema and improve VA in patients with BRVO
- The therapeutic effect of IVTA in BRVO is transient and once the steroids start to wear off, the macular edema recurs
- IVTA is significantly more effective in reducing macular edema and improving VA than repeated retrobulbar injections

21.3.8.2.4 Antiangiogenic Therapy

Essentials
- Increased VEGF levels were measured in patients with vein occlusion, and correlated with the extent of macular edema
- Anti-VEGF drugs could specifically reduce the permeability of retinal capillaries or might downregulate VEGF, but do not eliminate the cause of the disease

- Preliminary results in small case series are promising, but as with IVTA, the effect is transient and appears to be of shorter duration

21.3.8.2.5 Surgical Treatment

Essentials
- The separation of the retinal artery from the underlying vein is called arteriovenous adventitial sheathotomy (AAS) and addresses a theoretical pathogenic mechanism of BRVO
- AAS can lead to restoration of the blood flow in the occluded vein immediately after decompression
- Eyes without rest perfusion did not benefit from AAS
- In a considerable number of patients surgical separation of the vessels cannot be achieved
- Potential complications of the procedure include laceration of vein or artery with consecutive hemorrhage or arcuate scotoma due to incision of the nerve fiber layer
- Interestingly, also incomplete dissection of the vein may result in improved vision suggesting that freeing the vessels from their retinal bed or vitrectomy alone might be effective
- The role of AAS has not yet been clarified
- Removal of the posterior hyaloid alone and/ or additional removal of the internal limiting membrane may contribute to resolution of the macular edema and visual improvement
- Recently successful cannulations of branch retinal arterioles and branch retinal venules were described and may offer new future treatment options

21.3.8.2.6 Macular Grid-Laser Photocoagulation

Macular grid LC is an effective treatment for persisting, perfused macular edema associated with BRVO [9] and is still the gold standard (Fig. 21.3.19a–c). In the BVOS the mean vision in laser treated eyes improved from 20/40 to 20/50, compared with 20/70 in untreated eyes. The treatment effect was negligible when the initial vision was in the poorer range of 20/100 to 20/200. Other studies provide contrasting results.

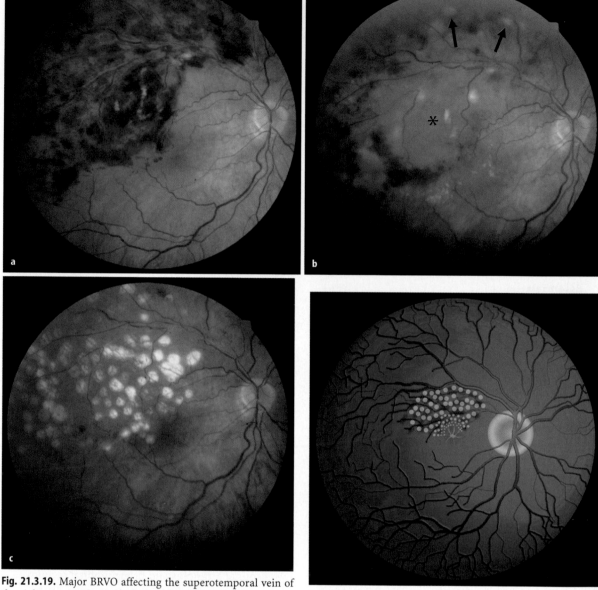

Fig. 21.3.19. Major BRVO affecting the superotemporal vein of the right eye. **a** Before treatment and **b** after scatter laser treatment (*arrow*) showing partial regression of hemorrhages but an increase in macular edema (*asterisk*). **c** Following macular grid LC complete resolution of macular edema can be seen

Fig. 21.3.20. Schematic drawing of the grid LC pattern in persisting macular edema associated with BRVO. The laser treatment covers the area of capillary leakage and spares the foveal avascular zone

In cases of persisting visual decrease to 20/40 or less and macular edema without capillary nonperfusion, the BRVO Study group recommended macular grid LC in the affected area [235] (Fig. 21.3.20). When macular nonperfusion is the cause of visual loss, no grid LC should be performed [235].

Argon green laser is probably the wavelength of choice in the routine macular grid LC treatment using 100 μm spot size, 100–200 mW power and 0.1–0.2 s duration depending on the patient's com-

pliance. Argon blue-green can also be used, but the blue wavelength is absorbed by the macular xanthophyll. The spots should be spaced approximately one burn width apart and directed at the area of capillary leakage identified by FLA, sparing the foveal avascular zone and not extending peripheral to the major vascular arcade. It is important to spare areas of collateral vessels during LC as well as areas with retinal hemorrhages which absorb the applied laser energy and lead to nerve fiber layer damage. Therefore, it is necessary to wait for sufficient clearing of retinal hemorrhages before LC is performed. It has to be considered that an increase of the central scotoma was found in 50% of patients after laser treatment [15].

In the BRVO Study, the effect of timing of the laser treatment was not examined. Since spontaneous remissions are commonly observed and no studies exist which have proven a less therapeutic effect of delayed therapy, early treatment does not seem justified. Most authors recommended waiting at least 2–3 months before considering grid LC. In cases of persisting macular edema, grid LC can be repeated after 3 months and visual improvement can be achieved despite multiple treatments [69]. The BVOS

recommendations, however, refer only to the occlusion of a major branch. No benefit of grid laser treatment versus the natural course could be proven [123, 187] in the subgroup of macular BRVO, but its therapeutic principle is also working (Figs. 21.3.21, 21.3.22). Recently, it was found that the presence of

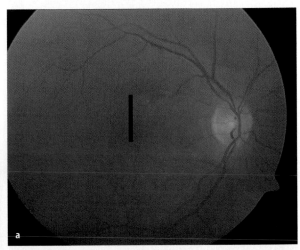

Fig. 21.3.22. The same patient as in Fig. 21 after grid laser treatment with **a** unchanged clinical features

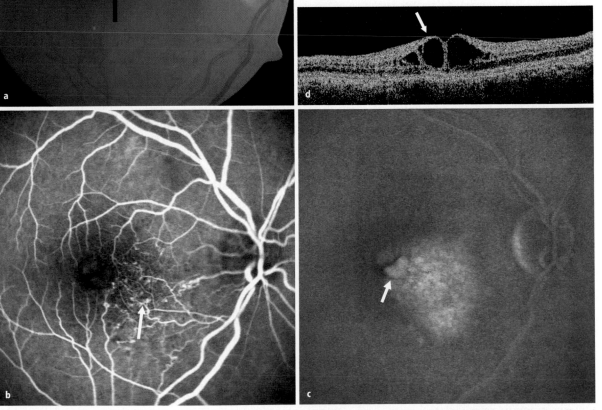

Fig. 21.3.21. a Macular BRVO with subtle clinical changes. **b** FLA shows microaneurysms (*arrow*) and **c** retinal edema in the affected area together with cystoid macular edema (*arrow*). **d** OCT (scan, see **a**) enables clear visualization of large localized foveal cysts (*arrow*)

21 III

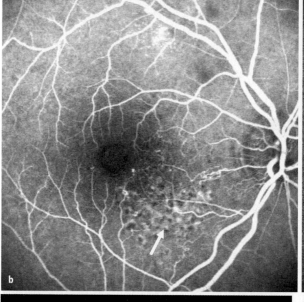
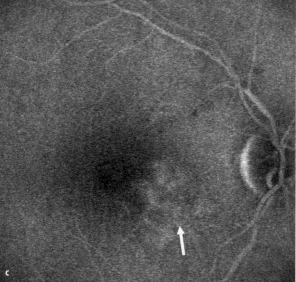

Fig. 21.3.22. b FLA shows focal laser burns (*arrow*) and **c** a reduction in edema (*arrow*). **d** Complete resolution of foveal cysts (*arrow*) can be seen with OCT (scan, see **a**)

21.3.8.2.7 Laser-Induced Chorioretinal Venous Anastomosis

Using an animal model of BRVO, MacAllister et al. successfully explored the possibility of creating an iatrogenic permanent anastomosis between the retinal and choroidal venous circulation [169] allowing for reperfusion of the affected retina. Collateral vessel formation in the early stages of the disease between a high pressure circulation, as in the circulation of an obstructed retinal vein, and a low pressure circulation, such as the choroidal venous system [26], could prevent closed loop circulation and its ischemic consequences.

Several lasers and settings were proposed to create a chorioretinal venous anastomosis as a therapeutic modality in CRVO and BRVO [70], directing the laser spot on the edge of or directly on top of a retinal vein adjacent to the occlusion site in order to achieve perforation of the retinal vein and rupture the underlying Bruch's membrane. When successfully treated, promising visual results could be achieved [70]. However, due to severe complications such as choroidal neovascular membrane formation, preretinal fibrosis, traction retinal detachment, and vitreous hemorrhage [16], this treatment has not gained general acceptance.

21.3.8.2.8 Corticosteroids

Although steroids have antiangiogenic, antifibrotic, and antipermeability properties, the principal effects of steroids are stabilization of the blood-retinal barrier, resorption of exudation, and downregulation of inflammatory stimuli [79]. Since they have been shown to reduce macular edema also in a wide variety of other conditions, their effect may be nonspecific. Whereas it is clear that there is a breakdown in the blood-retinal barrier in BRVO, studies have shown that there may also be an inflammatory component that would respond to steroids [133]. Increased levels of VEGF and interleukin (IL)-6 were found in vitreous samples of patients with BRVO [180]. It is known that inflammatory cell infiltrates are present in the area of the thrombotic occlusion including lymphocytic infiltration within the thrombus. Therefore steroids may have a therapeutic effect not only on macular edema, but also on the thrombus. Thus steroids may be useful in BRVO stabilizing the blood-retinal barrier, decreasing vascular permeability, and perivascular and intraluminal inflammation.

Systemic Corticosteroids

High dose oral steroid therapy was found to cause transient reduction in macular edema and improve vision over the short term, but these effects proved not to be long lasting and were associated with the potential adverse effects linked with systemic administration of steroids [216]. Assuming a vasculitic component, younger patients, especially, were treated with systemic corticosteroids. Currently, no controlled trials exist.

Intravitreal Corticosteroids

The intravitreal administration of steroids provides for a more concentrated dose of steroid delivery to the eye, together with the advantages of a potentially longer duration of action through its associated vehicle, and limitation of systemic effects. Intravitreal injection of triamcinolone acetate (IVTA), a long acting corticosteroid, was first described by the group of Machemer [110] and has been used in the treatment of macular edema of different origin [167]. Although the mode of action by which triamcinolone induces resolution of macular edema remains poorly understood, in vitro studies and clinical observations indicate that triamcinolone has the capacity to reduce the permeability of the outer blood-retinal barrier [191].

Due to the anti-edematous and anti-angiogenic effects shown in experimental investigations and clinical studies, IVTA has also been used in pilot studies on BRVO [35, 39, 52, 130, 131, 132, 185, 186, 260]. Case series have demonstrated that IVTA can reduce macular edema and improve VA in patients with BRVO (Fig. 21.3.23) [52, 130, 132, 134, 185, 186, 206]. In a recent randomized clinical trial IVTA was more effective than macular grid LC in patients with CME due to retinal vein occlusion or diabetic macular edema [11]. It is still controversial as to whether a combination of both is beneficial or not [11, 208]. The best point in time for grid LC is probably when the central macular thickness is maximally reduced between 6 and 12 weeks after IVTA [11, 113]. Jonas et al. reported that IVTA can significantly increase VA in nonischemic BRVO while no significant effect was found in the ischemic type [130, 131]. Jonas et al. used higher dosages of 20–25 mg triamcinolone, whereas in the majority of clinical studies between 4 and 8 mg

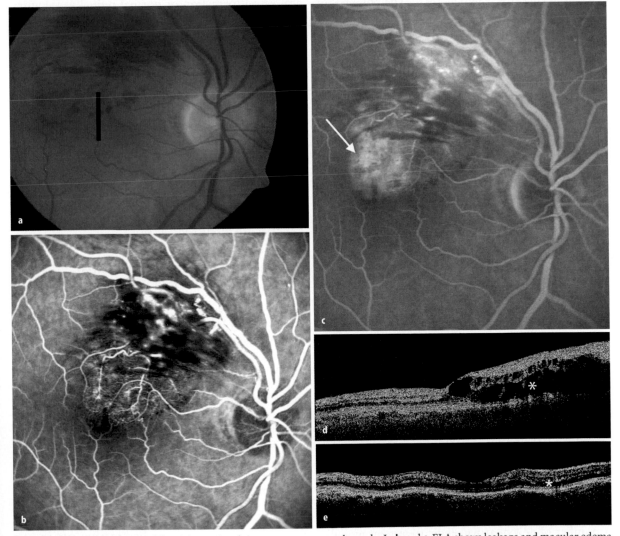

Fig. 21.3.23. a Macular BRVO of the right eye involving a superotemporal venule. In **b** and **c**, FLA shows leakage and macular edema (*arrow*) and **d** shows the typical OCT pattern (scan, see **a**) of cystoid spaces (*asterisk*). **e** Complete resolution of macular (*asterisk*) and retinal edema can be seen 6 weeks after intravitreal triamcinolone

21 III

is injected. However, Chen et al. found that IVTA is also beneficial in macular edema associated with BRVO and foveal capillary nonperfusion [40]. As a result, intravitreal injections of triamcinolone are now being performed as primary treatment of cystoid macular edema secondary to BRVO [155] with increasing frequency despite steroid-related adverse events such as cataract and glaucoma. Endophthalmitis, sterile endophthalmitis, pseudo-hypopyon, cataract, retinal tears, retinal detachment, and vitreous hemorrhage have all been described after IVTA [21, 87, 130, 176]. And, unfortunately, the therapeutic effect is not permanent [141, 260]. Because a single intravitreal injection should last roughly 3–6 months in a nonvitrectomized, phakic patient [18], it is not surprising that once the steroids start to wear off, the macular edema recurs. Therefore steroid injections need to be repeated and consequently the risk of side effects will increase. The potential of this intervention should be considered in the light of the small but real risk of endophthalmitis, which is in the range of 0.87% [21, 130].

It is unclear how long the effect of IVTA in the individual patient with a particular type of BRVO lasts and whether additional injections will continue to work. Only long-term studies evaluating IVTA will be able to answer these questions [78]. The Standard Care versus Corticosteroid for Retinal Vein Occlusion (SCORE) study aims to investigate the efficacy and safety of intravitreal triamcinolone for macular edema secondary to CRVO and BRVO [213]. Also the fact that intravitreal triamcinolone has never been approved by the Food and Drug Administration must be clearly weighed in the decision making.

Other forms of drug delivery with corticosteroid derivates such as the sub-Tenon capsule [241] or retrobulbar [135, 246] injection of triamcinolone and the parabulbar injection of anecortave acetate may have a beneficial effect as well. However, in a prospective randomized trial a single IVTA was significantly more effective in reducing macular edema and improving VA than repeated retrobulbar injections [105]. The use of a sustained release steroid implant would obviate the need for multiple injections, but may also magnify the risk of steroid related complications [65, 124]. A biodegradable, implantable, extended-release pellet has been developed that releases dexamethasone directly into the posterior segment. Clinical trials are currently being performed and preliminary anatomical and functional results are promising.

21.3.8.2.9 Antiangiogenic Therapy

A number of new pharmacologic treatment modalities are being developed and several substances are under current investigation. Application of vascular endothelial growth factor (VEGF) inhibitors represents a treatment option that targets the disease at the causal molecular level. Increased VEGF levels were measured in patients with vein occlusion, and correlated with the extent of macular edema [179, 180]. Preliminary studies using anti-VEGF drugs such as pegaptanib and bevacizumab show a promising effect on macular edema of vascular origin, especially venous occlusive disorders [121, 125, 203]. Both substances are potent anti-edematous agents, possibly specifically reducing the permeability of retinal capillaries or even downregulating VEGF, but they do not eliminate the cause of the disease. The exact duration of therapeutic effect is as yet unknown and seems to decrease within a few weeks. The need for repeat injections has been reported after 4 weeks. No information exists about the long term outcome and possible negative influences on the development of collateral vessel formation or normal retinal vasculature. Controlled studies are mandatory to evaluate the role of this off-label therapy.

21.3.8.2.10 Surgical Treatment

None of the aforementioned therapies is able to restore blood flow in BRVO. It is believed that the therapeutic goal must be directed to solving the obstructive process in the vascular lumen. The unique opportunity to visualize and reach the retinal vessels during vitreous surgery is leading to surgical approaches to either decompress or cannulate the affected vein. The surgical options of arteriovenous adventitial sheathotomy, vitrectomy and vessel cannulation are discussed below.

21.3.8.2.11 Arteriovenous Adventitial Sheathotomy

Pars plana vitrectomy with surgical separation of the retinal artery from the underlying vein at the site of the presumed pathologic AV crossing has been advocated as a potential treatment for BRVO. The separation addresses a theoretical pathogenic mechanism of BRVO and should effect reperfusion of the retina, rather than treating the BRVO sequelae. It is hypothesized that dissection of a common adventitia surrounding the artery and vein mitigates the obstruction. Mechanical relief of the retinal vessels may generate higher intraluminal pressure, resulting in better venous blood flow and improvement of visual function (Fig. 21.3.24).

Osterloh and Charles first described successful surgical decompression by dissection of the common adventitial sheath at the crossing site in one patient with BRVO [184]. Eleven years later, this therapeutic strategy was resurrected by Opremczak

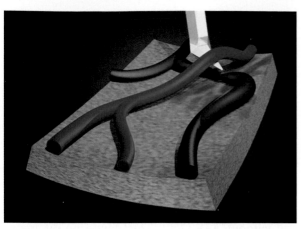

Fig. 21.3.24. Schematic drawing of surgical arteriovenous decompression with dissection of the common adventitial sheath at the occlusion site using a bent microvitreoretinal blade

and Bruce. They published favorable results with sheathotomy in 15 eyes with BRVO [182]. After 3 months, the vision of 67 % of patients improved by an average of four lines. Recently, successful arteriovenous adventitial sheathotomy (AAS) was reported using the 25 gauge vitrectomy technique [82, 150]. Kube et al. measured that AAS led to a significant decrease of AV passage time and can ameliorate retinal perfusion in the affected branch [142]. Yamaji et al. reported a reduced filling delay after AAS [258]. Fluorescein videoangiography with computer-assisted image analysis has shown improved retinal circulation [220]. Meanwhile, AAS has been reported in more than 250 patients. Pre- and postoperative images of a patient are shown in Figs. 21.3.31 and 21.3.32. There are studies claiming a favorable outcome [35, 85, 157, 158, 168, 172, 173, 215, 259] while others deny a treatment benefit [17, 32, 75, 83, 102]. In a long term follow-up (mean 6.5 years) of five patients with a poor visual prognosis, Shah et al. reported substantial visual improvement after AAS on a long term basis [215]. Mester and Dillinger reported significantly improved results in a larger series but in several patients the internal limiting membrane was also removed [172]. The aforementioned authors observed intraoperative restoration of the downstream blood flow in the occluded vein immediately after decompression. FLA revealed improved capillary perfusion in 93 % of patients 6 weeks after surgery. Functional improvement has been reported in patients with ischemic BRVO as well [157]. Following AAS a transient improvement of retinal blood flow was measured [116]. The addition of a thrombolytic over the crossing after its dissection may be helpful. Garcia-Arumi et al. performed additional fluid-air exchange and injected 25 µg of recombinant tissue plasminogen activator

(rTPA) in the affected area [85]. Using a bimanual technique they achieved successful unroofing of the vein in all 35 patients. In 28.6 % they observed thrombus release, an immediate restoration of downstream blood flow in FLA and in 74.3 % proximal venous widening. In most of these patients improvement of VA was four lines or more. Excepting the general risks of a vitrectomy, potential complications of the procedure include laceration of the vein or artery with consecutive hemorrhage, or arcuate scotoma due to incision on the nerve fiber layer [85, 182]. Different techniques and instruments such as vitreoretinal scissors, bent microvitreoretinal blades, ring forceps and microscissors have been suggested for separating the crossing and cutting the glial tag at the occlusion site [71, 73, 85, 182]. Despite these developments, in a considerable number of patients surgical separation of the vessels cannot be achieved [17, 72, 73]. In a clinicopathological correlation, Feltgen et al. showed surgically induced nerve fiber damage. Han et al. modified their approach to a more limited dissection [102] without separation of the retinal vessels. Interestingly, incomplete dissection or even iatrogenic perforation of the vein resulted in better visual results [52, 73, 102]. This supports the possibility that freeing the vessels from their retinal bed alone might be effective [73].

Although some authors speculate that early surgical decompression may reduce retinal hemorrhages and ischemia [85, 172, 182], it is questionable whether the procedure is worth the risks [52, 72]. Resolution of the stigmata of retinal venous insufficiency does not uniformly occur. The mechanism by which intraretinal hemorrhage may clear faster may be related to improved blood flow or to easier dissipation of blood into the vitreous cavity.

The role of AV dissection itself and the mechanisms by which AAS improves VA have not yet been clarified. The different results may depend on different inclusion criteria, surgical techniques and the time lapse between onset of the BRVO and surgery. Other confounding factors may also affect the outcome. Vitrectomy with AAS consists of several components, the relative contributions of which toward improvement of retinal venous blood flow (if any) remain undetermined. Other procedural factors include the vitrectomy method itself, stripping of the posterior hyaloid and the delamination of the internal limiting membrane (ILM). These components may play a decisive role as well and are discussed later [205, 224, 225]. Also the adjunctive use of triamcinolone at the end of vitrectomy for BRVO was reported [241].

Lakhanpal et al. proposed transvitreal limited arteriovenous crossing manipulation (LAM) without vitrectomy as a minimally invasive way to achieve clot dislodgement and reperfusion and to avoid the typical complications of vitrectomy such as cataract formation [149, 150].

In one trial comparing AAS and macular grid LC, no difference of the functional outcome could be detected [154], whereas another comparative study found significantly improved visual results following AAS [168]. However, both sample sizes were small, no randomization was performed or no Early Treatment Diabetic Retinopathy Study (ETDRS) VA was obtained. In a recent prospective case series a favorable functional outcome of AAS irrespective of successful dissection against isovolemic hemodilution was reported [73]. Irreversible fibrotic alterations of the vascular structure are probably due to the duration of vein occlusion [41, 50, 196] and as in most macular disorders a better functional outcome after AV decompression is reported after shorter duration of BRVO [172]. Also, youth had a beneficial influence on the functional results. Han et al. observed worse visual recovery in patients with unsuccessful preexisting macular grid LC [102]. Positive treatment results even after long-standing BRVO may be explained by the presence of a minimal rest-perfusion of the occluded vessel on FLA. Since eyes without rest-perfusion did not benefit from surgical decompression [172], this may serve as a predictive factor.

Unfortunately, the inclusion criteria and the postoperative results of all published studies are very heterogenous and hardly comparable. To date, it cannot be concluded that surgical AAS is more beneficial than other therapies for BRVO or even the natural course of the disease. To evaluate the effectiveness of this new treatment approach, several prospective randomized trials are currently underway.

21.3.8.2.12 Vitrectomy

Pars plana vitrectomy and posterior hyaloid peeling alone have been associated with visual and morphological improvement in eyes with BRVO [145, 147, 205, 212, 227]. Recent studies support the premise that vitrectomy improves macular edema and visual acuity whether or not AAS is performed [17, 38, 75, 83, 259]. In a small study, pars plana vitrectomy with posterior hyaloid removal and intraocular gas tamponade resulted in a reduction of macular edema and restoration of the normal foveal contour in 10 of 19 eyes with BRVO [205]. A recent study reported improved VA and new collateral vessel growth into previously nonperfused areas after vitrectomy in 75% of eyes with BRVO and ischemic macular edema [5].

Interestingly, eyes with BRVO and spontaneous posterior vitreous detachment have a significantly lower rate of macular edema [12, 228]. Eyes with vitreomacular separation, however, may still have macular edema [228], but vitrectomy may allow access of oxygenated aqueous to the inner retina, thereby improving macular edema and VA. Stefansson et al. demonstrated that inducing a BRVO in nonvitrectomized feline eyes resulted in retinal hypoxia, while inducing a BRVO in vitrectomized eyes produced no change in the retinal oxygen tension [224]. In seven eyes with BRVO and no PVD, Kurimoto et al. reported that vitrectomy and posterior hyaloid removal improved VA and decreased macular edema [147].

Yamamoto et al. reported a reduction in macular edema associated with BRVO after vitrectomy with and without AAS [259]. They found no difference between the treatment groups but the entry criteria were different, and in a considerable number of patients combined cataract surgery was performed. It is surprising that vitrectomy with posterior hyaloid removal alone could improve macular edema, because macular traction is rare in eyes with BRVO. It is possible that removal of the posterior hyaloid by vitrectomy improves oxygenation of the retina [224, 225]. Alternatively, vitrectomy may improve diffusion of harmful cytokines that promote increased vascular permeability, such as VEGF, leading them away from the retina, thereby influencing the oncotic pressure outside and the hydrostatic pressure within the vein. It is unknown whether vitrectomy also has a beneficial effect in eyes with preexisting posterior vitreous detachments.

Mandelcorn found a beneficial effect of ILM removal in BRVO patients with macular edema [166] as was seen in patients with diabetic macular edema. Different possible pathomechanisms are discussed including the removal of a diffusion barrier. Currently it is unknown whether an additional ILM removal during vitrectomy is advantageous.

21.3.8.2.13 Vessel Cannulation

Successful cannulations of branch retinal arterioles and branch retinal venules were described in porcine eyes [3, 240] and in a human cadaver eye model [233]. 10-0 monofilament nylon sutures were placed in the vessel lumen through an arteriotomy or phlebotomy opening [233]. Surgical penetration of retinal vessels was also accomplished in an allantois membrane model [156], and micromanipulator-assisted cannulation of retinal vessels could be achieved successfully in cat eyes [93]. Other researchers maximized the view of the cannulation site using a high resolution gradient index of refraction (GRIN) lens endoscope [117] allowing hand cannulation of canine retinal veins in vivo.

Retinal vein cannulation with prolonged intravascular injection of rTPA was found to be feasible and safe in an experimental BRVO in canine eyes [231]. Weiss et al. first described retinal endovascular surgery (REVL) injecting rTPA in a clinical setting [251] with favorable results. The method is difficult and has not gained general acceptance but may offer new future treatment options.

21.3.8.2.14 Fibrinolytic Therapy

Fibrinolytic agents such as streptokinase or rt-PA have been used in an attempt to cause dissolution of the venous obstruction. rt-PA has been applied as either an intravitreal injection [148, 252], as intravenous therapy [104], or directly injected into the obstructed vein [251]. Hattenbach reported encouraging results of systemic thrombolysis with low dose rt-PA (50 mg) in patients with hemorrhagic retinopathy in CRVO and BRVO [104]. However, the risk of life threatening hemorrhages is a major problem with intravenous fibrinolysis. Garcia-Arumi used rt-PA in BRVO in a combined approach with AAS. Recently, it was demonstrated that rt-PA administered intravitreally penetrates into the retinal vasculature in a porcine model of vascular occlusion [164]. A similar effect might be expected in humans. Thus, the fibrinolytic activity of rt-PA would help to restore venous flow.

21.3.8.3 Treatment of Neovascular Related Complications

Essentials
- Patients with areas of capillary nonperfusion exceeding five disk areas in fluorescein angiography should be closely followed for the development of neovascularization
- Only when preretinal neovascularization is present should scatter LC in the affected quadrant be performed
- Patients should be seen 4–6 weeks after scatter LC. If regression of neovascularizations is noted, further close observation is necessary
- If no regression is observed 4–6 weeks after treatments, additional scatter LC must be applied; after regression 6–12 months follow-up is sufficient
- The main indications for vitrectomy include nonclearing vitreous hemorrhage, tractional retinal detachment involving the macula, and epiretinal membrane formation

Concerning the prevention of both neovascularization and vitreous hemorrhage, the BVOS group concluded that there is no advantage of prophylactic laser photocoagulation in patients with ischemic BRVO. Regression of the new vessels and prevention of vitreous hemorrhage are likely to occur when scatter laser treatment of the involved quadrant is performed (Figs. 21.3.25a–d, 21.3.26a–d). As the patient in Figs. 21.3.26 and 21.3.27 is clearly showing, scatter LC should be performed also in the central avascular area using smaller spots to achieve regression of neovascularizations and to avoid aggravation of macular edema. Another option might be the combination with IVTA. Because only 25% of eyes with significant nonperfusion develop neovascularization and of these only 50% develop vitreous hemorrhage, laser is indicated only when capillary nonperfusion is accompanied by neovascularization, not in the setting of capillary nonperfusion alone. A peripheral scatter LC in the presence of neovascularization is as effective as a prophylactic LC, which would necessitate treating many patients who would never develop neovascularization [29]. As a result of the BRVO Study the following treatment lines were established:

Patients with areas of capillary nonperfusion exceeding five disk areas in fluorescein angiography should be closely followed at intervals of 4 months for the development of neovascularization. Only when preretinal neovascularization is present should scatter LC in the affected quadrant be performed. This restriction is important, since scatter LC can cause visual field loss and in superior major BRVO a loss of inferior peripheral visual

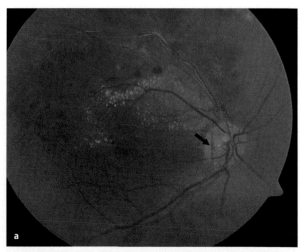

Fig. 21.3.25. a Major BRVO affecting the superotemporal quadrant of the right eye with lipid exudates, sheathed venules, partially regressed hemorrhages, a few laser burns and neovascularization of the disk (NVD) (*arrow*)

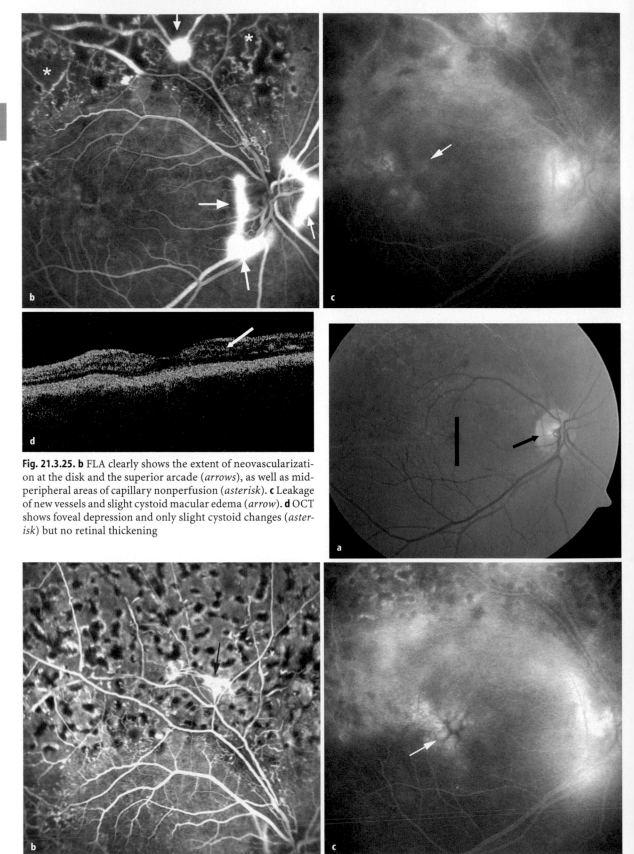

Fig. 21.3.25. b FLA clearly shows the extent of neovascularization at the disk and the superior arcade (*arrows*), as well as midperipheral areas of capillary nonperfusion (*asterisk*). **c** Leakage of new vessels and slight cystoid macular edema (*arrow*). **d** OCT shows foveal depression and only slight cystoid changes (*asterisk*) but no retinal thickening

Fig. 21.3.26a–c (legend see p. 497 ▷)

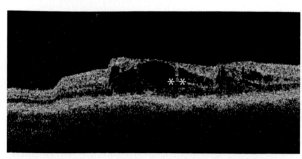

Fig. 21.3.26. d Large cystoid spaces (*asterisks*) in the corresponding OCT image can be seen (scan, see **a**)

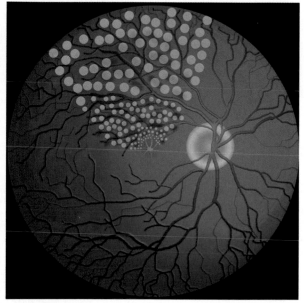

Fig. 21.3.27. Schematic drawing of scatter LC pattern covering the involved segment, sparing the area within two disk diameters of the fovea

field can be disabling for driving and navigating [109].

For scatter LC medium white burns with a size of 200–500 μm in diameter spaced one burn width apart and covering the entire involved segment sparing the area within two disk diameters of the fovea is recommended [29] (Fig. 21.3.27).

◁

Fig. 21.3.26. a Same patient as in Fig. 25 after first scatter LC treatment, with persisting NVD (*arrow*), and collateral vessels in the horizontal raphe. FLA shows **b** laser burns in the ischemic area and partial regression of the peripheral neovascularization (*arrow*) but **c** marked increase of cystoid macular edema can be seen (*arrow*).

21.3.8.3.1 Scatter Laser Photocoagulation in BRVO – Technique

Essentials

- Medium white burns 200–500 μm in diameter should be applied one burn width apart covering the entire involved ischemic segment and sparing the area within two disk diameters of the fovea
- The power should be increased in 50 mW increments starting at 200 mW until the desired burn intensity is achieved
- Duration may be variable from 0.1 s near the center to >0.2 s in the periphery
- A Goldmann – or wide angle – contact lens may be used
- The adjacent unaffected retina should not be treated
- Topical anesthesia is mostly sufficient; retrobulbar anesthesia is rare
- Neovascularizations and areas of vitreoretinal traction should be spared and not photocoagulated directly

In cases of reduced funduscopic view caused by corneal edema, progressed cataract or vitreous hemorrhage, sometimes LC is not possible. In these instances, peripheral retinal transscleral cryotherapy can be performed, but visual control of cryocoagulation effects is warranted to avoid overfreezing. Also an early vitrectomy can be considered, since usually transscleral cryotherapy provides no access to the avascular areas, which are mostly located posterior to the equator in BRVO.

Regular funduscopic checks should be performed in patients especially with the ischemic type of BRVO since neovascularization and vitreous hemorrhage may also develop years after the occlusion.

Also, in the presence of iris neovascularization scatter LC is essential to treat avascular zones as the underlying cause of the rubeosis iridis. If a neovascular glaucoma is present, cryo- or laser photocoagulation can be combined with cyclodestructive treatments such as cyclophotocoagulation (CPC) or cyclocryocoagulation. The main indications for vitrectomy include nonclearing vitreous hemorrhage, tractional retinal detachment involving the macula, and epiretinal membrane formation [4, 119].

21.3.9 Conclusions

21.3.9.1 Recommendations for Therapy

Based on the currently available evidenced data the following recommendations can be given:

- For patients with the mildest form of BRVO, careful observation seems reasonable.
- If neovascularization is present, prompt scatter photocoagulation is mandatory.
- In cases of persistent decreased vision of 20/40 or less and macular edema due to major BRVO, grid LC should be performed as the initial gold standard treatment.
- As long as no comparative data demonstrating the beneficial effect of other treatment modalities is available, intravitreal steroids or anti-VEGF drugs or surgical interventions should be considered only in cases of laser refractory macular edema or in patients in which grid LC has been shown to be ineffective, for example, patients with poor visual acuity in the range of 20/100 or foveal capillary nonperfusion.

Due to the associations with an increasing prevalence of diabetes mellitus and obesity [54], the prevalence of hypertension and retinal venous occlusive diseases is likely to increase in coming years. This makes it increasingly apparent from a public health standpoint that new and effective therapies are urgently needed.

A patient with an attack of BRVO may realize that the condition can be quite disabling and, with our present knowledge, several justifiable questions posed by the patient can barely be given a clear answer. Timely intervention can substantially reduce the incidence and severity of related complications and visual loss. Improved understanding of medical conditions associated with BRVO may have implications for completely new treatment strategies.

However, treatment benefits following focal laser are modest in most eyes. The BVOS shows that after 3 years, the mean number of lines of vision gained was 1.33 in the laser treated group and 0.23 in the control group [235]. While this improvement in vision was statistically significant, it offers little hope to patients with poor baseline VA or nonperfused macular edema.

In clinical practice, ophthalmologists are often confronted with patients having macular edema unresponsive to traditional treatment. Pilot studies, often without controls, with a variety of new treatment modalities suggest improvement over the natural history. Improved retinal perfusion of an occluded retinal vein by decompression at the AV crossing makes biomechanical sense, but posterior hyaloid removal alone showed similar results. In recent years, pharmatherapeutic approaches with intravitreal injection of steroids or anti-VEGF compounds appear promising. Unfortunately, the therapeutic effect is transient and the natural course of macular edema in BRVO is considerably longer. In fact, many of the patients within the published reports did have recurrent edema soon after the studies ended. Clearly, longer-term treatments are needed.

Currently, a substantial proportion of publications on BRVO belong to the class of case reports. They often do not differentiate between the different types of BRVO or between patients having a BRVO of short duration or a longer standing BRVO. This differentiation is urgently required, for the evaluation of new upcoming treatment strategies have to be evaluated. Further limitations of the available data including the small number of patients treated, the short duration of follow-up, and the lack of concurrent control patients preclude the drawing of definitive conclusions concerning the risk-benefit ratio compared with standard care. The treatment effect should not be judged only by analyzing retinal thickness but also by tortuosity and diameter of the affected vein.

There is significant lack of controlled treatment trials. BRVOs occur at high frequency allowing future prospective randomized controlled studies to be conducted to evaluate different therapies singly or in combination. But, until the results of such studies are available, what should be recommended to patients with vision loss due to BRVO? Although well-controlled clinical trials are not available to date, physicians are obliged to sort through the available evidence and make their best judgement to provide the best possible treatment for their patients. Given the relative perceived safety of steroids and anti-VEGF drugs weighed against the documented unfavorable natural history in patients with persistent macular edema with progressive visual loss and the limited benefit from macular grid LC, an argument for a low-risk intervention could be made. For eyes at higher levels of VA it would not seem prudent, in the absence of data to the contrary, to recommend initial treatment with unproven treatment modalities which may not be effective or safe in the long-term. However, the evidence-based, preferred method of therapy in eyes with poor visual prognosis remains unclear. In the absence of clear guidelines the ultimate decision as to whether to treat or not and the type of therapy to be used, is principally a decision made between the patient and his/her physician after thorough discussion of the known risks and benefits.

At the heart of it all, improved understanding of the pathophysiology of BRVO and its associated medical conditions may offer implications for completely new treatment strategies.

Acknowledgements. The author is grateful to PD Dr. L.-O. Hattenbach, the Eye Clinic Ludwigshafen, for proof reading of "associated risk factors" and "fibrinolytic therapy", Mrs. P. Hammermeister for typing and Mrs. A. McKenzie for translation assistance.

References

1. Abu el-Asrar AM, al-Momen AK, al-Amro S, Abdel Gader AG, Tabbara KF (1996) Prothrombotic states associated with venous retinal occlusion in young adults. Int Ophthalmol 20:197–204
2. Acheson JF, Gregson RMC, Merry P, Schulenburg WE (1991) Vaso-occlusive retinopathy in the primary antiphospholipid antibody syndrome. Eye 5:48–55
3. Allf BE, de Juan E Jr (1987) In vivo cannulation of retinal vessels. Graefes Arch Clin Exp Ophthalmol 225:221–225
4. Amirikia A, Scott IU, Murry TG, Flynn HW Jr, Smiddy WE, Fever WJ (2001) Outcomes of vitreoretinal surgery for complications of branch retinal vein occlusion. Ophthalmology 108:372–376
5. Ando N (2000) Vitrectomy for ischemic maculopathy associated with retinal vein occlusion. Vail Vitrectomy Meeting, Vail, 2000
6. Appiah AP, Greenidge KC (1987) Factors associated with retinal-vein occlusion in Hispanics. Ann Ophthalmol 19: 307–312
7. Appiah AP, Trempe CL (1989) Risk factors associated with branch vs. central retinal occlusion. Ann Ophthalmol 21: 153–157
8. Appiah AP, Trempe CL (1989) Differences in contributory factors among hemicentral, central and branch retinal vein occlusions. Ophthalmology 96:364–366
9. Arnarsson A, Stefansson E (2000) Laser treatment and the mechanism of edema reduction in branch retinal vein occlusion. Invest Ophthalmol Vis Sci 41:877–879
10. Asherson RA, Mery P, Acheson JF, Harris EN, Hughes GR (1989) Antiphospholipid antibodies: a risk factor for occlusive ocular vascular disease in systemic lupus erythematosus and the "primary" antiphospholipid syndrome. Ann Rheum Dis 48:358–361
11. Avitabile T, Longo A, Reibaldi A (2005) Intravitreal triamcinolone compared with macular laser grid photocoagulation for the treatment of cystoid macular edema. Am J Ophthalmol 140:695–702
12. Avunduk AM, Cetinkaya K, Kapicioglu Z, Kaya C (1997) The effect of posterior vitreous detachment on the prognosis of branch vein occlusion. Acta Ophthalomol Scand 75:441–442
13. Bandello F, Vigano D Angelo S, Parlavecchia M, et al. (1994) Hypercoagulability and high lipoprotein (a) levels in patients with central vein occlusion. Thromb Haemost 72: 39–43
14. Bandello F, Tavola A, Pierro L, Modorati GD, Azzolini C, Brancato R (1998) Axial length and refraction in retinal vein occlusions. Ophthalmologica 212:133–135
15. Barbazetto IA, Schmidt-Erfurth UM (2000) Evaluation of functional defects in branch retinal vein occlusion before and after laser treatment with scanning laser perimetry. Ophthalmology 107:1089–1098
16. Bavbek T, Yenice O, Toygar O (2005) Problems with attempted chorioretinal venous anastomosis by laser for nonischemic CRVO and BRVO. Ophthalmologica 219:267–271
17. Becquet F, Le Rouic JF, Zanlonghi X, Peronnet P, Hermouet-Leclair E, Pousset-Decre C, Ducournau D (2003) Efficiency of surgical treatment for chronic macular edema due to branch retinal vein occlusion. J Fr Ophtalmol 26:570–576
18. Beer PM, Bakri SJ, Singh RJ, Liu W, Peters GB III, Miller M (2003) Intraocular concentration and pharmacokinetics of triamcinolone acetonide after a single intravitreal injection. Ophthalmology 110:681–686
19. Behrman S (1962) Retinal vein occlusion. Br J Ophthalmol 46:336–346
20. Ben-Nun J (2001) Capillary blood flow in acute branch retinal vein occlusion. Retina 21:509–512
21. Benz MS, Murray TG, Dubovy SR, Katz RS, Eifrig CW (2003) Endophthalmitis caused by *Mycobacterium chelonae* abscessus after intravitreal injection of triamcinolone. Arch Ophthalmol 121:271–273
22. Bertelsen T (1950) The relationship between thrombosis in the retinal veins and primary glaucoma. Acta Ophthalmol 39:603–613
23. Bertram B, Haase G, Remkys A, Reim M (1994) Anti-Cardiolipin-Antikörper bei Gefäßverschlüssen am Auge. Ophthalmologe 91:768–771
24. Bertram B, Remky A, Arend O, Wolf S, Reim M (1995) Protein C, protein S, and antithrombin III in acute ocular occlusive diseases. Ger J Ophthalmol 4:332–335
25. Bick RL, Baker WF (1994) Antiphospholipid and thrombosis syndromes. Semin Thromb Hemost 20:3–15
26. Bill A (1963) The uveal venous pressure. Arch Ophthalmol 69:780–2
27. Blankenship GW, Ohun E (1973) Retinal tributary vein occlusion. Histology and management by photocoagulation. Arch Ophthalmol 89:363
28. Branch Vein Occlusion Study Group (1985) Argon laser photocoagulation for macular edema in branch vein occlusion. Am J Ophthalmol 99:218–219
29. Branch Vein Occlusion Study Group (1986) Argon laser scatter photocoagulation for prevention of neovascularization and vitreous hemorrhage in branch vein occlusion. A randomized clinical trial. Arch Ophthalmol 104:34–41
30. Brown BA, Marx JL, Ward TP, Hollifield RD, Dick JS, Brozetti JJ, Howard RS, Thach AB (2002) Homocysteine: A risk factor for retinal venous occlusion disease. Ophthalmology 109:287–290
31. Cahill MT, Stinnett SS, Fekrat S (2003) Meta-analysis of plasma homocysteine, serum folate, serum vitamin B12, and thermolabile MTHFR genotype as risk factors for retinal vascular occlusive disease. Am J Ophthalmol 136:1136–1150
32. Cahill MT, Kaiser PK, Sears JE, Fekrat S (2003) The effect of arteriovenous sheathotomy on cystoid macular oedema secondary to branch retinal vein occlusion. Br J Ophthalmol 87:1329–1332
33. Carbone J, Sanchez-Ramon S, Cobo-Soriano R, et al. Antiphospholipid antibodies: a risk factor for occlusive retinal vascular disorders. Comparison with ocular inflammatory diseases. J Rheumatol 128:2437–2441
34. Cassels-Brown A, Minford AMB, Chatfield SL, Bradbury JA (1994) Ophthalmic manifestation of neonatal protein C deficiency. Br J Ophthalmol 78:486–487
35. Cekic O, Chang S, Tseng JJ, Barile GR, Del Priore LV, Weis-

smann H, Schiff WM, Ober MD (2005) Intravitreal triamcinolone injection for treatment of macular edema secondary to branch retinal vein occlusion. Retina 25:851–5

36. Chabanel A, Glacet-Bernard A, Lelong F, Taccoen A, Coscas G, Samama MM (1990) Increased red blood cell aggregation in retinal vein occlusion. Br J Haematol 75:127–131

37. Chabanel A, Horellou MH, Conard J, Samama MM (1993) Retinal vein occlusion: hemorrheology and hemostasis. Thromb Haemost 69:996 Abstract

38. Charbonnel J, Glacet-Bernard A, Korobelnik JF, Nyouma-Moune E, Pournaras CJ, Colin J, Coscas G, Soubrane G (2004) Management of branch retinal vein occlusion with vitrectomy and arteriovenous adventitial sheathotomy, the possible role of surgical posterior vitreous detachment. Graefes Arch Clin Exp Ophthalmol 242:223–8

39. Chen SD, Lochhead J, Patel CK, Frith P (2004) Intravitreal triamcinolone acetonide for ischaemic macular oedema caused by branch retinal vein occlusion. Br J Ophthalmol 88:154–155

40. Chen SD, Sundaram Y, Lochhead J, Patel CK (2006) Intravitreal triamcinolone for the treatment of ischemic macular edema associated with branch retinal vein occlusion. Am J Ophthalmol 141:876–883

41. Christoffersen NL, Larsen M (1999) Pathophysiology and hemodynamics of branch retinal vein occlusion. Ophthalmology 106:2054–2062

42. Chua B, Kifley A, Wong TY, Mitchell P (2005) Homocysteine and retinal vein occlusion: A population-based study. Am J Ophthalmol 139:181–182

43. Chung MM, Trese MT, Hong YJ (1989) Protein C levels in retinal vein occlusions. Invest Ophthalmol 30:477

44. Ciardella AP, Yannuzzi LA, Freund KB, DiMichele D, Nejat M, De Rosa JT, Daly JR, Sisco L (1998) Factor V Leiden, activated protein C resistance, and retinal vein occlusion. Retina 18:308–315

45. Clement DB, Elsby JM, Smith WD (1968) Retinal vein occlusion. Br J Ophthalmol 52:111–117

46. Clemett RS, Kohner EM, Hamilton AM (1973) The visual prognosis in retinal branch vein occlusion. Trans Ophthalmol Soc U K 93:523–535

47. Clemett RS (1974) Retinal branch vein occlusion. Changes at the site of obstruction. Br J Ophthalmol 58:548–554

48. Cole MD, Dodson PM, Hendeles S (1989) Medical conditions underlying retinal vein occlusion in patients with glaucoma or ocular hypertension. Br J Ophthalmol 73:693–698

49. Coscas G, Dhermy P (1978) Occlusions veineuses retiniennes. Masson, Paris

50. Danis RP, Wallow HL (1987) Microvascular changes in experimental branch retinal vein occlusion. Ophthalmology 94:1213–1221

51. David R, Zangwill L, Badarna M, Yassur Y (1988) Epidemiology of retinal vein occlusion and its association with glaucoma and increased intraocular pressure. Ophthalmologica 197:69–74

52. Degenring RF, Kamppeter B, Kreissing I, Jonas JB (2003) Morphological and functional changes after intravitreal triamcinolone acetonide for retinal vein occlusion. Acta Ophthalmol Scand 81:399–401

53. Demirci FY, Guney DB, Akarcay K, Kir N, Ozbek U, Sirma S, Unaltuna N, Ongor E (1999) Prevalence of factor V Leiden in patients with retinal vein occlusion. Acta Ophthalmol Scand 77:631–633

54. Diabetes Public Health Resource page. Centers for Disease Control and Prevention Web site. Available at: http://www.cdc.gov/diabetes/index.htm.

55. Dieguez Millan JM, Suner Capo M, Olea Vallejo JL (2002) Intraoperatory rupture of the vein in an arteriovenous crossing sheathotomy. Arch Soc Esp Oftalmol 77:575–578

56. Disdier P, Harle JR, Mouly A, et al. (1992) Behçet's syndrome and Factor XII deficiency. Clin Rheumatol 11:422–423

57. Dithmar S, Hansen LL, Holz FG (2003) Venöse retinale Verschlüsse. Ophthalmologe 7:561–577

58. Dodson PM, Galton DJ, Winder AF (1981) Retinal vascular abnormalities in the hyperlipidemias. Trans Ophthalmol Soc UK 101:17–21

59. Dodson PM, Galton DJ, Hamilton AM, Blach RK (1982) Retinal vein occlusion and the prevalence of lipoprotein abnormalities. Br J Ophthalmol 66:161–164

60. Dodson PM, Westwick J, Marks G, Kakkar VV, Galton DJ (1983) β-Thromboglobulin and platelet factor 4 levels in retinal vein occlusion. Br J Ophthalmol 67:143–146

61. Dodson PM, Kritzinger EE (1985) Underlying medical conditions in young patients and ethnic differences in retinal vein occlusion. Trans Ophthalmol Soc UK 104:114–119

62. Dodson PM, Kritzinger EE (1987) Management of retinal vein occlusion. Br Med J 295:1434–1435

63. Dodson PM, Clough CG, Downes SM, Kritzinger EE (1993) Does type II diabetes predispose to retinal vein occlusion? Eur J Opthalmol 3:109–113

64. Dodson PM, Kubicki AJ, Taylor KG, Kritzinger EE (1985) Medical conditions underlying recurrence of retinal vein occlusion. Br J Ophthalmol 69:493–496

65. Driot JY, Nowak GD, Rittenhouse KD, Milazzo C, Pearson PA (2004) Ocular pharmacokinetics of fluocinolone acetonide after retisert intravitreal implantation in rabbits over a 1-year period. J Ocul Pharmacol Ther 20:269–275

66. Duker JS, Brown GC (1989) Anterior location of the crossing artery in branch retinal vein occlusion. Arch Ophthalmol 107:998

67. Durand F, Delamaire M, Carre V, et al. (1991) Study of rheological parameters in patients with retinal vein occlusion. Clin Hemorheol 11:533

68. Ellwein LB, Friedlin V, McBean AM, Lee PP (1996) Use of eye care services among the 1991 Medicare population. Ophthalmology 103:1732–1743

69. Esrick E, Subramanian ML, Heier JS, Devaiah AK, Topping TM, Frederick AR, Morley MG (2005) Multiple laser treatment for macular edema attributable to branch retinal vein occlusion. Am J Ophthalmol 139:653–657

70. Fekrat S, Goldberg MF, Finkelstein D (1999) Laser-induced chorioretinal venous anastomosis for nonischemic central or branch retinal vein occlusion. Arch Ophthalmol 116:43–52

71. Fekrat S (2005) Surgical management of branch retinal vein occlusion. Retina 71–73

72. Feltgen N, Auw-Haedrich C, Buchen R, Hansen LL (2005) Arteriovenous dissection in a living human eye: clinicopathologic correlation. Arch Ophthalmol 123:571–572

73. Feltgen N, Herrmann A, Agostini HJ, Sammain A, Hansen LL (2006) Arterio-venous dissection after isovolaemic haemodilution in branch retinal vein occlusion: a non-randomized prospective study. Graefes Arch Clin Exp Ophthalmol 244:829–35

74. Feist RM, Ticho BH, Shapiro MJ, Farber M (1992) Branch retinal vein occlusion and quadratic variation in arteriovenous crossings. Am J Ophthalmol 113:664–668

75. Figueroa MS, Torres R, Alvarez MT (2004) Comparative study of vitrectomy with and without vein decompression for branch retinal vein occlusion: a pilot study. Eur J Ophthalmol 14:40–47

76. Finkelstein D, Clarkson J, Diddie K, et al. (1982) Branch vein occlusion. Retinal neovascularization outside the involved segment. Ophthalmology 89:1357

77. Finkelstein D (1992) Ischemic macular edema. Recognition and favorable natural history in branch vein occlusion. Arch Ophthalmol 110:1427–1434

78. Flynn HW Jr, Scott IU (2005) Intravitreal triamcinolone acetonide for macular edema associated with diabetic retinopathy and venous occlusive disease. It's time for clinical trials. Arch Ophthalmol 123:258–259

79. Folkman J, Ingber DE (1987) Angiostatic steroids. Method of discovery and mechanism of action. Ann Surg 206:374–383

80. Frangieh GT, Green WR, Barraquer-Somers E, et al. (1982) Histopathologic study of nine branch retinal vein occlusions. Arch Ophthalmol 100:1132

81. Frucht J, Shapiro A, Merin S (1984) Intraocular pressure in retinal vein occlusion. Br J Ophthalmol 68:26–28

82. Fujii GY, de Juan E Jr, Humayun MS (2003) Improvements after sheathotomy for branch retinal vein occlusion documented by optical coherence tomography and scanning laser ophthalmoscope. Ophthalmic Surg Lasers Imaging 34:49–52

83. Fujimoto R, Ogino N, Kumagai K, Demizu S, Furukawa M (2004) The efficacy of arteriovenous adventitial sheathotomy for macular edema in branch retinal vein occlusion. Nippon Ganka Gakkai Zasshi 108:144–149

84. Fujino T, Curtin VT, Norton EW (1966) Experimental central retinal vein occlusion: a comparison of intraocular and extraocular occlusion. Trans Am Ophthalmol Soc 66:318

85. Garcia-Arumi J, Martinez-Castillo V, Boixadera A, Blasco H, Corcostegui B (2004) Management of macular edema in branch retinal vein occlusion with sheathotomy and recombinant tissue plasminogen activator. Retina 24:530–540

86. Gass JDM (1951) Arch Ophthalmol 46:550

87. Gillies MC, Simpson JM, Billson FA, et al. (2004) Safety of an intravitreal injection of triamcinolone: results from a randomized clinical trial. Arch Ophthalmol 122:336–340

88. Girolami A, Pellati D, Lombardi AM (2005) FXII deficiency is neither a cause of thrombosis nor a protection from thrombosis. Am J Ophthalmol 139:578–579

89. Glacet-Bernard A, Coscas G, Chabanel A, Zourdani A, Lelong F, Samama MM (1994) A randomized, double-masked study on the treatment of retinal vein occlusion with troxerutin. Am J Ophthalmol 118:421–429

90. Glacet-Bernard A, Chabanel A, Lelong F, et al. (1994) Elevated erythrocyte aggregation in patients with central retinal vein occlusion and without conventional risk factors. Ophthalmology 101:1483–1487

91. Glacet-Bernard A, Bayani N, Chretien P, Cochard C, Lelong F, Coscas G (1994) Antiphospholipid antibodies in retinal vascular occlusions. A prospective study of 75 patients. Arch Ophthalmol 112:790–795

92. Glacet-Bernard A, Coscas G, Chabanel A, Zourdani A, Lelong F, Samama MM (1996) Prognostic factors for retinal vein occlusion. A prospective study of 175 cases. Ophthalmology 103:551–560

93. Glucksberg MR, Dunn R, Giebs CP (1993) In vivo micropuncture of retinal vessels. Graefes Arch Clin Exp Ophthalmol 231:405–407

94. Glueck CJ, Bell H, Vadlamani L, Gupta A, Fontaine RN, Wang P, Stroop D, Gruppo R (1999) Heritable thrombophilia and hypofibrinolysis. Possible causes of retinal vein occlusion. Arch Ophthalmol 117:43–49

95. Greaves M (1997) Aging and the pathogenesis of retinal vein thrombosis (editorial). Br J Ophthalmol 81:810–811

96. Greiner K, Hafner G, Dick B, Peetz D, Prellwitz W, Pfeiffer N (1999) Retinal vascular occlusion and deficiencies in the protein C pathway. Am J Ophthalmol 128:69–74

97. Greven CM, Weaver RG, Owen J, Slusher MM (1991) Protein S deficiency and bilateral branch retinal artery occlusion. Ophthalmology 98:33–34

98. Gutman FA, Zegarra H (1974) The natural course of temporal retinal branch vein occlusion. Trans Am Acad Ophthalmol Otolaryngol 78:178–192

99. Gutman FA, Zegarra H (1976) Retinal detachment secondary to retinal branch vein occlusions. Trans Am Acad Opthalmol Otolaryngol 81:491

100. Hamilton AM, Marshall J, Kohner EM, et al. (1975) Retinal new vessel formation following experimental vein occlusion. Exp Eye Res 20:493

101. Hamilton AM, Kohner EM, Rosen D, Bird AC, Dollery CT (1979) Experimental retinal branch vein occlusion in rhesus monkeys. 1. Clinical appearances. Br J Ophthalmol 63:377–387

102. Han DP, Bennett SR, Williams DF, Dev S (2003) Arteriovenous crossing dissection without separation of the retina vessels for treatment of branch retinal vein occlusion. Retina 23:145–151

103. Hansen LL, Wiek J, Arntz R (1988) Randomized study of the effect of isovolemic hemodilution in retinal branch vein occlusion. Fortschr Ophthalmol 85:514–516

104. Hattenbach L-O, Steinkamp G, Scharrer I, Ohrloff C (1998) Fibrinolytic therapy with low-dose recombinant tissue plasminogen activator in retinal vein occlusion. Ophthalmologica 212:394–398

105. Hayashi K, Hayashi H (2005) Intravitreal versus retrobulbar injections of triamcinolone for macular edema associated with branch retinal vein occlusion. Am J Ophthalmol 139:972–82

106. Hayreh SS, March W, Phelps CD (1977) Ocular hypotony after retinal vascular occlusion. Trans Ophthalmol Soc UK 97:757–767

107. Hayreh SS, Zimmerman B, Podhajsky P (2001) Systemic diseases associated with various types of retinal vein occlusion. Am J Ophthalmol 131:61–77

108. Hayreh SS, Zimmermann MB, Podhajsky P (1994) Incidence of various types of retinal vein occlusion and their recurrence and demographic characteristics. Am J Ophthalmol 117:429–441

109. Hayreh SS (2005) Prevalent misconceptions about acute retinal vascular occlusive disorders. Prog Retinal Eye Res 24:493–519

110. Hida T, Chandler D, Arena JE, Machemer R (1986) Experimental and clinical observation of the intraocular toxicity of commercial corticosteroid preparations. Am J Ophthalmol 101:190–5

111. Hitchings RA, Spaeth GL (1976) Chronic RVO in glaucoma. Br J Ophthalmol 60:694–699

112. Hirota A, Mishima HK, Kiuchi Y (1997) Incidence of retinal vein occlusion at the Glaucoma Clinic of Hiroshima University. Ophthalmologica 211:288–291

113. Ho TC, Lai WW, Lam DSC (2006) Intravitreal triamcinolone compared with macular laser grid photocoagulation

for the treatment of cystoid macular edema. Am J Ophthalmol 141:786–787

114. Hochley DJ, Tripath RC, Ashton N (1976) Experimental retinal branch vein occlusion in the monkey. Histopathological and ultrastructural studies. Trans Ophthalmol Soc UK 96:202

115. Hochley DJ, Tripath RC, Ashton N (1979) Experimental retinal branch vein occlusion in rhesus monkeys. III. Histopathological and electron microscopical studies. Br J Ophthalmol 63:393

116. Horio N, Horiguchi M (2005) Effect of arteriovenous sheathotomy on retinal blood flow and macular edema in patients with branch vein occlusion. Am J Ophthalmol 139:739–740

117. Humayun MS, Hanza HS, Fujii G, et al. (1999) Endoscopic cannulation in central retinal vein occlusion. Subspecialty Day. American Academy of Ophthalmology Annual Meeting, October 22–23, Orlando, FL, 1999

118. Ikeda J, Hasegawa S, Suzuki K, Usui TE, Tanimoto N, Takagi M, Abe H, Kaiya T (2005) Evaluation of macula in patients with branch retinal vein occlusion using multifocal electroretinogram and optical coherence tomography. Nippon Ganka Gakkai Zasshi 109:142–147

119. Ikuno Y, Ikea T, Sato Y, Taro Y (1998) Tractional retinal detachment after branch retinal vein occlusion: influence of disc neovascularization on the outcome of vitreous surgery. Ophthalmology 105:417–423

120. Ingerslev J (1999) Thrombophilia: a feature of importance in retinal vein thrombosis? Acta Ophthalmol Scand 77:619–621

121. Iturralde D, Spaide RE, Meyerle CB, Klancnik JM, Yannuzzi LA, Fisher YL, Sorenson J, Slakter JS, Freund KB, Cooney M, Fine HF (2006) Intravitreal bevacizumab (Avastin) treatment of macular edema in central retinal vein occlusion: a short-term study. Retina 26:279–84

122. Jaeger EA (1981) Venous obstructive disease of the retinal. In: Duane TD (ed) Clinical ophthalmology, vol 3. Harper & Row, Philadelphia, revised 1983, chap 15, pp 1–22

123. Jaffe L, Goldberg RE, Magargel LE, et al. (1980) Macular branch vein occlusion. Ophthalmology 87:91–98

124. Jaffe GJ, Pearson PA, Ashton P (2000) Dexamethasone sustained drug delivery implant for the treatment of severe uveitis. Retina 20:402–403

125. Jaissle GB, Ziemssen F, Petermeier K, Szurman P, Ladewig M, Gelisken F, Völker M, Holz FG, Bartz-Schmidt KU (2006) Bevacizumab zur Therapie des sekundären Makulaödems nach venösen Gefäßverschlüssen. Ophthalmologe 103:471–75

126. Janssen MCH, den Heijer M, Cruysberg JRM, Wollersheim H, Bredie SJH (2005) Retinal vein occlusion: a form of venous thrombosis or a complication of atherosclerosis. Thromb Haemost 93:1021–1026

127. Jefferies P, Clemett R, Day T (1993) An anatomical study of retinal arteriovenous crossings: their role in the pathogenesis of retinal vein occlusion. Aust N Z J Ophthalmol 21:213–217

128. Jensen VA (1936) Clinical studies of tributary thrombosis in the central retina vein. Acta Ophthalmol 1 (Suppl X):1

129. Johnston RL, Bruckner AJ, Steinmann W, et al. (1985) Risk factors of branch retinal vein occlusion. Arch Ophthalmol 103:1831–1832

130. Jonas JB (2005) Intravitreal triamcinolone acetonide for treatment of intraocular oedematous and neovascular diseases. Acta Ophthalmol Scand 84:645–663

131. Jonas JB, Akkoyun I, Kamppeter B, Kreissig I, Degenring RE (2005) Branch retinal vein occlusion treated by intravitreal triamcinolone acetonide. Eye 19:65–71

132. Jonas JB, Akkoyun I, Kamppeter B, Kreissig I, Degenring RF (2005) Intravitreal triamcinolone acetonide as treatment of branch retinal vein occlusion. Eur J Ophthalmol 15:751–758

133. Kaiser PK (2005) Steroids for branch retinal vein occlusion. Am J Ophthalmol 139:1095–1096

134. Karacorlu M, Ozedmir H, Karacorlu SA (2005) Resolution of serous macular detachment after intravitreal triamcinolone acetonide treatment of patients with branch retinal vein occlusion. Retina 25:856–60

135. Kawaji T, Hirata A, Awai N, Takano A, Inomata Y, Fukushima M, Tanihara H (2005) Trans-tenon retrobulbar triamcinolone injection for macular edema associated with branch retinal vein occlusion remaining after vitrectomy. Am J Ophthalmol 140:540–2

136. Klein R, Klein BEK, Moss SE, et al. (2000) The epidemiology of retinal vein occlusion: the Beaver Dam Eye Study. Trans Am Ophthalmol Soc 98:133–139

137. Kohner EM (1964) Retinal vein occlusion. Proc R Soc Med 57:816

138. Kohner EM, Dollery CT, Shahib M, et al. (1970) Experimental retinal branch vein occlusive. Am J Ophthalmol 69:778

139. Koyanagi Y (1928) Die Bedeutung der Gefässkreuzung für die Entstehung der Astthrombose der retinalen Zentralvene. Klin Mbl Augenhk 81:219–231

140. Koyanagi Y (1937) Die pathologische Anatomie und Pathogenese des Kreuzungsphänomens der Netzhautgefasse bei Hochdruck. Graefes Arch Ophthalmol 137:619–635

141. Krepler K, Ergun E, Sacu S, Richter-Muksch S, Wagner J, Stur M, Wedrich A (2005) Intravitreal triamcinolone acetonide in patients with macular oedema due to branch retinal vein occlusion: a pilot study. Acta Ophthalmol Scand 83:600–4

142. Kube ET, Feltgen N, Pache M, Herrmann J, Hansen LL (2005) Angiographic findings in arteriovenous dissection (sheathotomy) for decompression of branch retinal vein occlusion. Graefes Arch Clin Exp Ophthalmol 243:334–338

143. Kuhli C, Hattenbach LO, Scharrer I, Koch F, Ohrloff C (2002) The high prevalence of resistance to APC in young patients with retinal vein occlusion. Graefes Arch Clin Exp Ophthalmol 240:163–168

144. Kuhli C, Scharrer I, Koch F, Ohrloff C, Hattenbach LO (2004) Factor XII deficiency: a thrombophilic risk factor for retinal vein occlusion. Am J Ophthalmol 137:459–464

145. Kumagai K, Ogino N, Furukawa M, Demizu S, Atsumi K, Kurihara H, Ishigooka H (2002) Vitreous surgery for macular edema in branch retinal vein occlusion. Nippon Ganka Gakkai Zasshi 106:701–707

146. Kumar B, Yu D, Morgan WH, et al. (1998) The distribution of angioarchitectural changes within the vicinity of arteriovenous crossing in branch retinal vein occlusion. Ophthalmology 105:424–427

147. Kurimoto M, Takagi H, Suzuma K, Oh H, Nonaka A, Kiryu J, Kita M, Ogura Y, Honda Y (1999) Vitrectomy for macular edema secondary to retinal vein occlusion: evaluation by retinal thickness analyzer. Jpn J Clin Ophthalmol 53:717–720

148. Lahey JM, Fong DS, Kearney J (1999) Intravitreal tissue

plasminogen activator for acute central retinal vein occlusion. Ophthalmic Surg Lasers 30:427–434

149. Lakhanpal RR, Javaheri M, Ruiz-Garcia H, De Juan E Jr, Humayun MS (2005) Transvitreal limited arteriovenous-crossing manipulation without vitrectomy for complicated branch retinal vein occlusion using 25-gauge instrumentation. Retina 25:272–280

150. Lakhanpal RR, Javaheri M, Equi RA, Humayum MS (2005) Improvement after transvitreal limited arteriovenous crossing manipulation without vitrectomy for complicated branch retinal vein occlusion using 25 gauge instrumentation. Br J Ophthalmol 89:922–923

151. Lang GE, Freißler K (1992) Klinische und fluoreszeinangiographische Befunde bei Patienten mit retinalen Venenastverschlüssen. Klin Monatsbl Augenheilkd 201:234–239

152. Larsson J, Olafsdottir E, Bauer B (1996) Activated protein C resistance in young adults with central vein occlusion. Br J Ophthalmol 80:200–202

153. Leber T (1877) Die Krankheiten der Netzhaut und des Sehnerven. In: Graefe A, Sämisch T (eds) Handbuch der gesamten Augenheilkunde, Pathologie und Therapie. Verlag von Wilhelm Engelmann, Leipzig, pp 521–535

154. Lee WH, Thompson JT, Sjaarda RN (2001) Visual acuity results in arteriovenous sheathotomy versus grid laser photocoagulation in branch retinal vein occlusion. Abstract no 3862. Poster presented at the annual meeting of the Association for Research in Vision and Ophthalmology, 2001

155. Lee H, Shah GK (2005) Intravitreal triamcinolone as primary treatment of cystoid macular edema secondary to branch retinal vein occlusion. Retina 25:551–5

156. Leng T, Miller JM, Bilbao KV, Palanker DV, Huie P, Blumenkranz MS (2004) The chick chorioallantoic membrane as a model tissue for surgical retinal research and simulation. Retina 24:427–434

157. Lerche RC, Richard G (2004) Arteriovenous sheathotomy in branch retinal vein occlusion. Klin Monatsbl Augenheilkd 221:479–484

158. Le Rouic JF, Bejjani RA, Rumen F, et al. (2001) Adventitial sheathotomy for decompression of recent onset branch vein occlusion. Graefes Arch Clin Exp Ophthalmol 239:747–751

159. Liebreich R (1855) Apoplexia retinae. Graefes Arch Ophthalmol 1:346–351

160. Lip PL, Blann AD, Jones AF, Lip GY (1998) Abnormalities in haemorheological factors and lipoprotein (a) in retinal vascular occlusion: implications for increased vascular risk. Eye 12:245–251

161. Lowe GDO, Trope G, McArdle BM, Douglas JT, Foulds W, Forbes CD, Prentice CRM (1982) Abnormal blood viscosity and haemostasis in chronic retinal vein thrombosis. Seventh International Congress on Thrombosis, Valencia, Spain, Oct. 13–16, 1982

162. Lowe GDO (1986) Blood rheology in arterial disease. Clin Sci 171:137–146

163. Luntz MH, Schenker HI (1980) Retinal vascular accidents in glaucoma and ocular hypertension. Surv Ophthalmol 25:163–167

164. Mahmoud TH, Peng YW, Proia A, Davidson M, Deramo VA, Fekrat S (2002) Intravitreal tissue plasminogen activator penetrates the retinal vessels in a porcine model of vascular occlusion. Invest Ophthalmol 43:3533. Abstract

165. Maiji AB, Janarthnan M, Naduvilath TJ (1997) Significance of refractive status in branch retinal vein occlusion: a case-control study. Retina 17:200–204

166. Mandelcorn MS, Nrusimhadevara RK (2004) Internal limiting membrane peeling for decompression of macular edema in retinal vein occlusion: a report of 14 cases. Retina 24:348–355

167. Martidis A, Duker JS, Greenberg PB, et al. (2002) Intravitreal triamcinolone for refractory diabetic macular edema. Ophthalmology 109:920–927

168. Mason J III, Feist R, White M Jr, Swanner J, McGwin G Jr, Emond T (2004) Sheathotomy to decompress branch retinal vein occlusion: a matched control study. Ophthalmology 111:540–545

169. McAllister IL, Yu D-Y, Vijayasekaran S, Barry C, Constable I (1992) Induced chorioretinal venous anastomosis in experimental retinal branch vein occlusion. Br J Ophthalmol 76:615–620

170. McGarth MA, Wechsler F, Hunyor ABL, Penny R (1978) Systemic factors contributory to retinal vein occlusion. Arch Intern Med 138:216–220

171. Merry P, Acheson JF (1988) Management of retinal occlusion. BMJ 296:294

172. Mester U, Dillinger P (2002) Vitrectomy with arteriovenous decompression and internal limiting membrane dissection in branch retinal vein occlusion. Retina 22:740–746

173. Mester U, Dillinger P (2001) Behandlung retinaler Venenastverschlüsse. Ophthalmologe 98:1104–1109

174. Michels RG, Gass JD (1974) The natural course of temporal retinal branch vein occlusion. Trans Am Acad Ophthalmol Otolaryngol 78:178–192

175. Mitchell P, Smith W, Chang A (1996) Prevalence and associations of retinal vein occlusion in Australia: The Blue Mountains Eye Study. Arch Ophthalmol 114:1243–1247

176. Moshfeghi DM, Kaiser PK, Scott IU, et al. (2003) Acute endophthalmitis following intravitreal triamcinolone acetonide injection. Am J Ophthalmol 136:791–96

177. Nakazato K, Watanabe H, Kawana K, Hiraoka T, Kiuchi TE, Oshika T (2005) Evaluation of arterial stiffness in patients with branch retinal vein occlusion. Ophthalmogica 219:334–337

178. Neetens A, Bocque G (1987) Bilateral retinal branch vein occlusion in protein C deficiency. Bull Soc Belge Ophtalmol 223:53–57

179. Noma H, Funatsu H, Yamasaki M, Tsukamoto H, Mimura T, Sone T, Jian K, Sakamoto I, Nakano K, Yamashita H, Minamoto A, Mishima HK (2005) Pathogenesis of macular edema with branch retinal vein occlusion and intraocular levels of vascular endothelial growth factor and interleukin-6. Am J Ophthalmol 140:256–216

180. Noma H, Minamoto A, Funatsu H, Tsukamoto H, Nakano K, Yamashita H, Mishima HK (2006) Intravitreal levels of vascular endothelial growth factor and interleukin-6 are correlated with macular edema in branch retinal vein occlusion. Graefes Arch Clin Exp Ophthalmol 244:309–315

181. Ohashi H, Oh H, Nishiwaki H, Nonaka A, Takagi H (2004) Delayed absorption of macular edema accompanying serous retinal detachment after grid laser treatment in patients with branch retinal vein occlusion. Ophthalmology 111:2050–2056

182. Opremcak EM, Bruce RA (1999) Surgical decompression of branch retinal vein occlusion via arteriovenous crossing sheathotomy. A prospective review of 15 cases. Retina 19:1–5

183. Orth DH, Patz A (1978) Retinal branch vein occlusion. Surv Ophthalmol 22:357–376

184. Osterloh MD, Charles S (1988) Surgical decompression of branch retinal vein occlusions. Arch Ophthalmol 106: 1469–1471

185. Ozkiris A, Evereklioglu C, Erkilic K, Ilhan O (2005) The efficacy of intravitreal triamcinolone acetonide on macular edema in branch retinal vein occlusion. Eur J Ophthalmol 15:96–101

186. Ozkiris A, Evereklioglu C, Erkilic K, Dogan H (2006) Intravitreal triamcinolone acetonide for treatment of persistent macular oedema in branch retinal vein occlusion. Eye 20:13–7

187. Parodi MB, Saviano S, Ravalico G (1999) Grid laser treatment in macular branch retinal vein occlusion. Graefes Arch Clin Exp Ophthalmol 237:1024–1027

188. Pasco M, Singer L, Romen M (1973) Chronic simple glaucoma and thrombosis of retinal vein. Ear Nose Throat J 52:294–297

189. Paton A, Rubenstein K, Smith VH (1964) Artificial insufficiency in retinal venous occlusion. Trans Ophthalmol Soc UK 84:559

190. Peduzzi M, Codeluppi L, Poggi M, Baraldi P (1983) Abnormal blood viscosity and erythrocyte deformability in retinal vein occlusion. Am J Ophthalmol 96:399–400

191. Penfold PL, Wen L, Madigan MC, et al. (2000) Triamcinolone acetonide modulates permeability and intercellular adhesion molecule-1 (ICAM-1) expression of the ECV304 cell line: implications for macular degeneration. Clin Exp Immunol 121:458–465

192. Peyman GA, Khoobehi B, Moshfeghi A, Moshfegi D (1998) Reversal of blood flow in experimental branch retinal vein occlusion. Ophthalmic Surg Lasers 29:595–597

193. Piermarocchi S, Segato T, Bertoja H, et al. (1990) Branch retinal vein occlusion: the pathogenetic role of blood viscosity. Ann Ophthalmol 22:303–311

194. Pulido JS, Lingua RM, Cristol S, Byrne SF (1987) Protein C deficiency associated with vitreous hemorrhage in a neonate. Am J Ophthalmol 104:546–547

195. Pulido JS, Ward LM, Fischman GA, et al. (1987) Antiphospholipid antibodies associated with retinal vascular disease. Retina 7:215–218

196. Rabinowicz IM, Litman S, Michaelson IC (1969) Branch venous thrombosis – A pathological report. Trans Ophthalmol Soc UK 88:191–210

197. Rabinowicz IM, Litman S, Michaelson IC (1968) Ibid 88:191

198. Rath EZ, Frank RN, Shin DH, Kim C (1992) Risk factors for retinal vein occlusions. A case-control study. Ophthalmology 99:509–514

199. Remky A, Wolf S, Hamid M, Bertram B, Schulte K, Arend O, Reim M (1994) Influence of hemodilution in retinal hemodynamics in branch vein occlusion. Ophthalmologe 91:288–292

200. Ring CP, Pearson TC, Sanders MD, Wetherley-Mein G (1976) Viscosity and retinal vein thrombosis. Br J Ophthalmol 60:397–410

201. Rosen DA, Marshall J, Kohner EM, et al. (1979) Experimental retinal branch vein occlusion in rhesus monkeys. II. Retinal blood flow studies. Br J Ophthalmol 63:388

202. Rosendaal FR (1997) Thrombosis in the young: epidemiology and risk factors: a focus on venous thrombosis. Thromb Haemost 78:1–6

203. Rosenfeld PJ, Fung AE, Puliafito CA (2005) Optical coherence tomography findings after an intravitreal injection of bevacizumab (Avastin) for macular edema from central retinal vein occlusion. Ophthalmic Surg Lasers Imaging 36:336

204. Rubenstein K, Jones EB (1976) Retinal vein occlusion: long term prospects. Br J Ophthalmol 60:148–150

205. Saika S, Tanaka T, Miyamoto T, Ohnishi Y (2001) Surgical posterior vitreous detachment combined with gas/air tamponade for treating macular edema associated with branch retinal vein occlusion: retinal tomography and visual outcome. Graefes Arch Clin Exp Ophthalmol 239: 729–732

206. Salinas-Alaman A, Garcia-Layana A, Sadaba-Echarri LM, Belzunce-Manterola A (2005) Branch retinal vein occlusion treated by intravitreal triamcinolone. Arch Soc Esp Oftalmol 80:463–5

207. Sallmann L (1937) Zur Anatomie der Gefässkreuzungen am Augenhintergrund: Zugleich ein Beitrag zur pathologischen Anatomie des Gunn und Salusschenzeichens. Graefes Arch Ophthalmol 137:619–635

208. Sarici A, Muftuoglu G (2006) Intravitreal triamcinolone compared with macular laser grid photocoagulation for the treatment of cystoid macular edema. Am J Ophthalmol 141:785–786

209. Scat Y, Morin Y, Morel C, Haut J (1995) Retinal vein occlusion and resistance to activated protein C (in French). J Fr Ophthalmol 18:758–762

210. Schatz H, Yannuzzi L, Stranky TJ (1976) Retinal detachment secondary to branch vein occlusion. Part I. Ann Ophthalmol 8:1437

211. Schatz H, Yannuzzi L, Stranky TJ (1976) Retinal detachment secondary to branch vein occlusion. Part II. Ann Ophthalmol 8:1461

212. Schepens CL, Avila MP, Jalkh AE, Trempe CL (1984) Role of the vitreous in cystoid macular edema. Surv Ophthalmol 28:499–504

213. Scott IU, Ip MS (2005) It's time for a clinical trial to investigate intravitreal triamcinolone for macular edema due to retinal vein occlusion: The SCORE study. Arch Ophthalmol 123:581–582

214. Seitz R (1962) Die Netzhautgefäße. Vergleichende ophthalmoskopische und histologische Studien an gesunden und kranken Augen. Bücherei des Augenarztes, Heft 40. Enke, Stuttgart

215. Shah GK, Sharma S, Fineman MS, Federman J, Brown MM, Brown GC (2000) Arteriovenous adventitial sheathotomy for the treatment of macular edema associated with branch retinal vein occlusion. Am J Ophthalmol 129: 104–106

216. Shaikh S, Blumenkranz MS (2001) Transient improvement in visual acuity and macular edema in central retinal vein occlusion accompanied by inflammatory features after pulse steroid and anti-inflammatory therapy. Retina 21: 176–78

217. Shahid H, Hossain P, Amoaku WM (2006) The management of retina vein occlusion: is interventional ophthalmology the way forward? Br J Ophthalmol 90:627–639

218. Shilling JS, Kohner EM (1976) New vessel formation in retinal branch vein occlusion. Br J Ophthalmol 60:810–15

219. Shilling JS, Jones CA (1984) Retinal branch vein occlusion. A study of argon laser photocoagulation in the treatment of macular oedema. Br J Ophthalmol 68:196–198

220. Shiraga F (2002) Evaluation of AV sheathotomy for branch retinal vein occlusion by means of fluorescein videoangi-

ography and a computer-assisted image analysis. Macular Society 25th Annual Meeting. Barcelona, Spain; June 13, 2002

221. Snyers B, Lambert M, Hardy JP (1990) Retinal and choroidal vasoocclusive disease in systemic lupus erythematosus associated with antiphospholipid antibodies. Retina 10:255–260

222. Soukiasian S, Lahav M, Snady-McCoy L (1989) Natural anticoagulants in retinal vein occlusions. Invest Ophthalmol Vis Sci (Suppl) 30:477

223. Sperduto RD, Hiller R, Chew E, Seigel D, Blair N, Burton TC, Farber MD, Gragoudas ES, Haller J, Seddon JM, Yannuzzi LA (1998) Risk factors for hemiretinal vein occlusion: comparison with risk factors for central and branch retinal vein occlusion: the eye disease case-control study. Ophthalmology 105:765–771

224. Stefansson E, Novack RL, Hatchell DL (1990) Vitrectomy prevents retinal hypoxia in branch retinal vein occlusion. Invest Ophthalmol Vis Sci 31:284–289

225. Stefansson E (2001) The therapeutic effect of retinal laser treatment and vitrectomy. A theory based on oxygen and vascular physiology. Acta Ophthalmol Scan 79:435–440

226. Staurenghi G, Lonati C, Aschero M, Orzalesi N (1994) Arteriovenous crossing as a risk factor in branch retinal vein occlusion. Am J Ophthalmol 117:211–213

227. Tachi N, Hashimoto Y, Ogino N (1999) Vitrectomy for macular edema combined with retinal vein occlusion. Doc Ophthalmol 97:465–469

228. Takahashi MK, Hikichi T, Akiba J, et al. (1997) Role of the vitreous and macular edema in branch retinal vein occlusion. Ophthalmic Surg Lasers 28:294–299

229. Takahashi K, Kashima T, Kishi S (2005) Massive macular hard exudates associated with branch retinal vein occlusion. Jpn J Ophthalmol 49:527–529

230. Talu S, Stefanut C (2004) Axial length and branch retinal vein occlusion. Oftalmologia 48:81–84

231. Tameesh MK, Lakhanpal RR, Fujii GY, Javaheri M, Shelley TH, D'Anna S, Barnes AC, Margalit E, Farah M, De Juan E Jr, Humayun MS (2004) Retinal vein cannulation with prolonged infusion of tissue plasminogen activator (t-PA) for the treatment of experimental retinal vein occlusion in dogs. Am J Ophthalmol 138:829–839

232. Tanaka TE, Muraoka K, Tokui K (1998) Retinal arteriovenous shunt at the arteriovenous crossing. Ophthalmology 105:1251–1258

233. Tang WM, Han DP (2000) A study of surgical approaches to retinal vascular occlusion. Arch Ophthalmol 118:138–143

234. Tavola A, D'Angelo SV, Bandello F, et al. (1995) Central retinal vein and branch artery occlusion associated with inherited plasminogen deficiency and high lipoprotein (a) levels: a case report. Thromb Res 80:327–331

235. The Branch Vein Occlusion Study Group. Argon laser photocoagulation for macular edema in branch vein occlusion. Am J Ophthalmol 98:271–282

236. The Eye Disease Case-Control Study Group. Risk factors for branch retinal vein occlusion. Am J Ophthalmol 116:286–96

237. Tekeli O, Gursel E, Buyurgan H (1999) Protein C, protein S and antithrombin III deficiencies in retinal vein occlusion. Acta Ophthalmol Scand 77:628–630

238. Tolosa-Villela C, Ordi-Ros J, Jordana-Comajuncosa R, et al. (1990) Occlusive ocular vascular disease and antiphospholipid antibodies. Ann Rheum Dis 49:203

239. Trope GE, Lowe GDO, McArdle BM, et al. (1983) Abnormal blood viscosity and haemostasis in long-standing retinal vein occlusion. Br J Ophthalmol 67:137–142

240. Tsilimbaris MK, Lit ES, Amico DJ (2004) Retinal microvascular surgery: a feasibility study. Invest Ophthalmol Vis Sci 45:1963–1968

241. Tsujikawa A, Fujihara M, Iwawaki T, Yamamoto K, Kurimoto Y (2005) Triamcinolone acetonide with vitrectomy for treatment of macular edema associated with branch retinal vein occlusion. Retina 25:861–7

242. US Census Bureau. Available at: http://eire.censusgov/popest/data

243. Vannas S, Tarkhanan A (1960) Retinal vein occlusion and glaucoma. Br J Ophthalmol 44:583–589

244. Vine AK, Samama MM (1993) The role of abnormalities in the anticoagulant and fibrinolytic systems in retinal vascular occlusions. Surv Ophthalmol 37:283–292

245. von Michel J (1878) Die spontane Thrombose der Vena Centralis des Opticus. Graefes Arch Ophthalmol 24:37–70

246. Wakabayashi T, Okada AA, Morimura Y, Kojima E, Asano Y, Hirakata A, Hida T (2004) Trans-tenon retrobulbar triamcinolone infusion for chronic macular edema in central and branch retinal vein occlusion. Retina 24:964–967

247. Weger M, Renner W, Steinbrugger I, Cichocki L, Temmel W, Stanger O, El-Shabrawi Y, Lechner H, Schmut O, Haas A (2005) Role of thrombophilic gene polymorphisms in branch retinal vein occlusion. Ophthalmology 112:1910–1915

248. Weinberg D, Dodwell DG, Fern SA (1990) Anatomy of arteriovenous crossings in branch retinal vein occlusion. Am J Ophthalmol 109:298–302

249. Weinberg DV, Egan KM, Seddon JM (1993) Asymmetric distribution of arteriovenous crossing in the normal retina. Ophthalmology 100:31–36

250. Weinberg DV, Seddon JM (1994) Venous occlusive diseases of the retina. In: Albert DM, Jakobiec FA (eds) Principles and practice of ophthalmology, 1st edn, vol 2. WB Saunders, Philadelphia, pp 735–746

251. Weiss JN (2001) Injection of tissue plasminogen activator into a branch retinal vein in eyes with central vein occlusion. Ophthalmology 108:2249–2257

252. Weizer JS, Fekrat S (2003) Intravitreal tissue plasminogen activator for the treatment of central retinal vein occlusion. Ophthalmic Surg Lasers Imaging 34:350–352

253. Wenzler EM, Rademakers AJ, Boers GH, Gryusberg JR, Webers CA, Deutman AF (1993) Hyperhomocysteinaemia in retinal artery and vein occlusion. Am J Ophthalmol 115:1162–1167

254. Wiechens B, Schroder JO, Potzsch B, Rochels R (1997) Primary antiphospholipid antibody syndrome and retinal occlusive vasculopathy. Am J Ophthalmol 123:848–850

255. Wiek J, Schade M, Wiederholt M, et al. (1990) Haemorheological changes in patients with retinal vein occlusion after isovolaemic haemodilution. Br J Ophthalmol 74:665–669

256. Williamson TH, Rumley A, Lowe GD (1996) Blood viscosity, coagulation, and activated protein C resistance in central retinal vein occlusion: a population controlled study. Br J Ophthalmol 80:203–208

257. Wong TY, Larsen EK, Klein R, Mitchell P, Couper DJ, Klein BE, Hubbard LD, Siscovick DS, Sharrett AR (2005) Cardiovascular risk factors for retinal vein occlusion and arteriolar emboli: the Atherosclerosis Risk in Communities & Cardiovascular Health Study. Ophthalmology 112:540–547

258. Yamaji H, Shiraga F, Tsuchida Y, Yamamoto Y, Ohtsuki H, Evaluation of arteriovenous crossing sheathotomy for branch retinal vein occlusion by fluorescein videoangiography and image analysis. Am J Ophthalmol 137:834–841

259. Yamamoto S, Saito W, Yagi F, Takeuchi S, Sato E, Mizunoya S (2004) Vitrectomy with or without arteriovenous adventitial sheathotomy for macular edema associated with branch retinal vein occlusion. Am J Ophthalmol 138: 907–914

260. Yepremyan M, Wertz FD, Tivnan T, Eversman L, Marx JL (2005) Early treatment of cystoid macular edema secondary to branch retinal vein occlusion with intravitreal triamcinolone acetonide. Ophthalmic Surg Lasers Imaging 36:30–36

261. Zhao J, Sastry SM, Sperduto RD, et al. (1993) Arteriovenous crossing patterns in branch retinal vein occlusion. Ophthalmology 100:423–428

21.4 Retinal Arterial Occlusion

M. Burton, Z. Gregor

Core Messages
- Retinal arterial occlusion is characterized by a sudden, painless loss of vision
- Occlusions can occur at several locations: ophthalmic artery, central retinal artery (CRAO), branch retinal arteries (BRAO) and retinal arterioles
- More proximal obstructions carry a worse visual prognosis
- Retinal arterial occlusion has many causes: emboli, thrombosis, hypercoagulability states, vasculitis, infections and trauma
- In the acute stages occlusion is characterized by marked reduction in retinal arterial blood flow, edematous whitening of the retina and a macular red spot. Intra-arterial emboli may be visible
- Many different treatments are used. However, these generally have a limited effect
- A thorough systemic assessment is necessary to identify any underlying cause of the retinal arterial occlusion. Additional treatment may be needed to reduce the risk of further retinal and cerebrovascular events

21.4.1 Epidemiology

The incidence of retinal arterial occlusion is estimated to be around 0.85/100,000 per year [55]. The condition mainly affects older people, with a mean age of 60 years at presentation. The pattern and causes of retinal arterial occlusion in younger age groups are quite distinct and will be discussed separately below. Men appear to be more susceptible than women, with a male to female ratio of 2:1. There may be some differences between racial groups. For example, affected African Americans tend to be younger than Caucasians, and in contrast to Caucasians, few have carotid artery disease as a cause of their obstruction [1]. In many cases retinal arterial obstruction is associated with general cardiovascular disease risk factors: hypertension (60%), smoking, diabetes (25%) and hypercholesterolemia.

21.4.2 Mechanisms of Retinal Arterial Obstruction

Many different causes of retinal arterial obstruction have been reported (Box 1). However, most cases are either due to an embolus from another site or to the development of a thrombosis often in association with atherosclerosis at the site of obstruction.

Box 1: Causes of retinal arterial occlusion
Emboli
Thrombosis ± atherosclerosis
Congenital thrombophilic states
Acquired thrombophilic states
Vasculitis
Infectious
Iatrogenic
Ocular abnormalities
Trauma
Vasospasm
Raised intraocular pressure

21.4.2.1 Emboli

Emboli arise from a more proximal part of the circulation, travel into the retinal arterial tree and obstruct the flow of blood. The most important source of retinal emboli is atherosclerotic disease of the carotid arteries. The point at which an embolus stops is determined by its size relative to the caliber of the vessel. Estimates of what proportion of retinal arterial occlusions is caused by emboli vary, but they probably account for about a third of all cases [33, 63]. Emboli are more commonly found in cases of branch retinal artery occlusions.

Various types of material have been found to embolize to the eye and cause arterial occlusion (Box 2). Most are cholesterol emboli (75%) [4, 46].

21 **III**

These are refractile with a yellow or orange color. The less common thrombus (15%) and calcific (10%) emboli are usually white in color. Cholesterol emboli are associated with atherosclerotic plaques in the aorta or carotid arteries (Box 3). Several cardiac abnormalities have been identified as a source of retinal arterial emboli: valve lesions, bacterial endocarditis, mural thrombus, arrhythmias and atrial myxoma [22, 54]. In individuals with a patent foramen ovale paradoxical emboli have been reported, which originate from the venous circulation [42, 48]. Rare cases of fat emboli from fractures and amniotic fluid emboli have also been reported [20, 72]. Multiple emboli are seen in 10–20% of cases where emboli are found, and these tend to be unilateral [39, 51].

Several large population based surveys have found asymptomatic retinal arterial emboli in about 1.4% of adults and in 3.1% of those over 75 years of age [39, 46]. The 10-year cumulative incidence rate has been estimated to be 1.5% [40]. Emboli are associated with being male, increasing age, hypertension, smoking and hypercholesterolemia. In patients with asymptomatic emboli 18% had a greater than 75% stenosis of the carotid artery [51].

The heart is probably the source of less than 10% of emboli [6]. In one study about half the patients with acute retinal arterial obstruction were found to have an abnormality by transthoracic echocardiogram. However, only about 10% required specific treatment such as anticoagulation or cardiac surgery [65]. This study subdivided individuals into low and high risk of cardiac emboli based on the history and examination. "High-risk" features included a history of myocardial infarction, rheumatic fever, valve disease, bacterial endocarditis or a cardiac murmur on auscultation. Individuals without any of these features very rarely had a cardiac defect requiring treatment. Therefore, echocardiography is recommended for any patient with a "high-risk" feature and in younger patients, but is not usually routine for "low-risk" patients.

Epidemiological studies indicate that individuals with asymptomatic retinal emboli have a relative risk of 2.6 for stroke [39, 76]. Among patients with symptomatic retinal arterial obstruction individuals with visible emboli had 4–10 times increased risk of stroke compared to controls, while those without emboli had no increased risk [58, 18].

Box 2: Types of embolic material
Cholesterol
Thrombus
Calcific
Fat
Bacterial vegetations
Tumor
Amniotic fluid

Box 3: Sources of emboli
Carotid arteries
   Atherosclerosis
Aorta
   Atherosclerosis
Heart
   Valvular abnormalities
   Mural thrombus (post-MI)
   Arrhythmias
   Atrial myxoma
   Infective endocarditis
   Septal defects
Paradoxical emboli

21.4.2.2 Thrombosis

Retinal arterial occlusion may result from the development of a thrombosis within the vessel. This frequently occurs at a site of atherosclerosis. The rupture of an atherosclerotic plaque triggers platelet aggregation and thrombus formation. It is possible that most central retinal arterial occlusions are due to a thrombosis at the level of the lamina cribrosa. The risk factors are those common to cardiovascular disease in general: hypertension, smoking, hypercholesterolemia and diabetes.

21.4.2.3 Thrombophilia

Thrombophilia is a term referring to a diverse group of disorders that predispose an individual to the development of thrombosis. Many of these conditions have been reported in association with retinal arterial occlusion, usually in younger patients [57]. However, the true significance of these associations is difficult to evaluate as these coagulation abnormalities are found in the asymptomatic general population and so associations could have arisen by chance in some cases.

Thrombophilia can be either congenital or acquired. Congenital thrombophilic states are inherited disorders, which result in disturbance of the coagulation system either through increased levels of procoagulants, deficiencies of anticoagulants or reduced fibrinolysis. Acquired causes are a diverse group of conditions. In the antiphospholipid syndrome autoantibodies to phospholipids form immune complexes resulting in thrombosis formation. Hyperhomocysteinemia damages the vascular endothelium promoting thrombus formation. Hyperviscosity states such as myeloproliferative disorders and pregnancy have also been associated with retinal arterial occlusion.

Box 4: Congenital and acquired thrombophilia
Congenital thrombophilic states:
Factor V Leiden [9, 44, 71]
Prothrombin 20210 [9, 71]
Protein C deficiency [49]
Protein S deficiency [31]
Antithrombin III deficiency [10]

Acquired thrombophilic states:
Antiphospholipid syndrome [21]
Hyperhomocysteinemia [19, 53]
Myeloproliferative disorders
Pregnancy
Oral contraceptive pill

21.4.2.4 Vasculitis

The systemic vasculitides are a group of conditions that produce inflammatory changes within the walls of blood vessels, leading to the obstruction of blood flow. Retinal arterial occlusions have been reported in association with a number of these disorders (Box 5). The most common of these is giant cell arteritis (temporal arteritis), which usually affects older people.

Box 5: Systemic vasculitis associated with retinal arterial occlusion
Giant cell arteritis
Polyarteritis nodosa
Wegner's granulomatosis [52]
Behçet's disease
Systemic lupus erythematosis
Susac's disease
Dermatomyositis

21.4.2.5 Infectious

A number of infectious conditions causing inflammatory problems in the retina can provoke occlusions of adjacent branch retinal arteries. These include toxoplasmosis, acute retinal necrosis, cat-scratch disease and loiasis [11, 23, 24, 45, 61, 67, 77].

21.4.2.6 Other Causes

Retinal arterial occlusion has been reported in association with a number of medical procedures, including vitreoretinal surgery, arterial angiography, retrobulbar injections and neck manipulations [38]. Structural abnormalities of the eye such as optic disk drusen and pre-papillary arterial loops have been associated with central retinal artery occlusions (Fig. 21.4.1) [50]. Traumatic injury to the orbit and its contents can lead to vascular occlusion, sometimes associated with a retrobulbar hemorrhage [35]. Vasospasm induced by either migraines or cocaine use has been linked to retinal arterial occlusion in a number of younger patients [15, 70].

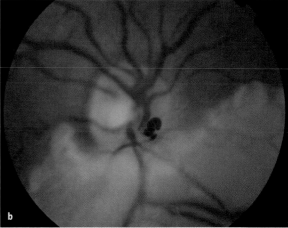

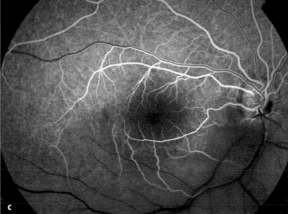

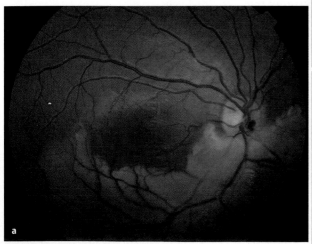

Fig. 21.4.1. Pre-papillary loop resulting in an inferior hemiretinal arterial occlusion (**a**). The twisted loop is in the inferior division of the central retinal artery that lies anterior to the optic disk (**b, c**)

21 III

21.4.2.7 Retinal Arterial Occlusion in the Young

Retinal arterial occlusion is a particularly rare event in younger people. Unlike older individuals they tend not to have atherosclerotic disease as the cause [32]. Instead a wide variety of underlying conditions have been reported including thrombophilic states, cardiac pathology, sickle cell disease, congenital abnormalities, trauma and migraine [15, 70].

21.4.3 Clinical Features

21.4.3.1 Anatomical Classification

The retinal arterial circulation can be obstructed at any point from the ophthalmic artery to the distal retinal arterioles (Box 6). The location of the obstruction determines the pattern of visual loss and clinical signs. The causes of the obstruction also vary by site with emboli being more commonly associated with branch (BRAO) rather than central retinal artery occlusion (CRAO).

Box 6: Sites of arterial obstruction
Ophthalmic artery
Central retinal artery
Branch retinal artery
Cilioretinal artery
Arterioles

21.4.3.2 History

Retinal arterial occlusion is characterized by a sudden, painless loss of vision. The visual defect is usually isolated to one eye and may be partial (BRAO and cilioretinal artery occlusion) or complete (CRAO and ophthalmic artery occlusion). Sometimes patients note the visual defect on waking up in the morning, which may be due to reduced retinal perfusion secondary to nocturnal hypotension. Some patients may report antecedent episodes of transient visual loss, amaurosis fugax. Patients who have giant cell arteritis as their underlying cause of arterial occlusion frequently have other symptoms such as temporal headache, jaw claudication, myalgia, weight loss and loss of appetite.

21.4.3.3 Examination

21.4.3.3.1 Ophthalmic Artery Occlusion

Ophthalmic artery occlusion interrupts the blood supply to both the retina and the choroid, resulting in profound loss of vision. An identical clinical phe-

notype develops if there are concurrent multiple occlusions affecting both circulations [14, 41]. Patients may experience ocular pain. The vision is usually reduced to no perception of light with a dense relative afferent pupillary defect (RAPD). There is extensive whitening of the retina, which is more pronounced than a CRAO and usually no cherry red spot, as there is reduced perfusion of the choroid. Severe posterior segment neovascularization has been reported in some eyes following ophthalmic arterial occlusion [41].

21.4.3.3.2 Central Retinal Artery Occlusion

In a central retinal artery occlusion the blockage may occur at any point between the origin at the ophthalmic artery and the optic disk head, although most probably occur at the level of the lamina cribrosa. Vision is usually reduced to the level of counting fingers or hand movements, unless there is a separate cilioretinal artery supplying the macula [13, 43]. An RAPD is present. The retina may initially appear normal. Within a few hours the nerve fiber layer becomes thickened with retinal whitening particularly in the macula (Fig. 21.4.2). A cherry red spot develops at the fovea where the choroid is still visible. Emboli may be seen in the CRA in about a quarter of cases. Retinal arteries become thin and attenuated and may have breaks in the column of blood (box-carring or cattle-trucking).

Over the course of several weeks the obstructed vessel may be recanalized allowing reperfusion of retinal vessels. The retinal edema resolves. The optic disk eventually becomes atrophic. Neovascularization of the iris occurs in 16–18% of eyes with CRAO (Fig. 21.4.3), while new vessels at the optic disk are quite rare [27, 28].

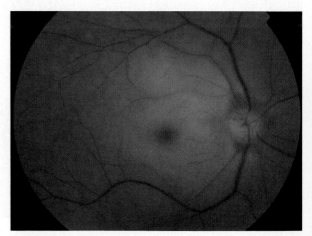

Fig. 21.4.2. Central retinal arterial occlusion. The retinal arteries are thin. The macula is pale with a cherry red spot at the center

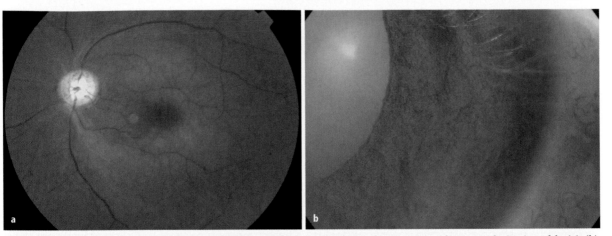

Fig. 21.4.3. Long-standing central retinal arterial occlusion with attenuated retinal arteries (**a**) and neovascularization of the iris (**b**)

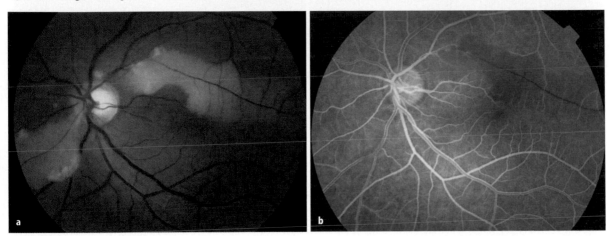

Fig. 21.4.4. Branch retinal arterial occlusion. A branch of the superotemporal retinal artery supplying the macula and the inferonasal retinal artery are occluded

21.4.3.3.3 Branch Retinal Artery Occlusion

Branch retinal artery occlusion tends to result in less severe visual loss than a CRAO (Fig. 21.4.4). Patients notice a partial loss of vision, often with an altitudinal visual field defect. An RAPD is usually present. The right eye is more commonly affected than the left, as cardiac emboli more readily travel up the right carotid artery. The temporal retina is more susceptible than the nasal [51]. Emboli are the cause of most BRAO. Visible emboli are found in up to 68% of cases [75]. The retina is pale in the affected sector. Neovascularization is a rare complication of BRAO.

21.4.3.3.4 Cilioretinal Artery Occlusion

Cilioretinal arteries arise from the short posterior ciliary arteries and can be found in about 32% of eyes [37]. There is marked variation in the number, size and distribution of these arteries. In about 19% of eyes they contribute to the macula blood supply. These vessels fill just before the central retinal artery

on fluorescein angiography. In cases of CRAO the presence of an additional cilioretinal blood supply to the macula can help to preserve central vision (Fig. 21.4.5) [17].

The cilioretinal arteries can become occluded leading to visual impairment. Three distinct patterns of cilioretinal occlusion have been described [16]. In the first group the occlusion is isolated. This is often associated with carotid artery atherosclerotic disease and the visual outcome is usually good (90% achieving 6/12). Secondly, cilioretinal artery occlusion can occur in eyes with central retinal vein occlusions (CRVO) [12, 16]. Various mechanisms may explain this association: back pressure in the cilioretinal arteries secondary to the CRVO, simultaneous reduction in perfusion pressure of the cilioretinal arteries and the retinal veins or a vasculitic process occurring in the optic disk. The third pattern of cilioretinal artery occlusion is associated with an anterior ischemic optic neuropathy, which has a very poor visual prognosis (usually less than 3/60).

21 **III**

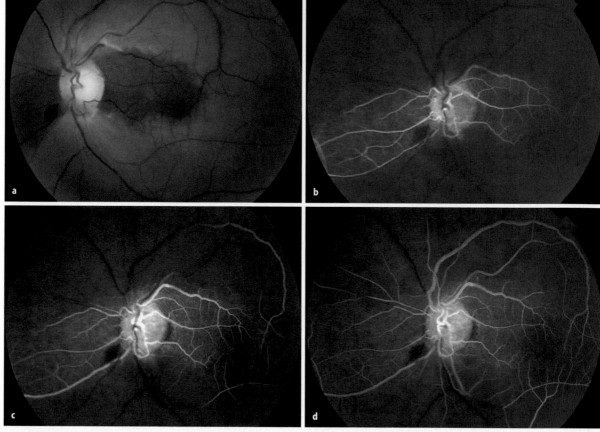

Fig. 21.4.5. Central retinal arterial occlusion with cilioretinal arteries to the macula and nasal retina (**a**). Fluorescein angiogram at **b** 6 s, **c** 14 s and **d** 27 s. This demonstrates rapid filling of the cilioretinal vessels and delayed minimal filling in a few branches of the central retinal artery

21.4.3.3.5 Retinal Arteriole Occlusion

Occlusion of the small retinal arterioles by microemboli can occur in a number of situations resulting in a characteristic clinical phenotype called Purtscher's retinopathy. Please refer to Chapter 24.1 for full details on this entity. The obstructions result in large patches of retinal whitening and some hemorrhage, usually centered on the optic disk. The condition is invariably bilateral and results in marked visual impairment. Talc retinopathy may develop following the repeated intravenous injection of crushed tablets. Multiple intra-arteriole emboli may be seen at the macula resulting in ischemic retinal changes.

21.4.4 Differential Diagnosis

The differential diagnosis of a CRAO is listed in Box 7 and that for BRAO is listed in Box 8. In addition to CRAO, a cherry red spot of the macula has a number of causes (Box 9). These are mostly congenital metabolic disorders, affecting both eyes from a young age.

Box 7: The differential diagnosis of central retinal artery occlusion
Ophthalmic artery occlusion
Multiple BRAO
Commotio retinae (Berlin's edema)
Macula hole
Acute retinal necrosis

Box 8: The differential diagnosis of branch retinal artery occlusion
CRAO
Cilioretinal artery occlusion
Myelinated nerve fibers
Acute retinal necrosis

Box 9: Causes of a cherry red spot at the macula
Central retinal artery occlusion
Tay-Sachs disease
Sandhoff disease
Niemann-Pick disease
Sialidosis Type 1 (Goldberg syndrome)
Gangliosidosis GM1
Farber's disease

21.4.5 Systemic Clinical Assessment

A systemic clinical assessment is necessary in cases of retinal arterial occlusion, as this may be the first presentation of a potentially serious medical condition [2, 58]. It is important to identify underlying medical conditions, which may be amenable to treatment to reduce the risk of further ocular or systemic morbidity. It is usually appropriate to refer patients with retinal arterial occlusion to a general physician for this medical assessment. A focused history should be taken to ascertain vascular risk factors such as those listed in Box 10. A general physical examination should be conducted for relevant clinical signs (Box 11).

Box 10: History relevant to retinal arterial occlusion
History of cerebrovascular disease
History of ischemic heart disease
Hypertension
Diabetes
Smoking history
Hypercholesterolemia
Family history of vascular disease
Symptoms of temporal arteritis
Thrombophilia (personal or family history)
Drug use (e.g., cocaine)
Trauma
Migraine

Box 11: Clinical examination of a patient with retinal arterial occlusion
Peripheral pulses – cardiac rhythm and presence of pulses
Blood pressure
Auscultation of carotid arteries for bruits
Cardiac auscultation for murmurs
Signs of temporal arteritis
Signs of infective endocarditis
Neurological signs of cerebrovascular disease

21.4.6 Investigations

A wide range of different ancillary investigations have been used in the context of retinal arterial occlusions to both confirm the diagnosis and to identify any underlying cause. The choice of tests needs to be tailored to the individual patient, guided by the history and clinical examination [62].

21.4.6.1 Investigations to Confirm the Diagnosis of Retinal Arterial Occlusion

21.4.6.1.1 Fluorescein Angiography

In cases of CRAO there is a marked delay in the transit time from the arm to the retina [25]. Often the front edge of the fluorescein is seen slowly progressing along the branch arteries. The arteriovenous transit time is prolonged. The choroid fills normally. In BRAO there is no fluorescein beyond the point of occlusion while other unaffected branches fill normally. In some cases of BRAO retrograde filling of the occluded vessel can be seen from adjacent vessels [59]. If there is occlusion of the ophthalmic artery in addition to poor perfusion of the retinal vessels there is reduced choroidal perfusion.

21.4.6.1.2 Visual Fields

In BRAO there is usually an altitudinal visual field defect corresponding to the affected portion of the retina [68]. Visual field testing is rarely attempted in cases of CRAO, although some have reported preservation of temporal peripheral vision.

21.4.6.1.3 Doppler Ultrasound

Colour Doppler ultrasonography allows the determination of blood flow direction and velocity [74]. It has been used to measure the blood flow in the ophthalmic artery, central retinal artery and short posterior ciliary arteries. In cases of CRAO occlusion there is reduced flow in the CRA while that in the short posterior ciliary arteries is maintained [73].

21.4.6.1.4 Electroretinography

Electroretinograms (ERG) from eyes with CRAO usually have an intact a-wave (derived from the photoreceptors) and a reduction or loss of the b-wave (from Müller and bipolar cells) [78]. In cases of BRAO multifocal ERG find similar abnormalities in the area of retina affected. In ophthalmic artery occlusion there is absence of both a- and b-waves [14].

21.4.6.2 Investigations of the Cause of Retinal Arterial Occlusion

21.4.6.2.1 Hematological Investigations

As many systemic conditions have been associated with retinal arterial occlusion numerous different blood tests could be requested (Box 12). However, a focused, stepwise approach is recommended. All patients should be tested for hyperglycemia and hyperlipidemia. In patients older than 50 years the ESR and CRP are measured as part of the evaluation for temporal arteritis. In younger patients and in older patients for whom no underlying cause for retinal arterial occlusion is initially found, various additional tests should be considered (Box 12).

Box 12: Hematological investigation of retinal arterial occlusion

All patients:
Glucose
Lipid profile

Older patients:
Erythrocyte sedimentation rate (ESR)
C-reactive protein (CRP)

Additional tests (young or second line):
Full blood count
Clotting screen
Thrombophilia screen (see Box 4)
Homocysteine
Plasma electrophoresis
Hemoglobin electrophoresis
Auto-antibody screen
   Anti-phospholipid antibodies
   Anti-nuclear antibodies (ANA)
   Anti-neutrophil cytoplasmic antibodies (ANCA)
Infection screen
   Toxoplasma gondii serology
   Bartonella henselae serology (Cat-scratch disease)
   Borrelia burgdorferi serology (Lyme disease)

21.4.6.2.2 Carotid Artery Assessment

The carotid arteries can be assessed in a number of ways: Doppler ultrasonography, magnetic resonance angiography and selective contrast angiography. Carotid ultrasonography is usually the first line investigation. It is non-invasive and provides information about the extracranial portion of the carotid artery. Among individuals with symptomatic retinal arterial occlusion, 19% had a significant carotid artery occlusion [64]. However, the presence of a visible embolus on retinal examination is a poor predictor of significant carotid stenosis [64]. Therefore, all adults with retinal arterial occlusion should have an ultrasound scan of the carotid arteries regardless of whether there is a visible embolus.

21.4.6.2.3 Cardiac Assessment

Patients are often referred to a physician for a cardiac assessment, which may involve various investigations. An electrocardiogram (ECG) is performed to screen for ischemic heart disease. A 24-h ambulatory ECG recording is made if arrhythmias are suspected. Transthoracic echocardiography is advised for all young patients and any with "high-risk" features [62, 66]. "High-risk" features include a history of myocardial infarction, rheumatic fever, valve disease, bacterial endocarditis or a cardiac murmur on auscultation. Transesophageal echocardiography may sometimes identify cardiac embolic sources not seen by transthoracic echocardiography [42].

21.4.7 Pathology

Animal models of central retinal artery occlusion have shown that irreversible damage of the retina develops after 100 min [34]. There is initially intracellular edema in the inner retina. Subsequently, necrosis of the inner retina develops with loss of the normal cellular architecture. Photoreceptors survive as they continue to receive nourishment from the choroidal circulation.

21.4.8 Management

The management of acute retinal arterial occlusion is difficult and the outcomes are often disappointing. There are no proven treatments. This is because retinal arterial occlusions are relatively rare events, so studies have been either retrospective or small case series without a randomly allocated control group.

Several different therapeutic maneuvers are advocated [47]. The aim of the treatment is to restore the retinal blood supply as soon as possible, increase oxygen delivery to the retina or limit the damage from hypoxia. Some of the interventions are simple and non-invasive while others are invasive, complex and have potentially serious side effects. For treatment to have any prospect of success it must be started immediately as irreversible visual loss develops after 100 min. If a patient with a CRAO presents within 24 h, most clinicians would attempt several of the non-invasive treatments described below. In some ophthalmic centers if a CRAO is diagnosed within a few hours of onset more invasive treatments may be performed. As the visual outcome for BRAO is relatively good invasive treatment is not indicated, although some clinicians would attempt to dislodge an embolus by ocular massage or a paracentesis.

21.4.8.1 Lie Patient Flat

Lie the patient flat to try to increase retinal perfusion pressure.

21.4.8.2 Ocular Massage

Ocular massage has been reported to occasionally help to dislodge an embolus [56]. Pressure is applied to the globe (through the eyelid or via a three-mirror contact lens) for 10 s and then suddenly released. This cycle is repeated for up to 15 min.

21.4.8.3 Anterior Chamber Paracentesis

Anterior chamber paracentesis is performed to rapidly reduce the intraocular pressure. The procedure is outlined in Box 13. Anecdotally this has been reported to help dislodge emboli. However, a retrospective review of paracentesis combined with carbogen treatment did not demonstrate any benefit [5]. Postparacentesis endophthalmitis has been reported.

Box 13: How to perform an anterior chamber paracentesis
Anesthetize the ocular surface with topical anesthetic
Instil a drop of povidone iodine 5% into the conjunctiva to reduce bacterial contamination
Support the eyelids with an eyelid speculum
Perform the paracentesis at the slit lamp or under an operating microscope
Use a tuberculin syringe with a 30-gauge needle to perform the paracentesis. Remove the plunger
The needle should enter the eye through the temporal limbus, ensuring that the tip remains over the iris at all times
Allow 0.2 ml of aqueous fluid to drain and then withdraw the needle
Instil a drop of antibiotic after the procedure

21.4.8.4 Pharmacological Reduction of Intraocular Pressure

Drugs to lower the intraocular pressure are given to try to augment the effect of massage or anterior chamber paracentesis in the hope of dislodging an embolus and improving the retinal perfusion (Box 14). Care should be taken to ascertain that the patient does not have a contraindication to any drug used.

Box 14: Drugs used to lower intraocular pressure in CRAO
Acetazolamide 500 mg i.v. or orally
Topical β-blocker (e.g., timolol)
Topical apraclonidine
Mannitol i.v.

21.4.8.5 Pharmacological Vasodilatation

Improved retinal blood flow may be achieved by vasodilatation of the vessels [47]. A number of drugs have been tried: pentoxifylline, glyceryl trinitrate and β-blockers. In one small study pentoxifylline was found to improve blood flow but not visual outcome [36].

21.4.8.6 Carbogen

Carbogen is a combination of carbon dioxide (5%) and oxygen (95%). It is used in some ophthalmic centers because it is thought that the carbon dioxide promotes dilation of the arterioles and the high concentration of inspired oxygen improves the oxygenation of ischemic retina. However, this effect is questionable, as vasodilatation was not found in healthy volunteers [26]. A retrospective study of carbogen and paracentesis did not find an improved outcome [5]. It is used for 10 min every 2 h for the first 2 days.

21.4.8.7 Hyperbaric Oxygen

Hyperbaric oxygen therapy produces a marked increase in the arterial oxygen tension. This results in an increase in the diffusion distance of oxygen from the choroid into the retina. This may be enough to sustain the retina until there is a spontaneous recanalization of the retinal artery. The patient is placed in 100% oxygen at 2.8 atmospheres absolute (ATA) for 90 min twice a day for 3 days and then once daily thereafter [8]. The treatment needs to be commenced within 8 h of the onset of the CRAO to be effective. In the largest retrospective study 83% of patients treated in hyperbaric oxygen had an improvement of three or more lines of Snellen acuity compared to 30% in those not treated with hyperbaric oxygen [8].

21.4.8.8 Steroids for Temporal Arteritis

If temporal arteritis is thought to be the cause of the retinal arterial occlusion, high-dose systemic corticosteroid treatment is indicated. Treatment schedules vary and need to be adjusted to the patient. An initial adult dose of 60–80 mg of oral prednisolone is usually used for several days. The dose is gradually reduced in line with improvement in symptoms and inflammatory markers (ESR and CRP). Intravenous methylprednisolone is sometimes used for the initial treatment followed by oral prednisolone. Patients need to be counselled about potential side effects and monitored for these during the course of the treatment. The patient is usually referred to a physician for the management of steroid treatment.

21.4.8.9 Thrombolysis

Thrombolysis has been used to treat both CRAO and BRAO. It has been delivered either directly into the ophthalmic artery via selective catheterization from the femoral artery [3, 60] or peripherally through an intravenous cannula [69]. Currently, the only data available are retrospective case series, some of which have found better outcomes among patients treated with thrombolysis [3, 60]. However, a meta-analysis of published studies found the effect to be only marginal [7]. These procedures can be complicated by stokes or hemorrhages. There are also major challenges in delivering the treatment within a time-frame for it to be effective. There is a need for a prospective randomized controlled trial to evaluate the role of thrombolysis in retinal arterial occlusion.

21 III

21.4.8.10 Secondary Prevention

The clinical assessment and investigations described above may identify underlying medical problems that require ongoing treatment by a physician. Effective treatment of these may reduce the risk of repeat vascular occlusion in the same or in the fellow eye, as well as preventing other co-morbidity such as a stroke. This will involve the control of hypertension, diabetes and hypercholesterolemia if present. The use of aspirin or warfarin may be indicated. Some patients with carotid artery disease may meet the criteria for consideration for carotid endarterectomy surgery.

21.4.9 Outcome and Follow-up

The prognosis for vision following a retinal arterial occlusion depends on the location and duration of the occlusion. Ophthalmic artery obstruction usually results in no perception of light [14]. In cases of CRAO about 10% of cases may show some improvement in vision; however, the vast majority will have a final visual acuity of counting fingers or worse [5, 13]. The prognosis for BRAO is much better with 80% having a final visual acuity of 6/12 or better [79].

Follow-up of patients with retinal vascular occlusion is important. Iris neovascularization (NVI) is common (18%) in CRAO but rare in BRAO. The onset of neovascularization has been documented to occur between 12 days and 15 weeks after the CRAO [30]. Panretinal photocoagulation is reported to be effective promoting regression of the NVI and preventing neovascular glaucoma [29].

References

1. Ahuja RM, Chaturvedi S, Eliott D, Joshi N, Puklin JE, Abrams GW (1999) Mechanisms of retinal arterial occlusive disease in African American and Caucasian patients. Stroke 30:1506–9
2. Appen RE, Wray SH, Cogan DG (1975) Central retinal artery occlusion. Am J Ophthalmol 79:374–81
3. Arnold M, Koerner U, Remonda L, Nedeltchev K, Mattle HP, Schroth G, et al. (2005) Comparison of intra-arterial thrombolysis with conventional treatment in patients with acute central retinal artery occlusion. J Neurol Neurosurg Psychiatry 76:196–9
4. Arruga J, Sanders MD (1982) Ophthalmologic findings in 70 patients with evidence of retinal embolism. Ophthalmology 89:1336–47
5. Atebara NH, Brown GC, Cater J (1995) Efficacy of anterior chamber paracentesis and Carbogen in treating acute nonarteritic central retinal artery occlusion. Ophthalmology 102:2029–34
6. Babikian V, Wijman CA, Koleini B, Malik SN, Goyal N, Matjucha IC (2001) Retinal ischemia and embolism. Etiologies and outcomes based on a prospective study. Cerebrovasc Dis 12:108–13
7. Beatty S, Au Eong KG (2000) Local intra-arterial fibrinolysis for acute occlusion of the central retinal artery: a meta-analysis of the published data. Br J Ophthalmol 84:914–6
8. Beiran I, Goldenberg I, Adir Y, Tamir A, Shupak A, Miller B (2001) Early hyperbaric oxygen therapy for retinal artery occlusion. Eur J Ophthalmol 11:345–50
9. Ben Ami R, Zeltser D, Leibowitz I, Berliner SA (2002) Retinal artery occlusion in a patient with factor V Leiden and prothrombin G20210A mutations. Blood Coagul Fibrinolysis 13:57–9
10. Bertram B, Remky A, Arend O, Wolf S, Reim M (1995) Protein C, protein S, and antithrombin III in acute ocular occlusive diseases. Ger J Ophthalmol 4:332–5
11. Braunstein RA, Gass JD (1980) Branch artery obstruction caused by acute toxoplasmosis. Arch Ophthalmol 98:512–3
12. Brazitikos PD, Pournaras CJ, Othenin-Girard P, Borruat FX (1993) Pathogenetic mechanisms in combined cilioretinal artery and retinal vein occlusion: a reappraisal. Int Ophthalmol 17:235–42
13. Brown GC, Magargal LE (1982) Central retinal artery obstruction and visual acuity. Ophthalmology 89:14–9
14. Brown GC, Magargal LE, Sergott R (1986) Acute obstruction of the retinal and choroidal circulations. Ophthalmology 93:1373–82
15. Brown GC, Magargal LE, Shields JA, Goldberg RE, Walsh PN (1981) Retinal arterial obstruction in children and young adults. Ophthalmology 88:18–25
16. Brown GC, Moffat K, Cruess A, Magargal LE, Goldberg RE (1983) Cilioretinal artery obstruction. Retina 3:182–7
17. Brown GC, Shields JA (1979) Cilioretinal arteries and retinal arterial occlusion. Arch Ophthalmol 97:84–92
18. Bruno A, Jones WL, Austin JK, Carter S, Qualls C (1995) Vascular outcome in men with asymptomatic retinal cholesterol emboli. A cohort study. Ann Intern Med 122:249–53
19. Cahill M, Karabatzaki M, Meleady R, Refsum H, Ueland P, Shields D, et al. (2000) Raised plasma homocysteine as a risk factor for retinal vascular occlusive disease. Br J Ophthalmol 84:154–7
20. Chang M, Herbert WN (1984) Retinal arteriolar occlusions following amniotic fluid embolism. Ophthalmology 91:1634–7
21. Cobo-Soriano R, Sanchez-Ramon S, Aparicio MJ, Teijeiro MA, Vidal P, Suarez-Leoz M, et al. (1999) Antiphospholipid antibodies and retinal thrombosis in patients without risk factors: a prospective case-control study. Am J Ophthalmol 128:725–32
22. Cogan DG, Wray SH (1975) Vascular occlusions in the eye from cardiac myxomas. Am J Ophthalmol 80:396–403
23. Cohen SM, Davis JL, Gass DM (1995) Branch retinal arterial occlusions in multifocal retinitis with optic nerve edema. Arch Ophthalmol 113:1271–6
24. Corrigan MJ, Hill DW (1968) Retinal artery occlusion in loiasis. Br J Ophthalmol 52:477–80
25. David NJ, Norton EW, Gass JD, Beauchamp J (1967) Fluorescein angiography in central retinal artery occlusion. Arch Ophthalmol 77:619–29
26. Deutsch TA, Read JS, Ernest JT, Goldstick TK (1983) Effects of oxygen and carbon dioxide on the retinal vasculature in humans. Arch Ophthalmol 101:1278–80
27. Duker JS, Brown GC (1988) Iris neovascularization associated with obstruction of the central retinal artery. Ophthalmology 95:1244–50
28. Duker JS, Brown GC (1989) Neovascularization of the optic

disc associated with obstruction of the central retinal artery. Ophthalmology 96:87–91

29. Duker JS, Brown GC (1989) The efficacy of panretinal photocoagulation for neovascularization of the iris after central retinal artery obstruction. Ophthalmology 96:92–5

30. Duker JS, Sivalingam A, Brown GC, Reber R (1991) A prospective study of acute central retinal artery obstruction. The incidence of secondary ocular neovascularization. Arch Ophthalmol 109:339–42

31. Golub BM, Sibony PA, Coller BS (1990) Protein S deficiency associated with central retinal artery occlusion. Arch Ophthalmol 108:918

32. Greven CM (1997) Retinal arterial occlusions in the young. Curr Opin Ophthalmol 8:3–7

33. Greven CM, Slusher MM, Weaver RG (1995) Retinal arterial occlusions in young adults. Am J Ophthalmol 120:776–83

34. Hayreh SS, Kolder HE, Weingeist TA (1980) Central retinal artery occlusion and retinal tolerance time. Ophthalmology 87:75–8

35. Hodes BL, Edelman D (1979) Central retinal artery occlusion after facial trauma. Ophthalmic Surg 10:21–3

36. Incandela L, Cesarone MR, Belcaro G, Steigerwalt R, De Sanctis MT, Nicolaides AN, et al. (2002) Treatment of vascular retinal disease with pentoxifylline: a controlled, randomized trial. Angiology 53 Suppl 1:S31–4

37. Justice J, Jr, Lehmann RP (1976) Cilioretinal arteries. A study based on review of stereo fundus photographs and fluorescein angiographic findings. Arch Ophthalmol 94:1355–8

38. Klein ML, Jampol LM, Condon PI, Rice TA, Serjeant GR (1982) Central retinal artery occlusion without retrobulbar hemorrhage after retrobulbar anesthesia. Am J Ophthalmol 93:573–7

39. Klein R, Klein BE, Jensen SC, Moss SE, Meuer SM (1999) Retinal emboli and stroke: the Beaver Dam Eye Study. Arch Ophthalmol 117:1063–8

40. Klein R, Klein BE, Moss SE, Meuer SM (2003) Retinal emboli and cardiovascular disease: the Beaver Dam Eye Study. Arch Ophthalmol 121:1446–51

41. Ko MK, Kim DS (2000) Posterior segment neovascularization associated with acute ophthalmic artery obstruction. Retina 20:384–8

42. Kramer M, Goldenberg-Cohen N, Shapira Y, Axer-Siegel R, Shmuely H, Adler Y, et al. (2001) Role of transesophageal echocardiography in the evaluation of patients with retinal artery occlusion. Ophthalmology 108:1461–4

43. Landa E, Rehany U, Rumelt S (2004) Visual functions following recovery from non-arteritic central retinal artery occlusion. Ophthalmic Surg Lasers Imaging 35:103–8

44. Larsson J (2000) Central retinal artery occlusion in a patient homozygous for factor V Leiden. Am J Ophthalmol 129:816–7

45. Lightman DA, Brod RD (1991) Branch retinal artery occlusion associated with Lyme disease. Arch Ophthalmol 109:1198–9

46. Mitchell P, Wang JJ, Smith W (2000) Risk factors and significance of finding asymptomatic retinal emboli. Clin Exp Ophthalmol 28:13–7

47. Mueller AJ, Neubauer AS, Schaller U, Kampik A (2003) Evaluation of minimally invasive therapies and rationale for a prospective randomized trial to evaluate selective intra-arterial lysis for clinically complete central retinal artery occlusion. Arch Ophthalmol 121:1377–81

48. Nakagawa T, Hirata A, Inoue N, Hashimoto Y, Tanihara H (2004) A case of bilateral central retinal artery obstruction with patent foramen ovale. Acta Ophthalmol Scand 82:111–2

49. Nelson ME, Talbot JF, Preston FE (1989) Recurrent multiple-branch retinal arteriolar occlusions in a patient with protein C deficiency. Graefes Arch Clin Exp Ophthalmol 227:443–7

50. Newman NJ, Lessell S, Brandt EM (1989) Bilateral central retinal artery occlusions, disk drusen, and migraine. Am J Ophthalmol 107:236–40

51. O'Donnell BA, Mitchell P (1992) The clinical features and associations of retinal emboli. Aust N Z J Ophthalmol 20:11–7

52. Peng YJ, Fang PC, Huang WT (2004) Central retinal artery occlusion in Wegener's granulomatosis: a case report and review of the literature. Can J Ophthalmol 39:785–9

53. Pianka P, Almog Y, Man O, Goldstein M, Sela BA, Loewenstein A (2000) Hyperhomocystinemia in patients with nonarteritic anterior ischemic optic neuropathy, central retinal artery occlusion, and central retinal vein occlusion. Ophthalmology 107:1588–92

54. Reimers CD, Williams RJ, Berger M, Wisnicki HJ, Tranbaugh RF (1996) Retinal artery embolization: a rare presentation of calcific aortic stenosis. Clin Cardiol 19:253–4

55. Rumelt S, Brown GC (2003) Update on treatment of retinal arterial occlusions. Curr Opin Ophthalmol 14:139–41

56. Rumelt S, Dorenboim Y, Rehany U (1999) Aggressive systematic treatment for central retinal artery occlusion. Am J Ophthalmol 128:733–8

57. Salomon O, Huna-Baron R, Moisseiev J, Rosenberg N, Rubovitz A, Steinberg DM, et al. (2001) Thrombophilia as a cause for central and branch retinal artery occlusion in patients without an apparent embolic source. Eye 15:511–4

58. Savino PJ, Glaser JS, Cassady J (1977) Retinal stroke. Is the patient at risk? Arch Ophthalmol 95:1185–9

59. Schmidt D (1999) A fluorescein angiographic study of branch retinal artery occlusion (BRAO) – the retrograde filling of occluded vessels. Eur J Med Res 4:491–506

60. Schmidt DP, Schulte-Monting J, Schumacher M (2002) Prognosis of central retinal artery occlusion: local intraarterial fibrinolysis versus conservative treatment. AJNR Am J Neuroradiol 23:1301–7

61. Shah SP, Hadid OH, Graham EM, Stanford MR (2005) Acute retinal necrosis presenting as central retinal artery occlusion with cilioretinal sparing. Eur J Ophthalmol 15:287–8

62. Sharma S (1998) The systemic evaluation of acute retinal artery occlusion. Curr Opin Ophthalmol 9:1–5

63. Sharma S, Brown GC, Cruess AF (1998) Accuracy of visible retinal emboli for the detection of cardioembolic lesions requiring anticoagulation or cardiac surgery. Retinal Emboli of Cardiac Origin Study Group. Br J Ophthalmol 82:655–8

64. Sharma S, Brown GC, Pater JL, Cruess AF (1998) Does a visible retinal embolus increase the likelihood of hemodynamically significant carotid artery stenosis in patients with acute retinal arterial occlusion? Arch Ophthalmol 116:1602–6

65. Sharma S, Naqvi A, Sharma SM, Cruess AF, Brown GC (1996) Transthoracic echocardiographic findings in patients with acute retinal arterial obstruction. A retrospective review. Retinal Emboli of Cardiac Origin Group. Arch Ophthalmol 114:1189–92

66. Sharma S, Sharma SM, Cruess AF, Brown GC (1997) Transthoracic echocardiography in young patients with acute retinal arterial obstruction. RECO Study Group. Retinal

Emboli of Cardiac Origin Group. Can J Ophthalmol 32: 38–41

67. Solley WA, Martin DF, Newman NJ, King R, Callanan DG, Zacchei T, et al. (1999) Cat scratch disease: posterior segment manifestations. Ophthalmology 106:1546–53

68. Tsumura T, Iijima H (1997) [Visual field defect in eyes with branch retinal artery occlusion]. Nippon Ganka Gakkai Zasshi 101:163–6

69. von Mach MA, Guz A, Wiechelt J, Pfeiffer N, Weilemann LS (2005) [Systemic fibrinolytic therapy using urokinase in central retinal artery occlusion. A case study]. Dtsch Med Wochenschr 130:1002–6

70. Wallace RT, Brown GC, Benson W, Sivalingham A (1992) Sudden retinal manifestations of intranasal cocaine and methamphetamine abuse. Am J Ophthalmol 114:158–60

71. Weger M, Renner W, Pinter O, Stanger O, Temmel W, Fellner P et al. (2003) Role of factor V Leiden and prothrombin 20210A in patients with retinal artery occlusion. Eye 17: 731–4

72. Williams DF, Mieler WF, Williams GA (1990) Posterior segment manifestations of ocular trauma. Retina 10 Suppl 1:S35–44

73. Williamson TH, Baxter GM, Dutton GN (1993) Color Doppler velocimetry of the optic nerve head in arterial occlusion. Ophthalmology 100:312–7

74. Williamson TH, Harris A (1996) Color Doppler ultrasound imaging of the eye and orbit. Surv Ophthalmol 40:255–67

75. Wilson LA, Warlow CP, Russell RW (1979) Cardiovascular disease in patients with retinal arterial occlusion. Lancet 1:292–4

76. Wong TY, Klein R (2002) Retinal arteriolar emboli: epidemiology and risk of stroke. Curr Opin Ophthalmol 13: 142–6

77. Yokoi M, Kase M (2004) Retinal vasculitis due to secondary syphilis. Jpn J Ophthalmol 48:65–7

78. Yotsukura J, Adachi-Usami E (1993) Correlation of electroretinographic changes with visual prognosis in central retinal artery occlusion. Ophthalmologica 207:13–8

79. Yuzurihara D, Iijima H (2004) Visual outcome in central retinal and branch retinal artery occlusion. Jpn J Ophthalmol 48:490–2

21.5 The Ocular Ischemic Syndrome

G.C. BROWN, M.M. BROWN

Core Messages

- Secondary to severe carotid artery disease, described as "venous stasis retinopathy," now mostly named as "ocular ischemic syndrome"
- Symptoms: abrupt or gradual loss of vision, dull aching pain, amaurosis fugax (15%)
- Signs: iris neovascularization, narrowed arteries, dilated veins, microaneurysms, neovascularization of the retina
- Therapy: panretinal photocoagulation (iris neovascularization), endarterectomy, aspirin (if carotid stenosis is <70%)

Kearns and Hollenhorst [23] described the ocular symptoms and signs occurring secondary to severe carotid artery obstructive disease in 1963 [23]. They named the entity "venous stasis retinopathy" and noted it was present in approximately 5% of patients with marked carotid artery insufficiency. Confusion has arisen using this term since it has also been used to refer to mild central retinal vein obstruction [17]. Other alternative names have been proposed, including ischemic ocular inflammation [24], ischemic coagulopathy [41], and the ocular ischemic syndrome [6, 7]. Histopathology of eyes with the disease generally does not reveal inflammation [20, 30], and thus the descriptive term we prefer is *ocular ischemic syndrome*.

The pathophysiologic, demographic, and clinical features of the ocular ischemic syndrome will be addressed. Ancillary diagnostic studies will be discussed, as will systemic abnormalities associated with the ocular ischemic syndrome, therapeutic modalities and the differential diagnosis.

21.5.1 Pathophysiology

Typically, a 90% or greater stenosis of the ipsilateral carotid arterial system is present in eyes with the ocular ischemic syndrome [6]. Flow abnormalities within the vessel are seen when the stenosis reaches 70%, and it has been demonstrated that a 90% carotid stenosis reduces the ipsilateral central retinal artery perfusion pressure by about 50% [22, 25]. The obstruction can be present within the common carotid or internal carotid artery (Figs. 21.5.1, 21.5.2). In approximately 50% of cases the affected vessel is 100% obstructed and in 10% of cases there is bilateral 100% carotid artery obstruction [6].

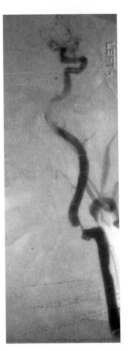 

Fig. 21.5.1. Carotid arteriogram reveals a left 95% internal carotid artery obstruction in a patient with the ocular ischemic syndrome in the left eye

Fig. 21.5.2. Carotid arteriogram in the same patient as shown in Fig. 1 demonstrates a right 100% common carotid artery obstruction

In select cases, obstruction of the ipsilateral ophthalmic artery can also be responsible for the ocular ischemic syndrome [6, 8, 26]. Rarely, a chronic central retinal artery obstruction alone can cause the dilated retinal veins and retinal hemorrhages seen in eyes with the ocular ischemic syndrome [27].

21 **III**

Atherosclerosis within the carotid artery is the cause of the majority of the ocular ischemic syndrome cases [6]. Dissecting aneurysm of the carotid artery has also been reported [13], as has giant cell arteritis [16]. In theory, entities such as Behçet's disease [11], fibromuscular dysplasia [14, 28], trauma [34], and inflammatory diseases that cause carotid artery obstruction could also produce the ocular ischemic syndrome.

21.5.2 Demography

The demographic features associated with the ocular ischemic syndrome are listed below:

Gender – male : female ratio is 2 : 1 [6]
Age range – 50's to the 80's [6]
Mean age – 65 years [6]
Bilaterality – 20 % [6]
Incidence – 7.5 cases/year/million population [38]
No ethnic predilection

21.5.3 Clinical Features

The clinical features and their associated frequencies of occurrence are shown below. In instances in which the incidence is uncertain, none is given.

A. Symptoms

Abrupt loss of vision – 12 % [6]
Often associated with a cherry red spot and iris neovascularization as the intraocular pressure exceeds that within the central retinal artery

Gradual visual loss over days to weeks – 80 % [6]

Ocular or periorbital pain – 40 % [6]
Dull aching pain
Referred to as "ocular angina"
Etiology: ischemia to the globe, increased intraocular pressure and/or ischemia to the ipsilateral meninges

Prolonged visual recovery after exposure to bright light [12, 40]

Amaurosis fugax – 15 % [31]

B. Signs [6]

Vision
Initial vision 20/20 – 20/50: 43 %; 20/800 or worse: 37 %
One year 20/20 – 20/50: 24 %: 20/800 or worse: 58 %

Head
Collateral vessels from the external to internal carotid arterial system (Fig. 21.5.3)

Anterior chamber
Iris neovascularization – 67 %
   Only half of such eyes develop increased intraocular pressure; poor ciliary body perfusion diminishes aqueous production in the others (Figs. 21.5.4 – 21.5.6)

Fluorescein angiography can demonstrate iris neovascularization (Fig. 21.5.5)
Flare – 50 %+ (most cases with iris neovascularization)
Cells – 18 %, no greater than 2+ on a 0 – 4+ classification [35]

Posterior segment (Figs. 21.5.7 – 21.5.18) [6]
Narrowed retinal arteries – 90 % (Figs. 21.5.7, 21.5.9)
Dilated (not tortuous) retinal veins – 90 % (Figs. 21.5.7, 21.5.9)
Retinal hemorrhages – 80 % (Fig. 21.5.8)
Microaneurysms – 80 % (Fig. 21.5.12)
Neovascularization of the optic disk – 35 % (Fig. 21.5.13)
Macular edema [5] – 14 % – often with less thickening clinically than evident with fluorescein angiography due to diminished retinal arterial perfusion pressure (Fig. 21.5.17)
Cherry red spot – 12 % – usually develops when neovascular glaucoma causes the intraocular pressure to exceed that in the central retinal artery
Neovascularization of the retina – 8 % (Fig. 21.5.14)
Cotton-wool spots – 6 %
Vitreous hemorrhage – 4 %
Spontaneous retinal arterial pulsations – 4 %
Retinal emboli (cholesterol) – 2 %
Anterior ischemic optic neuropathy [4] – 2 %
Acquired retinal arteriovenous communications [2]

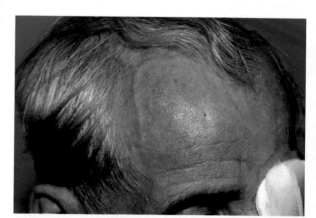

Fig. 21.5.3. Prominent collateral vessel from the external carotid system on the right side in a patient with a left 100 % common carotid artery obstruction

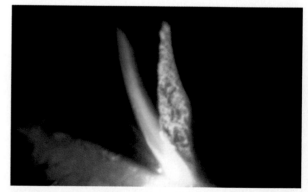

Fig. 21.5.4. Neovascularization on the brown iris of a patient with the ocular ischemic syndrome

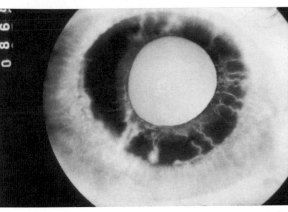

Fig. 21.5.5. Fluorescein angiogram at 86 s after injection discloses hyperfluorescence from iris neovascularization occurring secondary to the ocular ischemic syndrome

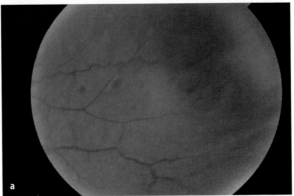

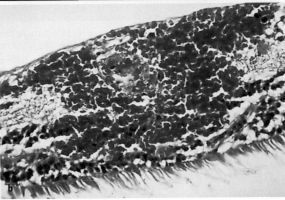

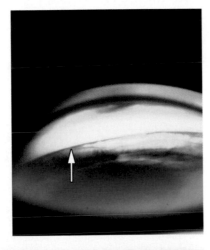

Fig. 21.5.6. Goni-oscopic view of iris neovascularization (*arrow*) closing off the anterior chamber angle in an eye with ocular ischemia

Fig. 21.5.8. a Mid-peripheral dot and blot retinal hemorrhages in the eye of a patient with the ocular ischemic syndrome (courtesy of Neal Atebara, MD). **b.** Histopathology of a retinal hemorrhage in an eye with the ocular ischemic syndrome. Blood is present throughout the entire thickness of the retina. (Courtesy of W. Richard Green, MD). H&E, ×40

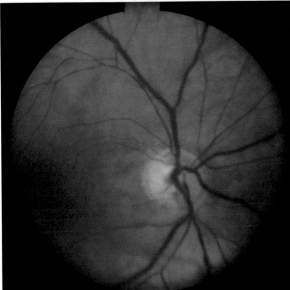

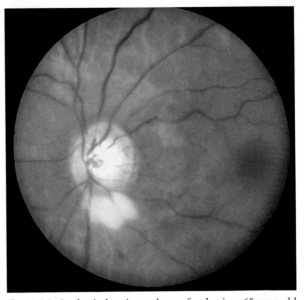

Fig. 21.5.7. Ocular ischemic syndrome in an eye with a 100% ipsilateral carotid artery obstruction demonstrates dilated, beaded (but not tortuous) retinal veins, while the retinal arteries are narrowed

Fig. 21.5.9. Ocular ischemic syndrome fundus in a 65-year-old man with a 100% left internal carotid obstruction. The retinal arteries are very narrowed and the veins dilated. The myelinated nerve fibers at the inferior border of the optic disk are unrelated to the ocular ischemia

21

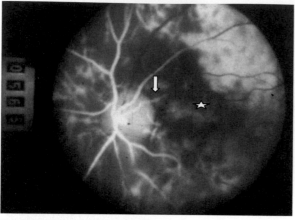

Fig. 21.5.10. Fluorescein angiogram at 56 s after injection in the ocular ischemic syndrome eye shown in Fig. 21.5.9. There is filling of the choroidal vasculature superotemporally, but a marked delay in choroidal filling elsewhere (*star*). Leading edges of fluorescein dye, distinctly abnormal phenomenona, can be seen within the retinal arteries (*arrow*)

Fig. 21.5.11. Histopathology of the poster segment in an eye with the ocular ischemic syndrome. Both the inner and outer retina are attenuated due to retinal vascular and choroidal ischemia, respectively. Note that the retinal pigment epithelium (*arrow*) appears to be intact. Retinal pigment epithelial changes are not a prominent feature of the ocular ischemic syndrome. (Courtesy of W. Richard Green, MD). H&E, ×20

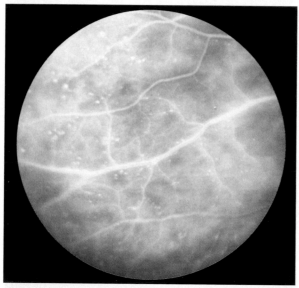

Fig. 21.5.12. Fluorescein angiography at 90 s after injection in an ocular ischemic syndrome eye reveals numerous, small punctate areas of hyperfluorescence which correspond to retinal microaneurysms. The retinal vessels are beginning to leak as well

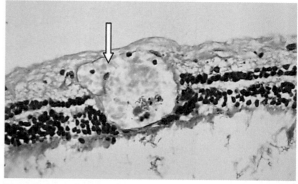

Fig. 21.5.13. Histopathologic specimen of a retinal microaneurysm in an eye with the ocular ischemic syndrome. Rupture (*arrow*) of these anomalies and damaged small retinal vessels result in the retinal hemorrhages seen in ocular ischemic syndrome eyes (Fig. 8)

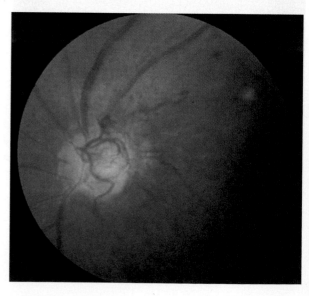

▷
Fig. 21.5.14. Neovascularization of the optic disk in a non-diabetic 80-year-old man with a 95% left internal carotid artery obstruction

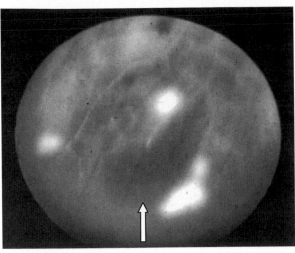

Fig. 21.5.15. Hyperfluorescent foci due to neovascularization of the retina occurring secondary to ocular ischemia in the eye of a non-diabetic man with severe ipsilateral carotid artery obstruction. Retinal capillary non-perfusion (*arrow*) can be seen immediately to the left of the largest area of retinal neovascularization

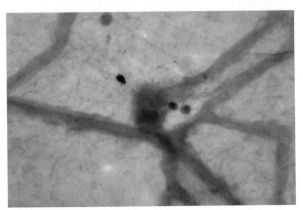

Fig. 21.5.16. Trypsin digest of a region of the retina with capillary non-perfusion in an ocular ischemic syndrome eye. The vessels are acellular tubules, thus explaining why reperfusion does not occur in non-perfused areas. (Courtesy of W. Richard Green, MD). ×160

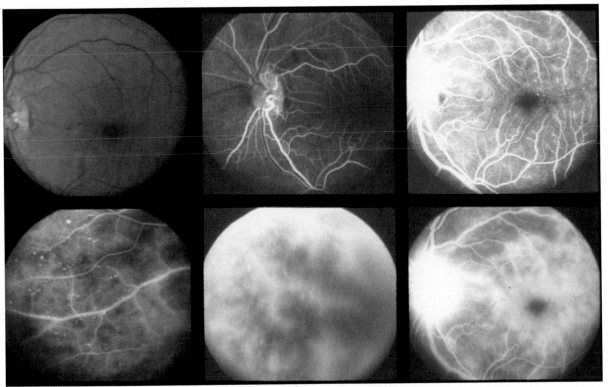

Fig. 21.5.17. Composite photographs of the left eye of a 60-year-old woman with a 100% left internal artery obstruction and a visual acuity of 20/40. *Top row, left:* The color photograph shows small retinal hemorrhages and venous dilation. *Top row, middle:* Fluorescein angiogram at 30 s after injection. Early retinal vascular filling is asymmetric. *Top row, right:* Fluorescein angiogram at 1 min after injection reveals hyperfluorescent foci in the macular retina corresponding to microaneurysms. *Bottom row, left:* Fluorescein angiogram of the midperipheral fundus at 90 s after injection retina discloses retinal microaneurysms, retinal capillary non-perfusion and early staining of the larger retinal vessels. *Bottom row, center:* At 5 min after injection, the mid-peripheral fundus demonstrates marked leakage of dye from the retinal vessels. *Bottom row, right:* At 6 min after injection, the optic disk is hyperfluorescent and intraretinal leakage of dye is present in a pattern consistent with macular edema. When macular edema is present, the optic disk is typically hyperfluorescent, despite the fact that it most often appears normal ophthalmoscopically. At follow-up 15 years after this series of pictures, the visual acuity was 20/30 with no treatment of the macular edema or carotid artery obstruction

21 **III**

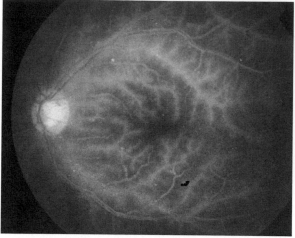

Fig. 21.5.18. Fluorescein angiogram at over 10 min after injection reveals staining of the larger retinal vessels, especially the retinal arteries. (Courtesy of Dr. Neal Atebara)

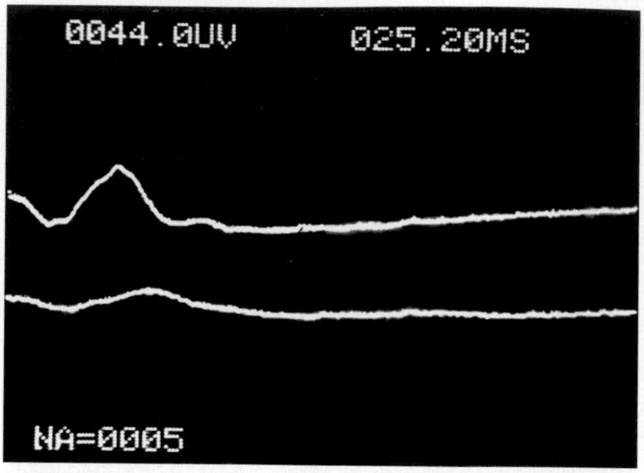

Fig. 21.5.19. Normal electroretinographic pattern (*upper tracing* in the normal right eye) and diminution of the amplitude of the a- and b-waves in the left eye (*lower tracing*) with the ocular ischemic syndrome. The a-wave is diminished due to choroidal vascular insufficiency and outer retinal ischemia, and the b-wave is diminished due to hypoperfusion within the retinal vessels supplying the inner retina

21.5.4 Ancillary Diagnostic Studies

A. Intravenous fluorescein angiography [6] – listed in order of specificity
Delayed choroidal filling – 60% (Fig. 21.5.10)
   Most specific
   > Five seconds from the first appearance of dye until complete choroidal filling
Late arterial staining – 85% (Fig. 21.5.18)
Delayed arteriovenous transit time – 95%
   Least specific
   > Eleven seconds from the first appearance of retinal arterial dye until complete retinal venous filling
Macular edema – 17% (Fig. 21.5.17)
Additional signs
   Retinal capillary non-perfusion (Fig. 21.5.15)
   Microaneurysmal hyperfluorescence (Fig. 21.5.17)
   Optic nerve head hyperfluorescence (Fig. 21.5.17)

B. Electroretinography [6, 7] (Fig. 21.5.19)
Diminished a-wave amplitude due to outer retinal ischemia
Diminished b-wave amplitude due to inner retinal ischemia

C. Ophthalmodynamometry
Positive in unilateral cases
Light digital pressure is a good substitute. Retinal arterial pulsations, if not already present, can be induced with light digital pressure on the lid

D. Color Doppler ultrasonography [18]
Diminished choroidal flow
Reversal of flow in the ophthalmic artery

E. Carotid non-invasive studies
Duplex ultrasonography and oculoplethysmography have an 88–95% chance of detecting a carotid stenosis of >75% [9, 18, 32]
MRA angiography is similar to Doppler ultrasonography in diagnosing carotid stenosis [10]

21.5.5 Systemic Associations [37]

A. Diseases
Systemic arterial hypertension – 73%
Diabetes mellitus – 56%
Atherosclerotic cardiac disease – 50%
Previous stroke – 25%
Peripheral arterial disease requiring bypass surgery – 20%
Stroke rate – 4%/year

B. 5-year mortality (Fig. 21.5.20)
40%
Cardiovascular disease is the leading cause of death
Referral to primary care physician or cardiologist is important

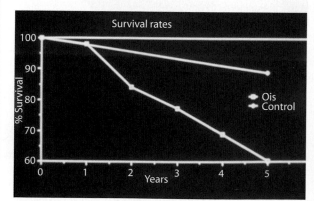

Fig. 21.5.20. Survival in patients with the ocular ischemic syndrome. The 5-year mortality is 40%, primarily due to cardiovascular death (*OIS* ocular ischemic syndrome)

21.5.6 Therapeutic Modalities

A. Ocular [36]
Panretinal photocoagulation if there is iris neovascularization and the anterior chamber angle is open or a glaucoma filtering procedure is under consideration (Fig. 21.5.21) 36 % success in eradicating iris neovascularization

B. Systemic
Endarterectomy [19, 21, 36] (Figs. 21.5.22, 21.5.23)
Improves or maintains vision in 1/3 of cases without a 100 % ipsilateral obstruction
100 % carotid obstruction: endarterectomy ineffective since clot propagation distally [39]

C. Carotid endarterectomy indications in general
a) Beneficial if ≥ 70 % stenosis and symptomatic (mild stroke, TIA and/or amaurosis fugax) [33]
  i) Endarterectomy 2-year stroke rate: 9 %
  ii) Aspirin only 2-year stroke rate: 26 %
b) Aspirin is the preferred treatment over surgery if the carotid stenosis is < 70 % [15, 1]
c) Angioplasty and stenting will have an increasing role [29]

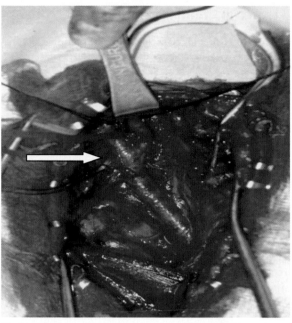

Fig. 21.5.22. Carotid endarterectomy surgery. The bifurcation of the common carotid artery into the internal carotid and external carotid arteries is shown by the *arrow*

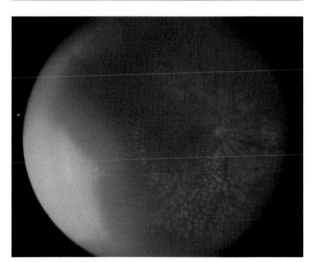

Fig. 21.5.21. Argon laser panretinal photocoagulation in an eye with iris neovascularization due to the ocular ischemic syndrome

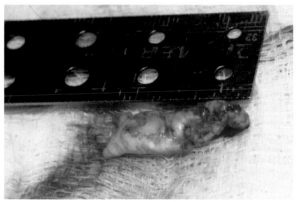

Fig. 21.5.23. Atherosclerotic plaque removed from a severely obstructed carotid artery. Fibrous, calcific and lipid components are present

21.5.7 Differential Diagnosis

Features differentiating the ocular ischemic syndrome (OIS), central retinal vein obstruction (CRVO) and diabetic retinopathy

	OIS	CRVO	Diabetic retinopathy
Laterality	80% unilateral	Usually unilateral	Bilateral
Age	50's to 80's	50's to 80's	Variable
Fundus signs			
Venous status	Dilated (not tortuous), beaded	Dilated, tortuous	Dilated, beaded
Hemorrhages	Peripheral, dot and blot	Nerve fiber layer, dot and blot	Posterior pole, dot and blot
Microaneurysms	Posterior pole and mid-periphery	Variable	Present in posterior pole
Exudate	Absent unless diabetic retinopathy present as well	Uncommon	Common
Optic disk	Normal	Swollen	Affected in papillopathy
Retinal arterial perfusion pressure	Decreased	Normal	Normal
Fluorescein angiography			
Choroidal filling	Delayed, patchy in 60%	Normal	Normal
Arteriovenous transit time	Prolonged – 95%	Prolonged – 95%	May be prolonged if proliferative
Retinal vessel staining	Arterial > venous	Venous	Usually absent

References

1. Barnett HJ, Taylor DW, Eliasziw M, Fox AJ, Ferguson GG, Haynes RB, Rankin RN, Clagett GP, Hachinski VC, Sackett DL, Thorpe KE, Meldrum HE, Spence JD (1998) Benefit of carotid endarterectomy in patients with symptomatic moderate or severe stenosis. North American Symptomatic Carotid Endarterectomy Trial Collaborators. N Engl J Med 339:1415–25

2. Bolling JP, Buettner H (1990) Acquired retinal arteriovenous communications in occlusive disease of the carotid artery. Ophthalmology 97:1148–1152

3. Bosley TM (1986) The role of carotid noninvasive tests in stroke prevention. Semin Neurol 6:194–203

4. Brown GC (1986) Anterior ischemic optic neuropathy occurring in association with carotid artery obstruction. J Clin Neuro-ophthalmol 6:39–42

5. Brown GC (1986) Macular edema in association with severe carotid artery obstruction. Am J Ophthalmol 102:442–448

6. Brown GC, Magargal LE (1988) The ocular ischemic syndrome. Clinical, fluorescein angiographic and carotid angiographic features. Int Ophthalmol 11:239–251

7. Brown GC, Magargal LE, Simeone FA, Goldberg RE, Federman JL, Benson WE (1982) Arterial obstruction and ocular neovascularization. Ophthalmology 89:139–146

8. Bullock J, Falter RT, Downing JE, Snyder H (1972) Ischemic ophthalmia secondary to an ophthalmic artery occlusion. Am J Ophthalmol 74:486–493

9. Castaldo JE, Nicholas GG, Gee W, Reed JF (1989) Duplex ultrasound and ocular pneumoplethysmography concordance in detecting severe carotid stenosis. Arch Neurol 46:518–522

10. D'Onofrio M, Mansueto G, Faccioli N, Guarise A, Tamellini P, Bogina G, Pozzi Mucelli R (2006) Doppler ultrasound and contrast-enhanced magnetic resonance angiography in assessing carotid artery stenosis. Radiol Med (Torino) 111:93–103

11. Dhobb M, Ammar F, Bensaid Y, Benjelloun A, Benabderra-zik T, Benyahia B (1986) Arterial manifestations in Behçet's disease: four new cases. Ann Vasc Surg 1:249–252

12. Donnan GA, Sharbrough FW (1982) Carotid occlusive disease. Effect of bright light on visual evoked response. Arch Neurol 39:687–689

13. Duker JS, Belmont JB (1988) Ocular ischemic syndrome secondary to carotid artery dissection. Am J Ophthalmol 106:750–752

14. Effeney DJ, Krupski WC, Stoney RJ, Ehrenfeld WK (1983) Fibromuscular dysplasia of the carotid artery. Austral N Z J Surg 53:527–531

15. European Carotid Surgery Trialists' Collaborative Group. MRC European Carotid Surgery Trial: interim results for symptomatic patients with severe (70–99%) or with mild (0–29%) carotid stenosis. European Carotid Surgery Trialists' Collaborative Group. Lancet 337:1235–43

16. Hamed LM, Guy JR, Moster ML, Bosley T (1992) Giant cell arteritis in the ocular ischemic syndrome. Am J Ophthalmol 113:702–705

17. Hayreh SS (1976) So-called – "central retinal vein occlusion." Venous-stasis retinopathy. Ophthalmologica 172:14–37

18. Ho AC, Lieb WE, Flaharty PM, Sergott RC, Brown GC, Bosley TM, Savino PJ (1992) Color Doppler imaging of the ocular ischemic syndrome. Ophthalmology 99:1453–62

19. Ishikawa K, Kimura I, Shinoda K, Eshita T, Kitamura S, Inoue M, Mashima Y (2002) In situ confirmation of retinal blood flow improvement after carotid endarterectomy in a patient with ocular ischemic syndrome. Am J Ophthalmol 134:295–7

20. Kahn M, Green WR, Knox DL, Miller NR (1986) Ocular features of carotid occlusive disease. Retina 6:239–252

21. Kawaguchi S, Okuno S, Sakaki T, Nishikawa N (2001) Effect of carotid endarterectomy on chronic ocular ischemic syndrome due to internal carotid artery stenosis. Neurosurgery 48:328–32

22. Kearns TP (1979) Ophthalmology and the carotid artery. Am J Ophthalmol 88:714–722

23. Kearns TP, Hollenhorst RW (1963) Venous stasis retinopa-

thy of occlusive disease of the carotid artery. Proc Mayo Clin 38:304–312

24. Knox DL (1965) Ischemic ocular inflammation. Am J Ophthalmol 60:995–1002

25. Kobayashi S, Hollenhorst RW, Sundt TM Jr (1971) Retinal arterial pressure before and after surgery for carotid artery stenosis. Stroke 2:569–575

26. Madsen PH (1965) Venous-stasis insufficiency of the ophthalmic artery. Acta Ophthalmol 40:940–947

27. Magargal LE, Sanborn GE, Zimmerman A (1982) Venous stasis retinopathy associated with embolic obstruction of the central retinal artery. J Clin Neuro-ophthalmol 2:113–118

28. Matonti F, Prost Magnin O, Galland F, Hoffart L, Coulibaly F, Conrath J, Ridings B (2006) Internal carotid artery dissection on arterial fibromuscular dysplasia causing a central retinal artery occlusion: a case report. J Fr Ophtalmol 29(7):e15

29. Mazighi M, Tanasescu R, Ducrocq X, Vicaut E, Bracard S, Houdart E, Woimant F (2006) Prospective study of symptomatic atherothrombotic intracranial stenoses: the GESICA study. Neurology 66:1187–91

30. Michelson PE, Knox DL, Green WR (1971) Ischemic ocular inflammation. A clinicopathologic case report. Arch Ophthalmol 86:274–280

31. Mizener JB, Podhajsky P, Hayreh SS (1997) Ocular ischemic syndrome. Ophthalmology 104(5):859–64

32. Neale ML, Chambers JL, Kelly AT, Connard S, et al. (1994) Reappraisal of duplex criteria to assess significant carotid artery stenosis with special reference to reports of the North American Symptomatic Carotid Endarterectomy Trial and the European Carotid Surgery Trial. J Vasc Surg 20:642–9

33. North American Symptomatic Carotid Endarterectomy Trial Collaborators (1991) Beneficial effect of carotid endarterectomy in symptomatic patients with high-grade carotid stenosis. N Engl J Med 325:445–453

34. Sadun AA, Sebag J, Bienfang DC (1983) Complete bilateral internal carotid artery occlusion in a young man. J Clin Neuro-ophthalmol 3:63–66

35. Schlaegel T (1983) Symptoms and signs of uveitis. In: Duane TD (ed) Clinical ophthalmology, vol 4. Harper & Row, Hagerstown, pp 1–7

36. Sivalingam A, Brown GC, Magargal LE (1991) The ocular ischemic syndrome. III. Visual prognosis and the effect of treatment. Int Ophthalmol 15:15–20

37. Sivalingham A, Brown GC, Magargal LE, Menduke H (1989) The ocular ischemic syndrome II. Mortality and systemic morbidity. Int Ophthalmol 13:187–191

38. Sturrock GD, Mueller HR (1984) Chronic ocular ischaemia. Br J Ophthalmol 68:716–7123

39. The EC/IC Bypass Study Group (1985) Failure of extracranial-intracranial arterial bypass to reduce the risk of ischemic stroke. Results of an international randomized trial. N Engl J Med 313:1191–1200

40. Wiebers DO, Swanson JW, Cascino TL, Whisnant JP (1989) Bilateral loss of vision in bright light. Stroke 20:554–558

41. Young LHY, Appen RE (1981) Ischemic oculopathy, a manifestation of carotid artery disease. Arch Neurol 38:358–361

22   Vascular Abnormalities
22.1 Idiopathic Juxtafoveolar Retinal Telangiectasis

D. Pauleikhoff, B. Padge

Core Messages

- Retinal telangiectasis is a group of rare retinal vascular anomalies affecting the retinal capillaries
- Characteristically irregular dilation, leakage and edema occur in the macula, rarely combined with vascular changes in the retinal periphery
- Telangiectasis may develop unilaterally or bilaterally and is commonly characterized by a slow decrease in visual acuity in adulthood
- The long-term prognosis for reading vision is usually good
- The pathogenesis of these changes is unknown

22.1.1 History

Idiopathic juxtafoveolar retinal telangiectases (IJRT) were classified by Gass and Oyakawa in 1982 using biomicroscopic and fluorescein angiographic data [6]. This classification was updated by Gass and Blodi in 1993 [5], who subdivided IJRT into three groups (Table 22.1.1). Group 1 IJRT are unilateral in most cases and characterized by dilated retinal capillaries and abnormal leakage leading to an easily visible exudation. Group 2 IJRT are mostly bilateral and characterized by late staining on fluorescein angiography with minimal exudation. Later in the disease process retinal pigment epithelial proliferation or secondary subretinal neovascularization may develop. In Group 3 changes are based on bilateral capillary occlusion. These changes lead to easily visible telangiectasis, parafoveolar capillary occlusion and minimal exudation.

22.1.2 Clinical Course of the Disease

Essentials

- Group 1: Unilateral telangiectasis with abnormal leakage of capillaries
  - Group 1A: Visible and exudative idiopathic juxtafoveolar retinal telangiectasis with lipid exudates, size > 1 disk diameter
  - Group 1B: Visible and exudative focal idiopathic juxtafoveolar retinal telangiectasis, size < 1 disk diameter
- Group 2: Bilateral idiopathic juxtafoveolar retinal telangiectasis
  - Group 2A: Nonexudative bilateral idiopathic juxtafoveolar retinal telangiectasis
  - Group 2B: Juvenile occult familial idiopathic juxtafoveolar retinal telangiectasis
- Group 3: Occlusive idiopathic juxtafoveolar retinal telangiectasis
  - Group 3A: Occlusive idiopathic juxtafoveolar retinal telangiectasis without central nervous system vasculopathy
  - Group 3B: Occlusive idiopathic juxtafoveolar retinal telangiectasis with central nervous system vasculopathy

Table 22.1.1. Summary of findings in idiopathic juxtafoveolar retinal telangiectasis [5]

Group	No.	Localization	Visual acuity	Mean age (years)	Gender	Characteristic
1A	31	Unilateral (97%)	20/40	37	Male	Leakage
1B	8	Unilateral (88%)	20/20	42	Male	Leakage
2A	92	Bilateral (98%)	20/40	55	No specificity	Diffusion
2B	2	Bilateral (100%)	20/70	11		Diffusion
3A	3	Bilateral (100%)	20/25	53		Occlusion
3B	4	Bilateral (100%)	20/50	42		Occlusion

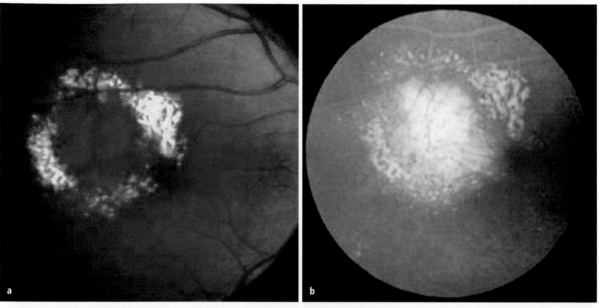

Fig. 22.1.1. Group 1A IJRT – biomicroscopic (**a**) and fluorescence angiographic changes (**b**): increased permeability of vessels with serous exudation (**b**) and a surrounding intraretinal lipid exudation is visible (**a**)

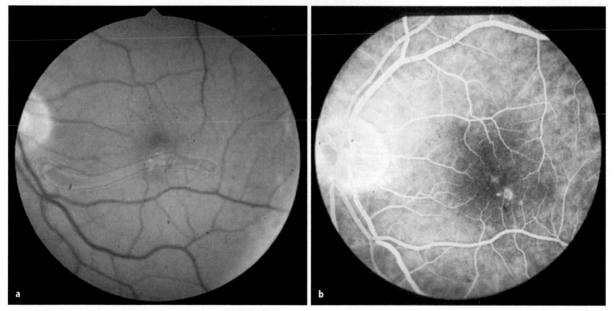

Fig. 22.1.2. Group 1B IJRT – biomicroscopic (**a**) and fluorescence angiographic changes (**b**): The Group 1B IJRT differs primarily in the size of lesion from Group 1A

In **Group 1A** primarily male patients are affected. Commonly this group is characterized by a unilateral development of retinal telangiectasis and in most cases the vascular changes are easily visible. Patients with unilateral IJRT may be asymptomatic or may experience a mild reduction of visual acuity. Biomicroscopic findings are prominent telangiectatic retinal capillaries in the temporal half of the fovea with a 2 DD involvement of the macula, mostly associated with surrounding intraretinal lipid exudates (Fig. 22.1.1a).

Group 1B telangiectasis differs primarily in the size of lesion from Group 1A. The extension of telangiectatic vessels is only 1 disk diameter temporal of the fovea and the changes are mostly not associated with lipid exudates (Fig. 22.1.2). Male predilection, unilaterality and biomicroscopic and fluorescein angiographic features are nearly the same. It may be a mild form of Group 1A telangiectasis [5].

Patients in **Group 2A** IJRT have the most common form of idiopathic juxtafoveolar retinal telangiecta-

22 III

sis, but the cause is still unknown. Unlike Group 1 there is no sex predilection found in Group 2A IJRT. Also the telangiectases are usually diagnosed in the late 5th life decade, nearly 20 years later than those in Group 1 (Table 22.1.1). Typically the changes demonstrate bilateral symmetry with telangiectatic vessels in the temporal half of the fovea, which are difficult to detect on biomicroscopy (Fig. 22.1.3a). Therefore the main diagnostic tool is fluorescein angiography (Fig. 22.1.3a, b). The biomicroscopic changes are microaneurysms, minimal exudation, increased reti-

nal thickness, small yellow crystalline exudate (Fig. 22.1.3a) and right-angled venules. In the course of the disease proliferation of retinal pigment epithelium and development of subretinal neovascularization may develop [5, 10].

The pattern on fluorescein angiography is characterized by symmetric bilateral rapidly fluorescein stained capillary walls (Fig. 22.1.3b). This staining is followed by a late diffuse staining of the middle and outer retina surrounding part but sparing the foveola itself (Fig. 22.1.3c). This disorder primarily affects

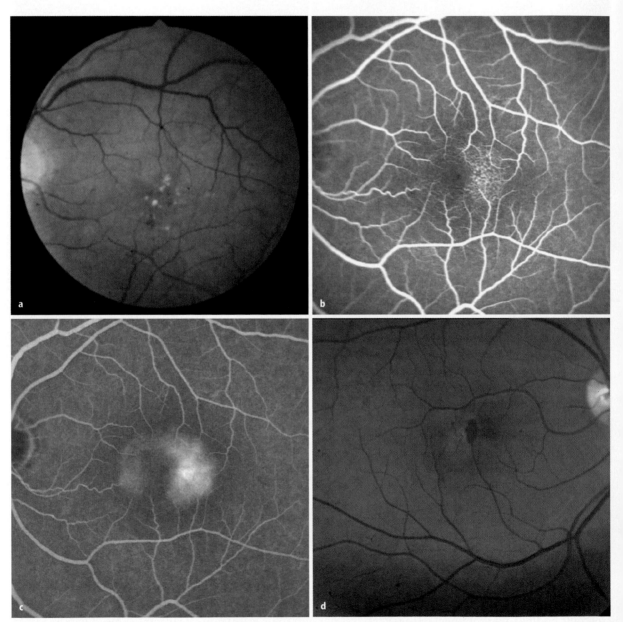

Fig. 22.1.3. Group 2A IJRT – biomicroscopic and fluorescence angiographic changes: In Stage 1 – 3 clinically only yellow crystalline deposits (**a**) and microaneurysms (**a**) could be found; on fluorescence angiography symmetric telangiectatic vessels are located temporal to the fovea (**b**), which develop a late diffuse hyperfluorescence (**c**). In Stage 4 a plaquoid pigment epithelial proliferation is visible (**d**).

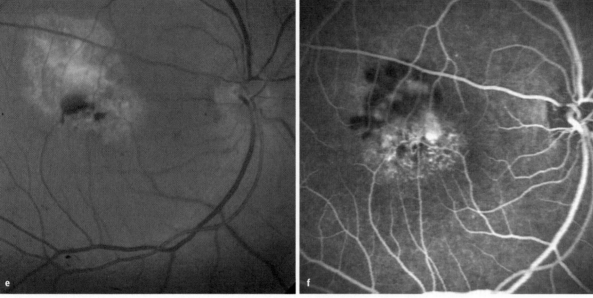

Fig. 22.1.3. In Stage 5 a subretinal neovascularization beside the juxtafoveolar telangiectass has developed (**e, f**)

the deep or outer juxtafoveolar capillary network in a zone within 1 DD of the center of the fovea.

Group 2A is further subdivided into five stages of development [5].

Stages 1–3 are characterized by intraretinal changes (Fig. 22.1.3). These stages include increased visibility of telangiectasis primarily affecting the outer capillary network temporal to the fovea. In Stage 3 there is evidence of one or several slightly dilated and blunted retinal venules that extend at right angles into the depth of the parafoveolar retina. Biomicroscopically there are often crystalline deposits visible in the temporal foveolar area (Fig. 22.1.3a) [10].

In **Stage 4** is seen a reactive pigment epithelial proliferation caused by the loss of outer receptor cells and consecutive retinal pigment epithelial proliferation (Fig. 22.1.3d). It leads to one or several foci of black plaques of retinal pigment epithelium within the retina, often beneath the blunted tips of the right angled venules. Sometimes these venules are enveloped by pigment epithelial cells.

Stage 5 is characterized by development of secondary subretinal neovascularization (Fig. 22.1.3e, f). This neovascularization usually develops in the temporal half of the fovea and in the vicinity of intraretinal pigment epithelial migration. The absence of pigment epithelial detachment and a limited size of the neovascular complex suggest the retinal rather than the choroidal vasculature is the primary source of the new vessels [5].

Group 2B IJRT are characterized by an early development of telangiectasis and subfoveal neovas-

cularization. It was described in two brothers, 9 and 12 years old, by Gass et al. [5]. There have been no further cases reported in the literature.

Group 3 IJRT are characterized by easily visible telangiectasis associated with parafoveolar capillary occlusion and minimal exudation. This group is further subdivided into two subgroups: Group 3A and Group 3B (Table 22.1.1).

The main characteristic in Group 3A IJRT is an extensive occlusion of the juxtafoveolar capillary network. This leads to a minimal exudation and to a remarkably good visual acuity despite marked enlargement of the capillary free zone. This zone can be greater than 1 DD in size [5]. The visual loss is primarily caused by capillary obstruction and occlusion but not by exudation. The usual range of visual acu-

Table 22.1.2. Group 2A IJRT – clinical findings

Stage	Characteristics
1	Occult, normal fundus, fluorescein staining, no exudation
2	Loss of retinal transparency, minimal telangiectasis, no exudation
3	Mild telangiectasis, blunted dilated right angle venules, capillary remodeling, and capillaries in the outer retina, no exudation
4	Intraretinal migration of retinal pigment epithelium (RPE), superficial stellate RPE plaques, less loss of retinal transparency, no exudation
5	Subretinal new vessels, exudation and hemorrhage

22 III

ity varies from 20/25 to 20/40. It cannot be determined whether the telangiectasis precedes, accompanies or follows the occlusive phenomenon.

In addition to telangiectasis of Group 3A, in Group 3B IJRT an association with central nervous system disease has been evidenced (Table 22.1.2).

22.1.3 Electron Microscopic and Light Microscopic Changes

In microscopic studies of Group 1A IJRT, changes in the superficial and deep juxtafoveolar retinal capillary plexus were observed. Deformed capillaries with endothelial decompensation lead to a serous exudation and retinal swelling primarily of the outer plexiform layer (Fig. 22.1.1B). The biomicroscopic picture of cystoid macula edema is caused by an extension of exudation into the foveal area. The often seen yellow exudates are accumulated large lipoproteins escaped from the insufficient vessels.

In electron microscopic and light microscopic studies of Group 2A IJRT [8], some narrowing of the capillary lumina has been observed. This is associated with thickening of the capillary wall by multilaminated basement membrane, focal endothelial defects and perithelial degeneration. A slight thickening of the retina by intra- and extracellular edema confined to the inner half of the retina has been noted. This suggests that the juxtafoveolar retinal capillaries are the primary tissue involved, and the late fluorescein staining is occurring within the sensory retina. Early fluorescein staining of the thickened capillary walls is responsible for the early angiographic appearance of telangiectatic vessels. The altered structure of the capillary wall is associated with decreased endothelial permeability, inciting chronic nutritional damages of retinal cells, particularly those at the level of the inner nuclear layer. The late staining in fluorescein angiography is probably caused by staining of extracellular matrix and intracellular diffusion into these damaged cells. Further changes and the nutritional deprivation of retinal cells in the median retinal layers lead to degeneration and atrophy of these cells and the connecting photoreceptor cells. This is responsible for the loss of visual acuity and a picture that may simulate a lamellar macular hole.

The yellow deposits are found in the inner surface of the retina, often anterior to the retinal blood vessels. The cause is unknown. It is suggested that they are composed of lipid. Their location in the region of the internal limiting membrane suggests that they might be a product of degenerating Müller cells whose nuclei are located in the inner nuclear layer at the site of the altered deep retinal capillary plexus.

Other groups of IJRT have not been investigated ultrastructurally in the literature.

22.1.4 Natural Course of IJRT

Essentials
- The natural course of most groups of IJRT demonstrates a very slow progression

The natural course of Group 2A IJRT is unknown, but slow progression from one stage to the next is suggested. Gass observed a progression from Stage 4 to Stage 5 in 24% of patients after a 6-year-follow up. The risk of disease progression in the early stages and future visual loss is unknown. Only a very slow rate of progression and the hope of possible stabilization at a specific stage of the disease can be postulated at the present time.

In order to obtain more insight into the pathogenesis, clinical presentations, genetic background and prognosis of Group 2A IJRT, an international prospective multicenter study (the Mac-Tel Study) has been initiated.

22.1.5 Association with Systemic Diseases and Differential Diagnosis

Idiopathic juxtafoveolar retinal telangiectasis must be differentiated from other systemic and ocular diseases associated with telangiectasis like diabetic retinopathy, retinal vascular occlusions, Eales disease, retinopathy of prematurity or sickle cell retinopathy.

Group 1A telangiectasis may be assigned to the great spectrum of congenital telangiectasis [3] and may be grouped with Leber's miliary aneurysms or central Coats' syndrome [12].

An association with systemic diseases seems to be typically in Group 3 IJRT. In his study of IJRT, Gass assigned only seven patients to Group 3A IJRT. Gass described in all seven patients other medical complications that may have contributed to the retinal changes: polycythemia, gouty arthritis, hypoglycemia and cardiovascular disease [5].

The evidence of atrophy of the juxtafoveolar retina and minimal exudation associated with capillary occlusion are similar to that found in patients with sickle cell retinopathy [14].

Group 3B IJRT are suspected to be an autosomally dominantly inherited cerebroretinal vasculopathy [5, 7]. This familial disorder is associated with central nervous system pseudotumor characterized by an unusual vasculopathy with fibrinoid necrosis and necrosis of white matter.

22.1.6 Therapy

Essentials

- In Group 1A a focal photocoagulation of the telangiectatic vessels is recommended, if lipid exudates progress toward the fovea
- In Group 1B a focal photocoagulation of the telangiectatic vessels is only very rarely recommended in case of progressive visual loss
- In Group 2A a focal photocoagulation can be suggested in Stage 1–3 in case of progressive visual loss. In case of subretinal neovascularization a photodynamic therapy or photocoagulation may be therapeutic options
- In Group 2B, 3A and 3B no therapy is recommend in the literature

Due to the different pathogenesis the therapy of IJRT is group and stage dependent.

In Group 1A IJRT photocoagulation of the telangiectatic vessels is effective by reducing the foveolar exudation and improves or preserves the visual function. Early treatment is recommended when visual loss combined with progression of lipid exudates toward the fovea is detected [1, 3, 4, 5, 6].

Focal photocoagulation of the telangiectatic capillaries is recommended only in a minority of Group 1B patients, because often no progression and visual loss will develop [5].

Different studies have shown that there are no proven therapeutic options available to treat Group 2A IJRT [5]. Many studies investigated photocoagulation as a possible treatment in Stage 1–3, but there were no significant differences between treated and untreated eyes with telangiectasis in Stage 1–3 [11] (Table 22.1.3). Photocoagulation leads to a stable visual acuity, but there has been no evidence that untreated eyes develop a decrease in visual acuity more rapidly. Probably intravitreal triamcinolone acetonide may reduce the intraretinal edema for a decent period [9]. Photodynamic therapy in bilateral parafoveal telangiectasis without subretinal neovascularization is not beneficial because there is no improvement of either the visual acuity or the macula edema [2]. In Stage 4 no therapy is recommended in the literature. In Stage 5 photocoagulation, photodynamic therapy or anecortave acetate could be therapeutic options, but only case reports or small series of treated patients with photocoagulation and consecutive stability of visual acuity have been reported (Table 22.1.3). Photodynamic therapy in combination with intravitreal triamcinolone acetonide has also demonstrated a regression of subfoveal neovascular membrane and improvement in visual acuity [13].

In the literature no therapy is recommended for treatment of the occlusive telangiectasis of Group 3A and 3B IJRT.

References

1. Chopdar A (1978) Retinal telangiectasis in adults: fluorescein angiographic findings and treatment by argon laser. Br J Ophthalmol 62:243–250
2. De Lahitte GD, Cohen SY, Gaudric A (2004) Lack of apparent short-term benefit of photodynamic therapy in bilateral, acquired, parafoveal telangiectasis without subretinal neovascularization. Am J Ophthalmol 138:892–894
3. Gass JD (1968) A fluorescein angiographic study of macular dysfunction secondary to retinal vascular disease. V. Retinal telangiectasis. Arch Ophthalmol 80:592–605
4. Gass JD (1987) Stereoscopic atlas of macular diseases: diagnosis and treatment, 3rd edn. Mosby, St. Louis, pp 390–396
5. Gass JD, Blodi BA (1993) Idiopathic juxtafoveolar retinal telangiectasis. Update of classification and follow-up study. Ophthalmology 100:1536–1546
6. Gass JD, Oyakawa RT (1982) Idiopathic juxtafoveolar retinal telangiectasis. Arch Ophthalmol 100:769–780
7. Grand MG, Kaine J, Fulling K, Atkinson J, Dowton SB, Farber M, Craver J, Rice K (1988) Cerebroretinal vasculopathy. A new hereditary syndrome. Ophthalmology 95:649–659
8. Green WR, Quigley HA, de la CZ, Cohen B (1980) Parafoveal retinal telangiectasis. Light and electron microscopy studies. Trans Ophthalmol Soc U K 100:162–170
9. Martinez JA (2003) Intravitreal triamcinolone acetonide for bilateral acquired parafoveal telangiectasis. Arch Ophthalmol 121:1658–1659
10. Moisseiev J, Lewis H, Bartov E, Fine SL, Murphy RP (1990) Superficial retinal refractile deposits in juxtafoveal telangiectasis. Am J Ophthalmol 109:604–605
11. Park DW, Schatz H, McDonald HR, Johnson RN (1997) Grid laser photocoagulation for macular edema in bilateral juxtafoveal telangiectasis. Ophthalmology 104:1838–1846

Table 22.1.3. Results of photocoagulation of Group 2A IJRT

| Visual loss | Stages 1–3 Follow-up 23.1 months | | Stage 4 Follow-up 25.1 months | | Stage 5 Follow-up 22.8 months | |
	Ø Treat. *n*=29	Treat. *n*=12	Ø Treat. *n*=12	Treat. *n*=0	Ø Treat. *n*=3	Treat. *n*=2
≤ 2 lines	29	10	11	0	1	1
≥ 3 lines	0	2	1	0	2	1

12. Pauleikhoff D, Wessing A (1989) Long-term results of the treatment of Coats' disease. Fortschr Ophthalmol 86:451–455

13. Smithen LM, Spaide RF (2004) Photodynamic therapy and intravitreal triamcinolone for a subretinal neovascularization in bilateral idiopathic juxtafoveal telangiectasis. Am J Ophthalmol 138:884–885

14. Stevens TS, Busse B, Lee CB, Woolf MB, Galinos SO, Goldberg MF (1974) Sickling hemoglobinopathies; macular and perimacular vascular abnormalities. Arch Ophthalmol 92: 455–463

22.2 Congenital Arteriovenous Communications and Wyburn-Mason Syndrome

A. WESSING

Core Messages
- Wyburn-Mason syndrome is a rare congenital oculocerebral syndrome consisting of retinal arteriovenous anastomoses and ipsilateral cerebral arteriovenous vascular abnormalities. It may, but does not always, also involve nevi and vascular malformations in the skin and mucosa in the area of the trigeminal nerve. Wyburn-Mason syndrome is one of the phacomatoses
- There is a wide range of arteriovenous anastomoses in the retina, from small uncomplicated short-circuits to hugely dilated ones. The arteries and veins fuse with each other directly, without a capillary plexus
- Small anastomoses are usually monosymptomatic; large ones are associated with cerebral anomalies in up to 90% of cases. Loss of vision is often the initial symptom, usually caused by central nervous damage
- Treatment for retinal arteriovenous anastomoses is not required, or not possible. Spontaneous remissions occur
- An interdisciplinary examination including neurological and neurosurgical consultation, with computed tomography (CT) or magnetic resonance imaging (MRI), is absolutely imperative, at least for patients with larger anastomoses

22.2.1 History

The first description of arteriovenous anastomoses dates back to the end of the 19th century [26, 33, 39, 52, 54]. In 1915, in Graefe-Saemisch's *Handbuch der gesamten Augenkrankheiten*, Leber presented the first comprehensive presentation of the condition [35]. The article includes highly instructive sketches of the cases reported by Schleich and Seydel. Further review articles were published by Weve in 1923 [67], Rentz in 1925 [48], and Junius in 1933 [29].

A 1903 report by Kreutz [33] is of particular interest, as it describes for the first time the combination of retinal arteriovenous anastomoses with arteriovenous vascular anomalies in the orbit. One of the earliest descriptions of cerebral arteriovenous vascular anomalies (cirsoid aneurysm) was provided by Heitmüller in 1904 [27]. From the mid-1920s onward, there were increasing numbers of reports on the simultaneous occurrence of retinal and cerebral arteriovenous anastomoses [10, 18, 34, 42, 43]. The first comprehensive presentation of the syndrome was published by P. Bonnet, J. Dechaume, and E. Blanc in 1937 with the title "*L'anévrysme cirsoide de la rétine (anévrysme racémeux), ses rélations avec lánévrysme cirsoide de la face et avec lánévrysme cirsoide du cerveau*" [8]. In 1943 followed an article by

R. Wyburn-Mason: "Arteriovenous aneurysm of mid-brain and retina, facial naevi and mental changes" [69]. This finally defined the disease as a clinical entity consisting of retinal and intracerebral vascular abnormalities. The complex of symptoms was later assigned to the phacomatoses [20, 21, 36].

It is perhaps of general biogenetic interest that arteriovenous anastomoses also occur in other primates. In rhesus monkeys, the typical vascular anomalies were observed ophthalmoscopically, imaged angiographically, and analyzed histologically by Bellhorn et al. in 1972 [5] and Horiuchi et al. in 1976 [28].

The nomenclature has changed over the course of time. Early authors speak of "*aneurysma arteriovenosum*" [39, 52] or "*varix aneurysmaticus* [54]. On the basis of Virchow's nomenclature, Leber in 1915 [35] recommended the term "*aneurysma racemosum*" or "*aneurysma racemosum arteriovenosum*" (Latin: *racemus* = bunch or cluster of grapes). Bonnet et al. [8] chose the term "*aneurysma cirsoides*" (Greek: *kirsoeides* = varicose). From the 1930s onwards, as familiarity with cerebral pathology increased, the condition was more and more frequently classified among hemangiomas and thus tumors [47]. Archer et al. use the term "arteriovenous communications" [1]. The terms "arteriove-

22 III

nous anastomoses" and "arteriovenous shunts" are used synonymously.

„*Neuroretino-angiomatosis Syndrom*," "*angiome encéphalo-rétino-faciale*," "*syndrome anévrismatique rétino-optico-mesencéphalic*," "arteriovenous cerebroretinal aneurysm," "retino-cephalic vascular malformation" and other localizing and descriptive terms represent the pattern of symptoms of retinal and cerebral vascular changes. However, it is more frequent for the condition to be named after the authors of the first comprehensive descriptions of it, as "*Syndrome de Bonnet, Dechaume et Blanc*" or "Wyburn-Mason syndrome" – the term most frequently used in the international literature.

22.2.2 Classification

Arteriovenous anastomoses are congenital vascular malformations (Fig. 22.2.1a). They arise due to disturbances in the maturation process of the retinal vascular system. Individual or multiple short-circuits of widely varying caliber develop between the arteries and veins in the retina. The surface extension of the malformation is also extremely variable, ranging from the involvement of small sectors to involvement of the entire retina [16, 19, 34, 24, 56, 57, 66].

Archer, Deutman, Ernest and Krill [1] distinguished two forms of arteriovenous anastomoses: firstly, arteriovenous anastomoses in which normal capillaries, or at least one more or less normal capillary plexus, are present between the arteries and veins; and, secondly, arteriovenous anastomoses with a direct transition from arteries to veins, without intermediate capillary or arteriolar links.

The arteriovenous anastomoses that occur in Wyburn-Mason syndrome, either monosymptomatic or as a retinocerebral symptom complex, belong exclusively to the second category. The artery and vein communicate directly, without any intervening capillary elements. Archer et al. [1] distinguish two more subgroups:

- Small and circumscribed arteriovenous anastomoses, completely without retinal complications, or with comparatively few. Vision is usually good. It is exceptional to find this combined with cerebral vascular anomalies (group 2 in Archer et al. [1]).
- The anastomoses are extremely extensive, immensely complex, and produce severe retinal complications. Cerebral involvement is usually found (group 3 in Archer et al. [1]).

The classification is based on the extreme variability of arteriovenous anastomoses and attempts to take into account the wide differences in the morphological appearance, the occurrence of retinal complications, and the association with cerebral, orbital, and facial vascular malformations.

Other classifications, such as that of Mansour et al. [40], are more detailed. But in view of the small numbers of cases they can only be used with difficulty.

22.2.3 Clinical Features

Arteriovenous anastomoses occur with equal frequency in both sexes. They are usually identified in adolescence, but can already be present in neonates as well. They occur in a strictly unilateral pattern. The few exceptions reported in the literature prove the rule [12, 40]. Smaller and medium-sized anastomoses (group 2 in the Archer classification) may be single or multiple, consisting of solitary channels or with a branching pattern (Fig. 22.2.2a). The location of preference is the central or temporal retina, and they are limited to individual sectors and quadrants.

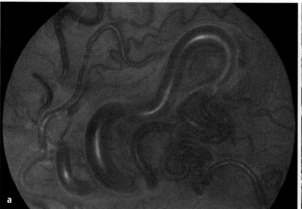 

Fig. 22.2.1. a Large congenital retinal arteriovenous communication originating from the optic disk (Archer classification group 3). **b** Same eye, fluorescein angiography

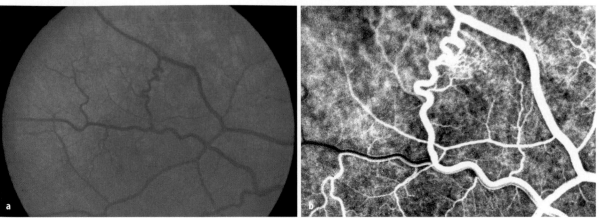

Fig. 22.2.2. a Isolated arteriovenous anastomosis in the mid-periphery of the retina (Archer classification group 2). **b** Same eye, fluorescein angiography

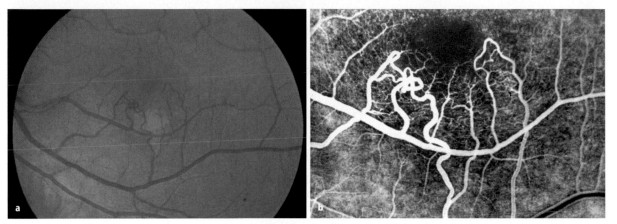

Fig. 22.2.3. a Parafoveal arteriovenous anastomoses (Archer classification group 2). **b** Same eye, fluorescein angiography

Individual vascular loops can reach as far as the perifoveal arcades, and sometimes even into the fovea itself (Fig. 22.2.3a). A cilioretinal artery is sometimes incorporated into the arteriovenous vascular loop. The anastomotic vessels can be 60 – 150 μm in diameter.

In severe cases (group 3 in the Archer classification), the anastomoses can develop huge, contorted shapes. The vessels are intertwined and convoluted. They can reach a diameter up to 10 – 12 times that of normal retinal vessels. They usually spread over the entire fundus. The dilated anastomotic vessels may hide the optic disk completely. Occasionally the anastomosis is limited to the disk area [64]. Even extremely dilated vessels do not show any pulsations.

Large vascular diameter creates a high flow rate, so that the venous side of the anastomosis also carries oxygenated blood. The arterial and venous vascular limbs are often only distinguishable angiographically. The intravascular pressure is increased [41, 51], and damage to the vascular walls consequently develops over time. The anastomoses devel-

op white-yellowish sheathings and are accompanied by serous exudates, lipid deposits, and reactive pigmentary hyperplasia (Fig. 22.2.4a, b). Bleeding can also occur [22, 41]. Large anastomoses cause extensive alterations in the surrounding capillary bed. The hyperoxia in the anastomoses is probably the reason why practically no reactive vascular proliferations develop or secondary glaucoma arises. There have only been two or three cases reported in the literature in which the patients with extremely massive retinal damage developed rubeosis iridis with glaucoma [7, 17]. Some of the complications appear to be due to central or branch vein occlusion or spontaneous thrombosis of the anastomosis (see p. 539). The observation reported by Tilanus et al. [60] is exceptional. They noted several bleeding macroaneurysms developing in the course of an arteriovenous vascular loop.

Loss of vision is rare in cases of simple anastomoses, and only arises when the macula or optic nerve is involved. In the severe forms, by contrast, visual field defects and loss of vision of varying degree, ranging

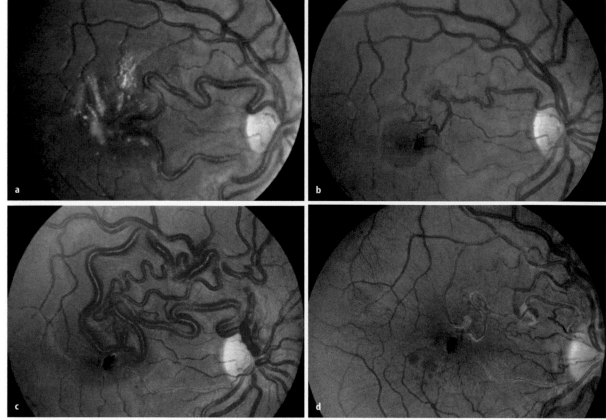

Fig. 22.2.4. a Arteriovenous anastomosis with follow-up of 17 years: first examination. **b** Three years later: partial regression of the vascular loop. **c** Nine years later: development of new anastomoses from primarily uninvolved retinal vessels. **d** Seventeen years later: complete regression of the arteriovenous communications in the central retina

up to complete blindness, usually develop. The functional deficits are usually due to the cerebral or orbital vascular changes. Overall, two-thirds of patients with anastomoses have functional defects (Meyer, cited in [57]).

22.2.4 Fluorescein Angiography

Fluorescein imaging (Figs. 22.2.1b, 22.2.2b, 22.2.3b) reveals that the anastomoses have a substantial influence on neighboring vascular areas and in some cases can also affect the entire vascular system of the retina. A particularly impressive finding on the fluorescein angiogram, however, is the extremely high blood flow velocity in the arteriovenous anastomoses. The larger the vascular diameter, the higher the flow rate.

The increase in blood flow is evidently the cause of severe damage to the surrounding capillary bed. Initially, broad avascular zones are observed along the anastomoses. Later, the capillaries are occluded even in distant areas of the retina. Large avascular zones can arise. The capillaries are obliterated either due to the hyperoxia or as a result of a steal effect. In con-

trast to other retinal vascular diseases, the avascular areas do not stimulate neovascularization. If damage to the vascular walls develops over time, the angiogram shows dye leakage from the dilated vessels, which can even extend to complete collapse of the blood-retina barrier.

22.2.5 Differential Diagnosis

Arteriovenous anastomoses do not give rise to any substantial differential-diagnostic difficulties and are relatively easy to distinguish from other vascular diseases in the retina.

- The anastomoses most resemble the dilated nutritive vessels of retinal capillary hemangiomas in von Hippel-Lindau's disease. In the latter conditions, however, arterial and venous elements are separated by the capillaries of the hemangioma. Fluorescein angiography is helpful to detect small or hidden angiomas.
- Congenital retinal macrovessels are obviously dilated vessels that lie between the optic disk and

the central retina, with end branches that extend beyond the horizontal raphe. The arteriovenous transition takes place via the capillary plexus.

- Primary congenital tortuosity of the retinal vessels is usually bilateral. Here again, the arteriovenous connection is via the capillary plexus.
- Acquired anastomoses, as in Coats' disease. Eales disease, after venous occlusion, in Takayasu's syndrome [59] and other forms of retinal angiopathy, are not difficult to identify on the basis of the underlying disease.

22.2.6 Natural Course

Arteriovenous anastomoses are regarded as being static and unchanging [9, 48, 49]. However, spontaneous remissions have been noted during long-term observations [6, 8, 9, 15, 25, 41, 45, 51, 66, 68]. The anastomotic vessels are transformed into ischemic, whitish bands or regress to such an extent that they are no longer distinguishable from their surroundings. Arteriovenous anastomoses represent a high-pressure system, with the corresponding hemodynamic stress on the vascular structures [51]. Sclerosis and thickening of the vascular walls develop. Increasing turbulence in the blood flow arises. The hemodynamic disturbances are intensified by mechanical compression of the vascular lumen within the optic nerve, at the optic disk, or at vascular crossings. Ultimately, the process leads to thrombosis and vascular obliteration.

There is evidence that the anastomosis formation itself is a dynamic process. Effron et al. [17] reported increasing size and increasing dilatation. In other cases, it has been observed that new arteriovenous connections develop after spontaneous occlusion of an anastomosis [45, 65, 66] (Fig. 22.2.4a–d). These develop from branches of the primary anastomosis, or at other points in the retina, including even previously uninvolved vessels. Augsburger et al. [2] reported obliteration of an anastomosis after ligation of the internal carotid artery and observed a new anastomosis arising subsequently.

22.2.7 Histopathology

The histological appearance of arteriovenous anastomoses is that of hypertrophic, normally matured blood vessels [70]. The vessels extend into all layers of the retina. Larger anastomoses bulge into the vitreous cavity and may also come into contact externally with Bruch's membrane. Segments of the vascular walls may be either thinned or thickened. Gigantically dilated arteries and veins are histopathologically barely distinguishable from one another. The vascular walls are fibrotic and interspersed with hyaline and lipid infiltrations. The surrounding retina shows cystoid changes and loss of ganglia cells and axons [14, 22, 34].

22.2.8 Systemic Involvement

Retinal arteriovenous anastomoses may be associated with analogous vascular malformations in the orbit and central nervous system, and more rarely with mucous and cutaneous vascular changes in the face [8, 69]. Cerebral malformations and retinal anastomoses are strictly ipsilateral in location and mainly appear to involve the complete optic tract as far as the optic cortex [16]. They only occur bilaterally very rarely [44]. In the cerebral region as well, the spectrum of arteriovenous aneurysms and angiomas ranges from small and medium-sized to extensive and voluminous malformations. Neurological symptoms due to compression and bleeding include corticospinal tract signs, cranial nerve pareses, epileptic fits, strabismus, visual field defects and visual loss ranging up to complete blindness ([8, 10, 13, 32, 38, 50, 53, 63, 69], etc.). Some 30% of Wyburn-Mason patients have homonymous hemianopia [19]. The more severe the changes in the retina are, the more likely it is that there will be cerebral involvement. Central aneurysms and angiomas are found in 90% of patients with huge, contorted anastomoses (group 3 in the Archer classification). According to Bech and Jensen [4], 17% of patients with retinal arteriovenous anastomoses have cerebral changes; this figure, however, dates from the period before computed tomography (CT) and magnetic resonance imaging (MRI). Wyburn-Mason originally reported an incidence of 81%. Conversely, he found retinal changes in 70% of patients with mesencephalic aneurysms.

Aneurysmal and angiomatous vascular malformations in the orbit can occur either in isolation or in various combinations with ocular and cerebral aneurysms and angiomas [30, 31, 33, 46, 50, 63, 69]. These are usually also located unilaterally, ipsilateral to the changes in the eye or brain. Bilateral findings are rare here as well [44]. Vascular anomalies in the orbit may lead to exophthalmos, with or without pulsation, vascular murmur, papilledema, optic atrophy, and loss of vision.

There are vascular malformations in the area of the maxilla, mandible, and pterygoid fossa, with epistaxis and hemorrhage [11, 32]. In the area of the face, telangiectasias and soft subcutaneous vascular tumors or nevi are found [8, 10]. These changes appear to be limited to the innervation area of the first trigeminal branch. In 1990, Patel and Gupta [44] reported a neonate with cerebral arteriovenous malformations combined with vascular malformations in both orbits and bilateral extensive cutaneous nevi in the innervation area of the trigeminal nerve.

22.2.9 Genetics

Congenital arteriovenous anastomoses are vascular malformations that are due to developmental disturbances in the early gestational period [46, 61, 62]. The embryological causes of the disturbances are unknown. A familial incidence has occasionally been discussed. However, MacDonald et al. clearly demonstrated in 1997 that Wyburn-Mason syndrome does not have genetic causes [37].

22.2.10 Therapy

Arteriovenous anastomoses do not initially have any therapeutic implications for the ophthalmologist. Basically, laser coagulation of smaller arteriovenous anastomoses is unnecessary, while coagulation of larger vessels is unpromising and hazardous. A few reports describing coagulation attempts do not alter this statement. To reduce the blood flow, Stucci and Höpping [58] attempted to narrow the lumen of dilated anastomoses by xenon arc photocoagulation burns. Baurmann et al. [3] attempted to arrest bleeding from damaged vessels, and Tilanus et al. [60] were able to stop hemorrhages from secondary aneurysms by laser photocoagulation. Shah et al. [55] recommended regular check-up examinations for eyes with secondary venous occlusion to allow timely coagulation in case of reactive neovascularizations. In general, however, laser coagulation is limited to individual selected cases with a specific pattern of symptoms.

It should be noted once again that the majority of patients with Wyburn-Mason syndrome are initially seen by ophthalmologists. One of the ophthalmologist's fundamental duties is therefore to refer the patient for neurological and neurosurgical work-up. This includes CT and MRI, as well as cerebral angiography when appropriate.

- Patients with smaller anastomoses (group 2 in the Archer classification) and with no serious suspicion of cerebral involvement need not necessarily undergo invasive diagnostic measures [19].
- In patients with extensive arteriovenous anastomoses (group 3 in the Archer classification), which are practically always associated with cerebral processes, comprehensive neurological and radiological diagnosis is absolutely imperative.

References

1. Archer DB, Deutman A, Ernest JT, Krill AE (1973) Arteriovenous communications of the retina. Am J Ophthalmol 75:224–241
2. Augsburger JJ, Goldberg RE, Shields JA, Mulberger RD, Magargal LE (1980) Changing appearance of retinal arteriovenous malformation. Graefes Arch Klin Ophthalmol 215:65–70
3. Baurmann H, Meyer F, Oberhoff P (1968) Komplikationen bei der arteriovenösen Anastomose der Netzhaut. Klin Monatsbl Augenheilkd 153:562–571
4. Bech K, Jensen OA (1961) On the frequency of co-existing racemose haemangiomata of the retina and brain. Acta Psychiat Neurol Scand 36:47–56
5. Bellhorn RW, Friedmann AH, Henkind P (1972) Racemose (cirsoid) hemangioma in rhesus monkey retina. Am J Ophthalmol 74:517–522
6. Bernth-Petersen P (1979) Racemose haemangioma of the retina; report of three cases with long term follow-up. Acta Ophthalmol 57:669–678
7. Bloom PA, Laidlaw A, Easty DL (1993) Spontaneous development of retinal ischaemia and rubeosis in eyes with retinal racemose angioma. Br J Ophthalmol 77:124–125
8. Bonnet P, Dechaume J, Blanc E (1973) L'anévrysma cirsoide de la rétine (anévrysme racémeux): ses relations avec l'anévrysma cirsoide de la face et avec l'anévrysma cirsoide du cerveau. J Méd Lyon 18:165–178
9. Brihaye M, Tassignon MJ, van Langen-Hove L, Demol S (1987) Anastomose artérioveineuse, isolée et unilatérale de la rétine avec follow-up de 25 ans. Bull Soc Belge Ophtalmol 225:71–78
10. Brock S, Dyke CG (1932) Venous and arteriovenous angioma of the brain. Bull Neurol Inst NY 2:247–293
11. Brower LE, Ditkowsky SP, Klien BA, Bronstein IP (1942) Arteriovenous angioma of mandible and retina with pronounced hematemesis and epistaxis. Am J Dis Child 64:1023–1029
12. Cagianut B (1962) Das arterio-venöse Aneurysma der Netzhaut. Klin Monatsbl Augenheilkd 140:180–191
13. Cameron ME (1958) Congenital arterio-venous aneurysm of the retina. Br J Ophthalmol 42:655–666
14. Cameron ME, Greer CH (1968) Congenital arteriovenous aneurysm of the retina: a post mortem report. Br J Ophthalmol 52:768–772
15. Dekking HM (1955) Arteriovenous aneurysm of the retina with spontaneous regression. Ophthalmologica 130:113–115
16. DeLaey JJ, Hanssens M (1990) Congenital arteriovenous communications in the retina. Bull Soc Belge Ophtalmol 225:85–99
17. Effron L, Zakov ZN, Tomsak RL (1985) Neovascular glaucoma as a complication of the Wyburn-Mason syndrome. J Clin Neuroophthalmol 5:95–98
18. Ehlers H (1924) Aneurysma racemosum arteriovenosum retinae. Acta Ophthalmol 2:374–387
19. Ferry AP (2001) Wyburn-Mason syndrome. In: Ryan JS (ed) Retina, vol 1, 3rd edn. Mosby, St Louis, pp 600–602
20. Font RL, Ferry AP (1972) The phacomatoses. Wyburn-Mason syndrome. Int Ophthalmol Clin 12:44–46
21. François J (1972) Ocular aspects of phacomatoses. In: Vinken PJ, Bruyn GW (eds) The phacomatoses. Handbook of clinical neurology, vol 14. North Holland Publishing, Amsterdam

22. François J, Rabaey M (1951) Hémosidérose prérétinienne et anévrysme artério-veineux. Ophthalmologica 122:348–356

23. Gass JDM (1978) Juxtapapillary hamartomas and choristomas. In: Jakobiec FA, Reese AB (eds) Ocular and adnexal tumors. Aesculapius Publishing, Birmingham Al, pp 208–210

24. Gass JDM (1997) Stereoscopic atlas of macular diseases. 4th edn. Mosby, St Louis, pp 440–443

25. Gregersen E (1961) Arteriovenous aneurysm of the retina; a case of spontaneous thrombosis and "healing". Acta Ophthalmol 39: 937–939

26. Gunn RM (1884) Direct arteriovenous communication on the retina. Trans Ophthalmol Soc UK 4:156–157

27. Heitmüller GH (1904) Cirsoid aneurysm of the branches of the internal carotid and basilar arteries. JAMA 42:648–649

28. Horiuchi T, Gass JDM, David NJ (1976) Arteriovenous malformation in the retina of a monkey. Am J Ophthalmol 82:896–904

29. Junius P (1933) Venöse und arterio-venöse Angiome im Bereich des Gehirns; ihre Beziehungen zum Sehorgan. Zentralbl Ges Ophthalmol 29:673–684

30. Kim J, Kim OH, Suh JH, Lew HM (1998) Wyburn-Mason syndrome: an unusual presentation of bilateral orbital and unilateral brain arteriovenous malformations. Pediatr Radiol 28:161

31. Kottow MH (1978) Congenital malformation of the retinal vessels with primary optic nerve involvement. Ophthalmologica 176:86–90

32. Krayenbühl H, Yasargil MG (1958) Das Hirnaneurysma. Doc Geigy Ser Chir 4

33. Kreutz (1903) Über einen Fall von Rankenaneurysma der A. ophth. dextra. Wien Med Wochenschr 53:1725

34. Krug EF, Samuels B (1932) Venous angioma of the retina, optic nerve, chiasm, and brain. A case report with postmortem observations. Arch Ophthalmol 8:871–879

35. Leber T (1915) Das Aneurysma racemosum der Netzhautgefäße. Graefe-Saemisch Handbuch der gesamten Augenkrankheiten, vol VII/A 1. Engelmann, Leipzig, pp 37–42

36. Lecuire J, Dechaume JP, Bret P (1972) Bonnet-Dechaume-Blanc syndrome. In: Vinken PJ, Bruyn GW (eds) The phakomatoses. Handbook of clinical neurology, vol 14. North Holland Publishing, Amsterdam

37. Mac Donald IM, Bech-Hansen NT, Britton WA et al (1997) The phacomatoses: recent advances in genetics. Can J Ophthalmol 32:4–11

38. Maeda H, Fujieda M, Morita H, Kurashige T (1992) Wyburn-Mason syndrome – a case report. No To Hattatsu 24:65–69

39. Magnus H (1874) Aneurysma arterio-venosum retinale. Virchows Arch Pathol Anat 60:38–45

40. Mansour AM, Walsh JB, Henkind P (1987) Arteriovenous anastomoses of the retina. Ophthalmology 94:35–40

41. Mansour AM, Wells CG, Jampol LM, Kalina RE (1989) Ocular complications of arteriovenous communications of the retina. Arch Ophthalmol 107:232–236

42. Mozetti M (1939) Sulle anomalie dei vasi retinici ed in particolar modo sull'aneurysma cirsoide della retina. Bull Oculist 18:455–468

43. Nicolato A (1933) Angioma retinico ed angioma cerebrale (maliorazione dei vasi cerebrali di Cushing e Bailey). Ann Ottal Clin Ocul 61:736–751

44. Patel U, Gupta SC (1990) Wyburn-Mason syndrome. A case report and review of the literature. Neuroradiology 31: 544–546

45. Pauleikhoff D, Wessing A (1991) Arteriovenous communications of the retina during a 17-year follow-up. Retina 11:433–436

46. Ponce FA, Han PP, Spetzler RF, Canady A, Feiz-Erfan I (2001) Associated arteriovenous malformation of the orbit and brain: a case of Wyburn-Mason syndrome without retinal involvement. J Neurosurg 95:346–349

47. Reese AB (1966) Racemose hemangiomas (arteriovenous aneurysms). Tumors of the eye, 3rd edn. Harper and Row, New York, pp 385–389

48. Rentz (1925) Aneurysma racemosum retinale. Arch Augenheilkd 95:84–91

49. Riffenburgh RS (1954) Arteriovenous aneurysm of the retina. Am J Ophthalmol 37:908–910

50. Rundless WZ Jr, Falls HF (1951) Congenital arteriovenous (racemose) aneurysm of the retina. Report of three cases. Arch Ophthalmol 46:408–418

51. Schatz H, Chang LF, Ober RR, McDonald HR, Johnson RN (1993) Central retinal vein occlusion associated with retinal arteriovenous malformation. Ophthalmology 100:24–30

52. Schleich (1884) Aneurysma arteriovenosum, Aneurysma circumscriptum et Varix (aneurysmat.) retinae. Mitt Ophthalmol Klin Tübingen 2:202

53. Schlieter F, Szepan B, Polenz B (1976) Über das Wyburn-Mason-Syndrom. Klin Monatsbl Augenheilkd 168:788–793

54. Seydel (1899) Ein Aneurysma arterio-venosum (Varix aneurysmaticus) der Netzhaut. Arch Augenheilkd 38:157–163

55. Shah GK, Shields JA, Lanning RC (1998) Branch retinal vein obstruction secondary to retinal arteriovenous communication. Am J Ophthalmol 126:446–448

56. Shields JA (1983) Diagnosis and management of intraocular tumors. Racemose hemangioma. Mosby, St Louis, pp 562–568

57. Speiser P (1978) Das arteriovenöse Hämangiom der Netzhaut. Adv Ophthalmol 36:90–101

58. Stucci CA, Höpping W (1966) Varix aneurysmaticus vicariens retinae et manifestations cérébrales. Traitment. Bull Mém Soc Fr Ophtalmol 79:90–101

59. Tanaka T, Shimizu K (1987) Retinal arteriovenous shunts in Takajasu disease. Ophthalmology 94:1380–1388

60. Tilanus MD, Hoyng C, Deutman AF, Cruysberg JRM, Aandekerk AL (1991) Congenital arteriovenous communications and the development of two types of leaking retinal macroaneurysms. Am J Ophthalmol 112:31–33

61. Tost M (1978) Kongenitale Anomalien an den Gefäßen des uvealen, retinalen und faszikulären Kreislaufs. In: Francois J (ed) Blutzirkilation in der Uvea, in der Netzhaut und im Sehnerven. Enke, Stuttgart

62. Tost F, Weidlich R, Tost M (1996) Zum Wyburn-Mason-Syndrom. Klin Monatsbl Augenheilkd 208:117–119

63. Unger HH, Umbach W (1966) Kongenitales okulocerebrales Rankenangiom. Klin Monatsbl Augenheilkd 148:672–682

64. Werneke T (1940) Angioma papillae nervi optici (caput medusae). Klin Monatsbl Augenheilkd 104:434–436

65. Wessing A (1979) Retinal vascular malformations: a follow-up study. In: Shimizu K, Oosterhuis JA (eds) International congress series 450. Excerpta Medica, Amsterdam, pp 798–801

66. Wessing A (1999) Racemöses Hämangiom der Netzhaut (arterio-venöse Anastomosen). In: Lommatzsch PK (ed) Ophthalmologische Onkologie. Enke, Stuttgart, pp 319–321

67. Weve H (1923) Varix aneurysmaticus vicariens retinae (Pseudoaneurysma arteriovenosum racemosum retinae) Arch Augenheilkd 93:1–13

68. Wiedersheim O (1942) Über zwei seltene angiomatöse Veränderungen des Augenhintergrundes und über Erweiterung des Begriffs Angiomatosis retinae. Klin Monatsbl Augenheilkd 108:205–213

69. Wyburn-Mason R (1943) Arteriovenous aneurysm of midbrain and retina, facial naevi, and mental changes. Brain 66:163–203

70. Yanoff M, Fine BS (1975) Ocular pathology. Arteriovenous communication. Harper and Row, New York, p 528

22.3 Retinal Arterial Macroaneurysms

S. BOPP

22.3.1 Clinical Presentation

22.3.1.1 Typical Clinical Findings

Retinal arterial macroaneurysms (RAMs) are characterized by unilateral, solitary round or fusiform dilatations of major retinal arteries within the first three orders of arteriolar bifurcation. Commonly, they are located at the site of an arteriolar bifurcation or an arteriovenous crossing at the posterior pole. The supratemporal artery is most commonly the site of RAMs; however, nasal arteries may be affected as well. Moosavi et al. [38] have analyzed the distribution of 34 RAMs and found 50% on superotemporal, 44.7% on inferotemporal and the remaining 5.2% on nasal vessels. Similar observations were reported by Tezel et al. [64], who investigated 21 symptomatic RAMs and found them to be located in 52.4% supe-

rotemporally, in 38% inferotemporally and in 9.6% nasally.

Simple RAMs consist of the vessel anomaly only. Narrowing of the distal artery may be observed, rarely branch arterial occlusion. In complex RAMs, exudation and bleeding into the adjacent retina occurs. These complications lead to macular edema or extrafoveal retinal edema, serous retinal detachment, circinate figure and intraretinal hemorrhages surrounding the aneurysm. Subtle microvascular changes may be present, in particular when using fluorescein angiography (FAG) for imaging (Fig. 22.3.4). Some RAMs show visible pulsations. The clinical significance is not clear, in particular, if this indicates a high risk of hemorrhagic complications [3]. Some eyes show focal yellow plaques (atheroma) in close proximity to the aneurysm. They are observed for retinal emboli, but according to the his-

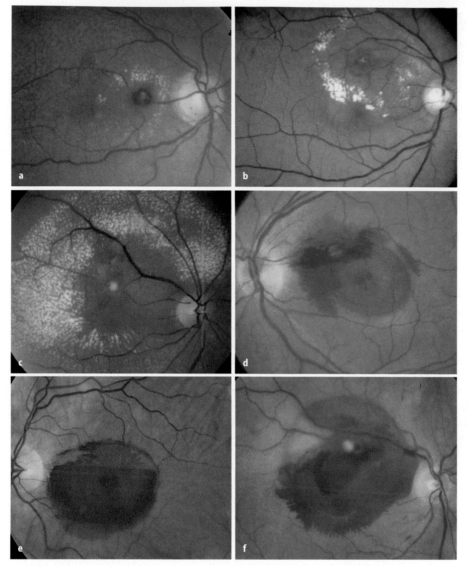

Fig. 22.3.1. Clinical spectrum of RAMs. **a–c** Primary exudative retinopathy and a characteristic lesion are surrounded by a circinate figure. A fleshy-appearing nodule bordered by small hemorrhages indicates an active lesion (**a, b**) whereas a whitish lesion reflects an inactive RAM that has undergone sclerosis (**c**). **d–f** Hemorrhagic complications due to RAM. Hemorrhages may be located primarily subretinally (**d**), preretinally (**e**) or may involve various layers ("mixed type," **f**)

topathologic findings they more likely represent arteriosclerotic vessel wall alterations. Moreover, the clinical spectrum of complex RAMs includes massive hemorrhage into the pre- and subretinal space and even into the vitreous cavity (Fig. 22.3.7). A "mixed-type" hemorrhage is highly suggestive of RAMs and usually does not involve retinal or choroidal vascular diseases of other origins (Fig. 22.3.13).

The spectrum of typical symptomatic clinical presentations is depicted in Fig. 22.3.1.

22.3.1.2 Special Clinical Findings

Multiple RAMs occur in approximately 15–20% of affected eyes. They can be observed in one eye along the same artery or elsewhere [18% in our series of hemorrhagic RAMs, up to 20% reported in the literature (Figs. 22.3.4, 22.3.11b, 22.3.13b)] [48]. Bilateral

disease is found in approximately 10% [31, 48] (Fig. 22.3.12).

Exceptional locations of a RAM are the cilioretinal artery [14] and the optic disk [5, 17, 30].

Some rare complications due to RAMs include macular hole formation [8, 36, 61], retinal detachment [62] and CNV [51]. To complete the clinical spectrum of unusual aneurysmal conditions, retinal venous macroaneurysms, RAMs associated with venous occlusive disease, have sporadically been reported [27, 29, 55].

22.3.1.3 Clinical Symptoms

Many RAMs remain asymptomatic and are being found by routine ophthalmoscopy. RAMs may become symptomatic for two major reasons: first, progressive and chronic exudative processes involving the macula, and, second, rupture of the aneurysm as a result of high intravasal arterial pressure leading to intra-, sub-internal limiting membrane (ILM) and vitreous hemorrhages (Figs. 22.3.1, 22.3.8). Typically, patients experience a slow or sudden visual loss with/without floaters. Complex RAMs indicate an active vision-threatening process, and therapy should be considered.

Asymptomatic active or asymptomatic non-leaking RAMs should be monitored regularly to allow early diagnosis in case the disease should progress and threaten the macula (Fig. 22.3.4).

22.3.2 Epidemiologic Data and Risk Factors

Large clinical series have shown that woman between 60 and 80 years of age are predominantly affected (approximately 70%) and mean age is 68–74 years [38, 41]. Sixty-four to 75% of patients have a history of systemic hypertension and clinical or ophthalmoscopic evidence of arteriosclerosis [32, 38, 42, 48]. Furthermore, an association with elevated serum lipid abnormalities was found [9]. As many diabetic patients suffer from hypertension as well, not diabetes mellitus, but hypertension is thought to be a major etiologic factor for RAMs. The association of hypertensive retinopathy and RAMs suggests a common pathophysiologic process [51].

The true incidence of RAMs is not known, as most often only symptomatic ones are diagnosed. Valsalva maneuvers are thought to increase the risk of bleeding.

22.3.3 Pathogenesis and Pathomorphology

Hypertensive and arteriosclerotic changes of arterial vessels (ageing symptoms) may explain the formation of RAMs, including medial muscle fiber replacement by collagen, hyaline degeneration of the vessel wall, endothelial cell damage with atheromas lining the inner wall leading to narrowing of the lumen, increased vessel rigidity, elevated intravasal pressure and transmural stress. Loss of autoregulation and blood flow turbulence are further circulatory disturbances. On the basis of these alterations, focal vessel dilatation and formation of RAMs occur. The pathogenetic process of aneurysm formation and rupture is not fully understood. There is histologic evidence suggesting focal thrombosis and localized embolic events to be inciting factors, until dissection of the inner vessel wall and full thickness rupture take place.

Histopathologic studies have shown a distended thickened vessel wall (hyalosis) that is more or less obstructed by a fibrinous substance (fresh or organized thrombus). Proteinaceous, lipid-rich material and hemosiderin deposits or blood can be found in the adjacent retina and subretinal space. Dilatated capillaries surround the lesion [11, 44].

22.3.4 Natural History

Weakening of the vessel wall leads to focal outpouching and aneurysm formation. Some RAMs can remain stationary over a long time period. Others show leakage or rupture. In these eyes, chronic exudative retinopathy or acute hemorrhages may cause visual symptoms, particularly if RAMs are located within the vascular arcades and adjacent to the macula. After such a period, most RAMs show spontaneous involution (Figs. 22.3.2, 22.3.8, 22.3.11). Focal thrombosis and sclerosis lead to restoration of the vessel wall, and perfusion is preserved in most cases (Fig. 22.3.2).

Similarly, exudative retinopathy and hemorrhages resolve. In cases involving the macula, visual restoration depends on the severity and duration of macular involvement. Structural damage from chronic cystoid edema, accumulation of lipid exudates and intraretinal hemorrhages is the most common cause of functional impairment (Fig. 22.3.2).

In cases of acute hemorrhage, functional recovery largely depends on the location of blood. Eyes with subfoveal hemorrhage are most likely to suffer from visual loss. Subfoveal retinal pigment epithelium (RPE) changes and fibrosis are frequent end stage alterations (Fig. 22.3.3).

Other late complications of complex RAMs are epiretinal membrane formation and macular holes [8, 61].

It is worth mentioning that in some eyes RAMs may eventually arise elsewhere in the same or other retinal vessels (Figs. 22.3.4, 22.3.11): patients should therefore be monitored regularly [1, 4, 38, 41, 48].

To summarize, functional impairment is frequent in eyes with RAMs affecting the macular area, despite spontaneous involution. In a large collaborative study of 142 eyes with macular involvement, Schatz et al. [56] reported that 95% suffered from incomplete functional recovery (20/30 or less) and 49% experienced severe visual loss (20/100 or less). Tonotsuka [65] reported on 65 eyes with symptomatic RAMs and demonstrated that spontaneous visual prognosis depends on macular pathology: a good outcome was found in eyes with vitreous and preretinal hemorrhage (mean 0.6–0.7), a moderate outcome in those with macular edema (0.5) and a poor outcome in cases with subretinal or mixed hemorrhages.

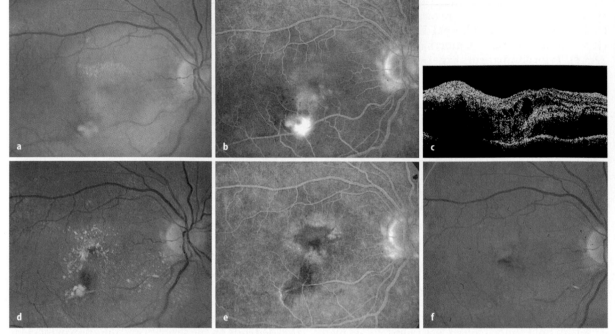

Fig. 22.3.2. Natural course of a RAM with exudative retinopathy. VA at initial presentation was 0.1 due to a serous macular edema (**a, c**). Late phase FAG shows significant leakage from the RAM (**b**). The RAM underwent spontaneous involution (**d–f**). Focal narrowing and sclerosis of the corresponding artery has occurred (**d**) and leakage has stopped (**e**). Lipid deposits indicate resorption of intra- and subretinal fluid (**d**). Finally, macular edema has resorbed, but pigmentary changes limit VA to 0.3 (**f**)

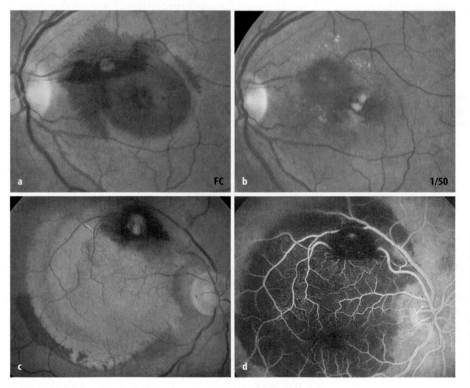

Fig. 22.3.3. Natural course of RAM with hemorrhagic retinopathy. **a, b** Subretinal blood finally resorbed, but function did not recover due to lipid and pigmentary changes. **c, d** Instead of liquefaction and resorption, the subretinal blood clot may undergo organization leading to marked scar formation

22

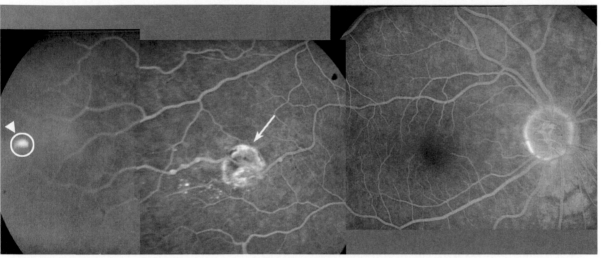

Fig. 22.3.4. Multiple RAM along a superotemporal artery. Multiple fusiform dilatations without significant leakage are visible. Laser therapy has been done to one RAM previously (*arrow*). In the equatorial region, another asymptomatic, but leaking saccular aneurysm was found (*arrowhead*). Note microvascular changes surrounding the area of grouped RAMs

22.3.5 Terminology and Classification

So far, no standardized classification for the variety of clinical presentations of RAMs has been recommended. As suggested before, RAMs can be described as *simple* (without associated retinal changes) and *complex* (associated with exudation and bleeding), which also correlates with an *active or inactive* process. Exudative complications with lipid deposits indicate a *chronic* stage, but significant hemorrhages are considered an *acute* stage of the disease. In terms of the predominant clinical finding, RAMs can also be classified into quiescent, exudative and hemorrhagic [31].

Palestine et al. [41] suggested categorizing RAMs on the basis of their anatomic location and vision affecting complications:

- RAMs within the vascular arcade and macular involvement due to complicating factors (e.g., edema, exudates, hemorrhages)
- RAMs within the vascular arcade ± complicating factors without macular involvement
- RAMs peripheral to the vascular arcade ± complicating factors without macular involvement

Visual prognosis was found to be strongly associated with macular complications. *Asymptomatic* RAMs (Groups 2, 3) were found to have a favorable prognosis, but visual prognosis in *symptomatic* disease (Group 1) was uncertain.

Furthermore, the shape of RAMs may influence the complication rate. *Round or saccular* RAMs were found to have fewer hemorrhagic complications compared to *fusiform* aneurysms [38].

22.3.6 Diagnosis and Imaging

Ophthalmoscopy reveals a focal fleshy saccular or fusiform thickening of the artery affected (Figs. 22.3.1, 22.3.3, 22.3.6, 22.3.9, 22.3.10, 22.3.13). Clear diagnosis may be difficult if secondary exudative changes dominate the picture or hemorrhages obscure the vessel anomaly. In those cases, angiography is indicated.

Fluorescein angiography shows a uniform filling in the early phase. Partial or incomplete filling (Figs. 22.3.4, 22.3.6b) is a sign of inner vessel wall thrombosis at the site of vessel dilatation. Blood flow may be restored completely and the vessel returns to normal. Residual findings can be subtle: a focal narrowing of the blood stream and irregular track or visible focal sclerosis of the artery (Figs. 22.3.2d, 22.3.6d, 22.3.12b).

Late phase angiography varies from little or no staining of the vessel wall to marked leakage into the adjacent retina. Unlike in age-related macular degeneration (ARMD), leakage only occurs at the site of the vessel anomaly, but does not affect the whole area showing exudative changes (Fig. 22.3.2b).

Microvascular anomalies surrounding RAMs can be found: a widening of the capillary-free zone around the vessel anomaly, adjacent capillary dilatation and non-perfusion, microaneurysms and small collateral vessels. They contribute to late phase leakage (Figs. 22.3.2b, 22.3.4).

In the case of RAMs with significant hemorrhages, FAG is useful for diagnosis. If the retinal vasculature is not completely obscured by blood, the presence of a characteristic hyperfluorescent nodule that is not visible on direct examination allows dif-

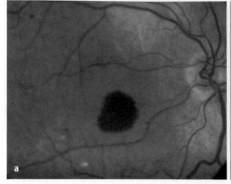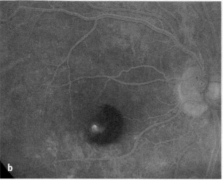

Fig. 22.3.5. Small preretinal hemorrhage of unknown origin with macular involvement. FAG allows diagnosis of a saccular RAM inferior to the macula

ferential diagnosis from other diseases with macular hemorrhage (Fig. 22.3.5).

Indocyanine angiography (ICG-A) allows better visualization of deeper retinal and choroidal structures, and near infrared light penetrates blood to a certain degree. Furthermore, the higher protein-binding capacity of ICG (98% to serum albumin) compared to fluorescein (60%) results in less dye leakage and more defined images. In case FAG shows no characteristic findings, ill-defined hyperfluorescence or dense hemorrhages, ICG-A may delineate the underlying pathology [16, 57, 66].

In about 50% of hemorrhagic RAMs, FAG leads to the correct diagnosis and ICG-A in about 75%. Thus, angiography contributes to the actual diagnosis at first examination and allows appropriate treatment or follow-up.

22.3.7 Differential Diagnosis

The broad spectrum of clinical presentations in RAMs and the similarity to other well-defined disorders often lead to misdiagnosis. Spalter has termed RAMs a "new masquerade syndrome" [60]. Depending on the leading symptom, RAMs mimic a variety of retinal disorders (Table 22.3.1).

Most often, complex RAMs with exudation and/or hemorrhage are misdiagnosed for wet ARMD, in particular when exudative maculopathy and subretinal hemorrhage dominate the clinical picture (Figs. 22.3.1a, d, 22.3.2d, 22.3.6). FAG then allows easy differentiation. Thick blood clots are sometimes misdiagnosed for malignant melanoma [44, 59].

Evaluation of our series with RAMs associated with severe hemorrhage showed that almost half of the cases were referred for treatment of hemorrhagic ARMD. Eccentric hemorrhage location surrounding an artery is highly suggestive of RAMs, even if the vessel anomaly itself is hidden by blood (Figs. 22.3.12c, 22.3.14a). Another important hint of RAMs is the fact that blood often involves various retinal layers. High pressure within the lumen of the artery

Table 22.3.1. Differential diagnosis of RAM

Leading symptom: atypical vessel structure
Angiomatosis retinae
Cavernous hemangioma
Vessel malformation in venous occlusive disease, diabetic retinopathy
Leading symptom: exudative retinopathy
Exudative ARMD
Disciform ARMD
Diabetic macular edema
Idiopathic parafoveal teleganiectasia
Leber's miliary aneurysm retinopathy/Coats' disease
Radiation retinopathy
Leading symptom: macular hemorrhage/dark mass in the macula
ARMD (subretinal/sub-RPE hemorrhage) and other conditions associated with SNVMs/hemorrhages
Proliferative diabetic retinopathy (subhyaloidal hemorrhage)
Chorioidal melanoma
Leading symptom: vitreous hemorrhage
Acute PVD, retinal tear
Ischemic retinal diseases (branch vein occlusion, central vein occlusion, PDRP, etc.)
Mass hemorrhage in ARMD

causes severe bleeding that does not respect the natural horizontal barriers of the retina, but extends into adjacent tissue sheets. Thus, a "mixed type" of hemorrhage is highly suggestive of RAMs (Figs. 22.3.1f, 22.3.8a, 22.3.13a). Similar findings may be present in Terson's syndrome, which is easy to exclude by the medical history.

Suspected underlying conditions in eyes that present with vitreous hemorrhage are venous occlusive disease, acute posterior vitreous detachment and diabetic retinopathy [34, 35].

Table 22.3.1 summarizes retinal disorders to be considered in the differential diagnosis of RAMs [13].

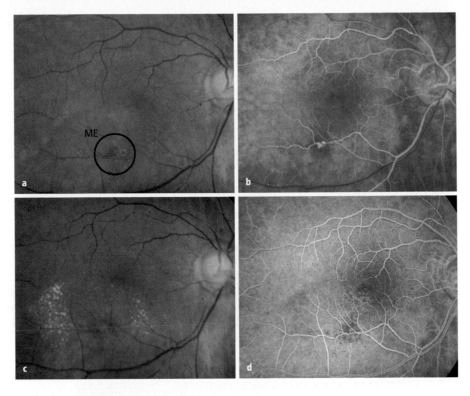

Fig. 22.3.6. Symptomatic RAM before (**a, b**) and 6 weeks after indirect laser treatment (**c, d**). Exudative maculopathy caused visual deterioration to 0.4. Careful laser was applied to the vessel wall. Fundus examination 6 weeks later revealed involution of the vessel anomaly. Temporarily hard exudates increase as a sign of increased phagocytotic processes in the retina (**c**). FAG shows minor irregularities in the area of RAM regression, normal perfusion, and no leakage (**d**). Final visual outcome was 0.7 after 6 months

22.3.8 Therapy

Treatment for RAMs is a controversial issue. In the face of the overall benign spontaneous prognosis, therapy is considered in symptomatic and active RAMs only. Indications may include [52]:

- Macular edema
- Macular exudates
- Macular hemorrhage (pre- and subretinal)
- Vitreous hemorrhage

A variety of therapeutic options for symptomatic RAMs have been suggested. Depending on the presenting symptoms, they range from laser therapy to subretinal surgery. An overview of current options will be given taking into account the leading symptom.

22.3.8.1 Exudative Complications

22.3.8.1.1 Laser for Exudative Retinopathy

Accumulation of fluid and lipids surrounding RAMs may originate from the leaking macroaneurysm itself and from microvascular anomalies surrounding the aneurysm. The main rationale for laser treatment is to induce thrombosis and sclerosis at the site of vessel dilatation, in other words, to enhance spontaneous involution and prevent retinal damage due to chronic edema and exudates, macular pigment changes and scarring. Although indications for laser treatment are poorly defined [4], most experts agree that laser should be performed, if edema and exudates involve or threaten the fovea [1, 12, 15, 17, 26, 31, 41, 46, 69].

Different laser techniques have been suggested:

- Direct laser of the RAMs (center)
- Indirect laser of the RAM (treating the vessel wall and bordering retina)
- Perianeurysmal laser (scatter laser in the adjacent area of microvascular changes)

Laser settings consist of low-power, medium-sized spots (200–500 μm) of longer duration (0.2–0.5 s) in order to avoid vessel rupture or complete vessel obstruction resulting in branch retinal artery occlusion. Both argon green and yellow krypton laser have been used, both with similar results.

Therapeutic efficacy is difficult to evaluate. There is clinical evidence that laser therapy actually induces rapid closure of the RAMs and this is followed by subsequent resolution of retinal edema and exudates (Figs. 22.3.4, 22.3.6, 22.3.9c, 22.3.14c). Which laser technique is more effective remains unclear. The risk of laser-induced complications, e.g., arteriolar occlusion, is considered a lower risk after indirect laser application [52]. Furthermore, final scientific verification of functional outcomes in laser-treated eyes compared to non-treated eyes (= natural course) is still lacking.

Adjuvant medical therapy may be taken into consideration as well. The rationale is to suppress edema and exudation in complex RAMs and to avoid retinal damage as a result of chronic leakage. Intravitreal triamcinolone and anti-VEGF drugs are the currently available drugs. Clinical experience with pharmacologic adjuncts in exudative retinopathy due to RAMs has not been presented so far.

22.3.8.2 Hemorrhagic Complications

Hemorrhage into the macula is the most serious visual threatening complication due to RAMs. Bleeding does not respect the natural horizontal barriers of the retina, spreading subretinally and possibly causing hemorrhagic cysts under the ILM and breaking into the subhyaloid space or the vitreous cavity (Fig. 22.3.7). Management of hemorrhages due to ruptured RAMs remains controversial. Goals of therapy are either to achieve early visual rehabilitation and/or to prevent blood-induced permanent foveal damage. Depending on the location of bleeding, therapeutic options may be different.

22.3.8.2.1 Laser for Preretinal Hemorrhage

Bleeding toward the inner retina and vitreous can split/separate the ILM or the posterior vitreous cortex and become encapsulated. Blood resorption is then often delayed. Since the central retina is obscured as well, evaluation and treatment of possibly associated exudative maculopathy or subfoveal hemorrhage is not possible. Drainage of blood into the vitreous cavity [43] will allow more rapid clearance.

For these reasons, photodisruption of the bordering/separating membranous tissue ("laser hyaloidotomy") was suggested [7, 10, 25, 28, 37, 47, 49, 50, 63, 67]. Neodymium-YAG, argon and krypton laser were applied in eyes presenting with encapsulated premacular hemorrhage with similar effect (Fig. 22.3.8). Most eyes treated showed marked clearing of hemorrhages and rapid improvement of function. Complications, in particular laser-induced retinal damage, were rarely reported. Park demonstrated that in case laser hyaloidotomy could not be achieved, pneumatic displacement is an option to induce posterior vit-

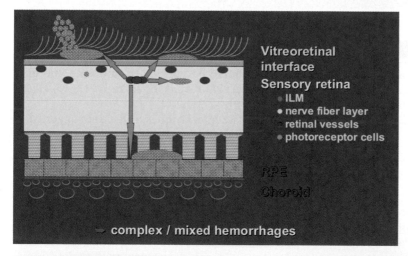

Fig. 22.3.7. Schematic drawing of pathways for blood in ruptured RAM

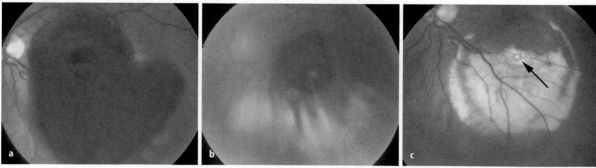

Fig. 22.3.8. Complex hemorrhagic RAM. An acute preretinal/subhyaloidal hemorrhage of unknown etiology was initially observed (**a**), then treated by YAG-laser hyaloidotomy. Blood escaped from the preretinal pocket allowing retinal examination 4 weeks later showing a dense hemorrhage extending from the macula to the inferior arcade (**b**). As further blood resorption was delayed, vitrectomy was performed. An occluded RAM (*arrow*) as the source of bleeding was identified (**c**). Visual outcome was 0.1 only due to macular pigment changes and blood remnants

reous detachment (PVD) and allow blood to disperse into the vitreous cavity [43].

22.3.8.2.2 Surgery for Premacular Hemorrhage

If laser fails to induce blood evacuation into the vitreous cavity or the blood pocket is too close to the macula, vitreous surgery may be considered. Vitrectomy with surgical PVD and incision of the ILM allows safe removal of encapsulated hemorrhages (Figs. 22.3.9, 22.3.13). Immediate visualization of RAM-related findings allows the decision of possible further measures in order to treat additional complications. In case of exudative maculopathy, additional laser may be considered. Subretinal hemorrhagic components can be treated by adjunctive pharmacologic therapy or surgical removal.

22.3.8.2.3 Surgery for Submacular Hemorrhage

Prognosis for restoration of vision in the presence of submacular hemorrhage is generally poor, regardless of the primary cause [21]. The duration and amount of blood deposition under the fovea is crucial for photoreceptor cell damage. Thus, surgical intervention is worth considering for those cases. However, subretinal blood removal has remained a challenge for some time. Direct and indirect approaches to submacular hemorrhages of various origins [ARMD, angioid streaks, presumed oscular

histoplasmosis syndrome (POHS), RAM, trauma] have been suggested.

Mechanical removal via retinotomy using forceps to grasp the clot was used in the 1990s, but includes a high risk of retinal trauma and RPE damage [22, 24]. Furthermore, we found that blood adheres firmly to the aneurysm, and attempts to remove clots often remain incomplete. We also observed that postoperative complications, such as marked scarring, pucker formation and PVR, frequently occur despite careful surgical manipulations (Fig. 22.3.12). Accordingly, this method is largely abandoned today.

A step toward a more atraumatic technique to remove submacular hemorrhages was achieved by injection of *recombinant tissue plasminogen activator (rt-PA) subretinally* into the blood coagulum [18, 23, 33, 39, 45, 53, 71]. The injected solution must be kept subretinally up to 1 h intraoperatively, to allow clot lysis. Blood evacuation is then performed using a fluid cannula, removing clot remnants with a forceps. Perfluorocarbon liquid (PFCL) may assist atraumatic blood displacement toward the retinotomy. Complications, as mentioned above, were reduced significantly (Figs. 22.3.12d, 22.3.13b, 22.3.14b).

As an alternative, techniques to dislocate subfoveal blood have been advocated, as direct elimination of subfoveal blood was thought not to be necessary. Removal of the hemorrhage from the macular area may be sufficient to protect foveal photoreceptors from toxic damage and nutritious deficiency by the

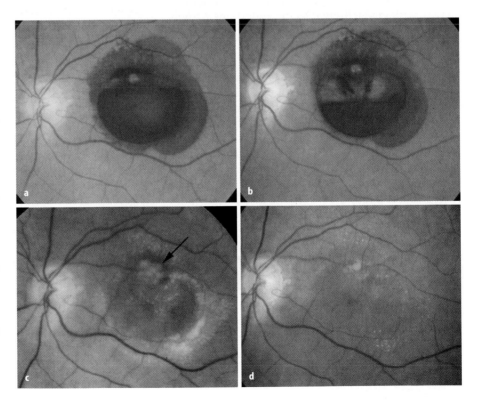

Fig. 22.3.9. a RAM with predominantly preretinal hemorrhage which is surrounded by a rim of subretinal blood. Intravitreal injection of SF$_6$ and rt-PA was performed, but sufficient blood dislocation was not achieved (**b**). During vitrectomy the major blood component was found to be encapsulated between the retina and ILM. After ILM peeling and removal of the liquefied blood remnants, severe macular edema and lipid deposits due to an active RAM became evident, and laser treatment to the RAM was performed (*arrow*) (**c**). Three months later, vision has improved to 0.6, the macula is dry and the RAM is occluded (**d**)

22 III

thick blood clot. Furthermore, the risk of surgically induced trauma was thought to be significantly lower.

Pneumatic displacement using an expansive gas was introduced in the late 1990s. The majority of eyes treated suffered from subretinal hemorrhage due to ARMD or were cases with hemorrhage of unknown origin. The rationale for gas-induced displacement was to prevent toxic photoreceptor cell damage and to allow early diagnosis, if additional therapy of the underlying disease was possible. Most reports stated that gas displacement was effective, but results varied considerably [40, 70]. We found that blood dislocation after gas injection alone was incomplete, most likely due to the fact that hemorrhages are already coagulated or partially organized and adhere firmly to the aneurysm. Attempts to dislocate blood with a gas bubble alone were therefore disappointing.

Improvement in the management of submacular hemorrhages occurred through the *combination of pneumatic displacement and fibrinolytic agents*, e.g., intravitreal rt-PA in conjunction with gas [6, 18, 20, 70]. Although most of the reports documented hemorrhagic ARMD, a few cases with RAMs were treated as well. Final visual outcomes were poor in ARMD eyes, but improved or remained stable in non-ARMD cases [18, 24, 39].

Another more invasive variant is to perform vitrectomy with *subretinal rt-PA application and subtotal gas tamponade*. A few clinical case series in ARMD patients have been reported. The studies showed successful displacements of macular hemorrhages, with vision limited by the underlying pathology [19, 33, 58]. We observed that subretinal rt-PA application using 40 gauge cannules was an atraumatic and effective approach that led to reliable clot lysis in fresh subretinal hemorrhages (Fig. 22.3.15b).

It has not yet been established whether subretinal rt-PA application with intraoperative blood removal or just pneumatic displacement is the safer and more efficient method.

To summarize: Surgical therapy for complex RAMs is still under debate. There are no clear recommendations on "when and how" to treat symptomatic RAMs. It is left to the surgeon to decide individual cases, taking into account the risk versus benefit for each patient.

22.3.8.3 Pearls and Pitfalls of Surgery for Hemorrhagic RAMs

Treatment concepts for RAMs have changed over the years. We investigated the impact on clinical practice at our clinic by retrospectively analyzing the medical records. A total of 205 patients with the diagnosis of RAMs were identified between 1995 and 2006, and 27 of them were operated on for hemorrhagic complications.

22.3.8.3.1 Patient Selection

As subretinal blood, with or without associated exudative maculopathy, is the most relevant symptom for prognosis in RAMs, these particular cases are potential candidates for surgical intervention. However, we have found that patient selection was difficult for the following reasons:

- In patients referred for vitreous hemorrhage of unknown origin approximately 5 – 10 % had complex RAMs as the source of bleeding [68]. Usually, clinicians rule out retinal detachment by ultrasound examination and wait for spontaneous blood resorption. Treatment is postponed until diagnosis becomes evident; complications arise or visual rehabilitation does not occur within a few months. With regards to subretinal bleeding, irreversible macular damage may already be present, when the right diagnosis is ultimately made (Fig. 22.3.8).
- In cases in which RAM is identified as the source of vitreous or preretinal hemorrhage, this may obscure a significant subretinal blood component. If the clinician decides to wait for spontaneous resolution, the same consequences may occur as mentioned above (Fig. 22.3.13a).
- Patients presenting with submacular hemorrhage may be diagnosed for exudative ARMD. Differential diagnosis is difficult, when other signs of ARMD are lacking and the characteristic aneurysmic nodule is covered by blood. In contrast to ARMD, hemorrhages due to RAM surround the RAM, and are located eccentrically. However, these characteristics are not always present (Fig. 22.3.1d, e).

22.3.8.3.2 Surgical Technique

The basic procedures are listed above. Surgical maneuvers varied according to the individual situation: In the case of preretinal and vitreous hemorrhage, vitrectomy was done, a PVD created and the ILM peeled off. In the case of submacular hemorrhage, careful subretinal balanced salt solution (BSS) injection via a small retinotomy was carried out and followed by drainage or extraction of the blood clot (early series) (Fig. 22.3.10). When rt-PA became available, additional clot lysis by injecting 12.5 µg/cc (0.1 – 0.2 cc) subretinally prior to drainage or extraction was performed (11 eyes) (cases depicted in Fig. 22.3.12 – 22.3.15). Initially, the aneurysm itself was not treated in view of the expected spontaneous prognosis, but later on, active appearing RAM received additional laser treatment (Figs. 22.3.9c, 22.3.14c).

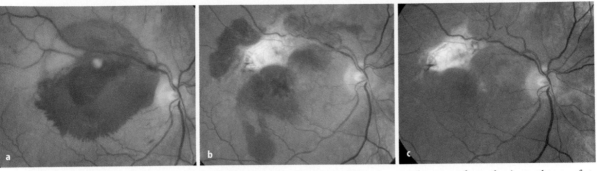

Fig. 22.3.10. RAM-related subretinal hemorrhage. Vitrectomy and subretinal blood removal were performed prior to the era of rt-PA. Initial VA was CF (**a**). After vitrectomy, subretinal BSS irrigation and blood clot removal via a retinotomy, recurrent bleeding occurred (**b**) that slowly resorbed (**c**). Severe scarring at the retinotomy site hints at significant surgical trauma. Both explain the poor visual outcome (1/35)

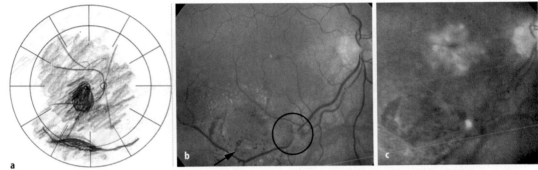

Fig. 22.3.11. RAM and vitreous hemorrhage. Macular involvement could be seen, but the location (pre- versus subretinal) remained unclear (**a**). Vitrectomy was carried out. Intraoperatively, subretinal hemorrhage originating from a nodular RAM (*circle*) was found to extend into the macula and was removed using subretinal rt-PA. Interestingly, a second, spontaneously occluded RAM was found at the same vessel (*arrow*) (**b**). Pigmentary changes surrounding the lesion were indicative of a previously active lesion with bleeding. Chronic cystoid macular edema resistant to medical therapy was the reason for limited functional recovery to 0.4 (**c**)

Fig. 22.3.12. Bilateral presentation of hemorrhagic RAMs. The patient experienced vitreous bleeding in her right eye (**a**). Vitrectomy revealed a "mixed type" of bleeding. A flat subretinal hemorrhage extended just into the macula, but appeared to be in resorption and was not removed. The RAM had already undergone involution (**b**). Postoperative vision remained 0.3 due to exudation-related macular changes. One year later, the patient returned because of vitreous and macular hemorrhage in her left eye. After clearance of some vitreous bleeding (VA=0.1), a dense subretinal hemorrhage extending into the macula became visible (**c**). rt-PA-assisted blood removal was performed. Despite a pucker formation at the retinotomy site, vision improved to 0.3 (**d**)

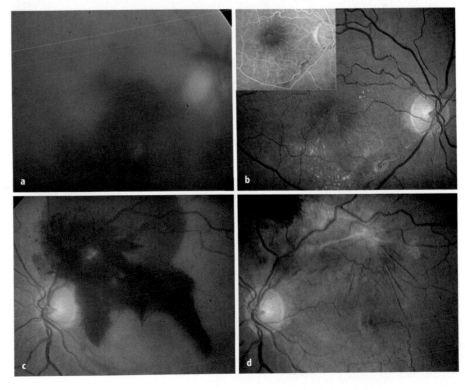

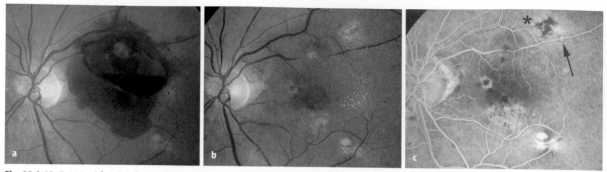

Fig. 22.3.13. RAM with mixed-type hemorrhage (sub-ILM and subretinal component). Initial VA was CF (**a**). Vitrectomy with ILM peeling and rt-PA assisted subretinal blood removal was performed and the active appearing RAM treated by laser 3 months postoperatively; vision had improved to 0.2 (**b**). The RAM appeared closed, but a second non-leaking RAM was detected distally from the primary one (*arrow*) (**c**). Note significant pigmentary alterations of the macula and a discrete scar at the retinotomy site (*asterisk*)

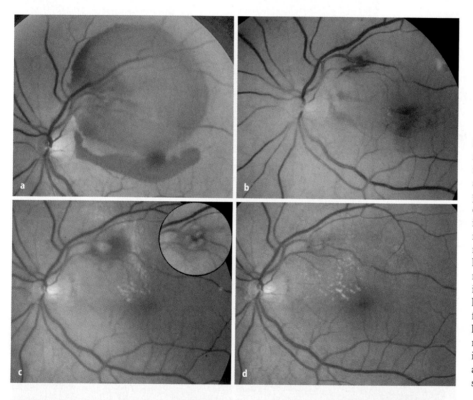

Fig. 22.3.14. RAM with subretinal hemorrhage (**a**). Vitrectomy with rt-PA assisted subretinal blood removal was performed. One week postoperatively, vision has improved to 0.5 (**b**). Two months later, vision deteriorated again as a result of leakage of the RAM. Note new hard exudates extending into the macula (**c**). Indirect laser to the RAM was performed (*inset*). One month later, the RAM showed sclerotic transformation, the retinal edema had disappeared and lipid exudates slowly resolved. VA increased to 0.9 (**d**)

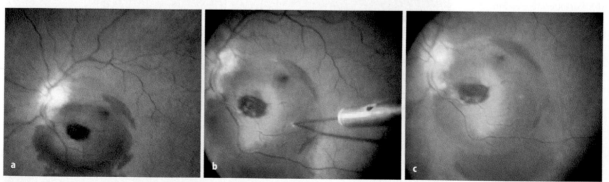

Fig. 22.3.15. Surgery with rt-PA assisted subretinal blood removal in a hemorrhagic RAM. Initial presentation (**a**) and injection of rt-PA using a 40 g cannula (**b**). Suggested incubation of rt-PA is 45 min to allow clot lysis (**c**)

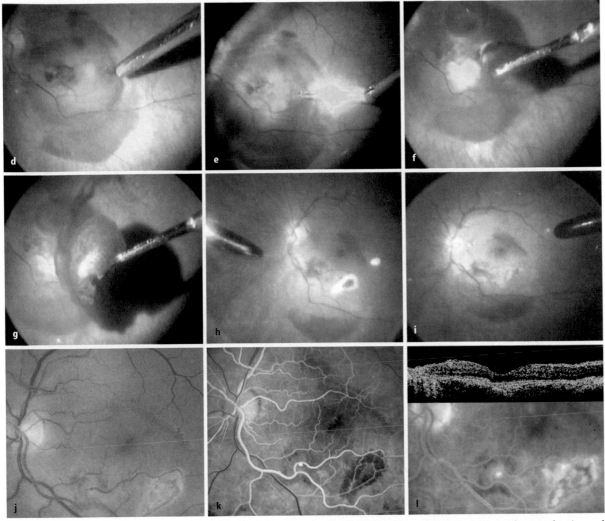

Fig. 22.3.15. Liquefied blood is removed using an extrusion cannula (**d**). Clotted remnants may remain dependent on the size and duration of hemorrhage. Extraction is performed with a subretinal forceps via the retinotomy (**e, f, g**). Usually the clot sticks firmly to the RAM. After removal of subretinal blood remnants (PFCL is useful to assist complete elimination), laser may be applied to the retinotomy site, if it appears large (**h**). Surgery is finished with air or gas tamponade (**i**). Postoperatively, vision improved to 0.4. The RAM is still visible, with minor leakage, and the macula shows a normal foveal depression without edema (**k, l**)

22.3.8.3.3 Preoperative and Intraoperative Findings

In our series of 26 eyes operated on for hemorrhagic RAMs, the leading symptom was macular hemorrhage in 70% and vitreous bleeding in the remainder. During initial examination, diagnosis of a RAM as the cause of bleeding was made in 48% only. Hemorrhagic ARMD was assumed in 35%, and vitreous hemorrhage of unknown origin was present in the remaining 20%.

In five eyes with preretinal or vitreous hemorrhage with an unknown source of bleeding, pretreatment was performed. One eye had intravitreal SF_6 injection, and another three eyes received a gas injection with rt-PA (50 µg) (see case in Fig. 22.3.9)

and one eye had Nd:YAG treatment to the surface of a dome-shaped preretinal blood pocket (see case in Fig. 22.3.8). Although consecutive blood dispersion into the vitreous cavity occurred, resorption was slow. In the other eyes, gas-induced blood displacement was insufficient. Thus, all five eyes underwent vitreous surgery. The time interval between onset of symptoms and surgery ranged from 2 days to 4 months.

The predominant location of hemorrhages, which is considered important for visual prognosis, was documented intraoperatively. Primary subretinal blood was present in 38% and mainly sub-ILM hemorrhage in 35%. Equal distribution of pre- and subretinal blood was observed in 23% and vitreous hemorrhage without macular involvement in one eye. Thus, almost two-thirds of eyes had a significant subretinal blood component. It is worth mentioning, in all cases – whether they were operated on early or delayed – blood was clotted or partially organized and did not show significant liquefaction. This may explain previous unsuccessful attempts to displace the blood pneumatically.

22.3.8.3.4 Postoperative Findings

Figure 22.3.16 shows the functional results and complications. Visual improvement of 2 lines or more was achieved in 76%. Subgroup analysis showed:

Eyes with preretinal/sub-ILM hemorrhage did best: Preoperative vision ranged from CF to 0.1 and increased from 0.2 to 0.9. Visual acuity (VA) was limited in three eyes as a result of chronic cystoid edema, amblyopia and ARMD.

Eyes with subretinal hemorrhage also improved, but not as favorably. Final VA correlated with the surgical technique and the timing of surgery. In the early series, surgery was postponed for the reasons

mentioned, and subretinal blood was already found to be organized. Furthermore, we mechanically removed clots using forceps, which led to significant trauma, subretinal RPE changes and pucker formation at the retinotomy site. Two eyes developed postoperative PVR. We learned that the use of subretinal rt-PA with earlier intervention was key to more favorable outcomes, e.g., better vision and fewer complications.

Other peculiarities were found in this particular patient group. Multiple RAMs in the same eye were identified in five cases (Figs. 22.3.11, 22.3.13). Four were located on the same vessel: Three showed focal vessel sclerosis and surrounding pigment changes after rupture and localized bleeding at an earlier time, but were asymptomatic. One eye developed another symptomatic RAM after subretinal surgery. Marked exudation with macular involvement required laser therapy, and vision was restored. Another eye had Nd:YAG hyaloidotomy years before and showed vitreous bleeding out of RAM elsewhere in the fundus.

Follow-up examinations revealed other unexpected findings: In four eyes of our early series RAM remained open after surgery and did not undergo spontaneous involution. Exudative retinopathy persisted and FAG disclosed prolonged abnormal leakage. Subsequent laser therapy was performed (Fig. 22.3.14c). Closure of the RAM was achieved and resulted in resolution of retinal edema.

The frequency of multiple, recurrent and persisting RAM is remarkable and indicates that in terms of RAM involution, the prognosis is not as good as expected from previous studies. One may speculate that the clinical course in the subgroup of hemorrhagic RAMs is more aggressive. Our conclusions from these observations were: Firstly, in case the RAM appears open during surgery, we perform additional soft laser treatment (Fig. 22.3.9c). This was done in 10/26 eyes, and all showed rapid regression of the vessel malformation. Secondly, patients should be monitored after surgery for hemorrhagic RAMs, as they may show persisting active or new RAM formation, and early treatment should be considered (Fig. 22.3.13b).

22.3.8.3.5 Discussion

In the face of the poor prognosis of subretinal hemorrhage in eyes with RAMs [38, 54, 65], functional results of the present case series are encouraging. Poor outcomes were associated with:

- Too late diagnosis
- Underestimated subretinal blood component
- Iatrogenic damage

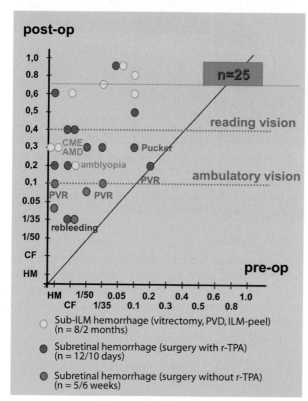

Fig. 22.3.16. Visual outcomes after surgery for hemorrhagic RAM. Intraoperatively, a classification according to the predominant location of blood was made. Furthermore, the median interval between the onset of symptoms and surgery is indicated. Reasons for poor or limited functional results in individual cases are indicated in the graph

By contrast, good function was achieved with:

- Early diagnosis
- Immediate therapeutic intervention
- Atraumatic surgical techniques

The awareness that vitreous and macular hemorrhages may result from a ruptured RAM and the fact that a total of 66 % of eyes had significant subretinal hemorrhage (half of them had complex bleedings with pre- and submacular components, the latter of which was often masked) motivated us to be less conservative in terms of early surgical intervention. Thus, to keep a balance between patient selection for surgery (those with vision-threatening alterations) and patient selection for observation (to avoid unnecessary surgery and complications) is a sensitive issue. Furthermore, multiple RAMs were not infrequent in our series, and they may threaten the macula again. Thus, we suggest continued monitoring of these eyes.

22.3.9 Conclusions and Practical Recommendations

The overall natural history of acquired RAMs is one of gradual spontaneous involution with good visual prognosis, with the majority of cases remaining asymptomatic. The usual benign course becomes vision-threatening when prolonged leakage with macular involvement or rupture with bleeding occurs, e.g., exudative maculopathy due to edema and lipid deposits or submacular hemorrhage. Visual prognosis largely depends on the degree and location of exudates or bleeding [41, 65].

The role of laser is not yet established, but clinical practice shows quick resolution of exudative complications due to chronic leakage from the aneurysm. In other words, laser shortens the active phase of the disease and reduces the risk of structural damage to the central retina. Retinal tissue recovers astonishingly well; at times, however, some visual loss is encountered in a significant number of cases.

The role of surgery is even more debatable. Attention has been paid mostly to subfoveal hemorrhagic complications. In view of the uncertain or poor prognosis and difficult treatment, most clinicians prefer a conservative approach [2, 38], whereas others recommend prompt surgical intervention for the reasons mentioned [6, 23, 70]. Doubtlessly, subretinal application of rt-PA is a big step toward an atraumatic technique for subretinal blood displacement or

Table 22.3.2

Clinical presentation	Technique	How	Why
Exudative RAM			
Macular involvement	Parafocal, touching the vessel wall	Low energy burns, long exposure time	Thermal-induced irritation of the vessel wall to accelerate involution of the leaking RAM
Progressive disease toward the macula	Ditto	Ditto	Ditto
Hemorrhagic RAM			
Preretinal bleeding: encapsulated by post. cortex or the ILM	Laser membranotomy/ hyaloidotomy	Nd:YAG photodisruption	To allow the blood to enter into the vitreous cavity for fast resorption
Preretinal bleeding and suspected subretinal component	Vitreous surgery	Vitrectomy, PVD, ILM peeling	For early diagnosis and treatment of subfoveal hemorrhage[a]
Preretinal bleeding and significant VR interface changes	Vitreous surgery	Vitrectomy, PVD, ILM peeling	To prevent additional retinal damage due to long-standing pucker
Subretinal bleeding	Early vitreous surgery	Vitrectomy, subretinal rt-PA ± blood removal, gas, positioning[a]	Blood removal or displacement to improve functional recovery
Vitreous bleeding			
Vitreous hemorrhage and known/suspected RAM	Early vitreous surgery	Vitrectomy + further measures according to the situation	For early diagnosis and treatment of subfoveal hemorrhage[a]
Vitreous hemorrhage of unknown origin	Early vitreous surgery	Ditto	Ditto Approx. 10 % are related to RAM [35]

[a] In our series, preretinal hemorrhage masked a significant subretinal blood component in one-third of eyes
[b] In case of predominantly liquefied blood > gas displacement may be effective. If a clotted hemorrhage is present > additional clot extraction may be more effective (not proven)

removal [24]. Hamayun et al. [23] reported no complications in a series of nine patients. We observed one eye with a macular hole (successfully repaired), one eye with late PVR development (final VA was 0.2 after subsequent silicone oil surgery) and one case with subretinal re-bleeding (the patient refused further treatment). Out of the 11 eyes treated with vitrectomy, subretinal rt-PA and gas, all but two (82%) improved and no eye lost vision.

Preretinal hemorrhages generally do not require treatment. Exceptions to this are cases of delayed visual rehabilitation. More importantly, however, preretinal blood can mask a significant subretinal blood component. Vitrectomy is a safe procedure in these cases. Surgery is hence preferable, in case of doubt.

22.3.10 Personal Approach to Symptomatic RAMs

There are still no clear recommendations "when and how" to treat RAMs. Having gone through all the therapeutic approaches mentioned, my current point of view can be summarized as follows. In general, treatment is restricted to active, symptomatic lesions only (Table 22.3.2).

In the particular situations listed, visual prognosis is either uncertain or poor and intervention justified. Risks of laser treatment, if carefully applied, appear irrelevant. Taking into account risks and complications due to surgery, the methods mentioned carry a reasonable risk/benefit ratio in the hands of an experienced surgeon.

Controlled randomized clinical trials to study the efficacy of various treatment options would be desirable. However, the relative benignancy and infrequent occurrence of complications due to RA will make it difficult to conduct such a trial. In the case of a patient with complex RAMs, we must keep in mind the patient's natural history, and the individual treatment will rely on logic-based considerations.

Acknowledgements. Surgery for RAMs was carried out by S. Bopp and K. Lucke (Augenklinik Universitätsallee).

References

1. Abdel-Khalek MN, Richardson J (1986) Retinal macroaneurysm: natural history and guidelines for treatment. Br J Ophthalmol 70(1):2–11
2. Berrocal MH, Lewis ML, Flynn HW Jr (1996) Variations in the clinical course of submacular hemorrhage. Am J Ophthalmol 122(4):486–93
3. Bleckmann H (1983) [Pulsating macroaneurysm of a retinal arterial branch] Pulsierendes Makroaneurysma einer retinalen Arterie. Klin Monatsbl Augenheilkd 182(1):91–3
4. Brown DM, et al. (1994) Retinal arteriolar macroaneurysms: long-term visual outcome. Br J Ophthalmol 78(7):534–8
5. Brown GC, Weinstock F (1985) Arterial macroaneurysm on the optic disk presenting as a mass lesion. Ann Ophthalmol 17(9):519–20
6. Buhl M, et al. (1999) Intravitreale rt-PA- und Gaseingabe bei submakularer Blutung [Intra-vitreal rt-PA and gas introduction in submacular hemorrhage]. Ophthalmologe 96(12):792–6
7. Chen YJ, Kou HK (2004) Krypton laser membranotomy in the treatment of dense premacular hemorrhage. Can J Ophthalmol 39(7):761–6
8. Ciardella AP, et al. (2003) Ruptured retinal arterial macroaneurysm associated with a stage IV macular hole. Am J Ophthalmol 135(6):907–9
9. Cleary PE, et al. (1975) Retinal macroaneurysms. Br J Ophthalmol 59(7):355–61
10. Dori D, et al. (1998) Nd:YAG laser treatment for premacular hemorrhage. Ophthalmic Surg Lasers 29(12):998–1000
11. Fichte C, Streeten BW, Friedman AH (1978) A histopathologic study of retinal arterial aneurysms. Am J Ophthalmol 85(4):509–18
12. Francois J (1979) Acquired macroaneurysms of the retinal arteries. Int Ophthalmol 1(3):153–61
13. Fritsche PL, Flipsen E, Polak BC (2000) Subretinal hemorrhage from retinal arterial macroaneurysm simulating malignancy. Arch Ophthalmol 118(12):1704–5
14. Giuffre G, Montalto FP, Amodei G (1987) Development of an isolated retinal macroaneurysm of the cilioretinal artery. Br J Ophthalmol 71(6):445–8
15. Godel V, Blumenthal M, Regenbogen L (1977) Arterial macroaneurysm of the retina. Ophthalmologica 175(3):125–9
16. Gomez-Ulla F, et al. (1998) Indocyanine green angiography in isolated primary retinal arterial macroaneurysms. Acta Ophthalmol Scand 76(6):671–4
17. Hannappel S, Gerke E (1989) Retinale Macroaneurysmen als Ursache fur Einblutungen in den Glaskorperraum [Retinal macroaneurysm as a cause of hemorrhage into the vitreous body]. Fortschr Ophthalmol 86(4):337–8
18. Hassan AS, et al. (1999) Management of submacular hemorrhage with intravitreous tissue plasminogen activator injection and pneumatic displacement. Ophthalmology 106(10):1900–6; discussion 1906–7
19. Haupert CL, et al. (2001) Pars plana vitrectomy, subretinal injection of tissue plasminogen activator, and fluid-gas exchange for displacement of thick submacular hemorrhage in age-related macular degeneration. Am J Ophthalmol 131(2):208–15
20. Hesse L, Schmidt J, Kroll P (1999) Management of acute submacular hemorrhage using recombinant tissue plasminogen activator and gas. Graefes Arch Clin Exp Ophthalmol 237(4):273–7
21. Hochman MA, Seery CM, Zarbin MA (1997) Pathophysiology and management of subretinal hemorrhage. Surv Ophthalmol 42(3):195–213
22. Hoh H, Khorsandian D, Ruprecht KW (1993) Frische submakulare Blutungen – Ergebnisse nach operativer Entfernung über Pars-plana-Zugang [Fresh submacular hemorrhage – results of surgical removal by pars plana approach]. Klin Monatsbl Augenheilkd 202(4):301–8
23. Humayun M, et al. (1998) Management of submacular hemorrhage associated with retinal arterial macroaneurysms. Am J Ophthalmol 126(3):358–61
24. Ibanez HE, et al. (1995) Surgical management of submacular hemorrhage. A series of 47 consecutive cases. Arch Ophthalmol 113(1):62–9

25. Iijima H, Satoh S, Tsukahara S (1998) Nd:YAG laser photo-disruption for preretinal hemorrhage due to retinal macroaneurysm. Retina 18(5):430–4

26. Joondeph BC, Joondeph HC, Blair NP (1989) Retinal macroaneurysms treated with the yellow dye laser. Retina 9(3):187–92

27. Kaiser HM, Pagani JM (2005) Isolated retinal venous macroaneurysm. Optometry 76(9):522–5

28. Khairallah M, Ladjimi A (2000) Dense premacular hemorrhage from a retinal macroaneurysm treated by argon laser. Retina 20(4):420–1

29. Khairallah M, et al. (1999) Retinal venous macroaneurysm associated with premacular hemorrhage. Ophthalmic Surg Lasers 30(3):226–8

30. Kowal L, Steiner H (1991) Arterial macroaneurysm of the optic disc. Aust N Z J Ophthalmol 19(1):75–7

31. Lavin MJ, et al. (1987) Retinal arterial macroaneurysms: a retrospective study of 40 patients. Br J Ophthalmol 71(11):817–25

32. Lewis RA, Norton EW, Gass JD (1976) Acquired arterial macroaneurysms of the retina. Br J Ophthalmol 60(1):21–30

33. Lim JI, et al. (1995) Submacular hemorrhage removal. Ophthalmology 102(9):1393–9

34. Lindgren G, Lindblom B (1996) Causes of vitreous hemorrhage. Curr Opin Ophthalmol 7(3):13–9

35. Lindgren G, Sjodell L, Lindblom B (1995) A prospective study of dense spontaneous vitreous hemorrhage. Am J Ophthalmol 119(4):458–65

36. Mitamura Y, Terashima H, Takeuchi S (2002) Macular hole formation following rupture of retinal arterial macroaneurysm. Retina 22(1):113–5

37. Moorthy RS (2000) Dense premacular hemorrhage from a retinal macroaneurysm treated by argon laser. Retina 20(1):96–8

38. Moosavi RA, Fong KC, Chopdar A (2005) Retinal artery macroaneurysms: clinical and fluorescein angiographic features in 34 patients. Eye 20(9):1011–20

39. Moriarty AP, McAllister IL, Constable IJ (1995) Initial clinical experience with tissue plasminogen activator (tPA) assisted removal of submacular haemorrhage. Eye 9(5):582–8

40. Ohji M, et al. (1998) Pneumatic displacement of subretinal hemorrhage without tissue plasminogen activator. Arch Ophthalmol 116(10):1326–32

41. Palestine AG, Robertson DM, Goldstein BG (1982) Macroaneurysms of the retinal arteries. Am J Ophthalmol 93(2):164–71

42. Panton RW, Goldberg MF, Farber MD (1990) Retinal arterial macroaneurysms: risk factors and natural history. Br J Ophthalmol 74(10):595–600

43. Park SW, Seo MS (2004) Subhyaloid hemorrhage treated with SF6 gas injection. Ophthalmic Surg Lasers Imaging 35(4):335–7

44. Perry HD, Zimerman LE, Benson WE (1977) Hemorrhage from isolated aneurysm of a retinal artery: report of two cases simulating malignant melanoma. Arch Ophthalmol 95(2):281–3

45. Peyman GA, et al. (1991) Tissue plasminogen activating factor assisted removal of subretinal hemorrhage. Ophthalmic Surg 22(10):575–82

46. Psinakis A, et al. (1989) Macroanevrysme arteriel retinien pulsatile: traitement par photocoagulation au laser argon [Pulsatile arterial macroaneurysm: management with argon laser photocoagulation]. J Fr Ophtalmol 12(10):673–6

47. Puthalath S, et al. (2003) Frequency-doubled Nd:YAG laser treatment for premacular hemorrhage. Ophthalmic Surg Lasers Imaging 34(4):284–90

48. Rabb MF, Gagliano DA, Teske MP (1988) Retinal arterial macroaneurysms. Surv Ophthalmol 33(2):73–96

49. Raymond LA (1995) Neodymium:YAG laser treatment for hemorrhages under the internal limiting membrane and posterior hyaloid face in the macula. Ophthalmology 102(3):406–11

50. Rennie CA, et al. (2001) Nd:YAG laser treatment for premacular subhyaloid haemorrhage. Eye 15(4):519–24

51. Ross RD, et al. (1996) Idiopathic polypoidal choroidal vasculopathy associated with retinal arterial macroaneurysm and hypertensive retinopathy. Retina 16(2):105–11

52. Russell SR, Folk JC (1987) Branch retinal artery occlusion after dye yellow photocoagulation of an arterial macroaneurysm. Am J Ophthalmol 104(2):186–7

53. Saika S, et al. (1998) Subretinal administration of tissue-type plasminogen activator to speed the drainage of subretinal hemorrhage. Graefes Arch Clin Exp Ophthalmol 236(3):196–201

54. Saito K, Iijima H (1997) [Visual prognosis and macular pathology in eyes with retinal macroaneurysms]. Nippon Ganka Gakkai Zasshi 101(2):148–51

55. Sanborn GE, Magargal LE (1984) Venous macroaneurysm associated with branch retinal vein obstruction. Ann Ophthalmol 16(5):464–8

56. Schatz H, Gitter K, Yannuzzi L, Irvine A (1990) Retinal arterial macroaneurysms: A large collaborative study. Presented at the American Academy of Ophthalmology, Annual Meeting, 1990, Chicago

57. Schneider U, Wagner AL, Kreissig I (1997) Indocyanine green videoangiography of hemorrhagic retinal arterial macroaneurysms. Ophthalmologica 211(2):115–8

58. Schulze SD, Hesse L (2002) Tissue plasminogen activator plus gas injection in patients with subretinal hemorrhage caused by age-related macular degeneration: predictive variables for visual outcome. Graefes Arch Clin Exp Ophthalmol 240(9):717–20

59. Shields CL, Shields JA (2001) Subretinal hemorrhage from a retinal arterial macroaneurysm simulating a choroidal melanoma. Ophthalmic Surg Lasers 32(1):86–7

60. Spalter HF (1982) Retinal macroaneurysms: a new masquerade syndrome. Trans Am Ophthalmol Soc 80:113–30

61. Tashimo A, et al. (2003) Macular hole formation following ruptured retinal arterial macroaneurysm. Am J Ophthalmol 135(4):487–92

62. Tashimo A, et al. (2003) Rhegmatogenous retinal detachment after rupture of retinal arterial macroaneurysm. Am J Ophthalmol 136(3):549–51

63. Tassignon MJ, Stempels N, Van Mulders L (1989) Retrohyaloid premacular hemorrhage treated by Q-switched Nd-YAG laser. A case report. Graefes Arch Clin Exp Ophthalmol 227(5):440–2

64. Tezel T, Gunalp I, Tezel G (1994) Morphometrical analysis of retinal arterial macroaneurysms. Doc Ophthalmol 88(2):113–25

65. Tonotsuka T, et al. (2003) Visual prognosis for symptomatic retinal arterial macroaneurysm. Jpn J Ophthalmol 47(5):498–502

66. Townsend-Pico WA, Meyers SM, Lewis H (2000) Indocyani-

ne green angiography in the diagnosis of retinal arterial macroaneurysms associated with submacular and preretinal hemorrhages: a case series. Am J Ophthalmol 129(1): 33–7

67. Ulbig MW, et al. (1998) Long-term results after drainage of premacular subhyaloid hemorrhage into the vitreous with a pulsed Nd:YAG laser. Arch Ophthalmol 116(11):1465–9

68. Verbraeken H, Van Egmond J (1999) Non-diabetic and non-oculotraumatic vitreous haemorrhage treated by pars plana vitrectomy. Bull Soc Belge Ophtalmol 272:83–9

69. Verougstraete C, Leroy CP (1991) [Arterial macro-aneurysm] Le macroanevrysme arteriel. Bull Soc Belge Ophtalmol 240:65–77

70. Wu TT, Sheu SJ (2005) Intravitreal tissue plasminogen activator and pneumatic displacement of submacular hemorrhage secondary to retinal artery macroaneurysm. J Ocul Pharmacol Ther 21(1):62–7

71. Zhao P, et al. (2000) Vitrectomy for macular hemorrhage associated with retinal arterial macroaneurysm. Ophthalmology 107(3):613–7

22.4 Coats' Disease

A. SCHUELER, N. BORNFELD

Core Messages
- Coats' disease presents with teleangiectasia and aneurysm-like abnormalities of the retinal vessels
- Massive subretinal lipid exudates develop in later stages of the disease
- Mostly unilateral, in healthy young boys
- Differential diagnosis is to be made toward retinoblastoma
- Treatment aims at control of exudation by the pathological retinal vessels

22.4.1 Introduction

Coats' disease is a rare disorder of the retinal vessels. Other terms are Coats' retinitis, Coats' syndrome, exudative retinitis, Leber's miliary aneurysms and retinal telangiectasis. It is a developmental retinal vascular disorder presenting with telangiectasia and aneurysm-like abnormalities of retinal vessels that may show significant leakage in fluorescein angiography. In more advanced stages of the disease, ophthalmoscopy reveals massive subretinal lipid exudates around the typical retinal vessel anomalies. Based on these primary pathologies, initially two different entities were described: a more severe form by Coats and a milder form by Leber [4, 11]. Today these two clinical presentations are thought to present variable expressions of the same disease and they are now subsumed under the term Coats' disease. Nevertheless, the cause of the disease is still unknown. Males are more often affected, but no genetic, racial or ethnic predispositions have been found. Furthermore some sporadic case reports of Coats' disease in patients with chromosomal deletions or in combination with other clinical syndromes have been reported [2, 3, 5, 16]. Retinal vessel anomalies, similar to those observed in Coats' disease, have been described in 3.6% of patients with retinitis pigmentosa [7].

Characteristically, Coats' disease is unilateral in about 90% of the cases and occurs in otherwise healthy, young boys [18]. The average age at the onset of symptoms is about 8 years, but it ranges between 4 months and the 7th decade of life. The majority of cases are diagnosed before the age of 20 years [6]. More than two-thirds of patients are males [18].

Presenting symptoms at the time of diagnosis are leukocoria, strabismus and reduced visual acuity. In more advanced stages of the disease, heterochromia and secondary glaucoma may be present.

In early stages, Coats' disease is normally detected during a routine examination. Localized areas with retinal telangiectasis, typically located in the periphery of the retina, are found (Fig. 22.4.1). Without treatment, the disease usually shows progression. Chronic leakage from the atypical vessels leads to pronounced retinal edema and lipid exudates, which dominates the clinical picture and may obscure the underlying pathology. Further progression of the disease shows the development of aneurysm-like abnormalities of retinal vessels and sometimes large areas of capillary non-perfusion and retinal hemorrhage. Ultimately a partial or total exudative retinal detachment occurs (Figs. 22.4.1, 22.4.3). Further late symptoms are rubeosis iridis and neovascular glaucoma, which leads to painful phthisis of the eye. Other complications during the course of the disease are vitreous hemorrhage, cataract, uveitis, neovascularization of the disk and elsewhere. Some rare cases develop a proliferative vitreoretinopathy or a rip of the retinal pigment epithelium [10, 19]. Visual deterioration is the result of macular edema, subretinal lipid exudates and the exudative retinal detachment. Subfoveal neovascularization may occur as a complication of the fibrous organization of long-standing subfoveal exudates.

An increased permeability of vascular endothelial cells in the area of vessel malformation is thought to be the principal pathomechanism in Coats' disease [12]. The clinical symptoms can be explained by an endothelial cell dysfunction. Telangiectasis and

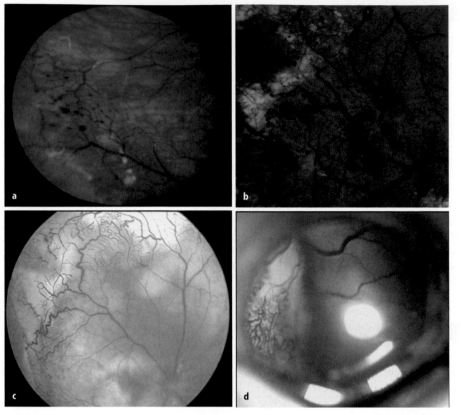

Fig. 22.4.1. Stages of Coats' disease: **a** Early stage with typical telangiectasis of retinal vessels without exudates. **b** Early stage with localized telangiectasis, aneurysm-like abnormalities and lipid exudates. **c** Advanced stage with diffuse retinal vessel anomalies in the periphery of the retina, massive subretinal lipid exudates and with exudative retinal detachment. **d** Late stage of Coats' disease with solid subretinal lipid exudates and total retinal detachment

development of microaneurysms and finally capillary occlusions may be the result of endothelial cell death. Breakdown of the blood-retina barrier with extravasion of blood components explains the intraretinal and subretinal fluid, protein and lipid deposits and development of exudative retinal detachment. Histopathological examination of enucleated eyes shows thickening and hyalinization of the retinal vessel walls. The subretinal space contains exudates, cholesterol and cholesterol-laden macrophages [21].

22.4.2 Classification

According to the variety of clinical pictures, various classifications have been suggested. Coats' disease is divided between type I disease with exudates without recognizable vascular abnormalities, type II disease with exudates and retinal telangiectasia and type III with local exudates around a solitary retinal angioma [4]. Coats' type III disease is consistent with the description of von Hippel's angiomatosis retinae. In type I disease fluorescein angiography also shows abnormal retinal vessels in the area of the exudates. Based on the proposal of Gomez-Morales, Shields published a more detailed clinical classification [6, 19]. Based on the clinical outcomes after

treatment of 124 eyes with Coats' disease, he proposed a classification that implies treatment selection and prognosis of the eye. Stage 1 shows telangiectasia only (Fig. 22.4.1a). Stage 2 presents with telangiectasia with extrafoveal (stage IIA) or foveal exudation (stage 2B). In stage 3 additional exudative retinal detachment occurs. Stage 3 is divided into localized retinal detachment (stage 3A) and total retinal detachment (stage 3B). Stage 4 is characterized by total detachment and secondary glaucoma. Stage 5 is the end-stage disease with phthisis and blindness.

With respect to functional prognosis, a poor visual outcome (20/200 or worse) was found in none of the eyes with stage 1. However, in stage 2 and 3 significant visual impairment was observed in 53% and 74%, respectively. Finally all eyes with stage 4 and 5 Coats' disease at presentation had a poor visual outcome.

This actual classification may be applicable for juvenile forms of Coats' disease. The clinical course in adults is normally less severe and shows a slower progression. Therefore an additional differentiation between juvenile (onset before 30 years) and adult (after age of 30 years) forms of the disease might be necessary.

22.4.3 Differential Diagnosis

The differential diagnoses of Coats' disease are listed in Table 22.4.1. In children, retinoblastoma is the most important differential diagnosis. Subretinal masses in Coats' disease normally appear more yellowish, in retinoblastoma more white, but this criterion is not reliable (Fig. 22.4.2). Subretinal masses with secondary calcifications are more common in retinoblastoma. Retinal telangiectasis over the solid masses is more typical for Coats' disease. A clear differentiation between these two entities is essential before the start of treatment. Vitreoretinal interventions are sometimes recommended in Coats' disease. In retinoblastoma, however, intraocular surgery may end up with a local spreading of tumor cells into the orbit, worsening the overall prognosis. Diagnostic tools should be performed in unclear cases to allow correct diagnosis. Additional ultrasound examination and fluorescein angiography (Fig. 22.4.3) might be helpful to detect typical vessel abnormalities. Computer tomography allows the detection of calcifications and magnetic resonance imaging allows a better staging of the tumor extension in the eye [15].

An intraocular medulloepithelioma is a rare disease. It is a unilateral solitary tumor of the non-pigmented epithelium of the ciliary body. It always presents in the first decade of life. Decreased vision, leukocoria, glaucoma, rubeosis iridis and hyphema are the most frequent findings at presentation [17]. A pathognomonic finding in medulloepithelioma is a cystic tumor, located in the ciliary body.

A subretinal mass in children may be caused by a choroidal hemangioma, sometimes associated with the Sturge-Weber syndrome. Clinical features of

Coats' disease may be mimicked by leukemic infiltrations of the choroid and the retina and by inflammatory diseases. Space occupying lesions like astrocy-

Table 22.4.1. Differential diagnosis of Coats' disease

Tumors	Retinoblastoma
	Leukemic infiltrates
	Medulloepithelioma
	Astrozytoma
	Choroidal hemangioma
	Pigment epithelium/retinal hamartoma
Hereditary conditions	Norrie disease
	Incontinentia pigmenti
	Familial exudative vitreoretinopathy, juvenile X-linked retinoschisis
	Retinal angiomatosis (von Hippel's)
Developmental anomalies	Persistent hyperplastic primary vitreous
	Congenital cataract
	Coloboma
	Myelinated nerve fibers
	Congenital retinal folds
	Morning glory syndrome
	Cavernous retinal hemangioma
Inflammatory conditions	Toxocariasis
	Toxoplasmosis
	Endophthalmitis
	Viral retinitis
	Uveitis
Others	Retinopathy of prematurity
	Rhegmatogenous retinal detachment
	Vitreous hemorrhage
	Eales disease
	Diabetic retinopathy
	Radiation retinopathy
	Retinal vasculitis

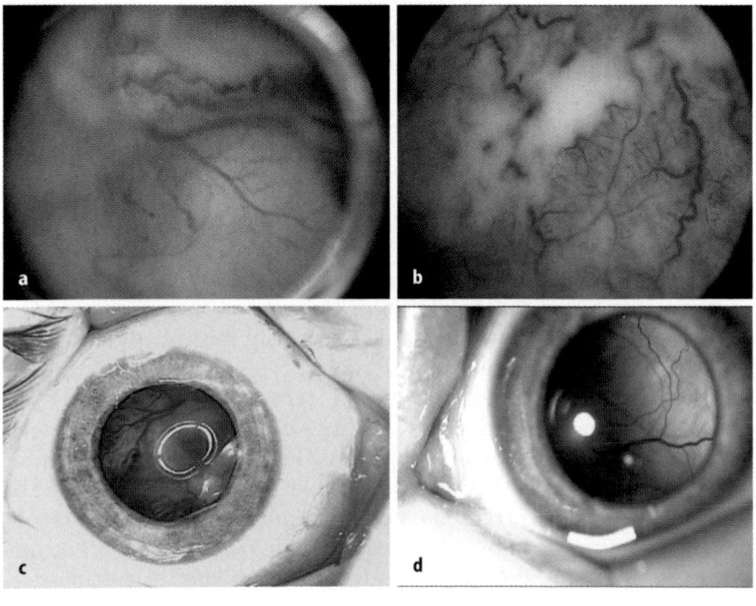

Fig. 22.4.2. In children, retinoblastoma is the most important differential diagnosis. Advanced Coats' disease with intra- and subretinal exudates (**a**) might be confused with a diffuse infiltrating retinoblastoma (**b**). Typical vessel anomalies and more yellow subretinal exudates are more common in Coats' disease (**c**). Differentiation to solid subretinal masses in retinoblastoma (**d**) might be difficult

22 III

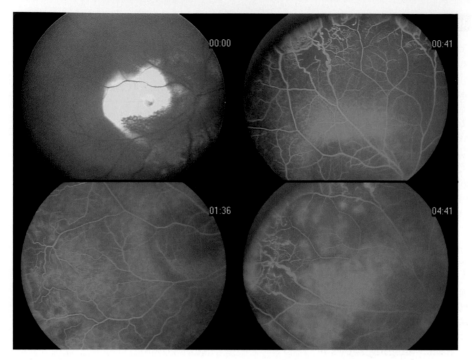

Fig. 22.4.3. Wide field fluorescein angiography (120°) of Coats' disease. Massive subretinal lipid exudates at the posterior pole and telangiectasis in the periphery of the retina. Retinal edema and lipid exudates can obscure the typical vessel anomaly in the periphery. Fluorescein angiography shows the extent and location of vessel abnormalities and capillary occlusions

tomic hamartoma may be misdiagnosed as Coats' disease. Astrocytomas are a common finding in tuberouse sclerosis and less frequent in neurofibromatosis. A careful search will reveal other stigmata of these diseases. Hereditary conditions that may simulate Coats' disease are incontinentia pigmenti and Norrie disease. Gender gives a hint in these two diseases: Norrie disease occurs only in boys and incontinentia pigmenti only in girls.

Despite some clinical similarities, specific findings or the medical history allow differential diagnosis. Retinopathy of prematurity has a typical medical history and is often bilateral. This is also valid for, e.g., familial exudative vitreoretinopathy and uveitis. Persistent hyperplastic primary vitreous is associated with microphthalmos in most cases. Blood analysis showing high levels of serum antibodies against specific antigens may be helpful for the diagnosis of uveitis and infections. In adults Coats' disease may be misdiagnosed as central retinal vein occlusion or branch retinal vein occlusion.

Using standard fluorescein angiography is of little diagnostic value in young children. Imaging of the peripherally located retinal changes can only rarely be achieved in children. Introduction of wide-field fluorescein angiography systems in recent years allows the viewing of peripheral findings under general anesthesia in very young children (Fig. 22.4.3). Typical findings in Coats' disease are relatively slow filling of the peripheral capillary changes in the early phase, sometimes combined with an early hypofluorescence due to the subretinal lipid exudates

(Fig. 22.4.3). Capillary occlusions and increased filling of aneurysms and capillary ectasia can be seen in the arteriovenous phase. An increasing focal or diffuse exudation can be seen in the late arteriovenous and in the late phase. Identification of retinal ischemia and areas of pathologically altered retinal vessel anomalies are important indicators toward considering treatment of the eye.

22.4.4 Treatment

The goal of treatment is the control of exudation by the pathological retinal vessels and to prevent further progression to visually threatening complications. Additional treatment of retinal ischemia is essential to prevent secondary complications. Effective therapy in stage 1 or 2 disease is destruction of the affected retinal vessels and ischemic retina by indirect laser photocoagulation or cryotherapy [13, 14]. Regression of the lipid deposits may take weeks to months, even after successful coagulation therapy. Repeated treatment may be successful even in more advanced stage 3 disease (Fig. 22.4.4). After initial coagulation therapy a partial regression of exudative changes can be achieved and may allow further coagulation therapy of the remaining retinal vessel anomalies several weeks later (Fig. 22.4.4). Finally a complete destruction of the abnormal vessels and a confluent coagulation of the ischemic retina should be achieved by thermal coagulation. It might be necessary to perform direct coagulation of the vessels or to use cryotherapy in triple freeze-thaw technique to control advanced

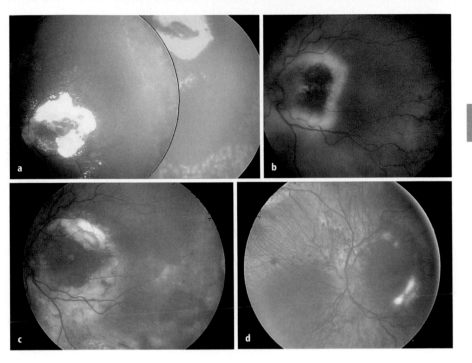

Fig. 22.4.4. a Advanced Coats' disease before and directly after laser coagulation of the retinal vessel abnormalities in the periphery. **b** Finding before a stepwise confluent laser coagulation of the affected retinal vessels. **c** Reattached retina and regression of the subretinal exudates 10 months after the first laser therapy. **d** Further regression and improved visual acuity 15 months after additional laser therapy

stages. External surgical drainage of the subretinal fluid and the exudates, followed by photo- or cryocoagulation, may be an option in cases showing retinal vessel abnormalities at the posterior pole or in cases with a bullous exudative retinal detachment [1]. Vitreoretinal procedures using maneuvers to reattach the retina by internal drainage of the subretinal fluid and lipid deposits, endocoagulation of the affected vessels and gas or silicone oil tamponade have been recommended [8, 9]. Cases with particular paramacular involvement showed an improvement of the visual acuity after vitrectomy and enucleation could be avoided in eight of nine eyes [8]. Nevertheless, the rank of vitrectomy in the therapy of Coats' disease is still controversial. In children, exclusion of retinoblastoma before surgery, attached vitreous as a surgical problem and the high rate of proliferative vitreoretinopathy remain a challenge and should be considered before a vitrectomy is taken into account.

22.4.5 Future Prospects

In future, anti-angiogenic drugs might play a role in the therapy of Coats' disease. At the moment it is not clear whether or not any of these drugs, currently used for the therapy of age related macular degeneration, would help patients with Coats' disease. The anti-angiogenic effect of these drugs might be a general problem for the use in children. The inclusion of children in controlled prospective studies on the efficacy of anti-angiogenic drugs in Coats' disease might be problematic.

Regular control examinations after successful therapy of Coats' disease are mandatory. New vessel abnormalities in prior unaffected retinal areas may develop and early treatment of these abnormalities is necessary to prevent further late complications [13, 20].

References

1. Adam R, Kertes P, Lam WC (2007) Observations on the management of coats' disease: less is more. Br J Ophthalmol 91:303–306
2. Beby F, Roche O, Burillon C, Denis P (2005) Coats' disease and bilateral cataract in a child with Turner syndrome: a case report. Graefes Arch Clin Exp Ophthalmol 243:1291–1293
3. Black GC, Perveen R, Bonshek R, Cahill M, Clayton-Smith J, Lloyd IC, McLeod D (1999) Coats' disease of the retina (unilateral retinal telangiectasis) caused by somatic mutation in the NDP gene: a role for norrin in retinal angiogenesis. Hum Mol Genet 8:2031–2035
4. Coats G (1908) Forms of retinal disease with massive exudation. R Lond Ophthalmic Hosp Rep 17:440
5. Genkova P, Toncheva D, Tzoneva M, Konstantinov I (1986) Deletion of 13q12.1 in a child with Coats disease. Acta Paediatr Hung 27:141–143
6. Gomez Morales A (1965) Coats' disease. Natural history and results of treatment. Am J Ophthalmol 60:855–865
7. Khan JA, Ide CH, Strickland MP (1988) Coats'-type retinitis pigmentosa. Surv Ophthalmol 32:317–332
8. Krause L, Kreusel KM, Jandeck C, Kellner U, Foerster MH (2001) Vitrektomie bei fortgeschrittenem Morbus Coats. [Vitrectomy in advanced Coats disease.] Ophthalmologe 98:387–390
9. Kreusel KM, Krause L, Broskamp G, Jandeck C, Foerster MH (2001) Pars plana vitrectomy and endocryocoagulation for paracentral Coats' disease. Retina 21:270–271

22 **III**

10. Kubota T, Kurihara K, Ishibashi T, Inomata H (1995) Proliferative vitreoretinopathy in Coats' disease. Clinicohistopathological case report. Ophthalmologica 209:44–46
11. Leber T (1912) Ueber ein durch Vorkommen multipler Miliaraneurismen characterisierte Form vor Retinlaer degeneration. Graefes Arch Clin Exp Ophthalmol 81:1–14
12. McGettrick PM, Loeffler KU (1987) Bilateral Coats' disease in an infant (a clinical, angiographic, light and electron microscopic study). Eye 1:136–145
13. Nucci P, Bandello F, Serafino M, Wilson ME (2002) Selective photocoagulation in Coats' disease: ten-year follow-up. Eur J Ophthalmol 12:501–505
14. Pauleikhoff D, Wessing A (1989) Langzeitergebnisse der Therapie bei Morbus Coats. [Long-term results of the treatment of Coats disease.] Fortschr Ophthalmol 86:451–455
15. Schueler AO, Hosten N, Bechrakis NE, Lemke AJ, Foerster P, Felix R, Foerster MH, Bornfeld N (2003) High resolution magnetic resonance imaging of retinoblastoma. Br J Ophthalmol 87:330–335
16. Schuman JS, Lieberman KV, Friedman AH, Berger M, Schoeneman MJ (1985) Senior-Loken syndrome (familial renal-retinal dystrophy) and Coats' disease. Am J Ophthalmol 100:822–827
17. Shields JA, Eagle RC Jr, Shields CL, Potter PD (1996) Congenital neoplasms of the nonpigmented ciliary epithelium (medulloepithelioma). Ophthalmology 103:1998–2006
18. Shields JA, Shields CL, Honavar SG, Demirci H (2001a) Clinical variations and complications of Coats disease in 150 cases: the 2000 Sanford Gifford Memorial Lecture. Am J Ophthalmol 131:561–571
19. Shields JA, Shields CL, Honavar SG, Demirci H, Cater J (2001b) Classification and management of Coats disease: the 2000 Proctor Lecture. Am J Ophthalmol 131:572–583
20. Shienbaum G, Tasman WS (2006) Coats disease: a lifetime disease. Retina 26:422–424
21. Tripathi R, Ashton N (1971) Electron microscopical study of Coat's disease. Br J Ophthalmol 55:289–301

22.5 Familial Exudative Vitreoretinopathy

A.M. JOUSSEN, B. KIRCHHOF

Core Messages

- Familial exudative vitreoretinopathy FEVR) is an inherited, bilateral peripheral vascular disease in children with no association to prematurity
- Unlike inherited diseases, expression of FEVR may be asymmetrical between the two eyes
- The clinical course is slowly progressive, and rarely stable. Its presentation may be confined to a peripheral avascular zone and a reduced angle kappa or it may lead to a falciform or a rhegmatogenous retinal detachment
- Retinal exudates and a peripheral fibrovascular mass result from the leakiness of peripheral vessels forming aneurysms, tubular dilatation, and neovascularization, indicating a progressive disease. Such abnormalities must be destroyed by laser or cryotherapy to stop exudation. Usually repetitive applications of retinopexy are necessary
- The main goal of treatment is the occlusion of abnormal retinal vessels to avoid complications such as tractional retinal detachment and regrowth of vitreous membranes
- The involvement of Wnt signaling in the pathogenesis of FEVR is currently under discussion. Frizzled-4 (Fz4), a presumptive Wnt receptor, and Norrin, the protein product of Norrie's disease gene, function as a ligand-receptor pair

22.5.1 History

Familial exudative vitreoretinopathy (FEVR) was first described by Criswick and Schepens in 1969 in six children as a peripheral vitreoretinopathy similar to retinopathy of prematurity (ROP) yet with familial occurrence [11].

Three stages of the disease were described by Gow and Oliver in 1971, who pointed out the inheritance and the vascular pathogenesis [17]. Later Canny and Oliver confirmed the vascular origin and described a peripheral avascular zone as demonstrated by fluorescein angiogram as a pathognomonic sign of the early stages of the disease [6]. Similarly, Ober and Bird emphasized the role of fluorescein angiography in identifying subclinical cases [27]. The gene was first identified by Li in 1992 [20, 21].

To date, a total of about 340 cases in more than 60 families have been reported in the literature.

22.5.2 Special Pathological Features

Essentials

- Failure of the (temporal) peripheral retina to vascularize
- Abnormal existing vessels and neovascularization
- Subretinal deposits as a result of increased vascular leakage
- Greaseproof-paper-like vitreous membranes

FEVR is a disorder of the peripheral retinal vessels often associated with vitreous traction. Systemic associations are absent and no association with prematurity is found.

Despite earlier theories that emphasized vitreoretinal changes, it is now clear that the fundamental abnormality in FEVR is the leakiness of the abnormal peripheral retinal vessels [6, 12, 19].

Preexisting vessels dilate and form bulb like and tubular dilatations at their peripheral ends, which later become leaky. Large peripheral avascular zones correlate with the presence of neovascularization at

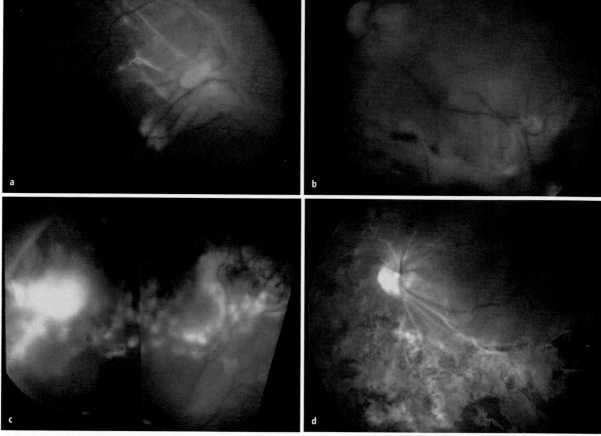

Fig. 22.5.1. a Peripheral vascular abnormalities with subretinal exudates indicating a requirement for photocoagulation. Radial vitreous strands are visible inserting into the retina without traction. **b** Persistent subretinal exudates and neovascularization. Scars are visible in the periphery. Supplemental photocoagulation is needed. **c** Regression of subretinal exudates after peripheral laser photocoagulation and cryopexy. **d** Extensive peripheral photocoagulation after second vitrectomy and recurrent re-proliferation of vitreous strands with persistent subretinal exudates (case shown in detail in Fig. 22.5.3)

its border and the development of peripheral fibrovascular masses [24].

Subretinal exudates signal a risk of disease progression. The abnormal peripheral vessels are leaky and considerable amounts of fluid and protein extravasate from these vessels, resulting in deposition of subretinal lipid and exudates. It is likely that the majority of leakage derives from the abnormally dilated and straightened second order vessels in the periphery and not from neovascularization. Figure 22.5.1a demonstrates that subretinal deposits can be found in the presence of abnormal vessels without formation of neovascularization.

Thus, peripheral ischemia, neovascularization and the abnormal vasculature combine to produce the pathology in FEVR.

Demonstration of hematoserological defects and abnormal platelet aggregation in affected members of two families [8] has not been confirmed by subsequent studies [4, 40].

Histopathology reveals similarities to other peripheral vasculopathies such as Coats' disease. Onion-skin-like vitreous membranes are typical of FEVR. Histopathological characteristics of FEVR comprise the following (Figs. 22.5.2, 22.5.3f):

- Thickened retina containing dilated, teleangiectatic blood vessels
- Vessel walls are thickened and may demonstrate a perivascular infiltrate [3, 12]
- Intraretinal [3] and subretinal [26] inflammation may be present
- Cellular and acellular vitreous membranes sprout from the retina [5]. They must be cut from the retina and cannot be peeled off the retina
- There is no retinal dysplasia

The failure of vascularization of the peripheral retina is the unifying feature seen in all affected individuals. A small avascular crescent usually remains asymptomatic. The visual problems in FEVR result

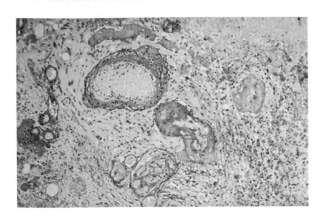

Fig. 22.5.2. Histological appearance of the affected peripheral retina. The retinal structure is lost. There are multiple vessels with thickened vessel walls which are infiltrated and surrounded by inflammatory cells and lymphocytes

from retinal round holes and from complications due to the development of hyperpermeable blood vessels, neovascularization, and vitreoretinal traction. Such appearances threaten the central vision in 20 % of cases, and lead to partial or total retinal detachment [40].

Fibrous proliferation may be the result of chronic peripheral vascular leakage [13]. In contrast to this theory, a „regrowth" of onion-skin-like vitreous membranes can be observed even after full treatment and regression of peripheral neovascularization. Macular traction or retinal detachment occurs with contraction of mesenchymal elements at the

Fig. 22.5.3. a Clinical course of a 10-year-old girl. **b** After lensectomy, fibrosed vitreous strands become visible, which are carefully dissected using the vitrectome. Care is taken not to pull the periphery as the strands insert into the retinal tissue and traction might cause retinal hole formation. **c** After further resection of the vitreous strands, abnormal peripheral vessels with subretinal exudates become visible. **d₁** View of the fundus

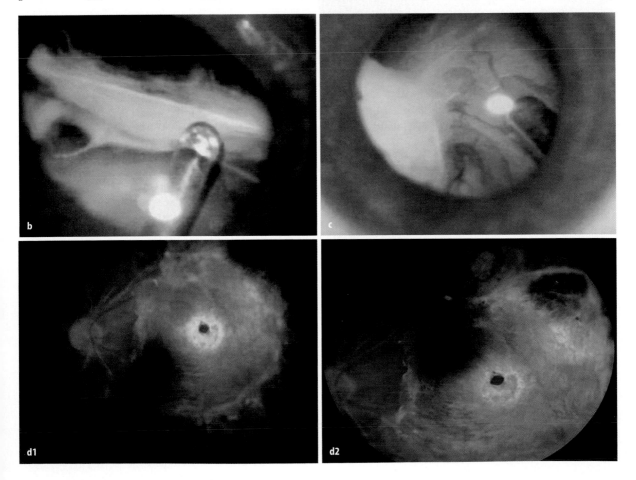

22 III

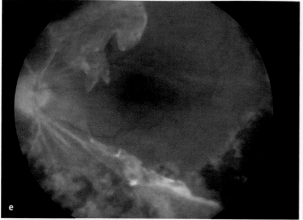

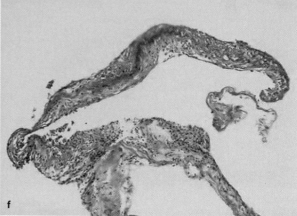

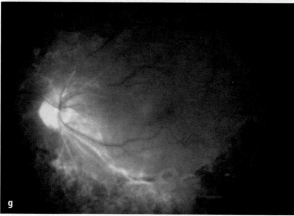

Fig. 22.5.3 (*Contin.*) after resection of the vitreous strands. The retinal surface is covered by a tightly adherent pucker forming a pseudomacular hole. **d₂** A closer view demonstrates insertion of the vitreous strands near the vessel arcades. Several thin membranes are piled up forming onion-skin-like structures. **e** After preparation of the membranes covering the macula, peripheral remnants have to be dissected but cannot be peeled off because of their insertion into the retinal tissue. Old laser scars are visible. **f** Histology section of the dissected membranes demonstrates fibrosed tissue intertwined with inflammatory cells. H&E. **g** Fundus view after extensive membrane peeling. The status remained stable without tamponade for several years

avascular border or of the fibrovascular mass that may occur just anterior to it. Mostly vitreoretinal traction is located in the temporal periphery, the area of the most apparent ischemia.

22.5.3 Genetics and Molecular Mechanisms

Essentials
- Autosomal dominant or recessive trait
- Complete penetrance
- Variable phenotypic expression
- Chr 11q13, principal locus EVR1
- Wnt signaling is demonstrated to be involved

FEVR is most commonly inherited as an autosomal dominant trait [18, 31]. However, a few families with an X-linked recessive form of FEVR [9, 10, 32] as well as one family with an autosomal recessive pattern of inheritance [33] have been reported. Isolated cases of sporadic or idiopathic cases of FEVR have been noted by Miyakubo [24].

The disease exhibits complete penetrance (almost 100%) with a highly variable phenotypic expression [27]. Many individuals demonstrate an asymptomatic expression of the disease, then only apparent on fluorescein angiography.

Figure 22.5.4 demonstrates two families from Syria of known consanguinity [1]. Different stages of the disease were seen in all six children of the families. Traction and distortion of the macula or optic disk was seen in 11 out of 12 eyes. The father demonstrated asymptomatic disease with temporal peripheral degenerations and atypical vessels.

The gene for autosomal dominant FEVR was mapped on chromosome 11q13-q23 with its principal locus (EVR1) in four northern European families [20, 25] and an Asian family [29]. To date, two autosomal dominant loci have been mapped. EVR1 on chromosome 11q was the first FEVR locus to be identified [20, 21] and verified in further families. The gene encoding Wnt receptor frizzled-4, FZD4 (MIM604579) was recently reported as the EVR1 gene [30]. The discovery that FEVR can be caused by mutations in the Wnt receptor FZD4 highlighted proteins involved in the Wnt-signaling pathway as candidate FEVR genes. Toomes et al. describe mutations in a second gene at the EVR1 locus, low-density-lipoprotein receptor-related protein 5 (LRP5), a Wnt coreceptor, which is one such candidate gene [36].

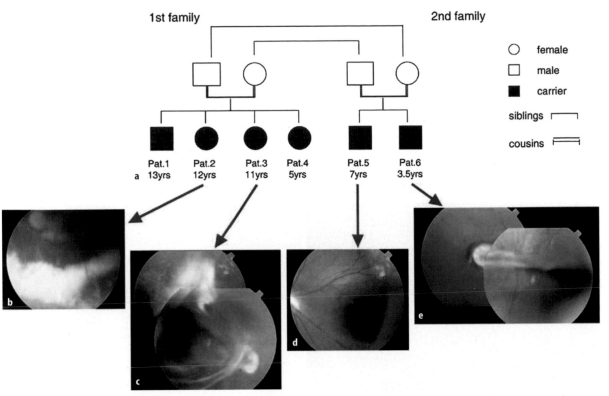

Fig. 22.5.4. a Family tree. **b** Patient 2 (female, 12 years). Subretinal exudates with abnormal vessels in the inferior hemisphere. VA OD: 20/600. **c** Patient 3 (female, 11 years). Temporal retinal folds, fibrovascular mass and subretinal exudates. VA OD: 20/400. **d** Patient 5 (male, 7 years). Temporally dragged vessels. VA OS: 20/60. **e** Patient 6 (male, 3.5 years). Temporally dragged disk, vitreo-retinal adhesion and retinal fold. VA OS: counting fingers

Xu et al. recently reported the vascular development in the retina and inner ear to be controlled by Norrin and Frizzled-4, a high-affinity ligand-receptor pair [41]. As described above, one form of FEVR is caused by defects in Frizzled-4 (Fz4), a presumptive Wnt receptor. Norrin, the protein product of Norrie's disease gene, is a secreted protein of unknown biochemical function. In a mouse model the authors were able to prove that Norrin and Fz4 function as a ligand-receptor pair based on: (1) the similarity in vascular phenotypes caused by Norrin and Fz4 mutations in humans and mice, (2) the specificity and high affinity of Norrin-Fz4 binding, (3) the high efficiency with which Norrin induces Fz4- and Lrp-dependent activation of the classical Wnt pathway, and (4) the signaling defects displayed by disease-associated variants of Norrin and Fz4. These data define a Norrin-Fz4 signaling system that plays a central role in vascular development in the eye and indicate that ligands unrelated to Wnts can act through Fz receptors.

At present, the cell types that express Norrin and Fz4 are not well defined in the retina. It is interesting, however, that linkage and candidate gene analysis have shown X-linked FEVR to be allelic to Norrie's disease [9, 15, 32]: A number of mutations in Norrie's

disease gene have been found in families with X-linked FEVR [15, 32].

Further effort is required to more closely define the cellular and subcellular localization of the proteins involved in the disease process leading to specific phenotypes, and more effort is required to develop new therapeutic targets.

22.5.4 Clinical Course of the Disease

Essentials
- Peripheral avascular zone (mostly temporal)
- Temporal dragging of disk and vessels (reduced angel kappa)
- Subretinal exudates indicate progressive disease
- Falciform tractional retinal detachment

The disorder is bilateral but often asymmetric. The majority of gene carriers suffer no visual impairment. Early manifestation (childhood) heralds progression. In the first years of life exotropia may be one of the first symptoms.

Symptoms of decreased visual acuity, often attributable to macular fold or retinal detachment, may occur in infancy and through adolescence.

Clinical severity is highly variable. Alsheikheh and coworkers recently presented two consanguine families with seven members affected at different stages (Fig. 22.5.4) [1]. While patients 5 and 6 show subclinical disease, patients 2 and 3 present with retinal exudates indicating progression of the disease.

The clinical signs always include abrupt termination of the temporal retinal vasculature with scall-oped borders (Fig. 22.5.5a, c). Fluorescein angiography confirms formation of a vascular-avascular border of variable distance from the ora serrata (Fig. 22.5.5b). The vessels just posterior to this zone may be tubularly dilated and straightened. At the avascular border, vessels end in arteriovenous anastomoses, which may leak fluorescein and from which neovascularization may develop. Vitreous hemorrhage is rare and signals the end stage of the natural course of the disease by retinal detachment and phthisis.

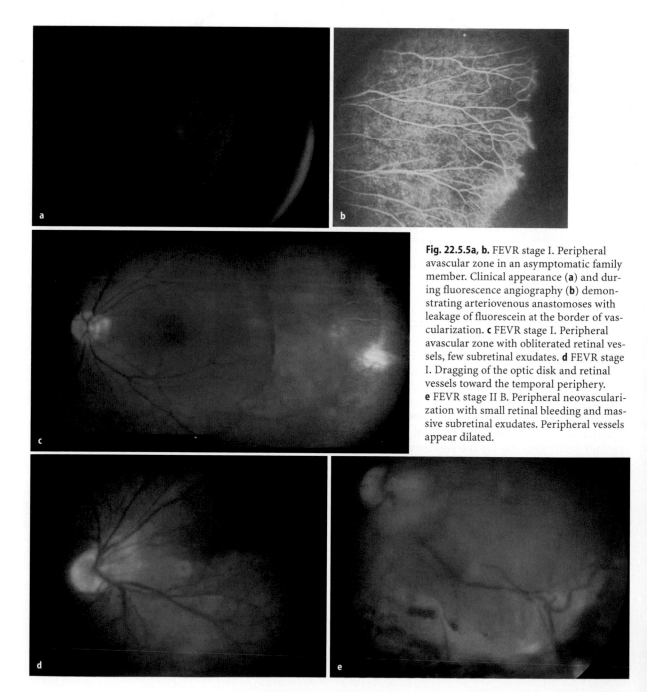

Fig. 22.5.5a, b. FEVR stage I. Peripheral avascular zone in an asymptomatic family member. Clinical appearance (**a**) and during fluorescence angiography (**b**) demonstrating arteriovenous anastomoses with leakage of fluorescein at the border of vascularization. **c** FEVR stage I. Peripheral avascular zone with obliterated retinal vessels, few subretinal exudates. **d** FEVR stage I. Dragging of the optic disk and retinal vessels toward the temporal periphery. **e** FEVR stage II B. Peripheral neovascularization with small retinal bleeding and massive subretinal exudates. Peripheral vessels appear dilated.

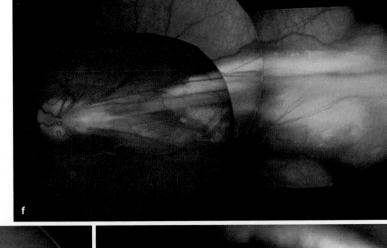

Fig. 22.5.5. f Falciform retinal detachment with macular involvement. Toward the periphery, massive subretinal exudates. **g** FEVR stage IV B. Vitreous veils obscuring fundus view. Retina attached. **h** FEVR stage V. Total retinal detachment, lens with pigment adhesion

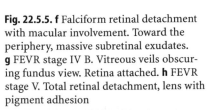

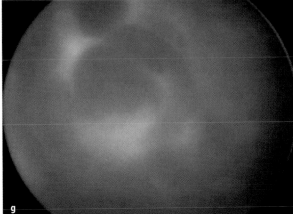

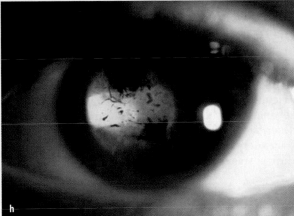

Just anterior to the avascular border, a fibrovascular mass (cyclic membrane) may develop with prominent, large-caliber arterial and venous feeders associated with marked retinal exudates. Exudates have been found in 9–22% of cases [2]. This mass may encompass the ciliary body and peripheral lens capsule.

Then, dragging of the macula, of the disk, and of retinal vessels can be seen in up to 50% of gene carriers. Tractional forces can result in retinal folds extending from the temporal quadrant through the macula or in tractional retinal detachment. Rhegmatogenous retinal detachment can be masked by epiretinal membranes covering up the retinal round hole.

Other clinical signs include myopia, peripheral white-with-pressure and white-without-pressure (Fig. 22.5.5c), peripheral cystoid degeneration [11, 14, 35], peripheral vitreous snowflakes [6, 11, 37], and condensed vitreous bands. Rarely, areas of coarse pigmentary changes or retinoschisis [11] are found.

In more advanced stages the subretinal exudates increase in size (Fig. 22.5.5e).

Advanced traction is followed by retinal detachment starting from the temporal periphery. Whereas visual function is frequently only minimally restricted when the macula is attached, a significant visual loss is associated with macular detachment. The typical form of the detachment is a falciform tractional detachment starting from the temporal periphery potentially involving the macula (Fig. 22.5.5f).

With increasing severity of vitreous strains a combined tractional-rhegmatogenous form with complete detachment and severe membrane formation can occur (Fig. 22.5.5g). Although there is a tractional component to the detachment, retinal holes may coexist, leading to a combined tractional-rhegmatogenous form of the detachment.

Nouhuys reports FEVR as a common cause of juvenile rhegmatogenous retinal detachment [39]. While falciform folds and total retinal detachments with retrolental organizations occur mostly during the 1st decade of life, rhegmatogenous retinal detachment was mostly observed during the 2nd and 3rd decades of life.

The anterior segment structures are usually uninvolved. However, in end stages of severely affected eyes with chronic retinal detachment, a cataract, band keratopathy, and glaucoma can be observed (Fig. 22.5.5h), all of which can be aggravated by previous surgical procedures.

There are two major classification systems according to the angiographic and funduscopic findings:

The Miyakabo and Hashimoto classification [24], according to the angiographic appearance, is as follows:

Type I: Avascular zone less than two disk diameters in width from the ora serrata, focal arteriovenous shunts, no neovascularization
Type II: Avascular zone greater than two diameters and more arteriovenous shunts
Type III: V-shaped notch in the avascular zone between the superior and inferior temporal arcades
Type IV: Additionally neovascularization including sea fans
Type V: Cicatricial disease

Miyakabo's classification focuses on early retinal vasculature changes. Fibrovascular proliferations of advanced FEVR and its complications are better described by Pendergast [28].

Pendergast and Trese [28] classify FEVR into five clinical stages as follows (Fig. 22.5.6a, b):

Pendergast's staging may be helpful for treatment considerations as all stages with exudates require treatment.

It is important to note that eyes without retinal exudates are only rarely affected by progressive disease.

Retinal folds manifesting in infancy or childhood may progress rapidly, slowly, or remain stable, depending on the presence of vascular leakage. Falciform retinal detachment unaccompanied by retinal exudation is usually stable.

Visual loss after the 2nd or 3rd decade of life is rare and usually relates to the development of rhegmatogenous retinal detachment or late peripheral exudation. Occasionally a fibrous macular pucker can cause traction and visual loss [13].

In a series of 170 eyes from 16 pedigrees [40], retinal exudates were found in 9%, retinal neovascularization in 11%, a peripheral fibrovascular mass in 6%, macular ectopia in 49%, retinal fold in 8%, retinal detachment in 21% and vitreous hemorrhage in 2%.

22.5.5 Differential Diagnosis

The differential diagnosis of FEVR includes a variety of peripheral vascular disease in childhood and differs as to the stage of the disease (Table 22.5.1, Fig. 22.5.7).

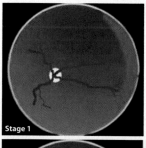

Stage 1:   Presence of an avascular zone, typically in the temporal periphery
Stage 2:   Peripheral avascular zone and extraretinal neovascularization
Stages 3 and 4: Eyes with subtotal retinal detachment
Stage 3:   Fovea attached
Stage 4:   Fovea detached
Further subdivision of stages 2–4: A. Without subretinal exudates
                                   B. With exudates
Stage 5:   Total retinal detachment with an open or closed funnel

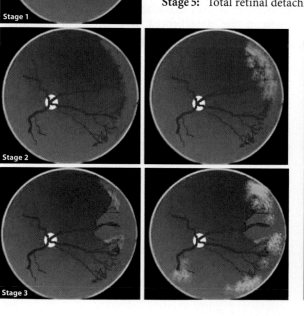
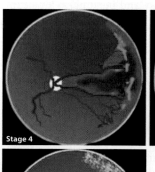
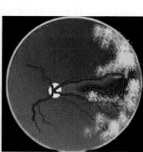
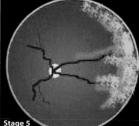

Fig. 22.5.6

Table 22.5.1. Differential diagnosis

	FEVR	ROP	IP	Norrie's disease	PHPV	Coats' disease
Sex	f=m	f=m	f	m	f=m	m
Inheritance	AD/X-R	-	X-D	X-R	-	-
Prematurity	-	+	-	-	-	-
Bilateral disease	Bilateral	Bilateral	Bilateral	Bilateral	Unilateral	Unilateral
Microphthalmus	-	-	-	+/-	+	-
Peripheral avascular zones	+	+	+	+	-	+
Retinal detachment	+	+	+	+	+/-	+
Peripheral retinal exudates	+	-/(+)	-	-	-	+
Systemic association	-	+/-	+	+	-	-

FEVR, familial exudative vitreoretinopathy; ROP, retinopathy of prematurity; IP, incontinentia pigmenti; PHPV, persistent fetal vitreous; f, female; m, male; AD, autosomal dominant; X-R, X-linked recessive; X-D, X-linked dominant.

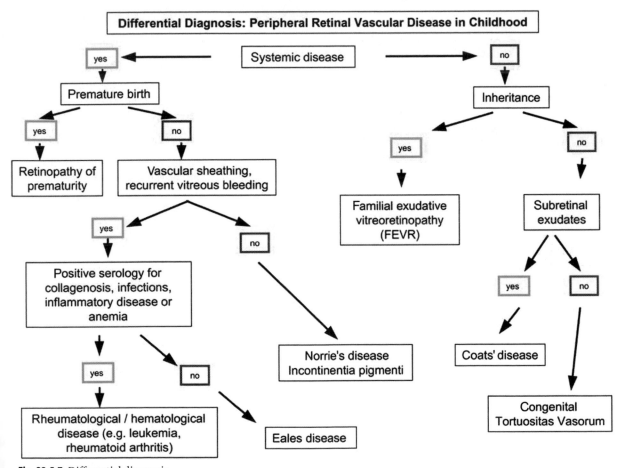

Fig. 22.5.7. Differential diagnosis

If a systemic association and premature birth are excluded, ROP, incontinentia pigmenti and Norrie's disease can be excluded with great likelihood.

This differentiation can be supplemented by other characteristics. Incontinentia pigmenti affects only female newborns. Norrie's disease demonstrates a primary retinal neuronal affection which can be shown by means of electrophysiology. Furthermore, subretinal exudates are not typically seen in Norrie's disease [7].

In contrast to ROP, changes in FEVR follow a different time course. In ROP there is the tendency to either progress to cicatricial stages or to abort and vascularize the periphery, whereas the avascular zone in FEVR and eventual retinal exudation remains a permanent feature throughout life.

22 III

Where there is peripheral vascular sheathing in the absence of a peripheral avascular zone and the temporal dragging of the disk or vasculature, Eales disease should be considered.

Coats' disease may resemble FEVR stages with subretinal deposits; however, it does not usually exhibit the vascular dragging or falciform detachment.

If the peripheral temporal traction in FEVR increases anterior-posteriorly, the clinical picture of a persistent fetal vitreous (PHPV) can be resembled. PHPV, however, is usually unilateral and the changes are predominantly in the lower part of the retina. Retinal folds in PHPV occur in conjunction with remnants of the hyaloid vascular system and may exhibit retinal dysplasia. No such association is found in FEVR.

Whereas PHPV results from abnormal development of the tunica vasculosa lentis and vitreous in the 7th or 8th weeks of gestation, the pathogenesis of FEVR appears to be a consequence of disturbed development of the retinal vasculature in the last months of gestation.

22.5.6 Treatment Recommendations Including Follow-up

Essentials
- Gene carriers suffering no visual impairment should be observed for progression of the disease into the 3rd decade of life
- There is no evidence for a prophylactic treatment of the avascular periphery
- Photocoagulation or cryotherapy are obligatory to regress peripheral subretinal exudates through destruction of abnormal vessels
- Retinotomies must be avoided as they are likely to result in proliferative vitreoretinopathy (PVR)

As discussed above a rapid progression of the disease is uncommon unless active leakage is involved.

The majority of gene carriers suffer no visual impairment and the condition is only diagnosed by chance. Out of this group, adult patients in which a peripheral avascular zone was found without other changes do not require treatment.

Similarly temporal dragging of the disk and the macula without detachment or peripheral traction may explain a reduction in macular function but should not be treated.

Children with symptomatic siblings, themselves exhibiting stage I or II disease, should be monitored carefully. Usually observation without treatment is justified in the absence of prominent exudates.

The effect of laser photocoagulation or cryotherapy in neonatal disease has not been investigated. Ablation of the avascular retinal zone in children as young as 4 years of age did not stimulate retinal vessels to advance beyond the avascular border [11].

Absolute indications for laser treatment include subretinal exudates, funduscopically visible peripheral neovascularization (Fig. 22.5.1b), and abnormal peripheral vessels with subretinal exudates (Fig. 22.5.1a).

Relative indications for treatment include treatments of fellow eyes which exhibit fewer severe peripheral changes, and stable falciform folds.

22.5.6.1 General Recommendations for Laser Photocoagulation in FEVR

Essentials
- Argon green or krypton red can be used according to the surgeon's preference
- Spot size and exposure time will differ according to the respective clinical appearance, usually 200–350 mW, 200 μm, 200–1,000 ms
- Subretinal exudates can only be addressed by direct laser coagulation of the leaky vessels
- Large subretinal exudates may take several months to regress. Six to 8 weeks after completion of the laser treatment, the treated area should be reexamined for persistence of neovascularization (Fig. 22.5.1c)
- Re-treatment is indicated in case of insufficient regression of neovascularization, persistent pathological peripheral vessels, or persistent subretinal deposits
- There is no rationale for panretinal photocoagulation as depicted in Fig. 22.5.1d
- If the retinal periphery cannot be inspected (e.g., media opacities, epiretinal membranes, or vitreous hemorrhage), it is necessary to perform a vitrectomy and decide intraoperatively on laser treatment (Fig. 22.5.8a–c)

Similar to laser photocoagulation, cryotherapy does induce regression of neovascularization, as evidenced by permanent replacement by a fibrous scar [40]. However, transscleral cryoapplication may be ineffective due to thick subretinal exudates. One may then consider endocryocoagulation. It is mandatory to freeze the abnormal vessels two to three times successively to reliably achieve tissue destruction.

Follow-up examinations are recommended since progression is reported in spite of laser coagulation: In a series of 15 eyes with active extraretinal vascularization and an attached retina, it was reported that

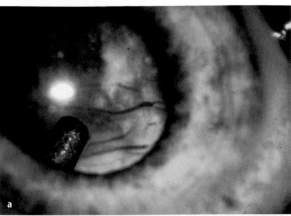

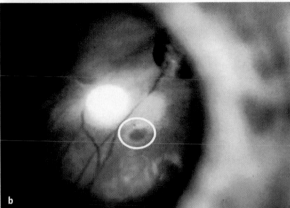

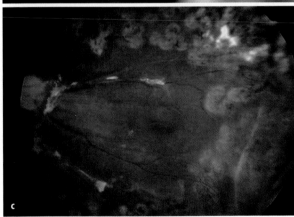

Fig. 22.5.8. a Vitrectomy of peripheral vitreous strands under indentation. Vitreous membranes are very adherent and cannot be peeled, but have to be sharply dissected. Subretinal exudates and abnormally dilated overlaying vessels are seen. **b** After removal of peripheral epiretinal membranes, a retinal round hole (*circle*) was detected associated with rhegmatogenous retinal detachment. **c** Fundus view after extensive peripheral photocoagulation. Remnants of vitreous strands can be seen after vitrectomy and peeling

8 (53%) of 15 eyes treated with peripheral laser required no further treatment [28]. Seven (47%) out of 15 eyes progressed to retinal detachment requiring vitreoretinal surgery.

A *surgical approach* becomes necessary if vitreous traction causes detachment involving the macula or if persistent vitreous hemorrhage prevents necessary laser treatment of peripheral neovascularization.

Vitreous hemorrhage is rare, but indicates a high risk of retinal detachment and phthisis in the end stage of the disease. Early vitrectomy should be considered. This allows for release of vitreous traction as well as treatment of the neovascularization triggering the hemorrhage.

III 22

22.5.6.2 Indications for Vitrectomy in FEVR

Essentials
- Progressive falciform detachment
- Progressing peripheral traction detachment threatening the macula
- Recent visual loss due to macular-off detachment
- Persistent vitreous hemorrhage (>8 weeks) or recent hemorrhage if retinal situation is unclear
- Vitreous onion-skin-like veils interfering with central vision
- Rhegmatogenous retinal detachment, when the retinal hole is obscured by peripheral epiretinal fibrosis
- PVR re-detachment

So far, surgical management of the complications has produced some encouraging results [12, 16, 23, 28, 34, 40].

Pars plana vitrectomy has been successfully applied in traction macular detachments, even in young patients [28]. In patients with vitreous hemorrhage and/or tractional retinal detachment a reattachment rate of 75% was achieved [34].

Nouhuys [38] describes the expected visual acuity after vitrectomy to be better than 20/80 in 72% of cases (Fig. 22.5.9). If an increase or at least stabilization of the initial visual acuity is considered a success, 11 out of 15 reported cases were successful in his series.

22.5.6.3 General Recommendations for Vitrectomy in FEVR

Essentials
- A three-port approach is generally used
- The lens must be removed in children to be able to address the most peripheral retinal changes. (This is in contrast to [23])

22 **III**

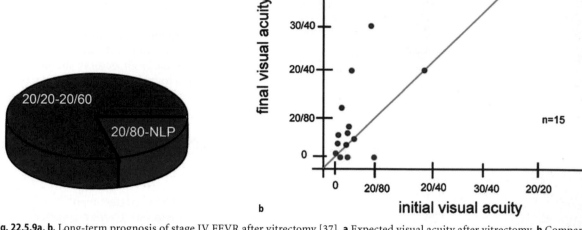

a   b

Fig. 22.5.9a, b. Long-term prognosis of stage IV FEVR after vitrectomy [37]. **a** Expected visual acuity after vitrectomy. **b** Comparison between initial and final visual acuity: *blue dots* can be considered successful, *red dots* as failure

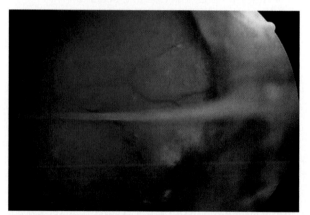

Fig. 22.5.10. Stable retinal fold after scleral buckle surgery (encircling band) and reduction of the peripheral traction

- Vitreous veils of FEVR are structured in an onion-skin-like fashion, parchment-like consistency and there is little retinal traction. Such vitreous veils cannot be peeled off the retina but must be cut from the retina
- Additional buckling procedures (encircling band) may be helpful in releasing peripheral traction (Fig. 22.5.10)
- Retinotomies should be avoided, since blood ocular barrier breakdown is typical and associated with a high risk of PVR (see below)
- Usually no vitreous substitute other than buffered saline solution is required. In cases with increased risk of PVR (e.g., in case of re-detachment or large retinal holes) or in eyes with aqueous insufficiency (cyclitic membrane), silicone oil is needed

As discussed earlier in this chapter, there is considerable recurrence of vitreous membranes forming an onion-skin-like pattern. Further, the role of inflammatory cells, which have been found in surgically removed membranes (Fig. 22.5.3f), is unknown.

It is an open question whether the recurrent greaseproof-paper-like vitreous membranes can be prevented by internal limiting membrane (ILM) peeling. However, these specific membranes seem to be anchored in or sprout deep from the retina. In this case ILM peeling would not be able to prevent regrowth of the membrane and could potentially even damage the retinal structure.

In tractional detachments, scleral buckling might be helpful in releasing some of the traction without performing a retinotomy (Figs. 22.5.10–22.5.12). Pendergast described an attachment rate of 62%, with 10 (35%) eyes achieving vision of 20/100 or better after pars plana vitrectomy and/or scleral buckling [28].

Buckle surgery or silicone tamponade alone [22] does not improve the vascular pathology and thus has to be supplemented by photocoagulation of peripheral neovascularization or abnormal vessels triggering the disease process [12].

Falciform folds and falciform retinal detachments are unspecific but common features of advanced FEVR. In the treatment, special considerations should be respected.

22.5.6.4 Surgical Approach to Falciform Detachment in FEVR (Figs. 22.5.11, 22.5.12)

Essentials

- Treatment of the falciform detachment involving the macular area causing visual deterioration should be not be attempted, since visual improvement is unlikely
- Relaxing retinotomies lead to formation of PVR membranes, PVR development and progression to re-detachment and potential loss of remaining vision
- Vitrectomy is performed to release peripheral vitreous traction and can be supported by an encircling band
- Stabilization is a more realistic aim and is to be addressed by occlusion of abnormal retinal vessels (Fig. 22.5.11e)
- Patients should be followed up for regression of subretinal fluid and retinal reattachment every 3–6 months during the first 2 years after surgery
- Ambulatory vision can improve despite persistent falciform fold, when subretinal exudates of the posterior retina are absorbed (Fig. 22.5.12c)

Although advanced stages of FEVR seem difficult to treat and success has been limited, it still is worthwhile trying if only to delay blindness. We report here on a girl who first reported to the clinic at age of 5 years with a severe visual loss after non-resolving vitreous bleeding of her „only eye" (Fig. 22.5.3a). She had a history of retinal detachment at an earlier age, but vision was reported to be better than 20/40 before the bleeding. After vitrectomy, vision improved up to 20/40 again, but deteriorated due to rebleeding and persistent neovascularization and subretinal exudates. We were unable to destroy peripheral leaking retinal vessels with the lens in place. Thus, 3 months after the initial vitrectomy we decided to remove the still clear lens and repeat coagulation of the now more readily accessible peripheral vessels. Figure 22.5.3b and c demonstrate the resection of tight vitreous strands. The retinal surface was covered with tight membranes (Fig. 22.5.3d I, II), which inserted near the vessel arcades and in the periphery. In the macular area the membranes formed a pseudohole. Membranes were carefully dissected (Fig. 22.5.3e, g). Without the need for a permanent tamponade the situation remained stable thereafter with a corrected visual acuity of 20/40.

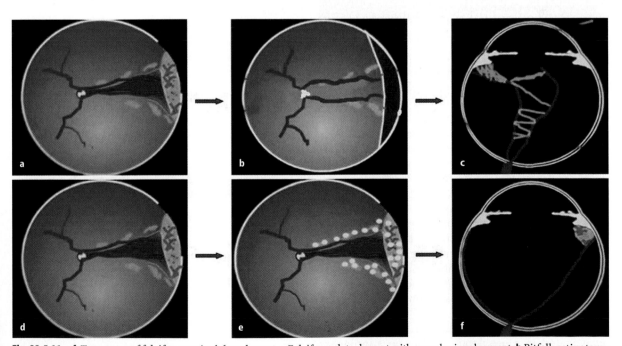

Fig. 22.5.11a–f. Treatment of falciform retinal detachment. **a** Falciform detachment with macular involvement. **b** Pitfall: retinotomy and retinal „reattachment." **c** PVR and progressing tractional detachment. **d** Falciform detachment involving macula. **e** Right: no retinotomy, coagulation of exudates and along falciform fold. **f** Stable situation with persistent falciform fold

22 III

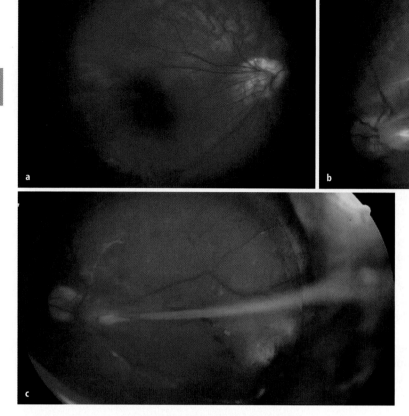

Fig. 22.5.12. a OD: VA 20/40, peripheral photocoagulation, stable for several years.
b OS: VA 20/800, stage IVA, falciform detachment involving the macula. **c** OS: VA 20/400, after encircling band and vitrectomy. Stable situation, falciform fold still present

22.5.7 Treatment Options Under Investigation

In contrast to several other vascular abnormalities resulting in increased vascular leakage, there are no current clinical trials on pharmacological treatment options. As indicated for the prophylactic laser treatment there is no evidence that the disease progress could be stopped. If the high rate of stable disease and the small risk of relentless progression are considered, it is unlikely that clinical trials will be initiated in the field and the necessity of a pathophysiology based treatment approach remains questionable.

Nevertheless, there could be a potential benefit from anti-inflammatory drugs not only by stabilizing the retinal vasculature and potentially reducing leakage, but also by regressing and preventing the perivascular inflammatory infiltrates, which are also seen in the vitreous membranes in histological examination. Still the benefit of such drugs in preventing regrowth of vitreous strains remains to be elucidated.

As the pathological mechanisms are still being investigated in more detail, to date no approaches have been reported to develop specific antibodies targeting Norrin or Fz4 or to selectively inhibit sub-sets of receptors and ligands in the Wnt/Fz signaling cascades.

Acknowledgements.   Figure 22.5.4 was generously provided by Drs. A. Alsheikheh, W. Lieb, and F. Grehn, Department of Ophthalmology, University of Würzburg, Josef-Schneider-Str. 11, 97080 Würzburg.

This chapter is devoted to the memory of Prof. Klaus Heimann, who has treated many FEVR patients and with his experience has greatly improved today's surgical approach. Most of the figures in this chapter are derived from Prof. Heimann's clinical collection.

References

1. Alsheikheh A, Lieb W, Grehn F (2004) Criswick-Schepens-Syndrom – Familiäre exsudative Vitreoretinopathie. Ophthalmologe 101:914–918
2. Benson WE (1995) Familial exudative vitreoretinopathy. Trans Am Ophthalmol Soc 93:473–521
3. Boldrey EE, Egbert P, Gass JD, Friberg T (1985) The histopathology of familial exudative vitreoretinopathy: a report of two cases. Arch Ophthalmol 103:238–241
4. Bopp S, Wagner T, Laqua H (1989) Keine Störung der arachidonsäureinduzierten Thrombozytenaggregation bei familiar exsudativer Vitreoretinopathie. Klin Monatsbl Augenheilkd 194:13–15
5. Brockhurst RJ, Albert DM (1981) Pathologic findings in

familial exudative vitreoretinopathy. Arch Ophthalmol 99: 2134–2146

6. Canny CLB, Oliver GL (1976) Fluorescein angiographic findings in familial exudative vitreoretinopathy. Arch Ophthalmol 74:1114–1120

7. Carl DR (2001) Familial exudative vitreoretinopathy. In Guyer DR, Yanuzzi LA, Chang S, Shield AS, Green WR (eds) Retina-Vitreous-Macula. Brave, Harcourt, pp 421–429

8. Chaudhuri PR, Rosenthal AR (1983) Familial exudative vitreoretinopathy associated with familial thrombocytopathy. Br J Ophthalmol 67:755–758

9. Chen ZY, Battinelli EM, Fielder A et al (1993) A mutation in the Norrie disease gene (NPD) associated with X-linked familiar exudative vitreoretinopathy. Nat Genet 5:180–183

10. Clement F, Backfort CA, Corral A, Jimenez R (1995) X-linked familial exudative vitreoretinopathy. Retina 15: 141–145

11. Criswick VG, Schepens CL (1969) Familial exudative vitreoretinopathy. Am J Ophthalmol 68:578–594

12. Culmann H (1993) Familiäre exsudative Vitreoretinopathie (Criswick-Schepens-Syndrom), Dissertation, Cologne

13. De Juan E, Lambert HM (1985) Recurrent proliferations in macular pucker, diabetic retinopathy, and retrolental fibroplasias-like disease after vitrectomy. Graefes Arch Clin Exp Ophthalmol 223:174–183

14. Dudgeon J (1979) Familial exudative vitreoretinopathy. Trans Ophthalmol Soc UK 99:45–49

15. Fuchs S, Kellner U, Wedemann H, Gal A (1995) A missense mutation in the Norrie disease gene associated with X-linked exudative vitreoretinopathy. Hum Mutat 6:257–259

16. Glazer LC, Maguire A, Blumenkranz MS et al. (1995) Improved surgical treatment of familial exsudative vitreoretinopathy in children. Am J Ophthalmol 120: 471–479

17. Gow J, Oliver GL (1971) Familial exudative vitreoretinopathy: an expanded view. Arch Ophthalmol 86:150–155

18. Kaufman SJ, Goldberg MF; Orth DH (1982) Autosomal dominant vitreoretinopathy. Arch Ophthalmol 100:272–278

19. Laqua H (1980) Familial Exudative Vitreoretinopathy. Graefes Arch Clin Exp Ophthalmol 213:121–133

20. Li Y, Fuhrmann C, Schwinger E, Gal A, Laqua H (1992) The gene for autosomal dominant familial exudative vitreoretinopathy (Criswick-Schepens) on the long arm of chromosome 11. Am J Ophthalmol 113:712–713

21. Li Y, Müller B, Fuhrmann C, van Nouhuys CE, Laqua H, Humphries P, Schwinger E, Glas A (1992b) The autosomal dominant familial exudative vitreoretinopathy locus maps on 11q and is closely linked to D11S533. Am J Hum Genet 51:749–754

22. Machemer R, Williams JM (1988) Pathogenesis and therapy of traction detachment in various retinal vascular diseases. Am J Ophthalmol 105:170–181

23. Maguire A, Trese MT (1993) Visual results of lens-sparing vitreoretinal surgery in infants. J Pediatr Ophthalmol Strabisus 30:28–32

24. Miyakabo H, Hashimoto K (1984) Retinal vascular pattern in familial exudative vitreoretinopathy. Ophthalmology 91:1524–1530

25. Mueller B, Orth U, van Mouhuys CE et al (1994) Mapping of the autosomal dominant exudative vitreoretinopathy locus (EVRI) by multipoint linkage analysis in four families. Genomics 20:317–319

26. Nicholson DH, Galvis V (1984) Criswick-Schepens syndromes (familial exudative vitreoretinopathy): a study of a Colombian kindred. Arch Ophthalmol 102:1519–1522

27. Ober RR, Bird AC, Hamilton AM, Sehmi K (1980) Autosomal dominant exudative vitreoretinopathy. Br J Ophthalmol 64:112–120

28. Pendergast SD, Trese MT (1998) Familial exudative vitreoretinopathy: results of surgical management. Ophthalmology 105:1015–1023

29. Price SM, Periam N, Humphries A, Woodruff G, Trembath RC (1996) Familial exudative vitreoretinopathy linked to D11S533 in a large Asian family with consanguinity. Ophthalmic Genet 17:53–57

30. Robitaille J, MacDonald ML, Kaykas A, Sheldahl LC, Zeisler J, Dube MP, Zhang LH, Singaraja RR, Guernsey DL, Zheng B, Siebert LF, Hoskin-Mott A, Trese MT, Pimstone SN, Shastry BS, Moon RT, Hayden MR, Goldberg YP, Samuels ME (2002) Mutant frizzled-4 disrupts retinal angiogenesis in familial exudative Vitreoretinopathy. Nat Genet 32:326–330

31. Saraux H, Laroche L, Koenig F (1985) Exudative retinopathy with dominant transmission. Report of a new pedigree. J Fr Ophthalmol 8:155–158

32. Shastry BS, Heijtmancik JF, Plager DA, Hartzer MK, Trese MT. 1995; Linkage and candidate gene analysis of X-linked familial exudative vitreoretinopathy. Genomics 27:341–344

33. Sharstry BS, Trese MT (1997) Familial exudative vitreoretinopathy: further evidence of genetic heterogeneity. Am J Med Genet 69:217–218

34. Shubert A, Tasman W (1997) Familial exudative vitreoretinopathy: surgical intervention and visual acuity outcomes. Graefes Arch Clin Exp Ophthalmol 235:490–493

35. Slusher MM, Hutton WE (1979) Familial exudative vitreoretinopathy. Am J Ophthalmol 87:152–156

36. Toomes C, Bottomley HM, Jackson RM, Towns KV, Scott S, Mackey DA, Craig JE, Jiang L, Yang Z, Trembath R, Woodruff G, Gregory-Evans CY, Gregory-Evans K, Parker MJ, Black GC, Downey LM, Zhang K, Inglehearn CF (2004) Mutations in LRP5 or FZD4 underlie the common familial exudative vitreoretinopathy locus on chromosome 11q. Am J Hum Genet 74:721–730

37. Van Nouhuys CE (1982) Dominant exudative vitreoretinopathy and other vascular developmental disorders of the peripheral retina. Doc Ophthalmol 54:1–414

38. Van Nouhuys CE (1981) Congenital retinal fold as a sign of dominant exudative vitreoretinopathy. Graefes Arch Clin Exp Ophthalmol 217:55–67

39. Van Nouhuys CE (1989) Juvenile retinal detachment as a complication of familial exudative vitreoretinopathy. Fortschr Ophthalmol 86:221–223

40. Van Nouhuys CE (1991) Signs, complications, and platelet aggregation in familial exudative vitreoretinopathy. Am J Ophthalmol 111:34–41

41. Xu Q, Wang Y, Dabdoub A, Smallwood PM, Williams J, Woods C, Kelley MW, Jiang L, Tasman W, Zhang K, Nathans J (2004) Vascular development in the retina and inner ear: control by Norrin and Frizzled-4, a high-affinity ligand-receptor pair. Cell 116:883–895

23 Vasculopathy After Treatment of Choroidal Melanoma

B. Damato

Core Messages

- Vasculopathy can develop after different forms of treatment for uveal melanoma
- After radiotherapy, macular edema and hard exudates can be caused by macular vasculopathy, extramacular retinopathy, or exudation from the irradiated tumor
- Maculopathy caused by exudation from persistent irradiated tumor can be treated by administering transpupillary thermotherapy to the tumor or by removing the tumor
- After radiotherapy of a large choroidal melanoma, severe exudative retinal detachment, rubeosis iridis and neovascular glaucoma can be treated by excising the irradiated tumor
- Photocoagulation of a choroidal melanoma can induce aggressive disk and retinal neovascularization if the major retinal vessels are occluded, as well as choroidal new vessels if Bruch's membrane is ruptured
- Choroidal new vessels can develop from a surgical coloboma of the choroid, to cause disciform macular degeneration

23.1 Uveal Melanoma

Essentials

- Most uveal melanomas arise in choroid and almost 50% of these extend close to optic nerve or fovea
- Treatment of choroidal melanoma is by radiotherapy, phototherapy, local resection or enucleation, individually or in combination

Uveal melanomas present in adulthood, their incidence peaking at around the age of 60 years [4]. More than 90% of uveal melanomas arise in choroid, about 40% extending close to optic disk or fovea. These tumors tend to cause visual loss from retinal pigment epithelial disease, macular edema, and exudative retinal detachment (Fig. 23.1). In advanced cases, the presence of a large intraocular tumor and extensive retinal detachment can result in rubeosis, neovascular glaucoma, phthisis, and a blind and painful eye. Approximately 50% of

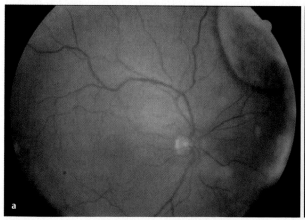
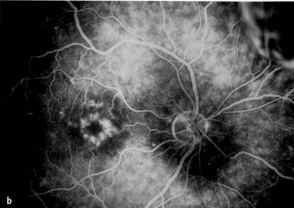

Fig. 23.1. Cystoid macular edema secondary to an untreated superonasal choroidal melanoma in the right eye of a 60-year-old woman: **a** color photograph; **b** fluorescein angiograph

patients develop metastatic disease, which usually involves the liver and which is usually fatal within a few months.

Treatment of uveal melanoma by enucleation has largely been superseded by a variety of methods aimed at conserving the eye with as much vision as possible [5]. Such "conservative" therapies consist of: (a) radiotherapy, which includes brachytherapy, proton beam radiotherapy, and stereotactic radiotherapy; (b) phototherapy, such as photocoagulation, transpupillary thermotherapy, and photodynamic therapy; and (c) tumor resection, performed by the transscleral or transretinal routes. Several of these treatments induce vascular changes in the tumor as well as in adjacent ocular tissues and these effects can cause visual loss.

The aims of this chapter are to describe vasculopathies caused by the different forms of treatment of uveal melanoma and to discuss their treatment.

23.2 Radiotherapy

23.2.1 Radiotherapy Techniques

In most centers, the first choice of treatment is brachytherapy, usually administered with a radioactive plaque containing ruthenium-106 or iodine-125. Less commonly used plaques include the binuclide plaque containing both ruthenium and iodine, palladium, iridium, and strontium [11, 12, 16, 18]. Proton beam radiotherapy is available in only a small number of centers around the world. Some oncologists use this treatment for all choroidal melanomas; others reserve it for tumors that cannot adequately be treated with brachytherapy, because of large size or close proximity to optic disk or fovea [14].

Stereotactic radiotherapy is generally used as an alternative to proton beam radiotherapy in centers where a cyclotron unit is not available [10].

23.2.2 Radiation Vasculopathy

Essentials
- Radiation retinopathy may be primary, as a result of retinal irradiation, or secondary, caused by the persistence of irradiated tumor in the eye
- Secondary radiation retinopathy can be treated by administering transpupillary thermotherapy to the irradiated tumor or by resecting the tumor

Ionizing radiation displaces electrons from water molecules, producing toxic free radicals, and from DNA and membranes, which are damaged reparably or irreversibly. Irradiated cells can die rapidly by apoptosis or they can lose their reproductive capacity, continuing to function normally until they reach the end of their normal life-span, becoming senescent or dying when trying to divide.

Radiation causes both acute and chronic effects. The acute effects are caused by breakdown of cell membranes, which result in optic disk swelling, macular edema, and exudative retinal detachment. These can resolve or may be followed by atrophy of the optic nerve, retina, retinal pigment epithelium and choroid. Chronic effects in the posterior segment of the eye are dominated by vascular degeneration, because of the relatively low cell turnover of cells in the retina and optic nerve. Gradual depletion of vascular endothelial cells and pericytes results in progressive circulatory decompensation, causing exudation, hemorrhage, occlusion, ischemia and neovascularization. These are manifest clinically as retinal telangiectasia, microaneurysms, macular edema, hard exudates, retinal hemorrhages, capillary closure, cotton wool spots, optic disk swelling, retinal and disk neovascularization, rubeosis, vitreous hemorrhage and neovascular glaucoma.

Visual loss after radiotherapy can be caused: (a) directly by cell death and vasculopathy in the optic nerve (Fig. 23.2) or macula (Fig. 23.3); or (b) indirectly, from exudation or hemorrhage arising from vasculopathy in another part of the eye, far from

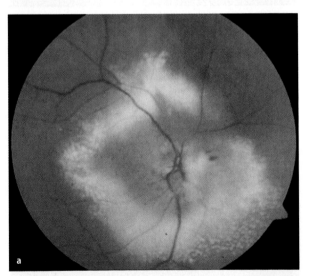

Fig. 23.2. Optic neuropathy after proton beam radiotherapy of a superonasal, juxtapapillary choroidal melanoma in the right eye: **a** exudative phase, 16 months after radiotherapy, with disk swelling and hard exudates

23 **III**

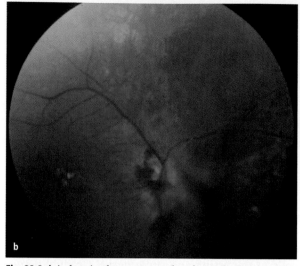

Fig. 23.2. b ischemic phase, 48 months after radiotherapy, with disk neovascularization and vitreous hemorrhage

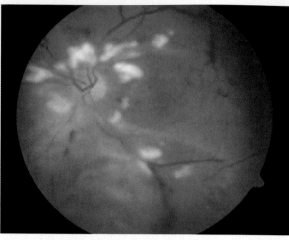

Fig. 23.5. Extensive radiation retinopathy after proton beam radiotherapy of an inferior choroidal melanoma in a 32-year-old, female, diabetic patient. Despite aggressive panretinal photocoagulation, enucleation was eventually necessary

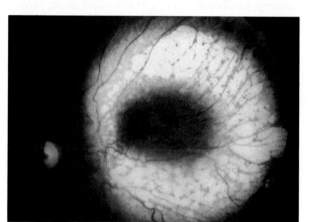

Fig. 23.3. Macular atrophy after ruthenium plaque radiotherapy of a posterior choroidal melanoma in the left eye

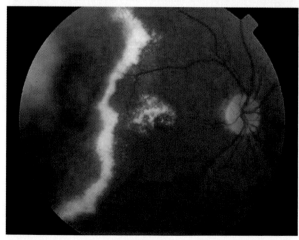

Fig. 23.4. Macular hard exudates from a temporal choroidal melanoma after proton beam radiotherapy

these structures (Fig. 23.4). The indirect effects of radiation diminish when there is complete vascular obliteration and atrophy of the irradiated tissues, because exudation ceases and because the ischemic stimulus to neovascularization is no longer present. This is why a low dose of radiation to a large part of the eye (e.g., from external beam radiotherapy of an intracranial tumor) can cause more ocular morbidity and visual loss than a high dose of radiation to a small area (e.g., brachytherapy of a choroidal melanoma). Irrespective of the type of radiation, ocular morbidity is more severe in the presence of diabetes (Fig. 23.5) or with concurrent chemotherapy.

23.2.3 Treatment of Radiation Vasculopathy

With iodine-125 brachytherapy, direct radiation damage to optic nerve and macula can be avoided by collimation (Fig. 23.6) [1]. With ruthenium-106 brachytherapy, such collateral damage can be prevented by positioning the plaque eccentrically, with its posterior edge aligned with the posterior tumor margin, using side-scatter of radiation to achieve the required safety margin (Fig. 23.7). This approach has improved conservation of vision, without any increase in local tumor recurrence rates (Fig. 23.8) [8, 9]. With proton beam radiotherapy, irradiation of optic nerve and macula is reduced by creating a notch in the beam (Fig. 23.9).

Direct radiation damage to optic nerve and macula is untreatable. In mild cases, spontaneous improvement can sometimes occur (Fig. 23.10). There is scope for investigating agents such as intraocular steroids (Fig. 23.11), and in the future radioprotective agents [17].

23

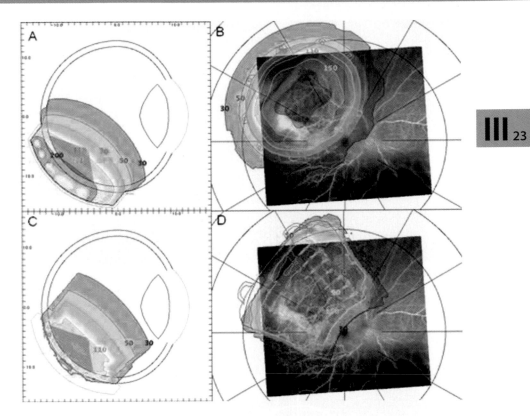

Fig. 23.6. Radiation dosimetry of iodine-125 brachytherapy with a COMS plaque (**A, B**) as compared with a collimated plaque (**C, D**). (Courtesy of M. Astrahan, University of Southern California Norris Cancer Hospital)

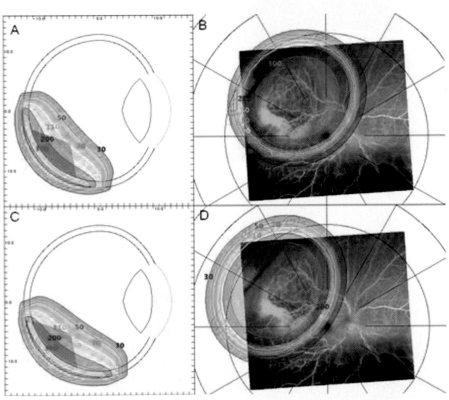

Fig. 23.7. Radiation dosimetry of ruthenium-106 brachytherapy with a ruthenium plaque placed centrally (**A, B**) as compared with one located eccentrically (**C, D**). (Courtesy of M. Astrahan, University of Southern California Norris Cancer Hospital)

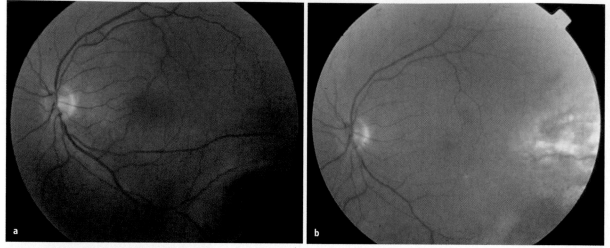

Fig. 23.8. Choroidal melanoma in the left eye of a 46-year-old woman: **a** before treatment, showing a 9.9×9.6×3.3 mm melanoma extending to within 3 mm of the fovea; and **b** 7 years after ruthenium-106 brachytherapy, with good tumor control and visual acuity of 6/6. The patient had also received argon laser photocoagulation to leaking juxtafoveal retinal vessels, which successfully led to disappearance of hard exudates threatening the fovea

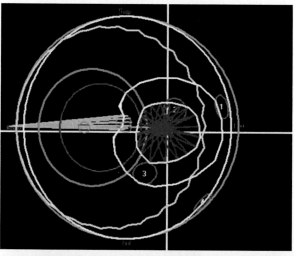

Neovascularization secondary to retinal ischemia can be treated by panretinal photocoagulation. The author has observed regression of radiation-induced rubeosis with Bevacizumab; however, further studies are indicated.

Exudation from the irradiated tumor is treated according to the size of the tumor. Small tumors can be treatable by transpupillary thermotherapy, both after brachytherapy (Fig. 23.12) and proton beam radiotherapy (Fig. 23.13). Thick, posterior tumors

◁
Fig. 23.9. Computerized 3D model showing a notch in the proton beam, designed to reduce optic nerve irradiation. (Courtesy of A. Kacperek, Clatterbridge Centre for Oncology)

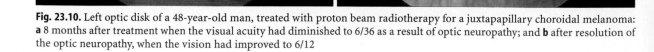

Fig. 23.10. Left optic disk of a 48-year-old man, treated with proton beam radiotherapy for a juxtapapillary choroidal melanoma: **a** 8 months after treatment when the visual acuity had diminished to 6/36 as a result of optic neuropathy; and **b** after resolution of the optic neuropathy, when the vision had improved to 6/12

can be treated by endoresection (Fig. 23.14). Extensive retinal detachment is treatable by transscleral local resection of the irradiated tumor, which also induces resolution of any neovascular glaucoma (Fig. 23.15).

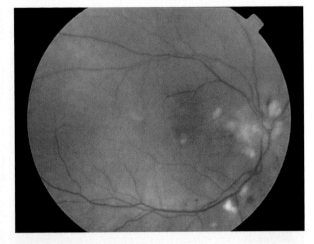

▷
Fig. 23.11. Right fundus of a 73-year-old diabetic man with an inferonasal choroidal melanoma treated with proton beam radiotherapy. Three years after treatment, he developed macular edema, which responded to intravitreal triamcinolone injection, with an improvement of visual acuity from 6/18 to 6/9

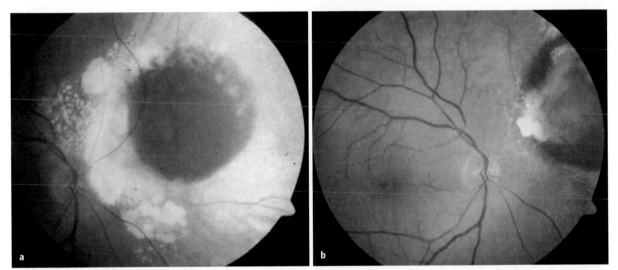

Fig. 23.12. Hard exudates arising from a choroidal melanoma after treatment with ruthenium-106 brachytherapy: **a** before; and **b** after transpupillary thermotherapy, which improved the vision from 6/18 to 6/9

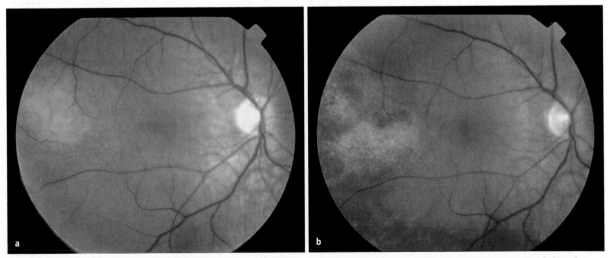

Fig. 23.13. Choroidal melanoma in the right eye of a 30-year-old woman: **a** at presentation, showing serous retinal detachment involving the fovea and reducing the vision to 6/9; and **b** 8 months after proton beam radiotherapy and transpupillary thermotherapy, with resolution of the retinal detachment and improvement in vision to 6/5

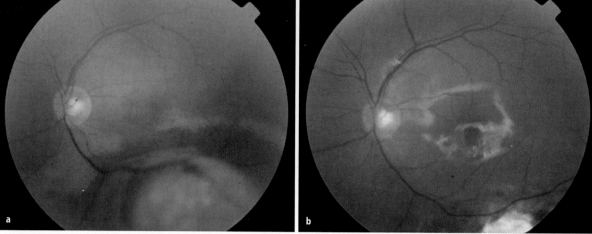

Fig. 23.14. Exudation from a choroidal melanoma in the left eye of a 53-year-old man: **a** 6 months after proton beam radiotherapy; and **b** after endoresection, showing resolution of the exudates, with improvement in vision from 6/24 to 6/12

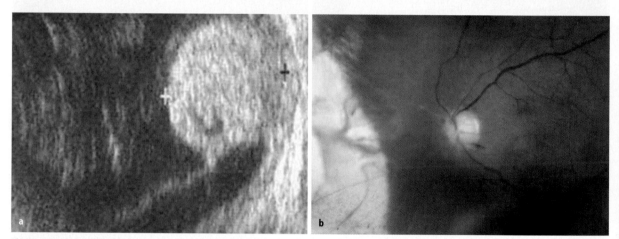

Fig. 23.15. Choroidal melanoma in the left eye of a 75-year-old man treated with proton beam radiotherapy. **a** Ocular ultrasound scan 16 months after this treatment, showing persistent tumor and exudative retinal detachment, which was associated with neovascular glaucoma of 50 mm Hg. **b** Fundus photograph 2 years after transscleral resection of the irradiated tumor, when the retina was flat, with regression of the rubeosis, good control of the intraocular pressure with latanoprost and conservation of vision of 6/36. The right eye was amblyopic with vision of 6/60

23.3 Phototherapy

23.3.1 Phototherapy Techniques

Photocoagulation heats the target tissue to approximately 60°C so that cellular proteins are coagulated. Transpupillary themotherapy heats the tumor by only a few degrees and is regarded as being safer and more effective than photocoagulation, which it supersedes [15]. There are anecdotal reports of photodynamic therapy of choroidal melanoma, but further evaluation is needed to determine the efficacy of this treatment [2].

23.3.2 Vasculopathy After Phototherapy

Photocoagulation obliterates not only all the tumor vasculature but also any overlying retinal vessels. Severe retinal ischemia caused by occlusion of the major retinal vessels can cause disk and retinal neovascularization. Furthermore, any breaks in Bruch's membrane can allow choroidal new vessels to develop (Fig. 23.16) [13]. These can be aggressive, growing as far forward as the lens (Fig. 23.17). As with radiation-induced neovascularization, any new vessels caused by phototherapy behave more aggressively if the patient suffers from diabetes mellitus. Neovascular complications are rarer with transpupillary thermotherapy than with photocoagulation.

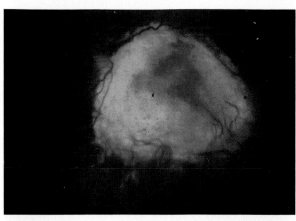

Fig. 23.16. Choroidal neovascularization with subhyaloid hemorrhage after photocoagulation of a choroidal melanoma in the left eye of a 50-year-old man

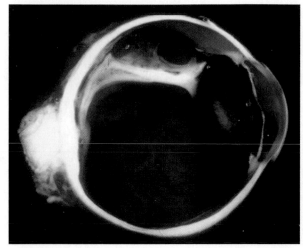

Fig. 23.17. Subluxation of the lens from traction by a choroidal fibrovascular membrane arising after photocoagulation of a superior, juxtapapillary choroidal melanoma in the right eye of a 70-year-old man. Neovascular glaucoma had also developed

23.3.3 Treatment of Vasculopathy After Phototherapy

The treatment of any new vessels induced by phototherapy is similar to that of neovascularization related to retinal vein occlusion, diabetic retinopathy and other causes of ischemia. Disk new vessels can regress after panretinal photocoagulation, but choroidal neovascularization may also require direct treatment.

23.4 Local Resection

23.4.1 Local Resection Techniques

Transscleral local resection involves the en bloc excision of the intact tumor, together with the deep scleral lamella to which it is attached, if possible without damaging the adjacent retina [7]. Endoresection is performed transvitreally, either through a hole in the retina over the tumor or after raising a retinal flap [6]. These operations are difficult and controversial and are therefore performed as primary procedures only by a few surgeons and for tumors considered unsuitable for radiotherapy. An alternative approach is to perform endoresection after radiotherapy [3].

23.4.2 Vasculopathy After Local Resection

After both types of local resection, choroidal new vessels can develop from the surgical margins. These vessels are subclinical if they grow into the surgical coloboma (Fig. 23.18). When the surgical excision extends far posteriorly, there is a tendency for any choroidal vessels to grow towards the fovea, resulting in a disciform scar, this condition arising both after transscleral resection (Fig. 23.19) and after transretinal resection (Fig. 23.20). Such disciform macular

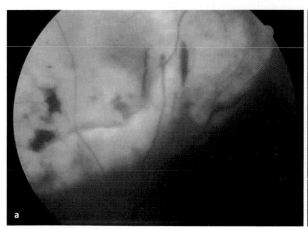
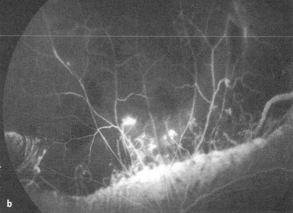

Fig. 23.18. Choroidal coloboma after local resection of a uveal melanoma in the left eye of a 63-year-old woman: **a** color photograph showing bare sclera; and **b** fluorescein angiogram demonstrating subclinical choroidal new vessels in the coloboma

23 **|||**

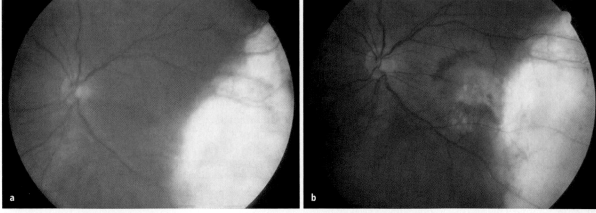

Fig. 23.19. Choroidal coloboma after transscleral local resection of a temporal uveal melanoma in the left eye of a 43-year-old man: **a** before and **b** after development of a macular disciform lesion

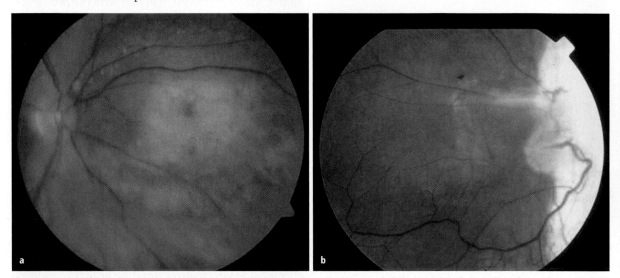

Fig. 23.20. Right fundus in a 46-year old woman: **a** at presentation, showing a nasal choroidal melanoma involving the optic disk; and **b** 25 months after endoresection, showing a disciform scar extending from the margin of the surgical coloboma to the fovea

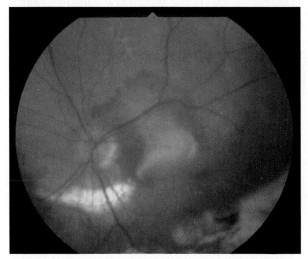

Fig. 23.21. Choroidal neovascular membrane arising from an iatrogenic choroidal tear following transscleral local resection of a choroidal melanoma in the left eye of a 46-year-old woman

degeneration can occur also as a result of an iatrogenic choroidal tear (Fig. 23.21).

23.4.3 Treatment of Vasculopathy After Local Resection

The treatment of choroidal new vessels threatening vision after local resection is the same as for age-related macular degeneration.

23.5 Conclusion

Vasculopathy is a common cause of visual loss after treatment of choroidal melanoma, resulting in ischemia and exudation. After radiotherapy, maculopathy and optic neuropathy can be the result of direct irradiation or can arise indirectly, because of exudation from irradiated retina or tumor as well as from neovascular complications. Phototherapy can induce retinal or disk neovascularization by obliterating large retinal vessels and, furthermore, choroidal neovascularization can occur if Bruch's membrane is ruptured. After transretinal or transscleral local resection, choroidal neovascularization can develop harmlessly into the coloboma or can extend posteriorly to cause a macular disciform scar. The principles of treating these neovascular and exudative complications are the same as for other diseases; however, when irradiated tumor is the cause of retinopathy, then phototherapy or excision of the residual tumor can be effective.

References

1. Astrahan MA, Luxton G, Pu Q, Petrovich Z (1997) Conformal episcleral plaque therapy. Int J Radiat Oncol Biol Phys 39:505–19
2. Barbazetto IA, Lee TC, Rollins IS, Chang S, Abramson DH (2003) Treatment of choroidal melanoma using photodynamic therapy. Am J Ophthalmol 135:898–9
3. Bornfeld N, Talies S, Anastassiou G, Schilling H, Schüler A, Horstmann GA (2002) Endoscopic resection of malignant melanomas of the uvea after preoperative stereotactic single dose convergence irradiation with the Leksell gamma knife. Ophthalmologe 99:338–44
4. Damato B (2000) Ocular tumours: diagnosis and treatment. Butterworth Heinemann, Oxford, pp 57–93
5. Damato B (2004) Developments in the management of uveal melanoma. Clin Exp Ophthalmol 32:639–47
6. Damato B, Groenewald C, McGalliard J, Wong D (1998) Endoresection of choroidal melanoma. Br J Ophthalmol 82:213–8
7. Damato B, Jones AG (2005) Uveal melanoma: resection techniques. Ophthalmol Clin North Am 18:119–28, ix
8. Damato B, Patel I, Campbell IR, Mayles HM, Errington RD (2005) Local tumor control after (106)Ru brachytherapy of choroidal melanoma. Int J Radiat Oncol Biol Phys 63:385–9
9. Damato B, Patel I, Campbell IR, Mayles HM, Errington RD (2005) Visual acuity after ruthenium(106) brachytherapy of choroidal melanomas. Int J Radiat Oncol Biol Phys 63:392–400
10. Dieckmann K, Georg D, Zehetmayer M, Bogner J, Georgopoulos M, Pötter R (2003) LINAC based stereotactic radiotherapy of uveal melanoma: 4 years clinical experience. Radiother Oncol 67:199–206
11. Finger PT, Berson A, Ng T, Szechter A (2002) Palladium-103 plaque radiotherapy for choroidal melanoma: an 11-year study. Int J Radiat Oncol Biol Phys 54:1438–45
12. Flühs D, Anastassiou G, Wening J, Sauerwein W, Bornfeld N (2004) The design and the dosimetry of bi-nuclide radioactive ophthalmic applicators. Med Phys 31:1481–8
13. Foulds WS, Damato BE (1986) Low-energy long-exposure laser therapy in the management of choroidal melanoma. Graefes Arch Clin Exp Ophthalmol 224:26–31
14. Gragoudas ES, Marie Lane A (2005) Uveal melanoma: proton beam irradiation. Ophthalmol Clin North Am 18:111–8, ix
15. Journée-de Korver HG, Midena E, Singh AD (2005) Infrared thermotherapy: from laboratory to clinic. Ophthalmol Clin North Am 18:99–110
16. Missotten L, Dirven W, Van der SA, Leys A, De Meester G, Van Limbergen E (1998) Results of treatment of choroidal malignant melanoma with high-dose-rate strontium-90 brachytherapy. A retrospective study of 46 patients treated between 1983 and 1995. Graefes Arch Clin Exp Ophthalmol 236:164–73
17. Nieder C, Andratschke NH, Wiedenmann N, Molls M (2004) Prevention of radiation-induced central nervous system toxicity: a role for amifostine? Anticancer Res 24:3803–9
18. Valcárcel F, Valverde S, Cardenes H, Cajigal C, de la Torre A, Magallón R, et al. (1994) Episcleral iridium-192 wire therapy for choroidal melanomas. Int J Radiat Oncol Biol Phys 30:1091–7

24   Vasculopathies with Acute Systemic Diseases
24.1 Purtscher's Retinopathy

D.V. Do, A.P. Schachat

Core Messages

- Purtscher's retinopathy is a traumatic retinal angiopathy and was originally described in individuals with severe head trauma
- Symptoms include rapid and painless vision loss
- Usually occurs in both eyes, but unilateral cases have been reported
- Acute funduscopic findings include cotton-wool spots and intraretinal hemorrhages primarily located in the posterior pole. Optic disk swelling may also occur
- Purtscher's-like retinopathy can occur in other traumatic conditions (compressive chest injuries, long bone fractures), collagen vascular diseases (systemic lupus erythematosus, sclerodermia, dermatomyositis), kidney diseases, and childbirth
- Treatment is supportive and directed at the underlying medical condition

24.1.1 History

Purtscher's retinopathy was first described by Otmar Purtscher in 1910 as a syndrome of multiple, white retinal patches, superficial retinal hemorrhages, and papillitis occurring in five patients with severe head trauma [11]. He originally named this condition *angiopathia retinae traumatica* and hypothesized that white, superficial retinal patches were lymphatic extravasations caused by a sudden increase in intracranial pressure secondary to massive head trauma.

The term Purtscher's retinopathy or Purtscher's-like retinopathy is currently used to describe similar fundus findings associated with various conditions: trauma (head trauma, compressive chest injuries, long bone fractures with fat embolism), collagen vascular diseases (systemic lupus erythematosus, sclerodermia, dermatomyositis), kidney disease, and childbirth (amniotic fluid embolism) [2, 4, 5, 8, 12].

Although associated with several systemic conditions, Purtscher's retinopathy is a relatively uncommon finding and published reports in the literature have been limited to case reports.

24.1.2 Special Pathologic Features

Essentials

- Complement-induced retinal leukoembolization is likely responsible for the ophthalmoscopic findings in Purtscher's retinopathy

Although Otmar Purtscher initially postulated that the white, superficial retinal patches (now recognized as cotton-wool spots) were lymphatic extravasations caused by the sudden increase in intracranial pressure seen in severe head trauma [11], the exact pathogenesis of Purtscher's retinopathy remains unknown. Several authors have hypothesized that microembolic events, such as fat embolization in long bone fractures and air embolization from compressive chest injuries, may be responsible [3]. Other scientists speculate that venous reflux with capillary engorgement of the upper body plays a role [10].

More recently, studies have shown that leukocyte aggregation by activated complement factor 5 (C5a) may lead to leukocyte emboli that can occlude the peripapillary retinal capillaries in conditions such as trauma, acute pancreatitis, and collagen vascular diseases [7]. Several authors have postulated that complement-induced retinal leukoembolization occurs in the setting of Purtscher's retinopathy, and animal models have shown evidence of multiple small retinal arteriolar occlusions [6, 9].

24.1.3 Clinical Course of the Disease

Essentials

- Painless, rapid vision loss
- Frequently bilateral involvement
- Acute funduscopic findings include cotton-wool spots and intraretinal hemorrhages located primarily in the posterior pole
- Fluorescein angiography may demonstrate retinal arteriolar occlusions, capillary non-perfusion, and leakage from vessels in the affected area

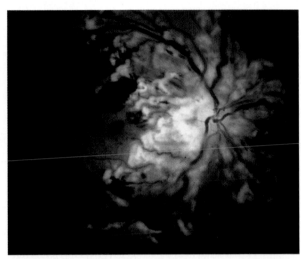

Fig. 24.1.1. Fundus photograph demonstrating Purtscher's retinopathy developing after an automobile accident. White, cotton-wool spots, intraretinal and preretinal hemorrhages, and arteriolar spasm are present primarily in the posterior pole

Purtscher's retinopathy is most often bilateral, but it has been reported to occur in unilateral cases [3]. It may occur after traumatic injuries or a Purtscher's-like retinopathy may be associated with other systemic conditions such as collagen vascular disease, kidney disease, and childbirth [2, 4, 5, 8, 12].

Acute funduscopic findings include cotton-wool spots and intraretinal hemorrhages, primarily located in the posterior pole (Fig. 24.1.1). Optic disk swelling may be present. Fluorescein angiography may reveal focal retinal arteriolar occlusions in the perifoveal area, capillary non-perfusion, and leakage from vessels in the areas of infarction (Fig. 24.1.2) [5].

Visual acuity is markedly decreased in the affected eye(s) and may range from 20/200 to light perception. An afferent papillary defect may be present if the condition is unilateral or asymmetric.

24.1.4 Differential Diagnosis

The differential diagnosis of Purtscher's retinopathy includes retinal vascular disorders that can result in intraretinal hemorrhages and cotton-wool spots. Central retinal vein occlusion, hypertensive retinopathy, diabetic retinopathy, Valsalva retinopathy, and lupus retinopathy may all present with intraretinal hemorrhages and nerve fiber layer infarcts.

Differentiating between Purtscher's retinopathy and other conditions requires a thorough medical history to document preexisting medical conditions, such as diabetes and hypertension, and possible precipitating events, such as trauma, that may suggest one disease over the other. In addition, funduscopic findings such as lipid exudates, tortuous and dilated veins, and retinal neovascularization are not typical-

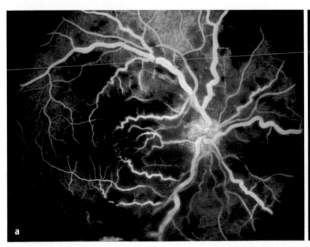

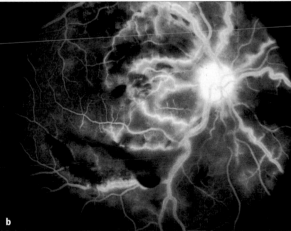

Fig. 24.1.2. a Early phase fluorescein angiogram from an eye with Purtscher's retinopathy demonstrating dilated vessels, blocked fluorescence from nerve-fiber layer infarcts, and capillary non-perfusion. **b** Late phase fluorescein angiogram shows leakage from the involved vessels. Optic disk staining and leakage is also present

ly seen in Purtscher's retinopathy, and these findings may suggest diabetic retinopathy or venous occlusive disease instead. Ultimately, Purtscher's retinopathy is a clinical diagnosis based on the medical history and examination findings.

24.1.5 Treatment Recommendations Including Follow-up

Visual outcome in Purtscher's retinopathy is variable depending on the location and amount of capillary non-perfusion, involvement of the macula, and degree of optic atrophy that can occur.

Currently there is no well-defined treatment for Purtscher's retinopathy. Initial treatment is supportive and aimed at resolving the underlying medical condition.

Although several published studies have cited steroid therapy for the treatment of Purtscher's retinopathy in hopes that steroids can stabilize damaged tissue and perhaps block the formation of complement-activated leukocyte aggregation [1, 13], no large scale studies have been conducted to determine the efficacy of this type of treatment. Large clinical trials would be difficult to plan given the infrequent occurrence of this condition and absent efficacy data; since there are safety concerns with steroids, the use of steroids to treat Purtscher's retinopathy is not currently advised.

References

1. Atabay C, Kansu T, Nurlu G (1993) Late visual recovery after intravenous methylprednisolone treatment of Purtscher's retinopathy. Ann Ophthalmol 25:330–333
2. Blodi B, Johnson MW, Gass JDM, Fine SL, Joffe LM (1990) Purtscher's-like retinopathy after childbirth. Ophthalmology 97:1654–1659
3. Burton TC (1980) Unilateral Purtscher's retinopathy. Ophthalmology 87:1096–1105
4. Cooper BA, Shah GK, Grand MG (2004) Purtscher's-like retinopathy in a patient with systemic lupus erythematosus. Ophthalmic Surg Lasers Imaging 35(5):438–439
5. Gass JDM (1987) Stereoscopic atlas of macular diseases: diagnosis and treatment. St. Louis, Mosby, pp 452–455
6. Jaco HS, Goldstein IM, Shapiro I, Craddock PR, Hammerschmidt DE, Weissmann G (1981) Sudden blindness in acute pancreatitis: possible role of complement-induced retinal leukoembolization. Arch Intern Med 141(1):134–136
7. Jacob HS, Craddock PR, Hammerschmidt DE, Moldow CF (1980) Complement-induced granulocyte aggregation: an unsuspected mechanism of disease. N Engl J Med 302:89–794
8. Kelley JS, Hartranft CD (2001) Traumatic chorioretinopathies. In: Ryan SJ, Schachat AP (eds) Retina. Mosby, St. Louis, pp 1815–1817
9. Lai JC, Johnson MW, Martonyi CL, Till GO (1997) Complement-induced retinal arteriolar occlusions in the cat. Retina 17(3):239–246
10. Marr WG, Marr EG (1962) Some observations on Purtscher's disease: traumatic retinal angiography. Am J Ophthalmol 54:693–705
11. Purtscher O (1910) Noch unbekannte Befunde nach Schadeltrauma. Berl Dtsch Ophthal Ges 36:294–301
12. Roden D, Fitzpatrick G, O'Donoghue H, Phelan D (1989) Purtscher's retinopathy and fat embolism. Br J Ophthalmol 73:677–679
13. Wang AG, Yen MY, Liu JH (1998) Pathogenesis and neuroprotective treatment in Purtscher's retinopathy. Jpn J Ophthalmol 42:318–322

24.2 Terson Syndrome

F. Kuhn, R. Morris, V. Mester

Core Messages

- Terson syndrome (Terson's syndrome, Tersons syndrome) is defined as vitreous hemorrhage secondary to subarachnoid or subdural hemorrhage (Fig. 24.2.1). Subarachnoid bleeding is the more common cause
- The less common subdural hemorrhage is often associated with brain trauma and permanent neurological or mental disability
- Screening of patients with subarachnoid or subdural hemorrhage through the dilated pupil should be part of these patients' management protocol
- Presence of intraocular hemorrhage in patients with subarachnoid hemorrhage is a statistically significant risk factor for coma as well as for death
- The vitreous hemorrhage may take months to spontaneously improve, but observation is not

without risks: proliferative vitreoretinopathy (PVR) can develop, leading to irreversible loss of vision
- In approximately 40% of patients with Terson syndrome, a hemorrhagic cyst (HMC) is found in the macular area. In approximately two-thirds of these cysts, the blood is located under the internal limiting membrane (ILM) (submembranous HMC)
- Treatment is aimed at removal of the intravitreal blood and evacuation of the cyst contents to hasten visual rehabilitation and to prevent blood-related complications
- The timing of vitrectomy is primarily determined by the patient's needs and wishes, but early intervention should be considered for eyes without spontaneous signs of blood resorption

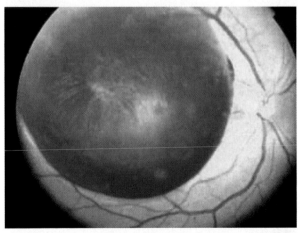

Fig. 24.2.1. Fresh macular hemorrhage (submembranous HMC) in an eye with Terson syndrome after subarachnoid bleeding; the vitreous hemorrhage has already been removed

24.2.1 History

Although named after Terson, who described vitreous hemorrhage in association with spontaneous subarachnoid hemorrhage in 1900 [29], intraocular hemorrhage resulting from subarachnoid hemorrhage was first reported 19 years earlier by Litten [10]. Terson published additional observations on the condition in 1926 [30]. Intraocular hemorrhage following subdural hemorrhage (SDH) was first reported by McDonald in 1931 [14]. Intracranial pressure rise from causes other than hemorrhage (e.g., epidural saline injection) has also been reported to lead to Terson syndrome [21].

24.2.2 Pathophysiology and Pathoanatomy

24.2.2.1 The Origin of the Intraocular Blood

As the most distal extension of the cerebrospinal space and its fluid include the orbital portion of the optic nerve, intraocular blood has been described to

24 III

directly originate from the subarachnoid space [1], entering the globe through the lamina cribrosa [31]. This could hardly explain, though, how subdural blood gets access to the globe.

A much more likely hypothesis is the sudden rise of intracranial pressure due to the extra fluid volume resulting from the extravasating blood. This in turn causes blockage and thus intraluminal pressure elevation in the venous channels draining blood from the globe. Consequent centrifugal rise in the venous pressure is immediate, and rupture of any of the veins inside the eye may then follow [33].

A third option would be partial or complete stasis in the central retinal vein, caused by transmission of the elevated intracranial pressure via the intravaginal spaces of the optic nerve sheath [5].

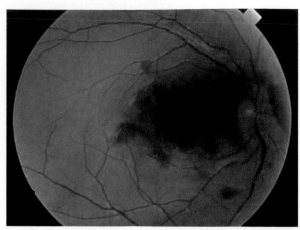

Fig. 24.2.2. Subretinal blood accumulation in an eye following spontaneous clearing of the vitreous hemorrhage after subarachnoid bleeding

24.2.2.2 The Type and Location of the Intraocular Blood

Since the pressure is elevated in each distal vein, hemorrhage can occur anywhere [12]: within the sheath of the optic nerve, in the orbit, and, most importantly, intraocularly. The blood inside the globe can appear in one or multiple locations:

- Subretinal (Fig. 24.2.2)
- Intraretinal
- Preretinal (subhyaloidal; Fig. 24.2.3)
- Intravitreal (Terson syndrome; Fig. 24.2.4)

The intraretinal hemorrhage can be in the deep layers or superficially, under the internal limiting membrane (ILM).

24.2.2.2.1 Blood Accumulation in the Macular Area: Hemorrhagic Macular Cyst

The rapidity of visual rehabilitation depends to a great extent on whether a *hemorrhagic macular cyst* (Fig. 24.2.5) also develops. Based on the exact location of the accumulated blood, two types of hemorrhagic cyst (HMC) are distinguished [19]:

- *Submembranous* HMC (Fig. 24.2.1): the blood is located under the ILM; therefore it is an intraretinal bleeding
- *Preretinal* HMC (Fig. 24.2.3): the blood is located in front of the ILM; therefore this is a subhyaloidal bleeding

The cyst can demonstrate various appearances, depending on its age:

- A fresh HMC contains blood; therefore it is *red* (Fig. 24.2.6)
- If degeneration of the blood has already started, the cyst appears *white* (Fig. 24.2.7)

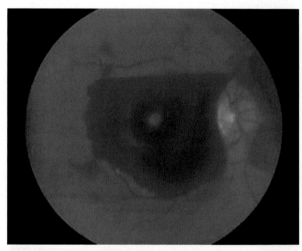

Fig. 24.2.3. Subhyaloidal hemorrhage (preretinal HMC) in a patient with Terson syndrome

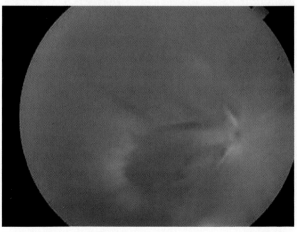

Fig. 24.2.4. Typical appearance of vitreous hemorrhage in a patient with Terson syndrome

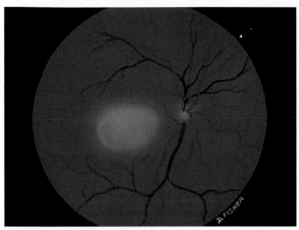

Fig. 24.2.5. Artist's rendering of an HMC in a patient with Terson syndrome. The blood is already in the process of degenerating; hence the HMC's whitish color. The rest of the fundus looks normal

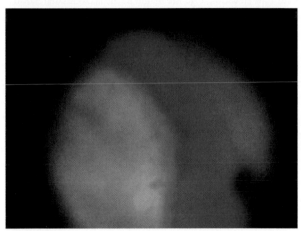

Fig. 24.2.6. Intraoperative image of a submembranous HMC in an eye with Terson syndrome. The condition is fresh, the blood is still red

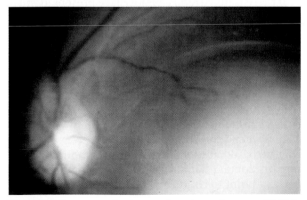

Fig. 24.2.7. Intraoperative image of a submembranous HMC in an eye with Terson syndrome. The condition is a few months old; the blood is partially absorbed already; the remaining blood has turned white

Fig. 24.2.8. Intraoperative image of a submembranous HMC in an eye with Terson syndrome. The condition is several months old; the blood has completely absorbed, leaving clear fluid behind. Within a few weeks this fluid also gets absorbed, leaving behind a collapsed anterior cyst wall (ILM) in the form of surface wrinkling

- If the blood breakdown process is complete, the cyst may still contain clear fluid and appear *transparent* (Fig. 24.2.8)
- The anterior wall of a submembranous cyst may collapse; its surface may be wrinkled and misdiagnosed as an epiretinal proliferation. In reality it is the nonelastic [7] ILM that has been permanently stretched by the blood

24.2.3 Epidemiology

The prevalence of intracranial aneurysm can be as high as 4.8% [22]; however, most of those affected will not develop actual subarachnoid bleeding [34]. The incidence among those 25 years and older was 13.3 per 100,000 in men and 24.4 per 100,000 in women in a Swedish study; interestingly, during the 16 years of observation, the age-standardized incidence in the 25–74 years group decreased significantly in men but remained essentially unchanged in women.

The largest prospective study [3] to date on patients surviving subarachnoid bleeding found that 17% developed some type of intraocular hemorrhage; the rate of Terson syndrome was 7%. Other studies found intraocular hemorrhage rates up to 40% and vitreous hemorrhage rates up to 27% [2, 3, 12, 23, 27, 35]. It must be noted that the figures underestimate the true incidence because they do not take into consideration people who die before they could have been examined, and mortality rates keep falling [28].

The largest study to date on eyes undergoing vitrectomy for Terson syndrome [9] found that 39% of them had HMC (Figs. 24.2.1, 24.2.3, 24.2.5–24.8).

24.2.4 Significance

24.2.4.1 Natural History of Vitreous Hemorrhage in Eyes with Terson Syndrome

Even without surgery, eyes with Terson syndrome generally have a good outcome: in most cases, the blood absorbs spontaneously. It must be noted, however, that complications are not rare with only observation – and today it appears that these complications may outweigh the risks of surgery. Complications include the following:

- Epiretinal membrane formation in 75% [25]
- Development of myopia (incidence unknown) [16]
- Amblyopia in the appropriate age group [4]
- Retinal detachment (incidence unknown) [15]
- Proliferative vitreoretinopathy (PVR) (incidence unknown) [32]

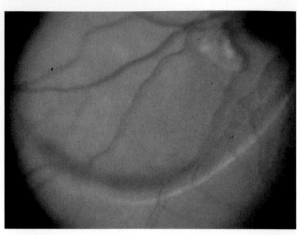

Fig. 24.2.9. Since the ILM is not elastic, its fluidic separation from the rest of the retina by the blood exerts traction along the circumference of the cyst as the ILM reinserts. Visibility of this fold ("macular ring"), appearing as a white line, is proportional with ILM elevation, e.g., the height of the accumulated blood

24.2.4.2 Systemic Significance

The prognosis for patients sustaining subarachnoid hemorrhage has significantly improved, with a 40% survival rate in the mid 1940s [11] to around 80% today [6]. Nevertheless, the presence of vitreous hemorrhage remains an important prognostic sign. In various studies, the mortality rate for patients with subarachnoid hemorrhage is shown to be two to four times higher if Terson syndrome was also present [13, 26]. In one prospective study [3], 100% of patients with Terson syndrome developed coma, as opposed to a 46% rate in patients without vitreous hemorrhage, a statistically significant difference.

24.2.4.3 Human Significance

For most patients with Terson syndrome, especially for those in whom extensive brain damage is also present, visual rehabilitation is a key in achieving improvement in their quality of life. Persons with bilateral vitreous hemorrhage – 22% of patients in Kuhn's study [9] – are completely deprived of sight. Early restoration of vision greatly contributes to the well-being of the individual, family, and society; this must be taken into account when the choice between observation versus surgical intervention is contemplated (see under Sect. 24.2.6.1).

24.2.5 Diagnosis*

Patients with intracranial hemorrhage typically are first seen by a neurologist or neurosurgeon, although occasionally no neurological symptoms are present [20]. Screening of all patients with subarachnoid or subdural bleeding should be the norm [9]. It must be noted that mental problems are not rare in patients with intracranial hemorrhage; therefore they may not be able to communicate that their vision has dropped.

An ophthalmoscope is all that is needed to recognize that vitreous hemorrhage is present. In case of media opacity or if the pupil cannot be adequately dilated, ultrasonography may be necessary.

If there is no vitreous hemorrhage present, it must be noted whether other symptoms such as *optic disc edema*, flamed and dot-like intraretinal, or subretinal hemorrhages have occurred.

Accumulation of *blood in the macular area* (HMC) is almost pathognomic.

A *retinal fold* ("macular ring," Fig. 24.2.9) may be seen at the reinsertion of the detached ILM in eyes with a submembranous HMC [8].

It is very important in young children to verify whether child abuse has caused the subdural bleeding. Intraocular hemorrhages, however, do not necessarily point to child abuse.

---

* Only the ophthalmic aspects are covered here; for the neurological aspects, the reader is referred to the appropriate textbooks.

24.2.5.1 Differential Diagnosis

Vitreous hemorrhage of other etiologies should be excluded.

24.2.6 Treatment*

24.2.6.1 Indication and Counseling

Typically, ophthalmologists use an artificially determined visual acuity value (e.g., 20/200) or duration (e.g., 3 months) to decide whether they recommend further observation or surgical intervention. Not only does this approach have no scientific basis, it completely ignores the patient's individual needs. Some people are more keen than others to hasten visual recovery because of personal or professional reasons; others would rather wait even if the condition is bilateral.

A careful, personalized approach is the most optimal option. The ophthalmologist should discuss the advantages and disadvantages of observation as well as those of surgery, and help the patient – or, in case of a minor or a mentally disabled patient, the family – to make the decision [17]. It is important to emphasize that delaying surgery has its own complications (see above) and that the risk of serious, vitrectomy-related complications is low. Early surgery should therefore definitely be considered [24] with the patient (guardian) having the final say.

---

* Only the ophthalmic aspects are covered here; for the neurosurgical treatment (Fig. 24.2.10), the reader is referred to the appropriate textbooks.

24.2.6.2 Surgery

The only intervention proven to be effective for vitreous hemorrhage is vitrectomy.

- Vitrectomy is standard (three-port)
- Most patients do not require general anesthesia; this is best discussed with the patient/family, and the anesthesiologist
- The posterior hyaloid is usually, but not always, detached. If still adherent, it should be carefully detached
- There is no need to perform extensive shaving at the vitreous base
- If an HMC is discovered, it should be excised in its entirety, including the ILM (Fig. 24.2.11) if the cyst is submembranous [9, 19]. Intentional removal of the still adherent ILM for treating conditions such as macular hole was first recommended based on the excellent long-term experience with ILM removal in eyes with submembranous HMC [18]
- Unless an intraoperative complication warrants it, there is no need to use gas or silicone oil tamponade, or to perform laser retinopexy
- A cataractous lens may be removed simultaneously; a clear lens should be not be extracted, unless the patient is expected to rapidly develop opacity because of advanced age, presence of mild cataract, or intraoperative lens injury

24.2.6.3 Complications

Surgery for vitreous hemorrhage is straightforward; serious intraoperative complications are rare. Delaying vitrectomy, however, increases the risk of retinal detachment, with [32] or without [15] PVR as well as epiretinal formation and amblyopia.

- Occasionally occurring *intraoperative* complications include peripheral retinal tears and lens touch

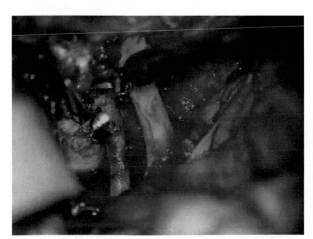

Fig. 24.2.10. Intraoperative image of clipping of the aneurysm during neurosurgery

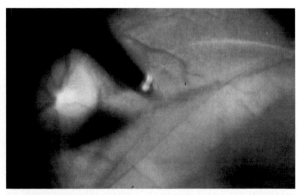

Fig. 24.2.11. Intraoperative image showing removal of the ILM in an eye with submembranous HMC. Degenerated blood products are still visible as is the retinal fold shown in Fig. 24.2.9

24 III

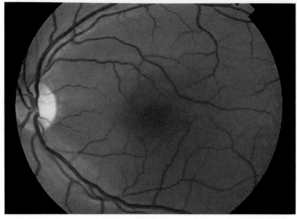

Fig. 24.2.12. Postoperative image of an eye that underwent ILM removal for submembranous HMC in a patient with Terson syndrome. The eye has maintained full vision after a follow-up exceeding 13 years. In this young individual, the shiny reflex from the area without ILM removal clearly delineates it from the central, ILM-deprived area

- Infrequent *postoperative* complications include retinal detachment with or without PVR, and epiretinal proliferation if the ILM has not been removed [36]. The rate of postoperative epiretinal proliferation occurrence is less than in non-vitrectomized eyes. Cataract is inevitable in phakic patients, although their development may be delayed for decades in young patients. No adverse effect caused by ILM removal is expected, even during long-term follow-up (Fig. 24.2.12)
- The prognosis is poor in children, due to concurrent brain damage [9]

24.2.6.4 Follow-up

Other than cataract, complications are not expected to occur once the patient is past the traditional 6-month follow-up. Long-term care, involving psychiatrists, neurologists, physical therapists, social workers, etc., may be necessary for people with nonophthalmological disabilities.

24.2.7 Summary

Screening of patients with acute intracranial bleeding should be performed to allow early recognition of vision-threatening intraocular hemorrhages. Once the diagnosis of Terson syndrome is made, surgery to remove the blood and the often present hemorrhagic macular cyst should be considered. Vitrectomy probably has fewer side effects than long-term observation, and hastens visual rehabilitation in these patients, who commonly have to face other consequences of the intracranial hemorrhage.

References

1. Castano-Duque C, Pons-Irazazabal L, Lopez-Moreno J (1997) Subarachnoid hemorrhage associated to subhyaloid hemorrhage: Terson syndrome. Rev Neurolog 25:1081–1083
2. Fahmy JA (1973) Fundal haemorrhages in ruptured intracranial aneurysms I. Material, frequency, and morphology. Acta Ophthalmol 51:289–298
3. Frizzel R, Kuhn F, Morris R, Quinn C, Fisher W (1997) Screening for ocular hemorrhages in patients with ruptured cerebral aneurysms: A prospective study on 99 patients. Neurosurgery 41:529–534
4. Greenwald MJ, Weiss A, Oesterle CS, Friendly DS (1986) Traumatic retinoschisis in battered babies. Ophthalmology 93:618–625
5. Hayreh S, Edwards J (1971) Ophthalmic arterial and venous pressures: effects of acute intracranial hypertension. Br J Ophthalmol 55:649–663
6. Ikawa F, Ohbayashi N, Imada Y, Matsushige T, Kajihara Y, Inagawa T, Kobayashi S (2004) Analysis of subarachnoid hemorrhage according to the Japanese Standard Stroke Registry Study – incidenc, outcome, and comparison with the International Subarachnoid Aneurysm Trial. Neurol Medico-Chirurg 44:275–276
7. Kuhn F (2003) Internal limiting membrane removal for macular detachment in highly myopic eyes. Am J Ophthalmol 135:547–549
8. Kuhn F, Morris R, Mester V (2000) Macular rings in Terson's syndrome. Acta Ophthalmol 78:721–722
9. Kuhn F, Morris R, Mester V, Witherspoon C (1998) Terson's syndrome. Results of vitrectomy and the significance of vitreous hemorrhage in patients with subarachnoid hemorrhage. Ophthalmology 105:472–477
10. Litten M (1881) Über einige vom allgemein-klinischen Standpunkt aus interessante Augenveränderungen. Berl Klin Wochnschr 18:23–27
11. Manschot W (1944) The fundus oculi in subarachnoid hemorrhage. Acta Ophthalmol 22:281–299
12. Manschot W (1954) Subarachnoid hemorrhage: intraocular symptoms and their pathogenesis. Am J Ophthalmol 38:501–505
13. McCarron M, Alberts M, McCarron P (2004) A systematic review of Terson's syndrome: frequency and prognosis after subarachnoid haemorrhage. J Neurol Neurosurg Psych 75:491–493
14. McDonald A (1931) Ocular lesions caused by intracranial hemorrhage. Trans Am Ophthalmol Soc 29:418–432
15. McRae M, Teasell R, Canny C (1994) Bilateral retinal detachments associated with Tersons syndrome. Retina 14:467–469
16. Miller-Meeks MJ, Bennett SR, Keech RV, Blodi CF (1990) Myopia induced by vitreous hemorrhage. Am J Ophthalmol 109:199–203
17. Morris R, Kuhn F, Witherspoon C (1988) Counseling the eye trauma victim. In: Alfaro V, Liggett P (eds) Vitrectomy in the management of the injured globe. Lippincott Raven, Philadelphia, pp 25–29
18. Morris R, Kuhn F, Witherspoon C (1994) Retinal folds and hemorrhagic macular cysts in Terson's syndrome. Ophthalmology 101:1
19. Morris R, Kuhn F, Witherspoon C, Mester V, Dooner J (1997) Hemorrhagic macular cysts in Terson's syndrome and its implications for macular surgery. Dev Ophthalmol 29:44–54

20. Murthy S, Salas D, Hirekataur S, Ram R (2002) Terson's syndrome presenting as an ophthalmic emergency. Acta Ophthalmol 80:665–666

21. Naseri A, Blumenkranz M, Horton J (2001) Terson's syndrome following epidural saline injection. Neurology 57:364

22. Ogungbo B, Mendelow A, Walker R (2004) The epidemiology, diagnosis and treatment of subarachnoid haemorrhage in Nigeria: what do we know and what do we need to know? Br J Neurosurg 18:362–366

23. Racz P, Bobest M, Szilvassy I (1977) Significance of fundal hemorrhage in predicting the state of the patient with ruptured subarachnoid aneurysm. Ophthalmologica 175:61–66

24. Ritland J, Syrdalen P, Eide N, Vatne H, Overgaard R (2002) Outcome of vitrectomy in patients with Terson syndrome. Acta Ophthalmol 80:172–175

25. Schultz PN, Sobol WM, Weingeist TA (1991) Long-term visual outcome in Terson syndrome. Ophthalmology 98:1814–1819

26. Shaw JHE, Landers M, Sydnor C (1977) The significance of intraocular hemorrhages due to subarachnoid hemorrhage. Ann Ophthalmol 9:1403–1405

27. Shaw JHE, Landers MB (1975) Vitreous hemorrhage after intracranial hemorrhage. Am J Ophthalmol 80:207–213

28. Stegmayr B, Eriksson M, Asplund K (2004) Declining mortality from subarachnoid hemorrhage: changes in incidence and case fatality from 1985 through 2000. Stroke 35:2059–2063

29. Terson A (1900) De l'hemorrhagie dans le corps vitre au cours de l'hemorrhagie cerebrale. La Clinique Ophthalmoloquie 6:309–312

30. Terson A (1926) Le syndrome de l'hematome du corps vitre et de l/hemorrhagie intracranienne spontanes. Ann d'Oculistque 163:666–673

31. Vanderlinden RG, Chisholm LD (1974) Vitreous hemorrhages and sudden increased intracranial pressure. J Neurosurg 41:167–176

32. Velikay M, Datlinger P, Stolba U, Wedrich A, Binder S, Hausmann N (1994) Retinal detachment with severe proliferative vitreoretinopathy in Terson syndrome. Ophthalmology 101:35–37

33. Walsh F, Hedges TJ (1951) Optic nerve sheath hemorrhage: The Jackson Memorial Lecture. Am J Ophthalmol 34:509–527

34. Wiebers D, Piepgras D, Meyer F, Kallmes D, Meissner I, Atkinson J, Link M, Brown RJ (2004) Pathogenesis, natural history, and treatment of unruptured intracranial aneurysms. Mayo Clinic Proc 79:1572–1583

35. Wietholter S, Steube D, Stotz H (1998) Terson syndrome: a frequently missed ophthalmologic complication in subarachnoid hemorrhage. Zentr Neurochirurg 59:166–170

36. Yokoi M, Kase M, Hyodo T, Horimoto M, Kitagawa F, Nagata R (1997) Epiretinal membrane formation in Terson syndrome. Jpn J Ophthalmol 41:168–173

24.3 Disseminated Intravascular Coagulopathy

M.M. Lai, A. Schachat

Core Messages

- Disseminated intravascular coagulopathy (DIC) is a pathologic condition associated with a severe underlying systemic illness characterized by simultaneous bleeding tendency and thrombosis
- The most common ocular manifestation of DIC is bilateral serous retinal detachment, submacular choroidal hemorrhage, or both

- The characteristic histopathologic feature in the eyes of patients with DIC is thrombotic occlusion of the choriocapillaris and adjacent choroidal vessels by fibrin-platelet clots
- Treatment is usually supportive and directed at the underlying systemic illness
- While the systemic prognosis of DIC is poor, the visual prognosis may be good in patients who survive the condition

24.3.1 Introduction

Disseminated intravascular coagulopathy (DIC) is not a primary disease, but rather a pathologic condition associated with a severe underlying systemic illness. It is characterized by simultaneous bleeding tendency and thrombosis. The mechanism of DIC is complex, but involves the simultaneous activation of fibrinolytic and coagulation systems. The process is thought to begin with the entrance of pro-coagulative substances such as thrombin, thromboplastin, or thromboplastic-like substances into the circulation, which leads to destruction of platelets, activation of plasma clotting factors, and activation of the plasminogen system. The secondary consumptive coagulopathy results in generalized hemorrhage. DIC may accompany many systemic illnesses, but is most commonly seen in association with sepsis, neoplasm, severe trauma, cardiopulmonary arrest, and obstetrical complications such as retained fetus, abruptio placentae, placenta previa, and septic abortion. The diagnosis of DIC can generally be made with laboratory studies that demonstrate severe thrombocytopenia, fragmented circulating red blood cells, hypofibrinogenemia, prolonged aPTT, and elevated levels of fibrin split products [2]. Treatment of DIC is directed at the underlying disease process. In addition, success of treatment is dependent on prompt recognition of its systemic manifestations, including characteristic findings in the eye.

24.3.2 History

Azar and colleagues first described ocular features of a neonate who developed DIC in 1974 [1]. This patient had bilateral hyphemas, retinal hemorrhages, and optic disk edema. The authors attributed these ocular findings to the patient's heparin therapy. The first comprehensive description of adult ocular manifestations of DIC was provided by Cogan in 1975 in seven autopsied eyes from patients who died of DIC [3]. Characteristic findings in the posterior segment of the eye including serous retinal detachment, disruption of the retinal pigment epithelium, and thrombotic occlusions of choroidal vessels were present in these cases. Similar ocular findings had been previously described by Yoshioka and Percival in patients with thrombotic thrombocytopenic purpura (TTP), a disease closely related to DIC [9, 10, 13]. Several subsequent studies provided additional clinical and histopathological descriptions of ocular involvement in DIC in a wide range of patients and clinical settings, including neonates, obstetrical complications, and patients with acquired immunodeficiency syndrome [5–8, 12].

24.3.3 Pathologic Features

Essentials
- Thrombotic occlusions of the choriocapillaris and choroidal vessels
- Thrombi consist of fibrin and platelets
- Submacular and peripapillary regions most commonly affected
- Choroidal hemorrhage
- Serous retinal detachment

The characteristic histopathologic finding in eyes of patients with DIC is thrombotic occlusion of the choriocapillaris and adjacent choroidal vessels by

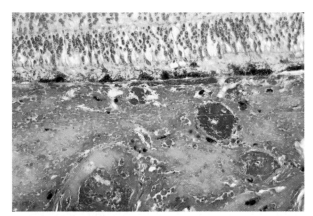

Fig. 24.3.1. Histopathological appearance of choroid and outer retina from a patient who died of DIC. Fibrinous thrombi (stained by phosphotungstic acid hematoxylin) in varying stages of organization and recanalization are seen in the choriocapillaris and large choroidal vessels. The photoreceptor inner and outer segments show marked degeneration and the retinal pigment epithelium exhibits patchy thinning and focal proliferation. (Courtesy of Helmut Buettner [3]). Original magnification × 320

fibrin-platelet clots (Fig. 24.3.1). This mirrors the histopathologic findings from elsewhere in the body.

In mild cases, fibrin-platelet clots may be the sole pathologic change in eyes with DIC. In more severe cases, choroidal hemorrhage can result from more extensive thrombosis of choroidal vessels. The retinal pigment epithelium may be variably disrupted as a result of non-perfusion of the underlying choriocapillaris. Disrupted or destroyed retinal pigment epithelium in turn can allow collection of serous fluid in the subretinal space and hence serous detachment of the overlying retina (Fig. 24.3.2).

Ocular features of DIC are usually confined to the submacular and peripapillary regions. Cogan postulated that this may relate to an unusual anatomic feature of the submacular choroidal vessels, where the posterior ciliary artery rapidly unloads its blood into a vast bed of choriocapillary sinusoids and veins. The rapid deceleration of blood flow in the area may favor the local precipitation of clots in a hypercoagulable condition such as DIC [4].

24.3.4 Clinical Course

Essentials
- Bilateral, usually symmetric process
- Serous retinal detachment
- Choroidal hemorrhage, usually submacular
- Cystoid macular edema
- Retinal pigment epithelium mottling
- Fluorescein angiography: dark central choroid, multifocal leakage from choroidal vessels
- Anterior segment findings in neonates

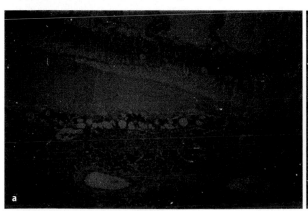
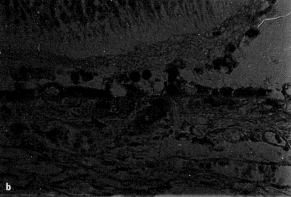

Fig. 24.3.2. a Thrombotic occlusion of the choriocapillaris by fibrinous material. The overlying retinal pigment epithelium is swollen and partially degenerated. Serous detachment of the overlying retina is present. H&E, original magnification ×100. **b** Phosphotungstic acid hematoxylin staining demonstrates the presence of fibrin thrombi in the choriocapillaris. (Courtesy of W. Richard Green, Wilmer Eye Pathology Laboratory). Original magnification ×160

24 III

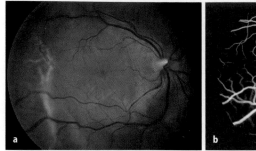 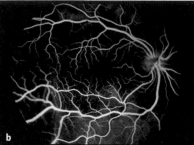 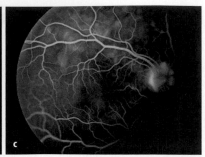

Fig. 24.3.3. a Fundus photograph of a patient with DIC secondary to abruptio placentae shows serous detachment of the posterior and temporal retina. **b** Fluorescein angiography demonstrates absence of fluorescence in large areas of the posterior choroid in the venous phase. **c** In the late phase of the study, multifocal leakage from choroidal vessels into the subretinal space is seen. The optic nerve is also hyperfluorescent. (Courtesy of Helmut Buettner [3])

As is the case in most ocular findings of systemic conditions, ocular features of DIC are bilateral and usually symmetric. The characteristic fundus manifestations consist of serous retinal detachment, hemorrhage in the submacular choroid, or both (Fig. 24.3.3a). However, a wide range of clinical findings may exist. In mild cases, retinal hemorrhages may be the only fundus finding on exam. In more severe cases, a dark fundus consisting of an elevated retina against a dark background of choroidal hemorrhage may be present. The latter appearance has rarely been described in conditions other than DIC and TTP [4]. Other fundus features reported in the literature include cystoid macular edema, optic nerve swelling, and mottling of the retinal pigment epithelium.

Clinical symptoms can vary from no visual disturbance to profound loss of vision. In some patients, blurred vision was posture-dependent secondary to the dependent nature of serous retinal detachment. In most clinical settings, patients with DIC may be too ill to report visual symptoms.

Fluorescein angiography typically reveals areas of non-perfusion in the posterior choriocapillaris in early phases of the study (Fig. 24.3.3b). Multifocal leakage from adjacent choroidal vessels with accumulation of dye in the subretinal space is seen in later phases of the study (Fig. 24.3.3c) [4, 5]. Choroidal hemorrhage, when present, causes blocked fluorescence. Retinal pigment epithelium mottling may cause patchy areas of increased fluorescence secondary to window defects [5].

Ophthalmic findings of DIC are usually confined to the posterior segment of the eye. However, anterior segment findings have been reported in a few cases of DIC in infants and neonates. Azar et al. reported an infant with DIC who developed bilateral hyphemas in addition to retinal hemorrhages and optic disk edema, although the authors attributed these ocular findings to the patient's heparin therapy [1]. Ortiz and colleagues reported two cases of neo-natal DIC with intravascular fibrin clots in the ciliary body and iris vessels [8].

24.3.5 Differential Diagnosis

The diagnosis of DIC is usually not in doubt by the time an ophthalmic examination has been requested. In the appropriate clinical setting, the diagnosis of systemic DIC is confirmed by laboratory tests that show severe thrombocytopenia, fragmented circulating red blood cells, hypofibrinogenemia, prolonged aPTT, and elevated levels of fibrin split products. A closely related hematologic condition, thrombotic thrombocytopenic purpura (TTP), may present with similar systemic and ocular manifestations.

The differential diagnosis for ocular complications of DIC includes other systemic conditions that produce serous retinal detachment (Table 24.3.1) [12].

Table 24.3.1. Systemic conditions associated with serous retinal detachment [12]

Systemic condition	Disease
Inflammatory	Sarcoidosis
	Vogt-Koyanagi-Harada
Infectious	Syphilis
	Cat-scratch disease
Choroidal ischemic vasculopathy	Disseminated intravascular coagulopathy
	Thrombotic thrombocytopenic purpura
	Systemic lupus erythematosus
	Eclampsia
Hematologic	Paraproteinemias
Renal	Goodpasture syndrome
	Type II glomerulonephritis
	Hemodialysis
	IgA nephropathy
Vascular malformation	Sturge-Weber
	Von Hippel-Lindau
Malignancy	Leukemia
	Myeloma

24.3.6 Treatment and Follow-up

Essentials
- Treat underlying condition
- Poor systemic prognosis
- Good visual recovery in patients who survive DIC

Treatment of DIC consists of supportive care and is directed at the underlying cause. Historically, the prognosis for DIC has been poor with a high mortality rate. This is supported by the fact that most published reports of ocular manifestations of DIC consist of autopsy studies.

Long-term follow-up data of eyes affected with DIC are not available in the literature, most likely due to the poor prognosis of this condition. In the few cases of surviving patients, visual acuity usually returns to the 20/20 to 20/30 range as serous retinal detachment resolves [5, 7]. However, this may be secondary to selection bias as patients who survive DIC may be more likely to have milder forms of the condition.

References

1. Azar P, Smith RS, Greenberg MH (1974) Ocular findings in disseminated intravascular coagulation. Am J Ophthalmol 78:493–496
2. Bell WR (1998). Bleeding disorders. In: Stobo JD, Hellmann DB, Ladenson PW, Petty BG, Traill TA (eds) The principle and practice of medicine. Appleton and Lange, Stamford, pp 748–750
3. Buettner H (1983) Department of Ophthalmology, Mayo Clinic. Ophthalmology 90:914–916
4. Cogan DG (1975) Ocular involvement in disseminated intravascular coagulopathy. Arch Ophthalmol 93:1–8
5. Hoines J, Buettner H (1989) Ocular complications of Disseminated intravascular coagulation (DIC) in abruption placentae. Retina 9:105–109
6. Lertsumitkul S, Whitcup SM, Chan C-C (1997) Ocular manifestations of disseminated intravascular coagulation in a patient with the acquired immunodeficiency syndrome. Arch Ophthalmol 115:676–677
7. Martin VAF (1978) Disseminated intravascular coagulopathy. Trans Ophthal Soc UK 98:506–507
8. Ortiz JM, Yanoff M, Cameron JD, Schaffer D (1982) Disseminated intravascular coagulation in infancy and in the neonate: ocular findings. Arch Ophthalmol 100:1413–1415
9. Percival SPB (1970) Ocular findings in thrombotic thrombocytopenic purpura (Moschcowitz's disease). Br J Ophthalmol 54:73–78
10. Percival SPB (1970) The eye in Moschcowitz's disease (Thrombotic thrombocytopenic purpura); a review of 182 cases. Trans Ophthalmol Soc UK 90:375–382
11. Samples JR, Buettner H (1983) Ocular involvement in disseminated intravascular coagulation (DIC). Ophthalmology 90:914–916
12. Wolfensberger TJ, Tufail A (2000) Systemic disorders associated with detachment of the neurosensory retina and retinal pigment epithelium. Curr Opin Ophthalmol 11:455–461
13. Yoshioka H, Yamanouchi U (1962) Case of retina and vitreous body hemorrhage due to thrombocytopenic purpura with special reference to pathohistological findings of the eye. Nagasaki Med J 37:234–237

24.4 Bone Marrow Transplant Associated Retinopathy

H. Tabandeh, N. Rafiei, A.P. Schachat

Core Messages

- Retinopathy occurs in 4.3–10% of patients with bone marrow transplants
- Bone marrow transplant associated retinopathy includes occlusive retinal microangiopathy, choroidopathy, and optic neuropathy
- Predisposing factors include capillary endothelial damage by drugs or irradiation, microemboli, hematological abnormalities and secondary diseases such as hypertension and septicemia
- Treatment is not required unless proliferative retinopathy develops

24.4.1 Introduction

Essentials
- Allogeneic BMT involves:
  - Conditioning by chemotherapy, total body irradiation, prophylactic cranial irradiation
  - Transfusion of human leukocyte antigen (HLA) matched donor hematopoietic progenitor cells
  - Immunosuppression
  - Supportive treatment until leukocyte, erythrocyte, and platelet population stabilize

Advances in immunosuppressive and antibacterial therapies during recent decades have significantly improved the success of bone marrow transplantation (BMT). BMT has become an important treatment modality for hematological and other forms of malignancies, and hereditary and acquired disorders such as sickle cell and aplastic anemias [1]. The international bone marrow transplant registry estimated that about 50,000 bone marrow transplants were performed during 1999. BMT involves intravenous infusion of hematopoietic progenitor cells in order to repopulate the dysfunctional host bone marrow, reestablishing the function.

Bone marrow transplant can be classified into allogeneic, autologous, and syngeneic transplant depending on the source of the donor tissue. Allogeneic transplant utilizes bone marrow obtained from an HLA-matched donor. Autologous transplant involves harvesting of the patient's own bone marrow that is subsequently transplanted back to the patient. Synergic bone marrow transplant is the least commonly performed type of BMT and involves the transplantation of bone marrow from one monozygotic twin to the other.

In most allogeneic BMTs a conditioning regimen is required before the transplantation can take place [7, 30, 38, 40]. Conditioning ablates the recipient's bone marrow in order to allow subsequent repopulation with donor cells. In the case of malignant disorders such as leukemia, conditioning also helps treat malignant cells that have spread throughout the body. By suppressing the immune system, conditioning reduces and may prevent the ability of the host to mount an immune reaction to the infused allogeneic bone marrow cells, therefore reducing the possibility of transplant rejection. Conditioning regimens include chemotherapeutic agents with or without total body irradiation (TBI) and prophylactic cranial irradiation. The most commonly used chemotherapeutic and immunosuppressive agents include cyclophosphamide, busulfan, and cytosine arabinoside. Following conditioning, the donor bone marrow is infused intravenously, allowing the cells to reach the host bone marrow. Prolonged immunosuppression is required in order to prevent the rejection of the transplanted tissue by the host, as well as avoid reaction of transplanted immunocompetent donor cells against the host, resulting in graft versus host disease (GVHD). Immunosuppressive drug therapies include corticosteroid, cyclosporine, and methotrexate in various combinations.

The major systemic complications of BMT include vaso-occlusive disease, infection, graft failure, GVHD, and hematologic abnormalities. GVHD mostly affects the skin, liver, gastrointestinal tract, and the eyes. In GVHD the transplanted bone marrow perceives the host as a foreign tissue and mounts a T-lymphocyte-mediated immunologic attack on the host [1, 7, 30, 41].

Ophthalmic complications of BMT may result from the underlying disease process, the conditioning regimen, immunosuppressive therapy, or secondary disorders such as infections, GVHD, hypertension, and hematologic abnormalities. Anterior segment complications of BMT are more common than those affecting the posterior segment and include keratoconjunctivitis sicca, pseudomembranous conjunctivitis, conjunctival graft-versus-host disease, bacterial and viral conjunctivitis, superficial punctate keratitis, sterile and infectious corneal ulceration, and cataracts [2, 6, 11, 14, 21–25]. Posterior segment manifestations of BMT include vitritis, retinopathy, choroidopathy, and optic neuropathy [3, 4, 11, 13, 15, 17, 18, 21, 27, 34].

Retinopathy associated with BMT was initially reported in 1983, consisting of multiple, bilateral cotton-wool spots, retinal hemorrhages, optic disk edema and ocular infections [18, 21, 22]. The term "BMT retinopathy" has been used to describe a number of retinal and posterior segment abnormalities that are encountered in patients who undergo BMT. The condition includes an occlusive retinal microvasculopathy, choroidopathy, optic neuropathy, retinal manifestations of hematological abnormalities, and infections. The retina findings are not specific to BMT and occur in other disorders that can predispose to occlusive retinopathy. The term "BMT associated retinopathy" will be used throughout the current chapter in an attempt to represent the condition more accurately.

24.4.2 Pathophysiology

A number of factors contribute to the development of BMT associated retinopathy (Table 24.4.1). Capillary endothelial damage results from composite insults related to drug toxicity, irradiation, or immune system abnormalities. Injury to retinal capillary endothelium leads to microvascular occlusion and incompetence causing ischemia, hemorrhage, vascular leakage, edema, hard exudates, microaneurysms, and telangiectasia. Webster and coworkers suggested that normal T-cell function may be necessary for the protection of retinal endothelial cells from insults such as radiation, which lower the threshold for subsequent endothelial loss and retinal ischemia [44]. Microvascular occlusion can also result from microembolization of small cellular aggregates during the administration of donor bone marrow as well as transfusion of platelet, red blood cells and other blood products in the period following transplantation. Septicemia complicating BMT may be associated with clinically significant septic embolization to the retina. Hematological abnormalities such as hypercoagulable states, hyperviscosity, thrombocytopenia, and anemia are often present in patients who undergo BMT and predispose to occlusive or hemorrhagic vasculopathy. Similarly, development of secondary conditions such as hypertension, or the presence of preexisting conditions such as diabetes, may promote microvasculopathy.

Reversible visual loss due to cyclosporine-induced retina toxicity has been reported in BMT patients [33]. Electrophysiological studies confirmed retinal dysfunction, with the improvement of visual function after withdrawal of cyclosporine.

Table 24.4.1. Factors predisposing to BMT associated retinopathy

Capillary endothelial damage
Drugs
Irradiation
Immune system abnormalities
Microemboli
Leukocyte microaggregates
Platelet transfusion
Blood transfusion
Septic emboli
Hematological abnormalities
Hypocoagulable state
Hyperviscosity
Anemia
Thrombocytopenia
Secondary diseases
Hypertension
Septicemia
Coexisting predisposing diseases
Diabetes mellitus
Sickle cell disease
Other disorder associated with microvascular retinopathy
Neurotoxicity
Drugs

24 III

24.4.3 Bone Marrow Transplant Associated Retinopathy

Essentials

- BMT associated retinopathy is usually associated with a good prognosis
- Following diagnosis, regular follow-up is recommended
- Treatment is not required in most cases
- Panretinal laser photocoagulation is considered in cases with proliferative retinopathy
- In cases with suspected drug toxicity, discontinuation of the offending medication should be considered
- Infections involving the posterior segment of the eye are often associated with systemic infection and should be treated in collaboration with the medical oncology team

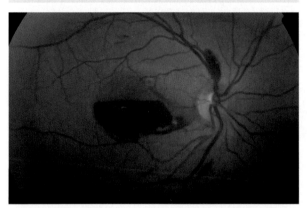

Fig. 24.4.1. Bone marrow transplant associated retinopathy in a 26-year-old patient with lymphoma. Retrohyaloid hemorrhage covering the central macular area reduced the vision to 20/400. Flame-shaped hemorrhages and white centered hemorrhages (pseudo-Roth hemorrhage), and cotton-wool spots were present. Visual acuity improved to 20/25 within 2 months

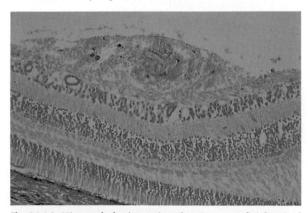

Fig. 24.4.2. Histopathologic section showing superficial retinal hemorrhage in a 19-year-old male following bone marrow transplantation for acute lymphocytic leukemia. (Courtesy of W. Richard Green, MD)

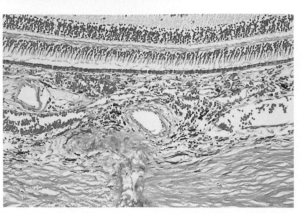

Fig. 24.4.3. Histopathologic section showing cellular infiltration of choroid in a patient with graft versus host disease. (Courtesy of W. Richard Green, MD)

BMT associated retinopathy comprises a number of retinal and posterior segment abnormalities that occur in patients who undergo BMT. The condition includes an occlusive retinal microvasculopathy, choroidopathy, optic neuropathy, retinal manifestations of hematological abnormalities, and infections.

In one report, 51 (12.8%) of 397 patients developed posterior segment complications after BMT [11]. The cause of these findings is likely to be multifactorial, resulting from the combined effects of high dose chemotherapy, TBI, infections, GVHD, immunosuppressive therapy, and recurrent malignancies.

24.4.3.1 Microvascular Retinopathy

BMT microvascular retinopathy is characterized by the presence of multiple superficial and intraretinal hemorrhages, microaneurysms, cotton-wool spots, telangiectasia, macular edema, and hard exudates. Rarely, severe capillary non-perfusion may result in retinal fibrovascular proliferation and neovascularization of the iris. BMT associated retinopathy is usually bilateral and symmetrical. Patients may be visually asymptomatic or present with decreased vision and scotoma. The reported occurrence of retinopathy ranges from 4.3% to 10% of patients with BMT. The variable rate likely represents variability in the diagnostic criteria, as well as differences in conditioning regimens and irradiation protocols.

BMT associated retinopathy usually occurs within 6 months of the transplantation [5, 21, 34] although there are reported cases of microvascular retinopathy occurring as late as 62 months after BMT [9]. Coexisting diseases such as diabetes mellitus and hypertension facilitate the development of BMT retinopathy by contributing to the microvasculopathy [19]. In patients that undergo BMT, hypertension may develop secondary to nephropathy or as a side effect of cyclosporine. Although mild to moderate ischemic retinopathy represented by retinal hemorrhages and cotton-wool spots is frequently seen in BMT retinopathy, severe ischemia resulting in retinal neovascularization is rare. Lopez et al. reported an occlusive microvascular retinopathy in five of eight (62%) long-term survivors of autologous and allogeneic transplant for acute leukemia [34]. Of this group, only one patient developed retinal neovascularization. Fluorescein angiography findings in patients with BMT associated retinopathy include microaneurysms, telangiectasia, intraretinal microvascular abnormalities, leakage consistent with macular edema, and a variable degree of capillary nonperfusion [5, 27].

Risk factors for development of BMT associated retinopathy include allogeneic BMT, TBI, prophylactic irradiation, shortened interval between TBI and BMT, acute lymphoblastic leukemia as the underlying disease, cyclosporine, prolonged immunosuppression, and the presence of other primary or secondary disorders that predispose to retinopathy.

The differential diagnosis for BMT associated retinopathy includes radiation retinopathy and other causes of microvascular retinopathy (Table 24.4.2). There are similarities and often an overlap between radiation and BMT associated retinopathies. Both conditions result in retinal hemorrhages, cotton-wool spots, telangiectasia, and other manifestations of an occlusive microvasculopathy. The retinopathy associated with BMT usually occurs within 6 months of transplantation and is typically reversible with a relatively benign course. The mean time for the development of radiation retinopathy is 18 months. Once it develops, radiation retinopathy often has a progressive course, often resulting in permanently decreased visual function [8].

Table 24.4.2. Differential diagnosis of BMT associated retinopathy

Radiation retinopathy
Hypertensive retinopathy
Leukemia retinopathy
Diabetic retinopathy
Retinal vein occlusion
Purtscher-like retinopathy
HIV retinopathy
Collagen vascular disorders
Ocular ischemic syndrome
Retinitis
Other forms of microvascular retinopathy

24.4.3.2 Retinal Pigment Epitheliopathy and Choroidopathy

Central serous chorioretinopathy (CSCR) has been reported to occur in association with bone marrow and solid organ transplantation. Friberg and Eller reported occurrence of CSCR in cardiac and renal transplant recipients [16]. Others have reported cases of CSCR after BMT [15, 26]. In a study of 270 patients with BMT, Karishma et al. described two patients with CSCR, who responded well to photocoagulation [26]. BMT associated CSCR has also been reported in the setting of GVHD [10, 15]. The combined effects of high doses of corticosteroid, emotional stress, systemic hypertension, cyclosporine, and other chemotherapeutic agents contribute to the pathogenesis of CSCR after bone marrow transplantation [10, 16, 21, 26]. Stewart et al. suggested that leukemic infiltration causes choriocapillary ischemia and reversible disruption of retinal pigment epithelium function, predisposing to serous retinal detachment [39]. The visual prognosis for CSCR in transplanted patients is generally good, and photocoagulation may be required only in persistent cases.

24.4.3.3 Hematologic Complications

Hematologic complications occur in the early postoperative period and are presumably related to the iatrogenic bone marrow aplasia, resulting in anemia, leukopenia, and thrombocytopenia. These changes lead to the development of intraretinal hemorrhages, microaneurysms, cotton-wool spots, and vitreous hemorrhage. In addition, during the early posttransplantation period, most patients receive frequent blood and platelet transfusions as well as other blood products. Blood and platelet transfusions are associated with formation of leukocyte, red blood cell, and platelet microaggregates. Embolization of these microaggregates to the retina vasculature results in an occlusive microvasculopathy with characteristic features of retinal hemorrhages, microaneurysms, and cotton-wool spots [31]. Cellular microaggregate embolization may also occur with the initial transfusion of the donor bone marrow. The leukopenic state and presence of multiple indwelling catheters predispose to the development of septicemia and subsequent septic microembolization that can potentially cause an occlusive microvascular retinopathy.

In a retrospective study of 397 patients, Coskuncan and coworkers reported 14 (3.5%) patients developed vitreous or intraretinal hemorrhages at a median of 51 days after BMT [11]. Over 90% of hemorrhagic complications were observed within 6 months after the transplantation. Visual loss due to vitreous and/or preretinal hemorrhages occurred in 1% of patients, and usually resolved spontaneously. In an autopsy study, 11 (28.9%) of 39 patients who were recipients of BMT had intraretinal hemorrhages [22].

Thrombocytopenia is an important factor in the development of retinal hemorrhages in patients who undergo BMT [42]. Hirst et al. reported two patients out of 45 patients that had transient visual loss as a result of subhyaloidal hemorrhages. The hemorrhages cleared with resolution of the underlying thrombocytopenia [21].

24.4.3.4 Optic Neuropathy

Bilateral optic disk edema has been described after BMT. One study observed optic disk edema in 11 of 397 patients after BMT [11]. In 8 patients, the disk edema resolved after discontinuation of cyclosporine, suggesting cyclosporine toxicity as the cause. Khawly and associates noted optic neuropathy in six of nine patients with advanced breast cancer who underwent autologous BMT without cyclosporine therapy [27]. All patients subsequently developed optic disk pallor.

24.4.3.5 Infectious Complications

Systemic infections account for most of the morbidity and mortality associated with bone marrow transplantation. Ocular infections are relatively uncommon and occur in approximately 2% of the patients [11]. Ocular infections are part of the spectrum of chorioretinopathies associated with BMT and are related to the recipient's iatrogenic immunodeficient state. The advent of effective antibacterial therapy has markedly decreased the occurrence and severity of both systemic and ocular bacterial infections in BMT patients. In the series of 397 patients reported by Coskuncan and associates, no cases of bacterial retinitis or endophthalmitis were encountered after BMT. Fungal infections, including *Candida* endophthalmitis and *Aspergillus* retinitis, were the most common intraocular infections, occurring in 1.5% of the patients [11]. Fungal infections generally occur within the first 4 months of BMT, with a median time to diagnosis of 2 months. The immunodeficient status and the increased frequency of invasive procedures such as intravenous lines during the first few months after BMT account for the higher incidence of infections. Although rare cases of *Aspergillus* endophthalmitis have been described, ocular aspergillosis is often limited to the subretinal or subretinal pigment epithelium space and is frequently associated with a negative yield for vitreous biopsy. *Fusarium* endophthalmitis, resulting in enucleation

of the affected eye, has been reported following autologous BMT for leukemia [37].

Viral retinitis caused by cytomegalovirus (CMV), herpes zoster (HZV) and herpes simplex (HSV) viruses is among the late ocular complications of BMT [11]. In a series of 785 patients who underwent BMT, 1% developed CMV retinitis [45]. In another study, only one patient out of 379 patients had CMV retinitis [11]. The incidence of CMV retinitis appears to be higher in HLA-matched unrelated donor transplants than in HLA-matched sibling transplants [29]. Repeated treatment with antithymocyte globulin and the use of certain immunosuppressive agents such as mycophenolate mofetil may increase the likelihood of CMV retinitis [20, 28]. CMV retinitis in the setting of BMT often responds to treatment with systemic and intravitreal administration of foscarnet and ganciclovir [35]. Severe, progressive CMV retinitis may result in poor visual outcome [29]. Prophylactic use of valacyclovir for prevention of CMV infection in BMT patients has been advocated by some authors.

Although cutaneous involvement by herpes zoster virus is a frequent complication following BMT and solid organ transplants, herpes zoster retinitis is an infrequent occurrence [43]. In a series of 397 patients, Coskuncan and coworkers reported one case of herpes zoster retinitis that occurred following disseminated herpes zoster infection 370 days after the BMT [11]. The clinical features of herpes zoster retinitis are similar to those of the acute retinal necrosis, and the progressive outer retinal necrosis. Management includes intravenous acyclovir therapy, as well as systemic or intravitreal ganciclovir or foscarnet for the refractory cases [32].

Toxoplasma retinochoroiditis may occur as an early or late complication of BMT, particularly in patients with GVHD who are chronically immunosuppressed [11]. It can be either unilateral or bilateral and is often due to reactivation of preexisting ocular toxoplasmosis [36], although acquired primary ocular disease may also occur. Many patients with old *Toxoplasma* chorioretinal scars do not experience reactivation of their disease after BMT. Ocular toxoplasmosis responds well to appropriate antibiotic therapy including pyrimethamine, sulfadiazine, and clindamycin. *Toxoplasma* retinochoroiditis in BMT patients can be associated with cerebral dissemination, resulting in significant morbidity and mortality. It is imperative to follow up these patients with serial CT scans of the brain [12].

24.4.3.6 Other Complications

Other posterior segment complications of BMT include posterior scleritis and serous retinal detachment.

24.4.4 Pathologic Features

In a light and electron microscopic study of a patient with BMT, Webster and coworkers noted infarction of the innermost layers of the retina with shrinkage of the inner plexiform layer, foci of cystoid bodies within the nerve fiber layer consistent with cotton-wool spots, and no gross abnormality of the outer layers of the neurosensory retina, retinal pigment epithelium, or choroid [44]. Electron microscopy confirmed the cystoid bodies as axonal swellings filled with degenerate organelles. Surrounding capillaries revealed loss of endothelial cells with relatively preserved pericytes. The authors suggested that normal T-cell function may be necessary for the protection of retinal endothelial cells from insults such as radiation, which may lower the threshold for subsequent endothelial loss and retinal ischemia. In an autopsy study of 39 patients with BMT, Jabs and coworkers noted infiltration of choroid by histiocyte-like cells, occasionally accompanied by chronic inflammatory cells in 80 % of cases and retinal hemorrhages in 29 % [22].

24.4.5 Management

BMT associated retinopathy generally has a good visual prognosis. The retinopathy typically resolves within 2–4 months after the cessation or lowering of the dosage of cyclosporine with or without use of systemic prednisone [5, 11]. Bernauer and coworkers noted complete resolution of the retinal findings in 69 % of patients, and full recovery of the baseline visual acuity in 46 % of patients [5]. In another study, the median visual acuity after BMT retinopathy was 20/50 [34]. Because of the relatively favorable prognosis and non-progressive nature of most cases of BMT associated retinopathy, aggressive treatment is usually not necessary. In severe cases resulting in proliferative retinopathy, panretinal laser photocoagulation may be indicated.

Ophthalmologic examination of patients after BMT with regular follow-up of those with retinopathy is recommended. In cases where drug toxicity is assumed to be a predisposing factor, discontinuation of the suspected medication should be considered. Ocular infections are treated in conjunction with the medical oncology team.

References

1. Armitage JO (1994) Bone marrow transplantation. N Engl J Med 330:827–838
2. Arocker-Mettinger E, Skorpik F, Grabner G, Hinterberger W, Gadner H (1991) Manifestations of graft-versus-host disease following allogenic bone marrow transplantation. Eur J Ophthalmol 1:28–32
3. Avery R, Jabs DA, Wingard JR, Vogelsang G, Saral R, Santos G (1991) Optic disc edema after bone marrow transplantation. Possible role of cyclosporine toxicity. Ophthalmology 98:1294–1301
4. Bernauer W, Gratwohl A (1992) Bone marrow transplant retinopathy. Am J Ophthalmol 113:604–605
5. Bernauer W, Gratwohl A, Keller A, Daicker B (1991) Microvasculopathy in the ocular fundus after bone marrow transplantation. Ann Intern Med 115:925–930
6. Bray LC, Carey PJ, Proctor SJ, Evans RG, Hamilton PJ (1991) Ocular complications of bone marrow transplantation. Br J Ophthalmol 75:611–614
7. Bron D (1994) Graft-versus-host disease. Curr Opin Oncol 6:358–364
8. Brown GC, Shields JA, Sanborn G, Augsburger JJ, Savino PJ, Schatz NJ (1982) Radiation retinopathy. Ophthalmology 89:1494–1501
9. Bylsma GW, Hall AJ, Szer J, West R (2001) Atypical retinal microvasculopathy after bone marrow transplantation. Clin Exp Ophthalmol 29:225–229
10. Cheng LL, Kwok AK, Wat NM, Neoh EL, Jon HC, Lam DS (2002) Graft-vs-host-disease-associated conjunctival chemosis and central serous chorioretinopathy after bone marrow transplant. Am J Ophthalmol 134:293–295
11. Coskuncan NM, Jabs DA, Dunn JP, Haller JA, Green WR, Vogelsang GB, Santos GW (1994) The eye in bone marrow transplantation. VI. Retinal complications. Arch Ophthalmol 112:372–379
12. Derouin F, Devergie A, Auber P, Gluckman E, Beauvais B, Garin YJ, Lariviere M (1992) Toxoplasmosis in bone marrow-transplant recipients: Report of seven cases and review. Clin Infect Dis 15:267–270
13. Dunn JP, Jabs DA (1992) Ocular microvasculopathy after bone marrow transplantation. Ann Intern Med 116:956–957
14. Dunn JP, Jabs DA, Wingard J, Enger C, Vogelsang G, Santos G (1993) Bone marrow transplantation and cataract development. Arch Ophthalmol 111:1367–1373
15. Fawzi AA, Cunningham ET, Jr (2001) Central serous chorioretinopathy after bone marrow transplantation. Am J Ophthalmol 131:804–805
16. Friberg TR, Eller AW (1990) Serous retinal detachment resembling central serous chorioretinopathy following organ transplantation. Graefes Arch Clin Exp Ophthalmol 228:305–309
17. Gloor B, Gratwohl A, Hahn H, Kretzschmar S, Robert Y, Speck B, Daicker B (1985) Multiple cotton wool spots following bone marrow transplantation for treatment of acute lymphatic leukaemia. Br J Ophthalmol 69:320–325
18. Gratwohl A, Gloor B, Hahn H, Speck B (1983) Retinal cotton-wool patches in bone-marrow-transplant recipients. N Engl J Med 308:1101
19. Gray RH, Tighe M, Russell NH (2000) Rapid onset retinopathy in a diabetic patient following bone marrow transplantation. Bone Marrow Transplant 26:695–696
20. Hambach L, Stadler M, Dammann E, Ganser A, Hertenstein B (2002) Increased risk of complicated CMV infection with the use of mycophenolate mofetil in allogeneic stem cell transplantation. Bone Marrow Transplant 29:903–906
21. Hirst LW, Jabs DA, Tutschka PJ, Green WR, Santos GW (1983) The eye in bone marrow transplantation. I. Clinical study. Arch Ophthalmol 101:580–584
22. Jabs DA, Hirst LW, Green WR, Tutschka PJ, Santos GW, Beschorner WE (1983) The eye in bone marrow transplantation. II. Histopathology. Arch Ophthalmol 101:585–590

24 III

23. Jabs DA, Wingard J, Green WR, Farmer ER, Vogelsang G, Saral R (1989) The eye in bone marrow transplantation. III. Conjunctival graft-vs-host disease. Arch Ophthalmol 107: 1343–1348

24. Jack MK, Hicks J (1981) Ocular complications in high-dose chemoradiotherapy and marrow transplantation. Ann Ophthalmol 13:709–711

25. Jack MK, Jack GM, Sale GE, Shulman HM, Sullivan KM (1983) Ocular manifestations of graft-v-host disease. Arch Ophthalmol 101:1080–1084

26. Karashima K, Fujioka S, Harino S (2002) Two cases of central serous chorioretinopathy treated with photocoagulation after bone marrow transplantation. Retina 22:651–653

27. Khawly JA, Rubin P, Petros W, Peters WP, Jaffe GJ (1996) Retinopathy and optic neuropathy in bone marrow transplantation for breast cancer. Ophthalmology 103:87–95

28. Kuriyama K, Todo S, Ikushima S, Fujii N, Yoshihara T, Tsunamoto K, Naya M, Hojo M, Hibi S, Morimoto A, Imashuku S (2001) Risk factors for cytomegalovirus retinitis following bone marrow transplantation from unrelated donors in patients with severe aplastic anemia or myelodysplasia. Int J Hematol 74:455–460

29. Larsson K, Lonnqvist B, Ringden O, Hedquist B, Ljungman P (2002) CMV retinitis after allogeneic bone marrow transplantation: a report of five cases. Transpl Infect Dis 4:75–79

30. Lazarus HM, Rowe JM (1995) New and experimental therapies for treating graft-versus-host disease. Blood Rev 9: 117–133

31. Leveille AS, Morse PH (1981) Platelet-induced retinal neovascularization in leukemia. Am J Ophthalmol 91:640–643

32. Lewis JM, Nagae Y, Tano Y (1996) Progressive outer retinal necrosis after bone marrow transplantation. Am J Ophthalmol 122:892–895

33. Lopez-Jimenez J, Sanchez A, Fernandez CS, Gutierrez C, Herrera P, Odriozola J (1997) Cyclosporine-induced retinal toxic blindness. Bone Marrow Transplant 20:243–245

34. Lopez PF, Sternberg P, Jr, Dabbs CK, Vogler WR, Crocker I, Kalin NS (1991) Bone marrow transplant retinopathy. Am J Ophthalmol 112:635–646

35. Okamoto T, Okada M, Mori A, Saheki K, Takatsuka H, Wada H, Tamura A, Fujimori Y, Takemoto Y, Kanamaru A, Kakishita E (1997) Successful treatment of severe cytomegalovirus retinitis with foscarnet and intraocular injection of ganciclovir in a myelosuppressed unrelated bone marrow transplant patient. Bone Marrow Transplant 20:801–803

36. Peacock JE, Jr, Greven CM, Cruz JM, Hurd DD (1995) Reactivation toxoplasmic retinochoroiditis in patients undergoing bone marrow transplantation: is there a role for chemoprophylaxis? Bone Marrow Transplant 15:983–987

37. Robertson MJ, Socinski MA, Soiffer RJ, Finberg RW, Wilson C, Anderson KC, Bosserman L, Sang DN, Salkin IF, Ritz J (1991) Successful treatment of disseminated Fusarium infection after autologous bone marrow transplantation for acute myeloid leukemia. Bone Marrow Transplant 8:143–145

38. Santos GW, Kaiser H (1982) Bone marrow transplantation in acute leukemia. Semin Hematol 19:227–239

39. Stewart MW, Gitter KA, Cohen G (1989) Acute leukemia presenting as a unilateral exudate retinal detachment. Retina 9:110–114

40. Storb R (1995) Bone marrow transplantation. Transplant Proc 27:2649–2652

41. Storek J, Saxon A (1992) Reconstitution of B cell immunity following bone marrow transplantation. Bone Marrow Transplant 9:395–408

42. Suh DW, Ruttum MS, Stuckenschneider BJ, Mieler WF, Kivlin JD (1999) Ocular findings after bone marrow transplantation in a pediatric population. Ophthalmology 106:1564–1570

43. Walton RC, Reed KL (1999) Herpes zoster ophthalmicus following bone marrow transplantation in children. Bone Marrow Transplant 23:1317–1320

44. Webster AR, Anderson JR, Richards EM, Moore AT (1995) Ischaemic retinopathy occurring in patients receiving bone marrow allografts and campath-1G: a clinicopathological study. Br J Ophthalmol 79:687–691

45. Wingard JR, Chen DY, Burns WH, Fuller DJ, Braine HG, Yeager AM, Kaiser H, Burke PJ, Graham ML, Santos GW, et al. (1988) Cytomegalovirus infection after autologous bone marrow transplantation with comparison to infection after allogeneic bone marrow transplantation. Blood 71:1432–1437

25 Inflammatory Vascular Disease

25.1 Eales' Disease

S. GADKARI

Core Messages
- Eales' disease is essentially a clinical entity, presenting as a recurrent vitreous hemorrhage in young adult males
- The clinical picture is one of an idiopathic retinal vasculitis (periphlebitis) and its sequelae
- Bilateral involvement in an economically productive age group makes it an important ophthalmic health care issue in South Asia
- Its exact etiopathology has evaded explanation, though the role of *Mycobacterium* tuberculosis, oxidative stress and immunological mechanisms have been studied
- Visual results with retinal laser treatment in patients with neovascularization and of vitrectomy in eyes having vitreous hemorrhage are good, if timely treatment is available
- Periodic follow-up is important given the progressive nature of the disease

25.1.1 History

Henry Eales, a British ophthalmologist, described this condition almost 125 years ago [14, 15]. Eales' original description was of recurring retinal and vitreous hemorrhages along with epistaxis, headaches, variation in peripheral circulation, dyspepsia, and chronic constipation in young men. He felt it was a vasomotor neurosis, wherein constriction of the alimentary vessels resulted in compensatory dilatation of the vessels in the head, leading to bleeding. Though Eales was honored with the eponym for this disease, Wadsworth was the first to describe the presence of retinal inflammation, 5 years later [37]. Duke Elder considered Eales' disease to be a clinical manifestation of many diseases.

25.1.2 Epidemiology

One in 200–250 general ophthalmic patients in India are affected, ten times more than in North America or Europe, where today it is a diagnosis of exclusion. The male preponderance is marked – almost 90% [10]. Bilateral affection has been reported in 50–90% of cases, depending on the study [26]. Most present in the 3rd decade of life (15–45 years). There is a definite predilection for the poor socioeconomic class and most patients come from rural communities. While the disease was reported globally at the beginning of the last century, improved sanitary conditions and standard of living have seen a gradual decrease in Europe and North America. Interestingly, Murphy et al. from the USA did not note a male preponderance in their series of 55 cases in America [27]. Today most cases are reported from Asia.

25.1.3 Clinical Features

Eales' disease is an idiopathic, usually peripheral, invariably bilateral retinal vasculitis resulting in peripheral non-perfusion and neovascularization in a young otherwise healthy male population.

The presenting symptoms in most patients are the presence of floaters and painless diminished vision of varying degrees. On most occasions patients have had many such episodes. These are usually self-limiting in the initial phase, as reabsorption takes place when the bleed is small. Visual loss usually takes place due to recurrent vitreous hemorrhage and rarely due to vascular ischemia.

Clinical signs are mainly restricted to the posterior segment; rarely a patient may show the presence of minimal anterior chamber reaction. The presence of anterior chamber flare or cells is usually the first indicator of the development of rubeosis iridis, as the presence of new vessels is difficult to visualize in dark brown irides.

Clinically, the various features of Eales' disease can be divided into signs of inflammation, signs of ischemia, and signs of neovascularization and its sequelae. In the normal course, they occur in this order; however, features from different stages may be present in the same patient. Given the bilateral tendency of this condition, careful examination of the fellow eye almost always shows peripheral sheathing.

It is important to emphasize that the diagnosis of Eales' disease is essentially clinical: no immunological, pathological or biochemical tests are available to make a diagnosis.

25.1.3.1 Signs of Inflammation

Peripheral periphlebitis or vasculitis is the hallmark of this disease. Perivascular exudates result in sheathing of the involved segment of the blood vessels. This can be localized, which can be referred to as cuffing (Fig. 25.1.1b) or involving a large segment of the vessels, called sleeving (Fig. 25.1.1a, c). Vessels with active vasculitis take up fluorescein (FA) dye, showing typical staining. It is important to note the extent of involvement of a vessel on FA far exceeds the ophthalmoscopic picture (Fig. 25.1.1d). Sometimes the vasculitis may involve one of the larger

venous branches, resulting in a secondary branch retinal vein occlusion (Fig. 25.1.2a, c). Such secondary venous occlusions are characterized by sheathing of the involved vessels. Associated superficial retinal hemorrhages and edema occurring in this condition are often seen to cross the horizontal midline raphae temporally, which is not seen normally in branch retinal vein occlusion (BRVO) (Fig. 25.1.2b, c) [27]. Though not common, some eyes do show the presence of patches of fresh or old chorioretinitis, against the backdrop of retinal vasculitis (Fig. 25.1.3a–c).

25.1.3.2 Signs of Ischemia

Retinal ischemia due to vessel closure manifests in the same way as in other retinal vascular diseases [16, 35]. Superficial retinal hemorrhages in the nerve fiber layer are often seen in the area of sheathed vessels (Fig. 25.1.4). Interestingly, dot blot hemorrhages are rarely seen. In response to ischemia, venous changes such as beading and reduplication are seen (Fig.

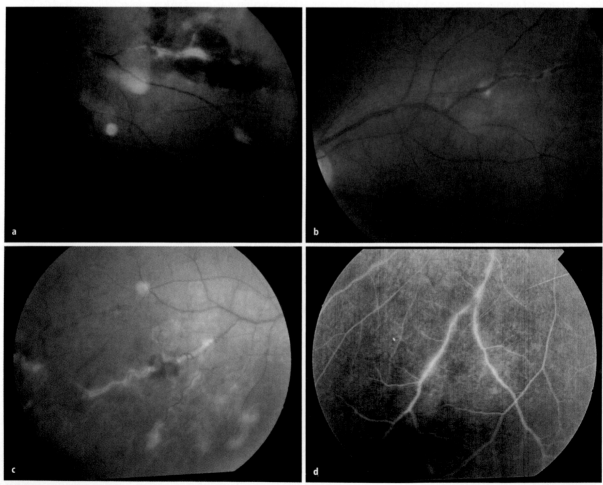

Fig. 25.1.1. a Multiple areas of active periphlebitis in the periphery, some associated with retinal hemorrhages. **b** Multiple cuffs of periphlebitis. **c** Sleeving of a large segment of vessels, laser marks seen. **d** FA of vein with active periphlebitis showing staining

III 25

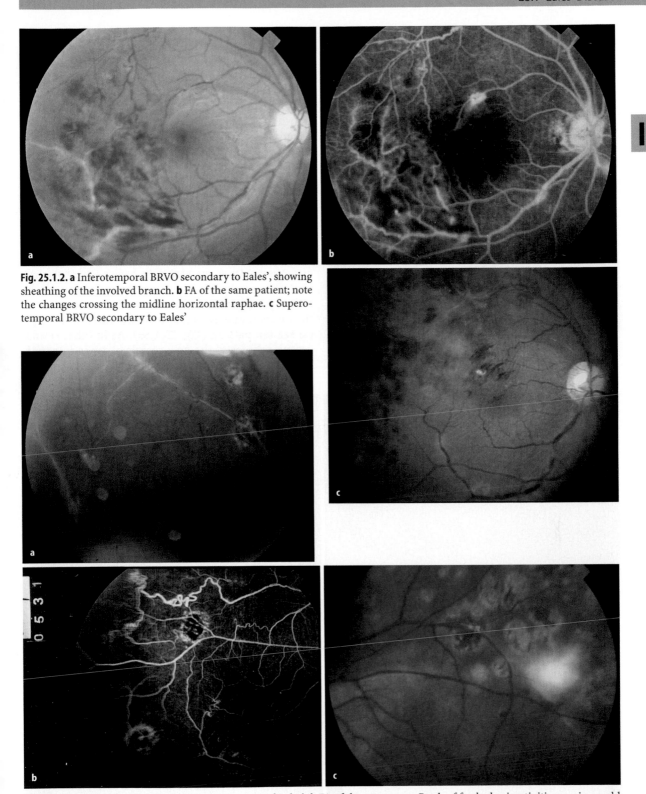

Fig. 25.1.2. a Inferotemporal BRVO secondary to Eales', showing sheathing of the involved branch. **b** FA of the same patient; note the changes crossing the midline horizontal raphae. **c** Supero-temporal BRVO secondary to Eales'

Fig. 25.1.3. a Patches of old chorioretinitis in a case of Eales'. **b** FA of the same eye. **c** Patch of fresh chorioretinitis seen in an old Eales' patient with previous laser marks

25.1.5a, b). In an attempt to correct the ischemic changes, collaterals or venovenous shunts are common (Fig. 25.1.5c). These represent a stabilization of the vasculature, and are not an indication for scatter laser treatment. These can be differentiated from

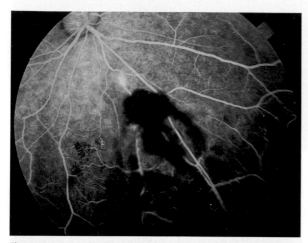

Fig. 25.1.4. Superficial retinal hemorrhages seen in the vicinity of a vessel with periphlebitis due to retinal ischemia

new vessels on FA as they do not leak dye (Fig. 25.1.5d). Areas of retinal edema are seen especially in the early stage of the disease because of the hemodynamic disturbance caused by the closure of inflamed vessels (Fig. 25.1.6). Macular edema, however, is not common as it is in diabetic retinopathy. Areas of non-perfusion can be well identified on FA and are peripherally located in the distribution of the involved vessels (Fig. 25.1.7a–d). The posterior pole and macula are rarely involved, except in cases where advanced disease and severe ischemia exist.

25.1.3.3 Signs of Neovascularization and Its Sequelae

Since most patients present after bleeding into the vitreous cavity, as many as 80 % have neovascularization. Since ischemia most commonly occurs in the retinal periphery, new vessels arise at the junction of the vascular and avascular retina, usually in the typical sea fan pattern (Fig. 25.1.8a). As in other retinal vascular diseases, the new vessels occur in intraretinal, preretinal and intravitreal locations progressive-

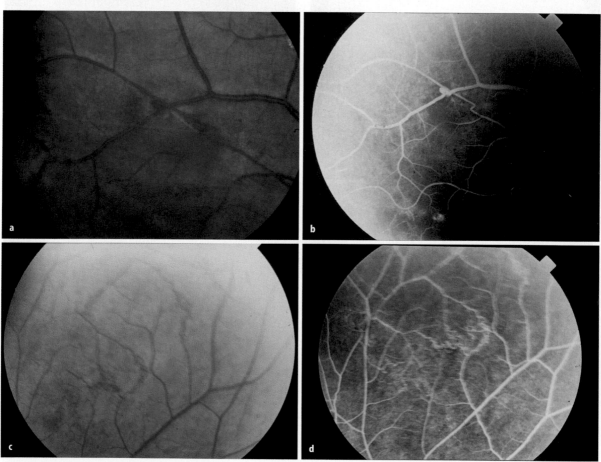

Fig. 25.1.5. a Reduplication of a segment of vein due to ischemia. **b** FA of the same lesion. **c** Venovenous shunt collaterals in the midperiphery. **d** FA of the collaterals; note the absence of leakage

25

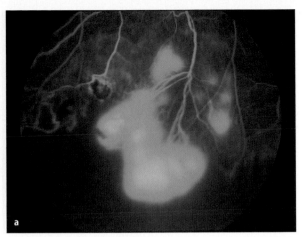

Fig. 25.1.8. a Typical peripheral sea fan pattern of neovascularization described in Eales' disease

Fig. 25.1.6. FA showing presence of retina and macular edema

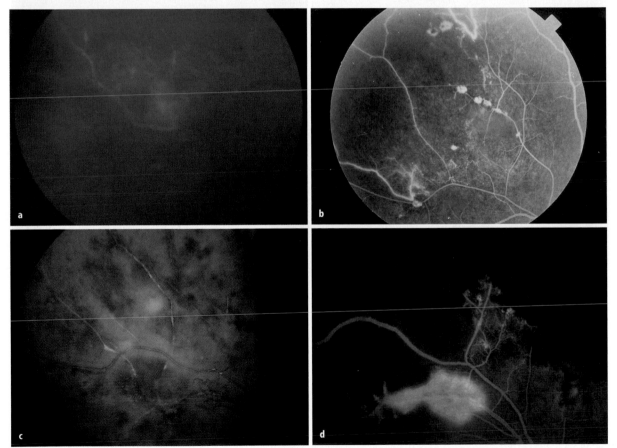

Fig. 25.1.7. a Slight difference in color of ischemic and non-ischemic retina. **b** FA of the same case highlighting the junction of perfused and non-perfused retina. **c** Ischemic retina with extensive occlusive vasculitis. **d** Severe retinal ischemia, extending far posteriorly. Note the stunted vessels

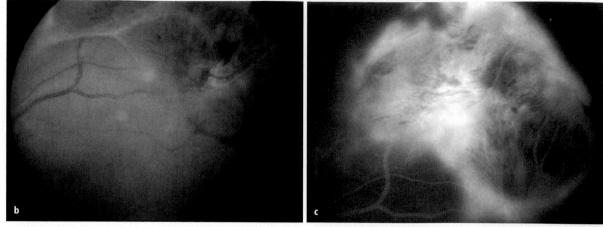

Fig. 25.1.8. b Neovascularization just outside the arcade growing into the vitreous. **c** Extensive leakage on FA seen from the same lesion

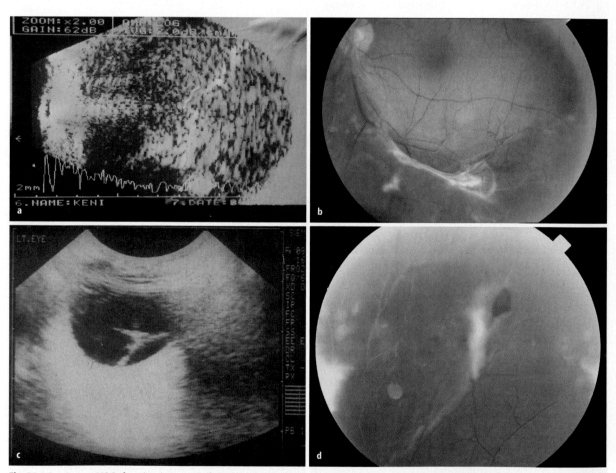

Fig. 25.1.9. a B scan USG showing in a case of vitreous hemorrhage due to Eales' disease. Note the intragel nature of the hemorrhage and the absence of PVD. **b** Inferotemporal TRD just outside the arcade. **c** Concave configuration of TRD on B scan, with vitreoretinal traction bands. **d** Linear tear due to contraction of fibrovascular glial tissue resulting in a combined retinal detachment

ly. Some workers have tried to show that the temporal periphery is more commonly affected. Sometimes the new vessels develop in the mid periphery, or just outside the arcades (Fig. 25.1.8b, c). They may radiate spider like from a common point. In more advanced disease with global ischemia, NVD may develop. The commonest sequela of neovascularization is vitreous hemorrhage (Fig. 25.1.9a), the others being traction retinal detachment (Fig. 25.1.9b, c) and combined retinal detachment. Vitreous hemorrhage occurs from new vessels and is rarely due to sudden occlusion of a large vessel. Most bleeds are intragel and small. Retrohyaloid hemorrhages, which are de rigueur in proliferative diabetic retinopathy, are rarely seen here. Most eyes show multiple bleeds in various stages of absorption, before they become dense enough to completely obscure vision. In Eales' disease, traction retinal detachments rarely affect the posterior pole and hence do not significantly affect vision. However, contraction of the fibrous component of the fibrovascular tissue laid down during neovascularization can cause retinal tears, resulting in a combined retinal detachment. These tears are usually linear and are related to contracted glial tissue (Fig. 25.1.9d).

25.1.3.4 Central Eales'

Central Eales' has been referred to inconsistently in the literature. Some studies place the prevalence of this at 6% in the case of Eales' disease [32]. This presents as a papillophlebitis or a non-ischemic central retinal vein occlusion (CRVO) (Fig. 25.1.10a, b). Good vision is usually present and a good response has been noted to corticosteroids or immune suppression (Fig. 25.1.11a–c).

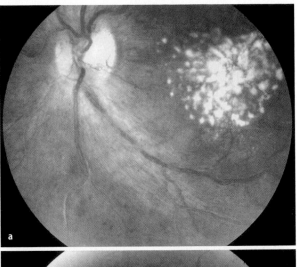
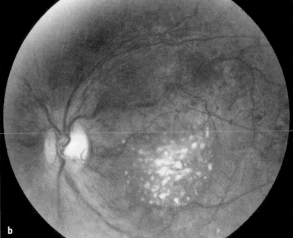

Fig. 25.1.11. a Central Eales' with disk edema, sheathing and extensive hard exudates over the macula. **b** Response to systemic steroids in a fortnight showing decreased venous fullness, disk edema and reduction of hard exudates.

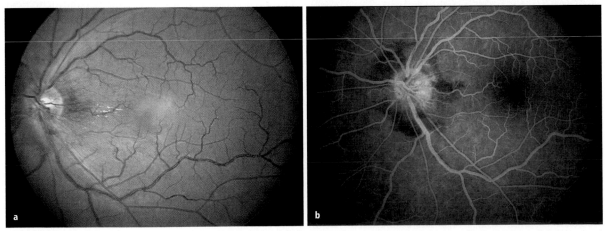

Fig. 25.1.10. a Central Eales': venous fullness disk edema and hemorrhages and hard exudates. VA 6/9. **b** Staining of the proximal segments of the retinal venous system, with disk edema. Note the good perfusion

25 III

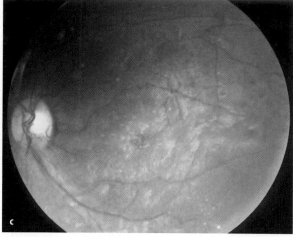

Fig. 25.1.11. c Marked improvement 8 weeks after systemic steroids. Lamellar macular hole noted in area of cleared exudates

25.1.4 Natural Course

The natural course of Eales' disease is variable. Some spontaneously remit after a single episode of bleeding while others gradually progress to end stage disease [20]. Retinal vasculitis causes retinal ischemia, which in turn promotes neovascularization and fibrovascular proliferation. This results in vitreous hemorrhage and/or retinal detachment. End stage eye disease (Fig. 25.1.12) is marked by neovascular glaucoma, proliferative vitreoretinopathy (PVR), cataract and ectropion uvea. Some patients may insidiously progress to a burnt out state with severe ischemia and disk pallor.

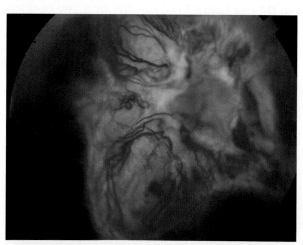

Fig. 25.1.12. End stage eye with severe neovascularization

25.1.5 Differential Diagnosis

Today, cytomegalovirus (CMV) retinitis is an important differential diagnosis due to the common patient profile.

Eales Disease has to be differentiated from other causes of retinal vasculitis/retinitis

CMV retinitis: immune compromised patients, clear vitreous, granular retinal opacification, brushfire like progress, frosted branch angiitis like sheathing

Behçet's: hypopyon, aphthous and genital ulcers

Sarcoid: female preponderance, candle wax dripping, hilar adenopathy, serum ACE levels

Leukemia: peripheral smear, cotton-wool spots (CWS) and preretinal hemorrhages

Syphilis: serological tests, genital ulcers/scars, lymphadenopathy

Tuberculosis: focal chorioretinitis, nodule formation, systemic features

Multiple sclerosis: optic neuritis, systemic features

Pars planitis: snow banking, vitritis, macular edema

Toxoplasma: necrotizing central retinochoroiditis, overlying vitritis, serology

Toxocara: usually unilateral, granuloma formation, ELISA

Systemic lupus erythematosus: systemic and skin signs, CWS

Lyme borreliosis: systemic disease, spirochetal, response to tetracycline

Eales Disease has to be differentiated from other retinal vascular diseases

Sickle cell disease: black sunburst appearance, salmon patch, sea fans with autoinfarction, Hb electrophoresis

Coats' disease: vascular abnormalities, lipid exudation, unilateral

Branch vein occlusion: no sheathing hemorrhages and exudates do not cross horizontal midline raphae

Central retinal occlusion: non-ischemic type difficult to differentiate from central Eales'

Diabetic retinopathy: hyperglycemia, dot blot, hemorrhages, maculopathy, no vasculitis

Retinopathy of prematurity: history, age of presentation, no vasculitis

25.1.6 Systemic Associations Described in Eales' Disease

The involvement of the CNS and vestibulocochlear apparatus [31] show that the pathology of Eales' is not confined to the eye. In the CNS it is thought to cause:

- Acute or subacute myelopathy [34]
- Multifocal white matter abnormality [25]
- Ischemic infarction of the brain [7]
  and therefore manifests as:
  - Focal neurological signs and demyelination
  - Abnormalities of the peripheral vestibular system
  - Bilateral sensorineural hearing loss
  - Hemiplegia and paraparesis
  - Internuclear ophthalmoplegia
  - Psychosis

25.1.7 Attempts at Classification

Over the years various attempts have been made at classification. Charmis et al. [9] suggested a classification system based on evolution and progress of the disease. The classification divided the disease into four stages:

- **Stage I**: mild periphlebitis of peripheral retinal capillaries
- **Stage II**: widespread periphlebitis of the venous system
- **Stage III**: new vessel formation and vitreous hemorrhage
- **Stage IV**: end result of multiple hemorrhages – retinitis proliferans

A more complete classification was suggested [11], which can be used to study the response to treatment. In addition to the lesions seen, it also noted the number of clock hours involved (Table 25.1.1).

In 2004, Saxena et al. suggested a new system of classification which first divides the disease into central and peripheral types [32], the latter being further subdivided into:

- **Stage 1** is periphlebitis of small (1a) and (1b) large caliber vessels with superficial retinal hemorrhages
- **Stage 2a** denotes capillary non-perfusion and **2b** neovascularization (NVE/NVD)
- **Stage 3a** is classified as fibrovascular proliferation and **3b** vitreous hemorrhage
- **Stage 4a** is traction/combined retinal detachment whereas **4b** is rubeosis iridis, NVG, cataract, optic atrophy

This classification addresses the issue of central and peripheral disease and also accommodates end stage disease.

25.1.8 Pathology

Histopathology of the epiretinal membranes (ERM) in Eales' disease can be performed after harvesting the sample during vitrectomy. Studies have been performed using light microscopy and even with immunohistochemical stains [4]. These show several neovascular channels with glial cells, macrophages, fibrocytes and lymphocytic infiltration. Majji et al. histologically compared ERM harvested from patients with vaso-occlusive conditions (other retinal vascular diseases) and from Eales' (vasoinflammatory group). The ERMs from both groups were similar except for the presence of inflammation in the latter. The presence of mast cells and eosinophils was also noted [23].

25.1.9 Etiopathology

More than 125 years since its description by Eales, this entity has evaded an exact explanation of its etiopathology. While it appears to be multifactorial -immununological, molecular biological, and biochemical studies have indicated the role of HLA antigens, autoimmunity, mycobacterial genome and oxidative stress mechanism.

The controversy of whether it is due to an immunological response to mycobacterial antigens, or due to the actual presence of the acid fast bacilli, seems to be getting unraveled using the latest investigative modalities. Laboratory studies have showed statistically significant higher phenotype frequencies of HLA B5, DR1 and DR4 among patients with Eales' disease compared to controls [6]. Since the most favored etiologies were tuberculosis or hypersensitivity to tuberculoprotein, Mantoux testing and lymphocyte proliferation assay to purified protein derivative (PPD) was performed on patients with Eales' against age and gender matched volunteers with normal fundus findings. No significant difference was found between the two groups [5]. Polymerase chain reaction (PCR) was applied using IS 6110 primers to

Table 25.1.1. Das and Namperumalsamy classification

Grades	I Mild 1/12=30% of a circle	II Moderate	III Advanced	IV Very advanced
Description of lesions				
Angiopathy				
Venous changes (tortuosity, periphlebitis)	<1/12	<2/12	<3/12	>3/12
Microaneurysms				
Retinal hemorrhages	<1/12	<2/12	<3/12	>3/12
Proliferative retinopathy				
New vessels		<1/12	<2/12	>2/12
Fibrous tissue proliferation	<1/12	<2/12	<3/12	>3/12
Vitreous hemorrhages	<2/12	<4/12	<8/12	>8/12

detect the bacterial genome in vitreous fluid and was found in statistically significant numbers in Eales' patients [19]. Statistically significant presence of *M. tuberculosis* DNA in the epiretinal membranes of Eales' patients was demonstrated, compared to the controls by nested PCR (nPCR) using primers for gene coding for MPB 64 protein of *M. tuberculosis* [21, 22]. Conversely the ocular morbidity pattern in 2,010 eyes of patients with active systemic TB was studied prospectively, and was found only in 1.39%. The commonest finding was bilateral healed focal choroiditis (50%). Significantly, none of the patients in this series had Eales' disease [3].

Some studies have drawn attention to the role of oxidative stress, also referred to as free radical induced damage. Biochemical studies show that protein carbonyl group content increases with severity of Eales' disease. This increase is correlated to decreased antioxidant status [30]. This fact has been corroborated by other studies, including those showing monocyte activation (MC) resulting in oxidant thrust and subsequent tissue damage. More studies need to place in perspective the role of antioxidant supplementation in Eales'. For now the etiology can be considered multifactorial [8].

25.1.10 Management

25.1.10.1 Fundus Fluorescein Angiography

Fundus fluorescein angiography (FFA) is recommended in Eales' disease to understand the actual extent of involvement, which is often more than the clinical picture. It is advised to examine patients with Eales' disease with an indirect ophthalmoscope after performing the FA. A good idea is obtained by using high end indirect ophthalmoscopes with blue colored interference filters. Some of these areas are difficult to visualize with a fundus camera. During the active phase of phlebitis, staining of the walls of the vein with dye and some leakage may be observed. Narrowing of the lumen is also noted in some cases. Areas of capillary non-perfusion are seen peripheral to the area of periphlebitis. Fortunately macular ischemia occurs only rarely. Since retinal hemorrhages are often seen in the vicinity of involved vessels, blocked fluorescence may be observed. Collateral channels are often seen. The margin of the perfused and non-perfused areas is usually characterized by the presence of stunted vessels, aneurysms and development of new vessels. The characteristic sea fans typically occur at this junction [24, 36].

25.1.10.2 Medical Treatment

Inflammatory vasculitis is the first step toward a chain of events that progress from vessel closure, to ischemia and subsequent neovascularization. Corticosteroids have been advocated during the phase of inflammation (active periphlebitis) to minimize and reverse the damage. Systemic steroids up to 2 mg/kg of prednisolone have been advocated in the active phase. It is advisable to give some time for the steroids to control the inflammatory component prior to starting retinal laser photocoagulation, where indicated. Needless to say, they have to be used keeping in mind their side effects and ruling out any systemic infection. These can be tapered over 6–8 weeks. Subtenon injections of triamcinolone have also been used in a dose of 0.5–1 ml (40 mg/ml). This route reduces the risk of systemic toxicity. Use of long-acting depot steroids poses problems in steroid responders and care is needed while administering this injection especially in high myopes. The use of antimetabolites has been reported in one eyed patients and in central disease. Low dose oral methotrexate pulse therapy (at a dose of 12.5 mg/week) for 12 weeks has been reported recently but needs more evaluation [2].

The use of antituberculous drugs in the absence of systemic tuberculosis is one of the controversies in the management of Eales' disease. A decade ago it was widespread. The treatment recommended is isoniazid (INH) 300 mg and rifampicin (RIF) 450 mg for a period of 9 months. Its indication for use today is restricted to patients with extensive exudative sheathing, nodule formation or a strongly positive Mantoux test (with induration and ulceration) in the presence of Eales' disease [10].

25.1.10.3 Role of Laser Treatment

The rationale for retinal laser is that regression of existing neovascularization can be achieved by scatter treatment of the ischemic retina. Prophylactic laser treatment to severely ischemic areas documented on FFA can prevent the development of new vessels [18]. Since photocoagulation has to be performed in the far periphery, delivery systems other than the slit lamp like the laser indirect ophthalmoscope are a useful option. In patients where some amount of hemorrhage obscures the view, transscleral diopexy or anterior retinal cryopexy can be used [29]. Laser treatment may be performed focally in areas of ischemia where the disease is localized (Fig. 25.1.13a–c). Panretinal photocoagulation may be performed in the presence of NVD or when extensive areas of ischemia exist (Fig. 25.1.14a, b). Retinal laser has also been used to perform "anchoring treat-

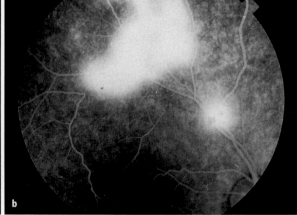

25

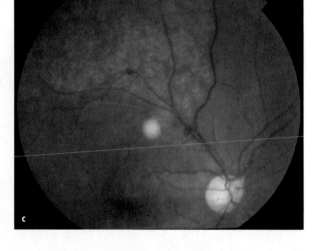

Fig. 25.1.13. a Spider shaped NVE just outside the superotemporal arcades. **b** FA showing leakage from the NVE. **c** Retinal laser treatment of the involved area (1 month post-treatment)

ment" (Fig. 25.1.15a, b). This is done by placing a barrage of laser at the base of the fibrovascular traction band. This helps to prevent the development of a combined retinal detachment should the band contract causing a retinal tear. It is important to do anchoring treatment prior to doing a scatter in such eyes, as regression is often associated with contraction of the fibrovascular bands [29]. Six monthly follow-up is advised to the patient as retinal non-perfusion may be progressive in this condition; hence new areas of neovascularization may develop which will require supplemental treatment. The patient is advised to report earlier if any new visual symptoms are noted.

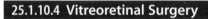

25.1.10.4 Vitreoretinal Surgery

Pars plana vitrectomy has an important role to play in the presence of media opacities or abnormal vitre-

oretinal relationships. Surgical anatomy of the vitreoretinal interface in Eales' disease has been studied [1]. Fibrous and fibrovascular proliferations have multiple areas of adhesions to the posterior vitreous cortex. The presence of type II collagen in

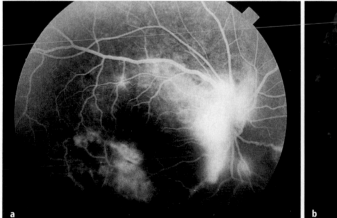

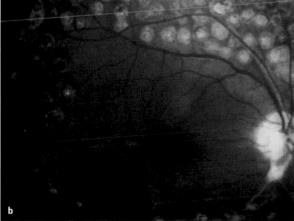

Fig. 25.1.14. a FA shows NVD in an eye with widespread ischemia due to vaso-occlusive damage in Eales' disease. **b** Complete regression noted after a tight laser PRPC

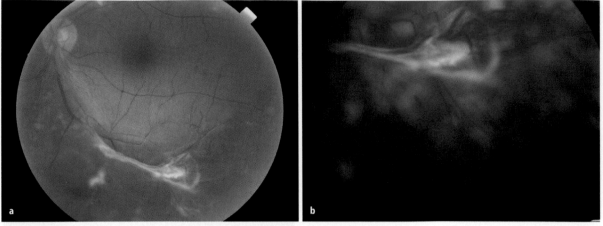

Fig. 25.1.15. a Fibrovascular tractional band with early TRD (pre-treatment). **b** Anchoring treatment applied at and around the base of the traction band

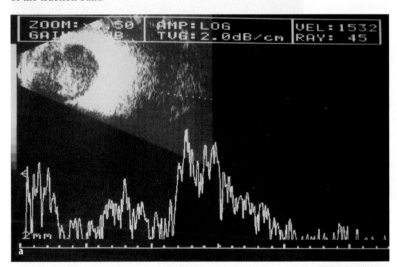

Fig. 25.1.16. a B scan USG with C vector of an eye with non-resolving vitreous hemorrhage

the ERM indicates a possible vitreous collagen component to the double layered membranes (vitreoschisis). Recognition of this double layered membranes aids in relief of traction during surgery by delamination. Also noteworthy of this condition is the fact that the hemorrhages are predominantly intragel.

Indications for surgery are the same as in other conditions: non-resolving vitreous hemorrhage, traction retinal detachment threatening the macula and combined retinal detachment. The recommended waiting period for the vitreous hemorrhage to clear spontaneously has reduced from 6 months previously to 6–8 weeks today. This is attributed to the higher safety margin of the vitrectomy procedure and to the fact that delay in dealing with the neovascularization causes progressive worsening.

Vitrectomy permits vitreous clearance, removal of the vitreous scaffolding and traction, and an opportunity to perform endolaser. Often, a large part of the inferior retina is otherwise not amenable to laser treatment, due to residual vitreous hemorrhage. Most vitrectomies except in long-standing disease are of a lower relative degree of difficulty. Wide angle fundus viewing systems permit a complete vitreous clearance with excellent visualization of the periphery – an asset in these cases. Endolaser is mandatory for these vitrectomies as vitreous rebleeds may not permit laser treatment postoperatively (Fig. 25.1.16a–d). Opinions on the use of encirclage routinely for vitrectomies in Eales' disease are divided. Care of the lens during vitreous surgery is especially important while performing any surgical maneuvers in the periphery. Vitrectomy for combined retinal detachment requires meticulous membrane peeling to relieve vitreoretinal traction and tamponade with gas or silicone oil (Figs. 25.1.17a, b, 25.1.18a, b). Delay in performing vitrectomy (repeated vitreous hemorrhages with only percep-

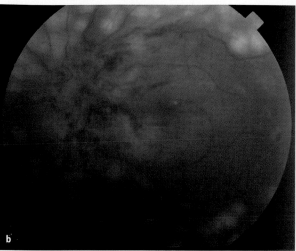

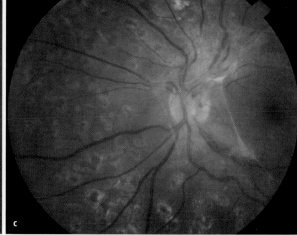

Fig. 25.1.16. b Post-PPV the next day. Note the fresh 500 u endolaser marks. **c** Post-PPV after 1 month showing complete regression of the NVD. VA 6/12. **d** Post-PPV after 1 week

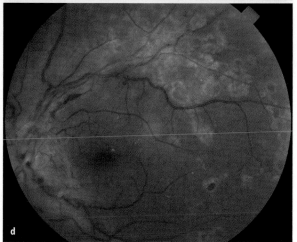

tion of light vision) is an important cause of poorer postoperative visual recovery [17, 28]. According to Shanmugham et al., visual acuities for individual cases were quite stable at the 60 month follow-up, with 50 eyes (78.5%) either maintaining or improving upon their 2 month postoperative visual acuity [33]. Response to surgical treatment is good opposed to conditions like sickle cell retinopathy. Visual results are good as the macula is usually unaffected.

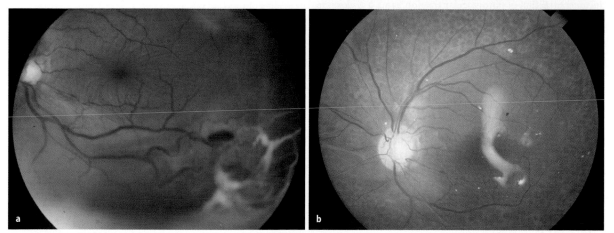

Fig. 25.1.17. a Combined retinal detachment with break seen near inferotemporally with the responsible fibrovascular tissue. **b** Post-PPV after endolaser and silicone oil showing flat retina with release of all VR traction

25 III

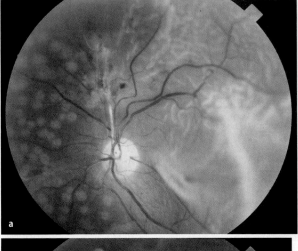
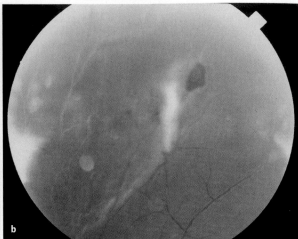
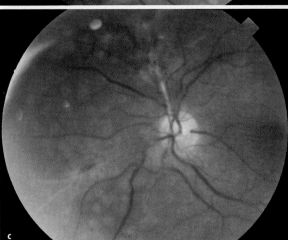

Fig. 25.1.18. a Combined retinal detachment after retinal laser due to contraction of fibrovascular traction band. **b** A close-up of the retinal tear and the offending fibrovascular tissue. **c** Re-attached retina after PPV + MP + endolaser + C3F8

References

1. Badrinath SS, Gopal L, Sharma T et al (1999) Vitreoschisis in Eales' disease: pathogenic role and significance in surgery. Retina 19:51–54
2. Bali T, Saxena S, Kumar D (2005) Response time and safety profile of pulsed oral methotrexate therapy in idiopathic retinal periphlebitis. Eur J Ophthalmol 15:374–378
3. Biswas J, Badrinath SS (1995/1996) Ocular morbidity in patients with active systemic tuberculosis. Int Ophthalmol 19:293–298
4. Biswas J, Rao NA (1990) Epiretinal membrane in Eales' disease and other vascular retinopathies. Invest Ophthalmol Vis Sci 31 [Suppl]:369
5. Biswas J, Narain S, Roy S, Madhavan HN (1997) Evaluation of lymphocyte proliferation assay to purified protein derivative, enzyme-linked immunosorbent assay and tuberculin hypersensitivity in Eales' disease. India J Ophthalmol 45: 93–97
6. Biswas J, Mukesh BN, Narain S, Roy S, Madhavan HN (1998) Profiling of human leukocyte antigen in Eales' disease. Int Ophthalmol 21:277–281
7. Biswas J, Raghavendran R et al (2001) Presumed Eales disease with neurologic involvement: report of three cases. Retina 21:141–145
8. Biswas J, Sharma T, Gopal L, Madhavan HN, Sulochana KN, Ramakrishnan S (2002) Eales disease – an update. Surv Ophthalmol 47:197–214
9. Charmis J (1965) On the classification and management of the evolutionary course of Eales' disease. Trans Ophthalmol Soc UK 85:187
10. Das T, Biswas J, Kumar A, Nagpal PN, Namperumalsamy P et al (1994) Eales' disease. Indian J Ophthalmol 42:3–18
11. Das TP, Namperumalsamy P (1987) Combined photocoagulation and cryotherapy in treatment of Eales' retinopathy. Proc All Ind Ophthalmol Soc 45:108
12. Das TP, Namperumalsamy P (1990) Photocoagulation in Eales' disease. Results of prospective randomised clinical study. Presented in XXVI Int Cong Ophthalmol, Singapore
13. Duke-Elder S, Dobree JH (1967) System of ophthalmology, vol X. Kimpton, London
14. Eales H (1880) Retinal haemorrhages associated with epistaxis and constipation. Brim Med Rev 9:262
15. Eales H (1882) Primary retinal haemorrhage in young men. Ophthalmic Rev 1:41
16. Elliot AJ (1975) Thirty years observation of patients with Eales' disease. Am J Ophthalmol 80:404
17. Gadkari SS, Kamdar P, Jehangir RP (1992) Pars plana Vitrectomy in Vitreous haemorrhage due to Eales' disease. Ind J Ophthalmol 40:35–37
18. Gopal L, Abraham C (1985) Efficacy of photocoagulation in Eales' disease. Trans Asia-pacific. Acad Ophthalmol 10:689
19. Gunisha P, Madhavan HN, Jayanthi U, Therese KL (2000) Polymerase chain reaction using IS6110 primer to detect Mycobacterium tuberculosis in clinical samples. Indian J Path Microbiol 43:395–402
20. Kalsi R, Patnaik B (1979) The developing features of phlebitis retinae (a vertical study). Ind J Ophthalmol 27:87
21. Madhavan, HN. Therese KL, Gunisha P, Jayanthi U, Biswas J (2000) Polymerase chain reaction for detection of Mycobacterium tuberculosis in epiretinal membrane in Eales' disease. Invest Ophthalmol Vis Sci 41:822–825
22. Madhavan HN, Therese KL, Dora Swamy K (2002) Further investigations on two association of Mycobacterium Tuberculosis with Eales disease. Ind J Ophthalmol 50:35–39
23. Majji AB, Vemuganti GK, Shah VA et al (2006) A compara-

tive study of epiretinal membranes associated with Eales' disease: a clinicopathologic evaluation. Eye 20:46–54

24. Malik SRK, Patnaik B (1973) Fluorescein angiography in Eales' disease. Ind J Ophthalmol 21:5
25. Masson C, Denis P, Prier S et al (1988) Eales' disease with neurologic disorders. Fre Revue Neulogeque 144:817–819
26. Murthy KR, Abrabam C, Baig SM et al (1977) Eales' disease. Proc All Ind Ophthalmol Soc 33:323
27. Murphy RP, Renie WA, Proctor LR, Shimuzu H, Lippmann SM, Anderson KC, Fine SL, Patz A, McKusick VA (1983) A survey of patients with Eales' disease. In: Fine SL, Owen SL (eds) Management of retinal vascular and macular disorders. Williams and Wilkin, Baltimore MD
28. Namperumalsamy P, Kelkar AR, Das TP (1990) Vitreous surgery in Eales' disease – when and why presented in XXVI Ophthalmol, Singapore
29. Patnaik B, Kalsi R, Chary P (1980) Cryopexy in the management of diabetic retinopathy. Preliminary report. Proc All Ind Ophthalmol Soc 39:197
30. Rajesh M, Sulochana KN et al (2004) Determination of carbonyl group content in plasma proteins as a useful marker to assess impairment in antioxidant defense in patients with Eales' disease. Indian J Ophthalmol 52:139–144
31. Renie WA, Murphy RP, Anderson KC et al (1983) The evaluation of patients with Eales' disease. Retina 3:243–248
32. Saxena S, Kumar D (2004) A new staging system for idiopathic retinal periphlebitis. Eur J Ophthalmol 14:236–239
33. Shanmugam MP, Badrinath SS, Gopal L, Sharma T (1998) Long term visual results of vitrectomy for Eales' disease complications. Int Ophthalmol 22:61–64
34. Singhal BS, Dastur DK (1976) Eales' disease with neurological involvement. Part I. Clinical Features in 9 patients. J Neurol Sci 27:313–321
35. Spitzans M, Meyer-Schwickerath GT, Stephen B (1975) Clinical Picture of Eales' disease. Graefes Arch Clin Exp Ophthalmol 194:73
36. Theodosisadis G (1970) Fluorescein angiography in Eales' disease. Am J Ophthalmol 69:271
37. Wadsworth OF (1887) Recurrent retinal haemorrhage followed by the development of blood vessels in the vitreous. Ophthalmic Rev 6:289

25.2 Ocular Manifestations of Systemic Lupus Erythematosus

J.T. Rosenbaum, F. Mackensen

25 III

Core Messages
- Sicca is the most common ocular manifestation of systemic lupus erythematosus (SLE)
- While sicca is commonly associated with many rheumatic diseases, cotton-wool spots and intraretinal hemorrhages are present in 7–8% of patients with SLE and are not commonly found in other rheumatic diseases
- SLE related retinopathy is an activity marker and a prognostic factor for poor outcome
- All ocular disease in patients with lupus is not due to inflammation related to SLE: infections, thrombosis related to antiphospholipid antibodies, and medication toxicities need to be considered
- Screening of all SLE patients for ophthalmic disease does not seem warranted; they should be seen as is recommended for their age group
- Patients with known SLE retinopathy or hydroxychloroquine treatment should be observed more closely

25.2.1 Epidemiology and Disease Criteria for SLE

Essentials
- Reported prevalence of systemic lupus erythematosus (SLE) ranges from 15 to 124 cases per 100,000 per year
- Mainly women of childbearing age are affected
- Socioeconomic factors influence course and outcome of SLE
- Four of 11 ACR (American College of Rheumatology) criteria must be fulfilled to make a certain diagnosis of SLE

The prevalence of SLE in the United States ranges from 15 to 124 cases per 100,000 per year [16, 18]. In Europe comparable prevalence rates have been reported in Sweden and Italy [6, 39]. The disease had been thought to be more common and more severe in blacks than in whites, but patients of Hispanic and Asian origin are also severely affected [17, 31]. In later studies socioeconomic and environmental factors have been shown to influence the course and outcome of SLE, which might have biased previous epidemiological studies assessing the contribution of race to disease severity [22, 30]. About 90% of affected persons are women, usually of child bearing age [17]. For men the incidence increases after the 7th decade of life.

The diagnosis of SLE is based on clinical and laboratory criteria proposed by the American College of Rheumatology (ACR) [40]. If 4 out of 11 criteria are present, a diagnosis of SLE can be made with 98% specificity and 97% sensitivity. Ocular disease is not included among the criteria. These criteria are to classify SLE for published studies and, thus, the criteria are stricter than what is sometimes used in practice. The main screening laboratory marker is the ANA (antinuclear antibody), but further findings must support the diagnosis. Systemic symptoms are usually prominent and include fatigue, fever, anorexia and arthralgia. Activity of SLE is measured by different clinical and laboratory criteria. Activity scores that are used, for example, include the Systemic Lupus Erythematosus Disease Activity Index (SLEDAI) or the European Consensus Lupus Activity Measurement (ECLAM) [15]. Only the former includes retinal changes as a criterion for activity.

25.2.2 Frequency and Prognostic Value of Ocular Findings in SLE

Essentials

- Ocular involvement in SLE is common in the form of dryness
- Retinal pathologies can be found in 7.5% of patients with SLE, mostly microangiopathic changes with cotton-wool spots
- Cotton-wool spots and retinal hemorrhages correlate with disease activity and are a negative prognostic sign
- SLE patients with antiphospholipid syndrome have an increased risk for retinal vaso-occlusive disease

A summary is given in Table 25.2.1. There are few prospective epidemiologic studies for ocular involvement in SLE. Generally vision threatening ocular manifestations of SLE are rare, perhaps thanks in part to the earlier diagnosis and better treatment the patients receive. Suspected uveitis in patients with SLE has been reported anecdotally, but not frequently [12]. Orbital disease is also rare [3]. Episcleritis and scleritis are slightly more frequent. In a case series of patients with scleritis, 1% had SLE [41]. Ocular involvement is correlated to the severity of disease and subtype of SLE [12, 38]. A study by Soo et al. on 52 randomly chosen patients from Malaysia with inactive SLE (SLEDAI score ≤4) and no ocular symptoms found pathologic Schirmer testing in 31% as the only abnormal ocular finding. Fifty

age matched controls had normal Schirmer tests [37]. Stafford-Brady and colleagues examined 550 patients with SLE of varying activity prospectively for retinopathy over a period of 16 years. They found retinal pathologies in 41 (7.5%) patients, most of which consisted of microangiopathic changes with cotton-wool spots and hemorrhages. Only 2 of the 41 patients presented with ischemic optic neuropathy, 2 more with retinal vaso-occlusive disease, and only one with choroidal vasculitis complicated by serous retinal detachment. These 41 patients had a reduced survival curve compared to the SLE patients without retinopathy. Of the patients with retinopathy, 88% had active systemic disease, with active central nervous system lupus in 73%, renal disease in 63.5% and detectable lupus anticoagulant in 38%. Many of the patients in this valuable series were evaluated during an era when anticardiolipin or antiphospholipid antibodies were not routinely measured. Twenty-seven percent of the patients with retinopathy also had mild to moderate hypertension. No data are given at all about the presence of ocular symptoms. Equally not mentioned is the severity of disease in the remaining 474 patients [38]. In patients with raised anticardiolipin antibodies (antiphospholipid syndrome) a higher prevalence of retinal vaso-occlusive disease was seen with 8% of 84 patients in one series [5] as well as retinopathy in general with 33% vs. 6% [28].

Table 25.2.1. Frequency of ocular disease in SLE

Type of ocular involvement	% of SLE patients	Sample size and selection	Studies performed	Reference
Sicca syndrome	31%	52 inactive Asian SLE patients	Complete ophthalmologic exam including color testing and Schirmer's	[37]
Retinopathy Microangiopathic changes Ischemic neuropathy Retinal vaso-occlusive disease Choroidal vasculitis	7.5% (41) 34 2 1	550 SLE patients of varying disease activity	Fundoscopy by rheumatology fellow, pathology confirmed by an ophthalmologist	[38]
Retinal vaso-occlusive disease	8% (7)	87 consecutive SLE patients with raised aCL antibodies		[5]
Retinal vascular disease	15% (13) 33% 6%	82 consecutive SLE patients; 49% with active disease With aPL Without aPL	Complete ophthalmologic exam	[28]

aCL anticardiolipin, *aPL* antiphospholipid syndrome

25.2.3 Special Pathological Features and Molecular Mechanisms

Essentials
- The pathogenetic mechanisms of SLE are complex and only partially understood
- A triggering event such as an infection might lead to activation of autoreactive T and B cells
- Retinal vasculitis is a misunderstood term and is reserved by internists generally for histological diagnosis
- There are few case reports about postmortem findings of the eyes of SLE patients. These are of note mainly for thrombosis of retinal arterioles

The variety of autoantibodies found in patients with SLE strongly suggests that it is an autoimmune disease. Although the mechanisms are not yet completely understood, a coincidence of several events such as an infection as a stimulus in a genetically susceptible organism and a subsequent activation and expansion of autoreactive T and B cells are a potential explanation for the loss of tolerance to autoantigens in SLE. Research in mice and humans has shown involvement of multiple cellular and humoral components of the immune system (summarized by [32]). Tissue damage has been shown to occur through immune-complex deposits or a direct effect of autoantibodies. A number of clinically different syndromes fulfill the ACR criteria for SLE, which makes it also likely that several pathogenetic mechanisms exist. SLE can affect nearly all organs including skin, kidney, brain, heart, lung and also the eyes [7, 8].

In some patients with SLE perivascular sheathing can be seen with the ophthalmoscope. This finding is usually termed retinal vasculitis. Sheathing is a non-specific sign of blood-retinal-barrier disruption and can have many causes. The term "vasculitis" should be reserved for histopathologically diagnosed cases, but this is a standard that is impossible to achieve with tissue that is not readily accessible to biopsy. As a clinical description microangiopathic retinopathy or retinal vaso-occlusive disease seem more apt [19, 33]. From animal studies we also know that leukocytes leave the vessel and enter the perivascular tissue if an intraocular inflammation exists, without the vessel wall itself being damaged. Ischemia can also produce secondary inflammatory changes.

In systemic vasculitis the diagnosis is commonly based on the Chapel Hill Consensus Conference relying mainly on the size of the affected vessels [21]. For the histological diagnosis infiltration of the vessel wall by inflammatory cells must be present and a destruction of the wall with necrosis and/or fibrinous exudates should be in evidence. Patients with SLE can evolve a secondary vasculitis which is mainly thought to be due to immune-complex deposits [27]. This is rare. More frequent systemic vascular involvement consists of thrombosis and/or endothelial damage and atherosclerotic changes [7, 8]. An additional vasodestructive influence of hypertension due to renal disease has to be taken into account. In mice with lupus-like syndrome, immune-complex deposits were found to produce non-inflammatory vascular lesions [1].

Histopathologic studies of affected eyes are rare due to the obvious difficulty of biopsies and most information stems from postmortem exams. Less than ten published cases report thrombosed retinal arterioles, but active vasculitis or foci of inflammatory cells have not been demonstrated. However, some authors describe choroidal cell infiltrates and choroidal vasculitis [10, 14] and others also detected immune-complex deposits in the choroidal vessels [4, 23, 29]. This is similar to changes detected in brain biopsies of SLE patients [14].

25.2.4 Retinopathy in SLE patients: Clinical Picture and Course of Disease

Case Report
An 18-year-old female presents to the Emergency Department with a history of recent, sudden bilateral vision loss left more than right. Systemically she reports having fever of 5 weeks duration of unknown origin, oral ulcers and multiple swollen lymph glands. A lymph gland biopsy shows non-specific inflammation. She has a history of pancytopenia. EBV reactivation has been suspected. Her only medication is roxithromycin. Visual acuity is 20/30 and 20/60. Intraocular pressure is normal. The slit lamp examination is unremarkable; fundus examination shows a single cotton-wool spot at the left optic disk with an intraretinal hemorrhage. Blood pressure is 140/85. MRI of the brain and orbits is normal except for enhancement in the orbital fat tissue. Fluorescence angiography was not performed.

25.2.5 Differential Diagnosis

Essentials
- Hypertensive retinopathy
- SLE related retinopathy
- Leukemic retinopathy

- Infection secondary to immunosuppression
- Bilateral optic neuropathy
- Choroidal vasculitis
- Orbital inflammatory disease

The patient was transferred to the Rheumatology Department with a suspected diagnosis of SLE, which was supported by ANA levels of 1:20,000, positive ds (double stranded) DNA antibodies and reduced complement levels. Proteinuria was detected and a renal biopsy showed active glomerulonephritis. With oral ulcers, nephritis, pancytopenia, positive dsDNA and an elevated ANA titer, she fulfilled 5 of the 11 ACR criteria for the classification of systemic lupus. Treatment with high dose corticosteroids and intravenous cyclophosphamide was initiated. Ocular symptoms responded quickly and visual acuity recovered. Three weeks after onset of therapy, visual acuity was 20/20 in both eyes. The left eye now showed some cotton-wool spots and small intraretinal hemorrhages; the right eye demonstrated only a few small cotton-wool spots. Three months after onset of therapy, visual acuity was 20/20 in both eyes and the retinal changes were slowly disappearing, but still showed cotton-wool spots and a dissolving retinal hemorrhage, sometimes termed a "Roth" spot (Fig. 25.2.1). The last pictures (Fig. 25.2.2) were taken 5 months after first presentation. Visual acuity was 20/20 and the fundus findings had nearly completely resolved. Further follow-up examinations were performed in 3 monthly intervals by the local ophthalmologist and the Rheumatology Department. Even though no further eye symptoms occurred, proteinuria increased again after azathioprine was substituted for cyclophosphamide, so that treatment with 15-deoxyspergualin was initiated. Deoxyspergualin (NKT-01) shows both in vitro and in vivo immunosuppressive activity affecting B-lymphocyte, T-lymphocyte and macrophage/monocyte function and is currently being tested in an open-label, multicenter Phase I/II pilot study for patients with SLE nephritis. It has been reported to be effec-

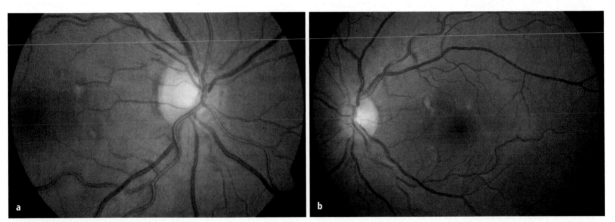

Fig. 25.2.1. a Right and **b** left eye of the patient 3 weeks after first examination. Several cotton-wool spots can be seen as well as vanishing intraretinal hemorrhages, forming a "Roth" spot

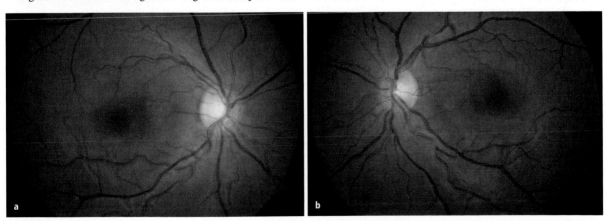

Fig. 25.2.2. a Right and **b** left eye of the patient 6 months after first examination. The cotton-wool spots have disappeared after systemic treatment

tive in a case series and has the advantage over cyclophosphamide of not being gonadotoxic [24]. Our patient responded favorably to this treatment, but about 1 year later, another flare of the nephritis occurred and rituximab was given.

Although a variety of problems could have explained this patient's visual loss, the precise cause was never determined with certainty. Several options within the differential diagnosis were excluded by the course and presentation. Hypertension, leukemia, and infection were considered, but these were not consistent with the presentation. The orbital disease was not considered to be clinically significant. It is most likely that this patient had either optic nerve disease and/or choroidal vasculitis. While an angiogram could have helped distinguish among these possibilities, the study was not obtained largely because of the acuteness of the patient's disease and the need to institute immunosuppression, empiric therapy for either underlying diagnosis.

In general it is more likely that a patient presenting to an ophthalmologist already has known SLE. In the case of suspicions that eye findings such as microangiopathic retinopathy, retinal vaso-occlusive events or neuropathy might be the first presentation of SLE, a detailed history asking specifically about symptoms such as fatigue, anorexia, fever, weight loss, pleuritic chest pain, skin rash, or arthritis should reveal further evidence for an associated systemic disease suspicious of SLE. Then laboratory investigations can be initiated. An ANA test is the best screening test for SLE, but should not be used unless an acceptable pretest likelihood of a positive result exists (Bayes' theorem) [35]. An ANA test has a sensitivity of roughly 97% for systemic lupus, so a negative test is very useful in excluding lupus as a likely diagnostic possibility. A positive ANA can occur in lower levels in normal individuals. This finding increases with age. It is frequently positive in patients with the subset of juvenile arthritis associated with uveitis and among patients with Sjogren's syndrome. In the case of retinal vaso-occlusive disease especially in younger patients with no previous cardiovascular risk factors, anticardiolipin antibodies and the lupus anticoagulant should be tested as these patients have increased risk for antiphospholipid syndrome [9].

25.2.6 Treatment Recommendations, Follow-up and Recommendations for Ophthalmologic Screening of SLE Patients

Essentials
- Treatment of microangiopathic changes should be chosen in the context of the systemic disease
- Remission is induced with intravenous cyclophosphamide and maintained with other immunosuppressive medications, e.g., azathioprine
- Emerging therapies for systemic lupus include mycophenolate mofetil and rituximab
- This treatment seems to be effective for SLE related neuropathy as well
- Patients with antiphospholipid syndrome are mainly treated with anticoagulation

Ocular findings can be seen as markers for disease activity and a bad prognostic sign [38]. Patients should be referred to a rheumatologist or nephrologist and treated in the context of systemic disease. High dose corticosteroids and intravenous cyclophosphamide in combination therapy are used in severe disease and have been shown to induce remission in a higher percentage of patients with lupus nephritis than prednisone alone [13]. Usually cyclophosphamide is used for up to 6 months to induce remission and then another immunosuppressive medication, e.g., azathioprine, is used to maintain that state. Several recent reports have supported the choice of mycophenolate mofetil orally as an alternative choice for mild or moderate renal disease and intravenous rituximab for severe renal disease [2, 42]. Another treatment under investigation is deoxyspergualin [24]. Systemic treatment also seems to be adequate for SLE related retinopathy. For optic neuropathy, an occlusive vasculopathy should be assessed by checking for antiphospholipid antibodies and the lupus anticoagulant. If the cause is thought to be inflammatory rather than thrombotic, high dose corticosteroids and/or intravenous cyclophosphamide should be administered promptly [11, 20, 34]. In patients with antiphospholipid syndrome without signs of inflammatory activity, anticoagulation alone can be sufficient [36].

Ophthalmologic examinations should be performed at regular intervals to monitor efficacy of treatment in patients with known retinopathy. These patients generally have a favorable outcome in terms of visual acuity [38]. Screening of all SLE patients

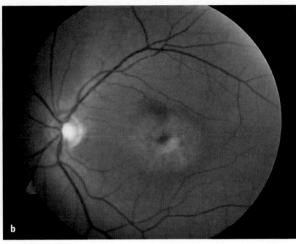

III 25

Fig. 25.2.3. a Right eye and **b** left eye: bilateral "bulls-eye" maculopathy as the typical ophthalmoscopic finding of retinal toxicity of hydroxychloroquine treatment in SLE

ophthalmologically does not seem warranted as vision threatening ocular manifestations are rare. AAO recommendations for ophthalmologic routine examinations for each age group are appropriate. The subgroup of SLE patients with antiphospholipid syndrome might benefit from regular fundus examinations. The finding of minor vascular occlusion might be a rationale for anticoagulation therapy or aspirin, but there are no studies confirming a benefit of this approach. Patients receiving chloroquine or hydroxychloroquine treatment should have a baseline examination and follow-up depending on the dose and duration of the drug [26]. Additional risk factors such as renal disease might prompt more frequent monitoring for antimalarial toxicity. The typical "bull's-eye maculopathy" is the clinical hallmark of retinal toxicity (Fig. 25.2.3). Unfortunately this can progress sometimes even after the treatment has been stopped [25].

References

1. Accinni L, Dixon FJ (1979) Degenerative vascular disease and myocardial infarction in mice with lupus-like syndrome. Am J Pathol 96:477–492
2. Anolik JH, Aringer M (2005) New treatments for SLE: cell-depleting and anti-cytokine therapies. Best Pract Res Clin Rheumatol 19:859–878
3. Arevalo JF, Lowder CY, Muci-Mendoza R (2002) Ocular manifestations of systemic lupus erythematosus. Curr Opin Ophthalmol 13:404–410
4. Aronson AJ, Ordonez NG, Diddie KR, Ernest JT (1979) Immune-complex deposition in the eye in systemic lupus erythematosus. Arch Intern Med 139:1312–1313
5. Asherson RA, Merry P, Acheson JF, Harris EN, Hughes GR (1989) Antiphospholipid antibodies: a risk factor for occlusive ocular vascular disease in systemic lupus erythematosus and the 'primary' antiphospholipid syndrome. Ann Rheum Dis 48:358–361
6. Benucci M, Del Rosso A, Li Gobbi F, Manfredi M, Cerinic MM, Salvarani C (2005) Systemic lupus erythematosus (SLE) in Italy: an Italian prevalence study based on a two-step strategy in an area of Florence (Scandicci-Le Signe). Med Sci Monit 11:CR420–CR425
7. Boumpas DT, Austin HA 3rd, Fessler BJ, Balow JE, Klippel JH, Lockshin MD (1995a) Systemic lupus erythematosus: emerging concepts, part 1. Renal, neuropsychiatric, cardiovascular, pulmonary, and hematologic disease. Ann Intern Med 122:940–950
8. Boumpas DT, Fessler BJ, Austin HA 3rd, Balow JE, Klippel JH, Lockshin MD (1995b) Systemic lupus erythematosus: emerging concepts. Part 2: Dermatologic and joint disease, the antiphospholipid antibody syndrome, pregnancy and hormonal therapy, morbidity and mortality, and pathogenesis. Ann Intern Med 123:42–53
9. Cobo-Soriano R, Sanchez-Ramon S, Aparicio MJ, Teijeiro MA, Vidal P, Suarez-Leoz M, Rodriguez-Mahou M, Rodriguez-Huerta A, Fernandez-Cruz E, Cortes C (1999) Antiphospholipid antibodies and retinal thrombosis in patients without risk factors: a prospective case-control study. Am J Ophthalmol 128:725–732
10. Cordes FC, Aiken SD (1947) Ocular changes in acute disseminated lupus erythematosus; report of a case with microscopic changes. Am J Ophthalmol 30:1541–1555
11. Galindo-Rodriguez G, Avina-Zubieta JA, Pizarro S, Diaz de Leon V, Saucedo N, Fuentes M, Lavalle C (1999) Cyclophosphamide pulse therapy in optic neuritis due to systemic lupus erythematosus: an open trial. Am J Med 106:65–69
12. Gold DH, Morris DA, Henkind P (1972) Ocular findings in systemic lupus erythematosus. Br J Ophthalmol 56:800–804
13. Gourley MF, Austin HA 3rd, Scott D, Yarboro CH, Vaughan EM, Muir J, Boumpas DT, Klippel JH, Balow JE, Steinberg AD (1996) Methylprednisolone and cyclophosphamide, alone or in combination, in patients with lupus nephritis. A randomized, controlled trial. Ann Intern Med 125:549–557
14. Graham EM, Spalton DJ, Barnard RO, Garner A, Russel RWR (1985) Cerebral and retinal vascular changes in systemic lupus erythematosus. Ophthalmology 92: 44–448
15. Griffiths B, Mosca M, Gordon C (2005) Assessment of patients with systemic lupus erythematosus and the use of

lupus disease activity indices. Best Pract Res Clin Rheumatol 19:685–708

16. Hahn BH (1997) Systemic lupus erythematosus. In: Harrison TR et al (eds) Harrison's principles of internal medicine. McGraw Hill, New York, pp 1874–1880

17. Hochberg MC, Boyd RE, Ahearn JM, Arnett FC, Bias WB, Provost TT, Stevens MB (1985) Systemic lupus erythematosus: a review of clinico-laboratory features and immunogenetic markers in 150 patients with emphasis on demographic subsets. Medicine (Baltimore) 64:285–295

18. Hochberg MC, Perlmutter DL, Medsger TA, Steen V, Weisman MH, White B, Wigley FM (1995) Prevalence of self-reported physician-diagnosed systemic lupus erythematosus in the USA. Lupus 4:454–456

19. Jabs DA, Fine SL, Hochberg MC, Newman SA et al (1986a) Severe retinal vaso-occlusive disease in systemic lupus erythematosus. Arch Ophthalmol 104:558–563

20. Jabs DA, Miller NR, Newman SA (1986b) Optic neuropathy in systemic lupus erythematosus. Arch Ophthalmol 104:564–568

21. Jennette JC, Falk RJ, Andrassy K, Bacon PA, Churg J, Gross WL, Hagen EC, Hoffman GS, Hunder GG, Kallenberg CG et al (1994) Nomenclature of systemic vasculitides. Proposal of an international consensus conference. Arthritis Rheum 37:187–192

22. Karlson EW, Daltroy LH, Lew RA, Wright EA, Partridge AJ, Fossel AH, Roberts WN, Stern SH, Straaton KV, Wacholtz MC, Kavanaugh AF, Grosflam JM, Liang MH (1997) The relationship of socioeconomic status, race, and modifiable risk factors to outcomes in patients with systemic lupus erythematosus. Arthritis Rheum 40:47–56

23. Karpik AG, Schwartz MM, Dickey LE, Streeten BW, Roberts JL (1985) Ocular immune reactants in patients dying with systemic lupus erythematosus. Clin Immunol Immunopathol 35:295–312

24. Lorenz HM, Grunke M, Wendler J, Heinzel PA, Kalden JR (2005) Safety of 15-deoxyspergualin in the treatment of glomerulonephritis associated with active systemic lupus erythematosus. Ann Rheum Dis 64:1517–1519

25. Marmor MF (2003) New American Academy of Ophthalmology recommendations on screening for hydroxychloroquine retinopathy. Arthritis Rheum 48:1764

26. Marmor MF, Carr RE, Easterbrook M, Farjo AA, Mieler WF (2002) Recommendations on screening for chloroquine and hydroxychloroquine retinopathy: a report by the American Academy of Ophthalmology. Ophthalmology 109:1377–1382

27. Meister P (2003) Vasculitides: classification, clinical aspects and pathology. A review. Pathologe 24:165–181

28. Montehermoso A, Cervera R, Font J, Ramos-Casals M, Garcia-carrasco M, Formiga F, Callejas JL, Jorfan M, Grino MC, Ingelmo M (1999) Association of antiphospholipid antibodies with retinal vascular disease in systemic lupus erythematosus. Semin Arthritis Rheum 28:326–332

29. Nag TC, Wadhwa S (2005) Histopathological changes in the eyes in systemic lupus erythematosus: an electron microscope and immunohistochemical study. Histol Histopathol 20:373–382

30. Petri M, Perez-Gutthann S, Longenecker JC, Hochberg M (1991) Morbidity of systemic lupus erythematosus: role of race and socioeconomic status. Am J Med 91:345–353

31. Ramal LM, Lopez-Nevot MA, Sabio JM, Jaimez L, Paco L, Sanchez J, de Ramon E, Fernandez-Nebro A, Ortego N, Ruiz-Cantero A, Rivera F, Martin J, Jimenez-Alonso J (2004) Systemic lupus erythematosus in southern Spain: a comparative clinical and genetic study between Caucasian and Gypsy patients. Lupus 13:934–940

32. Riemekasten G, Hahn BH (2005) Key autoantigens in SLE. Rheumatology (Oxf) 44:975–982

33. Rosenbaum JT, Robertson JE Jr, Watzke RC (1991) Retinal vasculitis – a primer. West J Med 154:182–185

34. Rosenbaum JT, Simpson J, Neuwelt CM (1997) Successful treatment of optic neuritis in association with systemic lupus erythematosus using intravenous cyclophosphamide. Br J Ophthalmol 81:130–132

35. Rosenbaum JT, Wernick R (1990) The utility of routine screening of patients with uveitis for systemic lupus erythematosus or tuberculosis. A Bayesian analysis. Arch Ophthalmol 108:1291–1293

36. Ruiz-Irastorza G, Khamashta MA, Castellino G, Hughes GR (2001) Systemic lupus erythematosus. Lancet 357:1027–1032

37. Soo MP, Chow SK, Tan CT, Nadior N, Yeap SS, Hoh HB (2000) The spectrum of ocular involvement in patients with systemic lupus erythematosus without ocular symptoms. Lupus 9:511–514

38. Stafford-Brady FJ, Urowitz MB, Gladman DD, Easterbrook M (1988) Lupus retinopathy. Patterns, associations, and prognosis. Arthritis Rheum 31:1105–1110

39. Stahl-Hallengren C, Jonsen A, Nived O, Sturfelt G (2000) Incidence studies of systemic lupus erythematosus in Southern Sweden: increasing age, decreasing frequency of renal manifestations and good prognosis. J Rheumatol 27:685–691

40. Tan EM, Cohen AS, Fries JF, Masi AT, McShane DJ, Rothfield NF, Schaller JG, Talal N, Winchester RJ (1982) The 1982 revised criteria for the classification of systemic lupus erythematosus. Arthritis Rheum 25:1271–1277

41. Tuft SJ, Watson PG (1991) Progression of scleral disease. Ophthalmology 98:467–471

42. Zimmerman R, Radhakrishnan J, Valeri A, Appel G (2001) Advances in the treatment of lupus nephritis. Annu Rev Med 52:63–78

25.3 Behçet's Disease

M. ZIERHUT, N. STÜBIGER, I. KÖTTER, C. DEUTER

Core Messages
- Behçet's disease is a systemic vasculitis, with a prevalence of 0.12–370/100,000 inhabitants
- Diagnosis is made according to the following criteria: recurrent oral aphthous ulcers in combination with two of the following lesions: eye lesions, genital ulcers, skin lesion, pathergy testing
- Immune mediated disorder (T, B, NK and neutrophilic cells), leading to inflammation and endothelial dysfunction
- HLA-B51 is the most important genetic marker
- Treatment is with systemic immunosuppression: corticosteroids, cyclosporin A, most recently successful biologicals such as interferon-α and tumor necrosis factor (TNF)-α antagonists

25.3.1 Epidemiology and Definition of Behçet's Disease

25.3.1.1 Definition

Behçet's disease (BD) is a systemic vasculitis [86, 94]; thus, almost any organ system can be affected. The diagnosis of BD is usually based on the criteria of the International Study Group from 1990 (Table 25.3.1) [39].

Table 25.3.1. Criteria of the International Study Group 1990 [39]

Recurrent oral aphthous ulcers	Small or large aphthous or herpetiform ulcerations, recurring at least 3 times in a 12 month period
Plus 2 of the following:	
Recurrent genital ulcers	Aphthous ulcerations or scarring
Eye lesions	Anterior uveitis, posterior uveitis or cells in vitreous on slit lamp examination or retinal vasculitis observed by an ophthalmologist
Skin lesions	Erythema nodosum, pseudofolliculitis, or papulopustulous lesions or acneiform papules in postadolescent patients without steroid treatment
Positive pathergy testing	Intracutaneous stick with 21G needle on the forearm (inside), read by a physician after 24–48 h

25.3.1.2 Epidemiology

Essentials
- Most common along the Silk Route from the Mediterranean to eastern Asian countries
- Incidence: 0.8/100,000 inhabitants in Japan
- Prevalence: 0.12 (USA)–370 (Turkey)/100,000 inhabitants
- Gender ratio: previously male:female 3:1; today probably 1:1

Behçet's disease has been reported from many countries all over the world. Especially along the ancient Silk Route from the Mediterranean to the eastern Asian countries, BD patients are endemic [44]. The disease is most common between the latitudes 30° and 45°N in Asian and European populations [113]. These geographic data and the fact that certain ethnic groups are affected hint at the transfer of genetic material and/or of an exogenous agent, which may have been responsible for the spread of the disease [117].

The prevalence of the disease (Table 25.3.2) varies around the world and seems to be strongly dependent on geographic area [44, 113, 117]. Single or small numbers of cases have been reported from all continents, but the highest prevalence has been described in Turks living in Anatolia (northwestern Turkey), with 370 patients per 100,000 inhabitants [112, 117]. The overall prevalence in Asia varies between 18 and 110 patients [44, 75, 112]. The preva-

Table 25.3.2. Prevalence of Behçet's disease in different populations

Population	Year of the study	Prevalence per 100,000 inhabitants
Turkish (northeastern Anatolia) [112]	1987	370
Chinese (Ningxiahei Province) [44]	1998	120
Japanese (Hokkaido region) [44]	1997	22
Northern Spanish [117]	1998	7.50
Greek [82]	1984	6.00
Italian [117]	1988	2.50
Swedish [117]	1993	1.18
German [118]	1994	0.55
US American [14]	1979	0.12

lence in southern Europe, especially in the Mediterranean countries, is 1.53–7.50 patients per 100,000 inhabitants; in northern Europe and in the USA the highest prevalence was recorded in the Swedish population with 1.18 patients per 100,000 inhabitants. Interestingly, the data from Germany exhibited a prevalence of 20.75 per 100,000 inhabitants in patients of Turkish origin compared to only 0.42 per 100,000 inhabitants in patients of German origin [118].

Data on the incidence of BD are available for only a few countries [113, 117]. From Japan, a country with well organized registration of patients with BD, it was reported that in the year 1990 0.8 new cases per 100,000 inhabitants were diagnosed with BD [73].

Average age of onset is recorded at 25–35 years worldwide and it seems to be independent of the origin of the patients and their gender [113, 117]. Variations in many studies may be based on different definitions. Most authors consider the onset of the disease to be the age at which the patient met the diagnostic criteria of the disease, while for others BD onset should be associated with the first symptom [113, 117].

Concerning the gender distribution of BD patients, Japanese and Turkish reports have found a predominance of males to females [68, 113, 117] of 3 to 1, but current epidemiological studies have registered an approximately equal rate [31, 73, 117].

25.3.2 Special Pathological Features

Essentials
- Immune mediated disorder with participation of T, B, NK and neutrophilic cells
- Hyperactive neutrophils can lead to increased chemotaxis, phagocytosis and superoxide generation
- Increase in gamma/delta T cells has been demonstrated

- Endothelial dysfunction with abnormalities of coagulation and fibrinolytic pathway may lead to venous and arterial occlusion
- Several microbial antigens have been shown to stimulate T cells of BD patients

BD is, at least in part, an immune mediated disorder, in which multiple cells are major and minor components. This includes cells of the innate immune system (neutrophils, NK cells), but also of the specific immune system, especially the T cells.

Hyperactive neutrophils, with a high expression of CD10, CD11a and CD14, infiltrate BD lesions, leading also to an increase in CD64+ neutrophils in the serum [103], and they seem to initiate an increased chemotaxis, phagocytosis and superoxide generation [11, 84]. Stimulation seems to be initiated by T cells which secret tumor necrosis factor (TNF)-α, interleukin (IL)-8 and granulocyte-macrophage colony-stimulating factor (GM-CSF) [74]. T cells in BD patients seem to be, at least in part, apoptosis resistant, e.g., due to NF-kappaB [100], and with a high expression of Fas ligand [107].

T-cell-mediated immune reactions seem to play a major role in BD, but how is not well understood. There is a strong TH1 response with expression of high amounts of IL-2, IFN-γ [24] and IL-18 [70, 80]. Excessive expression of Txk, a member of the Tec family of tyrosine kinases, seems to contribute to excessive Th1 cytokine production of T cells of BD patients [72]. A reduced CD4/CD8 ratio has been found, but there are indications for a defect of the CD8 suppressor cells [85]. Several microbial antigens have been shown to stimulate T cells in BD patients, e.g., staphylococcal antigens, streptococcal antigens [40, 57, 59], *E. coli* derived peptides, *Chlamydia pneumoniae* [6], and especially important heat-shock proteins (hsp) of various microbes [58]. Subcutaneous injection of hsp peptides (with adjuvant), but also oral and nasal administration (without adjuvant), can induce experimental autoimmune uveitis in rats [35]. The finding of cytotoxic T cells may show some evidence for a role of virus, e.g., HSV or others, in the pathogenesis of BD. Also unclear is the role of γδ-T cells, which are elevated in BD patients. In correlation to activity, a subset of these cells has a proliferative response to four mycobacterial hsp-derived peptides. This suggests a regulatory role of the γδ-T cells [33]. On the other hand, Vgamma9 Vdelta2 T cells have been isolated from eyes of BD patients, recognizing non-peptide prenyl pyrophosphate antigens [106].

The role of natural killer cells also remains unclear at the moment [114]. BD patients also show an increased concentration of antibodies against phosphatidylserine and ribosomal phosphoproteins, which may stress the role of NK cells in BD [9]. Recently, an abnormal KIR (killer inhibitory receptor) expression on NK cells of BD patients was found [98]. Kinectin [60] and alpha-tropomyosin [61] have also recently been suggested as important autoantigens in BD.

Vascular involvement is a major feature of BD, leading to venous or arterial occlusion. Additionally to the immunological reaction, it seems that endothelial dysfunction and abnormalities of the coagulation and fibrinolytic pathway facilitate BD. Endothelial dysfunction [53] in BD patients has been reported. Endothelial cells are characterized by a reduced flow-mediated dilation of arteries [15], impaired synthesis of endothelial cell-derived prostaglandins [45], increased levels of endothelium-derived antiaggregant PGF1α, and platelet-derived proaggregant TXB2 [34]. Additionally, antibodies against endothelial cells have been detected in up to 50% of BD patients, directed especially against microvascular cell antigens [7, 12, 21, 55], at least in part directed against alpha-enolase [54]. In most studies the incidence of anti-endothelial cells correlates with BD activity. An increase of soluble E-selectin in BD patients may be caused by activated endothelial cells and may serve as a control parameter in BD vasculitis [91].

Besides endothelial dysfunction, abnormalities of the coagulation and the fibrinolytic system have been detected. An elevation of the thrombin-anti-thrombin III complex, of thrombin activatable fibrinolysis inhibitor (TAFI) [22] and of prothrombin 1 and 2, but also of plasma plasminogen activator inhibitor-1 (PAI-1) antigens and PAI-1 activities [81], may cause intravascular thrombin formation. Further, also defects in protein C, protein S and factor V Leiden have been demonstrated in patients with BD after thrombosis, when they carry the factor V Leiden and prothrombin gene [13, 64, 89, 93]. Also mutations which increase the risk of thrombosis have been located in BD patients [27, 29, 62, 105]. The relevance of elevated anticardiolipin antibodies (25% in BD patients) [101], especially in the active state [69], remains unclear at present. All findings support an imbalance toward a prothrombotic state: endothelial dysfunction may facilitate immune-mediated vasculitis, leading to profound tissue damage.

As the genetic contribution to the pathogenesis of BD is estimated at only 20–30% [30], infectious agents, heat shock proteins, abnormalities in the innate immune system such as neutrophil hyperfunction or an increase in gamma delta T-cells together with pro-coagulatory factors play a major role. The present hypothesis for the pathogenesis of BD is depicted in Fig. 25.3.1.

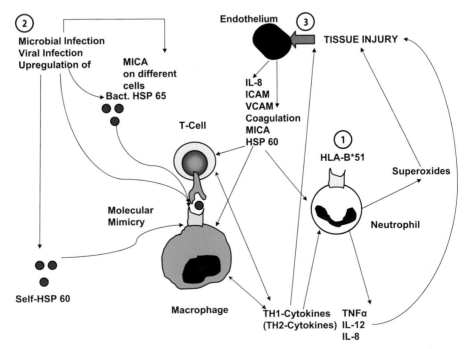

Fig. 25.3.1. Hypothesis for the pathogenesis of MB: (1) genetic factors, e.g., HLA-B*51, induce a general hyperactivity of the immune system (TH1 response, granulocytes). (2) Bacterial or viral infection stimulates the expression of HSP60 (self) and MICA on different cells (e.g., endothelial cells), upregulation of adhesion molecules, activation of coagulation, stimulation of T cells (especially γδ and NK cells), continuing elevation of the cytokine production; and finally (3) tissue damage (vasculitis). Additionally, molecular mimicry with HLA-B*51 may play a role. In parallel, B cells will be stimulated polyclonally (not shown on the figure) and may produce more antibody (e.g., against HSP)

25.3.3 Genetics and Molecular Mechanisms

Essentials

- The genetic contribution to the pathogenesis of BD is estimated to be 20–30%
- HLA-B51 is the most important genetic marker for BD
- Analysis of multicase families suggests a complex genetic inheritance model
- The most important allele for BD seems to be B*5101, while at least in Japanese B*5201 seems protective
- Other genes are in linkage disequilibrium with HLA-B51, like MICA
- The critical region for susceptibility antigens, coding for BD, lies between the HLA-B locus, the TNF locus and the locus for MICA, or possibly at the telomeric end of chromosome 6pi

Despite the fact that cause and pathogenesis of BD are still to a large extent unclear, important factors for the pathogenesis have been identified in the last approximately 30 years. With a strong genetic predisposition, primary or secondary abnormalities of the immune system, probably following a still unknown infection, seem to initiate BD.

It seems that HLA-B51, with HLA-B52 one of the split antigens of HLA-B5, is a very important genetic marker for this disease. HLA-B5 is present in 40–80% of all BD cases [67], while in the respective healthy population the prevalence is between 8% (northern Europe) and 24% (Turkey, Middle East) [104]. Table 25.3.3 shows the prevalence of HLA-B51 in different studies. This strong association has been confirmed in patients of different ethnic origin. The analysis of multicase families suggests a complex genetic inheritance model [29]. Hence, familial cases of BD are only rarely reported. This high association only exists with HLA-B51, but not HLA-B52. BD has not been reported yet from countries with an extremely low prevalence of HLA-B51.

Table 25.3.3. Prevalence of HLA-B*51 in patient groups, healthy controls and general population, relative risk for BD. (Modified from [104] and [116])

Country	Patients n	HLAB-51+ (%)	Controls n	HLA-B51+ (%)	RR	p value	Population HLA-B51 (%)
Asia							
Japan	91	52	140	20	7.9	<0.00005	18–22.3
Korea	113	51	112	16	4.0–6.8	<0.001	10–13
Taiwan	51	51	128	11	8.5		2.6–7.7
China	120	56	100	12	9.3		2.5–8.8
India	31	32	400	30	1.1		6.4–16
Iraq	52	62	175	29	3.9		2.3
Iran		53		33	2.3		
Turkey	520	77	1106	26	9.2		24
Saudi Arabia	85	72		26	9.0		26
Jordan	68	74	43	23	9.2		
Lebanon	100	54	100	34			6.1
Israel	126	75	790	21	11.5		
Africa							
Egypt	84	58	200	7	20.1	<0.0001	
Tunisia	55	62	80	24	5.2		
Morocco [59]	86	30.2	111	15.3	2.4	<0.015	15.8
Europe							
Russia	19	37	150	15	3.2		4.9
Great Britain	107	25	2032	9	3.3		1.9–4.4
Ireland [60]	24	25	96	3	6.3	<0.002	0
Germany	75	36	1415	14	3.5		6.1
Switzerland	8	38		17	3.0		9.5
Portugal	318	53	135	24	3.6		17
Spain	100	42	452	21	2.7		6.1–24.5[a]
France	105	51	591	13	6.7		6.5
Italy	57	75	304	22	10.9		17.4
Greece	170	79	670	28	9.7	<0.001	15.1
America							
USA	32	13	523	10	1.3		1.5–4.5%
Mexico	10	70	105	31	5.1		

Empty fields: no data available
[a] Southern Spain

The B51 antigen comprises 29 alleles at the amino acid level, B*5101-B*5124 [23]. The distribution of these alleles has been shown to be very different in ethnic groups. While B*5108 and B*5101 were found to be relatively high in patients from Greece, Iran, Italy and Saudi Arabia, this allele is unknown in Japan, where there is a dominance of 58.3% BD patients who are B*5101 positive. The fact that in Japanese patients B*5201 (13.2% of BD patients) seems protective for BD, in contrast to B*5101, has provoked the following hypothesis: B*5101 and B*5201 differ only by two amino acid substitutions at position 63 and 67. In B*5101 (also in B*5102 and B*5108) position 63 has asparagine, and position 67 has phenylalanine [66]. Both positions seem to be an important part of the pocket of the HLA groove. The difference in the allelic distribution between the Japanese patients and patients from the more Western located countries around the Silk Route may be due to different methods of distribution. It appears likely that BD spread among the various populations together with its associated HLA-B*51 allele prior to divergence into suballeles. Recently it has been shown for Japanese patients that HLA-A*2602 and B*3901 might also have some secondary influence on the onset of BD [41]. The analysis of sister chromatid exchange (SCE) has demonstrated a higher SCE frequency in HLA-B51 positive patients than in HLA-B51 negative ones, pointing to a higher rate of DNA damage in HLA-B51+ patients [37].

Besides the association with HLA-B51, the contribution of other genes has been investigated. Most research focuses on chromosome number 6, which also contains the genes for the major histocompatibility complex (MHC), in the human HLA system. Besides others, polymorphisms for the alleles encoding the intercellular adhesion molecule (ICAM-1) [10], the CTLA-4 gene [87] and the VEGF gene [90] have been found also for the endothelial nitric oxidase synthase [88], for IL-18 [42], and for IL-1 [46]. Recently it has been suggested that the gene responsible for the familial Mediterranean fever (MEFV) may also be involved in the pathogenesis of BD [38]. Some other protein alleles seemed to be associated with BD, like MICA [36], but were finally found to be in linkage disequilibrium with HLA-B*51. An additional susceptibility locus was mapped to 12p12–13 and 6p22–24, which is well beyond the strong linkage disequilibrium region [30, 47]. The TNF-α gene polymorphisms are unlikely to play an important role in the pathogenesis and severity of BD [4, 56] in contrast to the MBL2 gene (mannose-binding lectin gene-2), in which the HYPA haplotype seems to play a role in MBL levels and increases the susceptibility to MB [83], and also the endothelial nitric oxide synthase gene [49]. Identification of other susceptibility genes is under investigation.

Hence, at present HLA-B51 is the only major disease susceptibility antigen which is directly associated with BD. The critical region in which susceptibility antigens for BD are coded probably lies between the HLA-B locus, the TNF locus and the locus for MICA (MHC Class I chain related A), or possibly at the telomeric end of chromosome 6pi.

25.3.4 Course of the Disease

25.3.4.1 Extraocular Manifestations

Essentials
- BD is a multisystem vasculitis, affecting nearly all organ systems
- Common manifestations can be distinguished from rarer ones
- Predilection sites are mucous membranes with oral aphthous ulcers as the main feature
- Young and male patients may have shortened life expectancy due to more severe courses of the disease

Although BD as a multisystem vasculitis may affect nearly all organ systems, mucous membranes, the skin and the eye are preferred sites [86, 94]. Table 25.3.4 presents an overview of the extraocular manifestations and their frequency in patients with BD.

The main features of BD are oral aphthous ulcers, which affect nearly all patients. In contrast to habitual aphthosis, ulcerations in BD occur in uncommon locations (e.g., sublingually, hard and soft palate, epipharynx, pharynx and larynx), last a long time (2 weeks), relapse frequently and are very painful.

Genital ulcerations and skin lesions are also frequent manifestations of BD. The latter may present,

Table 25.3.4. Extraocular manifestations and their frequency in Behçet's disease

Manifestation	Frequency
Oral aphthous ulcers	90–100%
Genital ulcerations	60–80%
Skin lesions	41–94%
Pathergy phenomenon	19–53%
Skeletal system	47–69%
Gastrointestinal	3–30%
Neurological	8–31%
Vascular	28%
Cardiac	1–6%
Urogenital	4–31%
Renal	<1%

e.g., as papulopustules, acneiform pseudofolliculitis, erythema nodosa or superficial thrombophlebitis. Cutaneous hypersensitivity is demonstrated by the so-called pathergy phenomenon: 24–48 h after an intracutaneous stick with a sterile 21G needle a sterile papulopustule occurs [20, 26]. If the skeletal system is involved, arthritis is mostly oligoarticular (less than five joints affected), asymmetrical, affects the lower extremities and is non-erosive (thus usually not being visible on X-rays). Gastrointestinal, neurological and vascular (thromboses and arterial aneurysms) are less frequent but potentially very serious manifestations of BD. Cardiac (e.g., pericarditis), urogenital (mostly epididymitis) and renal manifestations rarely occur. Life expectancy may be shortened especially in young and male patients tending to more severe courses of BD. Involvement of the CNS, arterial aneurysms and gastrointestinal involvement are the most life threatening manifestations [109].

25.3.4.2 Ocular Manifestations

Essentials
- In 10–20% of patients intraocular inflammation may be the initial manifestation of BD
- Anterior uveitis (iritis with hypopyon), posterior or panuveitis with occlusive retinal vasculitis and cystoid macular edema are typical findings in ocular BD
- Ocular disease is mostly bilateral, running a chronic relapsing and progressive course

Ocular involvement in BD, often termed as "*ocular BD*," is usually characterized by the occurrence of a non-granulomatous inflammation, especially as diffuse uveitis with retinal vasculitis, and is found in 60–80% of patients [5, 16, 97, 99, 113]. The reported frequency of ocular manifestations is 83–95% in males and 67–73% in females, and several studies indicate that the disease is more severe in men [51, 97, 113].

The first signs of uveitis are reported on average 4 years after disease onset, but it could also be the initial BD symptom in 10–20% of patients [16, 97, 113].

Ocular BD demonstrates a chronic relapsing course and progress by successive attacks. As the healing of lesions is very slow, a new attack will occur before the healing is completed. As a result, lesions will accumulate from one attack to another, progressing gradually toward severe loss of vision [17].

The primary eye lesions in 50–87% of patients may be unilateral and may occur most often as an

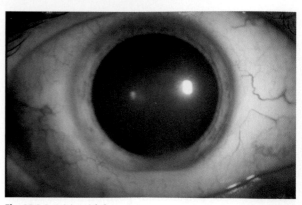

Fig. 25.3.2. Iritis with hypopyon

anterior uveitis, but later on, in 75% of cases bilateral panuveitis with a chronic relapsing course is reported [97, 115] – so an anterior segment type and a panuveitis type of inflammation is reported [65].

Less than 20% of patients present only with *anterior uveitis* [1, 65, 113]. The occurrence of hypopyon (Fig. 25.3.2), found in 16–39% of cases, is not necessarily a distinctive feature of the ocular manifestation in BD [1, 65]. The explanation for this phenomenon could be that nowadays, due to early and aggressive treatment, the inflammation will be dampened [97, 115].

Slit lamp examination will reveal conjunctival and ciliary injection, keratic precipitates, cells and flare in the anterior chamber, and vitreous opacity.

Interestingly, the acute inflammation of the anterior chamber may resolve spontaneously over 2–3 weeks even without therapeutic intervention. So, the visual prognosis for the anterior segment type is more favorable than for the panuveitis type.

Complications of recurrent anterior uveitis could be iris atrophy, and posterior and peripheral anterior synechiae, resulting in development of secondary glaucoma and secondary cataract formation [97, 113, 115].

Rarely described anterior segment manifestations in BD are conjunctivitis with or without subconjunctival hemorrhage, episcleritis or scleritis, keratitis, and, seldomly, extraocular muscle paralysis [16, 97, 115].

Inflammatory changes in the panuveitis type include vitritis, whereas an isolated vitreous inflammation is not characteristic of ocular BD. The most frequent finding is *retinal vasculitis*, resulting in intensive retinal edema, yellowish-white exudates and hemorrhages (Figs. 25.3.3, 25.3.4). This retinal vasculitis shows an occlusive course and ocular histopathology during the attack is characterized by severe angiitis with intensive infiltration of neutrophil leukocytes largely in the uveal tract and the retina; the latter is severely affected and loss of visual cells and other neural elements results [65].

In most patients retinal vasculitis occurs mainly affecting the retinal veins, which is pathognomonic for BD as it is the only systemic vasculitis affecting small and medium sized arteries and veins. So, perivascular sheathing has been observed frequently, whereas sheathing of the veins often precedes sheathing of the arteries [97, 115]. A typical complication is retinal vein branch occlusion, mentioned in a study by Özdal et al. [78] in about 6% of affected eyes.

Severe retinal vasculitis may lead to ischemic alterations due to the vascular occlusion (Fig. 25.3.5), which will cause tissue hypoxia and, subsequently, retinal neovascularizations due to stimulating growth factors. These neovascularizations, which could occur all over the retina, especially at the optic disk, have a high risk of bleeding in the vitreous capacity, which may induce membrane formation, causing retinal holes with secondary retinal detachment [97, 113, 115].

Also the optic nerve can be affected due to ischemic lesions and results in optic disk paleness in about 7% of cases [78], but more often, in at least

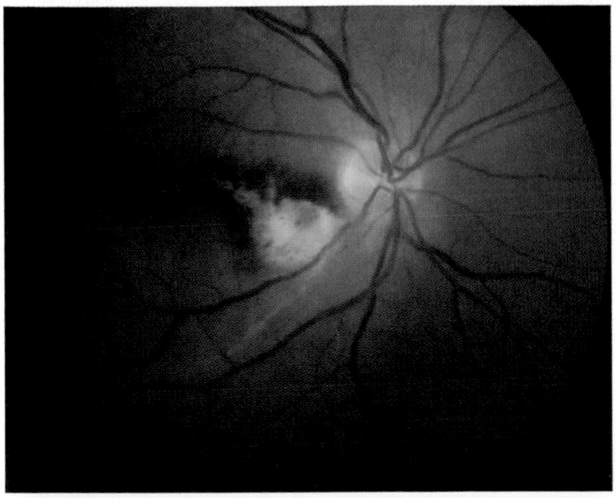

Fig. 25.3.3. Acute infiltration of the central retina

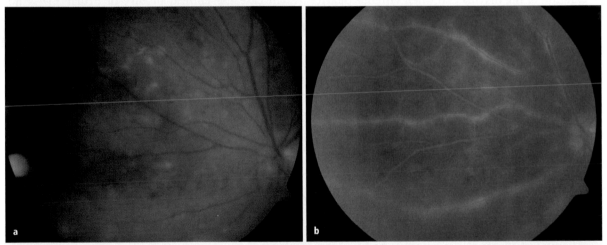

Fig. 25.3.4. Panuveitis with retinal vasculitis: **a** in funduscopy; **b** in fluorescein angiography

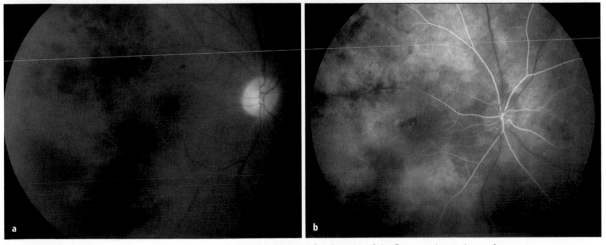

Fig. 25.3.5. Occlusive retinal vasculitis due to Behçet's disease: **a** in funduscopy; **b** in fluorescein angiography

one-fourth of BD patients, an affection of the optic nerve will be observed as papillitis (hyperemia of the optic disk with blurring of the margins). Papilledema is not frequent, but it may occur as a result of microvasculitis of the arterioles [97, 115].

Choroidal vascular involvement occurs as well, and choroidal infarcts are probably more common than generally suggested [97, 113, 115].

Another severe and sight threatening complication is retinal edema in 20–80% of cases, especially in the macular region [51, 97, 115]. At present, one of the most widely used investigations for confirming the presence of macular edema is fluorescein angiography (FA). Today optical coherence tomography (OCT) is a new method for high-resolution cross-sectional imaging of the retina. It is very useful in the objective evaluation of macular thickness and quantitative assessment of macular edema. Several studies found a specificity of 100% and a sensitivity of 96% on OCT when FA was taken as the reference standard. OCT is superior to FA in the quantitative evaluation of the macula and when vitreous opacity is present [5].

The most common fluorescein angiography findings are diffuse vascular leakage, hyperfluorescence of the optic disk and hyperfluorescence of the macula. These changes in the FA can occur even when eye involvement is not yet seen clinically. Hence, following patients closely, some authors recommend performing FA in all patients diagnosed as BD on a regular basis [78].

Severe posterior segment involvement with recurrent eye attacks is the most serious ocular complication of BD, leading to retinal changes, like narrowed and occluded silver-wired vessels, a variable degree of chorioretinal scars, retinal pigment epithelium alterations and optic nerve atrophy, which result in reduction of visual acuity up to blindness. Several studies indicate that more than three ocular attacks per year, and exudates within the retinal vascular arcade, are the risk and prognostic factors for a poor outcome for visual acuity in BD patients [99]. However, even in such cases, the prognosis will improve with early diagnosis and appropriate therapy [1].

25.3.5 Differential Diagnosis

The differential diagnosis of ocular BD includes various other causes of intraocular inflammation, as long as uveitis or vasculitis is the leading sign of ocular BD. Table 25.3.5 summarizes the most important disorders which may have similarities to ocular BD.

In case the clinical picture of ocular BD is not complete, multiple other systemic disorders can mimic BD. So, the combination of occlusive retinopathy with uveitis and genital and/or oral ulceration seems to be extraordinarily typical for BD.

In case of ocular vasculitis the most important differential diagnosis is intermediate uveitis (characterized by snowballs, probably associated with sarcoidosis or multiple sclerosis or without any systemic manifestations) or sarcoidosis without signs of inter-

Table 25.3.5. Most important disorders which may have similarities to ocular BD

	MB	Sarcoidosis	Sympathetic ophthalmia	AIDS ass. intra-ocular inflamm.	Arterial hypertension	Connective tissue disease	MS ass. inter-mediate uveitis	Syphilis	Colitis ulcerosa/Crohn's disease	Intermediate uveitis	Wegener's granulomatosis	Bacterial endo-phthalmitis	Eales disease	Tuberculosis
System manifestations														
a) Oral ulceration	+++	–	–	++	–	+	–	++	++	–	+	–	–	–
b) Genital ulceration	+++	–	–	++	–	–	–	++	–	–	–	–	–	–
c) Arthritis	++	+	–	+	–	++	–	+	+	–	++	–	–	–
d) Dermatitis	++	+++	–	+++	–	++	–	++	+	–	++	–	–	++
e) CNS manifestations	+	+	–	+	–	+	+++	++	–	–	++	–	–	+
f) Gastroint. manif.	++	+	–	+	–	+	–	+	+++	–	++	–	–	++
g) Lung manif.	–	+++	–	++	–	++	–	–	–	–	+++	–	–	+++
h) Renal manif.	+	+	–	+	+	++	–	+	–	–	+++	–	–	++
Ocular manifestations														
a) Vasculitis	+++	+++	+	+	–	++	++	++	++	+++	+++	+	++	++
b) Retinal vein occlusion	++	++	+	+	–	++	++	++	++	++	+++	+	++	++
c) Panuveitis	+++	+++	+++	–	–	++	–	+++	++	–	+++	+++	–	+++
d) Chorioretinitis	(+)	+++	+++	++	–	–	–	+++	(+)	–	+	++	–	+++
e) Snowballs	–	++	–	–	–	–	+++	+	+	+++	–	–	–	+
f) Granul. ocul. inflamm.	–	+++	+++	–	–	–	+	+	–	–	++	–	–	++
g) Cotton-wool spots	+++	–	–	+++	+++	+++	–	–	–	–	–	–	–	–

mediate uveitis. The definite diagnosis may become very difficult, when chest X-ray and even CT scan of the lung, besides angiotensin-converting enzyme, remain unremarkable.

25.3.6 Treatment

Essentials

- Due to poor prognosis, early and aggressive treatment of ocular BD is necessary
- Systemic immunosuppressive drugs are usually used as first-line therapy
- In case of inefficacy an early switch to novel therapeutic agents is recommended
- Biologicals seem to improve visual prognosis of the disease

Treatment of BD often remains a difficult task, because it should not only be effective for nearly all manifestations but should also produce few side effects while being inexpensive at the same time. In clinical practice, therapy will be provided with respect to the most serious and threatening manifestations of BD, which are usually represented by the involvement of the central nervous system and of the eyes.

Visual prognosis of ocular BD has shown to be poor. Without any treatment loss of vision will occur in more than 90 % of patients on average of 3.4 years after onset of ocular symptoms [63]. Thus, compared to other forms of intraocular inflammation, aggressive treatment should be initiated as soon as possible. This usually implies systemic immunosuppressive therapy as soon as inflammatory changes of the posterior eye segment occur. Although the introduction of modern immunosuppressive drugs doubtless initiated a great advance in the treatment of BD, these drugs have not been able to avoid the loss of useful vision in more than 50 % of patients after 5 – 10 years of disease [8]. Today, the preservation of good visual function for a long time is an adequate goal in the therapy of ocular BD, which often needs a switch to novel treatment options in time. Recently the so-called biologicals, interferon-α and TNF-α antagonists, have shown to be very promising in improving the visual prognosis for patients with BD.

This chapter aims to reflect the recent developments in the therapy of ocular BD.

25.3.6.1 Corticosteroids

Due to their high anti-inflammatory potential and their rapid onset of action, systemic corticosteroids, administered either orally or intravenously, are most effective in treating acute episodes of intraocular inflammation. However, because doses to maintain remission of ocular disease would be unacceptably high, they have been shown not to be suitable as long term monotherapy in BD [51]. Therefore a steroid-sparing immunosuppressive drug as described below has to be added early in most cases.

Topical corticosteroids (eyedrops and ointments) may be additionally used in case of inflammatory involvement of the anterior eye segment, but they have to be assumed to be ineffective in the treatment of posterior eye segment lesions. Intravitreally or periocularly administered corticosteroids may be useful as a temporary option in single cases, but usually they will not be able to substitute for systemic immunosuppressive therapy in severe cases of ocular BD and should be avoided.

25.3.6.2 Immunosuppressive and Cytotoxic Agents

Due to its rapid onset of action, **cyclosporin A** has been the most commonly used immunosuppressive agent for ocular BD so far. At doses of 3 – 5 mg/kg/day either as a monotherapy or in combination with low dose corticosteroids, it has been shown to be effective in controlling intraocular inflammation in ocular BD [79, 108]. However, the use of cyclosporin A often may be limited by its nephrotoxicity, particularly in doses higher than 5 mg/kg/day. It is also known that a rebound uveitis may occur after discontinuation of the drug [51]. It has been described moreover that cyclosporin A may exert neurotoxic effects in BD patients or trigger the onset of neuro-BD disease, respectively [50].

Tacrolimus (FK506) may be an alternative for cyclosporin A. Exerting a comparable immunosuppressive effect at lower doses, it has shown good efficacy in small groups of BD with sight-threatening posterior uveitis in whom previous cyclosporin A therapy had to be stopped due to a lack of response or unacceptable side effects [48, 95].

In a placebo-controlled double-blind study **azathioprine** was shown to be effective in treating ocular BD and, if treatment started less than 2 years after onset of the disease, in improving visual prognosis [32, 110, 111]. Because azathioprine, administered in doses of 2 – 2.5 mg/kg/day, is usually well tolerated, it might be a suitable alternative to cyclosporin A. However, one has to consider that azathioprine needs 2 – 3 months to become effective and therefore in the case of an acute inflammatory attack systemic corticosteroids have to be used to close this gap.

In contrast, **methotrexate, mycophenolate mofetil** and **colchicine** have been demonstrated or are believed to have only weak therapeutic effects on severe manifestations of BD and thus will not normally be in use for ocular involvement [2, 43, 77].

Due to their toxic side effects, **cyclophosphamide, chlorambucil** and **thalidomide** should be reserved for life-threatening manifestations of BD and should not be used any longer for ocular manifestations. Novel agents like interferon-α and TNF-α antagonists represent potent alternatives.

25.3.6.3 Novel Medical Treatment Approaches (Biologicals)

There is increasing evidence that the so-called biologicals, a novel generation of therapeutic agents, may be superior to conventional immunosuppressive drugs in the therapy of various chronic inflammatory conditions. On the other hand, depending on the drug, biologicals may be much more expensive and so far there has been only limited experience regarding their long term side effects.

Today there are two groups of biologicals, interferon (IFN)-α and TNF-α antagonists, in use for the treatment of BD especially with severe ocular involvement.

Interferon-α is a cytokine with strong antiviral, antiproliferative and various immunomodulatory effects. Since 1986 in open studies more than 330 BD patients have been treated with recombinant human interferon-α, more than 180 of them for acute ocular disease. IFN-α not only induced remission of intraocular inflammation in more than 90 % of patients who did not respond to conventional immunosuppressive treatment previously but also maintained it in more than 50 % of the patients after discontinuation of the drug [52]. IFN-α showed also to be effective in the resolution of cystoid macular edema as well as in regression of retinal neovascularizations and reperfusion of occluded vessels [51, 96]. Moreover, recent data suggest that IFN-α may lead to a significant improvement of visual prognosis in patients with ocular BD [19]. Although side effects will occur frequently (they are nearly all dose dependent and reversible), treatment with IFN-α is usually well tolerated. However, contraindications like depression, psoriasis and some autoimmune disorders such as sarcoidosis have to be ruled out before initiation of the drug. An IFN associated retinopathy which is known from patients with hepatitis C and skin melanoma, treated with inteferon-α, has not been seen yet in patients who received the drug for ocular BD or other intraocular inflammatory conditions [18]. Unfortunately there is no consensus so far on the optimal dosing schedule of IFN-α. In the light of our experience we recommend an initial dose of

3–6 million IU per day subcutaneously tapering subsequently to a maintenance dose of 3 million IU two or three times a week before discontinuation [51]. To avoid an antagonization of IFN-α, previous immunosuppressants should be stopped 1 day before initiation of IFN-α and corticosteroids have to be tapered to a maximum of 10 mg prednisolone equivalent per day as soon as possible [51].

In recent years **TNF-α antagonists** have been introduced more and more in the treatment regimens of various chronic inflammatory disorders, and also for ocular BD. TNF-α, a proinflammatory cytokine, is thought to play a major role in inducing such diseases. To block TNF-α, three different drugs are commercially available so far: etanercept (a soluble TNF-α receptor), infliximab (a mouse-human chimeric monoclonal TNF-α antibody) and adalimumab (a human monoclonal TNF-α antibody). In open clinical studies and in many case series, the most commonly used TNF-α antagonist infliximab, administered intravenously in doses of 3–5 mg/kg body weight at intervals of 2–8 weeks, has shown to be very effective in suppressing acute inflammatory attacks as well as in maintaining remission of ocular BD in patients who showed refractory to previous conventional immunosuppressive treatment. Regression of retinal neovascularization was also seen during infliximab treatment. However, there are few data existing so far regarding remission rates after discontinuation of the drug and no data about long term visual prognosis [25, 76, 92, 102]. Although infliximab as a chimeric molecule may induce allergic reactions in some cases, the drug is usually well tolerated. As TNF-α antagonists may lead to reactivation of previously existing tuberculosis, this infectious disease has to be ruled out before treatment initiation.

Although adalimumab as a human molecule probably will not induce allergy and can be self-administered by the patients (subcutaneous injection every 2 weeks), there is only one report which demonstrates that remission achieved with infliximab in ocular BD is able to continue by adalimumab [71]. In contrast, etanercept has not been reported in severe ocular BD.

25.3.6.4 Surgical Treatment

Ocular BD remains a domain of medical treatment. Surgery should be reserved to complications secondary to intraocular inflammation such as cataract or glaucoma in the anterior and retinal detachment or persistent vitreous hemorrhage in the posterior eye segment. As it has been shown that novel biologicals may induce regression of retinal and disk neovascularization (Fig. 25.3.6), one should be restrictive in

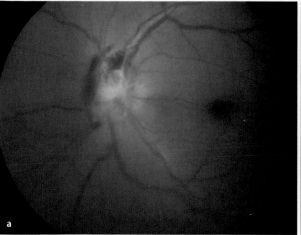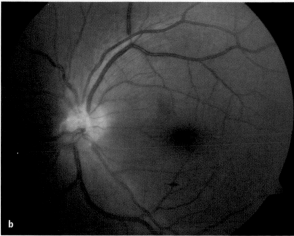

Fig. 25.3.6. a Occurrence of neovascularizations of the disk despite immunosuppressive treatment with azathioprine. **b** Complete regression of the neovascularizations after 3 months of treatment with IFN-α

performing laser photocoagulation in BD patients with retinal ischemia due to occlusive retinal vasculitis.

25.3.6.5 Recommendations for the Clinician

If a patient is suspected of having BD, it is strongly recommended to refer him to a specialized tertial medical center for diagnosis and treatment by an interdisciplinary approach. As initially mentioned, treatment has to be planned with respect to the most severe manifestation. If this is ocular disease with involvement of the posterior eye segment, a combination treatment of systemic steroids and an immunosuppressive drug, e.g., azathioprine or cyclosporin A, should be started without delay. It has to be kept in mind that immunosuppressive drugs may need weeks to months to develop their full anti-inflammatory effect. Therefore corticosteroids should be tapered very slowly to bridge this gap. If after adequate time a patient does not respond to or intraocular inflammation relapses under immunosuppressive treatment, a quick switch to a biological should be performed instead of trying several immunosuppressants one after another or using double- or triple-immunosuppressive regimens. In a BD patient who presents with an acute retinal vessel occlusion or with neovascularizations of the disk or the retina, we would suggest abstaining from conventional steroid and immunosuppressive treatment but initiating a first-line therapy with a biological immediately.

References

1. Accorinti M, Pirraglia MP, Corradi R, Marino M, Pivetti-Pezzi P (2003) Ocular lesions other than Behcet's disease in Behcet's disease-affected patients. In: Zouboulis CC (ed) Adamantiades-Behcet's disease. Kluwer Academic/Plenum Publishers, New York, pp 365–368
2. Adler YD, Mansmann U, Zouboulis CC (2001) Mycophenolate mofetil is ineffective in the treatment of mucocutaneous Adamantiades-Behcet's disease. Dermatology 203:322–324
3. Ambresin A, Tran T, Spertini F, Herbort C (2002) Behcet's disease in Western Switzerland: Epidemiology and analysis of ocular involvement. Ocul Immunol Inflamm 10:53–63
4. Ates A, Kinikli G, Duzgun N, Duman M (2006) Lack of association of tumor necrosis factor-alpha gene polymorphisms with disease susceptibility and severity in Behcet's disease. Rheumatol Int 26:348–353
5. Atmaca LS, Batioglu F, Müftüoglu O (2003) Fluorescein Angiography and optical Coherence Tomography in ocular Behcet's disease. In: Zouboulis CC (ed) Adamantiades-Behcet's disease. Kluwer Academic/Plenum Publishers, New York, pp 355–360
6. Ayaslioglu E, Duzgun N, Erkek E, Inal A (2004) Evidence of chronic Chlamydia pneumoniae infection in patients with Behcet's disease. Scand J Infect Dis 36:428–430
7. Aydintug AO, Tokgöz G, D'Cruz DP, Gürler A, Cervera R, Düzgün N, Atmaca LS, Khamashta MA, Hughes GR (1993) Antibodies to endothelial cells in patients with Behcet's disease. Clin Immunol Immunopathol 67:157–162
8. BenEzra D, Cohen E (1986) Treatment and visual prognosis in Behcet's disease. Br J Ophthalmol 70:589–592
9. Berlit P, Stueper B, Fink I, Rebmann V, Hoyer P, Kreuzfelder E, Grosse-Wilde H (2005) Behcet's disease is associated with increased concentrations of antibodies against phosphatidylserine and ribosomal phosphoproteins. Vasa 34:176–180
10. Boiardi L, Salvarani C, Casali B, Olivieri I, Ciancio G, Cantini F, Salvi F, Malatesta R, Govoni M, Trotta F, Filippini D, Paolazzi G, Nicoli D, Farnetti E, Macchioni L (2001) Intercellular adhesion molecule-1 gene polymorphisms in BD. J Rheumatol 28:1283–1287

11. Carletto A, Pacor ML, Biasi D, Caramaschi P, Zeminian S, Bellavite P, Bambara LM (1997) Changes of neutrophil migration without modification of in vitro metabolism and adhesion in Behçet's disease. J Rheumatol 24:1332–1336

12. Cervera R, Navarro M, Lopez-Soto A, Cid MC, Font J, Esparza J, Reverter JC, Monteagudo J, Ingelmo M, Urbano-Marquez A (1994) Antibodies to endothelial cells in BD: cell-binding heterogeneity and association with clinical activity. Ann Rheum Dis 53:265–267

13. Chafa O, Fischer AM, Meriane F, Chellali T, Sternberg C, Otmani F, Benabadji M (1992) Behçet syndrome associated with protein S deficiency. Thromb Haemost 67:1–3

14. Chajek T, Fainaru M (1975) Behcet's disease: Report of 41 cases and a review of the literature. Medicine 54:179–196

15. Chambers JC, Haskard DO, Kooner JS (2001) Vascular endothelial function and oxidative stress mechanisms in patients with Behçet's syndrome. J Am Coll Cardiol 37:517–520

16. Cochereau-Massin I, Wechsler B, Le Hoang P, Huong LeThi D, Girard B, Rousselie F (1992) Ocular prognosis in Behçet's disease. J Fr Ophthalmol 15:343–347

17. Davatchi F (2003) Treatment of ocular manifestations of Behcet's disease. Adv Exp Med Biol 528:487–491

18. Deuter CME, Zierhut M (2006) Interferons for intraocular inflammation. Ophthalmol Int (in press)

19. Deuter CME, Kötter I, Günaydin I, Zierhut M, Stübiger N (2004) Ocular involvement in Behçet's disease: first 5-year results for visual development after treatment with interferon alpha-2a. Ophthalmologe 101:129–134

20. Dilsen N, Konice M, Aral O, Ocal L, Inanc M, Gul A (1993) Comparative study of the skin pathergy test with blunt and sharp needles in BD: confirmed specificity but decreased sensitivity with sharp needles. Ann Rheum Dis 52:823–825

21. Direskeneli H, Keser G, D'Cruz D, Khamashta MA, Akoglu T, Yazici H, Yurdakul S (1995) Anti-endothelial cell antibodies, endothelial proliferation and von Willebrand factor antigen in Behçet's disease. Clin Rheumatol 14:55–61

22. Donmez A, Aksu K, Celik HA, Keser G Cagirgan S, Omay SB, Inal V Aydin HH, Tombuloglu M, Doganavsargil E (2005) Thrombin activatable fibrinolysis inhibitor in Behçet's disease. Thromb Res 115:287–297

23. Eksioglu-Demiralp E, Direskeneli H, Ergun T, Fresko I, Akoglu T (1999) Increased CD4+CD16+ and CD4+CD56+T cell subsets in Behçet's disease. Rheumatol Int 19:23–26

24. Freysdottir J, Farmer I, Lau SH, Hussain L, Verity D, Madanat W et al (1998) Cytokines in BD (P44). 8th International Congress on Behçet's Disease, Program and Abstracts. Italy, Reggio Emilia, p 143

25. Giansanti F, Barbera ML, Virgili G, Pieri B, Emmi L, Menchini U (2004) Infliximab for the treatment of posterior uveitis with retinal neovascularization in Behçet disease. Eur J Ophthalmol 14:445–448

26. Gül A, Esin S, Dilsen N, Konice M, Wigzell H, Biberfeld P (1995) Immunohistology of skin pathergy reaction in BD. Br J Dermatol 132:901–907

27. Gül A, Özbek U, Öztürk C, Inanç M, Koniçe M, Özçelik T (1996) Coagulation factor V gene mutation increases the risk of venous thrombosis in Behçet's disease. Br J Rheumatol 35:1178–1180

28. Gül A, Aslantas AB, Tekinay T, Koniçe M, Özçelik T (1999) Procoagulant mutations and venous thrombosis in Behçet's disease. Rheumatology (Oxford) 38:1298–1299

29. Gül A, Inanc M, Ocal L, Aral O, Konice M (2000) Familial aggregation of BD in Turkey. Ann Rheum Dis 59:622–625

30. Gül A, Hajeer AH, Worthington J, Ollier WE, Silman AJ (2001) Linkage mapping of a novel susceptibility locus for Behçet's disease to chromosome 6p22–23. Arthritis Rheum 44:2693–2696

31. Gürler A, Boyvat A, Türsen Ü (1997) Clinical manifestations of Behçet's disease: an analysis of 2147 patients. Yonsei Med J 38:423–427

32. Hamuryudan V, Özyazgan Y, Hizli N, Mat C, Yurdakul S, Tüzün Y, ᵃenocak M, Yazici H (1997) Azathioprine in Behçet's syndrome: effects on long-term prognosis. Arthritis Rheum 40:769–774

33. Hasan A, Fortune F, Wilson A, Warr K, Shinnick T, Mizushima Y, van der Zee R, Stanford MR, Sanderson J, Lehner T (1996) Role of γδT cells in pathogenesis and diagnosis of Behçet's disease. Lancet 347:789–794

34. Haznedaroglu IC, Dündar SV, Kirazli S (1995) Eicosanoids in the prethrombotic state of Behçet's disease. Thromb Res 80:445–446

35. Hu W, Hasan A, Wilson A, Stanford MR, Li-Yang Y, Todryk S, Whiston R, Shinnick T, Mizushima Y, van der Zee R, Lehner T (1998) Experimental mucosal induction of uveitis with 60-kDa heat shock protein-derived peptide 336–351. Eur J Immunol 28:2444–2455

36. Hughes EH, Collins RW, Kondeatis E, Wallace GR, Graham EM Vaughan RW, Stanford MR (2005) Associations of major histocompatibility complex class I chain-related molecule polymorphisms with Behçet's disease in Caucasian patients. Tissue Antigens 66:195–199

37. Ikbal, M, Atasoy M, Pirim I, Aliagaoglu C, Karatay S, Erdem F (2006) The alteration of sister chromatid exchange frequencies in Behçet's disease with and without HLA-B51. J Eur Acad Dermatol Venereol 20:149–152

38. Imirzalioglu N, Dursun A, Tastan B, Soysal Y, Yakicier MC (2005) MEFV gene is a probable susceptibility gene for Behçet's disease. Scand J Rheumatol 34:56–58

39. International Study Group for Behçet's Disease (1990) Criteria for diagnosis of Behçet's disease. Lancet 335:1078–1080

40. Isogai E, Ohno S, Kotake S, Isogai H, Tsurumizu T, Fujii N, Yokota K, Syuto B, Yamaguchi M, Matsuda H (1990) Chemiluminescence of neutrophils from patients with Behçet's disease and its correlation with an increased proportion of uncommon serotypes of Streptococcus sanguis in the oral flora. Arch Oral Biol 35:43–48

41. Itoh Y, Inoko H, Kulski J-K, Sasaki S, Meguro A, Takiyama N, Nishida T, Yuasa T, Ohno S, Mizuki N (2006) Four digit allele genotyping of the HLA-A and HLA-B genes in Japanese patients with Behçet's disease by a PCR-SSOP-Luminex method. Tissue Antigens 67:390–394

42. Jang WC, Park SB, Nam YH, Lee SS, Kim JW, Chang IS, Kim KT, Chang HK (2005) Interleukin-18 gene polymorphisms in Korean patients with BD. Clin Exp Rheumatol. 23:59–63

43. Kaklamani VG, Kaklamanis PG (2001) Treatment of Behçet's disease – an update. Semin Arthritis Rheum 30:299–312

44. Kaneko F, Nakamura K, Sato M, Tojo M, Zheng X, Zhang JZ (2003) Epidemiology of Behcet's disease in Asian countries and Japan. In: Zouboulis CC (ed) Adamantiades-Behcet's disease. Kluwer Academic/Plenum Publishers, New York, pp 25–29

45. Kansu E, Sahin G, Sahin F, Sivri B, Sayek I, Batman F (1986) Impaired prostacyclin synthesis by vessel walls in Behçet's disease. Lancet ii:1554

46. Karasneh J, Hajeer AH, Barrett J, Ollier WER, Thornhill M,

Gul A (2003) Association of specific interleukin 1 gene cluster polymorphisms with increased susceptibility for Behcet's disease. Rheumatology 42:860–864

47. Karasneh J, Gul A, Ollier WE, Silman AJ, Worthington J (2005) Whole-genome screening for susceptibility genes in multicase families with Behçet's disease. Arthritis Rheum 52:1836–1842

48. Kilmartin DJ, Forrester JV, Dick AD (1998) Tacrolimus (FK506) in failed cyclosporine A therapy in endogenous posterior uveitis. Ocul Immunol Inflamm 6:101–109

49. Kim JU, Chang HK, Lee SS, Kim JW, Kim KT, Lee SW, Chung WT (2003) Endothelial nitric oxide synthase gene polymorphisms in Behçet's disease and rheumatic disease with vasculitis. Ann Rheum Dis 62:1083–1087

50. Kotake S, Higashi K, Yoshikawa K, Sasamoto Y, Okamoto T, Matsuda H (1999) Central nervous system symptoms in patients with Behçet disease receiving cyclosporine therapy. Ophthalmology 106:586–589

51. Kötter I, Zierhut M, Eckstein A, Vonthein R, Ness T, Günaydin I, Grimmbacher P, Blaschke S, Peter HH, Kanz L, Stübiger N (2003) Human recombinant interferon-alpha2a (rhIFN alpha2a) for the treatment of Behçet's disease with sight-threatening retinal vasculitis. Br J Ophthalmol 87: 423–431

52. Kötter I, Günaydin I, Zierhut M, Stübiger N (2004) The use of interferon α in Behçet disease: review of the literature. Semin Arthritis Rheum 33:320–335

53. Kretschmann U, Seeliger MW, Ruether K, Usui T, Apfelstedt-Sylla E, Zrenner E (1998) Multifocal electroretinography in patients with Stargardt's macular dystrophy. Br J Ophthalmol 82:267–275

54. Lee EB, Kim JY, Lee YJ, Park MH, Song YW (2003a) TNF and TNF receptor polymorphisms in Korean Behçet's disease patients. Hum Immunol 64:614–620

55. Lee KH, Bang D, Choi ES, Chun WH, Lee ES, Lee S (1999) Presence of circulating antibodies to a disease-specific antigen on cultured human dermal microvascular endothelial cells in patients with Behçet's disease. Arch Dermatol Res 291:374–381

56. Lee KH, Chung HS, Kim HS, Oh SH, Ha MK, Baik JH, Lee S, Bang D (2003b) Human alpha-enolase from endothelial cells as a target antigen of anti-endothelial cell antibody in Behçet's disease. Arthritis Rheum 48:2025–2035

57. Lehner T (1997) The role of heat shock protein, microbial and autoimmune agents in aetiology of Behçet's disease. Int Rev Immunol 14:21–32

58. Lehner T (2000) Immunopathogenesis of Behçet's disease. In: Bang D, Lee ES, Lee S (eds) Behçet's disease. Design Mecca, Seoul, pp 3–18

59. Lehner T, Lavery E, Smith R, von der Zee R, Mizushima Y Shinnick T (1991) Association between the 65-kilodalton heat shock protein, Streptococcus sanguis, and the corresponding antibodies in Behçet's disease. Infect Immun 59:1434–1441

60. Lu Y, Ye P, Chen SL, Tan EM, Chan EK (2005) Identification of kinectin as a novel Behçet's disease autoantigen. Arthritis Res Ther 7:R1133–R1139

61. Mahesh SP, Li Z, Buggage R, Mor F, Cohen IR, Chew EY, Nussenblatt RB (2005) Alpha trompomyosin as a self-antigen in patients with Behçet's disease. Clin Exp Immunol 140:368–375

62. Mammo L, Al-Dalaan A, Bahabri SS, Saour JN. (1997) Association of factor V Leiden with Behçet's disease. J Rheumatol 24:2196–2198

63. Mamo JG (1970) The rate of visual loss in Behçet's disease. Arch Ophthalmol 84:451–452

64. Misgav M, Goldberg Y, Zeltser D, Eldor A, Berliner AS (2000) Fatal pulmonary artery thrombosis in a patient with Behçet's disease, activated protein C resistance and hyperhomocystinemia. Blood Coagul Fibrinolysis 11:421–423

65. Mishima S, Masuda K, Izawa Y, Mochizuki M, Namba K (1979) The eighth Frederick H. Verhoeff lecture presented by S. Mishima. Behçet's disease in Japan: ophthalmologic aspects. Trans Am Ophthalmol Soc 77:225–279

66. Mizuki N, Inoko H, Mizuki N, Tanaka H, Kera J, Tsuiji K, Ohno S (1992) Human leukocyte antigen serologic and DNA typing of Behçet's disease and its primary association with B51. Invest Ophthalmol Vis Sci 33:3332–3340

67. Mizuki N, Inoko H, Ohno S (1997) Molecular genetics (HLA) of BD. Yonsei Med J 38:333–349

68. Müftüoglu AÜ, Yazici H, Yurdakul S, Pazarli H, Ozyazgan Y, Tuzun Y, Altac M, Yalcin B (1981) Behçet's disease: lack of correlation of clinical manifestations with HLA antigens. Tissue Antigens 17:226–230

69. Musabak U, Baylan O, Cetin T, Yesilova Z, Sengul A, Saglam K, Inal A, Kocar IH (2005) Lipid profile and antivardiolipin antibodies in Behçet's disease. Arch Med Res 36:387–392

70. Musabak U, Pay S, Erdem H, Simsek I, Pekel A, Dinc A, Sengul A (2006) Serum interleukin-18 levels in patients with Behçet's disease. Is its expression associated with disease activity or clinical presentations? Rheumatol Int 26:545–550

71. Mushtaq B, Saeed T, Situnayake RD, Murray PI (2006) Adalimumab for sight-threatening uveitis in Behçet's disease. Eye (Epub ahead of print)

72. Nagafuchi H, Takeno M, Yoshikawa H, Kurokawa MS, Nara K, Takada E, Masuda C, Mizoguchi M, Suzuki N (2005) Excessive expression of Txk, a member of the Tec family of tyrosine kinases, contributes to excessive Th1 cytokine production by T lymphocytes in patients with Behçet's disease (2005). Clin Exp Immunol 139:363–370

73. Nakae K, Masaki F, Hashimoto T (1993) Behçet's disease. In: Wechsler B, Godeau P (eds) International congress series 1037. Excerpta Medica, Amsterdam, pp145–151

74. Niwa Y, Mizushima Y (1990) Neutrophil-potentiating factors released from stimulated lymphocytes; special reference to the increase in neutrophil-potentiating factors from streptococcus-stimulated lymphocytes of patients with Behçet's disease. Clin Exp Immunol 79:353–360

75. Ohno S, Ohguchi M, Hirose S, Matsuda H, Wakisaka A, Aizawa M (1982) Close association of HLA-Bw51 with Behçet's disease. Arch Ophthalmol 100:1455–1458

76. Ohno S, Nakamura S, Hori S, Shimakawa M, Kawashima H, Mochizuki M, Sugita S, Ueno S, Yoshizaki K, Inaba G (2004) Efficacy, safety, and pharmacokinetics of multiple administration of infliximab in Behçet's disease with refractory uveoretinitis. J Rheumatol 31:1362–1368

77. Okada AA (2000) Drug therapy in Behçet's disease. Ocul Immunol Inflamm 8:85–91

78. Özdal PC, Ortac S, Taskintuna I, Firat E (2002a) Posterior segment involvement in ocular Behçet's disease. Eur J Ophthalmol 12:424–431

79. Özdal PÇ, Ortaç S, Taskintuna I, Firat E (2002b) Long-term therapy with low dose cyclosporine A in ocular Behçet's disease. Doc Ophthalmol 105:301–312

80. Oztas MO, Onder M, Gurer MA, Bukan N, Sancak B (2005) Serum interleukin 18 an tumor necrosis factor-alpha levels are increased in Behçet's disease. Clin Exp Dermatol 30:61–63

25 III

81. Ozturk MA, Ertenli I, Kiraz S, C-Haznedaroglu I, Celik I, Kirazli S, Calguneri M (2004) Plasminogen activator inhibitor-1 as a link between pathological fibrinolysis and arthritis of BD. Rheumatol Int 24:98–102

82. Palimeris G, Papakonkonstantinou P, Mantas M (1984) The Adamantiades-Behcet's syndrome in Greek. In: Saari KM (ed) Uveitis update. Exerpta Medica, Amsterdam

83. Park KS, Min K, Nam JH, Bang D, Lee ES, Lee S (2005) Association of HYPA haplotype in the nannose-binding lectin gene-2 with Behçet's disease. Tissue Antigens 65:260–265

84. Pronai L, Ichikawa Y, Nakzawa H, Arimori S (1991) Enhanced superoxide generation and the decreased scavenging activity of peripheral blood leukocytes in Behçet's disease-effect of colchicines. Clin Exp Rheumatol 9:227–233

85. Sakane T, Kotani H, Takada S Tsunematsu T (1982) Functional aberration of T cell subsets in patients with Behçet's disease. Arthritis Rheum 25:1343–1351

86. Sakane T, Takeno M, Suzuki N, Inaba G (1999) Behçet's disease. N Engl J Med 341:1284–1291

87. Sallakci N, Bacanli A, Coskun M, Yavuzer U, Alpsoy E, Yegin O (2005) CTLA-4 gene 49A/G polymorphism in Turkish patients with BD. Clin Exp Dermatol 30:546–550

88. Salvarani C, Calamia K, Silingardi M, Ghirarduzzi A, Olivieri I (2000) Thrombosis associated with the prothrombin GA20210 mutation in Behçet's disease. J Rheumatol 27:515–516

89. Salvarani C, Boiardi L, Casali B, Olivieri I, Ciancio G, Cantini F, Salvi F, Malatesta R, Govoni M, Trotta F, Filippini D, Paolazzi G, Nicoli D, Farnetti E, Macchioni P (2002) Endothelial nitric oxide synthase gene polymorphisms in Behçet's disease. J Rheumatol 29:535–540

90. Salvarani C, Boiardi L, Casali B, Olivieri I, Cantini F, Salvi F, Malatesta R, La-Corte R, Triolo G, Ferrante A, Filippini D, Paolazzi G, Sarzi-Puttini P, Nicoli D, Farnetti E, Chen Q, Pulsatelli L (2004) Vascular endothelial growth factor gene polymorphisms in BD J Rheumatol 31:1785–1789

91. Sari RA, Kiziltunc A, Taysy S, Akdemyr S, Gundoglu M (2005) Levels of soluble E-selectin in patients with active Behçet's disease. Clin Rheumatol 24:55–59

92. Sfikakis PP, Theodossiadis PG, Katsiari CG, Kaklamanis P, Markomichelakis NN (2001) Effect of infliximab on sight-threatening panuveitis in Behçet's disease. Lancet 358:295–296

93. Shehto NM, Ghosh K, Abdul-Kader B, al Assad HS (1992) Extensive venous thrombosis in a case of Behçet's disease associated with heterozygous protein C deficiency. Thromb Haemost 67:283

94. Shimizu T, Matsumura N (1972) Behçet's disease. Nippon Rinsho 30:416–420

95. Sloper CML, Powell RJ, Dua HS (1999) Tacrolimus (FK506) in the treatment of posterior uveitis refractory to cyclosporine. Ophthalmology 106:723–728

96. Stübiger N, Kötter I, Zierhut M (2000) Complete regression of retinal neovascularisation after therapy with interferon alpha in Behçet's disease. Br J Ophthalmol 84:1437–1438

97. Stübiger N, Zierhut M, Kötter I (2003) Ocular manifestations in Behçet's disease. In: Zierhut M, Ohno S (eds) Immunology of Behçet's disease. Swets and Zeitlinger, Lisse, Tokyo, pp 36–45

98. Takeno M, Shimoyama Y, Kashiwakura J, Nagafuchi H, Sakane T, Suzuki N (2004) Abnormal killer inhibitory receptor expression on natural killer cells in patients with Behçet's disease. Rheumatol Int 24:212–216

99. Takeuchi M, Hokama H, Tsukahara R, Kezuka T, Goto H, Sakai J, Usui M (2005) Risk and prognostic factors of poor visual outcome in Behçet's disease with ocular involvement. Graefes Arch Clin Exp Ophthalmol 243(11):1147–1152

100. Todaro M, Zerilli M, Triolo G, Iovino F, Patti M, Accardo-Palumbo A, di Gaudio F, Turco MC, Petrella A, de Maria R, Stassi G 2005 (2005) NF-kappaB protects BD T cells against CD95-induced apoptosis up-regulating antiapoptotic proteins. Arthritis Rheum 52:2179–2191

101. Tokay S, Direskeneli H, Yurdakul S, Akoglu T (2001) Anti-cardiolipin antibodies in Behçet's disease: a reassessment. Rheumatology (Oxford) 40:192–195

102. Tugal-Tutkun I, Mudun A, Urgancioglu M, Kamali S, Kasapoglu E, Inanc M, Gül A (2005) Efficacy of infliximab in the treatment of uveitis that is resistant to treatment with the combination of azathioprine, cyclosporine, and corticosteroids in Behçet's disease. Arthritis Rheum 52:2478–2484

103. Ureten K, Ertenli I, Ozturk MA, Kiraz S, Onat AM, Tuncer M, Okur H, Akdogan A, Apras S, Calguneri M (2005) Neutrophil CD 64 expression in Behçet's disease. J Rheumatol 32:849–852

104. Verity DH, Marr JE, Ohno S, Wallace GR, Stanford MR (1999a) BD, the Silk Road and HLA-B51: historical and geographical perspectives. Tissue Antigens 54:213–220

105. Verity DH, Vaughan RW, Madanat W, Kondeatis E, Zureikat H, Fayyad F, Kanawati CA, Ayesh I, Stanford MR, Wallace GR (1999b) Factor V Leiden mutation is associated with ocular involvement in Behçet's disease. Am J Ophthalmol 128:352–356

106. Verjans GM, van Hagen PM, van der Kooi A, Osterhaus AD, Baarsma GS (2002) Vgamma9Vdelta2 T cells recovered from eyes of patients with Behçet's disease recognize non-peptide prenyl pyrophosphate antigens. J Neuroimmunol 130:46–54

107. Wakisaka S, Takeba Y, Mihara S, Takeno M, Yamamoto S, Sakane T, Suzuki N (2002) Aberrant Fas ligand expression in lymphocytes in patients with BD. Int Arch Allergy Immunol 129:175–180

108. Whitcup SM, Salvo EC, Nussenblatt RB (1994) Combined cyclosporine and corticosteroid therapy for sight-threatening uveitis in Behçet's disease. Am J Ophthalmol 118:39–45

109. Yazici H, Basaran G, Hamuryudan V, Hizli N, Yurdakul S, Mat C, Tuzun Y, Ozyazgan Y, Dimitriyadis I (1996) The ten-year mortality in Behcet's syndrome. Br J Rheumatol 35:139–141

110. Yazici H, Ozyagan Y (1999) Medical management of Behçet's syndrome. Dev Ophthalmol 31:118–131

111. Yazici H, Pazarli H, Barnes CG, Tüzün Y, Özyazgan Y, Silman A, Serdaroðlu S, Oðuz V, Yurdakul S, Lovatt GE, Yazici B, Somani S, Müftüoðlu A (1990) A controlled trial of azathioprine in Behçet's syndrome. N Engl J Med 322:281–285

112. Yurdakul S, Günaydin I, Tuzkun Y, Tankurt N, Pazarli H, Ozyazgan Y, Yasici H (1988) The prevalence of Behcet's syndrome in a rural area in northern Turkey. J Rheumatol 15:820–822

113. Zafirakis P, Foster CS (2002) Adamantiades-Behcet's disease. In: Foster CS, Vitale AT (eds) Diagnosis and treatment of uveitis. Saunders, Philadelphia, pp 632–652

114. Zierhut M, Mizuki N, Ohno S, Inoko H, Gul A, Onoe K, Isogai E (2003) Immunology and functional genomics of BD. Cell Mol Life Sci 60:1903–1922

115. Zierhut M, Stübiger N, Deuter CME (2004): Behçet's disease. In: Pleyer U, Mondino B (eds) Essentials in ophthalmology: uveitis and immunological disorders. Springer, Berlin Heidelberg New York, pp 173–195

116. Zouboulis CC (1999) Epidemiology of Adamantiades-Behcet's disease. Ann Med Interne (Paris) 150:488–498

117. Zouboulis CC (2003) Epidemiology of Adamantiades-Behcet's disease. In: Zierhut M, Ohno S (eds) Immunology of Behçet's disease. Swets and Zeitlinger, Lisse, Tokyo, pp 1–16

118. Zouboulis CC, Kötter I, Djawari D, Kirch W, Kohl PK, Ochsendorf FR, Keitel W, Stadler R, Wollina Uproksch E, Sohnchen R, Weber H, Gollnick HP, Holzle E, Fritz K, Licht T, Orfanos CE (1997) Epidemiological features of Adamantiades-Behcet's disease in Germany and in Europe. Yonsei Med J 38:411–422

25.4 Vasculitis in Multiple Sclerosis

M.D. Becker, U. Wiehler, D.W. Miller

Core Messages
- In addition to optic neuritis, multiple sclerosis (MS) can affect the eye with intraocular inflammation termed "uveitis." Uveitis in patients with MS is often a bilateral disease and manifests with vitritis and retinal vasculitis (called intermediate uveitis)

- Prognosis of visual acuity is often determined by optic atrophy or cystoid macular edema (CME) and the formation of retinal neovascularization
- In addition to treatment with high-dose corticosteroids and immunosuppressive drugs, therapy with interferon-β is evolving as a new treatment option for this condition

25.4.1 History

The association between MS and optic neuritis has been known since the late 19th century. However, intraocular inflammation (uveitis) was originally described in 1945 by Rucker [30].

25.4.2 Epidemiology

Multiple sclerosis (MS) is a chronic, inflammatory, demyelinating disease of the central nervous system mostly affecting young adults. The clinical picture is determined by the location of foci of demyelination within the CNS. The eyes are frequently affected with optic neuritis, extraocular muscle disturbances, and intraocular inflammation. In cooperation with consulting neurologists, the diagnosis of MS must be made based upon neurological examination and MRI imaging according to the guidelines described by Poser et al. [26]. The occurrence of the association of MS and uveitis varies widely, ranging from 0.4% to 26.9% in patients with MS [4, 9, 25] and from 0.8% to 14% in patients with uveitis [17, 20, 29]. These differences between reports are attributable to varying patient populations, examination techniques, and criteria for diagnosis. The longer the follow-up in these studies the higher the prevalence. The prevalence of MS in the uveitic population has been reported to be 1–2% [7, 29], with a higher prevalence among patients with intermediate uveitis, ranging from 7.8% to 14.8% [41]. In the database of the Interdisciplinary Uveitis Center Heidelberg (tertiary referral center), 50 out of 1,430 total uveitis patients (3%) had an association with MS. Seventy-

six percent of these patients were female with an age median of 43.5 years, and 72% had bilateral disease, with 54% patients having a chronic disease course, 68% intermediate uveitis, and 50% associated optic neuritis (papillitis, retrobulbar neuritis). Anterior uveitis is rare in patients with MS [19, 21]. It occurred in only 14% of MS patients in our database. Conversely, the prevalence of MS in patients with intermediate uveitis at the Interdisciplinary Uveitis Center in Heidelberg was 13%.

25.4.3 Pathogenesis

Experimental autoimmune encephalomyelitis (EAE) is a well-established rodent model of CNS-specific inflammatory disease and is known to be the best animal model for studying the etiology and pathogenesis of MS. It is mediated by T cells and results in progressive demyelination and paralysis. Anterior uveitis (AU) has been found to coincide with EAE in rabbits, monkeys, in the Lewis rat [2, 36] and more recently in (PL/J×SJL/J) F1 mice [11]. The encephalogenic T cells are specific for the antigen myelin basic protein (MBP), which is a component of the myelinated sheath surrounding nerve bundles. These myelinated nerve bundles are abundant in the spinal cord and in the iris; thus this "autoantigen" is located at sites of inflammation. This suggests shared antigenic determinants between neurological and ophthalmological manifestations of the disease with similar pathomechanisms [2]. In the rodent models of EAE, AU generally persists after the paralysis has subsided. EAE and AU can be induced actively by immunizing with MBP in the presence of adjuvant, or passively by using adoptive transfer of MBP-specific T cells that have been generated against the whole antigen or encephalogenic peptides. Treatment with IFN-β reduced ocular inflammation in this model [23].

Despite these findings, most EAE studies have focused on the problems of spinal cord and brain disease and there is little information on ocular pathologic events in the course of this experimental illness. As in MS, EAE is characterized by a breakdown of the blood-brain barrier. The inflammatory response is characterized by mononuclear infiltrates located around vessels of the CNS white matter, by activation of local microglia and astrocytes, and in the most severe cases it may be eventually followed by demyelination. Recently, several authors have examined the ophthalmologic manifestations of EAE in Lewis rats and all described the induction of anterior uveitis [2, 33, 36]. More recently, Hu et al. [14, 15] investigated the cellular responses in the optic nerve and retina of Lewis rats with MBP-induced EAE. Villarroya et al. induced relapsing EAE in rats to determine the phenotype and localization of immunocompetent cells as well as the MHC class I and class II antigen expression in various structures of the visual system [37]. Their findings suggest that ocular structures and optic pathways, in particular optic chiasma, are important targets in relapsing EAE, and the development of ocular findings shares similar pathogenic mechanisms with neurologic disease. Disruption of the cerebral endothelial barrier and transendothelial migration of inflammatory cells into the brain plays a significant role in the pathogenesis of MS.

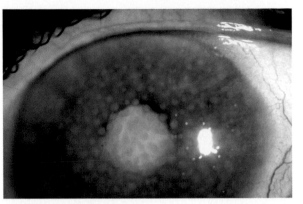

Fig. 25.4.1. Granulomatous "mutton-fat" keratic precipitates in a patient with MS-associated uveitis. The white pupil indicates the presence of cataract

25.4.4 Intermediate Uveitis and Retinal Vasculitis as the Clinical Manifestations in MS-Associated Uveitis

Optic neuritis (ON) is the most common ocular manifestation of MS, occurring in an estimated 30% of patients [24]; however, intraocular inflammation (uveitis) also occurs. Conversely, 30% of patients with ON will develop clinically definite MS, as reported by the Optic Neuritis Study Group [24]. The classic manifestation of ON includes unilateral, sudden onset reduction of visual acuity, relative afferent pupillary defect, and paracentral or centrocecal scotoma in a young female patient. In a recent retrospective multicenter case observational study [6] with 13 patients, in 7 patients MS started a median of 8.2 years (range 2.8–24.2 years) before uveitis, and in 6 patients uveitis was diagnosed a median of 5.5 years (0.3–20.6 years) before MS. Seven patients had suffered from ON in at least one eye during the previous course of their disease. In all 13 patients ON occurred before the onset of uveitis.

The most common type of uveitis is intermediate uveitis (according to Bloch-Michel and Nussenblatt [8]) with characteristic changes (iritis, pars planitis, vitritis, periphlebitis with or without neovascularization) and may present with "granulomatous changes" [1, 4, 7, 35]. Granulomatous changes in the anterior segment include large, "mutton-fat" keratic precipitates located on the corneal endothelium (Fig. 25.4.1), iris nodules (Koeppe nodules at the pupillary border, Bussacca nodules in the iris stroma). These changes are not only typical for histologically granulomatous diseases like sarcoid or tuberculosis but can also be seen in ocular structures in histologically non-granulomatous diseases like MS. In patients with this type of uveitis, secondary changes like the formation of CME or occlusive vasculitis with vasoproliferations often develop. These changes have a profound effect on visual prognosis. If they occur, the disease is often difficult to manage and requires high dosages of corticosteroids [38] or laser treatment, and often remains refractory to these standard therapies.

Retinal vasculitis in MS uveitis patients is considered to affect exclusively veins ("periphlebitis," retinal venous "sheathing"). It encompasses both active periphlebitis and chronic venous sclerosis [3]. The chronic form of sheathing appears typically as dense, white linear stripes following the course of several generations of the venous tree. In our MS patients vasculitis usually manifested in a continuous fashion over several branches of the vascular tree in contrast to sarcoidosis, which shows a multifocal pattern ("candle wax drippings"). There may be associated constriction and dilation of the veins, with occasional retinal hemorrhage. However, there are also focal lesions in MS-associated uveitis, especially in the acute phase. Venous sclerosis seems to be the result of persistent inflammation (Fig. 25.4.2). Retinal arteriolar involvement in MS-associated uveitis has not been described to our knowledge.

25 **III**

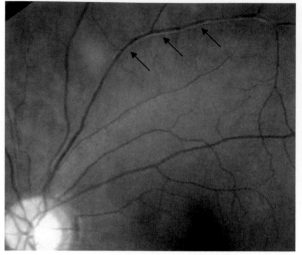

Fig. 25.4.2. Chronic venous sclerosis of the vena temporalis superior with a linear white opacity (*arrows*) in a patient with MS-associated uveitis with narrowing and irregularity of vein caliber. There is also optic atrophy due to recurrent retrobulbar neuritis

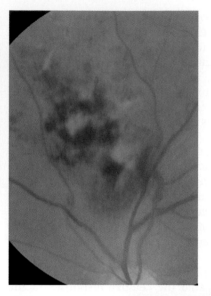

Fig. 25.4.4. Occlusive periphlebitis with branch vein occlusion indicated by intraretinal hemorrhage

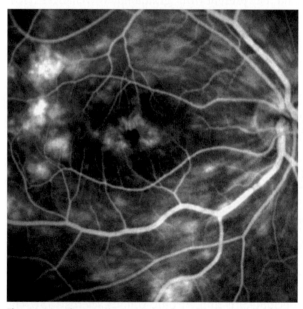

Fig. 25.4.3. Fluorescein angiography of active periphlebitis demonstrates delayed filling, persistence of dye, staining, and leakage from affected vessels, which indicates breakdown of the blood-retinal barrier

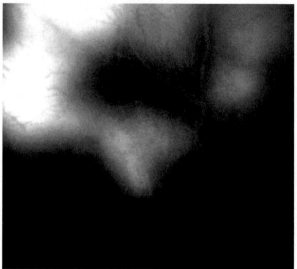

Fig. 25.4.5. Retinal neovascularization in patient with occlusive MS-associated vasculitis shown by fluorescein angiography

Activity of periphlebitis in MS is not correlated with optic neuritis, systemic exacerbations, or severity of disease. Schmidt et al. showed that MS patients with concomitant intraocular inflammation are not distinct from "classic" MS with regard to the clinical course of disease, disability, and neuroimaging features [32].

Fluorescein angiography of active periphlebitis demonstrates delayed filling, persistence of dye, staining, and leakage from affected vessels, which indicates breakdown of the blood-retinal barrier [40] (Fig. 25.4.3). Some vessels show staining of leakage in areas without clinically detectable periphlebitis. Venous sclerosis may show either late staining or normal appearance on angiography.

Complications of intermediate uveitis include cataract formation, CME, epiretinal membrane formation, glaucoma, retinal detachment, and occlusive vasculitis (Fig. 25.4.4) with subsequent neovascularization (Fig. 25.4.5) with and without vitreous hemorrhage. There seems to be no difference in the rate of complications in patients with primary or other secondary forms of intermediate uveitis.

Table 25.4.1 summarizes the characteristic findings of intraocular inflammation in patients with MS.

Table 25.4.1. Characteristic features of uveitis in multiple sclerosis

- Bilateral, intermediate uveitis with or without optic neuritis in the past
- „Granulomatous" changes in the anterior segment
- Linear periphlebitis
- Cystoid macular edema
- Occlusive vasculitis with subsequent formation of retinal neovascularization

Table 25.4.2. Differential diagnosis of MS-associated uveitis

Oculocerebral lymphoma	MRI imaging, CSF exam, diagnostic vitrectomy
CNS vasculitis	MRI imaging
Neurosyphilis	Syphilis serology, CSF exam
Neuroborreliosis	Erythema migrans, Lyme serology, CSF exam
Viral infections (herpes family viruses)	Retinal arteriolitis, diagnostic vitrectomy, retinal morphology
Behçet's disease	Occlusive retinal vasculitis
Sarcoidosis	Multifocal periphlebitis, ACE level, chest X-ray, granulomatous changes

25.4.5 Differential Diagnosis

Inflammatory systemic diseases that produce encephalomyelopathy, optic neuritis, extraocular muscle disturbances, intermediate uveitis with retinal vasculitis, and may mimic MS-associated uveitis include the diseases shown in Table 25.4.2.

25.4.6 Therapeutic Options

25.4.6.1 Corticosteroids and Immunosuppression

For patients with primary or non-infectious secondary forms of intermediate uveitis, systemic corticosteroids (initial dosage 1 mg/kg body weight) are still the mainstay of initial treatment. If the inflammatory activity (density of cells in the vitreous or anterior chamber) can be reduced by systemic corticosteroids and there is no relapse of disease above the Cushing threshold (about 7.5 mg prednisone equivalent per day) but a relapse after complete cessation, a low-dose prednisone scheme should be attempted. If the relapse is above the Cushing threshold, the use of corticosteroid-sparing drugs (like methotrexate or mycophenolate mofetil) should be used. The use of inhibitors of tumor necrosis factor (TNF) in patients with MS-associated uveitis is contraindicated due to the possibility of induction of MS. However, it should be considered in patients with severe primary or secondary (other than MS) intermediate uveitis and vasculitis; infliximab seems to provide a stronger anti-inflammatory potency than etanercept.

Periocular or intravitreal injections of triamcinolone should be used in cases of refractory CME. However, the formation of cataract or steroid-induced glaucoma has to be considered.

25.4.6.2 Immunomodulatory Therapy with Interferon for Cystoid Macular Edema

Interferon (IFN) has been shown to have beneficial effects in patients with MS [27, 28] and/or ON [10]. Jacobs et al. showed that initiating treatment with IFN-β1a at the time of the first demyelinating event, like optic neuritis, is beneficial for patients with brain lesions on MRI that indicate a high risk of clinically definite multiple sclerosis [16].

The biological effects of IFNs have been known since the early 1970s. Type I IFNs (IFN-α and -β) share about 30% homology in their amino acid sequence and use the same receptor (there is also an additional IFN-β receptor). Therefore, the therapeutic effects of type I IFNs are quite similar. Type II IFNs (IFN-γ) have different amino acid structures.

Several studies show that IFN-β treatment shifts the balance of cytokines in favor of a net anti-inflammatory response, by either suppressing Th1 or by increasing Th2 cytokine production or both [12, 39]. IFN-β also suppresses the IFN-γ-mediated expression of major histocompatibility complex (MHC) class II on astrocytes and microglia in vitro and reduces the antigen-presenting capacity of these cells [18, 31].

In the nomenclature of interferons, a "1" indicates a closely related protein within the β-family; "a" indicates that the synthetic product has the same primary structure as the natural substance. Minagar showed in a recent study that IFN-β1a and IFN-β1b blocked the IFN-γ-induced disintegration of endothelial junction integrity and therefore protected endothelial barriers. They also showed that the protective effects of IFN-β on occludin and VE-cadherin stability appear to represent molecular mechanisms for the therapeutic effects of the IFN-β on blood-brain barrier in MS [22]. Interestingly, the reduction of vascular leakage and hence reduction of edema of the posterior pole and increase in visual acuity seems to be an important effect of IFN in the treatment of patients with uveitis associated with MS in one of our recent studies [6]. Because of the different mode of action of IFNs ("immunomodulatory" rather than "immunosuppressive"), it is thought that IFNs need an unsuppressed immune system. Therefore, it is recommended to use additional corticosteroids only in low doses. The concurrent use of immunosuppressives should be avoided during IFN treatment.

The current approach to treating patients with optic neuritis has been modified by the results of the Controlled High-Risk Subjects Avonex Multiple Sclerosis Prevention Study (CHAMPS). Patients with an initial clinical episode of demyelination (optic neuritis, incomplete transverse myelitis, or brain-stem/cerebellar syndrome) and at least two characteristic demyelinating lesions within the brain were randomized to receive interferon β1a or placebo after initial treatment with intravenous corticosteroids. At a 3-year follow-up, patients treated with IFN-β1a showed a 50% reduced risk of progression to clinically definite MS (CDMS) [5, 10, 13]. The results of this study have set the standard of treatment for patients with a first bout of demyelinating optic neuritis.

25.4.6.3 Laser Photocoagulation

In cases of severe neovascularization, argon laser treatment might be an option if the disease is unresponsive to immunosuppressive drugs or treatment with IFN. Sectorial panretinal photocoagulation is the treatment of choice (Fig. 25.4.6). IFN-α has been shown to lead to regression of inflammatory retinal neovascularization in patients with occlusive periphlebitis associated with Behçet's disease [34]. In our experience, prophylactic treatment of occlusive periphlebitis in patients with MS-associated uveitis seems to shrink the total area of non-perfusion, inhibit the formation of retinal neovascularization and reduce the amount of vascular leakage (Figs. 25.4.7, 25.4.8).

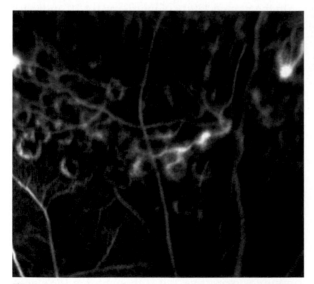

Fig. 25.4.6. Regression of neovascularization in the same patient of Fig. 25.4.5 after sectorial photocoagulation shown by fluorescein angiography

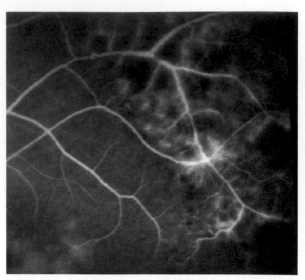

Fig. 25.4.7. Occlusive retinal vasculitis before treatment with interferon-β in a patient with MS-associated uveitis

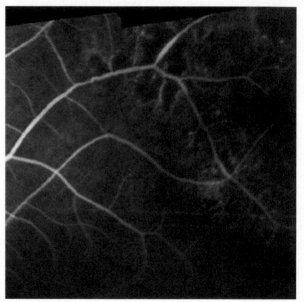

Fig. 25.4.8. Occlusive retinal vasculitis after treatment with interferon-β in the same patient as Fig. 7: regression of areas of non-perfusion, no formation of retinal neovascularization, reduction of vascular leakage

25.4.6.4 Treatment Strategy

Multiple factors influence the decision on how to treat patients with MS-associated uveitis. Table 25.4.3 delineates the personal treatment strategy of the authors. Due to the chronic course of MS-associated uveitis, patients should be on a regular follow-up according to clinical activity. In cases of mild inflammatory activity we recommend 3-monthly intervals of ophthalmologic examinations.

Table 25.4.3. The personal treatment strategy of the authors

Visual acuity	Complicating factors	Treatment option
≥ 0.8 without chronic complaints of floaters	≤1+ cells in the vitreous, no CME, mild periphlebitis	Observe
<0.8	Vitreous opacities, mild CME, 1–2+ cells, unilaterality	Periocular or intravitreal triamcinolone injection
0.1–0.4	VA mostly due to CME	Systemic corticosteroids, immunosuppression, treatment with interferon-β
	Occlusive vasculitis	Systemic corticosteroids, immunosuppression or treatment with interferon-beta
	Neovascularization	Treatment with interferon-β and/or sectorial argon laser photocoagulation

III 25

References

1. Acar MA, Birch MK, Abbott R, Rosenthal AR (1993) Chronic granulomatous anterior uveitis associated with multiple sclerosis. Graefes Arch Clin Exp Ophthalmol 231(3):166–8
2. Adamus G, Amundson D, Vainiene M, et al. (1996) Myelin basic protein specific T-helper cells induce experimental anterior uveitis. J Neurosci Res 44(6):513–8
3. Arnold AC, Pepose JS, Hepler RS, Foos RY (1984) Retinal periphlebitis and retinitis in multiple sclerosis. I. Pathologic characteristics. Ophthalmology 91(3):255–62
4. Bamford CR, Ganley JP, Sibley WA, Laguna JF (1978) Uveitis, perivenous sheathing and multiple sclerosis. Neurology 28(9):119–24
5. Beck RW, Chandler DL, Cole SR, et al. (2002) Interferon beta-1a for early multiple sclerosis: CHAMPS trial subgroup analyses. Ann Neurol 51(4):481–90
6. Becker MD, Heiligenhaus A, Hudde T, et al. (2005) Interferon as a treatment for uveitis associated with multiple sclerosis. Br J Ophthalmol (in press)
7. Biousse V, Trichet C, Bloch-Michel E, Roullet E (1999) Multiple sclerosis associated with uveitis in two large clinic-based series. Neurology 52(1):179–81
8. Bloch-Michel E, Nussenblatt RB (1987) International Uveitis Study Group recommendations for the evaluation of intraocular inflammatory disease. Am J Ophthalmol 103(2):234–5
9. Breger BC, Leopold IH (1966) The incidence of uveitis in multiple sclerosis. Am J Ophthalmol 62(3):540–5
10. CHAMPS Study Group (2001) Interferon beta-1a for optic neuritis patients at high risk for multiple sclerosis. Am J Ophthalmol 132(4):463–71
11. Constantinescu CS, Lavi E (2000) Anterior uveitis in murine relapsing experimental autoimmune encephalomyelitis (EAE), a mouse model of multiple sclerosis (MS). Curr Eye Res 20(1):71–6
12. Dhib-Jalbut S (1997) Mechanisms of interferon beta action in multiple sclerosis. Mult Scler 3(6):397–401
13. Galetta SL (2001) The controlled high risk Avonex multiple sclerosis trial (CHAMPS Study). J Neuroophthalmol 21(4):292–5
14. Hu P, Pollard J, Hunt N, Chan-Ling T (1998) Microvascular and cellular responses in the retina of rats with acute experimental allergic encephalomyelitis (EAE). Brain Pathol 8(3):487–98
15. Hu P, Pollard J, Hunt N, et al. (1998) Microvascular and cellular responses in the optic nerve of rats with acute experimental allergic encephalomyelitis (EAE). Brain Pathol 8(3):475–86
16. Jacobs LD, Beck RW, Simon JH, et al. (2000) Intramuscular interferon beta-1a therapy initiated during a first demyelinating event in multiple sclerosis. CHAMPS Study Group. N Engl J Med 343(13):898–904
17. James DG, Friedmann AI, Graham E (1976) Uveitis. A series of 368 patients. Trans Ophthalmol Soc U K 96(1):108–12
18. Jiang H, Milo R, Swoveland P, et al. (1995) Interferon beta-1b reduces interferon gamma-induced antigen-presenting capacity of human glial and B cells. J Neuroimmunol 61(1):17–25
19. Lim JI, Tessler HH, Goodwin JA (1991) Anterior granulomatous uveitis in patients with multiple sclerosis. Ophthalmology 98(2):142–5
20. Malinowski SM, Pulido JS, Folk JC (1993) Long-term visual outcome and complications associated with pars planitis. Ophthalmology 100(6):818–24; discussion 25
21. Meisler DM, Tomsak RL, Khoury S, et al. (1989) Anterior uveitis and multiple sclerosis. Cleve Clin J Med 56(5):535–8
22. Minagar A, Long A, Ma T, et al. (2003) Interferon (IFN)-beta1a and IFN-beta1b block IFN-gamma-induced disintegration of endothelial junction integrity and barrier. Endothelium 10(6):299–307
23. Okada AA, Keino H, Fukai T, et al. (1998) Effect of type I interferon on experimental autoimmune uveoretinitis in rats. Ocul Immunol Inflamm 6(4):215–26
24. Optic Neuritis Study Group (1997) The 5-year risk of MS after optic neuritis. Experience of the Optic Neuritis Treatment Trial. Neurology 49(1404–1418)
25. Porter R (1972) Uveitis in association with multiple sclerosis. Br J Ophthalmol 56(6):478–81
26. Poser CM, Paty DW, Scheinberg L, et al. (1983) New diagnostic criteria for multiple sclerosis: guidelines for research protocols. Ann Neurol 13(3):227–31
27. PRISMS-4 (2001) Long-term efficacy of interferon-beta-1a in relapsing MS. Neurology 56(12):1628–36
28. PRISMS Study Group (1998) Randomised double-blind placebo-controlled study of interferon beta-1a in relapsing/remitting multiple sclerosis. PRISMS (Prevention of Relapses and Disability by Interferon beta-1a Subcutaneously in Multiple Sclerosis) Study Group. Lancet 352(9139):1498–504
29. Rothova A, Buitenhuis HJ, Meenken C, et al. (1992) Uveitis and systemic disease. Br J Ophthalmol 76(3):137–41

25 III

30. Rucker CW (1945) Sheathing of the retinal veins in multiple sclerosis. JAMA 127:970–3

31. Satoh J, Paty DW, Kim SU (1995) Differential effects of beta and gamma interferons on expression of major histocompatibility complex antigens and intercellular adhesion molecule-1 in cultured fetal human astrocytes. Neurology 45(2):367–73

32. Schmidt S, Wessels L, Augustin A, Klockgether T (2001) Patients with multiple sclerosis and concomitant uveitis/periphlebitis retinae are not distinct from those without intraocular inflammation. J Neurol Sci 187(1–2):49–53

33. Shikishima K, Lee WR, Behan WM, Foulds WS (1993) Uveitis and retinal vasculitis in acute experimental allergic encephalomyelitis in the Lewis rat: an ultrastructural study. Exp Eye Res 56(2):167–75

34. Stübiger N, Koetter I, Zierhut M (2000) Complete regression of retinal neovascularization after therapy with interferon alfa in Behcet's disease. Br J Ophthalmol 84(12):1437–8

35. Towler HM, Lightman S (2000) Symptomatic intraocular inflammation in multiple sclerosis. Clin Exp Ophthalmol 28(2):97–102

36. Verhagen C, Mor F, Cohen IR (1994) T cell immunity to myelin basic protein induces anterior uveitis in Lewis rats. J Neuroimmunol 53(1):65–71

37. Villarroya H, Klein C, Thillaye-Goldenberg B, Eclancher F (2001) Distribution in ocular structures and optic pathways of immunocompetent and glial cells in an experimental allergic encephalomyelitis (EAE) relapsing model. J Neurosci Res 63(6):525–35

38. Wakefield D, Jennings A, McCluskey PJ (2000) Intravenous pulse methylprednisolone in the treatment of uveitis associated with multiple sclerosis. Clin Exp Ophthalmol 28(2):103–6

39. Yong VW, Chabot S, Stuve O, Williams G (1998) Interferon beta in the treatment of multiple sclerosis: mechanisms of action. Neurology 51(3):682–9

40. Younge BR (1976) Fluorescein angiography and retinal venous sheathing in multiple sclerosis. Can J Ophthalmol 11(1):31–6

41. Zierhut M, Foster CS (1992) Multiple sclerosis, sarcoidosis and other diseases in patients with pars planitis. Dev Ophthalmol 23:41–7

25.5 Sarcoidosis

S. Sivaprasad, N. Okhravi, S. Lightman

Core Messages

- Sarcoidosis is a multisystem granulomatous disease
- Ocular manifestations may precede or coexist with systemic disease
- Most common ocular features are dry eyes and uveitis
- Posterior segment disease may involve retinal vessels, retina and/or choroid
- Retinal vasculitis is a cardinal feature of posterior segment disease
- Retinal neovascularization may be due to inflammation and/or ischemia

25.5.1 Epidemiology

Sarcoidosis is a multisystem granulomatous disease with protean clinical manifestations and occurs worldwide. It affects all age groups, with two-thirds of patients being less than 40 years old at the time of diagnosis [43]. The clinical manifestations and severity of the disease vary widely and are strongly associated with racial and ethnic factors: acute and more severe disease is typical of black patients, whereas asymptomatic and chronic disease is more frequent in whites [35, 62]. Among the white patients, it is more prevalent among Scandinavians [55]. The disease also tends to affect certain organs in certain populations. For example, ocular and cardiac sarcoidosis appears to be more common in Japanese while erythema nodosum occurs more often in people of north European descent [32, 49].

25.5.2 Etiology

The etiology of sarcoidosis is unknown. Current theories suggest an abnormal immune response directed against one or more possible antigens in a susceptible individual [70]. Proposed antigens fall into three categories that include infectious, environmental, and autoantigens [44]. The most common infectious agents implicated are *Mycobacterium tuberculosis, Propionobacterium acne* and herpes virus 8 [16]. Specific HLA associations have also been connected with disease susceptibility and outcome [54]. Genetic polymorphisms of the ACE [24] and vitamin D receptor genes [45] may also be of potential relevance in the expression and outcome of the disease.

In addition, a promoter polymorphism of the TNF-α gene (TNF-308) has been associated with Löfgren's syndrome [63].

25.5.3 Molecular Mechanisms

Unknown antigen(s) or abnormal defense mechanisms triggered by various insults may be responsible for T cell activation in a genetically predisposed host. The compartmentalization of the T cells occurs as a result of cellular distribution from peripheral blood or in situ proliferation [15]. The CD4+ cells and macrophages are increased in areas of granuloma formation. Therefore, in areas of disease activity, the CD4:CD8 ratio is high while the CD4+ cells in the blood decrease, resulting in a reduced or inverted CD4:CD8 ratio [29]. B-cell activation also occurs, increasing the production of immunoglobulin. Increased serum vascular endothelial growth factor (VEGF) has also been observed in sarcoidosis and may be a useful prognostic indicator [56].

25.5.4 Pathology

The characteristic histopathologic findings are multiple non-caseating epithelioid granulomas. The center of the granulomas consists of histiocytes, epithelioid cells and multinucleated giant cells that may contain intracytoplasmic inclusions such as asteroid bodies and Schaumann bodies. The epithelioid cells are derived from monocytes and secrete several cytokines. They are surrounded by lymphocytes, plasma cells and fibroblasts. These granulomas secrete glucuronidase, collagenase, calcitriol and

25 III

angiotensin-converting enzyme (ACE). The granulomas may resolve spontaneously or with treatment.

25.5.5 Special Pathological Findings

25.5.5.1 Systemic Features

The disease predominantly affects the lungs and thoracic lymph nodes, skin, and eyes. It can have either a self-limited or chronic course. The majority of patients have constitutional symptoms such as fever, malaise, fatigue, and weight loss. Most of the patients will have some respiratory problems. Frequently the disease is asymptomatic and discovered by chest radiography (hilar adenopathy) performed for unrelated causes. Table 25.5.1 shows the common presentations specific to each organ involvement in sarcoidosis.

Specific manifestations of sarcoidosis include acute Löfgren's syndrome, which is a combination of erythema nodosum, arthritis, and hilar lymphadenopathy (sometimes associated with anterior uveitis) and Heerfordt's syndrome (uveoparotid fever), consisting of fever, parotid swelling, anterior uveitis, and sometimes facial palsy.

25.5.5.2 Ocular Sarcoidosis

Ocular involvement is the third commonest manifestation of sarcoidosis, preceded by pulmonary disease and hilar adenopathy [39]. It may precede or coexist with systemic disease. Ocular disease may be the presenting feature in 9% of cases. Twenty-five to 40% of patients with systemic sarcoidosis will at some point in the disease course show ocular involvement [46]. Sarcoidosis can present with a variety of ocular involvement. There are two peaks in age incidence for ocular sarcoid: 20–30 years and 50–60 years [52]. The most common manifestations are dry eyes and uveitis.

25.5.5.2.1 Anterior Segment Involvement

Eyelids can be involved as part of cutaneous or ocular sarcoidosis and it usually presents as painless skin or subcutaneous nodules [27]. Lacrimal gland involvement varies from 7% to 69% and is more common in black patients. It may manifest as bilateral or unilateral painless lacrimal gland enlargement with or without parotid gland involvement (Fig. 25.5.1). Lacrimal gland enlargement is usually asymptomatic, though it may lead to keratoconjunctivitis sicca or diplopia [14]. Orbital inflammatory syndrome can also occur in less than 1% of patients [11]. Orbital pseudotumor like reaction may manifest as the initial presentation of sarcoidosis with painful ophthalmoplegia and occasionally decreased vision [22]. The inflammation may also involve the extraocular muscles directly, resembling Graves' ophthalmopathy, or it may present as an orbital granuloma (Fig. 25.5.2).

Conjunctival involvement has been reported in up to 40% of patients with ocular sarcoidosis [46]. Mul-

Table 25.5.1. Organ involvement in sarcoidosis

Organ	Common manifestations	% of patients with specific organ involvement [2, 48, 57–59]
Lungs	Dyspnea, dry cough, chest pain	>90%
Mediastinal lymph nodes	Radiologic finding	95–98%
Liver	Abnormal liver function tests, hepatomegaly	50–80%
Spleen	Anemia, leukopenia, thrombocytopenia, pressure symptoms due to splenomegaly	40–80%
Musculoskeletal	Joint aches, tenosynovitis and myopathy	25–39%
Eyes	Dry eyes and uveitis	25–40%
Peripheral lymph nodes	Posterior triangle lymph nodes in the neck, axillary, epitrochlear and inguinal nodes	30%
Skin	Erythema nodosum, lupus pernio	25%
Neurosarcoidosis	Facial nerve palsy, other cranial nerve involvement, hypothalamic and pituitary lesions	10%
Heart	Benign arrhythmias to heart blocks	5%
Endocrine	Hypercalcemia, diabetes insipidus	2–10%
GIT	May mimic Crohn's disease	<1%

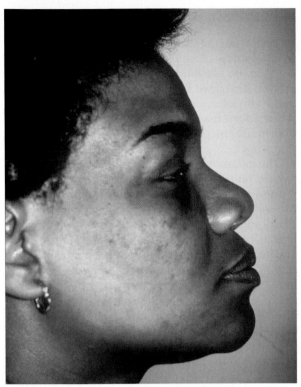

Fig. 25.5.1. Lacrimal gland enlargement in sarcoid

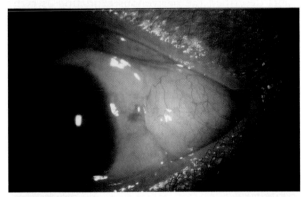

Fig. 25.5.2. Orbital granuloma

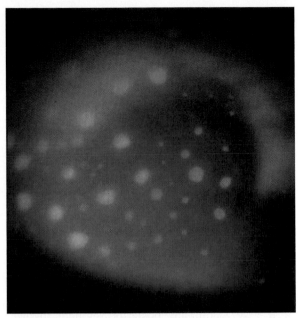

Fig. 25.5.3. Mutton fat keratic precipitates

tiple, translucent, pale conjunctival granulomas have frequently been described as the first clinical sign of sarcoidosis and they can be easily biopsied. Conjunctival biopsy is technically simple, cost-effective and gives a good positive yield when the granulomas are seen [18]. These granulomas usually respond well to topical steroid treatment. Chronic inflammation may lead to dry eyes, symblepharon, cicatricial conjunctivitis and even entropion [25]. Corneal involvement is extremely rare. Corneal band degeneration may develop as a consequence of long-standing anterior uveitis.

Scleritis is uncommon in sarcoidosis though a recent report suggests that sarcoidosis can present with scleritis [21].

Uveitis occurs in 30–70% of patients, with more than 80% presenting within 1 year of onset of systemic disease [30]. Sarcoidosis is the most frequently reported systemic disease associated with non-infectious uveitis in the Western world with a reported incidence of about 7%. The most common type of uveitis in sarcoidosis is anterior uveitis in black patients compared to posterior uveitis in elderly white females. If left untreated, the uveitis may cause severe visual loss due to cystoid macular edema, severe vitritis, cataract, secondary glaucoma and vitreous hemorrhage secondary to retinal neovascularization.

Sarcoid related uveitis may occur in five ways:

- Acute anterior uveitis
- Chronic anterior uveitis
- Intermediate uveitis
- Posterior uveitis
- Panuveitis

Acute anterior uveitis presents especially in Löfgren's syndrome, associated with erythema nodosum and bilateral hilar lymphadenopathy. This responds well to topical steroids but occasionally periocular steroids are also required. The prognosis for vision is good. Chronic uveitis is usually bilateral, with mutton fat keratic precipitates (Fig. 25.5.3) and granulomatous nodules in the iris or anterior chamber angle (Fig. 25.5.4). The iris nodules may be Koeppe' nodules and/or Busacca's nodules. Half of these patients have chronic and recurrent episodes with associated complications of iris bombé, cataract, glaucoma and

25 III

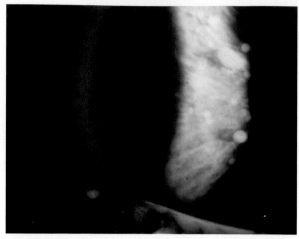

Fig. 25.5.4. Iris nodules

cystoid macular edema. Vitritis may result from a spill over from anterior uveitis, intermediate uveitis or it may accompany posterior segment involvement. Intermediate uveitis may present as a fine cellular reaction or run a more chronic course with snowball opacities that accumulate in the inferior vitreous and/or snowbank formation.

Posterior subcapsular lenticular opacities are a well-recognized complication of uveitis and both systemic and topical steroid therapy. Akova and Foster studied cataract extraction in patients with ocular sarcoidosis and found that posterior chamber lens implantation in patients with sarcoidosis-associated uveitis is well tolerated when absolute control of the inflammation is achieved. Preexisting retinal pathology and glaucoma as a result of uncontrolled inflammation resulting in permanent ocular structural damage were found to be the most important factors for determining the postoperative final visual acuity [1].

A major prognostic factor for visual morbidity in ocular sarcoidosis is the presence of secondary glaucoma. The main causes of secondary glaucoma include secondary open angle glaucoma due to inflammation, pupillary block, peripheral anterior synechiae, trabeculitis and neovascular glaucoma. Rubeosis may result from chronic inflammation but more commonly occurs due to ischemia from extensive retinal vascular closure. Anterior segment inflammation usually responds to topical steroids, but in some patients the intraocular pressure (IOP) rises on treatment with steroids (steroid response) and may require treatment to lower the IOP. Abnormalities noted in the endothelium of Schlemm canal suggest that Schlemm canalitis resembling microangiopathy seen in skin and lung biopsies of sarcoidosis may be another cause of secondary glaucoma [28].

25.5.5.2.2 Posterior Segment Involvement

Posterior segment involvement may be the only manifestation of ocular sarcoidosis [51]. Posterior segment manifestations also show demographic variations. American black patients have a higher incidence of posterior segment involvement compared to European black patients suggesting that genetic and environmental differences between populations may vary.

The inflammatory process may involve the retinal vessels, retina and/or choroid. Retinal vasculitis constitutes one of the cardinal features of posterior segment inflammation. It is typically characterized by perivenous sheathing in the equatorial retina, which is usually segmental (Fig. 25.5.5). The cuffing of the affected veins is due to the collection of infiltrating cells at sites of disruption of the inner blood-retinal barrier. The condition may remain asymptomatic at this stage and is often an incidental finding. It is nonspecific and non-diagnostic as it may occur in many types of uveitis.

However, vasculitis may also result in leakage or occlusion of the lumen of the veins resulting in more severe signs and symptoms. Leakage leads to retinal swelling, exudation and edema. Franceschetti and Babel described the perivenous exudation as tache de bougie (candle wax drippings) [23]. Though described as typical of sarcoid vasculitis, it is not a common finding (Fig. 25.5.6).

Retinal vasculitis can also lead to vascular occlusion resulting in intraretinal hemorrhage, cotton-wool spots and capillary non-perfusion.

Neovascularization occurs in approximately 11% of patients with posterior segment disease and is thought to be induced by a variety of angiogenic factors. Neovascularization may develop as a result of

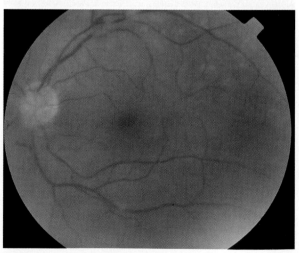

Fig. 25.5.5. Retinal vascular sheathing

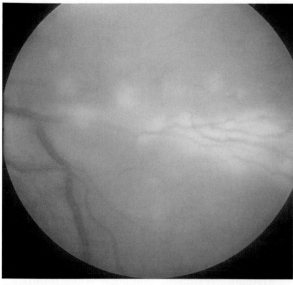

Fig. 25.5.6. Candle-wax drippings

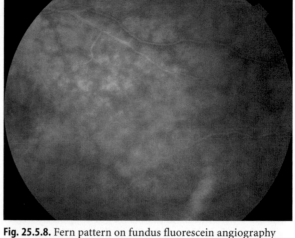

Fig. 25.5.8. Fern pattern on fundus fluorescein angiography

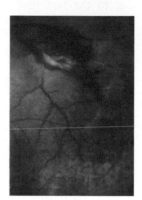

Fig. 25.5.7. Retinal neovascularization in ischemic retinal vasculitis

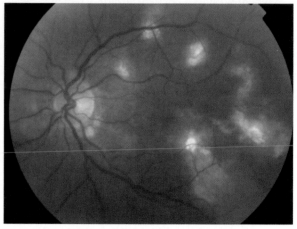

Fig. 25.5.9. Multifocal choroiditis

two mechanisms: (a) severe ocular inflammation and (b) retinal vascular occlusion and ischemia [20]. Neovascularization may involve the optic disk or peripheral retina (Fig. 25.5.7). A higher rate of vitreous hemorrhage has been noted in younger patients with periphlebitis than their older counterparts. The neovascularization usually responds to anti-inflammatory agents when the drive is inflammatory but laser photocoagulation may be required if there is any retinal ischemia present.

Fluorescein angiography is an important investigation for the assessment of the integrity of the vessels and the degree of ischemia. Vessel staining and leakage may indicate active vasculitis or retinal ischemia. Diffuse fern-like vascular leakage may be seen on angiography (Fig. 25.5.8). The areas of capillary closure are best delineated by fluorescein angiography. Likewise, macular edema may often be more evident angiographically than clinically.

Cystoid macular edema (CME) is a major cause of visual loss in ocular sarcoidosis. Females and older patients have a higher rate of CME and worse visual acuity [65]. CME usually responds to treatment with steroids but may require the use of additional immunosuppressive agents.

Multifocal chorioretinitis in sarcoidosis may be central, peripheral or both. It is often associated with CME (Fig. 25.5.9) [38]. Subfoveal choroidal neovascularization has also been reported in sarcoidosis secondary to multifocal choroiditis [31]. Peripapillary choroidal neovascularization that responds to systemic steroids has also been observed [10].

Retinal macroaneurysms are associated with multifocal choroiditis in women and are associated with severe cardiovascular disease. Elderly patients with multiple retinal macroaneurysms, uveitis, disk staining, and peripheral chorioretinitis should be thoroughly investigated for sarcoidosis [53].

Acute multifocal posterior placoid epitheliopathy (AMPPE) and bird shot chorioretinopathy like picture have also been associated with sarcoidosis [17, 37]. Extensive posterior pole choroiditis may resemble serpiginous choroiditis. Plaque-like yellowish choroidal infiltrates associated with systemic sarcoidosis may occur in eyes remarkably free of other signs of inflammation. The infiltrates tend to radiate from the region of the optic nerve in a confluent ameboid-like pattern. They generally respond to corticosteroids and may be the first recognized manifestation of systemic sarcoidosis [13].

Choroidal circulatory disturbances have been observed in sarcoidosis in the absence of visual symptoms or retinal lesions [41]. Indocyanine angiography in patients with ocular sarcoidosis may show features of choroidal ischemia [73].

Choroidal granulomas may exist without signs of anterior chamber or vitreous inflammation or intracranial granulomas [8]. They are observed in 5% of patients with ocular sarcoid. They are typically bilateral, pale yellow, elevated lesions that resolve spontaneously or with oral corticosteroids. They may recur but good vision is usually maintained. Optic disk granuloma may be the presenting sign of ocular sarcoidosis (Fig. 25.5.10) [9]. It may mimic papilledema or papillitis but is usually unilateral and responds well to steroids. Vision may or may not be affected. They respond to systemic steroids. Granuloma of the retina has rarely been reported in sarcoidosis [3].

Retinal pigment epithelial detachment has also been noted in sarcoidosis [6]. Serous retinal detachment at the macula may occur in association with pulmonary angiitis secondary to sarcoidosis [71]. The underlying mechanisms of subretinal exudation are thought to include choroidal vascular perfusion and permeability changes, which result in increased choroidal interstitial fluid with further extension into the subretinal space.

Other less frequent manifestations include posterior scleritis with annular ciliochoroidal detachment causing angle closure glaucoma [19].

Neurological manifestations of sarcoidosis are varied and are identified in 12% of patients with systemic sarcoidosis [67]. Posterior segment involvement is associated with neurological disease in 27% of cases. Neurological manifestations may include optic nerve disease, cranial nerve palsies, encephalopathy, and disorders of the hypothalamus and pituitary gland. Chiasmal syndromes and motility disorders may also occur. Facial nerve palsy is the most common manifestation of CNS involvement and is usually self limiting. Optic nerve involvement may be caused by direct sarcoid tissue infiltration or compression by cerebral mass or chronic meningitis or secondary to posterior segment involvement. Usually, the involvement of the optic nerve is an indication for systemic treatment. Diagnosis is often difficult due to the fact that the clinical presentation can mimic other disorders, such as multiple sclerosis. Sarcoidosis is also an important differential diagnosis in uveomeningeal syndromes [7].

25.5.5.3 Ocular Sarcoidosis in Children

Sarcoidosis is relatively rare in children and is an uncommon cause of childhood uveitis. The disease may run in families suggesting possible genetic and environmental etiology. There are two age groups of pediatric sarcoidosis. Children under 4 years of age are characterized by the triad of rash, uveitis, and tenosynovitis. The more frequent is the occurrence of sarcoidosis in older children (8–15 years of age),

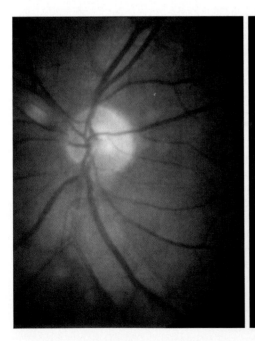 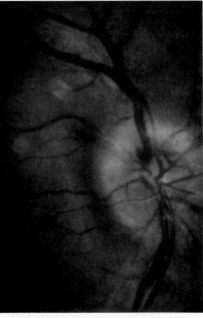

Fig. 25.5.10. Optic nerve granuloma

which is characterized by almost universal lung involvement and manifestations in eyes, skin, liver, and spleen. The diagnosis of sarcoidosis in children with uveitis should be suspected when the uveitis presents with red eyes and there are keratic precipitates of the mutton-fat type, when there is anterior and posterior segment involvement or when the tenosynovitis affects multiple joints. The prognosis for children is more favorable than for adults [33].

25.5.6 Diagnosis of Sarcoidosis

A fully directed clinical history and examination is mandatory. The diagnosis is established when clinical and radiographic findings are supported by histologic evidence of non-caseating epithelioid cell granulomas found on tissue biopsy. However, the initial investigations may be negative and it may not be possible to demonstrate a non-caseating granuloma. The common investigations include serum angiotensin converting enzyme (S-ACE) and chest radiography.

25.5.6.1 Serum Angiotensin Converting Enzyme

S-ACE is thought to correspond with macrophage activity. It catalyzes the conversion of angiotensin I to angiotensin II. S-ACE is used as both a diagnostic and prognostic indicator of sarcoidosis. It is elevated in 50–80% of patients with active sarcoidosis but can also be elevated in other disorders. In approximately 20–50% of patients, sarcoidosis can be present without elevated ACE levels [4]. This may be due to the fact that the disease is in its early phase, chronic stage or quiescent (burnt out) stage. The absolute value has no prognostic significance. The combination of raised S-ACE levels with abnormal gallium scanning is a specific and sensitive tool for diagnosing patients suspected of having ocular sarcoidosis who had normal chest radiographs [50].

25.5.6.2 Chest Radiographs

A staging system is used to classify chest X-rays taken to detect sarcoidosis (Table 25.5.2). These stages do not correlate with disease severity and patients up to stage 3 may be symptomless.

25.5.7 Clinical Course and Prognosis

25.5.7.1 Ocular Disease

Pure ocular sarcoidosis has a relatively benign course, with 72% maintaining good vision. In a

Table 25.5.2. Staging of X-ray chest in sarcoidosis

Stages	Findings
Stage 0	Normal chest X-ray
Stage 1	Enlarged lymph nodes with clear lungs
Stage 2	Enlarged lymph nodes plus infiltrates in the lungs
Stage 3	Lung infiltrates are present but the lymph nodes are no longer seen
Stage 4	Scarring in the lung

Table 25.5.3. Investigations that may be useful in diagnosing and assessing activity in sarcoidosis [2, 42, 60]

Chest radiography	Very useful in initial evaluation
Pulmonary function tests	Assess lung function at baseline and follow-up
Hematological evaluation	Non-specific test; anemia, lymphocytopenia, thrombocytopenia, hypergammaglobulinemia, slight elevation of ESR
Serum biochemistry	S-ACE; hypercalcemia; liver function tests and renal function tests to assess hepatic and renal involvement
Urinary calcium	Hypercalciuria due to increased sensitivity to vitamin D and may also indicate bone involvement
Cutaneous anergy	Negative tuberculin skin test
Bronchoalveolar lavage	High CD4/CD8 ratio indicates pulmonary activity
Kveim-Siltzbach tests	Not used nowadays; positive in 50–60% of patients
Gallium scan [74]	Lambda sign – symmetric uptake in mediastinal and hilar lymph nodes. Panda sign indicates increased uptake in lacrimal and parotid glands
Biopsy	Biopsy of conjunctival granuloma when present gives high positive yield; chorioretinal biopsy is only indicated when either infection or malignancy cannot be excluded and is relevant in differential diagnosis
MRI brain and orbit	Gadolinium enhanced T1-weighted MRI useful in neurosarcoidosis
CSF	Normal or elevated ACE in neurosarcoidosis

series of 75 patients with ocular sarcoidosis, panuveitis and peripheral multifocal choroiditis were reported to be the main causes of visual loss and accounted for 47% of patients [40]. Visual morbidity may be due to secondary glaucoma, CME and optic nerve damage. Poor visual prognosis is associated with advancing age, black race, female sex, chronicity of disease, posterior segment involvement and complications of uveitis [66].

25.5.7.2 Systemic Disease

The overall prognosis for systemic sarcoidosis is also relatively benign. Only about 10 % of patients develop serious disability from ocular, respiratory, or other organ damage, and mortality is less than 3 % [34]. Pulmonary fibrosis leading to cardiorespiratory failure is the most common cause of death, followed by pulmonary hemorrhage from a complicating aspergilloma.

The prognosis is better for patients who have radiologic evidence of hilar adenopathy without pulmonary disease. The most reliable indicator of a favorable outcome of sarcoidosis is onset with erythema nodosum. Factors associated with poor outcome include disease persisting for longer than 6 months, involvement of more than three organs, and stage III of pulmonary disease (see Table 25.5.2) [69].

25.5.8 Treatment of Ocular Sarcoidosis

Active ocular disease may be present without symptomatic systemic disease. Corticosteroids are the mainstay of therapy for ocular sarcoidosis. The need for topical, peribulbar or systemic steroids is dictated by the localization and severity of the disease. The indication for systemic treatment in ocular disease includes optic neuritis and severe vision threatening posterior uveitis. The initial dosage depends on the disease severity, but often high dosage is required to induce remission or quiescent stage of intraocular inflammation; thereafter slow tapering and prolonged low dose treatment is usually administered to maintain the disease quiescence. Blood pressure, blood sugar and weight should be monitored in all patients on steroid treatment. The main reason for failure of systemic steroids is often related to an inadequate initial dose. Periocular steroid injections and intravitreal triamcinolone have been found to be useful in refractory uveitis [64, 74]. The long acting analogue of somatostatin (octreotide) has been given in sarcoid related neurochorioretinitis that failed to respond to previous corticosteroid treatment [36].

For refractory disease, immunosuppressants are initiated. While these agents are clearly of value in selected patients, there are no randomized controlled trials on their effectivity in ocular sarcoidosis. Methotrexate, mycophenolate and infliximab have been found to be effective in ocular involvement [5, 61].

Anterior uveitis is treated with topical steroids and cycloplegics are needed to prevent synechiae when the inflammation is chronic.

Most rises in IOP can be controlled medically. Secondary glaucoma not responding to topical anti-glaucoma therapy may require trabeculectomy ±5-fluorouracil or mitomycin C.

The surgical interventions for the complications of ocular sarcoidosis are performed when the eyes are quiet. The inflammation must be controlled preoperatively and surgery performed under steroid cover in patients with posterior segment disease or history of CME. Intraocular lens (IOL) implantations are well tolerated and good visual results are achieved. The major causes of the decreased visual acuity following cataract surgery are CME, preexisting glaucomatous damage and posterior segment involvement [47]. Rarely, IOL placement after cataract extraction results in inflammatory membrane formation despite anti-inflammatory coverage.

In the absence of capillary closure, retinal neovascularization may regress spontaneously or with systemic anti-inflammatory treatment. Cases with persistent neovascularization benefit from laser treatment when ischemia is present [26]. Laser photocoagulation should be performed when the inflammation is quiescent to prevent the development or exacerbation of cystoid macular edema.

Vitrectomy may be required for cases with persistent vitreous hemorrhage, residual vitreous opacities or retinal detachment.

25.5.9 Treatment of Systemic Sarcoidosis

Corticosteroids are the mainstay of treatment to suppress severe symptoms (e.g., dyspnea, arthralgia, fever) or hepatic insufficiency, cardiac arrhythmias, CNS involvement, hypercalcemia and disfiguring skin infections. Patients with acute disease may need treatment for only a few weeks, but most of those requiring therapy have chronic sarcoidosis and need treatment for years. Maintenance doses of oral prednisolone as low as 5 mg/day may be needed indefinitely to control symptoms and radiological lesions.

About 10 % of patients requiring therapy are unresponsive to tolerable doses of corticosteroid and may need immunosuppressive drugs such as cyclosporin. Although these drugs are often more effective in refractory cases, relapse is frequent after their cessation.

The treatment of chronic neurological symptoms with methotrexate or cyclophosphamide was associated with a better therapeutic response than the treatment with corticosteroids alone. The long-term prognosis of neuro-ophthalmic sarcoid has not been studied in large patient populations, but the data that is available suggests that remission may occur in up to 47 % [12].

25.5.10 Conclusion

Ocular sarcoidosis may present with a wide variety of ocular signs in all parts of the eye and may be associated with either acute or chronic and progressive intraocular inflammation leading to visual deterioration. The diagnosis may be difficult owing to the absence of diagnostic criteria and the variety of presentations. Treatment is aimed at controlling the inflammatory process and it may require the use of topical, periocular or oral steroids plus immunosuppressive agents. In most patients the long term outlook for vision is good.

Case Reports

Case 1

A 70-year-old Caucasian lady presented with blurred vision and floaters associated with recent weight loss and lethargy of 3 months duration. Examination revealed vitreous cells and peripheral multifocal choroiditis. Vision was 6/12 mainly due to associated CME. A chest radiograph revealed bilateral hilar lymphadenopathy and investigations showed a raised S-ACE. Lung function tests showed reduced transfer factor. The patient received oral corticosteroids for both eye and systemic disease that resulted in partial resolution of her symptoms and signs with improvement of vision to 6/9. She is also under the joint care of the chest physicians and is on a maintenance dose of oral steroids.

Case 2

A 42-year-old black gentleman with cutaneous sarcoidosis on hydroxychloroquine developed floaters in both eyes for the previous 6 months. Examination revealed a visual acuity of 6/6 bilaterally. Examination showed perivenous sheathing, and small areas of equatorial retinal neovascularization and small choroidal lesions resembling Dalen Fuchs nodules. Fluorescein angiography showed no peripheral vascular closure. Prompt oral steroids led to involution of these vessels and laser photocoagulation was not indicated.

References

1. Akova YA, Foster CS (1994) Cataract surgery in patients with sarcoidosis-associated uveitis. Ophthalmology 101(3):473–9
2. American Thoracic Society (1999) Statement on sarcoidosis. Am J Respir Cri Care Med 160(2):736–755
3. Augustin AJ, Boker T, Seewald S, et al. (1994) Solitary retinal granuloma as a presenting sign of sarcoidosis. Ger J Ophthalmol 3:71–72
4. Baarsma GS, La Hey E, Glasius E (1987) The predictive value of serum angiotensin converting enzyme and lysozyme levels in the diagnosis of ocular sarcoidosis. Am J Ophthalmol 104(3):211–7
5. Baughman RP, Bradley DA, Lower EE (2005) Infliximab in chronic ocular inflammation. Int J Clin Pharmacol Ther 43(1):7–11
6. Bourcier T, Lumbroso L, Cassoux N (1998) Retinal pigment epithelial detachment: an unusual presentation in ocular sarcoidosis. Br J Ophthalmol 82(5):585
7. Brazis PW, Stewart M, Lee AG (2004) The uveo-meningeal syndromes. Neurologist 10(4):171–84
8. Campo RV, Aaberg TM (1984) Choroidal granuloma in sarcoidosis. Am J Ophthalmol 97(4):419–27
9. Castagna I, Salmeri G, Fama F, et al. (1994) Optic nerve granuloma as first sign of systemic sarcoidosis. Ophthalmologica 208(4):230–2
10. Cheung CM, Durrani OM, Stavrou P (2002) Peripapillary choroidal neovascularisation in sarcoidosis. Ocul Immunol Inflamm 10(1):69–73
11. Collison JMT, Miller NR, Green WR (1986) Involvement of orbital tissues by sarcoid. Am J Ophthalmol 102:302–307
12. Constantino T, Digre K, Zimmerman P (2000) Neuro-ophthalmic complications of sarcoidosis. Semin Neurol 20(1): 123–37
13. Cook BE Jr, Robertson DM (2000) Confluent choroidal infiltrates with sarcoidosis. Retina 29(1):1–7
14. Cook JR, Brubaker RF, Savell J (1972) Lacrimal sarcoidosis treated with corticosteroids. Arch Ophthalmol 88(5):513–7
15. Crystal RG, Bitterman PB, Rennard SI, et al. (1984) Interstitial lung diseases of unknown cause. Disorders characterized by chronic inflammation of the lower respiratory tract. N Engl J Med 310(4):235–44
16. Di Alberti L, Piattelli A, Favia G, et al. (1997) Human herpesvirus 8 variants in sarcoid tissues. Lancet 350:1655–1661
17. Dick DJ, Newman PK, Richardson J, et al. (1988) Acute posterior multifocal placoid pigment epitheliopathy and sarcoidosis. Br J Ophthalmol. 72(1):74–7
18. Dios E, Saornil MA, Herreras JM (2001) Conjunctival biopsy in the diagnosis of ocular sarcoidosis. Ocul Immunol Inflamm 9(1):59–64
19. Dodds EM, Lowder CY, Barnhorst DA (1995) Posterior scleritis with annular ciliochoroidal detachment. Am J Ophthalmol 120(5):677–9
20. Duker JS, Brown GC, McNamara JA (1988) Proliferative sarcoid retinopathy. Ophthalmology 95(12):1680–6
21. Dursun D, Akova YA, Bilezikci B (2004) Scleritis associated with sarcoidosis. Ocul Immunol Inflamm 12(2):143–8
22. Faller M, Purohit A, Kennel N, et al. (1995) Systemic sarcoidosis initially presenting as an orbital tumour. Eur Respir J 8(3):474–6
23. Franchescetti A, Babel J (1949) La chorio-retinie en 'taches de bougie', manifestation de la maladie de Besnier-Boeck. Ophthalmologica 118:701–710
24. Furuya K, Yamaguchi E, Itoh A, et al. (1996) Deletion polymorphism in the angiotensin I converting enzyme (ACE) gene as a genetic risk factor for sarcoidosis. Thorax 51(8): 777–80
25. Geggel HS, Mensher JH (1989) Cicatricial conjunctivitis in sarcoidosis: recognition and treatment. Ann Ophthalmol 21(3):92–4
26. Graham EM, Stanford MR, Shilling JS, et al. (1987) Neovascularisation associated with posterior uveitis. Br J Ophthalmol 71(11):826–33

III 25

25 III

27. Hall JG, Cohen KL (1995) Sarcoidosis of the eyelid skin. Am J Ophthalmol 119:100–101
28. Hamanaka T, Takei A, Takemura T (2002) Pathological study of cases with secondary open-angle glaucoma due to sarcoidosis. Am J Ophthalmol 134(1):17–26
29. Hunninghake GW, Crystal RG (1981) Pulmonary sarcoidosis: a disorder mediated by excess helper T-lymphocyte activity at sites of disease activity. N Engl J Med 305(8):429–34
30. Hunter DG, Foster CS (1994) Ocular manifestations of sarcoidosis. In: Albert DM, Jakobeiec FA (eds) Principles and practice of ophthalmology. Saunders, Philadelphia, pp 443–450
31. Inagaki M, Harada T, Kiribuchi T, et al. (1996) Subfoveal choroidal neovascularization in uveitis. Ophthalmologica 210(4):229–33
32. Iwai K, Sekiguti M, Hosoda Y, et al. (1994) Racial difference in cardiac sarcoidosis incidence observed at autopsy. Sarcoidosis 11:26–31
33. James DG, Kendig EL, Jr (1988) Childhood sarcoidosis. Sarcoidosis 5:57–59
34. James DG, Neville E, Siltzbach LE (1976) A worldwide review of sarcoidosis. Ann N Y Acad Sci 278:321–34
35. James GD, Hosada Y (1994) Epidemiology. In: Sarcoidosis and other granulomatous disorders. Marcel Dekker, New York, pp 729–743
36. Jurowski P, Gos R, Kunert-Radek J (2002) Long acting analogue of somatostatin (Octreotide) for treatment of patients with neurochorioretinitis due to ocular sarcoidosis who failed corticosteroids therapy. Klin Oczna 104(3–4):266–9
37. Kuboshiro T, Yoshioka H (1988) Birdshot retinochoroidopathy – a possible relationship to ocular sarcoidosis. Kurume Med J 35(4):193–9
38. Lardenoye CW, Van der Lelij A, de Loos WS, et al. (1997) Peripheral multifocal chorioretinitis: a distinct clinical entity? Ophthalmology 104(11):1820–6
39. Liggett PE (1986) Ocular sarcoidosis. Clin Dermatol 4(4):129–35
40. Lobo A, Barton K, Minassian D, et al. (2003) Visual loss in sarcoid-related uveitis. Clin Exp Ophthalmol 31(4):310–6
41. Machida S, Tanaka M, Murai K (2004) Choroidal circulatory disturbance in ocular sarcoidosis without the appearance of retinal lesions or loss of visual function. Jpn J Ophthalmol 48(4):392–6
42. Martin DF, Chan CC, de Smet MD, et al. (1993) The role of chorioretinal biopsy in the management of posterior uveitis. Ophthalmology 100(5):705–14
43. Maycock RL, Bertrand P, Morrison CE, et al. (1963) Manifestations of sarcoidosis. Analysis of 145 patients, with a review of nine series selected from the literature. Am J Med 35:67–89
44. McGrath DS, Goh N, Foley PJ, et al. (2001) Sarcoidosis genes and microbes: soil or seed. Sarcoidosis Vasc Diffuse Lung Dis 18(2):149–64
45. Niimi T, Tomita H, Sato S, et al. (1999) Vitamin D receptor gene polymorphism in patients with sarcoidosis. Am J Respir Crit Care Med 160:1107–1109
46. Obenauf CD, Shaw HE, Sydnor CF, et al. (1978) Sarcoidosis and its ophthalmic manifestations. Am J Ophthalmol 86:648–655
47. Okhravi N, Lightman SL, Towler HM (1999) Assessment of visual outcome after cataract surgery in patients with uveitis. Ophthalmology 106(4):710–22
48. Perez RL, Rivera-Marrero CA, Roman J (2003) Pulmonary granulomatous inflammation: From sarcoidosis to tuberculosis. Semin Respir Infect 18(1):23–32
49. Pietinalho A, Ohmichi M, Löfroos AB, et al. (2000) The prognosis of pulmonary sarcoidosis in Finland and Hokkaido, Japan. A comparative five year study of biopsy-proven cases. Sarcoidosis Vasc Diffuse Lung Dis 17:158–166
50. Power WJ, Neves RA, Rodriguez A, et al. (1995) The value of combined serum angiotensin-converting enzyme and gallium scan in diagnosing ocular sarcoidosis. Ophthalmology 102(12):2007–11
51. Rothova A (2000) Ocular involvement in sarcoidosis. Br J Ophthalmol 84:110–116
52. Rothova A, Alberts C, Glasius E, et al. (1989) Risk factors for ocular sarcoidosis. Doc Ophthalmol 72:287–296
53. Rothova A, Lardenoye C (1998) Arterial macroaneurysms in peripheral multifocal chorioretinitis associated with sarcoidosis. Ophthalmology 105(8):1393–7
54. Rybicki BA, Maliarik MJ, Major M, et al. (1997) Genetics of sarcoidosis. Clin Chest Med 18:707–717
55. Rybicki BA, Maliarik MJ, Poisson LM, et al. (1997) Racial differences in sarcoidosis incidence: a 5-year study in a health maintenance organization. Am J Epidemiol 145(3):234–41
56. Sekiya M, Ohwada A, Miura K, et al. (2003) Serum vascular endothelial growth factor as a possible prognostic indicator in sarcoidosis. Lung 181(5):259–65
57. Sharma OP (1972) Cutaneous sarcoidosis: clinical features and management. Chest 61(4):320–5
58. Sharma OP (1996) Vitamin D, calcium, and sarcoidosis. Chest 109:535–539
59. Sharma OP (1997) Cardiac and neurologic dysfunction in sarcoidosis. Clin Chest Med 18(4):813–25
60. Sherman JL, Stern BJ (1990) Sarcoidosis of the CNS: comparison of unenhanced and enhanced MR images. AJNR Am J Neuroradiol 11(5):915–23
61. Shetty AK, Zganjar BE, Ellis GS, Jr, et al. (1999) Low-dose methotrexate in the treatment of severe juvenile rheumatoid arthritis and sarcoid iritis. J Pediatr Ophthalmol Strabismus 36(3):125–8
62. Siltzbach LE, Geraint James D, Neville E, et al. (1974) Course and prognosis of sarcoidosis around the world. Am J Med 57:847–852
63. Somoskovi A, Zissel G, Seitzer U (1999) Polymorphisms at position –308 in the promoter region of the TNF-alpha and in the first intron of the TNF-beta genes and spontaneous and lipopolysaccharide-induced TNF-alpha release in sarcoidosis. Cytokine 11(11):882–7
64. Sonoda KH, Enaida H, Ueno A, et al. (2003) Pars plana vitrectomy assisted by triamcinolone acetonide for refractory uveitis: a case series study. Br J Ophthalmol 87(8):1010–4
65. Stanbury RM, Graham EM, Murray PI (1995) Sarcoidosis. Int Ophthalmol Clin 35(3):123–37
66. Stavrou P, Linton S, Young DW, et al. (1997) Clinical diagnosis of ocular sarcoidosis. Eye 11(3):365–70
67. Stern BJ, Krumholz A, Johns C, et al. (1985) Sarcoidosis and its neurological manifestations. Arch Neurol 42(9):909–17
68. Tahmoush AJ, Amir MS, Connor WW, Farry JK, Didato S, Ulhoa-Cintra A, Vasas JM, Schwartzman RJ, Israel HL, Patrick H (2002) CSF-ACE activity in probable CNS neurosarcoidosis. Sarcoidosis Vasc Diffuse Lung Dis 19(3):191–7
69. Takada K, Ina Y, Noda M, et al. (1993) The clinical course and prognosis of patients with severe, moderate or mild sarcoidosis. J Clin Epidemiol 46(4):359–66
70. Verleden GM, du Bois RM, Bouros D, et al. (2001) Genetic

predisposition and pathogenetic mechanisms of interstitial lung diseases of unknown origin. Eur Respir J 18:Suppl 32:17s–29s

71. Watts PO, Mantry S, Austin M (2000) Serous retinal detachment at the macula in sarcoidosis. Am J Ophthalmol 129(2):262–4

72. Winterbauer RH, Lammert J, Selland M, et al. (1993) Bronchoalveolar lavage cell populations in the diagnosis of sarcoidosis. Chest 104(2):352–61

73. Wolfensberger TJ, Herbort CP (1999) Indocyanine green angiographic features in ocular sarcoidosis. Ophthalmology 106 (2):285–9

74. Yoshikawa K, Kotake S, Ichiishi A, et al. (1995) Posterior sub-Tenon injections of repository corticosteroids in uveitis patients with cystoid macular edema. Jpn J Ophthalmol 39(1):71–6

25.6 Necrotizing Vasculitis

J.L. DAVIS

Core Messages

- Systemic necrotizing vasculitides are rare, affecting fewer than 1 in 10,000
- Wegener granulomatosis, polyarteritis nodosa (PAN), and Churg-Strauss syndrome (CSS) are histologically distinct, but overlap clinically
- Pathogenesis is linked to anti-neutrophil cytoplasmic antibodies plus environmental and genetic factors
- Ocular manifestations are more common in Wegener granulomatosis and may be localized; retinal vasculitis is rare in all three vasculitides
- Multidisciplinary management is the rule
- Treatment is usually initiated with corticosteroids and cyclophosphamide
- Mortality lessens with treatment but is higher than in the general population

25.6.1 Polyarteritis Nodosa and Microscopic Polyangiitis

25.6.1.1 Synonyms and Related Conditions

- Periarteritis nodosa
- Microscopic polyarteritis
- Small vessel vasculitis
- Kawasaki disease

Classic polyarteritis nodosa (PAN) can be further subdivided into that which is hepatitis B related (10%). Other viruses, such as HIV, CMV, or hepatitis C, have also been associated with PAN. Prevalence is about 4 cases per million.

25.6.1.2 Histopathology

- Focal necrotizing arteritis
- Mixed cellular infiltrate in the vessel wall
- Microaneurysm formation
- Intraorgan hemorrhage from ruptured microaneurysms

Classic PAN typically involves small and medium-sized muscular arteries with a focal necrotizing arterial inflammation and microaneurysm formation at arterial branch points; the relevance of the classic form of the disease to retinal vascular disease per se is therefore uncertain, although it is the diagnosis typically associated with non-granulomatous vasculitis affecting the eye. Microscopic polyangiitis (MPA), in contrast, denotes vasculitis in arterioles, venules, and capillaries, and is potentially more relevant to eye disease [22]. Older literature did not separate PAN and MPA and may be confusing.

25.6.1.3 Systemic Course of PAN

- Neuropathy (mononeuritis multiplex)
- Nephropathy
- Cutaneous ulcers
- Gastrointestinal thrombosis and infarction
- Musculoskeletal pain
- Coronary arteritis
- CNS involvement

PAN has a systemic onset with fever, malaise, and weight loss, followed by organ specific manifestations listed above in approximate order of frequency. Most patients are middle aged. PAN affects renal function by a vascular nephropathy. MPA uniquely causes glomerulonephritis and pulmonary disease by virtue of its involvement of smaller vessels [26]. Factors associated with a poor prognosis are increased serum creatinine, proteinuria, cardiomyopathy, CNS involvement, and gastrointestinal involvement. Presence of any of these factors is probably an indication for treatment with cyclophosphamide as well as corticosteroids [15]. With the use of cyclophosphamide, 5 year survival is about 80%.

Kawasaki disease (acute febrile mucocutaneous lymph node syndrome) in children is associated with coronary artery vasculitis and is indistinguishable from infantile PAN histologically; clinically, muco-

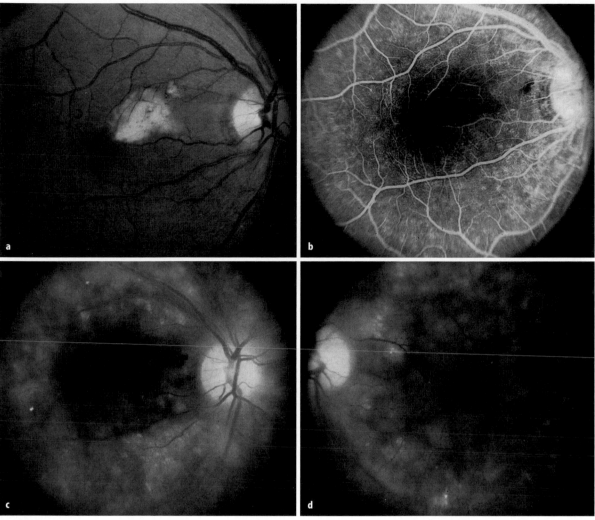

Fig. 25.6.1. Polyarteritis nodosa. **a** Red-free photograph. There is a large cotton-wool spot. **b** Capillary nonperfusion at the site of the cotton-wool spot. The capillary spacing in the perifoveal region is enlarged. **c, d** Late-phase angiogram of right and left eye. There is diffuse retinovascular leakage with areas of cystic retinal edema. (Courtesy of Robert B. Nussenblatt, Laboratory of Immunology, National Eye Institute, USA)

cutaneous involvement is necessary to diagnose Kawasaki disease.

25.6.1.3.1 Diagnosis

- General laboratory tests including complete blood count and C-reactive protein
- Chemistry and urinanalysis to detect glomerulonephritis (MPA)
- Hepatitis B surface antigen only in PAN (order cryoglobulins if HepB SAg+)
- P-ANCA/C-ANCA: anti-myeloperoxidase and anti-serine proteinase 3 (MPA)
- Chest X-ray to detect pulmonary infiltrate (MPA)
- Nerve or muscle biopsy of clinically affected tissue
- Abdominal angiography to detect microaneurysms in kidney, liver, and mesenteric artery

25.6.1.4 Ocular Manifestations of PAN

- Microangiopathy with cotton-wool patches
- Serous retinal detachment
- Retinal vasculitis
- Artery occlusion
- Ischemic optic neuropathy

There are no large series of ocular manifestations of PAN or MPA. Case reports indicate that retinal vasculitis, microangiopathy (cotton-wool spots), and retinal detachment can occur. Despite the rarity of involvement, screening eye examinations in patients with PAN or MPA have been recommended.

Ocular involvement has been estimated to occur in 10–20% of patients; some manifestations may be related solely to accompanying hypertension. Serous

25 **III**

retinal detachment occurred in association with scleritis in one case [24] and was attributed to choroidal vasculitis in two siblings [35]. Fibrinoid necrosis of choroidal vessels led to serous retinal detachment in one fatal infantile case [14]. Histologically confirmed involvement of the short ciliary arteries produced monocular blackouts of vision attributed to choroidal vascular insufficiency in one case; ocular ischemic syndrome was not present [31]. Central retinal artery occlusion, contralateral ischemic optic neuropathy, and multiple wedge-shaped choroidal occlusions occurred in a 70-year-old woman in the setting of symptomatic mononeuritis multiplex; sural nerve and muscle biopsy confirmed PAN [19]. More subtle manifestations of vascular involvement in retina, nerve and choroid were recorded in three case reports published in 2001 [33]. Rosen summarized angiographic findings in PAN in 1968 [32].

Despite the common association of macroaneurysms located at the retinal arteriolar branch points in idiopathic retinal arteriolar macroaneurysms and neuroretinitis (IRVAN) [6], there seems to be no association between IRVAN and PAN. There is a case report of bilateral retinal ischemia and unilateral ophthalmic artery microaneurysm in an infant who died of Kawasaki disease [12].

25.6.2 Churg-Strauss Syndrome (CSS)

25.6.2.1 Synonyms and Related Conditions

- Allergic granulomatous angiitis
- Granulomatous small-vessel vasculitis
- Eosinophilic vasculitis

25.6.2.2 Histopathology

Small necrotizing granulomas contain eosinophils. Necrotizing vasculitis is also typical. Glomerulonephritis, when it occurs, is pauci-immune. The role of the eosinophil in pathogenesis is accepted, but the cause of eosinophil activation is not known and may be T cell mediated [18].

25.6.2.3 Systemic Course of Disease

Asthma is the most common finding in CSS and may persist after clinical remission; cases diagnosed following administration of leukotriene receptor antagonists for asthma may be related to steroid withdrawal [23]. Polyneuropathy and constitutional symptoms of fever, malaise, and weight loss are also common [16]. Skin manifestations (palpable purpura, urticaria, nodules) are more frequent in CSS than PAN/MPA or Wegener granulomatosis. Morbidity

related to joints, sinus, GI, or pulmonary involvement is often seen. Ocular involvement is rare.

25.6.2.3.1 Diagnosis

- **ACR criteria** [29]
- **Four of six**
  - Asthma, often adult onset
  - Eosinophilia >10% in peripheral blood
  - Sinusitis
  - Pulmonary infiltrates
  - Neuropathy
  - Biopsy with vasculitis and eosinophils

Like MPA and Wegener granulomatosis, CSS is associated with anti-neutrophil cytoplasmic antibodies. Like MPA, CSS is more likely to display p-ANCA (anti-myeloperoxidase or MPO) than c-ANCA.

25.6.2.4 Ocular Manifestations

Orbital inflammatory disease led to the diagnosis of Churg-Strauss syndrome in two patients reported in 2001 [37]. Fifteen previously reported cases were cited that met the ACR criteria for diagnosis and had ocular involvement. The authors hypothesized that orbital inflammatory disease and ischemic vasculitis were distinct manifestation of CSS, with orbital inflammatory disease carrying a relatively good visual prognosis.

Of the nine reported cases of ischemic vascular involvement reported in the literature, the five tested for ANCA were all positive. Manifestations included three anterior ischemic optic neuropathies, one ischemic optic neuropathy, one branch retinal artery occlusion, one central retinal artery occlusion, one central retinal vein occlusion, and one case of retinal vasculitis. Three of the cases had amaurosis fugax [1, 2, 4, 9, 21, 40].

25.6.3 Wegener Granulomatosis

25.6.3.1 Synonyms and Related Conditions

- Wegener disease
- Systemic necrotizing angiitis
- Necrotizing glomerulonephritis

Wegener granulomatosis typically involves a triad of systemic necrotizing vasculitis (87%), necrotizing granulomatous involvement of the respiratory tract (69%), and necrotizing glomerulonephritis (48%) [27]. Localized forms of disease exist, for example, isolated disease of the eye or orbit. Prevalence is about 8 cases per million.

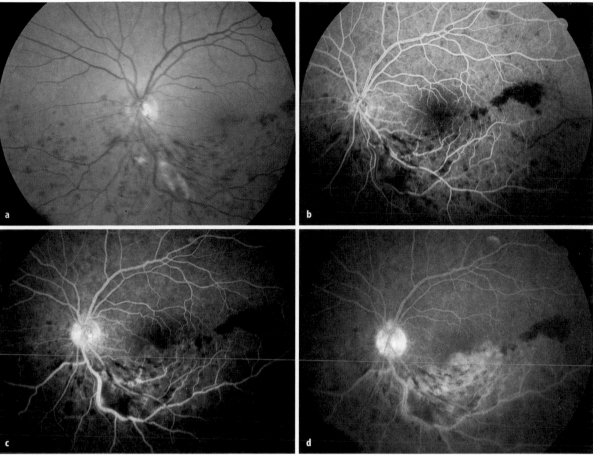

Fig. 25.6.2. Wegener granulomatosis. Hemiretinal vein occlusion. **a** Color photograph, left eye, showing intraretinal hemorrhages and cotton-wool patches. **b** Fluorescein angiogram at 20 s. There is a relative venous delay in the inferior hemiretinal venous branch. The retina is perfused with capillary closure in the region of the cotton-wool patches. Intraretinal hemorrhages block the underlying fluorescence. **c** Fluorescein angiogram at 102 s. Hyperfluorescence of the affected vein is noted. **d** Fluorescein angiogram at 780 s. There is staining of the walls of the affected vein and diffuse retinal leakage

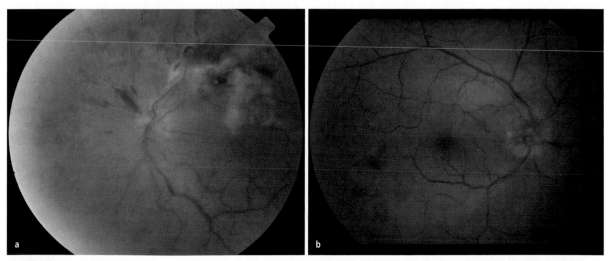

Fig. 25.6.3. Wegener granulomatosis. Inflammatory branch vein occlusion with vitreitis. **a** Color photograph, left eye. There is engorgement of the superotemporal vein with inflammatory sheathing, intraretinal hemorrhages and soft exudates. Sclerotic arterioles are present nasally. **b** Color photograph, right eye. Three years later, the right eye has an impending central retinal vein occlusion with optic nerve hyperemia, intraretinal hemorrhages, and vascular attenuation in the inferotemporal quadrant.

25 III

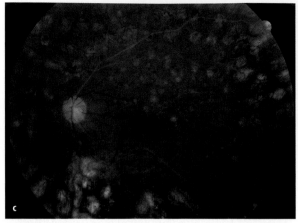

Fig. 25.6.3. c Fundus photo, left eye, 3 years after vein occlusion, at the time of presentation with visual symptoms in the right eye. Panretinal photocoagulation was applied because of neovascularization of the optic nerve head and macular edema. Multiple sclerotic vessels are visible. Vision is stable at 20/20

25.6.3.2 Histopathology

Granulomatous inflammation forms first, often with a typical palisade of histiocytes, followed by a necrotizing vasculitis involving small arteries, veins, and capillaries. The necrosis is basophilic, and collagenolytic, so-called "blue granuloma" [28]. Microabscesses and isolated multinucleated giant cells are also present [41]. The frequent association with antineutrophil cytoplasmic antibody (anti-serine proteinase-3) and the correlation of c-ANCA levels with disease activity indicate that these antibodies play a pathogenetic role. Passive transfer of ANCA antibodies can induce disease [8]. Neutrophils may be the principal effector of tissue damage, after being attracted to the site by endothelial bound c-ANCA. The pauci-immune glomerulonephritis of Wegener is characterized by focal necrosis and not by deposition of immunoglobulins or complement activation. In the lungs, necrotizing granulomatous inflammation is present in the lung fields and capillaritis in the alveoli.

25.6.3.3 Systemic Course of Disease

Organ involvement typically involves the "aerorespiratory" tract: upper airways and lungs. Sinusitis, nasal obstruction, subglottic stenosis occur. Hemoptysis, cough, dyspnea, or respiratory failure may occur or patients may be asymptomatic. On chest X-ray involvement may manifest as nodules, infiltrates, or cavities. Renal involvement is usually asymptomatic but can be detected on laboratory evaluation. Arthralgia, mononeuritis multiplex, purpuric rash also occur.

Diagnostic evaluation is directed toward the usual sites of disease and specific diagnostic criteria have been defined by the American College of Rheumatology (ACR). Anti-neutrophil cytoplasmic antibodies, especially c-ANCA (anti-serine proteinase 3), are highly useful; specificity increases in classic disease and may be accepted as proof of disease without biopsy with a compatible clinical presentation. C-ANCA may be negative and occasionally patients negative for c-ANCA are instead positive for p-ANCA, anti-myeloperoxidase.

25.6.3.3.1 Diagnosis

- **ACR Criteria** [20]
- **Two of Four**
  - Inflammation in the nose or mouth
  - Abnormal urinary sediment
  - Abnormal chest X-ray
  - Granulomatous inflammation of artery on biopsy

25.6.3.4 Ocular Manifestations [5]

- Orbit (13%)
- Eyelid or nasolacrimal duct (13%)
- Episcleritis or scleritis (11%)
- Keratitis (8%)
- Optic neuropathy or compression (6%)
- Conjunctivitis (4%)
- Retinal vasculopathy (5%)
- Uveitis (3%)

Orbital and external ocular involvement far exceeded intraocular involvement in a series of 140 biopsy-proven patients reported in 1983, as shown above. Forty patients had at least one ocular manifestation (29%). In the 1983 series, among the 7 patients considered to have retinal disease, 4 had findings consistent with vasculitis with retinal hemorrhages and cotton-wool spots; choroidal thickening resolved with retinal pigment epithelial changes after treatment. One patient had occlusion of a small branch retinal vein.

An earlier series reported ocular involvement in 47% of 29 patients with Wegener disease, one of whom had bilateral retinal artery occlusion and recovered vision in one eye after corticosteroid treatment [17]. A 2005 case report of bilateral central retinal artery occlusion cites 16 prior cases of central retinal artery occlusion (CRAO) in Wegener granulomatosis and 6 bilateral cases [7]. There is a case report of a hemiretinal vein occlusion in Wegener granulomatosis in 2003 [39].

25.6.4 Differential Diagnosis

Conclusions of a consensus congress on the nomenclature of systemic vasculitides are helpful in distinguishing these similar diseases from one another [22].

For Wegener granulomatosis, differential diagnosis includes Goodpasture syndrome, which has renal disease with immune complex deposition as well as pulmonary manifestations, and lethal midline granuloma, which occurs without renal disease.

Sarcoidosis produces granulomas, but is a T-cell-mediated disease of less severity than Wegener granulomatosis [11]. Giant cell arteritis is not associated with anti-neutrophil antibodies [3].

25.6.5 Treatment

Severe disease in all three systemic ANCA-associated necrotizing vasculitides is similar usually initiated with high-dose prednisone and cyclophosphamide [25]. Pulse cyclophosphamide may be inferior to oral cyclophosphamide, especially for preventing relapse [10]. Transition from cyclophosphamide to methotrexate or azathioprine after 3–6 months is usually successful and reduces the potential for toxicity [13]. Relapse with cessation of therapy may be reduced by maintenance therapy of at least 1 year after quiescence. Supportive therapy for end-organ damage is often needed and the involvement of multiple medical specialists, including ophthalmologists, is common.

Treatment of ocular manifestations in limited Wegener granulomatosis mimics that for systemic disease; limited disease is more likely to be treated initially with the alternative agents methotrexate or azathioprine. There is increasing interest in the use of tumor-necrosis alpha inhibitors for treatment of Wegener granulomatosis; a randomized controlled trial of etanercept showed no benefit [38]. Concomitant use of trimethoprim-sulfamethoxazole in Wegener granulomatosis may reduce relapse by an unknown mechanism [36]. Because of the mechanical effect of granulomatous masses in Wegener granulomatosis, surgical therapy can be required, for example, orbital decompression (which may spread disease into the sinuses), or tracheostomy for subglottic stenosis.

25.6.6 Prognosis

Systemic prognostic factors were assessed in PAN and CSS [15]. Risk of mortality increased with proteinuria, elevated serum creatinine, or GI tract involvement. Presence of more than three factors from a list of five (proteinuria, creatinemia, cardio-myopathy, GI tract involvement, and CNS signs) was associated with a statistically significant increase in mortality.

Death can ensue directly from complications of Wegener granulomatosis or its treatment, with infection being the leading cause of death; less than 80% survivorship at 6 years has been reported [30]. Churg-Strauss syndrome has a better prognosis than Wegener granulomatosis or PAN/MPA with mortality no greater than the general population in one study of 91 patients [23].

In a large prospective trial of etanercept in Wegener disease in which 5.2% of 180 patients had ocular involvement, visual impairment or diplopia was noted in 3.9% of patients either due to vasculitis or to the treatment of the disease and the same percentage were blind in one eye [34]. In Churg-Strauss syndrome, some vision loss can be reversible with treatment [37].

References

1. Acheson JF, Cockerell OC, Bentley CR, Sanders MD (1993) Churg-Strauss vasculitis presenting with severe visual loss due to bilateral sequential optic neuropathy. Br J Ophthalmol 77:118–119
2. Alberts AR, Lasonde R, Ackerman KR, Chartash EK, Susin M, Furie RA (1994) Reversible monocular blindness complicating Churg-Strauss syndrome. J Rheumatol 21:363–365
3. Baranger TA, Audrain MA, Castagne A, Barrier JH, Esnault VL (1994) Absence of antineutrophil cytoplasmic antibodies in giant cell arteritis. J Rheumatol 21:871–873
4. Bosch-Gil JA, Falga-Tirado C, Simeon-Aznar CP, Orriols-Martinez R (1995) Churg-Strauss syndrome with inflammatory orbital pseudotumour. Br J Rheumatol 34:485–486
5. Bullen CL, Liesegang TJ, McDonald TJ, DeRemee RA (1983) Ocular complications of Wegener's granulomatosis. Ophthalmology 90:279–290
6. Chang TS et al (1995) Idiopathic retinal vasculitis, aneurysms, and neuro-retinitis. Retinal Vasculitis Study. Ophthalmology 102:1089–1097
7. Costello F, Gilberg S, Karsh J, Burns B, Leonard B (2005) Bilateral simultaneous central retinal artery occlusions in Wegener granulomatosis. J Neuroophthalmol 25:29–32
8. Csernok E (2003) Anti-neutrophil cytoplasmic antibodies and pathogenesis of small vessel vasculitides. Autoimmun Rev 2:158–164
9. Dagi LR, Currie J (1985) Branch retinal artery occlusion in the Churg-Strauss syndrome. J Clin Neuroophthalmol 5:229–237
10. De Groot K, Adu D, Savage CO (2001) The value of pulse cyclophosphamide in ANCA-associated vasculitis: meta-analysis and critical review. Nephrol Dial Transplant 16:2018–2027
11. DeRemee RA (1994) Sarcoidosis and Wegener's granulomatosis: a comparative analysis. Sarcoidosis 11:7–18
12. Font RL, Mehta RS, Streusand SB, O'Boyle TE, Kretzer FL (1983) Bilateral retinal ischemia in Kawasaki disease. Postmortem findings and electron microscopic observations. Ophthalmology 90:569–577

25 III

13. Goek ON, Stone JH (2005) Randomized controlled trials in vasculitis associated with anti-neutrophil cytoplasmic antibodies. Curr Opin Rheumatol. 17:257–264
14. Googe JM Jr, Brady SE, Argyle JC, Apple DJ, Gooch WM III (1985) Choroiditis in infantile periarteritis nodosa. Arch Ophthalmol 103:81–83
15. Guillevin L et al. (1996) Prognostic factors in polyarteritis nodosa and Churg-Strauss syndrome. A prospective study in 342 patients. Medicine (Baltimore) 75:17–28
16. Guillevin L, Cohen P, Gayraud M, Lhote F, Jarrousse B, Casassus P (1999) Churg-Strauss syndrome. Clinical study and long-term follow-up of 96 patients. Medicine (Baltimore) 78:26–37
17. Haynes BF, Fishman ML, Fauci AS, Wolff SM (1977) The ocular manifestations of Wegener's granulomatosis. Fifteen years experience and review of the literature. Am J Med 63:131–141
18. Hellmich B, Ehlers S, Csernok E, Gross WL (2003) Update on the pathogenesis of Churg-Strauss syndrome. Clin Exp Rheumatol 21:S69–S77
19. Hsu CT, Kerrison JB, Miller NR, Goldberg MF (2001) Choroidal infarction, anterior ischemic optic neuropathy, and central retinal artery occlusion from polyarteritis nodosa. Retina 21:348–351
20. Hunder GG et al (1990) The American College of Rheumatology 1990 criteria for the classification of vasculitis. Introduction. Arthritis Rheum 33:1065–1067
21. Ibanez BF, Bonal-Pitz P, Fernandez-Torres C, Pulido F, de la RL (1983) A case of Churg-Strauss syndrome with ocular involvement. Med Clin(Barc) 81:769–771
22. Jennette JC et al (1994) Nomenclature of systemic vasculitides. Proposal of an international consensus conference. Arthritis Rheum 37:187–192
23. Keogh KA, Specks U (2003) Churg-Strauss syndrome: clinical presentation, antineutrophil cytoplasmic antibodies, and leukotriene receptor antagonists. Am J Med 115:284–290
24. Kielar RA (1976) Exudative retinal detachment and scleritis in polyarteritis. Am J Ophthalmol 82:694–698
25. Langford CA (2003) Treatment of ANCA-associated vasculitis. N Engl J Med 349:3–4
26. Lhote F, Cohen P, Guillevin L (1998) Polyarteritis nodosa, microscopic polyangiitis and Churg-Strauss syndrome. Lupus 7:238–258
27. Lie JT (1997) Wegener's granulomatosis: histological documentation of common and uncommon manifestations in 216 patients. Vasa 26:261–270
28. Lynch JM, Barrett TL (2004) Collagenolytic (necrobiotic) granulomas, part 1. The "blue" granulomas. J Cutan Pathol 31:353–361
29. Masi AT et al (1990) The American College of Rheumatology 1990 criteria for the classification of Churg-Strauss syndrome (allergic granulomatosis and angiitis). Arthritis Rheum 33:1094–1100
30. Matteson EL, Gold KN, Bloch DA, Hunder GG (1996) Long-term survival of patients with Wegener's granulomatosis from the American College of Rheumatology Wegener's Granulomatosis Classification Criteria Cohort. Am J Med 101:129–134
31. Newman NM, Hoyt WF, Spencer WH (1974) Macula-sparing monocular blackouts. Clinical and pathologic investigations of intermittent choroidal vascular insufficiency in a case of periarteritis nodosa. Arch Ophthalmol 91:367–370
32. Rosen ES (1968) The retinopathy in polyarteritis nodosa. Br J Ophthalmol 52:903–906
33. Schmidt D, Lagreze W, Vaith P (2001) Ophthalmoscopic findings in 3 patients with panarteritis nodosa and review of the literature. Klin Monatsbl Augenheilkd 218:44–50
34. Seo P et al (2005) Damage caused by Wegener's granulomatosis and its treatment: prospective data from the Wegener's Granulomatosis Etanercept Trial (WGET). Arthritis Rheum 52:2168–2178
35. Stefani FH, Brandt F, Pielsticker K (1978) Periarteritis nodosa and thrombotic thrombocytopenic purpura with serous retinal detachment in siblings. Br J Ophthalmol 62:402–407
36. Stegeman CA, Tervaert JW, de Jong PE, Kallenberg CG (1996) Trimethoprim-sulfamethoxazole (co-trimoxazole) for the prevention of relapses of Wegener's granulomatosis. Dutch Co-Trimoxazole Wegener Study Group. N Engl J Med 335:16–20
37. Takanashi T, Uchida S, Arita M, Okada M, Kashii S (2001) Orbital inflammatory pseudotumor and ischemic vasculitis in Churg-Strauss syndrome: report of two cases and review of the literature. Ophthalmology 108:1129–1133
38. The Wegener's Granulomatosis Etanercept Trial (WGET) Research Group (2005) Etanercept plus standard therapy for Wegener's granulomatosis. N Engl J Med 352:351–361
39. Venkatesh P, Chawla R, Tewari HK (2003) Hemiretinal vein occlusion in Wegener's granulomatosis. Eur J Ophthalmol 13:722–725
40. Vitali C, Genovesi-Ebert F, Romani A, Jeracitano G, Nardi M (1996) Ophthalmological and neuro-ophthalmological involvement in Churg-Strauss syndrome: a case report. Graefes Arch Clin Exp Ophthalmol 234:404–408
41. Yi ES, Colby TV (2001) Wegener's granulomatosis. Semin Diagn Pathol 18:34–46

25.7 Systemic Immunosuppression in Retinal Vasculitis and Rheumatic Diseases

J.J. HUANG, C.S. FOSTER

Core Messages

- Retinal vasculitis is often associated with underlying systemic diseases
- The diagnosis and the treatment of the underlying systemic diseases dictate the immunosuppressive regimen used for therapy
- Corticosteroids are often the first line therapy, offering rapid and effective control of most retinal and systemic vasculitis
- For patients with a chronic condition requiring long-term therapy or who are intolerant of corticosteroids therapy, steroid-sparing immunosuppressive agents are used, with the tapering of the corticosteroids
- Immunosuppressive treatments often require cooperation between the patient, the ophthalmologist, and the internist or rheumatologist
- A stepladder approach is used for the escalating vigor of treatment required for the severity of the disease process. Combination therapies are often required for additive and synergistic effects of each medication. At the lower dose of these medications, there is an associated reduction of toxicity and side effects of each immunosuppressive medication
- A new class of biologic drugs that inhibit the specific inflammatory mediators in the inflammatory cascade may offer high efficacy with reduced systemic toxicity compared to traditional immunosuppressive agents

25.7.1 Introduction

Retinal vasculitis represents a group of disorders with retinal vascular inflammation and associated ocular inflammation. Many of the patients with retinal vasculitis have been previously diagnosed with an associated systemic disease, but some may develop retinal vasculitis as the initial manifestation of an underlying systemic disorder. The critical aspect for the ophthalmologist's diagnosis of retinal vasculitis is the implication of possible associated systemic and central nervous system inflammation. The underlying etiology for the various causes of retinal vasculitis will ultimately determine the therapy for the inflammation.

Common clinical manifestations of retinal vasculitis include vascular sheathing, vitritis, intraretinal hemorrhage, macular edema, and vascular leakage. The retinal vascular changes are often demonstrated more prominently by fluorescein angiography. Untreated, retinal vasculitis may eventuate to severe ocular complications such as cystoid macular edema, macular ischemia, peripheral vascular occlusion, retinal neovascularization, optic nerve atrophy, and retinal detachment.

Table 25.7.1. Indications for immunosuppressive chemotherapy

Absolute
1. Adamantiades-Behçet disease with retinal involvement
2. Sympathetic ophthalmia
3. Vogt-Koyanagi-Harada syndrome
4. Rheumatoid arthritis with necrotizing scleritis/peripheral ulcerative keratitis
5. Wegener's granulomatosis
6. Polyarteritis nodosa
7. Relapsing polychondritis with scleritis
8. Juvenile idiopathic arthritis associated iridocyclitis
9. Ocular cicatricial pemphigoid
10. Bilateral Mooren's ulcer

Relative
1. Intermediate uveitis
2. Retinal vasculitis with vascular leakage
3. Severe chronic iridocyclitis
4. Sarcoid-associated uveitis

The goal of treatment for patients with retinal vasculitis is the suppression of intraocular inflammation, the prevention of visual loss and its long-term ocular complications. While therapy may not be required for some patients with mild disease with a good visual acuity, most patients with moderate to severe inflammation with associated visual loss will require long-term treatment. Candidates for corticosteroid-sparing immunosuppression include patients with severe associated co-morbidities or intolerance to corticosteroid therapy. Early and aggressive therapy can be critical for the prevention of irreversible visual loss.

The treatment of noninfectious uveitis often requires a thorough understanding of possible associated systemic conditions with risk and benefit for various systemic therapies. Every corticosteroid-sparing immunosuppressive regimen is associated with its own side effects and risks profile. A choice of therapy often involves an understanding between the patient, ophthalmologist and internist or rheumatologist. The risks and benefits of the therapy are weighed against the potential long-term ocular and systemic damages of the disease. While the acute phase of uveitis is commonly treated with corticosteroids in various preparations, the chronicity of the diseases dictates the use of steroid-sparing agents for long-term inflammation control.

Noninfectious uveitis is a significant cause of morbidities and visual loss. There are a limited number of clinical trials in the peer-reviewed literature on the matter of the treatment of ocular inflammatory diseases using immunosuppressive drugs. Thus, there continues to be a lack of "gold standard" for the treatment of uveitides, and for the comparison for future therapies. For most patients with retinal vasculitis and severe uveitis, initial therapy is with high dose corticosteroids. In patients with unilateral disease process, periocular or intravitreal corticosteroid injection offers rapid control of inflammation without the systemic side effects of corticosteroid therapy. Addition of a systemic nonsteroidal anti-inflammatory drug can be used to reduce the rate of recurrence and treat the inflammation associated cystoid macular edema (CME). Clinical improvement can be evident clinically 2 weeks after injection in 80% of patients. Some cases of retinal vasculitis and systemic vasculitis may respond poorly to corticosteroid therapy. And some patients are intolerant to the side effects of systemic corticosteroids and may require initiation of corticosteroid-sparing immunosuppressive agents. The process of immunosuppression is intended to downregulate the immune system. Current immunosuppressive agents interrupt the immune responses through blocking intracellular signaling, or immune cell division. Antimetabolites, T-cell inhibitors, and alkylating agents remain the most commonly used drugs for patients unresponsive to corticosteroids. Biologics, a new class of drugs that target specific inflammatory cytokines, may provide higher efficacy with fewer systemic side effects and toxicity.

Table 25.7.2. Disorders associated with retinal vasculitis

Ocular diseases
1. Idiopathic
2. Eales disease
3. Birdshot retinochoroidopathy
4. Intermediate uveitis
5. Frosted branch angiitis
6. Idiopathic retinal vasculitis aneurysms and neuroretinitis (IRVAN)
7. Acute multifocal hemorrhagic retinal vasculitis

Infectious diseases
1. Toxoplasmosis
2. Tuberculosis
3. Syphilis
4. Lyme disease
5. Cytomegalovirus
6. Herpes simplex
7. Varicella zoster
8. Whipple's disease
9. Human T-cell lymphotropic virus
10. Brucellosis
11. Hepatitis
12. Cat scratch disease
13. HIV

Systemic diseases
1. Admantiades-Behcet's disease
2. Sarcoidosis
3. Crohn's disease
4. Systemic lupus erythematosus
5. Wegener granulomatosis
6. Polyarteritis nodosa
7. Buerger disease
8. Relapsing polychronditis
9. Antiphospholipid syndrome
10. Churg-Strauss syndrome
11. Sjögren syndrome
12. Rheumatoid arthritis
13. Microscopic polyangiitis
14. Dermatomyositis
15. Takayasu disease
16. Primary central nervous system lymphoma
17. Acute leukemia
18. Cancer-associated retinopathy

25.7.2 Diagnosis, Imaging and Electrophysiology

A variety of diagnostic tools are commonly used for diagnosing and monitoring of retinal vasculitis. Intravenous fluorescein angiography (IVFA) is critical for the evaluation and the management of patients with retinal vasculitis. IVFA features commonly seen in patients with retinal vasculitis include inflammation or ischemia of the retinal vasculature with dye leakage demonstrating cystoid macular edema, optic disc edema, skip lesions, capillary non-perfusion, retinal neovascularization, and sclerosis of vessels. Evidence of macular ischemia may explain the poor vision associated with irreversible vision loss after control of the inflammation. For patients on systemic treatment for retinal vasculitis, significant improvement can be evident on clinical examination with subtle residual inflammation evident only on fluorescein angiography. Patients with retinal vasculitis on immunosuppressive therapy typically require two or more angiograms each year to determine the efficacy of the treatment and the persistence of mild active disease.

An additional angiography technique employed commonly for diagnosing and monitoring active inflammation is indocyanine green (ICG), a high molecular weight dye commonly used as an adjunct to fluorescein angiography. The protein bound ICG molecules remain intravascular; unlike fluorescein, ICG is used to study choroidal vasculature. Two common patterns of ICG findings can be seen: inflammation of the choriocapillaris or of the large choroidal stromal vessels. The white-dot uveitis syndromes commonly involve the choriocapillaris, while uveitides such as sarcoidosis, sympathetic ophthalmia, Vogt-Koyanagi-Harada disease, toxoplasmosis, and Admantiades-Behçet's disease commonly show deeper stromal vascular involvement.

Two additional useful tests for the evaluation of patients with ocular inflammatory diseases include optical coherence tomography (OCT) and electroretinogram (ERG) testing. Optical coherence tomography uses the difference in optical reflectivity of the various retinal layers to obtain detailed cross-sectional images of the macula and optic nerve with a spatial resolution of 10 μm. This resolution far exceeds ultrasound and scanning laser ophthalmoscope images. OCT has the same limitations as the other diagnostic instruments using the optical systems, difficulty obtaining images through media opacities such as dense cataract and vitreous hemorrhage. The OCT image provides a detailed anatomical structure of the vitreoretinal interface and foveal architecture with great correlation to the histologic section of the macula. The ability of OCT to detect subtle macular edema with associated intraretinal fluid is extraordinary, far more sensitive than the most expert contact lens examination of the retina. OCT has become the gold standard for diagnosing and evaluating the treatment of patients with macular edema and vitreoretinal pathology of any etiology. Electroretinograms (ERG) have been used for the study of various retinal degenerations and dystrophies. In certain uveitides such as birdshot retinochoroidopathy (BSRC), ERG changes are noted before clinical evidence of decreased vision, increased vitreous cells and macular edema. With early initiation of therapy for BSRC, the abnormal ERG findings normalize with the control of inflammation.

25.7.2.1 Laboratory Tests

In patients with noninfectious systemic diseases, evaluation for underlying collagen vascular diseases and systemic vasculitis syndromes is critical. Appropriate tests include erythrocyte sedimentation rate

Table 25.7.3. Immunosuppressive agents, doses and laboratory tests

Immunosuppressive agents	Dosing	Laboratory testing
Transcription factor inhibitors		
Corticosteroids	1–2 mg/kg/day prednisone 1 gm/day solumedrol	Blood pressure, CBC, glucose monitoring, weight, bone density
Cyclosporine (Sandimmune)	2–5 mg/kg/day	Blood pressure, CBC, BUN/creatinine, liver function tests
Tacrolimus (Prograf)	0.15–0.3 mg/kg/day	Blood pressure, CBC, BUN/creatinine, liver function tests
Sirolimus (Rapamycin)	2 mg/day	CBC, liver function test, lipid profile
Antimetabolite		
Methotrexate	5–25 mg/week	CBC, liver function test
Azathioprine	50–150 mg/week	CBC, liver function test
Mycophenolate mofetil	1,000–3,000 mg/week	CBC, liver function test
Alkylating agents		
Cyclophosphamide (Cytoxan)	2–3 mg/kg/day	CBC, liver function test (target of WBC 3,000–4,000), urine analysis
Chlorambucil (Leukeran)	0.1–0.2 mg/kg/day	CBC, liver function test (target of WBC 3,000–4,000)
Biologics		
Interferon-α2a	6 million IU/day	CBC, liver function test, electrolytes
Daclizumab (Zenapax)	1 mg/kg IV every 2–4 weeks	CBC, liver function test, electrolytes
Infliximab (Remicade)	3–10 mg/kg every 2–4 weeks	CBC, liver function test, electrolytes
Etanercept (Enbrel)	25 mg every 2–3 weeks	CBC, liver function test, electrolytes
Rituximab (Rituxin)	375 mg/m² IV every week 4 or 8 weeks dosing	CBC, liver function test, electrolytes CD20 B-lymphocytes

(ESR), C-reactive protein level (CRP), rheumatoid factor (RF), antinuclear antibody (ANA), antineutrophil cytoplasmic antibody (ANCA), anti-DNA, anti-Smith, anti-cardiolipin, anti-phospholipids, serum electrophoresis, serum cryoglobulins, complement levels and anti-hepatitis B antibodies.

25.7.3 Treatment

25.7.3.1 Corticosteroids

Corticosteroids, since their initial discovery in 1935 by Edward C. Kendall and their first clinical use in 1949 by Hench for the treatment of rheumatoid arthritis, have been the mainstay of therapy for patients with noninfectious inflammatory diseases. The drug is fast-acting, highly effective and generally well tolerated briefly, with limited short-term side effects. In an attempt to limit side effects, increase absorption, and efficacy, various corticosteroid preparations have been formulated for the treatment of ocular and systemic inflammation. All formulations of corticosteroids comprise 21 carbon molecules consisting of a cyclopentoperhydrophenathrene nucleus. The mechanism of action for corticosteroids is at the molecular level inside the cell nucleus. The steroid molecule, after entry into the cell, binds to the cytoplasmic steroid-receptor protein. The steroid-receptor protein complex then crosses the nuclear membrane and binds to sites known as glucocorticocoid response elements (GREs). The GREs directly control the transcription of various mRNAs and the translation of the protein end products. Anti-inflammatory and immunosuppressive effects of corticosteroids include lymphopenia, reduction of eosinophils and monocytes, inhibition of macrophage recruitment and migration, attenuation of bactericidal activity of macrophages, inhibition of prostaglandin synthesis and reduction of capillary permeability.

Orally administered corticosteroids are absorbed in the jejunum with a bioavailability of 90%. The most commonly employed form of the oral corticosteroid preparations is prednisone, which is typically initiated at a dose of 1 mg/kg/day. The regimen is tapered over a span of weeks to months. In patients with severe vision and life-threatening inflammation, intravenous methylprednisolone may be used at a dose of 250 mg/day to 1,000 mg/day for 3 days before the initiation of oral prednisone. In patients with monocular disease who are intolerant of oral prednisone, several regional applications of corticosteroid can be used as treatment option. These options include transeptal, subtenons, and intravitreal triamcinolone acetonide steroid injections. For patients who may require long-term regional steroid delivery, a slow release steroid implant may be a reasonable alternative. All these regional steroid delivery techniques are associated with risk of increased intraocular pressure and the development of cataract. While most steroid-related glaucoma can be controlled with topical ocular hypotensive medications, a small percentage of patients will require glaucoma surgery (trabeculectomy or a glaucoma drainage valve).

The side effects of systemic corticosteroid therapy are numerous. Chronic corticosteroid use is associated with adrenal suppression through its effect on the hypothalamic-pituitary-adrenal axis. Altered moods ranging from euphoria to depression are well known complications of steroid use. Mineralocorticoid activities of corticosteroids significantly alter the patient's intravascular fluid status, with sodium retention and concomitant potassium wasting. With the retention of sodium comes the risk of associated systemic hypertension. The alteration of protein synthesis and gluconeogenesis in the liver result in hyperglycemia, hyperlipidemia, and ketosis. Long-term use of corticosteroids should be avoided in all patients due to the guaranteed side effects of the chronic therapy. In children, growth retardation can occur rapidly. In patients where inflammation is steroid dependent and low dose steroid therapy is required long-term, all necessary steps should be taken to avoid bone loss through calcium and vitamin D supplementation. In other patients, bisphosphonate or calcitonin therapy may be indicated.

25.7.3.2 Nonsteroidal Anti-inflammatory Drugs

In the past 2 decades, many nonsteroidal anti-inflammatory medications have been developed for the treatment of pain and inflammation. Their application in ophthalmology has been extended to topical use for patients with postsurgical cystoid macular edema, prevention of intraoperative miosis, and of steroid-responsive postoperative inflammation. In patients with mild episodic uveitis, topical and systemic NSAID are routinely used in conjunction with other medications for the prevention of relapse and associated macular edema. Since the demonstration of the inhibition of prostaglandin production by aspirin and other NSAID in the 1970s, this class of medication has been one of the most prescribed for the treatment of inflammation and pain. Several different classes of NSAID currently exist.

The mechanism of action for all NSAID involves the inhibition of the cyclooxygenase conversion of arachidonic acid to endoperoxidase, the precursors of prostaglandin. All NSAID are readily absorbed in the gastrointestinal system and reach peak serum concentrations in 0.5–5 h. Over 90% of all NSAID in the serum are protein bound. The liver is the major site of NSAID metabolism.

NSAID therapy in patients with uveitis and retinal vasculitis is an important adjunct to other therapeutic approaches. Systemic use of oral NSAID may reduce the dose of systemic prednisone needed to control inflammation. In patients with acute anterior uveitis, chronic NSAID therapy can reduce the number of flare ups and the amount of topical steroids required for the control of inflammation. In patients with posterior uveitis and retinal vasculitis, the use of oral NSAID with transseptal steroid (triamcinolone acetonide 40 mg) is effective in eliminating cystoid macular edema (CME) and can prevent CME recurrence. While primary retinal vasculitis is not amenable to treatment with NSAID, NSAID are a key component of the multiple drug regimen including corticosteroids and steroid-sparing immunosuppressive agents.

The most common potential side effect of NSAID therapy is gastrointestinal irritation ranging from symptoms of nausea, vomiting, mucosal ulcers and frank bleeding. Other significant potential side effects can involve the central nervous system, hematologic, hepatic, dermatologic and cardiovascular systems. Recent clinical studies in high-risk cardiovascular patients postmyocardial infarction indicated a significantly increased risk of cardiovascular related death for all classes of NSAID therapy. The increased mortality in patients with postmyocardial infarction has raised significant concerns for all patients and prescribing physicians. The increased risk of cardiovascular disease may be related to renal and metabolic changes from NSAID therapy, including salt and fluid retention, edema, and associated systemic hypertension. The risks and benefits of chronic NSAID therapy for the treatment of uveitis and retinal vasculitis should be considered for all patients in a manner similar to all immunosuppressive steroid-sparing agents, and especially in elderly patients with cardiovascular disease or a family history of myocardial infarction.

25.7.3.3 Immunosuppressive Agents

As our knowledge of the cells and molecules involved in the inflammatory cascade have broadened in the past 2 decades, the number of medications used for immunosuppression have exponentially increased. The majority of the current immunosuppressive medications work through the inhibition of DNA and RNA synthesis, and protein translation. Lymphoid proliferative cells, due to their high mitotic activity, are extremely sensitive to this group of medications. A new class of medications, called "biolog-

ic," target the various cytokines of the inflammatory cascade for the suppression of inflammation.

Patients who are candidates for immunosuppressive therapy include patients intolerant or unresponsive to corticosteroids, and patients with chronic disease requiring prednisone doses greater than 6 mg/day. The selection of the proper immunosuppressive regimen is a contract between the treating physician and the patient. An understanding of the disease process along with the risks and benefits of various treatment options is critical. Mild to moderate chronic uveitis may be treated by monotherapy of a single immunosuppressive agent and the successive taper of the corticosteroids. In severe uveitis with a vision threatening disorder, combination therapy using agents with synergistic or additive effects is employed.

25.7.3.3.1 Transcription Factor Inhibitors

Cyclosporine A (Neoral, Sandimmune)

Cyclosporine A (CSA) is an 11 amino acid peptide formed by the fungus *Beauvaria nivea*. The drug exerts its effect on lymphocyte proliferation by blocking the T-cell receptor to genes that encode for multiple lymphokines and enzymes for the activation of resting T cells. CSA also inhibits the calcium dependent intracellular transcriptional signaling of the nuclear factor of activated T cells (NF-AT) for the production of interleukin-2 (IL-2), a potent T-cell mitogen. CSA binds to cyclophilin, a 17-kDa protein in the family of immunophilins. In short, CSA halts the progression of T-cell activation early in the cells cycle, significantly decreasing antibody production and decreasing cytotoxic T-cell activities. Its first application as an immunosuppressive agent in uveitis was reported by Nussenblatt, and CSA was subsequently used in various rheumatic diseases [26].

CSA is often administered as a single or combination agent for moderate to severe uveitis at a dose of 2–5 mg/kg/day. At higher levels, the drug is associated with unacceptable degrees of nephrotoxicity. CSA is formulated in 25 mg and 100 mg tablets. Absorption of CSA in the gastrointestinal tract is usually slow and poor, with bioavailability in the range of 20–50%. Peak serum levels are reached 3–4 h after ingestion. Over 90% of the drug is protein bound. Ocular penetration of the drug is normally poor, but in inflamed eyes, the concentration may reach up to 40% of serum levels. The majority of the drug is processed in the hepatic system through the cytochrome P-450 system. Most clinicians believe that CSA is most effective when used along with low dose prednisone. Side effects of CSA include nephrotoxicity and systemic hypertension, especially at doses above

5 mg/kg/day, such as those used for organ transplant patients. Nephrotoxicity is detected by increased serum creatinine with a disproportionate increase in BUN. Early reduction of CSA dose will reverse CSA induced nephrotoxicity. However, at chronic toxic levels, irreversible interstitial fibrosis of the renal tubules occurs. Hypertension is typically observed in a dose dependent fashion and is reversible in 15–25% of patients during the first weeks of CSA therapy. It is more commonly observed in patients on concomitant therapy with oral steroids and in patients with renal disease. Other common adverse reactions include paresthesia, temperature hypersensitivity, nausea, vomiting, hirsutism, gingival hyperplasia, neurotoxicity and increased risk of infections. Routine monitoring of blood pressure and serum creatinine levels are mandatory. If no clinical response is observed after 3 months of maximum therapy, the medication should be discontinued for an alternative immunosuppressive agent. In patients where a favorable response is observed, the drug is maintained for at least 2 years with subsequent slow gradual taper. In ours and other clinicians' experience, the reduction of CSA is associated with recurrent inflammation, at times necessitating the addition of a second steroid-sparing agent such as mycophenolate mofetil.

CSA have been widely used for therapy in a variety of ocular inflammatory diseases, including Adamantiades-Behçet disease, birdshot retinochoroidopathy, sarcoidosis, par planitis, Vogt-Koyangagi-Harada disease, multiple sclerosis associated uveitis, sympathetic ophthalmia and idiopathic vitritis. *Ozdal* in a large series of patients with Admantiades-Behçet's Disease demonstrated improvement in visual acuity in one-third of the patients, one-third with stable or decreased vision. Half of the patients did not have a flare up of uveitis during CSA treatment. Relapse after the discontinuation of CSA therapy *was typical* for patients in the study [28]. The use of CSA for the treatment of uveitis and retinal vasculitis is considered off-label.

Tacrolimus (FK-506, Prograf)

Tacrolimus is a macrolide antibiotic. A chemical product of the fungus *Streptomyces tsukubaensis,* first discovered in 1984, tacrolimus has a similar spectrum of immunosuppressive activity as CSA, inhibiting NF-AT signaling. It has been approved by the United States Food and Drug Administration for the prophylaxis of organ rejection in patients with liver transplant. Early transplant studies demonstrated that tacrolimus, while similar in action to CSA, allowed for a faster taper of steroid in transplant patients. Tacrolimus, like CSA, blocks the activation of lymphocytes through the suppression of lymphokines and the expression of IL-2 receptor on activated T cells. However, tacrolimus is at least 10 times more potent both in vivo and in vitro. Tacrolimus has a side effect profile similar to that of CSA, with significant effect on renal function and blood pressure. Common side effects include headache, dizziness, nausea, and electrolyte imbalance. The drug is given orally at 0.15–0.3 mg/kg/day. Tacrolimus is available in 1 mg or 5 mg anhydrous oral preparation or for intravenous injections. The drug is poorly absorbed from the GI tract after oral ingestion. Bioavailability ranges from 5% to 67% in studied transplant patients. The presence of food may decrease absorption of the tacrolimus. The drug is highly lipophilic and is strongly bound to the erythrocytes and plasma proteins, mainly albumin. Tacrolimus is metabolized in the liver, with two of the nine metabolites demonstrating persistent immunosuppressive activity. The half-life of the medication is highly variable, ranging from 3.5 to 40.5 h, and is further extended in patients with liver disease. Close monitoring of blood pressure and renal function is critical for patients on tacrolimus.

Mochizuki was the first to demonstrate the efficacy of tacrolimus in uveitis patients and as a monotherapy for patients with Admantiades-Behçet's disease. The majority of the treated patients had a reduction of inflammation as well as a decrease in the number and degree of flare ups during treatment. In patients refractory to CSA and prednisone therapy, a switch to tacrolimus therapy brought the inflammation under control [22].

Tacrolimus and CSA share similar side effect profiles, including nephrotoxicity, hypertension, neurotoxicity and hyperglycemia. However, hirsutism and gingival hyperplasia have not been reported in patients treated with tacrolimus. Neurotoxic symptoms include headache, paresthesia, tremors, aphasia, seizures, encephalopathy and coma. Systemic hypertension associated with tacrolimus in general occurs less frequently and requires less antihypertensive therapy than that occurring with CSA therapy. Opportunistic bacterial, viral and fungal infections are a potential risk with the use of the medication. Close monitoring of liver, renal function and blood pressure is essential for patients receiving tacrolimus. The use of tacrolimus for the treatment of uveitis and retinal vasculitis is considered off-label.

Sirolimus (Rapamycin)

Sirolimus macrolide isolated from the actinomycete *Streptomyces hygroscopicus* is commonly used for the prophylaxis of organ rejection. Although chemically similar in structure to tacrolimus and functionally similar to both CSA and tacrolimus in immunosuppression, rapamycin exerts its action through a separate mechanism. In contrast to CSA and tacrolimus, rapamycin does not act through a calcium dependent pathway. Its mechanism of action involves the binding of FK-binding proteins which target TORs (targets of rapamycin) or FRAPs (FK-rapamycin associated proteins). The drug blocks the signaling transduction of various proinflammatory cytokines, including IL-2 and IL-4. In short, rapamycin blunts the response of B cells and T cells to specific interleukins rather than inhibit their production. Both CSA and tacrolimus exert their effect on resting lymphocytes with no activity against activated lymphocytes, while the effects of rapamycin are independent of lymphocyte state.

Rapamycin is given orally with a loading dose of 6 mg, followed by a maintenance dose of 2 mg/day. Currently, rapamycin is available in 1 mg tablets with variable bioavailability. The pharmacokinetics of rapamycin is relatively unknown. The drug is commonly well tolerated. Common side effects of the drug include hyperlipidemia, thrombocytopenia, and leukopenia. All these side effects are reversible with the cessation of the drug. Routine laboratory monitoring of lipid profile and complete blood count (CBC) is required. Currently, rapamycin is FDA approved for kidney and liver transplant rejection prevention.

Due to the novel mechanism of action, rapamycin may be a good new alternative single agent or may be used in combination with other immunosuppressive agents for patients with moderate to severe ocular inflammation. The use of rapamycin for the treatment of uveitis and retinal vasculitis is considered off-label.

25.7.3.3.2 Antimetabolites

Methotrexate (Rheumatrex)

Methotrexate (MTX) was initially used for the treatment of leukemia in children. Today, MTX is widely used to treat acute lymphoblastic leukemia (ALL), central nervous system lymphoma, and a variety of inflammatory conditions including psoriasis, rheumatoid arthritis, juvenile idiopathic arthritis, Reiter's syndrome, and sarcoidosis. Due to its long track record of safety and efficacy, it has become the first line agent for various pediatric and adult inflammatory diseases.

MTX is a folic acid antagonist that inhibits DNA synthesis and repair, RNA transcription through the prevention of dihydrofolate to tetrahydrofolate conversion by competitively and irreversibly binding the enzyme dihydrofolate reductase (DHFR). Tetrahydrofolate is an essential cofactor in the synthesis of purine nucleotides and thymidylate. The inhibitory action of MTX is cell cycle specific, exerting its activity in the S-phase of the cell cycle. The inhibition of DHFR can be bypassed clinically by the use of leucovorin, a fully functional folate coenzyme. The "leucovorin rescue" allows for higher dose of MTX for clinical use. Methotrexate has no effect on resting nondividing cells. It exerts its cytotoxic effects on malignant cells, fetal cells, cells of gastrointestinal tract, and lymphoproliferative cells. Both B and T cells of the immune system are affected. Apparently, it has little to no affect on the cell-mediated immunity.

Methotrexate is available in 2.5 – 15 mg tablets and as a preparation for injection. Orally administered MTX is readily absorbed with peak serum levels in 1 – 4 h. Over 50 % of MTX is bound to plasma proteins with the remaining unbound fraction mediating its cytotoxic effects. It has been widely effective as a steroid-sparing agent in chronic juvenile and adult uveitis caused by juvenile idiopathic arthritis, sarcoidosis, and idiopathic uveitis at a dose of 5 – 25 mg/week. Hepatotoxicity is a potentially serious side effect of MTX; it is reversible on cessation of therapy. Routine liver function tests are used to monitor therapy. A rise in aspartate aminotransferase (AST) and alanine aminotransferase (ALT) to two times normal level warrants consideration for a reduction of dose or discontinuation of the therapy. Another critical side effect of therapy is interstitial pneumonia. This can present early or later during the course of therapy. Symptoms include cough, dyspnea, and exertional fatigue. Signs and symptoms are reversible if the problem is detected early and medication discontinued promptly.

While MTX has been demonstrated to be highly efficacious for the treatment of moderate to severe uveitis, its use is considered off-label. The great majority of our experience with MTX in uveitis has involved children. However, the drug can be safely used for patients of all ages. Samson published the largest series of patients treated with MTX with a variety of diagnoses, including idiopathic uveitis, HLA-B27 associated uveitis, juvenile idiopathic arthritis associated uveitis, and sarcoidosis. MTX has demonstrated efficacy in the treatment of all disease groups, with better results in sarcoidosis patients than in juvenile idiopathic arthritis patients [34]. Advantages of MTX as an immunosuppressive agent include weekly dosing, long track record of safety and efficacy in many uveitides.

Azathioprine (Imuran)

Azathioprine was first introduced in the 1960s. It is a pro-drug and is converted to 6-mercaptopurine after oral ingestion. It has a similar mechanism of action with prolonged duration of activity compared to 6-mercaptopurine. Azathioprine was first introduced into ophthalmic use in 1966.

Azathioprine is a prodrug that is quickly metabolized in the liver to 6-MP, which in turn interferes with purine metabolism of DNA, RNA and protein synthesis. 6-Mercaptopurine, through its conversion to thioinosine-5-phosphate, a purine analogue, impairs adenine and guanine nucleotide formation in actively dividing cells in a cell cycle-specific (S phase) fashion. Both B and T cells are suppressed.

Azathioprine is available in 50 mg tablets for oral administration, and approximately 50 % of the medication is absorbed in the gastrointestinal system within 2 h. A single or divided dose of azathioprine is administered at a concentration of 2–3 mg/kg/day. The dose is reduced by 25 % if allopurinol is concomitantly administered. One of the more common side effects of azathioprine is nausea and anorexia, common reasons for discontinuation of therapy. Ultimately, the frequency and severity of adverse effects of azathioprine therapy depend on the dose, duration of therapy and underlying hepatic and renal disease. Other less common potential side effects include hepatotoxicity, myelosuppression, pneumonitis, pancreatitis and alopecia. Ingestion of a large dose of azathioprine may lead to bone marrow hypoplasia, bleeding and even death. Azathioprine has been implicated in potentiating the risk of future neoplasia in transplant patients. Several clinical studies have demonstrated no statistical difference in the overall frequency of malignancy in patients treated with azathioprine when compared to the frequency of malignancy in the general population. Although azathioprine is well tolerated, routine hematologic monitoring is crucial and should include liver function testing and complete blood counts with differential.

Many reports document the use of azathioprine for treating ocular inflammatory disease both as a single or combination therapy. Newell demonstrated efficacy of azathioprine in patients with pars planitis [24]. Azathioprine has been effective in the treatment of JIA associated uveitis. Foster demonstrated efficacy of azathioprine in the treatment of patients of Adamantiades-Behçet disease, preventing the development of new lesions, reducing the frequency and intensity of ocular inflammation [10]. Currently, we are using azathioprine in our clinic as a steroid-sparing agent for disease entities such as multifocal choroiditis, sympathetic ophthalmia, Vogt-Koyanagi-Harada syndrome, sarcoidosis, par planitis, JIA associated uveitis, and Reiter's syndrome associated uveitis. The use of azathioprine for the treatment of uveitis and retinal vasculitis is considered off-label.

Mycophenolate Mofetil (Cellcept)

Mycophenolate mofetil is an immunosuppressive agent developed for use in the prevention of solid organ transplant rejection in 1995. Since then, its use in ocular inflammatory disease is increasing, and it is often used in place of various other drugs such as azathioprine, cyclosporine and methotrexate. Mycophenolate mofetil is a pro-drug of mycophenolic acid. The mechanism of action involves the selective, noncompetitive, reversible inhibition of proliferating T and B cells through inhibiting inosine monophosphate dehydrogenase, an enzyme critical in de novo purine synthesis. Unlike most cells that use the salvage pathway for growth and proliferation, rapidly dividing activated B and T cells cannot, and thus they are highly sensitive to mycophenolate mofetil.

The common side effects of mycophenolate mofetil therapy include diarrhea, nausea, vomiting, headache and fatigue. More severe and less common potential side effects include leukopenia, hepatic toxicity, risk of malignancy and sepsis. Drug dosing starts at 500 mg/day to a maximum dose of 3,000 mg/day. Routine liver function testing and complete blood cell count are critical for monitoring therapy. Drugs such as acyclovir and ganciclovir may compete for renal excretion. Concomitant use of antacids may reduce the absorption of mycophenolate.

Mycophenolate mofetil is often better tolerated than the aforementioned immunosuppressive drugs with less risk for renal and hepatic toxicity. The use of mycophenolate mofetil in the treatment of uveitis patients is considered off-label. Various small studies have demonstrated efficacy of mycophenolate mofetil as a monotherapy and also in a combination therapy for uveitides such as JIA associated uveitis, sarcoidosis, multifocal choroiditis, inflammatory bowel disease associated uveitis, orbital pseudotumor, and Admantiades-Behçet's disease.

25.7.3.3.3 Alkylating Agents

Cyclophosphamide (Cytoxan)

Cyclophosphamide is an alkylating agent derived from nitrogen mustard gas. The effect of leukopenia and aplasia with this class of agent was discovered during World War I after the use of sulfur mustard gas. The first medical application of cyclophosphamide was in the treatment of systemic lymphoma patients. Since then, it has been widely used for the treatment of various forms of systemic vasculitis.

Cyclophosphamide is a prodrug and is converted by hepatic cytochrome P-450 oxidase into the active metabolites phosphoramide mustard and 4-hydroxycyclophosphamide. The drug targets the 7-nitrogen atom of guanine to cause guanine-thymidine cross links in the DNA molecule, leading to miscoding, DNA strand breaks and formation of phosphodiester bond after repair of DNA breaks. Cyclophosphamide has profound effects on lymphoid cells. B-cell functions are affected more than T cells. It is the only immunosuppressive agent that can induce immunologic tolerance to the exposure of a particular antigen. But due to the potency and potential toxicity associated with all alkylating agents, cyclophosphamide is reserved for sight threatening refractory severe uveitis.

Cyclophosphamide is supplied in 25 and 50 mg tablets, and as a powder in 100, 200, 500 mg, 1 g and 3 g vials for injection. Effective dosing ranges from 1 to 3 mg/kg/day. Seventy-five percent of an oral dose is absorbed from the gastrointestinal tract reaching peak levels in 1 h. The half-life of the metabolized drug is 4–6 h. The oxidized inactive metabolite acrolein is the agent most responsible for the bladder toxicity associated with cyclophosphamide use. The most common side effect of cyclophosphamide is bone marrow suppression. The dose of cyclophosphamide is usually titrated to a white blood cell count between 3,000 to 4,000. Severe leukopenia significantly increases the risk of infection and severe sepsis. A portion of all patients receiving cyclophosphamide develop hemorrhagic cystitis. All patients on the drug should drink 3–4 L of fluids to reduce the bladder toxicity and risk for bladder cancer. Other associated potential side effects include gonadal toxicity, nausea, vomiting, and alopecia. Long-term therapy of greater than 1 year or a cumulative dose of greater than 76 g of cyclophosphamide is associated with hematopoietic malignancies. Routine monitoring of patient's CBC and urine analysis is critical.

Cyclophosphamide is the drug of choice for all patients with ocular manifestation of Wegener's granulomatosis or polyarteritis nodosa. Other applications of the drug include peripheral ulcerative keratitis (PUK) associated with rheumatoid arthritis, bilateral Mooren's ulcer, progressive ocular cicatricial pemphigoid, and Adamantiades-Behçet disease with retinal vascular involvement. While chlorambucil is a more commonly and safer agent used for retinal vascular ABD, intravenous cyclophosphamide is highly effective for providing rapid control of inflammation. Alkylating agents are most likely to induce long-term drug free remission in patients with chronic severe uveitis. Akpek reported a series of serpiginous choroiditis patients with persistent active disease on nonalkylating immunosuppressive medications. After switching to a regimen of cyclophosphamide or chlorambucil, no patient developed recurrence and seven patients had drug free remission [1].

Using the stepladder approach of treatment for various recalcitrant uveitides such as par planitis, sympathetic ophthalmia, Vogt-Koyanagi-Harada disease, in addition to the group of systemic vasculitis with associated retinal vasculitis, cyclophosphamide is used for these patients unresponsive to other immunosuppressive therapies. The use of cyclophosphamide is considered off-label for the treatment of uveitis and retinal vasculitis.

Chlorambucil (Leukeran)

Chlorambucil is another alkylating agent that crosslinks DNA in a mechanism similar to cyclophosphamide. The drug was first synthesized in the 1950s for the treatment of systemic lymphoma. Its use in ophthalmology was first reported for patients with Admantiades-Behçet's disease, and it is now commonly used for the treatment of this disease. Similar to cyclophosphamide, chlorambucil is a nitrogen mustard derivative with a common mechanism of action interfering with DNA replication and RNA transcription. The cytotoxic effects of the drug are cell cycle nonspecific. The normal efficacious dose for the drug is in the range of 4–18 mg/day with target white blood cell count in the range of 3,000–4,000. Long-term use of chlorambucil can be associated with unpredictable and sudden pancytopenia. Routine monitoring of white blood cell count is critical due to the risk of severe leukopenia, and the duration between laboratory testing is shortened as the therapy continues. Additional potential side effects include secondary malignancy, infertility, infection and gastrointestinal discomfort. The risk of secondary malignancy increases with treatment duration greater than a year. The advantage of chlorambucil therapy over cyclophosphamide is the reduced risk of hemorrhagic cystitis and bladder cancer.

Miserocchi reported 56 eyes of 28 patients with various uveitides unresponsive to corticosteroids and other immunosuppressive agents. Upon switching to chlorambucil, vision improved or stabilized in 82% of the patients, and 50% were in remission off medications [21]. Goldstein used chlorambucil in short-term high dose therapy to minimize cumulative drug toxicity. The treatment duration averaged 16 weeks with maximum daily dose of 30 mg. Seventy-seven percent of patients were in remission during follow-up [14]. The use of chlorambucil is considered off-label for the treatment of uveitis and retinal vasculitis.

25.7.3.3.4 Biologics

Recent research in the field of molecular biology has elucidated many of the pathways of the inflammatory cascade. Some of the key molecules involved in this pathway are the cytokines produced by lymphocytes and macrophages, including tumor necrosis factor-α (TNF-α), interferon-gamma (IFN-γ), interleukin-1 (IL-1), interleukin-2 (IL-2), and interleukin-10 (IL-10). A new class of drugs directed against the specific cytokines and its receptor are called "biologics," based on the fact that this class of drugs is produced by cultured cells rather than manufactured through a man-made chemical process. Biologics are usually very well tolerated compared to traditional immunosuppressive therapy, but are also far more expensive compared to traditional immunosuppressants.

Interferon

During the past 10 years, several studies have described the use of interferon for the treatment of retinal vasculitis, especially that associated with Adamantiades-Behçet's disease. Interferon-α has been used to treat ABD resistant to conventional immunosuppressive therapy. Wechsler reported efficacy of IFN-α treatment in eight patients with ABD with improvement of vision and reduction of oral prednisone [41]. Kotter, in a large study of 50 patients, reported a positive response in 92% of them, with significantly improved visual acuity and reduction in disease activity [19]. Dose dependent side effects of interferon therapy including reddening at the site of injection, flu-like symptoms, depression, and leukopenia were common. Sixteen percent of patients developed autoimmune phenomena, including antithyroid and antinuclear antibodies. The use of IFN-α for the treatment of uveitis and retinal vasculitis is considered off-label.

Daclizamab (Zenapax)

Daclizumab is an immunoglobulin G (IgG) monoclonal antibody directed against the CD25 subunit of human IL-2 receptor on activated T cells. The drug was first developed for the treatment and prevention of solid organ transplant rejection. Daclizumab is a humanized monoclonal antibody with its mechanism of action on the IL-2 receptor causing the blockade of the IL-2 mediated activation of the T cells. Daclizumab is supplied in 25 mg/5 ml concentrate for intravenous administration. The drug has an in vivo half-life of 20 days. Treatment using daclizumab is relatively safe, with main side effects being constipation, diarrhea, nausea, vomiting, and increased risk of infection during various multicenter studies for the treatment of acute renal allograft rejection. Daclizumab was highly effective for the treatment of renal transplant rejection in a combination therapy with prednisone, CSA and azathioprine, later with prednisone, CSA and mycophenolate

mofetil. The use of daclizumab for the treatment of uveitis is considered off-label use.

Nussenblatt reported the use of daclizumab in ten patients with bilateral uveitis. Patients treated included those with diagnoses of sarcoidosis, idiopathic intermediate uveitis, Vogt-Koyanagi-Harada disease, multifocal choroiditis and idiopathic panuveitis. Eight of the ten patients responded with stable or improvement in visual acuity. No therapy was stopped due to side effects or intolerance to therapy [27]. We reported favorable results in a nonrandomized trial of daclizumab in the treatment of 14 patients with scleritis, uveitis, or mucous membrane pemphigoid unresponsive to other immunosuppressive therapies. And we have recently treated seven patients with birdshot retinochoroidopathy recalcitrant or intolerant to conventional immunomodulatory therapy, and who were switched to intravenous daclizumab therapy. All seven patients responded clinically to daclizumab with stable or improved visual acuity and ERG abnormalities. As with all biologics used for the treatment of uveitis, no ideal dosing and frequency have yet been determined.

Infliximab (Remicade)

Infliximab, a chimeric human-mouse anti-TNF-α monoclonal antibody, is currently approved for use in patients with rheumatoid arthritis, Crohn's disease, psoriasis and ankylosing spondylitis. In animal and human studies, there is evidence for the efficacy of TNF-α antagonists in the treatment of uveitis. Hematologic studies have shown elevated levels of serum TNF-α in patients with chronic uveitis when compared to patients with a single flare up. Perez-Guijo found that serum levels of TNF-α are higher in patients with HLA-B27 associated uveitis compared to uveitis patients who were HLA-B27 negative [31]. Recent clinical trials of TNF-α antagonists have demonstrated significant efficacy in the treatment of patients with rheumatoid arthritis, ankylosing spondylitis, and Crohn's disease. TNF-α antagonists are usually well tolerated, with relatively few side effects such as fatigue, activation of latent tuberculosis, leukopenia, and serious infections. Other side effects include optic neuritis, worsening of multiple sclerosis, lupus-like reaction and anaphylaxis. Before initiation of therapy, prior tuberculosis exposure should be excluded by purified protein derivative intradermal skin testing and chest radiograph. Infliximab is typically given intravenously 5–10 mg/kg every 2–4 weeks.

Infliximab appears to have good efficacy in the short term treatment of patients with Adamantiades-Behçet disease including those with panuveitis and retinal vasculitis. Murphy showed that infliximab was efficacious for the treatment of various ocular inflammatory diseases including scleritis, retinal vasculitis, intermediate uveitis and idiopathic panuveitis [23]. An important advantage of infliximab therapy is the rapid onset of action compared to other medications used for immunomodulation therapy. Sfikakis treated five ABD patients with severe panuveitis. In all five patients, the inflammation improved in 24 h and completely resolved in 7 days [35].

Due to differences in the penetration of blood-retina barrier and in the mechanism of action for TNF-α blockade, clinically, infliximab has been more effective than etanercept for treatment of ocular inflammatory disease. Currently, the evidence is lacking whether infliximab is equal or superior to conventional immunosuppressive therapy in the treatment of uveitis and retinal vasculitis. In addition, the lack of knowledge for the ideal dose and frequency also hampers the treatment of chronic uveitis and retinal vasculitis patients. The use of infliximab for the treatment of uveitis and retinal vasculitis is considered off-label.

Etanercept (Enbrel)

Etanercept is a recombinant tumor necrosis factor receptor (p75)-Fc fusion protein that competitively inhibits tumor necrosis factor-alpha (TNF-α). It binds extracellular TNF-α, preventing the binding of native receptors and inhibiting downstream signaling. Etanercept has been successfully used in patients with rheumatoid arthritis, juvenile rheumatoid arthritis, ankylosing spondylitis, and psoriatic arthritis, reducing pain and inflammation. The drug is administered subcutaneously in 25 mg two or three times per week. In the rat model, Koizumi found that etanercept reduced leukocyte rolling, adhesion, and vascular leakage, resulting in decreased breakdown of the blood-retina barrier and associated apoptotic cell death [18]. A double-blind placebo controlled study of etanercept showed efficacy in suppressing mucocutaneous manifestations of Admantiades-Behçets disease. Reiff evaluated etanercept therapy in children with chronic recalcitrant uveitis. Within 3 months, 10 of 16 patients showed decreased inflammation, and 4 of 10 showed improvement in visual acuity. There were no reports of serious adverse reactions except local injection site reaction [32]. Foster reported a study of 20 patients with uveitis well controlled on MTX and later tapered to a regimen of etanercept versus placebo. Relapse occurred in three of ten on etanercept and five of ten on placebo. The authors concluded that etanercept was no better than placebo in preventing the relapse of uveitis previously controlled with MTX [13].

Side effects associated with etanercept therapy included risk of infection and injection site reactions. Routine liver function test and complete blood count should be performed. The use of etanercept in the treatment of uveitis and retinal vasculitis is considered off-label.

Rituximab (Rituxin)

The rituximab antibody is a genetically engineered chimeric murine/human monoclonal antibody directed against the CD20 glycoprotein found on the surface of normal B-lymphocytes. The antibody is an IgG$_1$ kappa immunoglobulin containing murine light- and heavy-chain variable region sequences and human constant region sequences. The drug is currently approved for the treatment of B-cell lymphoma. Due to the selectivity of the chimeric antibody, it allows for the direct targeting of B cells and the inflammatory responses that are antibody mediated. Rituximab has been highly successful for the treatment of systemic anti-neutrophil cytoplasmic antibody-positive vasculitis associated with Wegener's granulomatosis. In our clinic, it has been used successfully for a Wegener's granulomatosis patient refractory to traditional immunosuppressive agents with associated retinal vasculitis and panuveitis. The drug is administered intravenously every week for 4 weeks or for 8 weeks at a dose of 375 mg/m^2 surface area dosing.

Most side effects of rituximab therapy are encountered during the first infusion. Special attention should be paid to the rate of antibody infusion with a slow gradual increase in the hourly infusion rate. Fever, chills, respiratory symptoms, and occasionally hypertension are the most common effects. The side effects overall are much more mild compared to traditional immunosuppressive or chemotherapeutic agents. The use of rituximab for the treatment of uveitis and retinal vasculitis is considered off-label.

25.7.4 Immunosuppressive Therapy in Children

The use of corticosteroid-sparing immunosuppressive therapy for the treatment of children requires a different approach than that for adult patients. Prompt control of inflammation is crucial for the prevention of vision loss from inflammation as well as from amblyopia. Chronic use of corticosteroids in children may result in growth suppression, cataracts, weight gain and acne. Corticosteroids associated glaucoma is far more common in children than in adults.

Walton demonstrated safety and efficacy of cyclosporine therapy in children for the treatment of

25 **III**

severe vision threatening intermediate uveitis and panuveitis with follow-up of 4 years. Over 80 % of the children responded to the CSA therapy [40]. Additional studies have further confirmed the safety of CSA as an immunosuppressive agent for the treatment of pediatric uveitis. Methotrexate is one of the most commonly used medications for the treatment of juvenile idiopathic arthritis associated uveitis because of its long track record of safety in children. The medication can be used as a monotherapy or a combination therapy along with CSA. Over 60 % of patients with JRA associated uveitis will have clinical response to MTX as a single agent therapy. Currently, mycophenolate mofetil is gaining popularity for the treatment of children with JIA associated uveitis, pars planitis, sarcoid associated uveitis, and psoriatic arthritis associated uveitis.

25.7.5 Combination Therapy

There is strong evidence in our clinic and others that combination therapy for the treatment of uveitis and retina vasculitis can help to rapidly taper corticosteroids and reduce side effects associated with immunosuppressive therapy. Synergistic activities of combination therapy can help to reduce the dose needed for each individual immunosuppressive medication and the associated side effects. Transcription factor inhibitors such as CSA, tacrolimus and rapamycin are often used in combination with antimetabolites such as MTX, azathioprine and mycophenolate. Additional medications that can be incorporated as a part of the drug regimen include NSAID, low dose prednisone (less than 6 mg/day), and biologics. Diseases responsive to combination therapy include panuveitis, HLA-B27 associated uveitis, Admantiades-Behçet disease, serpiginous choroiditis, sympathetic ophthalmia, VKH, and multifocal choroiditis with panuveitis.

25.7.6 Conclusion

Although retinal vasculitis is a rare ophthalmic condition, the evaluation and the treatment are often complicated and require a thorough understanding of the ocular disease and underlying systemic associations. The long-term prognosis visually and systemically must be weighed against the risks and benefits of the treatment regimen. Fluorescein angiography is critical for assessing the full extent of the disease and for monitoring the efficacy of treatment. In patients with infectious retinal vasculitis, treatment with the appropriate antimicrobial therapy is essential and may cure the patient of the associated retinal vasculitis.

For patients with noninfectious retinal vasculitis associated with ocular and systemic disease, cortico-steroids are the mainstay of *initial* treatment. Traditional steroid-sparing immunosuppressive agents are used for patients who fail to respond to corticosteroids or who are intolerant to the side effects of chronic therapy. Due to the lack of clinical trials for the use of immunosuppressive therapy in the treatment of uveitis and retinal vasculitis, it is often difficult to make an evidence-based decision on which agent is best for which clinical situation. Studies of the efficacy for such drugs have been limited by the difficulty in enrolling large numbers of patients in clinical trials; hence the absence of a "gold standard" for comparison of all new immunosuppressive therapy. For all treating physicians and patients, a thorough understanding of the disease prognosis with the risks and benefits of treatment regimen is crucial. Collaboration with an internist and rheumatologist also can help to co-manage the disease and immunosuppressive regimen. It is especially important for those cases in which an ocular immunologist is not participating in treating the patient. In patients with mild to moderate uveitis and retinal vasculitis, single agent therapy of methotrexate and mycophenolate mofetil are commonly employed. In patients with severe retinal vasculitis, combination therapy or an alkylating agent is often required to control the inflammation. With the newly expanding class of biologics demonstrating efficacy against various recalcitrant uveitides, flexibility on the part of the treating physician is critical for formulating the proper treatment regimen.

References

1. Akpek EK, Jabs DA, Tessler HH, Joondeph BS, Foster CS (2002) Successful treatment of serpiginous choroiditis with alkylating agents. Ophthalmology 109:1508–1513
2. Antcliff RJ, Spalton DJ, Stanford MR, Graham EM, ffytche TJ, Marshall J (2001) Intravitreal triamcinolone for uveitic cystoid macular edema: an optical coherence tomography study. Ophthalmology 108:765–772
3. Becker MD, Rosenbaum JT (2000) Current and future trends in the use of immuno-suppressive agents in patients with uveitis. Curr Opin Ophthalmol 11:472–477
4. Berk PA, Goldberg JD, Silverman MN, Weinfeld A, Donovan PB, Ellis JT, Landaw SA, Laszlo J, Najean Y, Pisciotta AV, Wasserman LR (1981) Increased incidence of acute leukemia in polycythemia vera associated with chlorambucil therapy. N Engl J Med 304:441–447
5. Carnahan MC, Goldstein DA (2000) Ocular complications of topical, peri-ocular, and systemic corticosteroids. Curr Opin Ophthalmol 11:478–483
6. Ciulla TA, Walker JD, Fong DS, Criswell MH (2004) Corticosteroids in posterior segment disease: an update on new delivery systems and new indications. Curr Opin Ophthalmol 15:211–220
7. Cunningham ET (2002) Diagnosis and management of acute anterior uveitis. In Focal Points, AAO

8. Cunningham ET (2000) Uveitis in children. Ocul Immunol Inflamm 8:251–261
9. Eriksson P (2005) Nine patients with anti-neutrophil cytoplasmic antibody-positive vasculitis successfully treated with rituximab. J Intern Med 257:540–548
10. Foster CS, Baer JC, Raizman MB (1991) Therapeutic responses to systemic immunosuppressive chemotherapy agents in patients with Behcet's syndrome affecting eyes. In: O'Duffy JD, Kokmen E (eds) Behcet's disease: basic and clinical aspects. Dekker, New York, pp 581–588
11. Foster CS, Vitale AT (2002) Immunosuppressive chemotherapy. In: Foster CS, Vitale AT (eds) Diagnosis and treatment of uveitis. Saunders, Philadelphia, pp 141–214
12. Foster CS (2003) Diagnosis and treatment of juvenile idiopathic arthritis associated uveitis. Curr Opin Ophthalmol 14:395–398
13. Foster CS, Tufail F, Waheed NK, Chu D, Miserocchi E, Baltatzis S, Vredeveld CM (2003) Efficacy of etanercept in preventing relapse of uveitis controlled by methotrexate. Arch Ophthalmol 121:437–440
14. Goldstein DA, Fontanilla FA, Kaul S, Sahin O, Tessler HH (2002) Long-term follow-up of patients treated with short-term, high dose chlorambucil for sight-threatening ocular inflammation. Ophthalmology109:370–377
15. Jabs DA, Rosenbaum JT, Foster CS, Holland GN, Jaffe GJ, Louie JS, Nussenblatt RB, Stiehm ER, Tessler H, Van Gelder RN, Whitcup SM, Yocum D (2000) Guidelines for the use of immunosuppressive drugs in patients with ocular inflammatory disorders: recommendations of an expert panel. Am J Ophthalmol 130:492–513
16. Kilmartin DJ, Forrester JV, Dick AD (1998) Cyclosporin A therapy in refractory non-infectious childhood uveitis. Br J Ophthalmol 82:737–742
17. Kiss S, Letko E, Qamruddin S, Baltatzis S, Foster CS (2003) Long-term progression, prognosis, and treatment of patients with recurrent ocular manifestations of Reiter's syndrome. Ophthalmology 110:1764–1769
18. Koizumi K, Poulaki V, Doehmen S, Welsandt G, Radetzky S, Lappas A, Kociok N, Kirchhof B, Joussen AM (2003) Contribution of TNF-alpha to leukocytes adhesion, vascular leakage, and apoptotic cell death in endotoxin-induced uveitis in vivo. Invest Ophthalmol Vis Sci 44:2184–2191
19. Kotter I, Zierhut M, Eckstein AK, Vonthein R, Ness T, Gunaydin I, Grimbacher B, Blaschke S, Meyer-Riemann W, Peter HH, Stubiger N (2003) Human recombinant interferon alfa-2a for the treatment of Behcet's disease with sight threatening posterior or panuveitis. Br J Ophthalmol 87:423–431
20. McGhee CN, Dean S, Danesh-Meyer H (2002) Locally administered ocular corticosteroids: benefits and risks. Drug Saf 25:33–55
21. Miserocchi E, Baltatzis S, Ekong A, Roque M, Foster CS (2002) Efficacy and safety of chlorambucil in intractable noninfectious uveitis. Ophthalmology 109:137–142
22. Mochizuki M, Masuda K, Tuyoshi S, Ito K, Kogure M, Sugino N, Usui M, Mizushima Y, Ohno S, Inaba G (1993) A clinical trial of FK 506 in refractory uveitis. Am J Ophthalmol 115:763–769
23. Murphy CC, Ayliffe WH, Booth A, Makanjuola D, Andrews PA, Jayne D (2004) Tumor necrosis factor alpha blockade with infliximab for refractory uveitis and scleritis. Ophthalmology 111:352–356
24. Newell FW, Krill AE (1967) Treatment of uveitis with azathioprine. Trans Ophthalmol Soc 87:499–511
25. Nussenblatt RB, Palestine AG, Rook AH, Scher I, Wacker WB, Gery I (1983) Treatment of intraocular inflammation with cyclosporine A. Lancet 1:235–238
26. Nussenblatt RB, Palestine AG, Chan CC (1983) Cyclosporine A therapy in the treatment of intraocular inflammatory disease resistant to systemic corticosteroids and cytotoxic agents. Am J Ophthalmol 96:275–282
27. Nussenblatt RB, Thompson DJ, Li Z, Chan CC, Peterson JS, Robinson RR, Shames RS, Nagarajan S, Tang MT, Mailman M, Velez G, Roy C, Levy-Clarke GA, Suhler EB, Djalilian A, Sen HN, Al-Khatib S, Ursea R, Srivastava S, Bamji A, Mellow S, Sran P, Waldmann TA, Buggage RR (2003) Humanized anti-interleukin-2 (IL-2) receptor alpha therapy: long-term results in uveitis patients and preliminary safety and activity data for establishing parameters for subcutaneous administration. J Autoimmun 21:283–293
28. Ozdal PC, Ortac S, Taskintuna I, Firat E (2002) Long-term therapy with low dose cyclosporine A in ocular Behcet's disease. Doc Ophthalmol 105:301–312
29. Palestine AG, Nussenblatt RB, Chan CC (1985) Cyclosporine penetration into the anterior chamber and cerebrospinal fluid. Am J Ophthalmol 99:210–211
30. Papaliodis GN, Chu D, Foster CS (2003) Treatment of ocular inflammatory disorders with daclizumab. Ophthalmology 110:786–789
31. Perez-Guijo V, Santos-Lacomba M, Sanchez-Hernandez M, Castro-Villegas Mdel C, Gallardo-Galera JM, Collantes-Estevez E (2004) Tumor necrosis factor-alpha levels in aqueous humor and serum from patients with uveitis: the involvement of HLA-B27. Curr Med Res Opin 20: 155–157
32. Reiff A, Takei S, Sadeghi S, Stout A, Shaham B, Bernstein B, Gallagher K, Stout T (2001) Etanercept therapy in children with treatment-resistant uveitis. Arthritis Rheum 45:252–257
33. Rothova A (2002) Corticosteroids in uveitis. Ophthalmol Clin North Am 15:389–394
34. Samson CM, Waheed N, Baltatzis S, Foster CS (2001) Methotrexate therapy for chronic noninfectious uveitis: analysis of a case series of 160 patients. Ophthalmology 108:1134–1139
35. Sfikakis PP, Theodossiadia PG, Katsiari CG, Kaklamanis P, Markomichelakis NN (2001) Effect of infliximab on sight-threatening panuveitis in Behcet's disease. Lancet 358:295–296
36. Shetty AK, Zganjar BE, Ellis GS, Ludwig IH, Gedalia A (1998) Low-dose methotrexate in the treatment of severe juvenile rheumatoid arthritis. J Pediatr 133:266–268
37. Solomon SD, Cunningham ET (2000) Use of corticosteroids and noncorticosteroid immunosuppressive agents in patients with uveitis. Compr Ophthalmol Update 1:273–286
38. Suttorp-Schulten MSA, Rothova A (1996) The possible impact of uveitis in blindness: a literature survey. Br J Ophthalmol 80:844–848
39. Thall EH (2003) Dosage of intravitreal triamcinolone. Am J Ophthalmol 136:1192
40. Walton RC, Nussenblatt RB, Whitcup SM (1998) Cyclosporine therapy for severe sight-threatening uveitis in children and adolescents. Ophthalmology 105:2028–2034
41. Wechsler B, Bodaghi B, Huong DL, Fardeau C, Amoura Z, Cassoux N, Piette JC, LeHoang P (2000) Efficacy of interferon alfa-2a in severe and refractory uveitis associated with Behcet's disease. Ocul Immunol Inflamm 8:293–301
42. Yoshida A, Kawshima H, Motoyama Y, Shibui H, Kaburaki T, Shimizu K, Ando K, Hijikata K, Izawa Y, Hayashi K, Numaga J, Fujino Y, Masuda K, Araie M (2004) Comparison of patients with Behcet's disease in the 1980's and 1990's. Ophthalmology 111:810–815

26 Hypertensive Retinopathies
26.1 General Basics of Hypertensive Retinopathy

S. WOLF

Core Messages

- Hypertensive retinopathy is an acquired bilateral disease in patients with arterial hypertension
- The clinical course is often only recognized after onset of visual symptoms
- It is characterized by narrowing of the retinal arterioles, cotton-wool spots, hard exudates, retinal hemorrhages, and optic disk edema indicating severe parenchymal changes in the retina
- Treatment is the reduction of the elevated systemic blood pressure

The classification of fundus changes secondary to arterial hypertension has been used for monitoring the severity of vascular alterations. In the past, this was the only possibility to assess the status of the microcirculation. Therefore, very detailed classifications have been developed. These classifications are described below. The detailed classification of hypertensive vascular fundus changes may be still important for scientific reasons. However, for the clinical management of systemic hypertension today these detailed classifications are no longer necessary. Nevertheless, the separation between minor vascular changes in systemic hypertension (stage I and II) and hypertensive retinopathy (stage III and IV) is still very important for the management of arterial hypertension. Patients with hypertensive retinopathy have to be treated and monitored intensively by internal medicine since these patients are at high risk for cardiovascular and cerebrovascular complications.

26.1.1 Pathophysiology of the Retinal Vessels in Arterial Hypertension

- Constriction of the arterial vascular tract leads to hypertension due to increased vascular resistance.
- Increased cardiac output produces hypervolemic hypertension.
- Vasospastic arteriolar stenosis on the fundus is a characteristic finding in hypertension due to increased vascular resistance. Examples of this include hypertension in toxemia of pregnancy, acute glomerulonephritis, or pheochromocytoma.

- The normal caliber of the arterioles in the fundus is two-thirds that of the venules. Generalized vascular constriction is recognizable by a narrowed column of blood.
- Long-standing arterial hypertension over a period of months to years leads to organic vascular changes similar or identical to those in arteriosclerosis. Arterial hypertension is generally regarded as an important causative factor in arteriosclerosis. Findings in chronic hypertension include not only narrowed caliber of the arterioles but also wide, bright reflexes on the arteries.

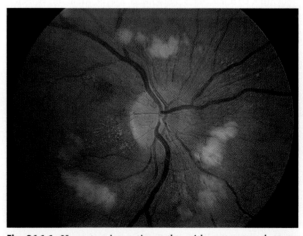

Fig. 26.1.1. Hypertensive retinopathy with cotton-wool spots, dilated capillaries, hard exudates, retinal hemorrhages, and optic disk swelling

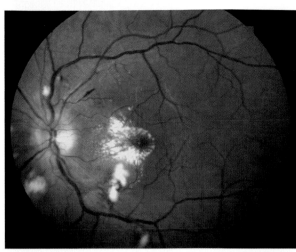

Fig. 26.1.2. Hypertensive retinopathy with macular star of lipid deposition in the macula, retinal hemorrhages, and cotton-wool spots

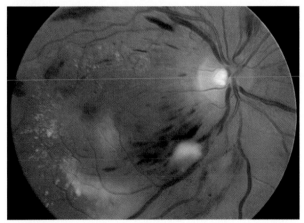

Fig. 26.1.3. Hypertensive retinopathy with macular edema, lipid deposition around the macula, retinal hemorrhages, and cotton-wool spots

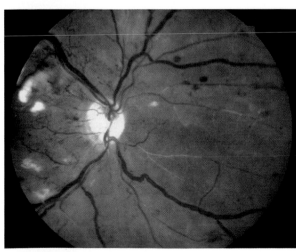

Fig. 26.1.4. Hypertensive retinopathy with occluded retinal arteries, retinal hemorrhages, capillary dilatation, and cotton-wool spots

26.1.2 Fundus Changes in Arterial Hypertension

- The fundus is well perfused, glistening, and moist
- The vasculature tends to be tortuous
- Arterioles will be narrow with bright reflexes; capillaries may be dilated on the optic disk and in the vascular arcades
- The margins of the optic disk become blurred and its parenchyma becomes somewhat hyperemic
- Other parenchymal changes include minor retinal bleeding, specifically spot hemorrhages or flame shaped hemorrhages in the nerve fiber layer

26.1.3 Fundus Changes in Hypertensive Retinopathy

Severe parenchymal changes in the retina in arterial hypertension often manifest themselves in the vascular arcades as cotton-wool spots and as hard exudates.

Hard exudates may appear like focal calcifications around the macula, forming a macular star. Severe hypertensive retinopathy may be characterized by edema in the central retina and optic disk edema, which may be as pronounced as papilledema. As hypertensive retinopathy is invariably bilateral, a differential diagnosis should consider a cerebral mass or primary elevated cerebrospinal fluid pressure (cerebral pseudotumor). Often patients are subjected to extensive examinations in an attempt to confirm the latter diagnosis before the physician considers measuring the blood pressure.

26.1.4 Clinical Diagnoses in Hypertensive Retinopathy

- Glomerulonephritis
- Chronic renal insufficiency
- Decompensated arterial hypertension
- Toxemia of pregnancy
- Pheochromocytoma
- Treatment: management of the arterial hypertension

26.1.5 Classification of Fundus Changes in Arterial Hypertension

- **Stage I**
  - Normal or widened arteriolar caliber
  - Brighter reflexes on arterioles
  - Vascular distension and tortuosity
  - No parenchymal changes
  - Arteriosclerosis of variable severity
  - Hypervolemic hypertension

- **Stage II**
  - Arterioles: generalized narrowing in caliber
  - Circumscribed areas of narrowed caliber
  - Increased and irregular reflexes
  - Paramacular venules distended and tortuous – sign of venous stasis
  - Isolated capillaries visible (capillary ectasia) in central retina and on the optic disk
  - Hyperemia of the optic disks
  - Fine retinal hemorrhages

- **Stage III: Hypertensive retinopathy**
  - Generalized narrowing of the arterioles
  - Arterioles appear thin and threadlike in places
  - Segmental or rosary-like constrictions
  - Bright and irregular reflexes
  - Siegrist's streaks and obliterated vessels
  - Cotton-wool spots
  - Hard exudates
  - Macular star
  - Retinal hemorrhages
  - Optic disk edema

- **Stage IV: Severe hypertensive retinopathy**
  - Changes as in stage III covering the entire fundus
  - Bilateral severe papilledema
  - Retinal edema
  - Exudative retinal detachment

26.2 Pregnancy-Induced Hypertension (Preeclampsia/Eclampsia)

T.R. Klesert, A.P. Schachat

Core Messages

- Pregnancy-induced hypertension (PIH) is a multisystem disorder affecting 6–8% of all pregnancies
- PIH is a major cause in both developed and developing countries of maternal and fetal morbidity and mortality
- Ocular involvement is common in PIH, but permanent visual loss is rare
- PIH affects both the retinal and choroidal vasculature, with clinical findings resembling hypertensive retinopathy
- No specific treatment is indicated for PIH-related retinopathy, which generally resolves soon after delivery of the fetus. Care consists of appropriate management of the underlying systemic disorder by a qualified obstetrician

26.2.1 History

Preeclampsia/eclampsia, also long known as "toxemia of pregnancy," has been recognized as a clinical entity for thousands of years. The danger of seizures during pregnancy was recorded in the literature as far back as ancient Egypt and China. Eclampsia is alluded to in the pre-Hippocratic *Coan Prognosis*, which reads "In pregnancy, drowsiness and headache accompanied by heaviness and convulsions, is generally bad." Indeed the word "eclampsia" derives from the Greek word "eklampsis," which means "shining forth." Prior to the 18th century, the term eclampsia was used only to refer to the visual phenomena that were known to accompany seizures during pregnancy. It was not until the 19th century – when the resemblance of eclamptic patients to those with nephritis prompted a London physician to check the urine of 14 pregnant patients suffering from blurred vision, seizures, edema and headache – that the full clinical syndrome was recognized (reviewed by Purkerson and Vekerdy [38]).

Retinal changes related to hypertension have been recognized since the early 19th century. The first report in the literature of retinal changes in association with preeclampsia was written by the great German ophthalmologist Albrecht von Graefe, who described serous retinal detachments in a patient with toxemia of pregnancy [52]. The year 1921 saw the first large ophthalmologic case series of toxemia of pregnancy published by Schiotz, who reported on 680 hospitalized patients with toxemia of pregnancy. Of these patients, 35 were found to have funduscopic changes, and 3 were found to have retinal detachments.

26.2.2 Hypertension in Pregnancy

Essentials

- Hypertension is the most common medical disorder during pregnancy
- Hypertension during pregnancy can be classified into one of four categories: chronic hypertension, gestational hypertension, preeclampsia/eclampsia, and superimposed preeclampsia

Hypertension is the most common medical disorder during pregnancy, affecting 6–8% of all pregnancies [39]. Until recently, the diagnostic criteria delineating the various subtypes of hypertensive disease in pregnancy had not been well defined or standardized [24]. Since 2000, uniform diagnostic criteria have emerged through the cooperation of international working groups [7]. Currently, hypertension during pregnancy is classified into 4 categories: chronic hypertension, gestational hypertension, preeclampsia-eclampsia, and preeclampsia superimposed on chronic hypertension (Table 26.2.1).

Chronic hypertension is defined as hypertension which was preexisting prior to pregnancy, or which

Table 26.2.1. Classification of hypertension during pregnancy. (Adapted from [10])

26 III

I. Gestational hypertension
Hypertension with onset after 20 weeks gestation in previously normotensive patient

Mild hypertension:
140 mm Hg ≤ systolic < 160 mm Hg or
90 mm Hg ≤ diastolic < 110 mm Hg

Severe hypertension:
Systolic ≥ 160 mm Hg or
Diastolic ≥ 110 mm Hg

II. Gestational proteinuria

Mild proteinuria: ≥ 1+ on dipstick and > 300 mg/24 h but < 5 g/24 h

Severe proteinuria: > 5 g/24 h

III. Preeclampsia (gestational hypertension + proteinuria)

Mild preeclampsia:
Mild hypertension + mild proteinuria

Severe preeclampsia:
Severe hypertension + mild proteinuria
Mild hypertension + severe proteinuria
Persistently severe cerebral or visual symptoms
Thrombocytopenia (< 100,000/mm³)
Pulmonary edema
Right upper quadrant pain or epigastric pain
Oliguria (< 500 cc/24 h)

Eclampsia:
Preeclampsia + convulsions

IV. Chronic hypertension
Hypertension before pregnancy
Hypertension before 20 weeks gestation

V. Superimposed preeclampsia
Chronic hypertension with new onset proteinuria

---

develops before 20 weeks gestation. Occasionally, chronic hypertension can be mistaken for **gestational hypertension** (new-onset hypertension which first develops *after* 20 weeks gestation) because the normal 10–15 mm Hg fall in physiologic blood pressure that occurs during the second trimester may temporarily bring pressures into the normotensive range. In gestational hypertension, blood pressure usually returns to normal soon after delivery. Hypertension that persists for more than 6 weeks postpartum therefore supports the diagnosis of chronic hypertension [10]. The diagnosis of hypertension requires two independent blood pressure measurements which are at least 6 h apart, but not more than 1 week apart.

Preeclampsia is gestational hypertension with an associated onset of proteinuria, defined as at least 300 mg/24 h. Approximately 46% of patients with gestational hypertension will develop preeclampsia. The severity of preeclampsia is determined based on the degree of hypertension, amount of proteinuria, and the involvement of other organ systems as manifested by associated signs and symptoms. Aside from ocular and visual manifestations, other signs and symptoms of severe preeclampsia include: convulsions (a condition called **eclampsia**), other CNS disturbances including stroke, anasarca, oliguria and renal failure, pulmonary edema, right upper quadrant or epigastric pain, liver failure, and thrombocytopenia. **HELLP** syndrome is a particularly severe form of preeclampsia with a rapidly progressive course. HELLP is an acronym for Hemolysis, Elevated Liver enzymes, and Low Platelets, and may be associated with normal or only minimally elevated blood pressure.

Women with chronic hypertension have an increased risk of developing preeclampsia, which is termed **superimposed preeclampsia** to distinguish it from traditional preeclampsia, although it is not a clearly distinct clinical entity. Superimposed preeclampsia is marked by the onset of proteinuria after 20 weeks gestation in a chronically hypertensive patient, and often involves an exacerbation of the preexisting hypertension.

26.2.3 Systemic Complications of PIH

Essentials
- PIH is a major cause of maternal and fetal morbidity and mortality
- Risk of complications for mother and fetus is related to disease severity

While mild gestational hypertension carries little increased risk to the health of the mother or fetus, severe gestational hypertension and preeclampsia cause significant morbidity and mortality. The risk of adverse maternal and fetal outcomes in developed countries (Table 26.2.2) is dependent on several factors, including: severity of disease, gestational age at the time of onset, presence of other systemic diseases, and quality of management. The worse outcomes are seen in those women developing severe disease before 33 weeks gestation. Mild disease developing after 36 weeks gestation generally portends a favorable prognosis for both mother and baby [46].

Although death from severe preeclampsia is rare, it remains a leading cause of maternal mortality in developed countries (10–15% of maternal deaths). Other serious complications for the mother include seizure, stroke, encephalopathy, disseminated intra-

Table 26.2.2. Maternal and fetal complications of severe preeclampsia. (Adapted from [46])

I. Fetal complications
Preterm delivery (15–67%)
Intrauterine growth restriction (10–25%)
Perinatal death (1–2%)
Hypoxia-induced CNS injury (<1%)
Long-term complications of preterm birth

II. Maternal complications
Disseminated intravascular coagulopathy (DIC)/HELLP
(10–20%)
Pulmonary edema/aspiration (2–5%)
Acute renal failure (1–5%)
Abruptio placentae (1–4%)
Liver failure or hemorrhage (1%)
Eclampsia (<1%)
Stroke (rare)
Death (rare)
Long-term cardiovascular complications

26.2.4 Ocular and Neurologic Manifestations

Essentials
- PIH affects both the retinal and choroidal circulation
- Retinal vascular manifestations of PIH resemble hypertensive retinopathy, and are rarely associated with permanent loss of vision
- Serous retinal detachments are a reflection of choroidal ischemia, and may occur in up to 1/3 of patients with severe preeclampsia or eclampsia.
- Cerebral vasospasm and edema in the occipital lobes can lead to transient cortical blindness in PIH.

III 26

vascular coagulopathy (DIC), pulmonary edema, left ventricular failure, acute renal failure, liver failure, placental abruption, and aortic dissection. In addition, recent studies have suggested an increased risk of coronary artery disease and cerebrovascular disease later in life for women with a history of PIH [25, 53].

Fetal complications of PIH include preterm delivery (with its associated short-term and long-term complications), intrauterine growth restriction, hypoxia-related CNS injury, and perinatal death.

Visual disturbances are common in PIH, and have long been associated with the disease. The most common complaint is blurred vision. Other less frequent symptoms include scotomata, diplopia, photopsia, amaurosis, and chromatopsia. Visual symptoms are present in up to 25% of patients with preeclampsia, and 50% of patients with eclampsia [12].

The most common ocular finding associated with PIH is a retinopathy resembling hypertensive retinopathy (Fig. 26.2.1a). The earliest changes are focal constriction or spasm of the retinal arterioles. As the

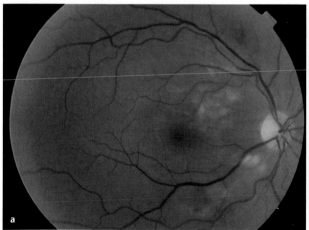
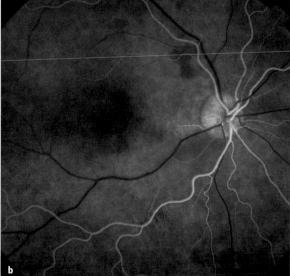

Fig. 26.2.1. Chorioretinopathy in a 22-year-old Caucasian female with preeclampsia, 1 week after child birth. (Courtesy of Homayoun Tabandeh, MD). **a** Shallow serous detachment involving the central macular region, extending toward the optic disk. Multiple cotton-wool spots, superficial hemorrhages, and yellow-white focal lesions at the level of the RPE are present. **b** Fundus fluorescein angiogram indicating patchy areas of choroidal non-perfusion, and abnormal early fluorescence at the level of the RPE

26 III

condition progresses, this narrowing may become generalized. In early studies of preeclampsia, the incidence of focal retinal arteriolar abnormalities was reported to be anywhere from 30% to 100%. More recent studies suggest a much lower incidence of 5% in preeclampsia and 30% in pregnant women with chronic hypertension [42, 43]. The lower incidence is most likely attributable to improvements in the treatment of PIH, leading to a reduced incidence of more severe systemic disease. Vision loss secondary to retinal arteriolar spasm in PIH is rare, but has been reported [15, 47].

Other retinal manifestations of systemic hypertension, such as cotton-wool spots, intraretinal hemorrhages, retinal edema and optic nerve edema are usually seen only in more severe cases of PIH, and their presence in a patient with only mild preeclampsia should immediately raise suspicion for other coexisting systemic conditions, such as chronic hypertension and diabetes [26].

Less commonly in PIH, the retinal vascular system may be affected by vascular occlusive events. A Purtscher's-like retinopathy developing after childbirth has been reported in patients with preeclampsia, and may be caused by complement-activated leukoembolus formation [4, 45]. Central retinal vein occlusion occurring 10 days post-partum in a patient with HELLP syndrome has been reported [18]. Transient unilateral vision loss, presumably from a thrombotic event, has also been reported in two patients with preeclampsia and antiphospholipid antibodies [6]. Finally, bilateral peripheral retinal neovascularization has been seen following PIH, which was attributed to microthrombus formation [5].

Choroidal involvement in PIH is marked by yellow-white focal lesions at the level of the retinal pigment epithelium (RPE), serous retinal detachment, and Elschnig's spots (small, isolated areas of hyperpigmentation with surrounding yellow or red halos). The serous detachments are often bullous, and usually bilateral (Fig. 26.2.1A). Although earlier studies estimated the incidence of serous retinal detachment at 1% in severe preeclampsia and 10% in eclampsia, a recent study by Saito and Tano [41] suggests a much higher incidence. The authors examined 71 women with severe preeclampsia or eclampsia within a few days of admission to the hospital, and found an incidence of serous retinal detachment of 32%. Because 72% of these detachments resolved within one week of initial examination, the authors suggested that prior studies underestimated the true incidence of retinal detachment, because patients were not examined within this short window of time.

Fluorescein angiography indicates that the choroidal manifestations of PIH are most likely a result of choroidal ischemia and infarction [31, 41]. Delayed choroidal filling is seen in areas corresponding to the yellow-white RPE lesions, the shape and distribution of which reflect the lobular pattern of the choriocapillaris (Fig. 26.2.1B). Subretinal leakage of fluorescein is also observed. The limited histopathologic evidence suggests that Elschnig's spots represent cicatricial change from infarction of the RPE and choriocapillaris [27].

Involvement of the occipital cortex in PIH can lead to transient cortical blindness. Complete recovery of vision usually occurs within one week, even in cases with complete loss of light perception. Computed tomography (CT) imaging typically shows multiple low density, nonenhancing lesions in the occipital lobe, which are felt to represent areas of decreased perfusion secondary to arterial vasospasm or cerebral edema. Magnetic resonance (MR) imaging will show focal areas of increased signal. Transient diplopia from ischemic cranial neuropathies have also been reported [36].

26.2.5 Pathophysiology and Epidemiology

Essentials
- PIH is a multisystem disorder of unknown etiology
- The angiogenic factors VEGF and PlGF may play important roles in disease pathogenesis through their effects on vascular endothelial function
- PIH-associated retinopathy results from the compensatory responses of the retinal and choroidal vasculature to elevated blood pressure, and ultimately, the failure of those compensatory systems.

PIH is a multisystem disorder of unknown etiology. At the simplest level, preeclampsia is a maternal response to placentation. A final common pathway in disease pathogenesis appears to be systemic endothelial dysfunction, manifested as increased vascular permeability, platelet aggregation, enhanced vascular sensitivity to angiotensin II and norepinephrine, and decreased production and activity of the vasodilators prostacyclin and nitric oxide [46].

Recent studies have suggested that soluble angiogenic factors may play an important role in the pathogenesis of endothelial cell dysfunction in PIH. In normal pregnancy, vascular endothelial growth factor (VEGF) and placental growth factor (PlGF) are released by uterine natural killer cells, resulting in significant increases in the maternal circulating levels of these factors. Increased levels of VEGF and

PlGF are thought to help maintain a quiescent endothelial state under the increased inflammatory and hydrodynamic stress of pregnancy [46]. Placental-derived sFlt1, a soluble antagonistic receptor of VEGF and PlGF, appears to be upregulated in preeclamptic patients, and elevated circulating sFlt1 levels correlate with lower circulating levels of VEGF and PlGF [31]. It is hypothesized that lower concentrations of VEGF and PlGF contribute to systemic endothelial dysfunction. Supporting this hypothesis, levels of circulating sFlt1 do correlate with disease severity in PIH [9, 30].

Numerous theories have been proposed to explain the root causes of PIH. Important clues may be provided by the unique epidemiologic features of the disease (Table 26.2.3). Two leading theories are the immune theory and the genetic-conflict theory. The immune theory considers PIH as a maternal immune maladaptation to foreign fetal antigens derived from the paternal sperm. In this theory, exposure to paternal sperm over time enhances maternal immune tolerance, and previous gestations with a single partner increase the tolerance to subsequent gestations from the same partner. This could explain why the incidence of preeclampsia is higher in teenage mothers (limited sperm exposure) and nulliparous mothers. It might also explain why there is an increased risk of preeclampsia in multiparous mothers who change partners [50]. Epidemiologic studies of the relationship between length of cohabitation (and thus sperm exposure) prior to pregnancy and risk of preeclampsia have had conflicting findings, however [13, 22, 35].

In the genetic-conflict theory, PIH is a consequence of the natural evolutionary conflict between the competing interests of fetal (paternal) genes and maternal genes during pregnancy. Evolution should select for fetal genes that maximize transfer of nutrients across the placenta, whereas the selection pressure for maternal genes is the limitation of transfer of

nutrients beyond some maternal optimum [21]. Thus, evolution would favor fetal genes that would raise maternal blood pressure, and thereby placental perfusion. These genes would be in conflict with maternal genes that act to limit the maternal blood pressure. Genomic imprinting is a well-established phenomenon in mammals by which certain genes are selectively expressed only from the maternally or paternally inherited chromosomes. Genomic imprinting is believed to have evolved as a manifestation of the maternal-fetal conflict, and many imprinted genes isolated thus far do play a role in growth and development. In the genetic-conflict theory, then, one might predict a role for imprinted genes in the pathogenesis of PIH. In this regard, Oudejans et al. [37] have identified a preeclampsia susceptibility locus on chromosome 10q22.1 containing genes expressed in the placenta that show evidence of imprinting.

The retinal vascular changes associated with PIH can be understood in terms of the regulatory mechanisms that control blood flow through the eye. When blood pressure becomes elevated, retinal arterioles constrict to increase vascular resistance, and thereby maintain relatively steady perfusion to the retinal tissue. Prolonged high blood pressure in chronic hypertension can lead to permanent arteriolar narrowing. Because the retinal vessels have no sympathetic innervation, this vasoconstriction is controlled by autoregulatory mechanisms. When the degree of hypertension exceeds the capacity of the vessels to autoregulate, the system fails and the capillary bed is exposed to elevated pressures. If prolonged, this leads to occlusion of terminal arterioles, capillary nonperfusion, retinal ischemia, cotton wool spots, hemorrhages and retinal edema [36].

Choroidal vessels do have sympathetic innervation, which likewise stimulates vasoconstriction in response to hypertension. If the blood pressure exceeds the capacity of the sympathetic system to regulate perfusion, damage to choroidal vascular bed may result, leading to choroidal occlusion and ischemia, ischemia to the overlying retinal pigment epithelium and outer retina, exudative retinal detachment, and long term pigmentary changes (Elschnig's spots).

Table 26.2.3. Risk factors for PIH. (Adapted from [2], American College of Obstetricians and Gynecologists, ACOG Tech Bull 219)

Factor	Relative risk ratio
Chronic renal disease	20:1
Chronic hypertension	10:1
Antiphospholipid syndrome	10:1
Family history	5:1
Twin gestation	4:1
Obesity	3:1
Nulliparity	3:1
Age < 15 years or > 40 years	3:1
Diabetes mellitus	2:1
Change in partners (multiparas)	1.5:1
African-American race	1.5:1

26.2.6 Diagnostic Evaluation

Essentials
- The role of ancillary diagnostic testing in PIH, such as indocyanine green (ICG) angiography and fluorescein angiography, is limited.

26 **III**

- Although fluorescein and ICG are probably safe for use during pregnancy, confirmatory epidemiologic evidence is lacking. Therefore, their use should be limited to conditions in which the results would affect treatment decisions. A careful discussion with the patient about the risks, benefits and alternatives is necessary and informed consent should be obtained.

PIH has a generally benign natural history. A complete resolution of visual symptoms and abnormal fundus findings soon after delivery of the fetus occurs in the vast majority of patients. For this reason, extensive diagnostic evaluation beyond simple documentation of the funduscopic findings by chart drawing or fundus photography is probably not warranted.

There are no data in the literature to suggest that fluorescein angiography poses a significant risk to the fetus: increased rates of birth defects or fetal loss have not been observed in animal studies or reported in humans. However, sodium fluorescein does cross the placenta and enter the fetus, and no large epidemiologic studies have been published to confirm the safety of fluorescein angiography during pregnancy. In a survey of 424 retinal specialists, Halperin et al. [23] found that 77% had never knowingly performed fluorescein angiography on a pregnant woman. As part of this study, the authors were able to gather outcome data from 116 pregnant women on whom fluorescein angiography had been performed. Four fetal or neonatal deaths were reported, 2 of which could clearly be attributed to other causes. Birth anomalies were reported in 2 children (undescended testicle and syndactyly). The authors concluded that there was no significantly increased risk of birth anomaly or fetal demise associated with the use of fluorescein angiography during pregnancy. Given the limited data, however, fluorescein angiography during pregnancy should probably be reserved for those conditions in which the results would affect treatment decisions, such as juxtafoveal choroidal neovascularization.

Evidence supporting the safety of indocyanine green (ICG) angiography during pregnancy is stronger. ICG does not cross the placenta at any measurable level, and ICG has been used extensively for non-ophthalmologic applications on pregnant women (in the study of hepatic blood flow and cardiac output), without any reported adverse maternal or fetal effects. Nonetheless, there remains widespread hesitation by retinal specialists to perform ICG angiography on pregnant patients. Of 520 retinal specialists surveyed by Fineman et al. [14], only 24% felt ICG was safe to use during pregnancy. Of the remaining respondents, 15% felt ICG was unsafe, and 60% were unsure.

It should be noted that both fluorescein and ICG are classified by the Food and Drug Administration as pregnancy category C, which means that safety during pregnancy is uncertain, due to insufficient data.

26.2.7 Treatment

Essentials
- Specific treatment for the ocular manifestations of PIH is not generally indicated.
- Systemic treatment of PIH consists of antihypertensive therapy, magnesium sulfate, and early delivery of the fetus when indicated

As permanent visual loss is rare in PIH, specific treatments for the ocular manifestations of PIH are not generally indicated. Rather, treatment is directed at the underlying systemic disease, and patients with PIH should be under the care of a qualified obstetrician. Perhaps the most important role of the ophthalmologist is to recognize the ocular manifestations of PIH, so that the rare undiagnosed patient can be referred in a timely manner to the appropriate specialists. More commonly, the ophthalmologist is consulted to evaluate the vision complaints of a patient with a preexisting diagnosis of PIH. In these cases, the primary role is to provide reassurance to the patient that the vision will likely return to normal after delivery of the fetus.

The cure for PIH (with rare exceptions – [32]) is delivery of the fetus. In severe preeclampsia, preterm delivery by induction of labor or C-section must be considered. The obstetrician must weigh the risks to the health of the mother and fetus of continuing the pregnancy, against the risk to the fetus from the complications of prematurity. In general, severe preeclampsia is an indication for preterm induction of labor once the pregnancy reaches 34 weeks of gestation. Those with mild preeclampsia are induced at 38 weeks [46].

In the past, the degree of retinal vascular changes associated with preeclampsia was felt to be predictive of fetal mortality [17, 40] and was sometimes used as an indication for early delivery of the fetus [12]. In a prospective, controlled and masked study of 31 patients with preeclampsia and 25 control patients, Jaffe and Schatz [26] found that it was indeed possible to distinguish patients with severe preeclampsia from those with no disease or mild disease based solely on abnormal fundoscopic findings, which consisted of focal arteriolar constrictions, and a reduction in the arteriole-to-venule ratio. Because the diagnosis of severe preeclampsia was already clinically obvious in these patients, however, the authors argued that the role of the ophthalmologist in guiding the management of preeclampsia is limited. Moreover, the study found that it was not possible reliably to distinguish patients with mild preeclampsia from those without disease. Therefore, the role of the ophthalmologist in routine screening for the disease is likewise limited. The authors have recommended that obstetricians measure Snellen acuity or perform Amsler grid testing as basic screening tests to identify those patients that need an ophthalmology consultation. Of course, the sensitivity and specificity of these screening measures is not clear, so perhaps the best advice is that visually symptomatic patients be examined and that patients who would have management altered if certain eye findings were seen be referred as well.

The mainstay of treatment for PIH is antihypertensive therapy. Because mild to moderate gestational hypertension has no clear effect on maternal or fetal outcomes, and because antihypertensive treatment does not reduce the risk of progression to preeclampsia, treatment is not generally indicated for mild to moderate gestational hypertension [1]. Treatment of mild gestational hypertension may also increase the risk for a small-for-gestational-age baby [51].

Severe gestational hypertension is associated with maternal and fetal morbidities that more closely resemble severe preeclampsia than mild preeclampsia. Therefore, it is recommended that both patients with severe gestational hypertension and patients with severe preeclampsia be admitted to the hospital for bedrest, monitoring and antihypertensive therapy (Table 26.2.4). Parenteral hydralazine, labetalol, sodium nitroprusside and short-acting oral nifedipine are the most common first-line medications for management of acute hypertension in pregnancy. For chronic management of PIH, oral methyldopa, labetalol, nifedipine, and thiazide diuretics are the drugs of choice. Furosemide, angiotensin-converting enzyme inhibitors (ACE inhibitors), and angiotensin receptor blockers (ARBs) are all contraindicated during pregnancy, but may be used in the postpartum period. Furosemide has been associated with hypospadias. ACE inhibitors and ARBs are associated with numerous fetal complications, including intrauterine demise, renal dysgenesis, oligohydramnios, pulmonary hypoplasia, fetal growth restriction and neonatal renal dysfunction [46].

Parenteral magnesium sulfate is used for the treatment of eclamptic seizures, as it has been demonstrated to be superior to both phenytoin and diazepam in preventing recurrent eclamptic seizures, and in preventing maternal and fetal death [49]. Patients

Table 26.2.4. Indications for antihypertensive therapy in pregnancy. (Adapted from [10])

I. Antepartum and intrapartum

Persistent elevations for at least 1 h:
SBP ≥ 180 mm Hg or
DBP ≥ 110 mm Hg or
MAP ≥ 130 mm Hg

Persistent elevations for at least 30 min:
SBP ≥ 200 mm Hg or
DBP ≥ 120 mm Hg or
MAP ≥ 140 mm Hg

With thrombocytopenia or congestive heart failure:
SBP ≥ 160 mm Hg or
DBP ≥ 105 mm Hg or
MAP ≥ 125 mm Hg

II. Postpartum

Persistent elevations for at least 1 h:
SBP ≥ 160 mm Hg or
DBP ≥ 105 mm Hg or
MAP ≥ 125 mm Hg

on magnesium sulfate need to be monitored closely for signs of magnesium toxicity, which include hyporeflexia, decreased mental status, slurred speech, muscular paralysis, respiratory distress and cardiac arrest.

Numerous studies have been conducted over the years to evaluate various medications and dietary modifications that might reduce the rate and/or severity of preeclampsia. The data are insufficient to recommend anything except low dose aspirin, which may provide modest reductions in the risks of preeclampsia and fetal death [28]; and calcium supplementation, which may reduce the risk for those women at high risk for preeclampsia, and for those with low dietary calcium intake [3].

26.2.8 Clinical Course and Outcomes

Essentials
- The visual symptoms and abnormal fundus findings associated with PIH generally resolve completely after delivery of the fetus
- Permanent loss of vision related to PIH is rare, but has been reported

Most women with ocular manifestations of PIH make a full recovery without permanent loss of vision or other ocular sequelae of the disease. Focal and generalized arteriolar narrowing, the most common ocular manifestations of the disease, generally

26 III

resolve after delivery of the fetus and normalization of blood pressure.

Saito and Tano [41] published a relatively large case series of 31 patients with severe preeclampsia or eclampsia, who were determined to have abnormalities in the choroidal circulation, based on the presence of yellowish opaque RPE lesions (47 eyes), and/or serous retinal detachments (40 eyes). Of the retinal detachments, 72% had completely resolved within 1 week of initial examination, and 97.5% within 3 weeks. The RPE lesions completely resolved without scarring within 3 weeks of initial examination in 83% of involved eyes. Persistent RPE mottling was observed in 8.5% of involved eyes, and localized chorioretinal atrophy was observed in 8.5% of eyes. None of the women in the study experienced any permanent measurable loss of vision. It should be noted that persistent RPE lesions from prior episodes of preeclampsia (Elschnig's spots) may be mistaken for an inherited macular dystrophy [16].

Permanent vision loss secondary to PIH has been observed in rare cases. One cause is optic atrophy secondary to retinal arterial occlusion or direct ischemia to the optic nerve [8, 15, 32, 47]. Vision loss secondary to a Purtscher's-like retinopathy developing after childbirth has also been reported in at least five patients with preeclampsia or gestational hypertension [4, 45]. This condition has also been seen in patients without preeclampsia, but with an underlying coagulopathy [19, 34].

In patients with cortical vision loss, full visual recovery is likewise the norm. Cunningham et al. [11] published a retrospective case series covering a 14-year period at Parkland Hospital, in which 15 patients with preeclampsia who developed associated cortical blindness were identified. In these patients, blindness persisted anywhere from 4 h to 8 days, but subsequently resolved in all. Persistent electroencephalographic abnormalities, however, have been measured in patients who have nonetheless experienced a full recovery of vision after cortical vision loss in PIH [20]. This suggests that persistent subclinical cortical dysfunction may occur.

Although abnormal fundoscopic, fluorescein angiographic and ICG angiographic findings usually resolve after delivery of the fetus and normalization of blood pressure, mild multifocal electroretinographic abnormalities may persist in areas with previous choroidal ischemia, despite otherwise normal measures of visual function [29]. This suggests that some degree of permanent damage to the retina can result from severe cases of PIH. The functional significance of this damage appears to be insignificant.

References

1. Abalos E, Duley L, Steyn DW, Henderson-Smart DJ (2001) Antihypertensive drug therapy for mild to moderate hypertension during pregnancy. Cochrane Database Syst Rev 1:CD002252
2. American College of Obstetricians and Gynecologists (1996) Hypertension in Pregnancy. ACOG Tech Bull 219
3. Atallah AN, Hofmeyr GJ, Duley L (2002) Calcium supplementation during pregnancy for preventing hypertensive disorders and related problems. Cochrane Database Syst Rev 1:CD001059
4. Blodi BA, Johnson MW, Gass JD et al (1990) Purtscher's-like retinopathy after childbirth. Ophthalmology 97:1654–1659
5. Brancato R, Menchini U, Bandell F (1987) Proliferative retinopathy and toxemia of pregnancy. Ann Ophthalmol 19:182–183
6. Branch DW, Andres R, Kigre KB et al (1989) The association of antiphospholipid antibodies with severe preeclampsia. Obstet Gynecol 73:541–545
7. Brown MA, Lindheimer MD, de Sweit M et al (2001) The classification and diagnosis of the hypertensive disorders in pregnancy: a statement from the International Society for the Study of Hypertension in Pregnancy (ISSHP). Hypertens Pregnancy 20:IX–XIV
8. Carpenter F, Kave HL, Plotkin D (1953) The development of total blindness as a complication of pregnancy. Am J Obstet Gynecol 66:641–647
9. Chaiworapongsa T, Romero R, Espinoza J et al (2004) Evidence supporting a role for blockade of the vascular endothelial growth factor system in the pathophysiology of preeclampsia. Am J Obstet Gynecol 190:1541–1547
10. Coppage KH, Sibai BM (2005) Treatment of hypertensive complications in pregnancy. Curr Pharm Design 11:749–757
11. Cunningham FG, Fernandez CO, Hernandez C (1995) Blindness associated with preeclampsia and eclampsia. Am J Obstet Gynecol 172:1291–1298
12. Dieckmann WJ (1952) The toxemias of pregnancy, 2nd edn. Mosby, St Louis
13. Einarsson JI, Sangi-Haghpeykar H, Gardner MO (2003) Sperm exposure and development of preeclampsia. Am J Obstet Gynecol 188:1241–1243
14. Fineman MS, Maguire JI, Fineman SW, Benson WE (2001) Safety of indocyanine green angiography during pregnancy: a survey of the retina, macula and vitreous societies. Am J Ophthalmol 119:353–355
15. Gandhi J, Ghosh S, Pillari VT (1978) Blindness and retinal changes with preeclamptic toxemia. NY State J Med 78:1930–1932
16. Gass JDM, Pautler SE (1985) Toxemia of pregnancy pigment epitheliopathy masquerading as a heredomacular dystrophy. Trans Am Ophthalmol Soc 83:114–130
17. Gibson GG (1938) The clinical significance of the retinal changes in the hypertensive toxemias of pregnancy. Am J Ophthalmol 21:22
18. Gonzalvo FJ, Abecia E, Pinilla I et al (2000) Central retinal vein occlusion and HELLP syndrome. Acta Ophthalmol Scand 78:596–598
19. Greven CM, Weaver RG, Owen J, Slusher MM (1991) Protein S deficiency and bilateral branch retinal artery occlusion. Ophthalmology 98:33–34
20. Grimes DA, Ekbladh LE, McCartney WH (1980) Cortical blindness in preeclampsia. Int J Gynecol Obstet 17:601–603

21. Haig D (1993) Genetic conflicts in human pregnancy. Q Rev Biol 68:495–532
22. Hall G, Noble W, Lindow S, Masson E (2001) Long-term sexual co-habitation offers no protection from hypertensive disease of pregnancy. Hum Reprod 16:349–352
23. Halperin LS, Olk RJ, Soubrane G, Coscas G (1990) Safety of fluorescein angiography during pregnancy. Am J Ophthalmol 110:323–325
24. Harlow FH, Brown MA (2001) The diversity of diagnosis of preeclampsia. Hypertens Pregnancy 20:57–67
25. Haukkamaa L, Salminen M, Laivuori H et al (2004) Risk for subsequent coronary artery disease after preeclampsia. Am J Cardiol 93:805–808
26. Jaffe G, Schatz H (1987) Ocular manifestations of preeclampsia. Am J Ophthalmol 103:309–315
27. Klein BA (1968) Ischemic infarcts of the choroid (Elschnig spots). A cause of retinal separation in hypertensive disease with renal insufficiency. A clinical and histopathologic study. Am J Ophthalmol 66:1069–1074
28. Knight M, Duley L, Henderson-Smart DJ, King JF (2002) Antiplatelet agents for preventing and treating pre-eclampsia. Cochrane Database Syst Rev 2:CD000492
29. Kwok AK, Li JZ, Lai TY et al (2001) Multifocal electroretinographic and angiographic changes in pre-eclampsia. Br J Ophthalmol 85:111–112
30. Levine RJ, Maynard SE, Qian C et al (2004) Circulating angiogenic factors and the risk of preeclampsia. N Engl J Med 350:672–683
31. Mabie WC, Ober RR (1980) Fluorescein angiography in toxaemia of pregnancy. Br J Ophthalmol 64:666–671
32. Matthys LA, Coppage KH, Lambers DS et al (2004) Delayed postpartum preeclampsia: an experience of 151 cases. Am J Obstet Gynecol 190:1464–1466
33. Maynard SE, Min JY, Merchan J et al (2003) Excess placental soluble fms-like tyrosine kinase 1 (sFlt1) may contribute to endothelial dysfunction, hypertension, and proteinuria in preeclampsia. Clin Invest 111:649–658
34. Nelson ME, Talbot JF, Preston FE (1989) Recurrent multiple-branch retinal arteriolar occlusions in a patient with protein C deficiency. Graefes Arch Clin Exp Ophthalmol 227:443–447
35. Ness RB, Markovic, Harger G, Day R (2004) Barrier methods, length of preconception intercourse, and preeclampsia. Hypertens Pregnancy 23:227–235
36. Ober RR (2001) Pregnancy-induced Hypertension (preeclampsia-eclampsia). In: Ryan SJ (ed) Retina, 3rd edn, vol 2. Mosby, St Louis, pp 1393–1403
37. Oudejans CB, Mulders J, Lachmeijer AM et al (2004) The parent-of-origin effect of 10q22 in pre-eclamptic females coincides with two regions clustered for genes with down-regulated expression in androgenetic placentas. Mol Hum Reprod 10:589–598
38. Purkerson ML and Vekerdy L (1999) A history of eclampsia, toxemia and the kidney in pregnancy. Am J Nephrol 19:313–319
39. Report of the National High Blood Pressure Education Program (2000) Working group report on high blood pressure in pregnancy. Am J Obstet Gynecol 183:S1–S22
40. Sadowsky A, Serr DM, Landau J (1956) Retinal changes and fetal prognosis in the toxemias of pregnancy. Obstet Gynecol 8:426–431
41. Saito Y, Tano Y (1998) Retinal pigment epithelial lesions associated with choroidal ischemia in preeclampsia. Retina 18:103–108
42. Saito Y, Omoto T, Kidoguchi K et al (1990) The relationship between ophthalmoscopic changes and classification of toxemia in pregnancy. Acta Soc Ophthalmol Jpn 94:870–874
43. Schiotz I (1921) Über retinitis gravidarum et amaurosis eclamptica. Klin Mbl Augenheilk 67:1–136
44. Schreyer P, Tzadok J, Sherman DJ et al (1990) Fluorescein angiography in hypertensive pregnancies. Int J Gynecol Obstet 34:127–132
45. Shaikh S, Ruby AJ, Piotrowski M (2003) Preeclampsia-related chorioretinopathy with Purtscher's-like findings and macular ischemia. Retina 23:247–250
46. Sibai B, Kuperminc M (2005) Pre-eclampsia. Lancet 365:785–799
47. Somerville-Large LB (1950) A case of permanent blindness due to toxemia of pregnancy. Br J Ophthalmol 34:431–434
48. Sunness JS, Santos A (1999) Retinal disease in pregnancy. In: Guyer DR, Yannijzzi LA, Chan GS, Shields JA, Green WR (eds) Retina-vitreous-macula. Saunders, Philadelphia, pp 498–513
49. The Eclampsia Trial Collaborative Group (1995) Which anticonvulsant for women with eclampsia? Evidence from the Collaborative Eclampsia Trial. Lancet 345:1455–1463
50. Trupin LS, Simon LP, Eskenazi B (1996) Change in paternity: a risk factor for preeclampsia in multiparas. Epidemiology 7:240–244
51. Von Dadelszen P, Magee LA (2000) Fall in mean arterial pressure and fetal growth restriction in pregnancy hypertension: a meta-analysis. Lancet 355:87–92
52. Von Graefe A (1855) Über eine Krebsablagerung im Innern des Auges, deren Ursprung Sitz zwischen Sclera und Choroidera war. Albrecht von Graefes Arch Klin Ophthalmol 2:214–224
53. Wilson BJ, Watson MS, Prescott GJ et al (2003) Hypertensive diseases of pregnancy and risk of hypertension and stroke later in life: results from cohort study. BMJ 326:1–7

27   Sickle Cell Retinopathy and Hemoglobinopathies

27.1 Histopathology of Sickle Cell Retinopathy

G.A. LUTTY

Core Messages
- Complications of sickle cell disease result from vaso-occlusive disease
- Packed sickle RBCs precede platelet fibrin thrombi and leukocytes
- Activation of endothelial cells is seen along with subsequent expression of adhesion molecules such as ICAM-1, VCAM-1, E-selectin, and P-selectin
- Inflammatory cytokines (TNF-α, IL1-β) are involved
- Black sunbursts represent hyperplastic retinal pigment epithelium (RPE)
- Vascular endothelial growth factor (VEGF) and pigment epithelium-derived factor (PEDF) are involved in proliferative disease

27.1.1 Introduction

Sickle cell hemoglobinopathies all share the common feature of an abnormal globin chain, which leads to sickling of erythrocytes and obstruction of the microcirculation. Sickle vaso-occlusive events are insidious and affect virtually every vascular bed in the eye, often with visually devastating consequences. Vaso-occlusion most profoundly affects the retina, the light-sensitive tissue that lines the inside wall of the posterior aspect of the eye, because it is exquisitely sensitive to deprivation of oxygen. Even temporary vaso-occlusion, if longer than about 1.5–2 h, can result in permanent infarction of the retina. Most, if not all, of the complications of sickle cell disease in retina originate from the vaso-occlusive processes. The pathological changes can be divided into nonproliferative and proliferative events.

27.1.2 Nonproliferative Changes

27.1.2.1 Retinal Vessel Occlusions and Remodeling

Retinal occlusive events occur first and most often in the peripheral retina and only rarely cause loss of peripheral or side vision because of the extremely far peripheral location of the obstructed vessels, where the retina has little important function. Occlusions of the peripheral retinal microvasculature have been documented in HbSS subjects as early as 20 months of age (Fig. 27.1.1) [29]. Although retinal capillaries

and precapillary arterioles appear to be the initial site of occlusion early in life, larger-caliber vessels eventually become nonperfused with age (Fig. 27.1.2). The sites of occlusion in larger vessels are often at arteriovenous crossings as in Fig. 27.1.3. We have documented and examined sites of occlusion in retina by incubating the retinas for adenosine diphosphatase (ADPase) activity and then flat-embedding them in the transparent polymer glycol methacrylate for serial sectioning. ADPase activity is only present in viable blood vessels, so sites of occlusion can be identified en bloc and sectioned (Figs. 27.1.1–27.1.4). Where vessels and ADPase activity end abruptly, sometimes hairpin-shaped loops will form. With repeated vaso-occlusive events in the peripheral retina over a prolonged time, centripetal recession of the most peripheral vascular arcades occurs away from the ora serrata and toward the equator. The end result is a totally ischemic peripheral retina (Fig. 27.1.2) [14].

Hairpin loops are short at sites of capillary and arteriolar occlusion (Fig. 27.1.1) and longer where major vessels end (Fig. 27.1.2). One channel of the loop is the original blood vessel lumen that became occluded and the second channel is a recanalization of the original wall of the occluded segment (Fig. 27.1.4). The recanalization appears to progress until the first viable branch of the original blood vessel is encountered and this branch then becomes the efferent channel or path for blood flow. Flow through the 360° turns must be awkward and one would imagine that hairpin loops might become subsequent sites of occlusion.

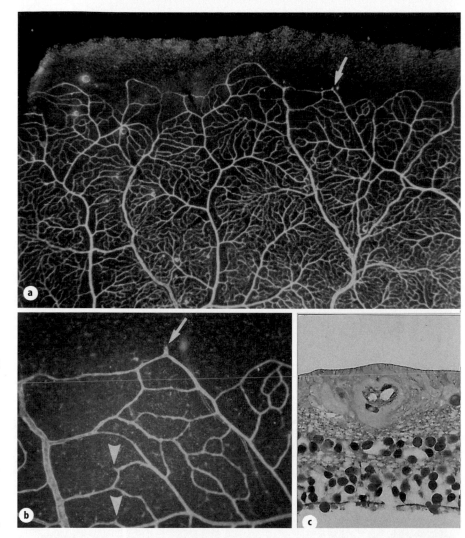

Fig. 27.1.1. ADPase incubated retina from a 20-month-old SS subject. **a** Low magnification of the far peripheral retina in which a small hairpin loop is present (*arrow*). **b** The area with the hairpin loop at higher magnification has abruptly ended capillary segments (loss in ADPase activity suggesting occlusions) (*arrowheads*). **c** Section through the hairpin loop

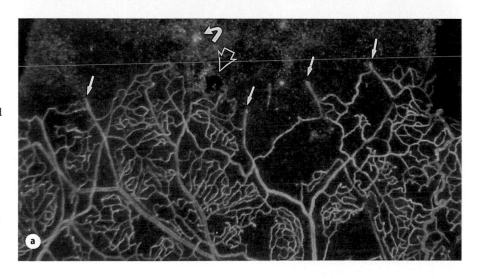

Fig. 27.1.2. Fifty-four-year-old SC disease subject with nonperfused peripheral retina (*top*). **a** Darkfield illumination of an area in peripheral retina viewed en bloc shows large blood vessels that end abruptly in hairpin loops (*straight arrows*) and arteriovenous (AV) anastomoses between occluded arteries and veins.

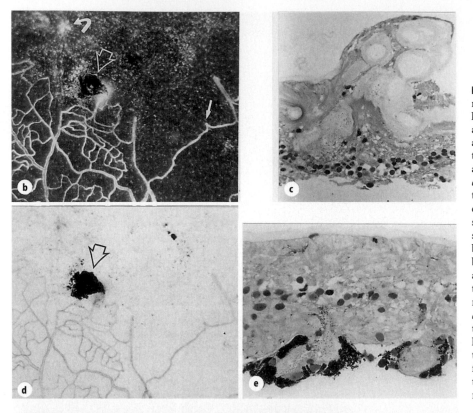

Fig. 27.1.2. b At higher magnification, a short hairpin loop (*straight arrow*) is apparent in one of the AV anastomoses. **c** A section through the *white area* in **a** and **b** indicated with a *curved arrow* shows collagenous tubes that represent the feeder vessels of an autoinfarcted sea fan. **d** When the area shown in **b** is viewed with bright field illumination, a black sunburst lesion is apparent just peripheral to the border of perfused and nonperfused retina (*open arrow* in **a, b**). **e** A section through the black sunburst lesion in **d** shows that hypertrophic RPE cells have formed acinar-like formations in this area of atrophic retina

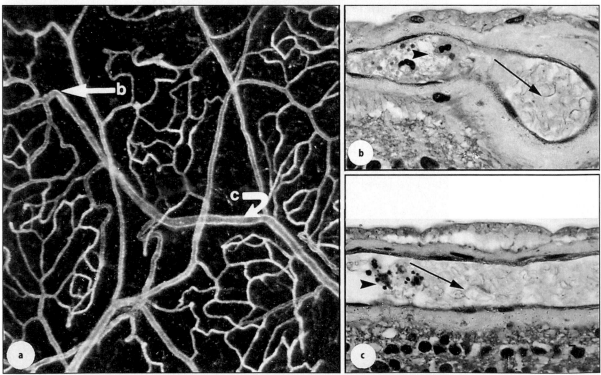

Fig. 27.1.3. Apparent occlusions in a vein of a 37-year-old HbSS subject. **a** Darkfield image of the area en bloc. Sections are shown from areas indicated. **b** Section taken at the kink in the blood vessel shows packed sickle erythrocytes downstream from a platelet fibrin thrombus. **c** Section just downstream from the AV crossing indicated in **a** also shows packed sickle erythrocytes downstream from a platelet fibrin thrombus with leukocytes within it. (**a** Darkfield illumination of ADPase incubated retina en bloc; **b, c** sections stained with toluidine blue-basic fuchsin) (Fig. 9 from [29])

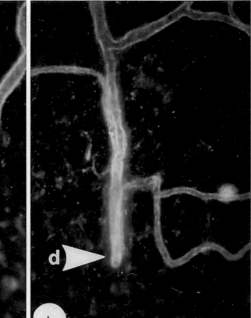

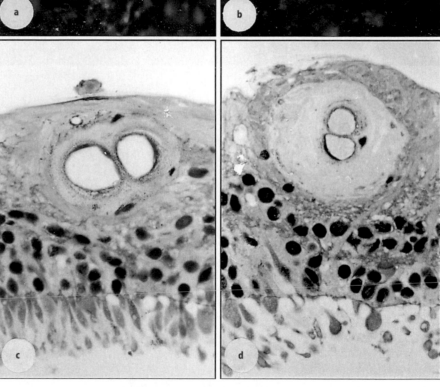

Fig. 27.1.4. Two retinal vascular hairpin loop formations in a 54-year-old SC sickle cell subject. In **a** and **b**, the ADPase activity in the loops is viewed with darkfield illumination showing that the artery (**a**) and the vein (**b**) abruptly terminate at the border of perfused and nonperfused peripheral retina. Subsequent sectioning peripheral to those structures demonstrated collagenous tubes where the original vessels were. When sectioned, it is apparent that a new channel has formed beside the original artery (**c**) in the original arterial wall. In sections of the venous loop (**d**), however, the new channel is below the original channel in the wall, which appears sclerotic (Fig. 5 from [29])

Connections form between occluded arterioles and adjacent terminal venules by way of preexisting capillaries, resulting in arteriovenous anastomoses at the border separating the vascularized and the ischemic peripheral retina (Fig. 27.1.2). The anastomoses do not show leakage on fluorescein angiography, confirming that they indeed represent enlargement of preexisting vessels with intact blood-retinal barrier properties, rather than true neovascularization [13, 31, 32].

Occlusion within a portion of a vein may exert substantial backpressure within the proximal vessel and result in focal vascular extrusion [29] (Fig. 27.1.5). This extruded vessel or loop may further enlarge as the elevated intraluminal pressure persists. Stretching of the vascular structures may lead to endothelial cell proliferation [7] and neovascularization [38].

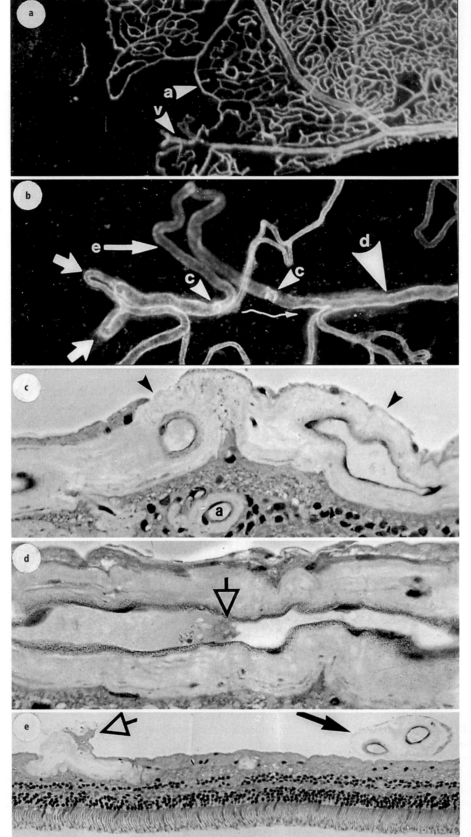

Fig. 27.1.5. Unusual vascular formation at the site of a peripheral arteriovenous anastomosis. **a** At low magnification in the flat perspective, the artery (*a*) and vein (*v*) are apparent at the border of nonperfused retina. **b** At higher magnification there are hairpin loops (*short arrow*) in both branches of a Y-shaped bifurcation. A tortuous loop appears out of focus. The original path of the vessel is shown by the *serpentine arrow*. The *arrowheads with letters* indicate where sections in **c–e** were cut from. **c** Section where the extruded loop breaches the internal limiting membrane (*arrowheads*). **d** A fibrin plug in the vein downstream from the extruded segment. **e** The extruded loop is shown on the right side of the section and another autoinfarcted preretinal formation on the left side. (**a, b** Darkfield illumination of ADPase activity; **c–e** sections stained with toluidine blue-basic fuchsin) (Fig. 6 from [29])

27.1.2.2 Causes of Vaso-occlusions

Density separation of sickle erythrocytes demonstrates that there is a heterogeneous population of erythrocytes ranging from dense irreversibly sickled cells to lightest density, younger reticulocytes [19]. Although dense cells have been traditionally implicated in vaso-occlusive events in the sickle hemoglobinopathies, there is emerging evidence that the pathophysiology of vaso-occlusion involves more than simple mechanical obstruction by rigid, dense, irreversibly sickled erythrocytes. At sites of occlusion like the arteriovenous crossing in Fig. 27.1.3, packed sickle RBCs often precede platelet fibrin thrombi and leukocytes. Leukocytes [23] and low-density circulating reticulocytes [18, 20, 28, 34] both express adhesion molecules that promote abnormal adherence to the vascular endothelium. Some young reticulocytes express integrin $\alpha_4\beta_1$ (VLA-4), enabling the cells to bind to vascular cell adhesion molecule-1 (VCAM-1) found on the surface of activated endothelial cells. Sickle reticulocytes also express the non-integrin glycoprotein IV (CD 36), which may mediate binding to vascular endothelial cells [18]. Red cell–endothelial adhesion is further promoted by the direct activation of endothelial cells, leading to the expression of adhesion molecules such as ICAM-1 (intracellular adhesion molecule-1),VCAM-1, E-selectin, and P-selectin [35]. We have observed a significant increase in polymorphonuclear leukocytes (PMNs) in sickle cell retina compared to control subjects and in the same sickle cell subjects a significant increase in ICAM-1 and P-selection, adhesion molecules responsible for PMN rolling and firm adherence respectively [23]. Sickle reticulocyte and leukocyte adherence to the vascular endothelium creates microvascular stasis, which in turn leads to increased red cell transit time, further polymerization of hemoglobin S, and complete vessel occlusion. Dense, irreversibly sickled cells do not adhere well to the vascular endothelium because they lack the adhesion molecules present on low-density reticulocytes and because their rigidity prohibits large areas of surface contact with endothelial cells; however, they can obstruct blood vessels in which reticulocytes and leukocytes are adherent.

Other research has examined the role of inflammatory cytokines, such as tumor necrosis factor-alpha (TNF-α) and interleukin-1-beta (IL1-β), which may contribute to vaso-occlusion by accelerating the production of adhesion molecules on the vascular endothelium and by activating polymorphonuclear leukocytes [10]. These cytokines may be released under conditions of stress, such as systemic infection or tissue hypoxia. Other investigators have theorized that some form of imbalance in the fibrinolytic system may also contribute to microvascular occlusions through enhanced deposition of fibrin and increased thrombin activity [9, 25]. Perhaps cytokine activation of the vascular endothelium is the critical event in leukocyte and reticulocyte adherence and in the initiation of the clotting cascade. Hematocrit also plays a role in vaso-occlusion by affecting the blood viscosity [15].

This may provide a partial explanation for the discrepancy in the severity of ocular and systemic manifestations in the various sickling hemoglobinopathies. HbSC and HbSThal subjects tend to have substantially higher hematocrits than HbSS subjects, contributing to higher viscosity with potentially more pronounced vaso-occlusion in the retinal microvasculature during any given sickling event. Even though HbSS patients have a larger number of circulating sickled red cells, their overall lower hematocrit may provide relative protection from vaso-occlusion in the small-caliber vessels of the retina [14]. An alternative theory proposes that the retinal vascular occlusions in HbSS disease may actually be so complete that total infarction and retinal necrosis occur, with no viable tissue remaining that is capable of initiating an angiogenic response. In contrast, the occlusions in HbSC disease may be less severe, resulting in chronic ischemia, but less complete infarction, and therefore with continuous secretion of angiogenic substances by the damaged tissues [12].

We propose a third theory. Using a rat model, we demonstrated that high-density HbSS erythrocytes (dehydrated dense discocytes and irreversibly sickled cells) are easily trapped in retinal capillaries and precapillary arterioles under hypoxic conditions, whereas HbSC cells (normal- and high-density cells) show very little retention in the retinal microvasculature, regardless of oxygen concentration [26]. Retention of HbSC cells does occur, however, after stimulation of the vascular endothelium with the cytokine TNF-α (unpublished data). We demonstrated further that TNF-α exposure causes SC and SS reticulocytes to adhere in retina and this is associated with VLA-4 on erythrocytes and fibronectin on retinal endothelial cells [27]. Perhaps vaso-occlusion in HbSC disease actually depends more on extra-erythrocytic factors, such as abnormalities in the fibrinolytic system, leukocyte interactions, activation of vascular endothelium, and the induction of adhesion molecule expression, than on the mechanical trapping of dense, rigid, sickled cells. This subtle difference in pathophysiology may very well provide an explanation for the discrepancy in the severity of systemic and ocular findings seen in subjects with HbSC and HbSS disease.

27.1.2.3 Hemorrhage, Schisis Cavities, Iridescent Spots, and Black Sunbursts

The occlusive processes are also associated with intraretinal hemorrhages. Salmon patch hemorrhages as they are called are round to oval areas of hemorrhage located within the superficial retina. The lesion may be up to 2 mm in diameter, with well-defined boundaries and either a flattened or a dome-shaped configuration. Such hemorrhages, which usually occur in the midperipheral retina adjacent to an intermediate arteriole, are thought to result from a blowout of vascular walls weakened by prior episodes of occlusion and ischemia. Although the hemorrhage is initially red, it may turn salmon-colored over time because of progressive hemolysis [11, 33].

After resorption of the salmon patch hemorrhage, the retina may appear entirely normal, without any evidence of residual blood. In the location of the hemorrhage, however, there may also be a faint indentation or depression representing thinning of the inner retina. This appears on ophthalmoscopy as a dimple, which may contain multiple glistening, refractile, yellowish granules, which are hemosiderin-laden macrophages. Histopathologic examination demonstrated that the granules and macrophages may be enclosed in retinoschisis cavities and both intracellular and extracellular iron may be present in the area [33].

The black sunburst lesion has also been associated with intraretinal hemorrhages [40]. These lesions appear as a flat, round to oval black patch, about 0.5–2 mm in size (Fig. 27.1.2D). Glistening, refractile granules, similar to those observed in iridescent spots, may be present. Histopathologic study discloses focal hypertrophy of the retinal pigment epithelium (RPE) along with areas of RPE hyperplasia and migration (Fig. 27.1.2E). Also present are diffuse iron deposits, hemosiderin-laden macrophages, and pigment deposition. Romayanada et al. hypothesized that the black sunburst represents intraretinal migration of hyperplastic RPE in response to blood that had dissected between the RPE and neurosensory retina [33] but the etiology of the black sunburst may be multifactorial. The black sunburst may evolve directly from a salmon patch hemorrhage depending on the plane of dissection of the resulting hemorrhage [1] as has been clearly demonstrated in a 17-year-old man with SC disease during a 6-year follow-up period [39]. However, it is interesting that hemorrhages occur often in diabetic retinopathy; yet no black sunburst-like lesion occurs in diabetic subjects. Others have observed choroidal neovascularization (CNV) within the black sunburst lesion [24, 25] and may develop following a localized choroidal occlusion [3, 41]. This suggests that RPE migration into retina may be associated with choroidal dysfunction in some cases. We have found that most black sunburst lesions are in atrophic, nonperfused regions of retina and associated with high levels of transforming growth factor beta (TGF-β), which is associated with fibrosis [25].

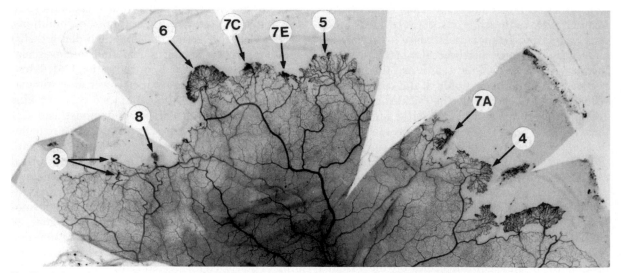

Fig. 27.1.6. Superior retina from 40-year-old sickle cell anemia subject (SS) who was incubated for ADPase activity. This image was taken when the retina was viewed with bright field illumination as a wet flat mount before embedding in glycol methacrylate, so the ADPase activity appears black. The entire peripheral retina is nonperfused (lacks ADPase+ vessels) and many neovascular structures (darkly stained blood vessels) are present at the border of perfused and nonperfused retina. The neovascular structure indicated by *arrow #6* has the appearance of a sea fan, which is the name ascribed to these structures. This structure is shown in detail in Fig. 27.1.7 (Fig. 1 from [30])

27.1.3 Proliferative Retinopathy

Peripheral retinal vascular occlusion is the initiating event in the pathogenesis of proliferative sickle retinopathy. Clinically, arteriolar closure appears to preferentially occur at or near Y-shaped bifurcations [37]. Occlusions also often occur at arteriovenous crossings [30] (Fig. 27.1.6). The occlusive process presumably produces local ischemia, which stimulates the production of vascular growth factors.

Peripheral retinal neovascularization (Fig. 27.1.7) often assumes a frond-like configuration, resembling the marine invertebrate *Gorgonia flabellum* (sea fan), but intraretinal neovascularization (IRMA) can also be found at sites of active angiogenesis (Fig. 27.1.8). The majority of the neovascular sea fan formations are found at the interface between perfused and nonperfused peripheral retina, growing toward the ischemic preequatorial retina (Fig. 27.1.6) [14, 32]. Sea fans may have multiple feeding arterioles and draining venules, probably due to their origin from multiple buds of angiogenesis that break through the internal limiting membrane of retina and grow along the surface of the retina at the vitreoretinal interface (Fig. 27.1.8) [30]. The sea fan is a dynamic neovascular formation in that the same formation may have actively growing blood vessels, established lumens ensheathed with pericytes, and autoinfarcted segments (Fig. 27.1.7). Sea fans have only vascular cells (endothelial cells and pericytes) in them and matrix between the vessels composed of collagen IV, heparan sulfate proteoglycan, and collagen II at the vitreo-sea fan border [2]. Statistically, sea fans are most commonly found in

III 27

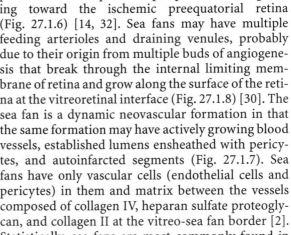
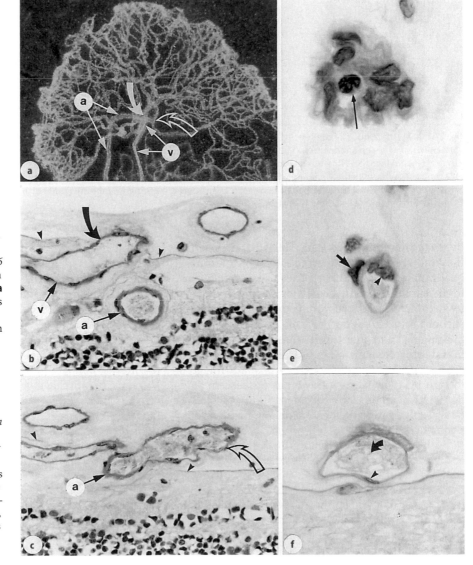

Fig. 27.1.7. Sea fan formation in ADPase-incubated retina and indicated by an *arrow #6* in Fig. 27.1.6. This formation shown in flat perspective in **a** occurred at an arteriovenous crossing. When sectioned where the vein passes through the ILM (**b**), the vein (*v*) passes internal to the artery (*a*) at the site shown by the *solid curved arrow* in **a**. The arterial feeder vessel (**c**) passes through the ILM at the site marked with an *open curved arrow* in **a**. This neovascular structure has newly forming blood vessels that resemble angioblastic masses (**d**), mature blood vessels (**e**) with endothelial cells (*arrowhead*) and pericytes (*arrow*), and autoinfarcted capillaries are only collagenous tubes (**f**) (Fig. 6 from [30])

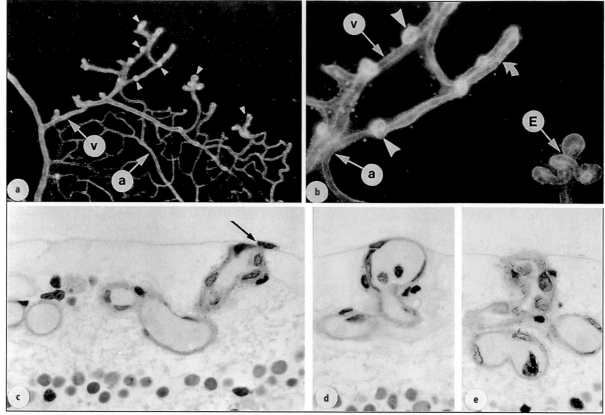

Fig. 27.1.8. Multiple buds of angiogenesis in close proximity at the border of perfused and nonperfused retina in a 20-year-old woman with sickle cell anemia (SS). The ADPase activity is always elevated in neovascularization, so the buds of neovascularization (*arrowheads*) and an IRMA formation (*arrow* in **e**) have more activity in the flat perspective (**a** and higher magnification **b**). Most buds are present in the venular part of the vasculature and both arteriole (*a*) and venule (*v*) terminate in hairpin loops (*curved arrow*). **c, d** When sectioned, the buds can be seen near internal limiting membrane (ILM) or breaching it (*arrow* in **c**). Even the IRMA-like formation (**e**) is at the ILM (Fig. 2 from [30])

the superotemporal quadrant, followed, in order, by the inferotemporal, superonasal, and inferonasal quadrants (Fig. 27.1.6) [14]. They usually are limited to the equatorial retina, rarely extending to the posterior pole. Sea fans represent true neovascular tissue and thus show profuse leakage of intravascular fluorescein dye and human serum albumen by immunohistochemistry, indicating loss of the blood-retinal barrier. The clinician can take advantage of this property by using intravenous fluorescein angiography or angioscopy to detect subtle patches of neovascularization not evident during standard ophthalmoscopy [14].

Both vascular endothelial growth factor (VEGF) and basic fibroblast growth factor (FGF-2) are associated with sea fans [2]. In a recent study, we ironically observed extremely high levels of the antiangiogenic factor pigment epithelial growth factor (PEDF) and VEGF associated with sea fans (Fig. 27.1.9) [22]. Using densitometric analysis of the reaction products, we found that the ratio of VEGF to PEDF was

reduced and shifted in the sea fans to 1:1, suggesting that the increase in VEGF alone shifted the balance toward angiogenesis [22].

As sea fans grow into the vitreous cavity, traction on their delicate vascular channels results in bleeding at irregular intervals for years. Traction-induced hemorrhage may occur in the setting of minor ocular trauma, normal vitreous movement, or contraction of vitreous bands induced by previous hemorrhage. Vitreous hemorrhage may be asymptomatic, if it remains localized to the region adjacent to the sea fan. The end result is the emergence of sea fan-shaped neovascular fronds that are predisposed to vitreous hemorrhage, subsequent tractional vitreous membrane formation, and ultimately retinal detachment [14, 17].

Vitreous bands and condensed contractile membranes are common sequelae of chronic vitreous hemorrhage and plasma transudation from incompetent neovascular tissue. They serve as key mechanical players in the progression to tractional and/or

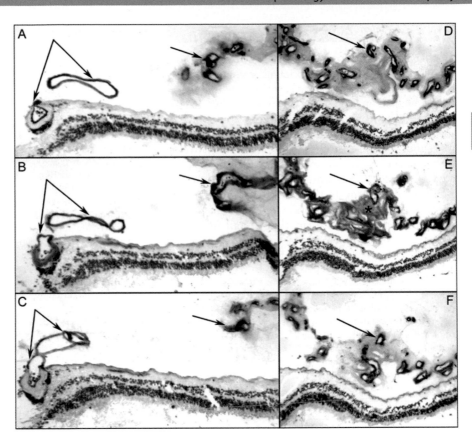

Fig. 27.1.9. PEDF and VEGF immunolocalization in retina and feeder vessel (**A–C**) and sea fan neovascular structure (**D–F**) in a 58-year-old SC disease subject. **A, B** Heparan sulfate proteoglycan perlecan immunolocalization demonstrates only viable retinal blood vessels and a feeder vessel (*double arrow*) that supplies the sea fans' viable vascular channels (*arrow* in **A, D**). PEDF immunoreactivity is present in the retinal vessels, feeder vessel, and sea fan but also is prominent in the matrix components of the sea fan (*asterisk*) (**B, E**). VEGF immunoreactivity is present predominantly in viable retinal blood vessels and the feeder vessel in **C** and viable vessels in the sea fan (**F**). (Red AEC immunoreaction reaction product and hematoxylin counterstain) (Fig. 3 from [22])

rhegmatogenous retinal detachment. As one would expect, retinal detachment is found most frequently in SC subjects [40], since these are the patients most plagued by proliferative disease. Retinal detachment is rarely present in HbAS [16] or HbSS [4, 21] subjects.

Sea fan fronds may regress over time with an eventual rate as high as 60% because of the process of autoinfarction [5, 11]. Although not completely understood, the pathophysiology of autoinfarction is probably multifactorial. In a milieu of chronic hypoxia and ischemia, multiple recurrent episodes of thromboses and sickling within sea fans may eventually lead to permanent infarction of the lesion. Kunz Matthews et al. found greatly elevated numbers of PMNs in sea fans, which was associated with elevated P-selectin, VCAM-1 and ICAM-1 [23]. Undoubtedly, PMNs contribute to autoinfarction of sea fans. We also observed high levels of PEDF in autoinfarcted sea fans while VEGF levels were very low, which could predispose the neovascular formation to regress [22]. The atrophic sea fans may slough off into vitreous leaving a stalk containing the major feeding and draining blood vessels (Fig. 27.1.2).

Major nutrient vessels may also become kinked or even avulsed by vitreous traction (Fig. 27.1.5), resulting in complete interruption of feeder vessel flow [29].

Although sea fans may autoinfarct in one region of the eye, they may continue to flourish in other regions of the same eye (Fig. 27.1.6). For this reason, most neovascular lesions are considered potentially dangerous and ordinarily should be treated and obliterated.

27.1.4 Choroidopathy

Described numerous times in patients with sickling hemoglobinopathies [6, 8, 29, 36], choroidal nonperfusion is thought to result from occlusive events in the posterior ciliary arterial circulation. As in retina, adhesion of reticulocytes with VLA-4 to endothelium may play a role [28]. Some investigators have suggested that vaso-occlusive events in the choroidal vasculature may be involved in the formation of the black sunburst lesion [3, 41]. Histopathologic features associated with choroidal vaso-occlusion include impacted erythrocytes, increased fibrin, and platelet-fibrin thrombi [25, 29].

Occlusions of the choroidal vasculature may also contribute to the orange-red or brown streaks that emanate radially in the fundus from the optic nerve called angioid streaks. These appear as crack lines in choroid and are thought to be due to calcification and brittleness of Bruch's membrane, the limiting membrane between retina and choroid.

27 III

27.1.5 Conclusions

Sickle cell vaso-occlusions are at the origin of most pathological changes in sickle cell retina and choroid. Sickle erythrocytes are certainly the major cell type involved in the occlusive process, but leukocytes and the fibrinolytic system may contribute to the process as well. Occlusions are common at arteriovenous crossings, as they are in branch vein occlusion, and they cause salmon hemorrhage and associated iridescent spots and retinoschisis cavities as well as contributing to the formation of black sunburst lesions. Vasoproliferation occurs in this ischemic retinopathy in the formation of hairpin loops, IRMA, and sea fan formations.

Acknowledgements.   The author acknowledges his collaborators D. Scott McLeod, Carol Merges, MES, Michaela Kunz-Matthews, and Jingtai Cao, who contributed substantially to the studies discussed in this manuscript. D. Scott McLeod is responsible for creating all of the figures shown and performed all of the ADPase analysis of the sickle cell retinas. This work was supported by NIH grants EY 01765 (Wilmer Institute) and HL45922 (GL), the Reginald F. Lewis Foundation (GL), and Research to Prevent Blindness (Wilmer). Gerard A. Lutty is an American Heart Association Established Investigator and the recipient of a Research to Prevent Blindness Lew Wasserman Merit Award.

References

1. Asdourian G, Nagpal KC, Godlbaum M, Patrianakos D, Goldberg MF, Rabb M (1975) Evolution of the retinal black sunburst in sickling haemoglobinopathies. Br J Ophthalmol 59:710–716
2. Cao J, Kunz Mathews M, McLeod DS, Merges C, Hjelmeland LM, Lutty GA (1999) Angiogenic factors in human proliferative sickle cell retinopathy. Br J Ophthalmol 83:838–846
3. Cogan DG (1974) Ophthalmic manifestations of systemic vascular disease. Saunders, Philadelphia, PA
4. Condon PI, Serjeant GR (1972) Ocular findings in homozygous sickle cell anemia in Jamaica. Am J Ophthalmol 73: 533–543
5. Condon PI, Serjeant GR (1980) Behaviour of untreated proliferative sickle retinopathy. Br J Ophthalmol 64:404–411
6. Condon PI, Serjeant GR, Ikeda H (1973) Unusual chorioretinal degeneration in sickle cell disease. Br J Ophthalmol 57:81–88
7. Curtis ASG, Seehar GM (1978) The control of cell division by tension or diffusion. Nature 274:52–53
8. Dizon RV, Jampol LM, Goldberg MF, Juarez C (1973) Choroidal occlusive disease in sickle cell hemoglobinopathies. Surv Ophthalmol 23:297–306
9. Francis RB Jr (1989) Elevated fibrin D-dimer fragment in sickle cell anemia: evidence for activation of coagulation during the steady state as well as in painful crisis. Haemostasis 19:105–111
10. Francis R Jr, Haywood LJ (1992) Elevated immunoreactive tumor necrosis factor and interleukin-1 in sickle cell disease. J Natl Med Assoc 84:611–615
11. Gagliano D, Goldberg MF (1989) Evolution of the salmon patch in sickle cell retinopathy. Arch Ophthalmol 107: 1814–1815
12. Gagliano DA, Jampol L, Rabb M (1996) Sickle cell disease. In: Tasman WS, Jaeger E (eds) Duane's clinical ophthalmology. Lippincott-Raven, Philadelphia
13. Goldberg MF (1971) Natural history of untreated proliferative sickle retinopathy. Arch Ophthalmol 85:428–437
14. Goldberg MF (1977) Retinal neovascularization in sickle cell retinopathy. Trans Am Acad Ophthalmol Otolaryngol 83:409–431
15. Horne MK 3rd (1981) Sickle cell anemia as a rheologic disease. Am J Med 70:288–298
16. Ishbey H, Clifforg G, Tanaka K (1958) Vitreous hemorrhage associated with sickle trait and sickle cell hemoglobin C disease. Am J Ophthalmol 45:870–879
17. Jampol LM, Green JL Jr, Goldberg MF, Peyman GA (1982) An update on vitrectomy surgery and retinal detachment repair in sickle cell disease. Arch Ophthalmol 100:591–593
18. Joneckis C, Ackley R, Orringer E, Wayner E, Parise L (1993) Integrin a₄b₁ and glycoprotein IV (CD36) are expressed on circulating reticulocytes in sickle cell anemia. Blood 82:3548–3555
19. Kaul DK, Fabry ME, Windisch P, Baez S, Nagel RL (1983) Erythrocytes in sickle cell anemia are heterogeneous in their rheological and hemodynamic characteristics. J Clin Invest 72:22–31
20. Kaul DK, Fabry ME, Nagel RL (1996) The pathophysiology of vascular obstruction in the sickle syndromes. Blood Rev 10:29–44
21. Kearney WF (1965) Sickle cell ophthalmology. NY State Med J 65:2677–2681
22. Kim SY, Mocanu C, McLeod DS, Bhutto IA, Merges C, Eid M, Tong P, Lutty GA (2003) Expression of pigment epithelium-derived factor (PEDF) and vascular endothelial growth factor (VEGF) in sickle cell retina and choroid. Exp Eye Res 77:433–445
23. Kunz Mathews M, McLeod D, Merges C, Cao J, Lutty G (2002) Neutrophils and leukocyte adhesion molecules in sickle cell retinopathy. Br J Ophthalmol 86:684–690
24. Liang JC, Jampol LM (1983) Spontaneous peripheral chorioretinal neovascularization in association with sickle cell anaemia. Br J Ophthalmol 67:107–110
25. Lutty G, Merges C, Crone S, McLeod D (1994) Immunohistochemical insights into sickle cell retinopathy. Curr Eye Res 13:125–138
26. Lutty GA, Phelan A, McLeod DS, Fabry ME, Nagel RL (1996) A rat model for sickle cell-mediated vaso-occlusion in retina. Microvascular Res 52:270–280
27. Lutty GA, Taomoto M, Cao J, McLeod DS, Vanderslice P, McIntyre BW, Fabry ME, Nagel RL (2001) Inhibition of TNFa-induced sickle RBC retention in retina by a VLA-4 antagonist. Invest Ophthalmol Vis Sci 42:1349–1355
28. Lutty GA, Otsuji T, Taomoto M, McLeod DS, Vanderslice P, McIntyre B, Fabry ME, Nagel RL (2002) Mechanisms for sickle RBC retention in choroid. Curr Eye Res 25:163–171
29. McLeod D, Goldberg M, Lutty G (1993) Dual perspective analysis of vascular formations in sickle cell retinopathy. Arch Ophthalmol 111:1234–1245
30. McLeod DS, Merges C, Fukushima A, Goldberg MF, Lutty GA (1997) Histopathological features of neovascularization in sickle cell retinopathy. Am J Ophthalmol 124:473–487
31. Nagpal KC, Goldberg MF, Rabb MF (1977) Ocular manifestations of sickle hemoglobinopathies. Surv Ophthalmol 21:391–411

32. Raichand M, Goldberg MF, Nagpal KC, Goldbaum MH, Asdourian GK (1977) Evolution of neovascularization in sickle cell retinopathy. Arch Ophthalmol 95:1543–1552

33. Romayananda N, Goldberg MF, Green WR (1973) Histopathology of sickle cell retinopathy. Trans Am Acad Ophthalmol Otolaryngol 77:652–676

34. Setty BN, Stuart MJ (1996) Vascular cell adhesion molecule-1 is involved in mediating hypoxia-induced sickle red blood cell adherence to endothelium: potential role in sickle cell disease. Blood 88:2311–2320

35. Solovey A, Lin Y, Browne P, Choong S, Wayner E, Hebbel RP (1997) Circulating activated endothelial cells in sickle cell anemia. N Engl J Med 337:1584–1590

36. Stein MR, Gay AJ (1970) Acute chorioretinal infarction in sickle cell trait. Arch Ophthalmol 84:485–490

37. Stevens TS, Busse B, Lee CB, Woolf MB, Galinos SO, Goldberg MF (1974) Sickling hemoglobinopathies; macular and perimacular vascular abnormalities. Arch Ophthalmol 92:455–463

38. Van Meurs JC (1990) Ocular findings in sickle cell disease on Curacao. Catholic University Nijmegen

39. Van Meurs JC (1995) Evolution of a retinal hemorrhage in a patient with sickle cell-hemoglobin C disease. Arch Ophthalmol 113:1074–1075

40. Welch RB, Goldberg MF (1966) Sickle-cell hemoglobin and its relation to fundus abnormality. Arch Ophthalmol 75:353–362

41. Wise GN, Dollery CT, Henkind P (1971) The retinal circulation. Harper and Row Publishers, New York, NY

27.2 Retinal Vascular Disease in Sickle Cell Patients

J.C. VAN MEURS, A.C. BIRD, S.M. DOWNES

Core Messages

- Sickle cell disease is a hemoglobinopathy caused by a hereditary mutation of the beta-chain of hemoglobin. Patients have the abnormal hemoglobin S (HbS), which makes their erythrocytes more rigid and causes sickling on deoxygenation. When patients are homozygous for HbS, we speak of HbSS disease, which carries the most severe prognosis in terms of general health; when double heterozygous for another abnormal hemoglobin (C), it is HbSC; when heterozygous for HbS and normal HbA it is called sickle cell trait (HbAS)
- Sickle cell disease can be combined with abnormalities in the synthesis of the alpha chains of hemoglobin, called the thalassemias
- The hallmark of sickle cell disease is vaso-occlusion, primarily by the abnormal erythrocytes, but also by secondary changes in endothelium and leukocytes
- Salmon patch hemorrhages, iridescent deposits and black sunbursts are early and late manifestations respectively of burst-out retinal hemorrhages and do not cause lasting visual complaints
- Proliferative sickle retinopathy (PSR) is the major cause of severe prolonged visual loss in sickle cell patients
- PSR is much more prevalent in patients with sickling disease of genotypes with a more benign general health course, such as patients with HbSC and HbS-beta-thal. It is not well understood why the peripheral retinal capillaries are particularly targeted in such patients, resulting in more extensive non-perfusion and ischemia
- Although feeder vessel treatment is effective in closing even elevated and large sea fans, it

may cause choroidal neovascularization and should only be considered with recurrent vitreous hemorrhages
- Extensive midperipheral scatter laser coagulation reduces the incidence of vitreous hemorrhage and may decrease the incidence of long-term visual loss. Evidence for the need for yearly fundus examinations and prophylactic laser coagulation in patients with PSR is not as strong, however, as in proliferative diabetic retinopathy
- Vitreoretinal surgery for non-clearing vitreous hemorrhage or retinal detachment in patients with PSR may be associated with per- and postoperative severe systemic complications (acute chest syndrome) as well as ocular ones (anterior segment ischemia). Empiric advice includes: local anesthesia instead of general anesthesia, no injury to the recti muscles, no encircling elements, prevention and early treatment of hyphema, and careful control of per- and postoperative intraocular pressure
- Retinal artery or vein occlusions may occur in patients with any sickling genotype, but remain rare and are approached as in non-sickling patients
- Sickle cell trait should not be associated with vaso-occlusive events in the retina; if so, a search for an explanatory co-morbidity should be made
- Hyphema, however, may run a severe clinical course even in HbAS carriers. Acetazolamide and mannitol may increase erythrocyte sickling; avoid these medications, but rely on topical medications such as atropine, timolol, apraclonidine, and latanopost and consider early and repeated surgical removal of the intracameral blood

27.2.1 Introduction to Sickle Cell Disease

Sickle cell disease (SCD) is clinically the most important hemoglobinopathy worldwide [97]. It is caused by a single point mutation in the β-globin gene that leads to the synthesis of sickle hemoglobin (HbS). This hemoglobin has the property of polymerizing when devoid of oxygen, leading to the formation of a rigid sickle-shaped erythrocyte [10] (Fig. 27.2.1). SCD is a heterogeneous disorder, with clinical manifestations including chronic hemolysis, increased susceptibility to infections and recurrent vaso-occlusive events that culminate in ischemic organ damage, a diminished quality of life and early death [93, 98].

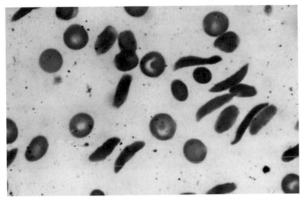

Fig. 27.2.1. Blood smear of a 20-year-old HbSS patient showing sickled erythrocytes with the sodium metabisulfite method

27.2.1.1 Normal Hemoglobin and Sickle Hemoglobin

The most important protein of the red blood cells (RBCs) is hemoglobin, which consists of four globin chains, each folded around a heme molecule. Hemoglobin delivers oxygen from the lungs to the tissues and carbon dioxide from the tissues to the lungs. The predominant hemoglobin in adulthood is HbA (±97%), which consists of two α- and two β-globin chains $(\alpha_2\beta_2)$. Other hemoglobins are HbA$_2$ (2–3.5%; $\alpha_2\delta_2$) and HbF (<2%; $\alpha_2\gamma_2$). During intrauterine development, several globin chains are synthesized (α, β, γ, δ, ε and ζ), with the predominant hemoglobin type during fetal life being HbF. In the first 12 weeks after birth, the HbF% quickly declines, leaving HbA and HbA$_2$ as the remaining hemoglobins [72]. The β-globin gene is found on chromosome 11p15.5. A single point mutation in the 6th codon leads to substitution of glutamic acid for valine, resulting in an abnormal globin: β^S. This results in the formation of "sickle hemoglobin," or HbS $(\alpha_2\beta^S_2)$. Upon deoxygenation, β^S forms hydrophobic interactions with adjacent β^S-globins, ulti-

mately resulting in the polymerization of HbS. This is the molecular hallmark of SCD [10]. As a consequence, the normally pliable RBC assumes a rigid sickled shape, with ensuing erythrocyte membrane damage and hemolysis.

Inheritance of two β^S genes leads to homozygous SCD, or sickle cell anemia (HbSS). Other genotypes that give rise to SCD include double heterozygous states in which the β^S gene is inherited together with other abnormal β genes, or with mutations that result in decreased synthesis of β-globin genes (β-thalassemias). In HbC a mutation in the β gene results in substitution of glutamic acid by lysine. The HbSC genotype is the most common compound heterozygous state, followed by HbS-β-thalassemia [67]. People that carry just one β^S mutation have the sickle cell trait (HbAS), and are generally asymptomatic [97]. Hence, the disease is recessive with respect to clinical manifestations, but the gene is dominant in its expression since sickled cells can always be visualized in deoxygenated blood of individuals with HbAS. Combinations of HbAS with β-thalassemias and of HbSS (or other forms of SCD) with α-thalassemias (mutations that result in decreased synthesis of α-globin genes) give rise to SCD with varying severity depending on the number and type of gene deletions [67]. The survival advantage of people with sickle cell trait with regard to infections with *Plasmodium falciparum* may explain the association of malaria distribution and the distribution of the sickle cell gene, as well as the balanced polymorphism of the β^S gene in the African population [2, 97, 98]. In some parts of Africa, 45% of the population is heterozygous for the β^S gene and in the United States and the Caribbean about 8% of blacks carry one β^S gene. The β^S gene also occurs in the Caribbean, the Mediterranean basin, Saudi Arabia and parts of India [113].

27.2.1.2 Determinants of HbS Polymerization and Patient Categories

The pathophysiological hallmark of SCD is intracellular polymerization of HbS upon deoxygenation [10]. Decreases in pH (which reduce the affinity of hemoglobin for oxygen) enhance HbS polymerization, as does a rise in temperature [64]. The concentration of HbS in the erythrocytes is also of great importance, with higher concentrations of HbS leading to more rapid polymerization. Another determinant of the tendency for polymerization is the presence of other hemoglobins, with HbF and HbA$_2$ limiting the HbS polymerization to a greater extent than HbC and HbA. HbSS patients have no HbA, an HbS% greater than 85%, a normal HbA$_2$% and an elevated HbF%. In people with HbAS the HbS% is approximately 40%. In HbSC patients the HbS% is about 10–15% higher and the mean corpuscular hemoglobin concentration (MCHC) is elevated (due to HbC induced erythrocyte potassium and water loss), explaining why people with HbSC can be severely affected (as opposed to the largely asymptomatic people with sickle cell trait) [55].

27.2.1.3 Combinations with Thalassemia

In combinations of HbS and β^0-thalassemia (no β-globin synthesis from the affected thalassemic allele), there is no normal β-globin production and hence no HbA. These patients show a HbS% similar to that of HbSS patients (>85%). Inheritance of HbS with β^+-thalassemia (reduced β-globin synthesis from the affected thalassemic allele) leads to a variable HbA% (ranging from 1% to 25%), and hence a variable HbS% [67]. Combinations of HbSS with α-thalassemia lead to a slight elevation of the HbA$_2$% with a concomitant reduction of the HbS%. In patients with HbS-β-thalassemias and HbSS with α-thalassemia, the mean corpuscular volume (MCV) and the MCHC are reduced, thereby lowering the HbS polymerization rate as compared to HbSS patients [33, 67].

27.2.1.4 Current Concepts in the Pathogenesis of Vaso-occlusion

Vaso-occlusion accounts for a major part of the clinical picture of SCD. Essential to the management of sickle cell patients is the increasing fundamental understanding of this complex process. Upon delivery of oxygen to the tissues, sickle hemoglobin polymerizes, resulting in the formation of a rigid, nondeformable sickle red blood cell (SRBC) that becomes trapped in the microcirculation, resulting in vaso-occlusion and tissue ischemia. Upon reoxygenation, sickled cells "unsickle," but with each sickling cycle red cells become less pliable, and ultimately irreversibly sickled dehydrated dense SRBCs are formed. Fortunately, the delay time (the time required for polymer formation) is generally longer than the transit time of red cells through the microcirculation and SRBCs do not usually form in the microcirculation. Furthermore, the number of irreversibly sickled cells is only related to the hemolytic component of the disease in most studies [7, 10, 64].

Therefore, it has become evident that other factors greatly influence the vaso-occlusive process. Factors that delay the transit time, such as the adhesion of leukocytes and sickle reticulocytes to cytokine activated vessel wall endothelium, are now recognized of importance in the initiation and propagation of sickle cell vaso-occlusion [10, 65].

Other factors, such as endothelial dysfunction, a hypercoagulable state, and cell dehydration (which shortens the delay time) also contribute [8, 35, 54, 55].

These events result in obstruction of the microcirculation, tissue ischemia and, with prolonged duration, organ damage. The resolution stage remains largely uncharacterized, but it is well established that reperfusion of ischemic areas also contributes to tissue injury [65, 80].

27.2.1.5 Pathophysiology of Ocular Vaso-occlusion

In the last decade, sophisticated histological and immunohistological studies in patients' eyes, as well in transgenic animal models, performed mainly at the Wilmer Eye Institute, have increased our insight into the pathophysiological events in the ocular circulation of sickle cell patients (please refer to Chapter 27.1 by G. Lutty).

27.2.2 Retinal Vascular Disease in Sickle Cell Patients

27.2.2.1 Introduction

Except for two occasions, the ocular manifestations of sickle cell disease are caused or initiated by vaso-occlusions, and those affecting the choroid and retina are described in this chapter.

The association with angioid streaks is the only manifestation not directly explained by vaso-occlusion, but may be a sequela of reperfusion injury after a vaso-occlusive event, with oxidative damage to elastin [1]. The other exception, hyphema, is clinically more important, particularly because it is the only ocular manifestation in which people with HbAS are also at risk.

Within the anterior chamber, erythrocytes sickle and compromise trabecular outflow, leading to a high and prolonged intraocular pressure rise; moreover, bloodflow of the disk and retina may be compromised earlier than in non-sickling persons. Early surgical evacuation of the blood is advised as well as the avoidance of acetazolamide, which may acidify the anterior chamber and encourage further sickling [112].

27.2.2.2 Database

The majority of the clinical reports on sickle cell retinopathy date from the two decades following the first detailed descriptions in 1971 and 1972 from the two major study centers, Chicago (Goldberg, Jampol) and Kingston, Jamaica (Serjeant, Condon, Talbot,

Bird). In the 1990s only three more modest prevalence studies were published (from Curacao, Miami and Togo: van Meurs, Clarcson and Balo, respectively).

The best information, however, is derived from the prospective sickle cell cohort study, initiated in 1973 by Serjeant, based on all sickle cell patients detected among 100,000 consecutive normal deliveries in Kingston, Jamaica. The cohort also included age matched controls. A comprehensive report with follow-up data up to 2000 was published in 2005 [30].

27.2.3 Choroid

Four sickle cell patients (two SC, one SS and one Sthal) with possible ciliary artery occlusions have been reported [25, 28, 29]. Dizon and coworkers were able to observe the course of these lesions: midperipheral wedge-shaped whitish areas deep to the retina in the acute stage were seen, which became mottled areas within weeks. The recurrent triangular lesions in Dizon's patient showed a striking resemblance to the ones produced by ligation of the lateral or all ciliary arteries in monkeys [53, 54]. Apart from choroidal occlusions, the HbSS patient studied by Dizon and colleagues had retinal arteriolar occlusions and occlusion of one internal carotid artery with amaurosis fugax. In this patient with an exceptional vaso-occlusive tendency, it is likely that emboli of erythrocyte plugs or thrombi from the carotid or ophthalmic artery occluded either a ciliary artery or simultaneously several short posterior ciliary or choroidal arteries.

As a complication of feeder-vessel photocoagulation for elevated sea fans (see Sect. 27.2.9.1 below), similar lesions were noted more frequently in sickle cell patients [43].

However, choroidal infarction remains a rare manifestation in sickle cell disease, probably because the flow rate of the choroid is very high (i.e., the transit time is considerably shorter than the delay time of polymerization; refer to Sect. 27.2.1.4).

27.2.3.1 Choriocapillary

Speculating that the uneven filling of the choriocapillaris in the early arteriovenous phase observed in some sickle cell patients (Fig. 27.2.2) might be related to impaired rheology, van Meurs studied choroidal filling in HbSS and HbSC patients as well as in age and race matched HbAS and HbAA controls [107]. He found no difference in choroidal filling pattern between sickle cell patients and controls; consequently, at least in African Caribbeans, a delayed filling pattern of the choriocapillaris in the early arteriovenous phase only cannot be regarded as pathologic.

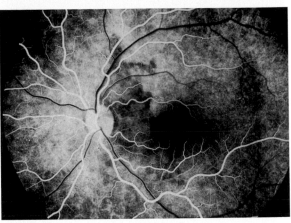

Fig. 27.2.2. Late filling in a geographic pattern, merging with the watershed zone, in a 24-year-old HbAS control patient

27.2.4 Retinal Vessel Occlusions

Vessel involvement at the posterior pole in sickle cell disease presents a spectrum ranging from central retinal artery occlusion to perimacular capillary drop-out, with the formation of perifoveal avascular zones. The smaller vessel occlusions are frequently asymptomatic and without reduced visual acuity, whereas arterial occlusions usually present with acute symptoms of visual loss.

Peripheral retinal vessel occlusions with capillary loss in large, often confluent areas extending from the ora serrata posteriorly are mostly described in relation to proliferative retinopathy. In many patients, however, these occlusions do not lead to proliferative changes.

As sequelae to arteriolar occlusion retinal hemorrhages, salmon patches, iridescent spots and black sunbursts are grouped together.

27.2.4.1 Major Retinal Vessel Occlusions

Central retinal artery occlusions have been reported in ten sickle cell patients. Symptomatic branch and macular artery occlusions have been reported in nine sickle cell patients; visual loss in the latter patients was often relatively mild (Table 27.2.1). Although major retinal arterial occlusions cause severe visual symptoms, only one central retinal occlusion (following retrobulbar anesthesia) and two minor branch occlusions were noted over a period of 15 years in approximately 800 HbSS and 300 HbSC patients in Jamaica [18, 19, 26]. Since an association with sickling disorders tends to hasten reporting, there may be underreporting of "idiopathic" young patients with these occlusions [3]. Given the plausibility of a causative relationship, however, we regard retinal arterial occlusion as a relatively rare ocular manifestation of sickle cell disease.

27 III

Author	Year	Hb	Age (years)	Site occl.	Retinal edema	Final VA	Pale disk	Remarks
Kabakov	1955	AS	36	ca	+	NLP		LE, active TBC
Goodman	1957	SS	18	ba	+	CF		
Conrad	1967	AS	32	ca	?	NLP	+	
Condon	1972	SS	13	ba	+	?		
Acacio	1973	SS	15	ca	+	0.1	+	
Knapp	1972	SS	10	ma	–	0.1		
Ryan	1974	SC	24	ma	–	1.0		Both eyes
Chopdar	1975	SC	22	ba	+	0.2		
Appen	1975	Sth	44	ca	?	?		
		AS	64	ca	?	?		
Weissman	1979	SS	25	ba	+	0.8		Both eyes
Klein	1982	SC	43	ca	–	1.0		After retrobulbar injection
		SC	33	ca	+	NLP		
Asdourian	1982	Sth	14	ma	+	0.5		
Condon	1985	SS	20	ma	+	1.0		Recurrent
Clarcson	1992	SC		ca				
Mansi	2000	SS						Both eyes
Al-Abdulla	2001	SS	9	ca				Albinism
Clarke	2001	SS	Child	Arteriole				Both eyes
Sangvi	2004	SS		ca				

Table 27.2.1. Symptomatic vessel occlusions in sickle cell patients

ca central retinal artery,
NLA no light perception,
ba branch retinal artery,
LE lupus erythematodes,
ma macular arteriole,
VA visual acuity

As in all arterial occlusions, treatment of the underlying disorder is important to decrease the risk of a recurrent occlusion; in sickle cell disease, unfortunately, few specific and proven measures are available.

27.2.4.2 Retinal Vein Occlusions

Interestingly, there is no mention in the literature of an increased occurrence of central or branch vein occlusion in sickle cell patients. Only Hasan et al. [50] reported one patient with HbSS, who was found to have a protein S deficiency as well. Lieb and coworkers [71] mentioned two cases with possible central vein occlusion. In both patients, who also had a subarachnoid hemorrhage, the drawings could have also represented severe disk edema. Recently, indeed, Henry [57] described the occurrence of pseudotumor cerebri in three pediatric patients (one HbSS, two HbSC).

27.2.5 Retinal Hemorrhages, Salmon Patches, Iridescent Spots and Black Sunbursts

Hannon [49] used the term "salmon-pink exudative plaque" for what has been shown to be a round or oval shaped intra- or pre-retinal hemorrhage, as distinguished from a more diffuse intravitreal hemorrhage or clearly subhyaloid hemorrhage, presumed due to an acute arteriolar obstruction.

Glistening deposits, refractile and lipid deposits [114], cholesterol deposits [71] and iridescent glistening deposits [19] are terms that have been used to

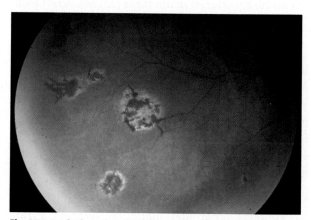

Fig. 27.2.3. Black sunbursts in a 45-year-old HbSC patient

describe the small crystal-like deposits seen superficially in the retina in sickling patients. They are often observed in association with mottled or pigmented round areas.

The deep retinal plaques of stellate or spiculate pigment dispersion, seen in the midperiphery in sickling patients, were termed "black sunbursts" by Welch et al. [114] (Fig. 27.2.3).

27.2.5.1 Reports

A review of the relevant reports reveals that iridescent spots, salmon patches, retinal hemorrhages, brown patches and black sunbursts (either one of these signs or a combination of them) are noted in two-thirds of either SS, SC or S-α-thalassemia patients (Table 27.2.2).

Table 27.2.2. Reported frequency retinal hemorrhages, salmon patches, iridescent spots and black sunbursts

Author	Year	*n*	Hb	Ret.	Sal.	Iri.	Mott.	BS
Cook	1930	1	SS	1				
Hannon	1952	52	SS		1			
Goodman	1956	5	SC	2				1
Lieb	1961	51	SS	23		12		23
		9	SC	5		3		5
Levine	1967	10	SC	2		4		
Welch	1967	35	SS	10		10		15
		22	SC	8		8		4
Condon	1972	74	SS	2		10	10	25
	1972	70	SC	2		23	11	33
	1972	50	Sth	7		9	13	10
Talbot	1982	59	SS				22	
		37	SC					
van Meurs	1990	81	SS		1			25
		97	SC		2			43

n number of patients, *Iri.* iridescent spots, *Hb* Hb type, *Mott.* mottled brown areas, *Ret.* retinal hemorrhage, *BS* black sunburst, *Sal.* salmon patch

27.2.5.2 Course

Asdourian et al. [5] observed a retinal hemorrhage gradually turn into a black sunburst lesion. Similarly described was the transformation of a "salmon patch" superficial retinal hemorrhage into a black sunburst-like lesion [110] (Fig. 27.2.4a–d).

Using light microscopy, Romayananda et al. [89] found hemosiderin-laden macrophages either preretinally, in "schisis cavities" formed by the internal limiting membrane and inner retina, or between the sensory retina and retinal pigment epithelium. In the latter case, there was focal pigment epithelial hypertrophy, hyperplasia and migration. Apparently, a hemorrhage breaking out from a retinal arteriole or capillary can give rise to different ophthalmoscopic signs after its absorption, depending upon its exact location and consequent plane of dissection: from preretinal iridescent spots, via a round or oval schisis cavity with or without iridescent spot or brownish pigment to the deeper retinal and pigment epithelial scar known as the black sunburst patch.

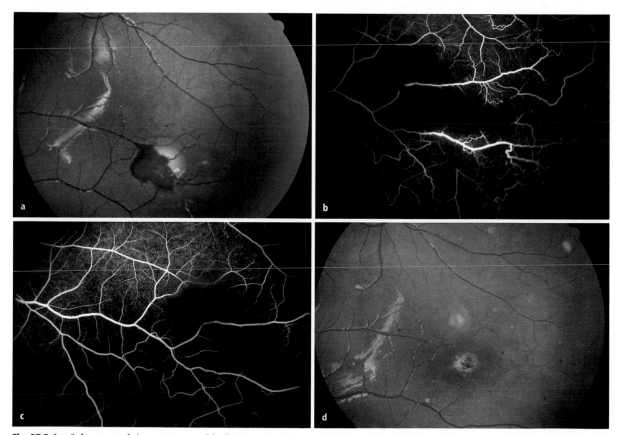

Fig. 27.2.4. a Salmon patch in an 18-year-old HbSC patient. The occluded arteriole is chalky white. A part of the blow-out, superficial retinal hemorrhage is hemolyzed and salmon pink. **b** Fluorescein angiogram of the same area as in **a**, showing blocking of fluorescein due to the salmon-patch hemorrhage and retinal capillary drop-out in the area supplied by the blocked arteriole. **c** Fluorescein angiogram of the same area as in **a** and **b**, demonstrating partial reperfusion of the capillary bed 3 weeks later. **d** Fundus picture of the same patient 4 years after the blow-out hemorrhage shown in **a–c**. A small scar at the level of the retinal pigment epithelium bears testimony of that episode

27 III

27.2.5.3 Relevance to Function

Even when these retinal hemorrhages occasionally break through the internal limiting membrane into the vitreous, they do not interfere with visual acuity, being both small and peripheral.

Although these patches do not by themselves cause visual loss, they are related to occlusions of small vessels, burst-out hemorrhages and capillary loss. Talbot et al. [105] did show a correlation between the patches and capillary closure in a series of 96 young children.

Interestingly, despite the fact that these different signs of vaso-occlusion are often accompanied by the loss of retinal perfusion and thus are a factor in the occurrence of proliferative sickle retinopathy (PSR), they are no more common in HbSC patients, where PSR is frequent.

27.2.6 Capillary Occlusions of the Posterior Pole

Stevens et al. [101] and Asdourian et al. [6] reported abnormalities in either or both the macula and temporal raphe in 32% of HbSS patients, 36% of HBSC patients and 20% of S-beta-thal patients in 100 consecutive cases. The macular abnormalities consisted of cotton-wool spots, microaneurysm-like dots, dark and enlarged segments of arterioles, hairpin-shaped venular loops, and on fluorescein angiography, of pathological avascular zones (PAZ) and widening and irregularities of the foveal avascular zone (FAZ). In children, Talbot et al. [105] found no pathologic avascular zones in the macular region in 59 HbSS and 37 HbSC patients with ages from 5 to 8 years.

Goldbaum [42] described the retinal depression sign on ophthalmoscopy, which was seen after the disappearance of cotton-wool spots and was thought to result from local thinning after necrosis of an infarcted inner retina.

Transient dark spots, apparently plugged deoxygenated erythrocytes within small surface vessels, were noted on the disk itself in 9 (7 HbSS) of 80 sickle cell patients [44] and in 17 of 74 HbSS patients [19].

27.2.6.1 Natural History

Some patients with cotton-wool spots were observed longitudinally [6]; at these sites, abrupt obstructions in the adjacent arteriole were seen on fluorescein angiography, with subsequent opening and reclosing of the initially non-perfused arteriole. In some cases, the dependent capillaries also reperfused. Venular loops were found to develop adjacent to areas that remained non-perfused.

Marsh et al. [73] observed an inverse relationship between the number of perimacular capillary abnormalities and the size of the FAZ; they also noted that a small FAZ and a high count of these capillary abnormalities were more common in younger than in older patients, suggesting that perimacular vessel abnormalities represent the early vaso-occlusive phase that may progress to enlargement of the foveal avascular zone.

27.2.6.2 Relevance of These Findings to Function

In only three of the patients studied by Stevens and Asdourian, a visual loss in the range of 20/25 to 20/30 could be attributed to a macular cause. One eye in Asdourian's series had a paracentral scotoma with a deutan defect on the Farnsworth D-15 panel. In only three cases could field defects be related to these lesions by static perimetry. However, using the Farnsworth-Munsell 100-hue test, Roy et al. [90] demonstrated mild to moderate yellow-blue defects in nearly all 61 eyes of sickle cell patients; there was no correlation with pathologic avascular zones or retinal depression signs. Mild defects were found in one-third of the control subjects.

These results have not shown that these chronic, remodeling small vessel changes cause significant functional loss. Furthermore, there is no mention in the reports concerning elderly sickle cell patients of major visual loss in association with ischemic maculopathy, which would be expected if these changes were progressive.

Interestingly, no specific mention is made in these papers of the coexistence of peripheral retinopathy: apparently, there is no strong correlation between the extent of peripheral non-perfusion and non-perfusion along the raphe, in spite of the similarities in vasculature in these two regions [101].

27.2.7 Peripheral Vessel Occlusions

Teleangiectasias, beading and dilatations are signs of the mild and moderate stages of peripheral microangiopathy according to Condon et al.'s [19–21] classification. Vessel sheathing and silver-wire appearance can be manifestations of obstruction: this has been classified by Condon as the severe stage, and by Goldberg [39] as stage I.

27.2.7.1 Reports

In Condon's series [19–21], there is no apparent difference between HbSS (24%) and HbSC (31%) patients in the prevalence of peripheral obstructions; indeed, Talbot [105] noted a greater number of obstructions in HbSS children (Table 27.2.3).

Table 27.2.3. Occurrence of peripheral vessel obstructions

Author	Year	n	Hb	Tele.	Silver.	Obstr.	Sheath.	A-V
Hannon	1956	3	SC		1			
Goodman	1957	1	SS		1			
Lieb	1961	51	SS				38	
		9	SC				5	
Munro	1960	9	SC				1	
Welch	1966	35	SS			17		
		23	SC			16		
Condon	1972a	74	SS	27	9	18		5
	1972b	70	SC	22		22		19
	1972c	50	Sth	10		7		
Talbot	1982	59	SS			14	30	0–3
	1982	37	SC			6	11	0–3

n number of patients, *Tele.* telangiectasia, *Silver.* silverwire vessels, *Obstr.* obstructions, *Sheath.* sheathing, *A-V* arteriovenous anastosmosis

27.2.7.2 Natural History

Galinos et al. [40] showed, prospectively, a continuous remodeling of the peripheral retinal vasculature due to successive closures and reopenings of equatorial retinal vessels (see Fig. 27.2.4b, c). Nevertheless, a centripetal (posterior) recession of the peripheral retinal vasculature usually resulted.

Raichand et al. [86], who followed 40 sickling patients with peripheral obstructions and arteriovenous anastosmoses, observed the formation of four sea fans from arteriovenous anastosmoses of these patients. These anastosmoses had been observed to exist for 7 months to 3 years prior to sea fan formation.

27.2.7.3 Relevance of These Findings to Function

No loss of visual acuity had resulted from these peripheral occlusions; moreover, no visual field loss had been detected [40]. However, peripheral vessel closure and capillary bed drop-out can be the setting for proliferative sickle retinopathy (see below).

27.2.8 Proliferative Sickle Retinopathy

Retinal and, in rare instances, disk neovascularization in sickling patients have been reported as retinitis proliferans [18, 19, 58] and as sea fans [114]. Some of these cases progressed to vitreous hemorrhage and retinal detachment, thus proving to be a sight-threatening variant of sickle retinopathy.

27.2.8.1 Classification of Sickle Cell Retinopathy

Lieb et al. [71] proposed that sickle cell retinopathy be classified as stage I when tortuosity and dilatation of the major retinal vessels were observed; and as grade II when accompanied by ischemic areas with retinal edema, sheathing of peripheral vessels, neovascularization, microaneurysms and peripheral telangiectasis. Grade III comprises retinal hemorrhages, iridescent spots and obstruction of small veins. Retinitis proliferans, vitreous hemorrhage and central artery or vein occlusion are grade IV. Lieb et al. felt that the higher the grade of retinopathy, the more severe the systemic clinical course of the patient had been.

Goldberg [45] proposed a classification of proliferative sickle retinopathy that was both clinically useful and pathogenetically logical. According to Goldberg, peripheral arteriolar obstruction constituted stage I. Peripheral arteriovenous anastosmoses, judged to be sequential to the obstructions, were designated stage II (Fig. 27.2.5). Peripheral retinal neovascularization originating from the anastosmoses was called stage III (Fig. 27.2.6). Vitreous hemorrhages from such neovascularizations were called stage IV. Partly tractional, partly rhegmatogenous (due to tears adjacent to sea fans with vitreous traction) detachments were called stage V. The extent of circumferential involvement was quantitated in substages. Condon et al. [19] proposed a subdivision of Goldberg's stage I, in which arteriolar obstruction constitutes the most advanced abnormality.

27.2.8.2 Reports of Proliferative Sickle Retinopathy

Reports pertaining to the occurrence of proliferative sickle retinopathy (PSR) are listed in Table 27.2.4. Only Condon's [19–21], van Meurs's [108] and obviously Downes's [30] series have not been selected specifically on the basis of ocular symptoms. PSR was present in 3–14% of the HbSS patients, in 33–43% of the HbSC patients and in 18% of the Hb-S-β-thalassemia patients. In the HbSC patients, up to 31% of these eyes progressed to vitreous hemorrhage and up to 15% to retinal detachment. With the exception of one case with an exudative detachment [31], all other reported cases of retinal detachment in PSR were either tractional or a combination of tractional and rhegmatogenous.

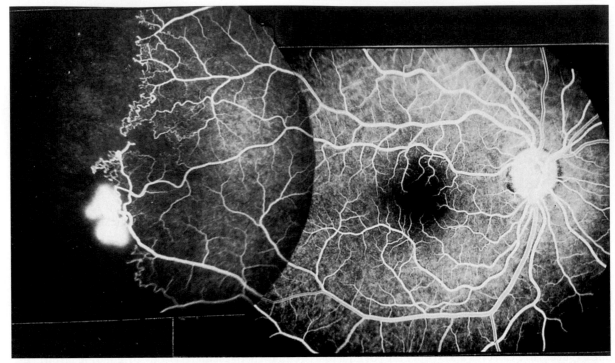

Fig. 27.2.5. Peripheral non-perfusion and arteriovenous anastosmoses in a 28-year-old HbSC female

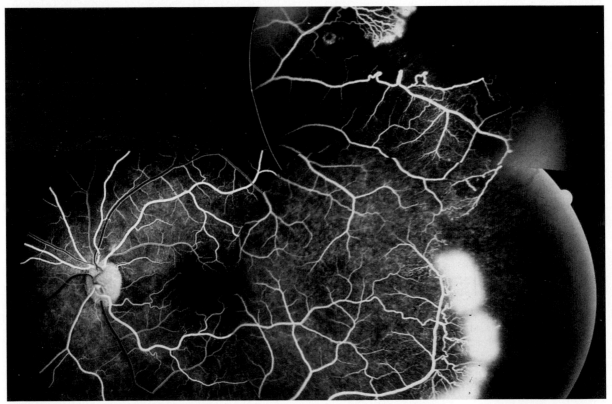

Fig. 27.2.6. Peripheral non-perfusion, arteriovenous anastosmoses and retinal neovascularization in a 23-year-old HbSC patient

Table 27.2.4. Reported frequency of proliferative retinopathy in non-treatment series

Author	Year	n	Hb	StIII	StIV	StV	St I or II (when indicated)
Edington	1952	1	SS		1		
Henry	1954	9	SS	2	1		
		1	SC			1	
Hannon	1954	20	SC	4	4		
Goodman	1956	5	SC	4		1	
Lieb	1961	51	SS	9			
		9	SC	1			
Munro	1960	9	SC	12	9	2	
Rubinstein	1967	9	SC	4	4	2	
Welch	1966	35	SS	3	1		
		22	SC	13	5		
Goldberg	1971	24	SC	39	14	4	
Condon	1972a	76	SS	3	1	2	
	1972b	70	SC	38	11	4	23
	1972c	50	Sth	9	2		9
	1972d	60	SS	12			8
Condon	1980	115	SS	29	5		23
		157	SC	75	11		63
		39	Sth	6	1		5
Hayes	1981a	261	SS	29			
	1981b	243	SC	90			
Lenne	1986	35	SS	7	2	1	
		24	SC	17	5	1	
Talbot	1983	59	SS	–			
	1988	37	SC	1			
Durant	1982	1	SC			1	
Kimmel	1987	135	SS	2		2	
			Or SC	5		5	
van Meurs	1990	81	SS	–			
		97	SC	23	18	9	
Clarcson	1992	59	SS	11			
		23	SC	11			
		3	Sthal	1			
Al-Hazzas	1995	61	SS				
Balo	1997	100	SS				
		90	SC	20			
Downes	1905	307	SS	14		1	
		166	SC	45		1	
Total			SS	121	11	6	8.4% PSR
Total			SC	392	81	31	39% PSR
Total		53	Sthal	16	3		–

n number of patients, *Hb* Hb type, *PSR* proliferative sickle retinopathy, *(partly) the same patients, *StIII* Goldberg stage III: retinal neovascularization, *StIV* Goldberg stage IV: vitreous hemorrhage, *StV* Goldberg stage V: retinal detachment

27.2.8.3 Non-peripheral Proliferative Lesions

Neovascularization on the disk has been reported in five HbSC patients [15, 66, 79, 86, 108] and one HbSS patient [19] (Fig. 27.2.7). More peripheral areas of PSR were also evident in these patients. In four patients efforts were made to exclude other causes for retinal or disk neovascularization.

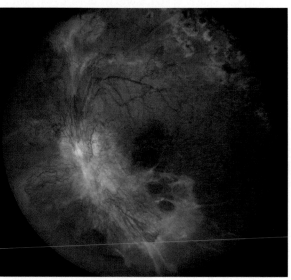

Fig. 27.2.7. A fortunately rare, very active fibrovascular proliferation located near the disk, more resembling proliferative diabetic retinopathy, but as a manifestation of PSR in a 38-year-old HbSC patient

27.2.8.4 Natural History of Proliferative Sickle Retinopathy

In patients with PSR, there is a tendency for a slow increase in the extent of neovascularization, as well as in the incidence of vitreous hemorrhage and retinal detachment (Table 27.2.5).

The youngest patient with reported PSR was an 8-year-old boy with HbSC [102]. The prevalence of PSR increased from the age of 15 years [51, 52] to the age of 35 years, after which time a plateau was reached.

Analyzing 114 eyes with PSR, Condon et al. [15, 22, 23] found autoinfarction (Figs. 27.2.8a, b, 27.2.9a, b) and increased neovascularization to occur simultaneously in the same eye in 32 eyes (28%). PSR was observed to develop (46 eyes) most commonly in patients under the age of 25 years. The most florid growth of neovascularization occurred also in this age group. The average age of patients with autoinfarction alone (11%) was higher than that of patients with an increase in number of sea fans (30%).

In Jampol and colleagues series [62], only four of the 22 vitreous hemorrhages interfered with visual acuity. Within this series risk factors for vitreous

Author	Year	Hb	n	More eyes	More sea fans	Ai.	Vit. hem.	Ret. det.	Av. FU
Goldberg	1971	SC	17		3	–	7 > 11	–	31
Condon	1975	PSR	45	12	3	1	12 > 14	3 > 5	27
Nagpal	1975	PSR	45		?	9			?
Condon	1980	PSR	114		34	13			48
					32	= 32			
Jampol	1983	PSR	80				22		26

Table 27.2.5. Natural course of proliferative sickle retinopathy (PSR)

Hb Hb type, *n* number of eyes, *More eyes* de novo PSR, *More sea fans* more sea fans per eye, *Ai.* autoinfarction of sea fans, *Vit. hem.* vitreous hemorrhage, *7 > 11* i.e., 4 more vitreous hemorrhages, *Ret. det.* retinal detachment in fellow eyes, *Av. FU* average follow-up in months

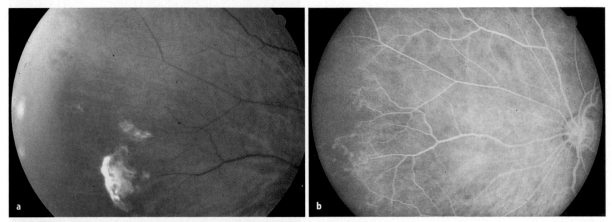

Fig. 27.2.8. a Autoinfarcted sea fan in a 53-year-old HbSC patient. **b** Autoinfarction confirmed by non-perfusion of the fibrotic look-ing sea fan with fluorescein angiography

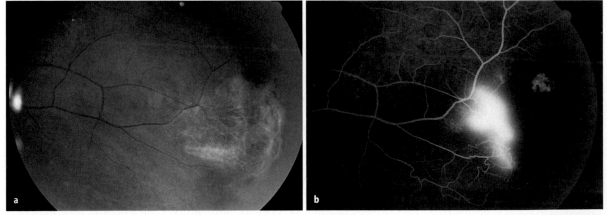

Fig. 27.2.9. a Fibrotic, involuted appearing sea fan in a 31-year-old HbSC patient. **b** Fluorescein angiography, however, reveals that the sea fan is well perfused

hemorrhage were the HbSC genotype, the initial presence of vitreous blood, and more than 60 degrees of perfused sea fans [13]. Clarkson [12] followed 11 HbSS (17 eyes) and 11 HbSC patients (19 eyes) with PSR over an average period of 7 years. Progression, either the increase in number or extent of sea fans or progression to vitreous hemorrhage, was more common in HbSC patients.

Downes [30] reported the 20-year longitudinal observation data of the Jamaican cohort of 307 HbSS

and 166 HbSC patients. Patients with unilateral PSR had a 16% (11% in SS, 17% in SC) probability of regressing to no PSR and a 14% probability (16% in SS, 13% in SC) of progressing to bilateral PSR. Those with bilateral PSR had a 8% (for SS and SC) probability of regressing to unilateral PSR and a 1% (0% in SS, 2% in SC) probability of regressing to a PSR-free state.

27.2.8.5 Clinical Correlations

Research that attempted to determine factors correlating with the clinical course was performed primarily in Jamaica by Serjeant and his coworkers. Investigating hematologic indices in a group of 261 HbSS patients, of whom 29 had developed PSR, Hayes et al. [51, 52] found a high Hb level (>9 g/dl) and a low HbF level (<5%) constituted a risk factor for PSR in males only. Among 243 HbSC patients of whom 90 had developed PSR, Hayes [51, 52] found a high mean cell hemoglobin concentration in males and a low fetal hemoglobin concentration in both sexes to be significantly more common in PSR.

PSR was more frequently (14%) found in HbSS patients 40 years of age and older. PSR was observed to develop most often between the ages of 20 and 30 years; 68% of HbSC patients aged 45 years or over were affected.

Talbot et al. [105] reporting on the 59 HbSS and 37 HbSC children ranging from 5 to 7 years old (cohort) found retinal closure to be consistently more common in HbSS children; however, this difference did not reach statistical significance. In the same cohort, Talbot et al. [103, 104] found that retinal closure was closely correlated with low total hemoglobin concentration and fetal hemoglobin concentration, high reticulocyte and irreversibly sickled cell (ISC) counts in HbSS disease and with high reticulocyte counts in HbSC disease. There was no apparent relationship between vessel closure and clinical events including dactylitis, pneumonia, sickle crisis and infarction, gastroenteritis, weight loss and fever in either genotype. In a follow-up report on the same cohort 6 years later, these observations held true [102].

In both SS and SC patients, BE Serjeant et al. [95, 96, 98] compared plasma and serum viscosity, whole blood viscosity and erythrocyte filterability in age- and sex-matched pairs with or without PSR (27 HbSS pairs, 31 HbSC pairs). HbSC patients with PSR showed significantly higher mean cell hemoglobin and lower HbF levels; however, the viscosity and erythrocyte filtration indices did not differ between the two groups. HbSS patients with PSR showed a higher Hb and lower HbF in males and a higher mean cell hemoglobin concentration in females. In males with PSR, significantly higher whole blood viscosity was measured at high shear and at the patient's own hematocrit.

Electroretinograms from patients with PSR showed reduced a-wave, b-wave and oscillatory potential amplitudes, possibly due to photoreceptor dysfunction secondary to choroidal ischemia or increased oxygen demands by the inner retina [81, 82]. Ischemia of the inner retina may also have contributed to the altered b-wave and oscillatory poten-tials. The same group [38] found a negative correlation between the ERG amplitude measurements and capillary non-perfusion.

In summary, although the most active phase of PSR was between 20 and 30 years of age, there appears to be a cumulative effect with age. Factors that increase sickling per cell (such as a high mean cell hemoglobin concentration and low HbF) and red cell numbers (such as HbSC) lead more frequently to PSR. The most prominent correlation is that between PSR and HbSC.

No reports have indicated that peripheral closure is clearly more extensive in HbSC than HbSS patients. Indeed, in the children's cohort the reverse was shown. To try and explain the different incidence of PSR in HbSS and HbSC patients, BE Serjeant et al. [95, 96, 98] and GR Serjeant [26] hypothesized that, due to the high obstructive tendency in HbSS, there would be a high prevalence of retinal infarction, in addition to early autoinfarction of developing neovascular tissue. In HbSC retinal infarction would develop; however, due to the moderate vaso-occlusive tendency in HbSC autoinfarction of neovascularizations would be rare. An alternative hypothesis by Serjeant [26] postulates that the greater vaso-occlusion in HbSS disease results in infarcted retina, which fails to release a vaso-proliferative substance, whereas the lower vaso-occlusion in HbSC disease might allow for the persistence of ischemic retina.

Van Meurs [109] showed in adult patients that the border between perfused and non-perfused retina was more peripheral in HbSS patients than in HbSC patients and suggested therefore that the greater area of non-perfused ischemic retina in HbSC patients explained the greater prevalence of PSR.

Penman [83] found that the development of PSR was dependent on the type of peripheral vascular border, i.e., patients without continuous arteriovenous loops and with capillary buds or stumps extending into non-perfused retina (type IIb, more prevalent in HbSC patients) were at greater risk for PSR. In a study in Saudi Arabia, Al-Hazza found that patients with the Benin haplotype of HbS (as in Jamaica) had such a type IIb border more frequently than patients with the Asiac haplotype [4].

27.2.8.6 Differential Diagnosis of Proliferative Sickle Retinopathy

The typical aspect of retinal neovascularizations in sickle cell patients renders the differential diagnosis list restricted. By "typical" we mean: an area of peripheral non-perfusion with neovascular formations bordering its posterior boundary, often already sea fan-like, either isolated or connected in ridges, some parts already partly fibrosed. Apart from

27 III

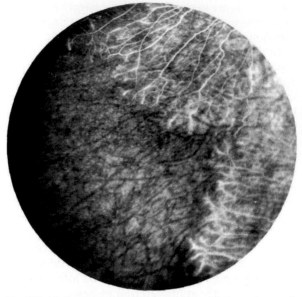

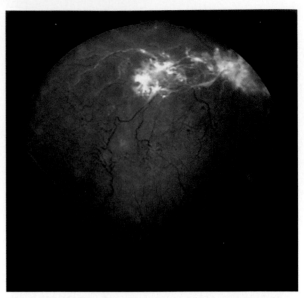

Fig. 27.2.10. Equatorial area in a 20-year-old Caucasian male with FEVR. Note the notch of extra non-perfusion between superior and inferior temporal vasculature and the abnormal parallel orientation of vessels due to the abnormal development of the retinal vascular network. (Courtesy of Eric van Nouhuys, MD)

Fig. 27.2.11. Peripheral capillary non-perfusion and retinal neovascularization with fibrotic parts in a 50-year-old Caucasian patient diagnosed with Eales' disease

branch vein occlusion, Eales' disease, retinopathy of prematurity and familial exudative vitreoretinopathy may be considered.

In regressed retinopathy of prematurity (ROP) [34a] and dominant exudative vitreoretinopathy (DEVR or FEVR) [111], the areas of non-perfusion are more anterior than in PSR. In PSR the perfused-nonperfused retina border is highly irregular, reflecting the ongoing centripetal process of vaso-occlusion; consequently, sea fans are generally not equidistant to the disk, as is usually the case in ROP and DEVR, where vascular development was impaired at one point in time. Similarly, there is no V-shaped border of perfused retina formed by the arrested growth of superior and inferior temporal arcades as in DEVR (Fig. 27.2.10). The features associated with retinopathy in incontinentia pigmenti are female sex, cutaneous manifestations, dental abnormalities and presentation at a generally younger age than PSR.

The most common diagnosis of exclusion is Eales' disease (Fig. 27.2.11). The clinical definition of Eales' disease is not yet well delineated. Elliot [32] and Renie et al. [88], in series of 30 cases each, include vasculitis or periphlebitis, together with peripheral occlusive vascular disease, within its definition. In these authors' view, vasculitis would be an early event in the course of Eales' disease, explaining the lack of inflammatory signs in many patients. Spitz-

nas [100], however, studying over 300 patients in Essen, considered periphlebitis and papillitis to be part of a separate entity: idiopathic periphlebitis, and reserved the term Eales' disease for the non-inflammatory peripheral obstructive vasculopathy, which resembles PSR in many ways. An important consideration was not to confuse the vascular sheathing (opacification of the vessel wall, hyaline degeneration [75]), as seen in Eales' disease and sickle cell disease, with the more irregular and wider inflammatory sheathing known to occur in vasculitis.

Obviously, in the diagnostic workup of each person of predominantly African (or Mediterranean or Middle Eastern) extraction with peripheral retinal vascular obstructions or neovascularization, a sickle cell test and hemoglobin electrophoresis are mandatory and conclusive. Any other positive finding revealed from the work-up might only suggest a joint pathological effect, which may be difficult to assess in terms of its relative impact. Interestingly, this issue has not been addressed in connection with HbSS, HbS-α-thal or HbSC disease. However, the combination of HbAS and diabetes [61, 78], active luetic disease, lupus erythematodes, or tuberculosis [78] has been reported in seven patients; the relative contribution of each disease to retinal neovascularization remains speculative. In the other reports on PSR and sickle cell trait patients (Rubinstein, 1967, two cases [91]; Welch et al. [114] and Treister et al. [106], one case each), no effort was made to exclude Hb S-α-thal-plus or other causes. In patients with HbAC (whose erythrocytes do not sickle, but have a higher MCHC) and diabetes, four patients with PSR-like fundus findings have been reported [59, 61].

27.2.9 Prophylactic Treatment of PSR

Hannon [49] successfully closed retinal neovascularizations with diathermy in two sickle cell patients. Condon et al. [21] used the xenon arc for this purpose, treating both the feeder vessels and the sea fan itself. Laser photocoagulation was initially used to treat the feeder vessels [13, 47, 48, 60], and later used in a scatter fashion around and peripheral to the neovascularizations [27, 34, 35, 87]. Cryocoagulation has rarely been administered [69]. Seibert mentioned the possibility of using transscleral diode laser treatment [94].

27.2.9.1 Reports

The feeder vessel technique has proven very successful in stopping perfusion of neovascularizations, as shown on fluorescein angiography [21, 47, 62] (Fig. 27.2.12a, b).

Apart from the retinal breaks [24, 43], a surprisingly high incidence of choroidal neovascularization has been reported after treatment of feeder vessels, for which high energy levels must be used to close the feeding artery. Choriovitreal neovascularizations often bled and proved very difficult to close with photocoagulation [16, 17, 29] (Fig. 27.2.13).

Due to the high energy settings used in the feeder vessel technique, choroidal arteries can become occluded, which results in choroidal ischemia. This shows initially as triangular grayish-white patches anterior to the treatment site, later fading into a similarly shaped region of granular hyperpigmentation

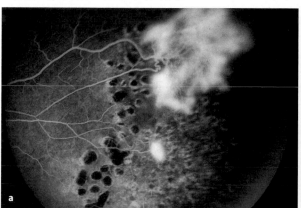
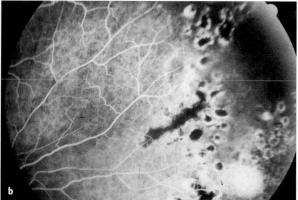

Fig. 27.2.12. a Large, elevated sea fan, which had caused a dense vitreous hemorrhage previously, in a 45-year-old HbSC patient. Scatter laser treatment had not been effective. **b** Successful closure of the feeder arteriole (feeder vessel technique). The smaller neovascular lesion just inferior to the larger sea fan has not been treated with this technique

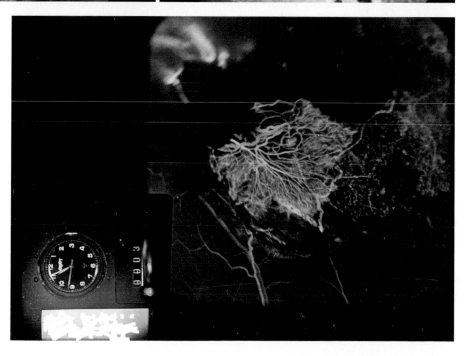

Fig. 27.2.13. Example of a chorioretinal neovascularization as a complication of a feeder vessel treatment of a sea fan in a 31-year-old HbSS patient

III 27

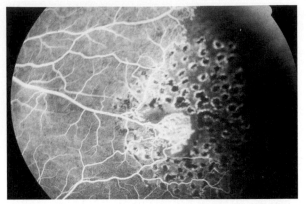

Fig. 27.2.14. Despite a fair amount of scatter laser coagulation, this sea fan remained perfused in a 41-year-old HbSC patient

[43]. Functionally, this resulted in subtle changes of the peripheral visual fields only.

Even with use of the highly effective feeder vessel technique, it was difficult to show a clinically significant effect in a prospective, randomized trial [14, 62]. This is caused partly by the introduction of a potentially sight-threatening complication, such as retinal breaks or choriovitreal neovascularization, and partly by the (relative to the period of observation) slowly progressive nature of sickle cell retinopathy.

The scatter technique caused few complications, but proved much less effective in the closure of sea fans, particularly elevated ones with fibrous tissue [87] (Fig. 27.2.14).

In a prospective randomized controlled study of scatter treatment of patients on Jamaica, Farber et al. [34] reported a statistically significant reduction in visual loss of more than 3 months duration as well as in the number of vitreous hemorrhages. Clarkson [12], however, points out that there is no difference in the occurrence of permanent reduction in vision between the treated group and controls. Therefore, because of the marginal benefit on visual function and the tendency of sea fans to autoinfarct, prophylactic laser is not advised by some experts [12, 30].

27.2.10 Epiretinal Membranes

Moriarty et al. [77] found epiretinal membranes (ERMs) in 25 eyes (22 HbSC, 3 HbSS patients) of the 699 examined eyes (3.6 %). All eyes with ERMs had PSR.

Carney et al. [11] detected epiretinal membranes in 55 of the 1,486 eyes for which fundus photographs were available (40 HbSC, 13 HbSS, 2 HbSthal) (3.7 %). The macula was involved in all but four eyes. In only six (HbSS) eyes there was no coexistent PSR. In one eye the pucker spontaneously peeled.

Hrisomalos et al. [60] reported on six eyes, in which vitreous bands exerted traction on the retina.

27.2.10.1 Etiology

Previous vitrectomy or detachment surgery had been performed in three of 55 eyes with ERMs [11]. In Carney's series there was an equal number of photocoagulated and untreated cases and in Hrisomanos' cases only one eye had been treated; the other five were control cases in treatment trials. In Moriarty's series epiretinal membranes were significantly less common in treated PSR eyes.

Thus, surgery and photocoagulation cannot be implicated in the genesis of these membranes, but rather the presence of neovascularization with its coexistent vitreoretinal changes. Transudation of plasma from PSR lesions may disorganize the cortical vitreous; the resulting posterior detachment may disrupt the internal limiting membrane, thus providing access for glial cell ingrowth [62] either along the retinal surface or onto the retracting, collapsing vitreous face, with secondary contraction of transformed retinal epithelial cells [68].

27.2.10.2 Relevance to Function

In 40 % of Moriarty's cases, visual acuity of less than 6/18 was due to epiretinal membranes; in two cases poor acuity was due to tractional retinal detachment. In nine of Carney's cases, moderat visual loss was caused specifically by the membranes; however, this was not the case in six patients demonstrating severe visual loss.

Carney et al. [11] reported visual improvement in one case following vitrectomy and membrane dissection. Hrisomalos et al. [60] reported usage of the neodymium-YAG laser in cutting vitreous bands. In three eyes, retinal traction appeared to be stabilized after treatment; in two of these, additional vitrectomy and membrane peeling were required.

Epiretinal membranes appear to be a common cause for a modest decrease of vision in patients with PSR. Given the uncertainty of visual improvement and the perioperative risk, an indication for vitrectomy has to be weighed against the possible benefits of epiretinal membrane removal, i.e., the restoration of binocular function.

27.2.11 Macular Holes

Few patients with macular holes in association with sickle cell disease have been reported [74, 85]. In these patients, macular holes were all associated with proliferative sickle retinopathy; the macular holes had a more distorted shape, more epiretinal tissue surrounding the macular hole and occurred at a much younger age than in patients with idiopathic macular holes. These macular holes are likely to be the result of more extended tangential traction as part of proliferative sickle retinopathy, very much like epiretinal membranes in these patients (Fig. 27.2.14).

Mason [74] described the successful closure of such a macular hole with an improvement in vision from 20/400 to 20/40.

27.2.12 Treatment of Vitreous Hemorrhages and Retinal Detachments

Non-resorbing dense vitreous hemorrhages and retinal detachments (tractional or rhegmatogenous, or both), sometimes associated with vitreous hemorrhage, have been treated either with ab externo detachment surgery (exoplants, encircling bands, cryocoagulation, external drainage), with intraocular surgery (vitrectomy, membranectomy, internal drainage), or by a combination of these. In patients with sickling disorders, these operations have proven to have a high percentage of intra- and postoperative complications.

27.2.12.1 Anterior Segment Ischemia

The anterior segment ischemia syndrome (ASI) can comprise keratopathy, intraocular inflammation with severe pain, cataract formation and hypotony. The syndrome in its most severe form can lead to phthisis bulbi and enucleation. Originally it was associated with detachment surgery involving diathermy, detachment of rectus muscles and tight encircling elements [8, 115] or with strabismus surgery on three or more muscles per eye [41]. Histological studies have revealed necrosis of parts of the iris and ciliary body [8, 115] resulting from interruption of either the long posterior ciliary arteries or the anterior ciliary arteries.

Ryan et al. [92] described the syndrome in six out of nine HbSC patients who underwent detachment surgery. Since he was unable to correlate this high incidence of ASI consistently with the previously cited risk factors, Ryan identified the hemoglobinopathy itself as one such factor. Obviously, any interference with blood flow and supply to the anterior segment can induce sickling and increase the risk of anterior segment ischemia. Leen [70] reported peroperative hypotension during general anesthesia as a possible cause, as well as laser coagulation over the long posterior ciliary arteries.

27.2.12.2 Surgical Reports

Surgical procedures, outcome and complications in surgical treatment reports are presented in Table 27.2.6. In 14 cases, eyes lost their function [light perception (LP) positive or negative, hand movements only (HM)]. In six of the 11 reported

Table 27.2.6. Results and complications of external retinal detachment surgery and vitrectomy in PSR

Authors	Year	Number	Ext.	Mu.	En.	Vit.	An./fu. success		TRD	RD	Tear	ASI	LP	HM
Ryan	1971	9	9	4	7		7	4	3			6		5
Eagle	1973	1	1	2	1		1				1			
Robertson	1975	1	1		1		1	1				1		
Ryan	1975	7				7	3	3		4	4			4
Freilich	1977	11	11	1	1		11	11						
Treister	1977	2	2		2	2	2	1					1	1
		1				1	0	1					1	1
Zinn	1981	1	1		1		1	1						
Jampol	1982	4	4				4	4	14					
		5				5	4	4			1		1	1
		10	10			10	6	5		4	5	1		
Brazier	1986	6	6				4	4	2	2				
		3			3		1			1				
		7	7		4	7	6		6			1		1
Morgan	1987	1	1				1	1						
		3			3		2	2		1	1			
		1	1				1	1	1			1		
Hrisomalo	1987	2					2	2	1					
Pulido	1988	11	5				11	10	10	5		1		
Leen	1902	1					1	1				1		
Downes	1905	2	1				1	2	2	1		1		
External										12	13			5
Vitrectomy									5	6			2	6
Combined										6	3			2
Total							64	58	26	6	13	11	2	13

Ext. external approach, *Mu.* rectus muscle disinsertion, *En.* encircling band, *Vit.* vitrectomy, *An./fu. success* anatomic or functional success, *TRD* traction retinal detachment, *RD* retinal detachment, *Tear* retinal tear, *ASI* anterior segment ischemia, *LP* light perception, *HM* hand movement, *SG* secondary glaucoma, *Tran.* exchange blood transfusion, *Combined* both external approach and vitrectomy

27 **III**

cases with anterior segment ischemia (Table 27.2.6), vision worsened to hand movements or no light perception.

At least 67% of the reported operations were successful in terms of restoring function. In two cases visual acuity was counting fingers only. The remaining cases had either attached retinas or non-progressive detachments, with visual acuity of 20/200 or better.

27.2.12.3 Comment

Even in relatively simple vitrectomies, the rate of iatrogenic breaks is rather high, which may be explained by the thinness of the avascular atrophic peripheral retina [46] as well as a tight vitreoretinal adherence in patients with PSR.

Anterior segment ischemia proved to be a major problem initially; in later series, however, this complication hardly ever arises. Using a hyperbaric oxygen chamber, Freilich et al. [37] were successful in external approach surgery without the complication of anterior segment ischemia.

In contrast to the many pre- and postoperative measures suggested by Ryan et al. [92] and Brazier [9], who particularly stressed the importance of exchange transfusions, Morgan and coworkers [76] used no additional preoperative steps. Their intraoperative complications, however, remain difficult to manage. Pulido et al. [84] feel that the risk of blood transfusion exceeds its benefits. Changes in general care of these patients, changes in anesthesia, more sophisticated surgery in the later papers and the relatively small series of a heterogeneous set of operations render interpretation of the influence of the different pre- and postoperative regimes speculative. Nevertheless, ASI still needs to be reckoned with [70]. Postoperatively hyphema and secondary sickle glaucoma may present a problem.

27.2.12.4 Measures to Consider in Vitreoretinal Surgery

As there is no evidence-based information available, the following perioperative measures appear rational to us.

- Local anesthesia with additional oxygen is preferable over general anesthesia, to decrease the chance of peroperative hypotension and postoperative sickle cell crises.
- Try to interfere as little as possible with the recti muscle and do not apply an encircling band, to decrease the risk of anterior segment ischemia.
- Try to avoid applying heavy laser coagulation over the long posterior ciliary arteries.

- As one of the best documented initiating factors for a sickle cell crisis is cold exposure, it has been proposed that the perfusion fluid for vitrectomy is warmed closer to body temperature than normal (Ramin Tadayaoni, verbal communication, 11 May 2005).
- Try to prevent hyphema, and if it occurs per- or postoperatively, be ready to remove the blood urgently.
- Monitor intraocular pressure during and after surgery and be careful to keep it in a normal range.
- Choose local anesthesia to decrease the incidence of an acute chest syndrome.
- PSR patients may have a severe and prolonged inflammatory reaction postoperatively, possibly because of a blood-retina barrier breakdown. Consider frequent and prolonged steroid drops or subconjunctival injections.

27.2.13 Blindness Caused by Sickle Cell Disease

We have tried to estimate the reported prevalence of visual loss in sickle cell patients that is attributable to sickle cell disease. All reported series on patients with sickle cell disease are listed, with, where stated, their cases of severe visual loss with probable cause. All reports are essentially cross-sectional surveys and the data provide information on the prevalence, but not on the incidence, of visual loss. Incidence figures may be derived from the Jamaican cohort [30] and to a lesser degree from the treatment series [34, 35].

We have paid special attention to the fact that in the publications from Jamaica, series may be reported several times, either because of further follow-up or to discuss another aspect of the same series. The papers from Chicago mostly appear to be separate groups, drawn from a larger screened population (Carney et al. [11] mentioned 769 patients screened), which has not been reported comprehensively.

27.2.13.1 Reports

We have included optic atrophy due to local vascular compromise as due to sickle cell disease, which may be controversial. In one of Asdourian et al.'s [6] patients, luetic serology was positive. One 27-year-old HbSS patient had no light perception in either eye, due to cortical blindness following severe circulatory shock and respiratory arrest after intra-abdominal blood loss following a cholecystectomy.

The complications of proliferative retinopathy, i.e., vitreous hemorrhage, retinal detachment, epiretinal membranes and anterior segment ischemia prove to be the major cause of visual loss: 82% (Tables 27.2.7, 27.2.8).

Table 27.2.7. Reported prevalence of severe visual loss in sickle cell patients

Author	Year	n	Hb	CAO	SG	VH	RD	ASI	AVODS	CF	HM	LP	0.1
Edington	1952	2	SC			1					1		
Henry	1954	10	SS			1					1		
		19	SC				1				1		
Smith	1954	15	SS										
		16	SC		3								
Kabakov	1955	1	?	2				1			2		
Hannon	1956	53	SS										
		20	SC			4					4		
Goodman	1957	5	SC	1		1			1		1		
Lieb	1959	51	SS	2 CVO		6							
		9	SC			1							
		5	AS										
Paton	1959	2	SS										
Munro	1960	9	SC				2			1		1	
Levine	1965	10	SC			1					1		
Welch	1966	35	SS			1							
		22	SC			5							
Conrad	1967	1	AS			1					1		
Rubinstein	1967	4	SC			1	1		2				
Ryan[a]	1971	9	SC				1	4			1	4	
Goldberg	1971	16	SC				Epi.						1
Goldberg	1971	14	Sth				1				1		
Goldberg[b]	1971	15	SC				Epi						1
Knapp	1972	1	SS	Mac. a.									1
Condon[c]	1972	76	SS	1			1			1			
[c]	1972	70	SC	Cil. a.		1	3			1		3	
[c]	1972	50	Sth										
Acacio	1973	1	?	1									1
Condon	1974	8	SS				1			1			
		6	Sbth				Epi.						1
Ryan	1974	1	SC										
Condon[c]	1975	88	SS							1			
[c]		6	SC				5			5			
[c]		50	Sth										
Lee[b]	1975	4	SC				1			1			
		2	SS										
Galinos[b]	1975	3	SC										
Galinos	1975	12	SS										
		10	SC										
Ryan	1975	7	SC				4			4			
Nagpal	1975	9	SS										
		29	SC		1	1	1				1	2	
		4	Sth										
Condon	1976	60	SS	Opt. a.			2				2		1
Asdourian	1976	41	SS	Opt. a.									1
[c]		36	SC										
		15	Sth										
Goldbaum[b]	1976	12	SC				1				1		
		4	SS										
		2	Sth										
Boase	1976	26	SC			4	1			1	1	3	
Goldbaum[a]	1976	5	SC		1		Epi.				1		1

n number of patients reported, *Hb* hemoglobin type, *CAO* central retinal artery occlusion, *Opt. a.* optic atrophy, *Mac. a.* macular arteriole, *Vitr. hem.* vitreous hemorrhage, *Cil. a.* ciliary artery occlusion, *CVO* central vein occlusion, *SG* secondary glaucoma, *VH* vitreous hemorrhage, *RD* retinal detachment, *Epi.* epiretinal macular pucker, *ASI* anterior segment ischemia, *0.1* visual acuity of 6/60 or 20/200, *CF* counting fingers, *HM* hand movements, *LP* light perception, *x* "loss of vision" or "reduced visual acuity" not specified, but in association with retinal detachment
[a] Surgery series, [b] Prophylactic treatment series, [c] Patients referred to in other reports as well

Table 27.2.7. (*Cont.*)

Author	Year	*n*	Hb	CAO	SG	VH	RD	ASI	AVODS	CF	HM	LP	0.1
Nagpal	1976	162	SS										
		101	SC										
		34	Sth				1					1	
Goldbaum[b]	1977	32	SC				Epi.,1			1			1
		3	Sth										
Raichand	1977	21	SS										
		14	SC										
		5	Sth										
Treister[a]	1977	1	SC					1		1			
Freilich[a]	1977	4	SC					2					2
		2	SS										
Ober	1978	1	SC										
Goldbaum	1978	13	SS	Opt. a.					1	1			1
		5	SC										
Goldbaum[b]	1979	7	SC										
		1	SS										
		1	AS										
Frank	1979	1	SC										
Condon	1980	115	SS				1					1	
		157	SC				4					4	
		25	Sth										
		14	Sth										
Condon[b,c]	1980	43	SC			1	3			2		2	
		8	SS			1					1		
		6	Sth										
Dizon-Moore[b,c]	1981	20	SC				1						1
		1	SS										
Condon[b,c]	1981	27	SC										
		2	SS										
		6	Sth										
Hayes[c]	1981	243	SC										
Hayes[c]	1981	261	SS										
Talbot[c]	1981	59	SS										
		37	SC										
Michels	1981	2	SC				1					1	
Zinn[a]	1981	1	SC										
Hamilton[c]	1981	17	SS			1					1		
Jampol[a,c]	1982	18											
Klein	1982	2	SC	1						1			
Asdourian	1982	1	Sbth										
Rednam[b]	1982	9	SC										
		3	SS										
		2	Sb										
		5	AS										
Hanscom[b]	1982	4	SC				1					1	
Jampol[b,c]	1983	122	PSR										
Cruess[b]	1983	23	SC										
		2	SS										
		2	Sb										
Talbot[c]	1983	54	SS										
		31	SC										
Condon[b,c]	1984	122	PSR										
Condon	1985	1	SS										
Lenne	1986	35	SS										
		24	SC										

n number of patients reported, *Hb* hemoglobin type, *CAO* central retinal artery occlusion, *Opt. a.* optic atrophy, *Mac. a.* macular arteriole, *Vitr. hem.* vitreous hemorrhage, *Cil. a.* ciliary artery occlusion, *CVO* central vein occlusion, *SG* secondary glaucoma, *VH* vitreous hemorrhage, *RD* retinal detachment, *Epi.* epiretinal macular pucker, *ASI* anterior segment ischemia, *0.1* visual acuity of 6/60 or 20/200, *CF* counting fingers, *HM* hand movements, *LP* light perception, *x* "loss of vision" or "reduced visual acuity" not specified, but in association with retinal detachment
[a] Surgery series, [b] Prophylactic treatment series, [c] Patients referred to in other reports as well

Table 27.2.7. (*Cont.*)

Author	Year	*n*	Hb	CAO	SG	VH	RD	ASI	AVODS	CF	HM	LP	0.1
Kimmel[b]	1986	1	SC										
Brazier[a]	1986	12	SC	1			5			1		5	
		1	SS										
Morgan[a]	1987	4					Epi., 1					1	1
Carney[(a)]	1987	33	SC	1		1	Epi., 3		2	1	3		
		11	SS										
		2	Sth										
Hrisomalo[a]	1987	6	SC?				Epi., 1		1		1		1
Kimmel	1987	135	SS										
			SC										
Van Meurs	1990	81	SS						1		2		
		97	SC			27	10		1		1	5	1
Clarcson	1992	59	SS										
		23	SC	1			1				2		
		3	Sthal										
Farber	1992	93	SC				3 RD 3 epi.				3		3
		21	SS										
		2	Sthal										
Fox	1993	66	PSR								2		
Al Hazza	1995	61	SS										
		10	Sthal										
Balo	1997	190	SS/SC										
Leen	2002	1	SC					1		1			
Downes	2005	307	SS				1 epi.						1
		166	SC										
Visual acuity Total		1498	SS	4	4	9	3		2	1	3	6	9
		1194	SC	4	4	52	56, 10 epi.	6	1	15	24	45	19
		204	Sbthal			2	1 epi.				2		
		2896	SCD	4	4	63	59, 11 epi.	6	3	16	29	51	26

n number of patients reported, *Hb* hemoglobin type, *CAO* central retinal artery occlusion, *Opt. a.* optic atrophy, *Mac. a.* macular arteriole, *Vitr. hem.* vitreous hemorrhage, *Cil. a.* ciliary artery occlusion, *CVO* central vein occlusion, *SG* secondary glaucoma, *VH* vitreous hemorrhage, *RD* retinal detachment, *Epi.* epiretinal macular pucker, *ASI* anterior segment ischemia, *0.1* visual acuity of 6/60 or 20/200, *CF* counting fingers, *HM* hand movements, *LP* light perception, *x* "loss of vision" or "reduced visual acuity" not specified, but in association with retinal detachment
[a] Surgery series, [b] Prophylactic treatment series, [c] Patients referred to in other reports as well

Table 27.2.8. Causes of visual loss possibly due to sickle cell disease

No. of patients	Diagnosis
1	Ciliary artery occlusion
1	Macular arteriolar occlusion
2	Macular hole (epi. membr.)
2	Cortical blindness (ODS)
4	Optic atrophy
4	Secondary glaucoma
5	Central retinal artery
5	Anterior segment ischemia
7	Epiretinal membrane
32	Vitreous hemorrhage
59	Retinal detachment
Total 122	

No. of eyes	Visual acuity
26	20/200
16	Counting fingers
29	Hand movements
51	Light perception
Total 122	

Severe visual loss related to sickle cell disease is almost entirely restricted to HbSC patients secondary to proliferative sickle retinopathy. The estimated prevalence in HbSC of a loss of vision of counting fingers or less in one eye and of hand movements or less was 9.5% and 7.4%, respectively. Fortunately, bilateral severe visual loss was rare and almost equally distributed among HbSS (cortical blindness, optical atrophy) and HbSC (PSR) alike.

As not all the reports clearly state severe visual loss and its cause, this compilation of available data represents only an estimation of visual loss related to sickle cell disease. The frequency of visual loss may be overestimated, because most series tended to be selected on visual symptoms or were treatment series. Again, the most unbiased data are from the Jamaican Cohort [30].

27 **III**

References

1. Aessopos A, Farmakis D, Loukopoulos D (2002) Elastic tissue abnormalities resembling pseudoxanthoma elasticum in beta thalassemia and the sickling syndromes. Blood 99:30–35

2. Aidoo M, Terlouw DJ, Kolczak MS, McElroy PD, ter Kuile FO, Kariuki S, Nahlen BL, Lal AA, Udhayakumar V (2002) Protective effects of the sickle cell gene against malaria morbidity and mortality. Lancet 3591311–1312

3. Al Abdulla NA, Haddock TA, Kerrison JB, Goldberg MF (2001) Sickle cell disease presenting with extensive perimacular arteriolar occlusions in a nine-year-old boy. Am J Ophthalmol 131:275–276

4. al Hazzaa S, Bird AC, Kulozik A, Serjeant BE, Serjeant GR, Thomas P, Padmos A (1995) Ocular findings in Saudi Arabian patients with sickle cell disease. Br J Ophthalmol 79:457–461

5. Asdourian G, Nagpal KC, Goldbaum M, Patrianakos D, Goldberg MF, Rabb M (1975) Evolution of the retinal black sunburst in sickling haemoglobinopathies. Br J Ophthalmol 59:710–716

6. Asdourian GK, Nagpal KC, Busse B, Goldbaum M, Patriankos D, Rabb MF, Goldberg MF (1976) Macular and perimacular vascular remodelling sickling haemoglobinopathies. Br J Ophthalmol60:431–453

7. Billett HH, Kim K, Fabry ME, Nagel RL (1986) The percentage of dense red cells does not predict incidence of sickle cell painful crisis. Blood 68:301–303

8. Boniuk M, Zimmerman LE (1961) Necrosis of the iris, ciliary body, lens and retina following scleral buckling operations with circling polyethylene tubes. Trans Am Acad Ophthalmol Otolaryngol 65:671–693

9. Brazier DJ, Gregor ZJ, Blach RK, Porter JB, Huehns ER (1986) Retinal detachment in patients with proliferative sickle cell retinopathy. Trans Ophthalmol Soc UK 105:100–105

10. Bunn HF (1997) Pathogenesis and treatment of sickle cell disease. N Engl J Med 337:762–769

11. Carney MD, Jampol LM (1987) Epiretinal membranes in sickle cell retinopathy. Arch Ophthalmol 105:214–217

12. Clarkson JG (1992) The ocular manifestations of sickle-cell disease: a prevalence and natural history study. Trans Am Ophthalmol Soc 90:481–504

13. Condon P, Jampol LM, Farber MD, Rabb M, Serjeant G (1984) A randomized clinical trial of feeder vessel photocoagulation of proliferative sickle cell retinopathy. II. Update and analysis of risk factors. Ophthalmology 91:1496–1498

14. Condon P, Jampol LM, Farber MD, Rabb M, Serjeant G (1984) A randomized clinical trial of feeder vessel photocoagulation of proliferative sickle cell retinopathy. II. Update and analysis of risk factors. Ophthalmology 91:1496–1498

15. Condon PI, Hayes RJ, Serjeant GR (1980) Retinal and choroidal neovascularization in sickle cell disease. Trans Ophthalmol Soc UK 100:434–439

16. Condon PI, Jampol LM, Ford SM, Serjeant GR (1981) Choroidal neovascularisation induced by photocoagulation in sickle cell disease. Br J Ophthalmol 65:192–197

17. Condon PI, Sergeant GR (1981) Choroid neovascularization. An important complication of photocoagulation for proliferative sickle cell retinopathy. Trans Ophthalmol Soc UK 101:429

18. Condon PI, Serjeant GR (1972) Ocular findings in hemoglobin SC disease in Jamaica. Am J Ophthalmol 74:921–931

19. Condon PI, Serjeant GR (1972) Ocular findings in homozygous sickle cell anemia in Jamaica. Am J Ophthalmol 73:533–543

20. Condon PI, Serjeant GR (1972) Ocular findings in sickle cell thalassemia in Jamaica. Am J Ophthalmol 74:1105–1109

21. Condon PI, Serjeant GR (1974) Photocoagulation and diathermy in the treatment of proliferative sickle retinopathy. Br J Ophthalmol 58:650–662

22. Condon PI, Serjeant GR (1980) Behaviour of untreated proliferative sickle retinopathy. Br J Ophthalmol 64:404–411

23. Condon PI, Serjeant GR (1980) Photocoagulation in proliferative sickle retinopathy: results of a 5-year study. Br J Ophthalmol 64:832–840

24. Condon PI, Serjeant GR (1980) Photocoagulation in proliferative sickle retinopathy: results of a 5-year study. Br J Ophthalmol 64:832–840

25. Condon PI, Serjeant GR, Ikeda H (1973) Unusual chorioretinal degeneration in sickle cell disease. Possible sequelae of posterior ciliary vessel occlusion. Br J Ophthalmol 57:81–88

26. Condon PI, Whitelocke RA, Bird AC, Talbot JF, Serjeant GR (1985) Recurrent visual loss in homozygous sickle cell disease. Br J Ophthalmol 69:700–706

27. Cruess AF, Stephens RF, Magargal LE, Brown GC (1983) Peripheral circumferential retinal scatter photocoagulation for treatment of proliferative sickle retinopathy. Ophthalmology 90:272–278

28. Dizon RV, Jampol LM, Goldberg MF, Juarez C (1979) Choroidal occlusive disease in sickle cell hemoglobinopathies. Surv Ophthalmol 23:297–306

29. Dizon-Moore RV, Jampol LM, Goldberg MF (1981) Chorioretinal and choriovitreal neovascularization. Their presence after photocoagulation of proliferative sickle cell retinopathy. Arch Ophthalmol 99:842–849

30. Downes SM, Hambleton IR, Chuang EL, Lois N, Serjeant GR, Bird AC (2005) Incidence and natural history of proliferative sickle cell retinopathy observations from a cohort study. Ophthalmology 112:1869–1875

31. Durant WJ, Jampol LM, Daily M (1982) Exudative retinal detachment in hemoglobin SC disease. Retina 2:152–154

32. Elliot AJ (1975) 30-year observation of patients with Eales' disease. Am J Ophthalmol 80(3 Pt 1):404–408

33. Embury SH, Steinberg MH (1994) Genetic modulators of disease. In: Embury SH, Hebbel RP, Mohandas N, Steinberg MH (eds) Sickle cell disease: basic principles and practice. Raven, New York, pp 279–298

34. Farber MD, Jampol LM, Fox P, Moriarty BJ, Acheson RW, Rabb MF, Serjeant GR (1991) A randomized clinical trial of scatter photocoagulation of proliferative sickle cell retinopathy. Arch Ophthalmol 109:363–367

34a. Ferrone PJ, Trese MT, Williams GA, Cox MS (1998) Good visual acuity in an adult population with marked posterior segment changes secondary to retinopathy of prematurity. Retina 18:335–338

35. Fox PD, Acheson RW, Serjeant GR (1990) Outcome of iatrogenic choroidal neovascularisation in sickle cell disease. Br J Ophthalmol 74:417–420

36. Francis RB Jr, Johnson CS (1991) Vascular occlusion in sickle cell disease: current concepts and unanswered questions. Blood 77:1405–1414

37. Freilich DB, Seelenfreund MH (1977) Long-term follow-up of scleral buckling procedures with sickle cell disease and retinal detachment treated with the use of hyperbaric oxygen. Mod Probl Ophthalmol 18:368–372

38. Gagliano DA, Goldberg MF (1989) The evolution of salmon-patch hemorrhages in sickle cell retinopathy. Arch Ophthalmol 107:1814–1815

39. Galinos S, Goldberg MF (1972) Photocoagulation therapy of sickle cell retinopathy. Sight Sav Rev 42:201–208

40. Galinos SO, Asdourian GK, Woolf MB, Stevens TS, Lee CB, Goldberg MF, Chow JC, Busse BJ (1975) Spontaneous remodeling of the peripheral retinal vasculature in sickling disorders. Am J Ophthalmol 79:853–870

41. Girard LJ, Beltranena F (1960) Early and late complications of extensive muscle surgery. Arch Ophthalmol 64:576–584

42. Goldbaum MH (1978) Retinal depression sign indicating a small retinal infarct. Am J Ophthalmol 86:45–55

43. Goldbaum MH, Galinos SO, Apple D, Asdourian GK, Nagpal K, Jampol L, Woolf MB, Busse B (1976) Acute choroidal ischemia as a complication of photocoagulation. Arch Ophthalmol 94:1025–1035

44. Goldbaum MH, Jampol LM, Goldberg MF (1978) The disc sign in sickling hemoglobinopathies. Arch Ophthalmol 96:1597–160.

45. Goldberg MF (1971) Classification and pathogenesis of proliferative sickle retinopathy. Am J Ophthalmol 71:649–665

46. Goldberg MF (1971) Natural history of untreated proliferative sickle retinopathy. Arch Ophthalmol 85:428–437

47. Goldberg MF (1971) Treatment of proliferative sickle retinopathy. Trans Am Acad Ophthalmol Otolaryngol 75:532–556

48. Goldberg MF, Jampol LM (1983) Treatment of neovascularization, vitreous hemorrhage, and retinal detachment in sickle cell retinopathy. Trans New Orleans Acad Ophthalmol 31:53–81

49. Hannon JF (1956) Vitreous hemorrhages associated with sickle cell-hemoglobin C disease. Am J Ophthalmol 42:707–712

50. Hasan S, Elbedawi M, Castro O, Gladwin M, Palestine A (2004) Central retinal vein occlusion in sickle cell disease. South Med J 97:202–204

51. Hayes RJ, Condon PI, Serjeant GR (1981) Haematological factors associated with proliferative retinopathy in homozygous sickle cell disease. Br J Ophthalmol 65:29–35

52. Hayes RJ, Condon PI, Serjeant GR (1981) Haematological factors associated with proliferative retinopathy in sickle cell-haemoglobin C disease. Br J Ophthalmol 65:712–717

53. Hayreh SS, Baines JA (1972) Occlusion of the posterior ciliary artery. I. Effects on choroidal circulation. Br J Ophthalmol 56:719–735

54. Hayreh SS, Baines JA (1972) Occlusion of the posterior ciliary artery. II. Chorio-retinal lesions. Br J Ophthalmol 56:736–753

55. Hebbel RP, Osarogiagbon R, Kaul D (2004) The endothelial biology of sickle cell disease: inflammation and a chronic vasculopathy. Microcirculation 11:129–151

56. Hebbel RP, Vercellotti GM (1997) The endothelial biology of sickle cell disease. J Lab Clin Med 129:288–293

57. Henry M, Driscoll MC, Miller M, Chang T, Minniti CP (2004) Pseudotumor cerebri in children with sickle cell disease: a case series. Pediatrics 113:e265–e269

58. Henry MD, Chapman AZ (1954) Vitreous hemorrhage and retinopathy associated with sickle-cell disease. Am J Ophthalmol 38:204–209

59. Hingorani M, Bentley CR, Jackson H, Betancourt F, Arya R, Aclimandos WA, Bird AC (1996) Retinopathy in haemoglobin C trait. Eye 10:338–342

60. Hrisomalos NF, Jampol LM, Moriarty BJ, Serjeant G, Acheson R, Goldberg MF (1987) Neodymium-YAG laser vitreolysis in sickle cell retinopathy. Arch Ophthalmol 105:1087–1091

61. Jackson H, Bentley CR, Hingorani M, Atkinson P, Aclimandos WA, Thompson GM (1995) Sickle retinopathy in patients with sickle trait. Eye 9:589–593

62. Jampol LM, Condon P, Farber M, Rabb M, Ford S, Serjeant G (1983) A randomized clinical trial of feeder vessel photocoagulation of proliferative sickle cell retinopathy. I. Preliminary results. Ophthalmology 90:540–545

63. Kampik A, Kenyon KR, Michels RG, Green WR, de la Cruz ZC (1981) Epiretinal and vitreous membranes. Comparative study of 56 cases. Arch Ophthalmol 99:1445–1454

64. Kaul DK, Fabry ME, Nagel RL (1996) The pathophysiology of vascular obstruction in the sickle syndromes. Blood Rev 10:29–44

65. Kaul DK, Hebbel RP (2000) Hypoxia/reoxygenation causes inflammatory response in transgenic sickle mice but not in normal mice. J Clin Invest 106:411–420

66. Kimmel AS, Magargal LE, Tasman WS (1986) Proliferative sickle retinopathy and neovascularization of the disc: regression following treatment with peripheral retinal scatter laser photocoagulation. Ophthalmic Surg 17:20–22

67. Kinney TR WR (1994) Compound heterozygous states. In: Embury H, Hebbel RP, Mohandas N, Steinberg NH (eds) Sickle cell disease: basic principles and practice. Raven, New York, pp 437–451

68. Kirchhof B, Kirchhof E, Ryan SJ, Sorgente N (1988) Human retinal pigment epithelial cell cultures: phenotypic modulation by vitreous and macrophages. Exp Eye Res 47:457–463

69. Lee CB, Woolf MB, Galinos SO, Goldbaum MH, Stevens TS, Goldberg MF (1975) Cryotherapy of proliferative sickle retinopathy. Part I. Single freeze-thaw cycle. Ann Ophthalmol 7:1299–1308

70. Leen JS, Ratnakaram R, Del Priore LV, Bhagat N, Zarbin MA (2002) Anterior segment ischemia after vitrectomy in sickle cell disease. Retina 22:216–219

71. Lieb WA GWGD (1951) Ocular and systemic manifestations of sickle cell disease. Acta Ophthalmol Scand 58 [Suppl]: 25–45

72. Ludvigsen FB (1998) Hemoglobin synthesis and function. In: Anne Stienne-Martin E, Lotspeich-Steininger CA, Koepke JA (eds) Clinical hematology: principles, procedures, correlations. Lippincott, Philadelphia, pp 73–87

73. Marsh RJ, Ford SM, Rabb MF, Hayes RJ, Serjeant GR (1982) Macular vasculature, visual acuity, and irreversibly sickled cells in homozygous sickle cell disease. Br J Ophthalmol 66:155–160

74. Mason JO III (2002) Surgical closure of macular hole in association with proliferative sickle cell retinopathy. Retina 22:501–502

75. Michaelson IC (1980) Textbook of the fundus of the eye. Michaelson IC, Ben-Ezra D (eds). Oxford University Press, Oxford

76. Morgan CM, D'Amico DJ (1987) Vitrectomy surgery in proliferative sickle retinopathy. Am J Ophthalmol 104:133–138

77. Moriarty BJ, Acheson RW, Serjeant GR (1987) Epiretinal membranes in sickle cell disease. Br J Ophthalmol 71:466–469

78. Nagpal KC, Asdourian GK, Patrianakos D, Goldberg MF, Rabb MF, Goldbaum M, Raichand M (1977) Proliferative retinopathy in sickle cell trait. Report of seven cases. Arch Intern Med 137:325–328

79. Ober RR, Michels RG (1978) Optic disk neovascularization in hemoglobin SC disease. Am J Ophthalmol 85:711–714

80. Osarogiagbon UR, Choong S, Belcher JD, Vercellotti GM, Paller MS, Hebbel RP (2000) Reperfusion injury pathophysiology in sickle transgenic mice. Blood 96:314–320

81. Peachey NS, Charles HC, Lee CM, Fishman GA, Cunha-Vaz JG, Smith RT (1987) Electroretinographic findings in sickle cell retinopathy. Arch Ophthalmol 105:934–938

82. Peachey NS, Gagliano DA, Jacobson MS, Derlacki DJ, Fishman GA, Cohen SB (1990) Correlation of electroretinographic findings and peripheral retinal nonperfusion in patients with sickle cell retinopathy. Arch Ophthalmol 108:1106–1109

83. Penman AD, Talbot JF, Chuang EL, Thomas P, Serjeant GR, Bird AC (1994) New classification of peripheral retinal vascular changes in sickle cell disease. Br J Ophthalmol 78: 681–689

84. Pulido JS, Flynn HW Jr, Clarkson JG, Blankenship GW (1988) Pars plana vitrectomy in the management of complications of proliferative sickle retinopathy. Arch Ophthalmol 106:1553–1557

85. Raichand M, Dizon RV, Nagpal KC, Goldberg MF, Rabb MF, Goldbaum MH (1978) Macular holes associated with proliferative sickle cell retinopathy. Arch Ophthalmol 96:1592–1596

86. Raichand M, Goldberg MF, Nagpal KC, Goldbaum MH, Asdourian GK (1977) Evolution of neovascularization in sickle cell retinopathy. A prospective fluorescein angiographic study. Arch Ophthalmol 95:1543–1552

87. Rednam KR, Jampol LM, Goldberg MF (1982) Scatter retinal photocoagulation for proliferative sickle cell retinopathy. Am J Ophthalmol 93:594–599

88. Renie WA, Murphy RP, Anderson KC, Lippman SM, McKusick VA, Proctor LR, Shimizu H, Patz A, Fine SL (1983) The evaluation of patients with Eales' disease. Retina 3:243–248

89. Romayananda N, Goldberg MF, Green WR (1973) Histopathology of sickle cell retinopathy. Trans Am Acad Ophthalmol Otolaryngol 77:642–676

90. Roy MS, Rodgers G, Gunkel R, Noguchi C, Schechter A (1987) Color vision defects in sickle cell anemia. Arch Ophthalmol 105:1676–1678

91. Rubinstein K (1967) Sicklaemia retinopathy. Br J Ophthalmol 51:217–221

92. Ryan SJ, Goldberg MF (1971) Anterior segment ischemia following scleral buckling in sickle cell hemoglobinopathy. Am J Ophthalmol 72:35–50

93. Schnog JB, Duits AJ, Muskiet FA, ten Cate H, Rojer RA, Brandjes DP (2004) Sickle cell disease; a general overview. Nether J Med 62:364–374

94. Seiberth V (1999) Trans-scleral diode laser photocoagulation in proliferative sickle cell retinopathy. Ophthalmology 106:1828–1829

95. Serjeant BE, Mason KP, Acheson RW, Maude GH, Stuart J, Serjeant GR (1986) Blood rheology and proliferative retinopathy in homozygous sickle cell disease. Br J Ophthalmol 70:522–525

96. Serjeant BE, Mason KP, Condon PI, Hayes RJ, Kenny MW, Stuart J, Serjeant GR (1984) Blood rheology and proliferative retinopathy in sickle cell-haemoglobin C disease. Br J Ophthalmol 68:325–328

97. Serjeant GR (1997) Sickle-cell disease. Lancet 350:725–730

98. Serjeant GR (2001) The emerging understanding of sickle cell disease. Br J Haematol 112:3–18

99. Serjeant GR, Serjeant BE, Forbes M, Hayes RJ, Higgs DR, Lehmann H (1986) Haemoglobin gene frequencies in the Jamaican population: a study in 100,000 newborns. Br J Haematol 64:253–262

100. Spitznas M, Meyer-Schwickerath G, Stephan B (1975) The clinical picture of Eales' disease. Albrecht Von Graefes Arch Klin Exp Ophthalmol 194:73–85

101. Stevens TS, Busse B, Lee CB, Woolf MB, Galinos SO, Goldberg MF (1974) Sickling hemoglobinopathies; macular and perimacular vascular abnormalities. Arch Ophthalmol 92:455–463

102. Talbot JF, Bird AC, Maude GH, Acheson RW, Moriarty BJ, Serjeant GR (1988) Sickle cell retinopathy in Jamaican children: further observations from a cohort study. Br J Ophthalmol 72:727–732

103. Talbot JF, Bird AC, Rabb LM, Maude GH, Serjeant GR (1983) Sickle cell retinopathy in Jamaican children: a search for prognostic factors. Br J Ophthalmol 67:782–785

104. Talbot JF, Bird AC, Serjeant GR (1983) Retinal changes in sickle cell/hereditary persistence of fetal haemoglobin syndrome. Br J Ophthalmol 67:777–778

105. Talbot JF, Bird AC, Serjeant GR, Hayes RJ (1982) Sickle cell retinopathy in young children in Jamaica. Br J Ophthalmol 66:149–154

106. Treister G, Machemer R (1977) Results of vitrectomy for rare proliferative and hemorrhagic diseases. Am J Ophthalmol 84:394–412

107. Van Meurs JC (1991) Choroidal filling patterns in sickle cell patients. Int Ophthalmol 15:49–52

108. Van Meurs JC (1991) Ocular findings in sickle cell patients on Curacao. Int Ophthalmol 15:53–59

109. Van Meurs JC (1991) Relationship between peripheral vascular closure and proliferative retinopathy in sickle cell disease. Graefes Arch Clin Exp Ophthalmol 229:543–548

110. Van Meurs JC (1995) Evolution of a retinal hemorrhage in a patient with sickle cell-hemoglobin C disease. Arch Ophthalmol 113:1074–1075

111. Van Nouhuys CE (1982) Dominant exudative vitreoretinopathy and other vascular developmental disorders of the peripheral retina. Doc Ophthalmol 54:1–414

112. Walton W, von Hagen S, Grigorian R, Zarbin M (2002) Management of traumatic hyphema. Surv Ophthalmol 47:297–334

113. Wang WC LJ (1998) Sickle cell anemia and other sickling syndromes. In: Lee R, Lukens J, Rodgers GM, Paraskevas F, Foerster J (eds) Wintrobe's clinical hematology. Williams and Wilkins, Baltimore

114. Welch RB, Goldberg MF (1966) Sickle-cell hemoglobin and its relation to fundus abnormality. Arch Ophthalmol 75:353–362

115. Wilson WA, Irvine SR (1955) Pathologic changes following disruption of blood supply to iris and ciliary body. Trans Am Acad Ophthalmol Otolaryngol 59:501–502

28 Vascular Tumors of the Retina

28.1 Histopathology of Retinal Vascular Tumors and Selected Vascular Lesions

M.A. CHANG, W.R. GREEN

Core Messages

- Vascular tumors of the retina are rare, benign lesions that may be associated with systemic disease, or may be an isolated finding
- Although decreases in visual acuity may be associated with certain types of lesions, initial observation is usually advocated
- It is important to distinguish these benign pro-

cesses from malignant tumors such as melanoma and metastatic tumors
- Characteristic fluorescein angiographic or echographic findings may be useful in securing the diagnosis
- Treatment may consist of laser photocoagulation, cryotherapy, radiotherapy, vitrectomy, or a combination of these

28.1.1 Cavernous Hemangioma

Essentials

- Benign, unilateral, congenital
- Cluster of slow-filling, vascular globules
- Vascular channels have normal walls and non-fenestrated endothelium and do not leak

These congenital, non-progressive lesions are benign, and often solitary and unilateral. Patients are usually asymptomatic, although some may experience decreased vision and neurologic symptoms, such as seizures and cranial nerve palsies [15, 87]. Cutaneous or cerebral hemangiomas may also be found, especially in cases of inherited lesions [2, 4, 15, 22, 23, 58]. Pedigree analysis suggests that this neurooculocutaneous syndrome may be inherited in an autosomal dominant pattern, with incomplete penetrance or variable expressivity [23, 58].

Ophthalmoscopy reveals a cluster of saccular aneurysms filled with venous blood one to two disk diameters in size, found in the mid-peripheral or peripheral retina. Occasional cases of posterior pole or peripapillary lesions have been reported [30, 42, 50, 60] (Fig. 28.1.1a). Overlying retinal hemorrhage or grayish-white preretinal membrane may be seen. No feeder vessels are present. The vascular globules have normal vascular permeability, and only rarely

lead to hard exudation. Fluorescein angiography (FA) reveals a normal arterial supply and slow, incomplete filling of the saccules, with characteristic layering of the erythrocytes in the inferior aspect of the saccules, and plasma in the superior aspect (Fig. 28.1.1b). The erythrocytes appear hypofluorescent, while the overlying plasma is hyperfluorescent. No leakage is seen.

Visual acuity may be impaired if the lesion is located near the macula, if a preretinal membrane overlies the lesion, or if vitreous hemorrhage from retinal traction occurs. Amblyopia has been reported in children with vitreous hemorrhage from cavernous hemangiomas [85]. It is relatively easy to distinguish cavernous hemangiomas from other retinal telangiectases due to Coats' disease, branch retinal vein occlusion, capillary hemangioma, and racemose hemangioma. However, they may be confused with Leber's miliary aneurysms. Cavernous hemangiomas tend to stay stable in size while Leber's is progressive and involves intrinsic retinal vasculature [22].

Histopathology of cavernous hemangioma
Histopathologically, the retina is thickened by multiple, large vascular channels with normal walls and non-fenestrated endothelium arising from the inner half of the retina [25] (Fig. 28.1.2a). In the area of the saccular aneurysms, the inner retinal layers are discontinuous. Thin stromal tissue separates the aneurysms. Vitreous condensates overlying the lesion may lead to hemorrhage from retinal traction in the case of a posterior vitreous detachment (Fig. 28.1.2b).

28 III

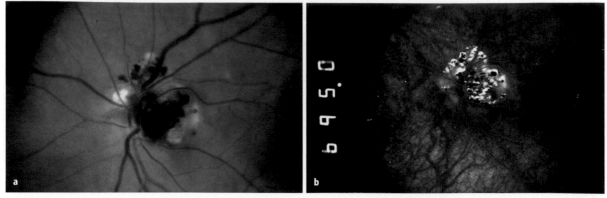

Fig. 28.1.1. Cavernous hemangioma of the optic nerve head and peripapillary retina. **a** Ophthalmoscopic appearance of a cluster of saccular globules filled with dark venous blood. **b** Fluorescein angiography of the lesion reveals incomplete filling of the saccules, and layering of the erythrocytes inferiorly with hypofluorescence. The overlying plasma is hyperfluorescent

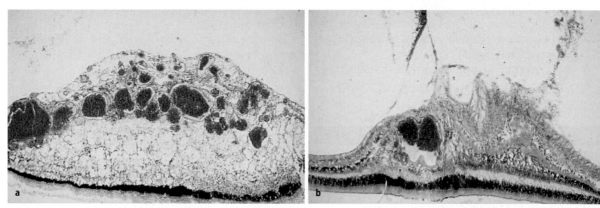

Fig. 28.1.2. Cavernous hemangioma of the retina. **a** The retina is edematous, and large, normal-appearing vessels are present in the inner retinal layers. H&E, original magnification ×40. **b** A different area has strands of vitreous with entrapped blood exerting traction on the cavernous hemangioma. H&E, original magnification ×40

Since most cavernous hemangiomas remain stable in size and rarely lead to severe vitreous hemorrhage or lipid exudate, observation is the mainstay of therapy. Cryotherapy, laser photocoagulation, or even vitrectomy may be performed in cases of severe vitreous hemorrhage [26].

28.1.2 Capillary Hemangioma

Essentials
- Benign, congenital, may be bilateral or multifocal
- May be associated with von Hippel-Lindau disease
- Dilated afferent and efferent vessels
- Abnormal capillary-like vasculature may have fenestrated endothelium and leak
- Lipidized fibrous astrocytes are present between vascular channels, and appear as vacuolated interstitial cells

Capillary hemangiomas are benign, congenital hamartomas which are typically found in the 2nd or 3rd decade of life, and have no gender or racial predilection. In up to 50% of cases, these tumors may be bilateral or multifocal [10]. The tumors are most commonly endophytic, involving the inner retinal layers, but may be exophytic, involving the outer layers of the retina. Both types may be seen in a peripheral or juxtapapillary location, though peripheral angiomas are more common [36, 49, 83]. Juxtapapillary hemangiomas tend to have worse visual prognosis, but are less likely to progress [54].

Both juxtapapillary and peripheral retinal capillary hemangiomas may be isolated findings, or may be associated with von Hippel-Lindau disease in about 20% of patients [48]. Von Hippel-Lindau disease is an autosomal dominant disorder linked to mutations in the VHL tumor suppressor gene on the short arm of chromosome 3 [64]. Along with retinal capillary hemangiomas, which are the most common and earliest manifestation in many cases, cerebellar and spinal cord hemangioblastomas, pheochromocytoma, and renal cell carcinomas are found. A suggestive family history warrants a thorough evaluation for other manifestations of von Hippel-Lindau disease, including magnetic resonance imaging of the brain and spinal cord, abdominal computed tomography, and urinary catecholamine studies [12]. In addition, genetic testing for von Hippel-Lindau disease, which is 99% sensitive, is available [73].

Early lesions may not have abnormal afferent and efferent vessels, and appear as a small red or gray nodule. With enlargement of the lesion, a more cluster-like appearance develops, and feeder vessels enlarge. The characteristic ophthalmoscopic appearance is that of a reddish or gray, round retinal lesion with prominent afferent and efferent vessels (Figs. 28.1.3, 28.1.4). Preretinal membranes or fibrovascular proliferation may be present on the surface of larger hemangiomas. Fluorescein angiography classically shows dilated feeder vessels, hyperfluorescence of the vascular lesion, and late continuous leakage.

Decreased vision may occur secondary to macular lipid exudation, vitreous hemorrhage, or exudative

retinal detachment. Additionally, glial proliferation on the surface of these lesions may lead to tractional retinal detachment.

Capillary hemangiomas have a characteristic clinical appearance, but associated findings may produce confusion. Associated exudation may simulate Coats' disease, choroidal neovascular membrane, familial exudative vitreoretinopathy, retinal macroaneurysm, and retinal angiomatous proliferation, while vitreous hemorrhage from neovascularization occurring on the surface of larger tumors may not be distinguishable from other causes of neovascularization such as diabetes. Exudative retinal detachment may mimic inflammatory causes such as sarcoid and toxoplasmosis, or other tumors such as melanoma and retinoblastoma.

If von Hippel-Lindau disease is not present, these lesions may be confused with vasoproliferative retinal tumors. However, the typical dilated feeder and draining vessels of capillary hemangiomas are not usually present in vasoproliferative retinal tumors. Juxtapapillary lesions may also be confused with papilledema, though papilledema is likely to be bilateral while juxtapapillary hemangiomas are usually unilateral.

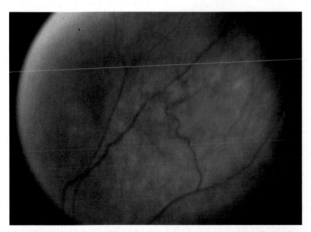

Fig. 28.1.3. Small capillary hemangioma in von Hippel-Lindau disease with dilated feeder vessels

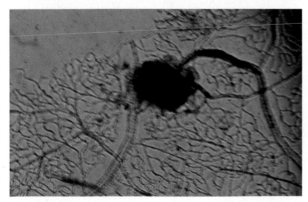

Fig. 28.1.4. Trypsin digest preparation of a small retinal hemangioma illustrates the feeder artery and efferent vein. H&E, periodic acid-Schiff, original magnification ×45

Histopathology of capillary hemangioma

Histopathologically, the tumor is composed of tortuous, capillary-like vessels lined by endothelium and delicate reticulum (Fig. 28.1.5), with larger feeding and draining vessels [36, 56] (Fig. 28.1.4). Pericytes, endothelial cells, and rare multilaminar pericytes with smooth muscle differentiation can be seen on electron microscopy [33]. In larger lesions, the abnormal capillary-like vessels may have fenestrated endothelium, and have a propensity to leak, leading to exudation that may lead to disciform scarring or exudative retinal detachment [33]. Cystic degeneration of surrounding retina may be present.

Collagenous fibrous tissue and vacuolated interstitial cells, representing lipidized fibrous astrocytes, are usually present between the vascular channels [33, 56] (Figs. 28.1.6, 28.1.7). The lipid in these cells is likely derived from leakage of plasma from the intravascular space. A study of patients with a history of von Hippel-Lindau (VHL) disease and retina angiomas found expression of vascular endothelial growth factor and loss of heterozygosity of the VHL gene in these vacuolated stromal cells [11]. Interestingly, this loss of heterozygosity was not found in vascular cells or glial tissue, suggesting that the vacuolated stromal cells may be the true neoplastic component of retinal angiomas.

Endophytic juxtapapillary tumors have minimal interstitial tissue, whereas exophytic tumors are more dispersed within retinal tissue, and have more prominent interstitium [56]. Fusiform thickening of the retina occurs in the areas of tumor. In some cases, the tumor may extend through a defect in the internal limiting membrane with overlying vitreous condensation; vitreous traction may then lead to hemorrhage [25] (Fig. 28.1.5).

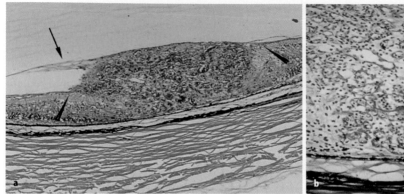

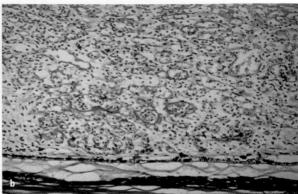

Fig. 28.1.5. Full-thickness retinal capillary hemangioma. **a** The tumor is extending through a defect in the internal limiting membrane (ILM) (*between arrowheads*) and onto the posterior surface of the detached vitreous (*arrow*). H&E, original magnification ×70. **b** Higher magnification reveals numerous capillary vessels in the hemangioma. Wilder reticulin, original magnification ×10

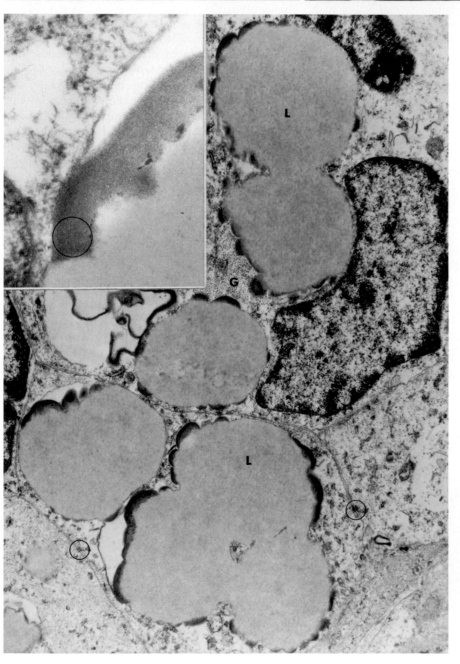

Fig. 28.1.6. Electron microscopy of retinal capillary hemangioma. Clusters of interstitial cells show large vacuolar intracytoplasmic lipid inclusions (*L*) that are homogenous, and of medium electron density. Glycogen granules (*G*) and desmosomes (*circles*) are also present. Original magnification ×17,000. At higher magnification (*inset*), the highly electron-dense material at the periphery of the lipid vacuoles had 6–7 nm periodicity, characteristic of complex lipids. Original magnification ×90,000. (From [56])

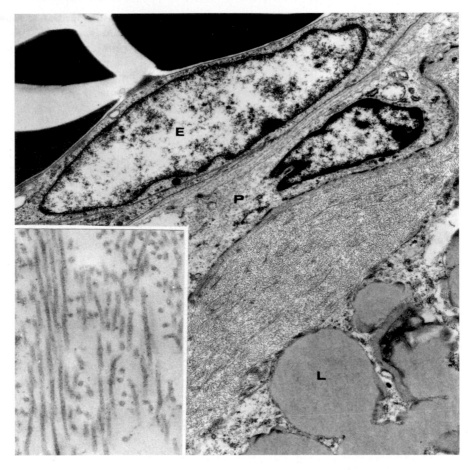

Fig. 28.1.7. Electron microscopy of retinal capillary angioma. Capillary endothelial (*E*) and perithelial (*P*) cells are present with intervening basement membrane, lipid-containing cells (*L*), and abundant fibrous interstitial tissue. Original magnification × 12,000. At higher magnification (*inset*), the fibrous tissue is composed of collagen fibrils of uniformly small diameter (22–24 nm). Original magnification × 50,000. (From [56])

Initially, most retinal capillary hemangiomas less than 500 μm in size may be carefully observed. Treatment should be considered for lesions which are larger than 500 μm, located in a vision threatening area, or associated with subretinal fluid or exudates. Laser photocoagulation is effective for lesions smaller than 1,500 μm, located in the posterior pole. Treatment is applied to the surface of the lesion and on the feeding artery to reduce blood flow through the tumor. For larger, more anterior tumors, cryotherapy may be required, and is generally applied until the lesion is completely enclosed by the ice ball [68]. Proton beam [57] and plaque radiotherapy [39, 68] have been used for capillary hemangiomas larger than 4 mm, since those lesions are poorly responsive to laser photocoagulation and cryotherapy. However, experience with radiotherapy has been limited to small numbers of patients. Transpupillary thermotherapy has been used for juxtapapillary tumors with mixed results [59, 68]. Capillary hemangiomas associated with exudative and tractional detachments often require vitrectomy and diathermy or endolaser, but outcomes are often poor [35, 47].

28.1.3 Retinal Vasoproliferative Tumors

Essentials
- Peripheral vascular tumors similar to capillary hemangiomas but with normal feeding and draining vessels
- Not associated with von Hippel-Lindau disease
- May be secondary to other ocular conditions
- Reactive retinal gliosis and hyalinization of vessel walls is common

These tumors were previously described as "retinal angiomas" [9], "angioma-like lesions" [21], "peripheral retinal telangiectasis" [46], and "retinal angiomatous masses" [24]. Retinal vasoproliferative tumor is a term coined by Shields et al. in 1995 [65] to describe peripheral vascular tumors without systemic or familial associations that are distinguished from capillary hemangiomas associated with von Hippel-Lindau disease [65]. Primary tumors are idiopathic, small, solitary, reddish-yellow lesions with normal feeding and draining vessels, found most often in the inferotemporal quadrant of the fundus near the ora serrata in middle-aged individuals. Secondary lesions are associated with other ocular conditions, most commonly retinitis pigmentosa, intermediate uveitis, retinal detachment, and toxoplasmosis [28, 45, 65]. Bilateral circumscribed tumors are rare [65]. A diffuse tumor has also been described. These tend to be bilateral, located posteriorly, and are more common in women [65].

These lesions are often associated with exudation, subretinal fluid, and vitreous hemorrhage. Though the tumor is often in the peripheral retina, macular changes such as epiretinal membrane, exudate, and edema may occur. Retinal pigment epithelial proliferation adjacent to tumors may be present. Fluorescein angiography demonstrates rapid early filling of the tumor through a non-dilated or minimally dilated feeder vessel, and late diffuse leakage and staining. Medium to high internal reflectivity on A-scan ultrasonography and echodensity on B scan ultrasonography are typical findings [65].

Exudation and telangiectatic dilation of retinal vessels associated with diffuse vasoproliferative tumors may be confused with Coats' disease. However, Coats' disease is usually unilateral, and found in young men, while diffuse vasoproliferative tumors are usually bilateral, and found in women. Circumscribed vasoproliferative tumors can be distinguished from capillary hemangiomas by the lack of dilated and tortuous feeding and draining vessels. Additionally, capillary hemangiomas associated with von Hippel-Lindau will have younger age of presentation, other systemic features of VHL, family history, and possibly multiple tumors. Choroidal melanomas and metastases are usually confined to the choroidal space, but should also be included in the differential, as these can involve the retina in later stages and can be associated with enlarged retinal vasculature mimicking feeder and draining vessels. Peripheral exudative hemorrhagic chorioretinopathy may also mimic retinal vasoproliferative tumors, with hemorrhage and exudates in subretinal, retinal, and vitreous spaces [3]. However, peripheral exudative hemorrhagic chorioretinopathy is usually subretinal in location, bilateral, associated with macular degeneration, and darkly pigmented [3].

Histopathology of retinal vasoproliferative tumors

Shields et al. hypothesized that these tumors develop secondary to underlying disease processes that stimulate reactive gliosis and vascular and pigment epithelial proliferation [65]. Histopathologic studies of vasoproliferative tumors have demonstrated marked spindle cell gliosis of the retina with cystic changes. These irregularly arranged, spindle-shaped cells stain positively with glial fibrillary acid protein [32, 34], and have no cellular atypia or mitotic figures. The vessels may be dilated or small and fibrotic, but hyalinization and lack of smooth muscle in vessel walls is usually found. In some vessels, thrombosis and intraluminal endothelial cell proliferation is present. In some instances, the lesion develops from choroidal neovascularization and the vascular complex is fed by an artery from the choroid (Fig. 28.1.8). Lipid, serous, and fibrinous exudates may be present in the connective tissue surrounding the abnormal vasculature [32]. Basal deposits on the inner surface of Bruch membrane, retinal pigment epithelium (RPE) proliferation and metaplasia leading to fibrosis, preretinal neovascularization, and associated retinal hemorrhage and exudate have also been described [28, 32, 65, 71].

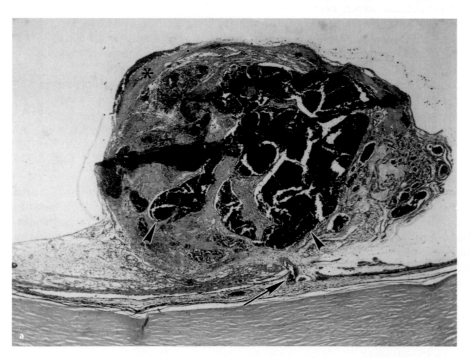

Fig. 28.1.8. Peripheral retinal vasoproliferative tumor. **a** Cavernous-like vascular channels (*arrowheads*) and hemorrhage (*asterisk*) are surrounded by fibrous connective tissue. A well-developed artery and vein traverse into the tumor through a defect in the ciliary epithelium (*arrow*). H&E, original magnification ×35.

Fig. 28.1.8. b Higher magnification shows the artery (*arrow*) and vein (*arrowhead*) extend from the choroid and into the tumor at the ora serrata. H&E, original magnification × 125. (From [25])

Most of these lesions can be observed, unless vision-threatening subretinal fluid or exudates, macular edema or epiretinal membrane is present. Laser photocoagulation, cryotherapy, plaque radiotherapy, and vitrectomy with endolaser and/or membranectomy are useful modalities that have been used in a limited number of patients [28, 34, 43, 65]. For patients in whom a malignant process cannot be excluded clinically, transscleral resection may prevent unnecessary enucleation by giving a tissue diagnosis [32, 71].

28.1.4 Combined Hamartoma of the Retinal Pigment Epithelium and Retina

Essentials
- Congenital, unilateral, solitary
- Elevated, often pigmented lesions most commonly found at optic nerve head
- Proliferation of retinal pigment epithelium around blood vessels
- Fluorescein angiography shows a spidery vascular pattern

These benign tumors containing vascular, glial, and retinal pigment epithelial components may be mistaken for choroidal melanoma or other malignant intraocular tumors. They are most likely congenital, as the diagnosis has been made in patients in the first few months of life [20, 62]. However, some cases have occurred following intraocular inflammation [13, 77], and may be acquired. Patients are usually diagnosed in the 2nd decade, but have presented as late as the 7th decade [20, 62]. There is no clear gender predilection, though it appears that Caucasians are more likely to be affected [20, 62]. Most cases are isolated ocular findings, though an association with neurofibromatosis types 1 and 2 has been found [6, 38, 69, 79, 81], especially for those with bilateral lesions.

Lesions are usually unilateral and solitary. Most involve the optic nerve head or juxtapapillary area, though macular lesions are also common. Midperipheral lesions have been noted, but comprise only 5% of cases [62]. Clinically, lesions of the posterior pole tend to be slightly elevated, black or charcoal gray, and have features of vascular, glial, and retinal pigment epithelial tissue, though one type usually predominates. If retinal pigment epithelial tissue is not prominent, the lesion may not be pigmented (Fig. 28.1.9a). Intralesional vascular tortuosity, contraction of the tumor's inner surface with juxtalesional tractional retinal distortion, vitreoretinal interface changes, and epiretinal membrane formation are common. The choroid is not involved. With follow-up examination, growth may be demonstrated. On early phases of fluorescein angiography, the lesion appears hypofluorescent. Vascular tortuosity and telangiectasis in a spidery configuration within the lesion, and traction of retinal vessels at the periphery of the hamartoma, are commonly detected

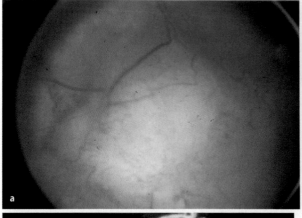

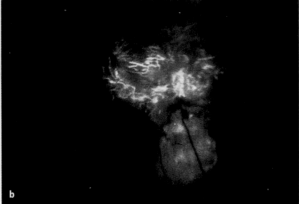

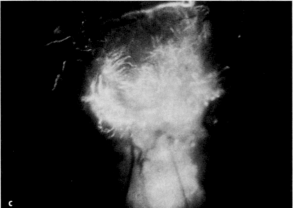

Fig. 28.1.9. Combined hamartoma of the RPE and retina. **a** Ophthalmoscopic appearance of an elevated, well-circumscribed amelanotic lesion adjacent to the optic nerve head. **b** Arterial phase of fluorescein angiography reveals spidery vasculature within the lesion. **c** Later frames reveal hyperfluorescence and leakage

[14, 37, 55, 83], choroidal neovascularization [18, 31, 62, 75], preretinal neovascularization [29], lipid exudation [62] or macular holes [53, 80].

Isolated epiretinal membrane may be considered in the differential diagnosis of combined hamartomas, as these may also be associated with vascular tortuosity and vitreoretinal interface changes. However, isolated epiretinal membranes are rarely pigmented. Other elevated, pigmented fundus lesions, such as congenital hypertrophy of the retinal pigment epithelium, choroidal nevi and choroidal melanoma, may also be confused with combined hamartomas, though these are usually not associated with vascular tortuosity and vitreoretinal interface changes, and typically do not involve the retinal layer. When choroidal melanomas are advanced enough to involve the retina, they are usually much thicker than the typical thickness of combined hamartomas.

during arterial and early venous phases (Fig. 28.1.9b). Late frames disclose leakage and hyperfluorescence of the tortuous intralesional vasculature (Fig. 28.1.9c). The lesion is hyperreflective and elevated on optical coherence tomography, with hyporeflective shadowing posteriorly [78].

Though approximately 10% of reported cases are found on routine examination of asymptomatic patients, the majority of patients present with painless visual loss. Thirty to 45% of cases have initial presenting visual acuities of 20/40 or better, but 30–40% had visual acuity of 20/200 or worse on first presentation [20, 62]. Decreased visual acuity leads to strabismus in 20–25% of patients, usually children [20, 62]. Uncommon presentations include leukocoria, floaters, and pain [62]. Visual loss may occur with direct involvement of the optic nerve, macular, or papillomacular area, epiretinal membrane formation, or retinal folds. Uncommonly, these hamartomas may cause vitreous hemorrhage

Histopathology of combined hamartoma of the retinal pigment epithelium and retina

Histopathologically, retinal glial proliferation and disorganization of retinal architecture is common [82] (Fig. 28.1.10a). In peripapillary lesions, vascular proliferation and hyperplastic RPE cause thickening of the optic nerve head and surrounding retina. Vascular tortuosity may occur as a result of retinal layer disturbances, but abnormal capillary proliferation is also present [45] (Fig. 28.1.10b). Proliferation of the retinal pigment epithelium is often found surrounding blood vessels (Fig. 28.1.10c). The proliferating RPE often extends into the inner retinal layers, and may be present in preretinal fibrovascular tissue [25]. Vitreous condensation often overlies the lesions. Infiltration of the optic nerve by retinal pigment epithelial cells has been reported in juxtapapillary hamartomas [74].

With an average 4-year follow-up period, approximately 65% of patients retained visual acuity within 2 lines of presenting visual acuity [62]. However, 20% lost 2 or more lines of vision. Epiretinal membranectomy may improve visual acuity [53, 61, 62,

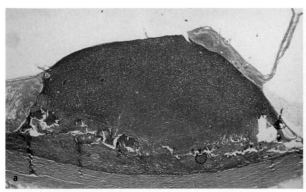

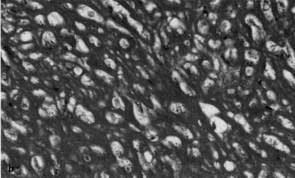

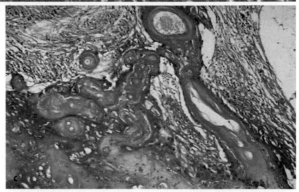

Fig. 28.1.10. Histopathology of combined hamartoma of the RPE and retina. **a** Low power view of the hamartoma reveals a mound-shaped lesion. H&E, original magnification × 10. **b** Higher power view discloses much of the lesion to be composed of layers and tubes of thickened basal lamina with only few remaining RPE cells. H&E, original magnification × 22. **c** High power reveals RPE proliferation surrounding blood vessels at the inner aspect of the lesion. H&E, original magnification × 25

72], with possible postoperative decrease in vascular tortuosity and leakage [61]. Vitrectomy for macular traction may also improve visual acuity [62], though in some cases anatomic success may not be associated with clinical improvement [62]. One case of choroidal neovascularization successfully treated with submacular surgical excision has been reported [31].

28.1.5 Racemose Hemangioma

Essentials
- Congenital retinal arteriovenous malformations
- May be associated with Wyburn-Mason syndrome
- Large vessels with variable wall thicknesses occupy the entire thickness of the retina

Racemose hemangiomas, also known as cirsoid aneurysms and congenital retinal arteriovenous communications, are enlarged vessels that connect and shunt blood from the retinal arterial to the retinal venous circulations (see Chapter 26.2). These are usually unilateral, are equally common in men and women, and are rarely hereditary [50]. Approximately 25–30% are associated with Wyburn-Mason syndrome, in which intracranial AV malformations are present, usually ipsilateral to the racemose hemangioma [50, 76] (see Chapter 26.2).

The retinal arteriovenous communications have a wide range of clinical manifestations. In the mildest form, the lesions may appear as small, tortuous, peripheral arteriovenous communications or can

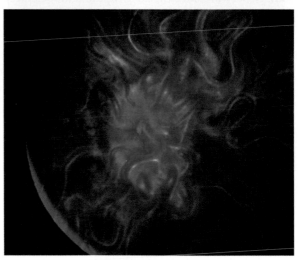

Fig. 28.1.11. Wyburn-Mason syndrome. Dilated, tortuous vessels emanate from the optic nerve head

appear as slightly enlarged retinal vessels that cross the horizontal midline. Patients may be asymptomatic at this stage. Larger lesions consist of dilated vessels emanating from the disk that follow a tortuous path through the fundus (Fig. 28.1.11). No spontaneous pulsation is seen [20]. Most lesions are found in the temporal quadrants or near the papillomacular bundle [50].

Visual loss is present in the majority of patients with large lesions, and may be secondary to optic nerve, retinal or intracranial involvement by these

28 III

vascular malformations, or from retinal capillary non-perfusion secondary to arteriovenous shunting [50]. In some cases, these lesions may lead to exudation, retinal or vitreous hemorrhage, macroaneurysms, neovascular glaucoma and central retinal vein occlusion [16, 51, 52, 63]. On fluorescein angiography, these enlarged vessels fill rapidly, with no leakage of dye. The arterial and venous circulations may appear to fill at the same time as a result of arteriovenous shunting.

When these lesions are small, the differential diagnosis should include other retinovascular diseases such as vein occlusion, sickle retinopathy, and ocular ischemic syndrome, as well as small capillary hemangiomas. Large lesions are unlikely to be confused with other entities.

Evaluation for Wyburn-Mason syndrome should be considered in patients with larger racemose hemangiomas, as these are more likely to be associated with Wyburn-Mason syndrome. Initial work-up may consist of magnetic resonance imaging and angiography of the brain, though conventional cerebral angiography may be warranted if the lesion is very large, or neurologic manifestations are present. A wide variety of neurological symptoms may be present, and depend on the areas of cerebral involvement. In addition to the pyramidal system, the visual pathways, including the optic tract and radiations, and midbrain structures tend to be involved. Cranial nerve palsies, internuclear ophthalmoplegia, nystagmus, and visual field loss have been described. Of patients with Wyburn-Mason syndrome, 25% also have facial AV malformations such as mandibular, maxillary, palatine, buccal mucosa, and nasopharynx lesions [76], with severe hemorrhage during dental procedures and recurrent epistaxis [7]. Another 25% have associated skin findings such as nevus flammeus [76]. Mild non-pulsatile proptosis due to the enlarged caliber of retrobulbar blood vessels may be present.

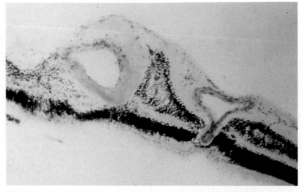

Fig. 28.1.12. Arteriovenous malformation of the retina. Large, thick walled vessels are present, and occupy the entire thickness of the retina. Masson trichrome, original magnification ×100. (From [8])

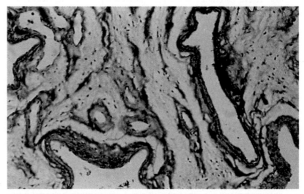

Fig. 28.1.13. Arteriovenous malformation in the optic nerve. The optic nerve is filled with thick-walled vessels. H&E, original magnification ×100. (From [8])

Histopathology of racemose hemangioma
Histopathologically, all retinal layers may be occupied by the enlarged vessels (Fig. 28.1.12), which adhere to Bruch membrane in some cases [8]. The distinction between arteries and veins may not be possible because all vessels appear to have walls of variable thickness, with fibrous hyalinization of the adventitial layers. Variable loss of ganglion cell and nerve fiber layers and cystoid changes of the retina may be present [8]. The optic nerve and tract may be distorted by abundant abnormal vasculature, consisting of thin venous channels or large, thick, muscular arteries [8] (Fig. 28.1.13).

These lesions are usually observed. Over time, venous thrombosis and vessel sclerosis may lead to a change in the appearance of the racemose hemangioma [53, 63]. Secondary findings such as venous

occlusion and macroaneurysm leakage may be treated accordingly.

28.1.6 Retinal Angiomatous Proliferation

Essentials
- Type of neovascular age-related macular degeneration with capillary proliferation within the retina and subretinal space
- May form retinal-choroidal anastomoses
- Intraretinal hemorrhages typically seen
- Fibrovascular membranes may be found in the subretinal space and within Bruch membrane

A small subset of patients with neovascular age-related macular degeneration (ARMD) have capillary proliferation within the inner retinal layers that can extend posteriorly to involve the subretinal spaces, termed "retinal angiomatous proliferation (RAP)" [86]. With progression, these lesions can become associated with sub-RPE and choroidal neovascularization, forming retinal-choroidal anastomoses [27, 40, 70]. Intraretinal hemorrhages seen at all stages may distinguish this form of neovascular ARMD [17, 86]. These lesions tend to be bilateral, with symmetric involvement [86]. Affected individuals are much more likely to be Caucasian and over the age of 80 [86]. Women are also more likely to be affected.

Clinically, RAP can be divided into three distinct stages. Stage I, or intraretinal neovascularization, is the earliest form of RAP [86]. The patient may be asymptomatic. A mass composed of angiomatous vasculature is present within the middle and inner retina, associated with intraretinal hemorrhages and edema. Occasional retinal-retinal anastomoses may be seen. Fluorescein angiography is notable for focal intraretinal leakage corresponding to the area of neovascularization and retinal edema [86].

Stage II, or subretinal neovascularization, occurs when the intraretinal neovascularization has extended to involve the space between the RPE and the photoreceptor layer. Clinically, these eyes have more tangential branching of the vasculature in the subretinal space, an increase in retinal edema, and a localized retinal detachment in the area of subretinal neovascularization. If neovascularization reaches or fuses with the RPE, a serous pigment epithelial detachment is usually seen. Subretinal, intraretinal, or preretinal hemorrhages may be seen. Retinal-retinal anastomoses are more common. Fluorescein angiography reveals diffuse intraretinal and subretinal leakage in the area of the lesion, making this difficult to differentiate from occult choroidal neovascularization; however, a hot spot on the indocyanine green (ICG) angiogram corresponding to the area of neovascularization can be seen [17].

Stage III, or choroidal neovascularization, occurs when both choroidal and retinal neovascularization are present. Retinal-choroidal anastomosis is presumed to have formed in such cases, though this may be difficult to detect clinically. Choroidal neovascularization usually appears more robust than the intraretinal or subretinal neovascular components. Disciform scarring may be present. Leakage in the sub-RPE and subretinal spaces occurs on fluorescein angiography. In the late phase of ICG angiography, RAP lesions become more hyperfluorescent since most of the dye remains intravascular. Unlike occult CNV, RAP does not exhibit a late washout effect [17]. Optical coherence tomography (OCT) can be used to demonstrate the intraretinal neovascular complex in cases of indistinct FA and ICG angiography [88].

RAP is most often misdiagnosed as occult or minimally classic choroidal neovascularization associated with ARMD, as it appears similar on fluorescein angiography. Unlike RAP, occult and minimally classic CNV are rarely associated with small intraretinal or preretinal hemorrhage and it is rare to see retinal perfusing or draining vessels. The characteristic "hot spot" on ICG angiography is well demarcated in RAP; in isolated CNV, ICG dye may leak extensively into subretinal or sub-RPE spaces. The ICG hot spot due to RAP may be difficult to differentiate from polypoidal CNV or focal occult CNV unless the CNV is associated with exudative detachment and the isolated choroidal nature of neovascularization in those lesions can be distinguished, or unless the RAP lesion has a distinct anastomotic connection between the retina and choroid. ICG hot spots may also be confused with retinal macroaneurysm or small capillary hemangiomas. In some cases, associated macular edema and small intraretinal hemorrhages may be confused with branch retinal vein occlusion [17].

Histopathology of retinal angiomatous proliferation
Histopathologic examination of six lesions obtained by submacular membranectomy [44] disclosed subretinal fibrovascular membrane with retinal pigment epithelium, basal deposits, and degenerated photoreceptor outer segments scattered in amorphous proteinaceous material (Fig. 28.1.14). A fibrocellular membrane within Bruch membrane was present in some lesions, associated with disruption of the basal deposit. In others, large vessels leaving the subretinal fibrovascular membrane extending toward the neurosensory retina were seen. The neurosensory retina was detached in most lesions, although outer neurosensory retina was firmly adherent to the subretinal fibrovascular tissue in some cases. The authors contrast this to occult and classic CNV lesions, which tend to remain adherent to the choroid. No CNV or retinal-choroidal anastomosis was found [44]. However, this study cannot confirm the histopathologic existence of RAP, as lesions were not sectioned in their entirety, and no intraretinal neovascularization extending into the subretinal space was documented. No other studies with confirmatory histologic evidence have been published. This will likely be possible only with detailed study of the whole eye.

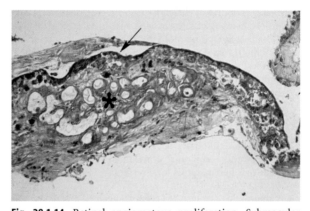

Fig. 28.1.14. Retinal angiomatous proliferation. Submacular membranectomy specimen with fibrovascular aggregate (*asterisk*) that is located internal to residual RPE and basal laminar deposit (*arrow*). Periodic acid-Schiff, original magnification ×100. (From [40])

28 III

Currently, no treatment modalities for RAP have been proven useful in large numbers of cases. Laser photocoagulation of extrafoveal lesions is often unsuccessful and requires repeated treatment due to communication with retinal blood vessels and revascularization [1]. Eyes with retinal-choroidal anastomosis are especially recalcitrant to laser photocoagulation due to the robust choroidal circulation, and have largely been unsuccessful in preventing visual loss [27, 40, 70]. Treatment of RAP with transpupillary thermotherapy has been reported to lead to rapidly progressive scarring [41], and should be avoided. Surgical ablation of Stage II RAP, performed by identifying feeder and draining vessels on ICG angiography, and surgically lysing those vessels, was shown in a small number of patients to improve visual acuity, and resolve intraretinal edema and pigment epithelial detachments [5]. A study of surgical excision of the neovascular lesions in Stage II patients found that visual acuity remained stable, but large defects of the RPE and choriocapillaris were formed [66]. Excision in Stage III cases also led to stabilization of visual acuity, but anatomic resolution of retinal hemorrhage and exudation occurred [66].

Supported in part by: Independent Order of Odd Fellows, Winston-Salem, North Carolina, and the Macula Foundation, New York, New York.

References

1. Arai K, Yuzawa M (2004) Therapeutic outcome in retinal angiomatous proliferation Jpn J Clin Ophthalmol 58:1423–1428
2. Backhouse O, O'Neill D (1998) Cavernous haemangioma of retina and skin. Eye 12:1027–1028
3. Bardenstein DS, Char DH, Irvine AR et al (1992). Extramacular disciform lesions simulating uveal tumors. Ophthalmology 99:944–951
4. Bell D, Yang HK, O'Brien C (1997) A case of bilateral cavernous hemangioma associated with intracerebral hemangioma. Arch Ophthalmol 115:818–819
5. Borrillo JL, Sivalingam A, Martidis A, Federman JL (2003) Surgical ablation of retinal angiomatous proliferation. Arch Ophthalmol 121:558–561
6. Bouzas EA, Parry DM, Eldridge R, Kaiser-Kupfer MI (1992) Familial occurrence of combined pigment epithelial and retinal hamartomas associated with neurofibromatosis 2. Retina 12:103–7
7. Bower LE, Ditkowsky SP, Klien BA, Bronstein IP (1942) Arteriovenous angioma of mandible and retina with pronounced hematemesis and epistaxis. Am J Dis Child 64:1023–1029
8. Cameron ME, Greer CH (1968) Congenital arterio-venous aneurysm of the retina. A post mortem report. Br J Ophthalmol 52:768–772
9. Cardoso RD, Brockhurst RJ (1976) Perforating diathermy coagulation for retinal angiomas. Arch Ophthalmol 94:1702–1715
10. Carr RE, Noble KG (1980) Retinal angiomatosis. Ophthalmology 87:956–961
11. Chan CC, Vortmeyer AO, Chew EY et al (1999) VHL gene deletion and enhanced VEGF gene expression detected in the stromal cells of retinal angioma. Arch Ophthalmol 117:625–630
12. Choyke PL, Glenn GM, Walther MM et al (1995) von Hippel-Lindau disease: genetic, clinical, and imaging features. Radiology 146:629–642
13. Dark AJ, Richardson J, and Howe JW (1978) Retinal hamartoma in childhood. J Pediatr Ophthalmol Strabismus 15:273–277
14. Destro M, D'Amico DJ, Gragoudas ES et al (1991) Retinal manifestations of neurofibromatosis: diagnosis and management. Arch Ophthalmol 109:662–666
15. Dobyns WB, Michels VV, Groover RV, Mokri B, Trautmann JC, Forbes GS, Laws ER Jr (1987) Familial cavernous malformations of the central nervous system and retina. Ann Neurol 21:578–583
16. Effron L, Zakov ZN, Tomsak RL (1985) Neovascular glaucoma as a complication of the Wyburn-Mason syndrome. J Clin Neuroophthalmol 5:95–8
17. Fernandes LH, Freund KB, Yannuzzi LA et al (2002) The nature of focal areas of hyperfluorescence or hot spots imaged with indocyanine green angiography. Retina 22:557–568
18. Flood TP, Orth DH, Aaberg TM, Marcus DF (1983) Macular hamartomas of the retinal pigment epithelium and retina. Retina 3:164–170
20. Font RL, Moura RA, Shetlar DJ, Martinez JA, McPherson AR (1989) Combined hamartoma of sensory retina and retinal pigment epithelium. Retina 9:302–311
21. Galinos SO, Smith TR, Brockhurst RJ (1979) Angioma-like lesion in hemoglobin sickle cell disease. Ann Ophthalmol 11:1549–1552
22. Gass JDM. Cavernous hemangioma of the retina: a neuro-oculocutaneous syndrome (1971) Am J Ophthalmol 799–814
23. Goldberg RE, Pheasant TR, Shield JA (1979) Cavernous hemangioma of the retina: A four-generation pedigree with neurocutaneous manifestations and an example of bilateral retinal involvement. Arch Ophthalmol 97:2321–2324
24. Gottlieb F, Fammartino JJ, Stratford TP, et al (1984) Retinal angiomatous mass. A complication of retinal detachment surgery. Retina 4:152–157
25. Green WR (1996) Retina: congenital variations and abnormalities. In: Spencer WH (ed) Ophthalmic pathology. Saunders, Harcourt Brace, Philadelphia, pp 682–720
26. Haller JA, Knox DL (1993) Vitrectomy for persistent vitreous hemorrhage from a cavernous hemangioma of the optic disc. Am J Ophthalmol 116:106–107
27. Hartnett ME, Weiter JJ, Staurenghi G, Elsner AE (1996) Deep retinal vascular anomalous complexes in advanced age-related macular degeneration. Ophthalmology 103:2042–2053
28. Heimann H, Bornfeld N, Vij O, Coupland SE, Bechrakis NE, Kellner U, Foerster macular hole (2000) Vasoproliferative tumours of the retina. Br J Ophthalmol 84:1162–1169
29. Helbig H, Niederberger H (2003) Presumed combined hamartoma of the retina and retinal pigment epithelium with preretinal neovascularization. Am J Ophthalmol 136:1157–1159
30. Hewick S, Lois N, Olson JA (2004) Circumferential peripheral retinal cavernous hemangioma. Arch Ophthalmol122:1557–1560

31. Inoue M, Noda K, Ishida S et al (2004) Successful treatment of subfoveal choroidal neovascularization associated with combined hamartoma of the retina and retinal pigment epithelium. Am J Ophthalmol 138:155–6

32. Irvine F, O'Donnell N, Kemp E, Lee WR (2000) Retinal vasoproliferative tumors: surgical management and histological findings. Arch Ophthalmol 118:563–569

33. Jakobiec FA, Font, RL, and Johnson FB (1976) Angiomatosis retinae: an ultrastructural study and lipid analysis. Cancer 38:2042–2056

34. Jain K, Berger AR, Yucil YH, McGowan HD (2003) Vasoproliferative tumours of the retina. Eye 17:364–368

35. Johnson MW, Flynn HW Jr, Gass JDM (1992) Pars plana vitrectomy and direct diathermy for complications of multiple retinal angiomas. Ophthalmic Surg 23:47–50

36. Joussen AM, Kirchhof B (2001) Solitary peripapillary hemangioblastoma. A histopathological case report. Acta Ophthalmol Scand 79:83–87

37. Kahn D, Goldberg MF, Jednock N (1984) Combined retinal-retina pigment epithelial hamartoma presenting as a vitreous hemorrhage. Retina 4:40–43

38. Kaye LD, Rothner AD, Beauchamp GR, Meyers SM, Estes ML (1992) Ocular findings associated with neurofibromatosis type II. Ophthalmology 99:1424–1429

39. Kreusel KM, Bornfeld N, Lommatzsch A et al (1998) Ruthenium-106 brachytherapy for peripheral retinal capillary hemangioma. Ophthalmology 105:1386–1392

40. Kuhn D, Meunier I, Soubrane G, Coscas G (1995) Imaging of chorioretinal anastomoses in vascularized retinal pigment epithelium detachments. Arch Ophthalmol 113:1392–1398

41. Kuroiwa S, Arai J, Gaun S, Iida T, Yoshimura N (2003) Rapidly progressive scar formation after transpupillary thermotherapy in retinal angiomatous proliferation. Retina 23:417–420

42. Kushner MS, Jampol LM, Haller JA (1994) Cavernous hemangioma of the optic nerve. Retina 14:359–361

43. Lafaut BA, Meire FM, Leys AM, Dralands G, De Laey JJ (1999) Vasoproliferative retinal tumors associated with peripheral chorioretinal scars in presumed congenital toxoplasmosis. Graefes Arch Clin Exp Ophthalmol 237: 1033–1038

44. Lafaut BA, Aisenbrey S, Vanden Broecke C, Bartz-Schmidt KU (2000) Clinicopathological correlation of deep retinal vascular anomalous complex in age related macular degeneration. Br J Ophthalmol 84:1269–1274

45. Laqua H, Wessing A (1979) Congenital retino-pigment epithelial malformation, previously described as hamartoma. Am J Ophthalmol 87:34–42

46. Laqua H, Wessing A (1983) Peripheral retinal telangiectasis in adults simulating a vascular tumor or melanoma. Ophthalmology 90:1284–1291

47. Machemer R, Williams JM Sr (1998) Pathogenesis and therapy of traction detachment in various retinal vascular diseases. Am J Ophthalmol 105:170–181

48. Maher ER, Webster AR, Moor AT (1995) Clinical features and molecular genetics of von Hippel-Lindau disease. Ophthalmol Genet 16:79–84

49. Malecha MA, Haik BG, Morris WR (2000) Capillary hemangioma of the optic nerve head and juxtapapillary retina. Arch Ophthalmol 118:289–291

50. Mansour AM, Jampol LM, Hrisomalos NF, Greenwald M (1998) Case report. Cavernous hemangioma of the optic disc. Arch Ophthalmol 106:22

51. Mansour AM, Walsh JB, Henkind P (1987) Arteriovenous anastomoses of the retina. Ophthalmology 94:35–40

52. Mansour AM, Wells CG, Jampol LM, Kalina RE (1989) Ocular complications of arteriovenous communications of the retina. Arch Ophthalmol 107:232–6

53. Mason JO 3rd, Kleiner R (1997) Combined hamartoma of the retina and retinal pigment epithelium associated with epiretinal membrane and macular hole. Retina 17:160–162

54. McCabe CM, Flynn HW Jr, Shields CL et al (2000) Juxtapapillary capillary hemangiomas. Clinical features and visual acuity outcomes. Ophthalmology 107:2240–2248

55. Moschos M, Ladas ID, Zafirakis PK, Kokolakis SN, Theodossiadis GP (2001) Recurrent vitreous hemorrhages due to combined pigment epithelial and retinal hamartoma: natural course and indocyanine green angiographic findings. Ophthalmologica 215:66–69

56. Nicholson DH, Green WR, Kenyon KR (1976) Light and electron microscopic study of early lesions in angiomatosis retinae. Am J Ophthalmol 82:193–204

57. Palmer JD, Gragoudas ES (1997) Advances in treatment of retinal angiomas. Int Ophthalmol Clin 37:150–170

58. Pancurak J, Goldberg MF, Frenkel M, Crowell RM (1985) Cavernous hemangioma of the retina. Genetic and central nervous system involvement. Retina 5:215–220

59. Parmar DN, Mireskandari K, McHugh D (2000) Transpupillary thermotherapy for retinal capillary hemangioma in von Hippel-Lindau disease. Ophthalmic Surg Lasers 3:334–336

60. Pierro L, Guarisco L, Zaganelli E, Freschi M, Brancato R (1992) Capillary and cavernous hemangioma of the optic disc. Echographic and histological findings. Acta Ophthalmol Suppl 204:102–106

61. Sappenfield DL, Gitter KA (1990) Surgical intervention for combined retinal-retinal pigment epithelial hamartoma. Retina 10:119–124

62. Schachat AP, Shields JA, Fine SL et al (1984) Combined hamartomas of the retina and retinal pigment epithelium. Ophthalmology 91:1609–1615

63. Schatz H, Chang LF, Ober RR, McDonald HR, Johnson RN (1993) Central retinal vein occlusion associated with retinal arteriovenous malformation. Ophthalmology 100:24–30

64. Seizinger BR, Rouleau GA, Ozelius LJ (1988) Von Hippel-Lindau disease maps to the region of chromosome 3 associated with renal cell carcinoma. Nature 332:268–269

65. Shields CL, Shields JA, Barrett J, De Potter P (1995) Vasoproliferative tumors of the ocular fundus. Classification and clinical manifestations in 103 patients. Arch Ophthalmol 113:615–23

66. Shimada H, Mori R, Arai K, Kawamura A, Yuzawa M (2004) Surgical excision of neovascularization in retinal angiomatous proliferation. Graefes Arch Clin Exp Ophthalmol Dec 17 [Epub ahead of print]

67. Singh AD, Nouri M, Shields CL, Shields JA, Smith AF (2001) Retinal capillary hemangioma: a comparison of sporadic cases and cases associated with von Hippel-Lindau disease. Ophthalmology 108:1907–1911

68. Singh AD, Nouri M, Shields CL, Shields JA, Perez N (2002) Treatment of retinal capillary hemangioma. Ophthalmology 109:1799–806

69. Sivalingam A, Augsburger J, Perilongo G, Zimmerman R, Barabas G (1991) Combined hamartoma of the retina and retinal pigment epithelium in a patient with neurofibromatosis type 2. J Pediatr Ophthalmol Strabismus 28:320–322

70. Slakter JS, Yannuzzi LA, Schneider U et al (2000) Retinal choroidal anastomoses and occult choroidal neovascularization in age-related macular degeneration. Ophthalmology 107:742–753

28 III

71. Smeets MH, Mooy CM, Baarsma GS, Mertens DE, Van Meurs JC (1988) Histopathology of a vasoproliferative tumor of the ocular fundus. Retina 18:470–472

72. Stallman JB (2002) Visual improvement after pars plana vitrectomy and membrane peeling for vitreoretinal traction associated with combined hamartoma of the retina and retinal pigment epithelium. Retina 22:101–104

73. Stolle C, Glenn G, Zbar B, et al (1998) Improved detection of germline mutations in the von Hippel-Lindau disease tumor suppressor gene. Hum Mutat 12:417–423

74. Theobald GD, Floyd G, and Krik HQ (1958) Hyperplasia of the retinal pigment epithelium simulating a neoplasm: report of two cases. Am J Ophthalmol 45:235–240

75. Theodossiadis PG, Panagiotidis DN, Baltatzis SG, Georgopoulos GT, Moschos MN (2001) Combined hamartoma of the sensory retina and retinal pigment epithelium involving the optic disk associated with choroidal neovascularization. Retina 21:267–270

76. Theron J, Newton TH, Hoyt WF (1974) Unilateral retinocephalic vascular malformations. Neuroradiology 7:185–96

77. Ticho BH, Egel RT, Jampol LM (1998) Acquired combined hamartoma of the retina and pigment epithelium following parainfectious meningoencephalitis with optic neuritis. J Pediatr Ophthalmol Strabismus 35:116–118

78. Ting TD, McCuen BW 2nd, Fekrat S (2002) Combined hamartoma of the retina and retinal pigment epithelium: optical coherence tomography. Retina 22:98–101

79. Tsai P, O'Brien JM (2000) Combined hamartoma of the retina and retinal pigment epithelium as the presenting sign of neurofibromatosis-1. Ophthalmic Surg Lasers 31:145–147

80. Verma L, Venkatesh P, Lakshmaiah CN, Tewari HK (2000) Combined hamartoma of the retina and retinal pigment epithelium with full thickness retinal hole and without retinoschisis. Ophthalmic Surg Lasers 31:423–426

81. Vianna RN, Pacheco DF, Vasconcelos MM, de Laey JJ (2001) Combined hamartoma of the retina and retinal pigment epithelium associated with neurofibromatosis type-1. Int Ophthalmol 24:63–66

82. Vogel MH, Zimmerman LE, Gass JDM (1969) Proliferation of the juxtapapillary retinal pigment epithelium simulating malignant melanoma. Doc Ophthalmol 26:461–481

83. Wang CL, Brucker AJ (1984) Vitreous hemorrhage secondary to juxtapapillary vascular hamartoma of the retina. Retina 4:44–47

84. Webster AR, Maher ER, Moore AT (1999) Clinical characteristics of ocular angiomatosis in von Hippel-Lindau disease and correlation with germline mutation. Arch Ophthalmol 117:371–378

85. Yamaguchi K, Yamaguchi K, Tamai M (1988) Cavernous hemangioma of the retina in the pediatric patient. Ophthalmologica 197:127–9

86. Yannuzzi LA, Negrao S, Iida T, et al (2001) Retinal angiomatous proliferation in age-related macular degeneration. Retina 21:416–34

87. Yen MY, Wu CC (1985) Cavernous hemangioma of the retina and agenesis of internal carotid artery with bilateral oculomotor palsies. J Clin Neuroophthalmol 5:258–262

88. Zacks DN, Johnson MW (2004) Retinal angiomatous proliferation: optical coherence tomographic confirmation of an intraretinal lesion. Arch Ophthalmol 122:932–933

28.2 Retinal Capillary Hemangioma

C.L. SHIELDS, J.A. SHIELDS

Core Messages
- Retinal capillary hemangioma is a benign vascular tumor of the ocular fundus
- Retinal capillary hemangioma can lead to retinal exudation, retinal detachment, vitreoretinal fibrosis, blindness, and loss of the eye
- This tumor can be associated with von Hippel-Lindau disease

- Von Hippel-Lindau disease is a condition with various combinations of retinal capillary hemangioma, cerebellar hemangioblastoma, pheochromocytoma, hypernephroma, pancreatic cysts, pancreatic islet cell tumors, endolymphatic sac tumor of the inner ear and cystadenoma of the epididymis

28.2.1 General Considerations

Retinal capillary hemangioma of the retina is a benign vascular hamartoma that can occur as an isolated tumor or as part of the spectrum of von Hippel-Lindau disease.

28.2.2 Definition and Incidence

The capillary hemangioma is a vascular hamartoma confined to the retina or optic disk [43, 44]. It is usually diagnosed in young patients between 10 and 30 years of age. The patient often presents with painless blurred vision and sometimes the mass is discovered on routine examination or screening of family members in those families with von Hippel-Lindau disease. There is no predisposition for sex or race although it appears that these tumors are more common in the white race. The tumors can be multiple in about one-third of cases [15]. Bilateral or multiple tumors imply the presence of underlying von Hippel-Lindau disease [13, 22, 23, 47]. Solitary tumors may or may not be associated with von Hippel-Lindau disease. A patient with a solitary hemangioma has a 45 % risk for developing von Hippel-Lindau if the patient in less than 10 years of age at diagnosis and 1 % risk if over age 60 years [45].

The von Hippel-Lindau disease [13, 22, 23, 47] is an autosomal dominant disorder with an estimated birth incidence of 1 per 40,000. In the United States, there are about 6,000 – 7,000 people with von Hippel-Lindau disease [47]. The features of von Hippel-Lindau disease include various combinations of:

Table 28.2.1. Tumors associated with von Hippel-Lindau disease

Tumor	Maher et al. series [22] ($n=152$)	Previously published studies ($n=554$)
Retinal capillary hemangioma	59%	57%
Cerebellar hemangioblastoma	59%	55%
Spinal cord hemangioblastoma	13%	14%
Renal cell carcinoma	28%	24%
Pheochromocytoma	7%	19%

- Retinal capillary hemangioma
- Cerebellar hemangioblastoma
- Renal cell carcinoma
- Pancreatic cysts and tumors
- Pheochromocytoma
- Epididymal cystadenoma
- Endolymphatic sac tumors
- Adnexal papillary cystadenoma of mesonephric origin (AMPO)
- Hemangioma of liver, lung, ovary
- Other manifestations (Table 28.2.1)

The phenotypic expression of the von Hippel-Lindau gene varies among families and individuals. Few patients manifest any feature of the disease before age 10 years. In general, the retinal capillary hemangioma is the first finding of the von Hippel-Lindau disease and occurs at a mean age of 25 years [22, 47]. The cerebellar hemangioblastoma occurs at a mean of 29 years and renal cell carcinoma at 44 years [22]. According to Maher and coworkers, the cumulative probability of a patient with von Hippel-Lindau disease developing a cerebellar hemangioblastoma by age 60 years was 0.84, for retinal capillary hemangioma it was 0.7, and for renal cell carcinoma it was 0.69 [22]. The mean age at death of a person with von Hippel-Lindau disease is 41 years, either of complications of cerebellar hemangioblastoma or metastatic renal cell carcinoma [22].

28.2.3 Clinical Features

The ophthalmoscopic appearance of a retinal capillary hemangioma varies with the size and location to the tumor. The earliest tumors may be detectable only as a blush of fluorescence on intravenous fluorescein angiography. Other early lesions may clinically be imperceptible but are suggested by slightly dilated retinal arteriole and venule feeding the tumor. Tumors greater than 50 μm in diameter are ophthalmoscopically visible to the experienced examiner and appear as a yellowish-red dot with minimally dilated afferent and efferent vessels. As the tumor enlarges, it assumes an orange-red color and the retinal feeder vessels are dilated (Fig. 28.2.1). The dilated blood vessels extend back to the optic disk and can be recognized in the posterior fundus remote from the tumor [43]. Twin vessels, described as a paired retinal arteriole and venule separated by no more than one venule's width from each other and coursing side by side, have also been found in association with retinal capillary hemangioma.

Retinal capillary hemangiomas in the periphery and at the equator of the eye tend to be well circumscribed. Those in an epipapillary or juxtapapillary location differ in that they may be less circumscribed and often do not have visible feeder vessels [7, 11, 18, 21, 28, 30, 33, 49, 54] (Fig. 28.2.2).

Retinal capillary hemangioma can assume an exudative or vitreoretinal form [43]. The exudative form is characterized by associated subretinal fluid and exudation (Fig. 28.2.2). Small tumors may only manifest a subtle rim of subretinal fluid or spotty adjacent exudation in the retina. Larger tumors accumulate more obvious subretinal fluid and yellow exudation. The exudation with larger tumors can be adjacent to the margin of the tumor in the retina or subretinal space or it can be remote from the tumor in the macula as stellate exudation.

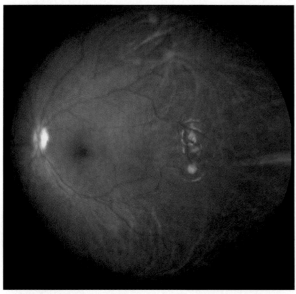

Fig. 28.2.1. Small retinal capillary hemangioma in a 25-year-old man

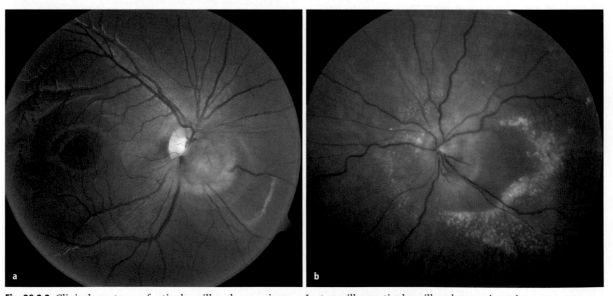

Fig. 28.2.2. Clinical spectrum of retinal capillary hemangioma. **a** Juxtapapillary retinal capillary hemangioma in a young woman. **b** Subtle sessile juxtapapillary retinal capillary hemangioma causing exudative retinal detachment in a child

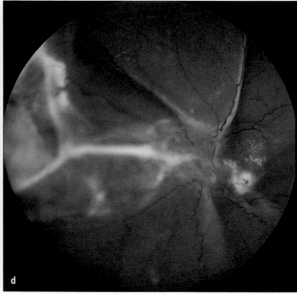

Fig. 28.2.2. c Vitreoretinal form of retinal capillary hemangioma. Peripheral retinal capillary hemangioma leading to preretinal fibrosis. **d** Exudative form of retinal capillary hemangioma. Total retinal detachment with fixed folds from peripheral retinal capillary hemangioma

The vitreoretinal form of retinal capillary hemangioma is characterized with more of a reactive phenomenon in the vitreous gel. Fibrosis of the overlying vitreous with traction bands elevating the retina may be visible [26] (Fig. 28.2.2). Flat preretinal fibrosis, especially in remote locations such as the macula, is typical of this form of retinal capillary hemangioma [17]. The two forms can sometimes overlap and we have observed the exudative type progress to the vitreoretinal type with or without treatment.

28.2.4 Differential Diagnosis

The differential diagnosis of retinal capillary hemangioma [37, 41, 42] includes:

- Retinal astrocytic hamartoma
- Retinoblastoma
- Retinal cavernous hemangioma
- Retinal racemose hemangioma
- Acquired retinal hemangioma (vasoproliferative retinal tumor)
- Choroidal melanoma
- Nematode granuloma
- Coats' disease
- Retinal artery macroaneurysm
- Sickle cell retinopathy
- Familial exudative vitreoretinopathy
- Age-related extramacular degeneration
- Age-related macular degeneration

The optic disk capillary hemangioma can resemble optic disk edema, papillitis, granuloma, glioma, metastasis, and others.

It is important to recognize the difference between a retinal capillary hemangioma and an acquired retinal hemangioma, more recently termed retinal vasoproliferative tumor [38, 41, 42]. The vasoproliferative retinal tumor can appear morphologically identical to the retinal capillary hemangioma, but it does not have the markedly dilated tortuous feeder vessels, shows a strong tendency to occur in the inferotemporal periphery, occurs in an older population, and carries no association with the von Hippel-Lindau disease (Fig. 28.2.3).

28.2.5 Pathology and Pathogenesis

Histopathologically, the retinal capillary hemangioma consists of a proliferation of retinal capillaries that replace the sensory retina architecture [27]. If the mass appears to grow inward toward the vitreous it is classified as endophytic and if outward toward the subretinal space it is classified as exophytic [11]. A proliferation of endothelial cells, pericytes, and vacuolated interstitial cell presumed of glial, vascular, or neuroectodermal origin is found [8]. The characteristic clear stromal cells seen in retinal hemangioblastoma are believed to be the cells of origin of this tumor [8]. However, their specific nature has not been identified and they do not appear to be vascular endothelial cells. Hence the term capillary hemangioma may not be accurate and this tumor

28 III

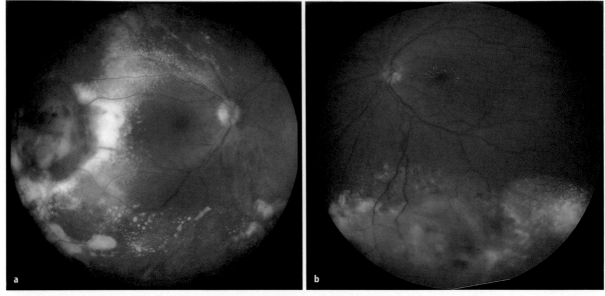

Fig. 28.2.3. Vasoproliferative tumor of the ocular fundus simulating retinal capillary hemangioma. **a** Peripheral vasoproliferative tumor with exudation and subretinal fluid. Note the lack of dilated feeder vessels. **b** Inferior vasoproliferative tumor with ill defined margins and macular star exudation. Note the lack of dilated feeder vessels

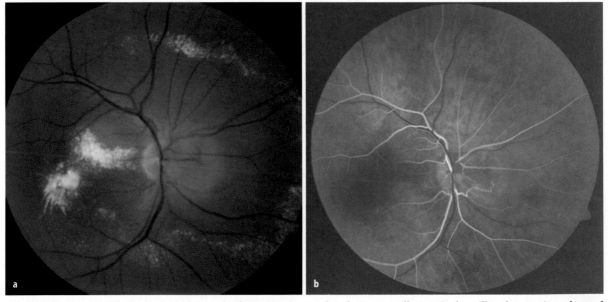

Fig. 28.2.4. Fluorescein angiography and optical coherence tomography of a juxtapapillary retinal capillary hemangioma located nasal to the optic disk in a young woman. **a** Subtle nasal juxtapapillary retinal capillary hemangioma with exudation and shallow subretinal fluid in the macula. **b** Early laminar venous phase of the fluorescein angiogram demonstrating retinal feeder vessel and early flush to the retinal tumor.

may be renamed in the future, once the histogenesis is better understood.

28.2.6 Diagnostic Approaches

The diagnosis of a retinal capillary hemangioma is most often made with careful indirect ophthalmoscopy. Ancillary studies such as fluorescein angiography, ocular ultrasonography, optical coherence tomography, color Doppler imaging, computed tomography, and magnetic resonance imaging assist in confirming the diagnosis.

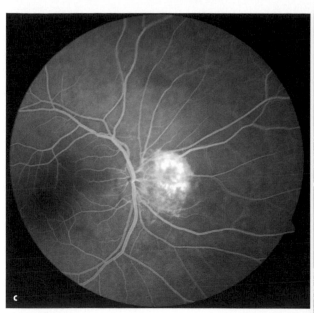

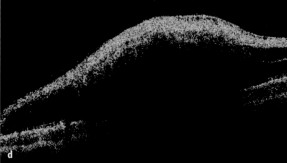

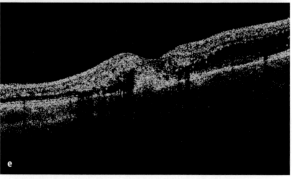

Fig. 28.2.4. c Full venous phase of the fluorescein angiogram demonstrating a nasal juxtapapillary mass with bright fluorescence, consistent with retinal capillary hemangioma. **d** Optical coherence tomography displaying an elevated dome-shaped retinal mass with bright optical surface and deep shadowing. **e** Optical coherence tomography of the fovea displaying subretinal fluid and optically dense subretinal debris, consistent with subretinal exudation

28.2.6.1 Fluorescein Angiography

Fluorescein angiography is the most helpful diagnostic test for recognizing retinal capillary hemangioma [43, 44] (Fig. 28.2.4). In the early arterial phase, the dilated retinal feeder arteriole appears prominent. Within seconds the retinal tumor is fluorescent as the fine capillaries that comprise the tumor fill with fluorescein. In the venous phase, the dilated draining vein fills with dye and the tumor maintains its bright fluorescence. In the late phase the tumor generally remains fluorescent and leaks dye into the vitreous. The intrinsic rapid fluorescence of the optic disk hemangioma assists in differentiating these tumors from other optic disk lesions.

28.2.6.2 Indocyanine Green Angiography

Indocyanine green angiography is used most often to visualize choroidal abnormalities as it is ideal for visualizing the choroidal vasculature [37]. It may be helpful in identifying a choroidal communication from the optic disk tumor to the adjacent choroid that is speculated to exist in some optic disk hemangiomas [27].

28.2.6.3 Ultrasonography

Ocular ultrasonography can detect small retinal capillary hemangiomas greater than 1 mm in thickness, but its sensitivity is best for those tumors larger than 2 mm in thickness. The A-scan demonstrates an initial high spike at the innermost apex of the tumor and high internal reflectivity throughout the mass. B-scan ultrasonography shows a dense echo at the inner apex of the mass and acoustic solidity throughout the mass with no choroidal component. The subretinal fluid and retinal detachment can be demonstrated on ultrasound.

28.2.6.4 Optical Coherence Tomography

Optical coherence tomography (OCT) is a method of cross sectional retinal imaging with high resolution to 10 µm. It is most useful for identifying subtle subretinal fluid, intraretinal edema, cystoid retinal edema, and retinal atrophy. The retinal layers can be appreciated on OCT, and atrophy or disorganization of the photoreceptor layer implies poor visual acuity. With regard to retinal capillary hemangioma, OCT can image the retinal mass, but it is most useful for monitoring related subretinal fluid and other retinal findings that threaten or cause poor visual acuity [36] (Fig. 28.2.4).

28.2.6.5 Color Doppler Imaging

Color Doppler imaging may be useful to demonstrate the blood flow within the mass while imaging the mass with an ultrasound cross sectional technique. In those eyes with opaque media, color Doppler may be an important imaging modality, but highly vascular retinal tumors such as retinoblastoma and choroidal tumors such as choroidal hemangioma and melanoma, especially those with a break in Bruch's membrane, may appear with similar features [20].

28.2.6.6 Computed Tomography

Computed tomography is generally reserved for larger retinal capillary hemangioma, especially those with total retinal detachment or those with opaque media. With contrast dye, the tumors enhance.

28.2.6.7 Magnetic Resonance Imaging

Magnetic resonance imaging is generally reserved for imaging larger tumors, especially those with total retinal detachment or opaque media. However, those patients with von Hippel-Lindau disease obtain yearly brain scans that may include views of the eyes. It is now realized that retinal capillary hemangioma greater than 2 mm in thickness can be detected by sensitive magnetic resonance imaging using a surface coil, contrast enhancement, and thin orbital views [10]. On T1-weighted images the retinal tumor appears with isointense to hyperintense signal compared to the vitreous and on T2-weighted images the tumor appears with isointense or hypointense signal compared to the bright vitreous. Gadolinium contrast provides moderate enhancement of the tumor and no enhancement of the subretinal fluid.

28.2.7 Systemic Evaluation

Analysis of the DNA of the patient and all family members can be performed in an attempt to identify the gene abnormality of von Hippel-Lindau disease. The gene for von Hippel-Lindau disease has been mapped to the short arm of chromosome 3 [23, 35]. All patients with von Hippel-Lindau disease should be followed carefully with yearly testing for systemic tumors as outlined in Table 28.2.2 [14, 22, 23, 47]. The relatives of patients with von Hippel-Lindau disease may benefit from a screening protocol (Table 28.2.2), depending on the results of DNA testing. The retinal capillary hemangioma is often the initial sign of von Hippel-Lindau disease and the various other systemic tumors found in this disease are best treated at an early stage; therefore it is important to routinely evaluate these patients systemically.

Table 28.2.2. Systemic evaluation for von Hippel-Lindau disease. Based on the Cambridge screening protocol [22]. The frequency of testing can be altered in relatives depending on the results of DNA analysis

Affected patient

Testing performed every 1 year:
Physical examination
Eye examination (indirect ophthalmoscopy)
Urine analysis
Urine 24 h collection for vanillylmandelic acid (VMA)
Renal ultrasound

Testing performed every 3 years:
Magnetic resonance (or computed tomography) of brain (after age 50 years, brain scan is performed every 5 years)
Computed tomography of kidneys

At risk relative

Testing performed every 1 year:
Physical examination
Eye examination (indirect ophthalmoscopy)
Urine analysis
Urine 24 h collection for VMA
Renal ultrasound

Testing performed every 3 years:
Magnetic resonance (or computed tomography) of brain (brain scan recommended every 3 years between age 15–40 years and then every 5 years until age 60 years)
Computed tomography of kidneys (abdominal scan recommended every 3 years between age 20–65 years)

28.2.8 Management

28.2.8.1 Ocular

Treatment of the retinal capillary hemangioma depends on the size and location of the tumor, clarity of media, and secondary features of the mass [3, 43, 46]. Some clinicians recommend treatment of all retinal capillary hemangiomas as these tumors tend to enlarge and produce subretinal fluid and exudation with visual loss. Others argue that these tumors may remain stable or even regress over a period of months to years and therefore recommend no treatment for small asymptomatic retinal capillary hemangiomas [52, 53]. Some tumors that have caused chronic retinal changes with poor visual potential are observed periodically (Fig. 28.2.5). Those small tumors with evidence of progression either in size or associated findings warrant treatment. Treatment of small tumors is generally limited to laser photocoagulation, cryotherapy, and diathermy [3–6, 9, 12, 19, 25, 40, 48, 50–52]. Larger tumors require techniques of photodynamic treatment, cryotherapy, radiotherapy, and retinal detachment surgery [2, 25, 29, 31, 39, 51, 52]. Enucleation is reserved for those eyes with advanced glaucoma and pain, usually from uncontrolled large capillary hemangiomas of the optic disk (Fig. 28.2.6).

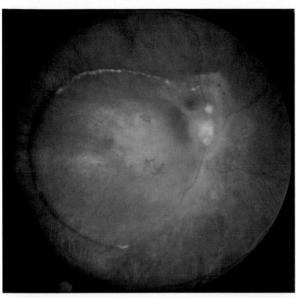

Fig. 28.2.5. Small juxtapapillary retinal capillary hemangioma with chronic macular retinoschisis managed with observation. There was no hope for visual improvement

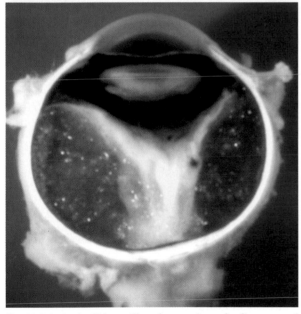

Fig. 28.2.6. Optic disk capillary hemangioma leading to total retinal detachment, neovascular glaucoma, and need for enucleation

A retrospective review of 68 patients with 174 retinal capillary hemangiomas from the Ocular Oncology Service at Wills Eye Hospital revealed initial management of observation (46%), laser photocoagulation (25%), or cryotherapy (23%) [46]. Small tumors (≤1.5 mm in size; 63 of 99; 64%) and those touching the optic disk (14 of 29; 48%) were more likely to be initially observed. Sixty-three (82%) of the 77 tumors that were initially observed remained stable

for a median follow-up of 12 years. The remaining 14 progressed and were successfully controlled with laser photocoagulation or cryotherapy. Either laser photocoagulation or cryotherapy was effective as the sole method of treatment in controlling 74% (26 of 35) and 72% (28 of 39) of extrapapillary tumors, with a mean number of 1.2 and 1.1 sessions, respectively. In a multivariate model, the only variables that were significantly related to final vision of ≤20/400 were poor initial vision [$P=0.01$, odds ratio (OR) 8.5] and the presence of retinal/vitreous hemorrhage ($P=0.024$, OR 5.7). Since the publication of this report, photodynamic therapy has assumed more of a role in the management of retinal capillary hemangioma.

Photocoagulation is delivered by the transpupillary route typically using argon or diode laser [43, 51]. Those tumors best suited for photocoagulation are less than 5 mm in diameter, without substantial subretinal fluid, located in the posterior pole of the eye. The goal of treatment is to occlude all feeder arterioles and avoid the draining venules. Small tumors can be obliterated by direct treatment to the tumor itself, but it is wisest avoid the tumor and treat the feeding arterioles if it is greater than 2 mm in diameter. The retinal arteriole leading to the tumor is first treated along its wall and then centrally for about 1–2 mm preceding entry into the tumor to induce spasm and decrease blood flow to the mass, and then the remaining peritumoral vessels are treated. A band of laser-induced ischemia is visible around the tumor after treatment. Those tumors near the optic disk typically show no feeder or draining vessels so treatment is directed on the tumor surface and surrounding the tumor in a double or triple row configuration of overlapping laser spots (Fig. 28.2.7). To improve laser uptake, some clinicians couple argon laser therapy with fluorescein dye potentiation [12] while others couple diode laser with indocyanine green dye [9].

Photodynamic therapy has been employed for medium sized retinal capillary hemangiomas that are too large to treat with laser photocoagulation or those tumors in the juxtapapillary and macular region [1, 34] (Fig. 28.2.8). Using verteporfin enhancing dye, a large spot laser light at 692 nm is directed to encompass the entire tumor, up to 7 mm diameter. Treatment results show resolution of subretinal fluid, but complications of retinal vascular occlusion, optic nerve ischemia, and vitreoretinal traction exist.

Cryotherapy is delivered by the transscleral route using a cryoprobe and indirect ophthalmoscopic guidance [2, 6, 39]. The tumor is elevated on the depressor tip of the cryoprobe and frozen completely and allowed to defrost. This freeze-thaw cycle is

28 **III**

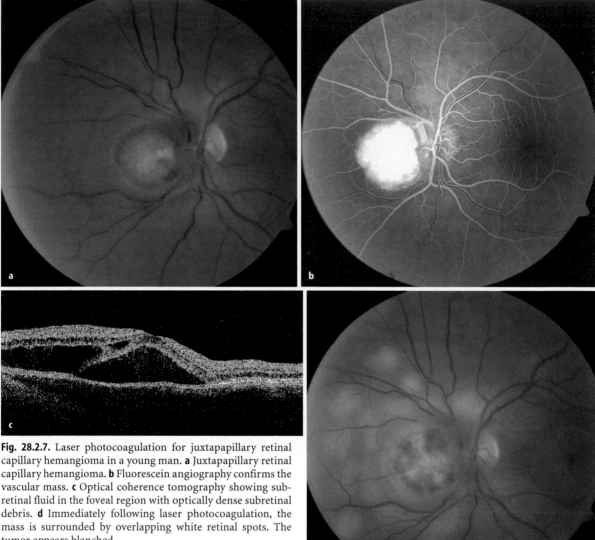

Fig. 28.2.7. Laser photocoagulation for juxtapapillary retinal capillary hemangioma in a young man. **a** Juxtapapillary retinal capillary hemangioma. **b** Fluorescein angiography confirms the vascular mass. **c** Optical coherence tomography showing subretinal fluid in the foveal region with optically dense subretinal debris. **d** Immediately following laser photocoagulation, the mass is surrounded by overlapping white retinal spots. The tumor appears blanched

repeated. Cryotherapy is generally reserved for peripheral retinal capillary hemangiomas, anterior to the equator of the eye and some posterior pole tumors that may be too large for photocoagulation. In some equatorial cases, a conjunctival incision will allow more accurate placement of the cryoprobe. Tumors less than 4 mm in thickness and less than 6 mm in base tend to respond adequately to cryotherapy. Larger tumors may require radiotherapy.

Plaque radiotherapy is a form of localized radiotherapy delivered by the transcleral route. Depending on tumor size, location, and associated features, the tumor is treated with an apex dose of 4,000 cGy over a 4- to 5-day period [16]. Larger tumors up to 8 mm thickness and 15 mm base can be treated with this method. External beam radiotherapy is a method of whole eye radiation and is reserved for very advanced tumors with substantial subretinal fluid or

those in the juxtapapillary region in which more conservative methods are expected to fail [31, 32]. In some advanced cases, especially those with the vitreoretinal form of retinal capillary hemangioma, vitreoretinal surgery with vitrectomy, repair of retinal detachment, endocryotherapy, and endophotocoagulation is employed.

The response to treatment with all of the above methods is slow and may need to be repeated. It is recommended to wait at least 1–2 months to assess the response to treatment and if there is a trend toward resolution, then further observation is indicated. If there remains minimal or no change or the disease worsens, then another session of treatment may be considered. The goal of treatment is to resolve associated subretinal fluid and exudation,

28

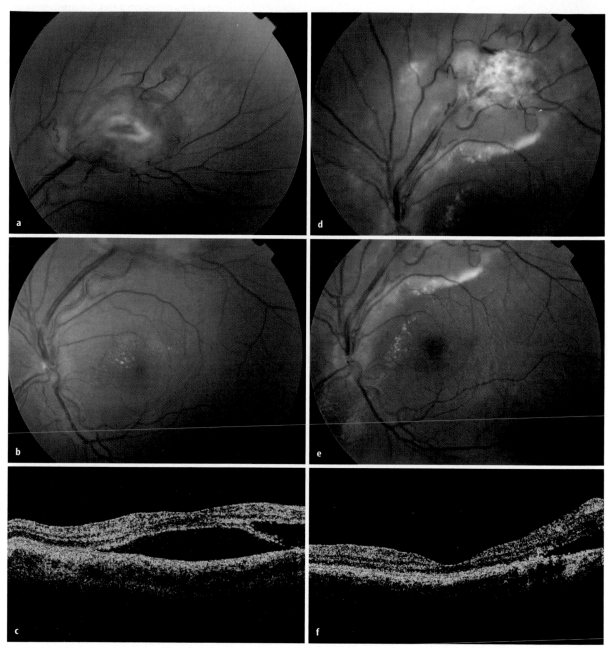

Fig. 28.2.8. Photodynamic therapy for four confluent retinal capillary hemangiomas in a 13-year-old girl. **a** One prominent and three small retinal capillary hemangiomas are visible. **b** Subretinal fluid and exudation in the macular region are noted with visual acuity of 20/80. **c** Optical coherence tomography of the foveal region reveals subretinal fluid and optically dense subretinal debris. **d** One month following photodynamic therapy, the retinal tumors are atrophic and fibrosed. **e** One month following photodynamic therapy, the subretinal fluid is partially resolved. **f** One month following photodynamic therapy, optical coherence tomography displays flat fovea with foveal thinning and persistent perifoveal subretinal fluid

but the residual tumor scar may remain as an elevated sclerosed mass.

Patients with juxtapapillary capillary hemangiomas are the most challenging to treat as the tumor, and related subretinal fluid and exudation can lead to profound loss of visual acuity. However, treatment itself can also lead to visual loss. A collaborative review of 60 eyes with juxtapapillary capillary hemangioma managed in ocular oncology or retina departments in Miami, Philadelphia, and San Francisco revealed poor long term vision (< 20/200) in 55 % of those treated with laser photocoagulation compared to 33 % poor vision in those not requiring or desiring laser treatment [24].

28 III

28.2.8.2 Systemic

Those patients with classic retinal capillary hemangioma, especially those of a young age or those with a family history of von Hippel-Lindau disease, should obtain a systemic evaluation as outlined in Table 28.2.2 [35, 39]. The availability of DNA analysis for patients with suspected von Hippel-Lindau has greatly increased detection of this disease and directed systemic monitoring.

28.2.9 Prognosis

The visual prognosis is quite variable depending upon tumor size and location as well as associated subretinal fluid, subfoveal gliosis, and preretinal fibrosis [24, 46]. The systemic prognosis is good if there is no associated von Hippel-Lindau disease. In those patients with von Hippel-Lindau disease, the projected median survival is 49 years [22].

28.2.10 Summary

Retinal capillary hemangioma is a benign vascular hamartoma of the retina. It can be associated with the von Hippel-Lindau disease and in those cases it is often the first finding of the disease. There is an exudative and vitreoretinal form of this tumor, leading to different clinical features. Treatment is directed toward early detection and obliteration of the tumors using methods of photocoagulation, photodynamic therapy, cryotherapy, radiotherapy, and others.

References

1. Aaberg TM Jr, Aaberg TM Sr, Martin DF, Gilman JP, Myles R (2005) Three cases of large retinal capillary hemangiomas treated with verteporfin and photodynamic therapy. Arch Ophthalmol 123:328 – 332
2. Amoils SP, Smith TR (1969) Cryotherapy of angiomatosis retinae. Arch Ophthalmol 81:689 – 691
3. Annesley WH, Leonard BC, Shields JA et al (1977) Fifteen-year review of treated cases of retinal angiomatosis. Trans Am Acad Ophthalmol Otolaryngol 83:446 – 453
4. Apple DJ, Goldberg MF, Wyhinny GH (1974) Argon laser treatment of von Hippel-Lindau retinal angiomas. II. Histopathology of treated lesions. Arch Ophthalmol 92:126 – 130
5. Balazs E, Berta A, Rozsa L, Kolozvari L, Rigo G (1990) Hemodynamic changes after ruthenium irradiation of Hippel's angiomatosis. Ophthalmologica 200:128 – 132
6. Blodi CF, Russell SR, Pulido JS, Folk JC (1990) Direct and feeder vessel photocoagulation of retinal angiomas with dye yellow laser. Ophthalmology 97:791 – 795
7. Brown GC, Shields JA (1985) Tumors of the optic nerve head. Surv Ophthalmol 29:239 – 264
8. Chan CC, Vortmeyer AO, Chew EY, Green WR, Matteson DM, Shen DF, Linehan WM, Lubensky IA, Zhuang Z (1999) VHL gene deletion and enhanced VEGF gene expression detected in the stromal cells of retinal angioma. Arch Ophthalmol 117:625 – 630
9. Costa RA, Meirelles RL, Cardillo JA, Abrantes ML, Farah ME (2003) Retinal capillary hemangioma treatment by indocyanine green-mediated photothrombosis. Am J Ophthalmol 135:395 – 398
10. DePotter P, Shields CL, Shields JA (1995) Disorders of the orbit. Tumors and pseudotumors of the retina. In: De Potter P, Shields JA, Shields CL (eds) MRI of the eye and orbit. Lippincott, Philadelphia, pp 93 – 116
11. Gass JDM, Braunstein R (1980) Sessile and exophytic capillary angiomas of the juxtapapillary retina and optic nerve head. Arch Ophthalmol 98:1790 – 1797
12. Gorin MB (1992) Von Hippel-Lindau disease: clinical considerations and the use of fluorescein-potentiated argon laser therapy for treatment of retinal angiomas. Semin Ophthalmol 7:182 – 191
13. Horton WA, Wong V, Eldridge R (1976) Von Hippel-Lindau disease: clinical and pathological manifestations in nine families with 50 affected members. Arch Intern Med 136: 769 – 777
14. Jennings AM, Smith C, Cole DR, Jennings C, Shortland JR, Williams JL, Brown CB (1988) Von Hippel-Lindau disease in a large British family: clinicopathological features and recommendations for screening and follow-up. Q J Med 66:233 – 249
15. Jesberg DO, Spencer WH, Hoyt WF (1968) Incipient lesions of von Hippel-Lindau disease. Arch Ophthalmol 80:632 – 640
16. Kreusel KM, Bornfeld N, Lommatzsch A, Wessing A, Foerster MH (1998) Ruthenium-106 brachytherapy for peripheral retinal capillary hemangioma. Ophthalmology 105: 1386 – 1392
17. Laatikainen L, Immonen I, Summanen P (1989) Peripheral retinal angiomalike lesion and macular pucker. Am J Ophthalmol 108:563 – 566
18. Landbo K (1972) A case of optic disc angioma. Acta Ophthalmol 50:431 – 435
19. Lane CM, Turner G, Gregor ZJ, Bird AC (1989) Laser treatment of retinal angiomatosis. Eye 3:33 – 38
20. Lieb WE, Shields JA, Cohen SM, Merton DA, Mitchell DG, Shields CL, Goldberg BB (1990) Color Doppler imaging in the management of intraocular tumors. Ophthalmology 97:1660 – 1664
21. Machmichael IM (1970) von Hippel-Lindau's disease of the optic disc. Trans Ophthalmol Soc UK 90:877 – 885
22. Maher ER, Yates JRW, Harris R, Benjamin C, Harris R, Moore AT, Ferguson-Smith MA (1990) Clinical features and natural history of von Hippel-Lindau disease. Q J Med 77:1151 – 1163
23. Maher RR, Iselius L, Yates JRW, Littler M, Benjamin C, Harris R, Sampson J, Williams A, Ferguson-Smith MA, Morton N (1991) Von Hippel-Lindau disease: a genetic study. J Med Genet 28:443 – 447
24. McCabe CM, Flynn HW Jr, Shields CL, Shields JA, Regillo CD, McDonald HR, Berrocal MH, Gass JD, Mieler WF (2000) Juxtapapillary capillary hemangiomas. Clinical features and visual acuity outcomes. Ophthalmology 107:2240 – 2248
25. Nicholson DH (1983) Induced ocular hypertension during photocoagulation of afferent artery in angiomatosis retinae. Retina 3:59 – 61
26. Nicholson DH, Anderson LS, Blodi C (1986) Rhegmatogenous retinal detachment in angiomatosis retinae. Am J Ophthalmol 101:187 – 189

27. Nicholson DH, Green WR, Kenyon KR (1976) Light and electron microscopic study of early lesions in angiomatosis retinae. Am J Ophthalmol 82:193–204
28. Oosterhuis JA, Rubinstein K (1972) Hemangioma at the optic disc. Ophthalmologica 164:362–374
29. Peyman GA, Rednam KRV, Mottow-Lippa L et al (1983) Treatment of large von Hippel tumors by eye wall resection. Ophthalmology 90:840–847
30. Pinkerton OD (1970) Papillary hemangioma (von Hippel's disease) of the optic papilla. A case report. J Ped Ophthalmol 7:157
31. Plowman PN, Harnett AN (1988) Radiotherapy in benign orbital disease I: Complicated ocular angiomas. Br J Ophthalmol 72:286–288
32. Raja D, Benz MS, Murray TG, Escalona-Benz EM, Markoe A (2004) Salvage external beam radiotherapy of retinal capillary hemangiomas secondary to von Hippel-Lindau disease: visual and anatomic outcomes. Ophthalmology 111:150–153
33. Schindler RF, Sarin LK, McDonald PR (1975) Hemangiomas of the optic disc. Can J Ophthalmol 10:305–317
34. Schmidt-Erfurth UM, Kusserow C, Barbazetto IA, Laqua H (2002) Benefits and complications of photodynamic therapy of papillary capillary hemangiomas. Ophthalmology 109:1256–1266
35. Seizinger BR, Smith DI, Filling-Katz MR et al (1991) Genetic flanking markers refine diagnostic criteria and provide insights into the genetics of von Hippel-Lindau disease. Proc Natl Acad Sci USA 88:2864–2868
36. Shields CL, Materin MA, Shields JA (2005) Review: optical coherence tomography of intraocular tumors. Curr Opin Ophthalmol 16:141–154
37. Shields CL, Shields CL, de Potter P (1995) Patterns of indocyanine green angiography of choroidal tumors. Br J Ophthalmol 79:237–245
38. Shields CL, Shields JA, Barrett J, de Potter P (1995) Vasoproliferative tumors of the ocular fundus. Classification and clinical manifestations in 103 patients. Arch Ophthalmol 113:615–623
39. Shields JA (1993) Response of retinal capillary hemangioma to cryotherapy. Arch Ophthalmol 111:551
40. Shields JA (1994) The expanding role of laser photocoagulation for intraocular tumors. The 1993 H. Christian Zweng Memorial Lecture. Retina 14:310–322
41. Shields JA, Decker WL, Sanborn GE, Augsburger JJ, Goldberg RE (1983) Presumed acquired retinal hemangiomas, Ophthalmology 90:1292–1300
42. Shields JA, Joffe L, Guibor P (1978) Choroidal melanoma clinically simulating a retinal angioma, Am J Ophthalmol 85:67–71
43. Shields JA, Shields CL (1992) Vascular tumors of the retinal and optic disc. Intraocular tumors: a text and atlas. Saunders, Philadelphia, pp 393–420
44. Shields JA, Shields CL (1999) Vascular tumors of the retinal and optic disc. Atlas of intraocular tumors. Lippincott Williams and Wilkins, Philadelphia, pp 244–253
45. Singh A, Shields J, Shields C (2001) Solitary retinal capillary hemangioma: hereditary (von Hippel-Lindau disease) or nonhereditary? Arch Ophthalmol 119:232–234
46. Singh AD, Nouri M, Shields CL, Shields JA, Perez N (2002) Treatment of retinal capillary hemangioma. Ophthalmology 109:1799–1806
47. Singh AD, Shields CL, Shields JA (2001) von Hippel-Lindau disease. Surv Ophthalmol 46:117–142
48. Straatsma BR (1954) Angiomatosis retinae. N Engl J Med 250:314–317
49. Takahashi T, Wada H, Tani E et al (1984) Capillary hemangioma of the optic disc. J Clin Neuroophthalmol 4:159–162
50. Vail D (1957) Angiomatosis retinae eleven years after diathermy coagulation. Trans Am Ophthalmol Soc 55:217–238
51. Watzke RC (1974) Cryotherapy of retinal angiomatosis. A clinicopathology report. Arch Ophthalmol 92:399–401
52. Welch RB (1970) Von Hippel-Lindau disease. The recognition of and treatment of early angiomatosis retinae and the use of cryosurgery as an adjunct to therapy. Trans Am Ophthalmol Soc 63:367–424
53. Whitson JT, Welch RB, Green WR (1986) Von Hippel-Lindau disease: case report of a patient with spontaneous regression of a retinal angioma. Retina 6:253–259
54. Yimoyines DJ, Topilow HW, Abedin S et al (1982) Bilateral peripapillary exophytic retinal hemangioblastomas. Ophthalmology 89:1388–1392

28.3 Cavernous Hemangioma

B. Jurklies, N. Bornfeld

Core Messages

- Cavernous hemangioma is a rare vascular hamartoma localized in the inner layers of the retina, which may involve the periphery of the retina and the optic nerve, respectively
- It is usually observed unilaterally, while bilateral manifestations have been reported
- It may be diagnosed more often in adults than in children
- The lesions of the retina are usually asymptomatic. Symptoms may rarely be caused due to a localization of the lesion involving the macula, and the optic nerve, and due to vitreous hemorrhages, respectively
- Histology detects thin walled, endothelium-lined, dilated blood vessels with non-fenestrated endothelium replacing the normal architecture of the inner retina. Preretinal membranes of glial origin on the surface of the lesion have been observed
- The typical clinical findings are represented by:
  - Clusters of saccular aneurysms within the inner retinal layers presenting a grape-like formation
  - A separation of the plasma and erythrocytic layers detected by fluorescein angiography. Usually there is no exudation observed
  - Thrombosis of the lesion, small hemorrhages and membranes on the surface may occur
- Therapy of the cavernous hemangioma is usually not necessary. Rarely vitrectomy may be considered in the case of persistent and severe vitreous hemorrhages
- Cavernous hemangioma of the retina may present a manifestation of cerebral cavernous malformations (CCM) with cavernomas of the central nervous system and the skin. Therefore an exclusion of this entity, involving the brain and the skin, is recommended
- Choroidal hemangiomas have been observed in patients of a family with autosomal dominant familial cavernous hemangiomas of the brain

28.3.1 Introduction

Cavernous hemangioma represents a rare vascular hamartoma of the retina involving the periphery of the retina and the optic nerve, respectively. The manifestation of this capillarovenous lesion is usually observed unilaterally. However, bilateral cases with a manifestation in both eyes have been reported [2]. The majority of patients are adult, while it is diagnosed rarely in children [28, 29, 44]. There are typical clinical findings of the tumor consisting of clusters of saccular aneurysms within the inner retinal layers, and presenting a grape-like formation [8]. Typical features with a separation of the plasma and erythrocytic layers may be detected by fluorescein angiography [8, 26]. Usually the lesions are diagnosed during a routine examination. Symptoms may rarely be caused by the tumor in the case of a localization involving the macula and the optic nerve and inducing hemorrhages, respectively [8, 19, 29].

In addition, cavernous hemangioma of the retina may present a manifestation of "neuro-oculocutaneous syndrome" with cavernous hemangiomas of the central nervous system and the skin [5, 11]. Due to their clinical and genetic characterization, cerebral hemangiomas have been defined as cerebral cavernous malformations (CCMs) and cavernomas [7, 16, 22].

This chapter reports on the clinical findings, characteristics, and pathological features of the retinal cavernous hemangioma, and the co-segregation which may at least in part present with cavernomas of the skin and the brain.

28.3.2 History

Cavernous hemangioma of the retina was first convincingly observed and published by Niccol and

Moore (1934) [30]. It was subsequently observed by others [32, 42]. However, it has been suggested that some reports published before 1971 described the clinical signs of cavernous hemangioma as telangiectasis, Coats' disease, congenital retinal angioma, angiomatosis retinae and a secondary vascular reaction to a previous exudative process [8]. Gass [8] compared the clinical findings observed in some of his patients with those of the literature and clearly defined the typical characteristics of this lesion. In addition, the term "neuro-oculocutaneous syndrome" has been used for cavernous hemangiomas of the retina associated with angiomatous lesions of the brain and the skin [8, 36]. The combination of clinical findings was first published by Weskamp and Cotlier (1940) [41]. They reported on a young female with a retinal vascular tumor. Vascular lesions of the skin and the brain suggested the presence of a cavernous hemangioma due to histological examinations.

28.3.3 Pathological Features

Thin walled, endothelium-lined and dilated blood vessels replacing the normal architecture of the inner retina have been detected histologically [8, 27]. The vascular lumens were at least in part connected to each other [8]. In addition, the telangiectatic retinal vessels were similar to normally encountered retinal vessels with a thin layer of non-fenestrated endothelial cells and a basement membrane [27]. The inner limiting membrane could not be separated over the entire area of the tumor [8]. However, preretinal

membranes have been observed in some cases. They consisted of spindle-shaped cells with glial filaments in the cytoplasm, the presence of glial-fibrillary acidic protein, and suggested a glial origin of the membranes [27].

28.3.4 Clinical Findings and Characteristics of Cavernous Hemangioma

Cavernous hemangioma may involve the retina and the optic nerve, respectively [8, 19, 23, 29, 31]. Usually, it is detected unilaterally, while bilateral cases have rarely been observed [2, 11].

Usually, there are no typical symptoms, except for hemangiomas involving the macula and the optic nerve, respectively [19, 29]. A location of the cavernous hemangioma within and beneath the macular area has been reported in up to 10% of cases [26]. Macular pucker [26] and retinal folds involving the macular area [8] have rarely been observed. Therefore, cavernous hemangiomas within the periphery of the retina are usually detected during a routine examination.

The clinical findings and the definition as a distinct entity have been characterized by Gass [8]: Ophthalmoscopy represents a grape-like tumor of the retina consisting of clusters of saccular aneurysms often localized beside a retinal vein (Fig. 28.3.1). It is localized within the inner layers of the retina [8, 26]. The vascular hamartoma may be associated with hemorrhages and fibrous (fibroglial) tissue on the surface of the tumor. The vascular pattern of the normal retina is not usually affected.

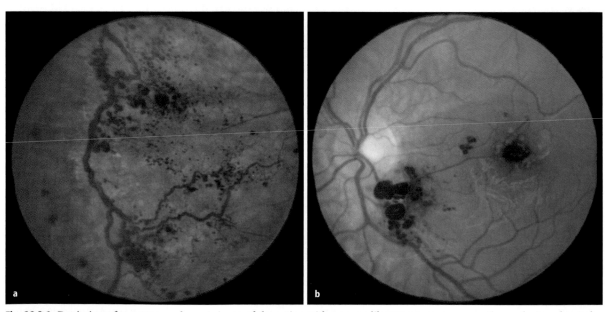

Fig. 28.3.1. Depiction of a cavernous hemangioma of the retina with a grape-like appearance representing a cluster of saccular aneurysms

28 III

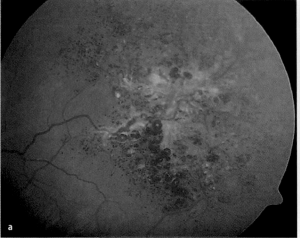
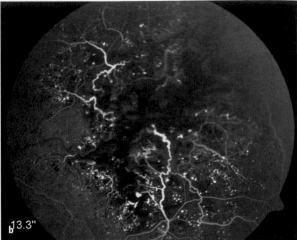

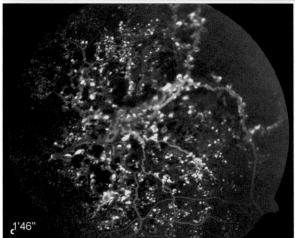

Fig. 28.3.2. Early (**a**) and late phase (**b**) of fluorescein angiography showing the features of a retinal vascular hemangioma: The telangiectatic lesions are demonstrated with hyperfluorescence of the upper half and hypofluorescence of the lower half without significant exudation in the late phase of fluorescein angiography

However, the presence of feeder vessels and a dilated feeder artery would be more typical for capillary hemangioma than for cavernous hemangioma, respectively [26].

Fluorescein angiography reveals a slow and incomplete filling of the cavernous hemangioma. Typically, separation of the plasma and the erythrocytes is detected during the late phases of fluorescein angiography, presenting hypofluorescence in the lower half and hyperfluorescence of the higher half of the vascular lesion (Fig. 28.3.2). This indicates a relatively isolated blood flow from the main stream of the retina [8]. In addition, extravascular leakage and arteriovenous shunts are typically not found in combination with cavernous hemangioma.

Cavernous hemangiomas involving the optic nerve usually represent similar clinical signs com-

pared to those of the periphery. Fluctuations in the size of the aneurysms have been suggested to be responsible for changes in visual function such as amaurosis [8]. Rarely a progression of the disease has been observed with growth and vitreous hemorrhages [19]. Echographic findings detected a dome shaped lesion with high reflectivity followed by signals with irregular reflectivity [31]. Optical coherence tomography revealed intraretinal cysts particularly of the inner retinal layer and signals with a high reflectivity [1].

The size of the aneurysms and the extent of the lesion have been reported to vary from case to case, and thrombosis with organization of the lesion has been observed during the follow-up [26]. However, the tumor usually only has minimal potential for changes during the follow-up [8].

Cavernous hemangioma of the retina may be associated with angiomatous vascular hamartomas of the brain and the skin, respectively [8, 33]. These cavernomas of the brain, CCMs, are defined as abnormally enlarged capillary cavities without intervening brain parenchyma. Intracranial hemorrhages, seizures, focal neurologic symptoms and sudden deaths may be the first signs of these localizations [10, 11]. In addition, choroidal hemangiomas have been diagnosed in two members of a family with autosomal dominant familial hemangiomas of the brain, skin and eye [33]. However, hereditary cerebral cavernous hemangiomas may vary according to the additional manifestation in the skin and eye [20, 33]. Examinations of 60 patients with a familial cerebral cavernoma revealed no manifestation on the skin in any out of 60 patients, while a retinal cavernous hemangioma was demonstrated in 3 out of 60 patients [20].

The incidence of solitary retinal cavernous hemangiomas is unknown due to the possibility of a follow-up without any symptoms. However, retinal cavernous hemangiomas have been found to be present in 5% of patients with familial cerebral cavernous malformations [20]. Familial cerebral cavernomas represent up to 10% of the cerebral cavernous malformations, with a prevalence of 0.5% in the general population [22]. Based on these findings, the incidence of retinal cavernous hemangioma has been estimated in up to 1 in 40,000 persons in the general population [20].

The age of patients with cerebral cavernous malformations and retinal cavernous hemangioma has been found similar to those without a retinal manifestation, with mean ages of 40.3 years (range 20.2–55.3 years) and 42.6 years (range 9.6–67.8 years), respectively [20].

Hereditary cases with an autosomal dominant pattern represented as CCMs have been associated with mutations in the CCM gene (OMIM 116860). They are classified as a distinct condition. Several genetic loci (CCM) have been identified. CCM1 has been shown to be associated with mutations in the KRIT1 gene on chromosome 7q11.2-q21 [12, 13, 16, 21, 25]. This encodes a microtubule-associated protein, which may play a role in endothelial cell function and vessel formation during angiogenesis [5, 14, 22]. CCM2 is caused by a mutation in the CCM2 (MGC 4607) gene (7p15-p13) normally encoding the protein malcavernin, while mutations in the PDCD10 (programmed cell death 10) gene (3q26.1) are responsible for CCM3 [4, 38–40]. Mutations of CCM2 and CCM3 have been suggested to induce disturbances in the angiogenesis (CCM2) and to be involved in the procedures of apoptosis (CCM3) [3, 6]. There is evidence for additional CCM loci, like the CCM4 locus on chromosome 3q26.3-27.2 [24].

A KRIT1/CCM1 truncating mutation [22] and a KRIT1/CCM1 splice-site mutation [17] have been found in a patient with cerebral and retinal angiomas. Recently, mutations in any of the three cerebral cavernous malformation genes, CCM/KRIT1, CCM2/MGC 4607, and PDCD10, have been detected in patients with cavernous hemangiomas of the retina and the brain [20].

28.3.5 Differential Diagnosis

The following diseases may play a role in differential diagnosis of the retinal cavernous hemangioma:

Parafoveal telangiectasis may present a similar picture compared to cavernous hemangiomas particularly in early stages of the disease. However, intraretinal and subretinal exudation and an increase in vascular dilatation may be observed during the follow-up [9]. In addition, due to its direct visualization of the structure of the retinal capillar-

ies, fluorescein angiography reveals a normal filling of the capillaries together with an atypical formation and dilatation of the capillaries. During the follow-up, leakage from the atypical capillaries is usually observed particularly in patients with symptoms and a decrease in visual acuity. In contrast, fluorescein angiography of cavernous hemangioma presents the typical separation of the plasma and erythrocytes with a low filling rate and without an exudation. As summarized by Gass [8], retinal telangiectasis represents a progressive disease which affects the retinal structure and the intrinsic retinal vasculature, while cavernous hemangioma is at least in part isolated from the normal capillary vessels.

Retinal capillary hemangioma may be differentiated typically from cavernous hemangioma due to the presence of feeder vessels, the supplying artery and the draining vein, the tortuosity of these vessels, the subretinal and intraretinal exudation, which may be associated with exudative retinal detachment, and the leakage during the fluorescein angiography. However, small retinal capillary hemangiomas may be difficult to detect and the feeder vessels may not be visible. In addition, retinal lesions of the retina may be associated with cerebral angiomas (von Hippel-Lindau syndrome) [34, 35, 37].

Racemous angioma may be associated with arteriovenous anomalies of the central nervous system (Wyborn-Mason syndrome). It represents dilated tortuous retinal vessels with a direct communication between the retinal veins and arteries, and typically without a capillary bed between the communicating vessels [34, 43].

28.3.6 Treatment

Cavernous hemangioma of the retina usually does not require any treatment. Rarely, pars plana vitrectomy may be indicated following vitreous hemorrhages [15]. However, usually the retinal cavernous hemangioma does not develop any progression. In addition, laser therapy and cryotherapy have not shown any significant benefits for the follow-up of cavernous hemangiomas [8, 18].

28.3.7 Conclusions

Cavernous hemangioma of the retina represents a benign vascular hamartoma. The typical clinical findings may help to differentiate this type of hamartoma from other retinal diseases. A significant progression of the disease and complications which may irreversibly threaten the visual function during the follow-up have been observed in only a few cases. Treatment is usually not required. Rarely pars plana vitrectomy may be taken into consideration if a severe vitreous hemorrhage does occur without any signs of regression. Follow-up examinations may help to exclude complications within the eye due to the cavernous hemangioma of the retina.

In addition, cavernous retinal hemangioma may be associated with vascular hamartomas of the brain and the skin, representing a sporadic or hereditary CCM, which may induce intracranial hemorrhages, seizures, focal neurologic symptoms and may be responsible for life threatening complications. Therefore, clinical examinations of the patient and the family members are recommended to exclude this condition with extraocular manifestations in the skin and central nervous system.

References

1. Andrade RE, Farah ME, Costa RA, Belfort R Jr (2005) Optical coherence tomography findings in macular cavernous haemangioma. Acta Ophthalmol Scand 83:267–269
2. Bell D, Yang HK, O'Brien C (1997) A case of bilateral cavernous hemangioma associated with intracerebral hemangioma. Arch Ophthalmol 115:818–819
3. Bergametti F, Denier C, Labauge P, Arnoult M, Boetto S, Clanet M, Coubes P, Echenne B, Ibrahim R, Irthum B, Jacquet G, Lonjon M, Moreau JJ, Neau JP, Parker F, Tremoulet M, Tournier-Lasserve E (2005) Mutations within the programmed cell death 10 gene cause cerebral cavernous malformations. Am J Hum Genet 76:42–51
4. Craig HD, Gunel M, Cepeda O, Johnson EW, Ptacek L, Steinberg GK, Ogilvy CS, Berg MJ, Crawford SC, Scott RM, Steichen-Gersdorf E, Saboroe R, Kennedy CTC, Mettler G, Beis MJ, Fryer A, Award IA, Lifton RP (1998) Multilocus linkage identifies two new loci for mendelian form of stroke, cerebral cavernous malformation at 7p15–13 and 3q25.2–27. Hum Mol Genet 7:1851–1858
5. Denier C, Gase JM, Chapon F, Domenga V, Lescoat C, Joutel A, Tournier-Lasserve E (2002) Ktit1/cerebral cavernous malformation 1 mRNA is preferentially expressed in neurons and epithelial cells in embryo and adult. Mech Dev 117:363–367
6. Denier C, Goutagny S, Labauge P, Krivosic V, Arnoult M, Cousin A, Benabid AL, Cornoy J, Frerebeau P, Gilbert B, Houtteville JP, Jan M, Lapierre F, Loiseau H, Menei P, Mercier P, Moreau JJ, Nivelon-Chevallier A, Parker F, Redondo AM, Scarabin JM, Tremoulet M, Zerah M, Maciazek J, Tournier-Lasserve E (2004) Mutations within the MGC4607 gene cause cerebral cavernous malformations. Am J Hum Genet 74:326–337
7. Dobyns WB, Michels VV, Groover RV, Mokri B, Trautmann JC, Forbes GS, Laws ER Jr (1987) Familial cavernous malformations of the central nervous system and retina. Ann Neurol 21:578–583
8. Gass JDM (1971) Cavernous hemangioma of the retina. Am J Ophthalmol, 71:799–814
9. Gass JDM, Blodi BA (1993) Idiopathic juxtafoveolar retinal teleangiectasis. Update of classification and follow-up study. Ophthalmology 100:1536–1546
10. Gislason I, Stenkulla S, Alm A, Wold E, Walinder PE (1979) Cavernous hemangioma of the retina. Acta Ophthalmol (Copenh) 57:709–717
11. Goldberg RE, Pheasant TR, Shields JA (1979) Cavernous hemangioma of the retina. A four generation pedigree with neurocutaneous manifestations and an example of bilateral retinal involvement. Arch Ophthalmol 97:2321–2324
12. Gunel M, Award IA, Finberg K, Anson J, Lifton RP (1995) Mapping a gene causing cerebral cavernous malformations. Proc Nat Acad Sci USA 92:6620–6624
13. Gunel M, Laurans MS, Shin D et al (2002) KRIT1 a gene mutated in cerebral cavernous malformation, encodes a microtubule-associated protein. Proc Natl Acad Sci USA 99:10677–10682
14. Guzeloglu-Kayisli O, Amankulor NM, Voorhees J, Luleci G, Lifton RP, Gunel M (2004) KRIT1/cerebral cavernous malformation 1 protein localizes to vascular endothelium, astrocytes and pyramidal cells of the adult human cerebral cortex. Neurosurgery 54:943–949
15. Haller JA, Knox DL (1993) Vitrectomy for persistent vitreous hemorrhage from cavernous hemangioma of the optic disc. Am J Ophthalmol 116:106–107
16. Johnson EW, Iyer LM, Rich SS, Ott HT, Gil-Nagel A, Kurth JH, Zabramski JM, Marchuk DA, Weissenbach J, Clericuzio CL, Davis LE, Hart BL, Gusella JF, Kosofsky BE, Louis DN, Morrison LA, Green DE, Weber JL (1995) Refined localization of the cerebral cavernous malformation gene (CCM1) to a 4 cM interval of chromosome 7q contained in a well defines YAC contig. Genome Res 5:368–380
17. Kitzmann AS, Pulido JS, Ferber MJ, Highsmith WE, Babovic-Vuksanovic D (2006) A splice-site mutation in CCM1/KRIT1 is associated with retinal and cerebral cavernous hemangioma. Ophthalmic Genet 27:157–159
18. Klein M, Goldberg MF, Cotlier E (1975) Cavernous hemangioma of the retina. Report of four cases. Ann Ophthalmol 7:1213–1221
19. Kushner MA, Jampol LM, Haller JA (1994) Cavernous hemangioma of the optic nerve. Retina 14:59–361
20. Labauge P, Krivosic V, Denier C, Tournier-Lasserve E, Gaudric A (2006) Frequency of retinal cavernomas in 60 patients with familial cerebral cavernomas. A clinical and genetic study. Arch Ophthalmol 124:885–886
21. Laberge-le Coteulx S, Jung HH, Labauge P, Houtteville J-P, Lescoat C, Cecillon M, Marechal E, Joutel A, Tournier-Lasserve E (1999) Truncating mutations in CCM1 encoding KRIT1 cause hereditary cavernous angiomas. Nature Genet 23:189–1993
22. Laberge-Le Couteulx S, Brezin AP, Fontaine B, Tournier-Lasserve E, Labauge P (2002) A novel KRIT/CCM1 truncating mutation in a patient with cerebral and retinal cavernous angiomas. Arch Ophthalmol 120:217–118
23. Lewis RA, Cohen MH, Wise GN (1975) Cavernous hemangioma of the retina and optic disc. A report of three cases an a review of the literature. Br J Ophthalmol 59:422–434
24. Liquori CL, Berg MJ, Squiteri F, Ottenbacher M, Sorlie M, Leedom TP, Cannella M, Maglione V, Ptacek L, Johnson EW, Marchuk DA (2006) Low frequency of PDCD10 mutations in a panel of CCM3 probands: potential for a fourth CCM locus. Hum Mutat 27:118
25. Marchuk Da, Gallione CJ, Morrison LA, Clericuzio CL, Hart BL, Kosofsky BE, Louis DN, Gusella JF, Davis LE, Prenger VL (1995) A locus for cerebral cavernous malformations maps to chromosome 7q in families. Genomics 28:311–314
26. Messmer E, Laqua H, Wessing A, Spitznas M, Weidle E, Ruprecht K, Naumann GO (1983) Nine cases of cavernous hemangioma of the retina. Am J Ophthalmol 95:383–390
27. Messmer E, Font RL, Laqua H, Höpping W, Naumann GO

(1984) Cavernous hemangioma of the retina. Immunohistochemical and ultrastructural observations. Arch Ophthalmol 102:413–418

28. Moffat KP, Lee MS, Ghosh M (1988) Retinal cavernous hemangioma. Can J Ophthalmol 23:133–135

29. Naftchi S, la Cour M (2002) A case of central visual loss in a child due to macular cavernous hemangioma of the retina. Acta Ophthalmol Scand 80:550–552

30. Niccol W, Moore RF (1934) A case of angiomatosis retinae. Br J Ophthalmol 18:454–457

31. Pierro L, Guarisco L, Zaganelli E, Freschi M, Brancato R (1992) Capillary and cavernous hemangioma of the optic disc. Acta Ophthalmol [Suppl] 204:102–106

32. Piper HF (1954) Über Cavernöse Angiome in der Netzhaut. Ophthalmologica 128:99–105

33. Sarraf D, Payne AM, Kitchen ND, Sehmi KS, Downes SM, Bird AC (2000) Familial cavernous hemangioma: an expanding ocular spectrum. Arch Ophthalmol 118:969–973

34. Shields JA, Shields CL (1999) Vascular tumors of the retina. In: Shields JA, Shields CL (eds) Atlas of intraocular tumors, vol 17. Lippincott Williams and Wilkins, Philadelphia, pp 244–267

35. Schmidt D (2005) Angiomatosis retinae. Klin Monatsbl Augenheilkd 222:90–109

36. Schwartz AC, Weaver RG Jr, Bloomfeld R, Tyler ME (1984) Cavernous hemangioma of the retina, cutaneous angiomas,

and intracranial vascular lesion by computed tomography and nuclear magnetic resonance imaging. Am J Ophthalmol 15:483–487

37. Singh AD, Nouri M, Shields CL, Shields JA, Perez N (2002) Treatment of retinal capillary hemangioma. Ophthalmology 109:1799–1806

38. Verlaan DJ, Davenport WJ, Stefan H, Sure U, Siegel AM, Rouleau GA (2002) Cerebral cavernous malformations: mutations in Krit1. Neurology 58:853–857

39. Verlaan et al (2004)

40. Verlaan DJ, Roussel J, Laurent SB, Elger CE, Siegel AM, Rouleau GA (2005) CCM3 mutations are uncommon in cerebral cavernous malformations. Neurology 65:1982–1983

41. Weskamp C, Cotlier I (1940) Angioma del cerebro de la retina con malformiciones capilares del la piel. Arch Oftalmol B Aires 15:1–10

42. Wessing A (1969) Fluorescein angiography of the retina: textbook and atlas. Mosby, St Louis, p 132

43. Wyborn-Mason R (1943) Arteriovenous aneurysm of midbrain and retina, facial naevi and mental changes. Brain 66:163–203

44. Yamaguchi K, Yamaguchi K, Tamai M (1988) Cavernous hemangioma of the retina in a pediatric patient. Ophthalmologica 197:127–129

28.4 Vasoproliferative Retinal Tumor

B. Damato, J. Elizalde, H. Heimann

Core Messages
- A nodular gliovascular proliferation causing exudation and fibrosis
- Inferotemporal, pre-equatorial, pink-yellow mass or masses
- Idiopathic or secondary to other ocular disease, such as uveitis
- Treatment with cryotherapy, brachytherapy photodynamic therapy are angiogenic agents
- Visual loss in most patients, as a result of maculopathy
- Enucleation may be necessary because of neovascular glaucoma

28.4.1 History

In 1982, Baines and associates described eight patients with inferotemporal, peripheral, retinal, telangiectatic nodules, associated with subretinal, exudates, macular edema, and epiretinal membranes [1]. Shields and colleagues initially called this condition "presumed acquired retinal hemangioma" but later renamed it "vasoproliferative retinal tumor" [7, 8]. Numerous single case reports and several case series have been published, with the disease described using a variety of terms [10].

28.4.2 Histological Features

Essentials
- Spindle-shaped glial cells with eosinophilic cytoplasm and small nucleus
- Dense capillary network and large, hyalinized blood vessels
- Foreign body giant cells, macrophages and cholesterol clefts

Vasoproliferative tumor consists mostly of glial cells, which are spindle shaped and eosinophilic, with no nuclear pleomorphism, and which stain positively for glial fibrillary acidic protein (GFAP, Dako) (Fig. 28.4.1) [3, 4, 11]. There is a fine capillary network throughout the lesion, with these blood vessels staining positively with CD31 antibody. There are also dilated and hyalinized blood vessels, some of which are occluded. Histology also shows exudates, macrophages and foreign body giant cells.

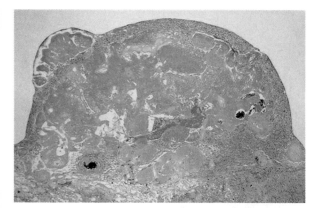

Fig. 28.4.1. Light micrograph of vasoproliferative retinal tumor showing eosinophilic glial cells, capillary network, hyalinized blood vessels, macrophages and giant cells. (Courtesy of S.E. Coupland, Berlin)

28.4.3 Pathogenesis

The pathogenesis is not known. Approximately 75% of cases are idiopathic and 25% are secondary to other ocular diseases, such as retinitis pigmentosa, uveitis, retinal detachment, congenital toxoplasmosis and Coats' disease (Fig. 28.4.2) [7, 9]. When such disease affects both eyes, multiple vasoproliferative lesions are more likely to develop. Some patients have bilateral vasoproliferative lesions in the absence of any apparent underlying ocular disease. Bilateral vasoproliferative tumors have been reported in a pair of monozygotic twins [12]. There is no sex preponderance. The condition can present at any age but is usually detected between the ages of 40 and 60 years. Diffuse tumors are rare, tending to arise in

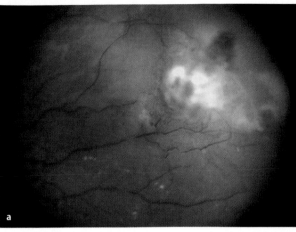 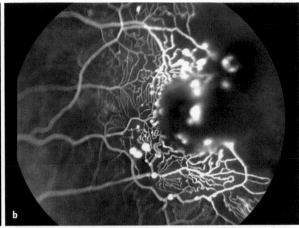

Fig. 28.4.2. Vasoproliferative tumor associated with Coats' disease: **a** color photograph and **b** fluorescein angiogram. (Courtesy of J. Elizalde, Barcelona)

young females and associated with more severe visual loss as well as neovascular glaucoma [7].

28.4.4 Clinical Features

Essentials
- Yellow-pink retinal mass
- Inferotemporal location in most cases
- Pre-equatorial or at equator
- Solitary, diffuse or multiple
- Small retinal feeder vessels
- Hard exudates, extending to macula
- Macular edema
- Exudative retinal detachment
- Vitreous hemorrhage
- Epiretinal membranes involving macula and causing retinal traction
- Anterior vitreous cells
- Retinal pigment epithelial hyperplasia
- Rubeosis and neovascular glaucoma

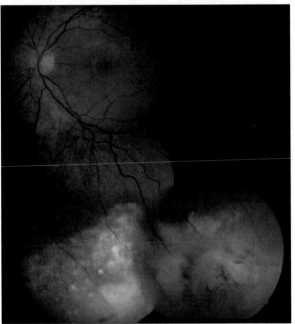

Fig. 28.4.3. Fundus photograph showing an inferotemporal vasoproliferative tumor associated with hard exudates. (Courtesy of C. Mosci, Genoa)

Patients tend to present with visual loss, floaters, and/or photopsia.

On ophthalmoscopy, the tumor has the appearance of a yellow or pink, intraretinal mass associated with adjacent hard exudates and occasionally retinal hemorrhages (Fig. 28.4.3). The hard exudates tend to extend posteriorly, eventually involving the fovea (Fig. 28.4.4). There may also be macular edema and exudative retinal detachment. In advanced stages, epiretinal membranes may develop, which can cause retinal distortion (Fig. 28.4.5). Solitary lesions predominate. Multiple lesions are more common in females when the disease is idiopathic and increase in incidence from about 6% to 41% in the presence of preexisting ocular disease [7]. A small minority of

patients have a diffuse vasoproliferative lesion and these tend to be relatively young, with an average age of 19 years [7].

Fluorescein angiography shows a rich capillary network and/or telangiectatic blood vessels within the tumor (Fig. 28.4.6). These leak profusely so that the entire lesion is hyperfluorescent in the late stages of the angiogram.

On ultrasonography, vasoproliferative tumors vary in size from 1.0 to more than 5 mm, averaging about 3 mm. The internal acoustic reflectivity varies from patient to patient and can be low, medium or high.

Tumor biopsy may be needed in some patients.

28 III

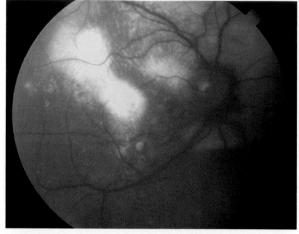

Fig. 28.4.4. Macular exudates from a peripheral vasoproliferative tumor. (Courtesy of H. Heimann, Berlin)

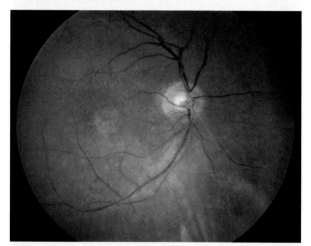

Fig. 28.4.5. Epiretinal membrane causing distortion of the retina. (Courtesy of H. Heimann, Berlin)

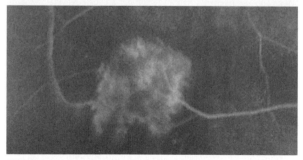

Fig. 28.4.6. Fluorescein angiogram showing patchy hyperfluorescence. (Courtesy of H. Heimann, Berlin)

28.4.5 Differential Diagnosis

Eccentric disciform lesions are usually associated with exudates and hemorrhages but are subretinal and tend to arise in older patients.

Retinal hemangioblastomas cause hard exudates, retinal hemorrhages, and fibrosis; however, these lesions are always associated with large retinal feeder vessels. There may also be extraocular manifestations of von Hippel-Lindau disease, a positive family history, or both.

Amelanotic melanomas can show dilated tumor vessels, particularly if the tumor has a collar-stud or mushroom shape. This tumor rarely causes exudates and hemorrhages.

Choroidal hemangiomas are pink, indistinct, nearly always located close to optic disk or fovea, and rarely associated with hemorrhages or hard exudates.

Coats' disease consists of intraretinal telangiectasia with intraretinal or subretinal hard exudates. It tends to occur in children, mostly males, and causes exudative retinal detachment without the formation of a mass. Rarely, it is associated with a vasoproliferative tumor.

28.4.6 Treatment

Essentials
- Observation for asymptomatic patients
- Cryotherapy for lesions up to 2 mm in thickness
- Brachytherapy for large lesions
- Photodynamic therapy for posterior lesions
- Vitrectomy for vitreous hemorrhage or epiretinal membranes
- Intravitreal antiangiogenic agents

Observation is indicated if a small, peripheral lesion is not causing much exudation so that there seems to be no threat to vision [6].

Cryotherapy using the triple free-thaw technique can conveniently be applied transconjunctivally in most cases, because these lesions are usually located anteriorly [7]. A single treatment session usually induces atrophy of the tumor and resorption of the hard exudates (Fig. 28.4.7). Some patients require several treatments, particularly if the tumor is large (i.e., >2 mm thick).

Plaque radiotherapy can be selected for tumors that do not respond to cryotherapy or for large lesions (Fig. 28.4.8) [3]. A dose sufficient to induce vascular closure is administered (e.g., more than 300 Gy to base of lesion). A ruthenium plaque may be preferable to an iodine applicator if it delivers a high dose of radiation to the tumor without excessive radiation to optic disk, macula and lens.

Photodynamic therapy has recently been shown to be effective (Fig. 28.4.9) [2]. With posterior lesions it may cause less visual loss than other methods. It may also be useful if it is important to avoid

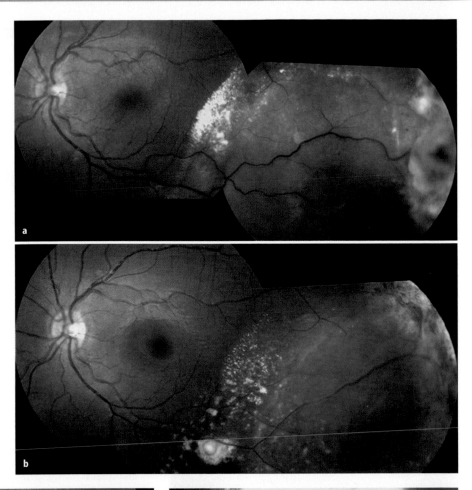

III 28

Fig. 28.4.7. Vasoproliferative tumor in the left eye of a 23-year-old female **a** at presentation, and **b** 1 year after cryotherapy. The tumor has become fibrotic and the exudates have regressed. (Courtesy of J. Elizalde, Barcelona)

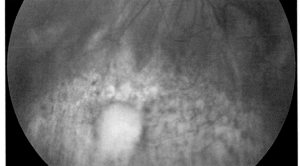

Fig. 28.4.8. Regressed vasoproliferative tumor surrounded by choroidal atrophy after ruthenium plaque radiotherapy. (Courtesy of H. Heimann, Berlin)

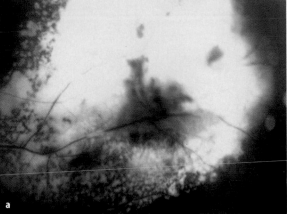

Fig. 28.4.9. Vasoproliferative tumor **a** at presentation and **b** after photodynamic therapy. (Courtesy of E. Balestrazzi and A. Tiberti, Rome)

inflammation, for example, if there is an epiretinal membrane or fibrosis with retinal traction.

Vitreoretinal surgery may be required to treat vitreous hemorrhage and epiretinal membranes. Hard exudates threatening the fovea have also been removed successfully, although this treatment requires further evaluation (Fig. 28.4.10).

In view of the fact that epiretinal membrane formation commonly causes visual loss, there is scope for investigating the scope of intravitreal triamcinolone injection as a means of preventing this complication. Intravitreal antiangiogenic agents such as bevacizumab (Avastin) may be useful [5].

28 III

Fig. 28.4.9b

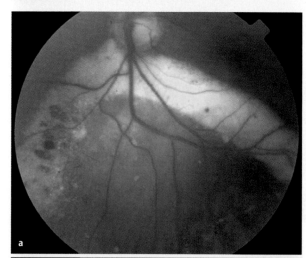

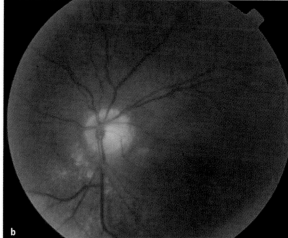

Fig. 28.4.10. Left fundus of a 23-year-old female with an inferior vasoproliferative tumor in the left eye showing **a** hard exudates threatening macula at presentation, and **b** conservation of macula after ruthenium brachytherapy of the tumor and surgical removal of the hard exudates. The retina was folded and the exudates were brushed away from its outer surface. (Courtesy of C. Groenewald, Liverpool)

28.4.7 Prognosis

The natural course of this disease varies from patient to patient, progressing slowly or not at all in some patients and causing severe exudation, retinal fibrosis, traction retinal detachment and/or rubeotic glaucoma in others. Visual loss often persists even when treatment successfully induces tumor atrophy and resorption of the hard exudates.

References

1. Baines PS, Hiscott PS, McLeod D (1982) Posterior non-vascularized proliferative extraretinopathy and peripheral nodular retinal telangiectasis. Trans Ophthalmol Soc U K 102:487–91
2. Barbezetto IA, Smith RT (2003) Vasoproliferative tumor of the retina treated with PDT. Retina 23:565–7
3. Heimann H, Bornfeld N, Vij O, Coupland SE, Bechrakis NE, Kellner U, et al. (2000) Vasoproliferative tumours of the retina. Br J Ophthalmol 84:1162–9
4. Irvine F, O'Donnell N, Kemp E, Lee WR (2000) Retinal vasoproliferative tumors: surgical management and histological findings. Arch Ophthalmol 118:563–9
5. Kenawy N, Groenewald C, Damato B (2007) Treatment of vasoproliferative tumour with intravitreal bevacizumab (Avastin). Eye 2007 (Epub ahead of print)
6. McCabe CM, Mieler WF (1996) Six-year follow-up of an idiopathic retinal vasoproliferative tumor. Arch Ophthalmol 114:617
7. Shields CL, Shields JA, Barrett J, De Potter P (1995) Vasoproliferative tumors of the ocular fundus. Classification and clinical manifestations in 103 patients. Arch Ophthalmol 113:615–23
8. Shields JA, Decker WL, Sanborn GE, Augsburger JJ, Goldberg RE (1983) Presumed acquired retinal hemangiomas. Ophthalmology 90:1292–300
9. Shields JA, Shields CL, Honavar SG, Demirci H (2001) Clinical variations and complications of Coats disease in 150 cases: the 2000 Sanford Gifford Memorial Lecture. Am J Ophthalmol 131:561–71
10. Singh AD, Rundle PA, Rennie I (2005) Retinal vascular tumors. Ophthalmol Clin North Am 18:167–76
11. Smeets MH, Mooy CM, Baarsma GS, Mertens DE, Van Meurs JC (1998) Histopathology of a vasoproliferative tumor of the ocular fundus. Retina 18:470–2
12. Wachtlin J, Heimann H, Jandeck C, Kreusel KM, Bechrakis NE, Kellner U, et al. (2002) Bilateral vasoproliferative retinal tumors with identical localization in a pair of monozygotic twins. 21:893–4

Subject Index

AC133   43
ACE, see angiotensin-converting
    enzyme
acellular capillary   131, 303
acetazolamide   714
activated protein C resistance
    (aPCR)   432, 435
acute multifocal posterior placoid epi-
    theliopathy (AMPPE)   661
acute retinal necrosis   512
adalimumab   644
adenosine   11
adenovirus   177
advanced
- diabetic eye disease   335
- glycation end-product (AGE)   66,
    108, 142, 213
- macular edema   335
AGE, see advanced glycation end-
    product
age-related
- extramacular degeneration   751
- macular degeneration (ARMD)   78,
    181, 194, 235, 547, 552, 751
- - hemorrhagic   548
- - neovascular   379
Ahmed glaucoma valve   278
AION, see anterior ischemic neuropa-
    thy
Airlie House classification   292
aldose reductase inhibitor   307, 309
Alzheimer's disease   110
amacrine cell   113
amaurosis fugax   670
amblyopia   598
aminoguanidine   305, 309
AMPPE, see acute multifocal posterior
    placoid epitheliopathy
amyloid-β   39
ANA, see antinuclear antibody
ANCHOR Trial   387
Ando iridectomy   265
anecortave   280
Ang-1, see angiopoietin-1
angioblasts   29
angiogenesis   24, 39, 76, 78
angiogenic factor   38
angioid streak   709, 714
angiomatosis retinae   548, 562
angiopoietin
- -1 (Ang-1)   43, 82, 99, 143
- -2   98, 99, 143

angiotensin II   98
angiotensin-converting enzyme
    (ACE)   128, 658
- elevated   663
- inhibitor   115
anterior
- ischemic optic neuropathy (AION)
    453, 511
- segment ischemia   727
antiangiogenic therapy   487
anticardiolipin antibodies   430
anticoagulation   457, 486
antinuclear antibody (ANA)   628
antioxidant   309
antiphospholipid antibody (APA)   430,
    628
- syndrome   440, 485, 629
antithrombin (AT) III   425, 436
anti-VEGF
- agent   419
- aptamer   395
- therapy   280
APC resistance   453, 484
aPCR, see activated protein C resistance
aphakia   265
apoptosis   304
aquaporin-1   141
argon laser   229, 409
ARMD, see age-related macular degen-
    eration   181
arterial
- hypertension   524
- macroaneurysm   10
- overcrossing   470
- oxygen tension   171
arteriolar narrowing   293
arteriovenous
- adventitial sheathotomy (AAS)   492
- anastomosis   536, 706
- early phase   195
- late phase   195
- passage time (AVP)   169
artery occlusion   669
aspirin   98, 307, 308, 525
asteroid body   657
astrocytes   26, 395
astrocytic hamartoma   751
atherosclerosis   425, 508
AVI contact lens system   261
azathioprine   643, 645, 681

B cell, autoreactive   630
Baerveldt implant   278
BAO, see branch retinal artery occlu-
    sion
basement membrane   5, 304
- capillary thickening   14
- tube   303
basic fibroblast growth factor (FGF-
    2)   708
BDP-MA, see benzoporphyrine deriva-
    tive monoacid ring A
Behçet's disease   430, 484, 635, 653,
    676
- anterior segment type   640
- panuveitis type of inflammation
    640
benfotiamine   310
benign vascular tumor   749
benzoporphyrine derivative monoacid
    ring A (BPD-MA)   244
Berlin's edema, see commotio retinae
bevacizumab   366
BIOM/SDI   261
birdshot chorioretinopathy   661, 676
black sunburst   706, 716
blood viscosity   447, 458, 485
blood-brain barrier   25
blood-retinal barrier   25, 139, 154,
    353, 377, 410, 562
- breakdown   444
BMT, see bone marrow transplantation
bone marrow
- niche   91
- transplantation (BMT)
- - retinopathy   606
bone morphogenic protein   82
bone-marrow derived cells   27, 32
brachytherapy   766
- vasoproliferative retinal tumor   766
brain development, IGF-1   399
branch
- retinal artery occlusion (BRAO)
    424, 510
- retinal vein occlusion (BRVO)   70,
    232, 424, 429, 435, 467, 477, 614, 724
- - study (BVOS)   482
- - surgical treatment   487
BRVO, see branch retinal vein occlu-
    sion
B-scan ultrasound   209
bull's eye maculopathy   633
buoyancy   264

Busacca's nodule   659
b-wave amplitude   452

calcitonin receptor-like receptor
   (CRLR)   231
Candida endophthalmitis   610
candle wax dripping   651
cannabidiol   115
capillary
– hemangioma   538, 736, 744
– microaneurysm   303
– non-perfusion   593, 622, 660, 676
carbogen   515
carbonic anhydrase inhibitor   364
cardiovascular risk factor   444
carotid
– artery insufficiency   519
– stenosis   514
cataract surgery   266, 338
– en bloc technique   339
– extracapsulat   339
– segmentation technique   339
cathepsin   87
caveolin   140
cavernoma   760
cavernous hemangioma   548, 735, 751,
   760
CCM, see cerebral cavernous malformation
CD4:CD8 ratio   657, 663
CD18   34, 97, 306, 307
CD45   101
CD117   84
cell
– fusion   82
– recruitment   76
central
– retinal artery occlusion (CRAO)
   181, 510, 715
– retinal vein occlusion (CRVO)   181,
   232, 369, 424, 429, 435, 443, 467, 670,
   716
– – classification   450
– – precipitating causes   446
– serous chorioretinopathy (CSCR)
   609
– serous retinopathy   235
cerebral cavernous malformation
   (CCM)   760
– differential diagnosis   763
– genetic loci   763
chemokine   48
chlorambucil   683
chloroquine   18, 633
chlorpromazine   280
chorioretinal venous anastomosis   460
choroid   167
choroidal
– granuloma   662
– hemangioma   248
– hemorrhage   602, 603
– melanoma   751
– neovascularization (CNV)   55, 240,
   241, 245, 544, 661, 706, 725, 742
choroidopathy   607
chromovitrectomy   262
chronic
– constipation   613

– neurodegeneration   108
Churg-Strauss syndrome (CSS)   670
ciliary
– artery occlusion   715
– body   167
cilioretinal
– anastomosis formation   456
– artery occlusion   511, 512
c-kit   84
– c-kit$^+$Sca-1$^+$ lineage (KSL)   79
clathrin   140
claudin   109, 146
clinically significant macular edema
   (CSME)   222, 293, 355
CMV retinitis   620
coagulation   432
Coat's disease   258, 413, 561, 568, 620,
   740, 751, 766
– differential diagnosis   563
collagen
– II   707
– IV   14, 707
color Doppler imaging   754
combined hamartoma of the retinal pigment epithelium an retina   741
commotio retinae (Berlin's edema)   512
complement factor   592
computed tomography (CT)   754
congenital
– arteriovenous communication   535
– retinal macrovessel   538
– vascular malformation   536
cortical blindness   694
corticosteroid   364, 459, 487, 490, 491,
   606, 664, 668, 675
– systemic   653
cotton wool spot   8, 291, 297, 447, 473,
   583, 592, 607–609, 629, 660, 669, 688,
   694, 718
COUP-TFII   45
COX-1/2, see cyclooxygenase-2
CRAO, see central retinal artery occlusion
C-reactive protein (CRP)   514
CRLR, see calcitonin receptor-like
   receptor
CRVO, see central retinal vein occlusion
cryocoagulation   409, 410
CRYO-ROP study   258
cryotherapy   256, 392, 421, 564, 739,
   755, 766
– panretinal   277
– vasoproliferative retinal tumor   766
CSCR, see central serous chorioretinopathy
CSME, see clinically significant macular
   edema
CSMT, see standardized change in macular thickness
cyclocryocoagulation   497
cyclohexadiene ring   244
cyclooxygenase
– -1 (COX-1)   98
– -2 (COX-2)   98, 307
– inhibitor   364
cyclophosphamide   664, 668, 682
cyclosporine   606
– A (CSA)   643, 645, 679

– cyclosporine-induced retina
   toxicity   607
cystoid macular edema (CME)   327,
   452, 532, 603, 650, 660, 661
– pseudo-hole   369
cytokine gradient   32

daclizamab   684
DAG, see diacylglycerol
daunomycin   269
db/db mice   306
deep
– retinal vascular plexus   32
– vascular plexus   26
delayed choroidal filling   524
deoxyspergualin (NKT-01)   631
dermatomyositis   676
diabetes   232, 425, 513
– mellitus   444, 524
diabetic
– macular edema   210, 235, 318, 548
– – pegaptanib   377
– – ranibizumab   386
– retinopathy   14, 161, 193, 258, 453,
   532, 609
– – animal model   303
– – EURODIAB grading system   294,
   299, 300
– – grading   291
– – growth hormone (GH)   325
– – insulin-deficient model   305
– – nonproliferative   303
– – severity scale   301
– – somatostatin analogues   324
– – study (DRS)   343
diacylglycerol (DAG)   66, 318
DIC, see disseminated intravascular
   coagulopathy
diminished a-/b-wave amplitude   524
diode laser   409
disseminated intravascular coagulopathy (DIC)   602, 604, 692, 693
drainage service   277
dye laser   229

EAE, see experimental autoimmune
   encephalomyelitis
Eales' disease   532, 613, 676, 724
Early Treatment of Diabetic Retinopathy
   Study (ETDRS)   99, 233, 292, 320
early vitrectomy   267
eclampsia   691
EGF, see epidermal growth factor
EIBOS   261
electroporation   179
electroretinogram (ERG)   513
Elschnig's spot   694
embolus   507, 511
endarterectomy   525
endocryocoagulation   257
endolaser coagulation   266, 277
endophthalmitis   160, 283, 382, 610
– non-infectious   287
endothelial
– cell (EC)   25, 38, 75, 78
– – proliferation   703
– – survival   82
– nitric oxide synthase   378

– progenitor cell (EPC) 40, 183
endothelin (ET) 131
– -1 11, 143
eosinophilic vasculitis 670
eph system 73
ephrin 51, 73
epidermal growth factor (EGF) 46
epiretinal membrane (ERM) 214, 218,
  598, 621, 726, 741
epistaxis 613
ERG, see electroretinogram
ERM, see epiretinal membrane
erythema nodosum 635
erythrocyte sedimentation rate (ESR)
  514
erythropoietin 48, 98
E-selectin 705
ESR, see erythrocyte sedimentation rate
ET, see endothelin
etanercept 99, 309, 644, 685
– protocol 350
ETDRS, see Early Treatment of Diabetic
  Retinopathy Study
EURODIAB grading system 293, 299
EVR1 570
experimental autoimmune encephalomyelitis (EA) 650
expression efficiency 175
extracapsular cataract surgery 339
extramacular degeneration, age-related
  751
exudation 530, 583, 660
exudative
– retinal detachment 582, 583, 739, 767
– retinopathy 545, 549
– vitreoretinopathy 724, 751
Eye Disease Case-Control Study 483

FA, see fluorescein angiography
Fabry's disease 14
factor V Leiden (FVL) 425, 432, 637
falciform
– retinal detachment 578
– tractional detachment 573
familial exudative vitreoretinopathy
  (FEVR) 258, 413, 567, 724, 751
– differential diagnosis 574
– falciform detachment 578
– laser photocoagulation 576
– vitrectomy 577
– X-linked recessive form 570
Fas ligand 357, 636
Fas/FasL 103, 307
Fas-L-mediated apoptosis 34
feeder vessel 373, 762
– photocoagulation 715
fenestration 378
fetal liver kinase 1 (Flk-1) 41
FEVR, see familial exudative vitreoretinopathy
FGF 68
FGF, see fibroblast growth factor
fibrillary acidic protein (GFAP) 114
fibrinolytic
– agent 495
– system 710
fibroblast growth factor (FGF) 30, 68,
  123, 231

fibronectin 14
fibrovascular membrane 330, 338, 403,
  405
filopodial-like process 31
final visual acuity (FVA) 454
FLA 469
Flk-1, see fetal liver kinase
Flt-1, see fms-like tyrosine kinase 1 42
fluorescein 193
– angiography (FA) 154, 193, 208, 234,
  244, 250, 347, 359, 448, 451, 469, 471,
  479, 675, 753
– – during pregnancy 696
– leakage 245
flux
– paracellular 140
– transcellular 140
fms-like tyrosine kinase 1 (Flt-1) 42
focal necrotizing arteritis 668
foreign body giant cell 766
FOXC2 45
frequency-doubled Nd:YLF laser 235
frizzled-4 (Fz4) 186
frosted branch angiitis 676
5-FU 269
fusiform aneurysm 547
FVL, see factor V Leiden
Fz4, see frizzled-4

gadolinium diethylenetriamine-pentaacetic acid (Gd-DTPA) 156
galactosemia 14
gap junction 5
gas-compression vitrectomy 159
G-CSF, see granulocyte colony-stimulating factor
Gd-DTPA, see gadolinium diethylenetriamine-pentaacetic acid
gene
– mutation 175
– therapy 175
– transfer 175
– – non-viral 170
genetically engineered virus 175
gestational age 403, 404
GFAP, see fibrillary acidic protein
GH, see growth hormone
giant cell arteritis 509
glaucoma 133, 218, 413, 439, 445, 484,
  663
– Ahmed valve 278
– rubeotic secondary 274
glial cell ingrowth 726
glomerulonephritis 688
glutamate 114, 133
Goodpasture syndrome 673
gradient index of refraction (GRIN)
  494
graft versus host disease (GVHD) 606
granulocyte colony-stimulating factor
  (G-CSF) 83
grid pattern 362
growth hormone (GH) 325
GVHD, see graft versus host disease

hairpin loop 700
hamartoma 741
– astrocytic 751

hard exudate 296, 582
HbSC 705, 723, 731
HbSS 700, 705
HbSThal 705
HCRVO, see hemicentral retinal vein
  occlusion
heavy silicon oil Densiron 264
Heerfordt's syndrome 658
Heidelberg retina flowmeter (HRF)
  168
HELLP syndrome 692
hemangioblast
– differentiation 41
– precursor 41
hemangioblastoma 749
hemangioma
– capillary 376, 744, 749
– cavernous 736, 751, 760
– racemose 743, 751
hematopoietic
– cells 24, 25
– progenitor cells 606
– stem cell (HSC) 44, 183
– – in vascular development 78
hematoporphyrin 244
hemicentral retinal vein occlusion
  (HCRVO) 468
hemodilution 458
hemoglobin 240
– S 705
hemoglobinopathy 700, 712, 713
hemorrhage 291, 583, 629
– ARMD 548
– choroidal 603
– cysts 595
– intraocular 597
– intraretinal 608, 660, 675, 694
– macular cyst 596
– premacular 551
– preretinal 550
– retinopathy 443
– Salmon patch 706
– subarachnoid 595
– subdural 595
– subhyaloidal 610
– submacular 551
– vitreous 595, 609, 613, 652, 708,
  727, 763
hemosiderin-laden macrophage 706,
  717
heparan sulfate proteoglycan 707
hepatocyte growth factor (HGF) 48
heterozygous thrombophilic disorder
  481
HGF, see hepatocyte growth factor
HIF-1, see hypoxia-inducible factor-1
HLA
– B27 684
– B51 638
homing 81
homocysteine 425, 426
homocystinuria 14
homonymous hemianopia 539
HRE, see hypoxia-response element
HRF, see Heidelberg retina flowmeter
HSC, see hematopoietic stem cell
hyaloidal vessel 27
hyaloidotomy 556

hydroxychloroquine 628, 633
hyperbaric oxygen 515
hypercholesterolemia 425, 507, 508
hyperfluorescence 203, 479
hyperglycemia 307, 318
hyperlipidemia 14, 471
hyperoxia 171, 395
hypertension 425, 444, 507, 508, 513
– pregnancy-induced 691
hypertensive retinopathy 210, 453,
    630, 688
hyperviscosity 444
hypofluorescence 203, 246
hypoxia 131, 709
– hypoxia-induced vascular prolifera-
    tion 393
– hypoxia-inducible factor-1 (HIF-1)
– – α 4, 122, 123
– – β 122
– – pathway 121
– hypoxia-inducible factor-2 (HIF-2)
– – α 46
– hypoxia-response element (HRE) 125
hypoxic insult 12

ICAM-1, see intercellular adhesion mol-
    ecule-1
idiopathic juxtafoveolar retinal telangi-
    ectasis 210, 252, 528
IGF-1, see insulin-like growth factor-1
IGFBP-3, see insulin-like growth factor-
    binding protein-3
immune-complex deposit 630
immunohistochemistry 116
in vitro experiment 182
indocyanine green angiography 753
inflammation in Eales' disease 613
infliximab 99, 644, 653, 664, 684
inner limiting membrane (ILM) 331,
    367, 578, 761
– peeling 353, 369
inner plexiform layer 111
Ins2^{Akita} diabetic mice 112
insulin 357
– resistance 305
– – type-2 diabetes 305
insulin-deficient model 305
insulin-like growth factor
– -1 (IGF-1) 30, 47, 324, 325, 326, 392,
    396, 397
– – brain development 399
– binding protein-3 (IGFBP-3) 86
integrin
– integrin-mediated adhesion 83
– α$_4$β$_1$ (VLA-4) 705
– αvβ$_{3/5}$ 48
intercellular adhesion molecule-1
    (ICAM-1) 34, 97, 129, 143, 305 – 307,
    639, 705
interfacial surface tension 264
interferon (IFN) 653, 684
– IFN-α 643
– IFN-β 650
– IFN-γ 129
interleukin (IL)
– IL-1
– – 1βm 231
– – β 47, 107, 129, 705

– IL-4 129
– IL-6 98
– IL-8 83, 636
intermediate uveitis 653, 740
intraocular
– hemorrhage 597
– pressure (IOP) 167, 275, 439
intraretinal
– edema 209
– hemorrhage 608, 660, 675, 694
– microvascular abnormality (IRMA)
    16, 233, 291, 298, 324, 303, 707
intravascular fibrin clot 604
intravitreal triamcinolone acetonide
    (IVTA) 216
– injection 491
intussusceptive growth 46
IOP, see intraocular pressure
iridectomy 413
iridescent spot 706
iris 167
– neovascularization 89, 510, 516
– nodule 659
IRMA, see intraretinal microvascular
    abnormality
Irvine-Gass syndrome 369
ischemia 110, 444, 583, 607, 709
ischemic
– index 451
– optic neuropathy 669
– retinopathy 609
isoniazid 622
isovolemic hemodilution 456, 459
IVTA, see intravitreal triamcinolone
    acetonide

JAM, see junction adhesion molecule
junction adhesion molecule (JAM)
    144

Kawasaki disease 668
kinase insert domain-containing recep-
    tor-1 (KD/flk-1) 232
Koeppe's nodule 659
krypton laser 229
KSL, see c-kit$^+$Sca-1$^+$ lineage

lack claudin-5 147
lactate 11
laser 353
– ablation 419
– coagulation 242, 360
– Doppler flowmetry 168
– Doppler velocimetry 168
– flare photometry 256
– hyaloidotomy 550
– laser-induced chorioretinal venous
    anastomosis 490
– photocoagulation 377, 392, 576, 654,
    725, 739, 755
leakage 203, 476
leakiness 26
Leber's miliary aneurysm 532, 561,
    735
lensectomy 413
– lensectomy-vitrectomy-membrane
    peeling 421
lens-sparing vitrectomy 419

lentivirus 177
leptin 48
leucovorin 681
leukocyte 710
lidocaine 286
lipoprotein A 438
lisinopril 310
Löfgren's syndrome 658
low-coherence interferometry 206
LRP5 570
lupus anticoagulant 430, 481
lymphangiogenesis 40
lymphatic system 38

macroaneurysm 744, 751
macrophage 53, 706, 766
– migration inhibitory factor (MIF)
    98
Macugen 69
macula hole 512
macular
– degeneration, age-related 751
– edema 9, 70, 337, 461, 583, 608,
    675, 767
– – clinically significant (CSME) 355
– inflammation 354
– grid laser coagulation 486, 487
– hole 544, 726
– pucker 256
– scar 448
– star 690
magnetic resonance imaging (MRI)
    155, 754
MAGUK, see membrane-associated
    guanylate kinase
mannose-binding lectin gene-2
    (MBL2) 639
MAPC, see multipotent adult progeni-
    tor cell
MAP-kinase, see mitogen-activated
    protein kinase
MARINA Trial 387
masquerade syndrome 543
matrix metalloproteinase 357, 378
MBL2, see mannose-binding lectin
    gene-2
mean
– corpuscular hemoglobin concentra-
    tion (MCHC) 714
– flow velocity (MFV) 170
– retinal circulation time (MRCT)
    168, 319
media opacity 330
melanin 240
meloxicam 309
membrane-associated guanylate kinase
    (MAGUK) 147
metalloprotease 9 (MMP-9) 84
methotrexate 606, 622, 653, 664, 681
methylenetetrahydrofolate reductase
    (MTHFR) 427
Meyer-Schwickerath 228
microaneurysm 292, 303, 386, 489,
    530, 608, 609, 621
microscopic polyangiitis (MPA) 668
microvasculature repair 13
MIF, see migration inhibitory factor
migration inhibitory factor (MIF) 98

Subject Index 775

mitogen-activated protein kinase (MAP-kinase) 66
monocyte 54, 99
– lineage cells 378
mouse model of RPO 393
MPA, see microscopic polyangiitis
MRCT, see mean retinal circulation time
MRI, see magnetic resonance imaging
MS, see multiple sclerosis
MTFR, see methylenetetrahydrofolate reductase
Müller cells 29, 67, 114, 395
multifocal chorioretinitis 661
multiple sclerosis (MS) 650
multipotent adult progenitor cell (MAPC) 43
multi-targeted receptor tyrosine kinase inhibitor (RTKI) 54
mural cells 25
Mycobacterium tuberculosis 613
mycophenolate mofetil 632, 653, 664, 682
myopia 412

Nd:YAG laser 229, 555
Ndp$^{y/-}$ mutant mice 188
necrotizing
– enterocolitis (NEC) 397
– vasculitis 668
neovascular
– disease 444
– glaucoma 218, 337, 447, 561, 582, 583, 767
neovascularization 131, 583, 703
– Eales' disease 613
– of the iris 510
nepafenac 309
netrin 51
neurodegeneration 108
neurofibromatosis 741
neuro-oculocutaneous syndrome 761
neurosarcoidosis, elevated ACE 663
neurovascular degeneration 108
neutrophil 99
NF-ϑB, see nuclear factor-ϑB
nitric oxide (NO) 99, 113, 307
– synthase 2 47, 123
NO, see nitric oxide
nodular gliovascular proliferation 766
nonenzymatic glycation 307
non-proliferative
– changes 700
– diabetic retinopathy (NPDR) 344
– – 4-2-1 rule 344
nonsteroidal anti-inflammatory drugs 678
non-viral gene transfer 179
Norrie's disease 186, 413, 571
Norrin 186, 571
– protein 186
Notch receptor 44
NPDR, see non-proliferative diabetic retinopathy
nuclear factor-ϑB (NF-ϑB) 98, 124, 232, 636

occludin 53, 109, 145, 653
occlusion 583

occlusive
– retinopathy 607
– vasculitis 653
OCT, see ocular coherence tomography
octreotide 326, 327, 664
– treatment recommendations 328
ocular
– coherence tomography (OCT) 155, 294, 370, 380, 387, 480, 753, 762
– – Fourier domain 206
– – hardware 222
– – software 223
– – title domain 206
– ischemia 453
– ischemic syndrome 519, 521, 609
– neovascularization 78, 88
oculocerebral
– lymphoma 653
– syndrome 535
open sky vitrectomy 422
optic
– disk
– – drusen 509
– – edema 688, 689
– – swelling 592
– nerve damage 663
– neuritis (ON) 651
– neuropathy 607
optical coherence tomography (OCT) 293, 364
– retinal vascular disease 205
oral aphtous ulcer 635
oxidative stress 110, 307
oxygen 393
– oxygen-induced retinopathy 35
– supplementation 132

p53 128
PAF 127
PAN, see polyarteritis nodosa
panretinal
– cryotherapy 277
– laser
– – coagulation 232
– – photocoagulation 611
– photocoagulation (PRP) 216, 275, 346, 525
Par6/Par3/atypical PKC (aPKC) polarity complex 144
paracellular flux 140
parafoveal retinal teleangiectasis 251
paraproteinemia 484
paroxysmal hemoglobinuria 484
PARP inhibitor 311
pars plana vitrectomy (PPV) 268, 277, 367, 494, 577, 623, 763
partial thromboplastin time (PPT) 433
PDGF, see platelet-derived growth factor
PDR, see proliferative diabetic retinopathy
PDT, see photodynamic therapy
peak systolic flow velocity (PSV) 170
pegaptanib 69, 280, 366, 377, 380
– sodium 283
pentoxifyllin 485
perfluorhexyloctane (F6H8) 264
perfluorocarbon
– gas 263

– liquid (PFCL) 262, 331
perfluorodecalin 262
perfluorooctane 262
peribulbar anesthesia 348
pericyte 16, 26, 304
peripheral fibrovascular mass 568
periphlebitis 613, 621, 651
perivascular
– mesenchymal cell 75
– sheathing 630
perivasculitis 20
perivenous
– exudation 660
– sheathing 660
persistent
– fetal vitreous (PHPV) 576
– hyperplastic primary vitreous (PHPV) 413
PFCL, see perfluorocarbon liquid
P-glycoprotein 141
pheochromocytoma 688
photocoagulation 345, 755
– panretinal 346
– protocol 348
– scattered lesions 346
photodisruption 240
photodynamic
– therapy (PDT) 239, 583, 755, 766
– – vasoproliferative retinal tumor 766
– vascular thrombosis 243
photosensitizer 240, 241, 243
photothrombosis 243
PHPV, see persistent hyperplastic primary vitreous phtocoagulation 332
PI-3 kinase 124
PI3K-AKT-mTOR pathway 124
pigment epithelial proliferation 528, 530, 740
pigment epithelium-derived factor (PEDF) 127, 183, 231, 246, 326, 414, 708
PIH, see pregnancy-induced hypertension
pinocytosis 140
PKC, see protein kinase C
placental growth factor (PlGF) 694
plasminogen activator inhibitor-1 (PAI-1) 637
plasticity 81, 90
platelet-derived growth factor (PDGF) 34
pluripotent cells 183
plus disease 404, 406, 408
polyarteritis nodosa (PAN) 668
polycythemia 484
polymyxin B 285
posterior
– hyaloid 422, 599
– vitreous detachment (PVD) 213
povidone-iodine 284
PPT, see partial thromboplastin time
pre-corneal lens 348
prednisolone 622
preeclampsia 691
pregnancy
– fluorescein angiography 696

- pregnancy-induced hypertension (PIH) 691
- – arteriolar narrowing
premacular hemorrhage 551
preretinal
- hemorrhage 550
- neovascularization 88, 742
pre-threshold 403
proliferative
- diabetic retinopathy (PDR) 70, 75, 78, 228, 330
- – laser coagulation 342
- retinopathy 396, 611
- sickle retinopathy (PSR) 718, 719, 726
- vitreoretinopathy (PVR) 256, 595, 598
- – severe progressive form 331
protein C 436, 637
- deficiency 509
- resistance 425, 432, 432
protein kinase C (PKC) 66, 70, 307, 356
- βI/II 317
- Diabetic Macular Edema Study 320
- Diabetic Retinopathy Study 320
- inhibitor 310, 317
- isoforms 318
protein S 637
- deficiency 509
prothrombin G20210A gene mutation 437
proton beam radiotherapy 587
PRP, see panretinal photocoagulation
pruning 46
P-selectin 705
PSR, see proliferative sickle retinopathy
Purtscher's retinopathy 512, 592, 609
PVD, see posterior vitreous detachment
PVR, see proliferative vitreoretinopathy
pyridoxamine 310

quinine 18

racemose hemangioma 743
radial optic neurotomy (RON) 219, 460
radiation
- retinopathy 17, 20, 548, 583, 609
- vasculopathy 584
radiotherapy 582, 756
Radner Reading Charts 370
RAM, see retinal arterial macroaneurysm
ranibizumab 283, 366, 386, 387
- safety 389
- visual acuity 389
RANTES 98
RAPD, see relative afferent pupillary defect
R-cadherin 31
RCH, see retinal capillary hemangioma
reacitve oxygen species (ROS) 66, 121
READ-1 Study 387
READ-2 Study 390
receptor tyrosine kinase 73
- inhibitor (RTKI) 54
recombinant tissue plasminogen activator (rt-PA) 485

relative afferent pupillary defect (RAPD) 452, 510
remodeling 46
respiratory distress syndrome (RDS) 411
reticulocyte 705
retinal
- angioma 249, 739
- angiomatous proliferation 744
- arterial macroaneurysm (RAM) 543, 751
- arterial occlusion 507
- capillary hemangioma (RCH) 248, 749
- detachment 218, 392, 393, 403, 405, 412, 418, 421, 544, 598, 675, 709, 719, 727, 740, 766
- – rhegmatogenous 422
- endovascular surgery (REVL) 495
- ganglion cell (RGC) 188
- hemorrhage 629, 688
- hypoxia 22, 187
- ischemia 27, 131, 275, 564
- ischemic disease 258
- neovascularization 476, 582, 653, 657, 675, 676, 694
- pigment epithelium/epithelial (RPE) 12, 212, 230, 244, 247, 602
- – cells 364
- – detachment 662
- – hyperplasia 767
- – hyperplastic 706
- racemose hemangioma 751
- revascularization 381
- telangiectasis 561
- thickness 377
- – measurement 221
- vascular
- development 35, 393
- –ischemia 661
- – occlusion 532, 661
- – tumors 735
- vascular disease
- – blood flow 173
- – optical coherence tomography (OCT) 205
- – photodynamic therapy 239
- – verteporfin 248
- – vitrectomy 260
- vasculature 6
- vasculitis 613, 640, 657, 669
- vasoocclusive disease 181, 629
- vasoproliferative tumor 739, 751
- vein
- – occlusion 210, 609
- – thrombosis 475
- venous sheathing 651
retinal-choroidal
- anastomosis 744
- tumor 20
retinectomy 289
retinitis pigmentosa 766
retinoblastoma 565
retinopathy 607
- bone marrow transplant-associated 606
- hypertensive 210, 630, 688
- occlusive 607

- of prematurity (ROP) 55, 75, 78, 126, 180, 258, 392, 403, 404, 532, 567, 575, 724
- – laser intervention 418
- – LIGHT-ROP study 411
- – mouse model 393
- – stage 4A 418
- – stage 4B (macula-off) 419
- – stage 5 (total) retinal detachment 419
- – surgical management 418
- – vascular endothelial growth factor 393
- oxygen-induced 35
- proliferative diabetic 78
retinopexy 228
retinoschisis 214
retinotomy 577
retrobulbar anesthesia 348
retrolental fibroplasia (RLF) 392, 403
retrovirus 177
RGC, see retinal ganglion cell
rhegmatogenous retinal detachment 422, 477, 573
rheological
- abnormalities 485
- treatment 458
rheumatic disease 675
Rheumatrex 681
rhexis hemorrhage 470
ridge 405
rifampicin 622
rituximab 632, 685
RLF, see retrolental fibroplasia
RNA aptamer 380
RON, see radial optic neurotomy
ROP, see retinopathy of prematurity
ROS, see reactive oxygen species
Roth spot 631
roxithromycin 630
RTKI, see receptor tyrosine kinase inhibitor
RTP-801 128
rt-PA 493, ETDRS 494
rubeosis 767
- iridis 408, 447, 537, 561, 582, 583, 621
rubeotic secondary glaucoma 274
ruboxistaurin 317, 318
- safety 322
- treatment-emergent adverse effects 322
ruby laser 228
ruthenium-106 583

salicylate 308
Salmon patches 716
- hemorrhage 706
Sandostatin LAR 327
- side effects 328
sarcoidosis 604, 642, 653, 657, 673, 676
Sca-1 84
scatter laser photocoagulation 497
Schaumann body 657
schisis cavity 706
scleral buckling 419
SCORE Study 42

SDF-1, see stromal cell-derived factor
sea fan   622, 707, 715
secondary glaucoma   413, 663
self-renewal capacity   80, 90
serine/threonine kinase   317
serous retinal detachment (SRD)   212, 602, 662, 669, 694
serum angiotensin converting enzyme   663
severe
– combined immunodeficiency syndrome (SCID)   175
– neovascular disease   444
– progressive proliferative retinopathy   331
sheathotomy   219
sickle
– cell
– retinopathy   432, 700, 751
– – disease   510, 620, 712, 713
– erythrocyte   710
– retinopathy   744
silicone   339
– oil
– – endotamponade   277
– – tamponade   265
silver-wired vessel   642
singlet oxygen   242
sirolimus   680
SLE, see systemic lupus erythematosus
small peripheral hemangioma   258
smooth muscle cells   25, 52
SMVL, see sustained moderate vision loss
SNAP-25   113
snowballs   642
– opacitiy   660
SOD, see superoxide dismutase
sodium iodate   159
soft exudate   292
soluble RAGE   311
somatostatin   664
– analogue   324
– receptor subtypes (SSTR1–5)   325
Sorafenib   54
sprouting   46
SRD, see serous retinal detachment
staining   203, 528
standardized change in macular thickness (CSMT)   359, 363
Starling's law   142
stereoscopic fundus photography   291
steroid response   660
strabismus   412
streptozotcin   305
stroma-derived factor-1 (SDF-1)   43, 83, 98
subarachnoid hemorrhage   595
subclinical retinal thickening   360
subdural hemorrhage   595
subhyaloidal hemorrhage   610
submacular hemorrhage   551
subretinal neovascularization   530
subthreshold laser coagulation   234
sulfur hexafluoride   263
superficial plexus   26
superluminescent diode (SLD)   222

superoxide
– anion   242
– dismutase (SOD)   127
suprachoridal seton implantation   280
sustained moderate vision loss (SMVL)   321
synaptophysin   113
systemic
– corticosteroids   653
– immunosuppression
– – in rheumatic diseases   675
– lupus erythematosus (SLE)   430, 509, 592, 604, 620, 628, 676
– – retinopathy   623, 630

T cell, autoreactive   630
tacrolimus (FK-506)   643, 680
Takayasu disease   676
TBI, see total body irradiation
telangiectasia   607, 608, 718
temporal arteritis   513
terminal dUTP nick end labelling (TUNEL)   114
Terson syndrome   595
tetradecapeptide   324
thermotherapy   582, 583
thrombocytopenia   607
thrombogenesis   447
thrombolysis   457, 515
thrombophilia   445, 508
thrombosis   428, 507
– risks   483
thrombospondin
– 1 (TSP-1)   47, 50
– 2 (TSP2)   86
thrombotic thrombocytopenic purpura (TTP)   602
Thy1.YFP mice   113
Tie2-angiopoietin system   82
tight junction   25, 46, 141, 378
– formation   139
TNF-α, see tumor necrosis factor α
total body irradiation (TBI)   606
toxoplasma   620
toxoplasmosis   677, 766
trabeculectomy   277
traction retinal detachment   619
tractional detachment   739
– rhegmatogenic retinal   331
transcellular flux   140
transdifferentiation   81
transforming growth factor (TGF)   68
– β (TGF-β)   47, 82, 706
transpupillary thermotherapy   582, 583
transscleral cyclophotocoagulation   278
trauma   445
triamcinolone   280, 331, 353, 365, 459, 491
– acetonide   161
trypsin   15, 19, 523
– digest retinae   113
TSP-1, see thrombospondin 1   47
tumor
– glial   21
– neuronal   21
– retinal-choroidal   20

tumor necrosis factor (TNF)   653
– α (TNF-α)   47, 98, 129, 307, 357, 636, 684, 705
– – antagonists   643
TUNEL, see terminal dUTP nick end labelling
tunica vasculosa lentis   408
tyrosine kinase   307

ultrasonography   753
uveal vascular system   167
uveitis   270, 653
– white-dot syndromes   677

VAMP2   113
vascular
– development
– – ephrins   74
– – hematopoietic stem cells   78
– dysfunction   108
– endothelial growth factor (VEGF)   3, 30, 357
– – retinopathy of prematurity (ROP)   393
– – inhibitor   353
– – mRNA levels   326
– – receptor-1 (VEGFR-1), see also fms-like tyrosine kinase 1 (Flt-1)   42, 66
– – receptor-2 (VEGFR-2), see also fetal liver kinase 1 (Flk-1)   41, 66
– – VEGF164/165   377
– hamartoma   761
– heterogeneity   25
– ischemia   613
– leakage   607, 675
– permeability   356, 394, 694
– permeability factor (VFP)   163
– permeability factor (VPF)   66
– progenitor cell   41
– resistance   167
– sheathing   675
– smooth muscle cells (VSMC)   14
– tortuosity   741
vasculitis   507, 650, 653
vasculogenesis   24, 29, 40, 41, 78
vasculopathy
– after phototherapy   588
– afterlocal resection   589
vaso-occlusive event   713
vasoproliferative retinal tumor   251, 751, 766
VCAM-1   129, 705
VE-cadherin   53, 653
VEGF, see vascular endothelial growth factor
vein occlusion   74
venous
– beading   293, 299
– occlusive disease   194
– overcrossing   470
– pressure   167
– sclerosis   651
– sheathing   293
– stasis retinopathy   443, 519
VEP, see visual evoked potential
verteporfin   239, 244, 247

vessel
– cannulation 494
– closure 614
– silver-wired 642
VFP, see vitreous fluorophotometry
VHL, see von Hippel-Lindau disease
Virchow's triad 485
viscosity 264
visual
– acuity (VA) 363, 467, 478
– loss 520
visually evoked potential (VEP) 418
vitamin E 411
vitreal hemorrhage 413
vitrectomy 260, 349, 599
– 25-gauge system 261
– early 267

– lens-sparing 419
– open sky 422
– pars plana 268, 277, 367, 494, 577, 623, 763
vitreomacular traction 209
vitreoretinal surgery 728
vitreous
– base 599
– bleeding 448
– detachment 334
– fluorophotometry (VFP) 155
– hemorrhage 160
– hemorrhage 218, 330, 331, 572, 595, 609, 613,652, 708, 719, 727, 742, 763, 767
– traction 737
vitritis 607, 650

Vogt-Koyanagi-Harada disease 677
Volk modified AVI system 261
von Hippel-Lindau disease (VHL) 124, 248, 538, 749
– HIF pathway 134
VPF, see vascular permeability factor

Wegener granulomatosis 509, 668, 670, 676
white-dot uveitis syndrome 677
wide-angle viewing system 331
Wnt pathway 186, 571
Wyburn-Mason syndrome 535, 743

xenon coagulation 228

zonula occludens-1 (ZO-1) 109, 144

See what your patients have been missing: improved vision in neovascular AMD[1]

Reference: 1. LUCENTIS® summary of product characteristics. Basel, Switzerland: Novartis Pharma AG; 2006.
Before prescribing LUCENTIS®, please consult the accompanying full national prescribing information approved in your country.